ILLUSTRATED DICTIONARY *and* RESOURCE DIRECTORY *of*

ENVIRONMENTAL & OCCUPATIONAL HEALTH

Second Edition

Herman Koren

Co-published with the
National Environmental Health
Association

CRC PRESS

Boca Raton London New York Washington, D.C.

Library of Congress Cataloging-in-Publication Data

Koren, Herman.
    Illustrated dictionary and resource directory of environmental and occupational health /
Herman Koren. — 2nd ed.
        p.   cm.
    Previous ed. published under the title: Illustrated dictionary of environmental health &
occupational safety.
    ISBN 1-56670-590-8 (alk. paper)
        1. Environmental health—Dictionaries. 2. Industrial safety—Dictionaries. I. Koren,
    Herman. Illustrated dictionary of environmental health & occupational safety. II. Title.

RA566.K59 2004
616.9′8′003—dc22   2003065998

Visit the CRC Press Web site at www.crcpress.com

Foreword

People can do some strange things in the course of a professional career. The exercise of writing a foreword for Dr. Koren's *Illustrated Dictionary and Resource Directory of Environmental and Occupational Health*, Second Edition, constitutes one of my top ten strangest things. I have now read the first edition of some 7500 entries and the second edition of over 16,000 entries. I can honestly say that I have read a dictionary as an interesting book. (I did hear once that if one was to read anything, read a dictionary. Why? All of the other books are in there!)

When I approached this task, I thought that I would be functioning more in the mode of a reviewer: Does that definition make sense? Why wasn't such and such a term included? After literally reading the first several pages of this dictionary, however, my sense of what I was doing began to change. In fact, I enjoyed it and, believe it or not, actually felt as if I were reading a fascinating book — complete with a story line! How in the world could I come to such a bizarre point of view? Actually, it was quite easy. Once into Dr. Koren's dictionary and once into the incredible details of all of the terms that he has included in this unusual and one-of-a-kind reference book, it quickly became apparent to me just how incredibly broad the fields of environmental and occupational health actually are, how many diverse and detailed sciences are involved, and how many organizations and other resources exist.

I have worked in the environmental field for my entire professional career, which now spans 30 years. Moreover, I have worked on, or have become familiar with, virtually every environmental issue imaginable. Yet, until I methodically proceeded through this dictionary, I confess that I really did not appreciate the enormous breadth of the field of practice and the underlying and supporting sciences.

That Dr. Koren illuminated me as to the distant horizons of this field, and the supporting sciences, is something that I shall be forever thankful for. Moving from word to word, thus reading it as a story, I was virtually spellbound by how encompassing and exciting this field of science actually is and the large body of knowledge needed by the practitioner and student.

Now, I wouldn't recommend that anyone other than a student or reviewer literally do what I did! I mentioned the impact that this book had on me only to emphasize to users of the dictionary that they have in their hands a comprehensive, environmental-health-focused publication that should cover any issue that might be of interest. For students at the various universities and colleges studying areas related to the environment, occupational health, medicine, law, engineering, and the specialties of air, water, soil, food, etc., it would be most helpful if they were assigned sections of the dictionary to read as a book, in order to gain an excellent understanding of the various environmental and scientific fields, the kinds of problems encountered, and the resources available to help to resolve problems.

If I might turn from the big picture that caught my attention to the details of the dictionary, I would like to share several specific observations with you here, as well. First, the cross-referencing is excellent. Often, within a definition, terms are used that also appear in the dictionary. I found it easy and instructive, for a fuller comprehension of the term, to move back and forth between all of the terms that were cross-referenced in the words that I was interested in.

I also liked the health-based orientation of the dictionary. The end result of most environmental problems is the effect on the health and welfare of people. The various chemicals and physical agents that may result in disease and injury are described, and the ensuing medical conditions are defined. Dr. Koren does justice to this special orientation, which makes the fields of environmental and occupational health so unique.

Representing a profession as I do, I am always interested in the application side of things. In other words, what is the usability index of this work? With that concern in mind, I asked questions as I read, such as, "Are the definitions expressed in such a way that the user can apply what is learned?" Consistently, I thought they were. In other words, this book will be useful to the practitioner as well as the student.

I also like that Dr. Koren did extensive research to make sure that all organizations and professionals related to the environment were included in his dictionary in the special Resource Directory section. How often have we all been frustrated by not knowing who the players are — or worse, knowing the initials of the players but not knowing what the initials stand for? You will find all the major players, and organizations, initials and all, in this material. Dr. Koren also includes many definitions pertaining to computer science, as well as environmental law. That's great! It also serves to make this a reference book that can be consulted for just about any issue the professional or student is studying.

The illustrations, which are computer generated, were devised and produced by Professor Alma Mary Anderson, C.S.C., Graphic Design Area Advisor, Department of Art,

Indiana State University. She is one of the most innovative graphic artists in the world today. She is the leader and foremost expert in the area of environmental and occupational health graphics. These graphics are sprinkled liberally throughout the text and even further enhance the comprehension of the terms being defined. An illustrated dictionary is many times better than one that lacks this useful learning tool. I recommend that serious students read the third paragraph of Dr. Koren's introduction, which just about summarizes the comprehensive nature of this piece of work.

In short, I think it is great that professionals in the environmental and occupational fields now have such a comprehensive, one-of-a-kind reference book that is so clearly health oriented. It is certainly going up on my reference shelf. I hope the cover is made of sturdy material because the book is going to be continually in use.

Nelson Fabian, M.E.
Executive Director
National Environmental Health Association
Denver, Colorado

Introduction

Environmental and occupational health is not a single topic, but rather a colorful, complex, and diversified range of interrelated subjects including all of the basic sciences, engineering, computer science, government, disease, injury identification, prevention, and control. Physical, chemical, microbiological, psychological, and social stresses are always present and can create or enhance the potential for disease or injury. People are exposed to these stresses in the home, community, workplace, and ambient environments.

The major environmental subject areas addressed include ecosystems, energy, transportation and alteration of chemicals in the environment, biological hazards, environmental impact statements, weather, geology and geography, food protection, foodborne disease, Hazard Analysis/Critical Control Point Inspection, food technology, insect control, rodent control, emerging diseases, pesticides, indoor environment, housing, indoor air pollution, hospitals, nursing homes and retirement homes, recreational environment, occupational environment, air quality, solid and hazardous waste management, private and public water supplies, swimming pools, plumbing, private and public sewage disposal, water pollution and water quality controls, environmental health emergencies, bioterrorism, instrumentation for evaluating indoor and outdoor environments, environmental law, and environmental programs.

The second edition of the *Illustrated Dictionary* has been increased from 7500 terms to over 16,000 terms because of the many comments from colleagues about the necessity of adding additional items to make this reference more comprehensive. It now represents as inclusive a list as possible. Thousands of references have been used to achieve this; however, it is recognized that some terms may still have been overlooked. The terms are drawn from varied, specialized, and technical fields and a variety of appropriate background areas but are related in such a manner as to be accessible to interested professionals in many different technical areas, students, and other specialized, as well as general, readers. They provide basic understanding in anatomy, biology, chemistry, genetics, geology, mathematics, microbiology, physics, and physiology to assist the reader in understanding environmental issues and consequences. Epidemiology, toxicology, risk assessment, statistics, computer science, geographic information systems, mapping, and instrumentation terminology are needed to understand how much and why environmental problems cause disease and injury. Medical terminology has been expanded to help explain the symptoms of a variety of exposures, and discussion of various diseases explains the negative endpoint of exposure, including death. Detailed computer-drawn illustrations by one of the leading experts in the world enhance these terms and provide needed visual assistance to make the terms more meaningful and understandable. Where applicable, definitions are supplemented with acronyms and abbreviations that are also alphabetized and cross-referenced.

The diseases selected were those most commonly related to the environment and industry. The chemicals selected were those most frequently encountered by the variety of environmental and occupational specialists, as well as the public. Further, all chemicals included are those considered to be hazardous by the U.S. Environmental Protection Agency and the National Institute of Occupational Safety and Health or potentially toxic or carcinogenic by the Agency for Toxic Substances and Disease Registry.

Considerable new terminology has been added in the areas of equipment, environmental controls, management communications, and other related areas that make environmental programs effective. New laws and terminology related to them have also been included.

The various resources within the first edition have been supplemented by hundreds more and have been put into a new section at the back of the book entitled Resource Directory. The new resources attempt to cover all governmental agencies and departments within them, civic groups concerned with the environment, professional organizations, abstract services, databases, and a variety of people working on environmental control.

The purpose of the second edition of the *Illustrated Dictionary* is to greatly enhance the quantity of high-quality definitions to fill the need created by the rapidly expanding field of environmental science for professionals in environmental health, environmental control, environmental engineering, occupational health, medicine, nursing, law, planning, loss prevention, safety management, insurance, government, industry, news organizations, teachers, and students. For students, it is recommended that sections of this one-of-a-kind reference be assigned for reading to help familiarize them with the breadth and depth of this extraordinary applied science. It also will provide an excellent understanding of the problems and resources.

As Nelson Fabian, Executive Director, National Environmental Health Association, said in the foreword, "That Dr. Koren illuminated me as to the distant horizons of this field and the supporting sciences is something that I shall be forever thankful for. Moving from word to word, thus, reading it as a story, I was virtually spellbound by how encompassing and exciting this field of science actually is and the large body of knowledge needed by the practitioner and student."

Acknowledgments

To Professor Alma Mary Anderson, C.S.C., Graphic Design Area Advisor, Department of Art, Indiana State University, for her extraordinarily fine computer-generated illustrations. The illustrations were based on extensive study in science areas in which she ordinarily does not work. She is considered to be the leader and foremost expert in the area of environmental health graphics today. She also incorporated into the illustrations for the original 7500 terms illustrations for an additional 8500 terms, for a total of 16,000 in this second edition of the dictionary. Her counsel and numerous suggestions for improvement have been of a high order of importance.

To the librarians and libraries of Indiana State University, Indiana University, Purdue University, University of South Florida, Vigo County Library, Largo Library, and the many other libraries where I was permitted to conduct my research for this book.

To Brian Lewis for coming up with the original idea for an illustrated dictionary and asking me to write it.

To Kathy Walters, former Associate Editor, Lewis Publishers, for her unique skills, her superb organizational work, and her dedication to a most difficult project. Without her help, innovative recommendations, guidance, and patience, the original work would never have been completed. Kathy dedicated her efforts to the memory of her mother, Jennie McInyk, for teaching her to be strong, compassionate, patient, and independent.

To Randi Cohen, Associate Acquisition Editor, CRC Press, for all of her time, encouragement, and good ideas for the second edition of the project. She has been a tremendous help in sorting through and resolving many of the problems that occur in a work of this nature.

To my wife, Donna, who at a critical time entered hundreds of new chemicals and hundreds of resources into the manuscript to meet publishing deadlines.

Dedication

To my wife, Donna,
who has brought to my life sweetness, beauty and music,
and who has provided a life for me and an environment
that helps me continue to be a highly productive teacher and writer.

A

Å — See angstrom unit.

a — (*indoor air quality*) See specific flow; SI symbol for the prefix *atto-*, representing a factor of 10^{-18}; symbol for arterial blood.

A — See ampere.

A horizon — (*soils*) Properties that reflect the influence of accumulating organic matter or eluviation alone or in combination.

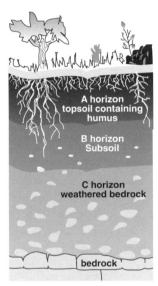

A Horizon
and other layers of soil

A&C — Acronym for abatement and control.

A&I — Acronym for alternative and innovative wastewater treatment system.

A&R — Acronym for air and radiation.

AA — Acronym for atomic absorption; adverse action.

AADI — Acronym for adjusted acceptable daily intake.

AAOE — Acronym for Airborne Antarctic Ozone Experiment.

AAQS — See ambient air quality standard.

AAR — See alkali–aggregate reaction; see Asbestos Analysts Registry.

AATSR — Acronym for advanced along-track scanning radiometer.

abacterial — A condition free of bacteria.

abandoned vehicle — Any unregistered motor vehicle deemed by applicable state laws to be an actual or potential safety hazard or which may be an actual or potential rodent harborage.

abandoned well — A well that has been permanently discontinued or is in such disrepair that it cannot be used.

abasia — The inability to walk because of a lack of motor coordination.

abate — (*law*) To put an end to; demolish; to do away with; to nullify or make void; a reduction in severity or degree.

abatement — (*asbestos*) The removal or other treatment of asbestos-containing materials to prevent asbestos contamination of a building or environment; the reduction in degree of intensity of a substance, such as a pollutant.

abatement debris — The waste from remediation activities.

abattoir — A slaughterhouse for animals intended for human consumption such as cattle, pigs, and sheep.

abdomen — The portion of the body between the thorax and the pelvis containing the lower part of the esophagus, the stomach, the intestines, the liver, the spleen, the pancreas, and other visceral organs.

abdominal aorta — The portion of the descending aorta that divides into the two common iliac arteries.

abdominal cavity — The space within the abdominal walls between the diaphragm and the pelvic area.

abdominal pain — An acute or a chronic localized or diffuse pain in the abdominal cavity signifying a need for immediate medical or surgical intervention.

abduct — To move away from the median plane of the body.

abduction — The movement away from something.

aberration — Deviation from the normal or usual.

ABG — Acronym for arterial blood gases.

abient — A tendency to move away from stimuli.

ability — The capacity to carry out specific tasks because of appropriate skills of mental and/or physical fitness.

abiotic — Referring to the non-living components of ecosystems.

A

abiotic factors — The physical and non-living chemical factors such as temperature, light, water, minerals, and climate that influence living organisms.

ABL — Acronym for atmospheric boundary layer.

ablation — Removal of a growth or harmful substance.

abnormal behavior — Acts or activities detrimental to the individual or to society.

ABO blood groups — A system for classifying human blood based on the antigenic components of red blood cells and their corresponding antibodies consisting of four blood types: A, B, AB, and O.

abortus fever — An infection in people caused by the ingestion of contaminated milk or meat from cows infected with *B. abortus*.

above grade — (*housing*) Any part of a structure or site feature that is above the adjacent finished ground level.

aboveground storage facility — A tank or other storage container, the bottom of which is on a plane not more than 6 inches below the surrounding surface.

ABP — Acronym for arterial blood pressure.

abrasion — Wearing away of any surface material by the scouring action of moving solids, liquids, gases, or combination thereof; a spot rubbed bare of skin or mucous membrane or superficial epithelium by rubbing or scraping.

abrasion hazard — (*swimming pools*) A sharp or rough surface that would scrape the skin by chance impact or normal use of a facility.

abrasive — Any of a wide variety of natural or manufactured substances of great hardness used to smooth, scour, rub away, polish, or scrub other materials; having qualities associated with abrasion.

abrasive blasting — The process of cleaning or finishing surfaces using sand, alumina, or steel in a blast of air.

ABS — See alkylbenzene sulfonate; black plastic pipe and fittings generally used in wastewater and drainage systems.

abscess — A localized collection of pus in a cavity formed by the disintegration of tissue. Abscesses are usually caused by specific microorganisms that invade the tissues through small wounds or breaks in the skin. An abscess is a natural defense mechanism by which the body attempts to localize an infection and isolate microorganisms to prevent spreading throughout the body.

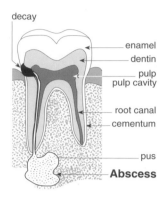

Abscess

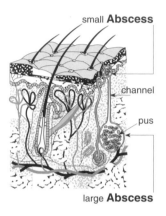

large Abscess

absenteeism — The absence from work of employees for health or related reasons, especially for influenza and occupationally related skin diseases.

absolute cavity radiometer — An instrument used for very accurate measurements of solar irradiance by adsorbing radiation on a blackened receiver, which is electrically self-calibrating.

absolute humidity — The mass (in grams) of water in a volume of air (in cubic meters); units are g/m^3.

absolute pressure — Gauge pressure plus atmospheric pressure.

absolute pressure transducer — A transducer that measures pressure in relationship to zero pressure where a vacuum exists on one side of the diaphragm.

absolute risk — An estimate of health risk by determining the rate of occurrence of a disease in the study population.

absolute temperature — The temperature in Celsius degrees relative to the absolute zero at –273.16°C or in Fahrenheit degrees relative to the absolute zero at –459.69°F.

absolute time — The geologic time expressed in years before the present.

absolute ventilation efficiency — A quantity that expresses the ability of a ventilation system to reduce pollution concentration relative to the feasible theoretical maximum performance.

absolute viscosity — A measure of the tendency of a fluid to resist flow regardless of its density.

absolute zero — The temperature (−273.16°C or −459.69°F) at which all gases theoretically contract to zero volume if Charles' law were obeyed exactly.

absorbance — The degree of absorption of light or other radiant energy by a medium through which the radiant energy passes.

absorbed dose — (*biology*) The amount of a substance penetrating across an absorption barrier of an organism by a physical or biological process; (*radiation*) the energy absorbed per unit mass in an irradiated medium; also known as gray (GY).

absorbent — The ability of attracting and absorbing substances.

absorber plate — A dark surface that absorbs solar radiation and converts it into heat.

absorbers — (*scrubbers*) A general classification of devices used in air pollution control to remove, treat, or modify one or more of the gaseous or vaporous constituents of a gas stream by bringing the stream into intimate contact with an absorbing liquid.

Countercurrent Packed Tower Absorber

absorbing media — The media in which the process of absorption and reflection of energy incident is completed.

absorptance — The ratio between the solar radiation absorbed by a surface and the total amount of solar radiation that strikes it.

absorption — A route of entry into the body through broken or unbroken skin. Any chemical can enter through a cut or abrasion; organic lead compounds, nitro compounds, and organic phosphate insecticides can enter through hair follicles; toluene and

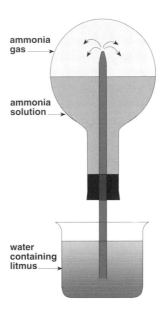

Absorption

xylene can be absorbed in fats and oils in the skin. (*radiation*) The phenomenon by which radiation imparts some or all of its energy to any material through which it passes; (*sound*) the ability of a material to absorb sound energy; (*physiology*) the process by which the end products of digestion, as well as other dissolved liquids and gases, enter the fluids and cells of an organism.

absorption barrier — Any of the body's barriers, such as skin, lung tissue, or gastrointestinal tract, allowing differential diffusion of substances across a boundary.

absorption capacity — The fraction of a radiant energy incident absorbed on a surface numerically equal to the emissivity for any given surface.

absorption coefficient — (*radiation*) The fractional decrease in the intensity of a beam of x- or gamma-radiation per unit thickness, per unit mass, or per atom of absorber due to deposition of energy in the absorber; (*acoustics*) the ratio of the sound energy absorbed by the surface of a medium or material to the sound energy incident on the surface; (*physics*) the constant for a material indicating the degree to which it absorbs radiation, atomic particles, rays of light, or sound; (*physiology*) the rate at which the human body absorbs a particular substance, usually expressed in milligrams per kilogram of body weight per hour.

absorption cross section — A measurement of the ability of an atom or molecule to absorb light at a specified wavelength measured in square centimeters per particle.

absorption field — An area through which septic tank effluent discharges through leaching or seepage into the surrounding ground by means of a series of perforated pipes laid in shallow trenches backfilled with gravel.

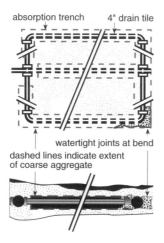

Absorption Field

absorption material — Any material that absorbs small amounts of sound energy, changing it into small amounts of heat energy, such as with the use of acoustical ceiling tile.

absorption of radiation — The uptake of radiation by a solid body, liquid, or gas; the energy may be transferred or re-emitted.

absorption spectrum — The range of electromagnetic energy that is used for spectral analysis including both visible light and ultraviolet radiation.

absorptivity — See absorption capacity.

abstract — A summary of a scientific article or other piece of writing.

abstraction reaction — (*chemistry*) A reaction that takes any atom away from another chemical species such as the gas phase removal of hydrogen from methane by hydroxyl radicals.

abuse — The improper use of equipment, materials, or services.

Ac — See altocumulus clouds.

AC — See alternating current; acronym for area conservationist.

AC pressurizing technique — A technique that allows building air-tightness to be examined at small pressure differentials with minimal interference from climatic forces. The air flow through a building can be evaluated by using a piston assembly to vary the effective volume of the structure and by measuring the amplitude of the pressure response inside the building and phase relationship between this pressure and the velocity of the piston, thus determining the air flow through the building.

AC scale sound level — A measurement of sound approximating the sensitivity of the human ear that is used to note the intensity or annoyance level of sound.

Acanthoscelides obtectus — See Bean Weevil.

acariasis — Any disease caused by a mite.

acaricide — A pesticide used to destroy mites on domestic animals, crops, and humans; also known as miticide.

acarid — One of the mites that are members of the order Acarina and may be parasitic and cause disease.

ACBM — Acronym for asbestos-containing building material.

accelerated bioremediation — A form of bioremediation within the subsurface at a given site that is accelerated beyond the normal actions of the naturally occurring microbial population and naturally occurring chemical, biological, and geological conditions.

accelerated erosion — The erosion of soil at a faster than natural rate.

acceleration — The rate of change of velocity with respect to time.

accelerator — (*chemistry*) A chemical additive that increases the speed of a chemical reaction; a device for providing a very high velocity to charged particles such as electrons or protons.

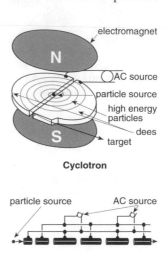

Cyclotron

Linear Accelerator

Accelerators

accelerogram — The record from an accelerometer showing acceleration as a function of time; the recording of the acceleration of the ground during an earthquake.

accelerograph — An instrument that records the acceleration of the ground during an earthquake; also called an accelerometer.

accelerometer — An instrument that measures the acceleration of a moving body; (*vibration*) an instrument that measures vibrations.

acceptable air quality — Air containing no known contaminants at harmful concentrations as determined by specialists and with which a substantial majority (80% or more) of the people exposed do not express dissatisfaction.

acceptable daily intake — The estimate of the largest amount of a chemical to which a person can be exposed on a daily basis that is not anticipated to cause adverse effects, usually expressed as milligrams per kilogram per day; the chemical ingestion level determined by combining the maximum no-observed-adverse-effect level with a safety factor where chemicals with these levels usually are not considered or suspected to be carcinogens.

acceptable intake for chronic exposure — An estimate similar in concept to the reference dose but derived using a less strictly defined methodology.

acceptable intake for subchronic exposure — An estimate similar in concept to the subchronic reference dose but derived using a less strictly defined methodology.

acceptable quality level (AQL) — The maximum percentage of regulated product that, for purposes of sampling inspection, can be considered as a process average.

acceptance test — (*geographic information system*) A test for evaluating the performance and conformity to specifications of a newly purchased computer system.

acceptor molecule — A molecule that accepts a phosphate.

accessible — Easily exposed for inspection and replacement of materials or parts; (*sewage*) a sewer located in a public right-of-way or easement contiguous to property; the service stub can be reached without excessive tunneling or boring under a roadway, building, or flowing stream, and the sewer or service stub is no farther than 200 feet from the residential building or 1000 feet from the commercial/industrial building to be served.

accessory cell — A cell required for, but not actually mediating, a specific immune response.

access time — (*computer science*) A measure of the time interval between the instant that data are called from storage and the instant that delivery is complete.

accident — An unplanned and usually unexpected occurrence that has the potential to cause damage to persons and/or property.

accident assessment — Evaluation of the nature, severity, and impact of an accident.

accident site — (*hazardous materials*) The location of an unexpected occurrence, failure, or loss resulting in a release of hazardous materials either at a plant or along a transport route.

accidental myiasis — The presence within a host of a fly not normally parasitic.

accidental parasite — A parasite found in animals or humans not normally hosts to the parasite.

acclimation — The process of biochemical adaptations enabling a species or population to withstand environmental changes such as changes in temperature, humidity, and pressure; also known as acclimatization.

acclimatization — See acclimation.

accommodation — (*optics*) The ability of the eye to adjust its focus for varying distances.

accordance agreement — The understood agreement by which a user of a pesticide agrees to follow labeled directions for proper application of the pesticide.

accreditation — The process by which an agency or organization evaluates a program of study or an institution as meeting certain predetermined standards.

accretion — The addition of air particles to hydrated drops, such as snow, rain, or sleet, by coagulation as the drops fall through the sky.

accumulated dose equivalent — An estimated lifetime maximum permissible dose of radiation for persons working with radioactive materials or x-rays (5 rems per year).

accumulated temperature difference — See degree day.

accumulation — The buildup of a chemical in the body due to long-term or repeated exposure.

accuracy — (*scientific technology*) The degree of agreement between a measured value and the true value of a calculation or reading; the extent to which a result is true when compared against a recognized standard.

accuracy of an input — (*engineering*) The maximum expected deviation of the indicated value of an input from the true value.

accuracy of an output — (*engineering*) The maximum expected deviation of the actual value of an output from the desired value.

Ace bandage — A woven elastic bandage used for control of injuries.

acenaphthylene ($C_{12}H_8$) — An odorless crystalline solid; MW: 152.20, BP: 509 to 527°F, Sol: in water, alcohol, ether, and benzene; sp. gr. 0.8988. A polycyclic aromatic hydrocarbon formed by catalytic dehydration of acenaphthene. It is a flammable/combustible material ignited by heat, sparks, or flames. Vapors may travel to a source of ignition and flash back. Exposure occurs in occupational settings from Coke ovens, coal tar, pitch, asphalt fumes, and carbon black and in a non-occupational setting from air, recreational activities, and contaminated waterways. It is hazardous to the skin, respiratory system, and eyes and is toxic by inhalation, ingestion, and contact. The symptoms of acute exposure include irritation of the eyes, nose, and throat; very high levels can cause headache, restlessness, lethargy, nausea, vomiting, anorexia, and anemia; permanent damage may occur in the liver and kidneys. Chronic exposure may cause headaches, fatigue, nausea, and skin rashes. It is a potential human carcinogen. NIOSH exposure limit: treat as a potential human carcinogen.

Acenaphtylene

acetaldehyde (CH₃CHO) — A colorless liquid or gas (above 69°F) with a pungent, fruity odor; MW: 44.1, BP: 69°F, Sol: miscible, Fl.P: –36°F, sp. gr. 0.79. It is used in the production of intermediates during the synthesis of acetic acid, acetic anhydride, and aldol compounds; during the manufacture of synthetic resins, in the synthesis of intermediates in the production of pesticides and pharmaceuticals, and in the synthesis of rubber processing chemicals; in coating operations in the manufacture of mirrors; as a hardening agent in photography and in the manufacture of gelatin, glue, and casein products; as a preservative in food products and leather. It is hazardous to the respiratory system, skin, and kidneys and is toxic by inhalation and ingestion; it is an eye, nose, and throat irritant. Symptoms of exposure include conjunctivitis, cough, central nervous system depression, eye and skin burns, dermatitis, delayed pulmonary edema; carcinogenic. OSHA exposure limit (TWA): 100 ppm [air].

acetate — A salt of acetic acid.

acetic — A substance having the sour properties of vinegar or acetic acid.

acetic acid (CH₃COOH) — A colorless liquid or crystal with a sour, vinegar-like odor; the pure compound is a solid below 62°F; it is often used in aqueous solution; MW: 60.1, BP: 244°F, Sol: miscible,

Fl.P: 102°F, sp. gr. 1.05. It is used in the production of acetic anhydride, vinyl acetate, and acetic esters; as an esterifying, acetylating, acidifying, and neutralizing agent; as a food additive or flavorant; in textile and dye industries as a solvent and intermediate, and in the production of nylon and acrylic fibers; in the manufacture of photographic chemicals; in the production of vitamins, antibiotics, and hormones; in the manufacture of rubber chemicals, accelerators, and as a coagulant of natural latex; in dry cleaning to remove rust; as a laboratory reagent in chemical and biochemical analysis; in field testing of lead fumes, aniline vapors, and separation of gases and in vinyl chloride determination; as a solvent for organic compounds. It is hazardous to the respiratory system, skin, eyes, and teeth and is toxic by inhalation. Symptoms of exposure include conjunctivitis, lacrimation, nose and throat irritation, pharyngeal edema, chronic bronchitis, eye and skin burns, skin sensitivity, dental erosion, black skin, and hyperkeratosis. OSHA exposure limit (TWA): 10 ppm [air].

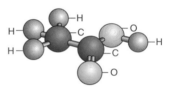

Acetic Acid

acetic anhydride ((CH₃CO)₂O) — A colorless liquid with a strong, pungent, vinegar-like odor; MW: 102.1, BP: 282°F, Sol: 12%, Fl.P: 120°F, sp. gr. 1.08. It is used in the manufacture of cellulose esters, fibers, plastics, lacquers, protective coating solutions, photographic films, cigarette filters, magnetic tape, and thermoplastic molding compositions; in the manufacture of pharmaceuticals and their intermediates; in organic synthesis as an acetylating, bleaching, and dehydrating agent; in synthesis of perfume chemicals, explosives, and weed killers. It is hazardous to the respiratory system, eyes, and skin and is toxic by inhalation, ingestion, and contact. Symptoms of exposure include conjunctivitis, lacrimation, corneal edema and opacity, nasal and pharyngeal irritation, cough, dyspnea, bronchitis, skin burns, vesiculation, sensitization, dermatitis. OSHA exposure limit (TWA): ceiling concentration 5 ppm [air].

acetogenic bacteria — A prokaryotic organism that uses carbonate as a terminal electron acceptor and produces acetic acid as a waste product.

acetone (CH₃COCH₃) — A colorless liquid with a fragrant, mint-like odor; MW: 58.1, BP: 133°F, Sol:

miscible, Fl.P: 0°F, sp. gr. 0.79. It is used during the application of lacquer, paints, and varnishes; during the use of solvents and cementing agents; during dip application of protective coatings; during fabric coating and dyeing processes. It is hazardous to the respiratory system and skin and is toxic by inhalation, ingestion, and contact. Symptoms of exposure include eye, nose, and throat irritation; headache; dizziness; and dermatitis. OSHA exposure limit (TWA): 750 ppm [air].

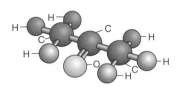

Acetone

acetone cyanohydrin (C₃H₆(OH)CN) — A colorless liquid; MW: 85.10, BP: 95°C, Sol: soluble in water and most organic solvents, Fl.P: 74°C (open cup); the vapor causes an explosive mixture in air; density: 0.932. It is used in the production of pharmaceuticals, foaming agents, and insecticides. It is hazardous to the respiratory tract, digestive tract, and skin. Inhalation of its vapors at high concentrations can cause instantaneous loss of consciousness and death. NIOSH exposure limit: 4 mg/15 min.

$$H_3C - \underset{\underset{\displaystyle OH}{|}}{\overset{\overset{\displaystyle CH_3}{|}}{C}} - C \equiv N$$

Acetone cyanohydrin

acetonitrile (CH₃CN) — A colorless liquid with an aromatic odor; MW: 41.1, BP: 179°F, Sol: miscible, Fl.P (open cup): 42°F, sp. gr. 0.78. It is used as a solvent; in organic synthesis in the preparation of vitamins, perfumes, water softeners, and plasticizers; as a catalyst; in low-temperature batteries, electrokinetic transducers, and angular accelerometers. It is hazardous to the kidneys, liver, cardiovascular system, central nervous system, lungs, skin, and eyes and is toxic by inhalation absorption, ingestion, and contact. Symptoms of exposure include asphyxia, nausea, vomiting, chest pain, weakness, stupor, convulsions, and eye irritation. OSHA exposure limit (TWA): 40 ppm [air].

acetophenone (C₈H₃O) — A colorless liquid; MW: 120.16; BP: 202°C; Sol: soluble in alcohol, ether, and chloroform; Fl.P: 82°C (closed cup); a com-bustible liquid; density: 1.033 at 15°C. It is used in making perfume and as a photosensitizer in organic synthesis. In rabbits, it causes severe eye irritation and can produce sleepiness. It is a mutagen. ACGIH: NA.

$$\langle\bigcirc\rangle - \underset{\underset{\displaystyle O}{\|}}{C} - CH_3$$

Acetophenone

acetylacetone (C₅H₈O₂) — A colorless liquid with a pleasant smell; MW: 100.12; BP: 140.5°C, Sol: soluble in alcohol, ether, acetone, and chloroform and moderately soluble in water; Fl.P: 34°C (closed cup); a flammable liquid; density: 0.972 at 25°C. It is used as a reagent for organic synthesis and as a transition metal chelating agent. It is hazardous to the eyes, mucous membranes, and skin. It causes irritation. ACGIH: NA.

$$CH_3 - \underset{\underset{\displaystyle O}{\|}}{C} - CH_2 - \underset{\underset{\displaystyle O}{\|}}{C} - CH_3$$

Acetyl acetone

2-acetylaminofluorene (C₁₅H₁₃NO) — A tan, crystalline powder; MW: 223.2, Sol: insoluble, BP: not known, Fl.P: not known. It is used in research and laboratory facilities. It is hazardous to the liver, bladder, kidneys, pancreas, skin, and lungs and is toxic by inhalation, absorption, ingestion, and contact. Symptoms of exposure include reduced function of liver, kidneys, bladder, and pancreas; an occupational carcinogen. OSHA exposure limit (TWA): lowest feasible concentration regulated use.

acetylation — A process for introducing acetyl groups into an organic compound containing –OH, –NH₂, or –SH groups.

acetylcholine (C₇H₁₇O₃N) — The acetic acid ester of choline normally present in many parts of the body and having important physiological functions. It is a neurotransmitter at the cholinergic synapses in the central sympathetic and parasympathetic nervous systems of many vertebrates.

acetylcholinesterase — An enzyme present in nerve tissue, muscles, and red blood cells that catalyzes the hydrolysis of the acetylcholine to choline and acetic acid, thus allowing neural transmission across synapses to occur.

acetylcholinesterase inhibitor — A compound or group of compounds (e.g., organphosphorus compounds) that block the action of the enzyme acetylcholinesterase, interfering with the transmission of impulses between nerve cells.

acetylene — See alkynes.

acetylene block assay — An estimate of denitrification by determining release of nitrous oxide from acetylene-treated soil.

acetylene tetrabromide ($CHBr_2CHBr_2$) — A pale yellow liquid with a pungent odor; a solid below 32°F; MW: 345.7, BP: 474°F (decomposes), Sol: 0.07%, sp. gr. 2.97. It is used as a catalyst or catalytic initiator in synthetic fibers; as a polymer additive for flameproofing and in flame-retardant polystyrenes, polyurethanes, and polyolefins; as a mercury substitute in the manufacture of gauges and balancing equipment; as a flotation agent in the processing and separation of mineral oils. It is hazardous to the eyes, upper respiratory system, and liver and is toxic by inhalation, ingestion, and contact; symptoms of exposure include eye and nose irritation, anorexia, nausea, severe headache, abdominal pain, jaundice, monocytosis. OSHA exposure limit (TWA): 1 ppm [air].

acetylene-reduction assay — An estimate of nitrogenase activity by measuring the rate of acetylene reduced to ethylene.

acetylene welding — See oxyacetylene welding.

ACFM — Acronym for actual cubic feet per minute.

ACH — A unit used for quantifying airflow; see air changes per hour.

ache — A pain characterized by a persistent, dull, moderate intensity.

AChE — See acetylcholinesterase.

achievement test — A standardized test for the measurement and comparison of knowledge or proficiency in various vocational or academic fields.

acid — A corrosive substance; any of a class of chemical compounds for which the aqueous solutions turn blue litmus paper red and which yields hydrogen ions, reacts with and dissolves certain metals to form salts, and reacts with bases to form water and salt.

acid aerosol — An acidic liquid or solid particle small enough to become airborne; high concentrations can be irritating to the lungs and are associated with respiratory diseases such as asthma.

acid–base balance — A state of equilibrium between acidity and alkalinity of the body fluids; also known as hydrogen ion balance.

acid–base metabolism — Metabolic process that maintains the balance of acids and bases essential in regulating the composition of body fluids.

acid bottom and lining — A melting furnace's inner bottom and lining composed of refractory materials that produce an acid reaction in the melting process.

acid burn — The damage to tissue caused by exposure to an acid, with the severity being determined by the strength of the acid and the duration and extent of exposure.

acid demand — A measure of the amount of acid required to reduce the pH to a predetermined level.

acid deposition — A serious environmental and economic problem that occurs when an emission of sulfur dioxide and/or nitrogen oxide interacts with sunlight, chemical oxidants, and water vapor in the upper atmosphere to form acidic substances which then fall to earth as acid rain, acid snow, or dry acid deposition.

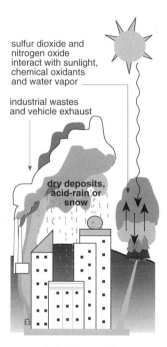

sulfur dioxide and nitrogen oxide interact with sunlight, chemical oxidants and water vapor

industrial wastes and vehicle exhaust

dry deposits, acid-rain or snow

Acid Deposition

acid detergent — A material made up of synthetic detergents and hydrochloric acid used for the removal of mineral deposits from surfaces containing water deposits, milk stone, and any other hardwater films. It is also used for deliming dishwashing machines and similar equipment and for sanitizing toilet bowls and urinals.

acid digestion — The first phase of biogas production in which complex molecules are broken down into smaller ones.

acid dust — An accumulation of highly acidic particles of dust that may contribute to respiratory illnesses and be part of acid rain.

acid equivalent — The amount of active ingredient expressed in terms of the parent acid.

acid extraction treatment system (AETS) — A process of soil washing that uses hydrochloric acid to extract contaminants from soils.

acid-fast bacillus — A type of bacteria that resists decolorizing by acid after accepting a stain.

acid-fast bacteria — Bacteria, especially *Mycobacterium tuberculosis*, that resist decolorizing by acids after being stained with basic dyes.

acid flush — A runoff of precipitation with a high acid content which may pollute rivers, reservoirs, and other bodies of water, killing fish and endangering ecosystems.

acid-forming bacteria — Organisms fermenting in the lactose of milk which produce mainly lactic acid. The lactic acid combines with the protein calcium caseinate to liberate free protein casein, which is insoluble in water and is precipitated in the form of smooth gelatinous curds.

acid hydrolysis — A chemical process that uses acid to convert starch to sugar.

acid mine drainage — Water low in dissolved oxygen containing dissolved iron pyrites (ferrous sulfide) left behind from coalmining operations that becomes acidic and deposits ferric hydroxide on stream bottoms.

acid mist — A mist containing a high concentration of acid or particles of any toxic chemical; chemicals often used by industry and stored in tanks that may leak their contents into residential areas and become especially dangerous if combined with fog.

acid-neutralizing capacity — The measure of ability of a base to resist changes in pH.

acid pickling — See pickling.

acid poisoning — A toxic condition caused by the ingestion of a toxic acid agent, such as hydrochloric, nitric, phosphoric, or sulfuric acids, which may be found in common household cleaning compounds.

acid rain — Rain that contains relatively high concentrations of acid-forming air pollutants, such as sulfur oxides and nitrogen oxides; acid rain may have a pH level as low as 2.8 as compared to normal rain pH of 6.

acid salt — A salt formed by replacing a part of the hydrogen ions of a polybasic acid with positive ions.

acid soil — A soil with a pH value of <6.6.

acidemia — An abnormally high level of acid in the blood.

acidic — The condition of water or soil that contains a sufficient amount of acid substances to lower the pH below 7.0.

acidic anhydride — An oxide that reacts with water to form an acid.

acidify — To convert into an acid.

acidity — The quality, state, or degree of being acid.

acidity control (food) — The adjustment of acidity in order to achieve desired flavor, texture, or cooking performance.

acidophile — An organism that grows best under acid conditions.

acidophilic — Referring to any substance, tissue, or organism easily stained with acid dyes; an organism showing a preference for an acid environment.

acidophilus milk — Milk that is inoculated with cultures of *Lactobacillus acidophilus* and is used in various enteric disorders to change the bacterial flora of the gastrointestinal tract.

acidosis — A pathological condition resulting from an accumulation of acid or depletion of alkaline reserves in the blood and body tissues; characterized by an increase in hydrogen ion concentration or a decrease in pH.

acid-pulse — Deposition of a powder-like substance over the ground surface which affects plant leaves and lowers the pH of water.

Acinetobacter — A genus of nonmotile aerobic bacteria of the family Neisseriaceae that often occurs in clinical specimens containing Gram-negative or Gram-variable cocci; does not produce spores.

ACL — Acronym for alternate concentration limit.

ACLF — Acronym for adult congregate living facility.

ACLS — See advanced cardiac life support.

ACM — Acronym for asbestos-containing material; acronym for advanced climate model.

acme — The peak or highest point.

acne — An inflammatory disease of sebaceous glands of the skin.

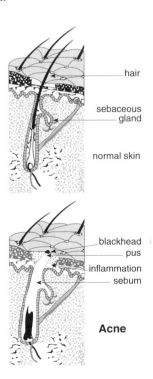

hair

sebaceous gland

normal skin

blackhead

pus

inflammation

sebum

Acne

A

acneiform — Resembling acne.

ACOP — Acronym for approved code of practice.

acoustic — Related to or associated with sound.

acoustical calibrator — A device compatible with the microphone of a personal noise dosimeter and which emits one or more sound pressure levels to provide a calibration check on the general operation of a personal noise dosimeter set.

acoustic center — The portion of the brain in the temporal lobe of the cerebrum in which the sense of hearing is located.

acoustic coupler — A device that enables a computer terminal to be linked to another over the telephone system via the handset of a conventional telephone.

acoustic hair cell — See auditory hair.

acoustic impedance — The effects of interference with the passage of sound waves (such as those generated by ultrasound equipment) by objects in the path of the sound waves.

acoustic meatus — The external or internal canal of the ear.

acoustic microscope — A microscope in which the object being viewed is scanned with sound waves and its image reconstructed with light waves, allowing a very close examination of cells and tissues without staining or damaging the specimen.

acoustic nerve — Either of a pair of cranial nerves composed of fibers from the cochlear nerve and the vestibular nerve in the inner ear containing impulses of the sense of hearing and the sense of balance.

acoustic trauma — Hearing loss caused by a sudden, loud noise in one ear or a sudden blow to the head resulting in temporary loss or permanent damage.

acoustics — The science of sound, including its generation, transmission, and effects on the environment; the characteristics found within a structure that determine the quality of sound within it relevant to hearing.

ACQR — Acronym for air quality control region.

acquired — A condition or disease caused by a reaction to environmental influences outside of the organism.

acquired immunity — Any type of immunity that is not inherited.

acquired immunodeficiency syndrome (AIDS) — (*disease*) A disease that is insidious and causes nonspecific symptoms such as anorexia, chronic diarrhea, weight loss, fever, and fatigue; incubation time is not known. It is caused by the retrovirus T-lymphotropic virus type III (HTLV-III) and occurs throughout the world, with the reservoir of infection being people. It is transmitted through sexual contact, through the use of unclean needles, and through blood transfusions; communicability, receptability, and resistance are not known. Its spread is controlled by the practice of safe sex, the screening of blood donors, and the avoidance of contact with the body fluids of an infected individual.

acquired trait — A physical characteristic that is not inherited and may be due to the environment or a mutation.

acre — A unit of measure equivalent to 43,560 square feet.

acre-foot — The volume of water required to cover 1 acre of land with 12 inches of water, equal to 325,851 gallons.

acrid — A sharp or pungent, bitter, and unpleasant smell or taste.

acrimony — A quality of bitterness, hardness, or sharpness.

ACRIMSAT — See Active Cavity Radiometer Irradiance Monitor Satellite Mission.

acrolein (CH_2CHCHO) — A colorless or yellow liquid with a piercing, disagreeable odor; MW: 56.1, BP: 127°F, Sol: 40%, Fl.P: –15°F, sp. gr. 0.84. It is used as a fungicide. It is hazardous to the heart, eyes, skin, and respiratory system and is toxic by inhalation, ingestion, and contact. Symptoms of exposure include eye, skin, and mucous membrane irritation; abnormal pulmonary function; delayed pulmonary edema; and chronic respiratory disease. OSHA exposure limit (TWA): 0.1 ppm [air].

acrophobia — A fear of high places that results in extreme anxiety.

acrylamide ($CH_2CHCONH_2$) — A white crystalline, odorless solid; MW: 71.1, BP: 347 to 572°F (decomposes), Sol. (86°F): soluble in water, Fl.P: 280°F, sp. gr. (86°F): 1.12. It is used in the manufacture of copolymers and polyacrylamides; as a grouting material in oil well drill holes, basements, tunnels, mine shafts, and dams. It is hazardous to the central nervous system, peripheral nervous system, skin, and eyes and is toxic by inhalation, absorption, ingestion, and contact. Symptoms of exposure include ataxia, numbness of the limbs, paresthesia, muscle weakness, absent deep tendon reflex, hand sweating, fatigue, lethargy, and eye and skin irritation; carcinogenic. OSHA exposure limit (TWA): 0.03 mg/m³ [skin].

acrylic — A family of synthetic resins made by polymerizing esters of acrylic acids.

acrylic acid ($C_3H_4O_2$) — A colorless liquid with an acrid odor; MW: 72.07, BP: 141°C, Sol: miscible with water, alcohol, ether, and other organic solvents, Fl.P: 130°F (closed cup); vapors form explosive mixtures with air; density: 1.052. Used in the manufacture of plastics, paints, polishes, adhesives and as a coating for leather. It is hazardous to the eyes, respiratory system, and skin. It damages vision and the respiratory tract and causes skin burns. ACGIH exposure limit (TLV-TWA): 10 ppm.

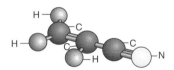

$$CH_2 = CH - \underset{\underset{O}{\|}}{C} - OH$$

Acrylic Acid

acrylic latex — An aqueous dispersion of acrylic resins in latex.

acrylonitrile (CH_2CHCN) — A colorless to pale yellow liquid with an unpleasant odor that can only be detected above the permissible exposure limit; MW: 53.1, BP: 171°F, Sol: 7%, Fl.P: 30°F, sp. gr. 0.81. It is used in the manufacture of ABS resin, SAN resin, plastic, or surface coating materials; in the manufacture and transfer of monomer to other reaction vessels or tank cars; as a chemical intermediate. It is hazardous to the cardiovascular system, liver, kidneys, central nervous system, skin, brain (tumor), and lung and bowel (cancer) and is toxic by inhalation absorption, ingestion, and contact. Symptoms of exposure include asphyxia, eye irritation, headache, sneezing, nausea, vomiting, weakness, lightheadedness, skin vesiculation, scaling dermatitis; carcinogenic. OSHA exposure limit (TWA): 2 ppm [air].

Acrylonitrile

Acrylonitrile Group — An organization of producers and users of acrylonitrile concerned with health, safety, environmental, and other regulatory matters regarding the manufacture, distribution, storage, and use of the chemical.

act — (*law*) The general term describing legislative enactment that provides the authority for a particular regulatory activity.

ACTD — Acronym for Advanced Concept Technology Demonstration.

ACTH — See adrenocorticotropic hormone.

act of God — An extraordinary, unpredicted, and unpreventable interruption, such as a flood or tornado, of normal activities.

Act to Require Aircraft Noise Abatement Regulation — A law passed in 1968 and amended several times until it became the Quiet Communities Act of 1978; it is used to abate aircraft noise and sonic booms.

actinic dermatitis — A skin inflammation or rash resulting from exposure to sunlight, x-ray, or atomic particles and which may lead to skin cancer.

actinometer — An instrument used to measure direct radiation from the sun.

actinomycetes — A large group of mold-like microorganisms that give off an odor characteristic of rich earth and are the significant organisms involved in the stabilization of solid wastes by composting; any of any order of filamentous or rod-shaped bacteria including actinomyces and streptomyces.

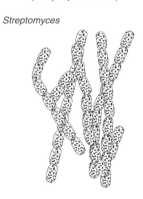

Actinomyces israelli
(causal agent of human actinomycosis)

Actinomycetes
(not proportionate)

Streptomyces

action current — See action potential.

action level — The level or concentration of chemical residue in food or feed above which adverse health effects are possible and above which corrective action is taken by the FDA; the level of toxicant that requires medical surveillance; (*pesticides*) the level set for inadvertent residues resulting from previous legal use or accidental contamination at which federal enforcement action is taken against the contaminated food or agricultural commodity.

action potential — The sequence of electrical changes occurring when a nerve cell membrane is exposed to a stimulus that exceeds its threshold.

A

activate — The ability to induce or prolong an activity or render an optimal action and result.

activated alumina — A granular adsorbent consisting mostly of highly porous aluminum oxide formed by heating aluminum hydroxide or most aluminum salts of oxy acids; it is especially useful for drying gases.

activated carbon — A specially treated and finely divided form of amorphous carbon that possesses a high degree of adsorption due to its large surface area per unit volume; also known as activated charcoal.

activated charcoal — See activated carbon.

activated sludge — A semi-liquid mass removed from a liquid flow of sewage that was subjected to aeration and aerobic microbial action.

activated sludge process — A system of biological treatment that utilizes bacteria and other organisms to consume impurities in sewage; activated sludge is mixed with sewage in aerators and impurities are absorbed by flocculent particles and consumed by the bacteria living in the floc.

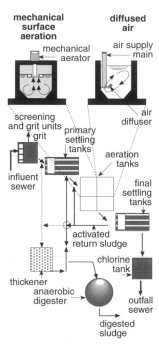

Activated Sludge Process

activation — The process of causing a substance to become artificially radioactive by bombarding it with protons or neutrons in a reactor.

activation energy (chemistry) — The amount of energy required to bring all molecules in 1 mole of a substance to their reactive state at a given temperature; the energy barrier that must be overcome during a collision of two potential reactants in order for a reaction to occur.

activation products — A material that has become radioactive through the process of activation.

activational effects — The effects of hormones that act as triggering influences for the expression of particular behavior patterns, with response being rapid (hours or days).

activator — A material added to a pesticide chemical formulation to increase its effect or toxicity.

active avoidance learning (rodents) — A conditioning response where the animal has to act in order to avoid some noxious consequence.

active carrier — A person without signs or symptoms of infection who carries the microorganism.

Active Cavity Radiometer Irradiance Monitor Satellite Mission — A NASA mission to measure total solar irradiance over a 5-year period.

active immunity — Disease resistance in an individual due to antibody production after exposure to a microbial antigen following disease.

active immunization — Stimulation with a specific antigen to promote antibody formation in the body.

active ingredient — (*pesticides*) The chemical or agent in a pesticide mixture that will destroy or prevent damage by pests.

active maintenance time — The time during which one or several persons work on a piece of equipment to carry out maintenance.

active material — (*pesticides*) See active ingredient.

active recreation — An area where moderate to high intensity-structured recreational use occurs.

active sensor — See active system.

active site — The place on the surface of an enzyme where catalytic action occurs.

active solar energy system — A system that requires the importation of energy from outside of the immediate environment, such as the energy needed to operate fans and pumps in solar heat.

active solar heating — The warming of an interior space with heat collected mechanically through a solar collector.

active system — A remote-sensing system that transmits its own radiation to detect an object or area for observation and receives the reflected or transmitted radiation. Radar is an example of an active system.

active transport — From a lower to a higher concentration by means of energy.

activities of daily living — Activities usually performed during a normal day, such as eating, dressing, washing, and household activities, which may be compromised by chronic illnesses and accidents.

activity — (*radiation*) The number of nuclear disintegrations occurring in a given quantity of material per unit time; see also curie.

activity coefficient — A proportionality constant relating activity to concentration.

activity plan — Written procedures of the asbestos-management plan of a school that detail the steps a local education agency will follow in performing initial and additional cleaning, operation and maintenance, periodic surveillance, and reinspection required by the Asbestos Hazard Emergency Response Act.

ACTM — Acronym for atmospheric chemical transport model.

ACTS — See Advanced Communications Technology Satellite.

actual dosage — The amount of a pure pesticide chemical in proper dilution used per unit area.

actuator — A mechanical device that moves or controls something.

acuity — The sensitivity of receptors in hearing or vision.

acupuncture — A method of producing analgesic effects or altering the function of a system of the body by inserting fine, wire-thin needles into the skin at specific sites on the body along a series of lines or channels.

acute abscess — A collection of pus in a body cavity accompanied by localized inflammation, pain, pyrexia, and swelling.

acute angle — Any angle of <90 degrees.

acute care — Care for individuals whose illnesses or health problems are of a serious nature; the care lasts until the patient is stabilized.

acute circulatory failure — A drop in heart output resulting from cardiac or noncardiac causes and leading to tissue hypoxia.

acute dermal LD$_{50}$ — A single dose of a substance, expressed as milligrams per kilogram of body weight, that is lethal to 50% of the test population of animals under specific test conditions.

acute diarrhea — A clinical syndrome of diverse etiology associated with loose or watery stools and often vomiting and fever.

acute disease — A disease with a relatively short duration of symptoms and which is usually severe.

acute dose — A quantity that is administered in a single day or in one dose.

acute effects — The effects that occur quickly and have a relatively short severe course.

acute exposure — Exposure to a chemical for a duration of 14 days or less.

acute health effect — A single exposure to a toxin or microorganism that results in an adverse effect on a human or animal and produces severe symptoms that develop rapidly and become critical quickly.

acute hemorrhagic conjunctivitis — A highly contagious eye disease usually caused by enterovirus type 70 and found primarily in densely populated human areas, especially in developing countries or places with a large immigrant population; produces symptoms of ocular pain, itching, redness, photophobia, edema of the eyelid, and profuse watery discharge.

acute hypoxia — A sudden or rapid depletion in available oxygen at the tissue level resulting from asphyxia, airway obstruction, acute hemorrhage, blockage of the alveoli by edema or infection, or abrupt cardio-respiratory failure.

acute infection — A relatively brief infection lasting a few days to a few weeks; afterwards, the microorganism is usually eliminated completely from the body by the immune system.

acute LC$_{50}$ — Concentration of a substance, expressed as parts per million parts of medium, that is lethal to 50% of the test population of animals under specified test conditions.

acute myocardial infarction — An early critical state of the necrosis of a portion of cardiac muscle caused by obstruction in a coronary artery.

acute nonbacterial infectious gastroenteritis — Generic term for an infection caused by astrovirus, calicivirus, enteric adenovirus, parvovirus, or viral gastroenteritis.

acute pain — Severe pain that may follow surgery, trauma, myocardial infarction, or other conditions and diseases.

acute radiation exposure — An exposure of short duration to intense ionizing radiation usually occurring as a result of an accident; exposure of the whole body to approximately 100 grays causes neurological and cardiovascular breakdown and fatality within 24 hours.

acute respiratory failure — A sudden inability of the lungs to maintain normal respiratory function caused by an obstruction in the airways or failure of the lungs to exchange gases in the alveoli.

acute toxicity — Relating to the acute effects resulting from a single dose of or exposure to a poisonous substance; the amount of a pesticide that will seriously affect or destroy a test animal in a single dose.

acutely hazardous waste — The wastes that are determined to be so dangerous in small amounts that they are regulated in the same way as large amounts of other hazardous wastes

acutely toxic chemicals — Chemicals that can cause both severe short- and long-term health effects after a single brief exposure; they can cause damage to living tissue, impairment of the central nervous system, severe illness, or, in extreme cases, death when injected, inhaled, or absorbed through the skin.

ACWM — Acronym for asbestos-containing waste material.

A/D converter — See analog-to-digital converter.

ad valorem **tax** — A tax on goods imposed at a rate percent of value.

ADALT — Acronym for Advanced Radar Altimeter.

adaptation — (*evolution*) The occurrence, by natural selection, of genetic or behavioral changes in a population or species in response to a new or altered environment.

adapter — A piece of equipment used for attaching parts that do not match.

ADC — See adult day care; see analog-to-digital converter.

additive — A substance introduced into a basic medium to enhance, modify, suppress, or otherwise alter some property of that medium; a substance added to a food to enhance flavor, taste, or quality.

additive effect — An effect in which the combined effect of two chemicals may be greater than the sum of the agents acting alone.

add-on control device — An air pollution control device such as a carbon absorber that reduces the pollution in an exhaust gas.

address (computer science) — A number or code that defines the position of information in the memory of a computer; the identification of a physical or virtual distinct entity in a network. On the Internet, this network address is referred to as a URL.

addressability — (*geographic information system*) The number of positions on the *x*- and *y*-axis on a visual display unit or graphics screen.

addressable point — (*geographic information system*) A position on a visual display unit that can be specified by absolute coordinates.

adduction — The movement toward something.

adenine — A purine base; a component of nucleotides and nucleic acids. A nitrogen base is found in both DNA and RNA.

adenitis — Inflammation of a gland or lymph node.

adenocanthoma — A carcinoma in which some or a majority of the cells show squamous differentiation.

adenocarcinoma — A malignant tumor with glandular elements.

adenofibrosis — A fibroid change in a gland.

adenoma — An epithelial tumor, usually benign, with a gland-like structure.

adenosine diphosphate (ADP) ($C_{10}H_{15}N_5O_{10}P_2$) — A low-energy compound, found in cells, that functions in energy storage and transfer; a nucleotide produced by the hydrolysis of adenosine triphosphate.

adenosine monophosphate (AMP) — A component of nucleic acid consisting of adenine, ribose, and phosphoric acid involved in energy metabolism and nucleotide synthesis; also known as adenylic acid.

adenosine triphosphate (ATP) ($C_{10}H_{16}N_5O_{13}P_3$) — A nucleotide in all cells that stores energy from oxidation and glucose in the form of high-energy phosphate bonds. The release of these phosphate bonds supplies free energy to drive metabolic reactions, especially those involving muscular activity, or to transport molecules against concentration gradients (active transport) when ATP is hydrolyzed to ADP, an inorganic phosphate, or AMP, an inorganic pyrophosphate. ATP is also used to produce high-energy phosphorylated intermediary metabolites, such as glucose 6-phosphate.

adenosis — Any disease of a gland, especially one involving the lymph nodes.

adenovirus — One of the 33 medium-sized viruses of the Adenoviridae family that are pathogenic to humans; causes conjunctivitis, upper respiratory infection, or gastrointestinal infection.

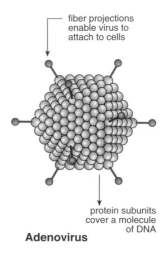

fiber projections enable virus to attach to cells

protein subunits cover a molecule of DNA

Adenovirus

adenylic acid — See adenosine monophosphate.

adequately wet — An asbestos-containing material that is sufficiently mixed or penetrated with liquid to prevent the release of particulates.

adequately wetted — Referring to a substance that has been sufficiently mixed or coated with water or an aqueous solution to prevent dust emissions.

ADH — See antidiuretic hormone.

adhere — To stick together or become fastened together.

adherence — The process in which a person follows rules, guidelines, or standards.

adherent — The tendency of a surface to cling to the surface of another substance.

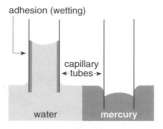

adhesion (wetting)

capillary tubes

water   mercury

Adhesion

adhesion — The force of molecular attraction that holds the surfaces of two or more substances in contact, such as water and rock particles; the ability of one substance to stick to another.

adhesive — A substance that causes close adherence of adjoining surfaces.

ADI — Acronym for acceptable daily intake of a substance, such as a contaminant in water or food.

adiabatic lapse rate — The −5.4°F theoretical change in air temperature for each 1000-foot increase in altitude. Also known as the dry-adiabatic lapse rate.

adiabatic — The cooling of air as it rises; the warming of air as it descends.

adient — A tendency to move toward rather than away from stimuli.

adipic — Fatty tissue.

adiponitrile ($C_4H_8(CN)_2$) — A colorless liquid; MW: 108.16; BP: 295°C; Sol: soluble in alcohol, chloroform, and benzene; Fl.P: greater than 94°C; forms an explosive mixture with air; density: 0.965 at 20°C. It is used to make amino resins. It is hazardous to the digestive tract and respiratory tract and causes symptoms of nausea, vomiting, respiratory tract irritation, and dizziness. NIOSH (TLV-TWA): 4 ppm.

$$N \equiv C - CH_2 - CH_2 +$$
$$+ CH_2 - CH_2 - C \equiv N$$

Adiponitrile

adipose — A collection of fat cells; fatty; of or relating to fat.

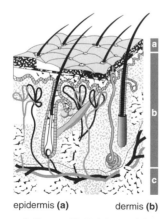

epidermis **(a)**   dermis **(b)**

Adipose (fatty) layer (c)

aditus — An opening or entrance.

adjacent — Neighboring; having a common border.

adjoining — See adjacent.

adjournment — (*law*) The putting off or postponing of business until another time or place; the act of a court; to defer; closing formally.

adjudicate — To carry a lawsuit to a conclusion.

adjunct — An additional substance, treatment, and procedure for increasing the efficiency or safety of a primary substance, treatment, or procedure.

adjusted decibel (dBA) — Sound level in decibels read on the A-scale of a sound-level meter.

adjusted rate — Rate calculated in such a way as to permit a more accurate comparison of the state of health of two or more populations that differ substantially in some important respect, such as distribution by age or sex; see also standard rate.

adjuvant — A chemical or agent added to a pesticide mixture that enhances the action of the active ingredient (e.g., wetting agent, spreader, adhesive, emulsifying agent, penetrant).

adjuvant therapy — A treatment that is given in addition to the main treatment to make it work better.

ADL — See activities of daily living.

ADME — Acronym for the process of absorption, distribution, metabolism, and excretion.

administered dose — The amount of a substance given to a test subject to determine dose–response relationships; administration for the replacement or supplementation of ions necessary for homeostasis.

administration of parenteral fluids — The intravenous infusion of various solutions to maintain adequate hydration, restore fluid volume, reestablish lost electrolytes, or provide partial nutrition.

administrative controls — (*ergonomics*) Changes in the way that work in a job is assigned or scheduled to reduce the magnitude, frequency, or duration of exposure to ergonomic risk factors; accomplished by use of employee rotation, job task enlargement, alternative tasks, and employer-authorized changes in work pace.

administrative hearing — A hearing held by an authorized agency in the executive branch of government to correct a violation of the rules and regulations of that agency.

administrative law — Rules and regulations in a given area established by an agency authorized by the legislature.

administrative order — A legal document signed by a governmental agency directing an individual, business, or other group to take corrective action or refrain from an activity.

administrative order on consent — A legal agreement signed by the U.S. EPA and an individual, business, or other entity in which the violator agrees to pay for correction of violations, take the required corrective or cleanup actions, or refrain from an activity.

administrative record — All of the documents that a governmental agency considers or relies on in selecting the remedy for a given environmental problem.

A

admittance (Y) — The ratio of current to voltage, which includes information on the relative phase of these quantities; the reciprocal of impedance; expressed in mhos.

admixture — A liquid or finely ground solid material combined into a particular concrete, mortar, or grout mix formula in predetermined, minute, and very controlled quantities, resulting in a major change in the behavior of the resulting product.

admonition — A reprimand from a judge to a person; warning of the consequences of conduct and punishment with greater severity if the same problem reoccurs.

adoptive immunity — The immune state created by inoculation of lymphocytes, not antibodies, from an immune animal rather than exposure to the antigen.

adrenal cortical steroids — The steroid hormones produced in the cortex of the adrenal gland.

adrenal glands — Pair of endocrine glands, located next to the kidneys in the abdomen, the cortex of which produces steroid hormones involved in water balance, glucose metabolism, and electrolyte balance. The adrenal medulla produces adrenaline and noradrenaline, which are involved in glucose metabolism, heart rate, and blood pressure.

adrenaline — See epinephrine.

adrenocorticotropic hormone (ACTH) — A pituitary hormone that stimulates the adrenal cortex to secrete its hormones. Also known as adrenotropic hormone.

adrenotropic hormone — See adrenocorticotropic hormone.

ADRV — See adult rotavirus.

adsorbent — A substance that is used to collect other substances, such as gaseous pollutants.

adsorption — The process by which gases, liquids, and dissolved substances are attracted to the surface of a solid and retained there; the process in which the molecular or ionic species are accumulated on a surface and bound to the surface by forces of molecular attraction.

adsorption coefficient (K_{oc}) — The ratio of the amount of a chemical adsorbed per unit weight of organic carbon in soil or sediment to the concentration of the chemical in solution at equilibrium.

adsorption ratio (K_d) — The amount of a chemical adsorbed by a sediment or soil in the solid phase divided by the amount of chemical in the solution phase which is in equilibrium with the solid phase at a fixed solid/solution ratio.

ADSS — See Advanced Decision Support System.

ADT — Acronym for average daily traffic.

adult — (*pests*) A full-grown, sexually mature insect, mite, nematode, or animal.

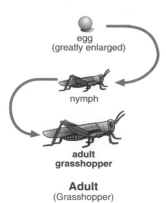

egg
(greatly enlarged)

nymph

**adult
grasshopper**

Adult
(Grasshopper)

adult day care (ADC) — A community-based, structured, comprehensive program for adults offered in a protective setting during daylight hours.

adult respiratory distress syndrome — A respiratory failure of sudden onset characterized by leakage of plasma into the lungs through damaged capillaries, resulting in fluid accumulation and depriving the lungs of their ability to expand.

adult rotavirus — A form of rotavirus that causes severe diarrhea in adults; it resembles the usual rotavirus and its genome but is not antigenically related to it and does not react against rotavirus antibodies.

adulterants — Chemicals or substances that by law do not belong in a food, plant, animal, or pesticide formulation.

adulterate — To debase or make impure by the addition of a foreign or inferior substance.

adulterated — Food that is made impure by the addition of a foreign or inferior substance.

adulterated and misbranded milk and milk products — Any milk or cream to which water has been added; milk or milk products that contain any unwholesome substance.

adulterated device — See adulterated drug.

adulterated drug — A drug or device consisting in whole or in part of any filthy, putrid, or decomposed substance. Also known as adulterated device.

adulterated food — Food bearing or containing any poisonous or deleterious substance that may render it injurious to health.

adulteration — The process of adulterating.

advance notice of proposed rule-making (ANPRM) — Preliminary notice that an agency is considering a regulatory action.

advanced cardiac life support — An emergency medical procedure in which basic life support efforts of cardiopulmonary resuscitation are augmented by establishment of an intravenous fluid line, drug administration, control of cardiac arrhythmias, and the use of ventilation equipment.

Advanced Communications Technology Satellite — The first all-digital-capacity communications satellite carrying digital communications with a fine quality of transmission using standard fiberoptic data rates. It pioneered advanced onboard traffic switching and processing.

Advanced Decision Support System — (*computer science*) Computer software designed to provide easy access, thus allowing efficient use of methods of analysis and information management.

Advanced Spaceborne Thermal Emission and Reflection Radiometer — An imaging instrument that is part of NASA's Earth Observing System and is used to obtain detailed maps of land surface, temperature, emissivity, reflectance, and elevation.

Advanced Very-High-Resolution Radiometer — A five-channel scanning instrument that quantitatively measures electromagnetic radiation. It is carried on NOAA environmental satellites and remotely determines cloud cover and surface temperature using the visible and infrared parts of the spectrum.

advanced waste treatment — Any biological, chemical, or physical treatment process used during any stage of treatment that employs unconventional, yet acceptable, techniques; also known as tertiary treatment.

advanced wastewater treatment (AWT) — A series of treatment techniques for the removal of additional contaminants from wastewater that has passed through a secondary treatment process to produce an effluent equivalent to potable water and which will meet drinking water standards; also known as tertiary wastewater treatment.

advect — A horizontal movement of a mass of fluid such as ocean or air currents; also refers to the horizontal transport of pollution, phytoplankton, ice, or even heat by such movement.

advection — (*meteorology*) The transfer of heat, cold, or other atmospheric properties by the horizontal motion of an air mass.

adventitious opening — An opening within the building envelope which in terms of ventilation is unintentional, such as cracks around doors and windows; also known as unintentional opening and fortuitous leakage.

adverse drug effect — A harmful, unintended reaction to a drug administered at a normal dosage.

adverse effect — A biochemical change, functional impairment, or pathological lesion that impairs performance and reduces the ability of a person to respond to additional challenges; an action that has an apparent direct or indirect negative effect on the conservation or recovery of an ecosystem component listed as threatened or endangered by the U.S. Forest Service.

adverse effect level — An exposure level at which statistically or biologically significant increases in frequency or severity of deleterious effects are observed between the exposed population and the control group.

adverse effects data — Required under the FIFRA law regarding pesticide registration, data that must be submitted to the U.S. EPA concerning any studies or other information regarding unreasonable adverse effects of a pesticide at any time after the registration.

adverse level — Measurement of the first discernible effects of air pollution likely to lead to symptoms of discomfort in humans.

advisory — A nonregulatory document that provides risk information to anyone who may have to make risk management decisions.

adynamia — Lack of strength due to a pathological condition.

AEA — Acronym for Atomic Energy Act.

AECD — See auxiliary emission control device.

Aedes — A genus of mosquito, one species of which transmits the virus that causes yellow fever in humans.

Eggs

Larva

Pupa

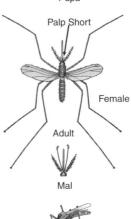

Palp Short

Female

Adult

Mal

Resting Position

Aedes **Life Cycle**

AEL — See adverse effect level; see airborne exposure limits.

aeolian — Materials carried, deposited, produced, or eroded by the wind; also spelled *eolian*.

aeolian deposits — Wind-deposited materials such as dune sands and loess deposits.

AER — Acronym for atmospheric and environmental research.

aerate — To bring in contact with air; to loosen or stir up soil mechanically for the purpose of introducing air; to expose to chemical action with oxygen; oxygenate by respiration.

aerated lagoon — See aerated pond.

aerated pond — A natural or artificial wastewater treatment pond or basin in which mechanical or diffused air aeration is used to supplement oxygen supply; also known as aerated lagoon.

aeration — Purifying of water by spraying it into the air or by passing air through it; the process of exposing a bulk material, such as compost, to air or of charging a liquid with a gas or a mixture of gases.

aeration tank — A chamber for injecting air into water.

aerator — (*hospitals*) A device for the removal of ethylene oxide from sterilized materials by exposing them to air circulation.

AERI — Acronym for Atmospheric Emitted Radiant Instrument.

aerial photography — High altitude photographs taken from an aircraft or satellite that are utilized for various environmental purposes.

aero filter — A filter bed for sewage treatment in which sewage is processed through continuous, rain-like application of sewage over the filter bed.

aeroallergen pollutants — Any airborne particulate matter that can induce allergic, infectious, and toxic responses in sensitive persons. Pollutants can include living organisms, such as animal dander, fungi, molds, infectious agents, bacteria, viruses, and pollen. Allergic reactions include watery eyes, runny nose and sneezing, nasal congestion, itching, coughing, wheezing, difficult breathing, and fatigue.

Aerobacter — In earlier classifications, a genus of the family Enterobacteriaceae consisting of Gram-negative facultative anaerobic motile rods.

aerobe — A microorganism that requires the presence of free oxygen to live and grow.

aerobic — A process that requires the presence of oxygen.

aerobic biological oxidation — A waste treatment process or other process using aerobic organisms in the presence of air or oxygen to reduce pollution loads, oxygen demand, or the amount of organic substances and waste.

aerobic digestion — The biological stabilization of sludge through partial conversion of putrescible matter into liquid, dissolved solids, and gaseous

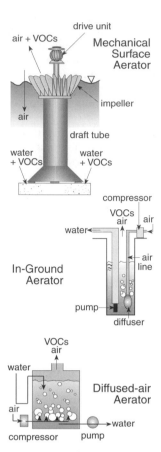

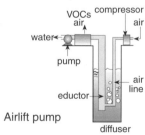

Aeration Devices
Aeration

byproducts with some destruction of pathogens in an oxygenated environment.

aerobic sewage treatment tank — Any unit incorporating as part of the treatment process the means of introducing air and oxygen into sewage held in a storage tank or tanks so as to provide aerobic biochemical stabilization during a detention period prior to its discharge.

aerobic treatment unit — A sewage treatment unit that incorporates a means of introducing air into sewage so as to provide aerobic biochemical stabilization during a detention period.

AEROCE — Acronym for Atmosphere/Ocean Chemistry Experiment.

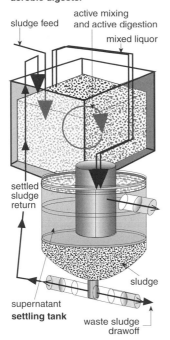

aerobic digester

sludge feed

active mixing
and active digestion

mixed liquor

settled
sludge
return

sludge

supernatant
settling tank

waste sludge
drawoff

**Continuous-Flow
Aerobic Digestion System**

aerodynamic diameter — The diameter of a unit density sphere having the same settling velocity as a particle no matter what its shape and density may be.

aerodynamic force — The force exerted on a particle in suspension by either the movement of air or gases around the particle or by the resistance of the gas or air to movement of the particle through the medium.

aeronomy — A study of the physical and chemical conditions of the upper atmosphere specifically concerning those regions upward of approximately 50 kilometers, where dissociation and ionization are fundamental properties.

aerosol — A particle of solid or liquid matter less than 1 micrometer in diameter that can remain suspended in the air because of its size; (*air pollution*) particles may cause irritation to the nose and throat, ulceration of the nasal passages, damage to the lungs, damage to other internal tissues (depending on the chemical present), and dermatitis; (*pesticides*) pesticide chemicals stored in a container under pressure; a propellant forces the mixture out of the can as tiny droplets of liquid or solid particles light enough to float in air. See also particulate matter.

aerosol, flammable — An aerosol that, when tested, yields a flame projection exceeding 18 inches at the valve opening or a flashback at any degree of valve opening.

aerosol photometry — A system of measurement employing a direct-reading physical instrument that uses an electrical impulse generated by a photocell to detect light scattered by a particle.

aerosol scavenger — A cloud droplet that attracts an aerosol into snow, rain, or some other water precipitate by coagulation and accretion.

aerosol transmission — A cloud or mist of solid or liquid particles containing pathogenic organisms released by sneezing or coughing.

aerotitus — An inflammation of the ear caused by changes in atmospheric pressure.

aerotitus media — An inflammation or bleeding in the middle ear caused by a difference between the air pressure in the middle ear and the atmosphere.

aerotolerant anaerobes — Microorganisms that grow under both aerobic and anaerobic conditions and do not shift from one mode of metabolism to another as conditions change, obtaining energy exclusively from fermentation.

AETS — See acid extraction treatment system.

a-f — See audio-frequency.

AFB — See acid-fast bacillus.

AFBC — Acronym for atmospheric fluidized bed combustor.

AFD — See air filtration device.

AFDM — Acronym for ash free dry mass.

AFEAS — Acronym for Alternative Fluorocarbons Environmental Acceptability Study.

afebrile — Without fever.

affect — To produce a material influence upon or alteration in something or someone.

affected environment — The existing biological, physical, social, and economic conditions of an area subject to change, directly and indirectly, as a result of a proposed human action.

affidavit — (*law*) A written sworn statement of persons having knowledge of certain facts.

affinity — The tendency of two substances to form strong or weak chemical bonds, thus forming molecules or complexes.

affinity chromatography — a highly specific separation technique that depends on the interaction between either enzyme and substrate or antibody and antigen.

aflagellar — Without flagella.

aflatoxin — Any of a group of carcinogenic mycotoxins that are produced by some strains of the fungus *Aspergillus flavus* in stored agricultural crops, such as peanuts.

AFO — Acronym for animal feeding operations.

African sleeping sickness — See African trypanosomiasis.

African tick fever — See relapsing fever.

African tick typhus — A rickettsial infection transmitted by ixodid ticks and characterized by fever, maculopapular rash, and swollen lymph nodes.

African trypanosomiasis — (*disease*) A parasitic disease caused by *Trypanosoma brucei gambiense* or *Trypanosoma brucei rhodesiense* and transmitted to humans by the bite of the tsetse fly, occurring only in the tropical areas of Africa where the flies are found. The disease progresses through three phases including localization at the site of invasion of the organism; systemic spread characterized by marked fever, chills, headache, anemia, edema of the hands and feet, and enlargement of the lymph glands; and, finally, neurological indications of central nervous system involvement, including lethargy, sleepiness, headache, convulsions, coma, and death if untreated. It may take years for the neurological phase to occur.

afterbay — The body of water immediately downstream from a power plant or pumping plant.

afterbirth — The placental and fetal membranes expelled from the uterus after childbirth.

afterburner — A device for augmenting the thrust of jet engines that includes an axillary fuel burner and combustion chamber to burn combustible gaseous substances; a device for burning or catalytically destroying unburned or partially burned carbon compounds in exhaust; see also secondary burner.

aftershocks — Earthquakes that follow the larger shock of an earthquake sequence.

agar — A dried hydrophilic colloidal substance extracted from various species of red algae. It is used in cultures for bacteria and other microorganisms.

agar culture plate — A culture plate, usually a Petri dish, with an agar medium.

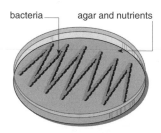

bacteria — agar and nutrients

Agar Culture Plate

AGC — See automatic generation control.

age-adjusted rate — The number of events occurring within a given age group divided by the population at risk of that age group multiplied by 100.

aged — A state of having grown older or more mature than others of the population group.

agency — An organization engaged in doing business for another person or persons.

agent — A physical, chemical, or biological entity that can be harmful to an organism.

agent activity/operation — Any operation involving chemical agents, including storage, shipping, handling, manufacturing, maintenance, test chamber activities, laboratory activities, surveillance, demilitarization, decontamination, disposal, and training.

agent orange — An herbicide, composed of 2-4-D and 2-5-T, that was used as a defoliant during the Vietnam War.

age-specific death rate — The mortality rate for a specific age class.

age-standardization, direct method — Standardization of age-specific rates observed in a studied population applied to an arbitrarily selected standard population to obtain the expected total number of events in the standard population. This expected total number is divided by the total number of persons in the standard population to obtain the age-standardized rates. Any difference between populations in age-adjusted rates based on the same standard population must be due to differences in age-specific rates and not due to differences in distribution by age.

age-standardization, indirect method — Age-specific rates obtained from a standard population are applied to age-appropriate groups of the studied population to obtain an expected number of events in the studied population. The observed number of events in the studied population is divided by the expected number to obtain a standardized mortality or morbidity ratio.

agglomerate — (*geology*) A rock composed of angular volcanic fragments fused by heat.

agglomeration — Consolidation of solid particles into larger forms by means of agitation or application of heat.

agglutination — Aggregation of separate particles into clumps or masses, especially the clumping together of bacteria by the action of a specific antibody directed against a surface antigen; the clumping together of blood cells in the presence of agglutinins.

agglutinin — Any substance causing agglutination of cells, particularly a specific antibody in the blood combining with its homologous antigen.

agglutinogen — A protein substance on a corpuscle surface responsible for blood types; an antigen that stimulates production of a specific antibody (agglutinin) when introduced into an animal body.

aggradation — A geologic process wherein streambeds, floodplains, sandbars, and the bottom of water bodies are raised in elevation by the deposition of sediment.

aggregate — Crushed rock or gravel, screened to sizes, which are mixed with cement and water for use in road surfaces, concretes, or bituminous mixes.

aggregate demand curve — A curve relating the total demand for the goods and services of an economy at each price level given the level of wages.

aggregated emissions data — Estimates of the amount of a substance emitted into the environment yearly from facilities that are not reporting facilities and anthropogenic sources other than facilities that emit a significant amount of that substance to the environment.

aggressive sampling — A sampling method using blowers and/or fans to keep particulates suspended during the sampling period.

aging — The process of growing old that results in part from a failure of body cells to function normally or to produce new body cells to replace those that are dead or malfunctioning; (*metals*) a change in the properties of certain metals and alloys that occurs at ambient and somewhat elevated temperatures after hot-working or heat treatment or after cold-working; involves a change in the properties but not in the chemical composition of the metal or alloy.

agitate — Keeping a solution from separating or settling by the use of mixing, stirring, or shaking; keeping a pesticide chemical mixed.

agitation — A state of chronic restlessness expressed as emotional tension.

agitator — A device for keeping liquids, or solids in liquids, in motion to prevent settling; a device using a paddle, air, or other object to keep a pesticide chemical mixed in a spray tank.

AGL — Acronym for above ground level.

agonal — Death and dying.

agribusiness — An agriculturally related business that supplies farm inputs such as fertilizer or equipment or is involved in the marketing of farm products.

agricultural pollution — The wastes, emissions, and discharges arising from farming activities, including runoff and leaching of pesticides and fertilizers, pesticide drift and volatilization, erosion and dust from cultivation, and improper disposal of animal manure and carcasses.

agricultural sewage — Waste produced through the agricultural processes of cultivating the soil, producing crops, or raising livestock.

agricultural solid waste — Solid waste that results from the rearing and slaughtering of animals and the processing of animal products and orchard and field crops.

agricultural waste — See agricultural pollution.

agrochemical — Synthetic chemicals, pesticides, and fertilizers used in agricultural production.

AGWAT — Acronym for Agriculture and Water Project.

AGWP — Acronym for absolute global warming potential.

AHF — Acronym for Argentine hemorrhagic fever virus.

AHH — See aryl hydrocarbon hydroxylase.

AHU — Acronym for air handling unit.

AI — See active ingredient; see artificial intelligence.

AIC — Acronym for acceptable intake for chronic exposure.

AIDS — See acquired immunodeficiency syndrome.

air — A mixture of gases that contains approximately 78% nitrogen, 21% oxygen, less than 1% carbon dioxide and inert gases, and varying amounts of water vapor, all forming the Earth's atmosphere.

air barrier — An intentionally designed and constructed barrier used to prevent both the exfiltration of indoor air to the outside and the infiltration of outdoor air into a building environment.

air basin — A land area with generally similar meteorological and geographic conditions throughout.

air binding — A condition where air enters a filter media and affects the filtration and backwash processes.

air change — A quantity of fresh air equal to the volume of the room or building being ventilated.

air change efficiency — A measure of how quickly the air in the room is replaced, which represents the ratio between the nominal time constant and the air change time for the room.

air change per hour — A unit that denotes the number of times a structure exchanges its entire volume of air with outside air in an hour, accomplished under natural conditions or forced air pressure.

air change rate — The volumetric rate at which air enters or leaves a building or zone expressed in units of building or zone volume; also called specific flow.

air classifier — A system that uses a forced-air stream to separate mixed material according to size, density, and aerodynamic drag of the pieces.

air cleaner — Any device used to remove atmospheric airborne impurities.

air cleaning — An indoor air quality control strategy to remove various airborne particulates and/or gases from the air; the three types of air cleaning used most often are particulate filtration, electrostatic precipitation, and gas sorption.

air compressor — A piece of equipment that compresses air for storage and use at later times.

air conditioning — A method of treating air to control temperature, humidity, cleanliness, and proper distribution within the space.

air conduction — The process by which sound is conducted to the inner ear through air in the outer ear canal.

air contaminant — Any smoke, soot, fly ash, dust, dirt, fume, non-naturally occurring gas, odor, toxin, or radioactive substance occurring within an environment.

air contaminant sources — All sources of emission of air contaminants, whether privately or publicly owned or operated.

air-cooled chillers — Chillers that consist of condensers and compressors that are usually located externally and are air cooled, with chilled water being provided.

air-cooled wall — A refractory wall with a lane directly behind it through which cool air flows.

air curtain — A method of containing oil spills in which air bubbles pass through a perforated pipe, thus causing an upward flow that slows the spread of oil; (*mechanical engineering*) a stream of high-velocity air directed downward to allow for air conditioning of a space with an open entrance; a means of keeping insects out of a structure.

air curtain destructor — A unit consisting of a combustion chamber pit and air blower designed to establish a curtain of high-velocity air above a fire burning in a pit, thus forcing the products of combustion up through the curtain before they reach the outside air.

air deficiency — A lack of air in an air–fuel mixture necessary to supply a quantity of oxygen stoichiometrically required to completely oxidize the fuel.

air distribution — The delivery of outdoor or conditioned air to various spaces in a building, usually by mechanical means.

air drying — The curing method of a film coating in which drying occurs by use of air without heat or the presence of a catalyst.

air embolism — See arterial gas embolism.

air emission — The release or discharge of a pollutant into the ambient air.

air entrapment — Intentional or accidental inclusion of air bubbles in materials such as concrete or liquid paint.

air exchange rate — The number of times that outdoor air replaces the volume of air in a building per unit time, typically expressed as air changes per hour.

air exfiltration — Uncontrolled outward leakage of indoor air through cracks, interstices, and other unintentional openings of a building caused by the pressure effects of the wind and/or stack effect.

air film — A layer of air next to a surface, such as a glass pane, which offers some resistance to heat flow.

air filter — A common air-cleaning device used to reduce the concentration of light particulate matter in the air to a level that can be tolerated.

air filtration device — A machine used to provide ventilation and a negative static pressure differential within a completely enclosed work area.

air flow — The mass/volume of air moved between two points.

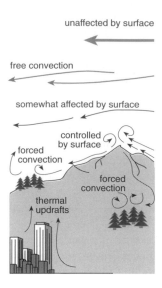

Air Flow

air flow rate — The mass/volume of air moved per unit of time through a space, opening, or duct; the mass flow rate is measured in kilograms per second, and the volume flow rate is measured in cubic meters per second.

air fluidization — The process of blowing warm air through a collection of microspheres to create a fluid-like environment.

air flush — Part of the sterilizer cycle in which the vacuum pump operates continuously and a valve opens, admitting filtered air into the sterilizer chamber.

air furnace — A reverberatory-type furnace in which metal is melted by a flame coming from fuel burning at one end of the hearth, passing over the bath, and exiting at the other end of the hearth.

air gap — The unobstructed vertical distance through the free atmosphere between the lowest opening from any pipe or faucet supplying water to a tank, plumbing fixture, or other device and the flood level rim of the receptacle. In all cases, the air gaps should be two times the diameter of the pipe.

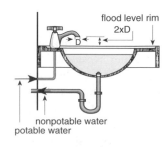

Air Gap
(Water Distribution System)

air–gas ratio — The ratio of air volume to gas volume which can be adjusted to change the character of combustion.

air hammer — A percussion-type pneumatic tool used for breaking through rock or concrete.

air heater — A heat exchanger through which air passes and is heated by a medium of higher temperature, such as hot combustion gases or heat filaments.

air impermeability — Airtightness caused by a high resistance to air flow.

air infiltration — The uncontrolled inward leakage of outdoor air through cracks, interstices, and other unintentional openings of a building caused by the pressure effects of wind and/or the stack effect.

air inlet — A deliberate opening in a room or duct wall for bringing outdoor or conditioned air into the room.

air jets — Streams of high-velocity air from nozzles in an incinerator enclosure that provide turbulence, combustion air, or a cooling effect.

air leakage — Uncontrolled flow of air through part of the building envelope that occurs when a pressure difference is applied across that part of the structure.

air leakage rate — The rate of air leakage into or out of a building or space.

air line respirator — Respirator connected to a compressed breathing air source by a hose of small inside diameter.

air mass — A vast body of air often covering hundreds of thousands of square miles in which the conditions of temperature and moisture are about the same at all points in a horizontal direction; also known as a high-pressure cell.

air monitoring — The process of continuous or periodic ambient air sampling to measure the concentration and type of pollutants present.

air mover — Any piece of equipment that is capable of moving air from one space to another.

air-moving device (AMD) — A piece of equipment, such as a fan, that moves air within a structure.

air outlet — A deliberate opening in a building envelope or duct through which air is expelled to the outside.

air padding — The pumping of dry air into a container to assist with the withdrawal of a liquid or to force a liquefied gas such as chlorine out of a container.

air plenum — Space used to convey air in a building, furnace, or structure.

air pollution — Mixture of solid, liquid, and gaseous contaminants discharged into the atmosphere by nature and the activities of humans and which may contribute to a variety of health or economic concerns and problems.

Air Pollution Control Act — A law passed in 1955 and updated through 1962 that states national policy to preserve and protect the primary responsibility and right of states and local governments in controlling air pollution. It initiated research by the U.S. Public Health Service and provided federal government technical assistance to states, established training programs in air pollution and in-house and external air pollution research, and directed the Surgeon General to conduct studies of motor vehicle exhausts related to effects on human health.

air pollution control district — An agency with authority to regulate stationery, indirect, and area sources of air pollution, including power plants, highway construction, and housing developments within a designated area.

air pollution episode — The occurrence of exceptionally high concentration levels of air pollution over a period of a few days or more.

air pressure — The weight of the atmosphere over a particular point; also called barometric pressure.

air-purifying respirator — A respirator that uses chemicals to remove specific gases and vapors from the air or that uses a mechanical filter to remove particulate matter; a respirator that reduces limited concentrations of air contaminants for breathing air but does not add oxygen and therefore cannot be used in an oxygen-deficient atmosphere.

Air Quality Act — A law passed in 1967 establishing eight air quality control regions in the United States and issuing air quality criteria for specific pollutants. The law provided for the development and issuance of information on recommended air pollution control techniques, required state and local agencies to establish air quality standards and state implementation of plans, and provided for federal action in cases of a state's noncompliance by giving the federal government emergency authority.

air quality assessment — The collection, handling, evaluation, analysis, and presentation of data required to understand the air pollution problem of a given area and its causes.

Air Quality Control Region (AQCR) — An area designated by the federal government in which communities share a common air pollution problem, sometimes involving several states.

air quality criteria — The levels of pollution and lengths of exposure at which specific adverse effects to health and welfare may occur; federal regulations reflect these levels.

Air Quality Index — A numerical index used for reporting the severity of air pollution to the public that incorporates the criteria pollutants of ozone, particulate matter, carbon monoxide, sulfur dioxide, and nitrogen dioxide into a single index; replaces the formally used Pollutant Standards Index.

index values	descriptor	color
0 50	good	green
51 100	moderate	yellow
101 150	unhealthy for sensitive groups	orange
151 200	unhealthy	red
201 300	very unhealthy	purple
301 500	hazardous	maroon

Air Quality Index

air quality manager — An individual employed by the local, state, or federal government to manage air quality.

air quality simulation model — A mathematical relationship between emissions and air quality that simulates on a computer the transport, dispersion, and transformation of compounds emitted into the air.

air quality specialist — An environmental health practitioner responsible for ensuring a quality of air that protects the health, welfare, and comfort of people and the community.

air quality standard — The prescribed concentration that a pollutant may reach in the outside air without exceeding the limits set forth by various federal

Pollutant	Averaging Period	Primary Standard	Secondary Standard
O_3	1 hr	0.125 ppm	0.125 ppm
	8 hr	0.085 ppm	0.085 ppm
CO	1 hr	35 ppm	
	8 hr	9 ppm	
SO_2	3 hr		0.5 ppm
	24 hr	0.145 ppm	
	annual	0.03 ppm	
NO_3	annual	0.053 ppm	0.053 ppm
Respirable Particulate Matter PM_{10}	24 hr	150 μg/m³	150 μg/m³
	annual	50 μg/m³	50 μg/m³
Respirable Particulate Matter $PM_{2.5}$	24 hr	65 μg/m³	65 μg/m³
	annual	15.1 μg/m³	15.1 μg/m³
Lead	quarter	1.55 μg/m³	1.55 μg/m³

National Ambient Air Quality Standards (NAAQS)

rules and regulations; the concentration represents the approximate level at which certain effects, clinical or subclinical, may be expected to begin to occur or the approximate level of concentration below which the effects defined should not ordinarily occur. In developing an air quality standard, it is necessary to identify the possible effects, to establish a criteria for the selection of the most sensitive group of receptors, to identify the existing burden of pollutants on the receptors, to specify a method for measurement of the concentration of the pollutant, and to specify an acceptable exposure limit to the pollutant.

Air Quality Working Group — An advisory group that provides a place for communication, cooperation, and coordination in the development and implementation of air quality control measures.

air-regulating valve — An adjustable valve used to regulate airflow to the facepiece, helmet, or hood of an air-line respirator.

air release valve — A valve (usually manually operated) that is used to release air from a pipe or fitting.

air return — A furnace duct through which interior cool air returns to the furnace and is then circulated through the heat exchanger, warmed, and distributed through the ducts.

air sacs — See alveoli.

air sampling — The process by which air samples or air pollution samples are collected.

air sampling professional — A person who does air monitoring sampling for the presence of various airborne contaminants, including asbestos.

air sealing — The practice of sealing unintentional gaps in the building envelope from the interior in order to reduce uncontrolled air leakage.

air setting — The characteristic of some materials, such as refractory cements, core pastes, binders, and plastics, to take permanent set at normal air temperatures of 20 to 25°C.

air slaking — The process of breaking up or sloughing when an indurated soil is exposed to air.

air spaces — Alveolar ducts, alveolar sacs, and alveoli of the respiratory system.

air sparging — Injection of air or oxygen into an aquifer to strip or flush volatile contaminants trapped in air bubbles up through the groundwater.

air speed — The speed of the air relative to its surroundings.

air stripping — A method of treatment that removes volatile organic compounds from contaminated groundwater or surface water by forcing an air stream through the water, causing the compounds to evaporate.

air-supplied respirator — A respirator that uses a filter or sorbent to remove harmful substances from the air.

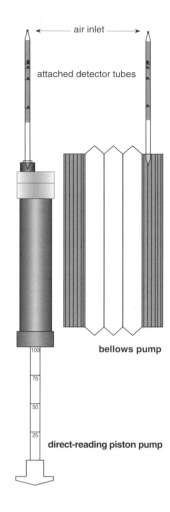

Air Sampling Devices

air-to-air heat pump — A piece of equipment that contains one fan to be connected to ductwork to cool or heat the indoor atmosphere and one fan to reject or reclaim heat to the external atmosphere.

air toxics — A generic term referring to a harmful chemical or group of chemicals in the air; any compound that is in the air and has the potential to produce adverse health effects.

air vapor barrier — A moisture-impervious layer that is applied to surfaces enclosing a space to limit moisture migration.

air velocity — The rate and direction of air movement; important when considering cooling effects and comfort criteria.

air vent — An air inlet or outlet.

air–void ratio — Ratio of the volume of air space to the total volume of voids in a soil mass.

air waves — Airborne vibrations caused by explosions.

airbone gap — The difference in decibels between hearing levels for a particular frequency determined by air conduction and bone conduction.

airborne chemical exposure monitoring — Sampling and measuring of the amount and duration of chemical contaminants in the air that can reach a person.

airborne contaminants — Materials in the atmosphere that can affect the tissues of the upper respiratory tract and lungs and therefore the health of individuals in the immediate or adjacent environments.

airborne exposure limits — Allowable concentrations in the air for occupational and general population exposures.

airborne particles — The liquid and solid aerosols as well as atmospheric particulates found in air.

airborne sound — Sound that reaches the point of measurement through the air.

airborne toxics — Toxic inorganic and organic chemicals released into the air that may cause short- or long-term health effects in humans.

airflow pattern — The pattern of movement of respiratory gases through the respiratory tract which may be affected by gas density and viscosity.

airglow — Continuous emission of upper atmosphere radiation during the day and at night over middle and low latitudes.

airless spray — A type of spraying system in which paint is atomized by using high hydraulic pressure rather than compressed air.

AIRS — Acronym for Aerometric Information Retrieval System; acronym for atmospheric infrared sounder.

AIRS/MHS — Acronym for Advanced Infrared Radiometer Sounder/Microwave Humidity Sounder.

AIRS–AMSU — Acronym for Atmospheric Infrared Radiometer Sounder–Advanced Microwave Sounding Unit.

airshed — A geographical area that shares the same air mass because of topography, meteorology, and climate.

airspace ratio — Ratio of the volume of water that can be drained from a saturated soil or rock under the action of force of gravity compared to the volume of voids.

airtightness — A general descriptive term for the leakage characteristics of the building.

airtightness standard — A standard value of building or component air leakage corresponding to a reference pressure difference across the building envelope or component.

A

airway — The passage by which air enters and leaves the lungs.

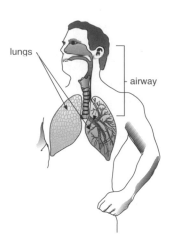

lungs

airway

Airway

airway resistance — Narrowing of the respiratory system air passages in response to the presence of irritating substances.

AIS — See acceptable intake for subchronic exposure; acronym for Automated Information System; acronym for airborne imaging spectrometer.

akinetic — A loss of ability to move a part or all of the body.

AL — Acronym for acceptable level.

ALADIN — Acronym for Atmospheric Laser Doppler Instrument.

Alar — Trade name for daminozide, a plant regulator classified as a pesticide that makes apples redder, firmer, and less likely to drop off trees before harvest; suspended by the U.S. EPA in 1989 following controversy raised regarding its cancer risk to children.

ALARA — Acronym for as low as reasonably achievable.

alarm — A signal that indicates a process has exceeded or fallen below the set or limit point.

alarm reaction — The first stage of the general adaptations syndrome causing mobilization of the various defense mechanisms of the body or mind to cope with a stressful physical or emotional situation.

albedo — A measure of the reflecting power of a surface that is the fraction of incident radiation reflected back in all directions.

albino — An organism showing deficient pigmentation in skin, eyes, and/or hair.

albumin — A protein material found in animal and vegetable fluids that is soluble in pure water.

albuminuria — The presence of albumin, serum globulin, or other proteins in the urine; usually indicative of renal disease.

alcohol (C$_2$H$_5$OH) — A colorless, volatile liquid; BP: 78.3°C, Sol: soluble in water, chloroform, and methyl alcohol. It is used as a solvent and in the manufacture of many chemicals and medicines and is an organic hydroxyl compound formed by replacing one or more hydrogen atoms of a hydrocarbon with an equal number of hydroxyl (OH) groups; primary alcohol contains a –CH$_2$OH group, secondary alcohol contains a –CHOH– group.

alcohol fuels — A class of liquid chemicals that have certain combinations of hydrogen, carbon, and oxygen and are capable of being used as fuel.

alcohol poisoning — Poisoning caused by the ingestion of any alcohol, including ethyl, isopropyl, and methyl, and leading to nausea, vomiting, abdominal pain, possible blindness, and death.

alcoholism — An extreme dependence on excessive amounts of alcohol that creates a chronic illness with slow insidious onset; it may occur at any age and results in patterns of deviant and destructive behaviors.

aldehyde — An organic compound formed by dehydrating oxidized alcohol; it contains the characteristic –CHO group.

aldicarb (C$_7$H$_{14}$N$_2$O$_2$S) — A white crystalline solid with a slightly sulfurous odor; MW: 190.25, BP: decomposes, Sol: 0.6%, Fl.P: NA (although it may burn, it does not readily ignite), sp. gr. 1.195. It is used as a restricted-use soil insecticide to control mites and nematodes. It is hazardous to the respiratory system and digestive system or through dermal or eye contact. Its primary target is the central nervous system, and in acute exposures it causes stupor, seizures, hypotension, hypertension, tachycardia, or cardiorespiratory depression, which may lead to death. It also causes blurred vision, sweating, nausea, and vomiting and in high exposures can cause pulmonary edema. Chronic exposure leads to occupational contact dermatitis and persistent central nervous system effects. No occupational exposure limit has been established, although exposure to aldicarb is hazardous.

aldoxycarb (C$_7$H$_{14}$N$_2$O$_4$S) — A crystalline solid; MW: 222.29, BP: NA, Sol: dissolves in most organic solvents, density: NA. It is used as a pesticide. It is highly toxic by all routes of exposure, is a cholinesterase inhibitor, and may cause death to adult humans.

$$CH_3-\underset{\underset{O}{\|}}{\overset{\overset{O}{\|}}{S}}-C(CH_3)_2-CH=N-O-\overset{\overset{O}{\|}}{C}-NH-CH_3$$

Aldoxycarb

aldrin (C$_{12}$H$_8$Cl$_6$) — A colorless to dark brown crystalline solid with a mild chemical odor that is used as an insecticide; MW: 364.9, BP: decomposes, Sol: 0.003%, sp. gr. 1.60. It is hazardous to the central nervous system, liver, kidneys, and skin and is toxic by inhalation, absorption, ingestion, and contact. Symptoms of exposure include headache, dizziness, nausea, vomiting, malaise, myoclonic jerks of the limbs, clonic and tonic convulsions, coma, hematopoietic, and azotemia; carcinogenic. OSHA exposure limit (TWA): 0.25 mg/m^3 [skin].

ALERT — See Automated Local Evaluation in Real Time.

alert level — That concentration of pollutants at which first stage control action is to begin.

algae — The general name for simple, single-cell or multicelled plants with chlorophyll that have no true roots, stems, or leaves and live in aquatic habitats and moist places on land.

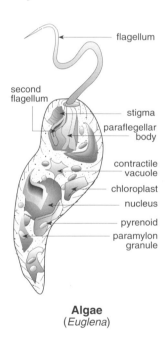

Algae
(*Euglena*)

algal blooms — A sudden, massive growth of microscopic and macroscopic plant life in lakes and reservoirs that affects water quality.

algicide — A chemical used for the control and prevention of algae growth in swimming pools.

algid malaria — A stage of malaria caused by the protozoan *Plasmodium falciparum* with symptoms of coldness of the skin, profound weakness, and severe diarrhea.

algorithm — A prescribed set of well-defined rules or processes for solving a problem in a given number of steps; frequently employed in electronic data processing.

alignment — The arrangement of a group of points or objects along a line.

alimentary — Referring to the digestive organs.

alimentary canal — Organs comprising the food tubes in animals and humans; they are divided into zones for ingestion, digestion, and absorption of food and for the elimination of indigestible material.

aliphatic — Of or pertaining to any open-chain carbon compound; three subgroups of such compounds are alkanes, alkenes, and alkynes.

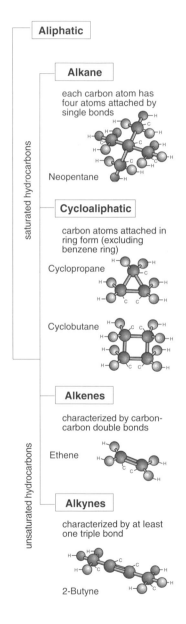

Aliphatic Hydrocarbons

aliquot — A sample that is representative of a larger quantity.

ALJ — Acronym for Administrative Law Judge.

ALK — Acronym for alkalinity.

alkali — Referring to any strong base, such as sodium hydroxide and potassium hydroxide.

alkali burn — Damage to tissue caused by exposure to an alkaline compound similar to lye.

alkali metal — Any of the highly reactive metals such as lithium, sodium, potassium, rubidium, cesium, and francium found in the first column of the periodic table and which act as bases to form strongly alkaline hydroxides.

alkali poisoning — A toxic condition caused by the ingestion of an alkaline agent such as ammonia, lye, or certain detergent powders.

alkali–aggregate reaction — A deterioration of concrete where the alkali in the cement paste in the concrete reacts chemically with the silica or carbon present in some aggregates to produce a gel, which, in the presence of free moisture, will expand and manifest into cracking and differential movement in structures as well as having other deleterious effects on durability and tensile strength.

alkalies — Caustic substances that dissolve in water to form a solution with a pH higher than 7. This includes either solid or concentrated liquid forms of ammonia, ammonium hydroxide, calcium hydroxide, potassium hydroxide, and sodium hydroxide. Alkalies are more hazardous and destructive to tissue than most acids and cause irritation of the eyes and respiratory tract and lesions in the nose.

alkaline — Referring to any substance having basic properties; having a pH greater than 7.

alkaline battery — A battery that uses sodium or potassium hydroxide as an electrolyte, and nickle–oxide flakes and powdered iron or nickel–cadmium for its active plates.

alkaline earths — Usually considered to be the oxides of magnesium, barium, calcium, strontium, and radium.

alkaline material — See alkaline substance.

alkaline phosphatase test — A tumor marker test that assists in diagnosis of bone and liver mestastases.

alkaline soil — A soil having a pH value of >7.3.

alkaline substance — A chemical compound in which the basic hydroxide ion is united with a metallic ion such as sodium hydroxide or potassium hydroxide, which causes alkalinity in water; it is used for neutralization of acids. Lime is the most commonly used alkaline material in wastewater treatment.

alkalinity — Capacity of a substance to neutralize acids.

alkali–silicate reaction — Reaction of an alkali with an aggregate that has the potential to cause deterioration of the concrete.

alkaloid — A group of nitrogen-containing compounds with diverse structures and having medicinal, hallucinogenic, or toxic properties.

alkalophile — An organism that grows best under alkaline conditions up to a pH of 10.5.

alkane — A series of aliphatic saturated hydrocarbons that have no carbon–carbon multiple bonds and have the general formula C_nH_{2n+2}.

alkapton — A class of substances with an affinity for alkali found in the urine and which causes the metabolic disorder known as alkaptonuria.

alkene — One of a class of unsaturated aliphatic hydrocarbons in which two carbon atoms are connected by a double bond, with the general formula C_nH_{2n}; also known as olefin.

alkyd — Oil-based paint.

alkyd resin — A modified form of resin prepared by combining alcohol and fatty acids used in general-purpose coatings such as lacquer paints, varnishes, and metal finishes.

alkyl — A monovalent radical C_nH_{2n+1} obtained from a saturated hydrocarbon by removing one hydrogen atom from an alkane (e.g., methyl CH_{3-}, derived from methane).

alkylating agents — A group of chemotherapy drugs among which chemical interactions interfere with cell growth and reproduction.

alkylation — The combination of a saturated and unsaturated hydrocarbon.

alkylbenzene sulfonate (ABS) — A major class of alkylaryl sulfonate surfactants used in detergents.

alkynes — A group of unsaturated hydrocarbons in which two carbon atoms are connected by a triple bond; also known as acetylene.

allegation — (law) An assertion undertaken to be proven with evidence by a party to a legal action.

allele — One of two or more alternative forms of a gene that exists at a specific gene location on a chromosome.

allergen — A substance, such as pollen or a drug, that activates the immune system and causes an allergic response in sensitized humans or animals.

allergic asthma — A form of asthma caused by exposure of bronchial mucosa to an inhaled airborne antigen, which leads to the production of antibodies that bind to mass cells in the bronchial tree, resulting in the release of histamine, which stimulates contraction of bronchial smooth muscle and causes mucosal edema.

allergic conjunctivitis — Increased blood in the conjunctiva caused by an allergy to pollen, grass, air pollutants, occupational irritants, and smoke.

allergic dermatitis — An acute inflammatory condition of the skin that occurs after exposure of a body part to an agent to which the person is hypersensitive.

allergic reaction — An abnormal physiological response by a sensitive person to a chemical or physical stimuli that causes no response in nonsensitive individuals.

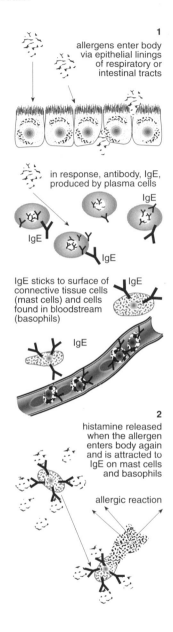

1

allergens enter body via epithelial linings of respiratory or intestinal tracts

in response, antibody, IgE, produced by plasma cells

IgE

IgE

IgE

IgE sticks to surface of connective tissue cells (mast cells) and cells found in bloodstream (basophils)

IgE

IgE

2

histamine released when the allergen enters body again and is attracted to IgE on mast cells and basophils

allergic reaction

Allergic Reaction

allergic rhinitis — An inflammation of the mucous membranes in the nose.

allergic sensitizer — Any chemical acting as a sensitizer, especially epoxy monomers and their amine hardeners, potassium dichromate, nickel, and formaldehyde.

allergy — A condition of abnormal sensitivity in certain individuals to contact with substances such as proteins, pollens, bacteria, and certain foods. This contact may result in exaggerated physiologic responses such as hay fever, asthma, hives, and, in severe enough situations, anaphylactic shock. Reexposure increases the capacity to react.

allethrin ($C_{19}H_{26}O_3$) — A clear or amber, fairly viscid liquid with a mild pleasant odor; highly miscible in kerosene, xylene, and other organic solvents but insoluble in water; an insecticide. Also known as synthetic pyrethrum.

allochthonous — Food material reaching an aquatic community that originated elsewhere in the form of organic detritus.

allotrope — One of the two or more forms of an element differing in either or both physical and chemical properties. In the case of oxygen, ozone (O_3) is an oxidant, molecular oxygen (O_2) is needed in respiratory processes, and elemental oxygen (O) changes into the other two forms.

allotropy — The existence of different forms of the same element in the same physical state with distinctly different properties due to differences in their energy content or arrangement of atoms.

allowable error — the amount of error that can be tolerated without invalidating the medical usefulness of the analytical result, usually considered to be the 95% limit.

allowable pile bearing load — The maximum load that can be permitted on a pile for adequate safety against movement of such magnitude that the structure is endangered.

alloy — Any of a large number of materials composed of two or more metals, or a metal and a nonmetal mixed together and having metallic properties.

alloying elements — Any of a number of chemical elements constituting an alloy.

alluvial — Transported and deposited by running water.

alluvial cones — See alluvial fans.

alluvial fans — The conical deposit of sediment laid down by a swift-flowing stream as it enters a plain or an open valley; also known as alluvial cones.

alluvial plane — A plane formed by the deposition of material in rivers and streams that periodically overflow.

alluvium — Surplus rock material consisting mainly of sand, silt, and gravel that a river or stream has eroded, carried in suspension, and deposited.

allyl alcohol (CH₂CHCH₂OH) — A colorless liquid with a pungent, mustard-like odor; MW: 58.1, BP: 205°F, Sol: miscible, Fl.P: 70°F, sp. gr. 0.85. It is used in the preparation of various allyl esters; in the preparation of chemical derivatives used in perfumes, flavorings, and pharmaceuticals; as a fungicide, herbicide, and nematicide; in refining and dewaxing of mineral oil. It is hazardous to the eyes, skin, and the respiratory system and is toxic

A

through inhalation, absorption, and ingestion. Symptoms of exposure include eye and skin irritation, tissue damage, upper respiratory system irritation, and pulmonary edema. OSHA exposure limit (TWA): 2 ppm [air].

allyl chloride (CH₂CHCH₂Cl) — A colorless, brown, yellow, or purple liquid with a pungent, unpleasant odor; MW: 76.5, BP: 113°F, Sol: 0.4%, Fl.P: –25°F, sp. gr. 0.94. It is used in the manufacture of glycerin, epichlorohydrin, allyl alcohol, allylamines, allyl ethers of starch, and pharmaceuticals. It is hazardous to the respiratory system, skin, eyes, liver, and kidneys and is toxic through inhalation, absorption, ingestion, and contact. Symptoms of exposure include eye, nose, and skin irritation; pulmonary edema; and liver and kidney damage in animals. OSHA exposure limit (TWA): 1 ppm [air].

allyl glycidyl ether (C₆H₁₀O₂) — A colorless liquid with a pleasant odor; MW: 114.2, BP: 309°F, Sol: 14%, Fl.P: 135°F, sp. gr. 0.97. It is used as a reactive diluent in the formulation of epoxy resins and as a copolymer for vulcanization of rubber, surface coatings, and epoxy resins. It is hazardous to the respiratory system and the skin and is toxic through inhalation, absorption, ingestion, and contact. Symptoms of exposure include dermatitis, eye and nose irritation, pulmonary irritation, edema, and narcosis. OSHA exposure limit (TWA): 5 ppm [air].

ALMS — Acronym for atomic line molecular spectroscopy.

ALOHA — Acronym for area location of hazardous atmospheres.

alopecia — The absence of hair from skin areas where it is normally found; a condition caused by trauma, cutaneous or systemic disease, drugs, chemical, or ionizing radiation.

alpha — The first letter of the Greek alphabet; often used in chemical nomenclature to distinguish one variation in chemical compound from others.

alpha decay — The process of radioactive decay in which the nucleus of an atom emits an alpha particle, creating a new atom with an atomic number lower by two and an atomic mass number reduced by four.

alpha-destructive distillation — The distillation of organic soils or liquids where the substance decomposes during distillation, leaving a solid or viscous liquid in the still.

alpha emitter — A radioactive substance that gives off alpha particles.

alpha particle — A positively charged particle emitted by several radioactive substances that has a massive charge identical to the nucleus of a helium atom and consists of two protons and two neutrons.

alpha-ray — A stream of fast-moving helium nuclei or alpha particles that is strongly ionizing and weakly penetrating.

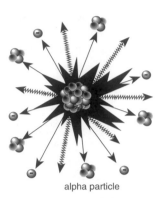

alpha particle

Alpha Particle
Alpha Ray

alphanumeric code — (*computer science*) Machine-processable letters, numbers, and special characters.

alphavirus — Any of a group of very small togaviruses consisting of a single molecule of single-stranded DNA within a lypoprotein capsule.

ALT — Acronym for altimeter.

alteration of generations — A reproductive cycle in which a sexual stage alternates with an asexual one, as in the case of a gametophyte alternating with a sporophyte.

alternating current (AC) — Current that reverses its flow with a constant frequency.

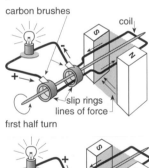

carbon brushes

coil

slip rings
lines of force

first half turn

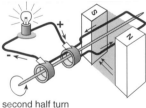

second half turn

ac generator

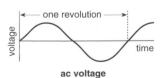

one revolution

voltage

time

ac voltage

Alternating Current

alternative fuels — Substitutes for traditional liquid-oil-derived motor vehicle fuels such as gasoline and diesel that are cleaner burning and help to meet mobile and stationery emissions standards. Common mixtures include alcohol-based fuels with gasoline, methanol, and ethanol, as well as compressed natural gas and hydrogen.

alternative system — Any approved on-site sewage disposal system used in place of, including modification to, a conventional subsurface system.

alternative technology — Any proven wastewater treatment process and technique that provides for the reclaiming and reuse of water, recycling wastewater constituents, or otherwise eliminating the discharge of pollutants or recovering of energy.

alternative wastewater treatment works — A wastewater conveyance and/or treatment system other than a conventional system, including small-diameter pressure and vacuum sewers and small-diameter gravity sewers carrying partially or fully treated wastewater.

alternator — A generator that changes mechanical energy into electrical energy by the rotation of its rotor.

altimeter — An instrument used to measure altitude with respect to a fixed level.

altitude — Any location on Earth with reference to a fixed surface point, which is usually sea level.

altocumulus clouds (Ac) — Patches or layers of puffy, roll-like gray or white clouds made up of water droplets; bases average 10,000 feet above the surface of the Earth.

Altocumulus Clouds

altostratus clouds (As) — Dense veils or sheets of gray or blue that very often totally cover the sky; bases average 10,000 feet above the surface of the Earth.

alum — Any group of double sulfates of a monovalent metal and a trivalent metal such as $K_2SO_4 \cdot Al_2(SO_4)_3 \cdot 24H_2O$. Alum is a common name for commercial aluminum sulfate, which is used in pools to form a gelatinous floc on sand filters or to coagulate and precipitate suspended particles in water. See also potassium alum.

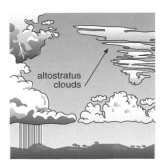

Altostratus Clouds

aluminosis — A form of pneumoconiosis caused by respiratory exposure to aluminum-bearing dust.

aluminum (Al) — A silver-like malleable and ductile metal found in soil; a flexible metal and a natural element in the earth; present in air, water, soil, and most foods; MW: 26.98, BP: 2307°C, Sol: insoluble in water, Fl.P: not known. The principle source is bauxite. It enters the environment through the weathering of rocks and minerals, and releases are associated with industrial processes such as smelting. It is used as a structural material in the construction, automotive, and aircraft industries; extensively in the production of metal alloys; in the electrical industry in overhead distribution and power lines, electrical conductors, and insulated cables and wiring; in auto parts, cooking utensils, decorations, highway signs, fences, cans, food packaging, foil, corrosion-resistant chemical equipment, dental crowns, and denture materials. Through inhalation exposure, the lungs receive Al^{3+} as particles of poorly soluble compounds; some of the particles stay within the lung tissue, while others are transported up the respiratory system and eventually swallowed. It may be distributed to the brain, bone, muscle, and kidneys and to some extent to the milk of lactating mothers and has been known to cross the placenta and accumulate in the fetus. It is not thought to be toxic to humans when ingested from cooking utensils and as part of the daily intake in food; however, sensitive subpopulations may include pregnant mothers and patients with Alzheimer's. A potential risk appears to exist in the breathing in of the dust and subsequent respiratory symptoms. It may interact with neuronal DNA to alter gene expression and protein formation. The nervous system may possibly be a target for aluminum; exposure limits in air for 8-hour periods for the aluminum metal are total dust, 15 mg/m³; respirable fraction, 5 mg/m³; pyro powders, 5 mg/m³; welding fumes, 5 mg/m³; soluble salts, 2 mg/m³; alkyls, 2 mg/m³. For aluminum oxide, the OSHA air regulation for total dust is 10 mg/m³, and the respirable fraction is 5 mg/m³.

aluminum potassium sulfate — See potassium alum.

alveolar air — Respiratory gases in an alveolus or air sack of the lung.

alveoli — The tiny sacks in the lungs where the actual exchange of oxygen and carbon dioxide takes place.

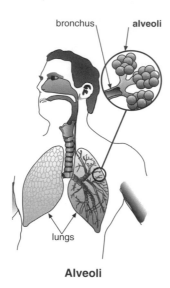

Alveoli

alveolitis — An allergic pulmonary reaction to the inhalation of antigenic substances with symptoms of acute episodes of dyspnea, cough, sweating, fever, weakness, and pain in the joints and muscles lasting from 12 to 18 hours.

amalgam — An alloy of mercury with some other metal or metals.

amalgamation — The process of alloying metals with mercury to separate them from ore.

ambient — Environmental or surrounding conditions.

ambient air — Air surrounding a particular area in which climate, meteorology, and topography influence its ability to dilute and disperse pollutants present.

Ambient Air Quality Standard (AAQS) — The limit on the amount of a given pollutant that will be permitted into the ambient air.

ambient compensation — An instrument designed in a manner so that changes in the ambient temperature will not affect readings of the instrument.

ambient concentration — Any air pollution concentration that occurs in the outdoor environment.

ambient noise level — The all-encompassing noise associated with a given environment, being a composite of sounds from all sources.

ambient pressure — The atmospheric pressure or pressure in the environment or surrounding area.

Pollutant	Averaging Period	Primary Standard	Secondary Standard
O_3	1 hr	0.125 ppm	0.125 ppm
	8 hr	0.085 ppm	0.085 ppm
CO	1 hr	35 ppm	
	8 hr	9 ppm	
SO_2	3 hr		0.5 ppm
	24 hr	0.145 ppm	
	annual	0.03 ppm	
NO_3	annual	0.053 ppm	0.053 ppm
Respirable Particulate Matter PM_{10}	24 hr	150 µg/m³	150 µg/m³
	annual	50 µg/m³	50 µg/m³
Respirable Particulate Matter $PM_{2.5}$	24 hr	65 µg/m³	65 µg/m³
	annual	15.1 µg/m³	15.1 µg/m³
Lead	quarter	1.55 µg/m³	1.55 µg/m³

National Ambient Air Quality Standards (NAAQS)

ambient temperature — The temperature of the surrounding air or other medium.

Ambu-bag — A breathing bag used to assist respiratory ventilation.

ambulance — An emergency vehicle usually used for the transport of a patient to a medical facility in the event of accident or severe illness.

ambulatory care — Health services provided on an outpatient basis to people who visit hospitals or other healthcare facilities.

AMD — See air-moving device.

ameba (amoeba) — Common name for a number of the simplest kinds of single-celled protozoans.

amebiasis — (*disease*) A parasitic infection of humans caused by the amoeba *Entamoeba histolytica*; symptoms include acute dysentery, fever, chills, and blood or mucous diarrhea. It occurs especially in areas of poor sanitation, and the reservoir of infection is humans. It is transmitted through the ingestion of water or food contaminated with feces containing amoebic cysts and is communicable as long as the cyst is passing with the feces; general susceptibility. It is prevented by the proper control of human feces, sewage protection of public and private water supplies, proper personal hygiene, and the control of flies. Incubation time is 2 to 4 weeks but can vary from a few days to several months.

amebic dysentery — An inflammation of the intestine caused by an infestation with *Entamoeba histolytica*;

symptoms include frequent loose stools flecked with blood and mucous; see also amebiasis.

amelioration — An improvement in a condition.

amended water — Water containing wetting agents, penetrants, and/or other agents to enhance the wetting of asbestos-containing materials to reduce the generation of dust.

amenorrhea — Absence of menstruation.

amensalism — The production of a substance by one organism that is inhibitory to one or more other organisms.

American Public Health Association–Public Health Service Housing Code — A code developed by the association and the federal government providing for local housing code requirements.

American Standard Code for Information Interchange (ASCII) — A widely used industry standard code for exchanging alphanumeric codes in terms of bit-signatures.

Ames assay — A test performed on bacteria to assess the capability of environmental chemicals to cause mutations; named for Bruce Ames.

ametryne ($C_9H_{17}N_5S$) — A crystalline solid; MW: 227.37, BP: NA, Sol: dissolves in methylene chloride, methanol, acetone, and toluene; density: NA. It is used as a herbicide and is toxic by ingestion; mild irritant to the skin and eyes.

$$C_2H_5NH-\!\!\!\!\bigg\langle\!\!\!\!\overset{N}{\underset{N}{\bigcirc}}\!\!\!\!\bigg\rangle\!\!\!\!-NHCH(CH_3)_2$$

SCH

Ametryne

AMI — See acute myocardial infarction; also, acronym for active microwave instrumentation.

amicus curiae — A friend of the court.

amide — One of a group of organic compounds formed by the reaction of any organic acid with ammonia or an amine and containing the $CONH_2$ radical, $RCONH_2$ (primary), $(RCO)_2NH$ (secondary), or $(RCO)_3N$ (tertiary), where R is a hydrocarbon group.

AMIE — Acronym for Automated Methane Instrument Evaluation.

amine — One of a class of organic compounds, such as CH_3NH_2, derived from ammonia by substituting one or more hydrocarbon radicals for hydrogen atoms.

amino acid — Any of the organic compounds containing an amino group and a carboxyl group that are polymerized to form peptides and proteins; the nitrogen-containing molecules that plants and animals combine in great variety in building cell proteins.

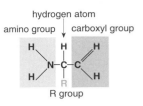

hydrogen atom
amino group | carboxyl group

R group

Amino Acid

4-aminodiphenyl ($C_6H_5C_6H_4NH_2$) — Colorless crystals that turn purple in contact with air and have a floral odor; MW: 169.2, BP: 576°F, Sol: slight, Fl.P: not known, sp. gr. 1.16. It is used in research and laboratory facilities. It is hazardous to the bladder and skin and is toxic by inhalation, absorption, ingestion, and contact. Symptoms of exposure include headache, dizziness, lethargy, dyspnea, ataxia, weakness, methemoglobinemia, urinary bleeding, and acute hemorrhagic cystitis; carcinogenic. OSHA exposure limit (TWA): regulated use through engineering controls, work practices, and personal protective equipment, including respirators.

aminoethane — See ethyl amine.

2-aminopyridine ($NH_2C_5H_4N$) — White powder or crystals with a characteristic odor; MW: 94.1, BP: 411°F, Sol: soluble, Fl.P: 154°F. It is used in the production of chemical intermediates for pharmaceuticals and in the manufacture of dyes, lubricant antioxidants, and herbicides. It is hazardous to the central nervous system and the respiratory system and is toxic by inhalation, absorption, ingestion, and contact. Symptoms of exposure include headache, dizziness, excitement, nausea, high blood pressure, respiratory distress, weakness, convulsions, and stupor. OSHA exposure limit (TWA): 0.5 ppm [air].

amiton ($C_{10}H_{24}O_3NPS$) — A colorless liquid; MW: 269.38, BP: 110°C. It is used as a pesticide and is highly toxic and hazardous to the digestive tract, eyes, and respiratory tract. It is a cholinestearase inhibitor; may cause nausea, vomiting, diarrhea, blurred vision, convulsions, coma, respiratory failure, and death.

$$\begin{array}{c} C_2H_5O \\ \searrow \overset{\displaystyle O}{\overset{\displaystyle \|}{P}}-S-CH_2-CH_2-N(C_2H_5)_2 \\ C_2H_5O \nearrow \end{array}$$

Amiton

amitosis — Direct cell division with simple fission of the nucleus and cytoplasm.

ammeter — An instrument used to measure the amount of the electric current in a circuit.

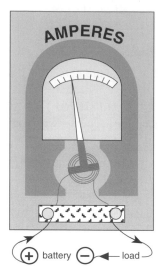

Ammeter
showing galvanometer

ammonia (NH₃) — A colorless gas with a very sharp odor; MW: 17.03, BP: –33.35°C, Sol: 42.8% at 0°C in water, 33.1% at 20°C in water, 34% at 25°C in water; flammability limits in air: 16 to 25%; density as a gas: 0.7710 g/l. A naturally occurring compound, ammonia is a key intermediate in the nitrogen cycle; it is essential for many biological processes, and a background concentration is present in most environmental media. It is synthesized commercially and released into the atmosphere by volatization from decaying organic matter; animal livestock excreta; fertilization of soil; venting of gas; leaks or spills during commercial synthesis; production or transportation from sewage or wastewater effluent; the burning of coal, wood, or other natural products; and through volcanic eruptions. The largest and most significant use is agricultural application of fertilizers; also used in production of fibers, plastics, explosives, urea, ammonium phosphates, nitric acid, ammonium nitrate, and ammonium sulfate; as a refrigerant or corrosion inhibitor; in purification of water supplies; as a component of household cleaners; in the pulp and paper, metallurgy, rubber, food and beverage, textile, and leather industries; in pharmaceuticals; and as chemical intermediates. It may cause death at short-term exposures of 5000 to 10,000 parts per million due to airway obstruction. Infections and other secondary complications may occur; chemical burns and edema of exposed tissues including the respiratory tract, eyes, and exposed skin occur after exposure to lethal levels. It is an upper respiratory irritant to humans; exposures to levels exceeding 50 parts per million result in immediate irritation to nose and throat, although a tolerance level may develop at much higher levels, especially when exposed to anhydrous ammonia, when nasopharyngeal and tracheal burns occur. Airway obstruction and respiratory distress as well as bronchiolar and alveolar edema occur; acute exposure to highly concentrated aerosols of ammonia may cause elevated pulse and blood pressure and cardiac arrest in humans. Bradycardia, hypertension, and cardiac arrhythmias leading to cardiovascular collapse follow acute exposure to concentrations exceeding 5000 parts per million. Highly concentrated aerosols of anhydrous ammonia can produce burns of the lips, oral cavity, and pharynx along with edema. Cyanosis, elevated white blood cell count, pulmonary artery thrombosis, spasms of the muscles of the extremities, hemorrhagic necrosis of the liver, and dermal and ocular irritations occur. Secondary infections as well as blurred vision, diffuse nonspecific encephalopathy, loss of consciousness, and decreased deep tendon reflexes also occur. OSHA exposure limit (TWA): 50 ppm [air].

ammonia oxidation — A test used during manufacturing processes to evaluate the ammonia oxidation rate for the nitrifiers.

ammonia stripping — A physical process in which ammonia is condensed and removed from alkaline aqueous waste solutions after contact with steam at atmospheric pressure.

ammonification — The release of ammonia, especially into the soil, from decaying protein by means of bacterial action.

ammonium — A form of nitrogen usable by plants.

ammonium ion — An NH₄⁺ ion formed by the reaction of ammonia (NH₃) with a hydrogen ion (H⁺).

ammonium nitrate (NH₄NO₃) — The most dangerous explosive among inorganic nitrates. In closed confinement, the mass causes a severe pressure buildup, resulting in an explosion. If mixed with oil, charcoal, or other combustible substances, it causes a violent reaction. Some of the worst industrial accidents, ship fires, and explosions of cargo vessels have been caused by ammonium nitrate.

ammonium sulfamate (NH₄SO₃NH₂) — A colorless to white crystalline, odorless solid; MW: 114.1, BP: 320°F (decomposes), Sol: soluble in water, Fl.P: 77°F, sp. gr. 0.88. It is used as an herbicide, in the manufacture of fire-retardant compositions, in the generation of nitrous oxide gas, and in the manufacture of electroplating solutions. It is hazardous to the upper respiratory system and eyes and is toxic by inhalation, ingestion, and contact. Symptoms of exposure include eye, nose, and throat irritation; coughing; and difficulty breathing. OSHA exposure limit (TWA): 10 mg/m³ [total] or 5 mg/m³ [respiratory].

AMMS — Acronym for Advanced Multichannel Microwave Sensor.

amnesia — A loss of memory caused by brain damage or severe emotional trauma.

amniocentesis — A procedure during pregnancy by which a transabdominal perforation of the amniotic sac is made for the purpose of obtaining a sample of the amniotic fluid to determine maturation and viability of the fetus; also used to determine the presence of certain genetic disorders.

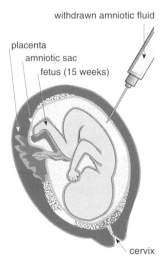

Amniocentesis

amniotic fluid — The watery fluid that surrounds the fetus or unborn child in the uterus to protect it from desiccation and shock.

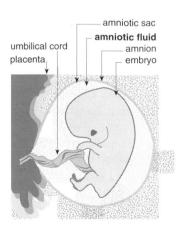

Amniotic Fluid

Amoeba (**ameba**) — A genus of single-celled ameboid protozoa, most of which are free-living; parasitic in humans.

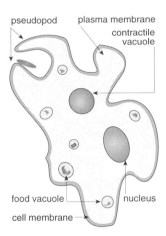

Amoeba

amorphous — Describing a noncrystalline solid having no orderly arrangement of molecules and neither definite form nor structure.

amorphous silica ($SiO_2 + 2H_2O = Si(OH)_4$) — A form of silica that does not exhibit a well-defined, characteristic diffraction pattern when an x-ray beam is passed through it.

AMOS — Acronym for air-management oversight system.

AMP — See adenosine monophosphate.

Amp — See ampere.

ampacity — The current-carrying capacity of electric conductors, expressed in amperes; used as a rating for power cables.

amperage — The strength in electric current measured in amperes.

ampere (A) or (Amp) — The meter-kilogram-second unit of electrical current equal to 6.25×10^{18} electrons (or 1 coulomb) per second; used for measuring the rate of flow of electricity.

amperometric titration — A means of measuring concentrations of certain substances in water based on the electric current that flows during a chemical reaction.

amperometry — A method of chemical analysis by techniques that involve measuring electric currents.

amphetamines — A group of nervous system stimulants that are subject to abuse because of their ability to produce wakefulness and euphoria; abuse leads to compulsive behavior, paranoia, hallucinations, and suicidal tendencies.

amphiboles — Grayish-brown minerals present in many igneous and metamorphic rocks and made up of

A

50% silica, 2% magnesium, and 40% iron. The formula typically is 5.5FeO, 1.5MgO, 8SiO$_2$, H$_2$O.

ampicillin — An antibiotic used in transformation of *Escherichia coli* through interference with formation of the bacterial cell wall.

ampicillin resistance — A property of bacterial cells that allows growth in the presence of the antibiotic ampicillin.

amplification — (*physics*) An increase in the magnitude or strength of an electric current or other physical quantity or force.

amplifier — A device consisting of one or more vacuum tubes or transistors and associated circuits used to increase the strength of an electrical signal where the weak signal is changed into a stronger signal with larger waves.

amplitude — (*physics*) The maximum vertical distance the particles in a wave are displaced from the normal position at rest; one half the extent of a vibration, oscillation, or wave where the amplitude depends upon its energy.

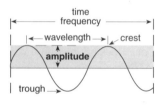

Amplitude

ampoule — A small glass vial containing ethylene oxide.

AMPS — Acronym for automatic mapping and planning system.

AMRIR — Acronym for advanced medium-resolution imaging radiometer.

AMTS — Acronym for advanced moisture and temperature sounder.

amu — See atomic mass unit.

amygdaloid — An almond shaped mass.

$$CH_3 - C - O - CH_2 - CH_2 - CH_2 - CH_2 - CH_3$$
$$\overset{\|}{O}$$

n-Amyl acetate

n-amyl acetate (CH$_3$COO[CH$_2$]$_4$CH$_3$) — A colorless liquid with a persistent banana-like odor; MW: 130.2, BP: 301°F, Sol: 0.2%, Fl.P: 77°F, sp. gr. 0.88. It is used as a vehicle solvent in the manufac-

ture of cellulose nitrate lacquers, lacquer thinners, adhesives, nail enamel, nail enamel removers, and paint; as an extractant of penicillin during production of antibiotics. It is hazardous to the eyes, skin, and respiratory system and is toxic through inhalation, ingestion, and contact. Symptoms of exposure include eye and nose irritation, narcosis, and dermatitis. OSHA exposure limit (TWA): 100 ppm [air].

sec-amyl acetate (CH$_3$COOCH(CH$_3$)C$_3$H$_7$) — A colorless liquid with a mild odor; MW: 130.2, BP: 249°F, Sol: slight, Fl.P: 89°F, sp. gr. 0.87. It is used during spray and dip application and manufacture of lacquers, varnishes, enamels, and metallic paints; in textile-sizing; as a solvent for chlorinated rubber; in the repair of motion picture film. It is hazardous to the respiratory system, eyes, and skin and is toxic through inhalation, ingestion, and contact. Symptoms of exposure include eye and nose irritation, narcosis, and dermatitis. OSHA exposure limit (TWA): 125 ppm [air].

amyl alcohols (C$_5$H$_{11}$OH) — Colorless liquids with the characteristic odor of alcohol; MW: 88.15, BP: NA, Sol: slightly soluble in water but mix readily with organic solvents. They are used as solvents for resins and gums and as diluents or lubricants, printing inks, and lacquers. They are hazardous to the respiratory tract, digestive tract, eyes, and skin. They may cause coughing, narcosis, headache, dizziness, nausea, and diarrhea, as well as eye, nose, and throat irritation. ACGIH exposure limit: (TLV-TWA) 100 ppm.

amylase — A class of enzymes that accelerates the hydrolysis of starches and other carbohydrates.

amyloid — A chemical structure accumulating in brain tissue as a result of various diseases; a waxy translucent substance consisting of proteins in combination with polysaccharides that is deposited in some animal organs and tissues under abnormal conditions.

amyloidosis — The buildup of amyloid in concentrations that cause damage to the tissues or actions of tissues in the body.

anabolic — Pertaining to anabolism.

anabolism — The constructive process of metabolism by which living cells convert simple substances into more complex compounds, especially into living matter; the method of synthesis of tissue structure.

anaerobe — A microorganism that grows and lives in complete or almost complete absence of oxygen.

anaerobic — A condition in which free dissolved oxygen is not present.

anaerobic bacteria — Any bacteria that live and grow in the absence of free oxygen; two types are facultative and obligate.

anaerobic biological treatment — A waste treatment process using anaerobic or facultative organisms in the absence of air to reduce organic matter in waste.

anaerobic contact process — An anaerobic waste treatment process in which the microorganisms responsible for waste stabilization are removed from the treated effluent stream by sedimentation or other means and held in or returned to the process to enhance the rate of treatment.

anaerobic decomposition — Reduction of the net energy level and changing chemical composition of organic matter caused by microorganisms acting in an environment free of oxygen.

anaerobic digestion — Biological stabilization of sludge through partial conversion of putrescible matter into liquid, dissolved solids, and gaseous byproducts with some destruction of pathogens in an oxygen-free environment.

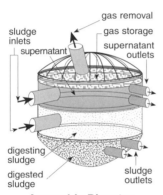

Anaerobic Digester

anaerobic infection — An infection caused by an anaerobic organism, usually occurring in puncture wounds that exclude air or in tissue that has diminished oxygen-reduction potential because of trauma, necrosis, or overgrowth with bacteria.

anaerobic respiration — The process by which an organism is capable of releasing energy without the use of oxygen.

analgesic — Any substance, such as salicylates, morphine, or opiates, used to relieve pain.

analog — A continuously varying electronic signal; (*biology*) an organ that is similar to another organ; (*chemistry*) a substance that resembles another substance structurally but differs from it in function, especially of concern when it interferes with a reaction because of its structural similarity. Also spelled analogue.

analog energy model — A research tool comparing the analogy that exists between electrical flow and heat flow; used to construct electrical devices for the study of complex heat flow phenomena.

analog-to-digital converter — A device for converting analog information such as temperature or ECG waveforms into digital form for processing by a digital computer; it converts analog signals to an equivalent digital form in either a binary code or a binary-coded decibel code.

analogous — Referring to structures that are similar in function and general appearance but not in origin.

analogue — See analog.

analogy — A resemblance in some particulars between things otherwise unlike.

analysis — (*law*) An examination of evidence; (*chemistry*) a determination of the nature or amount of one or more components of a substance.

analysis of variance — (*statistics*) A series of statistical procedures for determining whether the differences among two or more groups of scores are attributable to chance alone.

analyte — The chemical component of a sample to be determined or measured.

analytical air sampler — The sensor of a ventilating device that reacts directly with a contaminant or indirectly through a reagent; the change in chemical or physical properties of the sensor or the reagent is a measure of the concentration of the contamination in the air sample.

analytical epidemiological studies — Research used to test a specific hypothesis regarding the etiology of the disease being studied.

analytical epidemiology — The study of causes, determining the relatively high or low frequency of disease in specific groups by formulating hypotheses based on actual observations of existing diseases, conducting studies of outbreaks of disease, and evaluating the data to determine the accuracy of the original hypothesis. Age, sex, ethnicity, race, social class, occupation, and marital status are the principal variables.

analytical grade — The highest available purity of a chemical.

analytical methods — Methods of chemically analyzing the quantities of concentrations of air pollutants in an air sample.

analyzer — A device that conducts periodic or continuous measurement of some factor such as chlorine or fluoride.

anamorph — The asexual stage of fungal production in which cells are formed by the process of mitosis.

anaphase — The stage of mitosis or meiosis in which the chromosomes move from the equatorial plate toward the poles of the cell.

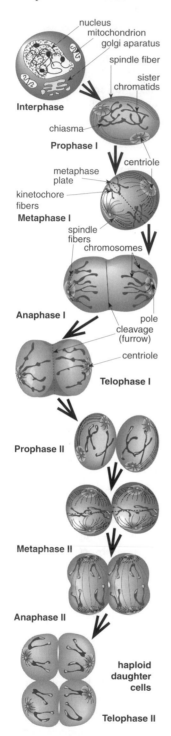

Anaphase

anaphylactic shock — A severe and sometimes fatal systemic hypersensitivity reaction to a sensitizing substance such as a drug, vaccine, type of food, insect venom, or chemical that, within seconds from the time of exposure, can cause respiratory distress and vascular collapse.

anaphylaxis — An unusual or exaggerated allergic reaction of an organism to foreign protein or other substances following a primary injection of the protein. It can produce reactions ranging from local swelling and redness to systemic collapse of the vascular network by permitting loss of fluid from blood vessels into the interstitial fluid area.

anaplastic — Cancer cells that divide rapidly and bear little or no resemblance to normal cells.

anatomy — The scientific study of the structures of plants and animals.

ANC — Acronym for acid-neutralizing capacity.

ancillary benefit — Additional benefits derived from environmental policy that is designed to control one type of pollution while reducing the admissions of other pollutants as well.

Ancylostomiasis — (*disease*) See hookworm disease.

androgen — Any of a class of steroid hormones that promotes male secondary sexual characteristics.

anecdotal — Knowledge of something based on isolated observations and not scientific study.

anecdotal evidence — Reports of individual cases of disease that are not scientifically validated but which may provide information that can be useful in prevention and/or treatment.

anemia — A condition marked by a low level of hemoglobin or circulating red blood cells; a symptom of a variety of disorders due to poor diet, blood loss, exposure to industrial poisons, diseases of the bone marrow, or any other upset in the balance between blood and blood production.

anemometer — An instrument to measure and indicate wind velocity.

Cup Anemometer

anemometry — The study of measuring and recording the direction and speed of wind.

aneroid barometer — A barometer that uses an aneroid capsule, a corrugated metal container from which the air has been removed. A spring inside of the container prevents air pressure from collapsing it completely. As air pressure increases, the top of the box bends in; as pressure decreases, the top bows out. Gears and levers transmit these changes to a pointer on a dial; the pull-down lever moves the pointer forward.

anesthesia — The entire or partial loss of sensation or consciousness following administration of an anesthetic.

anesthesia machine — A piece of equipment used for administering inhalant anesthetic agents.

anesthetic — A substance, such as ether, used to desensitize a reaction to pain or produce unconsciousness by inhibiting nerve activity.

aneurysm — A sac formed by localized, abnormal dilatation of a blood vessel that increases pressure against adjacent tissue or organs and creates the possibility of a rupture.

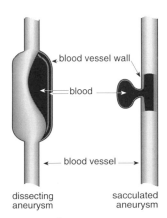

Aneurysm

anger — An emotional reaction characterized by extreme displeasure, rage, indignation, or hostility.

angina pectoris — Acute pain in the chest resulting from increased blood supply to the heart muscle. It may be brought on by physical activity or emotional stress that places added burden on the heart and increases the need for additional blood supply to the myocardium; may be accompanied by pain radiating down the arms, up into the jaw, or to other sites.

angina — See angina pectoris.

angiogenesis — A biological process by which cancerous growths send out chemical signals to promote the growth of blood vessels to feed the tumor.

angiography — A picture of a blood vessel that has been injected with a dye or radiopaque material.

angiosarcoma — A type of malignant tumor containing many fine blood vessels.

angiospasm — A sudden transient constriction of a blood vessel.

angiostrongyliasis — (*disease*) A disease of the central nervous system involving the meninges due to the nematode *Angiostrongylus contonensis*. Symptoms include severe headaches, stiffness of the neck and back, and occasionally temporary facial paralysis. Incubation time is usually 1 to 3 weeks; found in Hawaii, the Pacific Islands, the Philippines, Australia, and Asia. The reservoir of infection is the rat. It is transmitted through the ingestion of raw or improperly cooked snails, prawns, fish, and land crabs but is not transmitted from person to person. Those suffering from malnutrition or disease are especially susceptible. It is controlled through the destruction of rats and through proper cooking of snails, fish, crab, and prawn.

Angiostrongylus — A genus of nematode parasites.

angle — The space formed by two diverging lines emanating from a common point and measured by the number of degrees.

angle of abduction — The angle between the longitudinal axis of a limb and a sagittal plane.

angle of incidence — (*physics*) The angle formed between the direction of a ray or wave of sound emanating from a common point and striking a surface and the perpendicular to that surface at the point of arrival.

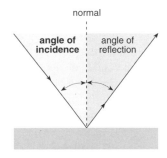

Angle of Incidence

angle of reflection — (*physics*) The angle formed between the direction of light or sound reflected by a surface and the line perpendicular to that surface at the point of reflection.

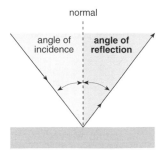

Angle of Reflection

A

angle of refraction — (*physics*) The angle that a ray of light makes with a line perpendicular to a surface separating two media.

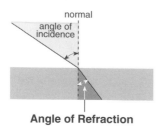

Angle of Refraction

angle of repose — The maximum angle at which loose material will come to rest when added to a pile of similar material.

angle stop — A shut-off valve arranged in a 90° configuration used to shut off the flow of water to a fixture.

Angoumois Grain Moth (*Sitotroga cerealella*) — A light grayish-brown or straw-colored moth with a satiny luster and wing expanse of 1/2 to 2/3 inches; the hind wings are fringed with long, dark setae and have a point at the tip like a finger which distinguishes the insect from the Crows Moth; a stored food product insect.

angstrom unit (Å) — A unit of measure of wavelength of optical spectra equal to 10^{-10} meters or 0.1 nanometers.

angular blocky — Cube-like soil structure where the aggregates have sides at nearly right angles and tend to overlap nearly block-like with six or more sides; all three dimensions are about the same; usually found in subsoil or B horizon.

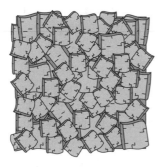

Angular Blocky Soil

angulation — An angular shape or formation.

anhidrosis — the absence of sweating.

anhydrase — An enzyme involved in the removal of water from a compound.

anhydride — A compound formed from an acid by the removal of water.

anhydrous — Containing no water.

aniline and homologs ($C_6H_5NH_2$) — A colorless to brown oily liquid with an aromatic amine-like odor; a solid below 21°F; MW: 93.1, BP: 363°F, Sol: 4%, Fl.P: 158°F, sp. gr. 1.02. This aromatic amine compound is a pale brown liquid at room temperature. One of the most important substances produced from coal tar, it is used in chemical synthesis and intermediates for rubber processing; in the synthesis of pharmaceuticals; in the manufacture of inks; in the synthesis of intermediates for artificial sweetening agents; in the synthesis of catalysts and stabilizers for hydrogen peroxide and cellulose. It is hazardous to the cardiovascular system, liver, kidneys, and blood and is toxic through inhalation, absorption, ingestion, and contact. Symptoms of exposure include headache, weakness, dizziness, cyanosis, ataxia, dyspnea on effort, tachycardia, and eye irritation; carcinogenic. OSHA exposure limit (TWA): 2 ppm [air] or 8 mg/m^3 [skin].

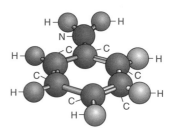

Aniline

animal bedding — Material, usually organic, that is placed on the floor of livestock quarters for animal comfort and to absorb excreta.

animal classification — A system of identification of animals and their natural relationships based on embryology, structure, or physiological chemistry. The system of classification in descending order is kingdom, phylum, class, order, family, genus, and species.

animal dander — See dander.

animal drugs — Drugs intended for use in the diagnosis, cure, mitigation, treatment, or prevention of disease in animals.

animal proteins — Proteins used in livestock feed that is derived from meat packing or rendering plants, surplus milk or milk products, and marine sources. These animal proteins can help contribute to the spread of disease in people.

animal studies — An investigation using animals instead of humans, with the expectation that the results in animals are pertinent to humans.

anion — A negatively charged ion; during electrolysis, it moves to the anode.

anion exchange — The chemical process where negative ions of one chemical are preferentially replaced by negative ions of another chemical, with the net effect in a water supply of removing unwanted ions.

anion exchange capacity — The sum total of exchangeable anions expressed as centimoles of negative charge per kilogram of soil.

anion exchange resin — A simple organic polymer with high molecular weight that exchanges anions with other ions in solution.

anionic — Characterized by an active, especially surface-active, anion.

anisidine (*o-, p*-isomers) ($NH_2C_6H_4OCH_3$) — The *o*-isomer is a red or yellow oily liquid with an amine-like odor and is solid below 41°F; the *p*-isomer is a yellow to brown crystalline solid with an amine-like odor; MW: 123.2, BP: 437°F for (*o*-isomer); 475°F for (*p*-isomer), Sol: insoluble for (*o*-isomer) and moderate for (*p*-isomer). Fl.P: 244°F (oc) for *o*-isomer and 86°F for *p*-isomer, sp. gr. 1.10 (*o*-isomer) and 107 for *p*-isomer. It is used in the synthesis of hair dyes; in the preparation of organic compounds; as a corrosion inhibitor for steel storage. It is hazardous to the cardiovascular system, blood, kidneys, and liver and is toxic through inhalation, absorption, ingestion, and contact. Symptoms of exposure include headache, dizziness, and cyanosis; carcinogenic. OSHA exposure limit (TWA): 0.5 mg/m³ [skin].

anisotrophic mass — A mass having different properties in different directions at any given point.

anisotropy — The characteristic of a substance for which a physical property depends upon direction with respect to the structure of the material; doubly refracting or having a double polarizing power.

ankylosing spondylitis — An inflammation of the vertebra accompanied by inflammation and degeneration of the connective tissue.

anneal — To treat a metal, alloy, or glass by heat with subsequent cooling to soften and render metals less brittle and make manipulation easier.

annealing — The process of cooling glass slowly to prevent brittleness.

Annelida — A diverse phylum of metazoan invertebrates comprised of segmented worms.

annihilation — (*physics*) The destruction of a particle and its antiparticle as a result of their collision; their energy is shared by lighter particles that emerge from the collision.

annual load factor — A factor equal to the energy generated in a year divided by the product of the peak demand for a year and the number of total hours in a year.

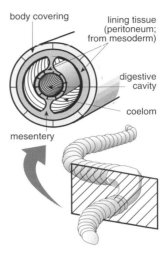

Annelid

annual operating expenses — The costs of operation and maintenance as well as capital-related charges, except interest or return on capital.

annual plant — A plant that completes its growth cycles in one growing season.

annualized cost — The equivalent annual cost that is equal to the revenue requirement, including annual operating expenses plus interest and return on capital.

annular — Ring shaped.

annular space — (*wells*) The space between the side of an excavation and the casing or wall of a well.

anode — The positive terminal of an electrolytic cell at which current enters and to which negative ions are attracted.

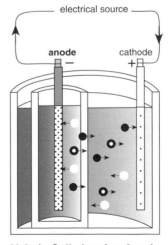

Voltaic Cell showing Anode

anomaly — Deviation of temperature or precipitation in a given region over a specified period from the normal value for the same region; the angular distance of an Earth satellite or planet from its perigee as seen from the center of the Earth.

A

anomie — A state of apathy, alienation, anxiety, personal disorientation, and distress resulting from the loss of social norms and goals.

Anopheles — A genus of mosquitoes, many species of which may transmit malaria parasites to humans.

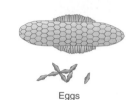

Eggs

Larva

Pupa

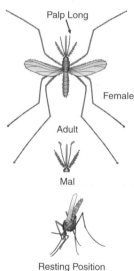

Palp Long

Female

Adult

Mal

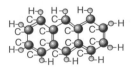

Resting Position

Anopheles **Life Cycle**

anorexia — Lack or loss of appetite for food.

ANOVA — See analysis of variance.

anoxia — Absence of oxygen in the tissues; the condition is accompanied by deep respirations, cyanosis, increased pulse rate, and impairment of coordination.

anoxic — Without oxygen.

anoxygenic photosynthesis — A type of photosynthesis in green and purple bacteria in which oxygen is not produced.

ANPR — Acronym for advance notice of proposed rule-making.

ANPRM — See advance notice of proposed rule-making.

ANS — See autonomic nervous system.

ANSIR — Acronym for Awareness of National Security Issues and Response.

answer — (*law*) A legal proceeding made after filing a claim of ownership in a seizure; claimants may file an answer in which they deny any or all of the allegations of the complaint for forfeiture. If this filing is made, a contest may ensue.

antagonism — A situation in which two chemicals, organisms, muscles, or physiologic actions, upon interaction, interfere in such a way that the action of one partially or completely inhibits the effects of the other; see amensalism.

antagonist — (*anatomy*) A muscle opposing the action of another muscle; necessary for control and stability of action.

antarctic — The South Pole or region near it.

antemetic — See antiemetic.

antemortem — Performed or occurring before death.

antenna — One of the sensory appendages on the head of many arthropods; a device for radiating or receiving radio waves.

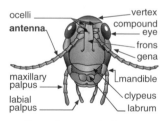

Grasshopper Antenna

anterior — Situated near the head or front end of an object.

anthelmintic — A substance that destroys or prevents the development of parasite worms.

anthracene ($C_{14}H_{10}$) — A polycyclic aromatic hydrocarbon formed during the incomplete combustion of coal, oil, gas, garbage, or other organic substances and found especially at hazardous waste sites; MW: 178.24. Carcinogenicity is not known; see polycyclic aromatic hydrocarbons.

Anthracene

anthracite — A hard, dense, metamorphic coal with a glassy texture formed from soft coal by heat and pressure within the Earth; cannot be used for making coke. It burns with a nonluminous flame and little smoke or odor; also known as hard coal, kilkenny coal, and stone coal.

anthrax — (*disease*) An acute bacterial disease of sheep and cattle usually affecting the skin; incubation time is within a 7-day period. Symptoms of exposure are mild and nonspecific initially and resemble a common upper respiratory infection. These are followed by acute symptoms of respiratory distress, fever, and shock; death may occur. It is caused by *Bacillus anthracis*; found in most industrial countries in occupational settings where people process hides, hair, bone and bone products, and wool. The reservoir of infection is soil contaminated with spores of *B. anthracis*, which come from the skins and hides of infected animals. It is transmitted by contact with the tissues of the animals or the skins; it is not communicable from person to person. Susceptibility is not known. It is prevented by immunization; dust control; thorough washing, disinfection, or sterilization of hair, wool, or hides; vaccination of animals; and control of waste.

anthrocosis — (*disease*) A disease of the lungs caused by prolonged inhalation of dust containing particles of carbon and coal.

anthropogenic — A substance made by people or resulting from human activities.

anthropogenic forcing — The influence exerted on a habitat or chemical environment by humans.

anthropogenic radionuclides — Radionuclides produced as a result of human activity.

anthropometry — (*ergonomics*) Measurement of the size and proportions of the human body to provide the dimensional data required for the positioning of controls and sizing of work spaces.

antibacterial — Having the capacity to destroy or suppress the growth or reproduction of bacteria.

antibiosis — The inhibition of growth of a microorganism by a substance produced by another microorganism.

antibiotic — A chemical substance produced synthetically or by a bacterium mold or other fungus plant and used to kill bacteria and stop the growth of disease microorganisms in the body.

antibiotic resistance — The tendency for bacteria not killed by a dose of antibiotics that has been used repetitively to develop a genetic resistance and pass the trait on to new generations of the bacteria.

antibiotic sensitivity tests — A laboratory method for determining the susceptibility of bacterial infections to therapy with antibiotics.

antibody — An immunoglobulin protein molecule, produced by certain white blood cells, that reacts with a specific antigen, inducing its synthesis with similar molecules. It is classified according to its mode of action as an agglutinin, bacteriolysin, hemolysin, opsonin, or precipitin. Antibodies are synthesized by B lymphocytes that have been activated by the binding of an antigen to a cell surface receptor.

antibody titer — A measure of the amount of antibodies present, usually shown in units per milliliter of serum.

anticaking agent — (*food*) An agent that prevents lumping by absorbing moisture.

anticoagulant — Any substance that suppresses, delays, or prevents coagulation of the blood; a chemical used in bait to destroy rodents. It affects the walls of the small blood vessels and keeps the blood from clotting, causing the animals to bleed to death.

anticonvulsant — A medicine to stop, prevent, or control seizures.

anticyclonic flow — Air flow produced about a high-pressure system by the combination of two forces: air flowing clockwise in the Northern Hemisphere and counterclockwise in the Southern Hemisphere.

antidegradation clause — A regulatory concept that limits deterioration of existing air or water quality by restricting the addition of pollutants.

antidepressant — A substance or technique that prevents or relieves depression.

antidiarrheal — A drug or other substance that relieves the symptoms of diarrhea.

antidiuretic — Any agent that prevents the excretion of urine.

antidiuretic hormone (ADH) — A polypeptide hormone from the posterior lobe of the pituitary gland that suppresses the secretion of urine; also known as vasopressin.

antidote — An agent to relieve, prevent, or counteract the effects of a poison.

antiemetic — A drug that reduces or eliminates nausea and vomiting.

antiestrogen — A chemical that interferes with the interaction between estrogen and its binding site in the cell.

antifatigue mats — (*ergonomics*) Padding on the floor designed to reduce musculoskeletal fatigue associated with static standing.

antifreeze — A substance added to a liquid to lower its freezing point.

antifungal — A substance that kills fungi or inhibits their growth or reproduction.

antigen — Any substance capable, under appropriate conditions, of inducing a specific immune response and of reacting with the products of that response. Antigens may be soluble substances such as toxins or foreign proteins or particulates such as bacteria and tissue cells.

A

antigen–antibody reaction — A process of the immune system in which immunoglobulin-coated B cells recognize an intruder or antigen and stimulate antibody production with the assistance of T cells.

antigen-binding site — The part of an immunoglobulin molecule that specifically binds antigen.

antigen determinant — A small area on the surface of an antigen molecule that fits a combining site of an antibody molecule and binds the antigen in the formation of an antigen–antibody complex.

antigenic draft — The tendency of a virus or other microorganism to alter its genetic makeup periodically, resulting in a mutant antigen requiring new antibodies and vaccines to change its effects.

antigenicity — The ability to cause the production of antibodies.

antiglobulin — An antibody occurring naturally or prepared in laboratory animals to be used against human globulin.

antiglobulin test — A test for the presence of antibodies that coat and damage red blood cells as a result of several diseases or conditions.

antihelminthic — See antihelmintic.

antihelmintic — A chemical agent used to destroy tapeworms in domestic animals; also spelled antihelminthic.

antihistamine — A drug that counteracts the effects of histamine; a normal body chemical that is believed to cause the symptoms of persons who are hypersensitive to various allergens; used to relieve the symptoms of allergic reactions, especially hay fever and other allergic disorders of the nasal passages.

antihypertensive — A substance or procedure that reduces high blood pressure.

antiinflammatory — A drug used to fight inflammation.

antiknock additives — Substances that allow fuel combustion to occur in an even manner in an internal combustion engine.

antimalarial — A substance that destroys or suppresses the development of malaria plasmodia; a procedure that destroys the mosquito vectors of the disease.

antimetabolite — A drug or other substance that resembles a normal human metabolite and interferes with its function in the body, usually by competing for the metabolites receptors or enzymes.

antimicrobial agent — A chemical compound that kills microorganisms or suppresses their multiplication or growth. Antimicrobial agents may interfere with the synthesis of the bacterial cell wall, which results in cell lysis and a weakening or rupturing of the cell wall, interferes with the synthesis of nucleic acids, changes the permeability of the cell membrane, and interferes with metabolic processes.

antimitotic — Inhibition of cell division.

antimony (Sb) and compounds — This metal occurs as a silver-white, lustrous, hard, brittle solid; as scale-like crystals; or as a dark gray, lustrous powder; MW: 121.8, BP: 2975°F, Sol: insoluble, Fl.P: N/A, sp. gr. 6.69. It is found during the crushing and transferring of antimony ore; during production of lead/antimony alloys; during machining, grinding, buffing, and polishing of metal products containing antimony; during the manufacture of paints, pigments, enamels, glazes, ceramics, and glass. It is hazardous to the respiratory system, cardiovascular system, skin, and eyes and is toxic by inhalation and contact. Symptoms of exposure include nose, throat, and mouth irritation; cough; dizziness; headache; nausea; vomiting; diarrhea; stomach cramps; insomnia; anorexia; skin irritation; inability to smell properly; and cardiac abnormalities. OSHA exposure limit (TWA): 0.5 mg/m^3 [air].

antineoplastic — Inhibiting the maturation and proliferation of malignant cells.

antineoplastic hormone — A chemical produced by an endocrine gland or a synthetic analog of the naturally occurring compound used to control certain types of cancer.

antinode — The position of maximum amplitude in a standing wave; also known as a loop.

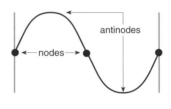

Antinode

antioxidant — A compound used to prevent deterioration by oxidation; used in food preservation as a retardant against rancidity of fats and to preserve the appearance and taste of fruits by preventing discoloration.

antiparasitic — A substance or procedure that kills parasites or inhibits their growth or reproduction.

antipruritic — A substance or procedure that helps prevent or relieve itching.

antipsychotic — A substance or procedure that counteracts or decreases symptoms of psychosis.

antipyretic — A substance or procedure that reduces fever.

antipyretic bath — A bath in which tepid water is used to reduce the temperature of the body.

antipyrotic — The treatment of burns or scalds.

antirheumatic — A substance or procedure used to relieve the symptoms of any painful or immobilizing disorder of the musculoskeletal system.

antiseptic — Any substance that inhibits the growth of infectious bacteria.

antiseptic dressing — A dressing containing an antiseptic, germicide, or bacteriostat applied to a wound or incision to prevent or treat infection.

antiserum — The serum of an animal or human that contains antibodies against a specific disease and which confers passive immunity.

antisiphon air gap — A device used to prevent the backflow of contaminated water into a potable water system.

antisiphon valve — A device installed on irrigation piping designed to prevent the drawing of contaminated groundwater into the domestic water supply system.

antiterrorism — Application of a full range of federal and state programs as well as other activities against terrorism, domestically and abroad.

antitoxin — A particular kind of antibody produced in the body in response to the presence of a toxin.

antitrust laws — Laws designed to promote open markets by limiting practices that reduce competition.

antiviral — A substance destructive to viruses.

ANTU — Alpha-naphthyl thiourea, a highly toxic rodenticide that causes death by inhibiting the clotting of blood and causing internal hermorrhaging; should not be used more than once a year.

anular — A ring-shaped lesion surrounding a clear, normal, unaffected part of skin.

anuria — Complete suppression of urine output by the kidney.

anxiety attack — An acute reaction manifested by intense anxiety and panic.

AO — Acronym for atmosphere–ocean.

AOAC Use Dilution Confirmation Test — See Association of Official Analytical Chemists Use Dilution Confirmation Test.

AOC — Acronym for abnormal operating conditions; see area of concern; see area of contamination.

AOL — Acronym for Airborne Oceanographic Lidar.

aorta — The large artery arising from the left ventricle that is the main trunk from which the systemic arterial system proceeds.

AP portable chest radiograph — A radiographic examination of the chest performed in the room of an immobilized patient with a portable x-ray machine.

APA — Acronym for Administrative Procedures Act; acronym for American Plastics Association.

apathy — Absence or suppression of emotion, feeling, or concern and an indifference to surroundings and life in general.

APCD — Acronym for Air Pollution Control District.

APE — Acronym for acute pulmonary edema.

APELL — Acronym for Awareness and Preparedness for Emergencies at Local Level.

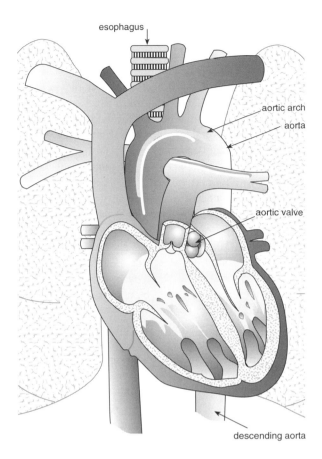

Aorta

aperture — An opening; (*optics*) the diameter of the opening through which light passes in a camera, telescope, or other optical instrument.

apex — The top or tip of a structure.

APF — See assigned protection factor.

APHA–PHS housing code — See American Public Health Association–Public Health Service housing code.

aphasia — Partial or total loss of the ability to articulate ideas in any form as a result of brain damage to the cerebral cortex.

API separator — A facility developed by the Committee on Disposal of Refinery Waste of the American Petroleum Institute for separation of oil from wastewater in a gravity differential; equipped with means for recovering the separated oil and removing sludge.

apical — Related to or situated at an apex.

APICE — See Association for Professionals in Infection Control and Epidemiology.

aplastic anemia — A condition in which the bone marrow fails to produce an adequate number of red blood corpuscles.

APMP — Acronym for Acid Precipitation Mitigation Program.

apnea — The absence of spontaneous respiration.

apogee — The point at which a satellite on an elliptical orbit path is farthest from the Earth.

apoptosis — The self-destruction of a cell.

apparatus — A device or a system made of different parts that act together to perform a function.

appeal — (*law*) A formal request for the review of a lawsuit and reversal of the verdict or decree.

appendix — A supplement.

applet — (*computer science*) A computer program that allows for animation or other interactive functions on a Web page.

appliance — A device or instrument designed for a specific purpose.

application — The placing of a substance on a plant, animal, building, or soil or the releasing of it into the air or water to destroy pests; (*computer science*) a task addressed by a computer system.

application efficiency — Ratio of the average depth of irrigation water infiltrated and stored in the root zone to the average depth of an irrigation water applied (expressed as a percent).

application program package — A set of computer programs designed for a specific task.

application rate — The quantity of coating material applied per unit area.

applicator — A person or piece of equipment that applies pesticides to destroy pests or prevent their damage.

applied anatomy — Study of the structure and morphology of the organs of the body as they relate to the diagnosis and treatment of disease.

applied dose — The amount of a substance in contact with the primary absorption boundaries of an organism and available for absorption.

applied psychology — A branch of psychology that emphasizes practical rather than theoretical approaches and objectives.

applied research — Research conducted to address a specific problem for the purpose of application, productivity, or commercial gain.

applied water — Water delivered to a user.

apply uniformly — To spread or distribute a pesticide chemical evenly.

apposition — The placing of objects in close proximity to each other.

appropriation — (*water*) The amount of water legally set apart or assigned to a particular purpose or use.

approved — Acceptable to a regulatory authority based on determination of conformity with widely accepted public health principles, practices, and generally recognized industry standards.

approved landfill — A site used for disposal of asbestos-containing and other hazardous waste that has been given approval by the U.S. EPA and state and local authorities.

approved source — A source acceptable to a regulatory authority based on determination of conformity with principles, practices, and generally recognized standards that protect public health.

APR — See air-purifying respirator.

aptitude — A natural ability, tendency, talent, or capability to learn, understand, or acquire a particular skill.

apyretic — An afebrile condition.

apyrexia — An absence or remission of fever.

AQCP — Acronym for Air Quality Control Program.

AQCR — See Air Quality Control Region.

AQDHS — Acronym for Air Quality Data Handling System.

AQDM — Acronym for Air Quality Display Model.

AQI — See Air Quality Index.

AQL — See acceptable quality level.

AQM — Acronym for air quality maintenance.

AQMA — Acronym for air quality maintenance area.

AQMD — Acronym for Air Quality Management District.

aqua — Latin for "water".

aquaculture — Propagation and rearing of aquatic species in controlled or selected environments, including ocean ranching, as defined by the National Aquaculture Act of 1980.

aquatic — The plants or animal life living in, growing in, or adapted to water.

aquatic habitat — An environment characterized by the presence of standing or flowing water.

aquatic life — The plants, animals, and microorganisms that spend all or part of their lives in water.

aquatic plants or leaves — Plants or leaves that grow on, in, or near water.

aquatic toxicity — Adverse effects to marine life that result from being exposed to a toxic substance.

aqueduct — Any artificial canal or passage for conveying water.

aqueous — Relating to, containing, or prepared with water.

aqueous humor — The transparent fluid filling the cavity between the cornea and lens of the eye which helps maintain the shape of the eye.

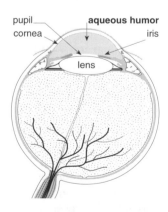

Aqueous Humor

aqueous solubility — The ability of a compound to dissolve in water; the maximum concentration of a chemical that will dissolve in pure water at a reference temperature.

aqueous solution — A solution in which water is the solvent.

aquiclude — See aquitard.

aquifer — The geological formation of a porous, water-bearing layer of sand, gravel, and rock below the Earth's surface that is capable of transmitting and storing economically important quantities of water for wells and springs.

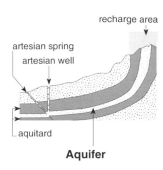

Aquifer

aquitard — A geological formation adjacent to an aquifer capable of transmitting and storing only negligible amounts of water; also known as an aquiclude.

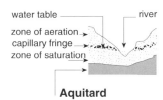

Aquitard

ARA — Acronym for Atmospheric Radiation Analysis.

arable — Soil or topographic features suitable for cultivation.

arachnid — Arthropod with four pairs of thoracic appendages.

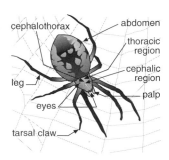

Arachnid
(Golden Orb Spider)

Arachnida — Common name for members of the class of arthropods to which spiders, mites, and ticks belong; characterized by four pairs of thoracic appendages.

arbovirus — Any of several viruses containing RNA spread by mosquitoes, ticks, and other arthropods. Approximately 90 viruses classified as arboviruses produce disease in people including yellow fever, dengue fever, and equine encephalitis.

Reservoir of infection: vertebrate hosts– birds, rodents, bats, reptiles

Vector: mosquito feeds and passes virus on–carries the virus for its lifetime.

Host: human or equine; accidental dead end host–transmission does not occur between hosts.

Arbovirus:Transmission of arboviral encephalitis

arc — A continuous part of the circumference of a circle or regularly curved line; a discharge of electricity through a gas; a line connecting a set of points that form one side of a polygon.

arc cutting — A type of thermal cutting of metal using temperature generated by the heat of an arc between an electrode and the base metal.

arc welding — Process that produces coalescence of metals by heating them with an arc; the application of pressure and the use of gases or filler metal are not necessary in this process.

arch furnace — A nearly horizontal structure that extends into a furnace and serves to deflect gases.

arch ignition — A refractory furnace, arch, or surface located over a fuel bed to radiate heat and to accelerate ignition.

archeology — The study of human cultures through the recovery and analysis of their material relics.

Archie — An information system offering an electronic directory service for locating information on the Internet by scanning the contents of more than 1000 sites around the world; contains more than 2 million file names.

architecture — Products created as a result of applying art and science to the design of buildings.

ARDS — Acronym for adult respiratory distress syndrome.

area — Any particular extent of a surface; a geographical region or tract.

area capacity table — A table giving reservoir storage capacity and sometimes surface areas in terms of elevation increments.

area monitoring — Routine monitoring of the level of radiation or radioactive contamination of any particular area, building, room, or equipment.

area of circle — Measurement of size found using the formula πr^2.

Area of Concern — (*CERCLA*) An area where releases of hazardous substances may have occurred or a location where there has been a release or threatened release into the environment of a hazardous substance, pollutant, or contaminant.

area of influence of a well — An area surrounding a well in which the piezometric surface has been lowered when pumping has produced a maximum steady rate of flow.

area of rectangle — Measurement of size found using the formula length × width.

area sanitary landfill method — A method in which waste is spread and compacted on the surface of the ground and a cover material is spread and compacted over the waste.

area source — Any source of air pollution from vehicles, other small engines, small businesses, and household activities released over a relatively small area but not classified as a point source.

area-wide sources — The sources of air pollution having emissions that are spread over a wide area, such as fireplaces, road dust, and farming operations.

arene — See aromatic hydrocarbon.

ARF — See acute respiratory failure.

argon — A colorless, and odorless, chemically inactive gas and one of six rare gases in the atmosphere

with an atomic weight of 39.95 and an atomic number of 18.

ARI — Acronym for Aquifer Risk Index.

arid — A climate or region in which precipitation is so deficient in quantity or occurs so infrequently that intensive agricultural production is not possible without irrigation; a relatively dry climate in which annual precipitation is less than 10 inches.

arithmetic growth — A pattern of growth that increases at a constant amount per unit time (e.g., 1, 2, 3, 4, 5, 6, 7).

arithmetic mean — The sum of a set of values or numbers divided by the quantity of those values or numbers; also known as mean.

ARMSAT — Acronym for Atmospheric Radiation Measurement Satellite.

Aroclor™ — See polychlorinated biphenyl.

AROM — Acronym for active range of motion.

aroma — An agreeable odor or pleasing fragrance.

aromatic — A strong but agreeable odor; an organic compound that contains a benzene ring or a ring with similar chemical characteristics.

aromatic alcohol — A fatty alcohol in which part of the hydrogen of the alcohol radical is replaced by a phenyl hydrocarbon.

aromatic compounds — Organic compounds that contain a benzene, naphthalene, or analogous ring.

aromatic hydrocarbon — Member of a class of organic compounds with a chemistry similar to that of benzene and characterized by large resonance energies; also known as arene.

Aromatic

benzene ring compounds of carbon and hydrogen

benzene ring

2 Nitrophenol

Aromatic Hydrocarbons

arousal — To awaken from sleep, to excite, or to create action or response to a sensory stimuli.

arraignment — (*law*) The appearance of a defendant in any criminal prosecution before the court to answer the allegations made against him or her and to enter a plea.

array — (*computer science*) A series of addressable data elements in the form of a grid or matrix; (*energy*) a set of photovoltaic modules or panels assembled for a specific application such as increased voltage or increased current.

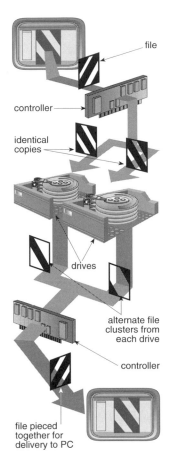

Array
(Mirrored Drive Array)

arrest — To inhibit, restrain, or stop, as over the course of a disease.

arrhythmia — Variation from the normal rhythm, especially of the heartbeat or respiration.

arroyo — A gully or channel cut by an intermittent stream.

arsenic (As) — A metal that is a silver-gray or tin-white, brittle, odorless solid; MW: 74.9, BP: sublimes, Sol: insoluble, Fl.P: NA, sp. gr. 5.73. It is used during the manufacture of insecticides, weed killers, and fungicides; as a wood preservative; in the manufacture of electrical semiconductors, diodes, and solar batteries; as an addition to alloys to increase hardening and heat resistance; during the smelting of ores. It is hazardous to the liver, kidneys, skin, lungs, and lymphatic system and is toxic through inhalation, absorption, ingestion, and contact. Symptoms of exposure include ulceration of the nasal septum, dermatitis, gastrointestinal disturbances, peripheral neuropathy, respiratory irrita-

tion, hyperpigmentation of the skin; carcinogenic. OSHA exposure limit (TWA): 0.010 mg/m^3 [air].

arsenic pentoxide (As$_2$O$_5$) — An odorless, white lumpy solid or powder that is corrosive; MW: 229.84, BP: decomposes, Sol: 150 g/100 ml at 16°C, Fl.P: nonflammable, sp. gr. 4.32 at 25°C. Arsenic is a naturally occurring element that is used to make glass, cloth, and electrical conductors and is also commonly used in the manufacture of insecticides, fungicides, and wood preservatives and in dyeing and printing. It is hazardous to the respiratory system, digestive system, skin, eyes, and other mucous membranes. Acute exposure results in nerve damage, weakness of arms and legs, poor appetite, nausea and stomach cramps, and, in high exposures, abnormal EKGs. Inhalation of inorganic arsenic compound products is the most common cause of industrial poisonings. Chronic exposure may cause reproductive damage in humans and cancer. OSHA exposure limit (8 hour TWA): 0.005 mg/m^3.

arsenic trichloride (AsCl$_3$) — A colorless or pale yellow, oily liquid with an acrid odor; MW: 181.28, BP: 130.21°C, Sol (water): 11 g/100 ml at 20°C, Fl.P: nonflammable, sp. gr. 2.149 at 25°C. Used to make other arsenic compounds such as pharmaceuticals, pesticides, ceramics, electrical semiconductors, fungicides, wood preservatives, and growth stimulants for plants and animals, as well as for veterinary purposes. It is hazardous to the respiratory system, digestive system, skin, eyes, and other mucous membranes. Acute exposure causes skin burns; nose, throat, and airway irritation; and severe irritation of the eyes. Chronic exposure causes genetic changes in living cells and becomes a cancer risk. OSHA exposure limit (8-hour TWA): 0.005 mg/m^3.

arsenic trioxide (As$_2$O$_3$) — An odorless white powder or colorless crystal; MW: NA, BP: 465°C, Sol (water): slight, Fl.P: nonflammable, sp. gr. 3.738 at 25°C. Used to manufacture glass, cloth, electrical semiconductors, fungicides, wood preservatives, and growth stimulants for plants and animals, as well as for veterinary purposes. It is hazardous to the respiratory system, digestive system, skin, eyes, and other mucous membranes. Acute exposure causes burning, itching, and a rash on the skin; eye damage; poor appetite; nausea and vomiting; muscle cramps; and, for high exposures, effects on the heart. Chronic exposure causes reproductive damage and skin and liver cancer. OSHA exposure limit (8-hour TWA): 0.005 mg/m^3.

arsenical — Containing or pertaining to arsenic.

arsine (AsH$_3$) — A colorless gas with a mild, garlic-like odor; MW: 78.0, BP: –81°F, Sol: 20%, Fl.P: NA

A

(gas). It is used during the manufacture of semiconductors and gallium arsenide; during the refining of metal ores containing arsenic; during the cleaning of metal equipment, electroplating of metals, metallic pickling, soldering and etching, photo-duplication. It is hazardous to the blood, kidneys, and liver and is toxic through inhalation. Symptoms of exposure include headache, malaise, weakness, dizziness, dyspnea, back and abdominal pain, nausea, vomiting, bronze skin, jaundice, and peripheral neuropathy; carcinogenic. OSHA exposure limit (TWA): 0.05 ppm [air] or 0.2 mg/m^3.

arterial gas embolism — A condition where bubbles of air from a ruptured lung segment under pressure enter the pulmonary circulation and travel to the arterial circulation, resulting in obstruction of the flow of blood through the vessel.

arteriole — A tiny artery that branches to become capillaries.

arteriosclerosis — Thickening of the lining of arterioles, usually due to hyalinization of fibromuscular hyperplasia.

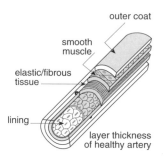

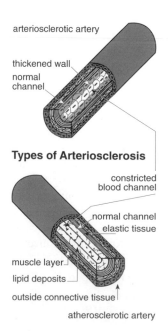

Types of Arteriosclerosis

artery — A large, muscular vessel that carries blood away from the heart toward other body tissue.

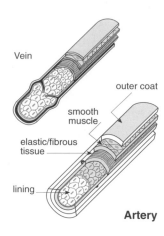

Artery

artesian — Water held under pressure in porous rock or soil confined by impermeable geological formations.

artesian well — A pressurized aquifer located between two nonpermeable rock layers penetrated by a pipe or conduit causing the water to come to the surface under its own internal pressure.

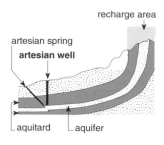

Artesian Well

arthritis — Any inflammation affecting joints or their component tissue. Arthritis and other rheumatic diseases are the major cause of chronic disability in the United States.

arthrogram — Radiograph of a joint after injection of a contrast medium.

arthrography — A method of radiographically visualizing the inside of a joint by injecting air or a contrast medium.

Arthropoda — The largest phylum in the animal kingdom, comprised of invertebrates with jointed legs and including insects, spiders, and crabs.

arthroscopy — Examination of the interior of a joint performed by inserting a specially designed endoscope through a small incision; permits biopsy of cartilage or synovia, diagnosis of a torn meniscus, and in some instances the removal of material.

artifactual association — A statistically significant association that is spurious because of some error in sampling, measurement, or analysis of data.

artificial fiber — A filament made from materials such as glass, rayon, or nylon.

artificial intelligence — A system that makes it possible for a machine to perform functions similar to human intelligence such as learning, reasoning, self-correcting, and adapting.

artificial radioactivity — Radioactivity produced by particle bombardment or electromagnetic irradiation as opposed to natural radioactivity; also known as induced radioactivity.

artificial recharge — An assortment of techniques used to increase the amount of water infiltrating an aquifer that has been depleted by abnormally large withdrawals.

artificial respiration — Any method of forcing air into and out of the lungs to maintain breathing in the absence of normal respiration.

artificial wetland — An area that under natural conditions was not a wetland but which now shows wetland characteristics because of human activities.

aryl — Chemical classification of hydrocarbon groups attached to compounds where the carbon atoms are arranged in aromatic rings.

aryl hydrocarbon hydroxylase — An enzyme that converts carcinogenic chemicals and tobacco smoke in polluted air into active carcinogens within the lungs.

As — See altostratus clouds; see arsenic.

ASAP — Acronym for as soon as possible.

ASAR — Acronym for Advanced Synthetic Aperture Radar.

asbestos — A generic term for a group of six naturally occurring impure magnesium silicate fibrous materials which have high tensile strength, are heat and chemical resistant and inert, and do not evaporate or dissolve. These materials include amosite: $((Mg,Fe)_7Si_8O_{22}(OH)_2)_n$, chrysotile: $Mg_3Si_2O_5(OH)_4$, tremolite: $(Ca_2Mg_5Si_8O_{22}(OH)_2)_n$, actinolite: $(Ca_2(Mg,Fe)_5Si_8O_{22}(OH)_2)_n$, anthophyllite: $((Mg,Fe)_7Si_8O_{22}(OH)_2)_n$, and crocidolite: $(NaFe_3^{2+}Fe_2^{3+}Si_8O_{22}(OH)_2)_n$. The basic unit of asbestos-class minerals is the silicate (SiO_4) group; this group can form a variety of polymeric structures through formation of Si–O–Si bonds. The amphibole class of asbestos has a characteristic polymeric structure consisting of a linear double chain that crystallizes into long, thin, straight fibers; this class includes amosite, crocidolite, tremolite, and anthophyllite. The serpentines class is the polymeric form of chrysotile, an extended sheet forming a tubular fiber structure. Asbestos, mainly chrysotile, has been used in over 3000 different products in the United States, including paper products; asbestos cement products; friction products; textiles; packaging and gaskets; coatings; asbestos-reinforced plastics for automobile clutchs, brakes, and transmission components; asbestos cement pipe; roofing products; coatings; and sealants. Disposal may occur only in special landfills approved and regulated by the federal government; currently, the standard method for measuring asbestos concentrations in workplace air uses phase contrast microscopy (PCM); a particle visible under PCM is counted as a fiber if it is >5 μm long and has a length-to-thickness ratio of 3:1 or more. Health hazards include respiratory effects, where inhalation exposure to asbestos fibers can lead to a characteristic pneumoconiosis called asbestosis, which is an inflammatory response in the lung caused by deposition of asbestos fibers; symptoms include shortness of breath, râles or cough, and, in severe cases, impairment of respiratory function which may ultimately lead to death. It indirectly affects the cardiovascular system by reducing the blood flow through the pulmonary capillary bed because of increased resistance. An increased risk of gastrointestinal cancer in humans is due to asbestos fibers being deposited in the lung during inhalation exposure and then being transported by mucociliary action to the pharynx, where they are swallowed and expose the gastrointestinal epithelium directly to the fibers. Cell-mediated immunity is depressed in workers who have radiological evidence of asbestosis. Concentrations of autoantibodies tend to be abnormally high, which may lead to rheumatoid arthritis; mesothelioma is rare in the overall population but more frequent in individuals who have been exposed to asbestos. Ingestion of asbestos in water causes little or no risk of noncarcinogenic injury; however, chronic oral exposure may lead to increased risk of gastrointestinal cancer. Dermal exposure may lead to the formation of small warts or corns from the asbestos fibers. OSHA exposure limit PEL (TWA): at the action level, 0.1 fiber per cc; at the excursion limit (30 minutes), 1 fiber per cc.

asbestos abatement — Procedures used to control fiber release from asbestos-containing materials in a building, including removal, encapsulation, repair, enclosure, encasement, and maintenance.

Asbestos Hazard Emergency Response Act (AHERA) — Law requiring the U.S. EPA to establish a comprehensive, regulatory system for controlling asbestos hazards in schools.

Asbestos School Hazard Act (ASHA) — Law authorizing the U.S. EPA to provide loans and grants to schools to help with a severe asbestos hazard.

A

asbestos tailing — Any solid waste product of asbestos mining or milling operations that contains asbestos.

asbestosis — Chronic lung inflammation caused by inhalation of asbestos fibers over an extended period of time. The onset of this disease is characterized by shortness of breath and some chest pain. Later, basal râles, clubbing of the fingers and occasionally the toes, and radiographic changes are found. Bronchitis occurs, and eventually a variety of pneumoconiosis may occur along with related heart problems.

A-scale sound level — A measurement of sound approximating the sensitivity of the human ear, used to note the intensity or annoyance of sound below 55 decibels.

ascariasis — (*disease*) A helminthic infection of the small intestine caused by *Ascaris lumbricoides*. Generally, few symptoms occur; however, live worms are passed in the stool or occasionally from mouth to nose, and pneumonitis can occur. It may aggravate nutritional deficiencies and may become fatal if bowel obstruction occurs. Incubation time is 2 months after ingestion of embryonated eggs; the disease is found worldwide. The reservoir of infection is humans and roundworm eggs in soil. It is transmitted by ingestion of the infected eggs from soil contaminated with human feces or from uncooked food contaminated with infected soil; it is not transmitted from person to person. It is communicable for 8 to 18 months, with general susceptibility. It is controlled through the proper disposal of feces and through good personal hygiene.

Ascaris lumbricoides
(in adult intestinal stage
of ascariasis)

Ascaris — A genus of large parasitic intestinal roundworms found in temperate and tropic regions.

ASCAT — Acronym for advanced scatterometer.

ASCEND — Acronym for Agenda of Science for Environment and Development into the 21st Century.

ASCII — See American Standard Code for Information Interchange.

byte   **ASCII Number**

01000001 ◄ **65** **A**

ascites — Accumulation of tissue fluid in the mesenteries and abdominal cavity.

Ascomycetes — Class of fungi including yeast and certain molds.

aseismic — A fault on which no earthquakes have been observed.

asepsis — Absence of septic matter; the state of being free from pathogenic microorganisms; freedom from bacteria or infectious materials.

aseptic processing and packaging — The filling of a commercially sterilized cooled product into presterilized containers followed by aseptic hermetical sealing with a presterilized closure in an atmosphere free of microorganisms.

aseptic technique — The performance of a procedure or operation in a manner that prevents infections from occurring.

asexual reproduction — Reproduction from a single parent without eggs and sperm.

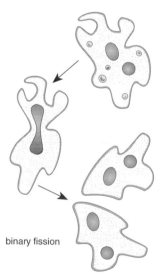

binary fission

Asexual Reproduction

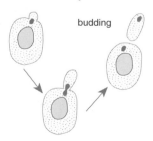

budding

ash — Inorganic residue less than 4 mm in diameter remaining after ignition of combustible substances, including cinders and fly ash.

ash pit — A pit or hopper located below a furnace or fireplace where residue is accumulated and from which it is removed.

ash sluice — A trench or channel in which water transports residue from an ash pit to a disposal or collection point.

ASHA — See Asbestos School Hazard Act.

ASHAA — Acronym for Asbestos in Schools Hazard Abatement Act.

ashlar — A type of masonry composed mostly of rectangular units of cut stones, cultured stones, burned clay, or shale and generally larger in size than regular brick.

askarel — A generic term for a group of nonflammable, synthetic, chlorinated hydrocarbons used as electrical insulating material.

ASP — See air sampling professional.

aspartame ($C_{14}H_{18}N_2O_5$) — A synthetic compound of two amino acids (L-aspartyl–L-phenylalanine methyl ester) used as a low-calorie sweetener. It is 180 times as sweet as sucrose. It degrades into diketopiperazine at certain temperatures. The metabolism of aspartame, a peptide, proceeds in the same way as the metabolism of proteins. After metabolism, aspartame proteins become aspartic acid, phenylalanine, and methanol.

aspect ratio — The ratio of the longer dimension to the shorter in any rectangular configuration; length-to-width ratio.

aspect — (*geographic information system*) The direction that a surface faces.

aspergillosis — (*disease*) A rare fungus infection in humans and animals causing a variety of clinical symptoms including bronchial effects; allergic effects; lung abscesses, possible damage to the brain, kidneys, and other organs; and invasion of blood vessels. It is caused by *Aspergillus fumigatus*, *A. niger*, and *A. flavus* and is possibly fatal. Incubation time is a few days to weeks. It is found worldwide and appears in compost piles, stored damp hay, decaying vegetation, cereal grains, and other food stocks beginning to heat up and undergoing fermentation and decay. It is transmitted by inhalation of airborne spores from grains, not people, and is controlled through prevention of growth of the fungus.

asperity — An area on a fault that is stuck.

asphalt — Black, solid, or semisolid bitumens that occur in nature or are obtained as residues during petroleum refining; used for paving and roofing and in paints and varnishes.

asphalt felt — A mat of organic or inorganic fibers impregnated with asphalt or coal tar pitch or impregnated and coated with asphalt.

asphenosphere — The region of the lithosphere in which the rock is less rigid than that above and below but rigid enough to transmit transverse seismic waves.

asphyxia — Suffocation due to a deficiency of oxygen in the blood and an increase in carbon dioxide in the blood and tissues.

asphyxiant — Any chemical that may be a vapor or gas that excludes oxygen or actively interferes with oxygen uptake and distribution.

asphyxiate — To cause an inability to breath due to circulatory congestion, chemical poisoning, electrical shock, or physical suffocation.

aspirant — Fluid, gases, or solid particles that are withdrawn from the body.

aspiration biopsy — The removal of living tissue for microscopic examination by suction through a fine needle attached to a syringe.

aspiration pneumonia — Inflammatory condition of the lungs and bronchi caused by the inhalation of foreign material or vomitus.

aspirator — An instrument that utilizes a vacuum to draw up substances.

ASR — See alkali–silicate reaction; acronym for aerosol solar radiation.

ASRL — Acronym for Atmospheric Sciences Research Laboratory.

ASSAS — Acronym for advanced solid-state array sensor.

assay — A qualitative or quantitative test for a particular chemical effect.

assembler — (*computer science*) A computer program that converts program-written instructions into computer-executable (binary) instructions.

assembler language — (*computer science*) Programming language that allows programmers to define labels and fix values and to use these labels with mnemomic instructions to produce machine-code computer programs.

assessment — The critical analysis, evaluation, or judgment of the status or quality of a particular condition, situation, or other subject of appraisal; evaluation and interpretation of measurements and other information to provide a basis for decision making.

assigned protection factor (APF) — The minimum anticipated protection provided by a properly functioning respirator or class of respirators to a given percentage of properly fitted and trained users.

assigned risk plan — A means for those who do not meet normal insurance standards to secure workers' compensation insurance.

assimilation — Conversion of nutritive material to living tissue; anabolism.

A

assimilation cycle — A natural process by which a body of water uses microbes to convert nonliving substances into protoplasm or cells to purify itself from pollutants.

assimilative capacity — The capacity of a water body to receive, dilute, and carry away wastes without harming water quality; in the case of organic matter, it also includes the capacity for natural biological oxidation (expressed in kilograms per day at a specific river flow rate and temperature); natural ability of soil and water to use and decompose potential pollutants without harmful effects to the environment.

assisted ventilation — The use of mechanical or other devices to help maintain respiration, usually by delivering air or oxygen under positive pressure.

associated gas — A gas occurring in combination with crude oil as distinct from gas occurring separately or manufactured from crude oil.

Association of Official Analytical Chemists Use Dilution Confirmation Test — A test carried out by placing sterile stainless-steel ring carriers in a broth culture of test organisms, drying the rings for 20 to 60 minutes, immersing them in a tub of germicide solution for 10 minutes at 20°C, immersing them in nutrient media, incubating them for 48 hours, and evaluating them for growth of microorganisms.

AST — Acronym for aboveground storage tank.

ASTER — See Advanced Spaceborne Thermal Emission and Reflection Radiometer.

asthenia — Loss of strength; weakness.

asthma — Pulmonary disease characterized by shortness of breath, wheezing, thick mucous secretions, and constricted bronchioles resulting from abnormal responsiveness of the airways to certain substances or emotional stresses.

asthma in children — An obstructive respiratory condition occurring between the ages of three and eight characterized by reoccurring attacks of paroxysmal dyspnea, wheezing, prolonged expiration, and an irritated cough that is chronic.

ASTM — Acronym for American Standards for Testing of Materials.

astringent — A substance that causes contraction of tissues upon application, usually used locally.

astrocyte — A common cell type in the brain that supports the functions of the neuron.

astrocytosis — The proliferation of astrocytes commonly found in CJD neuropathology.

astrovirus — (*disease*) An unclassified virus containing a single positive strand of RNA of about 7.5 kb surrounded by a protein capsid, 28 to 30 nm in diameter, which may cause sporadic viral gastroenteritis in children under 4 years of age and about 4% of hospitalized cases of diarrhea with additional symptoms of nausea, vomiting, malaise, abdominal pain, and fever. The mode of transmission is through the oral–fecal route, from person to person, or ingestion of contaminated food and water. The incubation period is 10 to 70 hours. The disease has been reported in England, Japan, and California. The reservoir of infection is people. Susceptibility is found in very young children and the elderly in institutional settings. Also known as acute nonbacterial infectious gastroenteritis; viral gastroenteritis.

asulam ($C_8H_{10}N_2O_4S$) — A crystalline solid used as a pesticide; Sol: highly soluble in acetone and methanol. It is hazardous to the digestive tract and may produce cholinergic effects.

asymmetric — Not similar in size, shape, form, or arrangement of parts on opposite sides of a line, point, or plane.

asymmetrical — The inability to be divided into like portions by hypothetical planes.

asymptomatic — Presenting no signs or symptoms of disease.

asystole — A life-threatening cardiac condition characterized by the absence of electrical and mechanical activity in the heart.

at risk — The state of being subject to some uncertain loss or difficulty; the situation of an individual or population being vulnerable to a particular disease or injury.

at. wt. — See atomic weight.

ata — See atmospheric absolute.

ataxia — Failure of muscular coordination or irregularity of muscular action due to some disease of the nervous system.

ATC — Acronym for automatic temperature compensation.

atelectasis — Incomplete expansion of the lung at birth; partial collapse of the lung later in life because of an occlusion of a bronchus or external compression due to a tumor or injury.

ATERIS — Acronym for Air Toxics Exposure and Risk Information System.

atgas — A synthetic gas produced by dissolving coal in a bed of molten iron.

atheroma — A mass or plaque of degenerated, thickened arterial intima occurring in atherosclerosis.

atherosclerosis — A form of arteriosclerosis in which atheromas-containing cholesterol, lipoid material, and lipophages are formed within the intima and inner medial of large and medium-sized arterias.

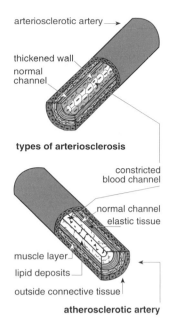

types of arteriosclerosis

arteriosclerotic artery
thickened wall
normal channel

constricted blood channel
normal channel
elastic tissue
muscle layer
lipid deposits
outside connective tissue

atherosclerotic artery

Atherosclerosis

Athrombin-K™ — Trademark for the anticoagulant warfarin.

ATLAS — Acronym for airborne turnable laser absorption spectrometer.

ATLID — Acronym for atmospheric lidar.

Atm — See atmosphere.

atmometry — The science of measuring the evaporation of water from solid or liquid surfaces; evapotranspiration.

atmosphere (Atm) — The gaseous envelope that surrounds a planet or celestial body and is held in place by gravity; a unit of pressure equal to 760 millimeters of mercury (mmHg) at sea level.

atmosphere convection — The process in which heat energy is transported through a medium, usually a gas or liquid, by warm air moving upward because of a lower density.

atmosphere supplying respirator — A respirator that provides breathing air from a source independent of the surrounding atmosphere.

atmospheric absolute — The atmospheric pressure at sea level as measured with a barometer (equal to 1 ata).

atmospheric deposition — The sedimentation of solids, liquids, or gaseous materials from the air containing acids, metals, and toxic organic chemicals transported by winds and deposited by snow, rain, or natural precipitation.

Atmospheric Infrared Sounder — An advanced sounding instrument used on satellites to retrieve vertical temperature and moisture profiles in the troposphere and stratosphere.

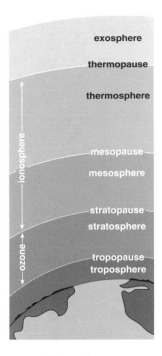

exosphere
thermopause
thermosphere
ionosphere
mesopause
mesosphere
stratopause
stratosphere
ozone
tropopause
troposphere

Atmosphere

atmospheric pressure — The pressure exerted by the weight of the air above any point on the Earth's surface; equal to 14.7 pounds per square inch at sea level. Also known as barometric pressure.

atmospheric radiation cloud station — Semiautonomous system of instruments, instruments systems, and communications systems for deployment at remote locations to measure meteorological properties.

atmospheric response variables — The variables that reflect the response of the atmosphere to external forces including temperature, pressure, circulation, and precipitation.

atmospheric survey — Examination of the air of a given geographical area to determine the nature, sources, extent, and effects of air pollution.

atmospheric turbidity — Haziness in the atmosphere due to aerosols, such as dust ranging from 0.1 to 1.0+ μm in diameter.

atmospheric turbulence — A state of the flow of air in which apparently random irregularities occur in the air, with instantaneous velocities often producing major deformations of the flow.

atmospheric windows — The range of wavelengths at which water vapor, carbon dioxide, or other atmospheric gases only slightly absorb radiation, thus allowing the Earth's radiation to escape into space.

atoll — A coral island consisting of a ring of coral surrounding a central lagoon; commonly found in the Indian and Pacific oceans.

atom — The smallest particle of a chemical that can take part in a chemical reaction without being permanently changed.

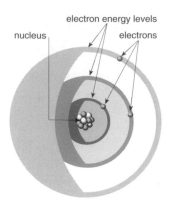

electron energy levels

nucleus   electrons

Structure of an Atom

atom smasher — An accelerator for atomic and subatomic particles so they can be used as projectiles to blast apart the nuclei of atoms.

atomic absorption — A method used to determine the composition of a sample; material is vaporized, and the amount of light it absorbs is measured.

atomic absorption spectrophotometry — A technique used in the analysis of wastewater that combines emission and absorption phenomena; resembles flame photometry.

atomic energy — The energy released in nuclear reactions; also known as nuclear energy.

atomic fluorescence spectrophotometry — A technique linked to atomic absorption spectrophotometry using atom fluorescence.

atomic mass — The mass of an atom expressed in terms of atomic mass units.

atomic mass number — The sum of the number of protons and the number of neutrons in an atom.

atomic mass unit (amu) — A unit of measure used to express relative atomic masses; equal to 1.660×10^{-27} kg, which is 1/12 of the mass of carbon-12 atom, the most abundant form of carbon atoms. Also known as Dalton.

atomic number — The total number of protons in an atomic nucleus.

atomic oxygen — An atom of elemental oxygen.

atomic power — See nuclear power plant.

atomic theory — A concept that all matter is composed of submicroscopic atoms that are in turn composed of protons, electrons, and neutrons; all of the atoms of the same kind are uniform in size, weight, and other properties.

atomic waste — The radioactive ash produced by the splitting of nuclear fuel in a nuclear reactor.

atomic weight (at. wt.) — The relative mass of an atom expressed in atomic weight units as compared with another element (carbon-12, which is assigned a mass value of 12); also known as relative atomic mass.

atomize — See nebulize.

atomizer — A device for reducing a liquid and ejecting it as a fine spray or vapor.

atoxic — See nontoxic.

ATP — See adenosine triphosphate.

atrazine ($C_8H_{14}ClN_5$) — An odorless, white crystalline powder; MW: 216.06, BP: decomposes, Sol (water): 2.8 mg/100 g water at 68°F, Fl.P: NA, sp. gr. 1.187. Used as an herbicide and a plant growth regulator. It is hazardous to the skin, eyes, mucous membranes, and nervous system (at very high concentrations) and causes muscle spasms and damage to the liver. Chronic exposure can cause mutations and may also result in cancer. OSHA exposure limit (8-hour TWA): 5 mg/m³; ACGIH (TLV-TWA): 5 mg/m³.

CH_3CH_2NH — N — $NHCH(CH_3)_2$

N   N

Cl

Atrazine

atria — Chambers of the heart.

atrial fibrillation — Cardiac arrhythmia marked by rapid, randomized contractions of the atrial myocardium causing an irregular ventricular response; also known as fibrillation. See also ventricular fibrillation.

Atropa belladonna — See belladonna.

atrophy — Wasting or diminution in structure or function of a cell, tissue, or organ that was once of normal size.

atropine ($C_{17}H_{23}O_3N$) — An anticholinergic alkaloid occurring in belladonna, hyoscyamus, and stramonium that acts as a competitive antagonist of acetylcholine at muscarinic receptors, blocking stimulation of muscles and glands by parasympathetic and cholinergic sympathetic nerves; used as a smooth muscle relaxant and as an antidote to organophosphate poisoning.

ATS — Acronym for Advanced Technology Satellite.

attached growth process — A wastewater treatment process in which the microorganisms treating the wastes are attached to the media in the reactor over which the wastes flow, resulting in removal of solids and reduction of BOD.

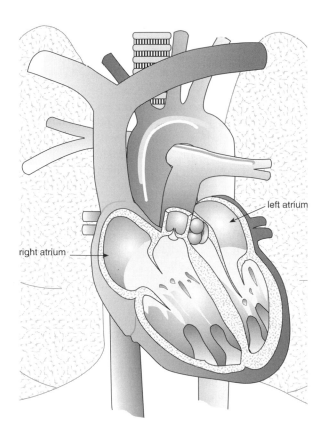

right atrium

left atrium

Atria

attachment — (*virus*) Binding of a virus particle to a specific receptor on the surface of a host cell.

attack — An episode or onset of illness.

attack rate — The incidence rate of an acute disease.

attainment area — (*air pollution*) Any area designated as having ambient air quality levels better than any national primary or secondary air quality standard for any pollutant.

attenuated — Something that is reduced in strength.

attenuated virus — Strain of virus for which the virulence has been lowered by physical or chemical processes.

attenuation — (*radiation*) The process by which a beam of radiation is reduced in intensity due to absorption and scattering when passing through some material; (*disease*) change in virulence of a pathogenic microorganism induced by passage through another host species which decreases its virulence, and how live vaccines are produced; (*earthquake*) decrease in amplitude of the seismic waves with distance due to geometric spreading, energy absorption, and scattering.

attenuator — A resistive network that absorbs part of a signal and transmits the rest with a minimum of distortion or delay.

attractant — A chemical or agent that lures insects or other pests by stimulating their sense of smell.

attraction — The force or influence on one object that is drawn toward another one.

attributable risk — The arithmetic difference between the incidence of disease in a group exposed to the factor and the incidence in an unexposed group — in other words, the number of new cases of the disease among persons exposed to the factor that can be attributed to the factor. It is the amount of a disease that can be attributed to a particular factor and is commonly used to assess the public health importance of that factor as the cause of a disease.

attribute — (*geographic information system*) Non-graphic information associated with a point, line, or area.

attribute survey — A survey to determine the important components of the recreational experience.

attrition — The wearing or grinding down of a substance by friction, which may lead to an increase in air pollutants.

atypical — Not typical.

atypical *Mycobacterium* — A group of mycobacteria, including pathogenic and nonpathogenic forms, classified according to their ability to produce pigments, their growth characteristics, and their reactions to chemical tests.

atypical pneumonia — A group of relatively mild symptoms of chills, headache, muscular pains, moderate fever, and coughing without evidence of a bacterial infection.

audible — Able to be heard.

audible range — (*sound*) The frequency range to which people with normal hearing respond; approximately 20 to 20,000 Hz.

audible sound — The sound that contains frequency components between 20 and 20,000 Hz.

audio frequency (a-f) — The frequency of sound to which the ear can respond.

audiogram — A graph or table obtained from an audiometric examination showing hearing level as a function of frequency (usually 500 to 6000 Hz).

audiologist — An individual who evaluates hearing functions to detect hearing impairment and/or hearing disorders.

audiology — The scientific study of hearing.

audiometer — An instrument for measuring in decibels the sensitivity of hearing; an apparatus used in audiometry.

audiometer setting — A setting on an audiometer corresponding to a specific combination of hearing level and sound frequency.

audiometric zero — The threshold of hearing; 0.0002 μbar of sound pressure.

audiometry — The measurement of the acuity of hearing at various frequencies of sound waves; testing of the sensitivity of the sense of hearing.

audit — Systematic review and evaluation of records and other data to determine the quality of services or products provided in a given situation.

audit gas — A primary standard bottle of calibrated gas used to calibrate a secondary standard.

auditory — Of or pertaining to the ear or sense of hearing.

auditory canal — One of two passageways for sound impulses traveling through the ear.

auditory hair — One of the cells with hairlike processes in the spinal organ of Corti.

auditory system assessment — Evaluation of a person's ears, hearing, and past and present diseases or conditions that affect hearing.

auditory threshold — The lowest intensity at which a sound may be heard.

auditory vertigo — A form of dizziness that is associated with ear disease; characterized by sensations of gyration and, when severe, with prostatation and vomiting.

auger — A rotating drill having a screw thread that carries cuttings away from the face.

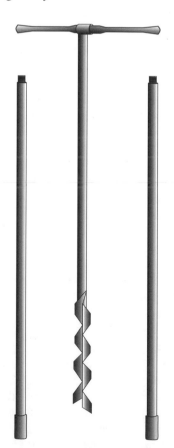

Auger with Extension Handles

auger effect — Spontaneous ejection of an electron by an excited positive ion to form a doubly charged ion.

augmentation — A process in which a substance or mechanism can produce an increased rate of biological activity such as a faster heartbeat.

auramine — A yellow aniline dye used in the manufacturer of paints, textiles, and rubber products; may cause bladder cancer in humans.

auricle — The part of the ear that projects from the head; one of the two upper, ear-shaped chambers of the heart.

aurora — A sporadic radiant emission from the upper atmosphere over middle and high altitudes seen most often along the outer areas of the Arctic and Antarctic.

ausculation — Listening for sounds within the body to evaluate the condition of the heart, blood vessels, lungs, intestines, or other organs.

authorization — An act by the U.S. Congress that authorizes use of public funds to carry out a prescribed action.

autoantigen — An endogenous body constituent that stimulates the production of autoantibody and the resulting autoimmune reaction against one or more tissues of the individual.

autochthonous malaria — An indigenous strain of malaria acquired by mosquito transmission in an area where malaria regularly occurs.

autochthonous — Something that was formed in the place where it is found.

autoclave — A piece of equipment used for sterilization of materials. Steam under pressure flows around articles placed in a chamber; the vapor penetrates the cloth or paper used to package the articles being sterilized.

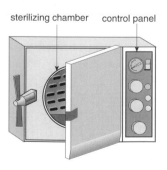

sterilizing chamber   control panel

Autoclave

autogenesis — A self-produced condition.

autogenous combustion — The burning of wet organic material where the moisture content is at such a level that the heat of combustion of the organic material is sufficient to vaporize the water and maintain combustion where no auxiliary fuel is required except for start-up of a process. Also known as autothermic combustion.

autogenous incineration — Combustion characteristic of sludge, having a composition (physical and chemical) such that no auxiliary fuel is required in incineration except at startup and shutdown.

autoignition temperature — The lowest temperature at which a gas–air mixture or vapor–air mixture will ignite from its own heat source or contact a heat source without a spark or flame.

autoimmune disease — A disease caused by the alteration of the function of the immune system of the body.

autoimmunity — Immune response to one's own proteins or other antigens.

autoinfection — Reinfection by a parasite juvenile without its leaving the host.

autointoxication — A condition of poisoning by substances generated by a person's own body.

automated cartography — The process of drawing maps with the aid of computer-driven display devices such as plotters and graphic screens.

automated filter tape — An air sampler that is a highly sensitive, portable instrument operating with a high degree of repeatability.

Automated Local Evaluation in Real Time — A flood warning system consisting of remote sensors, data transmission by radio, and a computer software package developed by the National Weather Service; a generic term used for a decision-making software package.

automatic chlorinator — A device that feeds a regulated amount of stabilized chlorine into pool water on a continuing basis.

automatic data processing — The use of any kind of automation in data processing.

automatic food vending machine — A machine designed and constructed so that, with insertion of a coin, food or drink is delivered to a customer in a package or in bulk.

automatic generation control — A computerized power system regulation to maintain scheduled generation within a prescribed area in response to changes in transmission system operational characteristics.

automatic rain gauge — An automatic instrument used for measuring all types of precipitation.

automatic safety controls — The devices designed and installed to protect systems and components from excessive high or low pressures, temperatures, electrical current, and water pressure, as well as loss of ignition, fuel leaks, fire, freezing, or other unsafe conditions.

automatic tape sampler — An air pollution sampler that collects ascended particulates by drawing air through a small area on a strip of filter paper at a rate of approximately 0.25 ft³/min.

autonomic imbalance — A disruption of a segment of the autonomic nervous system.

autonomic nerve — A nerve of the autonomic nervous system with the ability to function independently and spontaneously as needed to maintain optimum status of bodily activities.

autonomic nervous system (ANS) — A division of the nervous system in vertebrates that regulates the vital internal organs in an involuntary manner.

autonomic — The ability to function independently without outside influence.

autooxidation — An oxidation reaction that is self-catalyzed and spontaneous.

autopsy — Examination of a body after death to determine the actual cause of death; also known as postmortem or necropsy.

autoradiograph — A record of radiation from radioactive material produced by placing the object close to a photographic emulsion.

autosomal — Pertaining to an autosome.

autosome — Any of the 22 pairs of chromosomes found in an individual not concerned with determination of sex.

autothermic combustion — See autogenous combustion.

autotroph — An organism that synthesizes food molecules from inorganic molecules by using an external energy source; green plants that assimilate nutrient materials use energy to combine them into living organic substances.

autotrophic respiration — See photorespiration.

auto-zero — An automatic internal correction for offsets and/or drift at zero voltage input.

auxiliary emission control device (AECD) — Any element of design that senses temperature, vehicle speed, engine revolutions per minute, transmission gear, manifold vacuum, or any other parameter for the purpose of modulating, delaying, or deactivating the operation of any part of the emission control system.

auxiliary equipment — Accessory equipment necessary for operation of a plant.

auxiliary fuel-firing equipment — Equipment used in an incinerator to supply additional heat by burning an auxiliary fuel so that the resulting higher temperatures dry and ignite the waste material, maintain ignition, and combust the combustible solids, vapors, and gases.

auxochrome — (*chemistry*) Any group of atoms capable of making a chromogen into a dye or pigment.

available capacity — (*water*) The amount of water held in the soil and available to the plants.

available chlorine — Chlorine, either free or combined, which is active against bacteria in pool water.

available expansion — Vertical distance from the sand surface to the underside of a trough in a sand filter.

available heat — The gross quantity of heat released within a combustion chamber minus the sensible heat carried away by the dry flue gases, the latent heat, and the sensible heat carried away in water vapor contained in the flue gases; the quantity of useful heat produced per unit of fuel if it is completely burned.

average — (*statistics*) The arithmetic mean.

average annual runoff — The average value of annual runoff amounts calculated for a specific period of record that represents average hydrologic conditions for a specified area.

average energy — The total power generation produced by a power plant during all the years of its actual or simulated operation divided by the number of years of actual or simulated operation.

average head — Resistance of the flow of water in a pool recirculation system obtained by averaging the maximum and minimum resistance encountered in the course of a filter run.

average hot temperature — An average of the temperatures that occur during the hottest times of the year.

average lowest temperature — An average of the temperatures during the coldest times of the year.

average monthly discharge limitation — (*water*) The highest allowable average of "daily discharges" over a calendar month, calculated as the sum of all daily discharges measured during a calendar month divided by the number of daily discharges measured during that month.

average precipitation — An average of the amount of rain that falls in a given area at different times.

average temperature — An average of all the temperatures over the course of a year.

average year supply — The average annual supply of a water development system over a long period.

average year water demand — The demand for water under average hydrologic conditions for a defined level of development.

average-LET (linear energy transfer) — The average energy locally imparted to a medium by a charged particle of specified energy while traversing a short distance.

AVHRR — See Advanced Very-High-Resolution Radiometer.

avian — Referring to birds.

Aviation Safety and Noise Abatement Act — A 1979 law (enacted in 1980 and amended in 1982) providing assistance to airport operators to prepare and carry out noise compatibility programs; providing assistance to ensure continued safety in aviation; establishing a single reliable system for measuring noise; and establishing a single system for determining individuals' exposure to noise at airports.

AVIRIS — Acronym for Airborne Visible and Infrared Imaging Spectrometer.

avirulent — Nonpathogenic.

AVNIR — Acronym for Advanced Visible and Near-Infrared Radiometer.

Avogadro's constant — 6.022×10^{23}; the number of atoms in a gram atomic weight of any element. It is also the number of molecules in the gram molecular weight of any substance; 1 mole of any gas occupies 22.414 liters of volume; formerly called Avogadro's number.

Avogadro's hypothesis — See Avogadro's law.

Avogadro's law — The law stating that equal volumes of ideal gases under the same conditions of temperature and pressure contain an equal number of molecules; also known as Avogadro's hypothesis.

Avogadro's number — The number of molecules in a mole of any substance; equal to 6.02217×10^{23} at standard temperatures and pressures.

avoirdupois weight — System of units commonly used for measurement of the mass of any substance except medicines, precious metals, and precious stones; it is based on a pound (approximately 453.6 g).

avulsed fracture — A fracture that pulls the bone and other tissues away from the usual attachments.

Aw — See water activity.

AW — See artificial wetland.

awkward posture — (*ergonomics*) Any fixed or constrained body position that overloads muscles and tendons where loads are joined in an uneven or asymmetrical manner.

AWT — See advanced wastewater treatment.

axial to impeller — The direction in which material being pumped flows around the impeller or flows parallel to the impeller shaft.

axillary temperature — Body temperature recorded by a thermometer placed in the armpit; generally 1° less than the oral temperature.

axis — A straight line around which a shaft or body revolves.

axis of impeller — An imaginary line running along the center of a shaft.

axis of rotation — The true line about which angular motion takes place at any instance.

axis of thrust — The line along which thrust can be transmitted safely.

axon — A long, thread-like part of a nerve cell that carries an impulse away from the nerve body; also known as a neuraxon or neurite.

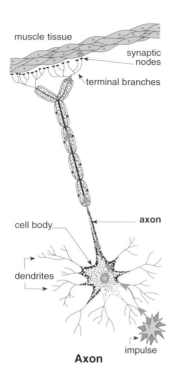

Axon

axoneme — The central strand of the eukaryotic cilium or flagellum.

azeotrope — A liquid mixture that has a constant minimum or maximum boiling point that is lower or higher than that of any of the components.

azide — A compound containing the group N_3 combined with an element or radical.

azimuth — The direction (in degrees) referenced to true north that an antenna must be pointed to receive a satellite signal, where the angular distance is measured in a clockwise direction.

azinphosethyl ($C_{12}H_{16}N_3O_3PS_2$) — A crystalline solid; MW: 345.38, BP: NA, Sol: soluble in most organic solvents, density: NA. It is used as a pesticide. It is highly toxic and hazardous to the digestive system, skin, and respiratory system. It may cause nausea, followed by vomiting, abdominal pain, diarrhea, headache, giddiness, weakness, slurring of speech, tightness in the chest, blurring of vision, breathing difficulty, convulsions, coma, and death.

Azinphos ethyl

azinphosmethyl ($C_{10}H_{12}N_3O_3PS_2$) — An odorless colorless crystal or brown waxy solid; MW: 317.3, BP: decomposes, Sol: 0.003 g/100 ml of water at 68°F, Fl.P: NA, sp. gr. 1.44. An organic phosphate chemical used to kill insects and parasites. It is hazardous to the skin, eyes, mouth, nose, and respiratory system. Acute exposure can cause central nervous system depression, agitation, confusion, delirium, coma, seizures, and death. Chronic exposure can cause dermal sensitization, damage to the nerves, poor coordination in arms and legs, personality changes, depression, anxiety, or irritability. OSHA exposure limit (8-hour TWA): 0.2 mg/m^3.

aziridine — See ethyleneimine.

azo- — A prefix meaning "containing nitrogen".

azo dyes — A group of widely used commercial dyes that carry the azo (–NN–) group in the molecular structure and usually contain impurities of 20% or more; synthesized by diazotization or tetrazotization of aromatic monomine or aromatic diamine compounds with sodium nitrite in an HCl medium; some of the dyes are carcinogenic to animals.

azotemia — An excess of urea or other nitrogenous compounds in the blood.

azurophilic — The quality of being stained with azure or similar metachromatic dye.

B

B — See magnetic field; see boron.

B cells — One of the two major classes of lymphocytes that mature into plasma cells producing antibodies directed at specific antigens.

B horizon — (*soils*) The horizon immediately beneath the A horizon characterized by a higher colloid clay or humus content or by a darker or brighter color than the soil immediately above or below, the color usually being associated with the colloidal materials. The colloids may be of alluvial origin as clay or humus; they may have been formed in place, such as clays including sesquioxide; or they may have been derived from a texturally layered parent material.

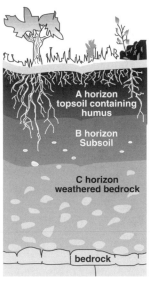

B Horizon
and other layers of soil

B2B exchanges — Business-to-business informational or transactional sites on the Internet.

Ba — See barium

b/cd — See barrels per calendar day.

b/d — See barrels per day.

B/L — See bill of lading.

B-lymphocytes — A class of cells producing antibodies independently from the thymus.

b.p. — before the present.

babbit — An alloy of tin, antimony, copper, and lead; used for bearings.

bacillary dysentery — (*disease*) See shigellosis.

bacilliform — A rod-shaped bacterium.

Bacillus — A genus of aerobic or facultatively anaerobic spore-forming rods, most of which are Gram-positive and motile; includes some pathogenic species.

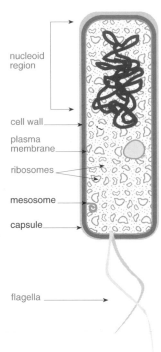

Bacillus

Bacillus anthracis — A Gram-positive, facultative anaerobe that causes anthrax.

***Bacillus cereus* food poisoning** — (*disease*) A gastrointestinal condition with symptoms that include the sudden onset of nausea and vomiting in some people and colic and diarrhea in others, usually lasting no more than 24 hours. Incubation time is 1 to 6 hours when vomiting occurs, 6 to 16 hours when diarrhea is most prominent. *Bacillus cerus*, an aerobic spore former, produces two enterotoxins; one is heat stable and causes vomiting, while the other is heat labile and causes diarrhea. It is found regularly in Europe and occasionally in the United States. The reservoir of infection is an organism found in the soil and in raw, dried, and processed foods. It is transmitted by ingestion of food that has been kept at room temperature after cooking, allowing multiplication of the bacteria. It is especially prominent in vegetable and meat dishes and is not communicable; susceptibility is not known. It is controlled with proper cooking of foods and rapid cooling in refrigerators.

Bacillus thuringiensis — A bacterium that produces a protein called Bt toxin; a gene that is inserted into the genetic material of several genetically engineered plants causes them to produce a natural protein that kills insect larvae after the protein is ingested.

bacitracin — A group of antibacterial substances useful in a wide range of infections.

back pressure — A pressure that can cause water to backflow into the water supply when a user's water system is at a higher pressure than the public water system; the pressure that builds in a vessel or a cavity as fluid is accumulated.

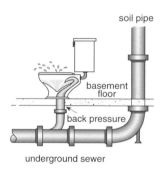

**Loss of trap seal as a result
of back-pressure**

backdrafting — The reverse flow of chimney gases into a building through the damper, draft hood, or burner unit; can be caused by chimney blockage or a pressure differential that is too high for the chimney to draw. Also known as flow reversal.

backend loader — A refuse truck that loads refuse from the rear and compacts it.

backfill — Material used to refill a ditch or other excavation, or the process of doing so.

backfire — See flashback.

backflow — Flow of water or other liquids, mixtures, or substances into the distribution pipes of a potable water supply from any source or sources other than the source of the potable water supply; backsiphonage is one form of backflow.

backflow connection — Any arrangement of plumbing whereby backflow can occur.

backflow preventer — A device or means to prevent backflow.

backflow prevention device — Any device, method, or type of construction to prevent backflow of water, liquids, mixtures, or substances into the distribution pipes of a potable supply of water from any source other than its intended source.

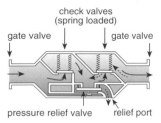

Backflow Prevention Device

backflow valve — A mechanical device installed in a waste pipe to prevent the reversal of flow under conditions of back pressure.

background concentrations — Concentrations of substances in the environment that are due to natural sources.

background leakage — The unidentified openings or gaps in a building envelope through which infiltration can take place.

background level — The amount of pollutant present in the ambient air due to natural sources such as marsh gases, pollen, volcanoes, and fire; may also refer to the amount of radiation or noise present, etc., so the term is used in a variety of situations, always as the constant or natural amount of whatever pollutant is present.

background noise — All of the sources of interference in a system used for the production, detection, measurement, or recording of a signal, independent of the presence of a signal; noise coming from sources other than the noise source being monitored.

background radiation — Low-intensity ionizing radiation arising from radioactive material other than that being directly considered. It may possibly be due to the presence of radioactive substances in other parts of the building; (*construction*) radiation due to building materials, radon gas, and naturally occurring radiation; (*physics*) low-level, natural radiation from cosmic rays and trace amounts of radioactive substances present in a given area.

backscattering — The scattering of radiation in a reverse direction from an irradiated substance; the process by which up to 25% of the radiant energy from the sun is reflected or scattered away from the surface by clouds.

back-siphonage — The backflow of used, contaminated, or polluted water from a plumbing fixture or vessel or other source in a potable water supply pipe as a result of negative pressure in the pipe.

backup — (*computer science*) Duplication of a copy of a file or disk for safekeeping.

backwash — Reverse flow of water through a system of granular media to dislodge and remove solids that have accumulated on the bed.

backwash rate — The rate of flow in gallons per minute per square foot of filter surface area required for efficient filter cleaning.

backwashing — Cleaning a filter or ion exchanger by reversing the flow of liquid and washing out the captured material.

backwater — A small generally shallow body of water with little or no current.

BACT — See Best Available Control Technology

bacteremia — The presence of bacteria in the blood.

bacteria — A group of microscopic, single-celled organisms without chlorophyll and lacking a distinct nuclear membrane.

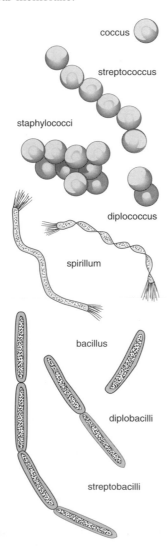

coccus

streptococcus

staphylococci

diplococcus

spirillum

bacillus

diplobacilli

streptobacilli

Bacteria

bacterial analysis — The examination of water and wastewater to determine the presence, number, and identification of bacteria.

bacterial endocarditis — An acute or subacute bacterial infection of the endocardium or the heart valves, or both.

bacterial examination — See bacterial analysis.

bacterial food poisoning — An illness due to the consumption of food containing a toxin created by bacterial growth within the food.

bacterial growth curve — A four-phase system, including a lag phase (lasting about 2 hours); a logarithmic or growth phase, when growth is at a maximum rate; stationary or resting phase, when bacteria die at the same rate as they are produced (spores are produced in this phase); and a death phase, which may occur 18 to 24 hours after the resting phase.

bacterial inflammation — An inflammation that is part of a body's response to a bacterial infection.

bacterial photosynthesis — A light-dependent anaerobic method of metabolism; carbon dioxide is reduced to glucose, which is used for both biosynthesis and energy production.

bacterial toxin — A poisonous substance produced by bacterium.

bactericide — An agent that kills bacteria.

bacteriocin — An agent produced by certain bacteria that inhibits or kills closely related species.

bacteriology — The scientific study of bacteria.

bacteriolysin — An antibacterial antibody that lyses bacterial cells.

bacteriophage — One of several kinds of viruses that can destroy bacteria.

bacteriophage typing — The process of identifying a species of bacteria according to the type of virus that attacks it.

bacteriostat — A product that prevents bacteria from multiplying without actually killing the bacteria; ineffective on hard surfaces such as walls, floors, and countertops.

baculoviruses — A family of DNA viruses infecting insects.

BADT — Acronym for Best Available Demonstrated Technology.

BaF — See benzo(*a*)pyrene; acronym for bioaccumulation factor.

baffle — A deflector vane, guide, grid, grading, or similar device constructed or placed in air- or gas-flow systems, flowing water, or slurry systems to effect a more uniform distribution of velocities; to absorb energy; to divert, guide, or agitate fluids; and to check eddies.

baffle chamber — A settling chamber in which baffles regulate the velocity or direction of the combustion

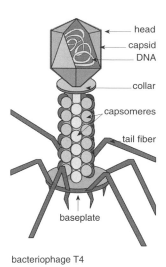

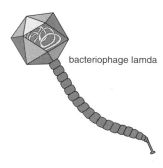

bacteriophage lamda

Bacteriophage

gases to promote the settling of fly ash or coarse particulate matter.

bag filter — A device containing one or more fabric bags for recovering particles from dust-laden gas or air.

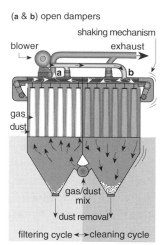

(**a & b**) open dampers

Bag Filter

bag sampling method (indoor air quality) — A method of measuring the air change rate by which a tracer is discharged into a sample volume and mixed, then sample bags are inflated with room air at intervals and the concentration of tracer is measured.

bagging — Artificial respiration performed with a ventilator or respirator bag.

baghouse collector — A device in which dust-laden gas is filtered through the use of bags made of porous cloth.

Bailer Sampler

bailer — (*plumbing*) A 10- to 20-foot-long pipe equipped with a valve at the lower end to remove slurry from the bottom or side of a well as it is being drilled or to collect groundwater samples from wells or open boreholes.

bailing — The compacting of solid waste into blocks to reduce volume and simplify handling.

bait — A food or other substance that will attract pests to a trap; a pesticide chemical that will destroy pests.

B

bake oven — A device that uses heat to dry or cure coatings.

baker's itch — A rash that may develop on the hands and forearms of bakery workers, probably caused by an allergic reaction to flour or other ingredients.

baking soda — The common name for sodium bicarbonate; a mild alkali that may be helpful in removing acid soils.

balance — An instrument for weighing a substance; a normal state of physiological equilibrium.

balanced diet — A diet containing all of the essential nutrients that cannot be synthesized in adequate quantities by the body.

balanced head condition — The condition in which the water pressure on the upstream and downstream sides of an object are equal.

balanced water — A water that will neither be corrosive or form a scale.

balancing reservoir — See equalizing basin.

balantidiasis — (*disease*) A protozoan infection of the colon producing diarrhea along with abdominal colic, nausea, and vomiting; stools may contain blood and mucous but little pus. Incubation time is not known but may be a few days. It is caused by *Balantidium coli*, a large ciliated protozoan, and occurs worldwide but especially as waterborne outbreaks where environmental water controls are poor. The reservoir of infection is pigs and possibly other animals. It is transmitted by ingestion of cysts from feces of the infected host, from contaminated water, and occasionally from hands contaminated with feces or particularly contaminated food. It is communicable as long as the infection persists; debilitated people may have a serious or fatal case. It is controlled through proper disposal of hog feces, protection of public water, adequate sewage systems, fly control, and good personal hygiene.

baler — A machine used to compress and bind solid waste or other materials into packages.

baling — A technique used to compact solid waste into blocks to reduce volume.

ball mill — A pulverizing machine consisting of a rotating drum that contains pebbles or metal balls as the pulverizing agent.

ball-milling — The repeated jarring action of stones, gravel, and sand caused by the force of water in a basin or other structure causing concrete abrasion.

ballast — An auxiliary piece of equipment designed to start and to properly control the flow of power to gas-discharge light sources, such as fluorescent and high-intensity discharge lamps.

ballistic separator — A machine used to sort organic from inorganic matter for composting by dropping them onto a high-speed rotary impeller and hurling them off at different velocities (because of their mass) to separate collecting bins.

band — (*radio*) A continuous sequence of broadcasting frequencies within given limits; (*radiometry*) a relatively narrow region of the electromagnetic spectrum to which a remote sensor responds; (*spectroscopy*) the spectral regions where atmosphere gases absorb and emit radiation: the 15-μm carbon dioxide absorption band, the 6.3-μm water vapor absorption band, and the 9.6-μm ozone absorption band.

band application or treatment — Placing a pesticide chemical on or in the soil in a narrow strip over or next to each row of plants in a field.

bandage — A strip or role of cloth or other material that may be wrapped around a part of the body to secure a dressing, maintain pressure, or immobilize a limb.

bandwidth — The total range of frequency required to pass a specific modulated signal without distortion or loss of data; (*computer science*) a term used to describe how much data can be sent through a connection to the Internet (measured in bits per second, kilobits per second, or megabits per second), where the larger the bandwidth the faster the information flows.

bank full — An established river stage at a given location along a river that is intended to represent the maximum safe water level that will not overflow the river banks or cause any significant damage to the river.

bank storage — The water that has infiltrated from a reservoir into the surrounding land, where it remains in storage until water levels in the reservoir are lowered.

bar — A measure of air pressure equal to 1 million dynes per square centimeter (dyn/cm^2).

bar code — A series of horizontal stripes or bars of varying widths that represent a string of characters that can be read by a bar code reader or scanner.

bar graph — (*statistics*) A graph in which frequencies are represented by bars extending from the ordinate or the abscissa, allowing the distribution of the entire sample to be seen at once.

bar screen — (*wastewater*) A filtering device that removes large solids at the beginning of the treatment process.

barban ($C_{11}H_9C_{12}NO_2$) — A crystalline solid; MW: 258.11, BP: NA, Sol: soluble in benzene and chlorinated hydrocarbons, density: NA. It is used as an herbicide. It is hazardous to the skin, digestive tract, and respiratory tract and may cause slow heart rate, blurred vision, lack of coordination, nausea, weakness, diarrhea, and abdominal pain.

$$\text{C}_6\text{H}_5-\text{NH}-\overset{\overset{\displaystyle O}{\|}}{\text{C}}-\text{O}-\text{CH}_2-\text{CH}=\text{CH}-\text{CH}_2\text{Cl}$$

Barban

baritosis — A benign form of pneumoconiosis caused by an accumulation of barium dust in the lungs.

barium (soluble compounds: Ba(NO$_3$)$_2$, BaO, BaCO$_3$, BaCl$_2$) — Barium nitrate and barium chloride are white odorless solids; MW: 261.4 and 208.3, respectively; BP: decomposes and 2840°F, respectively; Sol: 9 and 38%, respectively; Fl.P: NA; sp. gr. 3.24 and 3.86, respectively. They are used in the manufacture of pressed and blown glassware and flint and crown optical glass; in the manufacture of ceramic products; in the manufacture of magnets, vacuum tubes, cathodes, x-ray fluorescent screens, television picture tubes, and dry-cell vaporizers; in the manufacture of photographic papers, dyes, and chemicals; as pesticides, rodenticides, and disinfectants; in the manufacture of explosives, matches, and pyrotechnics; as catalysts, analytical reagents, and purifying agents; in the manufacture of pigments, paints, enamels, and printing inks; for treatment of textiles, leather, and rubber; as a smoke suppressant in diesel fuel. They are hazardous to the central nervous system, heart, respiratory system, skin, and eyes and are toxic through inhalation, ingestion, and contact. Symptoms of exposure include upper respiratory irritation, gastroenteritis, muscle spasm, slow pulse, extrasystoles, hypokalemia, eye and skin irritation, skin burns. OSHA exposure limit (TWA): 0.5 mg/m^3 [air].

barium cyanide (BaCN$_2$) — An odorless white crystalline solid; MW: 189.4, BP: 465°C, Sol: 80 g/100 ml water at 14°C, Fl.P: NA (however, when in contact with acids, acid salts, or carbon dioxide in air, highly flammable hydrogen cyanide gases may be produced), sp. gr. NA. Used in nuclear reactors and electronic tubes; as additives in lubricating oils; in the manufacture of pyrotechnics and explosives; in tanning and finishing leathers; as a mordant for fabrics and dyes; in electroplating, aluminum refining, or rubber manufacturing; and in the production of paints and enamels. It is hazardous to the skin, respiratory system, and digestive system. Acute exposure can cause cyanide poisoning with headaches, weakness, confusion, gasping for air, collapse, and death. Chronic exposure can cause loss of appetite, headaches, weakness, nausea, dizziness, irritation of the upper respiratory tract and eyes, skin rashes, and scars on the lungs. OSHA exposure limits (8-hour TWA): 0.5 mg/m^3 as barium, 5 mg/m^3 as cyanide.

barograph — An instrument that constantly monitors barometric pressure and records it.

barometer — An instrument used for measuring atmospheric pressure.

barometric pressure — The pressure at a given temperature and altitude due to the atmosphere; atmospheric pressure.

barometry — The science of pressure measurement.

barotrauma — An injury to the ear caused by a sudden alteration in barometric pressure.

barrel — A unit of liquid volume for petroleum equal to 42 U.S. gallons at 60°F.

barrel sampler — An open-ended steel tube used to collect soil samples.

barrels per calendar day — An oil industry measurement of actual refinery output compared to designed capacity.

barrels per day — A unit of measurement used in the oil industry for the production rates of oil fields, pipelines, and transportation.

barrier application or treatment — The use of a pesticide chemical or other agent to stop pests from entering a container, area, field, or building.

barrier island — A long narrow island that is built by waves along the coast.

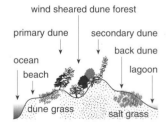

Barrier Island

barrier reef — A coral reef that runs approximately parallel to the shore and is separated from the shore by a lagoon.

barrier substances — (*industrial safety*) Creams and lotions applied to the skin in a thin film to act as a barrier to skin irritants; includes water-resistant, oil-resistant, film-forming, and specific-use substances.

BART — See Best Available Retrofit Technology.

barye — See microbar.

BAS — Acronym for Building Automation System.

basal application or treatment — The placing of an herbicide on stems or trunks of trees and brush just above the soil line.

basal cell carcinoma — A type of skin cancer that arises from basal cells.

basal cells — The small round cells found in the lower part or base of the epidermis, the outer layer of the skin.

basal metabolism — The activities required to maintain the body and to supply the energy necessary to support the basic life processes.

basalt — A fine-grained, dark-colored volcanic rock rich in iron bearing minerals.

basalt aquifers — Water stored within the layers or sheets of volcanic rock and unconsolidated sediments; found most often in areas where volcanic activity caused lava to flow across the surface.

basaltic domes — Mountain peaks consisting of dark, dense, igneous rock of a lava flow or minor intrusion composed essentially of labradorite and pyroxene; also known as shield volcanoes.

**Basaltic Dome
(Shield Volcano)**

base — A substance that turns litmus blue and produces hydroxyl ions in water solution; it reacts with acids to form water and salts.

base course — A layer of specified or selected material of planned thickness constructed on the subgrade or subbase to distribute the load, provide drainage, minimize frost action, etc.

base demand — A measure of the amount of alkali material required to raise pH to a predetermined level.

base exchange capacity — See cation-exchange capacity.

base flood plain — The flood plain inundated by a 100-year flood.

base flood — A flood for which there is a 1% chance of its being equaled or exceeded in any given year.

base flow — The groundwater that enters a stream channel, maintaining streamflow at times when it is not raining.

base map — (*geographic information system*) The most accurate spatial database within a data system serving as the point of reference when creating other spatial databases and having the highest level of accuracy.

base metal — A metal that reacts with dilute hydrochloric acid to form hydrogen.

base safety condition — The level of loading above which a failure does not contribute an incremental loss of life.

base temperature — An arbitrary reference temperature for determining liquid densities for adjusting the measured volume of a liquid quantity.

baseline — A condition that would prevail if no actions were taken.

baseline concentrations — See background concentrations.

baseline condition — An environmental situation in which a stable rate occurs until new factors are introduced.

baseline emissions — The emissions that would occur without policy intervention.

baseline profile — A survey of the environmental conditions and organisms existing in a region prior to unnatural disturbances.

baseload (energy) — A minimum load in a power system over a given period of time.

BASIC — See Beginner's All-Purpose Symbolic Instruction Code.

basic health services — The minimum amount of health care considered to be necessary to maintain adequate health and protection from disease.

basic life-support — The use of cardiopulmonary resuscitation and emergency cardiac care in restoring circulatory or respiratory function in the emergency treatment of a victim of cardiac or respiratory arrest.

basic oxygen furnace — A furnace in which molten metal, scrap steel, and a flux are the charge.

basin — A low-lying area in which water may accumulate as with rivers and their tributaries, streams, coastal waters, sounds, estuaries, bays, and lakes, as well as the lands drained by these waters.

basin runoff model — (*water quality*) One of the computer programs that mathematically models basin characteristics to forecast reservoir inflow from rainfall and/or streamflow data.

Basle Nomina Anatomica **(BNA)** — A system of anatomic nomenclature adopted at the annual meeting of the German Anatomic Society in 1895.

basophil — A granulocytic white blood cell characterized by cytoplasmic granules that stain blue when exposed to a basic dye.

BAT — See Best Available Technology.

batch-fed incinerator — An incinerator that is periodically charged with solid waste; one charge is allowed to burn down or burn out before another is added.

batch processing — (*computer science*) The processing of a group of similar jobs on the computer without operator intervention.

batch sample — Collection of substances or products of the same category, configuration, or subgroup drawn from a batch and from which test samples are taken.

BATEA — See best available technology economically achievable.

bather — Any person using a pool, spa, or other water area and the adjoining deck for the purpose of therapy, relaxation, recreation, and other activities.

bather load — The number of people, who are the major source of bacterial and organic contamination, using a pool or spa within a 24-hour period.

bathing place — A body of natural water, impounded or flowing, of such size in relation to the bathing load that the quality and quantity of water confined or flowing need be neither mechanically controlled for the purpose of purification nor contained in an impervious structure.

bathyal — Of or relating to the water from 600 to 6000 feet below the surface of the water.

bathymetry — The science of measuring ocean depths to determine the topography of the sea floor.

bathysphere — A round, heavy-walled vessel that can be lowered to great depths in the ocean by means of a tether.

battery — An apparatus of two or more electrolytic cells connected together to produce electricity.

baud — The binary units or bits per second used as a measure of data flow or the speed with which a computer device transmits information.

baud rate — (*computer science*) A measure of the speed of data transmission between a computer and other devices; equivalent to bits per second.

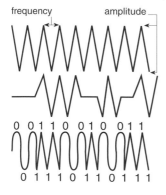

Baud Rate

bauxite — The impure mixture of aluminum, oxides, and hydroxides.

bay — A large estuarine system.

BBB — See blood–brain barrier.

bbl/d — See barrels per day.

BBSS — Acronym for balloon-borne sounding system.

bcf — Acronym for billion cubic feet.

BCF — See bioconcentration factor.

Bcfe — Acronym for 1 billion cubic feet of natural gas.

BCG vaccine (Bacillus Calmette–Guérin vaccine) — A tuberculosis vaccine containing a living, virulent, bovine-strain tubercle bacilli. It offers some protection against tuberculosis but cannot be relied on for total control of the disease.

BCPCT — Acronym for Best Conventional Pollution Control Technology.

BCT — See Best Conventional Pollutant Control Technology.

BDAT — See Best Demonstrated Available Technology.

BDCT — Acronym for Best Demonstrated Control Technology.

BDT — Acronym for Best Demonstrated Technology.

Be — See beryllium.

beach — A temporary accumulation of sediment that collects between low and high tide.

beach replenishment — The rebuilding of a beach by adding sand to it.

beaching — The action of water waves causing beach materials to settle into the water, due to removal of finer materials.

beam — Unidirectional or approximately unidirectional flow of electromagnetic radiation or particles.

beam axis — A line from the source through the centers of the x-ray fields.

beam-limiting device — A piece of equipment used to restrict the dimensions of the x-ray field.

Bean Weevil (*Acanthoscelides obtectus*) — A short-snout beetle that feeds upon stored beans and peas; the adult is 1/8-inch long with reddish legs and a light olive-brown color mottled with darker brown and gray; a stored food product insect.

bearing capacity — The maximum load per unit area that a material can safely support before failing.

bearing wall — (*construction*) A wall that supports any vertical load in addition to its own weight.

beats — Periodic variations that result from the superposition of waves having different frequencies.

becquerel (Bq) — The International System (SI) unit of activity equal to 1 transformation (disintegration) per second.

bed — Elevated areas in fields where seedlings, cuttings, or crops have grown; (*institutions*) a unit of measurement used to determine size and/or capacity of healthcare institutions.

B

bed elevation — The height of a streambed above a specified level.

bed layer — A flow layer, several grain diameters thick, that is immediately above the stream bed.

bed load — The sediment particles resting on or near the channel bottom that are pushed or rolled along by the flow of the water.

bed material — The unconsolidated material or sediment mixture of which a streambed is composed.

bed sore — See decubitus ulcer.

bedbug — A bloodsucking arthropod of the species *Cimex lectularius* or the species *C. hemipterus* that feeds on humans and other animals.

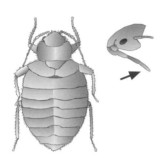

Bedbug
Order Hemiptera

bedding — The soil placed beneath and beside a pipe to support the load on the pipe.

bedrock — Unweathered solid rock of the Earth's crust overlaid by sand, gravel, or soil.

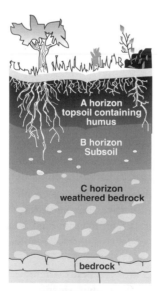

A horizon
topsoil containing
humus

B horizon
Subsoil

C horizon
weathered bedrock

bedrock

Bedrock
and layers of soil

bedroom — (*sewage*) Any room within a building that is used for sleeping purposes or could be converted into a room used for sleeping purposes, such as a den or sewing room. The number of bedrooms is used when calculating the amount of sewage coming from a home.

beef tapeworm disease — See taeniasis.

Beer–Lambert Law — See Beer's Law.

Beer's Law — For monochromatic radiation, absorbance $(A) = abc$, where a is absorptivity, b is the path length through the medium, and c is the concentration of the absorbing species.

beetle — The common name give to the hard-shelled insect of the order Coleoptera.

Beginner's All-Purpose Symbolic Instruction Code — A simple, high-level computer programming language originally designed for inexperienced computer operators.

behavior — The manner in which a person acts or performs.

BEIs — See biological exposure indices.

bel — (*physics*) The amount of energy in the form of sound transmitted to 1 square inch of the ear; a unit that expresses intensity of sound.

belladonna (*Atropa belladonna*) — A perennial, poisonous herb that is the source of various alkaloids, atrophine, and hyoscyamine; the leaf is used as an anticholinergic in the management of peptic ulcers and other gastrointestinal disorders. Also known as deadly nightshade.

belowground storage facility — A tank or other storage container for which the base is located more than 6 inches below the surrounding surface.

BEMS — See Building Energy Management System.

BEN — The U.S. EPA's computer model for analyzing the potential economic gain of violators who do not comply with environmental laws.

bench mark — A permanent or temporary monument of known elevation above sea level that is used as a reference in a survey or construction site.

bench research — Research done in a laboratory setting that does not involve the use of humans.

benchmark test — (*computer science*) A test to evaluate the capabilities of a computer system in terms of an individual's needs.

benchmarking — (*geographic information system*) The administration of a standardized test procedure that provides a systematic means of comparing the performance levels of competing systems.

bench-scale tests — The laboratory testing of potential cleanup technologies.

bend — A change in direction in piping.

bendiocarb (C₁₁H₁₃NO₄) — An odorless, white, crystalline powder; MW: 223.23, BP: NA, Sol (water): 40 ppm, Fl.P: NA (however, it is a flammable material and may be ignited by heat, sparks, or flames, and the vapors may travel to a source of ignition and flashback), sp. gr. NA. It is used as a contact insecticide. It is hazardous to the respiratory system and digestive system. Acute exposure results in central nervous system depression with symptoms of stupor, coma, seizures, hypertension, tachycardia, and cardiorespiratory depression. Death is frequently due to respiratory failure. Other symptoms include urinary incontinence, diarrhea, gastrointestinal cramping, blurred vision, nausea, and vomiting. Carcinogenicity, reproductive toxicity, and other chronic health effects are not known at this time. OSHA exposure limit (8-hour TWA): NA.

Bendiocarb

bends — A condition resulting from a too-rapid decrease in atmospheric pressure, such as when a deep-sea diver returns to the surface too rapidly; also known as caisson disease.

benefit — The sum of money specified in an insurance policy to be paid for certain types of loss under the terms of the policy.

benign — Not recurrent or not tending to progress; of no danger to life or health.

benign tumor — A noncancerous or nonmalignant growth.

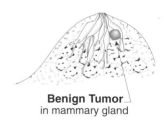

Benign Tumor
in mammary gland

malignant tumor

benomyl (C₁₄H₁₈N₄O₃) — A crystalline solid; MW: 290.36; BP: NA; Sol: soluble in chloroform, acetone, and benzene; density: NA. It is used as a pesticide. It is toxic to the digestive tract, respiratory tract, and skin and can cause headache, weakness, nausea, vomiting, and convulsions. It has shown teratogenic and mutagenic effects.

Benomyl

bent fracture — An incomplete greenstick fracture.

benthic — Aquatic bottom-dwelling organisms, including sponges, barnacles, muscles, oysters, insects, snails, certain clams, and worms.

benthic zone — The area of the aggregate of organisms living on or at the bottom of a body of water.

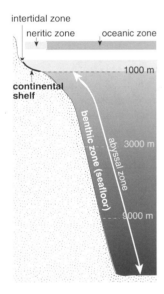

Benthic Zone

bentonite — A soft, absorbent, swelling, and colloidal clay formed from the alteration of volcanic ash; used widely in industry as a bonding agent, sealant, thickener, and filler and as a liner in landfills; added to ordinary natural clay-bonded sands where extra strength is required.

benz(*a*)anthracene (C₁₈H₁₂) — One of the polycyclic aromatic hydrocarbon compounds formed when gasoline, garbage, or any animal or plant material burns, usually found in smoke and soot; MW: 228.29; BP: 435°C; Sol: sublimes, virtually insoluble in water; Fl.P: not known; density: 1.274 at 20°C. It has some use as a research chemical; coal tar pitch is removed from the tar still as a residue.

It is used as a binder for electrodes in the aluminum reproduction process, as an adhesive in membrane groups, for wood preservation, and in the clinical treatment of skin disorders such as eczema, dermatitis, and psoriasis. It is a possible cause of cancer through inhalation when the chemical is part of a mixture of polycyclic aromatic hydrocarbons. OSHA exposure limit (8-hour TWA): 0.2 mg/m^3 [air].

benzene (C₆H₆) — A colorless, volatile, flammable, toxic, aromatic liquid hydrocarbon; MW: 78.11, BP: 80.1°C, Sol (water): 1780 mg/l, Fl.P: –11°C. Benzene has an acute narcotic effect on the individual and also acts as a local irritant to the skin and mucous membranes. It may injure the blood-forming tissues, even at low concentrations. Benzene is considered to be a hazardous air pollutant. Benzene recovered from petroleum and coal is primarily used as an intermediate in the manufacture of other chemicals and in products. It is used in the production of ethylbenzene, cumene, and cyclohexane. Ethylbenzene is an intermediate and a synthesis of styrene that is used in plastics and elastomers; cumene is used to produce phenol and acetone; phenol is used in the manufacture of phenolic resins and nylon intermediates; acetone is used as a solvent and in the manufacture of pharmaceuticals; cyclohexane is used to make nylon resins; benzene is also used to manufacture nitrobenzene, which is used in the production of aniline, urethanes, linear alkylbenzene sulphonates, chlorobenzene, and maleic anhydride, a solvent and a reactant. Benzene is a component of gasoline that occurs naturally in crude oil and is a byproduct of the oil refinery process. It is important for unleaded gasoline because of its antiknock characteristics; 2% of the benzene produced is used as a solvent in industrial plants, rubber cements, adhesives, paint removers, artificial leather, and rubber goods. It has also been used in pesticides, shoe manufacturing, and rotogravure printing industries; it may be detected in carpet glue, textured carpet liquid detergent, and furniture polish; inhalation and dermal exposure can occur through occupational or environmental exposure; however, the main route of entry to the body is through inhalation; in acute exposures, benzene can cause death in 5 to 10 minutes at 20,000 ppm; death is caused by asphyxiation, respiratory arrest, central nervous system depression, or cardiac collapse. Nonfatal symptoms may include headaches, nausea, staggering gait, paralysis, convulsions, and unconsciousness. When death does occur, cyanosis, hemolysis, and hemorrhage of the organs are observed. It may cause adverse immunological effects in humans, as it alters humoral immunity (the ability to produce changes in blood levels of antibodies). It also affects cellular immunity (changes in circulating leukocytes and lymphocytes). It can cross the human placenta and is present in the cord blood in amounts equal to or greater than those in maternal blood; it may impair fertility in women who are exposed to high levels. High concentrations of benzene may produce chromosome abnormalities; a cause–effect relationship exists between benzene and leukemia; it can cause intense toxic gastritis and later pyloric stenosis. Skin contact may cause swelling, edema, and central nervous system toxicity. OSHA exposure limits — PEL TWA: 1 ppm; action level (8-hour average): 0.5 ppm; STEL (15-minute average): 5 ppm. Also known as benzol.

benzene hexachloride (BHC) (C₆H₆Cl₆) — A crystalline solid not soluble in water but soluble to varying degrees in a wide variety of common solvents; used as an insecticide. See also hexachlorocyclohexane.

benzene ring — An arrangement of atoms in the shape of a hexagon; each of the points is made up of a carbon atom, and each carbon atom may be linked to other structural arrangements of atoms or to a single chemical.

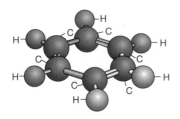

Benzene Ring

benzidine (NH₂C₆H₄C₆H₄NH₂) — A grayish-yellow, reddish-gray, or white crystalline powder; MW: 184.3, BP: 752°F, Sol (at 54°F): 0.04%, sp. gr. 1.25. It is used in the manufacture of azo dyes; as a hardener in the rubber industry; in research and analytical laboratories for detection of blood and inorganics. It is hazardous to the bladder, kidneys, liver, skin, and blood and is toxic through inhalation, absorption, ingestion, and contact. Symptoms of exposure include secondary anemia from hemolysis, acute cystitis, acute liver disorders, dermatitis, and painful and irregular urination; carcinogenic. OSHA exposure limit (TWA): regulated use.

benzo(a)anthracene (C₁₈H₁₂) — A crystalline compound that produces a greenish-yellow fluorescence; MW: 228.30, Sol: dissolves in most organic solvents, density: NA. It is an aromatic hydrocarbon that may be highly toxic by intravenous administration. It is a carcinogen in animals.

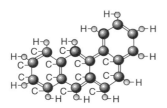

Benzo(*a*)anthracene

benzo(*a*)pyrene (C₂₀H₁₂) — One of the polycyclic aromatic hydrocarbons formed when gasoline, garbage, or any animal or plant material burns, usually in smoke and soot; pale-yellow but fluoresces yellow-green in ultraviolet light; MW: 252.3; BP: 310 to 312°C at 10 mmHg, 495°C at 760 mmHg; Sol: soluble in water (3.8 × 10–6 g/l); FL.P: not known. Some use as a research chemical; coal tar pitch is removed from the tar still as a residue. It is primarily used as a binder for electrodes in the aluminum reduction process and to bind carbon electrodes in reduction pots and as pitch, which is an adhesive in membrane roots. Creosote is produced and is sold for wood preservation. People may be exposed through environmental sources such as air, water, soil, and from cigarette smoke and cooked food. Studies in experimental animals have shown the ability of benzo(*a*)pyrene to induce respiratory tract tumors following long-term inhalation exposure; short-term or intermediate exposure to very high levels results in the death of experimental animals fed benzo(*a*)pyrene in the diet. Cancer appears to be induced in animals and possibly humans as a result of long-term dermal exposure. OSHA exposure limit (TWA): 0.2 mg/m³ [air].

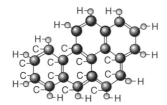

Benzo(*a*)pyrene

benzo(*b*)fluoranthene (C₂₀H₁₂) — A colorless aromatic hydrocarbon; odor: not known; MW: 252.32, BP: not known, Sol: in water 14 μg/l, Fl.P: not known. It is one of the polycyclic aromatic compounds formed when gasoline, garbage, or any animal or plant material burns, usually found in smoke and

soot. It is also found in coal tar pitch and is used to join electrical parts together and to preserve wood. No information is available regarding its effects on the various systems of the body; however, the chemical, when part of a mixture of PAHs, has been shown to have some relationship to cancer in the lungs and skin of animals. OSHA exposure limit: 0.2 mg/m³ [air].

Benzo(*b*)fluoranthene

benzo(*j*)fluoranthene (C₂₀H₁₂) — An aromatic hydrocarbon that is suspected of being a carcinogen (carcinogenic in animals); MW: 252.32.

Benzo(*j*)fluoranthene

benzo(*k*)fluoranthene (C₂₀H₁₂) — An aromatic hydrocarbon that causes lung and skin cancer in animals; MW: 252.32.

Benzo(*k*)fluoranthene

2,3-benzofuran (C₈H₆O) — A colorless, sweet-smelling oily liquid that does not mix with water and is formed when coal is processed to make coal oil; MW: 118.14, BP: 175°C, Sol: insoluble, Fl.P: not known. No studies are available on health effects on humans due to exposure, but some problems have been observed in laboratory mice and rats, such as mineralization of the pulmonary artery, chronic inflammation of the stomach, liver damage, kidney damage, and cancer. OSHA exposure limit: 200 ppm [food].

benzoic acid (C₇H₆O₂) — White powder or crystals with a faint, pleasant, slightly aromatic odor; MW: 122.12, BP: 480°F, Sol: 0.290 g/100 ml, Fl.P: 250°F, sp. gr. 1.27. It is used as a preservative in foods, juices, fats, and oils; in the manufacture of plasticizers, dyes, pharmaceuticals, and perfumes; as a laboratory reagent; as an ingredient in antiseptic ointments to treat fungal infections. It is hazardous to the respiratory system, skin, and digestive system and is toxic through inhalation, skin contact, or ingestion. Symptoms of exposure include irritation of the nose, throat, and upper respiratory system; reddening, swelling, and burning of skin; asthmatic sneezing; anaphylactic shock; violent cough; chest constriction; convulsions; collapse; and death. OSHA exposure limit: not established.

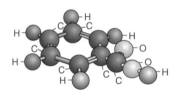

Benzoic Acid

benzol — See benzene.

benzo(g,h,i)perlene (C₂₂H₁₂) — An odorless, large, pale-yellow-green plate recrystallized from xylene; MW: 276.34, BP: 550°C, Sol: 0.0001 g/100 ml water at 20°C, Fl.P: NA, sp. gr. NA. Although small amounts are used for scientific research, no commercial production or commercial uses of this compound are known; however, it is found in crude oil and is a product of incomplete combustion associated with natural fires, lightning, volcanic activity, and spontaneous combustion. It may be produced and released into the environment through industrial effluents, wastewater treatment facilities, and waste incinerators. It is hazardous to the digestive system and respiratory system and through dermal contact. Acute exposure may irritate the eyes, nose, and throat, resulting in coughing and bronchitis. At very high levels headaches, nausea, and vomiting occur, and permanent damage may be found in the liver and kidneys. Chronic exposure may cause headaches, fatigue, and nausea, as well as itching and skin rashes. The exposure may increase the risk of cancers of skin, bladder, and lungs. OSHA exposure limit (8-hour TWA): NA.

benzo(e)pyrene (C₂₀H₁₂) — A crystalline solid, prism, or plate; MW: 252.32; BP: NA; Sol: soluble in methanol and dissolves in benzene, toluene, and methylene chloride; density: NA. It is an aromatic hydrocarbon that causes tumors in the stomach of animals. It is a mutagen.

Benzo(e)pyrene

1,4-benzoquinone (C₆H₄O₂) — A pale-yellow crystalline solid with an acrid odor similar to chlorine; MW: 108.09; BP: NA; Sol: soluble in alcohol, ether, and alkalies; Fl.P: a noncombustible solid but reacts violently with strong oxidizers; density: 1.318 at 20°C. It is used in the manufacture of dyes and fungicides and in photography. It is hazardous to the digestive tract, eyes, and skin. It causes irritation, ulceration, and necrosis. It is a mutagen and may cause cancer. ACGIH (TLV-TWA): 0.1 ppm.

1,4-Benzoquinone

benzoyl peroxide ((C₆H₅CO)₂O₂) — Colorless to white crystals or a granular powder with a faint benzaldehyde-like odor; MW: 242.2, Sol: <1%, sp. gr. 1.33 at 77°F. It is used in vulcanization of natural and synthetic rubber; as a bleaching agent for flour, cheese, fats, oils, and waxes; as an ingredient in skin creams for burns, dermatitis, poisoning, and external wounds; in the manufacture of fast-drying printing inks. It is hazardous to the skin, respiratory system, and eyes and is toxic through inhalation, ingestion, and contact. Symptoms of exposure include skin, eye, and mucous membrane irritation; sensitization dermatitis. OSHA exposure limit (TWA): 5 mg/m³ [air].

Benzaldehyde

benzyl alcohol (C₇H₈O) — A colorless liquid with a pleasant, fruity odor; MW: 108.13, BP: 402.8°F, Sol: 3.5 g/100 ml, Fl.P: 200°F (closed cup), sp. gr. 1.04535. It is used as a solvent for lacquers, cosmetics, and inks and in the manufacture of soaps, perfumes, pharmaceuticals, and dyes. It is hazardous to the respiratory and digestive systems. It is toxic by inhalation or ingestion. Symptoms of exposure include mild depression of the central nervous system, headache, nausea, dizziness, drowsiness, incoordination, and confusion; may be aspirated into the lungs. OSHA exposure limit: not established.

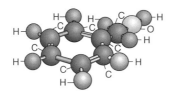

Benzyl Alcohol

benzyl chloride (C₇H₇Cl) — A colorless to slightly yellow liquid with a pungent aromatic odor; MW: 126.59, BP: 179°C, Sol: miscible with most organic solvents. It is used in the manufacture of dyes, artificial resins, tanning agents, pharmaceuticals, plasticizers, perfumes, and lubricants. It is a corrosive liquid that is hazardous to the eyes and respiratory system. It may cause injury to the cornea, lung edema, and depression of the central nervous system. ACGIH (TLV-TWA): 1 ppm.

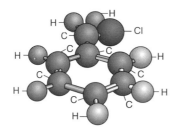

Benzyl Chloride

benzylaldehyde (C₆H₅CHO) — A colorless liquid; MW: 106.10, BP: 178°C, Sol: soluble in alcohol and ether, Fl.P: 74°C (open cup), density: 1.046 at 20°C. It is used as an intermediate in the production of flavoring chemicals, perfume, and soap. It shows low to moderate toxicity in test animals. ACGIH exposure limit: NA.

berm — A man-made mound or small hill of earth; a platform of wave-deposited sediment that is flat or slopes slightly landward.

Bernoulli's principle (physics) — The principle stating that the sum of the velocity and the kinetic energy of a fluid flowing through a tube is constant.

beryilla — See beryllium oxide.

beryl — A silicate of beryllium and aluminum.

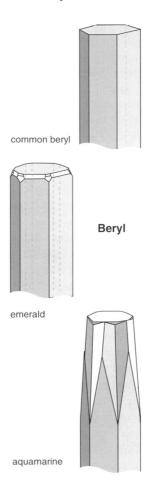

common beryl

Beryl

emerald

aquamarine

berylliosis — (*disease*) Beryllium poisoning, usually involving the lungs and less often the skin, subcutaneous tissues, lymph nodes, liver, and other structures; causes a toxic or allergic reaction.

beryllium (Be) — A rare, lightweight, gray metal. Beryllium–copper alloys are used in industry and science. It is hazardous to the respiratory system, lungs, liver, spleen, kidneys, and lymphatic system and is toxic through inhalation. Symptoms of exposure include shortness of breath, general weakness, weight loss, acute pneumonitis, immunological disorders, and death. OSHA exposure limit (TWA): 0.002 mg/m³ [air].

beryllium chloride (BeC₁₂) — A white to faintly yellow powder with a sharp odor; MW: 79.93, BP: 482°C, Sol: highly soluble in water, Fl.P: nonflammable (poisonous gases and fumes are produced in a fire, including beryllium and hydrogen chloride), sp. gr. 1.9 at 25°C. It is used in refining beryllium ores and as a chemical reagent. It is hazardous to the respiratory system, digestive system, skin, and eyes. Acute exposure can cause irritation of the airways and lungs, nasal discharge, tightness in the chest, cough, shortness of breath, fever, and death. Eye contact can cause redness, itching, and/or swelling of the eyelids. Skin contact can cause irritation, burns, and ulcers. Chronic exposure can cause chronic systemic disease primarily of the lungs but which can also involve the lymph nodes, liver, bones, and kidneys. It is a cancer-causing agent and may have the potential for causing reproductive damage in humans. OSHA exposure limit (8-hour TWA): 0.002 mg/m^3.

beryllium fluoride (BeF$_2$) — A colorless to white powder with no odor; MW: NA, BP: 1160°C, Sol: highly soluble in water, Fl.P: nonflammable (poisonous gases and fumes are produced in a fire, including beryllium oxide and fluoride fumes), sp. gr. 1.986 at 25°C. It is used in refining beryllium and manufacturing beryllium alloys and as a chemical reagent. It is hazardous to the respiratory system, digestive system, skin, and eyes. Acute exposure can cause irritation of the conjunctiva and mucous membrane and occasionally pneumonitis, severe bronchitis, and pneumonia leading to death. Chronic exposure can cause scarring of the lungs and other body organs; heart failure; systemic diseases of the liver, spleen, lymph nodes, bone, kidneys, and other organs; and severe skin rashes. It is a probable carcinogen. OSHA exposure limit (8-hour TWA): 0.002 mg/m^3.

beryllium nitrate (Be(NO$_3$)$_2$3H$_2$O) — An odorless, white to slightly yellow powder; MW: 133.03, BP: 142°C, Sol: highly soluble in water, Fl.P: nonflammable (although it will not burn, it will increase the intensity of a fire as it is an oxidizer and poisonous gases and fumes are produced, including nitrogen oxides and beryllium oxide), sp. gr. 1.56 at 25°C.

beryllium oxide (BeO•H$_2$O) — A white odorless powder; MW: 25.01, BP: 3900°C, Sol: 2 μg/100 ml at 30°C, Fl.P: nonflammable (fumes are produced in fires, including beryllium oxide fumes), sp. gr. 3.0 at 25°C. It is used in the manufacture of ceramics, glass, electronic tubes, electronics components, nuclear fuels, and nuclear moderators. It is hazardous to the respiratory system, digestive system, eyes, and skin. Acute exposure can cause irritation, pneumonitis, dyspnea, pulmonary edema, and dermatitis. High exposure can cause fever, shortness of breath, and death. Chronic exposure can cause scars in the lungs and other body organs; alteration of the skin; kidney stones; cyanosis; and tachycardia. OSHA exposure limit (8-hour TWA): 0.002 mg/m^3.

BEST — Acronym for Biomonitoring of Environmental Status and Trends.

Best Available Control Technology — An emission limitation based on the maximum degree of emission reduction achievable through application of production processes and available methods, systems, and techniques; the most stringent technology available for producing the greatest reduction of air pollutant emissions, taking into account energy, environmental, economic, and other costs.

Best Available Retrofit Technology (BART) — Emission limitations imposed on existing sources of air pollutants to protect visibility in certain areas where visibility is a particularly valuable resource, such as in national parks.

Best Available Technology (BAT) — Level of control described as "best of the best" technology in use, including controls on toxic pollutants.

Best Available Technology Economically Achievable — A level of control generally described as the best technology currently in use, including controls on toxic pollutants.

Best Conventional Pollutant Control Technology (BCT) — (*wastewater*) Technology-based effluent limitations for conventional pollutants that direct dischargers had to meet as of July 1, 1984, pursuant to the Federal Water Pollution Control Act.

Best Demonstrated Available Technology — The most effective commercially available means of treating specific types of hazardous wastes.

best estimate data — (*computer science*) Data stream of selected variables, the values of which have been assessed to be from a preferential source.

Best Management Practice (BMP) — The structural, nonstructural, and managerial techniques that are recognized to be the most effective practical means to control nonpoint source pollutants yet still be compatible with productive use of the resource to which they are applied.

Best Practical Control Technology Currently Available (BPT) — Technology-based effluent limitations that direct dischargers had to meet as of July 1, 1977, pursuant to the Federal Water Pollution Control Act.

beta-chlorprene (CH$_2$CClCHCH$_2$) — A colorless liquid with a pungent, ether-like odor; MW: 88.5, BP: 139°F, Sol: slight, Fl.P: –4°F, sp. gr. 0.96. It is used in the manufacture of a variety of neoprene elastomers and in the manufacture of neoprene. It is hazardous to the respiratory system, skin, and eyes and is toxic through inhalation, absorption, ingestion, and contact. Symptoms of exposure include eye and respiratory irritation, nervousness, irritability, dermatitis, alopecia; carcinogenic. OSHA exposure limit (TWA): 10 ppm [skin] or 35 mg/m^3.

beta decay — The process of radioactive decay in which a neutron loses a beta particle, thus increasing the atomic number of the atom by one while leaving the atomic mass number the same. This is the most common form of radioactive decay.

beta-hemolytic streptococci — The pyogenic streptococci that causes hemolysis in red blood cells in blood agar in the laboratory.

beta-naphthylamine ($C_{10}H_7NH_2$) — Odorless, white to red crystals with a faint aromatic odor; darkens in air to a reddish-purple color; MW: 143.2, BP: 583°F, Sol: miscible in hot water, Fl.P: 315°F, sp. gr. 1.06 at 208°F. It is used in the manufacture of dyes, acids, and rubber. It is hazardous to the bladder and skin and is toxic through inhalation, absorption, ingestion, and contact. Symptoms of exposure include dermatitis, hemorrhagic cystitis, dyspnea, ataxia, methemoglobinemia, and dysuria; carcinogenic. OSHA exposure limit (TWA): regulated use.

beta particle — A charged particle emitted from the nucleus of an atom during radioactive decay having mass and charge equal in magnitude to that of the electron.

beta-propiolactone ($C_3H_4O_2$) — A colorless liquid with a slightly sweet odor; MW: 72.1, BP: 323°F (decomposes), Sol: 37%, Fl.P: 165°F, sp. gr. 1.15. It is used in the manufacture of acrylic acids and esters; in the sterilization of blood plasma, tissue grafts, and surgical instruments; in research and laboratory facilities. It is hazardous to the kidneys, lungs, skin, and eyes and is toxic through inhalation, absorption, ingestion, and contact. Symptoms of exposure include corneal opacity, frequent urination, dysuria, skin irritation, blistering, and burns; carcinogenic. OSHA exposure limit (TWA): regulated use.

beta ratio — The ratio of the diameter of a pipeline constriction to the unconstricted pipe diameter.

beta ray — A stream of high-speed electrons or positrons of nuclear origin more penetrating and less ionizing than an alpha ray.

betatron — A cyclic accelerator that produces high-energy electrons for radiotherapy treatment.

Betz limit — The maximum power (theoretically) that can be captured by a wind turbine from the wind; equal to 59.3% of the wind energy.

BeV — Notation for one billion electronvolts (10^9).

bevel — Any angle other than a right angle between two planes or surfaces.

beverage — Any liquid intended for human consumption.

BGH — See bovine growth hormone.

BHC — See benzene hexachloride.

bias — (*statistics*) A systematic error that causes a population value to be under- or overestimated and may

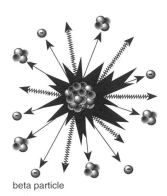

**Beta Particle
Beta Ray**

be due to improper sampling or improper measurement of the variable under study; a preference or inclination that inhibits impartial judgment.

biased sample (statistics) — A sample of a group in which all factors or participants were not equally balanced or objectively represented.

BIBEX — Acronym for biomass burning experiment.

bicarbonate — Any salt of carbonic acid in which only one of the hydrogen atoms has been replaced by a metal or radical.

biennial — A plant that completes its growth cycle in two growing seasons.

bifurcate — The division or branching into two forks.

bilateral — Affecting the right and left side of the body.

bile — A bitter-tasting, greenish-yellow alkaline liquid substance secreted by the liver and stored in the gall bladder and which flows into the small intestine, where it aids in the digestion and absorption of fat.

bill of lading (B/L) — The written order from a shipper to a carrier to move goods from one place to another; when available, this is the best source of shipping dates, origin, and name of shipper.

bill of materials — A list of components, ingredients, or materials required to manufacturer a product, including the proper ratios of quantities of each item.

bimodal distribution (statistics) — The distribution of quantitative data around two separate modes suggesting two separate normally distributed populations from which the data are drawn.

bin — (*statistics*) A range of values used for grouping purposes in order to make generalizations or to draw comparisons.

binary — Composed of or characterized by two elements or parts.

binary arithmetic — The mathematics of calculating in powers of two.

B

binary coded decimal (BCD) — (*computer science*) A system of number representation in which each digit of a decimal number is represented by a binary number.

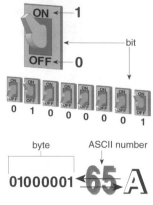

Binary Coded Decimal

binary fission — The division of cells into two approximately equal parts.

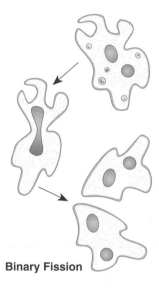

Binary Fission

binary number — A number represented in the binary system, used in computers, and represented by 0s and 1s.

binary system — A number system with a base of two.

binder — A resin or other cement-like material, other than clay, that is added to foundry sand to provide mechanical strength or ensure uniform consistency.

binding energy — The energy represented by the differences in mass between the sum of component parts and the actual mass of the nucleus.

binomial nomenclature — A system in which each organism is given a genus and species name.

bio- — A prefix meaning biologically based.

bioaccumulation — The uptake, retention, and concentration of environmental substances by an organism from its environment rather than through uptake from food; (*ecology*) the accumulation of toxic chemicals in living things through the consumption of food or water.

bioactive — A substance that has an effect on or causes a reaction in living tissue.

bioactivity — Any response from or reaction in living tissue.

bioaerosol — A microorganism suspended in the air as a free-floating particle either surrounded by a film of organic or inorganic material or attached to the surface of other suspended particulates.

bioassay — Determination of the concentration or dose of a given material necessary to affect the growth of a test organism under stated conditions; determination of the active power of a chemical by comparing its effects on a live animal or an isolated organ to a reference standard; (*pesticide*) the method of measuring the amount of pesticide chemical present in a test material by the action of a test plant or an animal exposed to it.

bioaugmentation — The process of using specifically cultured microorganisms to decontaminate polluted systems which speeds up the naturally occurring biodegradation processes by controlling the nature of the biomass.

bioavailability — The degree to which a chemical or other substance becomes available to a target tissue after being taken into the body.

biocatalysis — A chemical reaction mediated by biological systems, including microbial communities, whole organisms or cells, cell-free extracts, or purified enzymes.

biochelator — A biochemical compound synthesized by living organisms that binds and forms complexes with trace elements and polyvalent cations.

biochemical action system — A method of waste treatment that intensifies, stabilizes, oxidizes, and nitrifies the unstable organic matter; the systems include intermittent sand filters, contact beds, trickling filters, and activated sludge processes.

biochemical marker — Any hormone, enzyme, antibody, or other substance that is detected in the urine or other body fluids or tissues that may serve as a sign of disease or other abnormality.

biochemical oxygen demand (BOD) — The amount of oxygen reported in milligrams per liter that would be consumed by all the organics oxidized by bacteria and protozoa. The BOD is useful in predicting how low the levels of oxygen will go in a stream or river when organic wastes are oxidized in 1 liter of polluted water.

biochemistry — The study of the chemistry of living organisms and other chemical constituents and vital processes.

biocidal — Destructive to living organisms.

biocide — A substance that is destructive to many different organisms; also known as pesticide.

biocolloid — A colloid or colloidal mixture of biological components.

bioconcentration — The accumulation of a chemical in tissues of an organism to levels that are greater than the level in the medium, such as water, in which the organism lives.

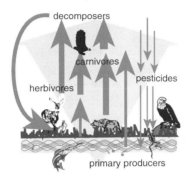

Bioconcentration

bioconcentration factor (BCF) — The quotient of the concentration of a chemical in aquatic organisms at a specific time or during a discrete time period of exposure, divided by the concentration in the surrounding water at the same time or during the same period.

bioconversion — A resource recovery method that uses the biological processes of living organisms to transform organic waste into usable material such as humus and compost.

biocurtain — A series of closely spaced wells spanning an aquifer designed to intercept polluted water with a colonized zone of biodegrading microbes.

biodegradability — The characteristic of organic matter to be broken down by microorganisms.

biodegradable — Susceptible to being degraded by biological processes.

biodegradation — The decomposition of a substance into more elementary compounds by the action of microorganisms such as bacteria.

biodegrade — To decompose, especially by bacterial action.

biodiversity — The variety and variability of living organisms and ecological complexities in which they occur; this encompasses different ecosystems, species, and genes.

bioelectricity — An electric current that is generated by living tissues such as nerves and muscles.

bioengineering — The application of engineering principles to biological or medical science.

bioequivalent — A chemical that has the same effect on the body as another chemical.

bioethics — A discipline dealing with the ethical applications of biological research and applications, especially in medicine.

biofeedback — A process providing a person with visual or auditory information about the autonomic physiological functions of his or her body.

biofilm — A slime layer that naturally develops when bacteria attach to an inert structure that is made of materials such as stone, metal, or wood in order to efficiently utilize nutrients that are concentrated on the solid surfaces, thereby saving energy. An extracellular polysaccharide is formed and acts as a natural glue to immobilize the cells; the microbial cells encased in an adhesive (usually a polysaccharide material) and attached to a surface.

biofiltration — A process that uses recirculation and a high rate of application to a shallow trickling filter to remove pollutants from effluent in a sewage treatment process.

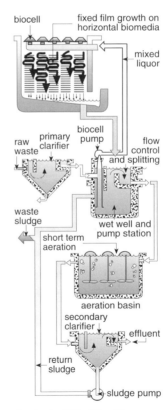

Activated Biofilter Process

bioflocculation — The clumping together of fine, dispersed organic particles by the action of certain bacteria and algae.

biogas — The methane that is produced from an anaerobic digester.

biogenesis — The belief that living material can originate only from preexisting life and not from inanimate material.

biogenic — Something produced by natural processes such as the emissions that are produced by plants and animals.

biogenic sediment — The sediment produced directly by the life processes of plants or animals.

biogeochemical cycle — The movement of matter within or between ecosystems caused by living organisms, geological forces, or chemical reactions (e.g., the cycling of nitrogen, carbon, sulfur, oxygen, phosphorus, and water); the chemical interactions that take place among the atmosphere, biosphere, hydrosphere, and geosphere.

biogeographical area — An entire self-contained natural ecosystem and its associated land, water, air, and wildlife resources.

biohazard area — Any area in which work has been or is being performed with biohazardous agents or materials.

Biological Hazard Symbol

biohazard control — Equipment and procedures utilized to prevent or minimize human and environmental exposure to biohazardous agents or materials.

biohazards — Biological substances that constitute a human health or environmental threat. These hazards include but are not limited to infectious and parasitic agents; noninfectious microorganisms such as some fungi, yeast, algae, plants, and plant products; and animals and animal products that contain infectious microorganisms, toxic biological substances, or biological allergens, or any combination of the three.

bioinsecticides — Bacteria and viruses causing diseases in insects.

biolarvicide — A delta endotoxin produced by the bacterium *Bacillus thuningiensis* subspecies *israelensis* which, when ingested by mosquito larvae, reacts with stomach secretions and causes gut paralysis and disruption of the ionic regulation

capacity of the gut epithelium, leading to death of the larvae.

biologic activity — The ability of a substance such as a drug or toxin to alter one or more of the chemical or physiological functions of a cell.

biologic dose — A measure of the concentration of a substance and its biologic effect.

biologic monitoring — A process of measuring the levels of various physiological substances, drugs, or metabolites within a patient during diagnosis or therapy; the measurement of toxic substances in the environment and the identification of health risks to the population.

biologic plausibility — A means of reasoning used to establish a cause-and-effect relationship between a biological factor and a disease.

biological — A medicinal preparation made from living organisms and their products; includes serums, antitoxins, and vaccines.

biological agent — Any living organisms or materials derived from them that cause harm to humans, animals, or plants or cause deterioration of material; may be found as liquid droplets, air cells, or dry powders.

biological amplification — An increase in the concentration of DDT, PCBs, and other slowly degradable fat-soluble chemicals in organisms as they proceed through higher levels of a food web.

biological contaminants — Living organisms or agents derived by viruses, bacteria, fungi, and mammal and bird antigens that can be inhaled and cause many health effects, including allergic reactions, respiratory disorders, hypersensitivity diseases, and infectious diseases.

biological control — Natural or applied methods using natural predators, parasites, or diseases to control populations of pest organisms.

biological diversity — Genetic and ecological differences in species.

biological exposure indices (BEIs) — Reference values intended as guidelines for the evaluation of potential health hazards in the practice of industrial hygiene.

biological growth — The activity and growth of any and all living organisms.

biological half-life — The time required for the body to eliminate one half of an administered dose of any substance during normal processes of elimination; this time is about the same for both stable substances and radioactive isotopes. Also known as effective half-life.

biological hazardous waste — Waste containing biohazardous materials.

biological incident — An event in which a biological agent is used as a terrorist weapon.

biological indicator — A vial containing bacterial spores used for determination of sterilization.

biological indicators of exposure study (ATSDR) — A study designed to use biomedical testing or the measurement of a chemical analyte, its metabolite, or another marker of exposure in human body fluids or tissues in order to validate human exposure to a hazardous substance.

biological integrity — The ability to support and maintain a balanced, integrated, functionality in a natural habitat of a given region; the condition of an aquatic community in unimpaired water bodies of a specified habitat as measured by community structure and function.

biological magnification — See biomagnification.

biological measurement — In exposure assessment, the measurement taken in a biological medium related to the established internal dose of a compound.

biological medium — One of the major components of an organism, including the blood, fatty tissue, lymph nodes, digestive system, and respiratory system, in which chemicals can be stored or be transformed.

biological monitoring — A technique for measuring the presence of a chemical or its metabolites in tissues or excreta or for measuring pathological effects of the toxin on the person.

biological oxidation — The process by which microorganisms decompose complex organic materials in activated sludge processes and self-purification of streams.

biological oxygen demand (BOD) — A standard test to determine the amount of dissolved oxygen utilized by organic material over a 5-day period.

biological residue — Any substance, including metabolites, remaining in livestock at the time of slaughter or in any of its tissues after slaughter as the result of treatment or exposure to a pesticide, organic or inorganic compound, hormone, hormone-like substance, growth promoter, antibiotic, anthelmintic, tranquilizer, or other therapeutic or prophylactic agent.

biological response modifiers — A substance that stimulates the body's response to infection and disease.

biological safety cabinet — An enclosure and ventilation system that prevents the release of infectious aerosols and may also prevent contact exposure to infectious agents used by the worker.

biological stressors — Organisms accidentally or intentionally dropped into habitats in which they do not evolve naturally.

biological survey — The collecting, processing, and analyzing of representative portions of a resident aquatic community to determine structure and function.

biological uptake (ATSDR) — The transfer of hazardous substances from the environment to plants, animals, and humans evaluated through environmental measurements (such as the amount of a substance in an organ known to be susceptible to that substance) to determine whether exposure has occurred. The presence of a contaminant, or its metabolite, in human biological specimens, such as blood, hair, or urine, is used to confirm exposure and can be an independent variable in evaluating the relationship between the exposure and any observed adverse health effects.

biological warfare — The use of bacteria or viruses or their toxins as weapons of mass destruction.

biological waste — Waste derived from living organisms.

biologically effective dose — The amount of a deposited or absorbed compound reaching the cells or target sites where adverse effects occur or where the chemical interacts with a membrane.

biologicals — See biologics.

biologics — The immunization vaccines, antigens, antitoxins, and other preparations made from living organisms and their products intended for use in diagnosing, immunizing, and treating humans and animals or in related research.

biology — A branch of the natural sciences dealing with the study of living organisms and their vital processes.

biomagnification — The process by which impurities found in water are concentrated in lower forms of life and reconcentrated substantially during their movement through the food chain; for example, whereas 0.001 µg of mercury might be found in the surrounding water, 30 to 50 µg of mercury could be found in fish that had bioconcentrated the mercury during their feeding process. Also known as biological magnification.

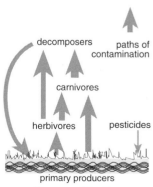

Biomagnification

biomarker — The measure of an individual's internal exposure or dose from a chemical; also known as biomonitoring.

biomass — Plant materials and animal waste used as a direct source of fuel which include forest products and residues, animal manure, sewage, municipal waste, agricultural crops, residues, and aquatic plants; the total mass or weight of living organisms expressed in terms of a given area of volume of the habitat. Also known as biomass fuel.

biomass fuel — See biomass.

biomaterial — The material used for, or suitable for use in, prostheses and coming in direct contact with living tissues.

biome — A major ecological community containing distinctive plant and animal life in a given climate usually extending over a large region of the Earth; examples are tundras and tropical rain forests.

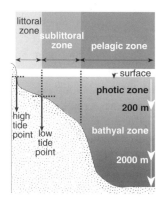

Marine Biome Strata

biomechanics — A discipline dedicated to the study of the living body as a structure that can function properly only within the confines of both the laws of Newtonian mechanics as well as the biological laws of life, including the muscular activity that occurs during movement or exercise; (*ergonomics*) the application of mechanical laws to human structures.

biomedical — Relating to or involving biological, medical, and physical science.

biomedical engineering — A system of techniques in which knowledge of biological processes is applied to solve practical medical problems and to answer questions in biomedical research.

biomedical testing (ATSDR) — The biological testing of persons to evaluate a qualitative or quantitative change in a physiological function that may be predictive of a health impairment resulting from exposure to a hazardous substance.

biomedicine — A branch of medical science concerned especially with the capacity of human beings to survive and function in abnormally stressing environments and with the protective modification of these environments.

biometeorology — The study of the relationship between living beings and atmospheric conditions.

biometrics — See biometry.

biometry — The statistical analysis of biological observations and phenomena; also known as biometrics.

biomimetic — The mimicking of a biological substance or process.

biomolecular engineering — The application of engineering technology to the solution of problems pertaining to organic molecules occurring in living organisms.

biomolecular materials — Complex biological macromolecules that can have an unusual combination of properties, such as being strong and supple yet lightweight at the same time.

biomonitoring — A means of surveillance of water quality by observing the biota from the field and laboratory standpoints; (*medicine*) see biomarker.

bionic potential — The inherent capacity of an organism to reproduce and survive despite limiting influences of the environment.

bionics — The science concerned with application of data about the functioning of biological systems to the solution of engineering problems.

biopesticide — A pesticide that is biological in origin (such as viruses, bacteria, pheromones, or natural plant compounds), in contrast to synthetic chemicals.

biophysics — The hybrid branch of science concerned with the application of physical principles and methods to biological problems.

biopolymer — The microbial products that can be manufactured commercially from renewable resources, are biodegradable, and provide alternatives to traditional plant and algal gums or to the plastics made from hydrocarbons.

biopotential — The electric charge produced by various tissues of the body, especially muscle tissue during contractions.

biopsy — The removal and examination of tissues, cells, or fluids from the living body for the purpose of diagnosis.

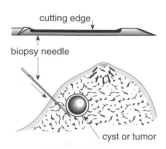

Breast Biopsy
in a closed operation

biopulping — An experimental way of using a fungus to pretreat wood chips before making paper pulp, thereby reducing energy use and water-polluting byproducts.

biorational control agents — See insect growth regulator.

bioreactor — A tank in which whole cells or cell-free enzymes transform raw materials into biochemical products and/or less undesirable byproducts.

bioremediation — A process of adding nutrients to groundwater to speed up the natural process in which bacteria break down chemicals into harmless compounds; the process by which living organisms act to degrade or transform hazardous organic contaminants.

biorhythm — An inherent rhythm that appears to control or initiate various biological processes.

biosafety level — Specific combinations of work practices, safety equipment, and facilities that are designed to minimize the exposure of workers and the environment to infectious agents — biosafety level 1 applies to agents that do not ordinarily cause human disease; biosafety level 2 is appropriate for agents that can cause human disease but whose potential for transmission is limited; biosafety level 3 applies to agents that may be transmitted by the respiratory route which can cause serious infection; and biosafety level 4 is used for the diagnosis of exotic agents that pose a high risk of life-threatening disease that may be transmitted by the aerosol route and for which there is no vaccine or therapy.

bioscience — The study of biology wherein all of the applicable sciences (physics, chemistry, and math) are applied.

bioscrubber technology — A system used to digest hazardous organic emissions from soil, water, and air decontamination processes.

biosensor — An analytical device comprised of a biological recognition element, such as an enzyme, receptor, DNA, antibody, or microorganism, in intimate contact with an electrochemical, optical, thermal, or acoustic signal transducer; together, these elements produce analyses of chemical properties or quantities.

biosequestration — The conversion of a compound through biological processes to a form that is chemically or physically isolated.

biosolid — The residues of wastewater treatment; formerly called sewage sludge.

biosorption — The property of the cell to absorb metals on its surface.

biosphere — The zone of air, land, and water at the surface of the Earth that is occupied by actively metabolizing matter and receives its supply of energy from an external source, which is ultimately the sun; however, spores may be found in areas outside the biosphere that are too dry, too cold, or too hot to support organisms that metabolize.

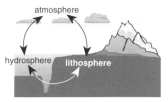

Structure of the Biosphere

biostabilizer — A machine that converts solid waste into compost by grinding and aeration.

biostatistics — Statistical processes and methods applied to the analysis of biological data.

biostimulation — A process that increases the rate of biological degradation, usually by the addition of nutrients, oxygen, or other electron donors and acceptors, to increase the number of indigenous microorganisms available for degradation of contaminants.

biosurfactant — A surface-active agent produced by microorganisms.

biosurvey — See biological survey.

biosynthesis — The production of a chemical compound and the organization of organic molecules by living organisms.

biot number — (*energy*) A dimensionless number: $Bi = (al_o)/l_a$, where a is the heat-transfer coefficient from the surface to the environment or from the environment to the surface, l_o is a specific dimension, and l_a is the thermal conductivity coefficient of the body.

biota — Animal and plant life characterizing a given area.

biotaxi — Systematic classification of living organisms according to their anatomic characteristics.

biotechnology — The use of genetic engineering to improve the quantity or quality of medical products, agricultural products, chemicals, and other products; applied biological science.

biotelemetry — The remote detection and measurement of a condition, activity, or function relating to the behavior and physiology of living organisms.

biotic — Of or pertaining to the living environment of a community made up of autotrophs (producer organisms that construct organic substances) and heterotrophs (consumer or reducer organisms that destroy organic substances).

biotic community — All living organisms comprising an ecosystem.

biotic factors — Those factors, such as weather and population, that affect the living organisms in an environment.

biotic index — A numerical index using various aquatic organisms to determine their degree of tolerance to differing water conditions.

biotic potential — All of the factors that contribute to increases in the number of a species, including reproduction, migration, adaptation, etc.

biotope — A specific biological habitat or place.

biotransformation — Alteration of the structure of a chemical compound by a living organism or enzyme; includes biodegradation.

biotransfusion — A collective term for various reactions (e.g., sewage broken down in soil by bacteria) that occur as a result of metabolism by organisms.

biotrophic — The nutritional relationship between two organisms in which one or both must associate with the other to obtain nutrients and grow.

biotype — A group of organisms sharing a specified genotype.

bioventing — The process of supplying oxygen in place to oxygen-deprived soil microbes by forcing air through unsaturated contaminated soil at low flow rates.

biphasic — Possessing both a sporophyte and a gametophyte generation in the life cycle; changing from positive to negative.

bipolar — Having or involving the use of two poles.

birth rate — See natality.

***bis*(2-chloroisopropyl) ether (C₃H₆Cl)₂₀)** — A colorless liquid with a characteristic odor; MW: 171.08; BP: 187°C; Sol: soluble in alcohols, ethers, and most organic solvents; Fl.P: 85°C (open cup), a combustible liquid; density: 1.4505. It is used as a solvent for resins, waxes, and oils and is hazardous to the respiratory tract and eyes; carcinogenic to animals. It causes irritation of the upper respiratory tract and eyes. ACGIH (TLV-TWA): NA.

$$CH_3 \qquad CH_3$$
$$| \qquad\quad |$$
$$ClCH_2 - CH - O - CH - CH_2Cl$$

***bis* (2-chloroisopropyl) ether**

***bis*(2-ethylhexyl) phthalate (C₂₄H₃₈O₄)** — A light-colored to colorless oily liquid with a slight odor; MW: 390.62, BP: 358°C, Sol (water): less than 0.01% at 25°C, Fl.P: 218°C, sp. gr. NA. It is the most commonly used phthalate ester in plasticizing, is a solvent and a component of dielectric fluids and electrical capacitors, and is used in pesticides. It is hazardous to the digestive system, eyes, and through dermal contact. Acute exposure may result in irritation of the eyes and mucous membranes, changes in liver function, central nervous system depression, testicular atrophy, chromosomal changes, and gastrointestinal irritation. It is potentially carcinogenic. OSHA exposure limit (8-hour TWA): 5 mg/m³.

***bis*(chloromethyl) ether (C₂H₄Cl₂O)** — A clear, colorless liquid with a suffocating odor detected at minimal levels; MW: 114.97; BP: 106°C; Sol: soluble in most organic solvents and decomposes in water; Fl.P: 35°C (closed cup), the vapor forms explosive mixtures with air; density: 1.315 at 20°C. It is used as an intermediate in anion-exchange quaternary resins. It is highly carcinogenic and toxic and is hazardous to the eyes, skin, digestive system, and respiratory system. It causes severe irritation and may rapidly lead to death. ACGIH (TLV-TWA): 0.001 ppm/m³.

$$Cl - CH_2 - O - CH_2 - Cl$$

***bis*(chloromethyl) ether**

bisect — Divide into two equal parts.

bit — The smallest unit of information that can be stored and processed in a computer; equivalent to the result of a choice between two alternatives: yes/no, true/false, or on/off.

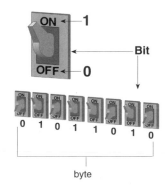

bit map — (*geographic information system*) A pattern of bits on the grid stored in memory that is used to generate an image on a raster scan display.

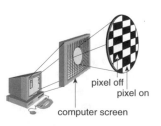

Bit Map

bit plane — (*geographic information system*) A memory grid in a graphics device used for storing information for display.

bits per inch (BPI) — (*computer science*) The density of bits recorded on a magnetic tape.

bitumen — Any of various mixtures of hydrocarbons (as tar), often together with their nonmetallic derivatives, occurring naturally or obtained as residue after heat-refining naturally occurring substances such as petroleum.

bituminous coal — A dark-brown to black, soft coal with high volatile gas and ash content.

bivalve — The common name for a number of bilaterally symmetrical animals, including mollusks and crustaceans; possessing a two-part valve or shell.

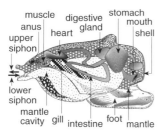
Bivalve (Clam)

BL — Acronym for biosafety level.

black body — An ideal emitter that radiates energy at the maximum possible rate per unit area at each wavelength for any given temperature. It also absorbs all of the radiant energy incident on it; none of the energy is reflected or transmitted.

black death — See plague.

black lead — See graphite.

black light — Ultraviolet light radiation (3000 to 4000 Å or 0.3 to 0.4 μm) that causes fluorescent materials to emit visible light.

black lung disease — An inflammation and fibrosis caused by accumulations and subsequent calcification of coal dust in the lungs or airways, resulting in permanent reduced lung capacity; also known as pneumoconiosis.

black oil — The crude oil or heavy fuel oil from the bottom of the refining process.

blackout — (*energy*) Disconnection of the source of electricity from all the electrical loads in a certain geographical area brought about by an emergency forced outage or other failure in the generation, transmission, or distribution system serving the area.

blackwater — Liquid and solid human body waste and carriage waters generated through toilet usage.

blackwater fever — A serious complication of chronic falciparum malaria with symptoms of jaundice, hemoglobinuria, acute renal failure, and the passage of bloody, dark-gray or black urine because of massive intravascular hemolysis; death rate is 20 to 30%.

blade — Part of an excavator that digs and pushes dirt or solid waste but does not carry it; part of a turbine that water or air reacts against, causing it to spin.

blanching — An operation in which raw food material is immersed in hot water or exposed to live steam. The product shrinks, and the respiratory gases contained in the plant cells are expelled. The release of gas prevents strain on the container during heat-processing and favors development of a higher vacuum in the finished product; also serves as an added cleaning measure and may remove raw flavors from foods.

blank — A bottle containing only dilution water or distilled water where the sample being tested has not been added. Tests are often run on a sample and a blank and the differences are compared.

blanket application or treatment — See broadcast application or treatment.

blast — A brief and rapid movement of air vapor away from a center of outward pressure, as in an explosion; a means of loosening or moving rock or soil by means of explosives or an explosion.

blast furnace — A furnace in which the cooked fuel is ignited by a blast of hot, compressed air to increase the combustion rate. The heat is used to reduce iron oxide ores to cast iron or pig iron; the raw materials are iron ore, coke, and limestone.

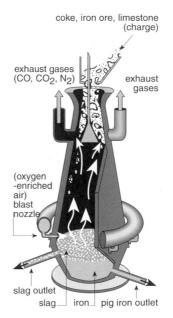
Blast Furnace

B

blast gate — A sliding metal damper in a duct usually used to regulate the flow of forced air.

blastocyst — An early stage of embryonic development consisting of a hollow ball of cells.

blastomycosis — An infectious disease caused by the yeast-like fungus *Blastomyces dermatitidis*.

bleach — A chlorine product that cleans, whitens, brightens, and removes stains from fabrics and other surfaces.

bleeder — An individual who is a hemophiliac or has any other vascular or hematological condition associated with a tendency to hemorrhage.

bleeders — (*food*) Openings used to remove air that enters retorts and steam chambers; promotes circulation of steam in such retorts and steam chambers and may serve as a means of removing condensation.

blende — See sphalerite.

blepharism — Spasm of the eyelids causing continuous, involuntary blinking.

blight — A general term for the sudden and wilting death of a plant or a plant part caused by fungi, bacteria, or viruses.

blister agent — A chemical, such as sulfur mustard, that produces local irritation and damage to skin and mucous membranes that progresses in severity to fluid-filled blisters on the skin.

blizzard — A severe weather condition characterized by low temperatures and strong winds greater than 35 mph and carrying a great amount of snow, either falling or blowing.

bloat — A swelling or filling up with gas.

blood — The fluid that circulates through the heart, arteries, capillaries, and veins. It is the body's chief means of transport of oxygen from the lungs to the body tissues and of carbon dioxide from the tissues to the lungs. It transports nutritive substances and metabolites to the tissues and removes waste products from the kidneys and other organs of excretion. Blood is composed of fluid plasma, suspended formed elements, blood cells, and platelets.

blood agent — A chemical agent that is absorbed into the general circulation system and carried to all body tissues.

blood bank — A place where whole blood or plasma is typed, processed, and stored for future use.

blood–brain barrier — A large group of capillaries in the central nervous system and surrounding membranes preventing or slowing the passage of some drugs, other chemical compounds, radioactive ions, and disease-causing organisms from the blood into the central nervous system; a thin membrane that protects the spinal fluid and brain from foreign substances.

blood cell — Any of the formed elements of the blood including red cells, white cells, and platelets.

blood circulation — The movement of blood through the body from the heart through the arteries, arterioles, capillaries, venules, and veins and back to the heart.

blood clot — A semisolid gelatinous mass made up of red cells, white cells, and platelets in a soluble fibrin network; produced to stop bleeding.

blood corpuscle — A blood cell.

blood count — A count of the number of corpuscles per cubic millimeter of blood.

blood fluke — See schistosome.

blood gas — A gas dissolved in the liquid part of the blood.

blood gas determination — An analysis of the pH of the blood and the concentration and pressure of oxygen, carbon dioxide, and hydrogen ions in the blood.

blood level — The concentration of a drug or other substance in a measured amount of plasma, serum, or whole blood.

blood pH — The hydrogen ion concentration of the blood that is a measure of blood acidity or alkalinity.

blood plasma — The liquid noncellular fraction of blood that includes dissolved substances.

blood poisoning — A common, non-medical term referring to the presence of infecting agents, such as bacteria or toxins, in the blood stream; see also septicemia.

blood pressure — The pressure of the blood against the walls of the blood vessels.

blood pressure monitor — A piece of equipment that automatically measures blood pressure and records the information on a continuous basis.

blood urea nitrogen — The urea concentration of serum or plasma; an important indicator of renal function.

blood vessel — A tube that carries blood.

bloodborne pathogens — Pathogenic microorganisms that are present in human blood and cause disease in humans.

bloom — (*phytoplankton*) A large and sometimes rapid growth of phytoplankton often associated with hypoxic conditions and giving a distinct color to the water.

blowdown — Discharge of water from a boiler or cooling tower by pressure to dispose of accumulated salts.

blower — A fan used to force air or gas under pressure.

blower door — A device that fits into a doorway of the building and contains a powerful fan to supply or extract a measured rate of airflow.

blowoff — A controlled outlet on a pipeline, tank, or conduit used to discharge water or accumulations of material.

blowout — A sudden, violent escape of gas and oil from an oil well when high-pressure gas is encountered.

blue-green algae — Any of the group of photosynthetic microorganisms classified as either plants or bacteria because they possess characteristics of both.

BM — See bench mark.

BMP — Acronym for Best Management Practice(s).

BNA — See *Basle Nomina Anatomica*.

BNICE — Acronym that identifies the five categories of terrorist incidents: biological, nuclear, incendiary, chemical, and explosives.

board certification — A process by which a professional is certified in a given specialty or subspecialty by peers.

Board of Health — An administrative body acting at a municipal, county, state, or national level to coordinate the projects and resources of a community and to set policy to satisfy areas of health needs, including prevention of disease, implementation of laws, and protection of the health of the public.

BOD — See biological oxygen demand.

BOD$_5$ — The amount of dissolved oxygen consumed in 5 days by biological processes breaking down organic matter.

body burden — The existing level of pollutant or radioactivity to which an individual has been subjected and the potential level of severity and seriousness of the effect of this pollutant; the amount of radioactive material in the body at a given time.

body cavity — Any of the spaces in the chest and abdomen that contain body organs.

body coat — (*swimming pools*) Diatomaceous earth that builds up on a filter element during the course of running water through a filter to help maintain filter porosity.

body feed — (*swimming pools*) Diatomaceous earth fed constantly or intermittently during the time water is running through a filter to produce a body coat.

body fluids — Fluids produced in the body such as semen, blood, vaginal secretions, and breast milk.

body language — Nonverbal signals, including body movements, postures, gestures, and facial expressions, that express various physical, mental, and emotional states of mind.

body odor — Odor originating from sweat and secretions from the skin, breath, and gases from the digestive tract, depending on diet, activity, and personal hygiene.

body scanner — A diagnostic, radiological instrument technique used for diagnosis of a patient.

body temperature — The level of heat produced and sustained by the body process, with variations and changes indicating disease and other abnormalities.

body wave — A seismic wave that travels through the body of the Earth rather than along its surface.

BOF — See basic oxygen furnace.

bog — A permanently wet, spongy land that is usually poorly drained, highly acidic, and rich in plant residue; also known as moor or quagmire.

Bohr effect — A characteristic of hemoglobin that causes it to disassociate from oxygen to a greater degree at higher concentrations of carbon dioxide.

boil — A painful nodule containing a central core and formed on the skin by inflammation of the dermis and subcutaneous tissue. It may develop wherever friction, irritation, or a scratch or break in the skin occurs and allows bacteria to penetrate the outer layer of the skin; most boils are caused by *Staphylococcus aureus*. Also known as a furuncle.

boiler — A vessel designed to transfer heat produced by combustion or electric resistance to water; boilers may provide hot water or steam.

boiler compensation — An operation that changes the operating temperature of a boiler, usually according to the outside air temperature.

boiler feedwater — Water provided to a boiler for conversion to steam and the steam generation process.

boiler optimization — An energy management function that acts to balance boiler operation to loads and control combustion air.

boiling — Movement that occurs in any liquid when pressure produced by the vapor within the liquid equals the pressure on its surface. The lower the pressure, the lower the temperature necessary to bring a liquid to a boil. Each pure liquid has its own definite boiling point.

boiling point (BP) — The temperature at which the vapor pressure of a liquid equals the atmospheric pressure.

boiling water reactor (BWR) — (*energy*) A reactor in which water, used as both a coolant and a moderator, is allowed to boil in the core, resulting in steam that can be used to drive a turbine.

BOLDER — Acronym for basic on-line disaster and emergency response.

bole — A variety of soft, friable clays of various colors, although usually red due to iron oxide.

bolide impacts — Asteroids or comets striking the Earth that are possible causes of major climatic changes and mass extinctions in the Earth's history.

Bolivian hemorrhagic fever — An infectious disease caused by an arenavirus transmitted to humans by infected rodents through contamination of food by rodent urine. The incubation period is 1 to 2 weeks, and symptoms include chills, fever, headache, muscle ache, anorexia, nausea, and vomiting, resulting in hypotension, dehydration, bradycardia, pulmonary edema, and internal hemorrhage; mortality rate is 30%.

BOM — See bill of materials.

B

bond — (*law*) A sum of money or other thing of value held by the court pending proper execution of a consent decree of condemnation in a seizure in which the goods are being reconditioned by a claimant or during import reconditioning, or while goods are being detained pending decision as to admission; (*chemistry*) see valence.

bonding — Interconnecting two objects by means of a clamp and bare wire to equalize the electrical potential between the objects to prevent a static discharge when transferring a flammable liquid from one container to another.

bone — A dense, hard, and slightly elastic connective tissue of the human skeleton composed of compact osseous tissue surrounding spongy tissue enervated by many blood vessels and nerves. The human skeleton has 206 bones.

bone marrow — The soft material that fills the cavities in most bones of vertebrates and manufactures most of the formed elements of the blood.

bone seeker — Any compound or ion that tends to migrate preferentially into bone when introduced in the body.

bonnet — (*plumbing*) The cover on a gate valve.

boom — Any heavy mechanical support beam that is hinged at one end and carries a weight-hooking device at the other.

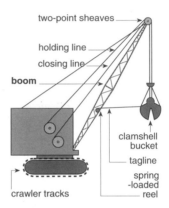

two-point sheaves

holding line

closing line

boom

clamshell bucket

tagline

spring-loaded reel

crawler tracks

Boom on Clamshell

booster injection — An antigen such as a vaccine or toxoid usually given in a smaller amount than the original immunization.

boot stage — The appearance of a grass or cereal seed head emerging from the sheaf covering it.

boot-up — (*computer science*) To start up a computer system.

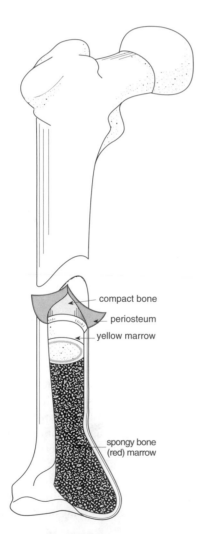

compact bone

periosteum

yellow marrow

spongy bone (red) marrow

Bone Marrow

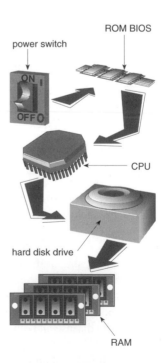

ROM BIOS

power switch

ON

OFF

CPU

hard disk drive

RAM

Boot-Up

B

borax (Na$_2$B$_4$O$_7$·10H$_2$O) — A white, crystalline, mildly alkaline, water-soluble salt (sodium borate) used as a laundry additive; provides moderate alkalinity buffering.

borderline disease — A state of health from which the patient has some of the signs and symptoms of a disease but not enough to have a definitive diagnosis.

Bordetella — A genus of Gram-negative coccobacilli, some of which are pathogens of the respiratory tract.

borehole — A hole made with drilling equipment.

boring — The act of rotary drilling.

boron (B) — A brown, amorphous, non-metallic powder that can be fused to a brittle mass; atomic number: 5; atomic weight: 10.811. It is used as borox, a cleaning material.

boron oxide (B$_2$O$_3$) — Colorless, semitransparent lumps or hard, white, odorless crystals; MW: 69.6, BP: 3380°F, Sol: 3%, sp. gr. 2.46. It is used in the manufacture of metal borates and in the preparation of fluxes; in the manufacture of glass for heat-resistance; as an herbicide; in the production of surface coatings and as fire-resistant additives in enamel and paints. It is hazardous to the skin and eyes and is toxic through inhalation. Symptoms of exposure include nasal irritation, conjunctivitis, and erythema. OSHA exposure limit (TWA): 10 mg/m^3 [air].

boron trifluoride (BF$_3$) — A colorless gas with a pungent, suffocating odor; forms dense white fumes in moist air; MW: 67.8, BP: –148°F, Sol: soluble in water. It is used as an acid catalyst for alkylation of aromatic compounds; in polymer technology; in the synthesis of other boron-containing organic and inorganic compounds; in the purification of hydrocarbons; in nuclear technology for separation of boron isotopes; as a filling gas for neutron counters; in metallurgy as a flux and antioxidant; as a flame-coloring agent for liquefied petroleum gas. It is hazardous to the respiratory system, kidneys, eyes, and skin and is toxic through inhalation and contact. Symptoms of exposure include nasal irritation, epistaxis, eye and skin burns, pneumonia, kidney damage, and epistaxis in animals. OSHA exposure limit (TWA): ceiling 1 ppm [air] or 3 mg/m^3.

boroscopic duct inspection — An inspection of the interior of a ventilation duct by drilling a hole in the duct wall and inserting a light and miniature video camera.

Borrelia — A genus of unevenly coiled helical spirochetes, several species of which cause tickborne and louseborne relapsing fever.

Borrelia burgdorferi — The etiological agent of Lyme disease transmitted to humans by tick vectors, primarily *Ixodes dammini*.

borrow (*soil*) — The material excavated from one area to be used as fill material in another area.

botanical pesticide — A pesticide chemical produced by a plant in nature.

bottled drinking water — Water, including mineral water, sealed in bottles or other containers and offered for sale for human consumption.

bottom ash — Ash that drops out of a furnace gas stream in the furnace and economizer sections.

botulinus toxin — A potent bacterial toxin produced by different strains of *Clostridum botulinum*; also called botulinum toxin.

botulism — (*disease*) A severe intoxication resulting from ingestion of toxin contained in contaminated food. Symptoms of exposure usually relate to the nervous system, including blurred or double vision, dry mouth, sore throat, paralysis, vomiting, diarrhea, and possible respiratory failure and death; neurological symptoms appear after an incubation time of 12 to 36 hours. It is caused by toxins produced by *Clostridium botulinum* types A, B, and E and occasionally F and G and is found worldwide. Reservoirs of infection include soil, sediment in the water, and the intestinal tracts of animals, including fish. It is transmitted by ingestion of food containing the toxin and is not communicable; general susceptibility. It is controlled by the careful processing of all foods and disposal of foods that appear to be off-color or have off-odors or where gas has been produced.

botulism antitoxin — An equine antitoxin administered intravenously and used against the toxins produced by type A, B, and E *Clostridium botulinum*.

boulder — A rock fragment with an average dimension of 12 inches or more.

boulder clay — Glacial drift that has not been subjected to the sorting action of water and therefore contains particles from boulder to clay sizes.

bounding estimate — An estimate of exposure, dose, or risk that is higher than that incurred by a person in the population with the currently highest exposure, dose, or risk.

boutonneuse fever — An infectious disease caused by *Rickettsia conorii* and transmitted to humans through the bite of the tick, with a fever lasting from a few days to 2 weeks, and a rash spreading over the body, including skin of the palms and soles of the feet.

bovine growth hormone — A naturally occurring protein that has been genetically engineered as a synthetic compound manufactured in large quantities and commercially available to farmers; causes cows to increase the efficiency of milk production per unit of feed consumed. As a result of public controversy concerning this hormone, some states require retail dairy product labels to identify use of the synthetic product.

bovine somatotropin — See bovine growth hormone.

bovine spongiform encephalopathy — A chronic progressive degenerative disease affecting the central nervous system of cattle and resulting in death. The causative agent has not been determined but may be a virino or prion. The etiological agent is extremely resistant to destruction. The incubation period in cattle is from 2 to 8 years.

bovine tuberculosis — A form of tuberculosis caused by *Mycobacterium tuberculosis* and primarily affecting cattle.

bowel — The intestine or part of the digestive tract from the stomach to the anus.

boxboard — Paperboard used to manufacture boxes and cartons.

Boyle's law — The volume of a confined gas at a constant temperature is inversely proportional to the pressure to which it is subjected; true only for ideal gases. The law states that the product of the volume and pressure of a gas contained at a constant temperature remains constant; also known as Mariotte's law.

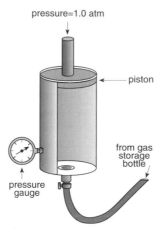

Boyle's Law
apparatus for investigation

BP — See boiling point.

bpd — See barrels per day.

BPI — See bits per inch.

BPJ — Acronym for best professional judgment.

bps — Acronym for bits per second.

BPT — See Best Practical Control Technology Currently Available.

Bq — See becquerel.

brachytherapy — The use of radiation sources in or on the body for treating certain types of cancer.

brackish water — A mixture of fresh- and saltwater; undrinkable.

bradycardia — (*physiology*) Abnormal slowness of the heart rate.

brain — The portion of the central nervous system contained within the cranium; composed of convoluted, soft, gray or white tissue that regulates the functions of the central nervous system.

brain concussion — A violent jarring or shaking or other nonpenetrating injury to the brain caused by a sudden change in momentum of the head.

brain death — An irreversible form of unconsciousness characterized by a complete loss of brain function while the heart continues to beat.

brain stem — Stemlike part of the brain connected to the spinal cord.

brake horsepower (plumbing) — Energy provided by a motor or other power source required at the top or end of the pump shaft.

branch — An addition to the main pipe in a piping system.

branching — (*physics*) The occurrence of competing decay processes in the disintegration of a particulate radionuclide; (*botany*) a shoot or secondary stem on the trunk or limb of a tree.

branching circuit — A portion of a wiring system in the interior of a structure that extends from a final overload protection device to a plug receptacle, motor, or heater.

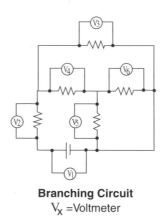

Branching Circuit
V_X = Voltmeter

brass — An alloy of copper and zinc that may contain a small portion of lead.

braze — To solder with a relatively infusible alloy such as brass.

breach — A gap, rift, hole, or rupture in a dam allowing stored water to flow through in an uncontrolled and unplanned manner.

breakbone fever — Another name for dengue fever.

breakdown product — A chemical that comes from a parent compound that has been altered, typically by heat, light, or enzymes.

breakpoint chlorination — Chlorine that reacts with ammonia in water to form chloramines; as additional chlorine is added, the total chlorine residual continues to rise until the concentration reaches a point that forces the reaction with ammonia to go rapidly to completion. The compounds of nitrogen and chlorine are released from the water, and the apparent residual chlorine decreases; the break-point is that point at which the residual suddenly drops off. Past the breakpoint, all combined chlorine compounds disappear, and the potential for eye irritation and chlorine odors disappear; the chlorine that remains in the water is all in the free state.

breakthrough — The localized penetration of unfiltered materials through a gas or vapor air-purifying element.

breakthrough time — Elapsed time between the initial contact of a chemical with the outside of the material and the time at which the chemical is first detected on the inside surface of the material by means of a chosen analytical instrument.

breakwater — An offshore structure used to protect a harbor or beach from the force of waves.

breast — Either of two protuberant milk-producing glands located on the chest of human females and some other mammals.

breast cancer — A malignant neoplastic disease of breast tissue.

breathalyzer — Piece of equipment used to analyze exhaled air, especially for blood alcohol levels.

breathing — Inhaling and exhaling movements of the chest and lungs to ventilate the alveoli.

breathing tube — A tube through which air or oxygen flows to the facepiece, helmet, or hood.

breathing zone — The zone of the ambient environment in which a person performs normal respiratory functions; the area of a room in which occupants breathe as they stand, sit, or lie down.

breathing zone sample — (*industrial*) An air sample collected in the breathing area of a worker to assess exposure to airborne contaminants.

breccia — A conglomerate-like rock made up of angular pieces of volcanic rock usually bound in volcanic ash.

brecciated — A rock made up of highly angular, coarse fragments.

breeching — A passage that conducts the products of combustion from a furnace to a stack or chimney.

breeder — A nuclear reactor that produces more fuel than it consumes.

breeder reactor — A nuclear reactor in which U_{238} or Th_{232}, neither of which is easily fissionable, absorbs neutrons and becomes atoms of Pu_{239} or U_{236}, which can later be used as fuels in fission reactors.

breeding density — The density of sexually mature organisms in a given area during the breeding period.

breeding potential — The maximum rate of increase in numbers of individuals of a species or population under optimum conditions.

breeding rate — The actual rate of increase in new individuals in a given population.

bremsstrahlung — Electromagnetic radiation (photons) emitted by the negative acceleration of a charged particle, usually an electron, after collision with the nucleus of an atom.

BRI — See building-related illness.

bridge crane — (*mechanical engineering*) A lifting unit in which the hoisting apparatus is carried by a bridge-like structure spanning the area of question.

brightness ratio — A comparison of two different brightnesses of light in an area.

brightness temperature — A measure of the intensity of radiation thermally emitted by an object given in units of temperature; a proportional correlation exists between the intensity of the radiation emitted and physical temperature of the radiating body.

brine — Water saturated with saline.

brine mud — Waste material associated with well-drilling or mining composed of mineral salts or other inorganic compounds.

briquet — Coal or ore dust pressed into an oval or brick-shaped block; also spelled briquette.

briquette — See briquet.

briquetter — A machine that compresses a material, such as metal turnings or coal dust, into small pellets.

British thermal unit (Btu) — Originally, the amount of heat required to raise the temperature of 1 pound of water 1°F; Btu is now defined as 1055.06 joules (British Standards Institution).

brittle — A structural behavior in which a material deforms permanently by fracturing.

brittle limit — The stress limit beyond which a material fractures.

broadband — (*computer science*) Computer access by means of channels, such as coaxial cable, that allow for greater bandwidth and faster connections to the Internet.

broadcast application or treatment — The placing of a pesticide chemical or agent over an entire field, lawn, or other vast area.

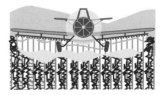

Broadcast Application
(as in crop dusting)

broadleaf plants or weeds — Any plants or weeds with wide, flat leaves having net veins, not the parallel veins of grasses or conifers.

broad-spectrum antibiotic — An antibiotic effective against a wide variety of pathogenic microbes.

B

brodifacoum (3-[3-(4¹-bromo-[1,1¹-biphenyl]-4-yl)-1,2,3,4-tetrahydro-1-naphthalenyl]-4-hydroxy-2H-1-benzopyran-2-one) — A potent anticoagulant poison used to exterminate warfarin-resistant Norway rats, roof rats, and house mice. It is hazardous to children, pets, domestic animals, and nontarget animals by reducing blood-clotting ability and causing hemorrhaging. It can be used around homes, industrial, commercial, agricultural, and public buildings, and modes of transportation but not in sewers.

bromates — General formula: $M(BRO_3)_x$, where M is a metal ion with valency x or the ammonium radical. These are salts of bromic acid ($HBrO_3$). Alkali metal bromates are used in analytical work for volumetric analysis, in extraction of gold, and as oxidizing agents. Bromates are powerful oxidizers and, on heating, decompose explosively. They react violently with combustible, organic, and other readily oxidizable substances. Potassium bromate and sodium bromate are moderately toxic compounds. They are hazardous to the digestive tract and respiratory tract. They may cause nausea, vomiting, diarrhea, respiratory stimulation, decrease in body temperature, methemoglobinemia, and renal injury. Bromates are carcinogenic in animals. ACGIH (TLV-TWA): 0.5 mg/m³ [air].

bromethalin (N-methyl-2,4-dinitro-N-(2,4,6-tribromophenyl)-6-(trifluoromethyl) benzenamine) — A rodenticide used for control of Norway rats, roof rats, and mice which works through the uncoupling of oxidative phosphorylation in the mitochondria and leads to fluid buildup between the myelin sheaths of the nerves, increased spinal fluid pressure, and pressure on the nerve axons. Acute exposure in humans causes symptoms of headache, confusion, personality changes, seizures, coma, and death.

bromide — A chemical compound containing bromine.

brominator — A device that allows fresh or recirculated water to be introduced into a bromine jar at atmospheric pressure to form a solution that is then conveyed to water.

bromine (Br₂) — A dark, reddish-brown, fuming liquid with suffocating, irritating fumes; MW: 159.8, BP: 139°F, Sol: 4%, sp. gr. 3.12. It is used during the synthesis of ethylene dibromide; in the manufacture of pesticides; as a laboratory reagent; as a disinfecting, sanitizing, and bleaching agent; in the preparation of flame retardants in plastics and fibers; in the manufacture of organic and inorganic compounds for use in photography, pharmaceuticals, fungicides, intermediates, dyes, and bleaching agents. It is hazardous to the respiratory system, eyes, and central nervous system and is toxic through inhalation, ingestion, and contact. Symptoms of exposure include dizziness, headache, lacrimation, epistaxis, cough, feeling of oppression, pulmonary edema, pneumonia, abdominal pain, diarrhea, measle-like eruptions, and severe skin and eye burns. OSHA exposure limit (TWA): 0.1 ppm [air] or 0.7 mg/m³.

bromine pentafluoride (Brf₅) — A colorless to pale-yellow liquid; MW: 174.91, BP: 40.8°C, Sol: NA (reacts with water), Fl.P: noncombustible liquid but highly reactive, density: 2.460 at 25°C. It is used as a fluorinating agent and an oxidizer. It is hazardous to the eyes, skin, and mucous membranes. It causes severe irritation of the eyelids, salivation, nephrosis, hepatosis, severe corrosion, and burns of the mouth. ACGIH (TLV-TWA): 0.1 ppm.

bromine trifluoride (BrF₃) — A colorless to pale-yellow liquid; MW: 136.91, BP: 125.7°C, Sol: miscible with water but reacts violently, Fl.P: a noncombustible liquid that reacts violently with water and any metal, density: 2.803 at 25°C. It is used as a solvent for fluorides. It is hazardous to the eyes, skin, and mucous membranes and causes severe burns. ACGIH (TLV-TWA): 2.5 mg/m³.

bromochlorodifluoromethane — A chemical belonging to the Freon family that is an effective gaseous fire suppression agent for use in the protection of computer-controlled rooms, museums, telecommunications switches, and other areas containing highly valuable materials; it causes ozone depletion.

bromoform (CHBr₃) — A colorless to yellow liquid with a chloroform-like odor; a solid below 24°F; MW: 252.8, BP: 301°F, Sol: 0.1%, sp. gr. 2.89. It is used as a heavy liquid flotation agent in mineral separation-sedimentary petrographical surveys; in chemical and pharmaceutical synthesis; as an industrial solvent in liquid-solvent extractions in nuclear magnetic resonance studies; as a flame retardant; as a catalyst, initiator, or sensitizer in polymer production, irradiation reactions, and vulcanization of rubber. It is hazardous to the skin, liver, kidneys, respiratory system, and central nervous system and is toxic through inhalation, absorption, and ingestion. Symptoms of exposure include respiratory system and eye irritation, central nervous system depression, and liver damage. OSHA exposure limit (TWA): 0.5 ppm [skin] or 5 mg/m³.

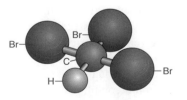

Bromoform

bromomethane (CH₃Br) — A colorless gas with little odor; MW: 94.95; BP: 3.6F°C; Sol: in water at 20°C, 0.9 g/l; Fl.P: not flammable; density: 3.97 at 20°C. It is primarily used as a soil or space fumigant for the control of insects, fungi, and rodents. It can cause accidental death through acute inhalation exposure during manufacturing and packaging operations, during the use of fire extinguishers containing bromomethane, or during fumigation activities. Death is not immediate, usually occurring within 1 to 2 days of exposure. The cause of death is not certain but is probably due to neurological and lung injury. Symptoms of exposure include edema accompanied by focal hemorrhagic lesions that severely impair respiratory function and lead to hypoxia, cyanosis, and complete respiratory failure. The liver may become swollen and tender in some cases. Renal effects include congestion, anuria or oliguria, and proteinuria. It is irritating to the skin and eyes and may cause conjunctivitis, arrhythmia, rashes, or blisters; it may also cause neurological problems with initial symptoms including headache, nausea, confusion, weakness, numbness, and visual disturbances, regressing to ataxia, tremor, seizures, paralysis, and coma. OSHA exposure limit: 5 ppm [air] or 20 mg/m³.

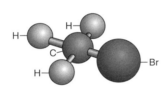

Bromomethane

bromthymol blue — A chemical dye sensitive to changes in pH; can be used to test pH over a range of 6.0 to 7.6. The color changes from yellow to blue as the pH increases.

bronchi — An air passage that leads from the windpipe to the lungs.

bronchial asthma — Abnormal responsiveness of the air passages to certain inhaled or ingested allergens. An attack of bronchial asthma consists of widespread narrowing of the bronchioles by muscle spasm, swelling of the mucous membranes, or thickening and increasing mucous secretion accompanied by wheezing, gasping, and sometimes coughing.

bronchial spasm — An excessive and prolonged contraction of the involuntary muscle fibers in the walls of the bronchi and bronchioles caused by irritation or injury to the respiratory mucosa, infections, or allergies.

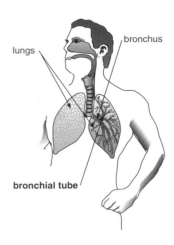

Bronchial Tube

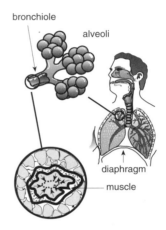

Bronchial Asthma

bronchiole — A small, thin-walled branch of the bronchial tubes within a lung.

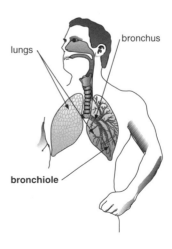

Bronchiole

bronchitis — An inflammation of the bronchial linings causing a persistent cough, copious amounts of sputum, and involuntary muscle spasms that constrict the airways.

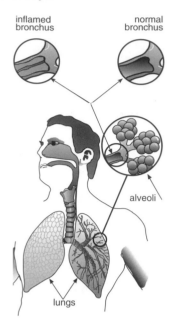

Bronchitis

bronchoconstrictor — An agent that causes a reduction in the diameter of a bronchus or bronchial tube.

bronchopneumonia — (*disease*) An inflammation of the lungs that usually begins in the terminal bronchioles; also known as locular pneumonia.

bronchoscope — A flexible, lighted instrument used to examine the trachea and bronchi, the air passages that lead into the lungs.

bronchus — One of the larger passages conveying air to and within the lungs.

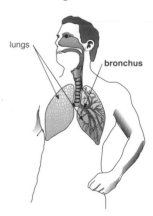

Bronchus

broth — A cultural medium used to support the growth of bacteria for laboratory analysis.

brownfields — Abandoned, lightly contaminated properties often found in economically distressed areas.

Brownian movement — The incessant and random movements of particles less than 1 μm in diameter in a fluid caused by collision of the particles with surrounding molecules.

brownout — (*energy*) Partial reduction of electrical voltages resulting in dimming of lights and slowing of motor-driven devices; a spot power outage.

browser — (*computer science*) A software program that allows users to visit sites on the World Wide Web.

BRP — Acronym for basic research plan.

Brucella — (*microbiology*) Genus of Gram-negative, aerobic, nonmotile, cocci, or rod-shaped bacteria, all of which are parasites and pathogens of mammals.

brucellosis — (*disease*) A systemic bacterial disease with acute or insidious onset of irregular fever, headache, weakness, profuse sweating, chills, depression, weight loss, and aching; may last for several months. Incubation time is highly variable from 5 to 30 days, or occasionally several months; caused by *Brucella abortus* biotypes 1 to 7 and 9, *B. melitensis* biotypes 1 to 3, *B. suis* biotypes 1 to 4, and *B. canis*. It is found worldwide. The reservoir of infection is cows, and it is transmitted by contact with tissues, blood, urine, vaginal discharges, aborted fetuses, and especially placentas through breaks in the skin or ingestion of raw milk or dairy products from infected animals; may also be an airborne infection. It is not communicable from person to person and has a wide variety of susceptibility. It is controlled through educating farmers, serological testing of animals, and pasteurization of milk.

bruise — See contusion.

BSA — Acronym for body surface area.

BSC — See base safety condition.

BSE — See bovine spongiform encephalopathy.

BST — See bovine somatotropin.

Bt — See *Bacillus thuringiensis*.

BT — Acronym for bleeding time.

BTA — Acronym for best technical approach.

BTPD — Acronym for body temperature, ambient pressure, dry.

BTPS — Acronym for body temperature, ambient pressure, saturated with water vapor.

Btu — See British thermal unit.

bu — Abbreviation for bushel.

bubble chamber — A chamber containing a liquefied gas, such as liquid hydrogen, and a charged particle that passes through the liquid and creates a visible path of bubbles.

bubble concept — A means of judging the amounts of air pollutants emitted from smoke stacks within a given area.

bubble policy — A situation where existing sources of air pollution in several facilities are better controlled than required at one emission point, making costs lower and allowing the company to have higher emissions at another area where costs are higher.

bubble tube — A glass tube filled with liquid and bubbles used to calibrate air-sampling pumps.

bubo — (*medicine*) An inflammatory swelling of a lymph gland.

bubonic plague — (*disease*) See plague.

bucket — An open container affixed to the moveable arms of a wheeled or track vehicle to spread solid waste and cover material and to excavate soil; part of an excavator that digs, lifts, and carries dirt or solid waste.

bucket-handle fracture — A double vertical fracture of the pelvis on the same side, resulting in dislocation of the pelvis.

budding — Asexual reproduction in which a portion of the cell body is thrust out and then becomes separated, forming a new cell.

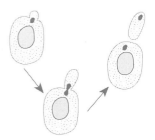

Budding

bufencarb ($C_{13}H_{19}NO_2$) — A solid at room temperature; MW: a mixture of several chemicals; BP: NA; Sol: soluble in methanol, benzene, and xylene; density: 1.024 at 26°C. It is used as a pesticide. It is moderately toxic to the digestive tract and skin, is a cholinesterase inhibitor, and may cause headaches, weakness, blurred vision, loss of muscle coordination, vomiting, diarrhea, and difficulty in breathing.

$$CH_3-NH-\overset{O}{\overset{\|}{C}}-O-\bigcirc-CH(C_2H_5)_2 \text{ or } -CH\overset{CH_2CH_2CH_3}{\underset{CH_3}{\diagup}}$$

Bufencarb

buffer — A substance that stabilizes pH; any substance in a fluid that tends to resist the change in pH when an acid or alkali is added.

buffer anions — The negatively charged bicarbonate, protein, and phosphate ions that make up the buffer systems of the body.

buffer capacity — (*chemistry*) The ability of a solution to maintain its pH when stressed chemically.

buffer cations — The positively charged ions of the body's electrolytes, including sodium, calcium, potassium, and magnesium.

buffer solution — (*chemistry*) A solution to which small amounts of an acid or base can be added without appreciably changing its hydrogen ion concentration.

bug — (*computer science*) An error in a computer program or in a piece of electronics that causes it to function improperly.

building cooling load — The hourly amount of heat removed from a building to maintain indoor comfort (measured in BTUs).

Building Energy Management System — A computerized system that operates to monitor and control energy usage in a building.

building envelope — Elements of a building, including all external building materials, windows, and walls that enclose the internal space.

building-related illness — A diagnosable illness with identifiable symptoms and causes that can be directly attributed to airborne building pollutants.

built environment — A human-modified environment that includes, for example, buildings, roads, and cities.

bulk density — The mass of powdered or granulated solid material per unit of volume.

bulk food — Unpackaged or unwrapped, processed or unprocessed food in aggregate containers from which quantities desired by the consumer are withdrawn.

bulk memory — (*computer science*) An electronic device, such as a disc or tape, that allows the storage of large amounts of data.

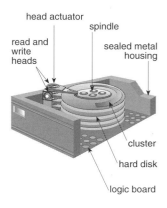

Bulk Memory

bulking agent — A fine, solid material added to a waste-water stream to produce clarification or coagulation by adding bulk to the solids.

bulking sludge — Sludge floating in the air that occurs throughout secondary clarifiers and sludge thickeners when the sludge becomes too light and will not settle properly.

bulking — The increasing volume of a material due to manipulation.

bulky waste — Items for which their large size complicates their handling by normal collection, processing, or disposal methods.

bull clam — A tracked vehicle with a hinged, curved bowl on the top of the front of the blade.

Bulldozer with Bull Clam

bulla — A blister; a circumscribed, fluid-containing, elevated lesion of the skin usually more than 5 mm in diameter.

bulldozer — (*mechanical engineering*) A track vehicle equipped with an earth blade.

Bulldozer

bullous — Of or pertaining to bulla.

BUN — See blood urea nitrogen.

bundle — (*U.S. EPA definition*) A structure composed of three or more fibers in a parallel arrangement, with each fiber closer than the diameter of one fiber; (*NIOSH definition*) a compact arrangement of parallel fibers in which separate fibers or fibrils may only be visible at the ends of the bundle; asbestos bundles having aspect ratios of 3:1 or greater and less than 3 nm in diameter are counted as fibers.

Bunyamwera arbovirus — An arthropod-borne virus that infects humans; carried by mosquitoes from rodents, causing California encephalitis, Rift Valley fever, and other diseases with symptoms of headaches, weakness, low-grade fever, myalgia, and rash.

bunyavirus — An arbovirus spread by mosquitos and ticks that causes fever, rashes, encephalitis, hemorrhage, meningitis, and hemorrhagic fever.

burden of proof — The necessity or responsibility of proving a fact or facts in dispute on an issue.

burettte — A piece of laboratory equipment used to deliver a wide range of volumes accurately.

burial ground — (*radiation*) A disposal site for unwanted radioactive materials using earth or water for a shield.

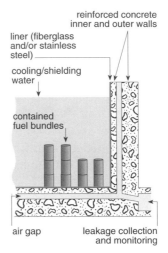

Burial Ground

burn — Injury to tissues caused by contact with heat, steam, chemicals, electricity, lightning, or radiation. A first-degree burn involves a reddening of the skin area; a second-degree burn involves blistering of the skin; a third-degree burn, the most serious type,

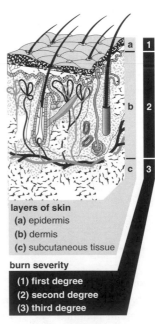

layers of skin
(a) epidermis
(b) dermis
(c) subcutaneous tissue

burn severity
(1) first degree
(2) second degree
(3) third degree

Burns

involves damage to the deeper layers of the skin, with necrosis throughout the entire skin.

burning area — A horizontal projection of a grate, a hearth, or both.

burning hearth — A solid surface without air openings to support the solid fuel or solid waste in a furnace during drying, ignition, or combustion.

burning rate — The quantity of solid waste incinerated or the amount of heat released during incineration; usually expressed in pounds of solid waste per square foot of burning area per hour or in BTUs per square foot of burning area per hour.

burr — A thin, rough edge of a machined piece of metal.

bursa — A sac filled with fluid and situated at a place in the tissues at which friction of movement has been reduced.

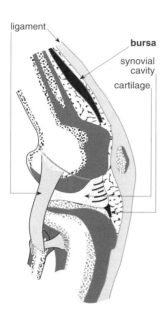

Bursa

bursitis — Inflammation of the bursa resulting in severe pain and limitation of motion of the affected joint.

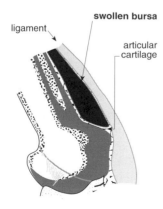

Bursitis in Knee

burst pressure — The maximum pressure applied to a transducer sensing element or case without causing leakage.

bus — (*computer science*) A circuit or group of circuits that provide a communication path between the computer and peripherals.

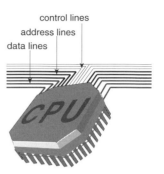

Bus

bushel — A container for grain, fruit, etc., equal to 4 pecks or 8 gallons.

1,3-butadiene (C$_4$H$_6$) — A colorless gas with a mild aromatic odor; MW: 54.10; BP: NA; Sol: soluble in organic solvents; Fl.P: a flammable gas that can travel a considerable distance to a source of ignition and flash back; density: 1.865. It is used to produce synthetic rubber, food-wrapping materials, butadiene, and styrene elastomer for tires. It is hazardous to the eyes, nose, throat, and respiratory tract. It may cause hallucinations, distorted perception, irritation, drowsiness, lightheadedness, and narcosis. It causes cancer in rats and mice and is a suspected human carcinogen. It is a mutagen and a teratogen. ACGIH (TLV-TWA): 10 ppm.

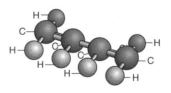

**1,3–Butadiene
(Butadiene)**

n-butane (C$_4$H$_{10}$) — A colorless gas; MW: 58.14; BP: NA; Sol: soluble in alcohol, ether, chloroform, and other organic solvents; Fl.P: –60°C (closed cup), a flammable gas; density: 2.04. It is used as a liquid

$$CH_3 - CH_2 - CH_2 - CH_3$$

n-Butane

fuel and is called liquified petroleum gas when mixed with propane. It is also used as a propellant for aerosols, production of synthetic rubber, and in organic synthesis. It is hazardous to the respiratory tract and causes asphyxia. ACGIH (TLV-TWA): 800 ppm.

1,4-butanediol (C₄H₈(OH)₂) — A colorless liquid; MW: 90.1; BP: 228°C; Sol: soluble in water, alcohol, and acetone; Fl.P: 121°C (open cup); density: 1.017 at 20°C. It is used to produce polybutylene terephthalate, a thermoplastic polyester, and polymeric plasticizers. It is hazardous to the digestive tract and central nervous system. It may cause excitement, depression of the central nervous system, nausea, and drowsiness. ACGIH: NA.

$$HO-CH_2-CH_2-CH_2-CH_2-OH$$

1,4 Butanediol

1-butanol (C₄H₉OH) — A colorless liquid with a wine-like odor; MW: 74.1; BP: 117.2°C; Sol: soluble in alcohol, ether, and acetone; Fl.P: 35°C; density: 0.810. It is used as a solvent in the coating industry; for extraction of oils, drugs, and cosmetic products; and as an ingredient in perfumes. It is hazardous to the skin, eyes, and respiratory system. It causes irritation of the eyes, nose, throat. In humans, chronic exposure at high concentrations causes photophobia, blurred vision, and lacrimation. In rats, at high concentrations it may cause teratogenetic effects. ACGIH (TLV-TWA): 50 ppm.

$$\begin{array}{c} OH \\ | \\ CH_3CH_2CH_2C-H \\ | \\ H \end{array}$$

1-Butanol

2-butanone (CH₃COCH₂CH₃) — A colorless liquid with a moderately sharp, fragrant, mint- or acetone-like odor; MW: 72.1, BP: 175°F, Sol: 28%, Fl.P: 16°F, sp. gr. 0.81. It is used in spray application of vinyl and acrylic coatings; during mixing

2-Butanone
(Methyl Ethyl Ketone)

of dye solutions; in laboratories; in the application of adhesives for artificial leather; during sponge and brush application of solvent for cleaning operations; during mixing of waterproofing compounds. It is hazardous to the central nervous system and lungs and is toxic through inhalation, ingestion, and contact. Symptoms of exposure include eye and nose irritation, headache, dizziness, and vomiting. OSHA exposure limit (TWA): 200 ppm [air] or 590 mg/m³.

2-butene-1,4-diol (C₄H₆(OH)₂) — A colorless liquid; MW: 88.1; BP: 234°C; Sol: soluble in water, alcohol, and acetone; Fl.P: 128°C (open cup); density: 1.070. It is used to make agricultural chemicals and the pesticide endosulfan. It is mildly toxic to the respiratory system and is a skin irritant. It causes depression of the central nervous system and skin problems. ACGIH: NA.

$$HO-CH_2-CH=CH-CH_2-OH$$

2-Butene-1,4-Diol

2-butoxyethanol (C₄H₉OCH₂CH₂OH) — A colorless mobile liquid with a mild ether-like odor; MW: 118.2, BP: 339°F, Sol: miscible, Fl.P: 143°F, sp. gr. 0.90. It is used in spray and brush application of varnishes, lacquers, and enamels; as an industrial solvent; in the production of plasticizers; as a stabilizing agent in metal cleaners and liquid household cleaners; in hydraulic fluids, insecticides, herbicides, and rust removers. It is hazardous to the liver, kidneys, lymphoid system, skin, blood, eyes, and respiratory system and is toxic through inhalation, absorption, ingestion, and contact. Symptoms of exposure include eye, nose, and throat irritation; hemolysis; hemoglobinuria. OSHA exposure limit (TWA): 25 ppm [skin] or 120 mg/m³.

butter — A fatty food product made from fresh or ripened, partially sour milk or cream and a small portion of other constituents natural to milk, with or without common salt and with or without additional harmless food coloring.

butter fat — See milk fat.

butterfly bandage — A narrow adhesive strip with broad, wing-like ends that is used to hold together the edges of a superficial wound while it heals.

buttermilk — A fermented fluid product resulting from the churning of milk or cream containing not less than 8.25% milk solids (not fat).

n-butyl acetate (CH₃COO(CH₂)₃CH₃) — A colorless liquid with a fruity odor; MW: 116.2, BP: 258°F, Sol: 1%, Fl.P: 72°F, sp. gr. 0.88. It is used during application of nitrocellulose by spraying, brushing, or dipping; in the production of lacquer thinner; as a solvent for oils, pitch, camphor, ethyl cellulose acetate, vinyl, polystyrene, methacrylate plastics, and chlorinated rubber; as a solvent in the production of artificial leather; in the manufacture of safety glass; in the production of flavorings, perfumes, cosmetics, adhesives, shoe polishes, and stain removers. It is hazardous to the eyes, skin, and respiratory system and is toxic through inhalation, ingestion, and contact. Symptoms of exposure include headache, drowsiness, and dryness and irritation of the eyes, upper respiratory system, and skin. OSHA exposure limit (TWA): 150 ppm [air] or 710 mg/m³.

***sec*-butyl acetate (CH₃COOCH(CH₃)CH₂CH₃)** — A colorless liquid with a pleasant, fruity odor; MW: 116.2, BP: 234°F, Sol: 0.8%, Fl.P: 62°F, sp. gr. 0.86. It is used as a solvent for coating paper, leather, and artificial leather material; as a lacquer and adhesive; in the manufacture of photographic film. It is hazardous to the eyes, skin, and respiratory system and is toxic through inhalation, ingestion, and contact. Symptoms of exposure include eye irritation, headache, drowsiness, and dryness of the upper respiratory system and skin. OSHA exposure limit (TWA): 200 ppm [air] or 950 mg/m³.

***tert*-butyl acetate (CH₃COOC(CH₃)₃)** — A colorless liquid with a fruity odor; MW: 116.2, BP: 208°F, Sol: insoluble, Fl.P: 62 to 72°F, sp. gr. 0.87. It is used as a drycleaning agent; in the formulation of paint systems; as an additive to improve antiknock properties of leaded aliphatic gasoline; as an activator in alkaline polymerization of caprolactam. It is hazardous to the respiratory system, eyes, and skin and is toxic through inhalation, ingestion, and contact. Symptoms of exposure include itching and inflamed eyes, irritation to the upper respiratory tract, headache, narcosis, and dermatitis. OSHA exposure limit (TWA): 200 ppm [air] or 950 mg/m³.

***n*-butyl acrylate (C₇H₁₂O₂)** — A colorless liquid; MW: 128.19, BP: 145°C, Sol: soluble in alcohol and ether, Fl.P: 48°C (open cup), density: 0.899 at 20°C. It is used to make polymers that are finished as resins for textile and leather and also in paints. It is hazardous to the skin and other mucous membranes. It is mildly irritating. ACGIH (TLV-TWA): 10 ppm.

$$CH_2 = CH - \underset{\underset{O}{\|}}{C} - O - CH_2 - CH_2 - CH_2 - CH_3$$

n-Butyl Acrylate

***n*-butyl alcohol (CH₃CH₂CH₂CH₂OH)** — A colorless liquid with a strong, characteristic, mildly alcoholic odor; MW: 74.1, BP: 243°F, Sol: 9%, Fl.P: 99°F, sp. gr. 0.81. It is used in the leather industry; in the manufacture of safety glass; as a solvent or diluent in the manufacture of brake fluids, perfumes, detergents, adhesives, denatured alcohol, and surface coatings; in photographic processing; in laboratory analysis. It is hazardous to the skin, eyes, and respiratory system and is toxic through inhalation, absorption, ingestion, and contact. Symptoms of exposure include eye, nose, and throat irritation, headache, vertigo, drowsiness, corneal inflammation, blurred vision, lacrimation, photophobia, and dry and cracked skin. OSHA exposure limit (TWA): ceiling 50 ppm [skin] or 150 mg/m³.

n-Butyl Alcohol
(1-Butanol)

***sec*-butyl alcohol (CH₃CHOHCH₂CH₃)** — A colorless liquid with a strong, pleasant odor; MW: 74.1, BP: 211°F, Sol: 16%, Fl.P: 75°F, sp. gr. 0.81. It is used in industrial cleaning compounds; in ethyl-cellulose-based surface coatings; in paint removers, perfumes, and hydraulic fluids. It is hazardous to the eyes and central nervous system and is toxic through inhalation, ingestion, and contact. Symptoms of exposure include narcosis and eye irritation. OSHA exposure limit (TWA): 100 ppm [air] or 305 mg/m³.

***tert*-butyl alcohol ((CH₃)₃COH)** — A colorless liquid with a camphor-like odor; pure compound is a liquid above 77°F; MW: 74.1, BP: 180°F, Sol: miscible, Fl.P: 52°F, sp. gr. 0.79. It is used in industrial cleaning compounds; in lacquer surface coatings; as a chemical intermediate; as a solvent for drug extraction, water removal, and wax solvent; in laboratory procedures. It is hazardous to the eyes and skin and is toxic through inhalation, ingestion, and contact. Symptoms of exposure include drowsiness and eye and skin irritation. OSHA exposure limit (TWA): 100 ppm [air] or 300 mg/m³.

***tert*-butyl chromate ((CH₃)₃CO₂CrO₂)** — A liquid that solidifies at 32 to 23°F; MW: 230.3, BP: not known, Sol: not known, Fl.P: not known. It is used in oxidation of steroids as a means of identification on paper chromatography; in the manufacture of catalysts used for polymerization of olefins. It is hazardous to the respiratory system, skin, eyes, and central nervous system and is toxic through inhalation, absorption, ingestion, and contact. Symptoms of exposure include lung and sinus cancer; carcinogenic. OSHA exposure limit (TWA): ceiling 0.1 mg/m³ [skin].

***n*-butyl glycidyl ether (C₇H₁₄O₂)** — A colorless liquid with an irritating odor; MW: 130.2, BP: 327°F, Sol: 2%, Fl.P: 130°F, sp. gr. 0.91 at 77°F. It is used as a reactive diluent of epoxy resins; as a chemical intermediate for preparation of ethers, surfactants, polymers, and resins; as a stabilizing agent for organic chemicals. It is hazardous to the eyes, skin, respiratory system, and central nervous system and is toxic through inhalation, ingestion, and contact. Symptoms of exposure include eye, nose, and skin irritation; sensitization; and narcosis. OSHA exposure limit (TWA): 25 ppm [air] or 135 mg/m³.

***tert*-butyl hydroperoxide (C₄H₁₀O₂)** — A colorless liquid; MW: 90.12; BP: 89°C; Sol: soluble in organic solvents; Fl.P: less than 27 to 54°C, a highly reactive, oxidizing, and flammable liquid; density: 0.896 at 20°C. It is used to initiate polymerization reactions and in organic synthesis. It is hazardous to the skin and respiratory tract. It is an irritant and may cause injury to the lungs; mutagenic. ACGIH exposure limit: 1.2 mg/m³.

$$H_3C - \overset{\overset{\displaystyle CH_3}{|}}{\underset{\underset{\displaystyle CH_3}{|}}{C}} - O - O - H$$

***tert*-Butyl Hydroperoxide**

***n*-butyl isocyanate (C₄H₉CNO)** — A colorless liquid with a mild odor; MW: 99.15; BP: 46°C; Sol: soluble in most organic solvents, decomposes in water and alcohol; Fl.P: 20°C (open cup), a flammable liquid that forms an explosive mixture with air; density: 1.4064 at 20°C. It is used as an acylating agent. It is hazardous to the respiratory tract and causes nausea, dyspnea, insomnia, coughing, and chest pain. ACGIH (TLV-TWA): 2 ppm.

$$H_3C - CH_2 - CH_2 - CH_2 - N = C = O$$

***n*-Butyl Isocyanate**

butyl mercaptan (CH₃CH₂CH₂CH₂SH) — A colorless liquid with a strong garlic-, cabbage-, or skunk-like odor; MW: 90.2, BP: 209°F, Sol: 0.06%, Fl.P: 35°F, sp. gr. 0.83. It is used as an odorant for natural gas; as a chemical intermediate in the manufacture of agricultural chemicals, herbicides, and defoliants; in the manufacture of polymerization

catalysts, stabilizers, modifiers, and chain transfer agents; as a solvent. It is hazardous to the respiratory system (in animals: the central nervous system, liver, and kidneys) and is toxic through inhalation, ingestion, and contact. Symptoms of exposure in animals include narcosis, incoordination, weakness, cyanosis, pulmonary irritation, and liver and kidney damage. OSHA exposure limit (TWA): 0.5 ppm [air] or 1.5 mg/m³.

***tert*-butyl peroxyacetate (C₆H₁₂O₃)** — MW: 132.18, Fl.P: a reactive, oxidizing, and flammable material. It is used to initiate polymerization and in organic synthesis. It is a mild irritant to the skin and eyes and toxic to the digestive tract and respiratory tract. ACGIH exposure limit: NA.

$$CH_3 - \overset{\overset{\displaystyle O}{\|}}{C} - O - O - \overset{\overset{\displaystyle CH_3}{|}}{\underset{\underset{\displaystyle CH_3}{|}}{C}} - CH_3$$

***tert*-Butyl Peroxyacetate**

butylamine (CH₃CH₂CH₂CH₂NH₂) — A colorless liquid with a fishy, ammonia-like odor; MW: 73.2, BP: 172°F, Sol: miscible, Fl.P: 10°F, sp. gr. 0.74. It is used as a vulcanizing accelerator and reaction initiator in rubber and polymer industries; as a chemical intermediate in the production of emulsifying agents; in chemical synthesis of developers for photography, pharmaceuticals, antioxidants for gasoline, synthetic tanning materials, curing agents, gum inhibitors; as a stabilizer in aviation fuels; on stored agricultural crops as a fungicide and to prevent decay. It is hazardous to the respiratory system, skin, and eyes and is toxic through inhalation, absorption, ingestion, and contact. Symptoms of exposure include eye, nose, and throat irritation, headache, and flushed and burned skin. OSHA exposure limit (TWA): ceiling 5 ppm [skin] or 15 mg/m³.

***n*-butylamine (C₄H₉NH₂)** — A colorless liquid with an ammonia odor; MW: 73.16; BP: 78°C; Sol: miscible with water, alcohol, and ether; Fl.P: –14°C (closed cup), flammable liquid; density: 0.733 at 25°C. It is used as an intermediate in dyestuffs, pharmaceuticals, and emulsifying agents. It is hazardous to the eyes, skin, and respiratory tract and produces severe burns of the eyes and skin and severe irritation of the nose and throat, as well as pulmonary edema. ACGIH (TLV-TWA): 5 ppm.

$$CH_3 - CH_2 - CH_2 - CH_2 - NH_2$$

***n*-Butylamine**

butylbenzyl phthalate ($C_{19}H_{20}O_4$) — A clear, oily liquid with a slight odor; MW: 312.39, BP: 370°C, Sol: 0.29 g/100 ml water at 20°C, Fl.P: 199°C, sp. gr. 1.113 to 1.121. It is used as a chemical intermediate and plasticizer. It is hazardous to the respiratory system, digestive system, skin, and eyes. Acute exposure may result in burning of the eyes, skin, and dizziness. Chronic exposure may damage a developing fetus and damage the testes. It is a possible carcinogen. OSHA exposure limit (8-hour TWA): NA.

butylene oxide (C_4H_8O) — A colorless liquid with a disagreeable odor; MW: 72.11; BP: 63.3°C; Sol: soluble in acetone and ether; Fl.P: –15°C (closed cup); density: 0.837 at 17°C. It is used as a fumigant and to stabilize sludge. It is hazardous to the skin, eyes, and respiratory tract and may cause severe burns. It may be mutagenic and carcinogenic in animals. ACGIH (TLV-TWA): NA.

$$H_2C\underset{O}{\diagdown}CH_2 - CH_2 - CH_3$$

Butylene Oxide

***p-tert*-butyltoluene ($(CH_3)_3CC_6H_4CH_3$)** — A colorless liquid with a distinct aromatic odor, somewhat like gasoline; MW: 148.3, BP: 379°F, Sol: insoluble, Fl.P: 155°F, sp. gr. 0.86. It is used as a primary intermediate in chemical and pharmaceutical industries. It is hazardous to the cardiovascular system, central nervous system, skin, bone marrow, eyes, and upper respiratory system and is toxic through inhalation, ingestion, and contact. Symptoms of exposure include eye and skin irritation, dry nose and throat, headache, low blood pressure, tachycardia, abnormal cardiovascular stress, central nervous system depression, and hematopoietic depression. OSHA exposure limit (TWA): 10 ppm [air] or 60 mg/m³.

2-butyne-1,4-diol ($C_4H_4(OH)_2$) — A crystalline solid; MW: 86; BP: 248°C; Sol: soluble in water, ethanol, acetone; Fl.P: 152°C (open cup), sp. gr. 1.114 at 60°C. It is used to produce butanediol and butenediol and in metal plating and pickling baths. It is a skin irritant and has moderate to high toxicity in test animals. ACGIH: NA.

$$HO - CH_2 - C \equiv C - CH_2 - OH$$

2-Butyne-1,4-Diol

***n*-butyraldehyde (C_3H_7CHO)** — A colorless liquid with a pungent odor; MW: 72.1; BP: 75.7°C; Sol: soluble in alcohol, ether, and acetone; Fl.P: –7°C (closed cup); density: 0.817 at 20°C. It is used to

make rubber accelerators, synthetic resins, and plasticizers and as a solvent. It is hazardous to the skin, eyes, and respiratory system and causes mild skin and eye irritation; can have a narcotic effect. ACGIH exposure limit: NA.

$$H_3C - CH_2 - CH_2 - \overset{\overset{\displaystyle H}{|}}{C} = O$$

***n*-Butyraldehyde**

butyronitrile (C_3H_7CN) — A colorless liquid; MW: 69.12; BP: 117.5°C; Sol: soluble in water, alcohol, and ether; Fl.P: 24°C (open cup); density: 0.795 at 15°C. It is used as a chemical intermediate. It is hazardous to the respiratory tract, digestive tract, and skin and affects the liver, kidney, central nervous system, lungs, sense organs, and peripheral nervous system. Symptoms include nausea, respiratory distress, ataxia, dyspnea, and spastic paralysis. NIOSH (TLV-TWA): 8 ppm.

$$H_3C - CH_2 - CH_2 - C \equiv N$$

Butyronitrile

BUV — Acronym for Backscatter Ultraviolet Radiometer.

bw — Acronym for body weight.

BW — Acronym for biological warfare.

BWR — See boiling water reactor.

by-metallic steamed thermometer — A food thermometer used to measure product temperatures.

byproduct — A useful or marketable product or service that is not the product or service being produced.

byssinosis — (*disease*) A pneumoconiosis disease occurring in workers from prolonged exposure to heavy air concentrations of cotton or flax dust.

byte — (*computer science*) A group of contiguous bits, usually eight, that represent a character and which are operated on as a unit.

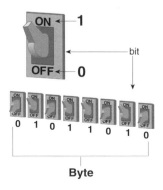

Byte

BZ — Acronym for breathing zone.

c — See capacitance; ceiling level.

C — (*chemistry*) See carbon and carbon black; (*computer science*) a high-level programming language used in graphics; (*electricity*) see coulomb, Celsius, and flow coefficient.

^{14}C method — A method of determining the age, in years, of organic matter by calculating the amount of radioactive carbon still remaining as compared to the stable isotope ^{12}C.

C factor — A factor of value used to indicate the smoothness of the interior of a pipe; the higher the C factor, the smoother the pipe and the greater its carrying capacity because of smaller friction or energy losses.

C horizon — (*soils*) The soil horizon that normally lies beneath the B horizon but may lie beneath the A horizon; it is unaltered or slightly altered parent material.

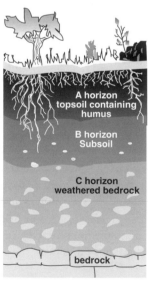

C Horizon
and other layers of soil

C/kg — A unit of radiation exposure in the SI system representing coulombs per kilogram of air.

C/N — See carbon–nitrogen ratio.

CA — Acronym for cloud-to-air lightning.

CAA — Acronym for Clean Air Act.

CAAA — Acronym for Clean Air Act Amendment.

cable-modem — (computer science) A device allowing users to access the Internet via coaxial cable.

cache — (*computer science*) A place on the hard drive of a computer where a Web browser saves information about Web pages recently visited to make return to those pages quicker and easier.

cachexia — A profound and marked state of wasting of the body resulting from general ill health, malnutrition, or terminal illnesses.

CAD — Acronym for coronary artery disease.

CAD/CAM — Acronym for computer-aided design/ computer-aided manufacturing.

cadastral — (*geographic information system*) The legally recognized registration of quantity, value and ownership of land parcels; relating to land boundaries and subdivisions, parcels of the land suitable for transfer title.

cadaver — A dead body used for dissection and study.

cadelle — A small black beetle destructive to stored grain.

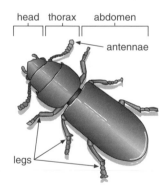

Cadelle
(*Tenebroides mauritanicus*)

cadmium (Cd) — A soft, silver-white metal in its pure state (rarely found this way in the environment); MW: 112.40, BP: 767°C, Sol: insoluble, density: 8.462. It is used in the production of nickel–cadmium batteries; in metal plating, plastics, and synthetics; for alloys and other miscellaneous uses. It is hazardous to the respiratory system, kidneys, and bones. Symptoms of exposure include severe tracheobronchitis, pneumonitis, pulmonary edema, calcium deficiency, osteoporosis, proteinuria, renal damage, and death. OSHA exposure limit (TWA): 0.1 mg/m³ [air].

cadmium acetate (Cd(C₂H₃O₂)₂) — A colorless solid/ crystal sugar- or sand-like substance; MW: 230.49, BP: decomposes, Sol: very soluble in cold water, Fl.P: nonflammable (poisonous gases and cadmium fumes are produced in a fire), sp. gr. 2.344 at 25°C. It is used in ceramics, process engraving, photoelectric cells, textile dyeing, printing, electroplating and to make other acetate compounds. It is hazardous to the respiratory system, digestive system, and eyes. Acute exposure may cause flu-like illness with chills, headaches, and/or fever, as well as severe lung damage, shortness of breath, chest pain, cough, buildup of fluid in the lungs, and, in severe cases, permanent lung damage or death. Chronic exposure may cause permanent kidney damage, kidney stones, lung scarring, emphysema, loss of sense of smell, and fatigue. It is a probable carcinogen and may also cause reproductive damage. OSHA exposure limit (8-hour TWA): 0.002 mg/m³.

cadmium bromide (CdBr₂) — An odorless white to yellowish crystalline powder; MW: 272.22, BP: decomposes, Sol: 57 g/100 ml water at 10°C, Fl.P: nonflammable (poisonous gases are produced in a fire), sp. gr. 2.192 at 25°C. It is used in photography, engraving, lithography, metal alloys, electroplating, photoelectric cells, and nickel–cadmium electrical storage batteries. It is hazardous to the respiratory system, digestive system, and eyes. Acute exposure may cause flu-like illness with chills, headaches, and fever. High exposures can cause rapid and severe lung damage, shortness of breath, chest pain, cough, buildup of fluid in the lungs, permanent lung damage, or death. Chronic exposure can cause permanent kidney damage, kidney stones, emphysema, and/or lung scarring. Because several related cadmium compounds are known cancer agents, it should be considered to be potentially carcinogenic. OSHA exposure limit (8-hour TWA): 0.2 mg/m³ (as Cd).

caecum — See cecum.

CAFE standards — See Corporate Average Fuel Economy.

caffeine — A substance found in coffee that is a central nervous system stimulant.

caffeine poisoning — A toxic condition caused by the chronic ingestion of excessive amounts of caffeine found in coffee, tea, cola beverages, or certain stimulant drugs with symptoms of restlessness, anxiety, general depression, tachycardia, tremors, nausea, diuresis, and insomnia.

CAFO — Acronym for concentrated animal feed operation.

CAH — Acronym for chronic active hepatitis.

CAI — See computer-assisted instruction.

cairn — A pile of stones used as a marker.

caisson — An underground cylindrical concrete and metal vault used during construction work under water.

caisson disease — See bends.

cake — (*wastewater*) Solids discharged from a dewatering apparatus.

cal — See calorie.

calcareous — Made of calcium carbonate.

calcareous soil — Soil containing sufficient calcium carbonate, often with magnesium carbonate, to effervesce visibly when treated with cold 0.1-*N* hydrochloric acid.

calcination — Heat treatment of solid materials to bring about thermal decomposition or a phase transition other than melting.

calcine — To heat to a high temperature without fusing in order to decompose, oxidize, and so on.

calcite (CaCO₃) — One of the most common minerals; usually colorless or white. It is a major constituent in limestones and marbles and is the major mineral in limestone. It is quite soluble, which accounts for its usual presence in water. It is one of three crystalline forms of calcium carbonate, the others being aragonite and vaterite; also known as calcspar.

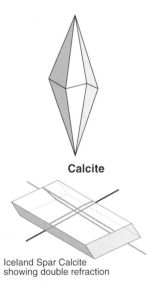

Calcite

Iceland Spar Calcite showing double refraction

calcium (Ca) — A silvery white metal that becomes bluish gray on exposure to moisture; atomic weight 40.08, BP: 1440°C, Sol: soluble in liquid ammonia, Fl.P: finely divided metal ignites in air at ordinary temperatures, density: NA. It is used as a deoxidizer for copper, steel, and beryllium and in making alloys. It is hazardous to the skin, eyes, and respiratory tract. Contact with the metal dust can cause burns and severe irritation. ACGIH exposure limit: NA.

calcium acetylide (CaC$_2$) — A colorless tetragonal crystal; MW: 64.10; BP: NA; Sol: decomposes in water; Fl.P: decomposes in water, producing acetylene, a highly flammable gas that also forms explosive mixtures in air; density: 2.22 g/cm^3. It is used to produce acetylene gas. It is hazardous to the skin because of its extreme corrosiveness. ACGIH exposure limit: NA

$$(-CaC \equiv C-)_n$$

Calcium acetylide

calcium arsenate (Ca$_3$(AsO$_4$)$_2$) — A colorless to white odorless solid; MW: 398.1, BP: decomposes, Sol: 0.01% at 77°C, sp. gr. 3.62. It is used as an insecticide and herbicide. It is hazardous to the eyes, respiratory system, liver, skin, lymphatics, and central nervous system and is toxic through inhalation, absorption, ingestion, and contact. Symptoms of exposure include weakness, gastrointestinal disturbance, skin hyperpigmentation, palmar planter hyperkeratoses, and dermatitis; carcinogenic. It causes liver damage in animals. OSHA exposure limit (TWA): 0.010 mg/m^3 [air].

calcium carbonate (CaCO$_3$) — A white powder occurring naturally as calcite. It is used in paint manufacture, as a dentrifice, as an anticaking medium for table salt, and in the manufacture of rubber tires.

calcium carbonate equivalent — An expression of the concentration of specified constituents in water in terms of equivalent value to calcium carbonate to help determine hardness.

calcium cyanide (Ca(CN)$_2$) — A white crystalline solid; MW: 92.12, BP: NA, Sol: slightly soluble in liquid ammonia, Fl.P: NA, density: NA. It is used for the extraction of gold and silver from ores and is a fumigant and rodenticide. It is a highly poisonous substance that is hazardous to the skin, respiratory tract, and digestive tract. It causes extreme intoxication and death. ACGIH exposure limit (TLV-TWA): 5mg/m^3 [skin].

calcium hardness — A measure of the calcium salts dissolved in water.

calcium hypochlorite (Ca(OCl)$_2$·4H$_2$O) — (*inorganic chemistry*) A compound of chlorine and calcium used in white granular or tablet form as a bacteri-

cide in pools; in water solution it releases 70% of its weight as available chlorine.

calcium oxide (CaO) — (*inorganic chemistry*) Caustic white or gray, odorless lumps or granular powder; MW: 56.1, BP: 5162°F, Sol: reacts, sp. gr. 3.34. It is used in the preparation of lime; in the manufacture of Portland cement, mortar, stucco, and plaster; in the manufacture of iron and steel in wire-drawing; as a flux in metal refining and smelting; as a softening, purifying, coagulating, and suspending compound in water treatment and purification; during food processing; in the manufacture of silicate and nonsilicate glass; in the manufacture of pesticides and fungicides. It is hazardous to the respiratory system, skin, and eyes and is toxic through inhalation, ingestion, and contact. Symptoms of exposure include eye and upper respiratory tract irritation, ulcer, perforated nasal septum, pneumonia, dermatitis. OSHA exposure limit (TWA): 5 mg/m^3 [air].

calcspar — See calcite.

calendar call — (*law*) A public calling of the docket or list of cases at the commencement of a term of court for setting a time for trial or entering orders of continuance, default, nonsuit, etc.; also known as calling the docket.

calibrate — (*science and technology*) To adjust mathematically the pollution concentration predicted by a model so as to match the results obtained from model validation field studies; to adjust instruments to a standard; to control the amount of a pesticide chemical applied by each nozzle or opening of a sprayer, duster, or granular applicator to a given area, plant, or animal; determination of variation from standards for accuracy of a measuring instrument to ascertain necessary correction.

calibrated visual area estimate — A semiquantitative method of estimating asbestos concentration using a stereomicroscope or polarized light microscope and comparing the sample with calibrated samples of known concentration.

calibrating solution — A solution containing a known amount of a substance having an effect equivalent to a pollutant concentration that has passed through the detection component during the static calibration of an analyzer.

calibration drift — Deviation in instrument responses from a reference value over time before recalibration.

calibration error — The difference between the pollution concentration indicated by the measurement system and the known concentration of the test gas mixture.

calibration gas — A gas used to calibrate an air-monitoring instrument.

calibration offset — An adjustment used to eliminate the difference between the indicated value and the actual process value.

caliche — See hardpan.

calicivirus — (*disease*) A virus classified in the family Caliciviridae and containing a single strand of RNA surrounded by a protein capsid 31 to 40 nm in diameter; may cause viral gastroenteritis in children 6 to 24 months of age, and accounts for approximately 3% of the hospitalized cases of diarrhea, with additional symptoms of nausea, vomiting, malaise, abdominal pain, and fever. The mode of transmission is the oral–fecal route, from person to person, or ingestion of contaminated food and water. The incubation period is 10 to 70 hours. The disease has been reported in England and Japan. The reservoir of infection is humans; susceptibility is found among the very young. Also known as acute nonbacterial infectious gastroenteritis and viral gastroenteritis.

California encephalitis — A common acute viral infection transmitted by a mosquito that affects the central nervous system and causes symptoms of headache, malaise, gastrointestinal distress, and a fever that may reach 104°F in its mild form. Its severe form causes high fevers, vomiting, headaches, lethargy, and neurological involvement such as loss of reflexes, disorientation, seizure, loss of consciousness, and paralysis.

California waste — A group of liquid hazardous wastes, including ones with polychlorinated biphenyls, heavy metals, and halogenated organic compounds that the U.S. EPA evaluated as of July 8, 1987, to determine if they should be banned from land disposal or if restrictions should be placed on the land disposal of these wastes.

calipers — An instrument with two hinged, adjustable, curved legs used to measure the thickness or the diameter of a convex or solid body.

calling the docket — See calendar call.

caloric — The heat value or calories.

calorie (cal) — (*thermodynamics*) The amount of heat necessary to raise the temperature of 1 gram of water 1°C, a measurement now largely replaced by the joule; 1 calorie = 4.1868 joules. See also gram-calorie.

calorigenic — A substance or process that produces heat or energy or that increases the consumption of oxygen.

calorimeter — An instrument used for the measurement of the amount of heat liberated or absorbed during a change in temperature caused by a chemical reaction, change of state, or formation of solution.

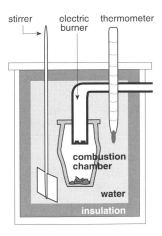

Calorimeter

calorimetry — Measurement of the amount of irradiated heat and amounts of heat absorbed.

Calvin cycle — The biochemical route of carbon dioxide fixation in many autotrophic organisms.

calyx — The outermost part of a flower that covers the bud.

CAM — See continuous air monitoring; acronym for chemical agent monitor.

CAMEO — Acronym for computer-aided management of emergency operations.

CAMP — Acronym for Continuous Air Monitoring Program.

camphor (synthetic) ($C_{10}H_{16}O$) — Colorless or white crystals with a penetrating, aromatic odor; MW: 152.3, BP: 399°F, Sol: insoluble, Fl.P: 150°F, sp. gr. 0.99. It is used in the preparation and loading of explosives; during embalming procedures; in mixing and blending of lacquers and varnishes; as a catalyst or chemical intermediate; in the packaging of moth and insect repellents. It is hazardous to the central nervous system, eyes, skin, and respiratory system and is toxic through inhalation, ingestion, and contact. Symptoms of exposure include eye, skin, and mucous membrane irritation; nausea; vomiting; diarrhea; headache; dizziness; excitement; and irrational epileptiform convulsions. OSHA exposure limit (TWA): 2 mg/m³ [air].

Campylobacter — A genus of Gram-negative, microaerophilic to anaerobic, motile, curved, or spiral rod-shaped bacteria found in the oral cavity, intestinal tract, and reproductive organs of humans and animals.

Campylobacter enteritis — An acute enteric disease of varying severity characterized by diarrhea, abdominal pain, malaise, fever, nausea, and vomiting. Incubation time is normally 3 to 5 days but can range from 1 to 10 days. It is caused by *Campylobacter jejuni* and *C. coli* and is found in all parts of the world. The reservoir of infection is animals, and it is transmitted by ingestion of organisms in food or in unpasteurized milk or contaminated water, by contact with infected pets, or from uncooked or poorly refrigerated food. It is communicable from several days to several weeks with universal susceptibility. It is controlled through isolation of patients and other individuals, as well as disinfection of articles soiled with feces and by control of sewage disposal.

campylobacteriosis — See *Campylobacter* enteritis.

canal — A channel that is usually open and conveys water by gravity to farms, municipalities, etc.

canal caps — Devices that rest on the ear canal opening and are supported by a headband.

cancellation — (*pesticides*) An act of disallowing the use of a pesticide by the U.S. Environmental Protection Agency when the chemical adversely affects the environment or does not comply with requirements of the Federal Insecticide, Fungicide, and Rodenticide Act.

cancer — Any malignant cellular tumor, including carcinoma and sarcoma. It encompasses a group of neoplastic diseases in which there is a transformation of normal body cells into malignant ones, probably involving some change in the genetic material of the cells (DNA), possibly as a result of faulty repair of damage to the cell caused by carcinogenic agents or ionizing radiation. The altered cells pass on inappropriate genetic information to their offspring and begin to proliferate in an abnormal and destructive manner. The mass of abnormal tissue enlarges, ulcerates, and begins to shed cells that spread the disease locally or to other sites by a process known as metastasis. Environmental, hereditary, and biological factors are important in the development of cancer.

cancer cell — A cell that devides and reproduces abnormally with uncontrolled growth and may spread to other parts of the body.

cancer effect level (CEL) — The lowest dose of chemical in a study or group of studies that produces significant increases in the incidence of cancer or tumors between the exposed population and its appropriate control; generally derived from animal studies and extrapolated to humans.

cancer risk score — Estimated cancer risk of a chemical compared with the cancer risk of other chemicals after each has been converted into a common unit for comparison.

cancroid — A lesion resembling cancer.

candela — The international reference standard for the unit of candle power, defined as 1/60 of the luminous intensity per square centimeter of a black body radiator operating at the temperature of freezing platinum; formerly known as standard candle.

Candida — A genus of yeast-like fungi that are commonly part of the normal flora of the mouth, skin, intestinal tract, and vagina and can cause a variety of infections.

candidiasis — A mycosis usually confined to the superficial layers of skin or mucous membrane; ulcers may be formed in the esophagus, gastrointestinal tract, or bladder; lesions may occur in the kidneys, spleen, lungs, liver, endocardium, eyes, meninges, or brain. Incubation time is 2 to 5 days. It is caused by *Candida albicans* and *C. tropicalis* and is found worldwide, as it is a part of normal human flora. The reservoir of infection is people. It is transmitted by excretions of the mouth, skin, vagina, and feces; patients are carriers, and it can also be carried from mother to infant during childbirth. It is communicable during the duration of the lesions; a widespread, low-level of pathogenicity occurs. It is controlled by disinfection of secretions and contaminated articles and specific treatment for the individuals.

Candiru fever — An arbovirus infection transmitted to humans by the bite of a sand fly with symptoms of acute fever, headache, and muscle aches.

candle — See candela.

candlepower (cp) — The unit of intensity or brightness of a standard candle at its source.

cane — The woody stem growing directly from the box of a plant as in raspberries, blackberries, and some vines.

canister — A container that holds the filter used in a gas mask.

canned meats — Meat products that are preserved and hermetically sealed in metal or glass containers.

canopy — The layer formed naturally by the leaves and branches of trees and plants.

cap — A fairly impermeable seal usually composed of a clay-type soil or a combination of clay soil and a synthetic liner placed over a landfill during closure to minimize leachate volume during biodegradation of waste by keeping precipitation from percolating through the landfill.

CAP — See community assistance panel.

capability — (*energy*) The maximum load that a generating unit, generating station, or other electrical apparatus can carry under specific conditions over a given period of time without exceeding approved limits of temperature and stress.

capacitance (c) — The property of a system of conductors and dielectrics which permits the storage of electric charge and electric energy when potential differences exist between the conductors.

capacitor — A device for holding and storing charges of electricity.

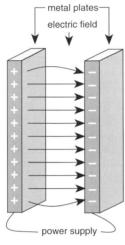

Capacitor

capacity — (*energy*) The load for which a generator, transmission line, or system is rated; expressed in kilowatts.

capillaries — The smallest blood vessels of both the circulation and lymphatic systems.

capillarity — (*physics*) The tendency of the surface of a liquid to rise or fall in contact with a solid; also known as capillary action.

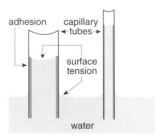

Capillarity

capillary action — See capillarity.

capillary attraction — The force of adhesion between a solid and a liquid in capillarity; movement of a liquid over or retention by a solid surface due to the interaction of adhesive or cohesive forces.

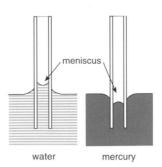

Capillary Attraction

capillary flow — See capillary migration.

capillary forces — The molecular forces that cause movement of water through very small places.

capillary fringe — A zone in the rocks just above the water table in which water is drawn upward into the spaces by capillary tension.

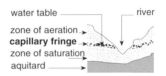

Capillary Fringe

capillary migration — The movement of water by capillary action.

capillary movement — The movement of underground water in response to capillary attraction.

capillary tension — The tendency for water to be drawn upward into very narrow spaces.

capillary water — The water held in the small pores of a soil by capillarity; usually with a tension greater than 60 cm of water.

capital costs — The long-term debt associated with financing construction and equipment.

capital investment — Investment for long-term use (over 1 year) to be treated as capital rather than as an expense.

capnometry — The measurement of carbon dioxide in a volume of gas; usually by infrared absorption or mass spectrometry.

capsid — The shell of protein that protects the nucleic acid of a virus.

capsomere — One of the building blocks of a viral capsid consisting of groups of identical protein molecules.

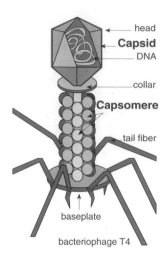

capture efficiency — The fraction of organic vapors generated by a process directly related to an abatement or recovery device.

capture system — The equipment, including hoods, ducts, fans, and dampers, used to capture or transport particulate matter generated by a pollutant source.

capture velocity — The air velocity at a point within or in front of an exhaust hood necessary to overcome opposing air current particle inertia and cause contaminated air to flow into the hood; the air velocity required to draw contaminants into the hood.

carbamate — A salt or ester of carbamic acid; a synthetic organic insecticide.

carbamic acid (H_2NCOOH) — An acid occurring only in the form of salts.

carbarl ($C_{12}H_{11}NO_2$) — A crystalline solid; MW: 201.22, BP: NA, Sol: soluble in acetone and cyclohexanone, density: NA. It is used as a pesticide. It is hazardous to the digestive tract and respiratory tract and may cause nausea, vomiting, diarrhea, abdominal cramps, excessive salivation, sweating, cyanosis, convulsions, and coma. ACGIH exposure limit (TLV-TWA): 5 mg/m³.

carbaryl (Sevin®) ($C_{10}H_7OOCNHCH_3$) — A white or gray odorless solid; MW: 201.2, BP: decomposes, Sol: 0.01%, sp. gr. 1.23. It is used as a pesticide and acaricide for field crops, fruit, vegetables, ornamentals, livestock, poultry, pets, domestic dwellings, medical facilities, schools, commercial and industrial areas, urban and rural outdoor areas, and sewage treatment plants. It is hazardous to the respiratory system, central nervous system, cardiovascular system, and skin and is toxic through inhalation, absorption, ingestion, and contact. Symptoms of exposure include miosis, blurred vision, tearing, nasal discharge, salivation, sweating, abdominal cramps, nausea, vomiting, diarrhea, tremors, cyanosis, convulsions, and skin irritation. OSHA exposure limit (TWA): 5 mg/m³ [air].

$$O-\overset{\overset{O}{\|}}{C}-NH-CH_3$$

Carbaryl

carbofenthion ($C_{11}H_{16}ClO_2PS_3$) — An amber liquid with a mild odor of sulfur; MW: 342.85, BP: 82°C, Sol: soluble in organic solvents, density: 1.271 at 25°C. It is used as a pesticide. It is a highly toxic cholinesterase inhibitor and is hazardous to the skin, digestive, and respiratory tracts. It may cause headaches, dizziness, blurred vision, vomiting, abdominal pain, diarrhea, chest pain, convulsions, coma, and death.

$$\begin{array}{c} C_2H_5O \\ C_2H_5O \end{array} \overset{\overset{S}{\|}}{P}-S-CH_2-S-\!\!\!\!\!\bigcirc\!\!\!\!\!-Cl$$

Carbofenthion

carbofuran ($C_{12}H_{15}NO_3$) — An odorless white crystalline solid; MW: 221.3, BP: NA, Sol (water): 0.07% at 25°C, Fl.P: NA (poisonous gases are produced in fire and containers may explode), sp. gr. 1.18. It is used as a carbonate pesticide sprayed on corn, rice, and other crops. It is hazardous to the respiratory system, digestive system, skin, and eyes. Acute exposure may cause central nervous system depression, including stupor, coma, seizures, hypertension, tachycardia, cardiorespiratory depression, respiratory failure, and death. It also causes urinary incontinence, diarrhea, gastrointestinal cramping, blurred vision, sweating, nausea, vomiting, and cyanosis. Chronic exposure symptoms are not known at this time. OSHA exposure limit (8-hour TWA): 0.1 mg/m³.

carbohydrate — Any group of organic compounds consisting of a ring or chain of carbon atoms with hydrogen and oxygen attached, usually in a ratio of 2:1. Sugars, starches, cellulose, and glycogen are carbohydrates.

carbolic acid — A poisonous, colorless to pale-pink crystalline compound obtained from coal-tar distillation and converted to a clear liquid, with a strong odor and burning taste, by adding 10% water.

carbon (C) — A nonmetallic element found either in its elemental state or as a constituent of coal, petroleum, limestone, and other organic or inorganic compounds; atomic number: 6; atomic weight: 12.01115. Its three main allotropic forms are diamond, graphite, and coal.

carbon-11 — A radioisotope of carbon with the half-life of 20 minutes produced by a cyclotron while emitting positrons.

carbon-14 — A beta-emitter with a half-life of 5760 years and which occurs naturally.

carbon-14 dating — See carbon dating.

carbon absorber — An add-on control device that uses activated carbon to absorb volatile organic compounds from the gas stream. The volatile organic compounds are later recovered from the carbon.

carbon adsorption/carbon treatment — A treatment system where contaminants are removed from groundwater, surface water, and air by forcing water or air through tanks containing activated carbon, which attract and hold the contaminants.

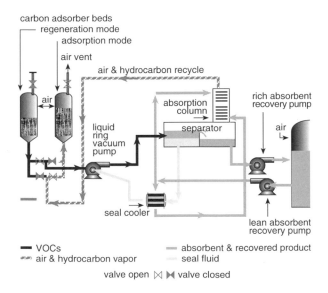

Carbon Absorption/Carbon Adsorption Treatment

carbon arc cutting — An arc-cutting process in which metals are severed by melting them with the heat of an arc between a carbon electrode and the base metal.

carbon arc welding — An arc-welding process that produces fusion of metals by heating them with an arc between a carbon electrode and the work.

carbon black (C) — A black, odorless solid; MW: 12.0, BP: sublimes, Sol: insoluble, Fl.P: not known, sp. gr. 1.8 to 2.1. It is used in the manufacture of natural and synthetic rubber, dry cells, explosives, plastics, and paper; in the manufacture of coating and printing inks; as a color pigment and source

of carbon. It is hazardous to the respiratory system and eyes and is toxic through inhalation and contact. Cough and irritated eyes occur; in the presence of polycyclic aromatic hydrocarbons it is carcinogenic. OSHA exposure limit (TWA): 3.5 mg/m^3 [air].

carbon cycle — (*geochemistry*) The natural sequence through which carbon circulates in the biosphere, beginning with the conversion of atmospheric carbon dioxide to carbohydrates by plants. The carbon stored in plants is returned to the atmosphere in the form of carbon dioxide by several routes, including respiration by animals that subsist on plants, the burning of fossil fuels, primarily coal and oil, or the eventual decomposition of plants, animals, and animal waste products.

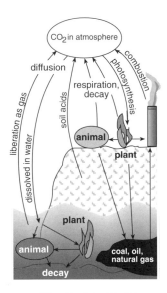

The Carbon Cycle

carbon damp — See damp.

carbon dating — The use of radioactive carbon-14 (C$_{14}$) isotope to estimate the approximate age and archeology of ancient organic materials; also known as radioactive carbon dating, radiocarbon dating, and carbon-14 dating.

carbon dioxide (CO$_2$) — Colorless, odorless gas that is a normal constituent of air (about 3000 ppm); MW: 44.0, BP: sublimes, Sol: 0.2% at 77°F. The solid form is utilized as dry ice. CO$_2$ is used in cooling and refrigeration for storage, preparation, and transfer of foods; as an inert gas in fire extinguishers; in shielded arc welding; in canned food production; as an inert pressure medium for aerosol packaging; for pressure spraying and spray painting; to inflate life rafts; in the manufacture of carbonated beverages; as a neutralizing agent in textile processing and treatment of leather hides; in the manufacture of urea, aspirin, carbonates and bicarbonates, beer, and sugar. It is hazardous to the lungs, skin, and cardiovascular system and is toxic through inhalation and contact. Symptoms of exposure include headache, dizziness, restlessness, paresthesia, dyspnea, sweating, malaise, increased heart rate and pulse, elevated blood pressure, coma, asphyxia convulsions (at high concentrations), and frostbite (from dry ice); associated with the greenhouse effect. OSHA exposure limit (TWA): 10,000 ppm [air] or 18,000 mg/m^3.

carbon dioxide cycle — A series of processes by which the balance of carbon dioxide and oxygen in the atmosphere is maintained.

carbon dioxide equivalent — A metric measure used to compare the emissions from various greenhouse gases based upon their global warming potential; derived by multiplying the tons of the gas by the associated global warming potential.

carbon dioxide toxicity — A buildup of carbon dioxide in the blood causing headaches, shortness of breath, and sudden blackouts.

carbon disulfide (**CS$_2$**) — Pure carbon disulfide is a colorless liquid with a pleasant chloroform-like odor, but impure carbon disulfide, used in most industrial processes, is a yellowish liquid with an unpleasant odor like that of rotting radishes; MW: 76.14, BP: 46.5°C, Sol (water): 0.294% at 20°C, Fl.P: 125 to 135°C. It is used in the viscose rayon industry and in the production of carbon tetrachloride, which is an intermediate in the production of fluorocarbon propellants and refrigerants. Increased levels of death occur to workers in viscose rayon plants where carbon disulfide as well as other chemicals are used; accidental inhalation may result in subtle and transient changes in pulmonary function with reduced vital capacity and decreased partial pressure of arterial oxygen; vascular atherosclerotic changes are a primary effect following long-term exposure; gastrointestinal effects may occur; inhibition of microsomal hepatic enzymes occur; severe ocular effects characterized by dot hemorrhages or microaneurysms of the retina have been observed; the primary target appears to be the nervous system with neuro, physiological, and behavioral effects as well as pathomorphology of peripheral nervous system structures. Symptoms of exposure include distal sensory shading, resting tremors, nerve conduction abnormalities, brain organic changes, and brain atrophy; systemic toxicity upon oral exposure usually occurs in the liver because of enzymatic disruptions. OSHA exposure limit: 4 ppm (12 mg/m^3) [air].

carbon equivalent — A metric measure used to compare the emissions of the different greenhouse gases based upon their global warming potential; to convert carbon to carbon dioxide multiply the carbon by 44/12, the ratio of the molecular weight of carbon dioxide to carbon.

carbon fiber — A material consisting of graphite fibers in a plastic matrix used in radiological devices to reduce patient exposure to x-rays.

carbon fixation — The conversion of inorganic carbon into energy-rich organic carbon, usually by photosynthesis.

carbon monoxide (**CO**) — A colorless, odorless, very toxic gas produced by the incomplete combustion of carbon-containing substances such as coal, oil, gasoline, and natural gas; MW: 28.0, BP: –313°F, Sol: 2%. It is produced on an industrial scale for use as a reducing agent in metallurgy. Carbon monoxide has an affinity for hemoglobin that is 200 to 300 times higher than that for oxygen, causing the body to become oxygen starved. Cigarette smokers' blood contains 2 to 10% carboxy hemoglobin, while nonsmokers have an average of 1% carboxy hemoglobin. Carbon monoxide can cause severe health symptoms or death when individuals are exposed to levels of 100 to 10,000 ppm. The low end of the scale results in slight headaches, and the upper end of the scale results in death. It is hazardous to the cardiovascular system, lungs, blood, and central nervous system and is toxic through inhalation and (liquid) contact. Symptoms of exposure include headache, tachypnea, nausea, weakness, dizziness, confusion, hallucinations, cyanosis, depressed S-T segment of electrocardiogram, angina, and syncope. OSHA exposure limit (TWA): ceiling 200 ppm [air] or 229 mg/m^3.

carbon monoxide toxicity — An illness caused by inhaling carbon monoxide resulting in headaches, unconsciousness, and potentially death.

carbon nitrogen ratio (**C/N**) — The ratio of the weight of carbon to the weight of nitrogen present in a compost or materials that are being composted.

carbon particulate matter — A product of incomplete combustion; also known as soot.

carbon sequestration — The uptake and storage of carbon, such as in trees and plants, where carbon dioxide is absorbed and oxygen is released while carbon is stored.

carbon sink — A place of carbon accumulation such as in large forests or ocean sediments; the carbon is removed from the carbon cycle for moderately long to very long periods of time.

carbon sinks — The carbon reservoirs and conditions that take in and store more carbon than they release.

carbon tax — A charge on fossil fuels based on their carbon content.

carbon tetrachloride (CCl₄) — A colorless liquid with a characteristic ether-like odor; MW: 153.8, BP: 170°F, Sol: 0.05%, sp. gr. 1.59. It is used in the manufacture of fluorocarbons for aerosols, refrigerants, and fire extinguishants; as an agricultural grain fumigant and pesticide; in polymer technology as a reaction medium, catalyst, chain transfer agent, and solvent for resins; in organic synthesis; as an industrial solvent for rubber cements and cable and semiconductor manufacture; as a laboratory solvent. It is hazardous to the central nervous system, eyes, lungs, liver, kidneys, and skin and is toxic through inhalation, absorption, ingestion, and contact. Symptoms of exposure include central nervous system depression, nausea, vomiting, liver and kidney damage, skin irritation; carcinogenic. OSHA exposure limit (TWA): 2 ppm [air] or 12.6 mg/m³.

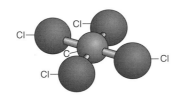

Carbon Tetrachloride

carbonaceous matter — Pure carbon or carbon compounds present in the fuel or residue of a combustion process.

carbonate aquifer — A type of aquifer made from limestone and dolomite rocks consisting of calcium and magnesium carbonate.

carbonate hardness — Temporary hardness due to soluble bicarbonates of calcium, magnesium, and iron, which can be removed by boiling.

carbonic acid — An unstable acid formed by dissolving carbon dioxide in water; used in carbonated beverages.

carbonization — A process by which the remains of living organisms are partially decomposed, leaving a residue of carbon.

carbonmonoxyhemoglobin — See carboxyhemoglobin.

carbonyl sulfide — A gas that is very stable and unreactive in the troposphere but is thought to photolyze to form carbon monoxide and sulfur in the stratosphere, where they are converted through atmospheric chemical reactions to SO₂ and H₂SO₄, which in turn form sulfate aerosols and cloud condensation nuclei, eventually settling into the troposphere and reacting to form sulfuric acid, a component of acid rain.

Carborundum™ — A trademarked name used for various manufactured abrasives.

carboxyhemoglobin — Hemoglobin combined with carbon monoxide; when the carbon monoxide is inhaled it occupies sites on the hemoglobin molecule that normally bind with oxygen. Also known as carbonmonoxyhemoglobin.

carboxyl group (COOH) — The acid group of organic molecules.

carboxylic acid — Any of a family of organic acids that contain the carboxyl group COOH.

carbuncle — (*bacteriology*) A necrotizing infection of skin and subcutaneous tissue composed of a cluster of boils usually due to *Staphylococcus aureus*.

carcass — All parts of a slaughtered animal that are capable of being used for human consumption, including viscera.

carcinogen — Any substance capable of increasing the incidence of neoplasms or decreasing the time it takes for them to develop.

carcinogenic — The ability of a substance to cause the development of cancer.

carcinogenicity — The power, ability, or tendency to produce cancerous tissue from normal tissue.

carcinogensis — The process by which normal cells are transformed into cancerous cells.

carcinoma — A malignant new growth made up of epithelial cells tending to infiltrate surrounding tissues, giving rise to metastases.

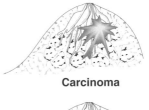

Carcinoma

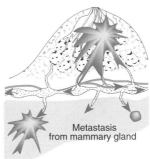

Metastasis from mammary gland

cardiac — Belonging or relating to the heart.

cardiac apnea — An abnormal temporary absence of ventilation.

cardiac arrest — A sudden cessation of cardiac output and effective circulation.

cardiac arrhythmia — An abnormal cardiac rate or rhythm.

cardiac catheterization — A diagnostic procedure in which a catheter is placed into a large vein and threaded through the circulatory system to the heart.

cardiac insufficiency — The inability of the heart to pump efficiently.

cardiac monitor — An instrument used for the continuous observation of cardiac function.

cardiogram — An electronically recorded tracing of cardiac activity.

cardiomyopathy — A disease that affects the structure and function of the heart.

cardiopulmonary — Of or pertaining to the heart and lungs.

cardiopulmonary resuscitation (CPR) — Reestablishment of heart and lung function, often applied in cases of cardiac arrest; the method of artificially oxygenating and circulating the blood of a person in cardiac arrest.

cardiotachometer — An instrument that continuously monitors and records the heartbeat.

cardiotoxic — Having a toxic or injurious effect on the heart.

cardiovascular and blood toxicity — Adverse effects on the heart or blood systems that result from exposure to toxic chemicals.

cardiovascular disease — Diseases of or involving the heart or blood vessels.

cardiovascular system — The heart and blood vessels.

carditis — Inflammation of the heart tissues.

caries — Progressive destruction of the hard tissues of teeth initiated by bacterially produced acids on the tooth surface.

carnivore — Any flesh-eating or predatory organism.

carotid pulse — The pulse of the carotid artery; found by gently pressing a finger in the groove between the larynx and the sternocleidomastoid muscle in the neck.

carpal — Of or pertaining to the carpus or eight bones of the human wrist.

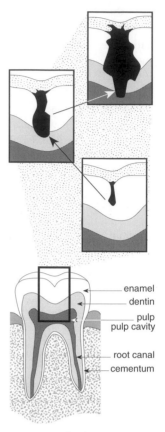

Caries

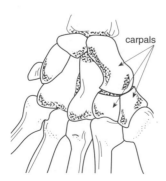

Carpals

carpal tunnel — The osteofibrous passage for the median nerve and flexor tendons formed by the flexor retinaculum and the carpal bones.

carpal tunnel syndrome — A painful syndrome resulting from compression of the median nerve in the carpal tunnel, with pain and burning or tingling paresthesia in the fingers and hand, sometimes extending to the elbow. The disorder is found most often in middle-aged women due to excessive wrist movements, arthritis, and swelling of the wrist.

carrier — (*disease*) An individual harboring specific infectious agents in the absence of discernible clinical symptoms and who serves as a potential source or reservoir of infection for other humans; (*pesticides*) a liquid, solid, or gas used to transport a pesticide chemical to the pest.

carrier gas — A gas or mixture of gases that contains and moves contaminant material but does not react with it.

carrying capacity — The maximum biomass that a system is capable of supporting continuously without deterioration.

Cartesian coordinate system — (*geographic information system*) Features on the Earth's surface that are referenced to map locations using an *x,y*-coordinate system.

C

cartilage — A specialized, elastic connective tissue attached to articular bone surfaces.

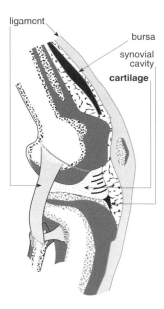

Cartilage

assembled view containing two culture dishes filled with semi-solid nutrient growth agar for sampling bioaerosols

Cascade Impactor (two-stage)

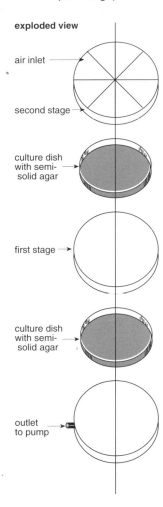

cartography — The production of maps, including construction of projections, design, compilation, drafting, and reproduction; also known as map.

cartridge disk — A type of magnetic-memory disk enclosed in a plastic cartridge and used in computer technology.

CAS registration number — A number assigned by the Chemical Abstract Service to identify a specific chemical.

cascade — Any process that develops in stages, with the later ones depending on the earlier ones.

cascade humidifier — A bubbling piece of respiratory care equipment in which gases travel down a tower and pass through a grid into a chamber of heated water, forming a liquid film that is converted to a froth and resulting in an airflow of relative humidity of up to 100%.

cascade impactor — A sampling device for particulates in air in which high-velocity, dust-laden air striking a flat surface at a 90° angle is forced to suddenly change direction and momentum, causing the particulates to settle out.

case — (*disease*) An instance of disease or injury that is studied to determine the cause of an outbreak.

case control study — (*epidemiology*) A study in which previously existing cases of a health condition are used in place of gathering new information from a random population and are compared to a control group of people who have not developed the same medical problem.

case fatality rate — A percentage of the number of persons diagnosed as having a specified disease and who die as a result of that disease.

case finding — The act of locating individuals with a specific disease.

case hardening — A process of hardening the surface of metals by raising the carbon or nitrogen content of the outer surface.

case history — Collected data concerning an individual, family, or environment that includes medical histories, environmental conditions, and other information useful in analyzing and diagnosing for teaching or research purposes.

case history study — An association between two variables such that a change in one is followed by a predictable change of the other; in the absence of experimental evidence, the following criteria are used to judge whether or not an association is causal: strength of association, consistency of association, correctness of temporal sequence, specificity of association, and coherence with existing information. Also known as causal association.

case management — A process that assesses, plans, implements, and monitors services required to meet an individual's health needs by using and creating resources to promote quality outcomes.

case study — (*ATSDR*) The medical or epidemiological evaluation of a single person or a small number of individuals to determine descriptive information about their health status or the potential for exposure through interview and/or biomedical testing.

casein — Solid or semisolid protein of milk obtained by precipitation of the milk solids by adding acids or whey.

cash flow — Annual cash receipts in the form of net profits plus the depreciation charge.

casing — (*water*) A pipe or tube of varying diameter and weight lowered into a borehole, during or after drilling, in order to support the sides of the hole or prevent contamination from water, gas, or other fluids.

cask — A thick-walled container, usually lead, used to transport radioactive material; also called a coffin.

CASNR — Acronym for Chemical Abstracts Service Registry Number.

casualty — Victim of a serious or fatal accident.

CAT — See computerized axial tomography.

CAT scan — See computed tomography.

cat scratch disease — See cat scratch fever.

cat scratch fever — A disease that results from the scratch or bite of a healthy cat that produces inflammation and pustules on the scratched skin and lymph nodes in the neck, head, groin, or axilla with symptoms of fever, headache, and malaise that may persist for months.

catabolic — Referring to any destructive process by which complex substances are converted by living cells into more simple compounds with release of energy.

catabolism — The decomposition process of metabolism in which relatively large food molecules are broken down to yield smaller molecules and energy. Catabolism releases energy as heat or chemical energy. The chemical energy released by catabolism is first converted into ATP molecules. The ATP molecules are then easily broken down and release the energy, which can be utilized in anabolism.

catalysis — The phenomenon in which a relatively small amount of a substance speeds up or slows down a chemical reaction but is regenerated at the end of the reaction in its initial form.

catalyst — A substance that facilitates or alters a chemical reaction without being changed itself.

catalytic — Of or pertaining to a process in which a catalyst is present in the incineration system and the chemicals are destroyed by a direct-flame oxidation process that proceeds at a lower temperature and in the absence of flame.

catalytic afterburner — An afterburner in which a catalyst induces incineration at lower temperatures than are possible in a direct fire-afterburner; a platinum alloy system that adsorbs the gases to be burned along with oxygen that has been injected in sizeable amounts so the gases may react with each other on the catalyst and oxidation products; carbon dioxide and water are the end products.

catalytic combustion system — A process in which a substance is introduced into an exhaust gas stream to burn or oxidize vaporized hydrocarbons or odorous contaminants while the substance itself remains intact.

catalytic converter — An air pollution abatement device that removes organic contaminants by oxidizing them into carbon dioxide and water through chemical reaction.

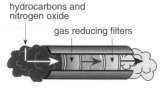

hydrocarbons and nitrogen oxide

gas reducing filters

Catalytic Converter

catalytic cracking — A refinery process that converts a high-boiling-range fraction of petroleum into gasoline, olefin feed for alkylation, distillate, fuel oil, and fuel gas by use of a catalyst and heat.

catalytic incinerator — A control device that oxidizes volatile organic compounds by using a catalyst to promote the combustion process. It requires lower temperatures than conventional thermal incinerators, thus saving fuel and other costs.

catalytic oxidation — A process to eliminate sulfur dioxide; flue gases are sent through a high-efficiency electrostatic precipitator to remove particulates; the gases then flow over a catalyst, vanadium pentoxide, which changes the sulfur dioxide to sulfur trioxide. The sulfur trioxide combines with water vapor in the flue gases to form sulfuric acid, which is condensed by cooling and can be used in a variety of applications.

catalyze — To act as a catalyst.

catalyzed — To be acted upon by a catalyst.

cataract — A clouding of the crystalline lens of the eye that obstructs the passage of light.

Eye with Cataract

catastrophe — A sudden and substantial disaster causing misfortune, destruction, or irreplaceable loss extensive enough to cripple activities in an area.

catastrophic disaster — An event or incident that produces large numbers of deaths and injury, and severe and widespread damages of a huge magnitude, resulting in the need for significant resources from outside of the affected area.

catastrophic illness — An illness that requires a lengthy hospitalization, extremely expensive therapies, or other special care beyond the financial resources of a family.

catastrophic reaction — The response to a drastic shock or sudden threatening condition, as in serious accidents or disasters.

cat-bite fever — See cat scratch fever.

catch basin — A sedimentation area designed to remove pollutants from runoff before it is discharged into a stream or pond.

catchment basin — An area from which all the drainage water passes into one stream or other body of water.

caterer — Any food-service establishment that prepares food or drink for service elsewhere.

caterpillar — The worm-like stage of moths and butterflies that usually feeds on plants.

catharsis — A cleaning or purging.

catheter — A tubular, flexible instrument passed through body channels for withdrawal or injection of fluids.

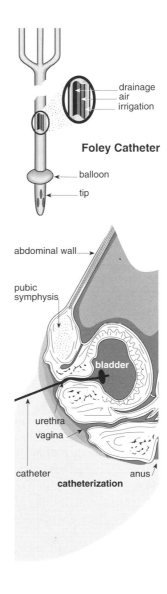

catheterization — The insertion of a catheter into a body cavity or organ to inject or remove a fluid.

cathode — The negative terminal of an electrolytic cell to which positive ions are attracted.

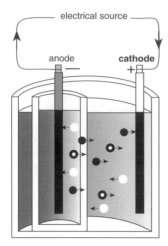

Voltaic Cell showing Cathode

cathode-ray oscilloscope — See oscilloscope.

cathode-ray tube (CRT) — (*electricity*) A device attached to a computer that can visually display information contained in magnetic disk files; an electron tube in which a beam of electrons can be focused on a small area on a surface. Also known as an electron-ray tube or visual display terminal (VDT).

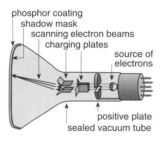

Cathode Ray Tube

cathode ray — (*electricity*) A stream of electrons emitted by the negative electrode of a gaseous discharge device when the cathode is bombarded by positive ions.

cathodic protection — An electrical system for prevention of rust, corrosion, and pitting of a metal surface by making the surface the cathode of an electrochemical cell.

cation — (*chemistry*) A positively charged ion that in solution is attracted to the negative electrode.

cation-adsorption capacity — See cation-exchange capacity.

cation exchange — (*chemistry*) The chemical reaction interchange between a cation and solution and another cation and the surface of any surface-active material such as clay or organic colloids.

cation exchange capacity — The sum total of exchangeable cations that a soil can adsorb; sometimes called total-exchange, base-exchange capacity, or cation-adsorption capacity. It is expressed in milliequivalents per hundred grams or per gram of soil. Also known as base exchange capacity or cation-adsorption capacity. The ability of the soil or other solid to exchange positive ions with a liquid.

cationic polymer — A polymer having positively charged groups of ions often used as a coagulant aid.

CAU — Acronym for carbon absorption unit.

caulking — A technique for making airtight joints by applying a sealing material.

causal association — See case history study.

causal organism — (*medicine*) A bacterium, fungus, nematode, or virus that can cause a given disease.

causation — A reasonable connection between occurrence of a problem and the resulting injury or damage suffered by an individual.

caustic — An alkali that strongly irritates, burns, corrodes, or destroys living tissue.

caustic poisoning — The accidental ingestion of strong acids or alkalies resulting in burns and tissue damage to the mouth, esophagus, and stomach; the victim experiences immediate pain, swelling, edema, irregular pulse, irregular respirations, and potential death.

caustic soda — See sodium hydroxide.

caution — (*pesticides*) A warning to the user of pesticide chemicals; a message placed on labels of pesticide containers alerting the user to special precautions.

CAV — Acronym for constant air volume.

caveat emptor — A principle of commerce where a warranty does not exist and the buyer takes the risk of receiving inadequate quality.

cavernous — Containing cavities or hollow spaces.

cavitation — (*fluid mechanics*) The formation of cavities in a fluid downstream from an object moving in it (e.g., behind the moving blades of a propeller); (*pathology*) the formation of cavities in any body structure, especially in tubercular lungs, or the formation of cavities within the body such as those occurring in the lung due to tuberculosis; the boiling of a liquid caused by a decrease in pressure rather than an increase in temperature.

cavitation corrosion — (*metallurgy*) The deterioration that occurs due to the production, formation, and collapse of vapor bubbles on a metal surface.

CAWM — Acronym for Chemical Agent Water Monitor.

CBA — See cost–benefit analysis.

CBC — See complete blood count.

CBD — Acronym for chemical and biological defense.

CBF — Acronym for cerebral blood flow.

CBM&S — Acronym for chemical/biological modeling and simulation.

CBMS — Acronym for chemical/biological mass spectrometer.

CBOD — Acronym for carbonaceous biochemical oxygen demand.

CBPS — Acronym for chemical and biological protected shelter.

CBR — Acronym for chemical, biological, and radiological.

CBRD — Acronym for chemical, biological, and radiological defense.

CBSD — Acronym for Chemical Biological Sustained Detector.

CBW — Acronym for chemical and biological warfare.

cc — See cubic centimeter.

CC — Acronym for cloud-to-cloud lightning.

CCEA — Acronym for conventional combustion environmental assessment.

CCL — See construction completion list.

CCN — See cloud condensation nuclei.

CCOs — Acronym for chemical control orders.

CCP — Acronym for composite correction plan.

CCU — See coronary care unit.

Cd — See cadmium.

CD — Acronym for climatological data; see criteria documents and compact disc.

CDBG — See Community Development Block Grant.

CDE — See carbon dioxide equivalent.

CDHS — Acronym for comprehensive data-handling system.

CDNS — Acronym for Climatological Data National Summary.

CDP — Acronym for census-designated places.

CD-ROM — (*computer science*) Acronym for compact disc read-only memory; a computer disc that can hold large amounts of information.

CDS — Acronym for compliance data system.

CDTA — Acronym for Chemical Diversion and Trafficking Act.

CE — See carbon equivalent; acronym for combustion efficiency.

cecum — (*anatomy*) The blind pouch-end of a cavity, duct, or tube; also spelled caecum.

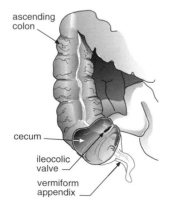

Cecum

CEGL — Acronym for continuous-exposure guidance level.

ceiling level (c) — Maximum allowable concentration of an airborne chemical in the workplace; also known as ceiling limit.

ceiling limit — See ceiling level.

ceiling plenum — The space below the flooring and above the suspended ceiling that accommodates the mechanical and electrical equipment that is used as part of the air distribution system; the space is kept under negative pressure.

ceiling value — The concentration of a substance that should not be exceeded even instantaneously.

CEL — See cancer effect level.

cell — (*biology*) The fundamental structural and functional unit of living organisms. It is any of the minute protoplasmic masses making up organized tissue consisting of a nucleus, surrounded by cytoplasm, and enclosed in a plasma membrane. The cell consists of the cytoplasmic membrane, endoplasmic reticulum (ER), Golgi apparatus, mitochondria, lysosomes, ribosomes, nucleus, and nucleoli; (*solid waste*) compacted solid wastes that are enclosed by natural soil or cover material in a sanitary landfill; (*geographic information system*) the basic element of spatial information on the roster (grid) description of spatial entities.

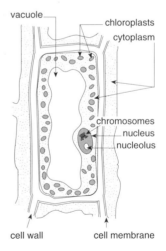

Plant Cell

cell biology — The science that deals with the structures, living processes, and functions of cells.

cell body — Part of the cell that contains the nucleus and surrounding cytoplasm exclusive of any projections or processes.

cell culture — Living cells that are maintained *in vitro* in artificial media of serum and nutrients for the study and growth of certain strains for controlling disease.

cell cycle — Sequence of events that occurs during growth and division of tissue cells.

cell death — Terminal failure of a cell to maintain essential life functions.

cell division — The continuous process by which a cell divides in four stages: prophase, metaphase, anaphase, and telophase.

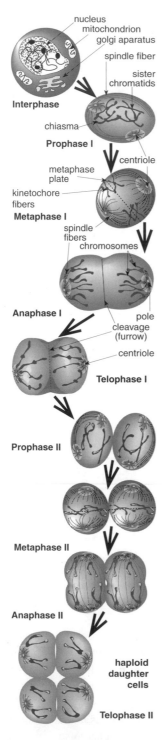

Interphase

Prophase I

Metaphase I

Anaphase I

Telophase I

Prophase II

Metaphase II

Anaphase II

Telophase II

Cell Division

cell height — (*solid waste*) Vertical distance between the top and bottom of compacted solid waste enclosed by natural soil or cover material in a sanitary landfill.

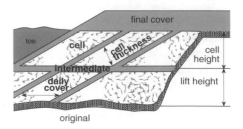

Cell (Solid Waste)
Cell Height
Cell Thickness

cell inclusion — Any foreign substance that is enclosed within a cell.

cell-mediated — Branch of the immune system in which specific defense cells, rather than antibodies, respond and act against a foreign antigen.

cell-mediated immune response — A delayed, type IV hypersensitivity reaction mediated primarily by sensitized T-cell lymphocytes instead of antibodies for defense against certain bacterial, fungal, viral pathogens, malignant cells, and other foreign proteins or tissues.

cell membrane — Thin, outer layer of cytoplasm of a cell consisting mainly of lipids and proteins.

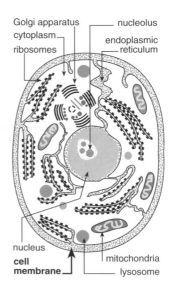

Animal Cell
Cell Membrane

cell residence time — The amount of time in days that an average microorganism remains in the process.

cell theory (biology) — A theory stating that the cell is a fundamental structural and functional unit of living matter and that the organism is composed of autonomous cells whose function is the essential process of life.

cell thickness — (*solid waste*) Perpendicular distance between cover materials placed over the last working faces of two successive cells in a sanitary landfill.

cell wall — The tough, nonliving outer layer of a plant cell consisting mainly of a cellulose secreted by the cytoplasm of the cell.

cell-type incinerator — An incinerator with grate areas divided into cells, each of which has its own ash drop, under fire, air control, and ash grate.

cellosolve — See 2-ethoxyethanol.

cellular — A wireless telephone technology that makes use of various radio frequencies to transmit telephone calls.

cellulitis — A diffuse, acute infection of the skin and subcutaneous tissue with symptoms of local heat, redness, pain, and swelling and occasionally fever, malaise, chills, and headache.

cellulose ($C_6H_{10}O_5$)$_n$ — A polymer of glucose with a very high molar mass and containing a type of linkage not easily split by hydrolysis; a carbohydrate is the chief substance contained in the cell walls of plants.

cellulose debris — Scrap wood buried after construction and found in substructure soil areas that leads to infestation by wood-destroying pests and other organisms.

Celsius — A unit of temperature where $1°C = 5/9 \times (°Fahrenheit - 32)$.

Celsius temperature scale — Temperature scale in which the freezing point of water at standard atmospheric pressure is nearly 0°C and the corresponding boiling point is nearly 100°C; formerly known as the Centigrade temperature scale.

CEM — Acronym for continuous emission monitoring.

cement — A substance made from limestone and clay that, when mixed with water, sets to a hard mass; used as an ingredient in concrete.

cementation — Consolidation of loose sediments or sand by injection of a chemical agent or binder.

CEMS — Acronym for Continuous Emission Monitoring System.

census days — The total number of days of patient care rendered by a healthcare facility in a given year or other specified period of time.

center — The middle point of a body or geometric entity that is equidistant from points on the periphery.

centi- (c) — Prefix meaning 1/100 or .01 in the metric system.

centiare — A metric unit of area equivalent to 1 m² or 10.76 ft².

Centigrade scale — See Celsius temperature scale.

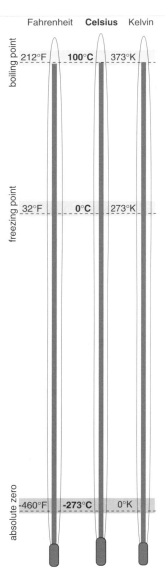

Celsius Temperature Scale
compared to K and F scales
(not proportionate)

centigram — A unit of metric measure equal to 0.01 grams.

centimeter (cm) — A metric measurement of length equal to 0.3937 inches.

centimeter–gram–second system — An internationally accepted scientific system of expressing length, mass, and time using SI units, or the International System of Units.

centimole — 100 moles.

central collection point — (*waste*) The location where a generator of regulated medical wastes, hazardous waste, or industrial waste consolidates the material; they are then treated on-site or transported elsewhere for treatment and/or disposal.

central heating — A single system supplying heat to one or more dwelling units or more than one rooming unit via pipes and ducts.

central nervous system (CNS) — Division of the vertebrate nervous system that is comprised of the brain and spinal cord and which coordinates all neural functions.

central nervous system depressants — Poisons that depress the central nervous system, resulting in headaches, vertigo, coma, and death; they include ethyl alcohol, wood alcohol, ether, chloroform, trichloroethylene, and benzene, among others.

central nervous system stimulant — A substance that increases the activity of the central nervous system by making neuronal discharges more rapid or by blocking inhibitory neurotransmitters.

central nervous system syndrome — A combination of neurological and emotional signs and symptoms that results from a massive, whole-body dosage of radiation; produces hysteria and disorientation, which increase during the final 24 to 48 hours before death.

central neurons — Those neurons in the brain and spinal cord that connect motor nerves to sensory nerves.

central processing unit (CPU) — (*computer science*) The part of the computer that operates the entire system.

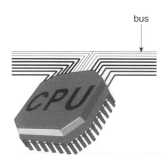

Central Processing Unit
CPU

central service (CS) — The department of a hospital responsible for sterilizing equipment and materials.

centrate — Water leaving a centrifuge after most of the solids have been removed.

centrifugal — Moving away from a center or axis of rotation.

centrifugal analyzer — A piece of equipment that uses centrifugal force to mix a portion of a sample with a reagent and a spinning rotor to pass the reaction mixture through a detector.

centrifugal collector — A mechanical system using centrifugal force to remove aerosols from a gas stream or to remove water from sludge.

centrifugal force — Force exerted by a revolving body in which movement extends away from the center of revolution.

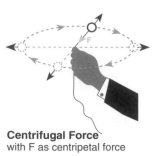

Centrifugal Force
with F as centripetal force

centrifugal pump — A pump that moves water by centrifugal force developed by rapid rotation of an impeller fixed on a rotating shaft that is enclosed in a casing and has an inlet and discharge.

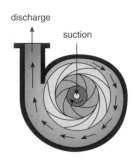

Centrifugal Pump

centrifuge — A device used to separate solid or liquid particles of different densities by spinning them at high speeds.

Simple Centrifuge

centriole — A minute cytoplasmic organelle usually found in the centrosome and considered to be the active division center of the animal cell.

centripetal force — Moving into the center toward the axis of rotation.

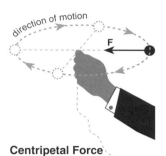

Centripetal Force

centromere — (*genetics*) Specialized area of a chromosome involved in its attachment to fibers of the spindle during cell division; also known as kinetochore.

ceramic — Calcined pottery, brick, and tile products molded from clay.

cercaria — A small, wormlike early developmental form of trematode that develops in a freshwater snail, is released into the water, and enters the body of the host by ingestion, such as in schistosomiasis.

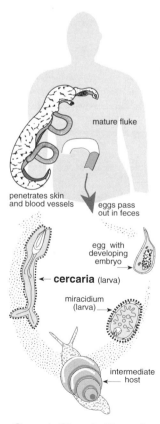

Cercaria Stage in life cycle of *S. mansoni*

CERCLA — See Comprehensive Environmental Response Compensation and Liability Act.

cerebellum — The part of the vertebrate brain region between the cerebrum and medulla concerned with equilibrium and muscular coordination and consisting of three lobes.

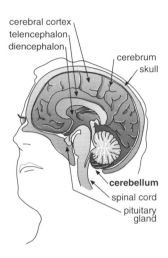

Cerebellum

cerebral cortex — A thin layer of gray matter on the surface of the cerebral hemisphere that integrates higher mental functions, general movement, visceral functions, perception, and behavioral reactions.

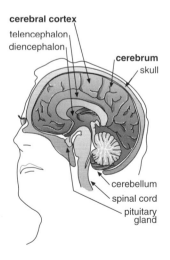

Cerebral Cortex

cerebral edema — An accumulation of fluid in the brain tissues.

cerebrospinal — Of or pertaining to the brain and spinal cord.

cerebrospinal fluid (CSF) — A clear fluid in the brain ventricles and surrounding the spinal cord that protects the central nervous system from mechanical injury.

cerebrovascular accident (CVA) — A disorder of the blood vessels serving the cerebrum resulting from an impaired blood supply to, and ischemia in, parts of the brain; also known as a stroke.

cerebrovascular disease — A disease of the blood vessels supplying the cerebrum.

cerebrum — The largest region of the vertebrate brain, considered to be the seat of emotions, intelligence, and other nervous system activities and consisting of two lateral hemispheres.

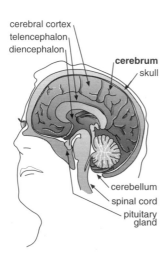

Cerebrum

certification — A process of testing and evaluating against specifications that is designed to document, verify, and recognize the competence of a person, organization, or other entity to perform a function or service, usually for a specific time, as specified by an association or agency.

certified applicator — (*pesticides*) A person who has been trained and certified as a pesticide applicator by a state agency authorized by the U.S. EPA.

certified industrial hygienist — An industrial hygienist certified by the American Board of Industrial Hygiene based on education, passing an examination, and active practice in the profession.

certified milk — Raw milk that is obtained, handled, and marketed in compliance with state health laws and produced by disease-free cows that are regularly inspected by a veterinarian and are milked by use of sterilized equipment in very clean surroundings containing less than a specified low bacterial count and delivered within 36 hours of milking.

certified safety professional (CSP) — A person who meets specific educational and experience criteria and passes the appropriate examinations to do the work of an occupational safety specialist.

certiorari — (*law*) To be certified; a writ commanding a court to certify and return records to a superior court.

cerumen — Earwax.

cervical — Of or pertaining to the neck.

cervix — The lower end of the uterus extending into the vagina.

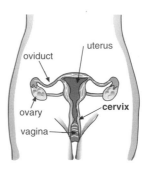

Cervix

cesium-137 — A radioactive material with a half-life of 30.2 years that is used in radiotherapy as a sealed source of gamma-rays for treatment of malignancies.

cesspool — A covered hole or pit for receiving untreated sewage.

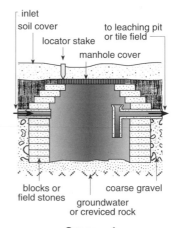

Cesspool

cesspool fever — An old term for typhoid fever.

Cestoda — A subclass of Cestoidea that includes the true tapeworm, which has a head (scolex) and segments (proglottides). The adults are endoparasitic in the alimentary tract and associated ducts of various vertebrate hosts; their larvae may be found in various organs and tissues.

cestode — See tapeworm.

CEU — See continuing education unit.

CFCl₃ — See trichlorofluorocarbon.

CFCs — See chlorofluorocarbons.

cfh — See cubic feet per hour.

cfm — See cubic feet per minute.

CFR — See Code of Federal Regulations; (*engineering*) see comprehensive facility review.

cfs — See cubic feet per second.

CFS — See chronic fatigue syndrome.

CFU — See colony-forming unit.

CG — Acronym for cloud-to-ground lightning.

CGI — See combustible gas indicator.

CGS — See centimeter–gram–second system.

CH₃CCl₃ — See methyl chloroform.

CH₃CL — See methyl chloride.

CH₄ — See methane.

chafe — An irritation of skin by friction.

chaff — Small strips of metal foil usually dropped in large quantities from aircraft or balloons to produce a radar echo that closely resembles precipitation and can be used for testing and calibration purposes.

Chagas' disease — A parasitic disease caused by *Trypanosoma cruzi* transmitted to humans by the bite of bloodsucking insects. It occurs in acute or chronic form. The acute form prevalent in children presents with a lesion at the site of the bite, fever, weakness, enlarged spleen and lymph nodes, edema of the face, legs, and tachycardia. Unless encephalitis develops, the disease runs its course within 4 months. The chronic form causes cardiomyopathy or dilation of the colon and esophogus.

chain — (*geographic information system*) A sequence of coordinates defining a complex line or boundary.

chain of custody — A legal method of documenting the history and possession of a sample from the time of collection through analysis and data reporting to its final disposition.

chain of infection — An organism leaving the reservoir of disease by means of a vector or vehicle and entering a host to cause disease.

chain reaction — A self-sustaining reaction in which the fission of nuclei produces subatomic particles that cause the fission of other nuclei.

chamber — An enclosed space inside an incinerator; (*mining engineering*) the workplace of a miner.

channel — (*computer science*) A path or circuit along which information travels; (*global positioning system*) circuitry necessary to receive a signal from a single GPS satellite.

channelization — The engineering of water courses by straightening, widening, or deepening them to permit water to move faster to reduce flooding and control mosquitoes.

char — A solid product of destructive distillation such as charcoal or coke; carbon.

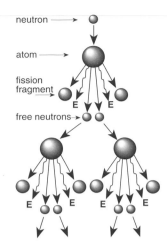

Chain Reaction

character — (*geographic information system*) An alphabetical, numerical, or special graphic symbol that is treated as a single unit of data.

characteristic radiation — Radiation originating from an atom after removal of an electron.

characteristics of hazardous waste — Current or potential hazards to human health and environment due to improper management; measured by quick, available, standardized test methods and reasonably detected by generators of solid waste through knowledge of waste including ignitability, corrosivity, reactivity, and extraction procedure toxicity.

characters per second (CPS) — (*geographic information system*) A measure of the speed with which a device, usually a printer or visual display unit, can process data in the form of characters.

charcoal filter — A filtering device made of activated carbon capable of eliminating certain materials, especially those organic in nature.

charge — A basic property of elementary particles of matter: positive, negative, or zero; the quantity of solid waste introduced into a furnace at one time.

charge sheet — Specifications for a violation attached to a notice of hearing; proposed statement of charges that the prosecution should use and which become the government's allegation in a criminal case.

charged density — The charge per unit area on a surface or per unit volume in space.

charged particle — A particle that possesses at least one unit electrical charge and will not disintegrate upon loss of the charge.

charging chute — An overhead passage through which waste materials drop into an incinerator.

charging gate — A horizontal movable cover that closes the opening on the top charge of the furnace.

Charles' law — When volume is held constant, the absolute pressure of a given mass of a perfect gas of a given composition varies directly with the absolute temperature; also known as the Gay–Lussac law.

CHB — Acronym for complete heart block.

CHC — Acronym for community health center.

CHD — Acronym for coronary heart disease.

check analysis — A second examination of a sample that has been examined and found to violate a law; usually conducted by an analyst other than the one performing the original analysis.

check valve — Any device that will allow fluid or air to pass through in only one direction; (*plumbing*) a special valve with hinged disk or flap that opens in the direction of normal flow and is forced shut when flows attempt to go in the reverse or opposite direction.

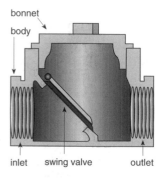

Check Valve

cheese — A food product made from separated curd obtained from the casein of milk, skim-milk, or milk enriched with cream; it is then cut, shaped, pressed, and salted or brined.

chelate — An inert complex compound or ion in which a metallic atom or ion is bound at two or more points to a molecule or ion to form a ring.

chelating agent — Any organic compound that will inactivate a metallic ion with the formation of an inner-ring structure in the molecule, where the metal ion becomes a member of the ring; a substance that removes organic material by forming a complex ion that does not have the chemical reactions of the ion that is removed.

chelation — A chemical complex of metallic cations with certain organic compounds.

chemical — A substance characterized by a specific molecular composition.

chemical action — A process in which natural elements and compounds react with each other to produce a chemical change or a different compound.

chemical affinity — An attraction that results in the formation of molecules from atoms.

chemical agent — A chemical substance intended to kill, seriously injure, or incapacitate people through its physiological effects; includes blood, nerve, choking, blistering, and incapacitating agents.

chemical agent casualty — An individual who has been affected sufficiently by a chemical agent to prevent or seriously degrade his or her ability to carry out work.

chemical agent monitor — A piece of equipment used to detect chemical agent vapors and provide a read-out of the concentration of the vapor present.

chemical antidote — A substance that reacts chemically with a poison to form a compound that is harmless.

chemical asphyxiant — A substance, such as carbon monoxide, in which rough, direct chemical action can prevent the uptake of oxygen by blood, interferes with the transportation of oxygen from the lungs to the tissues, or prevents normal oxygenation of tissues. Examples are hydrogen cyanide, which inhibits enzyme systems affecting utilization of molecular oxygen, and hydrogen sulfide, which paralyzes the respirator center of the brain and the olfactory nerve.

chemical bond — The force that holds atoms together in a molecule or crystal.

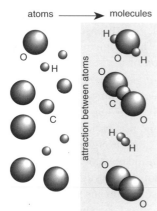

Chemical Bond

chemical burns — Burns similar to those caused by heat but are instead caused by chemical contact and require special treatment.

chemical carcinogen — A chemical agent that can induce the development of cancer in living tissue.

chemical cartridge — A type of absorption unit used with a respirator for removal of solvent vapors and certain gases.

chemical-cartridge respirator — A respirator, similar to a gas mask, that uses various chemical substances to purify inhaled air of certain gases and vapors; it is effective for concentrations up to ten times the threshold limit value of the contaminant.

C

chemical cauterization — The corroding or burning of tissue by a caustic substance.

chemical change — A transformation in which atoms are either rearranged to form new compounds or elements are separated from compounds.

chemical coagulation — The destabilization and initial aggregation of colloidal and finely divided suspended matter by the addition of a floc-forming chemical.

chemical compound — A distinct substance formed by the union of two or more elements in a specific proportion by weight.

chemical contamination — The presence of a chemical agent on a person, object, or area.

chemical decomposition indicators — Indicators that change due to the ionization produced by radiation in a chemical system and which indicates the amount of radiation received.

chemical dose–response curve — The relationship between the potential toxicity inherent in a given chemical and the kind of symptoms exhibited when the chemical interacts with the living, biological system.

chemical element — A substance composed of atoms that cannot be separated into similar parts by chemical means.

chemical energy — Energy contained in the chemical bond between atoms which can be released into the environment by a chemical reaction, such as combustion.

chemical engineering — The branch of engineering concerned with the development and application of manufacturing processes in which chemical or certain physical changes of materials are involved.

chemical equivalent — A drug or chemical having the same amounts of the same ingredients as another drug or chemical.

chemical family — A group of simple elements or compounds with a common general name.

chemical feeder — (*swimming pools*) A mechanism for the automatic addition of chemicals, such as pH control agents and disinfectants, to swimming pool water.

chemical food poisoning — An illness that develops rapidly and is due to chemicals being introduced accidentally into food or leached into food from a variety of containers; incubation period is less than 30 minutes.

chemical gastritis — An inflammation of the stomach caused by ingestion of a chemical compound.

chemical hazard — A chemical or biochemical substance that is toxic and can cause illness or death upon consumption.

Chemical Hazardous Information Profile — A profile of hazardous components found at an industrial site or disposal site, including all information listed on Material Safety Data Sheets; see also Material Safety Data Sheet.

chemical incident — Any event in which a chemical agent is used as a terrorist weapon.

chemical indicator — A substance that gives a visible change, usually of color, at a desired point in a chemical reaction, generally at a specified endpoint.

chemical interconversion — The conversion of one chemical into another within the cell.

chemical intermediate — A chemical formed or used during the process of producing another chemical.

chemical lifetime — The length of time a chemical species can survive without reacting, photolyzing, disassociating, or otherwise changing into another chemical species.

chemical mediator — A neurotransmitter chemical, such as acetylcholine.

chemical name — The name given to a chemical based on its structure in the nomenclature system developed by the International Union of Pure and Applied Chemistry (IUPAC) or the Chemical Abstract Service (CAS).

chemical oxygen demand (COD) — A measure of the amount of dissolved oxygen required in the receiving water to dispose of substances that can be oxidized chemically but cannot be removed biologically.

chemical peritonitis — An inflammation of the peritoneum resulting from chemicals.

chemical precipitation — Precipitation caused by the addition of chemicals, such as the process of removing acids by the addition of lime.

chemical process — A particular method of manufacturing or making a chemical that usually involves a number of steps or operations.

chemical protective clothing (CPC) — Items of clothing used to isolate parts of the body from direct contact with potentially hazardous chemicals.

chemical reaction — A change in the arrangement of atoms or molecules to yield substances of different composition and properties.

chemical sediment — The sediment formed by chemical precipitation from water.

chemical sensitivity — A sensitivity in some individuals to certain chemicals that results in health problems with symptoms of dizziness, eye and throat irritation, chest tightness, and nasal congestion.

chemical separation — (*nuclear energy*) A process for extracting uranium and plutonium from dissolved spent nuclear fuel and irradiated targets with fission products that are high-level waste; also known as reprocessing.

chemical sterilization — See cold sterilization.

chemical stressors — Chemicals released to the environment from industrial waste, automobile emissions, pesticides, and other sources that can cause illness and even death in plants and animals.

chemical toilet — A toilet in which sewage is retained and treated with chemicals.

chemical warfare — The use of poisonous chemicals and gases during war.

chemical waste — Waste generated by chemical, petrochemical, plastic, pharmaceutical, biochemical, or microbiological manufacturing processes.

chemical weathering — The decomposition of rocks and minerals by chemical reactions, usually through oxidation, carbonation, hydration, and solution in water above or below the surface.

chemically induced mutations — An error in the genetic code of a cell caused by a natural or synthetic chemical.

chemigation — Application of a pesticide and/or fertilizer through an irrigation system.

chemiluminescence — Emission of light as a result of a chemical reaction without an apparent change in temperature.

cheminosis — A disease caused by a chemical substance.

chemistry — The science that deals with the composition, structure, and properties of substances and their transformations.

chemoautotroph — An organism utilizing inorganic compounds as a source of energy.

chemolithotroph — An organism that obtains energy from the oxidation of inorganic compounds and uses inorganic compounds as electron donors.

chemoorganotroph — An organism that obtains energy and electrons from the oxidation of organic compounds.

chemoprophylaxis — The use of a chemical, including antibiotics, to prevent the development or progression of an infection.

chemoreceptor — A receptor that is stimulated by the presence of certain chemicals substances.

chemoresistance — Cancer cells that develop immunity to chemotherapy drugs.

chemosensitizers — Drugs or chemicals that enhance the effects of chemotherapy.

chemosis — An abnormal swelling of the mucous membrane covering the eyeball and lining the eyelid that is usually the result of local trauma or infection.

chemostat — A piece of equipment designed to grow bacteria indefinitely while keeping the conditions and colony size constant by providing a continuous flow of liquid nutrients that wash the colony and steadily remove bacteria.

chemosterilant — A pesticide chemical that controls populations by destroying their ability to reproduce.

chemosynthesis — Organization of carbohydrates of organisms by means of energy from inorganic chemical reactions rather than energy from light.

chemosynthetic bacteria — Nongreen plants that derive their energy for living from oxidizing inorganic compounds.

chemotaxis — The orientation or movement of a living organism in response to a chemical gradient.

chemotherapy — The use of chemicals of particular molecular structures in the treatment of specific disorders, where the structures exhibit an affinity for certain parts of the affected cell tissues or the invading bacterium or, in the case of cancer, the cancerous cells.

chemotroph — An organism that obtains its energy from oxidation of chemical compounds.

chemotropism — The orientation response of cells or organisms in relation to chemical stimuli.

chert — A microcrystalline form of silica.

chest tube — A tube inserted between the ribs and lungs to remove air and/or fluid that may be collapsing the lungs; may be used to reinflate a collapsed lung.

Cheyne–Stokes respiration — Breathing characterized by rhythmic waxing and waning of the depth of respiration; often observed in sleeping or unconscious people who seem to stop breathing for 5 to 40 seconds, then begin again and increase the intensity of their breathing, then stop breathing, and repeat the process; commonly found in healthy infants.

CHF — Acronym for congestive heart failure.

chigger — Six-legged mite larva of the family Trombiculidae that attaches to the host's skin and whose bite produces a wheal, usually accompanied by intense itching and severe dermatitis.

child nutrition programs — A group of programs funded by the federal government to support meal and milk service programs in schools, residential and daycare facilities, family and group daycare homes, and summer day camps for the benefit of low-income pregnant and postpartum women, infants, and children under age 5 visiting local WIC clinics.

child-resistant packaging — Any packaging specifically designed to protect children or adults from injury or illness resulting from accidental contact with or ingestion of pesticides, household chemicals, or pharmaceuticals.

childhood lead poisoning — Poisoning due to a variety of lead exposures but primarily through the consumption and digestion of lead paint or other inedible materials containing lead; prenatal exposure may cause decreased IQ and other forms of impaired learning in the fetus. It also causes increased blood pressure, brain and kidney damage, anemia, encephalopathy, and death.

chill — The sensation of cold caused by exposure to a cold environment.

chiller — A device that generates a cold liquid that is circulated through the cooling coil of an air-handling unit to cool the air supplied to the building.

chi-square — (*statistics*) A test of statistical significance that compares the observed number of cases falling into a particular category with the expected number of cases under the null hypothesis to determine whether the actual distribution differs significantly from what was expected.

Chironomid — A group of two-winged flying insects that live their larval stage underwater and emerge to fly about as adults.

chitin — A material present in the exoskeleton of insects and other arthropods.

Chlamydia — A genus of coccoid to spherical Gram-negative intracellular rickettsial parasites, including one that causes or is associated with various diseases of the eye, including trachoma and lymphogranuloma; also causes some forms of urethritis.

chloracne — An acneiform eruption caused by chlorinated naphthalenes and polyphenyls acting on sebaceous glands.

chloramine-T (sodium paratoluene sulfonchloramide; $CH_3C_6H_4SO_2NClNa·3H_2O$) — A chlorine compound that decomposes slowly in air; used for the sanitization of utensils and equipment.

chloramines — Compounds formed by the reaction of hypochlorous acid or aqueous chlorine with ammonia.

4-chloraniline (C_6H_6ClN) — A white to pale yellow to tan crystalline solid with a slightly sweetish characteristic amine odor; MW: 127.57, BP: 158 to 162.5°F, Sol: practically insoluble in cold water but soluble in hot water, Fl.P: 220°F, sp. gr. 1.427. Other information is not available. OSHA exposure limit (PEL-TWA): not established.

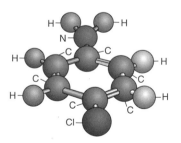

4–Chloroaniline

chlorasis — Discoloration of normally green plant parts that can be caused by disease, lack of nutrients, or various air pollutants.

chlorates — Metal or ammonium salts of chloric acid with the general formula $M(ClO_3)x$, where M is the metal ion with valency x or ammonium ion. The formula of chloric acid is $HClO_3$. Potassium and sodium chlorates are used in the production of explosives, fireworks, and matches; in dyeing cotton; for chemical analysis. Calcium chlorate is used as a disinfectant and in herbicides and insecticides. Sodium and potassium chlorates are low to moderately toxic in test animals, causing irritation of the gastrointestinal tract, anemia, and methemoglobinemia. Chlorates are powerful oxidizers and react explosively with organic or other readily oxidizable substances.

chlorbenside ($C_{13}H_{10}C_{12}S$) — A crystalline solid with an almond odor; MW: 269.20; BP: NA; Sol: dissolves in acetone, benzene, and petroleum ether; density: NA. It is used as a pesticide. It is hazardous to the digestive tract and may produce liver and kidney damage.

$$Cl\text{—}\bigcirc\text{—}CH_2\text{—}S\text{—}\bigcirc\text{—}Cl$$

Chlorbenside

chlordane ($C_{10}H_6Cl_8$) — An amber-colored, viscous liquid with a pungent, chlorine-like odor; MW: 409.8, BP: decomposes, Sol: insoluble, sp. gr. 1.56 at 77°F. It was used as an insecticide on soil before planting to treat fire ants and harvester ants but was banned by the U.S. EPA in 1976. It is hazardous to the central nervous system, eyes, lungs, liver, kidneys, and skin and is toxic through inhalation, absorption, ingestion, and contact. Symptoms of exposure include blurred vision, confusion, ataxia, delirium, cough, abdominal pain, nausea, vomiting, diarrhea, irritability, tremors, convulsions, and anuria; causes lung, liver, and kidney damage in animals. OSHA exposure limit (TWA): 0.5 mg/m³ [skin].

chlordecone ($C_{10}Cl_{10}O$) — An odorless tan-to-white crystalline solid; MW: 490.6, BP: sublimes, Sol (water): 0.5% at 100°C, Fl.P: NA (poisonous gases are produced in a fire and containers may explode), sp. gr. NA. A chlorinated hydrocarbon insecticide that was formerly used indoors or out for treating bananas, nonbearing citrus trees, tobacco plants, ornamental shrubs, lawns, turf, flowers, and buildings. It is produced during the synthesis of mirex and is a ketone analog of mirex. It is hazardous to the respiratory system and skin. Acute exposure may affect the central nervous system and cause symptoms of excitation, tremors, and seizures. Other symptoms include nausea, vomiting, diarrhea, respiratory depression, pneumonitis, acute renal failure, and dermal irritation. Chronic exposure may affect the reproductive and nervous systems, as well as cause harm to a developing fetus, decrease fertility rates, and damage the testes, liver, and kidneys. It may also induce liver cancer. OSHA exposure limit (8-hour TWA): NA. NIOSH-recommended airborne exposure limit: 0.001 mg/m³ averaged over a 10-hour work shift.

chlordimeform (C₁₀H₁₃ClN₂) — A colorless crystal or yellow liquid with a faint amine-like odor; MW: 196.67, BP: 156°C, Sol (water): 250 ppm at 25°C, Fl.P: NA (a flammable liquid that can be ignited by heat, sparks, or flames; vapors may travel to a source of ignition and flash back; containers may explode), sp. gr. 1.105. It is used as an insecticide and acaricide with ovicidal, larvicidal, and adulticidal properties against pests that are resistant to carbamates and organophosphorous compounds. It is hazardous to the respiratory system, digestive system, and skin. Acute exposure causes increased urination, acute hemorrhagic cystitis, abdominal pain, vomiting, anorexia, lethargy, muscular weakness, cyanosis, central nervous system excitation, seizures, coma, and death. Insufficient information is available to evaluate the chronic effects of the chemical on humans. OSHA exposure limit (8-hour TWA): NA.

chloride — A compound in which the negative element is chlorine.

chlorinate — To treat or combine with chlorine or a chlorine compound.

chlorinated camphene (C₁₀H₁₀Cl₈) — An amber, waxy solid with a mild chlorine- and camphor-like odor; MW: 413.8, BP: decomposes, Sol: 0.0003%, Fl.P: 275°F, sp. gr. 1.65 at 77°F. It is used in liquid, dust, powder, and granular insecticides. It is hazardous to the central nervous system and skin and is toxic through inhalation, absorption, ingestion, and contact. Symptoms of exposure include nausea, confusion, agitation, tremors, convulsions, unconsciousness, and dry, red skin; carcinogenic. OSHA exposure limit (TWA): 0.5 mg/m³ [skin].

chlorinated diphenyl oxide ((C₆H₂CL₃)₂O) — A light yellow, very viscous, waxy liquid; MW: 376.9, BP: 446 to 500°F, Sol: 0.1%, sp. gr. 1.60. It is used in organic synthesis; in corrosion inhibitors, drycleaning detergents, and thermal lubricants; as additives for soaps and lotions; in the manufacture of hydraulic fluids, pesticides, wood preservatives, and electric insulators. It is hazardous to the skin and liver and is toxic through inhalation, ingestion, and contact. Symptoms of exposure include acneform dermatitis and liver damage. OSHA exposure limit (TWA): 0.5 mg/m³ [air].

chlorinated hydrocarbons — A group of organic chemicals containing chlorine and used as pesticides.

chlorinated organic insecticide poisoning — Poisoning resulting from the inhalation, ingestion, or absorption of DDT or other insecticides, with symptoms of vomiting, weakness, and malaise.

chlorinated polyethylene (CPE) — A very flexible thermoplastic produced by a chemical reaction between chlorine and polyethylene.

chlorinated solvent — An organic solvent containing chlorine atoms used in aerosol spray containers, highway paint, and drycleaning fluids.

chlorination — The application of chlorine to water usually for the purpose of disinfection but frequently for aiding in coagulation and controlling tastes and odors.

chlorinator — A chemical feeder for the automatic addition of chlorine to water.

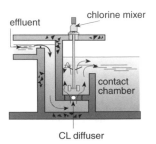

Mechanical Mixing Chamber

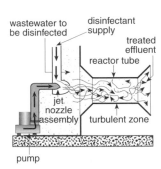

Chlorine Injector Mixer (Pump Type)

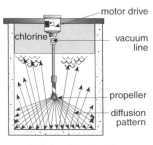

Chlorine Injector Mixer (Aspirating Type)

Chlorinator

chlorine (Cl$_2$) — A heavy, greenish-yellow gas with a pungent, irritating odor; shipped in steel cylinders; MW: 70.9, BP: –29°F, Sol: 0.7%. It is used as a chlorinating and oxidizing agent in organic synthesis; in the manufacture of solvents, automotive antifreeze, antiknock compounds, plastics, resins, elastomers, pesticides, refrigerants, bleaching, and inorganic chemicals; as a fluxing, purification, and extraction agent in metallurgy; as a bacteriostat, disinfectant, odor controller, and demulsifier in the treatment of sewage; as a bleaching and cleaning agent and disinfectant in laundries, dishwashers, and cleaning powders. It is hazardous to the respiratory system and is toxic through inhalation and contact. Symptoms of exposure include lacrimation, rhinorrhea, cough and choking, substernal pain, nausea, vomiting, headache, dizziness, syncope, pulmonary edema, pneumonia, hypoxemia, dermatitis, and burning of the eyes, skin, nose, and mouth. OSHA exposure limit (TWA): 0.5 ppm [air] or 1.5 mg/m^3.

chlorine combined available residual — That portion of the total residual chlorine remaining in water, sewage, or industrial waste at the end of a specified contact period that will react chemically and biologically as chloramines or organic chloramines.

chlorine-contact chamber — The part of a water treatment plant where effluent is disinfected by chlorine.

chlorine demand — The amount of chlorine necessary to oxidize all organic matter present in water at any given moment or over a period of time.

chlorine dioxide (ClO$_2$) — A yellow to red gas or a red-brown liquid (below 52°F) with an unpleasant odor similar to chlorine and nitric acid; MW: 67.5, BP: 52°F, Sol: 0.3% at 77°F, sp. gr. 1.6. It is used as a bleaching agent for wood pulp, paper, and textiles; in water treatment to remove tastes and odors by oxidation; as a bleaching and maturing agent in flour for baking; as a bleaching, cleaning, and dehairing agent in leather manufacture; in chemical synthesis as an oxidizing agent; as a bacteria and algae control agent in aquariums; and in packaged pharmaceuticals. It is hazardous to the respiratory system and eyes and is toxic through inhalation, ingestion, and contact. Symptoms of exposure include eye, nose, and throat irritation; cough; wheezing; bronchitis and chronic bronchitis; and pulmonary edema. OSHA exposure limit (TWA): 0.1 ppm [air] or 0.3 mg/m^3.

chlorine-free-available residual — The portion of total chlorine remaining in water, sewage, or industrial waste at the end of a specified contact period that would react chemically and biologically as hypochlorite ions.

chlorine monoxide — A radical species formed by the photolysis of CFCs in the stratosphere and subsequent destruction of an ozone molecule where the radicals act as a catalyst in the destruction of ozone while not being destroyed themselves; plays an important role in the breakdown of stratospheric ozone over Antarctica.

chlorine nitrate — A chemical formed from the reaction of chlorine monoxide and nitrogen dioxide reacting with HCl at low temperatures on the surfaces of polar stratospheric clouds; produces molecular chlorine and nitric acid, where the chlorine radicals rapidly catalyze the breakdown of ozone.

chlorine oxidation — A means of control of odor and reduction of pathogens without significantly reducing sludge solids through the use of chlorine.

chlorine pentafluoride (ClF$_5$) — A colorless gas with a suffocating odor; MW: 130.45; BP: NA; Sol: a violent reaction with water; Fl.P: a nonflammable gas but highly reactive and will cause paper, cloth, wood, and other organic matter to burst into flames; density: NA. It is used as a fluorinating and oxidizing agent. It is a highly toxic gas that is a severe irritant to the eyes, skin, and mucous membranes. It is hazardous to the respiratory tract, digestive tract, skin, and eyes. ACGIH exposure limit (TLV-TWA): 2.5 mg/m^3.

chlorine requirement — The amount of chlorine necessary for a particular purpose, such as reduction of coliform bacteria.

chlorine residual — The total amount of chlorine, combined and free available chlorine, remaining in water, sewage, or industrial waste at the end of a specified contact period following chlorination.

chlorine room — A separate room utilized for housing gas chlorine; it is adjacent to the filter room in swimming pool operations.

chlorine trifluoride (ClF$_3$) — A colorless gas or a greenish-yellow liquid (below 53°F) with a somewhat sweet, suffocating odor; MW: 92.5, BP: 53°F, Sol: reacts, sp. gr. 1.77. It is used as a fluorinating agent in organic and inorganic chemical synthesis; as a rocket fuel oxidizer; as an incendiary; as a cutting agent for well castings in oil well drilling. It is hazardous to the skin and eyes and is toxic through inhalation, ingestion, and contact. Symptoms of exposure include eye and skin burns; in animals, lacrimation, corneal ulcer, and pulmonary edema. OSHA exposure limit (TWA): ceiling 0.1 ppm [air] or 0.4 mg/m^3.

chlorites — A general formula, $M(ClO_2)x$, where M is a metal ion with valency x or the ammonium ion. These are the salts of chlorous acid. Chlorites are strong oxidizing substances. They are sensitive to heat or impact and will explode upon heating to 100 to 110°C. They are used as leaching agents for textiles and paper pulp, in water purification, and to produce chlorine dioxide. They are moderately toxic compounds when ingested and may affect the liver and kidneys.

1-chloro-1-nitropropane ($CH_3CH_2CHClNO_2$) — A colorless liquid with an unpleasant odor; MW: 123.6, BP: 289°F, Sol: 0.5%, Fl.P (oc): 144°F, sp. gr. 1.21. It is used as a solvent, an antigelling agent for rubber cements, and a fungicide. It is hazardous to the respiratory system, liver, kidneys, and cardiovascular system of animals and is toxic through inhalation, ingestion, and contact. Symptoms of exposure in animals include eye irritation, pulmonary edema, and liver, kidney, and heart damage. OSHA exposure limit (TWA): 2 ppm [air] or 10 mg/m³.

chloroacetaldehyde ($ClCH_2CHO$) — A colorless liquid with an acrid, penetrating odor; MW: 78.5, BP: 186°F, Sol: miscible, Fl.P: 190°F, sp. gr. 1.19. It is used during the manufacture of 2-aminothiazole; in the control of algae, bacteria, and fungi in water; during debarking operations; in acid media during chemical synthesis. It is hazardous to the eyes, skin, and respiratory system and is toxic through inhalation, ingestion, and contact. Symptoms of exposure include eye, skin, and mucous membrane irritation; skin burns; eye injury; pulmonary edema; skin and respiratory system sensitization; narcosis; and coma. OSHA exposure limit (TWA): ceiling 1 ppm [air] or 3 mg/m³.

α-chloroacetophenone ($C_6H_5COCH_2Cl$) — A colorless to gray crystalline solid with a sharp, irritating odor; MW: 154.6, BP: 472°F, Sol: insoluble, Fl.P: 244°F, sp. gr. 1.32 at 59°F. It is used during denaturing of industrial alcohol and in solutions for aerosols for law enforcement and civilian protective devices. It is hazardous to the eyes, skin, and respiratory system and is toxic through inhalation, ingestion, and contact. Symptoms of exposure include eye, skin, and respiratory system irritation and pulmonary edema. OSHA exposure limit (TWA): 0.3 mg/m³ [air].

chloroacne — A disease caused by chlorinated naphthalenes and polyphenols acting on sebaceous glands.

4-chloroaniline (C_6H_6ClN) — A colorless crystal with a slightly sweet characteristic amine odor; MW: 127.58, BP: 232°C, Sol: 0.39 g/100 g water at 20°C, Fl.P: 104°C, sp. gr. 1.169. It is used as a chemical intermediate in dyes, azoic coupling

agents, herbicides, agricultural chemicals, and pharmaceuticals. It is hazardous to the respiratory system, digestive system, skin, and eyes. Acute exposure may involve toxic response within several hours with symptoms of cyanosis, headache, weakness, lethargy, shortness of breath, nausea, confusion, drowsiness, loss of coordination, coma, and death. It may affect the heart, liver, and kidneys. Chronic exposure may cause central nervous system defects with symptoms of headaches, dizziness, vertigo, weakness, fatigue. It poses a significant risk to the fetus. OSHA exposure limit (8-hour TWA): NA.

chlorobenzene (C_6H_5Cl) — A colorless liquid with a faint, almond-like odor; MW: 112.56; BP: 131°C; Sol: miscible with organic solvents; Fl.P: 29°C (closed cup), the vapors form explosive mixtures with air; density: 1.107 at 20°C. It is used as a solvent for paints; as a heat transfer medium; in the manufacture of phenol and aniline. It is hazardous to the respiratory system, digestive system, and eyes. It causes irritation, drowsiness, incoordination, ataxia, and respiratory distress and may be carcinogenic to animals. ACGIH exposure limit (TLV-TWA): 75 ppm.

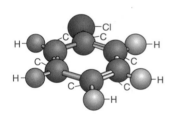

Chlorobenzene

o-chlorobenzylidene malononitrile ($ClC_6H_4CH C(CN)_2$) —A white crystalline solid with a pepper-like odor; MW: 188.6, BP: 590 to 599°F, Sol: insoluble. It is used as a harassing agent in civil disturbances and wars. It is hazardous to the respiratory system, skin, and eyes and is toxic through inhalation, absorption, ingestion, and contact. Symptoms of exposure include painful and burning eyes, lacrimation, conjunctivitis, erythema eyelids, blepharism, throat irritation, cough, chest constriction, headache, and vesiculation of the skin. OSHA exposure limit (TWA): ceiling 0.05 ppm [skin] or 0.4 mg/m³.

chlorobromomethane (CH_2BrCl) — A colorless to pale-yellow liquid with a chloroform-like odor; MW: 129.4, BP: 155°F, Sol: insoluble, sp. gr. 1.93. It is used as a fire extinguishing fluid in vaporizing fire extinguishers; in mineral and salt separations for flotation; as a grain fumigant. It is hazardous to the skin, liver, kidneys, respiratory system, and central nervous system and is toxic through inhalation, ingestion, and contact. Symptoms of exposure include disorientation, dizziness, pulmonary edema, and eye, skin, and throat irritation. OSHA exposure limit (TWA): 200 ppm [air] or 1050 mg/m³.

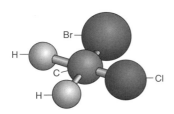

**Chlorobromomethane
(Bromochloromethane)**

chlorodibenzofurans ($C_{12}H_4ClO$, $C_{12}H_3Cl_5O$, $C_{12}H_2Cl_6O$, $C_{12}HCl_7O$, $C_{12}Cl_8O$) — A family containing 135 individual, colorless compounds known as congeners with varying harmful health and environmental effects. They are not commercially manufactured or used in the United States or abroad, except on a laboratory scale, but they are produced as unwanted compounds during the manufacture of several chemicals and consumer products such as wood treatment chemicals, some metals, and paper products. They are also produced from the burning of municipal and industrial waste in incinerators; from exhaust of leaded gasoline, heating, or production of electricity; from fires or breakdowns involving capacitors, transformers, and other electrical equipment containing polychlorinated biphenyls (PCBs). They are hazardous to the respiratory system, gastrointestinal system, liver, musculoskeletal system, skin, and nervous system and are toxic through inhalation, ingestion, and contact. Symptoms of exposure include frequent coughing, severe respiratory infections, chronic bronchitis, abdominal pain, muscle pain, acne rashes, skin color changes, unexpected weight loss, and nonmalignant or malignant liver disease. Listed as hazardous waste pollutants by the U.S. EPA; no exposure limits exist at this time.

chlorodibromomethane ($CHBr_2Cl$) — An odorless, colorless to pale-yellow liquid; MW: 208.28, BP: 119 to 120°C, Sol (water): 4400 ppm at 22°C, Fl.P: NA, sp. gr. 2.451. It is used as a chemical intermediate in organic synthesis for the manufacture of fire extinguishing agents, aerosol propellants, refrigerants, and pesticides. It is hazardous to the digestive system, skin, and eyes. Acute exposure may increase blood levels of methemoglobin and cause functional disturbances in the central nervous system,

adrenal glands, liver, and kidneys, which may be reversible when exposure ceases. Chronic exposure may lead to damage of the adrenal glands, liver, and kidneys. OSHA exposure limit (8-hour TWA): NA.

chlorodifluorobromomethane — See bromochlorodifluoromethane.

chlorodifluoromethane ($CHClF_2$) — A colorless gas or a liquid with a faint sweet odor under very cold conditions; MW: 86.47, BP: NA, Sol: 0.028 g/100 g water at 25°C, Fl.P: NA (poisonous gases are produced in a fire, including hydrofluoric acid, hydrochloric acid, phosgene, carbonyl fluoride, and chloride fumes), sp. gr. 1.194. It is used primarily as a refrigerant. It is hazardous to the respiratory system, digestive system, skin, and eyes. Acute exposure can cause symptoms of irritation to skin and eyes, dryness, dizziness, headaches, disorientation, cerebral edema, pulmonary edema, and irregular heart beat, which may become fatal. Chronic exposure can permanently impair psychomotor speed, memory and learning, emotional stability, liver, kidneys, and blood. OSHA exposure limit (8-hour TWA): 1000 ppm.

chlorodiphenyl (42% chlorine; $C_6H_4ClC_6H_3Cl_2$, approximately) — A colorless to light colored viscous liquid with a mild hydrocarbon odor; MW: 258 (approx.), BP: 617 to 691°F, Sol: insoluble, Fl.P: not known, sp. gr. 1.39 at 77°F. It is used as a high-temperature transfer medium in chemical and food processing vessels and drying ovens; as a dielectric in the manufacture of transformers, capacitors, resistors, and other electrical apparatus; in the manufacture of impregnants for cloth, paper, fiberboard, and wood; in the manufacture of natural and synthetic waxes and polishes; in the manufacture of hot-melt and other adhesives; in the manufacture of pesticides and fungicides; as a sealer for gaskets of natural rubber and synthetics; as a pigment carrier; as a pressure adhesive for sign backings; in compounding mastics and sealing and caulking materials; in compounding printing inks. It is hazardous to the skin, eyes, and liver and is toxic through inhalation, absorption, ingestion, and contact. Symptoms of exposure include eye irritation, chloracne, and liver damage; carcinogenic. OSHA exposure limit (TWA): 1 mg/m³ [skin].

chlorodiphenyl (54% chlorine — ; $C_6H_3Cl_2C_6H_2Cl_3$, approximately) — A colorless to pale-yellow viscous liquid (solid below 50°F) with a mild, hydrocarbon odor; MW: 326 (approx.), BP: 689 to 734°F, Sol: insoluble, Fl.P: not known, sp. gr. (77°F): 1.38. It is used as a high-temperature transfer medium in chemical and food processing vessels

C

and drying ovens; as a dielectric in the manufacture of transformers, capacitors, resistors, and other electrical apparatus; in the manufacture of impregnants for cloth, paper, fiberboard, and wood; in the manufacture of natural and synthetic waxes and polishes; in the manufacture of hot-melt and other adhesives; as a nonflammable working fluid in vacuum pumps, hydraulic systems, and expansion systems; in compounding and processing of plastics for flame retardancy; in the manufacture of pesticides and fungicides; as a sealer for gaskets of natural rubber and synthetics; as a pigment carrier; as a pressure adhesive; in compounding mastics and sealing and caulking materials; in compounding of printing inks. It is hazardous to the skin, eyes, and liver and is toxic through inhalation, absorption, ingestion, and contact. Symptoms of exposure include acneform dermatitis and eye and skin irritation; in animals, liver damage; carcinogenic. OSHA exposure limit (TWA): 0.5 mg/m³ [air].

chloroethane — See ethyl chloride; an anthropogenic volatile organic carbon that is highly reactive in the atmosphere.

***bis*-(2-chloroethyl) sulfide** — See mustard gas.

2-chloroethylvinyl ether (C₄H₇ClO) — A colorless liquid; MW: 106.56; BP: 109°C; Sol: soluble in alcohols, ethers, and most organic solvents; Fl.P: 27°C (open cup), a flammable liquid; density: 1.0525 at 15°C. It is used to produce sedatives, anesthetics, and cellulose ethers. It is hazardous to the respiratory tract, digestive tract, and skin and causes irritation, of the eyes, nose, lungs, and skin. It may exhibit carcinogenic properties in animals. ACGIH exposure limit (TLV-TWA).

$$Cl-CH_2-CH_2-O-CH=CH_2$$

2-Chloroethyl vinyl ether

chlorofluorocarbons (CFCs) — Relatively nontoxic, nonflammable, stable combinations of chlorine, fluorine, and carbon used in air conditioners, refrigerants, and freezers; in insulation and foam products; in cleaning solvents and medical sterilizing agents; as aerosol propellants. CFCs can exist for hundreds of years, eventually wafting up to the stratosphere, where bombardment by ultraviolet radiation releases chlorine atoms that consume ozone molecules and hinder formation of new ozone molecules; this process results in depletion of the ozonosphere.

chloroform (CHCl₃) — A colorless liquid with a pleasant nonirritating odor and a slightly sweet taste; MW: 119.38, BP: 61.3°C, Sol (water): 7.22 × 10³ mg/l at 25°C, Fl.P: none. It is used in producing fluorocarbon-22 (chlorodifluoromethane) and fluoropolymers; as a solvent or an extraction solvent for fats, oils, greases, resins, lacquers, rubber, alkaloids, gums, waxes, penicillin, vitamins, flavors, floor polishes, and adhesives; in artificial soot manufacturing; as a drycleaning spot remover; in fire extinguishers; as an intermediate; in the manufacture of dyes and pesticides; as a fumigant; in chlorination of wastewater. It may be transported for long distances before being degraded by reacting with photochemically generated hydroxyl radicals. Acute exposure can result in death. Changes in respiratory rates occur in individuals exposed to chloroform; respiration becomes depressed during sleep, and the effects of anesthesia are prolonged. It has cardiac effects in patients under anesthesia, producing bradycardia, cardiac arrhythmia, hypotension, first-degree atreoventricular block, or complete heart block; nausea and vomiting occur, as well as a fullness in the stomach. A major toxic effect is damage to the liver, as chloroform induces hepatotoxicity with impaired liver function, jaundice, liver enlargement, tenderness, delirium, coma, and death. Fatty degeneration of kidneys may also occur. The central nervous system is a target for chloroform toxicity, which produces headaches and light intoxication, exhaustion, hallucinations, delusions, and convulsions. OSHA exposure limits (TWA): 2 ppm [air] or 9.79 mg/m³. Also known as trichloromethane.

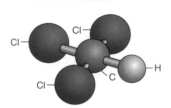

Chloroform

chloromethane (CH₃Cl) — A clear colorless gas or vapor that is difficult to smell but has a faintly sweet, nonirritating odor at high levels in the air; MW: 50.49, BP: –23.73°C, Sol (water): 0.5% at 25°C, Fl.P: less than 0°C. Cardiovascular effects include electrocardiogram abnormalities, tachycardia, increased pulse rate, and decreased blood pressure, with death occurring in extreme exposures. Gastrointestinal effects include nausea and vomiting accompanied by central nervous system toxicity, as well as clinical jaundice and cirrhosis of the liver. Renal effects include renal toxicity. Blurred, double vision; fatigue, drowsiness, staggering, headaches, mental confusion, tremor, vertigo, muscular cramping and rigidity, sleep disturbances, and ataxia may occur. See also methyl chloride. OSHA exposure limit (8-hour TWA): 50 ppm, (STL): 100 ppm.

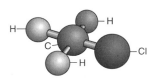

**Chloromethane
(Methyl Chloride)**

bis-**chloromethylether ((CH₂Cl)₂O)** — A colorless liquid with a suffocating odor; MW: 115.0, BP: 223°F, Sol: reacts, Fl.P: <66°F, sp. gr. 1.32. It is used in the manufacture of ion-exchange resins and polymers; as a solvent for polymerization reactions; as a chloromethylation agent in chemical synthesis; in the treatment of textiles. It is hazardous to the lungs, eyes, and skin and is toxic through inhalation, ingestion, and contact. Symptoms of exposure include pulmonary congestion and edema, corneal damage, necrosis, reduced pulmonary function, cough, dyspnea, wheezing, blood-stained sputum, bronchial secretions, and irritation of the eyes, skin, and mucous membrane of respiratory system; carcinogenic. OSHA exposure limit (TWA): regulated use.

chloromethylmethylether (CH₃OCH₂Cl) — A colorless liquid with an irritating odor; MW: 80.5, BP: 138°F, Sol: reacts, Fl.P (oc): 32°F, sp. gr. 1.06. It is used in the manufacture of ion-exchange resins and polymers; as a solvent for polymerization; as a chloromethylation agent in chemical synthesis; in the treatment of textiles. It is hazardous to the respiratory system, skin, eyes, and mucous membrane and is toxic through inhalation, absorption, and contact. Symptoms of exposure include pulmonary edema and congestion, pneumonia, burns, necrosis, cough, wheezing, blood-stained sputum, weight loss, bronchial secretions, irritation of the eyes, skin, and mucous membrane; carcinogenic. OSHA exposure limit (TWA): regulated use.

chlorophenoxy — A class of herbicides that may be found in domestic water supplies and cause adverse health effects; the two most important chemicals in this class are 2,4-D (2,4-dichlorophenoxyacetic acid) and 2,4,5-TP (2,4,5-trichlorophenoxypropionic acid, or silvex).

2-chlorophenol (C₆H₅–Cl–O) — A colorless to yellow-brown liquid with an unpleasant penetrating odor; MW: 128.56, BP: 347 to 348.8°F, Sol: 2.85 g/100 ml, Fl.P: 147°F, sp. gr. 1.241. Other information is not available. Short exposure could cause serious temporary or residual injury. OSHA exposure limit (PEL-TWA): not established.

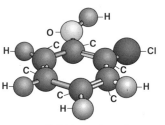

2-Chlorophenol

2,2-*bis*(*para*-chlorophenyl)-1,1-dichloroethane (DDD) (C₁₄H₁₀Cl₄) — A colorless, crystalline compound; MP: 109 to 111°C, Sol: insoluble. It is used as an insecticide on fruits and vegetables. An analog of DDT, it is equally as toxic as a mosquito larvacide and of lower toxicity to fish and shellfish.

chlorophyll — The generic name for several green pigmented substances in plants that capture and transform light energy into chemical energy that can be used by living organisms.

chloropicrin (CCl₃NO₂) — A colorless to faint yellow oily liquid with an intensely irritating odor; MW: 164.4, BP: 234°F, Sol: 0.2%, sp. gr. 1.66. It is used as a soil fumigant, disinfectant, and sterilizer for control of fungi, nematodes, and other injurious organisms; as a fumigant for stored cereals, grains, and fruits; as a rodenticide and insecticide; as a chemical intermediate in organic synthesis; as a warning agent in illuminating gas; as a nauseant in chemical warfare; as a chemical sterilant without high temperature. It is hazardous to the respiratory system, skin, and eyes and is toxic through inhalation, ingestion, and contact. Symptoms of exposure include eye irritation, lacrimation, cough, pulmonary edema, nausea, vomiting, and skin irritation. OSHA exposure limit (TWA): 0.1 ppm [air] or 0.7 mg/m³.

chloroplast — A type of cell plastid containing chlorophyll and functioning in photosynthesis.

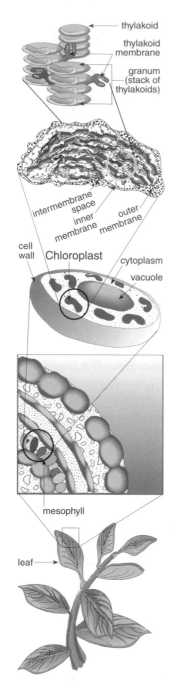

Chloroplast

chlororganic — An organic compound combined with chlorine.

chlorosis — Yellowing or whitening of normally green plant parts because of disease organisms, including viruses; lack of oxygen in water-logged soil; or lack of boron, iron, magnesium, manganese, nitrogen, or zinc.

chloroxuron (C₁₅H₁₅ClN₂O₂) — An odorless white crystal or colorless powder; MW: 290.77, BP: NA, Sol: insoluble in water, density: 1.34 at 20°C. It is used as an herbicide. It exhibits acute, delayed, and chronic toxicity and is hazardous to the digestive tract. It is listed by the U.S. EPA as an extremely toxic substance. It may form dimethylnitrosamine, under certain conditions, which is carcinogenic.

Chloroxuron

chlorpyrifos (C₉H₁₁Cl₃NO₃PS) — A white crystal with a mild odor; MW: 350.62, BP: NA, Sol: soluble in organic solvents, density: NA. It is used as a pesticide. It is a cholinesterase inhibitor and is hazardous to the skin and digestive and respiratory tracts. It produces acute, delayed, and chronic effects, including cholinergic symptoms in adult humans. It is highly toxic to birds.

Chlorpyrifos

chokes — A respiratory condition occurring in decompression sickness, with symptoms of shortness of breath, substernal pain, and a nonproductive cough caused by bubbles of gas in the blood vessels of the lungs.

choking — A condition in which a respiratory passage is blocked by constriction of the neck, obstruction in the trachea, or swelling of the larynx.

choking agent — A chemical agent that causes physical injury to the lungs.

cholecystitis — An inflammation of the gallbladder.

cholera — (*disease*) An acute, bacterial, enteric disease with sudden onset, profusely watery stools, occasional vomiting, rapid dehydration, acidosis, and circulatory collapse. Incubation time is a few hours to 5 days, but usually 2 to 3 days; it is caused by *Vibrio cholerae* and found worldwide. The reservoir of infection is people. It is transmitted through the ingestion of water contaminated with feces or vomitus or the ingestion of food contaminated with water or hands; it is also spread by flies. It is communicable during the stool-positive stage or in the carrier state for many months; susceptibility is variable. It is controlled through isolation using enteric precautions, handwashing, proper disposal of feces and vomitus and contaminated articles, and proper treatment and chlorination of water.

choleragen — An exotoxin produced by the cholera vibrio that stimulates the secretion of electrolyte and water into the small intestine, draining body fluids and weakening the patient.

cholesterol — A fat-soluble, crystalline, steroid alcohol found in animal fats, oils, and egg yolks and is widely distributed in the body, especially in the bile, blood, brain tissue, liver, kidneys, adrenal glands, and myelin sheaths of nerve fibers. It facilitates the absorption and transport of fatty acids and acts as a precursor for the synthesis of vitamin D at the surface of the skin as well as the synthesis of the various steroid hormones.

choline — A quaternary amine that occurs in the neurotransmitter acetylcholine and is an important methyl donor in intermediary metabolism.

cholinesterase — An enzyme found in blood and other tissues that splits acetylcholine into acetic acid and choline; helps regulate the activity of nerve impulses and is destroyed or damaged when organic phosphates or carbamates enter the body.

cholinesterase inhibitor — Any organophosphate, carbamate, or other pesticide chemical that can interrupt the action of enzymes that inactivate acetylcholine used in the nervous system.

chromatic — Relating to color.

chromatid — (*genetics*) The daughter strand of a chromosome that remains joined by the centromere to another chromatid following chromosomal duplication.

chromatograph — The apparatus used in chromatography to separate substances.

chromatography — A technique for the analysis of chemical substances. The substances to be analyzed are poured into a vertical glass tube containing an adsorbent; the various components of the substance move through the adsorbent at different rates of speed according to their degree of attraction to it, thereby producing bands of color at different levels of the adsorption column.

chromatolysis — The solution and disintegration of the chromatin of cell nuclei; disintegration of the Nissl bodies of a neuron as a result of injury, fatigue, or exhaustion.

chromic acid and chromates (H_2CrO_4) — Appearance and odor vary depending on specific compound; properties vary depending on specific compound. They are used in metal and leather finishing; as corrosion inhibitors in radiator coolants for internal combustion and gas turbine engines; in refrigerator and air conditioning systems and water-cooled nuclear reactors; in photoreproduction processes; as coloring pigments; in the dyeing of fur, leather, fabrics, wool, and nylon; in textile and paper printing; in the manufacture of some glues; as a fungicide; to increase the shelf-life of batteries; in the manufacture of safety matches and explosives; as a chemical reagent, oxidizing agent, and catalyst indicator; in chemical synthesis and analytical chemistry; in the manufacture of cement. They are hazardous to the blood, respiratory system, liver, kidneys, eyes, and skin and are toxic through inhalation, ingestion, and contact. Symptoms of exposure include respiratory system irritation, nasal septum perforation, liver and kidney damage, leukocytosis, leukopenia, monocytosis, eosinophilia, eye injury, conjunctivitis, skin ulcers, and sensitization dermatitis; carcinogenic. OSHA exposure limit (TWA): ceiling 0.1 mg CrO_3 per m^3 [air].

chromium (Cr) — A metallic element with oxidation states ranging from chromium II to chromium VI. The important balance states of chromium are 2, 3, and 4. Elemental chromium (O) does not occur naturally. Divalent state II is relatively unstable and is readily oxidized to trivalent state III, which is the chromic state. Hexavalent state VI is the chromate state. It is a naturally occurring element found in rocks, animals, plants, soil, and volcanic dust and gases; no known taste or odor. The metal is a steel-gray solid with a high melting point; MW: 51.996, BP: 2672°C, Sol: insoluble in water, Fl.P: not known. It is used in the metallurgical, refractory, and chemical industries to produce stainless steels, alloy cast irons, nonferrous alloys, and other miscellaneous materials; also used in drilling, muds, and water treatment; in rust and corrosion inhibitors; in chemical manufacturing and textiles; as toner for copying machines, magnetic tapes, and catalysts. Various and increased risk of death from noncancer respiratory disease is experienced by workers in chrome-plating plants. The respiratory tract is a major target of inhalation exposure of chromium compounds; chromate-sensitive workers acutely exposed to chromium VI compounds may develop asthma and other signs of respiratory distress accompanied by erythema of the face, nasopharyngeal pruritus,

C

nasal blocking, cough, wheezing, and dyspnea. Exposure to chromium trioxide causes coughing, expectoration, nasal irritation, sneezing, rhinorrhea, nose bleed, and ulceration and perforation of the nasal septum. Gastrointestinal effects have been associated with occupational exposure; hepatic effects of chromium VI include severe liver defects in workers, with derangement of the cells in the liver, necrosis, and lymphocytic and histiocytic infiltration. Acute systemic and dermal allergic reactions have been observed in chromium-sensitive individuals exposed to chromium via inhalation; it causes immunological effects with frequent skin eruptions, dyspnea, chest tightness, anaphylactic reactions, bronchospasms, and a tripling of the plasma histamine levels; neurological effects include dizziness, headache, anoxia, and weakness. Occupational exposure to chromium compounds has been associated with increased risk of respiratory system cancers, primarily bronchogenic and nasal. OSHA exposure limits (PEL, TWA) for chromium: 0; for insoluble salts: 1.0 mg/m^3; for chromium II: 0.5 mg/m^3; for chromium III: 0.5 mg/m^3. The ceiling concentration of chromium IV is 0.1 mg/m^3.

chromium hexacarbonyl (Cr(CO)$_6$) — A white crystalline solid; MW: 220.07, BP: NA, Sol: soluble in most organic solvents, Fl.P: 210°C, density: NA. It is used as a catalyst for polymerization and isomerization of olefins. It is also used as an additive to gasoline to increase octane. It is hazardous to the respiratory tract, digestive tract, and skin. It is highly toxic and causes headaches, dizziness, nausea, and vomiting. In its hexavalent form it is carcinogenic. ACGIH exposure limit (TLV-TWA): 0.05 mg Cr per m^3.

Chromium hexacarbonyl

chromium metal (Cr) — A blue-white to steel-gray, lustrous, brittle, hard solid; MW: 52.0, BP: 4788°F, Sol: insoluble, sp. gr. 7.14, atomic weight: 24. It is used in the fabrication of alloys; in the fabrication of plated products for decoration or increased wear resistance; as chemical intermediates; in dyeing, silk treating, printing, and wool mothproofing; in the leather industry; in photographic fixing baths; as catalysts for halogenation, alkylation, and catalytic cracking of hydrocarbons; as fuel additives and propellant additives. It is hazardous to the respiratory system and is toxic through inhalation and ingestion. Symptoms of exposure include histologic fibrosis of the lungs, systemic and dermal allergic reactions, and neurological effects. OSHA exposure limit (TWA): 1 mg/m^3 [air].

chromium oxychloride (CrO$_2$Cl$_2$) — A dark-red, fuming liquid with a musty burning odor; MW: 154.9, BP: 117°C, Sol: decomposes in hot or cold water, Fl.P: nonflammable (increases the intensity of a fire as it is an oxidizer; vapors may travel to a source of ignition and flash back), sp. gr. 1.91 at 25°C. It is used in making chromium complexes and dyes and in various organic oxidation and chlorination reactions. It is hazardous to the respiratory tract, digestive system, skin, and eyes. Acute exposure may cause severe damage with possible loss of vision from the corrosive chemical, a hole through the inner nose, irritation of the throat and bronchial tubes, skin burns, deep stomach ulcers, and an allergic reaction. Chronic exposure may cause irritation, discharge, bleeding, skin allergy, irritation of the bronchial tubes, shortness of breath, and kidney damage. OSHA exposure limit (8-hour TWA): 0.5 mg/m^3.

chromobacteriosis — A rare, usually fatal systemic infection caused by the Gram-negative bacillus *Chromobacterium violaceum* found in tropic and subtropic areas; enters the body through breaks in the skin. Symptoms include sepsis, multiple liver abscesses, severe prostration, and death.

chromoblastomycosis — A chronic infectious skin disease caused by a variety of fungi, with symptoms of pruritic warty nodules developing into a large ulcerated growth.

chromogen — (*chemistry*) An organic substance that forms a colored compound or becomes a pigment when exposed to air; a compound, not a dye, having color-forming groups and thus capable of being converted into a dye.

chromomere — A series of bead-like structures that lie along the coiled filament of a chromosome during the early stages of cell division.

chromosomal aberration — An irregularity in the number or constitution of chromosomes that may cause abnormalities in a developing embryo.

chromosomes — Thread-like structures visible during cell division that carry the hereditary traits and are contained in the nucleus of a cell; they are composed of chromatin.

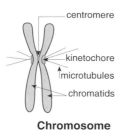

Chromosome

chronic — A condition or disease occurring over a long period of time either continuously or intermittently.

chronic alcoholism — A pathological condition resulting from the habitual use of alcohol in excessive amounts.

chronic bronchitis — A common debilitating pulmonary disease with symptoms of increased production of mucous by the glands of the trachea and bronchi and resulting in cough with expectoration for at least 3 months to a year or more.

chronic carrier — An individual who acts as a host to pathogenic organisms for an extended period of time without showing any signs of disease.

chronic condition — A condition that has one or more of the following characteristics: (1) it is permanent; (2) it results in residual disability; (3) it is caused by nonreversible pathological alteration; and (4) it requires special training for rehabilitation. Also known as chronic disease.

chronic daily intake — An exposure expressed as mass of a substance per unit body weight per unit time averaged over a long period of time.

chronic disease — See chronic condition.

chronic effect — A permanent adverse effect on a human or animal body with symptoms that develop slowly over a long period of time or which recur frequently as a result of exposure to a toxin or development of a disease.

chronic exposure — Exposure to a chemical or other substance for a period of 365 days or more.

chronic fatigue syndrome — The onset of disabling fatigue after an initial illness or an emotional situation.

chronic hepatitis — Condition in which the symptoms of liver disease continue for several months and may increase in severity.

chronic hypoxia — A slow, insidious reduction in tissue oxygenation resulting from gradual destructive or fibrotic lung diseases, congenital or acquired heart disorders, or chronic loss of blood.

chronic obstructive pulmonary disease (COPD) — A functional category designating a chronic condition of persistent obstruction of the bronchial air flow. It is the most significant chronic pulmonary disorder in the United States in regard to morbidity rate and is the second most common cause of hospital admissions in the country.

chronic pain — A pain that continues or recurs over a prolonged period caused by a variety of diseases or abnormal conditions.

chronic study — A toxicity study designed to measure the effects of chronic exposure to a chemical.

chronic toxicity — Adverse chronic effects resulting from repeated doses of or exposures to a substance over a relatively long period of time; the capacity of a substance to cause long-term poisonous health effects in humans, animals, fish, and other organisms.

chronobiology — The basic and applied study of temporal parameters of biological rhythms in plants, animals, and humans.

chronological age — The age of a person expressed as a period of time from birth.

chrysene (C$_{18}$H$_{12}$) — A polycyclic aromatic hydrocarbon formed when gasoline, garbage, or any animal or plant material burns, usually found in smoke and soot; MW: 228.3, BP: 448°C (sublimes *in vacuo*), Sol: 1.5 to 2.2 mg/l, Fl.P: not known. It is used as a research chemical, in coal tar pitch, as a binder for electrodes in the aluminum reduction process, and to bind carbon electrodes in reduction pots; also used as an adhesive in membrane roots. It is found in creosote wood preservative and is used in paving and sound and electrical insulation. It is a weak carcinogen through dermal exposure and a weak mutagen in experimental systems. OSHA exposure limit (TWA): 0.2 μg/m^3 [air].

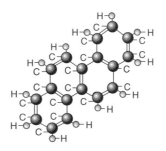

Chrysene

chrysotiles (Mg$_3$Si$_2$O$_5$(OH)$_4$) — A white-greenish hydrated magnesium silicate and a fibrous form of asbestos.

chute-fed incinerator — An incinerator that is charged through a chute that extends two or more floors above it; also spelled shute-fed incinerator.

chyme — Partially digested food as it leaves the stomach.

Ci — See curie.

cigarette beetle (*Lasioderma serricorne*) — Primarily a pest of tobacco but also a stored food product pest; a small, oval, light-brown beetle, 1/10-inch long, with smooth wing covers; the head is retracted to meet the thorax.

ciguatera poisoning — A food poisoning that results from eating fish contaminated with ciguatera toxin; symptoms include vomiting, diarrhea, and tingling or numbness of extremities and the skin around the mouth, itching, muscle weakness, pain, and respiratory paralysis.

cilia — Minute, short, hair-like cells capable of lashey movement to produce locomotion in unicellular organisms or a current in higher organisms; singular is cilium.

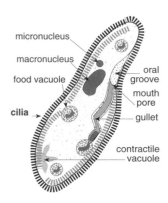

Cilia on *Paramecium*

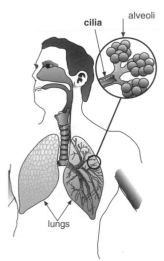

Cilia in Bronchiole

ciliary mucous transport — The movement of particles from the upper respiratory tract through the constant wave motion of microscopic ciliary lining the tract from the bronchioles to the trachea and the mucous layer that carries trapped particles upward to the lower pharynx.

CIM — See computer-integrated manufacturing.

cinnamaldehyde (C_8H_7CHO) — A yellowish liquid with a strong smell of cinnamon; MW: 132.17; BP: 253°C; Sol: soluble in alcohol, ether, and chloroform; Fl.P: 71°C; density: 1.0497 at 20°C. It is used in flavor and perfumes. It is hazardous to the skin, digestive tract, and respiratory tract and causes severe irritation to the skin and poisoning. In rats, mice, and guinea pigs, it causes convulsions, ataxia, coma, and diarrhea. It is a mutagen. ACGIH exposure limit: NA.

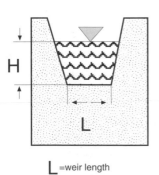

Cinnamaldehyde

CIP — See cleaning-in-place.

Cipolletti weir — A contracted weir of trapezoidal shape in which the sides of the notch are a given slope of 1 horizontal to 4 vertical.

Cipolletti Weir

circle of influence — The circular outer edge of a depression produced in the water table by the pumping of water from a well; see cone of depression.

circuit — *(electricity)* The complete path, or a part of it, over which an electrical current flows, usually including the generating apparatus.

Series Circuit

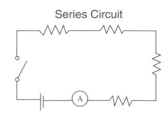

Parallel Circuit

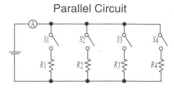

Branching Circuit

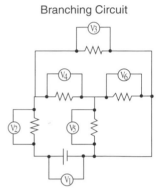

Circuit Diagrams

S=Switch   A=Ammeter
V_x=Voltmeter R_x =Resistor

circuit breaker — An electromagnetic switch that breaks an electric circuit when overloaded.

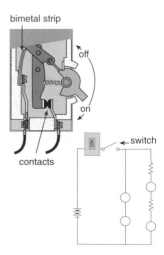

Circuit Breaker

circulation — The flow or motion of a fluid.

circulatory failure — The failure of the cardiovascular system to supply the cells with enough oxygenated blood to meet metabolic demands.

circulatory system — The series of organs comprising the lymphatic and cardiovascular systems in vertebrates that enable the blood to move throughout the body.

circumference of a circle — Distance around a circle equal to the diameter $\times$ 3.1416(π).

circumscribed — To be within a well-defined area with definite boundaries or limits.

cirrhosis — A progressive inflammatory disease of a tissue or organ, especially the liver, by excessive growth of hepatic connective tissue.

cirrocumulus clouds — A principal cloud type formed at 20,000 to 25,000 feet; appears as thin, patchy, clouds (often wave-like).

cirrostratus clouds — A principal cloud type formed at 20,000 to 25,000 feet; appears as thin sheets that look like fine veils or torn, wind-blown patches of gauze and is made up of ice crystals.

cirrus clouds — A principal high-altitude cloud type appearing as white patches or bands, usually formed 25,000 feet above the Earth and higher.

cis — A descriptive term meaning "on the same side" used to identify geometric isomers.

cistern — An underground box used for collecting water from the roof of a house for use within the home.

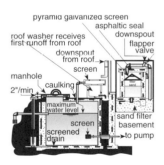

Cistern

cis-, *trans*-**isomers** — See geometric isomers.

cistron — The DNA sequence encoding a specific protein.

citation — See hearing.

citee — One who is cited; see also hearing.

city — (*housing*) A differentiated community with a sufficient population and resource base to allow residents to specialize in crafts, manufacturing, services, and professional occupations.

Civil Defense Agency — A state and/or local agency responsible for emergency operations, planning,

C

mitigation, preparedness, response, and recovery for all hazards.

civil suit — Generally any lawsuit other than a criminal prosecution.

Civilian–Military Contingency Hospital System (CMCHS) — A voluntary system of hospitals in the United States and other support services designed to make beds available to handle large numbers of military casualties or civilian injuries caused by a disaster.

CKD — Acronym for cement kiln dust.

Cl — Scientific notation for chlorine.

Cladophora — A filamentous green algae.

claim — A demand by an individual or corporation to recover losses under a policy of insurance; a statement of ownership under oath entered in response to seizure of a lot, which gives the right to contest.

clamshell bucket — A two-sided vessel used to hoist and convey materials; for excavating, it has two jaws that clamp together when it is lifted by specially attached cables.

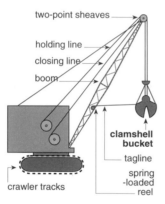

Clamshell Bucket

CLARET — Acronym for Cloud Lidar and Radar Exploratory Test

clarification — The process of removing suspended material from a liquid by sedimentation with or without chemicals or filtration.

clarifier — A settling tank used to remove settleable solids by gravity or colloidal solids by coagulation following chemical flocculation; also used to remove floating oil and scum through skimming.

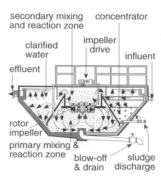

Suspended Solids Contact Clarifier

clarify (chemistry) — To clear a turbid liquid by allowing any suspended matter to settle through the addition of a substance that precipitates the material or by heating.

clarity — (*swimming pools*) The degree of transparency of pool water.

Clarke Belt — A belt 22,245 miles directly above the equator where a satellite orbits the Earth at the same speed at which the Earth is rotating.

class 1 area — An area specified under the Clean Air Act for which the visibility is protected more stringently than under the National Ambient Air Quality Standards; includes national parks, wilderness areas, monuments, and other areas of special national and cultural significance.

class I disposal facility — A sanitary landfill that services a municipal, institutional, and/or rural population and is used for disposal of domestic wastes, commercial wastes, institutional wastes, municipal wastes, bulky wastes, landscaping and land-clearing wastes, industrial wastes, construction/demolition wastes, farming wastes, discarded automotive tires, and dead animals.

class I injection well — An injection well regulated by federal and state permits used by generators of hazardous waste or owners/operators of hazardous waste management facilities to inject hazardous waste beneath the lowest formation containing an underground source of drinking water within the radius of 1 mile of the well bore of other industrial and municipal disposal wells that inject fluids beneath the lowest formation containing an underground source of drinking water within a radius of 1 mile of the well bore.

class II disposal facility — A landfill that receives wastes generated by one or more industrial or manufacturing plants and is used or is to be used for disposal of industrial wastes, commercial wastes, institutional wastes, farming wastes, bulky wastes, landscaping and land-clearing wastes, construction/demolition wastes, discarded automotive tires, and dead animals.

class III disposal facility — A landfill that is used or is to be used for disposal of farming wastes, landscaping and land-clearing wastes, and/or certain special wastes having similar characteristics.

class IV disposal facility — A landfill that is used or to be used for the disposal of demolition/construction wastes, certain special wastes having similar characteristics, and waste tires.

class V disposal facility — A land-farming facility.

class V injection well — An ejection well or system regulated by federal and state permits or rules and used for disposal of air conditioning return water from heating or cooling in a heat pump, drainage from the surface (primarily storm water runoff), water from cooling operations, and recharge well water.

class VI disposal facility — A surface impoundment used for disposal of solid wastes.

class A explosive — A material capable of detonation by means of a spark or flame or even with a small shock; includes nitroglycerine, lead azide, and black powder.

class A fire — A fire resulting from the combustion of ordinary cellulose materials that leaves embers or coals; generally involves wood, paper, and certain textiles. Water is an effective extinguishant.

class A poison — Department of Transportation term for extremely dangerous gas or liquid poisons; when mixed with air, very small amounts are dangerous to life. Examples include phosgene, cyanagene, hydrocyanic acid, and nitrogen peroxide.

class B explosive — Rapidly combustible materials, including photographic flash powder, that explode under extreme conditions of temperature.

class B fire — A fire resulting from the combustion of flammable liquids such as gasoline, kerosene, greases, oils, and flammable gases such as methane and hydrogen. Carbon dioxide and foam are effective extinguishants.

class B poison — A Department of Transportation classification for liquid, solid, paste, or semisolid substances other than class A poisons or irritating materials that are known or presumed on a basis of viable tests to be toxic to humans and considered a hazard to health during transportation.

class C explosive — Materials, such as flares, that do not ordinarily detonate in restricted quantities and are a minimum explosion hazard.

class C fire — A fire resulting from the combustion of materials occurring in or originating from live electrical circuits. Carbon dioxide is the effective extinguishant.

class D fire — A fire resulting from the combustion of certain metals that possess unique chemical properties, such as being reactive with water and carbon dioxide; includes titanium, magnesium, zirconium, and sodium. Special extinguishants such as graphite are effective.

classic emergency removal — The actions that are performed when a toxic release requires that the activities be initiated within hours of the determination that a removal action is appropriate.

classification code number — The identifying number for a classification listed in the *Workers' Compensation Manual*.

classification (epidemiology) — A process in data analysis in which data are grouped according to previously determined characteristics.

clay — Small mineral soil grains less than 0.0002 mm in diameter; a fine-textured soil material that breaks into very hard clods or lumps when dry and which is plastic and usually sticky when wet.

clay loam — A soil containing 27 to 40% clay and 20 to 45% sand, with the remaining portion being silt.

claypan — A layer of stiff, compact, relatively impervious clay that is not cemented.

CLD — Acronym for chronic lung disease.

Clean Air Act (CAA) — A law passed in 1963 that served as the basis of current air pollution laws. It provides matching grants for establishing and expanding air quality management programs, provides for developing air quality criteria, initiates efforts to control air pollution from federal facilities, encourages automotive companies and fuel industries to prevent air pollution, establishes federal authorities to abate interstate air pollution, and provides process for reviewing status of motor vehicle pollution. The initial Air Pollution Control Act was passed in 1955 and was amended in 1960 and 1962. The Motor Vehicle Air Pollution Control Act was passed in 1965, the Air Quality Act was passed in 1967, and the Clean Air Act Amendments were passed in 1970 and updated in 1977, 1983, and 1990.

Clean Air Act Amendments of 1970 — A law passed to stimulate additional air pollution research by authorizing state and regional grants, setting national ambient air quality standards, designating air quality control regions, setting fixed timetables for state implementation plans, setting new source performance standards, setting national emission standards for hazardous air pollutants, and requiring industry to monitor and maintain emission records.

Clean Air Act Amendments of 1977 — A law passed in 1977 requiring states with non-attainment areas to submit state implementation plan revisions; also established emission offset policy and banking policy and set modifications to emission standards for vehicles.

Clean Air Act Amendments of 1990 — A law passed in 1990 that recognizes existing severe ozone and carbon monoxide problems in certain metropolitan areas; creates tighter controls on tailpipe exhaust; calls for reduction of acid rain, nitrogen oxides, and air toxics and protection of the ozone layer by phasing out chlorofluorocarbons, carbon tetrachloride, methylchloroform, and hydrochlorofluorocarbons.

clean air standards — Any enforceable rule, regulation, guideline, standard, limitation order, control, or prohibition pursuant to the Clean Air Act.

clean-catch specimen — A urine specimen that is as free from contamination as possible.

clean coal technology — Any technology not in widespread use prior to the Clean Air Act Amendments of 1990 that will bring about significant reductions in pollutants associated with the burning of coal.

clean fuels — Blends or substitutes for gasoline fuels that include compressed natural gas, methanol, ethanol, and liquefied petroleum gas, among others.

clean-in-place — A system that uses cleaning and sanitizing applications for closed pipelines in food processing plants.

clean-out-of-place — A system providing for the disassembly of equipment and placement in recirculating tanks using high velocity physical and chemical cleaning followed by a sanitizing rinse before reassembling.

clean room — The area within a hospital where washed materials (such as instruments) are dried, inspected, and packaged; an area where elaborate precautions are taken to reduce contaminants in the air.

Clean Water Act (CWA) — A law passed in 1977 and amended numerous times through 1988 that emphasizes control of toxic pollutants in water; enhances the National Pollution Discharge Elimination System; establishes criteria for permits, including lists of pollutants regulated by federal or state law with permissible amounts of pollutants discharged per unit of time, monitoring requirements and schedules for implementing pollution concentration requirements; imposes special conditions on polluters by agencies; and establishes Best Available Technology (BAT) requirements for toxic substances.

Clean Water Standards — Any enforceable rule, regulation, guideline, standard, limitation order, control, or prohibition pursuant to the Clean Water Act.

cleaning — The act or process of the physical removal of residues of foods, ingredients, and other soiling materials.

cleaning-in-place (CIP) — A technique for cleaning pipelines in a milk plant dependent upon circulating appropriate cleaning solutions at relatively high velocities of not less than 5 feet per second; the milk pipelines go up from the pump to the outlet without any dips to provide self drainage of the solutions and for maintenance of full lines during circulation to ensure contact of solutions with all milk-contact surfaces.

CLEANS — Acronym for Clinical Laboratory for Evaluation and Assessment of Toxic Substances.

cleanup — The treatment, remediation, or destruction of contaminated material.

clear well — A reservoir for the storage of filtered water with sufficient capacity to prevent the need to vary the filtration rate with variations in demand.

clearance air monitoring — Air sample taken following completion of an asbestos abatement program to ensure that the activity is complete and that the area is safe to reoccupy.

Cleveland open-cup test (COC) — A laboratory flashpoint test method.

CLEVER — Acronym for Clinical Laboratory for Evaluation and Validation of Epidemiologic Research.

client — A person who is a recipient of a health service.

climate — The weather at a given place over a period of time; involves averages, totals, and extremes to set a picture of the weather pattern.

climate change — A significant change from one climatic condition to another, caused by either people or nature.

climate lag — The delay that occurs in climate change as a result of some factor that changes only very slowly, such as the effects of releasing more carbon dioxide into the atmosphere; it is not known for some time because a large fraction is dissolved in the ocean and only released to the atmosphere many years later.

climate model — A quantitative way of representing the interactions of the atmosphere, oceans, land surface, and ice.

climate system — The five physical components (atmosphere, hydrosphere, cryosphere, lithosphere, and biosphere) that are responsible for climate and its variations.

climatic — Having to do with weather conditions.

climatic analogue — A past climate situation in which changes similar to the present occurred; used when making climatic projections.

climatic variation — The change in one or more climatic variables over a specified time.

climatology — A quantitative description of climate showing the characteristic values of climate variables over a region; referring to the statistical collection of weather conditions over a specified period of time; the science dealing with climate and climate phenomena.

climax community — The stage of ecological development at which a community becomes mature, stable, self-perpetuating, and at equilibrium with the environment.

clinic — A department in a hospital or other medical facility where persons not requiring hospitalization may receive medical care.

clinical diagnosis — A diagnosis made on the basis of knowledge obtained from a medical history and physical examination without use of laboratory tests or x-rays.

clinical disease — A pathological condition that begins with an anatomical or physiological change that is sufficient to produce recognizable signs and symptoms of the disease.

clinical laboratory — A laboratory in which tests directly related to the care of patients are evaluated.

clinical studies — Scientific studies of humans under controlled conditions.

clinical thermometer — A thermometer designed for measuring the body temperature of patients.

clinical thermometry — A technique of determining temperature in heated tissue.

clinical trial — (*disease control*) An experiment to determine whether one procedure is more or less effective than another in preventing occurrence of a disease (prophylactic trial) or in treating an established disease process (therapeutic trial).

clinker — A hard, sintered, or fused piece of residue ejected by a volcano or formed in a furnace or by the agglomeration of ash, metals, glass, and ceramics.

clod — A small, hard mass of dirt formed by plowing or cultivating excessively wet or dry soil.

clone — An individual grown from a single somatic cell of its parent and genetically identical to it.

clonic — Alternate involuntary muscular contraction and relaxation in rapid succession.

cloning — The use of techniques of molecular biology to produce multiple copies of segments of DNA.

Clonorchis sinensis — The Chinese or oriental liver fluke, which affects people through the consumption of raw or imperfectly cooked fish; leads to chronic liver disease, diarrhea, edema, and eventually death.

close-coupled valve — (*milk*) A valve for which the seat is either flush with the inner wall of the holder or

so closely coupled that no fluid in the valve pocket is more than 1°F colder than the fluid at the center of the holder at any given time during the holding period.

closed chain — (*organic chemistry*) A compound in which the carbon atoms are bonded together to form a closed chain.

closed fracture — A bone fracture that is not accompanied by a break in the skin.

closed-loop recycling — The reclamation or reuse of wastewater for nonpotable purposes in an enclosed process.

closed-loop system — A system of heat exchange where the ground acts as a condenser for pipes from a house; the fluid within the closed loop consists of 20% glycol (antifreeze) and 80% water.

closed position — (*milk*) Any position of the valve seat that stops the flow of a liquid into or out of the holder.

closed system — A system that does not interact with its environment.

Clostridium — A genus of Gram-positive obligate anaerobic or microaerophilic, spore-forming, rod-shaped, motile bacteria.

Clostridium botulinum — See botulism.

Clostridium food poisoning — *Clostridium welchii* or *C. perfringens* food poisoning is an intestinal disorder with sudden onset of colic followed by diarrhea and nausea of short duration (1 day or less). Incubation time is 6 to 24 hours, usually 10 to 12 hours. It is caused by *C. welchii* or type A strains of *C. perfringens* or *C. welchii* and is produced by toxins from the microorganisms. It is found worldwide; reservoirs of infection include the soil and gastrointestinal tracts of healthy humans and animals. It is transmitted through ingestion of food in which a toxin has been produced by microorganisms found in the soil or in feces. It is not communicable; general susceptibility. It is controlled by proper cooking and rapid cooling of food, especially large cuts of meat.

Clostridium perfringens — A species of an aerobic Gram-positive bacteria causing gas gangrene in humans and various digestive and urinary tract diseases in livestock.

Clostridium welchii — See *Clostridium perfringens*.

closure — (*solid waste*) The procedure a landfill operator must follow when a landfill reaches its legal capacity for solid waste.

cloth filter collector — A dry collector working on the same principle as a vacuum cleaner bag; the gas stream carrying the particles to be filtered passes through woven fabric bags 30 feet or more in length, thus trapping the particles. Depending on the type of cloth filter, up to 100% of the particles down to 0.4 μm in size may be trapped.

cloud — A visible mass of particles of water or ice in the form of fog, mist, or haze suspended in the air.

cloud baby — A newborn child who appears well and healthy but is a carrier of infectious bacteria or viral organisms and spreads the organisms to the surrounding environment with airborne droplets.

cloud chamber — A device that detects nuclear particles through the formation of cloud tracks; also known as an expansion chamber or fog chamber.

cloud condensation nuclei — A tiny salt particle, sulfate or nitrate aerosol, or small particulate present in smoke that condenses extremely small quantities of water vapor floating in the air.

cloudburst — A torrential downpour of rain.

Clouds and the Earth's Radiant Energy System (CERES) — A system that measures solar-reflected and Earth-emitted radiation from the top of the atmosphere to the surface and also determines cloud properties, including the amount, height, thickness, particle size, and phase of clouds using simultaneous measurements by other instruments.

clouds of electrons — The distribution of electrons in space around an atomic nucleus.

cluster — (*U.S. EPA definition*) A structure with fibers in a random arrangement such that all fibers are intermixed and no single fiber is isolated from the group; groupings must have more than two intersections; (*NIOSH definition*) a network of randomly oriented interlocking fibers arranged so that no fiber is isolated from the group; dimensions of clusters can only be roughly estimated and clusters are defined arbitrarily to consist of more than four individual fibers.

cluster investigation — (*ATSDR*) A review of an unusual number, real or perceived, of health events grouped together in time and location, designed to confirm case reports and determine if an unusual disease outbreak has occurred.

cm — See centimeter.

CM — Acronym for corrective measures.

CMB — Acronym for chemical mass balance.

CMHC — Acronym for community mental health center.

CMI — See computer-managed instruction.

CMV — See cytomegalovirus.

CNEL — Acronym for Community Noisy Equivalent Level.

CNG — Acronym for compressed natural gas.

CNP — Acronym for community nurse practitioner.

CNS — See central nervous system.

CO — See carbon monoxide.

CO₂ — See carbon dioxide.

COAD — Acronym for chronic obstructive airway disease.

coagulant — A chemical that causes very fine particles to clump together into larger particles.

coagulant aid — A chemical or substance used to assist or modify coagulation.

coagulation — The combination of small particles into fewer, larger particles; a water treatment process in which chemicals are added to combine with or trap suspended and colloidal particles to form rapidly settling aggregates.

coal — Natural, rock-like, brown to black consolidated peat formed by the decomposition of woody plant debris.

coal conversion — Changing coal into a synthetic gas or liquid fuel.

coal distillation — Separation of components of coal by gradually increasing heat until they gasify at different boiling points.

coal dust — Carbon in amorphous form, used primarily as a fuel and to produce coal gas, water gas, coke, coal tar, synthetic rubber, and fertilizers. Chronic exposure to high concentrations may cause pneumoconiosis and bronchitis and impair the function of the lungs. ACGIH exposure limit (TLV-TWA): 2 mg/m^3 as respirable dust.

coal gasification — A process for converting coal, char, or coke to gas by reaction with hydrogen in the presence of steam, high temperature, and pressure; a process where coal is exposed to molecular oxygen and steam at temperatures of 900°C or higher. Carcinogenic compounds are reduced drastically in this process, but the potential exists for producing carbon monoxide, hydrogen sulfide, and other carbon–nitrogen products.

coal liquefaction — Any of a number of processes for converting coal to partially liquid form by heating with additives other than oxygen; occurs in a temperature range of 400 to 450°C. Some of the compounds found in this process are carcinogenic, such as benz(*a*)anthracene, chrysene, and benzopyrene.

coal oil — An oil that can be obtained by distilling bituminous coal.

coal seam — A mass of coal occurring naturally at a particular location and can be commercially mined.

coal tar — A byproduct obtained from destructive distillation of bituminous coal; see also naphtha.

coal tar color — Articles added or applied to a food, drug, cosmetic, the human body, or any part thereof and capable of imparting color and containing any substance derived from coal tar or any substance so related in its chemical structure to a constituent of coal tar.

coal tar pitch volatiles — The volatile components of dark brown amorphous residue obtained after distillation of coal tar pitch composed of polycyclic aromatics up to 10%, including benzopyrene to about 1.4%. These volatiles are used as binding agents in the manufacture of coal briquettes used for fuel; as dielectrics in the manufacture of battery electrodes; in the manufacture of roofing felts and papers; for protective coatings for pipes for underground conduits and drainage; in road paving and sealing; as coatings on concrete for waterproofing; in the manufacture of refractory brick and carbon ceramic items. They are hazardous to the respiratory system, bladder, kidneys, and skin and are toxic through inhalation and contact. Symptoms of exposure include dermatitis and bronchitis; carcinogenic. OSHA exposure limit (TWA): 0.2 mg/m³ [skin]; ACGIH exposure limit (TLV-TWA): 0.2 mg/m³ as a benzene-soluble fraction.

coal washing — The process of crushing coal and washing out soluble sulfur compounds with water or other solvents.

coalescence — The process by which small water droplets suspended in the air combine and grow into larger droplets.

coarct — The narrowing or constricting of a lumen.

coarse gravel protection — The gravel generally placed in a layer upon a finished surface to protect the finished surface from deterioration or erosion.

coarse texture — (*soils*) The texture exhibited by sands, loam sands, and sandy loams; being visibly crystalline.

coast — A narrow strip of land along the margin of the ocean extending inland for a variable distance from the low water mark.

Coastal Hazardous Waste Site Reviews — A series of documents produced by NOAA scientists describing their evaluations of more than 300 hazardous waste sites that the U.S. EPA has proposed for addition to the National Priority List; uncontrolled sites that could threaten natural resources for which NOAA is the natural resource trustee.

coastal shelf — The shallow region of the oceans surrounding a large land mass.

coastal waters — The waters of the coastal zone except for the Great Lakes and specified ports and harbors on inland rivers.

coastal zone — The lands and waters adjacent to the coast that exert an influence on the uses of the sea and the ecology of the area.

Coastal Zone Color Scanner — The first spacecraft instrument devoted to measurement of ocean color.

Coastal Zone Management Act — A law passed in 1972 and updated through 1986 that establishes effective management, beneficial use, protection, and development of the coastal zone, including

fish, shellfish, other living marine resources, and wildlife in an ecologically fragile area.

coating — Any organic material that is applied to a surface producing a continuous film.

coaxial cable — A cable that consists of electrically conducting material surrounding a central conductor held in place by insulators; used to transmit television and telephone signals of high frequency.

cobalt-60 — A radioactive isotope of the silver-white element cobalt with a mass number of 60 and a half-life of 5.2 years. It emits high energy and is used as a radioisotope in radiotherapy.

cobalt hydrocarbonyl ($HCo(CO)_4$) — A light-yellow gas or liquid; MW: 171.98, BP: 10°C, Sol: dissolves in organic solvents, Fl.P: a flammable gas that explodes when heated in a closed container, density: NA. It is used as a catalyst in organic synthesis. It is hazardous to the respiratory tract and can cause dizziness, giddiness, and headaches. ACGIH exposure limit (TLV-TWA): 0.1 mg/m³ as Co.

cobalt metal, dust, and fume (Co) — An odorless, silver-gray to black metal; MW: 58.9, BP: 5612°F, Sol: insoluble, sp. gr. (metal): 8.92, atomic weight: 27. It is used as a binder in the manufacture of cemented carbide items; during refining and concentration of ores; during manufacture of metal items from magnetic and super-temperature alloys; in the manufacture of dental prosthetics and osteosynthetic items; as pigments. It is hazardous to the respiratory system and skin and is toxic through inhalation, ingestion, and contact. Symptoms of exposure include cough, dyspnea, decreased pulmonary function, weight loss, dermatitis, diffuse nodular fibrosis, respiratory hypersensitivity. OSHA exposure limit (TWA): 0.05 mg/m³ [air].

COBOL — (*computer science*) Acronym for common business-oriented language.

COC — See Cleveland open-cup test.

cocarcinogen — A noncarcinogenic agent that increases the effect of a carcinogen by direct, concurrent, local effect on the tissue.

coccidioidomycosis — An infectious fungal disease caused by the inhalation of spores of the bacterium *Coccidioides immitis* carried on windborne dust particles in hot, dry regions of the southwestern United States and Central and South America. Initial symptoms include a cold or influenza; a secondary infection occurs after several weeks to years with symptoms of low-grade fever, anorexia, weight loss, cyanosis, dyspnea, focal skin lesions, and arthritic pain in the bones and joints.

coccidiosis — A parasitic disease of tropical and subtropical regions caused by the ingestion of oocysts of the protozoa *Isospora belli* or *I. hominis*; symptoms include fever, malaise, abdominal discomfort, and watery diarrhea.

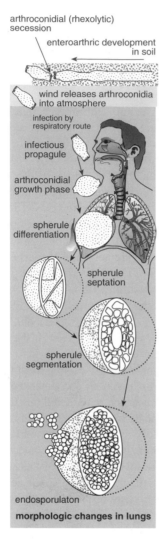

Dimorphic life cycle of *Coccidioides immitis*

coccus — A sphere-shaped bacterium.

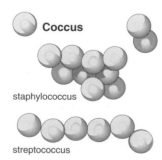

cochlea — The spiral-shaped structure that forms part of the inner ear of mammals and contains the essential organs of hearing.

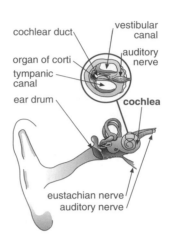

Cochlea

cockroach — See roach.

COD — See chemical oxygen demand.

***Code of Federal Regulations* (CFR)** — The compilation of general and permanent rules published in the *Federal Register* by the executive departments and agencies in the federal government; published annually by the U.S. Government Printing Office.

code (law) — A published body of statutes; a collection of standards and rules of behavior.

coding — The process of organizing information into categories for the purpose of sorting, storing, and retrieving the data.

codon — The nucleotide base triplet encoding a specific amino acid.

coefficient — Any factor in a product; a number placed in front of a formula to balance a chemical equation.

coefficient of entry — The ratio of the actual rate of air blown into an exhaust opening to the theoretical rate calculated by assuming that the negative static pressure in the exhaust opening is completely converted to velocity pressure.

coefficient of expansion — The ratio of increase in volume per degree of increase in temperature at constant pressure.

coefficient of friction — (*physics*) Ratio between the weight of an object being moved and the force pressing the surface together.

coefficient of haze (COH) — A measurement of visibility interference in the atmosphere.

coefficient of linear expansion — The fractional increase in length of a solid per degree rise in temperature at constant pressure.

coefficient of transmissibility — (*groundwater*) The rate of flow of water (gallons per day) at the prevailing water temperature for a vertical strip of aquifer 1 foot wide and extending the full saturated height of the aquifer under a hydraulic gradient of

100%. A hydraulic gradient of 100% reflects a 1-foot drop in head over 1 foot of flow distance.

coefficient of variation (CV) — The ratio of the standard deviation of a distribution to its arithmetic mean.

coefficient of volume expansion — The fractional increase in volume of a substance per degree rise in temperature at constant pressure.

coenocyte — A multinucleate mass of protoplasm resulting from repeated nuclear division unaccompanied by cell fission.

cofactor — A metal ion or inorganic ion with which an enzyme must unite in order to function.

coffee-ground vomitus — A dark-brown vomitus the color and consistency of coffee grounds and composed of gastric juices and old blood; indicates slow upper gastrointestinal bleeding.

cogeneration — The consecutive generation of useful thermal and electric energy from the same fuel source.

cognition — The mental process that includes knowing, thinking, learning, and judging.

cognitive learning — The learning process that is concerned with the acquisition of problem-solving abilities and with intelligence and conscious thought.

COGO — (*geographical information system*) See coordinate geometry.

COH — See coefficient of haze.

coherent waves — Waves whose crests and troughs are synchronized.

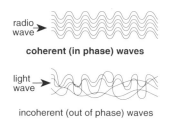

Coherent Waves

cohesion — The tendency of like molecules of a substance to hold together as a result of intermolecular attractive force; the act or state of sticking together tightly.

cohesive — Exhibiting cohesion or coherence.

cohesive soil — A predominantly clay and silt soil with fine-grain particles that sticks together whether wet or dry.

COHN — Acronym for Certified Occupational Health Nurse.

cohort — (*statistics*) The entire group of people who shared a similar experience or acquired characteristics in a defined period of time.

cohort effect — Systematic differences among two cohorts with respect to the distribution of some variable (e.g., height, weight at any specified attained age).

cohort study — A type of analytical epidemiologic study in which a cohort of unaffected persons is followed over time in order to investigate the incidence of a disease or health-related event in relation to a characteristic measured when the cohort was defined.

coke — Bituminous coal from which the volatile components have been driven by heat, leaving fixed carbon and ash fused together.

coke oven — An oven in which coal is converted to coke by destructive distillation.

coke-oven chemicals — Organic compounds derived from bituminous coal during its conversion to metallurgical coke; they are a major source of raw materials for a number of chemicals.

coke-oven emissions — Hazardous air pollutants emitted by coke ovens, including substantial quantities of carbon monoxide and all the byproducts of cleaning, grading, and combustion of coal. Polycyclic aromatic hydrocarbons may occur from smoke escaping when the ovens are filled or from leaks or from blow-off. Respiratory cancer and skin cancers have been found in individuals who have been exposed to coke-oven emissions.

cold — A condition of low temperature; the absence of heat.

COLD — Acronym for chronic obstructive lung disease.

cold air mass — A large, turbulent mass of unstable air with gusty winds; produces good surface visibility and the possibility of thundershowers.

cold front — An advancing edge of a cold air mass.

Cold Front

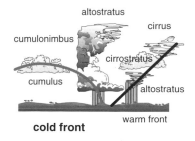

cold front

cold sterilization — A process conducted in a closed vessel using ethylene oxide gas for 2 hours to kill all bacteria and spores; also known as chemical sterilization.

cold vapor — A method of testing water for the presence of mercury.

cold welding — A solid-state welding process in which pressure is used at room temperature to produce coalescence of metals with substantial deformation at the weld.

colic — Acute abdominal pain usually caused by smooth muscle spasm, obstruction, or twisting.

coliform — Pertaining to fermentative Gram-negative enteric bacilli, sometimes restricted to those fermenting lactose (i.e., *Escherichia*, *Klebsiella*, *Enterobacter*, and *Citrobacter*).

coliform bacteria — A group of bacteria inhabiting the intestines of animals, including people; also found elsewhere. It includes all the aerobic, nonspore-forming, rod-shaped bacteria that produce from lactose, fermentation gas within 48 hours at 37°C; used to determine levels of pollution in water, soil, or milk.

coliform-group bacteria — A group of bacteria predominantly inhabiting the intestines of people or animals (but occasionally also found elsewhere) that may contaminate potable water.

coliform index — A rating of the purity of water based on a count of fecal bacteria.

coliform test — A bacteriological test involving the taking of a sample, inoculating it into a tube, incubating it, and measuring the number of samples that contain gas production. When run on raw milk, the test may indicate problems in the initial production of the raw milk; when run on pasteurized milk, the test may indicate problems in the pasteurization plant.

colitis — Inflammation of the colon.

collagen — Fibrous proteins found in vertebrates as the chief constituent of connective tissue.

collapse — An abnormal condition characterized by shock.

collateral — Secondary or accessory.

collection — The process of acquiring data from an instrument, instrument system, or external source.

collection efficiency — The amount of substance absorbed or detected divided by the amount sampled.

collection frequency — The number of times collection is performed in a given period of time.

collection stop — A stop made by a vehicle and crew to collect solid waste at one or more service sites.

collective effective dose equivalent — A measure of health risks to a population exposed to radiation; the sum of the effective dose equivalents of all individuals within an exposed population within 80 km from an environmental release point.

collective effective dose — The quantity obtained by multiplying the average effective dose by the number of people exposed to a given source of ionizing radiation.

collector — (*waste*) A public or private hauler that collects nonhazardous waste and recyclable materials from residential, commercial, institutional, and industrial sources.

collector sewers — The pipes used to collect and carry wastewater from individual sources to an interceptor sewer and on to a treatment facility.

collet — See cullet.

collimation — The process of restricting the useful beam of radiation to a predetermined cross-sectional area by attaching beam-limiting devices to the source of radiation.

collimator — A device that reduces the scatter of x-radiation while x-ray pictures are being taken of the body structures.

colloid — A chemical system composed of a continuous medium, such as protoplasm or albumen, throughout which are distributed small particles (1 to 1000 nm in size) that do not settle out under the influence of gravity.

colloid fraction — Organic and inorganic matter having a very small particle size and a large surface area per unit of mass.

colloidal state — A condition obtained by dispersing submicroscopic particles of gases, liquids, or solids through a second substance in a medium that may be gaseous, liquid, or solid; a gas dispersed in a gas is not colloidal.

colloidal suspension — A method of sediment transport in which water turbulence supports the weight of the sediment particles, keeping them from settling out or being deposited.

colluvial — (*soils*) Of or pertaining to material that has moved downhill by the force of gravity or action of frost and local wash and has accumulated on lower slopes or at the bottom of the hill.

colluvium — (*soils*) Material deposited in footslope positions by the action of gravity, soil creep, or local wash; frequently silty and loamy material.

colon — The portion of the human intestine exclusive of the cecum and the rectum; it is divided into four sections: ascending, transverse, descending, and sigmoid.

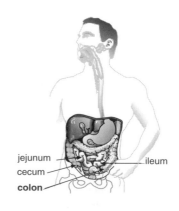

jejunum
cecum
ileum
colon

Colon

colonization — (*infection control*) The presence of methicillin-resistant *Staphylococcus aureus* on body tissue without symptoms or clinical manifestations of illness or infection.

colonoscope — A long instrument with a light and lens that allows examination of the interior of the colon.

colony counter — A piece of equipment used for counting colonies of bacteria growing in a culture and usually consisting of an illuminated transparent plate divided into sections.

colony-forming unit (CFU) — A unit of measurement for viable bacteria numbers.

color — A general term that refers to the property of light that is determined by its wavelength with particular reference to its visual appearance.

color additive — A dye pigment or other substance made by a process of synthesis or similar artifice or is extracted, isolated, or otherwise derived with or without intermediate or final change of identity from a vegetable, animal, mineral, or other source and, when added or applied to a food, drug, cosmetic, or to the human body, is capable of imparting color.

Colorado tick fever — (*disease*) An acute, febrile, two-phase disease where the initial fever subsides and a second fever occurs later, lasting 2 to 3 days; incubation time is 4 to 5 days. It is caused by viruses found in western Canada, Washington, Oregon, Idaho, California, and other western states. The reservoir of infection is small animals and ticks, primarily *Dermacentor andersoni*. It is transmitted by the bite of the tick and not person to person; general susceptibility. It is controlled by use of tick repellents and proper clothing in areas where ticks are prevalent.

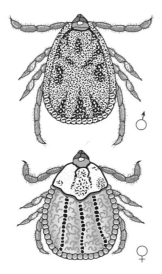

Dermacentor andersoni
(Rocky Mountain wood tick)
a vector for
Colorado Tick Fever

colorimeter — (*chemistry*) A device used in chemical analysis for comparing the color of a liquid with a standard color.

colorimetric analysis — A method of determining the concentration of a chemical substance by measuring the intensity of its color or of a color produced by it.

colostrum — Reddish-yellow, strongly odored, bitter-tasting milk with an increased number of proteins and decreased lactose; it is essential for calves' health, as it contains many antibodies. It should not be present in market milk.

column chromatography — A process of separating and analyzing a group of substances according to the differences in their absorption affinities for a given absorbent as shown by pigments deposited during filtration for the same absorbent contained in a glass cylinder or tube.

column ozone — The total amount of ozone that is found in a column of air, with the majority of it being typically in the stratosphere.

columnar soil — (*soils*) A soil structure with rounded caps, prism-like with the vertical axis greater than the horizontal, usually found in subsoil or the B horizon.

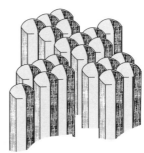

Columnar Soil

coma — A state of unconsciousness from which a patient cannot be aroused even by powerful stimuli.

comatose — Being in a state of abnormal deep sleep caused by illness or injury.

combined available residual chlorine — The concentration of residual chlorine that is combined with ammonia and/or organic nitrogen in water as a chloramine yet is still able to oxidize organic matter and utilize its bactericidal properties.

combined carbon dioxide — Portion of the total carbon dioxide that is contained in blood carbonate and can be calculated as a difference between the total and dissolved carbon dioxide.

combined chlorine — Chlorine that is available as a bactericide in water but which is combined with another substance, usually ammonia; it is less effective against bacteria.

combined oxygen — The oxygen that is physically bound to hemoglobin as oxyhemoglobin.

combined residual chlorination — The application of chlorine to water to produce a combined available residual chlorine that may be made up of monochloramines, dichloramines, and nitrogen trichloride.

combined sewer — A sewer receiving both surface-water runoff and sewage from homes, businesses, and/or industry.

combustible — Possessing the ability to catch fire and burn; material that can be ignited at a specific temperature in the presence of air to release heat energy; a term used by the National Fire Protection Association, Department of Transportation, and others to classify certain liquids that will burn on the basis of flashpoints.

combustible dust — Dust capable of undergoing combustion or of burning when subjected to a source of ignition.

combustible gas indicator (CGI) — An instrument used to determine the level of combustible gases present in a given environment before they reach their lowest explosive level.

combustible liquid — Any liquid having a flashpoint at or above 100°F but below 200°F, except any mixture having components with flashpoints of 200°F or higher.

combustion — The production of heat and light energy through a chemical process, usually oxidation of the gas, liquid, or solid fuel; (*air analysis*) a technique of analysis of combustible gases and vapors in which they are passed over a filament that is part of a Wheatstone bridge circuit; combustion of the chemicals alters the resistance of the filament, causing an imbalance in the circuit that is then used as a measure of concentration.

combustion air — The air used for burning the fuel.

combustion gases — The mixture of gases and vapors produced by combustion.

combustion product — Material produced or generated during the burning or oxidation of a material.

come-down time — (*milk*) The time that elapses between the processing or pasteurization temperature and the temperature required to cool the product.

come-up time — (*milk*) The time that elapses between the initial temperature of a product and reaching the temperature required to process or pasteurize the product.

cometabolism — The biodegradation of a pollutant by an organism using some other compound or compounds for growth and energy.

comfort conditioning — The provision of a environment comfortable for the majority of the occupants with a relative humidity range between 20 and 55% at normal confort temperatures.

comfort ventilation — Airflow intended to remove heat, odor, and cigarette smoke from an enclosure.

comfort zone — The temperature, humidity, wind velocity, and solar radiation for which a person is dressed and can still perform a task without discomfort.

command — (*computer science*) An instruction sent from the keyboard or other control device to execute a computer operation.

command language — (*computer science*) An English-like language for sending commands for complicated program sequences to the computer.

command post — (*emergencies*) A facility at a safe distance upwind from an accident site where the on-scene coordinator, responders, and technical representatives can make response decisions, deploy people and equipment, maintain liaison with the media, and handle communications.

commensal — Relating to or living in a state of commensalism.

commensalism — A relationship between two kinds of organisms in which one obtains food or other benefits and the other is neither damaged nor benefited.

commercial grade — A grade of chemical that is less than the purest available.

commercial processor — Any person engaged in commercial or custom processing of food.

commercial solid waste — Solid waste generated by stores, offices, or activities that do not actually produce a product.

commercial sterility — (*food*) The condition achieved by application of heat that renders food free of viable forms of microorganisms of public health significance; spores can exist during the commercial sterilization process.

commercial waste — A combination of garbage, refuse, ashes, demolition waste, urban renewal waste, construction waste, and remodeling waste.

commingled shipment — Two or more separate shipments of the same goods to one common location and bearing no distinguishing marks.

comminuted — (*food*) Reduced in size by chopping, flaking, grinding, or mincing; the food may be restructured or reformed, as in sausage or gyros.

comminution — See pulverization.

comminutor — A device for catching and shredding heavy solid matter.

commissary — A place where food, containers, or supplies are stored, prepared, or packaged for transit to, and sale or service at, other locations.

committed effective dose equivalent — The total effective dose equivalent received over a 50-year period following the internal deposition of a radionuclide; expressed in rems or sieverts.

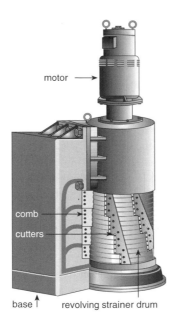

Comminutor

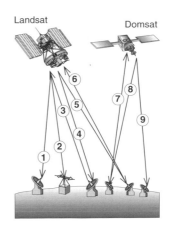

1 S-band (MSS/RBV data)
2 UHF (DCS data)
3 S-band (commands)
4 S-band/MSS/DCS/RBV
5 S-band/MSS/RBV/DCS
6 S-band VHF (commands)
7 Process MSS/RBV data
8 Video MSS/RBV data
9 Video MSS/RBV data

(Lansat) Communication
and Processing System

common business-oriented language — (*computer science*) A high-level compiler computer language used for programming.

common dining area — A central location in a group residence where people gather to eat.

common exposure route — A likely way by which a pesticide may reach and/or enter an organism; usually through oral, contact, or respiratory routes.

common law — A body of unwritten principles based on custom and the precedent of previous legal decisions.

common-mode rejection ratio — The difference in signal gain divided by the common-mode signal gain.

common name — Any identification such as code name, code number, trade name, brand name, or generic name, other than the chemical name, used to identify a chemical.

communicable disease — An illness due to an infectious agent or its toxic products being transmitted directly or indirectly to a well individual from an infected individual or animal or being transmitted through an intermediate animal host, vector, or inanimate environment.

Communicable Disease Center — Previous name for the Centers for Disease Control and Prevention.

communicable period — The time or times during which an etiologic agent may be transferred from an infected individual or animal to a non-infected individual.

communication — A process in which a message containing information is transferred from a person or persons.

communication means — A gesture, action, sound, written word, or visual image used to transmit a message.

communication system — Any combination of devices permitting the passage or exchange of messages.

communication theory — A system of communication consists of the sender, means of transmission, receiver, potential source of interference, purpose of the message, and final action taken by the receiver based on the nature of the communication.

communications superhighway — An electronic information system combining video, telephones, and computers in which all audio and video communications can be translated into digital information and stored and compressed to travel through existing telephone and cable lines; fiberoptic wiring provides the large capacity required for transmission of information over long distances, and coaxial cable wire carries this information effectively for about 1/4 mile to the user, thus eliminating the need to rewire. Also known as the electronic superhighway and the information superhighway.

community — A group of people who live in a designated geographical area and who share common interests or bonds.

community acquired infection — An infection spread through the community in contrast to one spread through a hospital or other healthcare setting where the close contact of individuals and use of certain medications may increase the problems.

Community Assistance Panel — A group of people utilized to facilitate constructive communication between a community and the ATSDR.

Community-Based Environmental Protection — A program that fosters implementation of integrated geographic approaches to environmental protection with an emphasis on ecological integrity, economic sustainability, and quality of life.

community-based residential facility — Any building or buildings, sections of a building, or distinct part of a building or other place, operated for profit or not, that undertakes through its ownership or management to provide (for a period exceeding 24-hour housing) food service and one or more personal services for persons not related to the owner or administrator by blood or marriage.

Community Health Investigation — (ATSDR) A medical or epidemiological evaluation of descriptive health information about individual persons or population of persons to evaluate and determine health concerns and to assess the likelihood that they may be linked to exposure to hazardous substances.

community health nursing — A combination of primary health care and nursing practice with public health nursing for the purpose of preventing, curing, and rehabilitating people who have various diseases and injuries.

community medicine — A branch of medicine with emphasis on early diagnosis of disease, recognition of environmental and occupational hazards, and prevention of disease in a community.

Community Noisy Equivalent Level — A 24-hour energy equivalent level derived from a variety of single noise events, with weighting factors of 5 dBA applied in the evening from 7 p.m. to 10 p.m. and 10 dBA applied to the nighttime from 10 p.m. to 7 a.m. to allow for the greater sensitivity to noise during these hours.

community right-to-know — The public's accessibility to information about toxic pollution.

Community Right-to-Know Act — See Emergency Planning and Community Right-to-Know Act.

community water system — A public water supply that serves at least 15 connections of 25 individuals on a regular basis.

compact disc — An optical disc on which computer data are recorded.

Compact Disc–Read Only Memory (CD-ROM) — A record-like storage medium that uses digital and optical laser technology to store about 600 to 800 Mb of text, images, and sound on a single disc.

compact fluorescent — A fluorescent light bulb small enough to fit into a standard light socket; more energy-efficient than standard incandescent bulbs.

compact fluorescent lamp — A small fluorescent lamp used as a more efficient alternative to incandescent lighting; also called PL, CFL, Twin-Tube, or BIAX lamps.

compaction — The consolidation and reduction in size of solid particles by rolling, tamping, or other means of applying mechanical pressure.

compaction curve — The curve showing the relationship between the dry density and water content of a soil for a given compacted effort.

compactor — A vehicle with an enclosed body containing mechanical devices that convey solid waste into the main compartment of the body and compress it; a machine designed to consolidate earth and paving materials.

comparative risk assessment — A process that generally uses the judgment of experts to predict effects and set priorities among a wide range of environmental problems.

comparison values (ATSDR) — The estimated contaminant concentrations in specific media that are not likely to cause adverse health effects, given a standard daily ingestion rate and standard body weight.

compatible pesticides — Two or more pesticide chemicals that can be safely mixed together without reducing their effectiveness or damaging the product, plant, or animal treated.

compensable injury — Accidental and not self-inflicted injury arising both out of and during the course of employment.

compensation point — The condition in a living system in which the uptake of carbon dioxide equals the release of carbon dioxide (i.e., photosynthesis equals respiration).

compensation — (*engineering*) Addition of specific materials or equipment to counteract a known error.

compiler — (*computer science*) A computer program that translates a high-level programming language into machine-readable code.

complainant — An individual registering a complaint, such as a consumer complaining about an article of merchandise or environmental situation.

complaint for forfeiture — A document furnished to the U.S. Attorney's Office for filing with the clerk of the court to initiate the seizure of a lot; formerly known as libel of information.

complete blood count (CBC) — A determination of the number of blood cells in a given blood sample expressed as the number of cells (red, white, platelet) in a cubic millimeter of blood.

complete fracture — A bone break that disrupts the continuity of osseous tissue across the entire width of the bone.

complete health history — A health history that includes a history of the current illness, past health issues, social history, occupational history, sexual history, and family health history.

complete metamorphosis — The four stages of development of certain insects: the egg, larva, pupa, and adult.

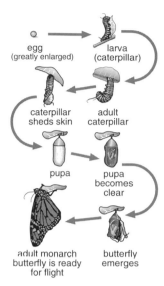

Complete Metamorphosis
(*Danaus plexippus*)

complete response — The total disappearance of a tumor.

completion — Sealing off of access by undesirable water to a well bore by proper casing and/or cementing procedures.

complex conductivity — A property of material that gives the ratio between current density and electric field and includes the phase difference between the two field quantities.

complex fracture — A closed fracture in which the soft tissues surrounding the bone is severely damaged.

complexing — The use of chelating or sequestering agents to form relatively loose chemical bonding as a means of treating certain pollutants such as nickel, copper, and cobalt.

compliance — A legally enforceable action taken to correct pollutant problems and meet the requirements of the law.

compliance monitoring — The type of monitoring done to ensure the meeting of immediate statutory requirements, the control of long-term water quality, the quality of receiving waters, or the maintenance of standards during any construction project.

component — A significant part of a larger unit.

composite map — A single map created by joining together several digitized maps.

composite power value — (*energy*) The value of power that includes both a capacity and energy component, usually expressed in mils per kilowatt-hour.

composite sample — A collection of individual samples obtained at regular intervals, usually every 1 or 2 hours over a 24-hour time span, forming a representative sample and analyzed to determine the average conditions during the sampling period; a series of water samples taken over a given period of time and weighted by flow rate.

compost — A mixture of decaying organic matter used for fertilizing and conditioning soil.

composting — A controlled process of degrading organic matter by microorganisms.

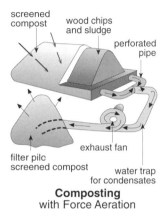

Composting
with Force Aeration

compound — (*chemistry*) A substance made up of two or more elements in union. The elements are united chemically, which means that each of the original elements loses its individual characteristics once it has combined with the other element or elements and cannot be separated by physical means. They combine in definite proportions by weight.

compound fracture — A fracture in which the broken end or ends of the bone have torn through the skin.

compound microscope — A microscope with two or more simple or complex lens systems.

comprehensive care facility — A licensed healthcare facility providing room, board, laundry, and administration of medication under supervision and responsibility of an attending medical staff.

Comprehensive Environmental Response Compensation and Liability Act (CERCLA) — A federal act passed in 1980 requiring the EPA to designate hazardous substances that can present substantial danger to the public, authorizes the cleanup of sites contaminated with these substances, and gives the federal government the power to respond to release or threatened release of any hazardous substance against the environment, as well as the release of any contaminant that may present imminent and substantial danger to public health or welfare. Also known as the Superfund Act.

Comprehensive Environmental Response, Compensation, and Liability Act Information System — A U.S. EPA database of information about Superfund sites intended for use by EPA employees to manage the Superfund program.

comprehensive facility review — (*engineering*) A review performed at a high-level or significant hazard facility every 6 years, including a field examination and a state-of-the-art review of the design assumptions of the structure, construction practices, and integrity under various loading conditions.

Comprehensive Health Planning and Public Health Services Amendments — A law passed by Congress in 1966 that emphasized regional planning and established for the first time the concept that each person had a right to health care.

Comprehensive Soil Classification System — A classification system commonly used by North American soil scientists based on the chemical and physical characteristics of a soil.

compressed air nebulizer — A device used to produce aerosols as from liquids.

compressed gas — A gas or mixture of gases having in a container an absolute pressure exceeding 40 psi at 70°F of gas; a mixture of gases having in a container an absolute pressure exceeding 104 psi at 130°F regardless of the pressure; a liquid having a vapor pressure exceeding 40 psi at 100°F.

compressibility — The property of a soil describing its susceptibility to decrease in volume when subjected to load.

compressible flow — A flow of high-pressure gas or air that undergoes a pressure drop, causing a significant reduction in density.

compression — Squeezing material from opposite directions; (*computer science*) shrinking digital image and video information to achieve smaller file sizes, making it possible to transmit lengthy information.

compression fracture — A bone break, especially in a short bone, that disrupts osseous tissue and causes the affected bone to collapse.

compromised host — A person who is less able to resist infection because of immunosuppressive therapy, immunological defects, severe anemia, concurrent diseases, or conditions such as AIDS, severe malnutrition, cancer, or specific medications.

Compton effect — An attenuation process observed for x- or gamma-radiation in which an incident photon interacts with an orbital electron of an atom to produce a recoil electron and a scattered photon of energy less than the incident photon.

computed tomography (CT) — Diagnostic, radiological technique that permits physicians to view cross-

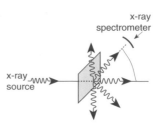

The Compton Effect

sectional images representing 1-cm-thick slices of internal body structures.

computer — A piece of electronic equipment used to process and store large amounts of information rapidly.

computer-assisted instruction — Teaching process using a computer and presentation of instructional materials that requires students to interact with the material.

computer graphics — A general term including any computer activity that results in a graphic image.

computer-integrated manufacturing — Relying upon a high level of automation in a manufacturing enterprise.

computer-managed instruction — A system in which a computer is used to manage several or many aspects of instruction, including learning assessment, pre-tests and post-tests, design and preparation of learning materials, and calculation, analysis, and storage of student scores.

computer operating system — A program that controls all the other parts of a computer, both the hardware and software.

computer program — A schedule or plan that specifies actions that may or may not be taken; expressed in the form of a set of instructions suitable for execution by computer.

computerized axial tomography (CAT) — Commonly called CAT scan; see computed tomography.

conc — See concentration.

concave lens — A lens with an inward curve that diverges parallel light rays.

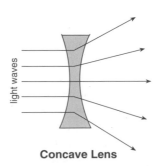

Concave Lens

concentrate — (*pesticides*) A pesticide chemical before dilution.

concentrated animal feeding operation — A facility where large numbers of farm animals are confined, fed, and raised, such as dairy and beef cattle feedlots, hog production facilities, and closed poultry houses. Individual operations that are potential sources of water pollution must obtain point-source discharge permits that specify the allowable levels of effluent from each of these places.

concentrated milk — An unsterilized and unsweetened fluid product resulting from removal of a large portion of water from milk.

concentration — The amount of active ingredient or pesticide equivalent in a quantity of diluent; expressed as pounds per gallon, milliliters per liter, etc.; the mass or volume of a substance present in a unit volume of a gas, solid, or liquid.

concentration factor — The amount of a compound accumulated in the tissue of an organism; calculated by dividing the concentration of the compound in the tissue by the concentration ingested in the diet or taken up from the surrounding medium.

concentration gradient — A gradient that exists across a membrane separating a high concentration of a special ion from a low concentration of the same ion; the difference in concentration in two parts of a system.

concentration threshold — The pollutant concentration below which no receptor experiences an ill effect.

concept — An abstract idea or thought that originates and is held within the mind.

concurrent disinfection — The disinfection of equipment and disposal of contaminated material that goes on routinely while a process or procedure is being carried out.

concurrent engineering — A program incorporating all the stages from research to sales where decisions have to be made jointly and may affect work at any of the other stages.

concurrent infection — A situation in which a person has two infections at the same time.

concussion — A cerebral trauma causing neuronal dysfunction and loss of consciousness, with no visible sign of brain tissue injury but a rise in intracranial pressure; (*explosives*) a shock or sharp airwave caused by an explosion or heavy blow.

condemned — (*food*) Food or any food product that has been determined by inspection or analysis to be unsound, unhealthful, unwholesome, or otherwise unfit for human consumption; (*housing*) a structure that has been determined by inspection and analysis to be unfit for human habitation.

condensate — A term used to describe light liquid hydrocarbons separated from crude oil after production and sold separately; liquid obtained by condensation of a gas or vapor, such as steam.

condensation — Formation of water upon aerosol particles acting as nuclei and causing the water temperature to pass through its dew point; a reduction to a denser form as from steam to water.

condensation nucleus — A tiny particle suspended in air upon which the condensation process begins.

condensed milk — Evaporated milk with sugar added.

condenser — A device in which the temperature of a gas is lowered to its condensation point, thereby converting it from the vapor to liquid stage.

condiment — Any food, such as ketchup, mayonnaise, mustard, and relish, that is used to enhance the flavor of other foods.

condition — The physical and mental health or well-being of an individual.

conditional maintenance — preventive maintenance that depends upon a predetermined schedule when a state of reduced function of the equipment may occur.

conditioned air — Air that has been heated, cooled, humidified, or dehumidified to maintain an interior space within the comfort zone; also known as tempered air.

conditioning — Pretreatment of a sludge to facilitate the removal of water in a thickening or dewatering process.

conditions to avoid — (*chemicals*) Conditions encountered during handling or storage that could cause the substance to become unstable.

conductance — The ratio of current to voltage in a circuit that absorbs but does not store electrical energy; a measure of the ability of a circuit to conduct electricity. Conductance is the reciprocal of resistance. A quick method for estimating the dissolved solids content of a water supply is to determine the capacity of the water sample to carry an electrical current.

conduction — The transfer of heat or electrical charge by physical contact between the molecules.

conductive flooring — Floor material that has the ability to allow electric energy to flow through it with minimum resistance, ranging between 25,000 and 1,000,000 ohms.

conductive hearing loss — A type of hearing loss due to any disorder in the middle or external ear that prevents sound from reaching the inner ear; not caused by noise.

conductivity — The capacity for conduction.

conductor — An object or substance that allows a current of electricity to pass continuously, such as a wire or combination of wires suitable for carrying current.

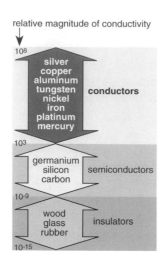

Conductor

conduit — A closed channel to convey water through, around, or under an area.

cone of depression — A natural depression in the water table around a well during pumping.

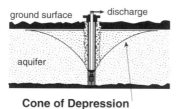

Cone of Depression

cone of influence — A depression, roughly conical in shape, produced in a water table by the pumping of water from a well.

cone penterometer testing — A direct push system used to measure lithology based on soil-penetrating resistance; the sensors in the tip of the cone of the direct push rod measure tip resistance and side-wall friction, transmitting electrical signals to digital processing equipment on the ground surface (see direct push).

confidence interval — (*epidemiology*) An interval that has the probability of including a given population being studied.

confidence limit — (*epidemiology*) Upper and lower limits of a confidence interval.

confidentiality — The nondisclosure of specific information to another person unless the individual is authorized to receive it.

configuration — The three-dimensional, spatial arrangement of electrons in atomic orbitals in an atom; (*computer science*) a combination of computer hardware and software used for a specific task.

confined aquifer — A geological formation in which water-bearing material is located between two aquitards and therefore is well protected against contamination percolating from the surface.

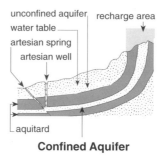

Confined Aquifer

confined areas — Rooms, buildings, and greenhouses with limited or inadequate ventilation.

confined space — Any area not intended or designated for continuous human occupancy that has limited openings for entry, exit, or ventilation or would otherwise make escape difficult in an emergency.

confluent growth — In coliform testing, a continuous bacterial growth covering the entire filtration area of the membrane filter, or a portion thereof, in which bacterial colonies are not discrete, thus making accurate measurement difficult.

conformal projection — (*geographical information system*) A projection that preserves the original shape of an area of interest but not the area or distance.

confounder — A factor, such as age, known to be associated with a health event; for example, age is not taken into account when determining if excess weight may be a cause of a heart attack.

confounding factors — Variables that can affect the incidence or degree of a parameter being measured.

Confused Flower Beetle (*Tribolium confusum*) — An elongated, reddish-brown, flightless beetle about 1/7-inch long with a distinct joint between the thorax and abdomen; its antennae gradually enlarge toward the tip. It is a stored food product insect.

confusion — A mental state of a person who is disoriented regarding time, place, or person, thus causing an inability to choose or act decisively and perform the activities of daily living.

congener — One of two or more things that are similar or closely related in structure, function, or origin.

congenital — Existing at or before birth.

congenital defect — An abnormality present at birth; it may be genetically determined or may result from some environmental insult during pregnancy.

congenital immunity — The immunity a child has at birth that has been acquired from the mother's antibodies that have passed through the placenta.

congestion — An abnormal accumulation of fluid in an organ or body area.

congestive heart failure — The failure of the heart to maintain adequate blood circulation.

conglomerate — A sedimentary rock composed of rounded stones including pebbles, cobbles, and boulders cemented together and usually found with sandstone.

conical burner — A hollow, cone-shaped combustion chamber that has an exhaust vent at its point and a space through which waste materials are charged; air is delivered to the burning solid waste inside the cone. Also known as a teepee burner.

conifers — Trees and shrubs with needle-like leaves, such as pine, cedar, spruce, and hemlock.

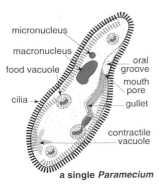

a single *Paramecium*

Eastern White Pine

Conifer

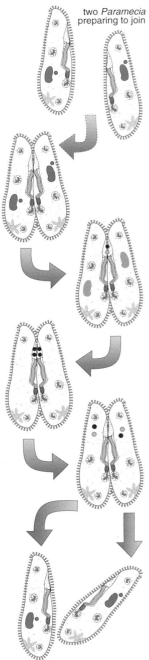

Conjugation

conjugation — A form of sexual reproduction in unicellular organisms in which two individuals join in temporary union to transfer genetic material; (*biochemistry*) the joining of a toxic substance with a natural substance of the body to form a detoxified product that can be eliminated.

conjunctiva — Delicate membrane lining the eyelids and covering the eyeball.

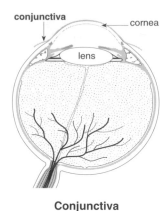

Conjunctiva

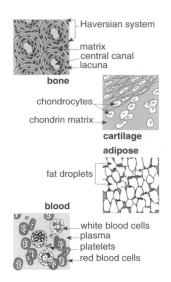

Connective Tissue

conjunctival burns — Chemical burns of the conjunctiva, which must be treated immediately.

conjunctive use — The coordinated use of surface water and groundwater resources.

conjunctivitis — An inflammation of the mucous membrane that lines the inner surface of the eyelid and the exposed surface of the eyeball; may be caused by bacteria, virus, allergic responses, or chemical or physical factors.

connate water — Highly mineralized water that has been trapped in igneous rock formations at the time the rocks were formed.

connection — (*energy*) Physical connection between two electric systems permitting the transfer of electric energy in one or both directions; the point where a rotating shaft provides energy to water and whirls it around, building up enough pressure to force the water through a discharge outlet.

connective tissue — Tissue that lies between groups of nerve, gland, and muscle cells and beneath epithelial cells; also includes bone, cartilage, blood, and lymph.

conscious — The ability to respond to sensory stimuli.

consciousness — A state of awareness of self and the environment.

consent decree of condemnation — A court order agreed to by a claimant in a seizure in which the individual admits that the article seized is in violation, as alleged in the complaint for forfeiture, and agrees to condemnation and either destruction or reconditioning as well as payment of costs. It commits the claimant to providing a bond and reconditioning the article under supervision of the Food and Drug Administration.

consequence management — (*FEMA*) An emergency management function used to protect public health and safety, restore essential government services, and provide emergency relief to governments, businesses, and individuals affected by the results of terrorism.

conservation of energy — The principle that the energy of the universe is constant and cannot be created or destroyed; no violations of this principle are known. Also known as law of conservation of energy and first law of thermodynamics.

conservation of mass–energy — The theory that energy and mass are interchangeable in accordance with Einstein's equation ($E = mc^2$), where E is energy, m is mass, and c is the velocity of light.

conservation of matter — The notion that matter can be neither created nor destroyed or that weight remains constant in an ordinary chemical change; theory has been violated by microscopic phenomena. Also known as law of conservation of matter.

conservation — Increasing the efficiency of energy use, water use, production, or distribution; the use, protection, preservation, and renewal of human and natural resources to ensure the highest economic or social benefits.

consistence — (*soils*) Resistance of a material to deformation or rupture; degree of cohesion or adhesion of a soil mass.

console — Main operating unit in which indicators and general controls of a radar or electronic group are installed; (*computer science*) a device that allows the operator to communicate with the computer.

consolidation grouting — Strengthening an area of ground by injecting grout.

consolidation — Reduction in particle spacing in a soil or solid waste and decrease in water content resulting from an increase in external pressure.

consortism — See symbiosis.

constancy — An absence of variation in quality.

constant air volume system — An air-handling system that provides a constant air flow of varying temperature to meet the heating and cooling needs of a specific area.

constant-flow sampler — A pump that overcomes variations in the rate of flow of air by use of sophisticated flow-rate sensors with feedback mechanisms permitting the maintenance of the preset flow rate during sampling.

constipation — Infrequent and difficult evacuation of feces that results in discomfort.

constriction — An abnormal closing or reduction in the size of an opening or passage of the body.

construction and demolition waste — Building materials and rubble resulting from construction, remodeling, repair, and demolition operations.

construction completion list — A list developed by the U.S. EPA to simplify its system of categorizing sites and better communicate about the successful completion of cleanup activities.

consultant — A person who by training and experience has developed a special knowledge in a subject area and has been recognized by a peer group.

consumer — (*ecology*) A plant or animal that cannot derive its energy from inorganic matter but must depend upon other plants or animals for its energy for living; also known as a heterotroph.

Consumer Price Index — A measure of the average amount paid for a market basket of goods and services by a typical U.S. consumer in comparison to the average paid for the same basket in an earlier base year.

Consumer Product Safety Commission Act (CPSCA) — A law passed in 1970 and updated in 1984 that established the Consumer Product Safety Commission as an independent regulatory agency, giving the commission the power to regulate consumer products and oppose unreasonable risk of injury or illness; to regulate consumer products except for foods, drugs, pesticides, tobacco, tobacco products, motor vehicles, aircraft, aircraft equipment, boats, and boat accessories regulated by other federal agencies; and to publish consumer product safety standards to reduce the level of unreasonable risk.

consumer protection legislation — Laws aimed at protecting consumers by ensuring that they have complete information about items they are considering buying and an understanding of potential hazards presented by consumer products.

consumptive water use — The total amount of water used by vegetation, activities of people, and evaporation of surface water; water removed from available supplies without direct return to a water resource system for such uses as manufacturing, agriculture, and food preparation.

contact aeration process — A secondary treatment process for sewage that depends on aerobic biological organisms breaking down the putrescible organic materials in sewage to simpler and more stable forms.

contact condensers — A device in which a vapor is condensed by being forced to give up its latent heat to a cooling liquid brought into contact with the vapor.

contact cooling water — The wastewater that has been in contact with machinery or equipment and is now contaminated.

contact dermatitis — A skin rash marked by itching, swelling, blistering, oozing, or scaling; caused by direct contact between the skin and a substance to which the individual is allergic or sensitive.

contact herbicide — (*pesticides*) A compound that kills primarily by contact with plant tissue rather than as a result of translocation.

contact load — Sediment particles that roll or slide along in almost continuous contact with a stream bed.

contact pesticide — A pesticide chemical that controls or destroys a pest that touches a treated surface or is touched by the pesticide.

contact poison — A poison that affects target organisms by physical contact rather than through ingestion or inhalation.

contact stabilization — A modification of the conventional activated sludge process in which two aeration tanks are used with one tank for separate aeration of the return sludge for at least 4 hours before flowing into the second one, where the sludge is mixed with primary effluent requiring treatment.

contactor — An electrical switch, usually magnetically operated.

contagion — The transmission of a disease by direct contact with the person who has the disease or by indirect contact through handling of contaminated inanimate objects.

contagious — Capable of being transmitted from one individual to another.

container — (*chemicals*) Any bag, barrel, box, bottle, can, cylinder, drum, reaction vessel, or storage tank that contains a hazardous chemical; any portable device in which material is stored, transported, treated, disposed of, or otherwise handled.

containment — The confinement of radioactive material in such a way that it is prevented from being dispersed into the environment or is released only at a specified rate.

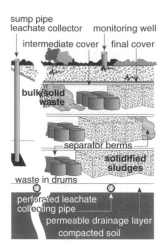

Containment
(Hazardous Waste Landfill)

contaminant — The agent of contamination.

contamination — (*radiation*) Deposition of radioactive material in any place where it is not desired, especially where it may be harmful or constitute a hazard; (*microbiology*) introduction of pathogenic organisms in or on the surface of an inanimate object.

contempt of court — Any act that is calculated to embarrass, hinder, or obstruct a court in the administration of justice.

contest — (*law*) To make defense to an adverse claim in a court of law; to impose, resist, or dispute the case made by a plaintiff.

contiguous — Actual contact with an object or surface or being near or adjacent to it.

continental drift — The slow movement of land masses caused by movement of molten rock underneath the Earth's crust.

continental shelf — Relatively shallow ocean floor bordering a continental land mass.

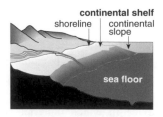

Continental Shelf

Continental Shelf Lands Act — A law passed in 1983 and updated through 1988 that states that the outer continental shelf, including all submerged lands lying seaward and outside of areas of lands beneath navigable waters, subsoil, and seabed attached to the United States, is of vital importance to the country as it is a national resource reserve of a variety of potential minerals, oils, etc. Exploration, development, and production of minerals are controlled by the federal government under this act.

contingency plan — A document setting out an organized plan and coordinated course of action to be followed in case of an unexpected event such as a fire, explosion, or release of hazardous waste that could threaten human life or the environment.

continuing education — A formal organized educational program designed to promote the knowledge, skills, and professional attitudes of a variety of individuals by means of a series of short courses.

continuing education unit — A credit awarded by a professional organization to someone who has attended 10 hours of specialized instruction.

continuous air monitoring (CAM) — A process in which air samples are collected continuously at a known rate.

continuous discharge — A discharge occurring without interruption throughout the operating hours of the facility.

continuous emission monitor — (*air pollution*) A type of air emission monitoring system installed to operate continuously inside of the smokestack or other emission source.

continuous feed incinerator — An incinerator into which solid waste is charged almost continuously to maintain a steady rate of burning.

continuous flow valve — A valve in a supplied air respirator that continuously supplies air to the facepiece at a constant flow rate.

continuous monitoring — The taking and recording of measurements at regular and frequent intervals during operation of a facility.

continuous positive pressure ventilation — A positive pressure above ambient pressure maintained in the upper airway throughout the breathing cycle.

continuous sampling — A process in which samples are collected continuously at a known rate.

continuous wave — A time-dependent function of constant amplitude.

continuous-wave laser (CW laser) — A laser in which the beam is generated continuously as required for communication or other application.

continuum — A continuous series.

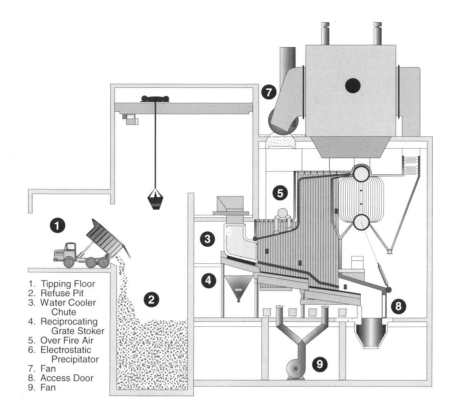

Continuous Feed Incinerator

1. Tipping Floor
2. Refuse Pit
3. Water Cooler Chute
4. Reciprocating Grate Stoker
5. Over Fire Air
6. Electrostatic Precipitator
7. Fan
8. Access Door
9. Fan

contour — A line connecting points of equal elevation.

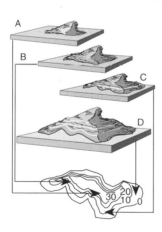

Contour Lines

contour map — A topographic map that shows relief or elevation differences by the use of lines indicating equal elevation.

contraband — Goods exported or imported to a country against the laws of that land.

contract — A binding agreement between two or more persons or parties that is legally enforceable.

contraction — A reduction in size, especially of muscle fibers.

contrails — Condensation trails that are artificial clouds made by the exhaust of jet aircraft.

contraindicate — A disease or physical condition that makes it impossible or undesirable to treat a particular person in the usual manner or to use certain medicines.

contrast — The difference in light between two surfaces or areas.

control — (*v.*) To take all necessary actions to ensure and maintain compliance with criteria established in the HACCP plan; (*n.*) the state where correct procedures are being followed and criteria are being met in a HACCP plan.

control efficiency — Ratio of the amount of pollutant removed from effluent gases by a control device to the total amount of pollutant without control.

control group — (*epidemiology*) An observed, unmanipulated set of test subjects compared to a manipulated test group to determine if a treatment or disease is related to special factors present in the second group.

control key — (*computer science*) A key used to access commands through the keyboard rather than the menus to utilize shortcuts.

control loop — A path through a control system between the sensor, which measures a process variable, and the controller, which controls or adjusts the process variable.

control measures — The actions and activities that can be used to prevent or eliminate a food safety hazard or reduce it to an acceptable level.

control of infectious diseases — Techniques directed against the reservoir of infection, such as isolation and quarantine; techniques used to interrupt the transmission of the agent, such as separation of food and water from fecal material or eradication of vectors; techniques used to reduce susceptibility of the host, such as immunization.

control panel — (*computer science*) A window that allows adjustment of various aspects of a computer, such as the volume, mouse speed, and clock.

control regulation — (*air pollution*) A rule of the EPA intended to limit the discharge of pollutants into the atmosphere and thereby achieve a desired degree of ambient air quality.

control rod — A rod, usually made of boron, used to control the power of a nuclear reactor.

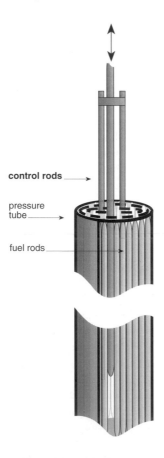

control rods

pressure tube

fuel rods

Control Rod

control system — A system that senses and controls its own operation on a continuous basis.

control technique guidelines — The U.S. EPA documents designed to assist state and local pollution authorities to achieve and maintain air quality standards for special sources through reasonably available control technologies.

control velocity — The air velocity required at the face of an enclosing hood to retain the contaminant within the hood.

controlled-air incinerator — An incinerator with two or more combustion chambers in which the amount of distribution of air is controlled; partial combustion takes place in the first zone and gases are burned in a subsequent zone or zones.

controlled area — An area where access is controlled to protect the individuals from exposure to radiation and radioactive materials.

controlled experiment — An experiment involving two or more similar groups; the control group is held as a standard for comparison while the other group is subjected to some procedure, treatment, measure, or disease for which the effect is being tested.

controlled reaction — A chemical reaction under temperature and pressure conditions maintained within safe limits to produce a desired product or process.

controller — A device that controls the starting, stopping, or operation of a device or a piece of equipment.

contusion — An injury caused by a blow to the body that does not disrupt the integrity of the skin and is characterized by swelling, discoloration, and pain.

convalescence — The time of recovery after an illness, injury, or surgery.

convection — The transfer of heat by the movement of molecules of liquids and gases.

convection cell — A cyclic pattern of movement in a fluid body such as the ocean, the atmosphere, or the Earth's mantle caused by density variations resulting from differences in temperature from one part of the fluid to another.

convection currents — Rising or sinking air currents that mix air in an atmosphere and transport heat from area to area.

convectional precipitation — Precipitation due to uneven heating of the ground that causes the air to rise and expand and the vapor to condense and precipitate.

convective — The transfer of heat, particularly through upwardly directed motion, within the atmosphere.

conventional filtration — A method of treating water to remove particulates that consists of the addition of

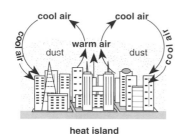

Convection Current

in earth's mantle

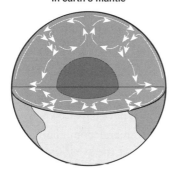

coagulant chemicals, flash mixing, coagulation, flocculation, sedimentation, and filtration.

conventional gas — A natural gas occurring in nature as opposed to a synthetic gas.

conventional septic system — See septic system.

convergence — The movement of two objects toward a common point.

convergence of information — (*geographical information system*) The principal of using multiple indicators to deduce information.

conversion — A process where a signal is changed from analog to digital or digital to analog.

conversion process — The process by which energy is converted from one type to another, such as radiant energy to heat or electric energy.

conversion rate — The rate at which sampled analog data are converted to digital data or digital data to analog data.

converted wetland — A wetland that has been drained, dredged, filled, leveled, or otherwise manipulated, including the removal of vegetation or any activity that results in impairing or reducing the flow, circulation, or reach of water that makes the production of an agricultural commodity possible.

converter — A technology that changes a potential energy in the fuel into a different form of energy such as heat or motion; a piece of equipment that changes the quantity or quality of electrical energy.

convex lens — A lens with an outward curve that converges parallel light rays.

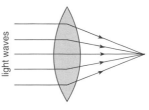

Convex Lens

conveyance — The process of water moving from one place to another.

conveyance loss — Water lost in movement through pipes, channels, conduits, ditches by leakage, evaporation, or transpiration, by plants growing in or near the channel.

conveyance system efficiency — The ratio of volume of water delivered to users to volume of water introduced into the conveyance system.

conveyor — A device that transports materials by belts, cables, or chains.

convulsant poisons — Poisons that act directly on the central nervous system; they include strychnine and camphor and produce convulsions, increased respiration, cardiac edema, vomiting, and death.

convulsion — A series of involuntary contractions of the voluntary muscles.

coolant — A liquid or gas used to reduce the heat generated by power production in nuclear reactors or electric generators; a transfer agent used in a flow system to convey heat from its source.

cooling air — Ambient air that is added to hot combustion gases to cool them.

cooling rate — The rate at which temperature decreases with time after completion of hypothermia treatment.

cooling sprays — Water sprays directed into flue gases to cool them and remove some fly ash.

cooling — Reduction of body temperature by the application of a hypothermia blanket, cold moist dressing, or ice pack; transfer of energy from a body of solid, liquid, or gas because of a temperature gradient from that body to the surroundings, which are at a lower temperature and may also be solid, liquid, or gas.

cooling tower — A piece of equipment used to cool water through a combination of sensing and evaporative heat transfer.

cooling water — The water that is used to cool machinery and equipment to remove heat from the metal.

Coombs' test — A test used to detect nonagglutinating antibodies on red blood cells by addition of an antiimmunoglobulin antibody.

cooperative — An organization owned by and operated for the benefit of those using its services.

cooperative extension system — A federal, state, or local cooperative education system that provides continuing adult education based on the academic programs of land grant colleges of agriculture and their affiliated state agricultural experimental stations.

coordinate — Advancing an exchange of information among essential people in a systematic manner in order to help them respond to a specified situation.

coordinate geometry — (*geographical information system*) A method of finding geometric features through the input of bearing and distance measurements; typically used by land surveyors.

coordinate system — (*geographical information system*) A method used to measure horizontal and vertical distances on a planimetric map.

coordinates — (*geographical information system*) Points on the *x,y*-axis that denote a particular spot on Earth.

coordination — The ability to perform smooth, voluntary movements by regulating muscle groups using the motor control areas of the brain and sensory nerves.

COPD — See chronic obstructive pulmonary disease.

coping — A process by which a person handles stress, solves problems, and makes decisions.

coping mechanism — An effort directed toward stress management that enables a person to regain emotional equilibrium after a stressful experience.

copolymer — A mixed polymer or heteropolymer that is a product of the polymerization of two or more substances.

copper dusts and mists (Cu) — A reddish, lustrous, malleable, odorless metal; MW: 63.5, BP: 4,703°F, Sol: insoluble, sp. gr. (metal): 8.94. It is used in the manufacture of copper rod, wire, piping, and tubing for electrical, plumbing, and construction; in the manufacture of domestic utensils; in fungicides, insecticides, and germicides; in paint pigments and coloring agents; in wood preservation, automotive emission controls, textile treatment, and organic synthesis. It is hazardous to the respiratory system, skin, liver, and kidneys and is toxic through inhalation, ingestion, and contact. Symptoms of exposure include nasal, mucous membrane, and pharynx irritation; nasal perforation; eye irritation; metallic taste; dermatitis; and increased risk of Wilson's disease. In animals, effects include lung, liver, and kidney damage and anemia. OSHA exposure limit (TWA): 1 mg/m³ [air].

copper fume (CuO/Cu) — Finely divided black particulate dispersed in air; MW: 79.5, BP: decomposes, Sol: insoluble, sp. gr. 6.4. It is liberated during the construction and installation of material fabricated from copper metal or copper alloys. It is hazardous to the respiratory system, skin, and eyes and is toxic through inhalation and contact. Symptoms of exposure include metal fume fever, chills, muscle ache, nausea, fever, dry throat, cough, weakness, lassitude, eye and upper respiratory tract irritation, metallic or sweet taste, discoloration of the skin and hair, and increased risk of Wilson's disease. OSHA exposure limit (TWA): 0.1 mg/m³ [air].

$$Cu(C \equiv N \longrightarrow O)_2$$

Copper(II)fulminate

copper sulfate (CuSO₄) — An effective algicide used infrequently in swimming pools because of its toxicity and incompatibility with some other compounds found in water; also known as cupric sulfate.

co-product — A mineral commodity that is recovered from mining at the same time as the desired mineral commodity.

core — The uranium-containing center of a nuclear reactor where energy is released; (*geology*) the central part of the Earth having a radius of approximately 2100 miles and physical properties different from those of the surrounding area.

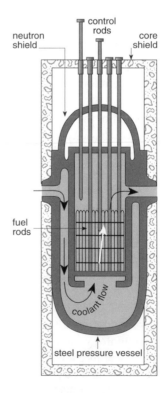

Reactor Core

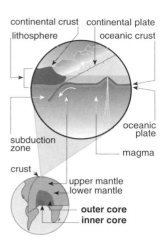

Core (geology)

core drill — A rotary drill, usually a diamond drill, equipped with a hollow bit and a core lifter.

Corer

Coriolis force — An apparent force that is a result of the Earth's rotation deflecting moving objects such as projectiles or air currents to the right in the Northern Hemisphere and to the left in the Southern Hemisphere.

corium — The deeper skin layer containing the fine endings of the nerves and the finest divisions of the blood vessels. Also known as derma.

cornea — A transparent bulge of the sclerotic layer of a vertebrate's eye which covers the iris and pupil and through which light rays pass.

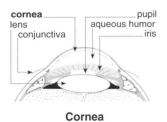

Cornea

corona — A luminous discharge due to ionization of the air surrounding a conductor that occurs when the local electric field exceeds the dielectric strength of air; a series of colored rings surrounding luminaries when veiled. It is due to diffraction by water droplets.

coronary artery — One of a pair of arteries (the left and right coronary arteries) that branch from the aorta and supply the heart with blood.

coronary artery disease — An abnormal condition that may affect the arteries of the heart and produce a variety of pathological conditions, especially a reduction in the flow of oxygen and nutrients to the myocardium.

coronary care unit — A specially equipped and staffed hospital area designed for the treatment of patients with sudden life-threatening cardiac conditions.

Corporate Average Fuel Economy — The overall average mileage required of a manufacturer for the fleet of automobiles and light trucks sold each year.

corpus — Dead body of a person or animal.

corpuscle — A red or white blood cell.

corrective action — A U.S. EPA requirement that treatment, storage, and disposal facilities handling hazardous waste undertake corrective actions to clean up spills resulting from failure to follow hazardous waste management procedures or through other mistakes.

Corrective Action Report — The information used to resolve a documented problem related to data quality.

corrective actions — Actions to be taken when the results of monitoring at a critical control point indicate a loss of control in a food operation.

corrective maintenance — Unscheduled maintenance carried out following an equipment failure.

correlation — (*statistics*) A relationship between variables that may be negative (inverse), positive, or curvilinear and is measured and expressed by using numeric scales; a statistical test that sets a numerical value to the amount of interdependence.

corridor — A narrow strip of land reserved for the location of transmission lines, pipelines, and service roads.

corrode — To weaken or destroy gradually; to deteriorate or erode by degrees.

corrosion — A chemical reaction at the surface of a substance, usually metal, causing an alteration or deterioration of the surface; the reactions approximately double with each 18°F increase in temperature.

corrosion inhibitor — A substance that slows the rate of corrosion of metal plumbing materials by water, especially lead and copper, by forming a protective film on the interior surface of these materials.

corrosion resistant — *(food)* Having the capacity to maintain original surface characteristics under the prolonged influence of the use environment, including expected food contact and normal use of cleaning compounds and sanitizing solutions; easily cleanable, readily accessible, and of such material and finish and so fabricated that cleaning can be accomplished by normal methods.

corrosive — *(food)* A substance that causes corrosion; *(biochemistry)* a chemical that causes visible destruction or irreversible alterations in living tissues by chemical action at the site of contact.

corrosive material — A liquid or solid that causes visible destruction or irreversible alterations in human skin tissue at the site of contact; a liquid that has a severe corrosive effect on steel or aluminum.

cortex — The outer layer of an organ or structure.

corticosteroid — A natural or synthetic hormone associated with the adrenal cortex which influences or controls such processes of the body as carbohydrate and protein metabolism and maintenance of serum glucose levels, balance of electrolytes and water, and functions of the cardiovascular system, skeletal muscle, kidneys, and other organs.

cortisone $(C_{21}H_{28}O_5)$ — A hormone secreted by the cortex of the adrenal gland of vertebrates.

corundum (Al_2O_3) — An impure mineral form of aluminum oxide; gem varieties are ruby and sapphire.

Corynebacterium — A genus of Gram-positive, nonmotile, straight to slightly curved, rod-shaped bacteria, including both pathogenic and nonpathogenic organisms that are widely distributed in nature and may cause diseases such as diphtheria.

coryza — Profuse discharge from the mucous membrane of the nose.

cosmic rays — High-energy particulate and electromagnetic radiations that originate outside the Earth's

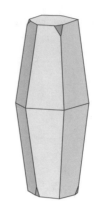

Corundum

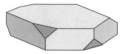

atmosphere and impinge upon the Earth from all directions of space with nearly the speed of light.

cost–benefit analysis (CBA) — A procedure in which the cost of a program is compared to its benefits, expressed in dollars, to determine the best investment; an evaluation of the costs and benefits of a proposed action that is used as a tool when trying to make decisions concerning specific kinds of problems or programs.

cost–benefit ratio — A comparison of the cost of an activity to the benefit of its outcome or product.

cost center — An accounting device whereby all related costs attributable to some center within an institution, such as an activity, department, or program, are segregated for accounting or reimbursement purposes.

cost containment — Control of overall cost.

cost-effective plan — A plan or combination of plans that produces the best results toward achieving stated objectives along with the most prudent use of resources.

cost-effectiveness analysis — The least expensive way to achieve a given environmental quality target or a way of achieving the greatest improvement in some environmental target for a given expenditure of resources.

cost of illness method — An approach that is used to estimate the societal costs of a particular illness or injury in a given time frame (usually a 1-year period); it focuses on direct medical and nonmedical costs and the indirect costs of lost productivity due to morbidity or premature mortality.

cost per ton per minute — A unit often used in cross comparisons between transfer and direct-haul operations.

cost sharing — A publicly financed program in which the community shares part of the cost of pollution control.

cottage cheese — Soft, uncured cheese prepared from curd obtained by adding harmless lactic-acid-producing bacteria with or without enzymatic action to pasteurized skim milk or pasteurized reconstituted skim milk.

cotton dust — A colorless, odorless solid; MW: not known, BP: decomposes, Sol: insoluble, Fl.P: not known. It is a result of the manufacture of cotton yarn. It is hazardous to the respiratory and cardiovascular systems and is toxic through inhalation. Symptoms of exposure include tight chest, cough, wheezing, dyspnea, decreased forced expiratory volume, bronchitis, malaise, fever, chills, and upper respiratory system symptoms after initial exposure. OSHA exposure limit (TWA): 1 mg/m^3 [air].

cough — The sudden audible expulsion of air from the lungs, which is an essential protective response that clears the lungs, bronchi, or trachea of irritants and secretions or prevents aspiration of foreign material into the lungs; chronic coughing is a symptom of disease or environmental problems.

coulomb (C) — (*electricity*) A unit of electrical charge equal to an ampere-second or 6.3×10^{18} electronic charges; the amount of electrical charge that crosses a surface in 1 second when a steady current of 1 absolute ampere is flowing across the surface.

Coulomb's law — The force between two charged bodies is proportional to the product of their charges and inversely proportional to the square of the distance between them; also known as the law of electrostatic attraction.

coulometry — Determination of the amount of an electrolyte released during electrolysis by measuring the number of coulombs used.

coumaphos ($C_{14}H_{16}ClO_5PS$) — A crystalline solid with a light brownish color and a slight sulfur-like odor; MW: 269.38, BP: NA, Sol: soluble in most organic solvents, Fl.P: NA (although difficult to ignite, irritating or poisonous gases such as oxides of sulfur and phosphorus may be produced in a fire), density: NA, sp. gr. 1.47. It is an organophosphate insecticide used to control cattle grubs, lice, and ectoparasites of sheep, goats, horses, swine, and poultry. It is a highly toxic compound and a choline stearase inhibitor that is hazardous to the respiratory system, digestive system, skin, and eyes.

Acute exposure may affect the central nervous system with agitation, confusion, delirium, coma, seizures, and death; it also causes bradycardia, bronchospasm, salivation, vomiting, and diarrhea. Chronic exposure may cause damage to the nerves, with weakness and poor coordination in arms and legs, depression, anxiety, and irritability and may affect vision. OSHA exposure limit (8-hour TWA): NA.

Coumaphos

count — (*radiation measurements*) External indication of a radiation-measuring device designed to enumerate ionizing events; (*law*) a statement of violation of an act in a criminal information or indictment; the number of counts depends on the number of interstate shipments involved in the case.

counter — An instrument or apparatus by which a numerical value is computed; (*radiology*) a device for enumerating ionizing events.

counterclaim — A claim presented by a defendant in opposition to the claim of the plaintiff; it is in effect a new suit in which the party named as the defendant under the bill becomes the plaintiff and the party named as plaintiff under the bill becomes the defendant.

counter-ion — The charge on the surface of a colloidal particle compensated for by an equal and opposite ionic charge in the liquid immediately in contact with the particle.

country-of-origin labeling — A section of the Tariff Act of 1930, as amended, declaring that products entering the United States to be used for food consumption must be clearly marked so the ultimate purchaser can identify the country of origin. Imported meat products are subject to this requirement.

coupled model — (*climate*) A numerical model that simulates both atmospheric and oceanic motions and temperatures and which takes into account the effects of each component on the other.

coupon — A steel specimen inserted into water to measure the corrosiveness of water; the rate of corrosion is measured as a loss of weight of the specimen in milligrams per surface area (expressed in square decimeters) exposed to the water per day.

covalent bond — A kind of chemical bond in which one or more pairs of electrons are shared by different atoms, with each electron in a pair coming from a different atom.

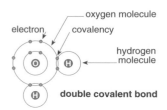

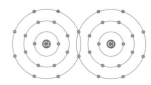

Covalent Bond

cover material — Soil used to cover compacted solid waste in a sanitary landfill.

coverage — (*geographical information system*) A spatial dataset containing a common feature type.

cow pox — See vaccinia.

coxsackie virus — One of a heterogeneous group of enteroviruses producing a disease in humans resembling poliomyelitis but without paralysis.

cp — See candlepower.

CPC — See chemical protective clothing.

CPD — Acronym for continuing professional development.

CPE — See chlorinated polyethylene.

CPF — Acronym for carcinogenic potency factor.

CPI — See Consumer Price Index.

CPO — Acronym for certified project officer.

CPPV — See continuous positive pressure ventilation.

CPR — See cardiopulmonary resuscitation.

cps — See cycles per second.

CPS — See characters per second.

CPSA — See Consumer Product Safety Act.

CPT — Acronym for current procedural terminology.

CPU — See central processing unit.

CQA — Acronym for construction quality assurance.

crab louse — A species of body louse, *Phthirus pubis*, that infests the genital area and is often transmitted between people during sexual intercourse.

cracking — The process of reducing large, complex hydrocarbon molecules into short-chain hydrocarbons, as in oil-refining processes.

cradle-to-grave — A management technique for hazardous waste disposal required by the Resource Conservation and Recovery Act; includes five basic elements: identification of generators and type of

waste; tracking of waste by a "uniform manifest"; requiring permits for all hazardous waste treatment, storage, and disposal facilities; restrictions and control on hazardous waste facilities; and enforcement of regulations and compliance by generators, transporters, and disposal facilities.

Crag™ herbicide ($C_6H_3Cl_2OCH_2CH_2OSO_3Na$) — A colorless to white crystalline, odorless solid; MW: 309.1, BP: decomposes, Sol (77°F): 26%, sp. gr. 1.70. It is used in the manufacture of weedicides. It is not known which organs are harmed by Crag™ herbicide, but it is toxic through inhalation, ingestion, and contact; no known symptoms of exposure. OSHA exposure limit (TWA): 10 mg/m^3 [total] or 5 mg/m^3 [resp.].

cramps — Painful muscular contractions that may affect almost any voluntary or involuntary muscle.

crash cart — A specially designed piece of equipment containing a variety of supplies and a defibrillator used to treat an individual when the heart stops.

crash (computer science) — A malfunction of computer hardware or software that may cause loss of all information and corruption of programs.

cream — A portion of milk that contains not less than 18% milk fat.

credentialing — Recognition of professional or technical competence.

creek — A small stream of water that serves as a natural drainage course for a drainage basin.

creep — The slow movement of rock debris or soil, usually imperceptible, except with observations of long duration.

creeping eruption — A skin lesion with irregular, wandering red lines on the foot made by the burrowing larvae of hookworms not able to mature in a given host and certain roundworms.

creosote — A translucent amber, brown, or blackish liquid; MW: not known, BP: >180°, Sol: practically insoluble in water, Fl.P: 75°C, sp. gr. not known. Type 1 results from high-temperature treatment of coal (coal-tar creosote); type 2 results from high-temperature treatment of beech and other woods (beechwood creosote); type 3 comes from the resin of the creosote bush (creosote-bush resin). Type 1 is the most widely used wood preservative in the United States; it is also used as a component of roofing pitch, fuel oil, and lamp black and as a restricted-use pesticide. Types 2 and 3 are used in industry. It is toxic by inhalation, absorption, ingestion, and contact; contains mutagenic polycyclic aromatic hydrocarbons, thus a possible relationship exists between creosote exposure and development of multiple myeloma. Symptoms of exposure include cancers of the nasal cavity, larynx, lungs, skin, and scrotum; ulceration of the oropharynx; petechial hemorrhages over the gastrointestinal serosal surfaces; degeneration and necrosis of hepatocytes; possible kidney failure; dermal and ocular irritation; and burns and warts. OSHA exposure limit (TWA): 0.2 mg/m³ [air].

cresol (all isomers) (CH₃C₆H₄OH) — A colorless, yellow, brown, or pinkish, oily liquid or solid with a sweet, tarry odor; MW: 108.2, BP: 376 to 397°F, Sol: 2%, Fl.P: 178 to 187°F, sp. gr. 1.03 at 77°F. It is used in the manufacture of antiseptics and disinfectants, phosphate ester, antioxidants, resins, herbicides, perfumes, explosives, and photographic developers; as a solvent and engine and metal cleaner. It is hazardous to the central nervous system, respiratory system, liver, kidneys, skin, and eyes and is toxic through inhalation, absorption, ingestion, and contact. Symptoms of exposure include central nervous system effects, confusion, depression, respiratory failure, dyspnea, irregular and rapid respiration, weak pulse, skin and eye burns, dermatitis, and lung, liver, and kidney damage. OSHA exposure limit (TWA): 5 ppm [skin] or 22 mg/m³.

o-**cresol (C₇H₈O)** — A white crystal that is a liquid above 88°F and has a sweet, tarry, or phenolic odor; MW: 108.2, BP: 191°C, Sol: 2% in water at 20°C, Fl.P: 178°F, sp. gr. 1.05. It is used in the production of disinfectants, herbicides, dyes, explosives, pharmaceuticals, phenolic resins. It is hazardous to the respiratory system, digestive system, skin, and eyes. Acute exposure may cause serious burns, blindness, confusion, central nervous system depression, seizures, dyspnea, irregular respirations, pulmonary edema, liver damage, heart damage, kidney damage, and death. Chronic exposure may produce allergic dermatitis, digestive disturbances, central nervous system effects, liver damage, kidney damage, vomiting, diarrhea, dizziness, mental disturbances, and death. It is a possible human carcinogen. OSHA exposure limit (8-hour TWA): 22 mg/m³.

p-**cresol (C₇H₈O)** — Colorless crystals with a phenolic odor; MW: 108.13, BP: 395.6°F, Sol: 2.4 g/100 ml, Fl.P: 187°F, sp. gr. 1.018; other information not available. Skin exposure may create health

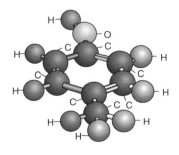

**p–Cresol
(4–Methylphenol)**

problems. OSHA exposure limit (PEL-TWA): 5 ppm or 22 mg/m³; also known as 4-methylphenol.

crest factor — The ratio of a peak voltage to the root-mean-square (rms) voltage of a wave form where both values are measured and referenced to the arithmetic mean value of the wave form.

Creutzfeldt–Jacob disease — A rare, progressive, fatal disease caused by a slow virus; symptoms include porous brain tissue, premature dementia in middle age, and gradual loss of muscular coordination.

Crimean–Congo hemorrhagic fever — An arbovirus infection transmitted to humans through the bite of a tick, with symptoms of fever, dizziness, muscle ache, vomiting, headache, and other neurological conditions leading to bleeding from the skin and mucous membranes and potentially death.

criminal law — The realm of law that addresses crimes and their punishments.

crisis — A turning point in the course of a disease usually indicated by a marked change in the intensity of signs and symptoms.

crisis intervention — (*psychiatry*) A means of psychological resolution of a person's immediate crisis and restoration to at least a level of functioning that existed before the situation occurred.

crisis management — (*FEMA*) A law-enforcement function including measures to identify, acquire, and plan the use of resources to anticipate, prevent, and/or resolve a threat or act of terrorism.

cristobalite — A crystalline, extremely hard, and chemically inert form of free silica created when refractory bricks and amorphous silica in diatomaceous earth are altered by exposure to high temperatures.

criteria — Plural of criterion.

Criteria Documents (CDs) — Recommended occupational safety and health standards for the Department of Labor; usually, part of the recommended standard is a recommended exposure limit.

criteria pollutants — Pollutants that are measured by ambient air quality standards and include total suspended particulates, sulfur dioxide, carbon monoxide, nitrogen oxide, photochemical oxidants (primarily ozone), gaseous hydrocarbons, and lead.

criterion — A standard on which a judgment or decision is made.

Criterion Continuous Concentration — An estimate of the highest concentration of a material in surface water to which an aquatic community can be exposed indefinitely without resulting in an unacceptable effect.

criterion standards — Measurable characteristics and the quantitative and/or qualitative value assigned to the particular characteristic.

critical care — See intensive care.

critical control point — (*food*) Any point or procedure in a specific food system at which a loss of control would result in an unacceptable health risk.

critical defect — A defect that may result in hazardous or unsafe conditions for individuals using and depending upon the product.

critical depth — The depth at which photosynthesis equals cell respiration.

critical factor — The single environmental factor closest to a tolerance limit for a given species at a given time.

critical habitat — (*Endangered Species Act*) An area essential to the conservation of a listed species, although the area need not actually be occupied by the species at the time it is designated.

critical item — A food that is in noncompliance with the Food Code of 1993 and is more likely than other violations to contribute to food contamination, illness, or environmental degradation.

critical limit — (*food*) One or more prescribed tolerances that must be met to ensure that a critical control point effectively eliminates or controls a microbiological hazard.

critical mass — The smallest amount of fissionable material that can sustain a chain reaction.

critical organ — An organ or tissue that is affected by physical, chemical, or microbiological factors and presents the greatest hazard to an individual or an individual's descendants.

critical orifice — A flow-rate meter pump that uses a critical or limiting orifice to regulate the rate of airflow.

critical pressure — The pressure exerted by a vapor in a closed system at the critical temperature.

critical soils — Soil materials that have been disturbed and/or have natural limitations enough to require alternative systems engineering or are perhaps so limited as to preclude the practicality of on-site wastewater treatment.

critical temperature — The highest temperature at which a substance can exist as a liquid regardless of the pressure.

CRM — Acronym for continuous radon monitoring.

crop residue — Portion of a plant, such as a corn stalk, left in the field after harvest.

cross-connection — Any physical connection or arrangement between two otherwise separate piping systems, one of which contains potable water and the other any fluid or suspension other than potable water; flow may occur from one system to another, with the direction depending on the pressure differential between the two systems.

cross-contamination — The movement of underground contaminants to one level or area to another due to invasive subsurface activities; the transfer of

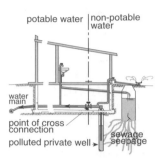

Direct Cross Connection

pathogens in an original source such as raw meat, poultry, or eggs to cooked food or other foods.

cross-flow turbine — A turbine where the flow of water is at right angles to the axis of rotation of the turbine.

cross-hatching — (*geographic information system*) The technique of shading areas on a map with a given pattern of lines or symbols.

cross-infection — An infection acquired by a patient during hospitalization; symptoms may appear while the patient is still hospitalized or may not appear until after discharge.

cross-reactivity — Ability of an antibody, specific for one antigen, to react with a second antigen; a measure of the relationship between two different antigenic substances.

cross-resistance — Resistance to a particular antibiotic or pesticide that results in resistance against a different antibiotic or pesticide from a similar chemical class to which the bacteria or insect may not have been exposed.

cross-sectional epidemiological study — An examination of a point in time of various health indices of a population.

cross-sectional study — (*epidemiology*) A one-time survey of a population generally intended to describe the current existing prevalence of a disease; this type of study can be markedly influenced by cohort effects.

cross-sensitivity — A sensitivity to one substance that predisposes a person to a sensitivity to other substances that are related chemically.

crotonaldehyde (CH₃CHCHCHO) — A water-white liquid with a suffocating odor; turns pale yellow on contact with air; MW: 70.1, BP: 219°F, Sol: 18%, Fl.P: 45°F, sp. gr. 0.87. It is used as an intermediate in the manufacture of butyl alcohol; in polymer technology; in organic synthesis in the manufacture of dyestuffs, sedatives, pesticides, and flavoring agents; as a solvent for the purification of mineral and lubricating oils; in the manufacture of surface-active agents and bactericides;

for petroleum well fluids, metal brighteners, and leather and paper sizing; in the manufacture of rubber; in the manufacture of chemical warfare agents; in photographic emulsions. It is hazardous to the respiratory system, eyes, and skin and is toxic through inhalation, ingestion, and contact. Symptoms of exposure include eye and respiratory irritation; in animals, dyspnea, pulmonary edema, and skin irritation. OSHA exposure limit (TWA): 2 ppm [air] or 6 mg/m³.

CRP — Acronym for child-resistant packaging.

CRT — See cathode-ray tube; see cell residence time.

crucible — A heat-resistant, barrel-shaped pot used to hold metal in a furnace.

Crucible

crude birth rate — Number of births in a year divided by the midyear population.

crude death rate — Number of deaths per thousand persons in a given year; also called crude mortality rate.

crude petroleum — A mixture of hydrocarbons having a flash point below 150°F and not processed in a refinery.

crude rates — Overall rates calculated from the actual number of events, such as deaths or births, in a total population during a specified period of time.

crumb soil — Soil structure in which the aggregates are 1 to 5 mm in diameter, soft, porous, and weakly held together (nearly spherical), with many irregular surfaces; usually found in surface soil or the A horizon.

crush syndrome — A severe, life-threatening condition caused by extensive crushing, characterized by destruction of muscle and bone tissue, hemorrhage, fluid loss, resulting in shock, renal failure, coma, and possibly death.

crushed gravel — Gravel that has been reduced in size by a machine.

crushed rock — Rock that has been reduced in size by a machine.

crusher — Machine that reduces rocks to smaller and more uniform sizes.

crust — (*earth science*) The outermost solid layer of the Earth, containing soils ranging in thickness from a few millimeters to perhaps as much as an inch and which is much more compact, hard, and brittle when dry than the material immediately beneath it; (*medicine*) an outer layer of solid matter formed by drying of a body exudate or secretion.

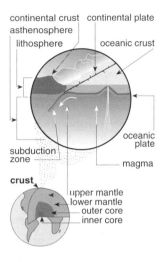

Earth's Crust

crustacean — Any edible, commercially distributed shrimp, crab, lobster, or other member of Crustacea and having jointed feet and mandibles, two pairs of antennae, and segmented bodies.

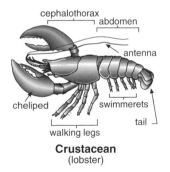

Crustacean
(lobster)

Cruz trypanosomiasis — See Chagas disease.

cryogenics — The science concerned with production of ultra-low temperatures and the study of properties of materials at these temperatures.

cryonics — The use of cold for a variety of therapeutic reasons.

cryosphere — One of the interrelated components of the Earth's system consisting of frozen water in the form of snow, permanently frozen ground (permafrost), floating ice, and glaciers for which fluctuations in volume cause changes in sea level.

cryptococcosis — An infectious decease caused by the fungus *Cryptococcus neoformans*, which spreads from the lungs to the brain and central nervous system, skin, skeletal system, and urinary tract; most likely to afflict persons with AIDS. Initial symptoms include coughing or other respiratory effects.

Cryptococcus — A genus of yeast-like fungi that reproduces by budding rather than by producing sporers; commonly found in the soil and on the skin and mucous membranes of people who are healthy.

Cryptococcus neoformans — A species of yeast-like fungi that causes cryptococcosis, a potentially fatal infection that affects the lungs, skin, and brain.

Cryptolestes pusillius — See Flat Grain Beetle.

cryptosporidiosis — (*disease*) A protozoan infection of the gastrointestinal tract associated with intestinal symptoms such as copious watery diarrhea, abdominal cramps, malabsorption, and weight loss. Incubation time is 2 to 14 days (average, 7 days) and duration of infection is 10 to 14 days, with watery diarrhea followed by continuing shedding of oocyst for an additional 14 to 21 days. It is caused by *Cryptosporidium*, an aflagellar intestinal protozoan with motile sporozoites that are released in the host intestinal tract upon dissolution of the oocyst's outer wall. Sporozoites implant on the host epithelium and undergo sexual and asexual development. The parasite–host relationship is unique in that it is intracellular; that is, the protozoa is enveloped by the host cell membrane but is extracytoplasmic to the cytoplasm. Sporozoites develop into trophozoites and subsequently go through asexual multiplication, formation of macrogametes and microgametes, fertilization, and oocyst formation. Newly formed oocysts that are expelled in the feces are immediately infective. It is found worldwide in daycare centers, among immunocompromised individuals, as a nosocomial infection, and in the general population through outbreaks of waterborne disease, with a reported level of 4 to 7% of the population (up to 65% of the population in areas of outbreaks of disease); 10 to 15% of asymptomatic individuals carry the parasite. It causes outbreaks of waterborne disease where the source may be surface water, water treated by rapid sand filtration and chlorination, water contaminated by heavy rainfall and runoff, or animals, especially cattle. The reservoir of infection includes humans and domestic animals. It is transmitted by ingestion of oocysts in water contaminated by feces, person to person, and by self-recontamination. It is communicable for the entire period of the infection and at least 2 to 3 additional weeks. Susceptibility is high, especially in target populations such as daycare centers and among the immunosuppresed. It is controlled through proper disposal of feces, vigorous personal hygiene, closing of contaminated water supplies, and superchlorination.

Cryptosporidium — An intestinal aflagellar protozoan, 4 to 5 nanometers in size, that is environmentally resistant and of the class of sporozoa, suborder Eimeriorina. It is parasitic in the intestines of humans and animals and may cause severe watery diarrhea; it can develop wholly in one host and therefore creates a tremendous potential for reinfection, especially for immunosuppressed individuals.

crystal — A mineral resulting from the arrangement of atoms, ions, or molecules in definite geometric patterns.

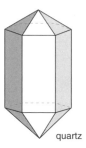

quartz

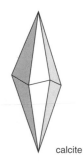

calcite

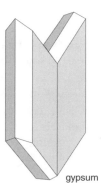

gypsum

Crystal

crystal structure — Regular and repeated three-dimensional arrangement of atoms or ions in a crystal.

crystalline — Having a definite molecular or ionic structure characteristic of crystals; crystalline materials need not exist as crystals.

crystalline aquifers — Geological formations of granites and various types of igneous and metamorphic rocks that store water in tiny fractures and cracks within the solid masses; this type of aquifer produces very low volumes of water.

crystallization — The process of forming definitive-shaped crystals when water has evaporated from a solution of the substance.

crystallography — The process of photographing the x-ray defraction pattern of a crystal.

CS — See central service.

CSF — See cerebrospinal fluid.

CSHO — Acronym for Compliance Safety and Health Officer.

CSI — Acronym for Compliance Sampling Inspection.

CSO — Acronym for combined sewer overflow.

CSP — See certified safety professional.

CSST — Acronym for chemical casualty site team.

Ct — A value for the concentration multiplied by time.

CT — See computed tomography and Ct_{calc}.

CT scan — See computed tomography.

CT scanner — An instrument designed for diagnostic radiological study of the head.

Ct value — A measure of vapor or gas exposure by inhalation.

Ct_{calc} — Product of the residual disinfection concentration (in mg/l) and corresponding disinfection contact time (in minutes).

CTD — Acronym for cumulative trauma disorder.

CTG — Acronym for Controlled Techniques Guidelines.

Ctrl — See control key.

CTSA — Acronym for Cleaner Technologies Substitutes and Assessment.

cu. ft. — See cubic foot.

cubic centimeter (cc) — A volumetric measurement that is equal to 1 milliliter.

cubic feet per hour (cfh) — Amount of cubic feet of air flowing past a specific point per hour of time.

cubic feet per minute (cfm) — Number of cubic feet of air drawn through a device per minute of time.

cubic feet per second (cfs) — Number of cubic feet of air drawn through a device per second of time.

cubic foot (cu. ft.) — One cubic foot is equal to 7.48 gallons, 28.3162 liters, or 1728 cubic inches.

cubic foot of water — Measurement equal to approximately 62.5 pounds.

cubic meter — The space inside of a box that is 1 meter wide, 1 meter high, and 1 meter deep; commonly used to measure a volume of air.

Culex — A genus of humpback mosquitoes that includes species that transmit viral encephalitis and filariasis.

Eggs

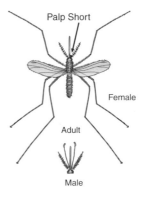

Larva

Pupa

Palp Short

Female

Adult

Male

Resting Position

***Culex* Life Cycle**

cullet — Clean, color-sorted, crushed glass that is used in glassmaking to speed up the melting of silica sand; also spelled collet.

culm — Waste material resulting from screening or removing coal from the surrounding materials.

culmination — The point at which a satellite reaches its highest position or elevation in the sky relative to an observer.

culture — The propagation of microorganisms or of living tissue cells in special media conducive to their growth.

culture medium — Any substance or preparation of material used for the growth of cultures and cultivation of microorganisms.

culvert — A pipe or small opening used for the free passage of surface drainage water under a highway, railroad, canal, or other embankment.

cumene ($C_6H_5CH(CH_3)_2$) — A colorless liquid with a sharp, penetrating, aromatic odor; MW: 120.2, BP: 306°F, Sol: insoluble, Fl.P: 96°F, sp. gr. 0.86. It is used as a constituent of solvents; in petroleum distillates; in the mixing and blending of aviation fuels. It is hazardous to the eye, upper respiratory system, skin, and central nervous system and is toxic through inhalation, absorption, ingestion, and contact. Symptoms of exposure include eye and mucous membrane irritation, headache, dermatitis, narcosis, and coma. OSHA exposure limit (TWA): 50 ppm [skin] or 245 mg/m³.

cumulative — Increasing by incremental steps.

cumulative dose — (*radiation*) Total dose resulting from repeated or continuous exposures to radiation.

cumulative effect — (*pesticides*) Buildup of pesticides and their storage in the body over a period of time that can sicken or kill an animal or human.

cumulative exposure — The sum of exposure of an organism to a chemical over a period of time.

cumulative incident rate — (*epidemiology*) Number of new cases of disease during a specified period divided by the population initially free of the disease.

cumulative trauma disorder — (*ergonomics*) An injury of the muscles, tendons, and nerves resulting from doing the same thing over and over without sufficient rest; a health disorder caused by repeated biomechanical stress due to ergonomic hazards.

cumulonimbus clouds — A principal cloud type, the bases of which almost touch the ground; thunderheads. Violent updrafts carry the tops to 75,000 feet; the most violent of these produce tornadoes.

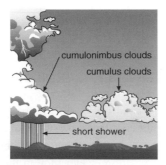

Cumulus and Cumulonimbus Clouds

cumulus clouds — A principal cloud type formed by rising air currents. They are puffy, resemble cauliflower at the head, and usually are a sign of good weather.

cup — A measurement equal to 16 tablespoons.

cupric acetylide (CuC_2) — A brownish-black powder; MW: NA, BP: NA, Sol: insoluble in water, Fl.P: extremely sensitive to impact and friction and can cause violent explosions, density: NA. It is used as a detonator and is hazardous due to its extreme explosiveness. ACGIH exposure limit: NA.

$$(-Cu \equiv C-)_n$$

Cupric acetylide

cupric sulfate — See copper sulfate.

cuprous acetylide (Cu_2C_2) — A red amorphous powder; MW: 151.11, BP: explodes on heating, Sol: soluble in acid, Fl.P: highly sensitive to shock and causes explosions on impact, density: NA. It is used to purify acetylene, in the preparation of pure copper powder, and in a diagnostic test. It is hazardous to the respiratory tract and nasal septum, causing irritation and ulceration. ACGIH exposure limit (TLV-TWA): 1 mg (Cu)/m³ [air].

$$\overset{+}{Cu} \; \overset{-}{C} \equiv \overset{-}{C} \; \overset{+}{Cu}$$

Cuprous acetylide

cuprous cyanide (CuCN) — A white powder that may also have dark-green or dark-red crystals; MW: 89.56, BP: NA, Sol: soluble in ammonium hydroxide, Fl.P: NA, density: 2.92. It is used in electroplating and as an insecticide and fungicide. It is hazardous to the respiratory tract, digestive tract, and skin. It is highly toxic and causes severe irritation. ACGIH exposure limit (TLV-TWA): 1 mg Cu/m³.

curb stop — A water service shut-off valve located in a water service pipe near a curb and between a water main and building.

cure — Restoration to health of a person who has a disease or other disorder.

cured meat — Meat preserved with chemicals or by drying, including salted, smoked, pickled, and dried meats.

curettage — The cleansing of a diseased surface.

curie (Ci) — Unit of radioactivity in which the number of disintegrations is 3.700×10^{10} per second.

curing — The addition of sodium chloride to meat mixed with sugar and spices. Curing was developed to preserve meat without the use of refrigeration; however, most processed meats require refrigeration.

current — Horizontal movement of water; (*energy*) the movement of electrons through a conductor (measured in amperes) or the movement or flow of electric charge.

current density — (*electricity*) Current per unit area of a conductor.

current dollars — Dollars at the most recent time.

current meter — A device used to measure the velocity of flowing water.

cursor — (*computer science*) A visible symbol guided by keyboard or mouse that indicates a position on a visual display device.

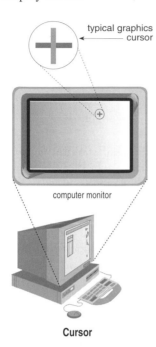

Cursor

curtain wall — A refractory, wall, construction, or baffle that deflects combustion gases downward; an external wall that is not load bearing.

curve — (*statistics*) A graphic method of demonstrating the distribution of data collected in a study or survey.

curvilinear — In a curved line.

curvilinear trend — (*statistics*) A trend in which a graphic representation of the data yields a curved line.

custodial care — The services of a nonmedical nature provided on a long-term basis for chronically ill individuals or older people.

cut — (*land*) Portion of a land surface or area from which earth or rock has been or will be excavated.

cutaneous — Pertaining to or affecting the skin.

cutaneous hazards — Chemicals that affect the dermal layers of the body by causing defatting of the skin, rashes, and irritation.

cutaneous larva migrans — (*disease*) A disease caused by an infective larva of dog and cat hookworm (*Ancylostoma braziliense* and *Ancylostoma caninum*) that causes dermatitis in individuals in contact with damp, sandy soil contaminated with cat and dog feces. The larvae enter the skin, migrate intracutaneously for long periods, and then penetrate to deeper tissues. A spontaneous cure occurs

after several weeks or months; also known as swimmer's itch.

cutaneous route — A route of entry into the body through the skin.

cutaneous toxicity — Adverse effects resulting from skin exposure to a substance.

cutaneous toxins — Chemicals that affect the dermal layer of the body; symptoms include defatting of the skin, rashes, and irritation; chemicals include ketones and chlorinated compounds.

cutdown — An incision into a vein with insertion of a catheter to supply intravenous fluid.

cutie pie — A portable instrument equipped with a direct-reading meter used to determine the level of ionizing radiation in an area; a radiation dose-rate meter mounted on a pistol-type grip.

cutting fluid — An oil or oil-in-water emulsion used to cool and lubricate a cutting tool.

cuttings — The spoils left by conventional drilling with hollow stem auger or rotary drilling equipment.

CV — See coefficient of variation; acronym for chemical vocabulary.

CVA — See cerebrovascular accident.

CVS — Acronym for constant volume sampler.

CW — Acronym for continuous working level monitoring.

CW laser — See continuous-wave laser.

CWA — See Clean Water Act.

CWC — Acronym for Chemical Weapons Convention.

cyanazine ($C_9H_{13}ClN_6$) — A white crystalline solid; MW: 240.73, BP: NA, Sol: moderately soluble in hydrocarbon solvents, density, NA. It is used as an herbicide and is hazardous to the digestive tract and skin. It shows moderate to high toxicity and is teratogenic in experimental animals.

$$C_2H_5NH - \underset{\underset{Cl}{\overset{N}{\underset{N}{\bigcirc}}}}{} - NH - \underset{\overset{CN}{|}}{C}(CH_3)_2$$

Cyanazine

cyanide (KCN and NaCN) — Potassium cyanide and sodium cyanide are white granular or crystalline solids with a faint almond-like odor; MW: 65.1 and 49.0, respectively; BP: not known and 2725°F, respectively; Sol: 72 and 58%, respectively, at 77°F; sp. gr. 1.55 and 1.60, respectively. They are used as fumigants and pesticides in greenhouses, ships, mills, and warehouses; in metal treatment; in the manufacture of intermediates in pharmaceuticals, dyes, vitamins, plastics, and sequestering agents; in cellulose technology, paper manufacture, and dyeing; as photographic fixatives. They are hazardous to the cardiovascular and central nervous systems and the liver, skin, and kidneys and are toxic through inhalation, absorption, ingestion, and contact. Symptoms of exposure include weakness, headache, confusion, nausea, vomiting, increased rate of respiration, slow gasping respiration, eye and skin irritation, asphyxia, and death. OSHA exposure limit (TWA): 5 mg/m³ [air].

cyanogen bromide — A crystalline solid; MW: 105.93; BP: volatile at ordinary temperature; Sol: soluble in water, alcohol, and ether; Fl.P: NA (may react violently with acid and decomposes in water, releasing toxic gases), density: 2.015 at 20°C. It is used in organic synthesis and as a reagent in bioanalysis. It is a highly toxic substance that is hazardous to the respiratory tract. It causes symptoms of nausea and chronic pulmonary edema and may be fatal. ACGIH exposure limit (TLV-TWA): NA; however, recommended is 0.5 ppm.

$$Br - C \equiv N$$

Cyanogen bromide

cyanogen (C_2N_2) — A colorless gas with an almond-like smell that is very acrid and pungent; MW: 52.04; BP: –21.1°C; Sol: highly soluble in water, alcohol, and ether; Fl.P: highly flammable and forms an explosive mixture with air; density: 0.87 at 20°C. It is used as a fumigant and propellant and in organic synthesis. It is a highly toxic gas that affects the skin, respiratory tract, and digestive tract. It causes symptoms of nausea, vomiting, headache, confusion, weakness, and death by asphyxiation. ACGIH exposure limit (TLV-TWA): 10 ppm.

$$N \equiv C - C \equiv N$$

Cyanogen

cyanogen chloride (CNCl) — A colorless liquid or gas with a pungent smell; MW: 61.47; BP: 13.8°C; Sol: soluble in water, alcohol, ether, and organic solvents; Fl.P: NA, density: 1.186. It is used in organic synthesis and as a military poisonous gas. It is a highly toxic substance that is hazardous to the eyes, nose, and respiratory tract. It causes severe irritation, hemorrhage of the bronchi and trachea, pulmonary edema, and death. ACGIH exposure limit (TLV-TWA): 0.3 ppm.

$$Cl - C \equiv N$$

Cyanogen chloride

cyanophos ($C_9H_{10}NO_3PS$) — A yellow to reddish brown transparent liquid; MW: 243.23, BP: NA, Sol: readily dissolves in most organic solvents, sp. gr. 1.26 at 25°C. It is used as a pesticide. It is a highly toxic cholinesterase inhibitor and is hazardous to the skin, digestive tract, and respiratory tract. It exhibits acute, delayed, and chronic symptoms and may cause headaches, dizziness, blurred vision, vomiting, abdominal pain, seizures, respiratory depression, respiratory paralysis, and death.

Cyanophos

cyanosis — A bluish discoloration of the skin and mucous membranes due to excessive concentrations of carbon monoxide and reduced hemoglobin in the blood.

cyanuric acid — A crystalline, weak acid that yields cyanic acid when heated and is used to prevent the decomposition of chlorine by ultraviolet light.

cyberspace — (*computer science*) A general term used to describe the space inhabited by computer networks, such as on the Internet.

cyclamate — A nonnutritive sweetener banned in the United States when animal testing showed it may be carcinogenic.

cycle — A completed set of regularly recurring events or phenomena; one complete execution of a repeatable process.

cycles per second (cps) — See hertz.

cycling — Periodic change in the controlled variable from one value to another.

cyclohexane (C_6H_{12}) — A colorless liquid with a sweet, chloroform-like odor; a solid below 44°F; MW: 84.2, BP: 177°F, Sol: insoluble, Fl.P: 0°F, sp. gr. 0.78. It is used as a solvent to dissolve cellulose, ethers, resins, fats, waxes, oils, bitumen, and crude rubber; in the manufacture of perfumes; in the synthesis of adipic acid for production of nylon. It is hazardous to the eyes, respiratory system, skin, and central nervous system and is toxic through inhalation, ingestion, and contact. Symptoms of exposure include eye and respiratory system irritation, drowsiness, dermatitis, narcosis, and coma. OSHA exposure limit (TWA): 300 ppm [air] or 1050 mg/m³.

cyclohexanol ($C_6H_{11}OH$) — A sticky solid or colorless to light-yellow liquid (above 77°F) with a camphor-like odor; MW: 100.2, BP: 322°F, Sol: 4%, Fl.P: 154°F, sp. gr. 0.96. It is used in the synthesis of adipic acid and caprolactam; in surface coatings of natural and synthetic textile dyes, cotton mercerizing, paints, varnishes, lacquers, and shellacs; in the production of soaps, synthetic detergents, rubber cement, dicyclohexyl phthalate, and other cyclohexyl esters. It is hazardous to the eyes, respiratory system, and skin and is toxic through inhalation, absorption, ingestion, and contact. Symptoms of exposure include eye, nose, and throat irritation and narcosis. OSHA exposure limit (TWA): 50 ppm [skin] or 200 mg/m^3.

Cyclohexanol

cyclohexanone (C$_6$H$_{10}$O) — A water-white to pale-yellow liquid with a peppermint- or acetone-like odor; MW: 98.2, BP: 312°F, Sol (50°F): 15%, Fl.P: 146°F, sp. gr. 0.95. It is used in the surface coating of fabrics and plastics; in the cleaning of leather and textiles; as a solvent in crude rubber, insecticides and epoxy resins; as a sludge solvent in lubricating oils. It is hazardous to the respiratory system, eyes, skin, and central nervous system and is toxic through inhalation, absorption, ingestion, and contact. Symptoms of exposure include eye and mucous membrane irritation, headache, narcosis, coma, dermatitis. OSHA exposure limit (TWA): 25 ppm [skin] or 100 mg/m^3.

cyclohexene (C$_6$H$_{10}$) — A colorless liquid with a sweet odor; MW: 82.2, BP: 181°F, Sol: insoluble, Fl.P: 11°F, sp. gr. 0.81. It is used in organic synthesis as a starting material or chemical intermediate; in the synthesis of polymers as polymer modifiers to control molecular weight; as a stabilizing agent. It is hazardous to the skin, eyes, and respiratory system and is toxic through inhalation, ingestion, and contact. Symptoms of exposure include eye, skin, and respiratory system irritation and drowsiness. OSHA exposure limit (TWA): 300 ppm [air] or 1015 mg/m^3.

Cyclohexene

cyclone — A cylindrical-shaped, air-cleaning apparatus through which a gas enters, swirls, and throws particles to the outside, where they slide down into a hopper.

cyclone collector — A device in which the direction of the gas flow is changed suddenly, causing the particles to flow straight ahead because of inertia.

clean gas

gas and dust

collected dust

dust out

Cyclone Collector

cyclonite (C$_3$H$_6$N$_6$O$_6$) — A white solid; MW: 222.15; BP: NA; Sol: slightly soluble in alcohol and ether and soluble in acetone; Fl.P: a highly powerful explosive; density: 1.82 at 20°C. It is highly explosive and the principal constituant of plastic bombs. It is a highly toxic substance that causes nausea, vomiting, and convulsions. ACGIH exposure limit (TLV-TWA): 1.5 mg/m^3 [skin].

NO$_2$

O$_2$N   NO$_2$

Cyclonite

cyclopentadiene (C$_5$H$_6$) — A colorless liquid with an irritating, terpene-like odor; MW: 66.1, BP: 107°F, Sol: insoluble, Fl.P (oc): 77°F, sp. gr. 0.80. It is used in thermal cracking and chemical synthesis; in the production of modified oil; in the preparation of epoxy resins, polymers, and varnishes; as a drying oil for varnishes; in the production of fire-resistant polyurethane foams; in foundry-core bindings and epoxidized olefins. It is hazardous to the eyes and respiratory system and is toxic through inhalation, ingestion, and contact. Symptoms of exposure include eye and nose irritation. OSHA exposure limit (TWA): 75 ppm [air] or 200 mg/m^3.

cyclotron — A device for accelerating charged particles to high energy by means of an alternating electric field between electrodes placed in a constant spiraling magnetic field; used for bombarding the nuclei of atoms.

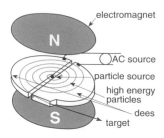

electromagnet

N

AC source

particle source

high energy particles

dees

S

target

Cyclotron

cyclotron D — See dee.

cyst — A closed epithelium-lined sac or capsule containing a liquid or semisolid substance; a stage in the life cycle of certain parasites during which they are contained in a protective wall; (*physiology*) a small, abnormal saclike growth in animals or plants usually containing liquid and diseased matter produced by inflammation.

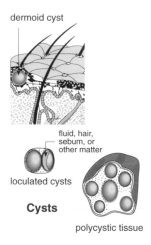

dermoid cyst

fluid, hair, sebum, or other matter

loculated cysts

Cysts

polycystic tissue

cysticercosis — Infection with larval forms of *Taenia solium*.

cystitis — Inflammation of the urinary bladder that may result from an infection descending from the kidney or coming from the exterior of the body by way of the urethra.

cystoscope — An instrument that allows a doctor to see inside the bladder and remove tissue samples or small tumors.

cytochemistry — The study of various chemicals within a living cell and their actions and functions.

cytochrome — Several iron-containing pigments that serve as electron carriers in aerobic respiration.

cytogenetics — The study of the structure and function of chromosomes as units of heredity and evolution.

cytokine — A molecule secreted by an activated or stimulated cell, such as a macrophage, that causes physiological changes in certain other cells; a natural substance made by the body that is important in host defenses against infections and malignant tumors.

cytokinesis — Division of the cytoplasm of a cell into two parts, as distinct from the division of the nucleus (mitosis).

cytokinin — A group of hormones that promote growth by stimulating cell division.

cytologic map — A graphic representation of the location of genes on the chromosomes.

cytological aberration — Deviation from normal cell structure or function.

cytology — The science that deals with cell function, growth, reproduction, and structure.

cytomegalovirus (CMV) — A viral infection that may occur without symptoms or result in mild flu-like symptoms; severe infections can result in hepatitis, mononucleosis, or pneumonia; the virus is shed in body fluids and is of particular concern for individuals suffering from some form of immune deficiency.

cytometer — A piece of equipment used for counting and measuring the number of cells within a given amount of blood, urine, or cerebrospinal fluid.

cytophotometer — An instrument for measuring light density through stained portions of cytoplasm; used for locating and identifying chemical substances within cells.

cytoplasm — The jelly-like protoplasm lying outside the nucleus of an animal or plant cell.

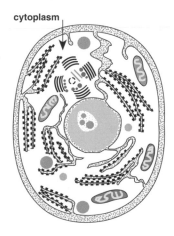

cytoplasm

Cytoplasm in Animal Cell

cytoplasmic membrane — The cell boundary that maintains the integrity of the cell.

cytoscopy — The diagnostic study of cells from a person.

cytotoxicity — The capacity of a material to interfere with cell metabolism.

cytotoxin — Substance having a specific toxic effect on cells of special organs.

CZARA — Acronym for Coastal Zone Management Act Reauthorization Amendments.

CZCS — See Coastal Zone Color Scanner.

CZMA — See Coastal Zone Management Act.

D

d — Notation for Dalton; an SI symbol for the prefix *deci-*, representing a factor of 10^{-1} power.

D — See deuterium; electric displacement.

2,4-D — See 2,4-dichlorophenoxyacetic acid.

da — SI symbol for the prefix *deka-*, representing a factor of 10^1 power.

DA — Acronym for developmental age.

D/A — See digital-to-analog converter.

DAC — See digital-to-analog converter.

Dacthal — An herbicide used especially on vegetables whose breakdown products are environmentally significant and are among the most commonly detected pesticide residues in drinking water wells.

dactylitis — Inflammation of a finger or toe.

DAF — Acronym for dissolved air flotation.

daily average temperature — The mean of the maximum and minimum temperature for a 24-hour period.

daily cover — Material that is spread and compacted on the top and sides of solid waste at the end of each operating day.

Dalton — See atomic mass unit.

Dalton's law — When gases or vapors having no chemical interaction are present as a mixture in a given space, the pressure exerted by a component of the gas mixture at a given temperature is the same as it would exert if it filled the whole space alone; also known as the law of partial pressures.

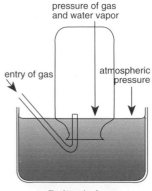

Dalton's Law
in which the total pressure is
equal to the barometric pressure

dam — A barrier built across a watercourse to impound or divert water.

dam failure — A catastrophic failure characterized by the sudden, rapid, and uncontrolled release of impounded water.

dam safety deficiency — A physical condition capable of causing the sudden uncontrolled release of reservoir water; partial or complete failure of a dam, structure, or facility.

damage risk criterion — Suggested baseline of noise tolerance that should not result in hearing loss.

damp — (*noise*) To reduce the generation of oscillatory or vibrational energy of an electrical or mechanical system; a harmful gas or mixture of gases found in coal mining.

damper — A manually or automatically controlled valve or plate in a breeching duct or stack that is used to regulate a draft or the rate of flow of air or other gases.

damping — Reduction in amplitude of a wave due to dissipation of the wavelength; soundproofing.

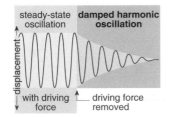

Damping

damping ratio — Ratio of actual damping to critical damping, where critical damping is the minimal amount of damping that prevents free oscillatory vibration.

dander — Small scales from the hair or feathers of animals that may cause an allergic reaction in sensitive persons.

Dano biostabilizer system — An aerobic, thermophilic, composting process in which optimum conditions of moisture, air, and temperature are maintained in a single, slowly revolving cylinder that retains the compostable solid waste for 1 to 5 days.

darkfield microscopy — Examination of organisms with a darkfield microscope where the specimen is illuminated by a peripheral light source; used primarily to identify the syphilis spirochete.

Darwinian theory — Evolution of a species results from the process of natural selection of variations in plants and animals that are best suited to promote survival in their environmental surroundings.

DAS — See data acquisition system.

DASHO — See Designated Agency Safety and Health Official.

data — Numerical or qualitative information for analysis used as the basis for decision making; singular is datum.

data acquisition — (*computer science*) A system of equipment or software that gathers data from devices through a communications channel.

data acquisition system — A radiation detection system that measures the amount of radiation passing through a patient.

data analysis — (*statistics*) Portion of a study that includes classifying, coding, and tabulating information necessary to perform analyses utilizing a specific research design.

data call-in — A process of developing key necessary test data, especially on long-term effects of existing pesticides, in advance of scheduled registration standard reviews; required by the Office of Pesticide Programs of the U.S. EPA.

data clustering — Grouping of related information in order to make an appropriate analysis.

data collection — Phase of a study that includes gathering information and identifying sampling units to meet a specific research design.

data collection and evaluation — A step in the risk assessment process where site data relevant to human and ecological health evaluations are gathered and analyzed.

data collection platform — (*computer science*) An electronic equipment device that gathers digital and analog data from monitoring devices and remotely relates that data to a central computer for interpretation and archiving.

data comparability — Characteristics that allow information from many sources to be of equivalent quality to be used to resolve problems and create programs.

data dictionary — (*geographical information system*) A document containing information about a series of datasets.

data integrity — (*geographical information system*) The consistency and accuracy of data stored in a database.

data level — (*computer science*) A standard for identifying the amount of processing applied to a file.

data link — Communication lines and related hardware and software systems necessary to send data between two or more computers over telephone lines, fiberoptics, satellite, or cable.

data message — (*global positioning satellite*) A message included in the GPS signal that reports the location of the satellite, clock corrections, and potential problems.

data mining — (*geographical information system*) Searching through datasets to find patterns and trends.

data model — (*computer science*) A standard structure that associates metadata to a file.

data object description — (*computer science*) The basic information, definitions, and metadata required to process raw measurement data into standard files.

data processing — (*computer science*) The techniques and practices used in the manipulation of information by a computer.

data quality report — (*computer science*) A report submitted and recorded by the data system that documents information about the quality of the data stream.

data retrieval — (*computer science*) A recovery of information from a computer database.

data source — The origin of information.

data stream — (*computer science*) Timed sequence series of like files.

data validation — The process of determining if information gathered is complete and accurate.

database — (*computer science*) A collection of interrelated information stored on a magnetic tape or disc; also known as a tuple.

database management system (DBMS) — A set of computer programs for organizing information on a database.

dataset — A minimum aggregation of uniformly defined and classified statistics that describe an element of a situation.

dateometer — A small calendar disc attached to motors and equipment to indicate the year in which the last maintenance service was performed.

datum — See data; (*geology*) any level surface taken as a plane of reference from which to measure elevations.

daughter — (*nuclear fission*) A decay product of a radioactive substance. See also decay product.

daughter cell — Newly formed cells resulting from the division of a previously existing cell (called a mother cell). The two daughter cells receive nuclear materials that are identical.

DAVES — See desorption and vapor extraction system.

day tank — A tank used to store a chemical solution of known concentration to be used for a chemical feeder.

daycare — A specialized program or facility that provides care for preschool children, usually within a group setting and including some form of educational input.

dB — See decibel.

2,4-DB ($C_{10}H_{10}Cl_2O_3$) — A crystalline solid; MW: 248.98; BP: NA; Sol: dissolves in acetone, ethanol, and ether; density: NA. It is used as an herbicide. It is moderately toxic by the oral route, skin, eyes, digestive tract, and respiratory tract and is irritating to mucous membranes.

$$O-CH_2-CH_2-CH_2-COOH$$

2,4-DB

dBA — See adjusted decibel.

dBC — See decibels on C scale.

DBMS — See database management system.

DBPs — Acronym for disinfection byproducts.

dBZ — A nondimensional unit of radar reflectivity that represents a logarithmic power ratio in decibels with respect to radar reflectivity factor Z. The value of Z is a function of the amount of radar beam energy that is backscattered by a target and detected as a signal or echo. The amount of backscattered energy generally is related to precipitation intensity; however, the width of the radar beam, precipitation type, drop size, and presence of ground clutter can affect the results.

dc — See direct current.

dcf — Notation for dry cubic feet of air.

DCG — See derived concentration guide.

DCS — Acronym for data collection system.

DD — See developmental disability.

DDD — See 2,2-*bis*(*para*-chlorophenyl)-1,1-dichloroethane.

DDE — See dynamic data exchange.

***p, p′*-DDE ($C_{14}H_8C_{14}$)** — A crystalline solid; MW: 318.02, BP: NA, Sol: soluble in most organic solvents, density: NA. It is used as a pesticide. It is hazardous to the digestive tract, is teratogenic in humans, and is carcinogenic in experimental animals.

p,p'-DDE

DDT ($(C_6H_4Cl)_2CHCCl_3$) — Common name for dichlorodiphenyltrichloroethane; colorless crystals or off-white powder with a slight aromatic odor; MW: 354.5, BP: 230°F (decomposes), Sol: insoluble, Fl.P: 162 to 171°F, sp. gr. 0.99. It is used as an insecticide but has been banned in the United States by the EPA. It is hazardous to the central nervous system, kidneys, liver, skin, and peripheral nervous system and is toxic through inhalation, absorption, ingestion, and contact. Symptoms of exposure include tremors, apprehension, dizziness, confusion, malaise, headache, fatigue, convulsions, paresis of hands, vomiting, eye and skin irritation, and paresthesia of the tongue, lips, and face; carcinogenic. OSHA exposure limit (TWA): 1 mg/m^3 [skin].

DE — See dose equivalent.

de minimus — (*law*) Not enough to be considered.

De Quervain's disease — An inflammation of the tendon sheath of the thumb attributed to excessive friction between two thumb tendons and their common sheath, usually caused by twisting and forceful gripping motions with the hands.

deactivation — The process of making something inactive or inoperable.

dead capacity — The reservoir capacity of stored water that cannot be removed by gravity.

dead end — The end of a water main that is not connected to other parts of the distribution system by means of a connecting loop of pipe.

dead-end host — Any animal or human from which a parasite cannot escape to further its life cycle.

dead storage — See dead capacity.

dead time — The time interval between a change in a signal and initiation of a perceptible response to that change.

deadly nightshade — See belladonna.

deafness — Temporary or permanent impairment or loss of hearing.

death — Cessation of life as shown by the absence of a heartbeat or respiration.

death rate — The ratio of total deaths to total population; usually expressed per 1000, 10,000, or 100,000 population. See also mortality rate.

debilitating — Producing weakness, such as a disease, injury, or surgical procedure.

debridement — A medical procedure for removing all foreign, contaminated, and devitalized tissues from or adjacent to a traumatic or infected lesion.

debris — The remains of something broken or destroyed.

debris flow — A flash flood consisting of a mixture of rocks and sediment and containing less than 40% water by volume.

debug — Remove errors from a program or hardware.

debugger — A program that is used to remove programming errors.

decaborane ($B_{10}H_{14}$) — A colorless orthorhombic crystal with a pungent smell; MW: 122.24; BP: 213°C; Sol: decomposes in hot water and dissolves in carbon disulfide, alcohol, benzene, ethel acetate, and acetic acid; Fl.P: 80°C (closed cup), density: 0.94 g/cm³ at 25°C. It is used in rocket propellants and as a catalyst in olefin polymerization. It is a highly toxic compound and hazardous to the respiratory tract, digestive tract, and skin. It causes headaches, nausea, vomiting, dizziness, lightheadedness, muscle spasms, convulsions, and loss of coordination. ACGIH exposure limit (TLV-TWA): skin 0.05 ppm.

decalcification — The loss or removal of calcium or calcium compounds from calcified material, such as bone or soil.

decant — To draw off the upper layer of liquid after the heavier material has settled.

decay — Reduction of the organic substances of a plant or animal body to simple inorganic compounds, usually by the action of bacteria or fungi; to undergo decomposition; (*radioactivity*) change in the nuclei of radioactive isotopes that spontaneously emits high-energy electromagnetic radiation and/or subatomic particles, while gradually changing into a different isotope or element.

decay chain — A sequence of radioactive decays (transformations) beginning with one nucleus; that initial nucleus (the parent) decays into a daughter nucleus that differs from the first by whatever particles were emitted during decay. Also known as decay series.

decay constant — The fraction of the number of atoms of the radioactive nuclide that decay in unit time, represented by c in the following equation: $I = I_0 e^{-ct}$, where I is the number of disintegrations per unit time.

decay product — A new isotope formed as a result of radioactive decay; also known as a daughter.

decay rate — The rate at which a group of radioactive atoms decays into stable daughter atoms; often expressed in terms of half-life.

decay series — See decay chain.

deceleration — The rate at which the velocity of a moving object decreases; the opposite of acceleration.

dechlorination — The deliberate removal of chlorine from a solution.

deci- — Prefix meaning 1/10 (.1) in the metric system.

decibel (dB) — (*physics*) A unit for measuring the relative intensity of sounds, equal to 1/10 of a bel. One decibel is approximately the smallest difference in acoustic power the human ear can detect, where an increase of 10 dB doubles the loudness of sound; the sound pressure level equal to 20 micropascal.

decibel meter — An instrument for measuring sound waves as they strike a microphone, producing an electric current for which the strength depends on the loudness. The current is then amplified and registered on a meter, which is calibrated in decibels.

decibels on C scale (dBC) — The sound level in decibels read on the C-scale of a sound-level meter, which discriminates very little against very low frequencies.

deciduous — Describes leaves that fall off or are shed seasonally.

decigram — A unit of metric measure equal to 0.1 g.

decimal — A numbering system subdivided into tenths of units.

deciview — (*air pollution*) A measurement of visibility; 1 deciview represents the minimal perceptible change in visibility to the human eye.

deck — Any structure that is on top of or adjacent to the outer edges of a pool, spa, or hot tub wall and supports people in a sitting or upright position.

declination — Angular distance from the equator to a satellite measured as positive north and negative south.

declining growth — A growth phase in which the availability of food begins to limit the cell growth.

decode — To put coded information into a form that is readily usable.

decolonization — (*infection control*) Elimination of methicillin-resistant *Staphyloccus aureus* from humans by use of infection control measures and/or antibiotics.

decommissioning — The process of closing down a nuclear reactor, removing the spent fuel, dismantling some of the other components, and preparing them for disposal.

decompose — To rot or decay.

decomposer — A heterotrophic organism that breaks down organic compounds.

decomposition — The breaking down of a compound into simple substances or into its constituent elements; (*food*) any food, whole or in part, that is filthy, putrid, or decomposed and unfit for human consumption; breakdown of a material or substance by heat, chemical reaction, electrolysis, decay, or other processes into parts, elements, or simpler compounds.

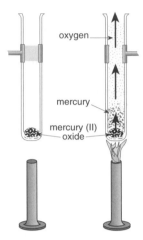

Decomposition Reaction

decomposition product — Material produced or generated by the physical or chemical degradation of a parent material.

decompression sickness — A painful, sometimes fatal syndrome caused by the formation of nitrogen bubbles in the tissues of divers who rise too rapidly from areas of greater pressure.

Decon — See decontamination unit.

decongestant — An agent that relieves nasal or respiratory congestion.

decontaminate — To break down, neutralize, or remove a radioactive, chemical, or biological substance that poses a hazard to personnel or equipment.

decontaminating material — Any substance used to chemically destroy, physically remove, seal, or otherwise make harmless a chemical, biological, or radiological agent.

decontamination — The removing of biological, radiological, or chemical contamination, making the environment or specific object safe to handle.

decontamination room — An area within a hospital where used materials are washed prior to sterilization.

decontamination unit — A five-chamber unit with a shower that abatement workers utilize to remove contaminating material.

decubitus ulcer — An ulcer of the skin and subcutaneous tissues due to local interference with the circulation; also known as a bed sore; pressure sore.

dedicated exhaust — An exhaust system serving only the aerator sterilizer and/or the immediate area.

deduction — A system of reasoning leading to the drawing of conclusions from known principles or data; from the general to the specific.

dee — A hollow accelerating cyclotron electrode shaped like the letter D; also known as cyclotron D.

deep mining — An extraction of coal or minerals at depths >1000 feet with a maximum for coal of approximately 1500 feet.

deep percolation — The movement of water by gravity downward through the soil profile below the maximum effective plant root zone.

deep percolation — The movement of water.

Deep-Water Port Act — A law passed in 1974 and amended through 1984 that authorizes and regulates location, ownership, construction, and operation of deep-water ports in waters beyond the territorial limits of the United States.

deep well injection — A means of disposal of liquid hazardous waste underground by injecting the waste with force and under pressure into steel-and-concrete-encased shafts placed deep into the Earth.

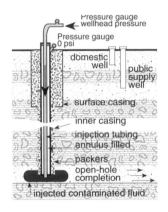

Deep Well Injection

deer mouse (*Peromyscus maniculatus*) — Member of the family Cricetidae, which is native to North America and occupies every type of habitat from forests to grasslands. It may be found in rural, semi-rural, and even urban areas. Characteristics include a weight of 2/3 to 1-1/4 ounces; total length, including tail, of 4-4/5" to 9"; white feet, usually white underside, and brownish upper surfaces. Gestation time is 21 to 23 days; litter size is 2 to 4 with 3 to 5 young each. It nests anywhere and is a burrowing animal with a range of 0.13 to 1.6 hecacres. It is primarily a seed eater but will feed on nuts, acorns, fruits, insects, insect larva, fungi, and green vegetation. It is the reservoir for hantavirus and the cause of outbreaks of hantavirus pulmonary syndrome.

Deerfly fever — See tularemia.

deet (C₁₂H₁₇NO) — A nearly odorless white to amber-colored liquid; MW: 191.26, BP: 160°C, Sol: NA, Fl.P: NA, sp. gr. 0.996. It is used as an insect repellant on humans for mosquitoes, biting flies, gnats, and biting crawling arthropods such as ticks, fleas, and chigger mites. It is hazardous to the digestive system, skin, and eyes. Acute exposure may cause severe central nervous system problems, including confusion, ataxia, coma, seizures, and death. It may also cause abdominal pain, nausea, vomiting, bradycardia, hypotension, toxic hepatitis, painful skin necrosis, and burning sensation of the eyes, tongue, and mouth. Chronic exposure may cause encephalopathy, which may lead to death. OSHA exposure limit (8-hour TWA): NA.

DEF (C$_{12}$H$_{27}$OPS$_3$) — A colorless to pale-yellow liquid with a mercaptan-like odor; MW: 314.5, BP: 150°C, Sol: NA, Fl.P: NA, sp. gr. 1.06. It is an organophosphate chemical used as a chemical defoliant, especially for cotton. It is hazardous to the respiratory system, digestive system, skin, and eyes. Acute exposure can cause central nervous system depression, agitation, confusion, delirium, coma, seizures, and death. Chronic exposure may result in dermal sensitization, damage to the nerves, poor coordination in the arms and legs, personality changes, depression, anxiety, and irritability. It is mutagenic in laboratory animals. OSHA exposure limit (8-hour TWA); NA.

default — (*computer science*) A selection automatically used by a computer program when a choice is not made by the user.

default decree of condemnation — An order requested of the court when lots under seizure are not claimed or defended; it provides for destruction, donation to charity, sale, or disposal as the court may elect to decree.

defecate — To void excrement or waste through the anus.

defendant — The opponent of either the plaintiff or prosecution; the party or parties named in an information, indictment, or complaint for injunction and against whom the government is proceeding.

defense meteorological satellite (DMS) — A satellite using infrared sensors to map global and spectral analysis of the ozone layer.

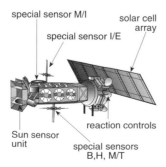

Defense Meteorological Satellite DMSP 5D-2

defibrillation — Delivering an electric shock to a patient's heart to terminate ventricular fibrillation.

defibrillator — A piece of equipment that delivers an electric shock at a preset voltage to the myocardium through the chest wall; used to restore normal cardiac rhythm.

deficit — A deficiency or difference from that which is normal.

definitive host — A host in which a parasite achieves sexual maturity.

deflation — The force of wind erosion.

deflocculation — The action in which groups or clumps of particles are broken up into individual particles and spread out and suspended in the solution; dispersion.

defluoridation — The removal of excess fluoride in drinking water to prevent the mottling of teeth.

defoaming agents — Chemicals that are added to wastewater discharges to prevent the water from foaming when it is released into a receiving water body.

defoliant — A pesticide chemical that causes the leaves of a plant to drop off prematurely.

deformation — A change in shape or size.

degasification — A water treatment process that removes dissolved gases from water by either mechanical or chemical treatment methods or a combination of both.

degeneration — Death or functional impairment of cells connected to neurons that have been destroyed or are severely damaged.

degradable — The ability to decompose chemically or biologically.

degradation — Breakdown of a complex chemical by the action of microbes, water, air, sunlight, or other agents; the wearing down of a land surface by the process of erosion; (*protective clothing*) a deleterious change of one or more physical properties of protective clothing material due to contact with a chemical.

degrease — To use an organic solvent as a surface cleaning agent or to remove grease, oil, or fatty material from a surface.

degree — A unit of angular measure represented by the symbol °.

degree day — A unit that represents 1 degree of difference in the mean daily outdoor temperature; used to measure heat requirements. It is a rough measure used to estimate the amount of heating or cooling required in a given area based on the difference between the mean daily temperature and 65°F.

degree of freedom — Statistical term that is a measure of stability related to the number of independent equivalent terms entering into a distribution.

dehalogenation — Removal of chlorine or bromine ions from organic contaminants to reduce toxicity or improve degradation.

dehumidification — The process of reducing the moisture content of the air.

dehydrating agent — A substance that is able to dry another substance by withdrawing water.

dehydration — Removal or loss of water from any substance.

deionized water — Water free of inorganic chemicals.

deionizing — Purifying water of dissolved salts by passing the water through synthetic materials.

DEIS — See Draft Environmental Impact Statement.

deka- — A prefix meaning 10 in the metric system.

dekagram — A unit of metric measure equal to 10 g.

delamination — Separation of one layer from another.

Delaney clause — Food additive amendment of 1958 to the Food, Drug, and Cosmetic Act; it states that known additives will be deemed to be unsafe if found to induce cancer when ingested by humans and/or animals or are found, after tests that are appropriate for the evaluation of food additives, to induce cancer in humans or animals.

delay time — Downtime; the amount of time that a signal is slowed after transmission.

delayed toxicity — The effects that occur after the lapse of some time from the administration of a substance.

delayed-type hypersensitivity — A manifestation of cell-mediated immunity, distinguished from immediate hypersensitivity in that maximal response is reached about 24 hours or more after intradermal injection of the antigen. The lesion site is infiltrated primarily by monocytes and macrophages.

deleterious — Causing harm or injury.

delirium — A state of frenzied excitement with symptoms of confusion, disorientation, restlessness, incoherent behavior, fear, anxiety, excitement, and hallucinations.

delist — Use of the petition process to rescind the toxic designation of a chemical; to remove a site from the National Priority List; to exclude a particular waste from regulation even though it is a listed hazardous waste.

delisting — The process by which a Superfund site is removed from the National Priorities List after it has been completely cleaned up.

delousing — A process of destroying an infestation of lice on a person.

delta — A mass of sediment usually triangular in shape and deposited where rivers lose their velocity as they enter lakes or seas.

delta-agent hepatitis — An infection caused by an RNA virus associated with the hepatitis B surface antigen in cases of chronic hepatitis and progressive liver damage.

delta-endotoxin — A toxic molecule produced within the cell of a bacterium.

delta-ray — An energetic or swiftly moving electron ejected from an atom during the process of ionization.

DEM — (*geographical information system*) See digital elevation model.

demand — The number of funded positions in a given occupation, whether filled or unfilled; (*energy*) the rate at which electric energy is used, expressed in kilowatts at a given instant or over a specific period of time.

demarcation — Setting limits or boundaries.

dementia — A progressive organic mental disorder with symptoms of chronic personality disintegration, confusion, disorientation, stupor, and a deterioration of intellectual capacity and function, control of memory, judgment, and impulses.

demeton (C$_8$H$_{19}$O$_3$PS$_2$) — An amber-colored oily liquid with a mild odor; MW: 258.34, BP: 134°C, Sol: soluble in most organic solvents, sp. gr. 1.183. It is a cholinesterase inhibitor and is used as a pesticide. It is extremely toxic and hazardous to the skin, digestive tract, and respiratory tract. It may cause acute, delayed, and chronic effects with symptoms of headache, dizziness, muscle spasms, blurred vision, abdominal pain, vomiting, nausea, diarrhea, seizures, low or high blood pressure, chest pains, coma, and death. ACGIH exposure limit (TLV-TWA): 0.1 mg/m³ [skin].

$$C_2H_5O \diagdown \overset{\displaystyle O}{\underset{\displaystyle \parallel}{P}} - S - CH_2 - CH_2 - S - CH_2 - CH_3$$
$$C_2H_5O \diagup$$

and

$$C_2H_5O \diagdown \overset{\displaystyle S}{\underset{\displaystyle \parallel}{P}} - O - CH_2 - CH_2 - S - CH_2 - CH_3$$
$$C_2H_5O \diagup$$

Demeton

demeton-*s*-methyl (C$_6$H$_{15}$O$_3$PS$_2$) — A pale-yellow, oily liquid; MW: 230.30, BP: 118°C, Sol: miscible with most organic solvents, sp. gr. 1.207 at 20°C. It is a cholinesterase inhibitor and is used as a pesticide. It is highly toxic and hazardous to the skin, digestive tract, and respiratory tract and may cause headaches, dizziness, blurred vision, vomiting, diarrhea, and abdominal pains, sweating, anorexia and loss of ability to concentrate. ACGIH exposure limit (TLV-TWA); 0.5 mg/m³.

$$CH_3O \diagdown \overset{\displaystyle O}{\underset{\displaystyle \parallel}{P}} - S - CH_2 - CH_2 - S - CH_2 - CH_3$$
$$CH_3O \diagup$$

Demeton-S-methyl

demineralization — A treatment process that removes dissolved minerals from water.

demiset — Death or destruction.

demodulation — Process of extracting the encoded intelligence from a modulated signal.

demographic transition — A pattern of falling death and birth rates in response to improving living conditions; a pattern of rising death and birth rates in response to deteriorating conditions.

demography — Statistical study of populations with reference to natality, mortality, migration, age, sex, and other socionomic, ethnonymic, and economic factors.

demolition order — A legal process that gives governmental agencies the authority to enter and investigate an abandoned site.

denature — Decharacterization of an article through the addition of some foreign substance in sufficient quantity so as to preclude its original purpose for future use.

dendrite — Any nerve cell process that conducts impulses toward the cell body.

dengue fever — (*disease*) An acute, febrile disease with sudden onset of fever for approximately 5 to 7 days, intense headache, joint and muscle pain, and rash; incubation time is 3 to 15 days. It is caused by the viruses of dengue fever immunological types 1, 2, 3, and 4, which are flaviviruses. It is found in most parts of Asia, Central and South America, Cuba, and Mexico. The reservoir of infection includes people and mosquitos; it is transmitted by the bite of infected mosquitos, *Aedes aegypti*, as well as other Aedes. It is not directly transmitted from person to person and is universally susceptible; it is controlled through community surveys and mosquito control efforts.

dengue hemorrhagic fever shock syndrome — An extremely severe form of dengue fever with symptoms of shock, collapse, prostration, weak thready pulse, and respiratory distress, as well as all of the symptoms of dengue fever.

denitrification — The biochemical reduction of nitrate or nitrite to gaseous molecular nitrogen or an oxide of nitrogen; brought about by denitrifying bacteria.

denitrifying bacteria — The nitrosomonas and nitrobacteria that facilitate reactions in the process of nitrification of wastewater. Nitrosomonas remove three pairs of electrons from ammonia, thus facilitating the formation of nitrite. Nitrobacteria remove electrons from nitrite to form nitrate.

dense nonaqueous-phase liquid — Liquids such as chlorinated hydrocarbon solvents or petroleum fractions with a specific gravity of >1.0; they sink through a water column until they reach a confining layer. Because they are at the bottom of aquifers instead of floating on the water table, their presence is not indicated in typical monitoring wells.

densitometer — A piece of equipment that uses a photoelectric cell to detect differences in the density of light transmitted through a liquid.

density — The weight of a solid or liquid (in grams per milliliter); ratio of the mass of a substance to its volume; also known as specific weight; (*statistics*) components of the birth and death rates of a natural population with stable age distribution.

density gradient — A variation in the density of a solution due to variation of the concentration of a solute in a confined solution.

dental fluorosis — A disorder caused by excessive absorption of fluorine and characterized by brown staining of teeth.

denudation — Wearing away of the land by sun, wind, rain, frost, running water, moving ice, and intrusion of the sea.

deossification — Loss of mineral matter from bones.

deoxygenation — Removal of oxygen from a substance, such as blood or tissue.

deoxyribonucleic acid (DNA) — A long, double-helix molecule in the nucleus of cells that consists of

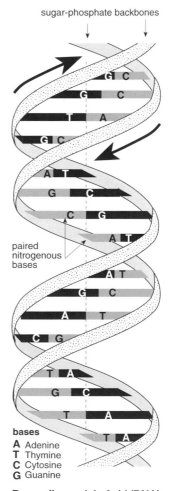

sugar-phosphate backbones

paired nitrogenous bases

bases
A Adenine
T Thymine
C Cytosine
G Guanine

Deoxyribonucleic Acid (DNA)
Double Helix

alternating units of deoxyribose sugar, phosphates, and organic bases; the nucleic acid in chromosomes that contains the master code for the hereditary traits of the organism in the arrangement of its units. DNA transmits the hereditary information and controls cellular activities.

Department of Agriculture Reorganization Act of 1994 — A law giving the Secretary of Agriculture broad authority to reorganize the USDA to achieve greater efficiency, effectiveness, and economy.

dependent variable (statistics) — A variable that may change in response to manipulation of an independent variable by the experimenter.

depleted — Exhausted; reduced.

depletion — The mineral exhaustion of a soil through continued planting of agricultural crops without proper fertilizing; using a resource faster than it can be replenished.

depletion curve — (*hydraulics*) Graphical representation of water depletion from storage-stream channels, surface soil, and groundwater; it can be drawn for base flow, direct runoff, or total flow.

depolarization — Loss of an electrical charge on the surface of a membrane.

depolarized — A reduction in the difference in charge between the outside and inside of a membrane.

depolarizer — A device that removes the effects of polarization.

deposit — A coating of solid matter that is gradually laid down on a surface by a natural process or agent.

deposition — Settling of any material, either liquid or solid, on a surface; (*law*) the sworn testimony of a witness in affidavit form that is obtained outside the courtroom through examination and cross-examination by attorneys for the government and the defense.

depot — Any area of the body, such as fat, where chemicals can be stored for later distribution.

depreciation — Decline in value of capital assets over time with use.

depressant — A substance that reduces a bodily function, activity, or instinctive desire, such as appetite.

depression — (*weather*) See low.

depressor — An agent, such as a chemical, that reduces the activities of the body.

depressurization — A condition that occurs when the air pressure inside a structure is lower than the air pressure outside; it occurs when fireplaces or furnaces consume or exhaust air and are not supplied with enough to make it up; it may be the cause of radon being drawn more rapidly into houses.

derived concentration guide — Concentration of a radionuclide in air or water under conditions of continuous exposure for 1 year by a single pathway,

resulting in an effective dose equivalent of 100 mrem (1 mSv).

derived data — (*computer science*) A new data stream generated by applying an algorithm or model to existing data.

derma — See corium.

Dermacentor — A genus of ticks that includes the species that transmits Rocky Mountain spotted fever.

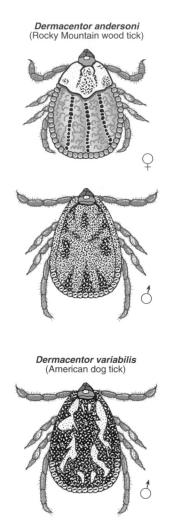

Dermacentor andersoni
(Rocky Mountain wood tick)

♀

♂

Dermacentor variabilis
(American dog tick)

♂

Dermacentor

dermal — Of or referring to the skin or dermis.

dermal absorption and penetration — A process by which a chemical penetrates the skin and enters the body as an internal dose.

dermal absorption factor — A chemical-specific value that accounts for the desorption of a chemical from the soil matrix and absorption of the chemical across the skin.

dermal exposure — Exposure to a chemical by absorption through the skin.

dermal penetration — Entering the body through layers of skin.

dermal toxicity — Adverse effects resulting from exposure of skin to a pesticide, chemical, or other substance that is absorbed through the skin.

dermatitis — Inflammation of the skin resulting from various animal, vegetable, and chemical substances; heat; cold; mechanical irritation; malnutrition; infectious diseases; and psychological problems. Symptoms may include itching, redness, crustiness, blisters, water discharges, fissures, or other changes in the normal condition of the skin.

dermatological injury — Injury to the skin resulting from instantaneous trauma or brief exposure to toxic agents involving a single incident in the work environment.

dermatologist — A doctor who specializes in the diagnosis and treatment of skin problems.

dermatology — The study of the anatomy, physiology, and pathology of skin and the diagnosis and treatment of skin disorders.

Dermatophagoides farinae — A species of household dust mite responsible for allergic reactions in sensitive individuals.

dermatophyte — Pathogenic fungi that attack the epidermis and cause ringworm.

Dermestes lardarius — See larder beetle.

dermis — Thick living layer of skin underneath the thinner epidermis in vertebrates; it contains blood vessels, nerves, and sensory organs.

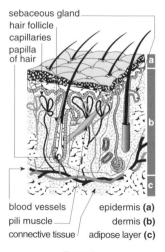

sebaceous gland
hair follicle
capillaries
papilla of hair

blood vessels
pili muscle
connective tissue

epidermis **(a)**
dermis **(b)**
adipose layer **(c)**

Dermis

derrick — A nonmobile tower equipped with a hoist; a crane.

DES — See diethylstilbestrol.

desalination — Purification of saline or brackish water to remove salts by the process of freezing, evaporation, or electrodialysis.

desalinization — The removal of salt from water to make it potable.

desaturation — The formation of an unsaturated chemical compound from a saturated one.

descriptive biochemistry — The precise structural chemistry of animate matter.

descriptive epidemiology — (*statistics*) Mathematical or graphical techniques to study events, frequency, time, place, and population to describe various phenomena and the occurrence of disease.

desensitize — To make an individual insensitive to a stimulant.

desert — A type of biome characterized by low moisture levels and infrequent and unpredictable precipitation so that little or no plant or animal life can survive.

desertification — The spread of desert to previously fertile areas; a process involving the rapid loss of top soil and depletion of plant life that is frequently caused by overgrazing of animals and misuse of the land.

desiccant — A chemical agent used to remove moisture and promote drying.

desiccate — To dry up or remove moisture from a substance.

desiccation — The process of drying; dehydration.

design capacity — The number of tons of solid waste that a designer anticipates an incinerator will be able to process in a 24-hour period.

design ground-level concentration — Maximum acceptable average exposure to a substance at ground level; used to calculate allowable emissions of substances from industries.

design heat load — Total heat loss of a building during the most severe winter conditions.

Designated Agency Safety and Health Official (DASHO) — The executive official of a federal department or agency who is responsible for safety and occupational health matters within the federal agency and is designated or appointed by the head of the agency.

designated area — A geographical area that, according to a presidential major disaster declaration, is eligible to receive disaster assistance in accordance with the law.

designated health planning areas — Geographical areas composed of counties for the purpose of state health planning.

designated uses of water — Water uses identified in state water quality standards that must be achieved and maintained as required under the Clean Water Act.

designer bugs — A non-technical term used for microbes developed through biotechnology that can degrade specific toxic chemicals at their source, in toxic waste dumps or in groundwater.

desorption — Removing a substance from the state of absorption or adsorption.

desorption and vapor extraction system (DAVES) — A low-temperature, fluidized bed to remove organic and volatile inorganic compounds from soils, sediments, and sludges.

desquamation — Shedding skin in scales or sheets.

destratification — The development of vertical mixing within a lake or reservoir to eliminate either totally or partially separated layers of temperature, plant, or animal life.

destruction and removal efficiency (DRE) — A percentage that represents the number of molecules of a compound removed or destroyed in an incinerator relative to the number of molecules entering the system; a DRE of 99.99% means that 9999 molecules are destroyed for every 10,000 that enter the incinerator.

destructive distillation — The process of heating an organic substance, such as coal, in the absence of air to break it down into volatile products and a solid char consisting of fixed carbon and ash.

desulfurization — The removal of sulfur from fossil fuels to reduce pollution.

detectable leak rate — The smallest leak from a storage tank expressed in terms of gallons or liters per hour; a test can reliably determine this level of leakage without a false alarm.

detection — The action of identifying by means of close, continuous or noncontinuous monitoring the appearance of a failure or the existence of a failing element; determination of the presence of a chemical or radiological agent.

detection limit — The analytical capability based on the amount of sample and sensitivity of the analytical method; the lowest concentration of a chemical that can be reliably distinguished from zero concentration.

detector — A device that detects the presence of an entity of interest and indicates its magnitude as the deviation from a reference and converts these indications into a signal; also known as sensor.

detention — The act of detaining an imported product to allow the owner to come forth and defend the lot, much in the same manner as a hearing prior to prosecution; the detention impedes entry into the country with any article that appears to be in violation of any federal food or product law.

detention time — The time drinking water or wastewater is kept in a process vessel during treatment.

detergent — A synthetic cleaning agent containing surfactants and added to water to improve its cleaning properties.

detergent disinfectants — Disinfectants with cleaning ability.

detergent sanitizers — Materials used in highly contaminated areas to first clean the area and then act as a sanitizer to kill bacteria that may cause disease.

deterioration — A condition of gradual worsening.

detonation — The practically instantaneous decomposition or combustion of an unstable compound with a tremendous increase in volume.

detoxification — Reduction of poisonous or toxic properties of a substance in the body.

detoxify — To make harmless by removing toxic properties.

detritus — Unconsolidated sediments comprised of both inorganic and dead and decaying organic material removed from rocks and minerals by mechanical means; remains of any broken-down tissue.

deuterium (D) — The isotope of the element hydrogen with one neutron and one proton in its nucleus and an atomic weight of 2.0144; see also heavy hydrogen.

deuterium oxide — See heavy water.

development location — Place where a development well can be drilled.

development toxicology — Field of study dealing with induction of adverse effects in humans up to the time of puberty.

development well — A well drilled within an approved area of an oil or gas reservoir to the depth of a horizon known to be productive in an attempt to recover proved undeveloped reserves.

developmental disability (DD) — Federally defined as a severe, chronic disability caused by mental or physical impairment, occurring before the age of 22 and likely to continue indefinitely, resulting in substantial functional limitations.

developmental toxicant — An agent affecting growth, development, or acquisition of normal organ function between conception and puberty; chemical, physical, or biological agents that adversely affect the growth or development of an embryo, fetus, or immature individual.

developmental toxicity — The occurrence of adverse effects on a developing organism that may result from exposure to a chemical in either parent prior to conception, during prenatal development, or postnatally to the time of sexual maturation.

deviation — Failure to meet a required critical limit for a critical control point; the difference between the actual value of a controlled variable and the desired or expected value.

deviation — (*ergonomics*) Movement of a body part toward its extreme in its range of motion.

deviation signal — Difference between a setpoint and the measured value.

device — Piece of equipment used for a specific function.

dew — Water vapor condensed on solid surfaces that have cooled below the condensation point of the air in contact with them because of radiation cooling during the night.

dew point — The temperature and pressure at which condensation occurs for a given concentration of water vapor in the air.

dewar — A double chamber used to maintain the temperature of superconducting magnetic coils at near absolute zero in nuclear magnetic resonance imaging.

dewatering — (*sludge*) The removal of water from sludge so that its physical form is changed from essentially that of a fluid to that of a damp solid.

dextrose — A glucose solution used for intravenous administration.

DFO — See disaster field office.

DFP ($C_6H_{14}FO_3P$) — An odorless, clear to pale-yellow oily liquid; MW: 184.2, BP: 62°C, Sol: 1.54 g/100 g water at 20°C, Fl.P: NA (a fire may produce irritating or poisonous gases), sp. gr. 1.055. An organophosphate chemical formally used as an insecticide and now considered as a chemical warfare agent. It is hazardous to the respiratory system, digestive system, skin, and eyes. Acute exposure may result in central nervous system depression, agitation, confusion, delirium, coma, seizures, and death. It also causes bradycardia, vomiting, and diarrhea. Chronic exposure causes dermal sensitization, poor coordination in arms and legs, personality changes, depression, anxiety, and irritability. OSHA exposure limit (8-hour TWA): NA.

DHFS — See dengue hemorrhagic fever shock syndrome.

D/I — Acronym for deionizing unit.

diabetes mellitus — A metabolic condition resulting from lack of insulin production in the pancreas such that the body loses its ability to store or oxidize sugar efficiently.

diacetone alcohol ($CH_3COCH_2C(CH_3)_2OH$) — A colorless liquid with a faint, minty odor; MW: 116.2, BP: 334°F, Sol: miscible, Fl.P: 125°F, sp. gr. 0.94. It is used in the manufacture of artificial silk and leather, quick-drying inks, photographic film, antifreeze preparations, and hydraulic fluids; as a solvent for cellulose esters, epoxy resins, hydrocarbons, oils, fats, resin gums, dyes, tars, cements, and waxes; in paper and textile coatings, wood stains, and preservatives. It is hazardous to the eyes, skin, and respiratory system and is toxic through inhalation, ingestion, and contact. Symptoms of exposure include corneal tissue damage, narcosis, and eye, nose, throat, and skin irritation. OSHA exposure limit (TWA): 50 ppm [air] or 240 mg/m^3.

diacetyl peroxide ($C_4H_6O_4$) — A colorless liquid with a pungent odor; MW: 118.10, BP: 63°C, Sol: soluble in alcohol and ether, Fl.P: a flammable and highly dangerous compound that is shock sensitive and will explode, density: NA. It is used in dilute solutions as a free-radical source to initiate polymerization and organic synthesis. It is hazardous to the eyes and skin and causes severe irritation. ACGIH exposure limit: NA.

Diacetyl peroxide

diagnosis — A concise, technical description of the cause, nature, or manifestations of a condition, situation, or problem by identifying its signs and symptoms.

diagnostic radiology — The use of x-rays in medicine for identifying disease or injury of patients.

dialifor ($C_{14}H_{17}ClNO_{14}PS_2$) — A white crystalline solid or colorless oily liquid; MW: 393.84, BP: NA, Sol: dissolves in most organic solvents, density: NA. It is a cholinesterase inhibitor and is used as a pesticide. It is highly toxic and hazardous to the skin, digestive tract, and respiratory tract and may cause weakness, headache, dizziness, tightness in the chest, blurred vision, slurred speech, mental confusion, drowsiness, nausea, vomiting, diarrhea, difficulty in breathing, convulsions, coma, and death.

Dialifor

dialog — (*computer science*) A complex form of computer-assisted instruction in which the student is actively involved with a computer in the learning process.

dialysis — The diffusion of solute molecules through a semipermeable membrane passing from the side of higher concentration to that of lower concentration; the process of separating a true solution from a colloidal dispersion by means of a membrane.

D

diameter of a circle — A line segment that passes through the center of a circle and which is equal to 2× the radius.

Diameter-Index Safety System — A system of standardized connections between cylinders of medical gases and flow meters or pressure regulators.

diapause — A period of arrested development in the life-cycle of insects and certain other animals in which physiological activity is very low and the animal is highly resistant to unfavorable external conditions.

diapause phase — A period of dormancy, common in insect species, that occurs during the more rigorous portions of the annual climatic cycle.

diaphoresis — Profuse sweating that occurs due to elevated body temperature, physical exertion, exposure to heat, mental or emotional stress, and heart attack.

diaphragm — A sensing element consisting of a membrane that is deformed by a pressure differential applied across it; (*anatomy*) the thick sheet of muscle that assists in breathing and separates the chest and the abdomen in mammals. See also membrane.

diaphragm pump — A pump that moves water by reciprocating motion of a diaphragm in a chamber having both inlet and outlet check valves.

diarrhea — A diffuse and abnormal discharge from the bowels.

diastase — An enzyme controlling the digestion of starch to maltose.

diastole — The period of time between contractions of the atria or ventricles during which blood enters the relaxed chambers from the circulatory system and the lungs.

diastolic blood pressure — The minimum level of blood pressure measured between contractions of the heart.

diastolic pressure — Blood pressure reading taken during the time between heartbeats when the heart chamber wall is relaxed.

diathermy — The generation of heat in tissues by electric currents for medical, therapeutic, or surgical purposes.

diatomaceous earth — A white, yellow, or light-gray powder composed of the fossilized skeletons of one-celled organisms called diatoms; it is highly porous and contains microscopic holes. It is used as a filter and paint filler, adsorbent, abrasive, and thermal insulator. It is also used to purify liquids, in the manufacture of fire brick and heat insulators, and in metal polishers. Under normal conditions of occupational exposure, fibrogenic or toxic effects are insignificant. ACGIH exposure limit (TLV-TWA): 10 mg/m³ as total dust.

diatomaceous earth filter — A filtering system that uses a hydrous form of silica or opal composed of the shells of diatoms; a filter designed to use diatoma-

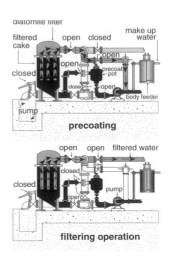

Diatomaceous Earth Filter
(Diatomite Filter)

ceous earth as a filter medium and which may be either pressure or vacuum tight.

diatomite — A shortened name for diatomaceous earth.

diatoms — Microscopic, unicellular algae whose silicious skeletons are often found in seafloor sediments.

diazinon ($C_{12}H_{21}N_2O_3PS$) — A colorless liquid with a faint fruity or ester-like odor; MW: 304.36, BP: decomposes, Sol (water): 0.004% at 20°C, Fl.P: 83°C, sp. gr. 1.12. It is an organophosphate insecticide for non-agricultural applications but is used for home and garden insect control. It is a cholinesterase inhibitor and is hazardous to the respiratory system, digestive system, skin, and eyes. Acute exposure may result in central nervous system depression, agitation, confusion, delirium, coma, seizures, and death. It also causes bradycardia and diarrhea. Chronic exposure may cause dermal sensitization, poor coordination in the arms and legs, personality changes, depression, anxiety, and irritability. It is mutagenic and damages a developing fetus. OSHA exposure limit (8-hour TWA): is 0.1 mg/m³.

Diazinon

diazomethane (CH_2N_2) — A yellow gas with a musty odor; shipped as a liquefied compressed gas; MW: 42.1, BP: –9°F, Sol: reacts. It is used as a methylating agent in chemical analysis and laboratory organic synthesis. It is hazardous to the respiratory system, skin, and eyes and is toxic through inhalation, ingestion, and contact. Symptoms of exposure include cough, shortness of breath, headache, flushed skin, fever, chest pain, pulmonary edema, pneumonitis, eye irritation, asthma, and fatigue. OSHA exposure limit (TWA): 0.2 ppm [air] or 0.4 mg/m³.

dibenz(*a,h*)anthracene (C₂₂H₁₄) — One of the polycyclic aromatic hydrocarbon compounds formed when gasoline, garbage, or any animal or plant material burns, usually found in smoke and soot; MW: 278.35, BP: sublimes, Sol: 0.5 mg/l, Fl.P: not known. It is used as a research material; coal-tar pitch is used primarily as a binder for electrodes in the aluminum reduction process to bind the carbon electrodes in the reduction pots; as an adhesive in membrane roofs; as a wood preservative; in clinical treatment of skin disorders such as eczema, dermatitis, and psoriasis. It is potentially carcinogenic. OSHA exposure limits (TWA): 0.2 mg/m³ [air].

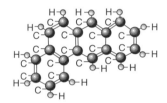

Dibenzo (a,h) anthracene

dibenz(*a, j*)anthracene (C₂₂H₁₄) — An aromatic hydrocarbon that has been shown to cause skin cancer in animals; MW: 278.36.

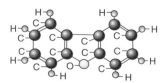

Dibenz[*a,j*]anthracene

dibenzofuran — An organic compound that contains two benzene rings fused to a central furan ring. Chloro-dibenzofurans are serious contaminants of some hazardous waste sites; see also chlorodibenzofurans.

Dibenzofuran

diborane (B₂H₆) — A colorless gas with a repulsive, sweet odor; usually shipped in pressurized cylinders; MW: 27.7, BP: –135°F, Sol: reacts. It is used as a reducing agent in the synthesis of organic chemical intermediates; as a component or additive for high-energy fuels; as a catalyst in olefin polymerization; in the electronics industry to improve crystal growth or impart electrical properties in pure crystals. It is hazardous to the respiratory and central nervous systems and is toxic through inhalation and contact. Symptoms of exposure include tight chest, precordial pain, shortness of breath, nonproductive cough, nausea, headache, lightheadedness, vertigo, chills, fever, fatigue, weakness, tremors, and muscle fasciculation; in animals, liver and kidney damage, pulmonary edema, and hemorrhage are observed. OSHA exposure limit (TWA): 0.1 ppm [air] or 0.1 mg/m³.

1,2-dibromo-3-chloropropane (CH₂BrCHBrCH₂Cl) — A dense yellow or amber liquid with a pungent odor at high concentrations; a solid below 43°F; MW: 236.4, BP: 384°F, Sol (70°F): 0.1%, Fl.P (oc): 170°F, sp. gr. 2.05. It is used as a pesticide and fumigant. It is hazardous to the central nervous system, skin, liver, kidneys, spleen, reproductive system, and digestive system and is toxic through inhalation, absorption, and contact. Symptoms of exposure include drowsiness; nausea; vomiting; irritation of the eyes, nose, skin, and throat; and pulmonary edema; carcinogenic. OSHA exposure limit (TWA): 0.001 ppm [air].

1,2-Dibromo-3-chloropropane (Dibromochloropropane)

1,2-dibromoethane (BrCH₂CH₂Br) — A colorless liquid with a mild, sweet odor; MW: 187.86, BP: 131 to 132°C, Sol (20°C): slightly soluble in water and miscible with most organic solvents, Fl.P: not flammable. It is used as an additive in leaded gasoline where it acts as a scavenger that converts lead oxides in cars to lead halides; as a soil fumigant to protect against insects, pests, and nematodes in citrus, vegetable, and grain crops; as a fumigant for turf, especially on golf courses. It is currently banned as a soil and grain fumigant. The respiratory tract, particularly the nasal cavity, is the point-of-contact target organ affected by inhalation. Symptoms of exposure include pharyngitis, bronchitis, conjunctivitis, anorexia, headache, and depression;

the liver is the target organ for toxic effects in experimental animals; renal effects reported also in experimental animals. The reproductive effects are antispermatogenic, producing a change in sperm velocity and a possible decrease in male fertility. Nasal carcinomas are found in laboratory animals. Renal lesions have been recorded in humans dying after acute oral exposure; endocrine lesions have also been found. Fatal cases of occupational exposure have been reported. OSHA exposure limits (PEL TWA): 20 ppm; ceiling, 30 ppm. Acceptable maximum peak above ceiling level (5 minutes per 8-hour shift): 50 ppm. STEL (15 minutes): 0.5 ppm [air].

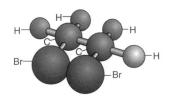

**1,2–Dibromoethane
(Ethylene Dibromide)**

dibutyl phosphate (C₄H₉O₂(OH)PO) — A pale-amber, odorless liquid; MW: 210.2, BP: 212°F (decomposes), Sol: insoluble, Fl.P: not known, sp. gr. 1.06. It is used as a catalyst in the manufacture of phenolic and urea resins and in metal separation and extraction. It is hazardous to the respiratory system and skin and is toxic through inhalation, ingestion, and contact. Symptoms of exposure include headache and irritation of the respiratory system and skin. OSHA exposure limit (TWA): 1 ppm [air] or 5 mg/m³.

dibutyl phthalate (C₆H₄(COOC₄H₉)₂) — A colorless to faintly yellow, oily liquid with a slight, aromatic odor; MW: 278.3, BP: 644°F, Sol: 0.5% at 77°F, Fl.P: 315°F, sp. gr. 1.05. It is used in the manufacture of nitrile rubber; in the manufacture of polyvinyl acetate surface coatings; in the production of cellulose acetate butyrate and polyvinyl acetate adhesives; in the manufacture of polyester and epoxy resins and nitrocellulose surface coatings; in the manufacture of explosives and propellants. It is

hazardous to the respiratory system and gastrointestinal system and is toxic through inhalation, ingestion, and contact. Symptoms of exposure include upper respiratory tract and stomach irritation. OSHA exposure limit (TWA): 5 mg/m³ [air].

dibutyl sulfate (C₈H₁₈O₄S) — Moderately toxic by ingestion; MW: 210.32.

dicapthon (C₈H₉ClNO₅PS) — An odorless white crystalline solid; MW: 297.66, BP: NA, Sol: insoluble, Fl.P: NA (a flammable, combustible material that may be ignited by heat, sparks, or flames; vapors may travel to a source of ignition and flash back; containers may explode in a fire), sp. gr. NA. It is an organophosphate chemical primarily used as an insecticide, aphicide, and acaricide. It is hazardous to the respiratory system, digestive system, skin, and eyes. Acute exposure may cause central nervous system depression, agitation, confusion, delirium, coma, seizures, and death, as well as bradycardia, vomiting, and diarrhea. Chronic exposure may result in dermal sensitization, poor coordination in arms and legs, personality changes, depression, anxiety, and irritability. OSHA exposure limit (8-hour TWA); NA.

1,3-dichloro-5,5-dimethylhydantion (C₅H₆Cl₂N₂O₂) — A white powder with a chlorine-like odor; MW: 197.0, BP: not known, Sol: 0.2%, Fl.P: 346°F, sp. gr. 1.5. It is used as a bactericide, sporicide, or sanitizer in swimming pools, dairies, laundries, restaurants, cutting oils, and the pharmaceutical industry; as a general cleanser; as a bleaching agent; as a stabilizer, discoloration preventer, and catalyst in the polymer industry; to emboss or texturize resinous sheet preparations. It is hazardous to the respiratory system and eyes and is toxic through inhalation, ingestion, and contact. Symptoms of exposure include eye, mucous membrane, and respiratory system irritation. OSHA exposure limit (TWA): 0.2 mg/m³ [air].

1,1-dichloro-1-nitroethane (CH₃CCl₂NO₂) — A colorless liquid with an unpleasant odor; MW: 143.9, BP: 255°F, Sol: 0.3%, Fl.P: 136°F, sp. gr. 1.43. It is used as an insecticidal fumigant for grain. It is

$$CH_3-CH_2-CH_2-CH_2-O-\overset{\displaystyle O}{\underset{\displaystyle O}{\overset{\|}{\underset{\|}{S}}}}-O-CH_2-CH_2-CH_2-CH_3$$

Dibutyl sulfate

hazardous to the lungs and is toxic through inhalation, ingestion, and contact. Symptoms of exposure in animals include eye and skin irritation; liver, heart, and kidney damage; pulmonary edema; and hemorrhage. OSHA exposure limit (TWA): 2 ppm [air] or 10 mg/m^3.

1,2-dichlorobenzene (C₆H₄Cl₂) — A colorless to pale-yellow liquid with a faint aromatic odor; MW: 147.00; BP: 180.5°C; Sol: miscible with organic solvents; Fl.P: 66°C (closed cup), vapors form explosive mixtures with air; density: 1.306 at 30°C. It is used as a solvent, fumigant, insecticide for termites, and a degreasing agent for metals, wool, and leather. It is hazardous to the respiratory tract, digestive tract, and skin. It causes lacrimation, depression of the central nervous system, anesthesia, and liver damage. OSHA exposure limit: 50 ppm.

1,2-Dichlorobenzene

1,3-dichlorobenzene (C₆H₄Cl₂) — A colorless, odorless liquid; MW: 147, BP: 173.53°C, Sol: 123 mg/l water at 25°C, Fl.P: 151°F, sp. gr. 1.29 at 20°C. It is used as a fumigant and an insecticide. It is hazardous to the respiratory system, digestive system, skin, and eyes. Acute exposure may cause central nervous system excitation, depression, tremors, seizures, nausea, vomiting, diarrhea, respiratory depression, pneumonitis, acute renal failure, nervousness, amnesia, and skin irritation. Chronic exposure may cause ocular toxicity, systemic toxicity, anorexia, nausea, fatigue, headaches, gastrointestinal difficulties, memory deficits, personality changes, numbness, loss of coordination, and seizures. OSHA exposure limit (8-hour TWA): 50 ppm.

1,4-dichlorobenzene (C₆H₄Cl₂) — A colorless or white crystalline solid with a mothball-like odor; MW: 147, BP: 173.53°C, Sol: 65.3 mg/l water at 25°C, Fl.P: 65.5°C, sp. gr. 1.45 at 20°C. It is used in liquid form as a deodorant; in the control of mildew and molds; in toilet blocks as a disinfectant. It is hazardous to the respiratory system, skin, and eyes. Acute exposure can cause headaches; dizziness; nausea; swelling around the eyes, hands, and feet; and skin burns. Chronic exposure can cause damage to the nervous system, weakness, trem-

bling, and numbness in the arms and legs. OSHA exposure limit (8-hour TWA): 75 ppm.

Cl
|

Cl

1,4-Dichlorobenzene

o-dichlorobenzene (C₆H₄Cl₂) — A colorless to pale-yellow liquid with a pleasant, aromatic odor; MW: 147.0, BP: 357°F, Sol: 0.01%, Fl.P: 151°F, sp. gr. 1.30. It is used in cleaning and degreasing metals, leather, paper, wool, drycleaning, brick, and upholstery; as a fumigant for poultry houses and stockyards; in organic synthesis in pesticides, herbicides, dyestuffs, and pharmaceuticals; as a chemical intermediate in the manufacture of toluene-diisocynate; as a deodorizing agent; in textile dyeing operation. It is hazardous to the liver, kidneys, skin, and eyes and is toxic through inhalation, absorption, ingestion, and contact. Symptoms of exposure include eye and nose irritation, liver and kidney damage, skin blistering. OSHA exposure limit (TWA): ceiling 50 ppm [air] or 300 mg/m^3.

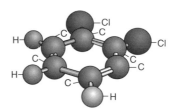

**O–Dichlorobenzene
(1,2–Dichlorobenzene)**

p-dichlorobenzene (C₆H₄Cl₂) — A colorless or white crystalline solid with a mothball-like odor; MW: 147.0, BP: 345°F, Sol: 0.008%, Fl.P: 150°F, sp. gr. 1.25. It is used in moth control; as a deodorant for garbage and restrooms; as an insecticide; in organic synthesis for preparation of dye intermediates. It is hazardous to the liver, respiratory system, eyes, kidneys, and skin and is toxic through inhalation, ingestion, and contact. Symptoms of exposure include headache, eye irritation, periorbital swelling, profuse rhinitis, anorexia, nausea, vomiting, low weight, jaundice, and cirrhosis; carcinogenic. In animals, it causes liver and kidney damage. OSHA exposure limit (TWA): 75 ppm [air] or 450 mg/m³.

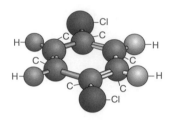

**p–Dichlorobenzene
(1,4–Dichlorobenzene)**

3,3′-dichlorobenzidine (C₆H₃ClNH₂C₆H₃ClNH₂) — A gray to purplish crystalline solid; MW: 253.1, BP: 788°F, Sol: almost insoluble, Fl.P: not known. It is used as an intermediate in the manufacture of dyes, pigments, and isocyanate-containing polymers. It is hazardous to the bladder, liver, lungs, skin, and gastrointestinal tract and is toxic through inhalation, absorption, ingestion, and contact. Symptoms of exposure include allergic skin reaction, sensitization, dermatitis, headache, dizziness, caustic burns, frequent urination, dysuria, gastrointestinal upsets, and upper respiratory infection; carcinogenic. OSHA exposure limit (TWA): regulated use.

dichlorodifluoromethane (CCl₂F₂) — A colorless gas with an ether-like odor at extremely high concentrations; shipped as a liquefied compressed gas; MW: 120.9, BP: –22°F, Sol: 0.03% at 77°F. It is used as a propellant and refrigerant; in the manufacture of aerosols for cosmetics, pharmaceuticals, insecticides, paints, adhesives, and cleaners; as a blowing agent for cellular polymers; as a foaming agent in fire-extinguishing aerosols; in water purification and copper and aluminum purification; in the manufacture of glass bottles; in regulating devices for leak detection; in thermal expansion valves; in organic synthesis of Freon. It is hazardous to the cardiovascular system and peripheral nervous system and is toxic through inhalation and contact. Symptoms of exposure include dizziness, tremors, unconsciousness, cardiac arrhythmias, and cardiac arrest. OSHA exposure limit (TWA): 1000 ppm [air] or 4950 mg/m³.

dichlorodiphenyltrichloroethane — See DDT.

1,1-dichloroethane (CHCl₂CH₃) — A colorless, oily liquid with a chloroform-like odor; MW: 99.0, BP: 135°F, Sol: 0.6%, Fl.P (oc): 22°F, sp. gr. 1.18. It is used as a dewaxer of mineral oils; as an extractant for heat-sensitive substances; as a fumigant; in the manufacture of vinyl chloride by vapor-

phase cracking; in the manufacture of high-vacuum rubber and silicon grease; as a chemical intermediate. It is hazardous to the skin, liver, and kidneys and is toxic through inhalation, ingestion, and contact. Symptoms of exposure include central nervous system depression, skin irritation, and liver and kidney damage. OSHA exposure limit (TWA): 100 ppm [air] or 400 mg/m³.

1,1–Dichloroethane

1,2-dichloroethane (C₂H₄Cl₂) — A colorless liquid with a pleasant odor and sweet taste; MW: 98.96; BP: 83.5°C; Sol: miscible in organic solvents; Fl.P: 13°C (closed cup), a flammable liquid; density: 1.257 at 20°C. It is used in paint removers, as a fumigant, as a degreaser, as a wetting agent, and as a solvent. It is hazardous to the respiratory system, digestive system, skin, and eyes. It causes central nervous system depression, irritation of the eyes, nausea, vomiting, diarrhea, cyanosis, pulmonary edema, and coma; may produce injury to the kidney and liver; and may cause death. ACGIH exposure limit: 10 ppm.

**Ethylene Dichloride
(1,2–Dichloroethane)**

1,2-dichloroethene (C₂H₂Cl₂) — A highly flammable, colorless liquid with a sharp, harsh odor; MW: 96.94, BP: 60.3°C; Sol (25°C): 3500 mg/l, Fl.P: 6°C, sp. gr. 1.2837. It is used as a chemical intermediate in the synthesis of chlorinated solvents and compounds; as a low-temperature extraction solvent for organic materials such as dyes, perfumes, lacquers, and thermoplastics. Fatality results from inhalation of the vapor in a small enclosure; inhalation of high concentrations depresses the central nervous system, causing nausea, drowsiness, fatigue, vertigo, and intracranial pressure. Biochemical changes in the liver have been reported in mice and rats. OSHA exposure limits PEL (8-hour TWA): 200 ppm [air].

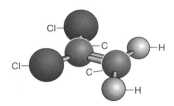

1,2–Dichloroethene

dichloroethyl ether ((ClCH₂CH₂)₂O) — A colorless liquid with a chlorinated, solvent-like odor; MW: 143.0, BP: 352°F, Sol: 1%, Fl.P: 131°F, sp. gr. 1.22. It is used as a solvent and dewaxing agent in the petroleum industry; in the manufacture of oils, fats, waxes, gums, tars, resins, soaps, ethyl cellulose, paints, varnishes, and lacquers; as a scouring and penetrating agent in the textile industry; as a fumigant; as a chemical intermediate in synthesis during the manufacture of pharmaceuticals, rubber chemicals, resins, plasticizers, and chemicals; as a lead scavenger during production of gasoline engine anti-knock compounds. It is hazardous to the respiratory system, skin, and eyes and is toxic through inhalation, absorption, ingestion, and contact. Symptoms of exposure include lacrimation, nose and throat irritation, cough, nausea, and vomiting; carcinogenic; in animals, pulmonary irritation and edema and liver damage have been observed. OSHA exposure limit (TWA): 5 ppm [air] or 30 mg/m³.

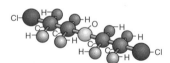

Dichloroethyl Ether
(*Bis* (2-Chloroethyl) ether)

1,1-dichloroethylene (C₂H₂Cl₂) — A colorless liquid with a milky, chloroform-like odor, MW: 96.94; BP: 31.7°C; Sol: slightly soluble in water and miscible with organic solvents; Fl.P: –18°C (closed cup), a flammable liquid; density: 1.213 at 20°C.

$$\begin{array}{c} Cl \\ \diagdown \\ \\ \diagup \\ Cl \end{array} C = CH$$

1,1-Dichloroethylene

It is used to produce copolymers for films and coatings. It is hazardous to the respiratory tract and eyes and causes irritation of mucous membranes and narcotic effects. ACGIH exposure limit (TLV-TWA): 5 ppm.

1,2-dichloroethylene (ClCHCHCl) — A colorless liquid (usually a mixture of *cis*- and *trans*-isomers) with a slightly acrid, chloroform-like odor; MW: 97.0, BP: 118 to 140°F, Sol: 0.4%, Fl.P: 36°F, sp. gr. 1.27 at 77°F. It is used as a low-temperature solvent for heat-sensitive substances in extraction of caffeine, perfume oils, and fats from the flesh of animals; in rubber and dye industries; as a direct solvent in gums, waxes, oils, camphor, and phenol; in solvent mixtures for esters and ether derivatives, lacquers, resins, thermoplastics, and artificial fibers; in organic synthesis; in miscellaneous applications such as liquid drycleaning agents, cleaning solutions for circuit boards, food packaging adhesives, and germicidal fumigants. It is hazardous to the respiratory system, eyes, and central nervous system and is toxic through inhalation, ingestion, and contact. Symptoms of exposure include central nervous system depression and irritation of the eyes and respiratory system. OSHA exposure limit (TWA): 200 ppm [air] or 790 mg/m³.

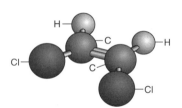

1,2–Dichloroethylene

dichloromonofluoromethane (CHCl₂F) — A colorless gas with an ether-like odor (liquid below 48°F) that is shipped as a liquefied compressed gas; MW: 102.9, BP: 48°F, Sol: 0.7% at 86°F. It is used as an aerosol propellant in pharmaceuticals; as a refrigerating agent; as an industrial solvent in the separation of fatty acids or polymers; as a blowing agent for rigid foams and insulating foams; as a working fluid in Rankine cycle turbogenerators; in water purification; in organic synthesis; in the preparation of other types of Freon. It is hazardous to the respiratory system and cardiovascular system and is toxic through inhalation, ingestion, and contact. Symptoms of exposure include asphyxia, cardiac arrhythmias, and cardiac arrest. OSHA exposure limit (TWA): 10 ppm [air] or 40 mg/m³.

D

2,4-dichlorophenol (C$_6$H$_4$Cl$_2$O) — A white solid with a medicinal smell; MW: 163.00, BP: 210°C, Sol (20°C): soluble, Fl.P: 93.3°C. It is used mostly to make other organic compounds and as an intermediate in production of herbicides, disinfectants, and mothproofing compounds. Through inhalation or dermal contact, symptoms of exposure include liver injury, chloracne, porphyria, hyperpigmentation, and hirsutism. Little evidence is available to link this chemical with cancer; although human cancer cases have been reported, the individuals were also exposed to pesticides, including volatile compounds produced as intermediates. No OSHA standards are available; U.S. EPA guidelines: 3.09 mg/l [water].

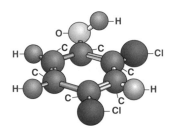

2,4–Dichlorophenol

2,4-dichlorophenoxyacetic acid (2,4-D) (Cl$_2$C$_6$H$_3$OCH$_2$ COOH) — A white to yellow crystalline, odorless powder; MW: 221.0, BP: decomposes, Sol: 0.05%, sp. gr. (86°F): 1.57. It is used as an herbicide and plant hormone. It is hazardous to the skin and central nervous system and is toxic through inhalation, absorption, ingestion, and contact. Symptoms of exposure include weakness, stupor, hyperflexia, muscle twitch, convulsions, dermatitis; in animals, liver and kidney damage has been observed. OSHA exposure limit (TWA): 10 mg/m^3 [air].

1,2-dichloropropane (C$_3$H$_6$Cl$_2$) — A colorless liquid with a chloroform-like odor; MW: 112.99, BP: 96.4°C, Sol: 0.26 g/100 g water at 20°C, Fl.P: 21°C, sp. gr. 1.159. It is used as an insecticide for stored grain and in a mixture with other chlorinated hydrocarbons as an insecticide on crops and for livestock, as well as other agricultural applications. It is hazardous to the respiratory system, digestive system, skin, and eyes. Acute exposure can cause irritation of the eyes, nose, throat, and air passages and can cause damage to the liver, kidneys, adrenal glands, and brain. Symptoms include headaches, lightheadedness, nausea, a feeling of drunkenness, changes in vision and speech, poor muscle coordination, shakes, seizures, life-threatening heart rhythms, coma, and death. Chronic exposure can cause skin irritation and rash and damage to the

liver, kidneys, adrenal glands, and brain. OSHA exposure limit (8-hour TWA): 75 ppm.

1,3-dichloropropene (C$_3$H$_4$Cl$_2$) — A colorless liquid with a sweet smell; MW: 110.98, BP: 104°C at 1 atm, Sol: 2700 ppm at 25°C, Fl.P: 35°C. It is used predominantly as a component of several formulations in agriculture as soil fumigants for parasitic nematodes. Symptoms of exposure include respiratory effects, mucous membrane irritation, chest pain, cough, and breathing difficulties. No OSHA exposure limits; ACGIH guidelines (TWA-TLV): 1 ppm [air].

dichlorotetrafluoroethane (CClF$_2$CClF$_2$) — A colorless gas with a faint, ether-like odor at high concentrations; a liquid below 38°F; shipped as a liquefied compressed gas; MW: 170.9, BP: 38°F, Sol: 0.01%. It is used in the manufacture of aerosols with other types of Freon to lower vapor pressure and produce nonflammable aerosol propellants; as a refrigerant in industrial cooling and air conditioning systems; as a blowing agent for cellular polymers; in the preparation of explosives; in the extraction of volatile substances; as a foaming agent in fire extinguishers and aerosols; in organic synthesis; in strengthening glass bottles; in magnesium refining. It is hazardous to the respiratory system and cardiovascular system and is toxic through inhalation, ingestion, and contact. Symptoms of exposure include asphyxia, cardiac arrhythmias, cardiac arrest, and respiratory system irritation. OSHA exposure limit (TWA): 1000 ppm [air] or 7000 mg/m^3.

dichlorvos ((CH$_3$O)$_2$P(O)OCHCCl$_2$) — A colorless to amber liquid with a mild, chemical odor; MW: 221.0, BP: decomposes, Sol: 0.5%, Fl.P (oc): >175°F, sp. gr. (77°F): 1.42. It is used as an insecticide. It is hazardous to the respiratory system, cardiovascular system, central nervous system, eyes, skin, and blood cholinesterase and is toxic through inhalation, absorption, ingestion, and contact. Symptoms of exposure include miosis, eye aches, rhinorrhea, headache, tight chest, wheezing, laryngeal spasms, salivation, cyanosis, anorexia, nausea, vomiting, diarrhea, sweating, muscle fasciculation, paralysis, giddiness, ataxia, convulsions, low blood pressure, cardiac irregularities, eye, and skin irritation. OSHA exposure limit (TWA): 1 mg/m^3 [skin].

dichromates — Sodium dichromate (Na$_2$Cr$_2$O$_7$) and potassium dichromate (K$_2$Cr$_2$O$_7$) are powerful oxidizing materials that decompose on heating and may ignite readily in a violent manner. They are used for bleaching resins, oils, and waxes; in pyrotechniques and safety matches; and as corrosion inhibitors. They are corrosive substances causing ulceration of the hands and injury to mucous membranes. Ingestion may be lethal to humans. ACGIH exposure limit (TLV-TWA): 0.05 mg/m^3 [air].

Dick test — A skin test to determine immunity to scarlet fever; *Streptococcus pyogenes* toxin is injected intracutaneously and produces a reaction if no circulatory antitoxin is present.

dicobalt octacarbonyl (Co₂(CO)₈) — An orange crystalline solid; MW: 391.94, BP: NA, Sol: soluble in organic solvents, Fl.P: NA, density: NA. It is used as a catalyst in organic conversion reactions. It is hazardous to the respiratory tract and digestive tract and exhibits moderate toxicity. ACGIH exposure limit (TLV-TWA): 0.1 mg/m^3 as Co.

Dicobalt octacarbonyl

die — A hard metal or plastic form used to shape material to a particular contour.

dieldrin (C₁₂H₈Cl₆O) — Colorless to light-tan crystals obtained by oxidation of aldrin and having a mild, chemical odor; MW: 380.9, BP: decomposes, Sol: 0.02%, sp. gr. 1.75. It is used as an insecticide and to mothproof carpets and other furnishings. It is hazardous to the central nervous system, liver, kidneys, and skin and is toxic through inhalation, absorption, ingestion, and contact. Symptoms of exposure include headache, dizziness, nausea, vomiting, malaise, sweating, myoclonic limb jerks, clonic tonic convulsions, and coma; carcinogenic. In animals, liver and kidney damage are observed. OSHA exposure limit (TWA): 0.25 mg/m^3 [skin].

dielectric — A nonconductor of direct electric current.

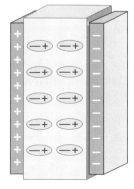

Dielectric
inserted in charged capacitor

dielectric constant — A property of material that, when multiplied by the permitivity of free space, determines the electric energy stored per unit volume per unit electric field.

diesel engine — A type of internal-combustion engine that uses a fuel injector and produces combustion temperatures by compression.

diesel oil — An oil obtained from distillation of petroleum and composed mainly of aliphatic hydrocarbons.

***N,N*-diethyl-*p*-phenylene-diamine** — Chemical used for the measurement of the chlorine residual in water by either titrating or comparing a developed color with color standards.

diethyl phthalate (C₁₂H₁₄O₄) — An odorless, colorless, oily liquid; MW: NA, BP: 295°C, Sol: 1.08 g/100 g water at 25°C, Fl.P: 163°C, sp. gr. 1.32. It is used as a solvent, as a vehicle for pesticide sprays, and in perfume manufacture. It is hazardous to the respiratory system, digestive system, and skin. Acute exposure may cause irritation of the eyes, nose, and throat. Chronic exposure may cause nerve damage, pain, numbness, or weakness in the arms and legs. It should be handled as a teratogen. OSHA exposure limit (8-hour TWA): NA.

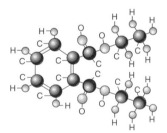

Diethyl Phthalate

diethyl sulfate (C₄H₁₀O₄S) — A colorless, oily liquid with an odor of peppermint; MW: 154.19, BP: 208°C, Sol: miscible with alcohol and ether and decomposes rapidly in hot water, density: 1.172 at 25°C. It is used as an ethylating agent in organic synthesis. It is a severe skin irritant and may cause cancer.

$$CH_3-CH_2-O-\overset{\overset{\textstyle O}{\|}}{\underset{\underset{\textstyle O}{\|}}{S}}-O-CH_2-CH_3$$

Diethyl sulfate

diethylamine ((C₂H₅)₂NH) — A colorless liquid with a fishy, ammonia-like odor; MW: 73.1, BP: 132°F, Sol: miscible, Fl.P: –15°F, sp. gr. 0.71. It is used in the preparation of textile finishing agents, surfactants, rubber processing chemicals, agricultural

D

chemicals, and pharmaceuticals; as a polymerization inhibitor and catalyst; as an intermediate in the dye industry; as a depilatory of animal skins; in electroplating solutions. It is hazardous to the respiratory system, skin, and eyes and is toxic through inhalation, absorption, ingestion, and contact. Symptoms of exposure include eye, skin, and respiratory system irritation. OSHA exposure limit (TWA): 10 ppm [air] or 30 mg/m^3.

2-diethylaminoethanol ((C$_2$H$_5$)$_2$NCH$_2$CH$_2$OH) — A colorless liquid with a nauseating, ammonia-like odor; MW: 117.2, BP: 325°F, Sol: miscible, Fl.P: 126°F, sp. gr. 0.89. It is used in the preparation of medicinals, pharmaceuticals, pesticides, protective surface coatings for metals, emulsifying agents for polishes, resinous materials for treating fiber surfaces, and fluorescent brightening agents; in polymer production; in organic synthesis to prepare compounds for surfactants, detergents, and wetting agents; for yarn-treating; in synthetic-fiber dyeing; as a photographic stabilizing solution. It is hazardous to the respiratory system, skin, and eyes and is toxic through inhalation, absorption, ingestion, and contact. Symptoms of exposure include nausea, vomiting, and irritation of the respiratory system, skin, and eyes. OSHA exposure limit (TWA): 10 ppm [skin] or 50 mg/m^3.

diethylberyllium (Be(C$_2$H$_5$)$_2$) — A colorless liquid; MW: 67.13; BP: NA; Sol: soluble in ether, hexane, and benzene but reacts violently with water; Fl.P: ignites spontaneously in air, producing dense white fumes of beryllium oxide; density: NA. It is used as an intermediate in organic synthesis. It is hazardous to the skin and causes severe burns. ACGIH exposure limit: NA.

diethylene glycol monoethyl ether (C$_6$H$_{14}$O$_3$) — A colorless liquid; MW: 120.15; BP: 193°C; Sol: miscible with water and organic solvents; Fl.P: 93°C (open cup), a noncombustible liquid; density: 1.035. It is used as a thermally stable solvent for lacquers, varnishes, and dyes. It is hazardous to the digestive tract, skin, and eyes. It may lower hemoglobin levels, cause renal damage, and eye irritation. It may be teratogenic in mice. ACGIH exposure limit: NA.

$$HO - CH_2 - CH_2 - O - CH_2 - CH_2 - O - CH_3$$

Diethylene glycol monomethyl ether

diethylmagnesium (Mg(C$_2$H$_5$)$_2$) — A colorless liquid; MW: 82.44, BP: NA, Sol: soluble in ether, Fl.P: catches fire spontaneously with air, density: NA. It is used as an intermediate in organic synthesis. It is hazardous to the skin and causes severe burns. ACGIH exposure limit: NA.

diethylstilbestrol (C$_{18}$H$_{20}$O$_2$) — A white, crystalline, nonsteroid estrogen used as a substitute for natural estrogenic hormones.

diethylzinc (Zn(C$_2$H$_5$)$_2$) — A colorless liquid; MW: 123.50; BP: 118°C; Sol: soluble in ether, hexane, and benzene but reacts violently with water; Fl.P: ignites spontaneously in air; density: 1.21 g/ml. It is used in organic synthesis. It is hazardous to the skin and causes severe burns when reacting with moisture. ACGIH exposure limit: NA.

dietitian — A specialist in proper nutrition.

differential pressure — The difference in static pressure between two locations.

differentiation — Chemical and physical changes associated with the developmental process of an organism or cell.

diffraction — The bending of waves as they pass through an aperture or around the edge of a barrier.

diffuse — To become widely spread through a membrane or fluid.

diffuse radiation — Radiation that has traveled an indirect path from the sun; it has been scattered by particles in the atmosphere, such as air molecules, dust, and water vapor.

diffused air — (*sewage*) A technique by which air under pressure is forced into sewage in an aeration tank; the air is pumped down into the sewage through a pipe and escapes out through holes in the side of the pipe.

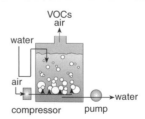

Diffused-Air Aerator

diffuser — A component of the ventilation system that supplies, distributes, and diffuses air to promote air circulation in the occupied space.

diffuser plate — A plate used for capturing incoming solar radiation for measurement and intercepting radiation from a mercury–argon calibration lamp.

Dome Diffuser

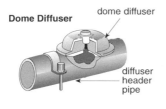

Sparger

Disc Diffuser

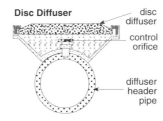

Static Tube Aerator

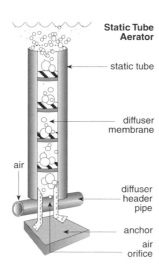

Diffuser

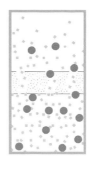

a)

Diffusion

b)

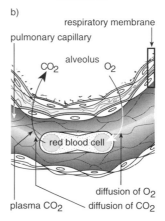

diffusion — The tendency of gases or liquids to mix freely, spreading out their component molecules, atoms, and ions in a given space away from the area of greatest concentration; the random movement of molecules from one location to another because of random thermal molecular motion, always occurring from a region of higher concentration to a region of lower concentration.

diffusion badge — A small, clip-on unit worn in the breathing zone using neither pump nor batteries for the testing of a gas or vapor in the air by relying on the tendency of the gas or vapor to disperse based on its density compared to air.

diffusion rate — A measure of the rate at which one gas or vapor disperses into or mixes with another gas or vapor.

difluorodibromomethane (CBr$_2$F$_2$) — A colorless, heavy liquid or gas (above 76°F) with a characteristic odor; MW: 209.8, BP: 76°F, Sol: insoluble, sp. gr. (59°F) 2.29. It is used as a fire-extinguishing agent; in organic synthesis; in processing cutting tools such as razor blades, hypodermic needles, scalpels, culinary knives, and garden tools; as a special solvent for preparation of explosive mixtures. It is hazardous to the respiratory system and is toxic through inhalation, ingestion, and contact. Symptoms of exposure in animals include respiratory system irritation, central nervous system symptoms, and liver damage. OSHA exposure limit (TWA): 100 ppm [air] or 860 mg/m^3.

difluorodichlormethane (CCl$_2$F$_2$) — A clear, colorless gas with a sweetish odor at high concentrations and practically odorless at low concentrations; liquid under pressure; MW: 120.92, BP: –29.8°C, Sol: practically insoluble in water, Fl.P: not combustible, sp. gr. (liquid) 1.452 at –17.8°C. It is used as a refrigerant, blowing agent, aerosol propellant, solvent, degreasing agent, monomer for resins, and leak-detecting agent and in the preparation of frozen tissue samples. It is hazardous to the central nervous system, lungs, skin, and eyes and is toxic through inhalation and contact. Symptoms of exposure include mild irritation of the nose, throat, and upper airways, lightheadedness, giddiness, dizziness, drowsiness, slurred speech, tingling sensations, humming in the ears, and frostbite. If abused through aerosol sniffing, it can cause an irregular heartbeat, leading to death from cardiac arrest. OSHA exposure limit (TWA): 1000 ppm [air] or 4950 mg/m³. Also known as Freon-12.

digest — To break into smaller parts and simpler compounds by mastication, hydrolysis, and the action of intestinal secretions and enzymes.

digester — (*wastewater*) A closed tank, sometimes heated to 95°F, where sludge is subjected to intense bacterial action.

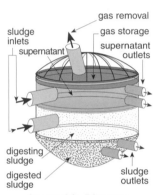

Anaerobic Digester

Aerobic Digester

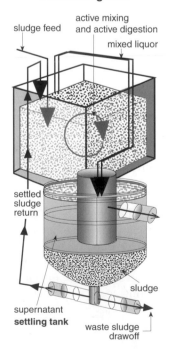

digestion — The process by which foods are chemically simplified and made soluble so that they can be used by the cells; (*sewage*) a reduction in volume by decomposition of highly putrescible organic matter to relatively stable or inert, organic, or inorganic compounds carried out by anaerobic organisms in the absence of free oxygen, resulting in partial gasification, liquefaction, and mineralization.

digestive system — The series of organs taking part in the digestion and absorption of food.

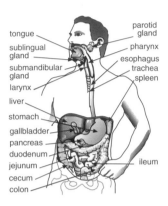

Digestive System

digit — (*mathematics*) An Arabic numeral from 0 to 9.

digital — (*computer science*) The ability to represent data in discrete units or digits.

digital elevation model — (*geographical information system*) A topographic surface arranged in a data file as a set of regularly spaced x-, y-, z-coordinates, where z represents elevation.

digital line graph — (*geographical information system*) Digital format standards published by USGS for exchange of cartographic data files.

digital output — An output signal representing the size of an input in the form of a series of discrete quantities.

digital readout — The use of numbers to indicate the value or measurement of a variable by using an instrument with a direct numerical reading of the measured value.

digital-to-analog converter — (*computer science*) A hardware device used to convert a digital signal into a voltage or current proportional to the digital input.

digitize — (*geographic information system*) To encode map coordinates in digital form.

digitizer — (*computer science*) A device that converts analog information into a digital format.

digitizing — (*geographic information system*) A method of converting information from one format to another using a trace methodology and creating a spatial dataset from a hard-copy source such as a paper map or a plan.

diglycidyl ether (C$_6$H$_{10}$O$_3$) — A colorless liquid with a strong, irritating odor; MW: 130.2, BP: 500°F, Sol: not known, Fl.P: 147°F, sp. gr. 1.26 at 77°F. Diglycidyl ether is not generally used outside research laboratories; it has no common industrial uses. It is hazardous to the skin, eyes, and respiratory system and is toxic through inhalation, ingestion, and contact. Symptoms of exposure include eye and respiratory system irritation, skin burns; carcinogenic. In animals, hematopoietic system, lung, liver, and kidney damage are observed. OSHA exposure limit (TWA): 0.1 ppm [air] or 0.5 mg/m^3.

diisobutyl ketone ((CH$_3$)$_2$(CHCH$_2$)$_2$CO) — A colorless liquid with a mild, sweet odor; MW: 142.3, BP: 334°F, Sol: 0.05%, Fl.P: 120°F, sp. gr. 0.81. It is used as a paint thinner; as a solvent in the production of synthetic coatings, soap, or nitrocellulose; as an intermediate in organic synthesis; as an extractant in the pharmaceutical industry; as a separating agent in the chemical industry. It is hazardous to the respiratory system, skin, and eyes and is toxic through inhalation, ingestion, and contact. Symptoms of exposure include eye, nose, and throat irritation; headache; dizziness; dermatitis. OSHA exposure limit (TWA): 25 ppm [air] or 150 mg/m^3.

Diisobutyl ketone

diisopropyl peroxydicarbonate (C$_8$H$_{14}$O$_6$) — A white crystalline solid with a mild characteristic odor; MW: 206.18; BP: NA; Sol: soluble in ether, chloroform, benzene, and hexane; Fl.P: 12°C. It is a

Diisopropyl peroxydicarbonate

highly reactive oxidizing and combustible substance that explodes on heating and can decompose violently below room temperature. It is used as a catalyst to initiate polymerization reactions. It is a moderate skin irritant and can cause conjunctivitis and corneal ulcerations when it comes in contact with the eyes. ACGIH exposure limit: NA.

diisopropyl sulfate (C$_6$H$_{14}$O$_4$S) — A colorless liquid with a sharp odor; MW: 182.26, BP: 120°C, Sol: miscible with alcohol and ether, density: 1.11. It is hazardous to the skin and digestive tract. It is a possible human carcinogen.

Diisopropyl sulfate

diisopropylamine ((CH$_3$)$_2$CH)$_2$NH) — A colorless liquid with an ammonia- or fish-like odor; MW: 101.2, BP: 183°F, Sol: slight, Fl.P: 20°F, sp. gr. 0.72. It is used in the synthesis of corrosion inhibitors in iron and steel; in the synthesis of herbicides; as a delayed-action vulcanization accelerator for sulfur-cured rubbers; as a catalyst for chemical synthesis of alkylene cyanohydrin; as a component in gels for cosmetic and medical applications; in deodorants and aftershave solutions. It is hazardous to the respiratory system, skin, and eyes and is toxic through inhalation, absorption, ingestion, and contact. Symptoms of exposure include nausea, vomiting, headache, eye irritation, visual disturbance, pulmonary irritation. OSHA exposure limit (TWA): 5 ppm [skin] or 20 mg/m^3.

dike — (*hazardous materials*) A barrier constructed to control or confine hazardous substances and prevent them from entering sewers, ditches, streams, or other flowing waters.

dilapidated — No longer adequate for the purpose or use for which it was originally intended.

dilate — An increase in diameter of a body opening, blood vessel, or tube.

diluent — A liquid or dust material used to weaken a concentrated chemical so it can be safely and economically used.

diluent gas — Any gas used to dilute the air pollution concentration in an air sample.

dilute — To make less concentrated by adding more solvent, such as water.

dilute solution — A solution that has been made weaker, usually by the addition of water.

dilution — Reduction of the concentration of a substance in air or water.

dilution air — The air required by some combustion heating systems in order to isolate the furnace from outside pressure fluctuations and maintain an effectively constant chimney draft.

dilution plate-count method — A technique for estimating the numbers of viable organisms in the sample by serially diluting the material and then transferring it to agar plates and determining the quantity of colony-forming units.

dilution rate — The amount of a diluent that must be added to a unit of a chemical to obtain the desired dosage; (*pesticides*) the amount of concentrated product mixed with water, oil, or inert carrier to achieve the proper application dilution in accordance with label usage.

dilution ratio — The relationship between the volume of water in a stream and volume of incoming water; it affects the ability of the stream to assimilate waste.

dilution ventilation — Air flow designed to dilute contaminants to acceptable levels.

dimanganese decacarbonyl (Mn$_2$(CO)$_{10}$) — A golden-yellow, crystalline solid; MW: 389.99, BP: NA, Sol: soluble in most organic solvents, Fl.P: may ignite when heated above 150°C, density: 1.75 g/ml at 25°C. It is used as a catalyst and a fuel additive to increase octane number. Chronic exposure to the dusts may be hazardous to the pulmonary system, central nervous system, skin, and eyes. ACGIH exposure limit: NA.

Dimanganese decacarbonyl

dimefox (C$_{14}$H$_{12}$FN$_2$OP) — A colorless liquid with a fishy odor; MW: 154.13, BP: 86°C, Sol: readily soluble in water and most organic solvents, sp. gr. 1.115 at 20°C. It is used as a pesticide. It is extremely toxic by all routes of exposure including

Dimefox

skin, digestive tract, and respiratory tract and is a cholinesterase inhibitor. It exhibits acute, delayed, and chronic effects and may cause headaches, dizziness, blurred vision, vomiting, abdominal pains, seizures, respiratory paralysis, and death.

dimethyl acetamide (CH$_3$CON(CH$_3$)$_2$) — A colorless liquid with a weak ammonia- or fish-like odor; MW: 87.1, BP: 329°F, Sol: miscible, Fl.P (oc): 158°F, sp. gr. 0.94. It is used as a spinning solvent for synthetic fibers; as a solvent for film-casting and top-coating resins; as a reaction medium and catalyst; as an extraction solvent for recovery and purification of butadiene; as a paint-stripping solvent. It is hazardous to the liver and skin and is toxic through inhalation, absorption, ingestion, and contact. Symptoms of exposure include jaundice, liver damage, depression, lethargy, hallucinations, delusions, and skin irritation. OSHA exposure limit (TWA): 10 ppm [skin] or 35 mg/m^3.

dimethyl fluorphosphate (C$_2$H$_6$FO$_3$P) — A strong acetylcholinesterase inhibitor; MW: 128.05.

Dimethyl fluorophosphate

dimethyl mercury (Hg(CH$_3$)$_2$) — A colorless liquid with a faint sweet odor; MW: 230.66, BP: 94°C, Sol: soluble in ether and alcohols, Fl.P: 101°F, density: 3.19 g/ml at 20°C. It is used as a reagent in inorganic synthesis and is a reference standard for mercury and nuclear magnetic resonance. It is a highly toxic chemical that is hazardous to the skin, digestive tract, and respiratory tract. It causes a severe neurological disease known as Minamata disease, which has resulted in hundreds of deaths. It causes nausea, vomiting, ataxia, slurred speech, and loss of vision and hearing. ACGIH exposure limit (TLV-TWA): 0.01 mg Hg/m^3.

dimethyl sulfate ((CH$_3$)$_2$SO$_4$) — A colorless, oily liquid with a faint, onion-like odor; MW: 126.1, BP: 370°F (decomposes), Sol (64°F): 3%, Fl.P: 182°F, sp. gr. 1.33. It is used as a methylating agent in the organic chemical industry for the manufacture of esters, ethers, and amines; in the manufacture of dyestuffs, dyes, coloring agents, and perfumes; in the pharmaceutical industry as a solvent for separation and preparation of mineral oils; during analysis of automobile fluids. It is hazardous to the eyes, skin, respiratory system, liver, kidneys, and central nervous system and is toxic through inhalation, absorption, ingestion, and contact. Symptoms of exposure include eye and nose irritation, headache, giddiness, conjunctivitis, photophobia, periorbital edema, dysphagia, productive cough, chest pain, dyspnea, cyanosis, vomiting, diarrhea, dysuria, analgesia, fever, icterus, albuminuria, skin and eye burns, and delirium; carcinogenic. OSHA exposure limit (TWA): 0.1 ppm [skin] or 0.5 mg/m^3.

$$CH_3-O-\overset{\overset{O}{\|}}{\underset{\underset{O}{\|}}{S}}-O-CH_3$$

Dimethyl sulfate

dimethyl-1,2-dibromo-2,2-dichloroethyl phosphate ($C_4H_7O_4PBr_2Cl_2$) — A colorless to white solid or straw-colored liquid (above 80°F) with a slightly pungent odor; MW: 380.8, BP: decomposes, Sol: insoluble, sp. gr. 1.96 at 77°F. It is used as an agricultural insecticide and acaricide that is applied to vegetables, fruit, and agronomic crops and in compounding waterbase paints and floor polishes. It is hazardous to the respiratory system, central nervous system, cardiovascular system, skin, eyes, and blood cholinesterase and is toxic through inhalation, absorption, ingestion, and contact. Symptoms of exposure include miosis, lacrimation, headache, tight chest, wheezing, laryngeal spasms, salivation, cyanosis, anorexia, nausea, vomiting, abdominal cramps, diarrhea, weakness, twitching, paralysis, giddiness, ataxia, convulsions, low blood pressure, cardiac irregularities, and skin and eye irritation. OSHA exposure limit (TWA): 3 mg/m^3 [skin].

dimethylamine (($CH_3)_2NH$) — A colorless gas with an ammonia- or fish-like odor (liquid below 44°F) that is shipped as a liquefied compressed gas; MW: 45.1, BP: 44°F, Sol: 24% at 140°F, Fl.P (liquid): 20°F, sp. gr. 0.67. It is used in the preparation of spinning solvents for acrylic and polymeric fibers; as a raw material in the synthesis of agricultural chemicals; in vulcanization accelerators for sulfur-cured rubber; in textile waterproofing agents; as a photographic chemical; as a stabilizer in natural rubber latex and certain types of resins; as a component of rocket propellants; as an antiknock agent in other fuels. It is hazardous to the respiratory system, skin, and eyes and is toxic through inhalation, ingestion, and contact. Symptoms of exposure include nose and throat irritation, sneezing, cough, dyspnea, pulmonary edema, conjunctivitis, skin and mucous membrane burns, and dermatitis. OSHA exposure limit (TWA): 10 ppm [air] or 18 mg/m^3.

4-dimethylaminoazobenzene ($C_6H_5NNC_6H_4N(CH_3)_2$) — Yellow, leaf-shaped crystals; MW: 25.3, BP: sublimes, Sol: insoluble, Fl.P: not known. It is used in research and laboratory facilities and as a coloring agent for polishes, waxes, polystyrene, petroleum, and soap. It is hazardous to the liver, skin, and bladder and is toxic through inhalation, absorption, ingestion, and contact. Symptoms of exposure include enlarged liver, hepatic and renal dysfunction, contact dermatitis, cough, wheezing, difficulty breathing, bloody sputum, bronchial secretions, frequent urination, dysuria; carcinogenic. OSHA exposure limit (TWA): regulated use.

dimethylaniline ($C_6H_5N(CH_3)_2$) — A pale-yellow, oily liquid with an amine-like odor; a solid below 36°F; MW: 121.2, BP: 378°F, Sol: 2%, Fl.P: 142°F, sp. gr. 0.96. It is used in the synthesis of dye and dye intermediates; in the synthesis of explosives; in pharmaceuticals; in intermediates for vanillin; in absorption of sulfur dioxide. It is hazardous to the blood, kidneys, liver, and central nervous system and is toxic through inhalation, absorption, ingestion, and contact. Symptoms of exposure include anoxia symptoms, cyanosis, weakness, dizziness, and ataxia. OSHA exposure limit (TWA): 5 ppm [skin] or 25 mg/m^3.

9,10-dimethylanthracene ($C_{16}H_{14}$) — An aromatic hydrocarbon that causes tumors in the lungs and skin of mice; MW: 206.30.

9,10-Dimethylanthracene

dimethylformamide (C_3H_7NO) — A colorless liquid with a faint ammonia-like or fishy odor; MW: 73.1, BP: 153°C, Sol: miscible in water, Fl.P: 58°C, sp. gr. 0.95. It is used as a solvent for pesticides and polyacrylic fabrics and in the metal-working, dyeing, and construction industries. It is hazardous to the respiratory system, digestive system, and skin. Acute exposure can irritate the eyes and skin, cause liver injury, vertigo, sleep disorders, upset stomach, abdominal pain, loss of appetite, nausea, vomiting, renal disorders, hematologic disorders, cardiovascular disorders, and dermatological injuries. It may lead to testicular cancer, damage to a developing fetus, and teratogenic disorders. Chronic exposure may cause serious kidney and liver damage, poor appetite, upset stomach, and cancer. OSHA exposure limit (8-hour TWA): 10 ppm.

dimethylformamide (HCON(CH₃))₂ — A colorless to pale-yellow liquid with a faint amine-like odor; MW: 73.1, BP: 307°F, Sol: miscible, Fl.P: 136°F, sp. gr. 0.95. It is used as a spinning solvent for acrylic fibers; as a booster solvent in coating, printing, and adhesive formulations; as a chemical intermediate, catalyst, and reaction medium in chemical manufacturing; as a selective absorption and extraction solvent for recovery, purification, absorption, separation, and desulfurization of nonparaffinics from paraffin hydrocarbons; in the manufacture of paint stripper; in the pigment and dye industries; as a crystallization solvent in the pharmaceutical industry; in miscellaneous applications for high-voltage capacitors. It is hazardous to the liver, kidneys, skin, and cardiovascular system and is toxic through inhalation, absorption, ingestion, and contact. Symptoms of exposure include nausea, vomiting, colic, liver damage, hepatomegaly, high blood pressure, flushed face, and dermatitis; in animals, kidney and heart damage has been observed. OSHA exposure limit (TWA): 10 ppm [skin] or 30 mg/m³.

1,1-dimethylhydrazine ((CH₃)₂NNH₂) — A colorless liquid with an ammonia- or fish-like odor; MW: 60.1, BP: 147°F, Sol: miscible, Fl.P: 5°F, sp. gr. 0.79. It is used in the formulation of jet and rocket propellants; in the chemical synthesis of catalysts, automotive antifreeze, pharmaceuticals, dyestuffs, and stabilizing agents; in the formulation of photographic developers. It is hazardous to the central nervous system, liver, gastrointestinal tract, blood respiratory system, eyes, and skin and is toxic through inhalation, absorption, ingestion, and contact. Symptoms of exposure include eye and skin irritation, choking, chest pain, dyspnea, lethargy, nausea, anoxia, convulsions, and liver injury; carcinogenic. OSHA exposure limit (TWA): 0.5 ppm [skin] or 1 mg/m³.

2,4-dimethylphenol (C₈H₁₀O) — A colorless needle shape or liquid with a strong medicinal-type odor. MW: 122.16, BP: 211.5°C, Sol: NA, Fl.P: >112°C,

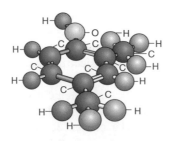

2,4 Dimethylphenol

sp. gr. 0.965. It is used in the manufacture of pharmaceuticals, plastics, insecticides, fungicides, rubber chemicals, wetting agents, and dyestuffs. It is hazardous to the respiratory system, digestive system, skin, and eyes. Acute exposure may cause headaches, nausea, weakness, fainting, shortness of breath, severe burns of the eyes and skin, arrhythmias, pulmonary edema, seizures, coma, and death. Chronic exposure may causes damage to the kidneys, liver, brain, pancreas, lungs, and heart muscle. OSHA exposure limit (8-hour TWA): 10 ppm.

dimethylphthalate (C₆H₄(COOCH₃)₂) — A colorless, oily liquid with a slight, aromatic odor (solid below 42°F); MW: 194.2, BP: 543°F, Sol: 0.4%, Fl.P: 295°F, sp. gr. 1.19. It is used as a plasticizer in the compounding of plastics; in the manufacture of surface coatings containing plasticized resins and polymers, including furniture lacquers, printing inks, textile and paper coatings, and moisture-proof coatings for cellophane; in insect repellents. It is hazardous to the respiratory system and gastrointestinal tract and is toxic through inhalation, ingestion, and contact. Symptoms of exposure include stomach pain and irritation of the upper respiratory system. OSHA exposure limit (TWA): 5 mg/m³ [air].

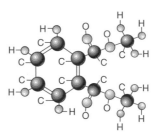

Dimethyl Phthalate

dimictic — Lakes and reservoirs that freeze over and normally go through two stratifications and two mixing cycles within a year.

dinitrobenzene (*ortho-, meta-,* and *para-*isomers) (C₆H₄(NO₂)₂) — A pale white or yellow solid; MW: 168.1; BP: 606, 572, and 570°F, respectively; Sol: 0.05, 0.02, and 0.01%, respectively; Fl.P: 302°F, 302°F, and not known, respectively; sp. gr. 1.57, 1.58, and 1.63, respectively, at 64°F. It is used in the preparation of dyes and dye intermediates; in organic synthesis in the chemical industry as photographic developers and as explosives; in the plastic industry as a substitute for camphor; as a polymerization inhibitor; as an explosive by substituting for TNT; in explosive shells. It is hazardous to the blood, liver, eyes, skin, cardiovascular system, and central nervous system and is toxic through inhalation, absorption, ingestion, and contact. Symptoms of exposure include anoxia, cyanosis, visual disturbance, central scotomas, bad taste, burning mouth, dry throat, thirst, anemia, liver damage, and yellowing of the hair, eyes, and skin. OSHA exposure limit (TWA): 1 mg/m³ [skin].

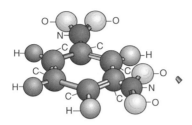

Dinitrobenzene
(1,3–Dinitrobenzene)

dinitrogen fixation — Conversion of molecular dinitrogen (N_2) to ammonia and subsequently to organic combinations or to forms useful in biological processes.

dinitro-*o*-cresol ($CH_3C_6H_2OH(NO_2)_2$) — A yellow, odorless solid; MW: 198.1, BP: 594°F, Sol: 0.01%, sp. gr. 1.1 (estimated). It is used as an herbicide, insecticide, and fungicide. It is hazardous to the cardiovascular system, endocrine system, and eyes and is toxic through inhalation, absorption, ingestion, and contact. Symptoms of exposure include a sense of well-being, headache, fever, lassitude, profuse sweating, excessive thirst, tachycardia, hyperpnea, cough, shortness of breath, and coma. OSHA exposure limit (TWA): 0.2 mg/m³ [skin].

2,4-dinitrophenol ($C_6H_4N_2O_5$) — A yellowish crystal with a sweet, musty odor. MW: 184.1, BP: NA, Sol: 5600 mg/l water at 18°C, Fl.P: NA (may be ignited by heat, sparks, or flames and produces poisonous oxides of nitrogen in a fire), sp. gr. 1.6 at 20°C. It is hazardous to the respiratory system, digestive system, skin, and eyes. Acute exposure causes seizures, coma, cyanosis, pulmonary edema, arrhythmias, malaise, and renal and hepatic damage. Chronic exposure may cause cataracts, secondary glaucoma, reproductive toxicity in animals, and permanent kidney and liver damage. OSHA exposure limit (8-hour TWA): NA.

dinitrotoluene ($C_6H_3CH_3(NO_2)_2$) — An orange-yellow, crystalline solid with a characteristic odor (often shipped molten); MW: 182.2, BP: 572°F, Sol: insoluble, Fl.P: 404°F, sp. gr. 1.32. It is used in the manufacture of toluene diisocyanate for production of polyurethane plastics. It is hazardous to the blood, liver, and cardiovascular system and is toxic through inhalation, absorption, ingestion, and contact. Symptoms of exposure include anoxia, cyanosis, anemia, and jaundice; carcinogenic. OSHA exposure limit (TWA): 1.5 mg/m³ [skin].

dinoseb ($C_{10}H_{12}N_2O_5$) — A yellow to orange crystalline solid with a pungent odor (a red-brown liquid in the technical grade); MW: 240.24, BP: NA, Sol: 0.00532 g/10 g water, Fl.P: 27°C, sp. gr. 1.26. It is used as an herbicide on soybeans, vegetables, deciduous fruits and nuts, peanuts, citrus, grain crops, and other field crops. It was formerly used as an insecticide. It is hazardous to the respiratory system, digestive system, skin, and eyes. Acute exposure may cause seizures, coma, cyanosis, pulmonary edema, arrhythmias, fever, increased respiration, extreme diaphoresis, headache, malaise, thirst, jaundice, nausea, vomiting, abdominal pain, and renal and hepatic damage. Chronic exposure may cause cataracts, secondary glaucoma, and permanent damage to the kidney and liver. OSHA exposure limit (8-hour TWA): NA.

diode — An electronic tube containing only two electrodes: a cathode and an anode.

diopters — A measure of the power of a lens or prism equal to the reciprocal of its focal length in meters.

dioxane ($C_4H_8O_2$) — A colorless liquid or solid (below 53°F) with a mild, ether-like odor; MW: 88.1, BP: 214°F, Sol: miscible, Fl.P: 55°F, sp. gr. 1.03. It is used as a solvent for fats, oils, waxes, greases, and natural and synthetic resins; as a wetting agent in textile processing, dye baths, and stain and printing composition; in the manufacture of detergents and cleaning preparations; as a dehydrating agent in the preparation of histological slides; in the preparation of cosmetics and deodorants; as a working fluid for scintillation counter samples; as a solvent in pulping wood. It is hazardous to the liver, kidneys, skin, and eyes and is toxic through inhalation, absorption, ingestion, and contact. Symptoms of exposure include drowsiness, headache, nausea, vomiting, liver damage, kidney failure, and irritation of the skin, eyes, nose, and throat; carcinogenic. OSHA exposure limit (TWA): 25 ppm [skin] or 90 mg/m³.

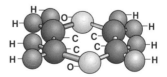

Dioxane
(1,4–Dioxane)

1,4-dioxane ($C_4H_8O_2$) — A colorless liquid with a mild ether-like odor; MW: 88.1, BP: 101°C, Sol: Miscible with water, Fl.P: 65.5°C, sp. gr. 1.034 at 20°C. It is used as a solvent for cellulose compounds, resins, oils, waxes, and dyes, textile processes, printing processes, and detergent preparation. It is hazardous to the respiratory system, skin, and causes irritation of the nose, throat, and air passages. Acute exposure may cause eye, nose, and throat irritation, headaches, nausea, dizziness, lightheadedness, loss of consciousness, serious kidney and liver damage, and death. Chronic exposure may cause kidney and liver damage, reproductive damage, upset stomach, and tenderness in the abdomen. It is a probable carcinogen in humans. OSHA exposure limit (8-hour TWA): 100 ppm.

1,4-Dioxane

dioxathion (C₁₂H₂₆O₆P₂S₄) — An odorless viscous tan to brown liquid: MW: 456.54, BP: NA, Sol: insoluble in water, sp. gr. 1.257. It is used as an insecticide and acaricide for deciduous fruit or as a spray or dip for horn fly, lice, and ticks on cattle, goats, pigs, and sheep. It is hazardous to the respiratory system, digestive system, skin, and eyes. Acute exposure may affect the central nervous system causing agitation, confusion, delirium, coma, seizures, and death. It may also cause bradycardia, bronchospasm, diaphoresis, vomiting, and diarrhea. Chronic exposure may cause nerve damage, poor coordination in arms and legs, personality changes, depression, anxiety, and irritability. OSHA exposure limit (8-hour TWA): 0.2 mg/m³.

dioxide — An oxide that contains two oxygen atoms.

dioxin — See 2,3,7,8-tetrachlorodibenzo-*p*-dioxin.

1,3-dioxolane (C₃H₆O₂) — A colorless liquid; MW: 74.09; BP: 74°C; Sol: miscible with water and organic solvents; Fl.P: 2°C (open cup), highly flammable liquid, vapor is heavier than air and can travel a considerable distance to a source of ignition and flash back; density: 1.061 at 20°C. It is used as an intermediate in organic synthesis. It is hazardous to the eyes and respiratory tract and causes severe irritation. ACGIH exposure limit: NA.

1,3-Dioxolane

diphasic — Something that occurs in two stages or phases.

diphenyl (C₆H₅C₆H₅) — A colorless to pale-yellow solid with a pleasant, characteristic odor; MW: 154.2,

BP: 489°F, Sol: insoluble, Fl.P: 235°F, sp. gr. 1.04. It is used as a heat-transfer medium; as a dye carrier for plastics and synthetic resin dyeing; in the impregnation of wrapping and packaging papers. It is hazardous to the liver, skin, central nervous system, upper respiratory system, and eyes and is toxic through inhalation, absorption, ingestion, and contact. Symptoms of exposure include headache, nausea, fatigue, numb limbs, liver damage, and irritation of the throat and eyes. OSHA exposure limit (TWA): 1 mg/m³ [air].

1,2-diphenylhydrazine (C₁₂H₁₂N₂) — A white solid; MW: 814.24, BP: 309°C, Sol: 66.9 mg/l at 20°C, Fl.P: not known, sp. gr. 1.158. It is used in the dye industry for the production of benzidine-based dyes, including many of the direct dyes, such as direct red 28, direct black 4, and direct blue 2. It is also used as a starting material in the production of benzidine; for the production of the drugs phenylbutazone (antiinflammatory agent) and sulfinpyrazone (uricosuric agent for the treatment of gouty arthritis). No health effects in humans are known by any route of exposure; however, in animals, significantly increased incidences of hepatocellular carcinomas, neoplasm, liver nodules, mammary adenocarcinomas, Zymbal's gland carcinomas, and adrenal pheochromocytomas have been observed. No OSHA exposure limits. U.S. EPA guidelines as an inhalation unit risk 2.2 × 10⁻⁴ µg/m³ [air].

diphenylmethane-4,4′-diisocyanate ((C₆H₄)₂CH₂(NCO)₂) — A light-yellow to white crystal; MW: 250.27, BP: 172°C, Sol: slightly soluble in water, Fl.P: noncombustible (open cup), 202°C, density: 1.197 at 7°C. It is used in the production of rigid urethane foam, coatings, and elastomers. It is hazardous to the respiratory tract. It causes symptoms of bronchitis, coughing, fever, shortness of breath, chest pains, nausea, and insomnia. ACGIH exposure limit (TLV-TWA): 0.051 mg/m³.

O = C = N — ⬡ —

CH₂ — ⬡ — N = C = O

Diphenylmethane-4,4′-diisocyanate

diphtheria — (*disease*) An acute bacterial disease of the tonsils, pharynx, larynx, nose, and sometimes mucous membranes of the skin. Incubation time is 2 to 5 days; it is caused by *Corynebacterium* *diphtheriae*. It is found in temperate zones among unimmunized children under 15 years of age or among adults who have not received immunization. The reservoir of infection is people. It is transmitted by contact with a carrier or through raw milk and is communicable until virulent bacteria have disappeared, usually 2 weeks or less; infants and those unprotected by immunization are most susceptible. It is controlled by proper immunization.

diphyllobothriasis — (*disease*) An intestinal fish tapeworm or broadworm infection of long duration, with symptoms ranging from nonexistent to vitamin B$_{12}$-deficiency anemia. Massive infections cause diarrhea, obstruction of the bowel duct or intestine, and toxic symptoms. Incubation time is 3 to 6 weeks from ingestion of eggs. It is caused by *Diphyllobothrium latum* cestodes and is found in lake regions and all areas of consumption of raw or partially cooked freshwater fish. Reservoirs of infection include infected people, dogs, bears, and other fish-eating mammals. It is transmitted by eating raw or inadequately cooked fish and is not communicable from person to person; general susceptibility. It is controlled by proper cooking of all freshwater fish.

Diphyllobothrium — A genus of large, parasitic, intestinal flatworms having a scolex with two slitlike grooves.

diploid number — The full set of chromosomes in a nucleus with both members of each pair present; designated as 2N.

diplopia — The condition of seeing single objects as two; double vision.

dipole — A positive and a negative charge bound together in such a way that the center of gravity of the positive charge is separated in space from the center of gravity of the negative charge, as with a magnet.

dipropylene glycol methyl ether (CH$_3$OC$_3$H$_6$O C$_3$H$_6$OH) — A colorless liquid with a mild, ether-like odor; MW: 148.2, BP: 374°F, Sol: miscible, Fl.P: 180°F, sp. gr. 0.95. It is used as a general solvent for oils and greases; as a coupling and dispersing agent in the manufacture and application of printing pastes, dyes, and inks; in the manufacture of cosmetics; in the manufacture of latex paints, lacquers, and leather protective coatings; as a slimicide in food packaging and adhesives in the food industry; as a heat-transfer agent in hydraulic brake fluid; in glass and metal cleaning and antifogging compounds. It is hazardous to the respiratory system and eyes and is toxic through inhalation, absorption, ingestion, and contact. Symptoms of exposure include weakness, lightheadedness, headache, and irritation of the eyes

and nose. OSHA exposure limit (TWA): 100 ppm [skin] or 600 mg/m^3.

dipstick — A chemically treated piece of paper used in the analysis of fluids.

Diptera — An order of insects that includes flies and mosquitoes.

Diquat (diquat dibromide; C$_{12}$H$_{12}$N$_2$Br$_2$) — An odorless, colorless to pale-yellow, crystalline solid for which aqueous solutions are a dark reddish-brown; MW: 344.1, BP: decomposes, Sol: 70 g/100 g water at 20°C, Fl.P: NA (poisonous gases produced in a fire include nitrogen oxides and bromine), sp. gr. 1.22 to 1.27. It is used as a contact herbicide, plant growth regulator, and desiccant. It is hazardous to the respiratory system, digestive system, and skin. Acute exposure may cause severe poisoning with nausea, vomiting, diarrhea, cerebral hemorrhage, lung, kidney and liver damage, and death. Chronic exposure may cause cataracts, damage to the lungs, and cracked skin and may damage the developing fetus. OSHA exposure limit (8-hour TWA): 0.5 mg/m^3.

direct count — A method of estimating the total number of microorganisms in a sample by direct microscopic examination.

direct current (dc) — Electric current that flows in one direction only.

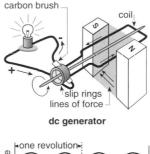

dc generator

dc voltage

Direct Current

direct-dump transfer system — The process of unloading solid waste directly from a collection vehicle to an open-top transfer trailer or container.

direct feed incinerator — An incinerator that accepts solid waste directly into its combustion chamber.

direct filtration — A filtration method of treating water that consists of the addition of coagulant chemicals, flash mixing, coagulation, minimal flocculation, and filtration.

direct-fire afterburner — An afterburner in which the fuel burner supplies the necessary heat; a device whereby a polluted gas stream goes through a fire box before the steam is vented to the atmosphere.

direct-flame incineration — A process by which organic emissions in concentrations well below the lower explosive limit are destroyed under the proper conditions by exposure to temperatures of 900 to 1400°F in the presence of a flame.

direct push — A technology used for performing subsurface investigations by driving, pushing, and/or vibrating small-diameter hollow steel rods into the ground.

direct-reading colorimetric devices — An instrument in which the gas reacts directly with a reagent to produce a color, the value of which is directly measured.

direct-reading instrument — An instrument that gives an immediate indication of the concentration and magnitude of aerosols, gases, vapors, radiation, noise, light, and heat.

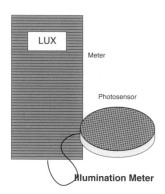

LUX

Meter

Photosensor

Illumination Meter

Direct-Reading Instrument

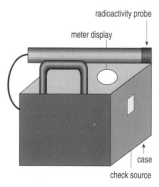

radioactivity probe

meter display

case

check source

Geiger-Müller Meter

direct runoff — Water that flows over the ground surface or through the ground directly into streams, rivers, or lakes.

directory — (*computer science*) A list of files or other directories at an Internet site.

dirofilariasis — Human infestation of the dog heartworm *Dirofilaria immitis*, transmitted through the bite of a mosquito.

dirt — Displaced soil.

disability — Limitation of an individual's physical, mental, or social activity when compared with others of similar age, sex, and occupation; inability to carry on one's normal occupation due to personal injury or illness.

disability benefits — Benefits that compensate disabled employees for loss of income or earning capacity due directly to their disabilities.

disability days — The total number of days individuals are disabled during a year.

disabling injury — An injury causing death, permanent disability, or any level of temporary total disability beyond the day of the accident.

disaccharide — Any sugar, such as $C_{12}H_{22}O_{11}$, that hydrolyzes to form two molecules of simpler sugars or monosaccharides.

disassociation — (*chemistry*) The breakup of a molecule into ions when placed in water or other solvents.

disaster — An event that demands a crisis response and presents a threat to a community or larger area.

Disaster Field Office — (*FEMA*) An office established in or near a designated area to support federal and state response and recovery operations.

disaster plan — A plan outlining individual and departmental responsibilities for natural disasters and multiple casualty incidence.

disc — See disk.

discharge — The release of stored charge from a capacitor or a battery; the flow of surface water into a stream or canal or the outflow of groundwater from a flowing artesian well, ditch, or spring; the release of liquid effluent from a facility; the release of chemical emissions into the air through designated venting mechanisms.

discharge area — An area where groundwater moves toward or is delivered to the soil surface.

discharge coefficient — A dimensionless coefficient relating to mean flow rate for an opening to an area and a corresponding pressure difference across the opening.

discharge head — The pressure (in psi) measured at the center line of a pump discharge and very close to the discharge flange which is then converted into feet.

Discharge Monitoring Report — The U.S. EPA's uniform and national form, including any subsequent additions, revisions, or modifications, for the reporting of self-monitoring results by National Pollutant Discharge Elimination System permitees.

discovery — (*law*) The process of using interrogatories to obtain facts or documents from a party to an action which are in the individual's exclusive knowledge or possession.

disease — A definite pathological process having a characteristic set of signs and symptoms that are detrimental to the well-being of the individual.

disease and symptom prevalence study — (*ATSDR*) A study designed to measure the occurrence of self-reported disease that may be validated through medical records or physical examination and to determine those adverse health conditions that may require further investigation.

disease prevention — Techniques used to protect patients or the public from actual or potential health threats and their consequences.

disease registry — (*ATSDR*) A system for collecting and maintaining in a structured record information on persons having a common illness or adverse health condition.

di-*sec*-octyl phthalate (C$_{24}$H$_{38}$O$_4$) — A colorless, oily liquid with a slight odor; MW: 390.5, BP: 727°F, Sol: insoluble, Fl.P (oc): 420°F, sp. gr. 0.99. It is used as a plasticizer in the production of polyvinyl chloride (PVC) and vinyl chloride resins. It is hazardous to the eyes, upper respiratory system, and gastrointestinal tract and is toxic through inhalation, ingestion, and contact. Symptoms of exposure include eye and mucous membrane irritation; carcinogenic. OSHA exposure limit (TWA): 5 mg/m^3 [air].

disequilibrium — Loss of balance or adjustment, physically or mentally.

disinfectant — A product that kills all vegetative bacteria but does not kill spores; a germicide or bactericide.

disinfectant byproduct — A compound formed by the reaction of a disinfectant such as chlorine with organic material in the water supply.

disinfectant time — The time it takes water to move from the point of disinfectant application to a point where the residual disinfectant is measured. In pipelines, the time is calculated by dividing the internal volume of the pipe by the maximum hourly flow rate; in mixing basins or storage reservoirs, it is determined by using tracer studies.

disinfection — A process designed to kill most microorganisms in water or on objects or people that include essentially all pathogenic bacteria.

disinfestation — The elimination of an infestation by insects or rodents.

disintegration — The separation or breaking up of a substance into its parts; (*nuclear*) a transformation or change involving the nuclei in which particles or photons are emitted.

disintegration constant — The fraction of the number of atoms of a radioactive nuclide that decay per unit interval of time.

disintegration rate — The absolute rate at which radiation is emitted from a radioactive source.

disk — (*soils*) An implement used to break up, turn, and loosen the top few inches of the soil; (*computer science*) a storage system for recording digital information. Also spelled disc.

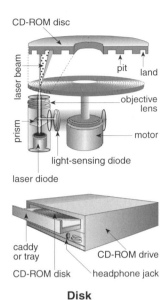

Disk

disk drive — (*computer science*) A piece of computer equipment containing a disk that spins at a high speed with a head that allows electric impulses to be written on it and to be read from the electromagnetic surface.

disk operating system — (*computer science*) A program used to control the transfer of information to and from a disk.

diskette — An inexpensive, low-capacity storage medium; also known as floppy disk.

Diskette
(Floppy Disk)

dislocation — Movement of any part of the body from its normal position.

dismissal — An order or judgment finally disposing of an action, suit, motion, or other proceeding by sending it out of court without a trial of the issues involved.

disorder — A disruption of or interference with normal functions of established systems.

disorientation — A state of mental confusion characterized by inadequate or incorrect perceptions of place, time, or identity; may be due to mental disorders, advanced age, drug and alcohol intoxication, severe stress, environmental chemicals, or other environmental stresses.

dispersant — A chemical agent used to break up concentrations of organic materials such as spilled oil.

disperse — To spread out or scatter; (*sewage*) to break up compound particles, such as aggregates, into individual component parts.

dispersing agent — A material that reduces the forces attracting like particles of a chemical mixture to each other so they will mix better with unlike particles (e.g., wetting agents, detergents).

dispersion — (*light*) Scattering of the values of a frequency distribution from an average; (*physics*) separation of light or other electromagnetic radiation into its different wavelengths; separation of nonhomogeneous radiation into components in accordance with some characteristics; a distribution of finely divided particles in a medium; the dilution or removal of a substance by diffusion or turbulence; the distribution of spilled oil into the upper layer of a water column by the natural action of waves or through the action of chemical dispersants. See also deflocculation.

dispersion force — See van der Waals force.

displaced fracture — A fracture where one or both pieces of the bone are out of normal alignment.

displacement — (*geology*) The difference between the initial position of a reference point and any later position; the distance by which portions of the same geological layer are offset from each other by a fault.

displacement flow — The displacement of internal room air by incoming outdoor or conditioned air without appreciable mixing of the two.

displacement flow ventilation — Air introduced into an air-conditioned space at low level and at low velocity.

displacement pump — A reciprocating type of pump where a piston draws water into a closed chamber and then expels it under pressure.

disposal — The removal of waste material to a site or facility that is specifically designed and permitted to receive such wastes; (*hazardous waste*) the improper discharge, deposit, injection, dumping, spilling, leaking, or placing of any solid waste or hazardous waste into or on any land or water so that any constituent thereof may enter the environment, be emitted into the air, or discharged into any waters, including groundwater.

disposal container — A plastic or paper sack designed for storing solid waste.

disposal facility — A facility or part of a facility in which hazardous waste is intentionally placed into or on any land or water and at which waste will remain after closure.

disposal well — A well used for the disposal of waste.

DISS — See Diameter-Index Safety System.

dissent — A statement written by a judge who disagrees with the decision of the majority of the court.

dissipate — To cause to lessen by scattering or spreading out.

dissipation constant — The ratio for a thermistor that relates a change in the internal power being dissipated to a resultant change in body temperature.

dissociation constant — A value that quantitatively expresses the amount a substance disassociates in solution varying with temperature, ionic strength, and nature of the solvent.

dissolution — (*chemistry*) The passing of particles into the liquid phase; dissolving; liquefaction.

dissolve — To cause a substance to pass into solution; turning a solid into a liquid and then going into a solution.

dissolved gas — A natural gas found mixed with oil in underground formations; (*chemistry*) a gas in a simple physical solution compared to one that has reacted chemically with a solvent or other solute to form a new compound.

dissolved oxygen (DO) — A measurement of the amount of oxygen dissolved in a given volume of water at a given temperature and atmospheric pressure.

dissolved solids — (*sewage*) A sewage term (rather than a technical definition) that includes solids in colloidal as well as those in true solution, both organic and inorganic; theoretically, the anhydrous residues of dissolved constituents in water.

dissolving — A chemical action of detergents that liquefies water-soluble materials or soils.

distal — Away from the central axis of the body.

distance/velocity lag — The delay time between an alteration in the value of a signal and its unchanged manifestation at a latter part of the system arising solely from the finite speed of propagation of the signal.

distillate — A liquid product condensed from vapor during distillation.

distillation — (*oil processing*) The first stage in the refining process in which crude oil is heated and unfinished petroleum products are initially separated.

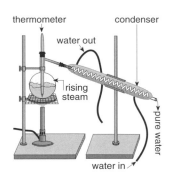

Distillation

distilled water — Water that has been treated by boiling and condensation to remove solids, inorganics, and some organics.

distress — An emotional or physical state of pain, sorrow, misery, suffering, or discomfort.

distribution — Movement of a chemical from blood to other tissues.

distribution box — A structure designed to distribute effluent by gravity from a septic tank equally into the pipes of an absorption system connected to it.

distributor — A person or business enterprise responsible for delivering goods in commerce.

disulfoton ($C_8H_{19}O_2PS_3$) — A pale-yellow to brown oily liquid with a disagreeable sulfur odor; MW: 274.38, BP: 132°C, Sol (water): 25 ppm at 23°C, Fl.P: less than 82°C, sp. gr. 1.144. It is used as an insecticide in liquid mixtures. It is hazardous to the respiratory system, digestive system, skin, and eyes. Acute exposure may cause central nervous system depression, agitation, confusion, delirium, coma, seizures, and death. It also causes bradycardia, bronchospasm, diaphoresis, vomiting, and diarrhea. Chronic exposure may damage the nerves and cause poor coordination in the arms and legs, as well as depression, anxiety, and irritability. OSHA exposure limit (8-hour TWA): 0.1 mg/m³.

diuresis — Increased excretion of urine.

diuretic — Any agent that increases the volume of urine excreted.

diurnal — Active during daylight hours; having a daily cycle.

diuron ($C_9H_{10}C_{12}N_2O$) — A crystalline solid; MW: 233.10, BP: NA, Sol: moderately soluble in alcohol and acetone, density: NA. It is used as an herbicide. It has low toxicity in experimental animals but adverse reproductive effects from chronic exposure to high concentrations have been observed in animals. ACGIH exposure limit (TLV-TWA): 10 mg/m³.

$$Cl - \overset{\overset{\displaystyle O}{\|}}{NH - C - N(CH_3)_2}$$

Diuron

divalent — Having a valence of 2.

diversion rate — The percentage of waste materials diverted from landfilling or incineration to be recycled, composted, or reused.

diversity — (*ecosystems*) The distribution and abundance of different kinds of plant and animal species and communities in a specified area.

divert — To direct a flow away from its natural course.

diverted flow position — (*milk*) In high-temperature, short-time pasteurization, the position in which the diaphragm remains in place across the forward flow line when the temperature is less than 161°F and the microswitch does not activate the solenoid in the flow diversion value.

divide — (*geology*) A ridge or high area of land that separates one drainage basin from another.

dizziness — A sensation of unsteadiness and feeling of movement within the head.

DMDS — Acronym for dimethyl disulfide.

DMM — An electronic instrument that measures voltage, current, resistance, or other electrical parameters by converting the analog signal to digital information and display.

DMS — See defense meteorological satellite; acronym for data management system.

DNA — See deoxyribonucleic acid.

DNA fingerprint — Profile of an organism's genetic material as determined from DNA segments; provides the ability to distinguish among different individuals.

DNAPL — Acronym for dense nonaqueous-phase liquid.

DNR — Acronym for do not resuscitate.

DO — See dissolved oxygen.

DOA — Acronym for dead on arrival.

Dobson unit — A unit of measurement of total ozone equal to 2.69×10^{16} molecules per square centimeter. An equivalent amount of ozone at one atmosphere and 273 K would form a layer 0.001 cm thick.

DOC — Acronym for dissolved organic carbon.

docket — A book containing an entry and brief of all the important acts done in court and conduct of each case from its inception to its conclusion.

documentation — A collection of documents to support a certain element of proof (e.g., the copying of a bill of lading to demonstrate that substances and goods have gone through interstate commerce).

doldrums — A region near the equator characterized by low pressure and light, shifting winds.

dolomite ($CaMg(CO_3)_2 = Ca^{2+} + Mg^{2+} + 2CO_3^{2-}$) — A usually white or colorless carbonate mineral consisting of a mixed calcium–magnesium carbonate crystallizing in the rhombohedral system.

domain name — (*computer science*) A website address usually followed by .com, .org, .gov, or .edu.

domestic or other non-distribution-system plumbing problem — A coliform contamination problem in a public water system with more than one service connection that is limited to the specific service connection from which the coliform positive sample was taken.

domestic waste — Wastewater derived principally from plumbing fixture drains in dwellings, business or office buildings, institutions, food-service establishments, and similar facilities and not including industrial or commercial processing waste.

domestic water use — Water used for household purposes such as drinking, food preparation, bathing, washing clothes and dishes, flushing toilets, and watering lawns and gardens.

dominant — A gene that is expressed at a high rate.

dominant lethal assay — A mutagenic bioassay used in assessing the ability of a chemical to penetrate gonadal tissue and produce genetic damage.

DOMSAT — A domestic satellite regulated by the Federal Communications Commission.

donor — A human or other organism that gives living tissue to another body.

DOPLID — See Doppler lidar.

Doppler effect — Apparent change in the frequency of acoustic or electromagnetic waves due to the relative

Doppler Effect
(not in proportion)

motion of the source of the waves and the observer; the frequency increases as the two approach each other and decreases as they move apart.

Doppler radar — A weather radar system that uses the Doppler shift of radio waves to detect air motion that can result in tornadoes and precipitation; it also measures the speed and direction of rain and ice.

Doppler scanning — A technique of ultrasound imaging monitoring the flow of blood or a beating heart.

Doppler shift — An apparent change in the frequency of a signal caused by the relative motion of a transmitter and receiver.

dormancy — A period of inactivity.

dormant spray — (*pesticides*) A chemical applied to plants to kill overwintering insects and diseases.

dorsal — The back or posterior.

DOS — (*computer science*) See disk operating system.

dosage — The quantity of a given substance associated with a given measurable or observable effect; the amount or rate of a chemical used per unit area, volume, or individual.

dose — (*radiation*) A general term denoting the quantity of radiation or energy absorbed; (*chemical*) the quantity of a chemical absorbed and available for interaction with the metabolic processes.

dose assessment — An estimate of the radiation dose to an individual or a population group usually made by means of predictive modeling techniques and sometimes supplemented by results of measurement.

dose-distribution factor — A factor that accounts for modification of dose effectiveness in cases where the radionuclei distribution is nonuniform.

dose equivalent (DE) — A quantity used in radiation protection expressing all radiation on a common scale for calculating the effective absorbed dose. The unit of dose equivalent is the rem, which is numerically equal to the absorbed dose in rads multiplied by certain modifying factors such as the quality factor, the distribution factor, etc.; in SI units, dose equivalent is the sievert, which is equal to 100 rem.

dose rate — The absorbed dose delivered per unit of time.

dose-rate meter — An instrument used to measure radiation dose rates.

dose–response — The mathematical relationship between the dose of the chemical and the physiological response.

dose–response assessment — The relationship between the level of exposure or the dose and the incidence of disease it describes.

dose–response curve — A figure that quantitatively defines the relationship between the degree of exposure to a chemical (dose) and the observed biological effect or response; the extent or incidence of the observed biological effect.

dose–response evaluation — A component of risk assessment that describes the quantitative relationship between the amount of exposure to a substance and the extent of the toxic injury or disease.

dose–response ratio — A concept that treatment effectiveness increases with higher doses of chemotherapy drugs.

dose–response relationship — The relationship that exists between the degree of exposure to a chemical dose and the magnitude of the effect or response in the exposed organism; projected on a dose–response curve.

dose threshold — (*radiation therapy*) The minimum amount of absorbed radiation that produces a detectable degree of a given effect.

dose to skin — (*radiation therapy*) The amount of absorbed radiation at the center of an irradiation field on the skin.

dosimeter — An electroscope used in personal monitoring to detect and measure an accumulated dosage of radiation.

Pocket Dosimeter

dosimetry — Scientific determination of amount, rate, and distribution of radiation emitted from a source of ionizing radiation; the process of measuring radiation dose.

dot-matrix plotter — A plotter in which the printing head consists of many closely spaced wire points that can print dots on paper to produce a map; also known as electrostatic plotter.

double-blind design — An experiment in which neither the participants nor the investigators or observers are aware of the allocation of actual and placebo treatments.

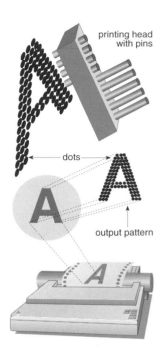

Dot-Matrix Plotter

double bond — (*chemistry*) A covalent bond sharing two pairs of electrons.

double glazing — A window that has two sheets of glass with an air space in between.

doubling time — The time required for a population of microorganisms to increase their number by a factor of 2.

DOUR — Acronym for dissolved oxygen uptake ratio.

down — (*computer science*) Describes a computer that cannot be operated because of a malfunction or maintenance problem.

downdraft — An air stream with a significant downward directional component of velocity occurring adjacent to cold surfaces; usually accompanied by precipitation, such as a shower or thunderstorm.

downer — An animal that is disabled due to illness or injury and is ready for slaughter.

downgradient — The downstream flow of groundwater.

download — (*computer science*) The transfer of a file from one computer to another user's computer.

downpass — A chamber or gas passage placed between two combustion chambers to carry the products of combustion downward.

downspout — A conduit used to carry water from the gutter to the ground.

downstream — (*oil processing*) Relating to all activities associated with the processing of crude oil, from refining to petroleum products to distribution, marketing, and shipping of the products.

downstream processing — The recovery, concentration, and purification of products or byproducts of a substance.

downtime — The time needed to locate a problem and then repair it; (*computer science*) a period during which a computer system is inoperable.

downwelling — The process of accumulation and sinking of warm surface waters along a coastline.

DPA — Acronym for Deepwater Ports Act.

DPD — Acronym for method of measuring chlorine residual in water; see *N,N*-diethyl-*p*-phenylene-diamine.

dr — See dram.

draft — Difference between the pressure in an incinerator or any component part and that in the atmosphere, causing air or the product of combustion to flow from the incinerator into the atmosphere; actively drawing or removing water from a tank or reservoir; water that is drawn or removed from a tank or reservoir; excessive air movement in an occupied enclosure that causes discomfort.

Draft Environmental Impact Statement (DEIS) — A document prepared by the U.S. EPA, or under its guidance, that identifies and analyzes the environmental impacts of a proposed EPA action and feasible alternatives and is circulated for public comment prior to preparation of the final environmental impact statement.

drag — Resistance to movement through water or any other medium.

drag conveyor — A conveyor that uses vertical steel plates fastened between two continuous chains to drag material across a smooth surface in solid waste disposal.

dragline — A revolving shovel that carries a bucket attached only by cables and digs by pulling the bucket toward itself.

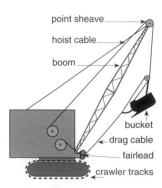

point sheave
hoist cable
boom
bucket
drag cable
fairlead
crawler tracks

Dragline

drainage — A technique to improve the productivity of some agricultural land or the usefulness of soil for sewage disposal by removing excess water from the soil; the process of transporting surface water over a land area to a river, lake, or ocean or the removal of water from soil using buried and regularly spaced perforated pipelines.

drainage area — An area that drains to a special point on a river or stream.

drainage basin — An area of land that drains water, sediment, and dissolved materials to a common outlet at some point along the stream channel; all the area drained by a river system.

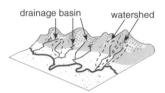

drainage basin   watershed

Drainage Basin

drainage blanket — (*construction*) A layer of pervious material placed directly on foundation material to facilitate drainage over a foundation and/or embankment.

drainage curtain — A line of vertical wells or boreholes to facilitate drainage of the foundation and to reduce water pressure.

drainage district — A legally established entity that has responsibility for developing, installing, and maintaining a drainage program for a specified land area that encompasses multiple ownership.

drainage system — A collection of surface and/or subsurface drains together with structures and pumps used to remove surface or groundwater.

drainage way — Channel portion of the landscape in which surface-water or rainwater runoff gathers intermittently to flow to a lower elevation.

drainage well — A well drilled to carry excess water off of the land that might cause groundwater contamination.

dram (dr) — A unit of weight formerly used in the avoirdupois system equal to 27.3437 grains or 1/16 ounce; a unit of weight used in the apothecaries' system equal to 60 grains or 1/8 ounce.

draping — (*geographical information system*) The process of applying another dataset over a shaded relief image.

draw — (*geology*) A small valley or gully.

drawdown — A drop in the water table or level of water in the ground when water is being pumped from a well; the amount of water used from a tank or reservoir; a drop in water level of a tank or reservoir.

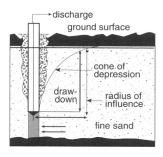

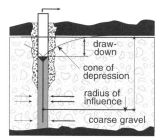

Drawdown

d-RDF — A refuse-derived fuel that has been compressed to improve certain handling or burning characteristics.

DRE — See destruction and removal efficiency.

dredge — To dig underwater; a machine that digs underwater.

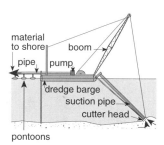

Suction Dredge

dredge spoils — Material removed from the bottom during dredging operations.

dredging — The removal of earth from the bottom of water bodies by using a scooping apparatus.

drench treatment (pesticides) — The application of a liquid chemical to an area until the area is completely soaked.

DRES — Acronym for Dietary Risk Evaluation System.

dress code — Standards of dress set by an institution for its members.

dressing — Any of various materials used for covering and protecting a wound.

drier — Any catalytic material that accelerates drying or hardening of a film when added to a drying oil.

drift — The movement of droplets, particles, sprays, or dust by air current from the target area to an area not intended for treatment.

Drift
(from crop dusting)

driller's well log — A log produced at the time of drilling showing the depth, thickness, and character of the different strata penetrated to reach water, as well as the depth, size, and type of casing installed.

drilling mud — A heavy suspension used in drilling an injection well introduced down the drill pipe and through the drill bit to lubricate and cool the bit.

drinking-water equivalent level — The estimated exposure (in mg/l) that is interpreted to be protective for noncarcinogenic endpoints of toxicity over a lifetime of exposure, developed for chemicals that have a significant carcinogenic potential and inferring that carcinogenicity should be considered the toxic effect of greatest concern.

drinking water supply — Any raw or finished water source that is or may be used by a public water system or as drinking water.

drip — The process of a liquid or moisture forming and falling in drops.

drip irrigation — The use of a pipe or tubing perforated with very small holes to deliver water slowly to the soil.

drive-point profiler — An exposed groundwater direct push system used to collect multiple-depth but discrete groundwater samples.

drop — A small spherical mass of liquid having a volume of 0.06 to 0.1 ml.

droplet — A tiny liquid particle suspended in a gas.

droplet infection — A disease spread through contact with cough or sneeze droplets carrying infectious microorganisms.

Drosophila — A genus of fly that includes the Mediterranean fruit fly and is sensitive to environmental effects; useful in genetic experiments because of the large chromosomes found in the salivary glands.

dross — A molten or solidified substance that is the non-metallic fused byproduct of a furnace operation; a scum that forms on the surface of molten metal. Also known as slag.

drought — A period of dryness that, when prolonged, causes extensive damage to crops or hinders their successful growth.

drug — A chemical substance taken orally; injected into a muscle, the skin, a blood vessel, or a body cavity; or applied to the skin or eyes to treat or prevent a disease or condition; an illegal narcotic substance.

drug absorption — The process by which a drug moves from its site of entry in the body to the target organ or tissue.

drug metabolism — A chemical process in body tissues that transforms a drug into its metabolites.

drug potency — The effect of a drug compared to another one of the same dosage.

drug reaction — An adverse effect in the body as a result of taking medications or illegal drugs or the interaction of two or more pharmacologically active chemicals within a short time.

drug tolerance — An adaptation of body cells to a pharmacologically active substance requiring increasingly larger doses to produce the same physiological or psychological effects as obtained earlier with smaller doses.

drum mill — A long, inclined steel drum that rotates and grinds solid waste in its rough interior; smaller material falls through the holes near the end of the drum, and the larger material drops out the end.

drum scanner — (*geographic information system*) A device for converting maps automatically to digital form.

dry acid deposition — The process by which particles such as fly ash or gases such as sulfur dioxide (SO_2) or nitric oxide (NO) are deposited onto surfaces.

dry adiabatic lapse rate — See adiabatic lapse rate.

dry alkali injection — Spraying dry sodium bicarbonate into flue gas to absorb and neutralize acidic sulfur compounds.

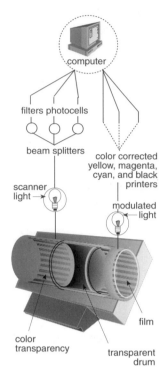

Drum Scanner

dry bulb temperature — The temperature of a gas or mixture of gases shown by a dry bulb thermometer after correcting for radiation.

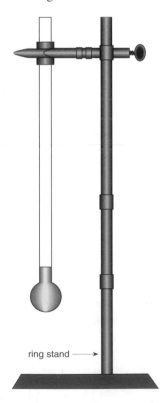

Dry-Bulb Thermometer

dry bulb thermometer — An ordinary thermometer with an unmoistened bulb that is not dependent upon atmospheric humidity.

dry-casket storage (nuclear energy) — A method for storing spend fuel in steel or steel-reinforced concrete and lead 18 or more inches thick.

dry cell — An electric cell that has a moist paste instead of a liquid electrolyte.

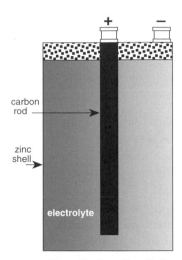

Zinc–Carbon Dry Cell

dry chemical — A powdered, fire-extinguishing agent usually composed of sodium bicarbonate or potassium bicarbonate and used especially for electrical fires.

drycleaning operation — Use of an organic solvent in the commercial cleaning of garments and other fabric materials.

dry deposition — The transfer of gases or particles from the atmosphere to surfaces because of molecular diffusion, Brownian diffusion, or gravitational settling in the absence of active precipitation.

dry gas — A gas that contains no water vapor.

dry gas meter — A device that measures total volume of a gas passed through it without the use of volatile liquids.

dry heat sterilization — A process conducted in a closed vessel at 160°C for 1 hour to kill all bacteria and spores.

dry hole — A well found to be incapable of producing either oil or gas in sufficient quantities to justify completion as an oil or gas well.

dry ice — Solid carbon dioxide, formed into blocks, that is usually used as a refrigerant.

dry limestone process — An air pollution control method that uses limestone to absorb the sulfur oxides in furnaces and stack gases.

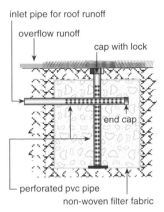

Dry Well

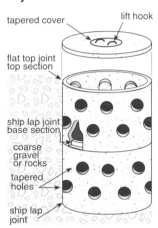

Precast Dry Well

dry milk — Milk from which at least 95% of the water has been removed.

dry rot — A fungus that decays wood in the presence of moisture and warm conditions and in the absence of light.

dry skin — An epidermis lacking moisture or sebum and exhibiting a pattern of fine lines and symptoms of scaling and itching.

dry soil — Any soil from which moisture has been removed.

dry well — A covered deep hole usually lined or filled with rocks that holds drainage water until it soaks into the ground.

DSCF — Acronym for dry standard cubic feet.

DSCM — Acronym for dry standard cubic meter.

DSP — Acronym for digital signal processing.

DSR — See dynamic spatial reconstructor.

DU — See Dobson unit.

dual-phase extraction — The active withdrawal of both liquid and gas phases from a well, usually involving the use of a vacuum pump.

duct — A passage with well-defined walls for the conveyance of air in a ventilation or other system; a tubular structure that delivers glandular secretions to parts of the body through tubes.

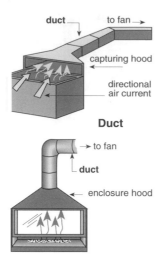

Duct

duct silencers — Cylindrical or rectangular structures fitted to the intake or discharge of air-moving equipment designed to reduce noise levels.

duct velocity — Velocity of air through a duct cross-section.

ductile — Capable of being elongated or deformed without fracture.

ductless gland — A body organ or tissue that delivers its products (known as hormones) directly into the blood stream; also known as an endocrine gland.

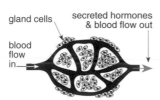

Ductless (Endocrine) Gland

due diligence — (*environment*) The process of evaluting the environmental condition of a parcel of land, usually in connection with a real estate transaction.

dull pain — A mildly throbbing acute or chronic pain.

dump — A site at which solid waste is disposed in a manner that may contaminate the environment.

dumped — Refers to a method of compacting soil by dumping it into place with no compactive effort.

dumping — An indiscriminate method of disposing of solid waste.

duodenitis — Condition of inflammation of the duodenum.

duodenum — The region of the small intestine immediately below the stomach; in humans, it is approximately 10 inches long.

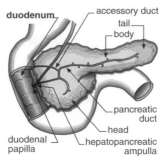

Duodenum

duplicates — Two separate samples taken at the same time and at the same place and placed in separate containers for comparative analysis.

duration — (*ergonomics*) The continuous time a task is performed without a rest period.

dust — A fine-grain particulate matter that is capable of being temporarily suspended in air, ranging in size from 0.1 to 25 μm and generated by physical processes such as the handling, crushing, or grinding of solids; (*pesticides*) a dry mixture of a finely ground material containing a pesticide chemical.

dust arrester — A piece of equipment for removing dust from air handled by ventilation or exhaust systems.

dust collector — An air-cleaning device used to remove heavy particles from exhaust systems before release to the outside air.

dust devil — A small atmospheric vortex, not associated with a thunderstorm, that is made visible by a rotating cloud of dust and debris; it forms in response to surface heating during fair, hot weather.

dust diseases of the lungs — See pneumoconiosis.

dustfall bucket — (*air pollution*) An open-mouthed container used to collect and roughly estimate the kinds and quantities of materials falling out in a given location.

dust loading — The amount of dust in a gas, usually expressed in grains per cubic foot or pounds per thousand pounds of gas.

Dutch liquid — See ethylene dichloride.

dwarf tapeworm infestation — An intestinal parasitic disease caused by an infestation of *Hymenolepsia nana* occurring mainly in the southern United States and usually affecting children who ingest the eggs by placing contaminated objects in their mouths.

DWEL — Acronym for drinking water equivalent level.

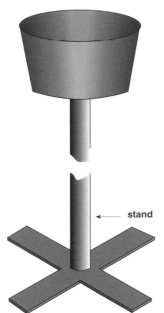

Dustfall Bucket

dwell period — That part of the sterilizer cycle during which sterilization takes place.

dwelling — Any enclosed space wholly or partially used or intended to be used for living, sleeping, cooking, and eating.

dwelling unit — A room or group of rooms arranged for use by one or more individuals living together as a single household and whom share living, sleeping, cooking, and eating facilities.

DWS — Acronym for drinking water standard; acronym for Doppler wind sensor.

dye — Any of various colored substances containing auxochromes and capable of imparting color to the other materials.

dynamic — Tending to change or encourage change.

dynamic biochemistry — Chemical or metabolic changes occurring in living systems.

dynamic calibration — Calibration of an instrument by means of sampling either a gas of known concentration or an artificial atmosphere containing a pollutant of known concentration.

dynamic compaction — A method of compacting soil by dropping a heavy weight onto loose soil.

dynamic data exchange — (*computer science*) A standard Microsoft Windows protocol that defines how Windows applications share information with one another.

dynamic head — The sum of all the pressure due to resistance in a complete system in operation.

dynamic imaging — The rapid sequential imaging of vascular blood flow to a body organ or system utilizing an imaging device either in nuclear medicine or by ultrasound.

dynamic loss — The loss of energy in air encountered during air flow resulting from turbulence caused by a change in direction or velocity within a duct.

dynamic pressure — Vertical distance (in feet) from a reference point to a hydraulic grade line (dynamic head) when a pump is operating.

dynamic response — The accuracy with which an instrument such as an electrocardiograph will simulate the actual event being recorded.

dynamic spatial reconstructor — A radiographic machine used in research that provides moving, three-dimensional images of human organs that can be examined visually from all directions.

dynamics — The scientific study of mechanics, which deals with forces in action.

dynamics of stressed ecosystems — Changes that occur in ecosystems due to stresses caused by humans.

dynamite — A substance composed of nitroglycerine and silicious earth; the nitroglycerine is absorbed into the porous material to make it much safer to handle than in its liquid form. It is highly explosive but can be transported in this manner without risk of spontaneous decomposition.

dyne — Metric unit of force; the amount of force that produces an acceleration of 1 cm/sec/sec in a particle of 1 gram mass.

dynometer — An apparatus used to measure force or work output external to a subject.

dysentery — (*disease*) Any of a number of disorders marked by inflammation of the intestine, especially the colon, with abdominal pain and frequent stools often containing blood and mucous.

dysfunction — Disturbance, impairment, or abnormality in the functioning of an organ.

dysmenorrhea — Painful menstruation.

dyspepsia — Impairment of the power or function of digestion.

dysphagia — Difficulty in swallowing.

dysplasia — An abnormality of development or growth, especially of the cells.

dyspnea — Labored breathing, which is a symptom of a variety of disorders and an indication of inadequate ventilation or insufficient amounts of oxygen in the circulating blood; a shortness of breath or a difficulty in breathing that may be caused by certain heart conditions, extraneous exercise, or anxiety; respiratory depression, paralysis, seizures, and loss of consciousness.

dystrophic — A term related to groundwater, lakes, and streams, usually with a low lime content and a high organic content; pertaining to an environment lacking in nutrients.

dysuria — Painful or difficult urination.

E

E — See electric field; an SI symbol for the prefix *exa-*, representing a factor of 10^{18}.

E&GW — Acronym for Economic and General Welfare.

E. coli — Abbreviation for *Escherichia coli.*

E. coli 015:H7 — See hemorrhagic colitis.

EA — Acronym for endangerment assessment, enforcement agreement, environmental action, environmental assessment, and environmental audit.

EAA — Acronym for Export Administration Act.

EAF — Acronym for electric arc furnaces; see Environmental Action Foundation.

EAG — Acronym for exposure assessment group.

EAP — Acronym for environmental action plan; see emergency action plan.

EAP program — See employee assistance program.

ear — The organ of hearing and balance consisting of the internal, middle, and external ear.

ear drum — Thin, tympanic membrane that vibrates; three small, hinged bones in the middle ear pass the vibration along to the inner ear.

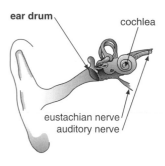

Ear Drum

ear muffs — Devices that fit over the ears and are supported either by a hard hat or headband that connects the two individual muffs; used to protect the ear from excessive sound.

ear plugs — Devices that fit within the ear canal to protect the ear from water or excessive noise.

earache — A pain in the ear that may be sharp, dull, burning, intermittent, or constant; caused by either a disease or a disorder of the nose, oral cavity, or larynx.

**Ear Plugs
(3–flange design)**

early effects — (*radiation*) Effects that appear within the first 60 days of an acute exposure.

early warning system — A system that will ensure timely recognition of a threatening event and provide a reliable and timely warning and evacuation of the population at risk.

earth blade — A heavy, broad plate that is connected to the front of a tractor and used to push and spread soil or other materials.

earth ground — A metal rod, usually copper, that provides an electrical path to the earth to prevent or reduce the risk of electric shock.

Earth Observing System — A series of small to intermediate-size spacecraft that are the core of NASA's Earth Science Enterprise; they carry a suite of instruments designed to study global climate change.

Earth system — The Earth regarded as a unified system of interacting components, including geosphere, atmosphere, hydrosphere, and biosphere.

Earth system science — An integrated approach to the study of the Earth that stresses investigations of interactions among the components of the Earth in order to explain Earth dynamics, evolution, and global change.

earthquake risk — Probable building damage and number of people that are expected to be injured or killed if an earthquake occurs on a particular fault.

earthquakes — A series of elastic waves traveling through the crust of the Earth; they are caused by movements along fault lines and emanate from an epicenter.

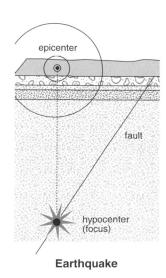

Earthquake

easement — The right to use land owned by another person for some specific purpose.

easily cleanable — A surface that is readily accessible and of such material, finish, and fabrication that residue may be effectively removed by normal cleaning methods.

easily movable — (*food service*) Weighing 30 pounds or less and mounted on casters, gliders, or rollers or able to be moved with a force of 30 pounds or less.

East African sleeping sickness — See Rhodesian trypanosomiasis.

eastern equine encephalitis — (*disease*) An acute, mosquito-borne, viral, inflammatory disease of short duration affecting the brain, spinal cord, and meninges with symptoms ranging from mild to severe headaches, high fevers, meningeal signs, stupor, coma, convulsions, and possible death; incubation time is 5 to 15 days. It is caused by an alpha virus and is found in eastern and northcentral United States and adjacent Canada, Central and South America, and the Caribbean islands. Possible reservoirs of infection include birds, rodents, bats, reptiles, amphibians, surviving mosquito eggs, or adult mosquitos. It is transmitted by the bite of infected mosquitos to horses as an intermediate, but not person to person; infants and the elderly are most susceptible. It is controlled by the destruction of mosquito larvae, breeding places, and adult mosquitos and by the use of mechanical mosquito control techniques.

EB — Acronym for emissions balancing.

Ebert–Fastie monochromatic spectrometer — An instrument used to measure energy intensity within the ultraviolet region of the electromagnetic spectrum.

Ebola — A virus transmitted through direct contact with the blood or bodily fluids of an infected person, unsterilized needles, or an infected animal. Symptoms include high fever, headaches, muscle aches, stomach pain, fatigue, and diarrhea, occurring 4 to 16 days after infection. If not treated, it results in failure of all major organs and death.

EBS — See Emergency Broadcasting System.

EBT — An acronym for electronic benefit transfer.

ECARP — An acronym for Environmental Conservation Acreage Reserve Program.

ECC — Acronym for emergency cardiac care.

ECD — Acronym for electronic capture detector.

ECG — See electrocardiogram; also called EKG.

echinococcosis — An infestation, usually of the liver, caused by the larval stage of a tapeworm of the genus *Echinococcus*. Dogs are the principal hosts of adult worms, and sheep, cattle, rodents, and deer are intermediate hosts for the larvae.

Echinococcus — A genus of small tapeworms that infest primarily canines.

echo — Reflection of sound waves from a surface or object causing a weaker version to be detected shortly after the original is emitted.

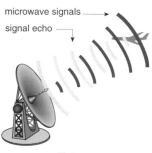

Echo

echocardiogram — A graphic outline of movement of the heart structures provided by the echoes of ultrasound vibrations moving through the heart.

echocardiography — A diagnostic procedure for studying the structure and motion of the heart by using ultrasonic waves directed through the heart and reflected backward or echoed.

echovirus — Any of a group of viruses (genus *Enterovirus*) isolated from people; may produce different types of human disease, especially aseptic meningitis, diarrhea, and various respiratory diseases. The name of the genus is derived from the group designation enteric cytopathogenic.

ECIS — Acronym for Economic Impact Statement.

eclampsia — A sudden attack of convulsions, edema, or elevated blood pressure during the latter half of pregnancy.

ECLIPS — Acronym for Experimental Cloud Lidar Pilot Study.

eclosion — The process whereby the adult form of an insect emerges from the pupa.

ecological adaptation — The short-term changing responses expressed by an individual organism to its environment.

ecological chemistry — The study of chemical compounds synthesized by plants that affect the ecosystem because of their toxicity.

ecological development — A gradual process of environmental modification by organisms.

ecological diversity — A number of different species coexisting in a community of organisms.

ecological fallacy — A false assumption that the presence of a pathogenic agent and disease within a given population is proof of the cause of the disease in any particular individual.

ecological health ranking — A chemical's adverse effect on plants and animals compared with that of other chemicals in a relative ranking system.

ecological impact — The effect that people or natural activities have on living organisms and their nonliving environment.

ecological indicator — A collective term for response, exposure, habitat, and stressor indicators.

ecological marker — An early indicator of potential ecological change, such as measurement of the addition of phosphates to a stream.

ecological niche — The combination of function and habitat of each of the approximately 1.5 million species of animals and the 0.5 million species of plants on the Earth; many interactions occur among species in an ecosystem, and a balance is determined by nature.

ecological pyramid of energy — The total energy content of all organisms at each trophic level in a food chain; the energy content declines at each higher level.

ecological pyramid of numbers — The number of organisms supported at each trophic level in a food chain; fewer organisms are supported at higher levels.

ecological risk assessment — The use of an analytical process or model to estimate the effects of human actions on a natural resource and to interpret the significance of the effects; a process used to estimate how likely it is that there will be adverse effects on plants or animals from exposure to chemicals or to other potential stresses.

ecology — The study of the interaction between organisms and their environments.

economic analysis — A procedure that includes tangible and intangible factors to evaluate various alternatives.

economic geology — The search for and exploitation of mineral resources and water.

economic poisons — Chemicals used as insecticides, rodenticides, fungicides, herbicides, etc. that may inadvertently or purposely be introduced into the environment.

economics — A social science concerned chiefly with the prediction and analysis of production, distribution, and consumption of goods and services.

ecosphere — The layer of Earth and troposphere inhabited by or suitable for the existence of living organisms.

ecosystem — The complex of a community and its abiotic and biotic environment functioning as an ecological system in nature.

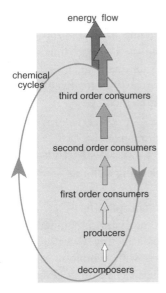

Ecosystem

ecosystem management — The integration of ecological, economic, and social goals in a unified system approach to resource management.

ecosystem restoration — The reinstating of an entire community of organisms to as near its natural condition as possible.

ecotone — The transition zone between two adjacent communities.

ecotopy — A condition in which an organ or substance is not in its natural or proper place.

ecotox thresholds — The amounts of media-specific contaminated concentrations that indicate further site investigations are needed.

ecotoxicity — The ability to be poisonous or harmful to plants or animals to some degree.

ecotoxicological study — A measurement of the effects of environmental toxicants on indigenous populations of organisms.

ECRA — Acronym for Economic Cleanup Responsibility Act.

ECt$_{50}$ — The dosage causing a specifically defined effect in 50% of a given population when the route of exposure is either percutaneous or through inhalation.

ectoderm — The outer layer of skin cells.

ectoparasite — A parasite that lives outside the host, such as a tick or mite.

ectopic — An infection in a location other than normal or expected.

ectopic beat — A heartbeat that appears someplace other than in the sinoatrial node.

ectopic pregnancy — Gestation outside the uterus, often in a fallopian tube.

ectotherm — An organism whose internal temperature varies with that of the environment.

ECU — Acronym for extended care unit.

eczema — An inflammatory skin condition, usually caused by an allergy, with symptoms of redness, itching, and scaly crusts.

ED — Acronym for effective dose.

EDA — Acronym for emergency declaration area.

edaphic — Relating to the soil.

EDB — See ethylene dibromide.

EDD — Acronym for enforcement decision document.

eddy — An irregularly patterned motion of turbulent air or fluid.

EDE — See effective dose equivalent.

edema — An excessive, abnormal accumulation of fluid in the intercellular spaces of the body causing a swelling of tissues.

edible — Intended for use as food.

edit — To remove errors and unwanted material from a work.

EDP — Acronym for electronic data processing.

EDTA titration — The use of ethylenediaminetetraacetic acid or its salts as a standard method of measuring the hardness of a solution.

eductor — A hydraulic device used to create a negative pressure (suction) by forcing a liquid through a restriction such as a Venturi tube.

EDXA — Acronym for energy-dispersive x-ray analysis.

EDXS — Acronym for energy-dispersive x-ray spectroscopy.

EEA — Acronym for energy and environmental analysis.

EEE — See eastern equine encephalitis.

EEG — See electroencephalogram.

EEGL — Acronym for Emergency Exposure Guidance Level.

EENT — Acronym for eyes, ears, nose, and throat.

EER — Acronym for excess emission report.

EERU — Acronym for environmental emergency response unit.

EF — See ejection fraction; acronym for emission factor.

effective — Something that follows a cause or agent.

effective corrosion inhibitor residual — Concentration of a corrosion inhibitor sufficient to form a protective coating on the interior walls of a pipe, thus reducing its corrosion.

effective dose — The quantity obtained by multiplying the equivalent dose to various tissues and organs by a weighting factor appropriate to each and summing the products.

effective dose equivalent — A value used to express health risk from radiation exposure to a tissue or tissues in terms of an equivalent whole body exposure.

effective dose$_{50}$ (ED$_{50}$) — A dose of a chemical that creates a specific effect in 50% of the test subjects.

effective half-life — The time required for a radioisotope in a biological system to diminish 50% as a result of the combined action of radioactive decay and biological elimination. The effective half-life = (biological half-life × radioactive half-life)/(biological half-life + radioactive half-life). See also biological half-life.

effective homogeneous conductivity — The conductivity of an object viewed as though its dielectric properties were uniform throughout.

effective porosity — The portion of pore space in saturated permeable material where the movement of water takes place.

effective range — The portion of a design range, usually the upper 90%, in which an instrument has acceptable accuracy.

effective size — The diameter of particles in a granular sample of filter media for which 10% of the total grains are smaller and 90% are larger on a weight basis.

effective soil depth — The depth of slightly or moderately limited soil material lying above a restricted soil layer such as clay, hardpan, or bedrock.

effective sound pressure — Root mean square (rms) value of instantaneous sound pressure over a timed interval at a point during a complete cycle, expressed in dynes per square centimeter.

effective temperature — A combination of globe temperature, wet bulb temperature, air speed, and metabolic rate.

effective temperature index — An empirically determined index of the degree of warmth perceived on exposure to different combinations of temperature, humidity, and air movement.

effector — An organ, tissue, or cell that becomes active in response to stimulation.

effervescence — The rapid escape of excess gas bubbles in a liquid by chemical reaction without the application of heat.

efficacy — The probability of benefit to individuals in a defined population from a program applied under ideal conditions; (*electricity*) the rate at which a lamp is able to convert electrical power (watts) into light (lumens) expressed in terms of lumens per watt (LPW); a watt is the amount of electricity being put in, and a lumen is the amount of power being put out.

efficiency — Relationship between the quantity of inputs or resources and the quantity of outputs produced.

effluent — An air flow, discharge, or emission of a liquid or gas that flows outward or away from a main area; the material that comes out of a treatment plant after completion of the treatment process.

effluent guidelines — Technical U.S. EPA documents that establish effluent limitations for industries and pollutants.

effluent limitation — The maximum amount of a specific substance or characteristic that can be present in effluent discharge without violating water quality standards in receiving waters.

effluent sewerage — A low-cost alternative sewage system that combines some features of septic tanks and centralized municipal treatment systems, where the septic tank is used by a residence and is pumped periodically to a central treatment plant.

effusion — The escape of fluid from blood vessels because of rupture or seepage.

EFNEP — See Expanded Food and Nutrition Education Program.

egestion — The disposal of indigestible or waste matter from the body by any normal route.

EGL — See energy grade line.

EGR — Acronym for exhaust gas recirculation.

egress — Exits that ensure a safe means of escape from buildings.

EHC — Acronym for Environmental Health Criteria; acronym for Environmental Health Center.

ehv — See extra-high-voltage.

EI — Acronym for emissions inventory; acronym for erosion index.

EIA — Acronym for Environmental Impact Assessment; acronym for Economic Impact Assessment

EIL — Acronym for environmental impairment liability.

EIP — Acronym for Emerging Infections Program.

EIR — Acronym for environmental impact report.

EIS — See Environmental Impact Statement.

EJ — See environmental justice.

ejection fraction — The proportion of blood that is ejected during each ventricular contraction compared with the total ventricular filling volume, where the normal fraction is 65%.

ejector — An air mover consisting of a two-flow system wherein a primary source of compressed gas is passed through a Venturi tube; the vacuum that develops at the throat of the Venturi tube is used to create a secondary flow of fluid.

EKG — See electrocardiogram.

Ekman spiral — The spiral change of water movement in a water column when the water is pushed by wind.

EL — Acronym for exposure level.

El Niño — A warming of surface waters of the eastern equatorial Pacific that occurs at irregular intervals of 2 to 7 years and usually lasting 1 to 2 years. Along the west coast of South America this warming creates southerly winds that promote the upswelling of cold, nutrition-rich water that sustains large fish populations and abundant seabirds. Where the conditions last for many months, more extensive ocean warming occurs and economic results can be serious. It has been linked to wetter, colder winters in the United States; drier, hotter summers in South America and Europe; and drought in Africa.

El Niño southern oscillation — A large-scale change in the normal weather patterns of the Pacific basin and adjacent regions resulting in a warming of the surface currents in the eastern Pacific.

elaborate — (*physiology*) To build up complex organic compounds from simple ingredients.

elastic tissue — A type of connective tissue containing elastic fibers.

elasticity — The property of matter that causes it to resume its original shape after a distorting force is employed and removed.

elastomer — A synthetic polymer with rubber-like characteristics.

elbow — A pipe fitting with two openings that creates a 90° change of direction in a run of pipe.

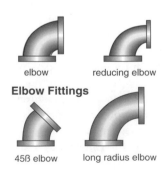

elbow   reducing elbow

Elbow Fittings

45β elbow   long radius elbow

elective — A procedure that is performed by choice but which is not essential at that moment.

electric arc furnace — A steel-making furnace in which the heat is obtained by an electric arc formed between the material to be heated and electrodes.

electric charge — A fundamental quantity required to explain measurable forces and the fundamental unit from which other electromagnetic quantities are developed.

electric circuit — The complete conducting path of an electric current.

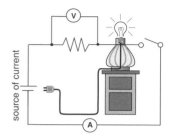

Electric Circuit
diagram of closed circuit with
battery as emf

electric current — The flow of electrons along a conductor due to potential difference in a circuit.

electric dipole movement — The product of either of the charges and the distance between them.

electric displacement (D) — A field quantity whose source is a net-free charge that has been created by separation of charge.

electric field (E) — Any space in which force due to an electric charge exists.

electric field strength — A measure of the intensity of an electric field.

electric furnace — A furnace heated by the passage of an electric current, such as in the arc-type furnace, where an electric arc jumps the gap between carbon electrodes.

electric generator — A device that changes mechanical energy to electrical energy by electromagnetic induction.

electric potential — The work required to transport a unit of electric charge from a reference position to another position.

electric potential gradient — The net difference in electric charge across the membrane of a cell.

electric shock — A traumatic physical event caused by the passage of electric current through the body with duration, type of current, frequency, and intensity determining the specific symptoms and

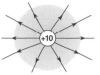

positively charged object

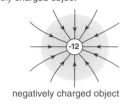

negatively charged object

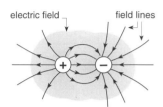

Electric Field

results, such as burns, coagulation of blood, necrosis of affected body parts, and death.

electric vehicle — A motor vehicle that uses an electric motor as the basis of its operation and emits virtually no air pollutants.

electrical conductivity — A measurement of the ability of a solution to conduct an electrical current; a measure of the salt content of water

electrical control — (*energy*) A control system that operates on line or low voltage and uses a mechanical means such as a temperature-sensitive bimetal to perform the control functions of actuating switches or positioning potentiometers.

electrical interaction — A force of repulsion acting between electric charges of like sign and a force of attraction between electric charges of unlike sign.

electrical interference — Electrical noises that obscure desired information signals being transmitted.

electrical plant — A facility containing electric generators and other equipment for converting mechanical, chemical, and/or fission energy into electric energy.

electrical precipitator — See electrostatic precipitator.

electrical resonance — An effect in which resistance to the flow of electrical current becomes very small over a narrow frequency range.

electrical synapse — A synapse at which local currents resulting from electrical activity flow between the two neurons through gap junctions joining them.

electroanalytical chemistry — The analysis of compounds by using an electric current to produce characteristic, observable change in a substance being studied.

electrocardiogram (ECG, EKG) — A graphic tracing and record of the electric current produced by contractions of the heart muscle.

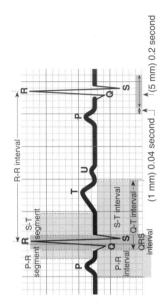

Electrocardiogram

electrochemical analysis — A system of electrodes and electrochemical techniques used in on-site analysis and continuous monitoring of wastewater effluent.

electrochemical reaction — Chemical changes produced by electricity (electrolysis) or the production of electricity by chemical changes (galvanic action).

electrochemical series — A list of metals with the standard electrode potentials given in volts. The size of the electrode potential indicates how easily these elements will take on or give up electrons or corrode. Hydrogen is assigned a value of 0.

electrode — An electric conductor for the transfer of charge between an external circuit and a medium; a terminal of an electrolytic cell that emits or collects electrons.

electrode potential — The difference in potential between an electrode and a solution in which it is immersed.

electrodiagnosis — The diagnosis of disease or injury by applying electrical stimulation to various nerves and muscles.

electrodialysis — Dialysis accelerated by applying an electromotive force to electrodes adjacent to the membranes; a dialysis that utilizes direct current and the arrangement of permeable-active membranes to achieve separation of soluble minerals from the water.

electrodynamics — The study of electrostatic charges in motion such as the flow of electrons in an electric current.

electroencephalogram (EEG) — A graphic record of the electric currents developed in the brain as detected by electrodes on the scalp.

electrokinetic soil processing — An *in situ* separation and removal technique for extracting heavy metals and organic contaminants from soils.

electrolysis — The production of chemical changes in matter by means of an electric current passing through an electrolyte.

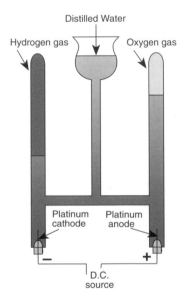

Electrolysis of Water

electrolyte — A nonmetallic chemical compound that will conduct an electric current through the movement of ions when dissolved in certain solvents or when fused by heat; a common example is salt.

electrolyte balance — An equilibrium among electrolytes in the body.

electrolyte solution — A solution containing electrolytes prepared for oral, parenteral, or rectal use.

electrolytic cell — A device in which the chemical decomposition of material causes an electric current to flow; a device in which a chemical reaction occurs as a result of the flow of electric current.

electrolytic process — A process that causes the decomposition of a chemical compound by the use of electricity.

electromagnet — An iron core that becomes a magnet when current is passed through a coil wrapped around it and becomes demagnetized when the current is interrupted.

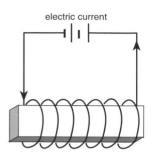

Electromagnet

electromagnetic compatibility — The ability of equipment or a system to function in its electromagnetic environment without causing electromagnetic disturbances or being affected by those that are present.

electromagnetic field — (*physics*) A field created by interaction of an electric field and a magnetic field that occurs when an electric current passes through a wire.

electromagnetic geophysical method — A way to measure subsurface conductivity by means of low-frequency electromagnetic induction.

electromagnetic induction — The production of electric current in a circuit when a current passes through a changing magnetic field.

electromagnetic interference (EMI) — A disturbance in the operation of an electronic device that results from the presence of undesired local electric or magnetic fields.

electromagnetic radiation — The propagation of varying electric and magnetic fields through space at the speed of light; it exhibits the characteristics of wave motion.

electromagnetic spectrum — A range of electromagnetic radiations including radio, infrared, visible light, ultraviolet, gamma-ray, and cosmic ray.

electromagnetic waves — Waves of radiant energy classified according to wave frequency and length.

electromotive force (emf) — The voltage of electricity due to the potential difference between two dissimilar electrodes.

electromotive series — A list of metals and alloys presented in the order of their tendency to go into solution.

electron — A negatively charged alimentary particle that is a constituent of the shell of every neutral atom. Its unit of negative electricity equals 4.8×10^{-10} electrostatic units or 1.6×10^{-19} coulombs, and its mass equals 0.00549 atomic mass units.

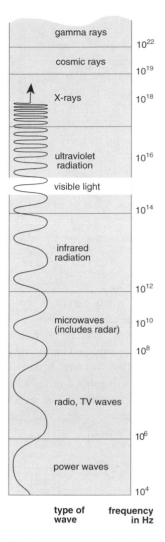

Electromagnetic Spectrum
(not to scale)

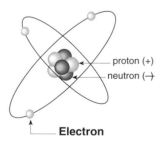

electron acceptor — A small inorganic or organic compound that is reduced to complete an electronic transport chain; a compound that is reduced in a metabolic redux reaction.

electron beam imagers — A device that exposes a view on a photosensitive surface by means of a shutter or similar mechanism and then uses a beam of electrons to provide a means of reading the photosensitive surface after the shutter has closed.

electron beam therapy — The delivery of external beam radiation therapy by means of an electronic beam.

electron beam welding — A welding process used in outer space that produces a coalescence of metals with heat obtained from a concentrated beam composed primarily of high-velocity electrons impinging on the joint to be welded.

electron capture — A mode of radioactive decay involving the capture of an orbital electron by its nucleus, where capture from the particular electron shell is designated as K-electron capture, L-electron capture, etc.

electron capture detector — An extremely sensitive gas chromatographic detector for halogen-containing compounds such as chlorofluorocarbons.

electron cloud — An average region around the nucleus of an atom where electrons are predicted to be at various states of excitement.

electron donor — A small inorganic or organic compound that is oxidized to initiate an electron transport chain; a compound from which electrons are derived in a metabolic redox reaction.

electron gun — (*electronics*) The part of a cathode-ray tube that guides the flow and greatly increases the speed of electrons.

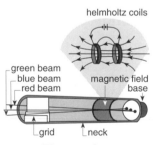

Electron Gun

electron microscope — A microscope in which high degrees of magnification are produced by an electron beam and magnetic field projecting enlarged images upon a fluorescent surface of photographic plate.

electron-ray tube — See cathode-ray tube.

electron shell — An energy level surrounding the nucleus of an atom in which an electron may be found.

electron shells — The positions of increasing energy levels occupied by electrons in an atom. They are designated as K (closest to the nucleus); L (second closest); M (third closest); N (fourth closest), etc.

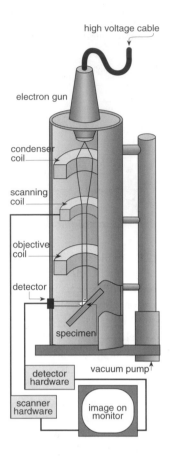

**Scanning
Electron Microscope**

electron transfer — (*biology*) The process of energy transfer in biological systems, usually in small steps with only slight changes in energy levels.

electron transport chain — The final sequence of reactions in biological oxidations made up of a series of oxidizing agents arranged in order of increasing strength and ending in oxygen.

electron transport system — The final stage of respiration, where a series of enzymes and coenzymes on the inner membrane of the mitochondria function in the transfer of electrons and the resulting synthesis of ATP.

electron volt (eV) — A unit of electrical energy used in radiation physics equal to the amount of energy gained by an electron passing through a potential difference of 1 volt, which equals 1.6×10^{-19} joules.

electronic air sampler — An instrument that measures contaminants through changes of an electronic signal.

electronic bubble flow meter — An electronic device used for calibrating air-sampling pumps from a specific manufacturer who produces those pumps.

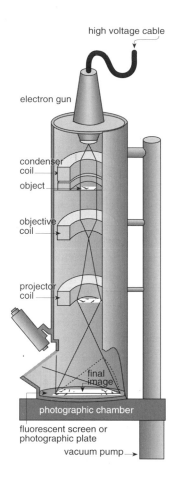

high voltage cable

electron gun

condenser coil

object

objective coil

projector coil

final image

photographic chamber

fluorescent screen or photographic plate

vacuum pump

Transmission Electron Microscope

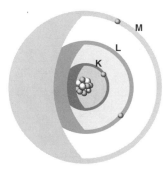

M

L

K

Electron Shells

electronic control — (*energy*) A control circuit that operates at low voltage and uses solid-state components to amplify input signals and perform control functions such as operating a relay or providing an output signal to position an actuator.

electronic mail (e-mail) — (*computer science*) Correspondence among individuals via computers, with each person having some form of network address.

electronic measurement — Detection of pollutants in a gaseous phase through spectroscopic techniques.

electronic product — Any manufactured or assembled product that, when in operation, continues or acts as part of any electronic circuit or emits electronic product radiation; any manufactured or assembled article that is intended for use as a component, part, or accessory of an electronic product.

electronic product radiation — Any ionizing or non-ionizing electromagnetic or particulate radiation or any sonic, infrasonic, or ultrasonic wave emitted from an electronic product as a result of operation of an electronic circuit in such product.

electronic superhighway — See communications super-highway.

electronic thermometer — A thermometer that rapidly registers temperatures electronically.

electroosmosis — The movement of a charged fluid as a result of the application of an electric field.

electrophilic compounds — Chemicals drawn to regions in other chemicals in which electrons are readily available.

electrophoresis — The movement of suspended particles through a fluid under the action of electromotive force, especially applying to colloids; an electro-chemical process in which colloidal ions or cells with a net electric surface density move in an electric field.

electrophysiology — The science that studies the relationship between living organisms and electricity.

electroplating — The depositing of a thin layer of a metallic element by electrodeposition.

electroreceptor — A receptor that senses changes in an electrical current, usually in the surrounding water.

electroscope — An instrument for detecting the presence of electric charges by the deflection of charged bodies.

electroshock — A condition caused by accidental contact with an electrical current, resulting in a variety of health problems and even death, depending on the amount of voltage and time of contact.

electrostatic — Pertaining to the charge on a substance at rest.

electrostatic discharge — An electrical discharge, usually of high voltage and low current.

electrostatic field — An electric field such as that produced by a stationary charge.

electrostatic plotter — See dot-matrix plotter.

electrostatic precipitator (ESP) — A dry collector that captures particles electrically through the use of electrodes; the particles go to the oppositely charged plate and then either fall or are shaken into a collection hopper. Also known as electrical precipitator.

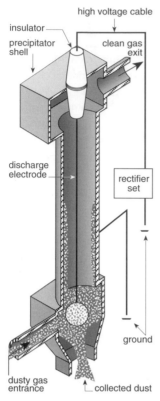

high voltage cable

insulator

precipitator shell

clean gas exit

discharge electrode

rectifier set

ground

dusty gas entrance

collected dust

Electrostatic Precipitator

electrostatic unit of charge (esu) — The quantity of electric charge that, when placed in a vacuum 1-cm distance from an equal and like charge, will repel it with a force of 1 dyne; also known as a statcoulomb.

electrotelluric survey — A low-frequency telluric instrument used to map hydrocarbons in subsurface strata; the instrument incorporates digital signal processing software and special hardware.

electrovalent bond — See ionic bond.

element — A molecule composed of atoms of the same atomic number and which cannot be broken down into simpler units by chemical reactions; common examples are hydrogen, gold, and iron; (*geographic information system*) a fundamental geographical unit of information, such as a point, line, or pixel. Also known as entity.

elevated sand mound — A mound of sandy fill material placed on the surface of the ground at a site approved for its use; the fill material serves as a physical and biological medium in which sewage effluent is filtered and treated before it is absorbed by the natural soil.

elevation — Height of a point above a plane of reference; the height above mean sea level.

elevator — A tall warehouse facility that uses vertical conveyors to raise or elevate and store grain.

ELF — See extremely low frequency.

elimination — The removal of a chemical from the body, primarily in urine, feces, or exhaled air.

ELISA test — See enzyme-linked immunosorbent assay test.

elliptical orbit — An orbit of a spacecraft that is elongated rather than circular because of gravity and drag.

elongation — Being lengthened or extended.

eluate — Solution that results from the elution process.

elution — (*chemistry*) Separation of material by washing.

elutriate — To purify, separate, or remove by washing.

elutriation — Purification of a substance by dissolving it in a solvent and pouring off the solution, thereby separating it from the undissolved foreign material; (*sewage*) the process of washing the alkalinity out of anaerobically digested sludge to decrease the demand for acidic chemical conditioning and to improve settling and dewatering characteristics.

eluviation — The removal of soil material from a layer of soil into a suspension.

EM — Acronym for electromagnetic conductivity.

e-mail — (*computer science*) See electronic mail.

embargo — A legal prohibition on commerce.

embolism — Sudden blocking of an artery or vein by a clot or embolus carried in the bloodstream.

embolus — A clot or plug foreign to the bloodstream that is forced into a smaller blood vessel, causing obstruction.

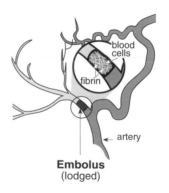

blood cells

fibrin

artery

Embolus
(lodged)

embryo — A multicellular organism in the earliest stage of development; from the time of fertilization in the ovum until the beginning of the third month of human pregnancy.

embryology — The study of the origin, development, and function of an organism from fertilization to birth.

embryotoxicity — Any toxic effect on the embryo as a result of prenatal exposure to a chemical.

EMC — See electromagnetic compatibility.

emergence — The action of a young plant breaking through the surface of the soil; an insect coming out of an egg or pupa; dry land that was part of the ocean floor.

emergency — A situation created by an accidental release or spill of hazardous chemicals and which poses a threat to the safety of workers, residents, the environment, or property; any occasion, as determined by the President of the United States, requiring federal assistance to supplement state and local efforts to save lives and protect property and public health and safety.

emergency action plan — A formal plan of procedures designed to minimize an emergency situation or unusual occurrence.

Emergency Broadcasting System (EBS) — A system of broadcast stations and interconnecting facilities authorized by the U.S. Federal Communications Commission to inform the public about the nature of a hazardous state of public peril or national emergency.

emergency department — See emergency room.

emergency exercise — An activity designed to promote emergency preparedness; evaluate emergency operations, policies, plans, procedures, and facilities; train personnel in emergency management and response duties; and demonstrate operational capability.

emergency level — The level of concentrations of single pollutants or combinations of pollutants and meteorological factors likely to lead to acute illness or death.

emergency management — The system by which mitigation, preparedness, response, and recovery activities are undertaken to save lives and protect property from all hazards.

emergency medical service (EMS) — Emergency ambulance services and other services used to help individuals in need of immediate medical care to prevent loss of life or aggravation of physiological or psychological illness or injury.

emergency medical system (EMS) — The entire emergency care system in a state or locality.

emergency medical technician (EMT) — An individual certified by the Emergency Medical Services Commission and responsible for the administration of emergency care to patients and for their handling and transportation.

emergency medicine — A branch of medicine concerned with the diagnosis and treatment of people with trauma or sudden illnesses.

emergency operations center — A facility where responsible officials gather during an emergency to direct and coordinate emergency operations, to communicate with other jurisdictions and field emergency forces, and to formulate protective action decisions and recommendations during an emergency.

emergency operations plan — A plan that clearly and concisely describes a jurisdiction's emergency organization, means of coordination with other jurisdictions, and approach to protecting people and property from disasters and emergencies caused by any of the hazards to which the community is particularly vulnerable.

Emergency Planning and Community Right-to-Know Act — A federal act passed in 1980 requiring states to establish state emergency response commissions, emergency planning districts, emergency planning committees, and comprehensive emergency plans and to develop programs to respond to hazardous chemical releases; also requires industries to report the presence of hazardous chemicals and the release of hazardous chemicals. Also known as the Community Right-to-Know Act.

emergency preparedness plan — Predecessor to the term "emergency action plan".

emergency public information — (*FEMA*) Information given out in anticipation of an emergency or at the actual time of an emergency; includes the transmission of direct orders and instructions that results in specific actions.

emergency response personnel — People authorized to respond to fire, rescue, police, and hazardous materials emergencies within a given jurisdiction.

emergency response team — (*FEMA*) A group of people made up of federal program and support personnel activated and deployed into an area affected by a major disaster or emergency.

emergency room — A hospital department specially designed to receive and initially treat patients suffering from sudden trauma or medical problems.

emergency support function — (*FEMA*) An activity used to facilitate coordinated federal delivery of assistance during the response phase of a disaster to save lives, protect property, protect health, and maintain public safety.

emergent beam diameter — Diameter of a laser beam at the exit aperture of the system.

emery — A natural or synthetic abrasive made of aluminum oxide.

emesis — The act of vomiting.

emetic — Any agent that causes vomiting.

emf — See electromotive force.

EMI — See electromagnetic interference.

emigrate — To move from one area to another to take up residence.

eminent domain — The authority of the state to seize, appropriate, or limit the use of property in the best interest of the community.

E

emission — Material released into the air, either by a primary source or a secondary source, as a result of a photochemical reaction or chain of reactions.

emission factor — A statistical average of the mass of pollutant emitted from each source of pollution per unit or per quantity of material handled, processed, or burned.

emission inventory — A list of primary air pollutants emitted into the atmosphere of a given community in certain amounts per day by type of source.

emission rate — The amount of pollutant emitted per unit time.

emission spectroscopy — The detection of radiation emissions of characteristic wavelengths from metallic elements (either oxides or salts) when they are excited within an electric arc or spark.

emission standard — Regulations restricting the maximum amounts of pollutants that can be released into the air from a specific polluting source.

emission taxes — Taxes levied on air or water emissions, usually on a per-ton basis.

emissions trading — A program in which the sources of a particular pollutant, usually air pollutants, are given permits to release a specified number of tons of the pollutant. Because only a limited number of permits consistent with the desired levels of emissions are issued, the owners of the permits may keep them and release the pollutants or reduce their emissions and sell the permits, thereby providing an economic incentive for reducing pollutants. Through the accumulation of surplus emission units where pollution requirements have been met and trading these units with other facilities or industries having greater problems an overall constant emission level is maintained.

emissive — Adjective for emission.

emissivity — The ratio of radiant energy emitted by a body to that emitted by a perfect black body. A perfect black body has an emissivity of 1; a perfect reflector has an emissivity of 0.

emittance — Rating of the ability of a material to give off heat as radiant energy.

emollient — A softening agent, especially for use on the skin.

emphysema — An irreversible, obstructive pulmonary disease in which airways become permanently constricted and alveoli are damaged or destroyed; heart action is frequently impaired.

employee — One who performs services for another under contract of hire, who acts under the direction and control of that person, and who performs the service for valuable consideration.

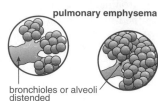
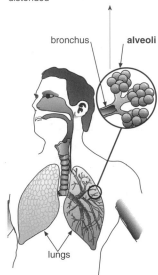

Emphysema

employee assistance program (EAP) — A program set up by a business or agency to help employees with problems such as stress, difficulty at work or home, alcoholism, or drugs; presents various wellness programs to employees.

employer — A person, firm, partnership, association, corporation, or other group that engages the services of another under a contract of hire.

EMR — Acronym for Environmental Management Report.

EMS — See emergency medical service and emergency medical system.

EMT — See emergency medical technician.

EMTD — See estimated maximum tolerated dose.

emulsifiable concentrate — A pesticide chemical mixture dissolved in a liquid solvent; an emulsifier allows dilution of a chemical with oil or water.

emulsification — The process of dispersing one liquid into another liquid with which it is immiscible.

emulsifier — A colloidal chemical that holds in suspension two immiscible liquids by forming a film around the particles so that they remain suspended in solution; also known as emulsifying agent.

emulsifying agent — See emulsifier.

emulsion — A stable dispersion of one liquid in a second, immiscible liquid.

enamel — An oil-based, paint-like coating that produces a glossy finish when applied to a surface.

encapsulate — To seal a substance in an impervious material to prevent physical or chemical degradation.

encephalitis — Inflammation of the brain.

encephalomyelitis — An inflammatory condition of the brain and spinal cord that may result in seizures, paralysis, personality changes, coma, or death; symptoms include fever, headache, stiff neck, back pain, and vomiting.

encephalopathy — Any degenerative disease of the brain.

encode — (*computer science*) To rewrite a message, signal, or stimulus by a computer program into a form that can be interpreted by a computer.

encoder — A feedback device that converts mechanical motion into electronic signals.

end-stage disease — A disease that is terminal because of damage to vital tissues or organs.

end use — The final intended use for a chemical or a substance.

Endameba — See *Entamoeba*.

endamebiasis — See amebiasis.

endangered species — Animals, birds, fish, plants, or other living organisms threatened with extinction by humans or natural changes in the environment.

Endangered Species Act — A federal law protecting species and the ecosystems on which they depend; administered primarily by the Fish and Wildlife Service and the National Marine Fisheries Service.

endangerment assessment — A site-specific risk assessment of the actual or potential danger to human health or welfare and the environment presented by the release of hazardous substances or waste.

endangerment assessment document — A document based on a special study in support of enforcement actions under CERCLA or RCRA.

endemic — The regular occurrence of a fairly constant number of human cases of a disease within an area.

endemic typhus — See murine typhus fever.

endergonic reaction — A chemical reaction that requires energy.

endocarditis — Inflammation of the inner lining of the heart or endocardium.

endocrine disruptor — A chemical agent that interferes with natural hormones in the body.

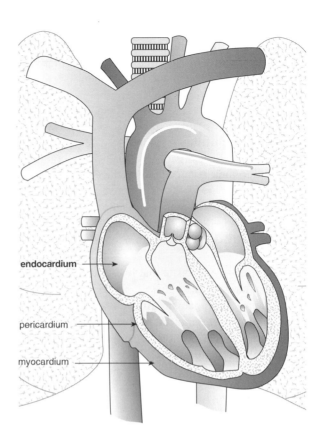

Endocardium

endocrine gland — Ductless gland that secretes hormones directly into the blood; also known as ductless gland.

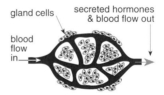

Endocrine Gland

endocrine system — Network of ductless glands and other structures that produce and secrete hormones directly into the bloodstream to affect the function of specific target organs.

endocrine toxicity — Any adverse structural and/or functional changes to the endocrine system resulting from exposure to chemicals.

endocrinologist — A physician who specializes in the diagnoses and treatment of disorders of the glands of internal secretion such as pituitary, thyroid, adrenal glands, and pancreas.

endoderm — The layer of cells in which the linings of the digestive system, liver, and lungs develop.

endogenous factors — Internal factors regulating an organism's growth and development.

endogenous infection — An infection caused by the reactivation of organisms lying dormant in the body, such as tuberculosis.

endogenous respiration — A reduced level of respiration in which organisms break down compounds within their own cells to produce the oxygen they need.

endometrium — The mucous membrane lining of the uterus.

endomyocarditis — Inflammation of the lining of the heart.

endoparasite — A parasite that lives within a host, such as a tapeworm.

endophyte — An organism growing within a plant in a symbiotic or parasitic manner.

endoplasm — Inner portion of cytoplasm.

endoplasmic reticulum (ER) — A vascular system of the cytoplasm that serves as the circulatory system of cells.

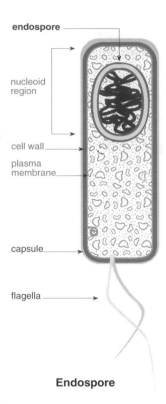

Endospore

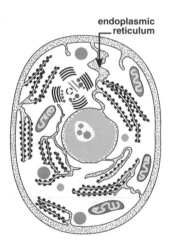

Endoplasmic Reticulum in Animal Cell

endorphin — A neuropeptide composed of many amino acids, elaborated by the pituitary gland and acting on the central and peripheral nervous systems to reduce pain.

endoscope — A thin, lighted tube that allows examination of tissues inside the body.

endospore — A thick-walled body formed within the vegetative cells of certain bacteria that is able to withstand adverse environmental conditions for prolonged periods; under favorable conditions, it can germinate to form the vegetative bacteria.

endosulfan ($C_9H_6Cl_6O_3S$) — Cream to brown-colored solid crystalline flakes with a distinct terpene-like odor or an odor similar to sulfur dioxide; MW: 406.95, sp. gr. (20°C): 1.745. It is used as an insecticide and miticide on vegetable and forage crops and on ornamental flowers and to control termites and tsetse flies. It is hazardous to the nervous system, cardiovascular system, and respiratory system; ingestion can be fatal in humans. Symptoms of exposure include gagging, vomiting, diarrhea, agitation, writhing, loss of consciousness, cyanosis, dyspnea, foaming at the mouth, and noisy breathing; autopsies show edema of the brain and lungs, hemorrhage of the medullary layer of the kidneys, acute lung emphysema, and chromatolysis of the neurons. OSHA exposure limit (8-hour skin PEL TWA): 0.1 mg/m^3 [air].

endothelium — Epithelial layer of cells lining the serous cavities, lymphatic cavities, joint cavities, and other closed cavities of the body.

endothermic — Referring to a chemical change or reaction in which heat is absorbed.

endotoxin — A heat-stable toxin associated with the outer membranes of certain Gram-negative bacteria including *Brucella*, *Neisseria*, and *Vibrio* species. Endotoxins are not secreted and are released only when the cells are disrupted. They are less potent and less specific than the exotoxins and do not form toxoids.

endotoxin shock — A septic shock in response to the release of endotoxins produced by the death of Gram-negative bacteria.

endotracheal — Within or through the trachea.

endpoint — The final result, positive or negative, of a health event such as recovery or death; (*titration*) that stage in titration indicating, usually with a color change, that a desired point in titration has been reached.

endrin ($C_{12}H_8CL_6O$) — A colorless to tan crystalline solid with a mild, chemical odor; MW: 380.9, BP: decomposes, Sol: insoluble, sp. gr. 1.70. It is used as an insecticide on cotton fields and vegetable crops and as a rodenticide against mice and chipmunks in orchards. It is hazardous to the central nervous system and liver and is toxic through inhalation, absorption, ingestion, and contact. Symptoms of exposure include epileptiform convulsions, stupor, headache, dizziness, abdominal discomfort, nausea, vomiting, insomnia, aggressiveness, confusion, lethargy, weakness, and anorexia; in animals, liver damage. OSHA exposure limit (TWA): 0.1 mg/m³ [skin].

endurance limit — The number of cycles that may occur without failure at a particular level of stress.

endurance — The ability to continue an activity despite increasing physical or psychological stress.

enema — A liquid injected into the rectum to cause movement of the feces.

energy — The capacity to do work; classified as either potential or kinetic energy.

energy budget — A quantitative description of the energy exchange for a given physical or ecological system including radiation, conduction, convection, latent heat, and sources and sinks of energy.

energy conservation — The principle that energy cannot be created or destroyed but can be changed from one form to another, such as heat energy to light energy.

energy content — The amount of energy available for doing work; the amount of energy in fuel available for powering a motor vehicle.

energy density — The intensity of electromagnetic radiation per unit area per pulse; expressed as joules per square centimeter.

energy efficiency — A measure of energy produced compared to energy consumed.

energy grade line — A line that represents the elevation of energy head (in feet) of water flowing in a pipe, conduit, or channel.

Energy Policy Act — Comprehensive federal legislation enacted in 1992 that resulted in fundamental changes in the electric utility industry by promoting competition in wholesale electricity markets.

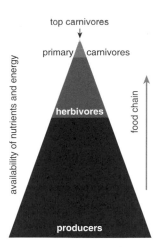

Energy Pyramid

energy pyramid — A representation of the loss of useful energy in each level of a food chain.

energy recovery — The capture of energy from waste through any of a variety of processes.

energy recovery ventilation system — A device or combination of devices applied to provide the outdoor air for ventilation in which energies are transferred between the intake and exhaust airstreams.

energy source — The primary source of power converted to electricity through chemical, mechanical, or other means using coal, petroleum, petroleum products, gas, water, uranium, wind, sunlight, or geothermal and other sources.

Energy Supply and Environmental Act — A law passed in 1974 and updated in 1978 providing for a way to assist in meeting the needs of the United States for fuels in a consistent, practical manner and protecting and improving the environment; it allows for coal conversion or coal derivatives in place of oil in power plants.

engineering control — A method of controlling employee exposure to contaminants or hazards by modifying the work environment; a system in which exposure is reduced through mechanical means such as ventilation, acoustic materials, or clean air control booths.

enhanced greenhouse effect — The natural greenhouse effect that has been intensified by anthropogenic emissions of greenhouse gases.

enhanced rhizosphere degradation — Enhanced biodegradation of contaminants near plant roots, where compounds exuded by the roots increase microbial biodegradation activity; water uptake by plant roots can enhance biodegradation by drawing contaminants to the root zone.

Enhanced Thematic Mapper Plus — An 8-band, multispectral scanning radiometer aboard the Landsat 7 satellite that is capable of providing high-resolution imaging of the surface of the Earth.

enhancement — The improvement of a product, facility, or environment beyond its original condition.

enjoin — (*law*) To direct or impose by court order or with urgent admonition.

enriched flour — Flour to which has been added the vitamins and other ingredients necessary to conform to the definition and standard of identity of enriched flour.

enriched material — Material in which the relative amount of one or more isotopes of a constituent has been increased.

enriched uranium — Uranium in which the content of the uranium-235 isotope has been increased above its natural level of 0.7% by weight.

enrichment — The addition of nutrients from sewage effluent or agricultural runoff to surface water that enhances the growth potential for algae and other aquatic plants; the chemical and mineral nutrients found in soil that results in an increase in the population of one organism compared to other organisms. It is an important aspect in the removal of chemical pollutants.

enrichment culture — A technique in which environmental conditions are controlled to enhance the development of a specific organism or group of organisms.

ENT — Acronym for ear, nose, and throat.

entamebiasis — See amebiasis.

Entamoeba — A genus of amebas parasitic in invertebrates and vertebrates, including humans.

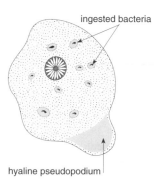

ingested bacteria

hyaline pseudopodium

Entamoeba histolytica
(cause of amebic dysentery)

enteric — Pertaining to the small intestine.

enteric adenovirus — (*disease*) A virus representing serotypes 40 and 41 of the family Adenoviridae and containing a double-stranded DNA surrounded by a distinctive protein capsid approximately 70 nm in diameter, which may cause viral gastroenteritis in 5 to 20% of young children, with symptoms of diarrhea, nausea, vomiting, malaise, abdominal pain, and fever. The mode of transmission is the oral–fecal route, from person to person, or from ingestion of contaminated food or water; it may also be transported by the respiratory route. The incubation period is 10 to 70 hours. The disease has been reported in England and Japan in children in hospitals or daycare centers. The reservoir of infection is people. Susceptibility is found in young children. Also known as acute nonbacterial infectious gastroenteritis and viral gastroenteritis.

enteritis — Inflammation of the intestine, especially the small intestine; a general condition produced by a variety of causes including bacteria and certain viruses and producing symptoms of abdominal pain, nausea, vomiting, and diarrhea.

Enterobacter cloacae — A common species of bacteria found in human and animal feces, dairy products, sewage, soil, and water and which rarely causes disease.

Enterobacteriaceae — A family of aerobic and anaerobic bacteria that includes both normal and pathogenic enteric microorganisms including the genera *Escherichia*, *Klebsiella*, *Proteus*, and *Salmonella*.

enterobacterial — A species of bacteria found in the digestive tract.

enterobiasis — (*disease*) Pinworm disease; a common intestinal infestation that may cause no symptoms or anal itching, disturbed sleep, irritability, and sometimes secondary infections. Incubation time is 4 to 6 weeks. It is caused by an intestinal nematode, *Enterobius vermicularis*, found worldwide. The reservoir of infection is people. It is transmitted by infected eggs, by hands, from anus to mouth, and from clothing, bedding, food, or other contaminated articles or dust; it is communicable for approximately 2 weeks with general susceptibility. It is controlled by proper disinfection, cleaning, and good personal hygiene.

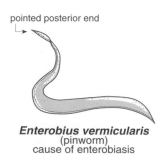

pointed posterior end

Enterobius vermicularis
(pinworm)
cause of enterobiasis

Enterobius vermicularis — A common parasitic nematode that resembles a white thread between 0.5 and 1 cm long; also called pinworm, seat worm, threadworm.

enterococcus — A streptococcus that inhabits the intestinal tract.

enterocolitis — An inflammation involving the large and small intestines.

enteropathy — A disease or disorder of the intestines.

enterotoxigenic — An organism or other agent that produces a toxin causing an adverse reaction on the cells of the intestinal mucosa.

enterotoxin — A toxin produced by *Micrococcus pyogenes* var. *aureus* (*Staphylococcus aureus*) that specifically affects cells of the intestinal mucosa, causing vomiting and diarrhea.

enterovirus — Any of a subgroup of the picornaviruses infecting the gastrointestinal tract and discharged in feces, including coxsackieviruses, echoviruses, and polioviruses; may be involved in respiratory disease, meningitis, and neurological disease.

enterprise zone — An area designated by the city and certified by the state to receive various tax incentives and other benefits to stimulate economic activity and revitalize neighborhoods.

enthalpy — The sum of the internal energy of a body and the product of its volume multiplied by the pressure.

entities — (*geographical information system*) Individual spatial features that include trees, lakes, grasslands, etc.

entity — See element.

entomology — The study of insects.

entozoic — An internal parasite, especially a parasitic worm.

entrain — The trappings of bubbles in water either mechanically through turbulence or chemically through a reaction.

entrainment — A process in which suspended droplets of liquid, particles, or gas are drawn into the flow of a fluid and transported.

entrainment velocity — The gas flow velocity that tends to keep particles suspended and become airborne.

entrance loss — Loss in static pressure of a fluid that flows from an area into and through a hood or duct opening.

entrapment — To engage in artifice and deceit to afford an individual the opportunity to commit a voluntary criminal act.

entropy — The tendency of a system to go from a state of order to a state of disorder.

entry loss — A loss in static pressure caused by air flowing into a duct or hood opening.

entry loss factor — A factor derived from the coefficients of entry which, when multiplied by the velocity pressure at the hood, yields the entry loss (in inches) of water gauged.

entry of decree — An indication that the court has entered a final decree or order of judgment, usually signifying the end to a particular proceeding; also known as entry of judgment.

entry of judgment — See entry of decree.

enucleation — The process of removing an organ or tumor in its cavity without cutting into it.

environment — The sum of all surrounding physical conditions in which an organism or cell lives, including the available energy in living and non-living materials.

environmental analysis — A systematic process for consideration of environmental factors in land, air, water, or facility management actions.

environmental aspect — The elements of an organization's activities, products, or services that can interact with the environment.

environmental assessment — A National Environmental Policy Act compliance document used to determine whether a federal action would significantly impact the environment and require a more detailed environmental impact statement.

environmental audit — An independent assessment by the U.S. EPA of the current status of compliance with environmental requirements, policies, practices, and controls.

environmental contamination — Release into the environment of radioactive, hazardous, and toxic materials.

environmental design — (*ergonomics*) A design ensuring that the lighting, heating, ventilation, noise, vibration, and level of chemical exposure are appropriate to prevent disease and injury.

environmental epidemiologist — An environmental health professional who develops, implements, and evaluates environmental epidemiological investigations, including observational and experimental studies of the relationships among etiological, statistical, and mathematical tools, to identify the causes of disease in human populations.

environmental epidemiology — The study of environmental factors that influence the distribution and determinants of disease in human populations; can either identify disease or injury and the characterization of the exposed population or identify the exposure and determine the health endpoint of the exposure to the population.

environmental equity — The equal protection from environmental hazards for individuals, groups, or communities regardless of race, ethnicity, or economic status.

environmental equity/justice — The development, implementation, and enforcement of environmental laws, regulations, and policies to provide equal protection from environmental hazards for individuals, groups, or communities regardless of race, ethnicity, or economic status.

E

E

environmental ethics — The search for moral values and ethical principles in the manner in which humans relate to the natural world.

environmental exposure — The level of human exposure to pollutants originating from facility emissions; the threshold levels are not necessarily surpassed, but low-level chronic pollutant exposure is present.

environmental factors — Conditions other than indoor air contaminants that cause stress, discomfort, and/or health problems to individuals residing in that environment.

environmental fate — The fate of a chemical or biological pollutant following release into the environment based on time, transport, transfer, storage, and transformation.

environmental health — The art and science of the protection of good health, the promotion of aesthetic values, the prevention of disease and injury through the control of positive environmental factors, and the reduction of potential physical, biological, chemical, and radiological hazards.

environmental health practitioner — An applied scientist and educator who uses his or her knowledge and skills in regard to the natural, behavioral, and environmental sciences to prevent disease and injury and to promote human well-being. Also known as an environmental health specialist; formerly known as a sanitarian.

environmental health professional — An environmental health practitioner whose educational background includes, at a minimum, a baccalaureate degree in the major components of environmental health and the basic public health sciences of epidemiology, biostatistics, and toxicology and whose primary focus is on protecting human health.

environmental health specialist — See environmental health practitioner.

environmental hormones — The chemical pollutants that substitute for, or interfere with, naturally occurring hormones in our bodies and may trigger reproductive failure, developmental abnormalities, or tumors.

Environmental Impact Statement (EIS) — An analysis required by the National Environmental Policy Act of 1970 before major programs affecting the environment are undertaken.

environmental income tax — A tax on corporations imposed on their modified alternative minimum taxable income over $2 million, with the proceeds going to the Hazardous Substance Superfund Trust Fund.

Environmental Index — See National Environmental Data Index.

environmental indicator — A measurement, statistic, or value that provides an approximate gauge or evidence of the effects of environmental management programs or the state or condition of the environment; a measurable feature or features providing managerial and scientifically useful evidence of environmental and ecosystem quality or reliable evidence of trends in quality.

environmental insults — Any factor in the physical environment that inhibits the growth and/or development of an organism.

environmental justice — The fair treatment and meaningful involvement of all people regardless of race, national origin, or income in the development, implementation, and enforcement of environmental laws, regulations, and policies; the combination of civil rights with environmental protection to develop a safe, healthy, life-sustaining environment for everyone.

environmental management specialist — See occupational safety specialist.

environmental media — Surroundings, including air, groundwater, surface water, soil, flora, and fauna.

environmental monitoring — Sampling for contaminants in air, water, sediments, soils, food, plants, and animals by direct measurement or by collecting and analyzing samples.

environmental noise — The intensity, duration, and character of sounds from all sources.

environmental pathway — The route of transport by which a toxicant can travel from its release site to human populations through air, land, water, and food.

environmental physiology — A measure of the stress due to light, noise, vibration, temperature, radiation, and chemicals on the functions of living organisms, their parts, and their physical and chemical processes.

environmental planning — Planning for the use of environmental resources on a sustainable basis.

Environmental Protection Agency number — A number assigned to a chemical regulated by the U.S. EPA; also known as EPA number.

environmental resistance — All of the limiting factors that tend to reduce population growth rates and set the maximum allowable population size or carrying capacity of an ecosystem.

environmental resources — The resources of land, water, air, flora, and fauna.

environmental response team (ERT) — A group of professionals, representing various agencies, who are trained and equipped to determine the extent of an environmental emergency and how to deal with it safely.

environmental restoration — (*nuclear energy*) An environmental cleanup including stabilizing contaminated soil; treating groundwater; decommissioning process buildings, nuclear reactors, chemical separation plants, and other facilities; and exhuming sludge and other buried drums of waste.

environmental safety specialist — See occupational safety specialist.

environmental science — Study of the complex interactions of human populations with their environments and energy resources.

environmental services — A department within a hospital or other healthcare facility that has responsibility for laundry, liquid and solid waste control, safe disposal of materials contaminated by radiation or pathogenic organisms, general maintenance of facilities, and safety.

environmental site assessment — A process of determining whether or not contamination is present on a piece of property and the level of contamination present.

environmental surveillance — See environmental monitoring.

environmental technology — The hardware and/or methods or techniques used in pollution control devices and systems for waste prevention, waste treatment, waste control, and site remediation.

environmental terrorism — Conducting acts of terrorism through destruction or pollution of the environment.

environmental threshold carrying capacity — An environmental standard necessary to maintain a significant scenic, recreational, educational, scientific, or natural value of the region or to maintain public health and safety within the region; includes, but is not limited to, standards for air quality, water quality, soil conservation, vegetation preservation, and noise.

environmental tobacco smoke (ETS) — Tobacco smoke that poses a risk of lung cancer in nonsmokers and which contains a variety of pollutants including inorganic gases, heavy metals, particulates, volatile organic compounds, and products of incomplete burning, such as polynuclear aromatic hydrocarbons.

environmental toxicologist — An environmental health professional who determines the adverse health effects and the mechanisms of those effects resulting from exposure to physical, chemical, and biological aspects of the human environment.

environmental toxicology — A study of potential poisons in the environment as they relate to individuals

affected by exposure, time, dose, and biological effect injuries.

environmental/ecological risk — The potential for adverse effects on living organisms related to pollution of the environment by effluents, emissions, wastes, or accidental chemical releases.

ENVISAT — Acronym for environmental satellite.

enzootic — A disease that is constantly present in the animal community but occurs in only a small number of cases.

enzyme — Any of a group of organic compounds, frequently proteins, found in plants and animals that accelerate by the catalytic action of specific transformations of material without being destroyed or altered, as in the digestion of food, and serve to regulate certain physiological functions.

enzyme-linked immunosorbent assay test (ELISA test) — A diagnostic test that detects the presence of antibodies to human immunodeficiency virus (HIV) in human blood components.

eolian — Windblown.

EOS — Acronym for Earth Observing System.

eosin ($C_{20}H_8Br_4O_5$) — A red fluorescent dye obtained by the action of bromine on fluorescein.

eosinophil — A leukocyte or other granulocyte with cytoplasmic inclusions readily stained by eosin.

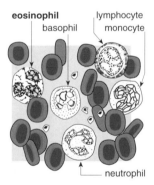

Eosinophil

eosinophilia — An increase of certain white cells in the blood or tissues.

EOSP — Acronym for Earth Observing Scanning Polarimeter.

EP — See extraction procedure.

EP toxicity test — An extraction procedure designed to identify waste likely to leach hazardous concentrations of particular toxic constituents into the groundwater as a result of improper management.

EPA health advisory — An estimate of acceptable drinking water levels for a chemical substance based on health effects information and not a legally enforceable federal standard.

EPA I.D. number — A unique code assigned by regulating agencies to each generator, transporter, and treatment, storage, or disposal facility to facilitate identification and tracking of chemicals or hazardous waste.

EPA number — See Environmental Protection Agency number.

EPACT — Acronym for Environmental Policy Act and for the Energy Policy Act of 1992, which includes comprehensive energy legislation.

EPCA — Acronym for Energy Policy and Conservation Act.

EPCRA — See Emergency Planning and Community Right-to-Know Act.

EPDC — See expected peak day concentration.

EPDM — See ethylene–propylene terpolymer.

epicenter — Focal point on the surface of the Earth directly above the origin of a seismic disturbance.

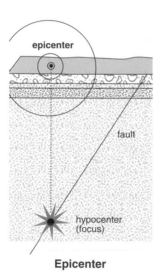

Epicenter

epichlorohydrin (C_3H_5OCl) — A colorless liquid with a slightly irritating, chloroform-like odor; MW: 92.5, BP: 242°F, Sol: 7%, Fl.P: 93°F, sp. gr. 1.18. It is used as an intermediate in the manufacture of DGEBA resins, epoxy novolac resins, phenoxyl resins, and wet-strength resins for paper; in the manufacture of glycerol; as a stabilizer in the manufacture of chlorinated rubber and chlorinated insecticides; in coating fibers and textile surface coatings. It is hazardous to the respiratory system, skin, and kidneys and is toxic through inhalation, absorption, ingestion, and contact. Symptoms of exposure include nausea, vomiting, abdominal pain, respiratory distress, coughing, cyanosis, and eye and skin irritation with deep pain; carcinogenic. OSHA exposure limit (TWA): 2 ppm [skin] or 8 mg/m³.

epidemic — The occurrence in a community or region of a disease at a rate substantially in excess of normal expectancy and derived from a common or propagated source.

epidemic hemorrhagic fever — A severe viral infection with symptoms of fever and bleeding that develops rapidly and results in hemorrhage, peripheral vascular collapse, hypovolemic shock, acute kidney failure, and death; caused by an arbovirus or other pathogenic organism.

epidemic myalgia — A disease caused by coxsackie B virus with symptoms of sudden acute chest and epigastric pain and fever lasting a few days followed by complete spontaneous recovery.

epidemic typhus — An acute severe rickettsial infection with symptoms of prolonged high fever, headache, and a dark maculopapular rash that covers most the body; the causative agent is *Rickettsia prowazekii*, which is transmitted when the human body louse bites and defecates on the skin of a person who then scratches and thereby inserts the feces containing the pathogens into the body. The incubation period is between 10 and 14 days and is followed by the rash. Complications may include vascular collapse, renal failure, pneumonia, gangrene, and death.

epidemic viral gastroenteritis — See Norwalk type disease.

epidemiological reasoning — A three-step process consisting of determining an association between exposure and endpoint, formulating a hypothesis of the relationship, and testing the hypothesis.

epidemiological study — A descriptive study to determine whether correlations exist between the frequency of disease in human populations and some specific factor such as a microorganism or concentration of a toxic chemical in the environment.

epidemiological surveillance — The ongoing systematic collection, analysis, and interpretation of health data essential to the planning, implementation, and evaluation of public health practice, closely integrated with the timely dissemination of the data to people who need to know; a surveillance system that includes a functional capacity for data collection, analysis, and dissemination linked to public health programs.

epidemiologist — A medical scientist who studies the incidence, prevalence, spread, prevention, and control of disease in the community or a specific group of individuals.

epidemiology — The study of the occurrence and distribution of disease and injury specific to person, place, and time.

epidermis — The outer protective cell layers of an organism.

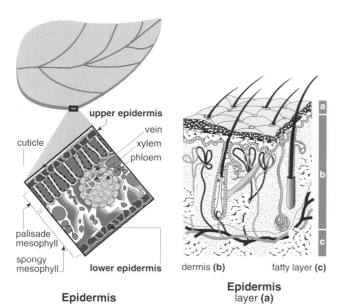

Epidermis

Epidermis layer (a)

dermis (b)   fatty layer (c)

epigastric pain — A pain in the epigastric region of the abdomen, which is located beneath the cartilage of the lower ribs.

epiglottis — A thin, leaf-like structure at the opening of the trachea that helps food go from the mouth to the esophagus without blocking the airway.

epihydrin alcohol — See glycidol.

epilepsy — Neurological disease with symptoms of seizures involving convulsions and loss of consciousness.

epileptiform — Resembling epilepsy or its manifestations.

epilimnion — A freshwater zone of thermally stratified water which is mixed as a result of wind action and convection currents.

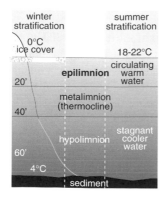

Epilimnion

epinephrine ($C_3H_{13}O_3N$) — A hormone produced by the medulla of the adrenal glands that acts to increase blood pressure; also known as adrenaline.

epiphytic — Of or pertaining to a plant that grows nonparasitically on another plant or nonliving structure.

episode — (*pollution*) An air pollution incident in a given area caused by a concentration of atmospheric pollution reacting with meteorological conditions; may result in a significant increase in illness or death.

episome — A plasmid that replicates itself by insertion into the bacterial chromosome.

epistaxis — Hemorrhage from the nose usually due to rupture of small vessels overlying the anterior part of the cartilaginous nasal septum.

epithelial tissue — Layers of cells covering the inside or outside surfaces of an animal's body that form glands and parts of sense organs; they serve as protective, absorptive, secretory, and sensory functions.

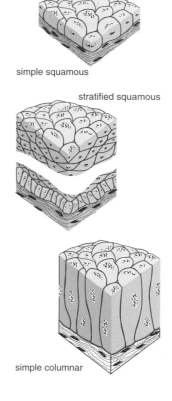

simple squamous

stratified squamous

simple columnar

Epithelial Tissue

epitope — The region of an antigen to which the antibody binds.

epizootic — Affecting many of one kind of animal in a region simultaneously, as with an epidemic disease.

EPN — See O-ethyl-O-*para*-nitrophenyl phenylphospho-nothioate.

EPNL — Acronym for effective perceived noise level.

epoxide — (*chemistry*) An epoxy compound.

epoxy — A large group of compounds containing a triangular structure in which oxygen acts as a bridge between two atoms already bound together.

epoxy resin — A synthetic resin composed of epoxy polymers and characterized by adhesiveness, flexibility, and resistance to chemical actions.

EPP — See Emergency Preparedness Plan.

EPTC — Acronym for extraction procedure toxicity characteristic.

EQIP — Acronym for Environmental Quality Incentive Programs.

equal-area projection — (*geographical information system*) A projection that preserves the area of interest shown but does not preserve the real-world distance or shape.

equalizing basin — A holding basin in which variations in flow and composition of liquid are averaged to provide a reasonably uniform volume and composition for treatment purposes.

equilibrium — (*radiation*) The state at which the radioactivity of consecutive elements within a radioactive series is neither increasing nor decreasing.

equilibrium fraction (*F*) — A measure of the equilibrium/disequilibrium in radon–radon daughters, where the parents and daughters have equal radioactivity; that is, as many decay into a specific nuclide as decay out.

equine — Pertaining to, characteristic of, or derived from the horse.

equine encephalitis — An arbovirus infection with symptoms of inflammation of the nerve tissues of the brain and spinal cord, high fever, headache, nausea, vomiting, and neurological symptoms; the virus is transmitted by the bite of an infected mosquito. The horse is the primary host.

equipment — (*food*) Items other than utensils used in the storage, preparation, display, and transportation of food, such as stoves, ovens, hoods, slicers, grinders, mixers, scales, meat blocks, tables, shelving, reach-in refrigerators and freezers, sinks, icemakers, and similar paraphernalia found in retail food stores.

equipment room — (*swimming pools*) An area provided for the storage of safety and deck equipment.

equipotential — A line or surface characterized by a single electric potential or voltage.

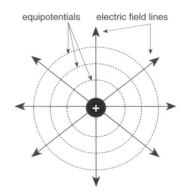

Equipotential

equitransference — The equal diffusion rates of positively and negatively charged ions of an electrolyte across a liquid junction without charge separation.

equity law — A decision made by a judge in a given situation where rules and regulations have been violated and the judge decides the penalty rather than deciding if the law is valid.

equivalency — A group of procedures and techniques of sample collection and/or analysis that have been demonstrated in specific cases to produce results not statistically different from a reference method.

equivalent aerodynamic diameter — The diameter of a sphere of unit density (1.0 g/cm^3) that exhibits the same aerodynamic behavior of a particle of any shape or density being measured.

equivalent dose — Quantity obtained by multiplying the absorbed dose by a factor that allows for the varying effectiveness of ionizing radiations to cause harm to tissue.

equivalent method — A method of sampling and analyzing ambient air for an air pollutant that has been demonstrated as equivalent to a reference sampling technique in accordance with the Code of Federal Regulations.

equivalent weight — (*chemistry*) The weight of an oxidizing or reducing agent that will produce or accept one electron in a chemical reaction.

equivocal symptom — A symptom that may be attributed to more than one cause or that may occur in several diseases.

ER — Acronym for emergency room and for emergency response program; see endoplasmic reticulum.

eradication — Elimination of pests from an area; elimination of infections from a facility.

ERAP — Acronym for emergency response action plan.

erg — A unit of work equal to a force of 1 dyne acting over a distance of 1 cm in the direction of the force; 10^{-7} joule, or 2.4×10^{11} kcal.

ergometry — The study of physical work activities, including work performed by specific muscles or muscle groups.

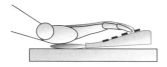

Ergonomic Keyboard

ergonomics — A multidisciplinary activity dealing with interactions between people and the total working environment and stresses related to such environmental elements as atmosphere, heat, light, and sound, as well as all tools and equipment in the workplace; literally, the study of measurement; organization of work by making purposeful human activities more effective.

ergonomics hazard — A workplace condition that poses a biomechanical stress to the worker.

ergonomics program — The application of ergonomics in a structured system including health and risk factor surveillance, job analysis and design, medical management, and training.

ergot — The spore-producing reproductive body of a fungus that infects grain crops.

ergotism — A rare, chronic poisoning associated with the consumption of rye contaminated with the fungus *Claviceps purpurea* or ergot fungus.

EROC — Acronym for ecological rates of change.

EROS — Acronym for Earth Resources Observation Satellite and for Earth Resources Observation System.

erosion — The removal of soil and rock fragments by natural agents such as wind, water, organisms, and gravity.

Glacial Erosion

Stream Erosion

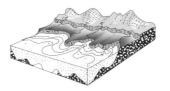

Wave Erosion

Erosion

erosion–corrosion — Deterioration occurring from movement of a corrosive over a surface, thereby increasing the rate of attack due to mechanical wear and corrosion.

erratic — Deviating from normal with no apparent purpose.

ERP — Acronym for emergency response team.

error — The difference between an observed or calculated value and a true value; variation in measurements, calculations, or observations of a quantity due to mistakes or uncontrollable factors; (*epidemiology*) a defect in the design of a study, in the development of measurements or instruments, or in the interpretation of findings.

error band — The allowable deviations of output from a specific reference norm.

error message — (*computer science*) A short statement on the computer screen indicating an incorrect procedure or malfunction.

ERT — See environmental response team.

ERTS — Acronym for Earth Resources Technology Satellite (now Landsat).

eruption — Rapid development of a skin lesion.

E

eruptive fever — Any disease with symptoms of fever and rash.

erysipeloid — (*disease*) A bacterial cellulitis with symptoms of fever, leukocytosis, and red, tender, edematous, spreading lesions of the skin with a definite raised border. Three forms exist: a usually self-limited, mild, localized form manifested by an erythemous and painful swelling at the site of inoculation which spreads peripherally with central clearing; a generalized or diffuse form, which may be accompanied by fever and arthritis symptoms and resolves spontaneously; and a rare and sometimes fatal systemic form associated with endocarditis. Incubation time is 1 to 7 days, longer if the infection enters through an unobserved breech of the skin. It is caused by *Erysipelothrix rhusiopathiae*, which is found in occupational food-processing settings. The reservoirs of infection are meat, poultry, and fish. It is transmitted by contact with the food source and is not communicable with unknown susceptibility. It is controlled by avoiding direct contact between the food source and skin.

erythema — An abnormal redness of the skin due to distention of the capillaries with blood caused by physical or chemical agents and due to skin injury, infection, or inflammation.

erythemal region — The ultraviolet light radiation between 280 and 320 nm as absorbed by the eye.

erythremia — An increase in the number of circulating erythrocytes.

erythroblast — A polychromatic nucleated cell of red bone marrow that synthesizes hemoglobin and is an intermediate in the initial stage of red blood cell formation.

erythrocyte — Red blood cell or corpuscle.

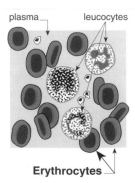

Erythrocytes

ES — See effective size.
ESA — Acronym for the Endangered Species Act
ESCA — Acronym for electron spectroscopy for chemical analysis.
eschar — The crust formed after injury by a caustic chemical or heat.

Escherichia coli — See coliform.

Escherichia coli **diarrhea** — (*disease*) Three types exist; invasive strains cause disease localized primarily in the colon with fever, mucoid, and occasional bloody diarrhea; enterotoxigenic (toxin-producing) strains are similar to *Vibrio cholerae*, producing profuse watery diarrhea without blood or mucous, abdominal cramping, vomiting, acidosis, prostration, and dehydration with symptoms lasting 3 to 5 days; enteropathogenic strains are usually associated with outbreaks of acute diarrheal disease in newborn nurseries. Incubation time is 12 to 72 hours. It is caused by enterotoxigenic (heat-labile) toxin-LT, heat stabile toxin-ST, or both toxin-LT/-ST invasive or enteropathogenic strains of *E. coli*. It is found in nurseries and institutions and as outbreaks of foodborne or waterborne disease in communities, especially those with poor water treatment and handling. The reservoir of infection is people. It is transmitted through fecal contamination of food and water, from mites, and from person to person. It is communicable, probably for the duration of the excretion of feces, but may last several weeks. Infants and malnourished children are most susceptible. Travelers may also be affected, in which case it is sometimes referred to as "travelers diarrhea." It is controlled through the proper handling of food, water, and fecal material and by breaking the oral–fecal route of disease. It is also important to have appropriate handwashing and other preventative procedures in institutions.

ESD — See electrostatic discharge.

ESDIM — Acronym for Environmental Services Data and Information Management.

ESECA — Acronym for Energy Supply and Environmental Coordination Act.

esophagus — The tube connecting the mouth and the stomach in the digestive system.

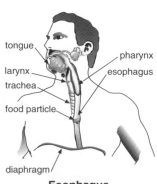

Esophagus

ESP — See electrostatic precipitator.

essential fever — A fever of unknown origin.

essential hypertension — An elevated systemic arterial pressure for which no cause can be found.

essential nutrient — A nutrient that cannot be synthesized by an organism and must be provided by diet for proper health.

essential oil — Any of a class of volatile, odoriferous oils found in plants and imparting to the plant its odor and often other characteristic properties; used in perfume.

ESTAR — Acronym for electronically scanned thinned array radiometer.

ester — The organic compound formed by the action between an alcohol and organic or inorganic acid; a compound of the type RCOOR, where each R is a hydrocarbon group.

esthetics — A judgment about what is beautiful or pleasing; also spelled aesthetics.

estimated maximum tolerated dose (EMTD) — The estimated maximum dose of a chemical to which individuals can be exposed without causing disease or injury.

estrogen — Female sex hormone that includes estradiol, estriol, and estrone, which are formed in the ovary, adrenal cortex, testis, and fetoplacental unit. Estrogen is responsible for secondary sex characteristic development; during the menstrual cycle, it acts on the female genitalia to produce an environment suitable for fertilization, implantation, and nutrition of an early embryo.

estuary — An enclosed or partially enclosed shallow body of water that forms where a river enters the ocean and creates an area that mixes fresh and ocean water.

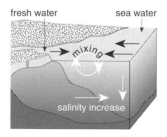

Estuary

ESU — See electrostatic unit of charge.

ET — Acronym for emissions trading and for ecotox threshold.

etch — To cut or eat away material with acid or other corrosive substance.

ETF — Acronym for environmental task force.

ethamine — See ethyl amine.

ethane (CH₃·CH₃) — A colorless, odorless gas used as a fuel and refrigerant; forms an explosive mixture with air.

1,2-ethanediol (C₂H₄(OH)₂) — A colorless, odorless, syrupy liquid with a sweet taste; MW: 62.07; BP: 197.6°C; Sol: soluble in water, alcohol, acetone, and acetic acid; Fl.P: 115°C (open cup); density: 1.1135 at 20°C. It is used as an antifreeze in heating and cooling systems, as a hydraulic brake fluid, and as a solvent for paints, plastics, and inks. It is hazardous to the respiratory system and digestive system. It may cause lacrimation, irritation of the throat and upper respiratory tract, headache and burning cough, stimulation followed by depression of the central nervous system, nausea, vomiting, drowsiness, coma, and death. ACGIH exposure limit (TLV-TWA): 50 ppm.

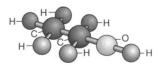

1,2-Ethanediol

ethanol (C₂H₅OH) — A colorless, clear, volatile liquid with an alcohol smell and taste, MW: 46; BP: 78.3°C; Sol: soluble in water, acetone, ether, and chloroform; density: 0.789 at 20°C. It is used as a solvent for resins, lacquers, pharmaceuticals, and cleaning agents and in the production of raw materials for cosmetics, perfumes, drugs, and plasticizers. It is hazardous to the digestive system, heart, liver, and central nervous system. It causes symptoms of excitation, intoxication, stupor, hypoglycemia, and coma and may result in death. ACGIH exposure limit: NA (however, ingestion of 250 to 500 ml can be fatal).

Ethanol

ethanolamine (NH₂CH₂CH₂OH) — A colorless, viscous liquid (or solid, below 51°F) with an unpleasant, ammonia-like odor; MW: 61.1, BP: 339°F, Sol: miscible, Fl.P: 185°F, sp. gr. 1.02. It is used in the production of monoalkanolamides; in fuel-oil additives, pharmaceuticals, agricultural chemicals, cosmetics, emulsion paints, polishers, and cleansers; in the manufacture of inks, paper, glues, textiles, and polishes; in the synthesis of 2-mercaptothiazole in rubber vulcanization acceleration;

in the recovery and removal of acid gases from natural fuel, and process gas in the synthesis of ammonia; in the manufacture of dry ice. It is hazardous to the skin, eyes, and respiratory system and is toxic through inhalation, ingestion, and contact. Symptoms of exposure include lethargy and skin, eye, and respiratory system irritation. OSHA exposure limit (TWA): 3 ppm [air] or 8 mg/m^3.

ethene — See ethylene.

ether — See ethyl ether.

ethion (C$_9$H$_{22}$O$_4$P$_2$S$_4$) — An odorless (technical product has a disagreeable odor), colorless to amber-colored liquid; MW: 384.48, BP: decomposes, Sol: 1.1 ppm in water at 20°C, Fl.P: 176°C, sp. gr. 1.22. It is an organophosphate used as an insecticide and acaricide for citrus fruit, apples, nuts, other fruit, and cotton. It is hazardous to the respiratory tract, digestive tract, skin, and eyes. Acute exposure may cause central nervous system depression, agitation, confusion, delirium, coma, seizures, and death. It also causes bradycardia, bronchospasm, vomiting, and diarrhea. Chronic exposure may damage the nerves and cause poor coordination in the arms and legs, depression, anxiety, and irritability. OSHA exposure limit (8-hour TWA): 0.4 mg/m^3.

2-ethoxyethanol (C$_2$H$_5$OCH$_2$CH$_2$OH) — Colorless liquid with a sweet, pleasant, ether-like odor; MW: 90.1, BP: 313°F, Sol: miscible, Fl.P: 110°F, sp. gr. 0.93. It is used as a solvent in the manufacture of lacquers, lacquer thinners, nitrocellulose lacquers, alkyd resins, printing ink solvents, varnish removers, cleaning compounds, soaps, cosmetics, pesticides, pharmaceuticals, adhesives, and detergents; in textile dyeing and printing; in the manufacture of leather finishes; as an anti-icing additive; in the manufacture of brake fluids, aviation fuels, and automotive antistall additives. In animals, it is hazardous to the lungs, eyes, blood, kidneys, and liver and is toxic through inhalation, absorption, ingestion, and contact. Symptoms of exposure in animals include pulmonary irritation, hematologic effects, eye irritation, and damage to the liver, kidneys, and lungs. OSHA exposure limit (TWA): 200 ppm [skin] or 740 mg/m^3. Also known as cellosolve.

2-ethoxyethyl acetate (CH$_3$COOCH$_2$CH$_2$OC$_2$H$_5$) — A colorless liquid with a mild odor; MW: 132.2, BP: 313°F, Sol: 23%, Fl.P: 124°F, sp. gr. 0.98. It is used as a solvent vehicle during hot and cold spray applications of lacquers and wood stains; in the manufacture of varnishes, thinners, lacquers, wood stains, adhesives, varnish removers, fabric and paper coatings, and textile sizing; in the manufacture of cellulose acetate films; as a solvent for many oils and gums. It is hazardous to the respiratory system, eyes, and gastrointestinal tract and

is toxic through inhalation, absorption, ingestion, and contact. Symptoms of exposure include vomiting, kidney damage, paralysis, and irritation of the eyes and nose. OSHA exposure limit (TWA): 100 ppm [skin] or 540 mg/m^3.

ethrane (C$_3$H$_2$ClF$_5$O) — A clean and colorless liquid with a pleasant odor; MW: 184.50, BP: 56.5°C, Sol: soluble in most organic solvents and mixes with oils and fats, Fl.P: greater than 94°C (a noncombustible liquid), density: 1.5167 at 25°C. It is used as an anesthesia by inhalation. It is hazardous to the respiratory tract, digestive tract, and skin in concentrations above 1.5 to 2% by volume of air. It is an anesthetic that can affect the central nervous system, cardiovascular system, kidney, and bladder and produce symptoms of respiratory depression and seizure. ACGIH exposure limit (TLV-TWA): 75 ppm.

$$
\begin{array}{ccccccc}
 & F & & F & & & F \\
 & | & & | & & & | \\
Cl - & C & - & C & - O - & C & - H \\
 & | & & \| & & & | \\
 & H & & F & & & F
\end{array}
$$

Ethrane

ethyl acetate (CH$_3$COOC$_2$H$_5$) — A colorless liquid with an ether-like, fruity odor; MW: 88.1, BP: 171°F, Sol: 10% at 77°F, Fl.P: 24°F, sp. gr. 0.90. It is used in the manufacture of smokeless powder and artificial leather; in the preparation of photographic films; as a cleaning agent in the textile industry; in the manufacture of linoleum and plastic wood; in the manufacture of dyes and drug intermediates. It is hazardous to the eyes, skin, and respiratory system and is toxic through inhalation, ingestion, and contact. Symptoms of exposure include narcosis, dermatitis, and irritation of the eyes, nose, and throat. OSHA exposure limit (TWA): 400 ppm [air] or 1400 mg/m^3.

$$
\begin{array}{c}
O \\
\| \\
CH_3COC_2H_5
\end{array}
$$

Ethyl acetate

ethyl acrylate (CH$_2$CHCOOC$_2$H$_5$) — A colorless liquid with an acrid odor; MW: 100.1, BP: 211°F, Sol: 2%, Fl.P: 48°F, sp. gr. 0.92. It is used in the manufacture of acrylic resins for use in paint formulations, industrial coatings, and latexes; in the manufacture of plastics such as ethylene ethyl acrylate; in the manufacture of polyacrylate elastomers and acrylic rubber; in forming denture materials. It is hazardous to the respiratory system, eyes, and skin and is toxic through inhalation, absorption, ingestion, and contact. Symptoms of exposure include irritation of the eyes, respiratory system, and skin; carcinogenic. OSHA exposure limit (TWA): 5 ppm [skin] or 20 mg/m^3.

ethyl alcohol — See ethanol.

ethyl amine (CH$_3$CH$_2$NH$_2$) — A colorless gas or water-white liquid (below 62°F) with an ammonia-like odor; shipped as a liquefied compressed gas; MW: 45.1, BP: 62°F, Sol: miscible, Fl.P: 1°F, sp. gr. (liquid) 0.69. It is used in the synthesis of agricultural chemicals for herbicides and of chemical intermediates and solvents; as a dyestuff intermediate; as a solvent for dyes, resins, and oils; as a catalyst for polyurethane foams and for curing epoxy resins; in pharmaceuticals, as emulsifying agents; as a vulcanization accelerator for sulfur-cured rubber; as a stabilizer for rubber latex; as a selective solvent in the refining of petroleum and vegetable oils; in the synthesis of rhodamine dyes; as a deflocculating agent in the ceramics industry; in the manufacture of detergents. It is hazardous to the respiratory system, eyes, and skin and is toxic through inhalation, absorption, ingestion, and contact. Symptoms of exposure include skin burns, dermatitis, and eye and respiratory irritation. OSHA exposure limit (TWA): 10 ppm [air] or 18 mg/m^3. Also known as aminoethane; ethamine.

ethyl amyl ketone (C$_8$H$_{16}$O) — A colorless liquid with a mild fruity odor; MW: 128.21; BP: 160°C; Sol: mixes with alcohols, ethers, ketones, and other organic solvents; Fl.P: 59°C (open cup), a combustible liquid; density: 0.8184 at 20°C. It is hazardous to the eyes, nose, skin, and respiratory tract. It can cause irritation, ataxia, prostration, respiratory pain, and narcosis. ACGIH exposure limit (TLV-TWA): 25 ppm.

CH$_3$—CH$_2$—C—CH$_2$—CH—CH$_2$—CH$_2$—CH$_3$
　　　　　‖　　　　　　‖
　　　　　O　　　　　CH$_3$

Ethyl amyl ketone

ethyl benzene (CH$_3$CH$_2$C$_6$H$_5$) — A colorless liquid with an aromatic odor; MW: 106.2, BP: 277°F, Sol: 0.01%, Fl.P: 55°F, sp. gr. 0.87. It is used in the manufacture and application of rubber adhe-

sives; during electroplating of aluminum on copper or steel; as a heat-transfer medium; as a dielectric; as an intermediate in dye manufacture. It is hazardous to the eyes, upper respiratory system, skin, and central nervous system and is toxic through inhalation, ingestion, and contact. Symptoms of exposure include headache, dermatitis, narcosis, coma, and irritation of the eyes and mucous membrane. OSHA exposure limit (TWA): 100 ppm [air] or 435 mg/m^3.

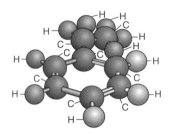

Ethyl benzene

ethyl bromide (C$_2$H$_5$Br) — A colorless, volatile substance that turns yellowish on exposure to air or light, with a burning taste and an ether-like odor; MW: 108.98; BP: 38.4°C; Sol: miscible with organic solvents; Fl.P: –23°C (closed cup), the vapor, which is heavier than air, can travel a considerable distance to a source of ignition and flash back; density: 1.461 at 20°C. It is used as a refrigerant and as an ethylating agent. It is hazardous to the eyes, respiratory tract, and digestive tract. It causes central nervous system depression, narcosis, severe irritation, and damage to the liver, kidneys, and intestine. ACGIH exposure limit (TLV-TWA): 200 ppm.

```
      H   H
      |   |
  H — C — C — Br
      |   |
      H   H
```

Ethyl bromide

ethyl butyl ketone (CH$_3$CH$_2$CO(CH$_2$)$_3$CH$_3$) — A colorless liquid with a powerful, fruity odor; MW: 114.2, BP: 298°F, Sol: 1%, Fl.P (open cup): 115°F, sp. gr. 0.82. It is used in lacquers, varnishes, epoxies, vinyl coatings, finishes, and adhesives; in cellulosic, acetate, and rubber adhesives in shoe manufacture, book binding, and packaging; in mastics and other natural gum-based adhesives for floor and wall tiles and other floor coverings; in

E

automobile silencers and lining pads. It is hazardous to the eyes, skin, and respiratory system and is toxic through inhalation, ingestion, and contact. Symptoms of exposure include headache, narcosis, coma, dermatitis, and irritation of the eyes and mucous membrane. OSHA exposure limit (TWA): 50 ppm [air] or 230 mg/m³. Also known as 3-heptanone.

ethyl chloride (CH₃CH₂Cl) — A colorless gas or liquid (below 54°F) with a pungent, ether-like odor that is shipped as a liquefied compressed gas; MW: 64.5, BP: 54°F, Sol: 0.6%, sp. gr. 0.92. It is used in the production of tetraethyl lead and ethyl cellulose; as a local general anesthetic; as a refrigeration compound; as a solvent for fats, oils, waxes, phosphorus, acetylene, and many resins; in organic synthesis of perchloroethane, esters, and Grignard reagents; in the manufacture of dyes, drugs, and perfumes; as a propellant in aerosols. It is hazardous to the liver, kidneys, respiratory system, and cardiovascular system and is toxic through inhalation, absorption, ingestion, and contact. Symptoms of exposure include incoordination, inebriation, abdominal cramps, cardiac arrhythmias, cardiac arrest, and damage to the liver and kidneys. OSHA exposure limit (TWA): 1000 ppm [air] or 2600 mg/m³. Also known as chloroethane.

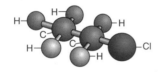

**Ethyl Chloride
(Chloroethane)**

ethyl chlorohydrin (CH₂ClCH₂OH) — A colorless liquid with a faint, ether-like odor; MW: 80.5, BP: 262°F, Sol: miscible, Fl.P: 140°F, sp. gr. 1.20. It is used as a fumigant and in treatment of seed potatoes; in dyeing operations during color printing of textiles; in cleaning, degreasing; in organic synthesis and synthesis of dyes, pharmaceuticals, and resin processing; in the extraction of pine lignin and oil of rose. It is hazardous to the respiratory system, liver, kidneys, skin, cardiovascular system, and central nervous system and is toxic through inhalation, absorption, ingestion, and contact. Symptoms of exposure include nausea, vomiting, vertigo, incoordination, numbness, visual disturbance, headache, thirst, delirium, low blood pressure, collapse, shock, coma, and irritation of the

mucous membrane. OSHA exposure limit (TWA): ceiling 1 ppm [skin] or 3 mg/m³.

ethyl ether (C₂H₅OC₂H₅) — A colorless liquid (gas above 84°F) with a pungent, sweetish odor; MW: 74.1, BP: 94°F, Sol: 8%, Fl.P: –49°F, sp. gr. 0.71. It is used as a solvent in the manufacture of smokeless powder; as a laboratory solvent and chemical extractant; as a solvent and cleaning agent in the shoe and textile industries; in the manufacture of warm-process and cold-process pharmaceuticals; as an additive in motor fuels, perfumes, and denatured alcohol; as an anesthetic by animal handlers. It is hazardous to the central nervous system, skin, respiratory system, and eyes and is toxic through inhalation, ingestion, and contact. Symptoms of exposure include dizziness, drowsiness, headache, excitedness, narcosis, nausea, vomiting, and irritation of the eyes, skin, and upper respiratory system. OSHA exposure limit (TWA): 400 ppm [air] or 1200 mg/m³. Also known as ether.

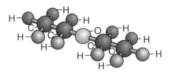

**Ethyl Ether
(Diethyl Ether)**

ethyl formate (CH₃CH₂OCHO) — A colorless liquid with a fruity odor; MW: 74.1, BP: 130°F, Sol (64°F): 9%, Fl.P: –4°F, sp. gr. 0.92. It is used in the shoe industry to dissolve celluloid heel coverings; as a solvent in the manufacture of artificial leather and in the manufacture of cellulose acetate for artificial silk; in the manufacture of safety glass; as a fumigant and larvacide for tobacco, cereals, and dried fruits; as an intermediate in organic synthesis; during the formulation of synthetic flavors. It is hazardous to the eyes and respiratory system and is toxic through inhalation, ingestion, and contact. Symptoms of exposure include narcosis and irritation of the eyes and upper respiratory system. OSHA exposure limit (TWA): 100 ppm [air] or 300 mg/m³.

$$\overset{\displaystyle O}{\overset{\displaystyle \|}{HCOC_2H_5}}$$

Ethyl formate

ethyl hydrogen sulfate ($C_2H_6O_4S$) — A colorless, oily liquid; MW: 126.13; BP: 280°C; Sol: soluble in water, forming sulfur acid; density: 1.367. It is used as an intermediate in the synthesis of ethanol from ethylene. It is hazardous to the skin and respiratory tract because reaction with moisture in the skin produces sulfuric acid.

$$CH_3-CH_2-O-\overset{\overset{O}{\|}}{\underset{\underset{O}{\|}}{S}}-OH$$

Ethyl hydrogen sulfate

ethyl mercaptan (CH_3CH_2SH) — A colorless liquid (gas above 95°F) with a strong, skunk-like odor; MW: 62.1, BP: 95°F, Sol: 0.7%, Fl.P: –55°F, sp. gr. 0.84. It is used as a fuel gas odorant; as a chemical intermediate; as raw material in the production of insecticides, antioxidants, plasticizers, and pharmaceuticals; as a stabilizer; as a solvent for elastomeric polymers and oil-soluble dyes in industrial stains and quick-setting rubber cements. It is hazardous to the respiratory system and in animals causes damage to the liver and kidneys; it is toxic through inhalation, ingestion, and contact. Symptoms of exposure include headache, nausea, and irritation of the mucous membrane; in animals, incoordination, weakness, pulmonary irritation, cyanosis, and liver and kidney damage. OSHA exposure limit (TWA): 0.5 ppm [air] or 1 mg/m³.

n-**ethyl morpholine ($C_4H_8ONCH_2CH_3$)** — A colorless liquid with an ammonia-like odor; MW: 115.2, BP: 281°F, Sol: miscible, Fl.P (open cup): 90°F, sp. gr. 0.90. It is used as a catalyst for flexible, semiflexible, and rigid polyurethane foam production; in polymer technology as a promoter for resin surface curing; as a stabilizer for fiber spinning solutions; in the manufacture of vat dyes; in the manufacture of pharmaceuticals; in the purification of penicillin G; in organic synthesis as a special solvent and pH regulator; for the preparation of chemical intermediates. It is hazardous to the respiratory system, skin, and eyes and is toxic through inhalation, absorption, ingestion, and contact. Symptoms of exposure include visual disturbance; irritation of the eyes, nose, and throat; and severe eye irritation from splashes. OSHA exposure limit (TWA): 5 ppm [skin] or 23 mg/m³.

O-ethyl-O-*para*-nitrophenyl phenylphosphonothioate (EPN; $C_2H_5O(C_6H_5)P(S)OC_6H_4NO_2$) — A yellow solid with an aromatic odor; a brown liquid above 97°F; MW: 323.3, BP: not known, Sol: insoluble, sp. gr. 1.27 at 77°F. It is used as an insecticide and acaricide. It is hazardous to the respiratory system, cardiovascular system, central nervous system, eyes, skin, and blood and is toxic through inhalation, absorption, ingestion, and contact. Symptoms of exposure include miosis, lacrimation, rhinorrhea, headache, tight chest, wheezing, laryngeal spasms, salivation, cyanosis, anorexia, nausea, abdominal cramps, diarrhea, paralysis, convulsions, low blood pressure, cardiac irregularities, eye and skin irritation. OSHA exposure limit (TWA): 0.5 mg/m³ [skin].

ethyl pyrophosphate — See TEPP.

ethyl silicate (($C_2H_5)_4SiO_4$) — A colorless liquid with a sharp, alcohol-like odor; MW: 208.3, BP: 336°F, Sol: reacts, Fl.P: 99°F. It is used as a bonding agent for investment castings, ceramic shells, crucibles, and types of refractory shapes; as a protective coating for paints, lacquers, and films; in the impregnation of porous materials to increase strength, hardness, stiffness, and abrasion; as a waterproofing and weatherproofing agent in porous rocks; as a chemical intermediate; as a jelling agent in organic liquids; as a coating agent inside electric lamp bulbs; in the synthesis of fused quartz; in the textile industry. It is hazardous to the respiratory system, liver, kidneys, blood, and skin and is toxic through inhalation, ingestion, and contact. Symptoms of exposure include eye and nose irritation; in animals, lacrimation, dyspnea, pulmonary edema, tremors, narcosis, liver and kidney damage, anemia. OSHA exposure limit (TWA): 10 ppm [air] or 85 mg/m³.

ethylene (C_2H_4) — A colorless, flammable, gaseous organic compound resulting from the cracking of either selective petroleum fractions or natural gas. It is widely used as a petrochemical base for numerous chemical reactions, especially for the preparation of polyethylene and related plastics. Also known as ethene and olefiant gas.

ethylene bromide — See ethylene dibromide.

ethylene chloride — See ethylene dichloride.

ethylene chlorohydrin ($CICH_2CHO_2H$) — A colorless liquid with a faint smell of ether; MW: 80.52; BP: 129°C; Sol: soluble in water, alcohol, and ether; Fl.P: 60°C (closed cup); density: 1.197 at 20°C. It is used in the manufacture of insecticides and in making ethylene glycol and ethylene oxide. It is hazardous to the respiratory system, gastrointestinal system, and skin. It is a severe, acute poison that affects the central nervous system, cardiovascular system, kidneys, liver, and gastrointestinal system; symptoms include respiratory distress, paralysis, brain damage, headaches, chest pain, stupor, and death. ACGIH exposure limit (TLV-TWA): 5 ppm [air] and 1 ppm [skin].

$$Cl-CH_2-CH_2-OH$$

Ethylene Chlorohydrin

ethylene dibromide (BrCH₂CH₂Br) — A colorless liquid or solid (below 50°F) with a sweet odor; MW: 187.9, BP: 268°F, Sol: 0.4%, sp. gr. 2.17. It is used in fumigation operations; in preplanting and on grains, fruits, and vegetables; in the production of antiknock fluids in fuels; in the production of waterproofing agents and fire-extinguishing agents; in gauge fluids during the manufacture of measuring instruments; in organic synthesis; in the production of dyes, pharmaceuticals, and ethylene oxide; as a specialty solvent for resins, gums, and waxes. It is hazardous to the respiratory system, liver, kidneys, skin, and eyes and is toxic through inhalation, absorption, ingestion, and contact. Symptoms of exposure include dermatitis with vesiculation and irritation of the eyes and respiratory system; carcinogenic. OSHA exposure limit (TWA): 20 ppm [air]. Also known as ethylene bromide and 1,2-dibromomethane.

$$
\begin{array}{ccc}
 & H & H \\
 & | & | \\
Br- & C- & C-Br \\
 & | & | \\
 & H & H
\end{array}
$$

Ethylene dibromide

ethylene dichloride (ClCH₂CH₂Cl) — A colorless liquid with a pleasant, chloroform-like odor that decomposes slowly, becomes acidic, and darkens in color; MW: 99.0, BP: 182°F, Sol: 0.9%, Fl.P: 56°F, sp. gr. 1.24. It is used as a chemical intermediate in the manufacture of vinyl chloride; as an intermediate in the production of chlorinated solvents and ethyleneamines; in the production of gasoline using tetraethyl lead as an antiknock agent and ethylene dichloride as a lead scavenger; as a fumigant or industrial solvent. It is hazardous to the kidneys, liver, eyes, skin, and central nervous system and is toxic through inhalation, absorption, ingestion, and contact. Symptoms of exposure include central nervous system depression, nausea, vomiting, dermatitis, corneal opacity, and irritation of the eyes; carcinogenic. OSHA exposure limit

(TWA): 1 ppm [air] or 4 mg/m³. Also known as ethylene chloride and Dutch liquid.

ethylene glycol (HOCH₂CH₂OH) — A colorless organic liquid used as an absorbent because it absorbs twice its weight of water at room temperature and 100% humidity. It is used as an antifreeze; as a solvent for dyes, resins, drugs, or other organic chemicals; and in the manufacture of dynamites. Also known as glycol.

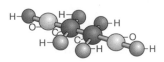

Ethylene Glycol

ethylene glycol dinitrate (O₂NOCH₂CH₂ONO₂) — A colorless to yellow, oily, odorless liquid; MW: 152.1, BP: 387°F, Sol: insoluble, Fl.P: 419°F, sp. gr. 1.49. It is used as an explosive ingredient (60 to 80%) in dynamite along with nitroglycerine (20 to 40%). It is hazardous to the cardiovascular system, blood, and skin and is toxic through inhalation, absorption, ingestion, and contact. Symptoms of exposure include throbbing headache, dizziness, nausea, vomiting, abdominal pain, hypotension, flush, palpitations, methemoglobinemia, delirium, depressed central nervous system, angina, skin irritation; in animals, it causes anemia, and mild liver and kidney damage. OSHA exposure limit (TWA): 0.1 mg/m³ [skin].

ethylene glycol monobutyl ether (C₆H₁₄O₂) — A colorless liquid with a mild odor of ether; MW: 118.20; BP: 172°C; Sol: miscible with water and most organic solvents; FL.P: 60°C (closed cup), a combustible liquid; density: 0.902 at 20°C. It is used as a solvent for oil and grease and in drycleaning. It is hazardous to the respiratory tract, digestive tract, and skin. It may produce nausea, vomiting, and headache. ACGIH exposure limit (TLV-TWA): 25 ppm [skin].

$$HO-CH_2-CH_2-O-CH_2-CH_2-CH_2CH_3$$

Ethylene glycol monobutyl ether

ethylene glycol monoethyl ether (C₄H₁₀O₂) — A colorless and odorless liquid; MW: 90.12, BP: 135°C, Sol: miscible with water and most organic solvents, Fl.P: 44°C (closed cup). It is used as a solvent for nitrocellulose, lacquers, and varnishes. It is hazardous to the respiratory tract and digestive tract and affects the lungs, kidneys, liver, spleen, and skin. It is a teratogen. ACGIH exposure limit (TLV-TWA): 5ppm [skin].

$$HO-CH_2-CH_2-O-CH_2-CH_3$$

**Ethylene glycol
monoethyl ether**

ethylene glycol monoisopropyl ether (C₅H₁₂O₂) — A colorless liquid; MW: 104.17; BP: 142°C; Sol: soluble in water and organic solvents; Fl.P: 45°C (closed cup), a combustible liquid; density: 0.903. It is used as a solvent for resins, dyes, and cellulose esters. It is hazardous to the respiratory tract, digestive tract, and skin. It may cause nausea, vomiting, headache, and anemia. ACGIH exposure limit (TLV-TWA): 25 ppm [skin].

$$HO-CH_2-CH_2-O-CH \begin{array}{c} \diagup CH_3 \\ \diagdown CH_3 \end{array}$$

Ethylene glycol monoisopropyl ether

ethylene glycol monomethyl ether (C₃H₄O₂) — A colorless liquid; MW: 76.11; BP: 124°C; Sol: miscible with water, alcohol, ether, and acetone; Fl.P: 42°C (closed cup); density: 0.965. It is used as a solvent for cellulose acetate, synthetic and natural resins, and dyes. It is hazardous to the respiratory tract, skin, digestive tract, blood, kidneys, and the central nervous system. It causes headaches, drowsiness, weakness, and indigestion and may be fatal. It is a teratogen. ACGIH exposure limit (TLV-TWA): 5 ppm [skin].

$$HO-CH_2-CH_2-O-CH_3$$

**Ethylene glycol
monomethyl ether**

ethylene oxide (ETO) (C₂H₄O) — A colorless, flammable gas with a somewhat sweet odor; MW: 44.05, BP: 11°C, Sol: 1 × 1⁻⁶ mg/l at 20°C, Fl.P: less than −18°C. It is used in organic synthesis, for sterilizing, and for fumigating. It is hazardous to the respiratory system, nervous system, and reproductive system and is toxic through inhalation, ingestion, absorption, and contact. Symptoms of exposure include bronchitis, pulmonary edema, emphy-

sema, nausea and vomiting, cataracts, headache, peripheral neuropathy, impaired eye coordination, memory loss, lethargy, and numbness and weakness in the extremities. OSHA exposure limit (PEL TWA): 1 ppm [air]. Also known as oxirane.

ethylene–propylene terpolymer (EPDM) — A high-strength, flexible compound designed especially for contact with potable water.

ethylenediamine (NH₂CH₂CH₂NH₂) — A colorless, viscous liquid with an ammonia-like odor; a solid below 47°F; MW: 60.1, BP: 241°F, Sol: miscible, Fl.P: 93°F, sp. gr. 0.91. It is used in the manufacture of carbamate fungicides, pesticides, and weed killers; in the manufacture of chelating agents; in dyes, soaps, and cleaning compounds; in water treatment; in agriculture; in rubber, pulp, and paper processing; as an activator for epoxy resin menders; in the preparation of spandex, adhesives, inks, surface coatings, cross-linking agents, plasticizers, resin-curing compounds; as a dye assist; in the manufacture of surfactants, emulsifying agents, dispersants, corrosion inhibitors, detergents, and textile surface treatments; in organic synthesis for preparation of dye intermediates, heterocyclic compounds, pharmaceuticals, and salts. It is hazardous to the respiratory system, liver, kidneys, and skin and is toxic through inhalation, absorption, ingestion, and contact. Symptoms of exposure include nasal irritation, primary irritation, sensitization dermatitis, asthma, damage to the liver and kidneys, and irritation of the respiratory system. OSHA exposure limit (TWA): 10 ppm [air] or 25 mg/m³.

ethyleneimine (C₂H₅N) — A colorless liquid with an ammonia-like odor; MW: 43.1, BP: 133°F, Sol: miscible, Fl.P: 12°F, sp. gr. 0.83. It is used in the manufacture of paper, textiles, adhesives, binders, petroleum refining products, rocket and jet fuels, lubricants, chemosterilant chemicals, chemotherapeutic agents, coating resins, varnishes, lacquers, agricultural chemicals, cosmetics, ion-exchange resins, photographic chemicals, colloid flocculants, and surfactants. It is hazardous to the eyes, lungs, skin, liver, and kidneys and is toxic through inhalation, absorption, ingestion, and contact. Symptoms of exposure include nausea, vomiting, headache, dizziness, pulmonary edema, damage to the liver and kidneys, eye burns, skin sensitization, and irritation of the nose and throat; carcinogenic. OSHA exposure limit (TWA): regulated use. Also known as aziridine.

di(2-ethylhexyl)phthalate (C₂₄H₃₈O₄) — A colorless liquid with almost no odor; MW: 390.57, BP: 385°C, Sol: 0.285 mg/l at 24°C, Fl.P: 196°C. It is used principally as a plasticizer in the production of polyvinyl chloride (PVC) and vinyl chloride resins; PVC is used in toys, vinyl, upholstery, shower curtains, adhesives, coatings, components of paper and paperboard, disposable medical examination and surgical gloves, flexible tubing used to administer parenteral solutions, and in hemodialysis treatment. Single oral doses up to 10 g are not lethal to humans, and no cases of death in humans after oral exposure are known; higher doses are lethal to rats, rabbits, and guinea pigs. Acute exposures to large oral doses can cause gastrointestinal distress with mild abdominal pain and diarrhea; oral exposures to rats and mice result in marked increase in liver mass. Hepatic hyperplasia appears to be the initial physiological response in rats; an autopsy of the liver shows fat deposits, a decline in centrilovular glycogen deposits, and structural changes in the bile ducts. Developmental toxicity including teratogenic effects have occurred in both rats and mice. OSHA exposure limits (PEL TWA): 5 mg/m^3 [air]; STL: 10 mg/m^3 [air].

etiological agent — A pathogenic organism or chemical causing a specific disease in a living body.

etiology — The science dealing with causes of disease.

ETM+ — See Enhanced Thematic Mapper Plus.

ETO — See ethylene oxide (also known as oxirane).

ETP — Acronym for emissions trading policy.

ETS — Acronym for emergency temporary standards; see environmental tobacco smoke.

eugenics — The science of human heredity.

eukaryotic cells — (*genetics*) The cells of higher animals and plants that contain genetic material in the nuclei.

euphoria — The absence of pain or distress; may be induced by chemical exposure.

euphotic zone — The lighted region that extends vertically from the water surface to the level at which photosynthesis fails to occur because of ineffective light penetration.

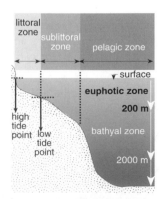

Euphotic Zone

eurythermal — The bodies of water located at the lower end of a river that are subjected to tidal fluctuations.

eurytopic — An organism with a wide environmental range.

eustachian tube — A tube connecting the nasopharynx with the middle ear cavity.

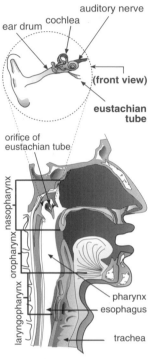

Eustachian Tube

eutectic chemicals — Phase-changing chemicals used in heat storage systems to store a large amount of energy in a small volume.

eutectic temperature — The lowest possible melting point of a mixture of alloys.

eutrophic — Pertaining to a lake abundant in nutrients and having high rates of productivity that frequently result in oxygen depletion below the surface layer.

eutrophication — A process by which pollutants cause a body of water to become overly rich in organic and mineral nutrients so that algae grows rapidly and depletes the oxygen supply.

eV — See electron volt.

evacuation — Removal of residents from an area of danger; taking the protective action of leaving an area of risk until the hazard has passed and it is safe to return to the area.

evacuation/exhaust — Part of the sterilizer cycle when a vacuum pump operates to remove the bulk of ethylene oxide from a chamber, followed by the opening of a valve to admit filtered air into the sterilizer chamber to return it to atmospheric pressure.

evaluation — A critical appraisal or assessment.

evaporated milk — A homogenized whole milk from which 50 to 60% of the water content has been removed prior to canning and sterilization.

evaporation — The physical transformation of a liquid to a gaseous state at any temperature below its boiling point.

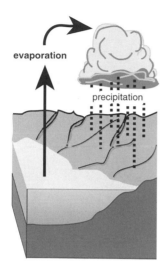

Evaporation

evaporation ponds — Areas where sewage, sludge, or wastes are dumped and allowed to dry.

evaporation rate — The rate at which a material will vaporize or evaporate compared to the known rate of vaporization of a standard material.

evaporative cooler — An unit that cools air by saturating it with water vapor.

evaporative emissions — Emissions from the evaporation of gasoline that occurs during vehicle refueling and operation and when the vehicle is parked, accounting for two thirds of the hydrocarbon emissions from gasoline-fueled vehicles on hot summer days.

evapotranspiration — The combined loss of water in a given area during a specified period of time by evaporation from the soil or water surface and by transpiration from plants.

evolution — A slow process of change by which organisms acquire their distinguishing characteristics and come to differ from their ancestors.

evolutionary adaptation — The changes occurring in a population of individuals over many generations by processes of natural selection.

EWG — An acronym for environmental working group.

EWP — See Emergency Watershed Protection.

EWRP — See Emergency Wetlands Reserve Program.

EWS — See early warning system.

ex parte — (*law*) On one side only; referring to cases in which only one side is represented.

ex situ — Out of the original position.

exacerbation — An increase in severity.

EXAMS — Acronym for Exposure Analysis Modeling System.

exceedance — Measured level of an air pollutant that is higher than national or state air quality standards.

excess combustion air — Air supplied in excess of theoretical air, usually expressed as a percentage of the theoretical air.

excess deaths — (*epidemiology*) The number of deaths over the statistically expected number in a population within a given time span.

excessive perspiration — A level of sweat greater than normal for the ambient environment, indicating potentially serious health problems including septic fever, pulmonary tuberculosis, chronic renal disease, malaria, extreme nervousness, or an approaching heart attack.

exchange capacity — Total ionic charge of the adsorption complex active in the adsorption of ions.

excise — To remove completely, as in the act of surgery.

excitation — The process by which the addition of energy to a system transfers it from its ground state to an excited state; an act of irritation or stimulation.

excitation energy — The minimum energy required to change a system from its ground state to an excited state.

exclusion area — The area immediately surrounding one or more receptacles in which chemical agents are contained.

excreta — Waste material eliminated from the body, including feces and urine.

excretion — The process by which waste materials are removed by living cells from the body.

excretory system — The series of organs adapted to removing waste from an organism.

execute — (*computer science*) To follow a set of instructions to perform a specified function.

execution of decree — Completion of a decree ordering an action to be taken — for example, destroying goods seized by the U.S. Marshall in response to a default decree of condemnation.

exergonic reaction — A chemical reaction that releases energy.

exfiltration — The air leakage outward through cracks and interstices and through the ceilings, floors, and walls of a space or building.

exhalation valve — A device that allows exhaled air to leave a respirator but prevents outside air from entering.

exhaust emissions — Substances emitted to the atmosphere from any opening downstream from the exhaust port of a motor vehicle.

exhaust gas recirculation — An emission control method that involves recirculating exhaust gases from an engine back into the intake and combustion chambers, thus lowering the combustion temperatures and reducing the level of nitrogen oxides.

exhaust rate — The volumetric rate at which air is removed.

exhaust ventilation — Mechanical removal of air from a portion of a building.

exhaustion — Extreme loss of physical or mental capabilities caused by fatigue or illness.

exocrine — A type of gland that releases its secretion through a duct.

exocrine gland — A gland that secretes its product to an epithelial surface directly through ducts.

exogenous — A chemical substance formed outside of the body.

exogenous factors — External factors regulating growth and development of an organism.

exogenous infection — An infection that develops from bacteria normally found outside the body but which are now present in the body.

exoskeleton — Stiff, tough external covering on the body of certain invertebrates, especially arthropods, that provides support and protection to the softer inner parts.

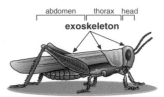

Grasshopper Exoskeleton

exosphere — The region of the atmosphere beyond 400 km that fades into interplanetary space.

exothermic — Referring to a chemical reaction in which heat is given off, such as in the burning of materials.

exothermic reaction — A chemical reaction in which the energy released is greater than the energy required to start the reaction.

exotoxin — A soluble toxin excreted by certain microbacteria and absorbed by the tissues of host organisms.

expansion chamber — See cloud chamber.

expansion factor — A correction factor for the change in density between two pressure measurement areas in a constricted flow.

expansion joint — A joint that is purposely left open to allow for movement of its parts.

expected deaths — (*epidemiology*) The number of deaths statistically expected to occur in a population over a given time from all causes.

expected peak day concentration — (*air pollution*) A calculated value that represents the concentration of a pollutant expected to occur at a particular site

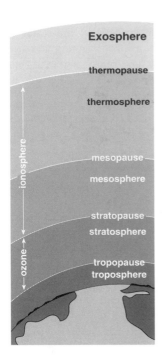

once each year on average; the calculation procedure uses measured data collected at the site during a 3-year period.

expectorant — A substance that reduces viscosity and promotes the ejection of mucous or other substances from the lung, bronchi, and trachea.

expenses — All payments transferred or paid to entities outside the operating organization for costs incurred for and related to the operation.

experience — A record of premiums and losses relating to insurance and used as the basis upon which future rates or costs are predicted.

experience modification — A provision designed to recognize the merits or demerits of individual risks as they relate to insurance.

experiment — An artificially created situation used to test a hypothesis, where variables that cause an effect are prevented from doing so.

experimental animal — An animal used in any research investigation involving the feeding or other administration of or subjection to an experimental or nonexperimental biological product, drug, or chemical used in a manner for which it was not intended.

experimental design — The use of groups, dependent and independent variables, and sample sizes for the testing of a particular hypothesis.

experimental epidemiology — An experimental model used for special studies to confirm a causal relationship seen during observational data gathering.

experimental study — Research in which the investigator has complete control over allocation of subjects to treatment groups.

expert witness — A person who has special knowledge and experience in a specific area and who presents testimony on this area in a court of law.

expertise — The special skills or knowledge a person gains from education, training, and experience.

expiration — The discharge of air from the lungs.

expiration date — The date beyond which a drug, chemical, or food should not be used.

expiratory flow rate — The maximum rate at which air can be expelled from the lungs.

expired gas — Any gas that is exhaled from the lungs.

explosion-proof enclosure — An enclosure that can withstand an explosion of gases within it and prevent the explosion of gases due to sparks, flashes, or external temperature.

explosive — Any chemical compound, mixture, or device for which the primary or common purpose is to produce a substantial instantaneous release of gas or heat when subjected to a sudden shock, pressure, or high temperature.

explosive incident — An event in which explosive devices are used as terrorist weapons.

explosive limit — See flammable limit.

explosive range — See flammable range.

exponent — (*mathematics*) Superscript on a number that indicates how many times that number is to be multiplied by itself.

exponential growth — The growth of an organism at a constant rate of increase per unit of time; see geometric growth.

exposure — A measure of the ionization produced in air by x- or gamma-rays. It is determined by dividing the sum of the electrical charges of all ions of one sign produced in air when all electrons liberated by photons in a volume element of air are completely stopped by the mass of air in the volume element. The unit of measurement of exposure is the roentgen; (*medicine*) the act or condition of coming in contact with, but not necessarily being infected by, a disease-causing agent; (*hazardous materials*) the state of being open and vulnerable to a hazardous chemical in the course of employment by inhalation, ingestion, skin contact, absorption or any other method, including potential and accidental exposure.

exposure assessment — Determination or estimation, either qualitative or quantitative, of the magnitude, frequency, duration, and route of human exposure to a particular agent; a combination of field monitoring, mathematical modeling, measurement of actual concentrations of chemicals in tissue, and laboratory monitoring data; the process of determining how much of a chemical is in the environment and capable of contact with organisms.

exposure coefficient — A term that combines information on the frequency, mode, and magnitude of contact of a contaminated medium to yield a quantitative value of the amount of contaminated medium contacted per day.

exposure concentration — The concentration of a chemical or other pollutant representing a health threat in a given environment.

exposure dose reconstruction — (*ATSDR*) An approach that uses computational models and other approximation techniques to estimate cumulative amounts of hazardous substances internalized by people at presumed or actual risk of contact with substances associated with hazardous waste sites.

exposure indicator — An environmental characteristic measured to provide evidence of the occurrence or magnitude of contact with a physical, chemical, or biological stressor.

exposure investigation — (*ATSDR*) The collection and analysis of site-specific information to determine if human populations have been exposed to hazardous substances; environmental sampling, exposure–dose reconstruction, biologic or biomedical testing, and evaluation of medical information are used.

exposure level — (*chemical*) Concentration of a chemical at the absorptive surfaces of an organism.

exposure pathway — The course a chemical or physical agent takes from a source to an exposed organism; a unique mechanism by which an individual or population is exposed to chemicals or physical agents at or originating from a hazardous waste site.

exposure period — The length of time an object or individual is exposed to a chemical or microorganism.

exposure point — Any hypothetical event that would lead to human contact with a toxin.

exposure point concentration — An estimate of the arithmetic average concentration for a contaminant based on a set of site-sampling results.

Exposure Registry — (*ATSDR*) A system for collecting and maintaining in a structured record information on persons with documented environmental exposures. The system evolved from the need for fundamental information concerning the potential impact on human health of long-term exposure to low and moderate levels of hazardous substances.

exposure route — The path a chemical or pollutant uses to enter an organism, such as by ingestion, inhalation, or dermal absorption.

exposure scenario — A set of conditions or assumptions about sources, exposure pathways, concentrations of toxic chemicals, and populations that aid investigators in evaluating and quantifying exposure in a given situation.

exposure standard — An airborne concentration of a particular substance in the worker's breathing zone that should cause neither adverse health effects nor undue discomfort to nearly all workers, measured as a time-weighted average, peak reading, or short-term exposure limit.

exposure unit — One electrostatic unit (ESU) of charge in 1 cm^3 of air at standard temperature and pressure; formerly called the roentgen.

exposure–response relationship — The relationship between exposure levels and the incidence of adverse effects.

exsanguination — A loss of blood.

extended aeration — The modification of the activated sludge process that provides for aerobic sludge digestion within the aeration system.

extended care facility — A residential facility for patients who require 24-hour nursing care or rehabilitative therapy on a less-intensive basis than is provided by a comprehensive rehabilitation center.

extension service — A nationwide continuing-education program that is based on the academic programs of land-grant colleges of agriculture.

extermination — The control and elimination of insects, rodents, and other pests by eliminating harborage and food and by the use of poisoning or trapping techniques.

external absorbtion — Taking up substances through the mucous membranes of the skin.

external auditory canal — The sound channel connecting the external and middle ears.

external ear — The outer structure of the ear, which funnels sound waves to the middle ear.

external forcing — The influence on the Earth system or one of its components by an external agent such as solar radiation or the impact of extraterrestrial bodies such as meteorites

external radiation — Radiation that penetrates the skin from a source outside the body.

external respiration — The exchange of oxygen and carbon dioxide between the alveolar air and blood.

external — The outside or exterior of a body or an organ.

extinction coefficient — A measure of the rate of reduction of transmitted light through a substance.

extinguishing media — Firefighting substances used to control a material during a fire.

extra high voltage (EHV) — A voltage above 345 kilovolts (kV) that is used for power transmission.

extracellular — Outside of a cell or cell tissue or any cavities or spaces between cell layers.

extracellular digestion — The digestion that takes place outside cells, usually in the gut or digestive cavity.

extract — A substance prepared by the use of solvents or evaporation to separate it from the original material.

extract air — The air that is removed from a building or space, a portion of which is often used for recirculation and added to incoming air.

extraction procedure (EP) — A procedure used to identify waste likely to leach hazardous concentrations of particular toxic constituents into the groundwater from landfills.

extraocular — Outside the eye.

extrapolation — Estimation of unknown values by extension or projection from known values; projecting, extending, or expanding known data or experience into an area not known or experienced to arrive at an estimation.

extrasystole — A premature cardiac contraction that is independent of the normal rhythm and arises in response to a sinoatrial node.

extremely hazardous substances — Any of several hundred chemicals identified by the U.S. EPA as toxic and listed under SARA Title 3; the list is periodically revised.

extremely hazardous waste — Any dangerous waste persisting in a hazardous form for several years or more at a disposal site with a potential for being concentrated by living organisms through the food chain or affecting the genetic makeup of humans or wildlife.

extremely low frequency (ELF) — The frequency range below 300 hertz (Hz) in the radio spectrum.

extremophile — An organism that is adapted to living in extreme conditions including high salt, ice, or thermal springs.

extrinsic — Related to some region other than the organ with which it is associated.

extrusion — The forcing of raw material through a die or form in either a hot or cold state in a solid or partial solution.

extubation — The removal of a tube from an orifice or cavity of the body.

exudate — The product of diffusion.

exudation — A discharge of fluid, pus, or serum.

eye — One of a pair of organs of sight, contained in a bony orbit at the front of the skull, embedded in orbital fat, and innervated by one of a pair of optic nerves from the forebrain.

eye cup — A small vessel that is shaped to fit over the eyeball and is used to bathe the eye.

eye hazards — Chemicals that affect the eye or visual capacity by causing conjunctivitis and corneal damage.

eye protection — Recommended safety glasses, chemical splash goggles, and faceshields to be utilized when handling a hazardous material.

eye toxins — Chemicals, including organic solvents and acids, that produce symptoms of conjunctivitis and corneal damage.

eyepiece — A gas-tight, transparent window in a full facepiece through which a person can see.

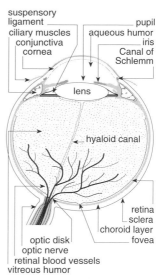

Eye

E

F

f — See frequency; an SI symbol for the prefix *femto-*, representing a factor of 10^{-15}.

F — See Fahrenheit scale; equilibrium fraction; farad.

F₁ — Notation for the first filial generation resulting from a cross.

FAA — Acronym for Federal Aviation Administration.

fabric filter — A device designed to remove particles from a carrier gas by passage of the gas through a porous fabric medium.

face velocity — Average air velocity into an exhaust system measured at the opening into the hood or booth.

facepiece — That portion of a respirator that covers the wearer's nose and mouth (half-mask facepiece) or nose, mouth, and eyes (full facepiece).

facial — Referring to the face.

facilitate — To make an action easier or promote better understanding.

facility — Geographical sublocation of a site.

facing — A coating of a different material for architectural or protection purposes.

factor of safety — Ratio of the ultimate strength of a material to an allowable or working stress.

facultative — Able to adapt to more than one condition.

facultative aerobe — A microorganism that may grow under anaerobic conditions but develops most rapidly in an aerobic environment.

facultative anaerobe — A microorganism able to grow under aerobic conditions but that develops most rapidly in an anaerobic environment.

facultative pond — A pond used for animal waste disposal wherein the upper portion is aerobic, with oxygen supplied by algae, while the bottom layer is anaerobic.

Fahrenheit scale (F) — Temperature scale where 32° is freezing and 212° is boiling.

failed — Something that no longer functions as designed or intended.

failure — An incident resulting in the uncontrolled release of water or a chemical from a facility.

failure potential assessment — A judgment of the potential for failure of an essential element within the expected life of the project or facility.

faint — See syncope.

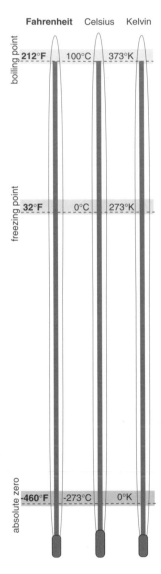

Fahrenheit Temperature Scale
compared to K and C scales
(not proportionate)

falciparum — The most severe form of malaria, caused by the protozoan *Plasmodium falciparum*.

fall — (*plumbing*) The amount of slope given to horizontal runs of pipe.

fallout — The dust particles that contain radioactive fission products resulting from a nuclear explosion.

false color — An imaging process that produces a color that does not correspond to the true or natural color of the subject as seen by the eyes.

false galena — See sphalerite.

false negative — Negative result of a test for a disease when the subject does have the disease.

false negative rate — The number of inaccurate negative test results compared to the true negative test results.

false positive — Positive result of a test for a disease when the subject does not have the disease.

false positive rate — The number of inaccurate positive test results compared to the true positive test results.

FAM — See fibrous aerosol monitor.

familial — A characteristic, condition, or disease that is present in some families and not in others in greater numbers than expected by chance.

family health — A description of health and diseases of the immediate family.

family history — A series of questions asked of an individual concerning the health and diseases of various members of the family in order to determine potential health problems for the individual, such as diabetes, heart disease, or cancer.

fan rating table — Data describing the volumetric output of a fan at different static pressures.

fan static pressure — Pressure added to a system by a fan; equal to the amount of pressure lost in the system minus the velocity pressure in the air at the fan outlet.

FAO — Acronym for the Food and Agricultural Organization of the United Nations.

FAQ — (*computer science*) Acronym for frequently asked questions, which allow computer users to find answers to questions about new products or services.

Far Eastern hemorrhagic fever — A form of epidemic hemorrhagic fever found in Asia and transmitted by a virus carried by an Asian rodent. Symptoms include chills, fever, headache, abdominal pain, nausea, vomiting, anorexia, and extreme thirst, leading to hypotensive shock; the mortality rate is as high as 33%.

far infrared — Electromagnetic radiation longer than the thermal infrared, with wavelengths between about 25 and 1000 μm.

farad (F) — The unit of capacitance in the meter–kilogram–second system.

faraday — The unit of electric charge required to deposit by electrolysis one equivalent weight of an element; a measurement equal to 96,500 coulombs; also known as Faraday constant.

Faraday constant — See faraday.

Faraday's law — During electrolysis, the weight of a liberated element is proportional to the quantity of electricity passing through the cell and equivalent to the weight of the element.

farm — A tract of land used for agricultural purposes.

farm pond — An impoundment used to provide water for agricultural and domestic needs, recreation, and conservation.

farmer's lung — A respiratory disorder caused by inhalation of actinomycetes or other organic dusts from old hay; symptoms include coughing, dyspnea, cyanosis, tachycardia, nausea, chills, and fever.

fasciculation — A small, local, involuntary, muscular contraction visible under the skin representing spontaneous discharge of a number of fibers innervated by a single motor nerve filament.

fascioliasis — An infection with the liver fluke, *Fasciola hepatica*, with symptoms of epigastric pain, fever, jaundice, eosinophilia, diarrhea, and fibrosis of the liver as a consequence of prolonged infection; acquired by ingestion of the cysts of the fluke found on aquatic plants such as raw watercress.

fat — Any glycerol ester of a saturated fatty acid that form a class of neutral organic compounds; a substance composed of lipids or fatty acids occurring in various forms or consistencies; a type of body tissue composed of cells storing fatty material.

fat free — See non-fat.

fatal — Having the capacity to cause death.

fatal flaw — Any problem or conflict, real or perceived, that will destroy a solution or process.

fatality rate — The number of persons diagnosed as having a specific disease who die as a result of the illness.

fate — The combined transport and transformation of a pollutant.

fate and exposure modeling — The scientific process used to predict where chemicals go after being released into the environment.

fatigue — A state of increased discomfort and decreased efficiency resulting from prolonged exertion; a generalized feeling of tiredness or exhaustion; a phenomenon due to overexertion leading to a decrease in physical performance and a buildup of lactic acid in the body.

fatigue fever — A period of benign fever after overexertion when the accumulation of metabolic wastes has not been removed.

fatty acid — One of the soluble products into which fats are broken down in the body by digestion.

faults — Fractures where once-continuous rocks have suffered relative displacement.

F

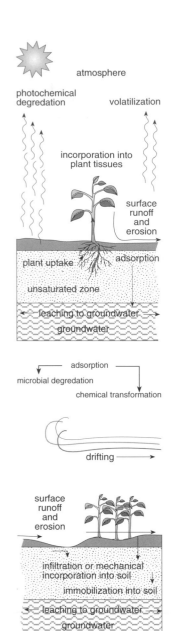

Fate of pesticides in soil

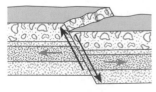
normal thrust fault

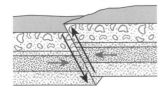
reverse thrust fault

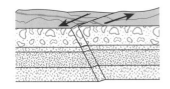
strike-slip fault

Faults

fauna — Animals characteristic of a region, period, or special environment.

favism — Poisoning from eating the fava bean (*Viciafava*) or being around the plants when they are in bloom.

fax — A method of transmitting images or printed material by electronic means.

F/B — See freight bill.

FBC — Acronym for fluidized bed combustion.

FBI — Acronym for fluidized bed incinerator.

fc — See footcandle.

FC — Acronym for fuel consumption.

f/cc — See fibers per cubic centimeter.

FCC — Acronym for fluid catalytic converter.

FCO — See federal coordinating officer.

FDA monitoring — The collection and analysis of samples for pesticide residues carried out by the Food and Drug Administration.

FDCA — See Federal Food, Drug, and Cosmetic Act.

FDECU — Acronym for field-deployable environmental control unit.

FDL — Acronym for final determination letter.

FDV — See flow diversion valve.

FE — See fugitive emissions.

feasibility estimate — An estimate used for determining the economic feasibility of a project, for defining the probable sequence and costs of construction of a project, and for making a choice among alternative locations or plans.

feasibility study — The acquisition of information to determine if a proposed project is practical and/or capable of being accomplished from the standpoint of advantages vs. disadvantages.

febrile — Associated or characterized by fever with a body temperature over 100°F.

febrile delirium — A symptom of improper nervous system operation, with excitement, restlessness, and disorientation due to acute fevers.

febrile seizure — A seizure caused by high fevers.

febrile state — A significant increase in body temperature accompanied by elevated pulse, respiration, headache, and pain.

fecal — The feces or excrement of people or animals.

fecal coliform bacteria — Bacteria of the coliform group and of fecal origin from the intestines of warm-blooded animals as opposed to coliforms of non-fecal sources.

feces — Solid intestinal waste materials.

Federal Advisory Committee Act — A law passed by Congress seeking consultation and public involvement as a critical step in making the federal pesticide programs work. The advisers are referred to as stakeholders, as they are involved in the results of the pesticide problems and programs and provide advice on strategic approaches for pest management planning, transition, and tolerance reassessment, as well as feedback on various topics. The Act provides a forum for stakeholders to discuss pesticide regulation, policies, and implementation and offer advice to the EPA on the screening and testing of endocrine disruptors.

federal coordinating officer — (*FEMA*) A person appointed by the FEMA director, deputy director, or associate director for response and recovery following a declaration of a major disaster or an emergency by the President of the United States to coordinate federal assistance.

Federal Facilities Compliance Act — A federal act that requires the Department of Energy to develop and submit to the states or the U.S. EPA plans for developing mixed waste treatment capacities and technologies.

Federal Facility Agreement and Consent Order — A 1991 agreement between the Department of Energy, Environmental Protection Agency, and the state of Idaho that establishes a plan for the cleanup of the Idaho National Engineering and Environmental Laboratory in accordance with the Comprehensive Environmental Response, Compensation, and Liability Act (CERCLA), Resource Conservation and Recovery Act (RCRA), and Idaho Hazardous Waste Management Act, which establishes the legally enforceable actions to be taken in the cleanup of the facility and the aquifer below it.

Federal Food, Drug, and Cosmetic Act (FDCA) — An amendment to the Food and Drug Act of 1906, passed in 1938 and amended through the 1980s, establishing the authority of the Food and Drug Administration to regulate food additives, contaminants, drugs, cosmetics, and potential carcinogens related to food, drugs, and cosmetics.

Federal Implementation Plan — A plan prepared by the U.S. EPA providing measures that nonattainment areas must use to meet requirements of the Federal Clean Air Act in the absence of an approved state implementation plan.

Federal Insecticide, Fungicide, and Rodenticide Act (FIFRA) — A 1947 law, updated through 1988, that authorizes the U.S. EPA to register and reregister pesticides; to review, cancel, and suspend registered pesticides; and to regulate the production, storage, transportation, use, and disposal of pesticides, including areas of research and monitoring.

Federal Land Policy and Management Act — This act passed by Congress updated all laws from colonial times to the present in the area of federal land ownership and management and provided that the United States receive fair market value for the use of the public lands and their resources that public lands be managed in a manner to protect the quality of scientific, scenic, historical, ecological, environmental, air and atmospheric, water resource, and archeological values and, where appropriate, provide feed and habitat for fish and wildlife and domestic animals; and that lands be provided for outdoor recreation and human occupancy and use.

federal on-scene commander — An FBI official who ensures appropriate coordination of the overall U.S. government response with federal, state, and local authorities until such time as the Attorney General transfers responsibility to FEMA for a disaster or emergency.

Federal Radiological Emergency Response Plan — A comprehensive, coordinated plan broadly describing the entire federal government's response to radiological emergencies in support of federal, state, and local government agencies.

Federal Response Plan — A plan developed to help expedite federal support to disaster areas.

Federal Water Pollution Control Act (FWPCA) — A 1956 law funding water pollution research and training, establishing construction grants for municipalities, and establishing a three-stage enforcement process.

Federal Water Pollution Control Act Amendments (FWPCAA) — A 1972 amendment to the Federal Water Pollution Control Act setting national water quality goals, establishing a zero discharge goal by 1985, setting technology-based effluent limitations, establishing National Pollution Discharge Elimination System permits, providing federal enforcement based on permit violations, and increasing federal funding to water treatment facilities.

F

fee for service — A fee charged by a governmental agency to license or approve an environmental service, such as a food establishment, septic tank system, waste site, etc.; the fee is comprised of the time, materials, and other expenses necessary to perform the service and can be partially subsidized by taxes.

feed — Animal food; the act of giving food to a plant or animal.

feedback — Opinions or information regarding an action or process provided to the originator of the action or process.

feedlot — A concentrated, confined, animal- or poultry-growing operation for meat, milk, or egg production; stabling in pens or houses wherein the animals or poultry are fed at the place of confinement and the waste material is concentrated.

feedstock — Raw materials supplied to manufacturing or processing plants for use in the production of goods or for treatment.

feedstock tax — An excise tax that is levied on 42 chemical raw materials, with the proceeds going to the Hazardous Substance Superfund Trust Fund.

feelers — See antenna.

FEF — Acronym for forced expiratory flow rate.

FEIS — Acronym for Fugitive Emissions Information System.

felony — A crime punishable by more than 1 year in prison.

femtogram — A unit of mass equal to 1 billionth of a microgram.

femtogram per cubic meter — One billionth of a microgram of a substance in a cubic meter of air, soil, or water.

femur — The longest and strongest bone in the body which extends from the pelvis to the knee.

fen — A type of wetland that accumulates peat deposits and is less acidic than bogs, deriving much of the water from groundwater containing calcium and magnesium.

fenestrated — Having numerous small holes or openings, as in a membrane or other object.

fenthion $C_{10}H_{15}O_3PS_2$) — A colorless to brown liquid with a faint garlic-like odor; MW: 278.3, BP: NA, Sol: 0.006 g/100 g water at 20° C, Fl.P: NA (may be ignited by heat, sparks, or flames and produces poisonous gases, including oxides of phosphorous and sulfur, in a fire), sp. gr. 1.25. It is an organophosphate chemical used as an insecticide, especially for mosquitoes. It is hazardous to the respiratory system, digestive system, skin, and eyes. Acute exposure may cause central nervous system depression, agitation, confusion, delirium, coma, seizures, and death. It also causes bradycar-

dia, bronchospasm, vomiting, and diarrhea. Chronic exposure may damage the nerves and cause poor coordination in arms and legs, depression, anxiety, and irritability. OSHA exposure limit (8-hour TWA): 0.2 mg/m³.

fenuron ($C_9H_{12}N_2O$) — A white crystal; MW: 164.23, BP: NA, Sol: slightly soluble in water, density: NA. It is used as an herbicide, is moderately toxic by the intraperitoneal route, and may hydrolyze to aniline in the body.

Fenuron

FEPCA — Acronym for Federal Environmental Pesticide Control Act.

ferbam (($CH_3)_2NCS_2\ _3Fe$) — A dark brown to black, odorless solid; MW: 416.5, BP: decomposes, Sol: 0.01%, Fl.P: not known. It is used as a fungicide for the control of scab and rust diseases. It is hazardous to the skin, respiratory system, and gastrointestinal tract and is toxic through inhalation, ingestion, and contact. Symptoms of exposure include gastrointestinal distress, dermatitis, and irritation of the eyes and respiratory system. OSHA exposure limit (TWA): 10 mg/m³ [air].

fermentation — The glucose oxidation process in which an anaerobic chemical change is brought about by enzymes, bacteria, or other living organisms, resulting in the formation of lactic acid or alcohol and generally accompanied by the evolution of gas; also known as zymosis.

ferric iron — A compound of iron that is insoluble in water and will precipitate.

ferrihemoglobin — See methemoglobin.

ferromagnetism — A property exhibited by certain metals, alloys, and compounds in which the internal magnetic moments spontaneously organize in a common direction; gives rise to permeability somewhat greater than air.

ferrous iron — A compound of iron that is soluble in water and which will impart a clear green color.

ferrovandium dust (FeV) — A dark, odorless particulate dispersed in air; ferrovandium metal is an alloy usually containing 35 to 80% vanadium; MW: 106.8, BP: not known, Sol: insoluble. It is used in the production of steel as an additive to produce refinement and hardenability. It is hazardous to the respiratory system and eyes and is toxic through inhalation and contact. Symptoms of exposure include irritation of the eyes and respiratory system; in animals, bronchitis and pneumonitis. OSHA exposure limit (TWA): 1 mg/m³ [air].

ferrule — A metal sleeve used for joining or binding one part to another.

fertilization — The physiochemical union of sperm and egg to form a zygote.

fertilizer — A plant food, usually sold in a mixed formula, that contains basic plant nutrients, including nitrogen, potassium, soluble phosphorous, sulfur, and sometimes mineral compounds.

fetal — Pertaining to the fetus.

fetal death — The death of a product of conception prior to complete expulsion or extraction from the mother in at least the 20th week of gestation.

fetal death rate — The number of fetal deaths per 1000 total births for a specified time period.

fetal dose — The estimated amount of radiation received by a fetus during an x-ray examination of a pregnant woman.

fetid — Something that has a foul or putrid odor.

fetotoxicity — Any toxic or damaging effect on a fetus as a result of prenatal exposure to a chemical.

fetus — A mammalian embryo after the main body features become present, which, in humans, is from 7 to 8 weeks after fertilization to birth.

FeV — See ferrovandium dust.

FEV₁ — See forced expiratory volume measure at 1 second.

fever — An elevation of the central body temperature caused by abnormal functioning of the thermoregulatory mechanisms.

FF — See full-face respirator.

FFA — Acronym for Federal Facility Agreement.

FFA/CO — See Federal Facility Agreement and Consent Order.

FFV — See flexible fuel vehicle.

FGD — See flue gas desulfurization.

FHSA — Acronym for Federal Hazardous Substances Act.

fiat — A command or order directing that some legal act be done.

fiber — (*U.S. EPA definition*) A structure having a minimum length equal to 0.5 μm and an aspect ratio (length to width) of 5:1 or greater with substantially parallel sides; (*NIOSH definition, A rules*) thread-like structure longer than 5 μm measured along the curve, if applicable, with a length-to-width ratio equal to or greater than 3:1; (*NIOSH definition, B rules*) thread-like structure longer than 5 μm and less than 3 μm in diameter with a length-to-width ratio equal to or greater than 5:1.

fiberoptic — An optical fiber composed of thin glass wire designed for light transmission; the fiber is capable of transmitting billions of bits per second and is not affected by random radiation in the environment.

fiberoptics — The technical process designed to view an internal organ or cavity using glass or plastic fibers to transmit light through a specially designed tube and to reflect a magnified image.

fibers per cubic centimeter (f/cc) — The number of fibers present in a cubic centimeter of air.

fibers per cubic meter (f/m³) — The number of fibers present in a cubic meter of air.

fibrillation — See atrial fibrillation and ventricular fibrillation.

fibrin — An insoluble protein formed during blood clotting by the union of thrombin and fibrinogen.

fibrinogen — A plasma protein involved in clotting.

fibrocarcinoma — A carcinoma with a hard structure due to the formation of dense connective tissue.

fibrosis — A condition marked by a relative increase in formation of fibrous tissue in any organ or region of the body, usually in the lungs.

fibrous aerosol monitor (FAM) — A direct-reading instrument used to measure airborne fibers, such as asbestos, by use of a helium–neon laser and electrooptical sensor that detects fiber oscillations as the aerosol passes through a rapidly oscillating, high-intensity electrical field.

Fick's law — (*chemistry*) The rate at which one substance diffuses through another is directly proportional to the concentration gradient of the diffusing substance.

FID — See flame ionization detector.

field — The set of influences (electricity, magnetism, gravity) that extend throughout space.

field blank — A clean filter cassette assembly that is taken to a sampling site and handled in every way the same as for the air samples, except that no air is drawn through it.

field capacity — The maximum amount of water that a soil can retain after excess water from saturated conditions has been rapidly drained by the force of gravity; the amount of water or sewage held in soil against the pull of gravity.

field fever — A form of leptospirosis caused by *Leptospira grippotyphosa* and primarily affecting agricultural workers with symptoms of fever, abdominal pain, diarrhea, vomiting, stupor, and conjunctivitis.

field of view — The range of angles scanned or sensed by a system or instrument, measured in degrees of an arc.

field of vision — The area in which objects are visible.

field tile — The short lengths of clay pipe that are installed as subsurface drains.

FIFRA — See Federal Insecticide, Fungicide, and Rodenticide Act.

filament — A fine, thread-like fiber; a single thread or a thin flexible thread-like object.

filamentous organisms — Organisms that grow in a thread or filamentous form and are a common cause of sludge bulking in the activated sludge process; includes *Thiothrix*, *Actinomycetes*, and *Cyanobacteria*.

filaria — A parasitic nematode worm of the superfamily Filarioidea.

Filariasis
(*Wuchereria bancrofti,* cause
of filariasis and elephantiasis)

filariasis — (*disease*) A human infection by a nematode that normally develops in the lymph glands; the female worms produce microfilariae which reach the bloodstream and circulate. Symptoms of exposure include fever, lymph node disease, and swelling of the breasts and genitalia. The incubation time is 1 to 12 months, depending on the organism. It is caused by *Wuchereria bancrofti*, *Brugia malayi*, and *B. timori* and is found in Latin America, Africa, Asia, and the Pacific and among some American veterans of foreign wars. The reservoir of infection is humans. It is transmitted by the bite of a mosquito and is not communicable from person to person; general susceptibility. It is controlled by elimination of mosquitos.

file — (*computer science*) A discrete collection of related data records arranged in order for preservation and easy reference; a body of information stored on a computer disk.

filiform — Thread-like.

fill — (*solid waste*) Soil transported and deposited intentionally, as well as soil transported and deposited by natural forces and used as backfill for sanitary landfill operations.

filling — Depositing dirt, mud, or other materials into aquatic areas to create more dry land for agriculture or commercial development, resulting in serious ecological consequences.

film badge — A personal radiation monitor where radiation interacts with the silver atoms in a photographic film to expose the film the same way as light rays do; the amount of darkening of a film is then compared to a control film not exposed to radiation to determine the amount of radiation exposure.

filter — A porous mesh of spun-glass fibers, asbestos cellulose, or other materials that allow air or liquid to pass through but retain solid particles; a mechanical device for straining suspended particles from water; (*swimming pools*) any of a variety of devices through which water continuously flows for removal of particulate matter from the water; a device for removing extraneous signals or radiation.

filter aid — A powder-like substance, such as diatomaceous earth or volcanic ash, used to coat a septum pipe filter; can also be used as an aid on a sand filter.

filter background level — The concentration of structures per square millimeter of filter that is considered indistinguishable from the concentration measured on a blank filter through which no air has been drawn.

filter cake — (*wells*) A deposit of mud on the walls of a drill hole.

filter cartridge — A disposal element, usually a fibrous material, used as a filter septum in some pool filters.

filter collector — A mechanical filtration system for removing particulate matter from a gas stream for measurement, analysis, or control.

filter cycle — The time of filter operation between backwash procedures.

filter efficiency — The efficiency of a filter based on the amount of particles collected compared to the amount of particles that could be collected.

filter element — A filter cartridge or that part of a diatomite filter on which the filter aid is deposited.

filter media — Any fine-grain material carefully graded to size which entraps suspended particles as water passes through.

filter rate — The rate of flow of water through a filter during the filtering cycle; expressed in gallons per square foot of effective filter area.

filter rock — Graded, rounded rock, and/or gravel used to support filter media.

filter room — An area that houses recirculation and filtration equipment and chemical treatment units, with the exception of gas chlorinators, chemicals, and supplies for repairs.

filter run — The time of filter operation between backwash procedures.

filter sand — A type of filter media composed of hard, sharp silica, quartz, or similar particles graded for size and uniformity.

filter septum — The part of the filter in which diatomaceous earth or filter media is deposited.

filter strip — A strip of area vegetation used for removing sediment, organic matter, and other pollutants from runoff and wastewater.

filterable virus — A virus that passes through the extremely small pores of unglazed porcelain filters used to separate bacteria from fluid.

filtering — A means of reducing noise errors in a signal.

filtrate — The liquid that is passed through a filter.

filtration — The process in which suspended matter is removed from a liquid by passing the liquid through a porous material.

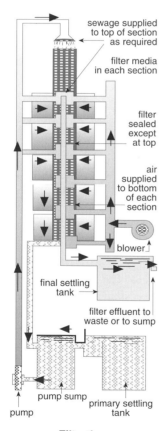

Filtration
(showing controlled filtration)

final cover — Material at a landfill that is used to permanently seal the surface.

final order — (*law*) A court order stating the final judicial action.

final sedimentation — A technique for thickening sludge.

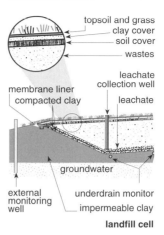

Final Soil Cover

landfill cell

final soil cover — The upper 2 feet of compacted, graded earth on top of buried solid waste; used to prevent insect and rodent problems or as a base for topsoil and grass.

financial analysis — A procedure that considers only tangible factors when evaluating various alternatives.

financial impact analysis — A study of the probable changes in the financial structure of a system that may result from a proposed or impending action.

financing — Acquiring monies for a specified use.

finding of no significant impact — A National Environmental Policy Act compliance document affirming that an environmental assessment was evaluated and that a proposed plan would not have a significant impact on human environment.

fine texture — The texture exhibited by soils having clay as a part of their textural class name.

fines — Fine particulates.

finfish — The portion of the aquatic community made up of the true fishes as opposed to invertebrate shellfish.

finished grade — The elevation or surface of the earth after all work has been completed after construction.

finished water — The water that has passed through a water treatment plant and is ready to be delivered to customers.

FIP — Acronym for final implementation plan; see Federal Implementation Plan.

fire — The phenomenon of combustion manifested in light, flame, and heat.

fire point — The lowest temperature at which a volatile, combustible material can evolve enough vapors to support combustion.

fire retardant — A treatment or coating applied to interior-finished building materials to reduce the rate of flame spread compared to that of untreated material.

fire-resistant rating — The time in hours or fraction of an hour that a material or group of materials will withstand the passage of flame and transmission of heat when exposed to fire under specified test conditions and subject to specific performance criteria.

firebox — A chamber that contains a fire; a furnace.

firebrick — Refractory brick made from fireclay and able to withstand high temperatures.

firedamp — The accumulation of an explosive gas, usually methane, in a mine.

firewall — (*computer science*) A security device that makes a computer or computer network inaccessible to the general public.

firing rate — The rate of fuel usage in boilers or other direct-fired process equipment expressed in terms of energy equivalent as BTU/hr.

first draw — The water that comes out when a faucet is first opened and which is likely to have the highest level of lead contamination from plumbing material.

first-draw sample — A 1-liter sample of tap water that has been standing in plumbing pipes at least 6 hours and is collected without flushing the tap.

first law of thermodynamics — Energy can be neither created nor destroyed; it can only be converted from one form into another — for example, mechanical energy can be converted into heat and heat into mechanical energy. Also known as conservation of energy.

first responder — The individual who is first to arrive at the scene of a medical emergency and is capable of rendering basic life support.

fiscal year — Any 12-month period for which annual accounts are kept.

fish — Any edible, commercially distributed fresh- or saltwater member of the animal kingdom classed as fish (*Pisces*).

Fish and Wildlife Coordination Act of 1958 — An act of Congress that ensures that wildlife conservation receives equal consideration and is coordinated with other features of water resource development programs. It authorizes the U.S. Fish and Wildlife Service to survey, investigate, prepare reports, make recommendations, and develop techniques to prevent the loss of, or damage to, wildlife resources.

fish poisoning — Toxic effects caused by the ingestion of fish containing substances that may produce symptoms ranging from nausea and vomiting to respiratory paralysis; see scrombroid poisoning as an example.

fish processing plant — Any food establishment or portion thereof in which fish or shellfish are handled, processed, or packed for market or shipment as fresh fish or fresh or perishable shellfish.

fish tapeworm infection — An infection caused by the tapeworm *Diphyllobothrium latum* that is transmitted to humans who consume raw or undercooked freshwater fish.

fishery — The aquatic region in which certain species of fish live.

fishing — The process of catching fish; (*wells*) the recovery of an object left or dropped in a drill hole.

fissile — Capable of being split by a low-energy neutron; relating to radioactive materials that are capable of undergoing or sustaining nuclear fission and requiring controls to ensure nuclear safety during transport. Plutonium-238, plutonium-239, plutonium-241, uranium-233, and uranium-235 are fissile.

fissile material — A specific set of nuclear materials such as uranium-235 and plutonium-239 that may be used in making a nuclear explosive.

fission — (*biology*) A method of asexual reproduction in single-celled organisms in which the cell divides in half and each half develops into a new organism exactly like the original organism; (*physics*) splitting of an atom nuclei accompanied by the release of great quantities of energy.

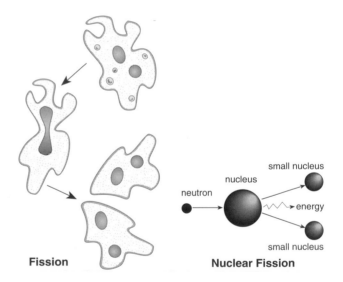

Fission   **Nuclear Fission**

fission products — The large variety of smaller atoms, including cesium and strontium, left after the splitting of uranium and plutonium.

fissure — A narrow opening or crack of considerable length and depth.

FIT — Acronym for field investigation team.

fixative — An agent used in preserving a histological or pathologic specimen so as to maintain the normal structure of its constituent elements.

fixed carbon — The ash-free carbonaceous material that remains after volatile matter is driven off during the proximate analysis of a dry solid waste sample.

fixed grate — A grate that has no moving parts; also known as a stationary grate.

fixed groundwater — Water that is held in saturated material and is not available for pumping.

fixed location monitoring — The sampling of an environmental or ambient medium for pollutants at one location continuously or repeatedly over time.

fixed noise source — A stationary device that creates sounds while fixed or motionless, including but not limited to residential, agricultural, industrial, and commercial machinery and equipment; pumps; fans; air-conditioner compressors; and refrigeration equipment.

fixed pupils — An abnormal condition in which the pupils fail to dilate or contract when stimulated.

fixer — A chemical used in processing photographic or x-ray film.

fl. oz. — See fluid ounce.

Fl.P — See flash point.

flagellates — Microorganisms that move by the action of tail-like projections.

flagellum — A flexible whip-like appendage on motile cells that is used as an organ of locomotion.

flame ionization detector (FID) — A direct-reading physical instrument in which gas is fed into a stainless steel burner mixed with hydrogen and then burned with air or oxygen; a loop of platinum is set about 6 mm above the tip of the burner. The current carried across the electrode gap is proportional to the number of ions generated during burning of the gas.

Flagella →

flame photometry — The spectrochemical measurement of light emitted from certain chemicals excited by a flame.

flame polished — The sharp or broken edges of glass rotated in a flame until the edge melts slightly and becomes smooth.

flame reactor process — A hydrocarbon-fueled flash smelting system that treats residues and wastes containing metals.

flameproofing material — A chemical that catalytically controls the decomposition of material at the flame temperature.

flammable — The capability of a substance to be set on fire or support combustion easily.

flammable aerosol — An aerosol that, when tested by federally regulated methods, yields a flame projection exceeding 18 inches at full valve opening or flashback at any degree of valve opening.

flammable chemical — Any aerosol, liquid, gas, or solid capable of combustion.

flammable gas — A gas that, at ambient temperature and pressure, forms a flammable mixture with air at a concentration of 13% or less by volume.

flammable limit — A minimum concentration of a substance below which propagation of flame does not occur on contact with a source of ignition; also known as explosive limit.

flammable liquid — Any liquid having a flash point below 100°F except any mixture having components with flash points of 100°F or higher, the total of which make up 99% or more of the total volume of mixture; a liquid that gives off combustible vapors.

flammable range — Difference between the lower and upper flammable limits, expressed in terms of percentage of vapor or gas by volume range; also known as explosive range.

flammable solid — A solid other than a federally defined blasting agent or explosive that is liable to cause fire through friction, absorption of moisture, spontaneous chemical change, retention of heat from manufacturing, or processing or which can be ignited readily and burns so vigorously and consistently as to create a serious hazard.

flange — A projecting rim of an organism or mechanical part.

flare — A fire primarily used to control hydrocarbon gases released intermittently in certain refinery and petrochemical processes; causes smokeless combustion of gases.

flash blindness — Temporary loss of sight caused by an intense light source.

flash flood — A flood of short duration with a relatively high peak rate of flow usually resulting from a high-intensity rainfall over a small area.

flash flood warning — A flash flood has been reported or is imminent.

flash flood watch — A flash flood is possible within the designated watch area.

flash point (Fl.P) — The minimum temperature at which a liquid gives off a vapor in sufficient concentration to ignite when tested by specific tests established by the American National Standard Methods; test conditions can be either open or closed cup.

flash point minimum — The lowest temperature at which a chemical will give off fumes able to burst into flames.

flashback — A flame from a torch that burns back into the tip, the torch, or the hose; also known as backfire.

flashover — The sudden spread of flame over an area when it becomes heated to the flash point; an abnormal electrical discharge.

flat — A tin can for which both ends are concave; the can remains in this condition even when it is brought down sharply on its end on a solid, flat surface; possibly caused by a loss of vacuum or in storage.

flat grain beetle (*Cryptolestes pusillius*) — One of the smallest beetles found in stored grain; a tiny reddish-brown beetle about 1/16 inch long with antennae nearly as long as the insect.

flatulence — Distention of the stomach or intestines due to gas.

flatware — Eating and serving utensils that are more or less flat such as forks, knives, and spoons.

flatworm — See fluke.

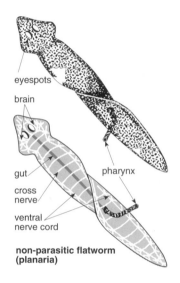

non-parasitic flatworm
(planaria)

Flatworm

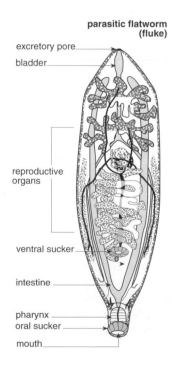

parasitic flatworm
(fluke)

flavivirus — An arbovirus spread by mosquitos and ticks which may cause fever, encephalitis, aseptic meningitis, rashes, hemorrhagic fever, and hepatitis.

flavored milk — A beverage consisting of milk to which has been added syrup or flavor made from wholesome ingredients.

FLD — Acronym for field hospital.

flea — A small, wingless, bloodsucking ectoparasite that may act as a disease carrier.

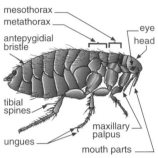

Rat Flea
(Xenopsylla cheopis)

flea bite — A small puncture wound produced by a bloodsucking flea.

fleaborne typhus — See murine typhus.

flexible fuel vehicle — A vehicle that can use a combination of fuels such as alcohol and unleaded gasoline.

flextime — A management technique that allows staff members to have individualized work schedules with different work hours and different work days.

flipper — A tin can that normally appears flat; however, when brought down sharply on its end on a flat surface, one end flips out. When pressure is applied to this end, it flips in again and can appear flat.

FLM — Acronym for federal land manager.

floating matter — Matter that passes through a 2000-μm sieve and separates by use of flotation for an hour.

floc — (*sewage*) A clump of solid formed in sewage by biological or chemical action. Also known as flocculent.

flocculation — The process by which suspended colloidal or very fine particles are assembled into large masses or floccules, which eventually settle out of suspension; the stirring of water after coagulant chemicals have been added to promote the formation of particles that will settle; (*sewage*) the process by which clumps of solids in sewage are enlarged by chemical, physical, or biological action.

flocculent — A compound, usually some type of alum, used with sand-type filters to form a thin layer of gelatinous substance on top of the sand; also known as floc.

flood — The peak flow that goes over the banks of a stream channel; an overflow of water on lands not normally covered by water for which the inundation is tem-

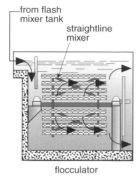

Flocculation

porary and comes from the overflow of a river, stream, lake, ocean, or blocked stormwater drains.

flood base elevation — The elevation, in terms of the U.S. Geological Survey mean sea level, that establishes the 100-year flood limit.

flood gate — A gate to control flood releases from a reservoir.

flood hazard area — Any area composed of floodplain land.

floodplain — The area contiguous to a lake, water course, stream or stream bed, or depressional pocket or area, the elevation of which is greater than the normal water level or pool elevation but equal to or lower than the flood base elevation; the part of a stream valley that is covered with water during the flood stage.

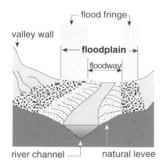

Floodplain

flood stage — An established gauge height within a given river above which a rising water surface level is defined as a flood.

floodway — A channel built to carry excess water from a stream.

floppy disk — See diskette.

flora — Plant life present in a given environment.

flotation — Process used to separate certain ores from rock by wetting.

flow — Volume of water that passes a given point within a given period of time.

flow chart — A graphic presentation of the components of a system and how the process flows.

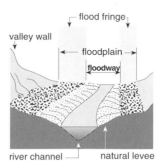

Floodway

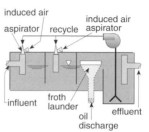

induced air flotation

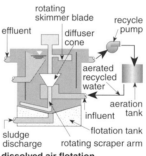

dissolved air flotation

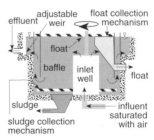

circular dissolved air flotation unit

Flotation

flow coefficient — A correction factor used for calculating volume flow rate of a fluid through an orifice, including the effects of contraction and turbulence loss, the compressibility effect, and the effect of an upstream velocity other than zero; (*indoor air quality*) a parameter used in conjunction with the flow exponent in a flow equation.

flow diversion valve (FDV) — (*milk*) A three-way valve automatically controlling product flow where forward flow position occurs when a recorder controller microswitch preset to operate at 161°F energizes a solenoid-operated air valve, permitting compressed air to flow across a diaphragm and forcing it to sit securely on the diverted flow line.

flow equalization system — A device or tank designed to store a portion of the peak flow of a liquid for release during low flow periods.

flow equation — (*indoor air quality*) An equation describing the air flow rate through a building or component in response to the pressure difference across the building or component. The equation is $Q = C\Delta pn$, where Δp is the change in pressure over the component or envelope, n is the flow exponent, and Q represents the resulting volume flow rate expressed in m³/hr.

flow exponent — (*indoor air quality*) A parameter that characterizes the type of flow through a building or component and is used in conjunction with the flow coefficient in a flow equation; when the flow exponent $n = 1$ the flow is laminar, and when $n = 0.5$ the flow is assumed turbulent. Typically, for most openings, the n value is between these two extremes.

flow hood — A device that measures air flow quantity.

flow network — (*indoor air quality*) A network of zones or cells of different pressures connected by a series of flow paths; a passage or duct for smoke or fumes from a boiler or fire, etc.

flow rate — The volume per unit time of a flow of gas or other fluid substance that emerges from an orifice, pump, or turbine or passes along a conduit or channel.

flow technique — The flow of a substance from the point of delivery to the point of production of the product and byproducts, as well as waste products and ultimate disposal. The importance of the flow technique is to determine potential problems that may occur at each point of flow and how to deal with them in an appropriate manner.

flowing spring — A spring created by water under pressure under the ground which rises above the surface.

flowing well — A well that taps groundwater under pressure so that water rises without pumping.

flowmeter — An instrument used for measuring the flow or quantity of moving liquids or gases.

FLPMA — See Federal Land Policy and Management Act.

flue — A conduit made of masonry material or other approved, noncombustible, heat-resistant material that is used to remove the products of combustion from solid, liquid, or gaseous fuel.

flue-fed incinerator — An incinerator that is charged through a shaft that functions as a chute for waste and has a flue to carry away the products of combustion.

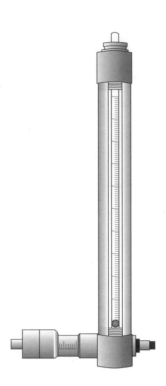

Flowmeter

flue gas — Waste gas from a combustion process.

flue gas desulfurization (FGD) — Any pollution control process that treats stationary source combustion flue gas to remove sulfur oxides.

flue gas scrubber — A piece of equipment that removes fly ash and other objectionable materials from flue gas by the use of sprays, wet baffles, or other means that require water as the primary separation medium.

fluid — The gas or liquid phase of matter in which molecules are able to flow past each other without limit.

fluid balance — A state in which the volume of body water and its solutes, both electrolytes and non-electrolytes, is within normal limits and fluids are distributed normally within the intracellular and extracellular compartments.

fluid ounce (fl. oz.) — A measure of liquid volume equal to 29 ml.

fluidize — To cause to flow like a fluid; to suspend solid particles in a rapidly moving stream of gas or vapor to induce flowing motion of the whole.

fluidized — A mass of solid particles that is made to flow like a liquid by injection of water or gas.

fluidized bed combustion — A method of burning particulate fuel, such as coal, in which high-pressure air is forced through a mixture of crushed coal and limestone particles, causing the burning fuel to move like a boiling liquid; the limestone combines

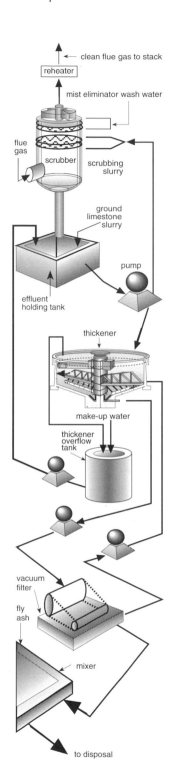

Flue Gas Desulfurization
(lime-limestone process)

with the sulfur oxide released in combustion to form calcium sulfate.

fluidized bed technique — A combustion process in which heat is transferred from finely divided particles, such as sand, to noncombustible materials in a combustion chamber; the materials are supported and fluidized by calm or moving air.

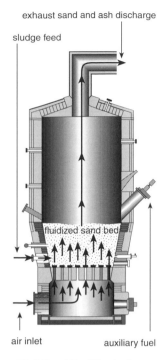

Fluidized Bed Technique

fluke — Common name for more than 40,000 parasitic organisms of the class Trematoda, characterized by a body that is usually flat and often leaf like, that can infect the blood, liver, intestines, and lungs; also known as parasitic flatworm.

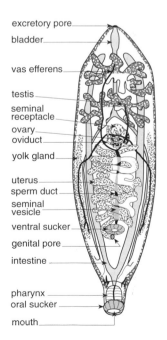

Fluke

flume — A natural or synthetic channel that diverts water from a natural course for power, industry, and so on.

fluoranthene ($C_{16}H_{10}$) — An odorless, pale-yellow needle-like structure or plate; MW: 202.3, BP: 375°C, Sol: 0.20 to 0.26 mg/100 g water at 20°C, Fl.P: NA (may be decomposed by heat and emit acrid smoke and fumes), sp. gr. 1.25. It is used as lining material to protect the interior of steel and ductile-iron potable water pipes and storage tanks. It is hazardous to the skin and the eyes. Acute exposure may cause severe irritation and burns to the skin, eyes, and mucous membranes. It also causes acute poisoning with dizziness, lightheadedness, headache, weakness, and nausea. Chronic exposure may cause skin and eye disorders and damage kidneys. It has been associated with increased incidents of lung, bladder, and kidney cancer. OSHA exposure limit (8-hour TWA): 0.2 mg/m^3.

Fluoranthene

fluorescein ($C_{20}H_{12}O_5$) — A yellow or red crystalline dye with a bright yellow-green fluorescence in alkaline solution.

fluorescence — The emission of electromagnetic radiation, usually visible, as the immediate result of absorption of energy from another source.

fluorescent lamp — A high-efficiency lamp utilizing an electric discharge through low-pressure mercury vapor to produce ultraviolet energy, which excites phosphor materials applied as a thin layer on the inside of the glass tube. The phosphors transform the ultraviolet light to visible light.

fluorescent light — A form of lighting that uses long, thin tubes of glass containing mercury vapor and various phosphoric powders to produce white light; considered to be the most efficient form of home lighting.

fluorescent screen — A screen coated with a fluorescent substance that emits light when irradiated with x-rays.

fluoridation — The addition of fluoride to the municipal water supply at a rate of 1 part per million as part of the public health program to prevent or reduce the incidence of dental caries and strengthen brittle bones.

fluoride (NaF and Na_3AlF_6) — NaF is an odorless white powder or colorless crystal, and the pesticide grade is often dyed blue; MW: 42.0, BP: 3099°F, Sol: 4%, sp. gr. 2.78. Na_3AlF_6 is a colorless to dark, odorless solid that loses color on heating; MW: 209.9, BP: decomposes, Sol: 0.4%, sp. gr. 2.90. They are used in the manufacture of cryolite from phosphate rock processing; in the formulation of insecticides; in the chemical industry; in the treatment of water; in the manufacture of fiberglass, optical glass, lenses, and pottery; in the manufacture of welding rods and welding fluxes. They are hazardous to the eyes, respiratory system, central nervous system, skeleton, kidneys, and skin and are toxic through inhalation, ingestion, and contact. Symptoms of exposure include nausea, abdominal pain, diarrhea, excessive salivation, thirst, sweating, stiff spine, dermatitis, calcification of ligaments of the ribs and pelvis, and irritation of the eyes and respiratory system. OSHA exposure limit (TWA): 2.5 mg/m^3 [air].

fluorine (F_2) — A pale-yellow to water-white, nearly odorless liquid or gas (above 75°F) and member of the halide family; MW: 38.0, BP: –307°F, Sol: reacts, atomic weight: 9. It is used during organic and inorganic synthesis in the production of fluorine compounds, refrigerants, plastics, incendiary devices, and electrolytic solvents; as a rocket fuel oxidizer for ammonia, hydrazine, hydrogen, JP-4, and diborane fuels. It is hazardous to the respiratory system, skin, and eyes; in animals, it is hazardous to the liver and kidneys and is toxic through inhalation and contact. Symptoms of exposure include laryngeal spasms; bronchial spasms; pulmonary edema; eye and skin burns; irritation of the eyes, nose, and respiratory system. In animals, it causes liver and kidney damage. OSHA exposure limit (TWA): 0.1 ppm [air] or 0.2 mg/m^3.

fluoroacetamide — (*rodenticide*) A rodenticide that causes convulsions, cyanosis, ventricular fibrillation, and death; also known as 1081.

fluorocarbon — A hydrocarbon, such as Freon, in which all or part of the hydrogenations have been replaced by fluorine; used as a propellant in aerosols, refrigerant, and solvent. It is thought to be modifying the ozone layer by allowing increased levels of harmful solar radiation to reach the surface of the Earth.

fluorochrome stain — A technique used to stain a clinical specimen with fluorescent dyes to perform a microscopic examination for mycobacterium.

fluorography — Photographic examination by means of a device that exposes deep body structures by using x-rays produced on a fluorescent screen.

fluorophotometry — The measurement of light given off by fluorescent substances.

fluoroscope — A fluorescent screen mounted with respect to an x-ray tube for ease of direct observation of internal organs or other internal structures in materials; a device for examining deep tissues by means of roentgen rays.

fluorosis — An abnormal condition caused by excessive intake of fluorine, characterized chiefly by mottling of the teeth.

fluorotrichloromethane (CCl₃F) — A colorless to waterwhite, nearly odorless liquid or gas (above 75°F); MW: 137.4, BP: 75°F, Sol: 0.1% at 77°F, sp. gr. 1.47. It is used as a propellant in aerosols for insecticides, floor waxes, paint, cosmetics, and perfumes; as a refrigerant; as a blowing agent in foam plastics; as a solvent and degreaser; in the production of polymeric resins; as a dielectric fluid in bubble chambers and in wind tunnels; as a sulfonation solvent in chemical synthesis. It is hazardous to the cardiovascular system and skin and is toxic through inhalation and contact. Symptoms of exposure include incoordination, tremors, dermatitis, frostbite, cardiac arrhythmia, and cardiac arrest. OSHA exposure limit (TWA): ceiling 1000 ppm [air] or 0.2 mg/m³.

flush — To open a coldwater tap to clear out all the water that may have been sitting for a long time in pipes; to force large amounts of water or other liquid through piping, tubing, or tanks in order to clean them; (*physiology*) a sudden reddening of the face and neck.

flush valve — A device located at the bottom of the tank of flushing water closets and similar fixtures.

flushing — A method used to clean water distribution lines by passing a large amount of water through the system.

flushing water closet — A toilet bowl that is flushed with water supplied under pressure and equipped with a water-sealed trap above floor level.

flushometer valve — A device activated by direct water pressure that discharges a predetermined quantity of water to fixtures for flushing purposes.

fluvial — Referring to streams and stream processes; produced by the action of the stream or river.

flux — A material or mixture of materials used in soldering and welding that causes other compounds with which it comes into contact to fuse at a temperature lower than their normal fusion temperatures and to prevent the formation of oxides; a substance that helps to melt and remove the solid impurities as a slag; the flow of a liquid.

flux-cored arc welding — An arc welding process that produces coalescence of metals by heating them with an arc between a continuous filler metal electrode and the work.

flux density — An expression of the number of neutrons entering a sphere of unit cross-sectional area in unit time.

fly — See *Musca domestica*.

fly ash — Fine, solid-particulate refuse including ash, charred paper, cinders, dust, soot, or other partially incinerated matter carried in a gas stream from a furnace.

fly bite — A bite that may be caused by a species of deer, horse, or sand fly and produces a small, painful wound with swelling due to substances in the saliva that are injected beneath the surface of the skin.

f/m³ — See fibers per cubic meter.

FM — See frequency modulation.

F/M — A ratio of the amount of food to the number of organisms used to control an activated sludge process.

FMAP — Acronym for Financial Management Assistance Project.

FML — Acronym for flexible membrane liner.

FMLA — Acronym for Family and Medical Leave Act.

FMP — Acronym for facility management plan.

foam — A colloidal dispersion of gas bubbles on the surface of a liquid.

foam resins — Plastic materials rendered light and porous by the inclusion of minute gas bubbles.

focal point — The point at which parallel rays of light converge after passing through a lens or mirror.

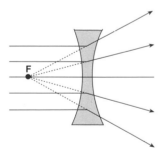

Focal Point

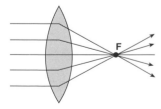

focus — The site of a disease process; (*optics*) to move an optical lens toward or away from an object to obtain the sharpest image.

fog — Suspended liquid particles formed by condensation of vapor in sufficient quantities to reduce visibility.

fog chamber — See cloud chamber.

fogging — (*pesticides*) Application of a pesticide chemical by rapidly heating the liquid form, resulting in very fine droplets with the appearance of smoke.

FOIA — Acronym for Freedom of Information Act.

Foley catheter — A tube inserted into the bladder for the purpose of draining and measuring urine output.

foliage — The leaves, needles, and blades of plants.

folic acid — A yellow, crystalline, water-soluble vitamin of the B-complex group, essential for cell growth and development.

follicle — A sac or pouch-like depression or cavity.

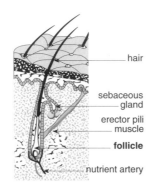

Hair Follicle

folliculitis — An infection of a hair follicle often caused by a natural or industrial oil obstruction.

fomite — An inanimate object that can harbor or transmit pathogenic organisms.

fonofos ($C_{10}H_{15}OPS_2$) — A pale-yellow liquid; MW: 246.32, BP: 130°C, Sol: readily miscible with organic solvents, density: 1.16 at 25°C. It is a cholinesterase inhibitor that is used as a pesticide. It is highly toxic and hazardous to the skin, digestive tract, and respiratory tract. It produces acute, delayed, and chronic symptoms that include headache, blurred vision, muscle spasms, vomiting, abdominal pain, diarrhea, seizures, respiratory arrest, and death. ACGIH exposure limit (TLV-TWA): 0.1 mg/m³ skin.

$$C_2H_5O \quad S \qquad \qquad$$
$$\text{P} - \text{O} - CH_2 - CH_3$$
$$\text{S}$$

Fonofos

FONSI — Acronym for finding of no significant impact.

font — A set of type of one size and design.

food — Any substance ingested by an organism that yields material for energy, growth, and repair of tissue and regulation of the life processes.

food additive — A substance, usually a chemical, added to the production, processing, packaging, or cooking of food to preserve, protect, or enhance flavor, nutritive content, or appearance.

food allergy — A hypersensitive state resulting from the ingestion of a specific food antigen.

food and drug interaction — An adverse health effect resulting from the combination of certain foods and medications.

food chain — The feeding habits of organisms in an ecological community based on trophic levels; results in the transfer of food energy from plants or organic detritus organisms.

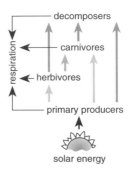

Food Chain

food contact surface — Surface of equipment, utensils, containers, or wrappings that comes in direct contact with food.

food contaminants — Any substance that makes food unfit for human consumption including bacteria, toxic chemicals, carcinogens, radioactive material, and debris.

food handler — Any individual who comes into contact with food or utensils used in connection with the storage, manufacture, preparation, handling, sale, or service of food.

food hypersensitivity reaction — See food allergy.

food infection — The introduction of infectious organisms into the body by means of food.

food poisoning — A group of acute illnesses due to ingestion of contaminated food carrying toxins inherent in the food or formed by bacteria; usually causes inflammation of the gastrointestinal tract, pain, cramps in the abdomen, nausea, vomiting, diarrhea, weakness, and dizziness; the disease is relieved when the toxin leaves the system.

food preservation — Processing designed to keep food clean, fresh, and wholesome in such a way that it will not lose nutritive value or acquire injurious properties.

food processing establishment — A commercial establishment in which food is manufactured or packaged for human consumption.

food processing waste — The combination of wastes resulting from the growing, harvesting, processing, and packaging of fruits, vegetables, and other food crops.

food protection — A technique used to prevent foodborne disease consisting of proper plan review and development, use of adequate facilities, procurement and use of correct equipment, necessary cleanliness, proper storage, proper temperatures, safe preparation and serving of food, removal of infected individuals, and adequate disposal of food remnants.

Food Quality Protection Act — Pesticide food safety legislation passed in 1996 by the U.S. Congress. It is meant to update and resolve inconsistencies in the Federal Insecticide, Fungicide, and Rodenticide Act and the Federal Food, Drug, and Cosmetic Act. It makes it mandatory to use a single, health-based standard for all pesticides and all foods; provides special protection for infants and children; expedites approval of safer pesticides; creates incentives for the development and maintenance of effective crop protection tools for American farmers; and requires periodic reevaluation of pesticide registrations and tolerances to ensure that the scientific data supporting the pesticide registrations will remain up to date into the future.

food safety — Protection of the food supply from microbial, chemical, and physical hazards or contamination at all stages of food production, handling, growing, harvesting, processing, transporting, preparing, distributing, and storing.

food service establishment — Any place where food is prepared and intended for individual portion service.

food spoilage — Deterioration of the color, texture, odor, or flavor of a food caused by chemicals including enzymes, bacteria, molds, and yeast.

food-vending vehicle — Any conveyance used by a food vendor offering food for human consumption.

food vendor — Any person who sells, offers for sale, or gives away from any vehicle, container, or public market any article of food for human consumption.

food web — The many connected food chains by which the organisms of a particular ecological community obtain resources of energy.

foodborne bacterial poisoning — Bacterial poisoning resulting from the ingestion of food on or in which bacterial organisms have grown and produced toxins.

foodborne chemical poisoning — Chemical poisoning characterized by a short interval (varying from 10 minutes to 2 hours) between ingestion of the food containing a specific chemical or chemicals and the onset of symptoms.

foodborne disease outbreak — An incident in which two or more persons have a similar illness after ingesting a common food and epidemiological analysis implicates the food as the source of the illness; a single case of illness, such as one case of botulism or chemical food poisoning.

foodborne infection — A disease caused by the ingestion of food in which large numbers of specific microorganisms have grown.

fool's gold — See pyrite.

foot (ft) — A unit of measurement equivalent to 12 inches or 0.3048 meters.

foot bath — An area of shallow water between bathhouse showers and a pool deck through which pool patrons must walk.

foot-pound (ft-lb) — The unit of work done by a force of 1 pound acting through a distance of 1 foot; equal to approximately 1.355818 joules.

foot-pound of torque — A measurement of the physiological stress exerted upon any joint during the performance of a task; also known as pound-foot.

foot spray — A device for spraying bathers' feet with water or a disinfectant in public bathing areas.

foot valve — (*plumbing*) A special type of check valve located at the bottom end of a suction pipe that opens when the pump operates to allow water to enter the suction pipe and closes when the pump shuts off to prevent water from flowing out of the suction pipe.

footcandle (fc) — A unit of illumination equal to 1 lumen/ft^2; it is a measure of the brightness of light 1 foot away from a 1-candela light source.

footing — The part of the foundation of a structure that transmits loads directly to the soil.

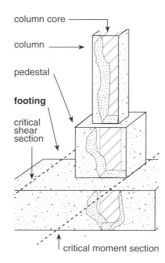

Footing

footlambert (fL) — A unit of brightness; 1 lumen uniformly emitted or reflected by an area of 1 square foot.

forage — Plants grown to feed dairy animals, beef cattle, horses, sheep, and goats.

forb — A weed or a broadleaf plant.

force — (*physics*) Any influence that causes a body to move or accelerate.

forced-air cooling — The rapid movement of refrigerated air over food in special rooms.

forced draft — Positive pressure created by the action of a fan or blower that supplies primary or secondary combustion air in an incinerator.

forced expiratory volume measure at 1 second (FEV$_1$) — A test of pulmonary function used for routine monitoring.

forced outage — Unscheduled shutdown of a generating unit or other facility for emergency or other unforeseen circumstances.

forced vital capacity (FVC) — The maximum gas volume that can be expired immediately after a maximum expiration; used as a test of pulmonary function.

forecast year — Any 12-month period in the future for which an estimation or calculation of something of interest is made.

foreign body — Any object or substance found in the body or an organ or tissue in which it does not belong under normal circumstances.

foreseeable emergency — Any potential occurrence such as, but not limited to, equipment failure, rupture of containers, or failure of control equipment, which could result in an uncontrolled release of a hazardous chemical into the workplace.

foreshocks — The relatively smaller earthquakes that precede the largest earthquake in a series, called the main shock.

forfeiture of bond — Failure to perform the conditions for which an obligor was to be excused of the penalty in the bond.

formaldehyde (HCHO) — A nearly colorless gas with a pungent, suffocating odor that is often used in an aqueous solution; MW: 30.0, BP: 207 to 214°F, Sol: miscible, Fl.P: 140 to 185°F, sp. gr. 1.08 to 1.10 at 77°F. It is used in the synthesis of chelating agents and dyes; in textile manufacturing and handling; in tanning operations; in the manufacture of particleboard, softwood plywood, sandpaper, and grinding wheels; as an embalming fluid. It is hazardous to the respiratory system, skin, and eyes and is toxic through inhalation, ingestion, and contact. Children, the elderly, and individuals with lung disease or poorly functioning immune systems are particularly affected. Symptoms of exposure include lacrimation, burning nose and cough, bronchial spasms, pulmonary irritation, dermatitis, and irritation of the eyes, nose, and throat; carcinogenic. OSHA exposure limit (TWA): 1 ppm [air].

formalin — A clear, 37% solution of formaldehyde and water used for fixing and preserving biological specimens for pathological and histological examination.

format — The arrangement of data in a systematic manner.

formazin — A polymer suspension used as a standard for turbidity.

formazin turbidity unit — A measure of water turbidity equivilant to but not equal to Jackson turbidity units.

formic acid (HCOOH) — A colorless liquid with a pungent, penetrating odor that is often used in an aqueous solution; the 90% solution freezes at 20°F; MW: 46.0, BP: 224°F, Sol: miscible, Fl.P (open cup): 122°F, sp. gr. 1.22. It is used in the textile dyeing and finishing industries as a dye-exhausting agent; in chrome dyeing; to impart finishes on cotton; as an acidifying agent; as a shrink- and wrinkleproofing agent; as a chemical intermediate for acids, salts, dyes, fumigants, refrigerants, pharmaceuticals, and solvents; in the leather processing industry as a deliming agent and neutralizer; in the rubber industry as a coagulant for natural rubber latex; in electroplating to control particle size and plating thickness; as an antiseptic in wine and beer brewing; as a preservative in animal feed additives; in miscellaneous operations. It is hazardous to the respiratory system, skin, kidneys, liver, and eyes and is toxic through inhalation, ingestion, and contact. Symptoms of exposure include eye and throat irritation, lacrimation, nasal discharge, cough, dyspnea, nausea, skin burns, dermatitis. OSHA exposure limit (TWA): 5 ppm [air] or 9 mg/m^3. Also known as methanoic acid.

formula — (*chemistry*) A set of numerals and other symbols expressing the constituents of a chemical compound.

formula translation — See FORTRAN.

formulation — (*pesticides*) A mixture of one or more pesticides plus other materials required to make it safe and easy to store, dilute, and apply.

Fort Bragg fever — See pretibial fever.

fortified milk — A pasteurized milk containing one or more added nutrients.

FORTRAN (formula translation) — (*computer science*) A high-level programming language used in computer graphics and computer-aided design/computer-aided manufacturing.

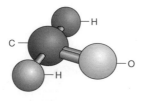

Formaldehyde

fossil — The hardened remains or traces of plant or animal life from a previous geological period preserved in the Earth's crust.

fossil fuels — Coal, oil, and natural gas; remnants of ancient plants and animals having high carbon or hydrogen contents and typically containing varying levels of sulfur as a contaminant.

FOT — Acronym for Fourier optical transform.

fouling — Impedance to the flow of gas or heat that results when material accumulates in gas passages or on heat-absorbing surfaces in an incinerator or other combustion chamber; (*naval architecture*) the adhesion of marine organisms to the underwater parts of ships, causing ships to lose speed.

foundation — The lower part of a structure which transmits the load to the earth.

foundation drain — The portion of a residential drainage system provided to drain only groundwater from outside the foundation of the house or from under the basement floor; tile or pipe for collecting seepage within a foundation.

Fourier analysis — (*mathematics*) a method of dissociating time series or spatial data into sets of sine and cosine waves.

Fourier number — A dimensionless number with the equation of $FO = (at)/lO2$, where a is the thermal diffusivity, lO is a specific dimension, and t is time.

Fourier transform — A mathematical procedure that separates out the frequency components of a signal from its amplitudes as a function of time or vice versa.

FOV — Acronym for field of view; field of vision.

fp — See freezing point.

FP — Acronym for fine particles.

FPAR — See fraction of photosynthetically active radiation.

FPD — Acronym for flame photometric detector.

FPEIS — Acronym for Fine Particulate Emissions Information System.

FPLA — Acronym for Fair Packaging and Labeling Act.

FPM — Acronym for federal personnel manual or flow velocity in feet per minute.

FPPA — Acronym for Federal Pollution Prevention Act.

FPRS — Acronym for Federal Program Resources Statement.

fps — Acronym for feet per second.

FQPA — See Food Quality Protection Act.

FR — Acronym for *Federal Register*, final rulemaking, or flame resistant.

FRA — Acronym for Federal Register Act.

fraction of photosynthetically active radiation — Radiation between 400 and 700 nm used by the green canopy in the photosynthetic process.

fractional distillation — The process of separating a mixture of liquids having different boiling points.

fractional efficiency — The percentage of particles of a specific size that are removed and retained by a particular type of collector or sampler.

fractionation — The separation of a mixture into two different portions.

fractionation dose — A method of administering radiation in which relatively small doses are given daily or at longer intervals.

fracto clouds — Fragmented or wind-blown clouds.

fracture — A break in a rock formation due to structural stresses; also known as faults, shears, joints, and planes of fracture cleavage; a crack or break in rocks along which no movement has occurred; a traumatic injury to a bone where the continuity of the tissue of the bone is disrupted.

franchise collection — Collection made by a private firm that has been given the right by a governmental agency to collect solid waste for a fee.

FRC — See functional residual capacity.

FREDS — Acronym for Flexible Regional Emissions Data System.

free available residual chlorine — The portion of the total available residual chlorine composed of dissolved chlorine gas, hypochlorous acid, and/or hydrochlorite ions remaining in water after chlorination.

free energy — The energy available for doing work in a chemical system.

free field — (*noise*) Distance from a noise source where the sound pressure level decreases 6 decibels on the A scale for each doubling of distance.

free groundwater — Groundwater in aquifers not bounded by or confined in impervious strata.

free radical — An electrically neutral fragment of a molecule having at least one impaired electron or an unused chemical bond.

free residual chlorination — Chlorination that maintains the presence of hypochlorous acid (HOCl) or hypochlorite ion (OCl^-) in water.

freeze-out sampling — The process by which an air sample is drawn through an extremely cold chamber and condensed contaminants such as hydrocarbons are collected.

freeze–thaw damage — The damage to concrete caused by extreme temperature variations, as evidenced by random crack patterns.

freezing condensation — A process that occurs in the clouds when ice crystals trap water vapor, become larger and heavier, and fall as rain or snow.

freezing point (fp) — The temperature at which a liquid changes to a solid at normal pressure.

freight bill (F/B) — The document stating the transportation charges by the carrier for a shipment; next to the bill of lading, the freight bill is considered the most authentic documention of movement of a shipment in interstate commerce.

F

French drain — A covered ditch containing a layer of loose stone or other pervious material; (*hazardous waste*) a chemical disposal well.

Freon — The commercial name for chlorofluorocarbon.

Freon-12 — See difluorodichloromethane.

Freon-12B1 — See bromochlorodifluoromethane.

frequency (*f*) — The number of waves, vibrations, or cycles that pass any given point in a certain period of time (usually 1 second) expressed in hertz.

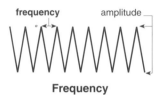

Frequency

frequency band — A particular range of frequencies.

frequency modulation (FM) — (*electronics*) A deliberate modulation of the frequency of the transmitting wave in broadcasting in order to agree with the sounds or images being transmitted to reduce static.

frequency spectrum — The distribution of signal amplitudes as a function of frequency.

fresh meat — Regular retail cuts of beef, veal, lamb, and pork.

fresh water — Water that generally contains less than 1000 mg dissolved solids per liter water; however, more than 500 mg dissolved solids per liter water is considered undesirable for drinking and for many industrial uses. Fresh water covers about 2% of the surface of the Earth and includes streams, rivers, lakes, ponds, and water wells.

friable — Easily crumbled or pulverized.

friable asbestos — A material containing more than 1% asbestos that can be crumbled or reduced to powder by hand.

friction — The force opposing the motion of an object relative to another with which it is in contact.

friction burn — A tissue injury caused by abrasion of the skin.

friction factor — A factor used in calculating loss of pressure due to friction of a fluid flowing through a pipe or duct.

friction head — Pressure lost by the flow in a stream.

friction loss — The loss in energy of a substance when it encounters resistance along any surface.

FRM — Acronym for Federal Reference Methods.

FRN — Acronym for Federal Register Notice; Final Rulemaking Notice.

front — (*weather*) The boundary at which air masses of different temperature and density collide to produce very unstable weather.

front-end loader — A waste collection vehicle with forward arms that engage a detachable container, move it up over the cab and into the body of the vehicle, and then return it to the ground.

frost — A covering of ice caused by the sublimation of water on objects colder than 32°F.

frost line — The greatest depth at which ground may be expected to freeze.

frostbite — Injury to tissues due to exposure to extreme cold.

Froude number — (*geology*) The ratio of inertial forces to gravitational forces in flow; also, the ratio of flow velocity to the velocity of a small gravity wave in the flow. When the Froude number is <1, the flow is tranquil; when the flow is >1, the flow is rapid.

frozen dessert — Any frozen or partially frozen combination of two or more milk or milk products, eggs or egg products, sugar, water, fruit or fruit juices, candy, nut meats, flavors, color, and harmless stabilizers; includes ice cream, frozen custard, ice milk, milk sherbet, ices, and other products.

frozen food processing plant — Any food establishment or portion thereof where food is prepared, processed, packed, stored for sale or shipment, or held in storage as a frozen product.

FRP — Acronym for fiberglass-reinforced plastic; see Federal Response Plan.

FRS — Acronym for formal reporting system.

fructose ($C_6H_{12}O_6$) — A sugar in which the molecules contain the same number and kind of atoms as the glucose molecule but in different arrangements; the sweetest of sugars, found in the free state in fruit juices, honey, and nectar.

fruit sherbets — Foods that are prepared by freezing while stirring a mix composed of one or more of the optional fruit ingredients and one or more of the pasteurized dairy ingredients, sweetened with one or more of the optional saccharin ingredients: sugar, dextrose, or corn syrup, with or without added water.

FS — Acronym for a feasibility study.

FSA — Acronym for Food Security Act.

ft — See foot.

ft-lb — See foot-pound.

FTE — See full-time equivalency.

FTIR — Acronym for Fourier transform infrared.

FTP — Acronym for federal test procedure; (*computer science*) acronym for file transfer protocol, an efficient way to transfer files from one computer to another over the Internet.

FTS — Acronym for Fourier transform spectrometer, acronym for file transfer service.

FTTS — Acronym for FIFRA/TSCA Tracking System.

FTU — See formazin turbidity unit.

FUA — Acronym for Fuel Use Act.

fuel — Any material that is capable of releasing energy or power by combustion or other chemical or physical reaction.

fuel cell — A device for converting chemical energy into electrical energy.

fuel efficiency — Proportion of the energy released on combustion of a fuel that is converted into useful energy.

fuel oil — An oil that is heavy-distilled in the refining process and is used to supply energy to power stations and factories.

fuel pellets — The uranium fuel for nuclear reactors in the form of ceramic cylinders about 1/2 inch long and 3/8 inch in diameter; they are stacked in long tubes to form fuel rods.

fuel rod — A long, slender tube that holds fuel pellets assembled into bundles called fuel elements which are loaded into the reactor core.

fuel switching — A precombustion process whereby a low sulfur fuel is used in place of a high sulfur fuel to reduce sulfur dioxide emissions.

fugitive dust — Particulate matter composed of soil that is uncontaminated by pollutants resulting from industrial activity.

fugitive emissions — Any air pollutants emitted to the atmosphere other than from a stack.

fulcrum — A pivot for a lever.

full hydraulic capacity — The designed capacity of a pipe or conduit.

full-face respirator — A respirator with a facepiece that covers the entire face from under the chin to the forehead and can be attached to either an air-purifying or air-supplied system.

full-time equivalency (FTE) — A standard of hours per year used to determine the standard number of employees; the actual time spent delivering direct services in the field by one person over the course of a year.

Fuller's earth — A hydrated silica-alumina compound used as a filter medium.

fully closed position — Closed position of a valve seat that requires maximum movement of the valve to reach the fully open position.

fully open position — Position of a valve seat that permits maximum flow into or out of the holder.

fulminate — To come suddenly and follow a severe, intense, or rapid course.

fume fever — See metal fume fever.

fumes — Fine, solid particles (under 1 μm in diameter) derived from condensing the volatization of molten metals from the gaseous state; they cause metal fume fever, allergic responses, and other respiratory-type problems.

fumigant — A pesticide chemical compound that burns or evaporates to form a gas or vapor that destroys pests.

fumigation — The use of a fumigant to destroy pests.

function — An act, process, or series of processes that serve a special purpose.

functional antagonism — An effect occurring when two chemicals counterbalance each other by producing opposite reactions to the same physiological function.

functional assessment — Sex health history.

functional disease — A disease that affects performance.

functional group — A relatively reactive group of atoms responsible for the characteristic reactions of a compound such as –COOH, –OH, or –NH$_2$. These may be attached to a relatively inert hydrocarbon group, and their presence imparts characteristic properties to the molecule.

functional residual capacity (FRC) — The volume of gas remaining in the lungs at the end of a normal expiration.

fungal — Having the characteristics of fungi.

fungal infection — An inflammatory condition caused by a fungus.

fungal septicemia — A blood poisoning caused by a fungus.

fungemia — The presence of fungi in the blood.

fungi — Spore-bearing plants lacking chlorophyll, such as mushrooms, molds, and yeast.

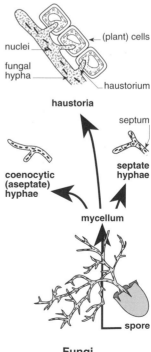

Fungi

fungicide — A chemical that kills the non-green microscopic plants, fungi.

fungistat — A chemical that stops fungi from growing.

fungus — Singular of fungi.

furan (C_4H_4O) — A colorless liquid; MW: 68.08; BP: 31.5°C; Sol: soluble in alcohols and ether; Fl.P: –35° C (closed cup), the vapor is heavier than air and can travel a long distance to a source of ignition and flash back, and it is a highly flammable liquid; density: 0.937 at 20°C. It is used as an intermediate in organic synthesis. It is a highly toxic compound and is hazardous to the respiratory tract. It causes acute pulmonary edema and lung damage. ACGIH exposure limit: NA.

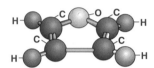

Furan

furfural (C_4H_3OCHO) — A colorless to amber liquid that darkens in light and air and has an almond-like odor; MW: 96.1, BP: 323°F, Sol: 8%, Fl.P: 140°F, sp gr: 1.16. It is used in the manufacture of reinforced plastic products and surface coatings; during *in situ* desludging and decarbonizing of internal combustion engines; during cold molding of abrasive grinding wheels; during the molding of friction materials such as brake linings and clutch facings; as an accelerator in the vulcanization of rubber. It is hazardous to the eyes, respiratory system, and skin and is toxic through inhalation, absorption, ingestion, and contact. Symptoms of exposure include headache, dermatitis, and irritation of the eyes and upper respiratory system. OSHA exposure limit (TWA): 2 ppm [skin] or 8 mg/m³.

furfuryl alcohol ($C_4H_3OCH_2OH$) — An amber liquid with a faint, burning odor; MW: 98.1, BP: 338°F, Sol: miscible, Fl.P: 149°F, sp. gr. 1.13. It is used during the manufacture of cements, molded high-density carbon, and graphite articles; in sand consolidation for oil and gas recovery operations. It is hazardous to the respiratory system and is toxic through inhalation, absorption, ingestion, and contact. Symptoms of exposure include dizziness, nausea, diarrhea, diuresis, respiratory and body temperature depression, and vomiting; in animals, drowsiness and eye irritation. OSHA exposure limit (TWA): 10 ppm [skin] or 40 mg/m³.

furnace — An enclosed combustion chamber in which heat is produced for the purpose of effecting a physical or chemical change.

furnace brazing — A process in which the parts to be joined are placed in a furnace heated to a suitable temperature.

furuncle — A localized inflammation caused by a bacterium in a hair follicle or skin gland that discharges pus and has a central core of dead tissue; also known as a boil.

fuse — A device that protects electrical circuits by interrupting power when an overload occurs.

fused silica — An amorphous substance formed by heating silica or quartz, which is insoluble in water or acids. It is used in rockets as an ablative material and for reinforcing plastics. The dusts are fibrogenic and impair the functioning of the lung. ACGIH exposure limit (TLV-TWA): 0.1 mg/m³ as respirable dust.

fusion — Changing the state of a substance from solid to liquid by heating; also known as melting.

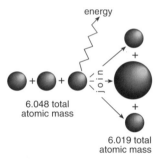

energy

6.048 total atomic mass

6.019 total atomic mass

join

+

+

+

Nuclear Fusion

fusion point — The temperature of a plasma above which the rate of energy generation by nuclear fusion reactions exceeds the rate of energy loss from the plasma, so that the fusion reaction can be self-sustaining.

fuzzy boundary — (*geographical information system*) A boundary that is treated as a band of uncertainty.

FVC — See forced vital capacity.

FWCA — See Fish and Wildlife Coordination Act of 1958.

FWPCA — See Federal Water Pollution Control Act.

FWWTA — Acronym for fraction of wastewater treated anaerobically.

fx — See salt effect.

FY — Acronym for fiscal year.

G

g — See gram; the force of acceleration due to gravity equal to 32.1739 ft/sec² or 386 in/sec².

G — See gauss; SI symbol for the prefix *giga-*, representing a factor of 10^9.

GAAP — Acronym for Generally Accepted Accounting Principles.

gabbro — (*geology*) Any of a group of fine- to coarse-grained, greenish-gray to black, plutonic, igneous rock containing pyroxene.

GABHS — See group A beta-hemolytic streptococcal skin disease.

GAC — Acronym for granular activated carbon.

GACT — Acronym for granular activated carbon treatment.

GAEG — See Graphic Approach to Environmental Guidance.

Gaia hypothesis — Living organisms of the biosphere form a single, complex, interacting system that creates and maintains a habitable Earth.

gaining stream — A stream or portion of the stream where flow increases because of discharge from groundwater.

gal — See gallon.

galaxy — A large-scale concentration of stars, gas, and dust.

galena (PbS) — A bluish-gray lead sulfide mineral that is heavy and brittle and breaks into cubes.

Galena

Galileo spacecraft — A spacecraft launched October 18, 1989, sent deep into space, and designed to orbit the planet Jupiter in December 1995 while sending probes into its atmosphere to determine the nature and compostition of the planet and atmosphere.

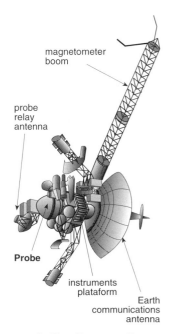

Galileo Spacecraft

gallbladder — A pear-shaped reservoir for bile.

gallon — A liquid measurement equal to 231 cubic inches, 128 fluid ounces, 3.7853 liters, or 4 quarts.

galvanic action — The creation of an electrical current by electrochemical action.

galvanic cell — An electrolytic cell capable of producing electrical energy by electrochemical action.

galvanic corrosion — Electrochemical corrosion of metals that occurs when two or more dissimilar metals are immersed in an electrolyte.

galvanic series — A list of metals and alloys presented in the order of their tendency to corrode or to go into solution.

galvanize — To coat a metal, especially iron or steel, with zinc.

galvanizing — Depositing a protective coating on metals by dipping them into a bath of molten zinc.

galvanometer — An instrument used to detect small electric currents by means of a coil of wire that pivots between the poles of a magnet.

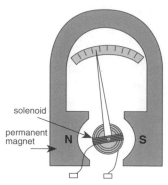

Weston Galvanometer

Gambian trypanosomiasis — A usually chronic form of African trypanosomiasis caused by the bite of the tsetse fly.

gamete — A reproductive cell that fuses with another gamete to form a zygote.

gamma — One nanotesla; a unit of magnetic field strength.

gamma-globulin — A blood protein sometimes used to impart temporary immunity to people for certain infectious diseases.

gamma-radiation — A very-high-frequency electromagnetic emission of photons from certain radioactive elements.

gamma ray — Radiant electromagnetic energy of a very short wavelength (about 10^{-8} to 10^{-11} cm) capable of penetrating most substances; it is given off by radioactive substances and during fission and fusion reactions.

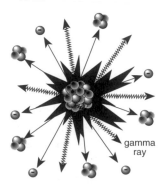

Gamma Ray

gamma-spectroscopy — An analysis technique used to identify specific radionuclides by measuring their radiation emissions, which are unique for each nuclide.

ganglion — A mass of nerve tissue containing cell bodies and synapses lying outside of the central nervous system in vertebrates.

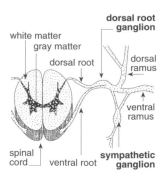

Ganglion

gangrene — The death of body tissue, usually in considerable mass, caused by loss of blood supply and followed by bacterial invasion and putrefaction.

gantry — An overhead structure that supports machines or operating parts.

gap — (*geographic information system*) The distance between two graphic entities on a digitized map.

gap analysis — A biogeographical technique of mapping biological diversity and endemic species to find gaps between protected areas that leave endangered habitats vulnerable to disruption.

garbage — All putrescible animal and vegetable waste, except sewage and body waste.

gas — (*physics*) A substance characterized by very low density and viscosity compared with liquids and solids; a fluid that has neither independent shape nor volume but tends to expand indefinitely. See also gasoline.

gas absorption — (*chemistry*) The solution of one component of a gaseous mixture in a liquid or the absorption onto a surface; selective absorption of gases from a mixture is a means of separating them for testing or collection.

gas bacillus — A species of bacteria that produces a gas as a byproduct of metabolism.

gas barrier — Any device or material used to divert the flow of gases produced in a sanitary landfill or by other land disposal techniques.

gas chromatograph/mass spectrometer — An instrument that identifies the molecular composition and concentration of various chemicals in water and soil samples.

gas chromatography (GC) — A type of automated chromatography in which the mobile phase is separated in columns and measured by a detector; a direct-reading physical instrument based on the principle that components of a complex mixture can be separated physically because of the varying infinities of the components for the different packing materials used.

gas embolism — See arterial gas embolism.

gas exchange — The movement of oxygen and other gases between the atmosphere and the ocean or between the water or atmosphere and living organisms.

gas flammable — A gas that at ambient temperatures and pressures forms a flammable mixture with air at a concentration of 13% by volume or less.

gas-forming bacteria — (*milk*) Microorganisms that ferment lactose, producing both acid and gas and a smooth, gelatinous curd that often contains gas bubbles.

gas gangrene — A death of tissue accompanied by bubbles occurring after surgery or trauma; caused by anaerobic organisms. Symptoms include pain, swelling, tenderness in the wound area, fever, tachycardia, and hypotension.

gas ionization instruments — Instruments based on collecting ions formed by the action of ionizing radiation in a gas.

gas laws — Scientific laws that predict how gases will behave with changes in pressure, temperature, and volume.

gas–liquid chromatography (GLC) — A form of gas chromatography in which the substances to be separated are moved by an inert gas along a tube filled with a finely divided inert solid coated with a nonvolatile oil. Each component migrates at a rate determined by its solubility in oil and its vapor pressure.

gas metal arc welding — A process that produces coalescence of metals by heating them with an arc between a continuous filler metal electrode and the work, using an inert gas to prevent oxidation of the weld.

gas oil — A medium-distilled oil from the refining process used in diesel fuel.

gas sorption — A piece of equipment used to reduce levels of airborne gaseous compounds by passing the air through materials that extract the gases.

gas sterilization — The use of a gas such as ethylene oxide to sterilize material or equipment.

gas stream — Air, clean or polluted, that is present during a production process or combustion and is eventually vented to the atmosphere.

gas–tungsten arc welding — A process that produces coalescence of metals by heating them with an arc between a tungsten electrode and the work.

gas turbine — An engine that uses a compressor to draw in air, compresses it, mixes it with fuel, and produces hot combustion gases to turn a turbine that turns the compressor.

gaseous — The form of or being a gas.

gaseous supersaturation — A condition of higher levels of dissolved gases in water due to entrapment, pressure increases, or heating.

gasification — Any chemical or heat process of converting a solid or liquid fuel into a gaseous fuel.

gasify — To convert into a gas.

gasohol — Mixture of gasoline and ethanol used as a fuel.

gasoline — Any petroleum distillate having a Reid vapor pressure of 4 pounds or greater and which is produced for use as a motor fuel; derived from crude petroleum and commonly known as gas.

gasoline volatility — Evaporative properties of a volatile organic compound.

gasometric method — An analytic technique in which a substance is isolated in gaseous form or is adsorbed from a gaseous mixture.

gastric acid — Digestive acid in the stomach secreted by glands in the stomach wall.

gastric analysis — Examination of the contents of the stomach to determine the quantity of acid present and also the presence of blood, bile, bacteria, and abnormal cells.

gastric intubation — Insertion of a tube through the mouth to the stomach to administer a substance.

gastritis — Inflammation of the lining of the stomach.

gastroenteritis — Inflammation of the lining of the stomach and intestine due to psychological or emotional upset, allergic reactions to certain foods, irritation caused by alcohol, and microorganisms.

gastroenterologist — An internist specializing in diagnosis and treatment of the stomach and gastrointestinal tract.

gastrointestinal (GI) — Pertaining to the stomach and intestine.

gastrointestinal allergy — An immediate hypersensitive reaction after ingesting certain foods or drugs including itching and swelling of the mouth and oral passages, nausea, vomiting, diarrhea, severe abdominal pain, and possibly anaphylactic shock.

gastrointestinal bleeding — Any bleeding from the gastrointestinal tract that may be due to a peptic ulcer, colitis, or cancer and is considered to be a potential emergency.

gastrointestinal infection — An infection of the digestive tract caused by bacteria, viruses, or parasites; symptoms include nausea, vomiting, diarrhea, and dehydration.

G

gastrointestinal tract — Part of the digestive tract where the body processes food and eliminates waste.

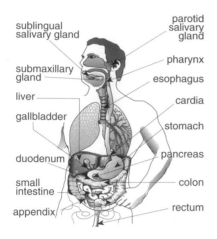

The Gastrointestinal Tract

gastrointestinal tube — A tube inserted through a surgical opening into the stomach for introducing liquids, food, or medication when a patient is unable to take these substances by mouth.

gate — A device that controls the flow in a conduit, pipe, or tunnel without obstructing any portion of the waterway when in the fully open position.

gate valve — (*plumbing*) A shut-off valve using a rising disc or gate to control liquid flow.

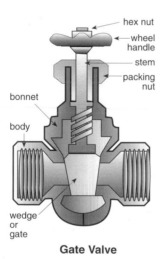

Gate Valve

gauge — A device for registering water level, discharge, velocity, pressure, etc.; the thickness of wire or sheet metal.

gauge height — Elevation of a water surface as measured by a gauge.

gauge pressure (psig) — The pressure measured within a piece of equipment compared to the atmospheric pressure.

gauging station — Specific location on a stream where systematic observations of hydrological data are obtained through mechanical or electrical means.

gauss (G) — The CGS (centimeter–gram–second) unit of magnetic field replaced by the MKS (meter–kilo-gram–second) tesla, where 1 tesla = 10^4 gauss.

gavage — Forced feeding, especially through a tube passed into the stomach; also known as tube feeding.

Gay–Lussac law — See Charles' law.

GBS — See Guillain–Barré syndrome.

GC — See gas chromatography.

GCM — Acronym for Global Climate Model; see general circulation model.

GCSS — Acronym for Global Cloud Systems Study.

GCTM — Acronym for global chemical transport model.

Geiger counter — A radiation counter using a gas-filled tube that indicates the presence of ionizing particles.

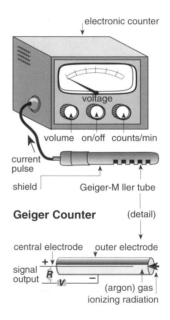

Geiger Counter   (detail)

Geiger–Müeller (GM) counter — A highly sensitive, gas-filled, radiation-measuring device that operates at voltages sufficiently high to produce ionization.

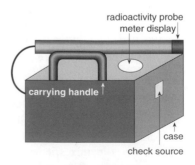

Geiger-Müller Counter

gel — A colloidal dispersion of a solid and a liquid; the dispersion coalesces in thin films to form a jelly-like material.

gel chromatography — Techniques used for the separation of mixtures of organic substances.

gene — A segment of deoxyribonucleic acid nucleus of a cell that contains information for the synthesis of a specific protein or for determining the inheritance of a particular trait.

gene expression — Transcription and translation of a gene into a protein.

gene insertion — The process by which one or more genes from one organism are incorporated into the genetic makeup of a second organism.

gene pool — The total genetic material of a freely interbreeding population at a given time.

gene regulation — The process by which genes are turned on and off to regulate growth and development of an organism.

gene splicing — The insertion of a strand of DNA into another segment of DNA.

gene therapy — The injection of healthy genes into the bloodstream of a patient to cure or treat a hereditary disease or other illness.

general circulation model — A global, three-dimensional computer model of a climate system that can be used to simulate human-induced climate change.

general exhaust — A system for exhausting contaminated air in a general work area.

general obligation bond — A bond issued by a municipality backed by its basic taxing authority for the purpose of raising capital for programs such as wastewater treatment; the bond has a fixed rate of interest and fixed maturities.

general-use pesticide — A pesticide that may be purchased or used by individuals without any special certification or licensing.

general ventilation — Mechanical ventilation applied to a room or an area for the purpose of climate control and dilution of hazardous chemical concentration to safe levels.

generalization (*GIS*) — (*geographical information system*) The removal of detail from a data layer to make processing or visualization easier at smaller scales.

generalized anaphylaxis — A severe, immediate reaction to an allergen with symptoms of itching, edema, wheezing respirations, apprehension, cyanosis, rapid weak pulse, and falling blood pressure; results in shock and death.

generalized peritonitis — Bacterial infection of the peritoneum subsequent to infection in another organ; has acute and severe results.

generation rate — The quantity of solid waste that originates from a defined activity.

generation time — The interval between receipt of infection by a host and maximal communicability of the host; an important factor in determining the rapidity of person-to-person spread of a disease.

generator — (*electricity*) A device that changes mechanical energy into electrical energy; (*waste*) a facility that produces hazardous waste.

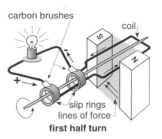

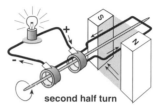

Alternating Current Generator

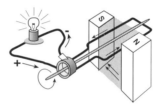

Direct Current Generator

generator status — Status determined by the quantity of hazardous wastes generated by a facility in a calendar month or accumulated prior to shipping off-site.

generic — A term relating to or descriptive of an entire group, class, or general category.

genetic code — Information carried by DNA molecules that determines the specific amino acids and their arrangement in the polypeptide chain of each protein synthesized by the cell.

genetic diversity — The amount of genetic variability among individuals of a single species.

genetic effects of radiation — Inheritable changes, chiefly mutations, produced by the absorption of ionizing radiation by male and female germ cells.

genetic engineering — A process of inserting new genetic information into existing cells in order to modify an organism for the purpose of changing particular characteristics.

genetic mutation — A chemical change in the genes of a cell that causes it to function differently.

genetic screening — Testing to identify individuals at risk of having children with specific genetic disorders.

genetically engineered microorganisms — Bacteria that have been altered by the insertion or deletion of DNA segments.

genetically significant dose (GSD) — The dose of radiation that, if received by every member of the population, would be expected to produce the same genetic injury to the population as would the actual doses received by individuals.

genetics — A branch of biology dealing with the phenomenon of heredity and variation.

genitourinary system — The parts of the body that are involved in reproduction and eliminating waste products in the form of urine.

genome — The complete set of hereditary factors contained in the haploid set of chromosomes.

genotoxic — Pertaining to a chemical that causes adverse effects in the genetic material of living organisms.

genotoxic carcinogens — Those carcinogens that affect genetic material only.

genotoxicity — The amount of damage that specific toxins can cause to a chromosome.

genotype — All or part of the genetic constituents of an organism.

genus — (*classification system*) A taxonomic subgrouping of plants or animals, crust, salts, and other lithospheric properties.

geocoding — (*geographic information system*) The activity of defining the position of geographic objects relative to a standard reference grid.

geodesy — (*geology*) A science related to determining the size and shape of the Earth and the precise location of points on its surface by direct measurements.

geodetic — Of or determined by geodesy.

geodynamics — Study of the motion of the Earth, including rotation, tectonics, and ocean tides, and its structure, core, and mantle.

Geographic Information Manipulation and Mapping Systems (GIMMS) — A low-cost polygon mapping geographic information system.

geographic information system (GIS) — A computer-based system used to store and manipulate geographic information linking the location of objects to characteristics of the objects such as soil types, presence of hazardous waste disposal sites, and groundwater supplies.

geohydrology — Geological study of the character, source, and mode of groundwater.

geoid — A surface of constant gravitational potential around the Earth; a hypothetical surface everywhere perpendicular to the force of gravity.

geological log — A detailed description of all underground features discovered during the drilling of a well.

geological repository — A mined facility for disposal of radioactive waste that uses waste packages and the natural geology to provide waste isolation.

geology — The science that deals with the study of planet Earth, the materials of which it is made, the processes that act to change these materials from one form to another, and the history recorded by these materials.

geomagnetism — (*geophysics*) Science concerned with the magnetism of the Earth.

geometric isomers — Molecules having the same types of bond in the same order but having atoms with different spatial relationships; also known as *cis-/trans*-isomers.

geomorphology — The geological study of the configuration, characteristics, origin, and evolution of land forms and earth features.

geophysical focus area — A delineation of science support activities centered around geophysical datasets related to a common set of scientific issues.

geophysical log — A record of the structure and composition of the Earth encountered when drilling a well or similar type of test hole or boring.

geophysics — A branch of earth science dealing with the physical processes and phenomena occurring especially in the Earth and in its vicinity.

geopressured brines — Hot, pressurized saltwater that contains dissolved methane at depths of 10,000 to 20,000 feet; the best-known reservoirs are along the Gulf Coasts of Texas and Louisiana.

geopressured reservoir — A geothermal reservoir consisting of porous sands that contain water or brine under conditions of high temperatures and pressure.

georeference — (*geographical information system*) Establishment of a relationship between page coordinates on a planar map and real-world coordinates.

GEOS — Acronym for Geodynamics Experimental Ocean Satellite.

Geos-3 — Abbreviation for NASA's Geodetic Satellite Mission.

Geosat — Abbreviation for Geologic Satellite.

geosphere — The soils, sediments, and rock layers of the Earth's crust, continental shelf, and area beneath the ocean floor.

geostationary — (*geographical information system*) Being or having an equatorial orbit; having an angular velocity the same as that of the Earth (e.g., the position of a satellite in such an orbit is fixed with respect to the Earth).

Geostationary Operational Environmental Satellite (GOES) — A satellite that remains fixed at 22,300 miles above a point on the Equator, takes photographs once every 30 minutes day and night, and provides up-to-the-minute data used by weather forecasters around the world; it moves at a speed synchronous with the Earth's rotation.

Geostationary Satellite Server — A server that provides retrospective imagery for the past 21 days; it cannot be guaranteed to be available or timely for use in supporting emergency disaster mitigation operations or forecasting of weather.

geosynchronous — See geostationary.

geotechnical — The use of scientific methods and engineering principles to acquire, interpret, and apply knowledge of earth materials for solving engineering problems.

geothermal — The heat of the interior of the Earth.

geothermal energy — An energy produced by tapping the internal heat of the Earth. At present, the only available technologies to do this are those that extract heat from hydrothermal convection systems, where water or steam transfers heat from deeper part of the Earth to areas where the energy can be tapped. The amounts of pollutants found in such geothermals vary from area to area but may contain arsenic, boron, selenium, lead, cadmium, and fluorides. They may also contain hydrogen sulfide, mercury, ammonia, radon, carbon dioxide, and methane. They may cause contamination of surface water from the wastewater, produce elliptical, dish-shaped depressions in the ground, enhance seismic activity, and create noise levels as high as 120 dBA.

geothermal field — A geographical region with known geothermal power sources that might be used to produce energy.

geothermal gradient — The rate at which temperature increases with depth below the surface.

geothermal heat pump system — A system that uses a condenser circuit that either is in the Earth, in the case of a closed-loop system, or is a water supply, in the case of an open-loop system, that provides or removes heat from a dwelling.

geothermal power — Energy obtained from heat under the surface of the Earth.

geothermal steam — Steam drawn from sources within the Earth.

geothermometry — The science of measuring temperatures below the surface of the Earth.

geotrichosis — A condition associated with the fungus *Geotrichum candidum* that causes oral, bronchial, pharyngeal, and intestinal disorders; commonly occurs in immunosuppressed patients.

GEP — An acronym for Good Engineering Practice.

g-eq — See gram equivalent.

geriatrics — The study and treatment of biological and physical changes and diseases of old age.

germ — A pathogenic microorganism.

germ cells — Reproductive cells of the body, either egg or sperm cells.

germ theory — The concept that all infections and contagious diseases are caused by living microorganisms.

germicide — An agent capable of killing microorganisms.

germination — The process of development of a seed or spore.

gerontology — Scientific study of the aging process.

gestation period — Period of intrauterine development in mammals from fertilization to birth.

GEWEX — Acronym for Global Energy and Water Cycle Experiment.

geyser — A natural, thermal spring that periodically discharges its water or steam with enormous power.

GF ($C_7H_{14}FO_2P$) — A highly toxic nerve agent that is a potent inhibitor of acetylcholinesterase and a neurotoxicant; MW: 180.18.

GFCI — See ground-fault circuit interrupter.

GH — Acronym for growth hormone.

GHP — See ground-source heat pump.

GI — See gastrointestinal.

GI tube — See gastrointestinal tube.

Giardia — A genus of flagellate protozoa that is parasitic in the intestines of people and animals and may cause protracted, intermittent diarrhea with symptoms suggesting malabsorption.

giardiasis — (*disease*) A protozoan infection, usually of the upper small intestine, that may be asymptomatic or associated with intestinal symptoms such as chronic diarrhea, abdominal cramps, fatigue, weight loss, and poor absorption of fats or fat-soluble vitamins. Incubation time is 5 to 25 days or longer, usually 7 to 10 days. It is caused by *Giardia lamblia*, a flagellate protozoan found worldwide, with children being infected more frequently than adults, especially in areas of poor environmental control and in institutions such as daycare centers. It may be the cause of waterborne outbreaks of disease in the United States. The reservoir of infection includes people and possibly wild or domestic animals. It is transmitted by ingestion of cysts in water or food contaminated with feces or by person-to-person transmission through the oral–fecal route. It is communicable for the entire period of the infection; the asymptomatic carrier rate is high so susceptibility may be sizable. It is controlled by proper protection and processing of water and food, proper sanitary disposal of feces, and good personal hygiene.

gigabit — One billion bits.

gigajoule — One billion joules.

gigawatt (GW) — A unit of power equal to 1 billion watts.

gigawatt-hour — One hour of electricity consumed at a constant rate of 1 gigawatt.

gigayears — Time measured in billions of years commonly used to measure geologic time.

GIMMS — See Geographic Information Manipulation and Mapping Systems.

gingivitis — An inflammation of the gums.

GIPME — Acronym for Global Investigation on Pollution in the Marine Environment.

GIS — See geographic information system; acronym for Global Indexing System.

g/kg — Grams per kilogram.

GL — Acronym for geophysic laboratory.

glacial drift — The unconsolidated mixture of gravel and partly weathered rock fragments left by glaciers.

glacial drift aquifers — Geological formations found in those parts of the United States that were covered by the advance of glaciers during the Ice Age. As the climate cooled and the glaciers advanced, the ice packed aerobic soil and bedrock. When temperatures warmed, the glaciers melted, the material was redeposited, and some of the water was trapped from the melted glacier.

glacial erosion — The covering or alteration of an area by an ice sheet or glaciers; also known as glaciation.

Glacial Erosion

glacial moraine — A massive mound of loose rock, soil, and earth deposited by the retreating edge of a glacier.

glacial outwash — The partially sorted mixture of gravel, sand, and silt left from ice melting in a preexisting valley bottom or broadcast in a form similar to an alluvial plain.

glacial till — Unsorted mixture of clay, silt, sand, gravel, and boulders deposited by an ice sheet.

glaciation — See glacial erosion.

glacier — A huge mass of ice formed on land by the compaction and recrystallization of snow; it moves very slowly downslope or outward due to its own weight.

gland — An organ in animals secreting a particular substance or substances vital to existence.

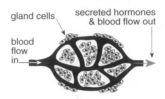

ductless (endocrine) gland

gland cells — secreted hormones & blood flow out

blood flow in

Gland

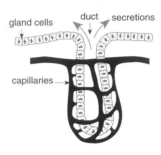

gland cells — duct — secretions

capillaries

Duct (Exocrine Gland)

glare index — An index used in modeling and simulation by environmental engineers to predict visual discomfort in order to avoid it.

glass — An inorganic solid having no crystalline structure.

glaucoma — A condition of the eye where the fluid that normally fills the eyeball fails to drain properly, thereby causing a buildup of excess pressure and damage to the optic nerve; it is caused by aging, infection, injury, and congenital defect and is a leading cause of blindness.

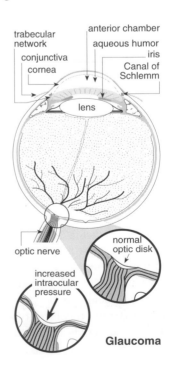

trabecular network
anterior chamber
conjunctiva
aqueous humor
cornea
iris
Canal of Schlemm
lens
optic nerve
normal optic disk
increased intraocular pressure

Glaucoma

GLC — See ground-level concentration; gas–liquid chromatography.

global carbon budget — Balance of exchanges (gains and losses) of carbon between carbon reservoirs or between one specific loop of the carbon cycle.

global climate change — Long-term fluctuations in temperature, precipitation, wind, and all other aspects of the climate of the Earth.

Global Climate Model — A computer program that is used by atmospheric scientists to study and predict worldwide trends in weather patterns and the forces and physical laws that affect climate.

global measurement — All of the activities required to specify a global variable, such as ozone.

Global Positioning System (GPS) — A method for identifying locations on Earth using triangulation calculations of satellite positions; originally created by the U.S. military and now used commercially. The U.S. Department of Defense Global Positioning System consists of a constellation of 24 satellites orbiting the Earth at a very high altitude. The GPS satellites transmit signals that allow determination, with great accuracy, of locations on the Earth. The GPS system is used in air, land and sea navigation, mapping, surveying, and other applications where precise positioning is necessary.

global variables — Functions of space and time that describe the large-scale state and evolution of the Earth system.

global warming — An increase in the near-surface temperature of the Earth; referring to the warming predicted to occur as a result of increased emissions of greenhouse gases; can also occur naturally.

global warming potential — The ratio of warming caused by a substance such as chlorofluorocarbon-12 to warming caused by a similar mass of carbon dioxide.

globe thermometer — A thermometer placed in the center of a metal sphere that has been painted black to measure radiant heat.

globule — A small spherical mass.

globulin — A general term for proteins that are insoluble in water and in highly soluble, concentrated salt solutions but are soluble in moderately concentrated salt solutions; all plasma proteins except albumin and prealbumin are globulins.

glove box — A sealed enclosure having long, impervious gloves sealed to the openings that are used to handle all items in the box; (*radiation*) a sealed box with gloves attached to the wall, filled with an inert gas, and fitted with a filter ventilation system to handle radioactive materials.

GLP — Acronym for Good Laboratory Practices.

GLRS — Acronym for Geoscience Laser Ranging System.

glucose ($C_6H_{12}O_6$) — A white, crystalline, monosaccharide; a product of photosynthesis and an end-product of digestion; the most common sugar.

glutaraldehyde (($CH_2)_3(CHO)_2$) — A colorless crystal; MW: 100.1; BP: 187°C (decomposes); Sol: soluble in water, alcohol, ether, other organic solvents; Fl.P: not flammable; density: 1.062. It is used as a cold sterilizing disinfectant, in tanning, and as a fixative for tissues. It is hazardous to the nose, eyes, skin, and digestive tract and causes strong irritation of the nose, eyes, and skin. It can cause upper respiratory tract irritation, headache, and nervousness. ACGIH exposure limit: (TLV-TWA): 0.2 ppm

$$O = C - CH_2 - CH_2 - C = O$$
with H attached above each terminal C

Glutaraldehyde

GLWQA — See Great Lakes Water Quality Agreement.

glycerin — See glycerol.

glycerol ($CH_2OHCHOHCH_2OH$) — A clear, sweet, colorless, syrup-like liquid that may be solid at lower temperatures. It is used in the manufacture of medicines, explosives, soaps, antifreeze, inks, perfumes, and cosmetics, finishes or as a solvent. It is one of the soluble products into which fats are broken down in the body by digestion; also known as glycerin or glycyl alcohol.

glycidol ($C_3H_6O_2$) — A colorless liquid; MW: 74.1, BP: 320°F (decomposes), Sol: miscible, Fl.P: 162°F, sp gr: 1.12. It is used in surface coatings; in chemical synthesis; as a chemical stabilizer for natural oils and vinyl polymers; as a demulsifying agent; as a dye leveling agent. It is hazardous to the eyes, skin, respiratory system, and central nervous system and is toxic through inhalation, ingestion, and contact. Symptoms of exposure include narcosis and irritation of the eyes, nose, skin, and throat. Also known as epihydrin alcohol. OSHA exposure limit (TWA): 25 ppm [air] or 75 mg/m³.

$$H_2C \overset{\diagdown}{\underset{O}{\diagup}} CH_2 - CHO$$

Glycidaldehyde

glycogen — A starch-like reserve carbohydrate formed and stored in the liver cells and muscles of all higher animals.

glycol — See ethylene glycol.

glycyl alcohol — See glycerol.

glyoxal (C₂H₂O₂) — A yellowish liquid; MW: 58.04; BP: 50.4°C; Sol: soluble in water, alcohol, and ether; Fl.P: NA; density: 1.14. It is used in the production of textiles and glues and in organic synthesis. It is hazardous to the skin, eyes, and respiratory tract and causes an irritation to the skin, eyes, respiratory tract, and digestive tract. It may cause gastrointestinal pain if ingested. ACGIH exposure limit: NA.

Glyoxal

GM counter — See Geiger–Müeller counter.

g/mi — Grams per mile.

GMP — See Good Manufacturing Practices.

GNP — See Gross National Product.

GNSS — Acronym for Global Navigation Satellite System.

goal — Community expectations of desired obtainable levels of program and system performance; a goal differs from an objective in that it does not have a deadline and it is usually for a longer term.

goat milk — Lacteal secretion practically free of colostrum that is obtained by milking of healthy goats; the generic term "milk" includes goat milk.

GOCO — Acronym for government-owned/contractor-operated.

GOES — See Geostationary Operational Environmental Satellite; Geostationary Satellite Server.

GOGO — Acronym for government-owned/government-operated.

goiter — An enlargement of the thyroid gland.

goitrogen — A natural product found in plant foods causing hypothyroidism.

Golgi apparatus — A complex cellular organella consisting mainly of a number of flattened sacs and associated vesicles and which is involved in the synthesis of glycoproteins, lipoproteins, membrane-bound proteins, and lysosomal enzymes.

GOME — Acronym for Global Ozone Monitoring Experiment.

GOMI — Acronym for Global Ozone Monitoring Instrument.

GOMR — Acronym for global ozone monitoring radiometer.

gonads — Male and female primary sex glands.

Gonyaulax catanella — A species of planktonic protozoa that produces a toxin ingested by shellfish and results in seafood poisoning; also called red tide.

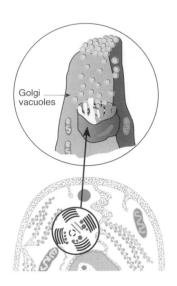

**Golgi Apparatus
in Animal Cell**

Good Manufacturing Practices — (*food*) Manufacturing standards for food processing used to ensure safety and wholesomeness to minimize the risk of exposing a product and its ingredients to unsanitary conditions that could lead to injury or disease.

good soil suitability — The quality of cover soil characterized by favorable properties for use.

GOOS — Acronym for Global Ocean Observing System.

gooseneck — (*plumbing*) A portion of a service connection between the distribution system water main and a meter.

GOP — Acronym for General Operating Procedures.

Gopher — (*computer science*) A menu-based system for browsing, searching, and retrieving files on the Internet.

GOPO — Acronym for government-owned/privately operated.

gouging — Removal of material by forming a bevel or groove.

gpad — Gallons per acre per day.

gpd — Gallons per day.

gpg — Grams per gallon.

GPH — The volumetric flow rate in gallons per hour.

gpm — Gallons per minute.

GPR — Acronym for ground-penetrating radar.

GPS — Acronym for groundwater protection strategy; see global positioning system.

gr — See grain.

GR — Acronym for grab radon sampling.

grab sample — A sample representing a larger gross sample taken at a particular place over a brief period of time and produced when a container emptied of liquid or air is used to quickly draw in a small representative sample of a gas.

gradation — The proportion of material of each grain size present in a given soil.

grade — Inclination or slope of a pipeline, conduit, stream channel, or natural ground surface usually expressed in terms of the ratio or percentage of number of units of vertical rise or fall per unit of horizontal distance (e.g., a 1% grade would be a drop of 1 foot per 100 feet of pipe); the elevation or slope of a surface or a surface slope; (*construction*) the ground level around a structure utilized to control the flow of water.

grader — A machine with a centrally located blade that can be angled to either side.

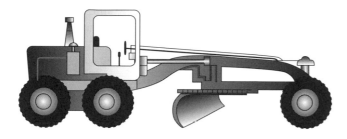

Motor Grader

gradient — The degree of slope or a rate of descent or ascent.

grading — Any excavating, filling, or leveling of earth material conducted at a site in preparation for construction, landfilling, or improvements.

Graham's law of diffusion — The rate of diffusion of a gas through porous membranes varies inversely with the square root of its density.

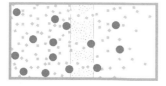

Graham's Law

grain (gr) — A unit of weight equal to 1/7000 pound, or 6.479891×10^{-5} kilograms.

grain alcohol — See ethanol.

grain dust — Dust of grains, wheat, barley, and oats that, through inhalation at high concentrations, can cause coughing, wheezing, breathlessness, dyspnea, and bronchial asthma. ACGIH exposure limit (TLV-TWA): 4 mg/m³ as total particulates.

grain per gallon — A measure of solution concentration equal to 17.1 parts per million.

gram (g) — The basic unit of mass (weight) in the metric system equivalent to 15.432 grains or 0.0353 ounces; 454 grams = 1 pound.

gram-calorie — The amount of heat necessary to raise the temperature of 1 gram of water 1°C in the range of 14.5 to 15.5°C; also known as calorie.

gram equivalent (g-eq) — The equivalent weight of a substance expressed in grams.

gram formula — See mole.

gram molecular weight — A mass in grams numerically equal to the molecular weight of a substance, or the sum of all the atomic weights in its molecular formula.

Gram-negative — A method of characterizing organisms in microbiology by staining that produces a pink color.

Gram-positive — A method of characterizing organisms in microbiology by staining that produces a violet color.

grams per kilogram (g/kg) — A unit used in the measurement of toxicological effects on humans or animals.

Gram's stain — A differential bacteriological staining procedure in which a smear is stained with crystal violet, treated with strong iodine solution, decolorized with ethanol or ethanol–acetone, and counterstained with a contrasting dye. Those retaining the primary stain are Gram-positive and those with the counterstain are Gram-negative.

Granary Weevil (*Sitophilus granarius*) — A stored food product insect similar to the Rice Weevil in appearance but with oval pits on the thorax and with winged covers uniformly dark brown.

grand jury — A jury of inquiry whose duty is to receive complaints and accusations in criminal cases, hear the evidence on the part of the federal or state government, and find bills of indictment in cases where it is satisfied a trial should commence.

grand name — See trade name.

granite — Light-colored, crystalline, coarse-grained rock containing approximately 30% quartz and 60% alkali feldspar with plagioclase, biotile, and hornblende.

granular activated carbon treatment — A filtering system often used in small water systems and individual homes to remove organics.

granular formulation — A dry, ready-to-use pesticide product that consists of an active ingredient mixed with or impregnated into a carrier.

granular pesticide — A pesticide chemical mixed with or coated with small pellets or a sand-like material.

granular soil — A soil structure in which the aggregates are small and nonporous and are held strongly together; nearly spherical with many irregular surfaces; usually found in surface soil or the A horizon.

granuloma — A mass or nodule of chronically inflamed tissue with granulations, usually associated with an infection.

Graphic Approach to Environmental Guidance — A unique method of presenting guidance for complying with environmental requirements employing flowcharts, checklists, and reference materials that are easily understandable.

graphical user interface (GUI) — (*computer science*) A software program designed to allow a user to execute commands by pointing and clicking on icons or text.

graphing — An organization of data with two or more variables plotted along horizontal and vertical axes to show the relationship between them.

graphite (natural) — Steel gray to black, greasy-feeling, odorless mineral; MW: 12.0, BP: not known, Sol: insoluble, Fl.P: not known, sp. gr. 2.0 to 2.25. It is used in foundry facings; in the manufacture of refractories in crucibles and retorts; in the manufacture of lubricants and adhesives in bearings, slides, gears, engines, and mold-release agents; in the manufacture of writing and drafting agents; in the manufacture of industrial paints, ink, and polishes; in the manufacture of motor and generator brushes, batteries, carbon resistors, and electrodes; in the manufacture of tungsten carbide cutting tools, and friction materials; in electroplating; in the manufacture of ammunition; for miscellaneous uses in hard rubber, engine packing, cord, rope, twine, floor coverings, and coatings for fertilizers. It is hazardous to the respiratory system and cardiovascular system and is toxic through inhalation, and contact. Symptoms of exposure include cough, dyspnea, black sputum, pulmonary function impairment, and lung fibrosis. OSHA exposure limit (TWA): 2.5 mg/m^3 [resp]. Also known as black lead or plumbago.

GRAS — (*food additives*) Acronym for Generally Recognized As Safe.

grass — Common name for all plants with jointed stems; narrow leaves having parallel veins that encircle the stem and produce a seed-like grain as their fruit.

grassland — A region characterized by grasses and other erect herbs, usually without trees or shrubs; usually the climate is dry for long periods of the summer and it freezes in the winter.

grate — A device used to support the solid fuel or solid waste in a furnace during drying, ignition, or combustion; openings permit combustion air to pass through it.

gravel — Rock fragments from 2 to 64 mm in diameter; an unconsolidated mix with sand, cobbles, and boulders.

gravimetric method — A method of chemical analysis in which a substance is isolated in the pure state or as one of its compounds; its weight is determined with an analytical balance.

gravimetry — A procedure dependent upon the formation or use of a precipitate or residue that is weighed to determine a concentration of a specific contaminant in a previously collected sample.

gravity — Heaviness or weight of an object resulting from the universal effect of the attraction between any body of matter and any planetary body; the force of the attraction depends on the relative masses of the bodies and on the distance between them.

gravity spray tower — An air pollution control device in which particles in the gas stream are trapped by much larger drops of water sprayed from above, causing the particles to be removed from the gas stream.

gray (Gy) — The International System unit of absorbed dose that replaces the term rad (1 Gy = 1 joule/kg – 100 rads); see also absorbed dose.

gray scale — (*geographic information system*) The level of brightness used to display information on a monochrome display device.

graywater — Wastewater generated by water-using fixtures and appliances, excluding the toilet and possibly the garbage disposal.

grease skimmer — A device for removing floating grease or scum from the surface of wastewater in a tank.

Great Lakes Water Quality Agreement — An agreement between the United States and Canada recognizing the urgent need to improve environmental conditions in the Great Lakes, by restoring and enhancing water quality, reducing nuisance conditions, eliminating the discharge of toxic substances, and setting numerical targets for the reduction of loading of phosphorus.

greenfields — The land outside the urban area that has been untouched by development and is considered to be free from environmental contamination.

greenhouse effect — A phenomenon in which energy from the sun, in the form of light waves, passes through the atmosphere and is absorbed by the Earth. It is then reradiated as heat waves, which are absorbed by the air. The air then acts in a similar manner as if it were trapped under glass in a greenhouse.

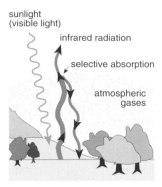

Greenhouse Effect

greenhouse gas — A gas, such as carbon dioxide or methane, that contributes to potential climate change.

Greenwich mean time — A world standard time implemented in Greenwich, England, in the 1840s and based on the motion of the Earth.

GRGL — Acronym for groundwater residue guidance level.

grid — An open lattice for mounting a sample to aid in its examination by transmission electron microscope or microscopy; the term is used by the U.S. EPA to denote a 200-mesh copper lattice approximately 3 mm in diameter. Alternatively, a network of uniformly spaced points or lines on a visual display device for locating positions; a set of regularly spaced sample points; the distribution network of utility resources; (*cartography*) an exact set of reference lines over the surface of the Earth.

Grignard reagents — Compounds of the type RMgX, where R represents an alkyl group and X is an aryl halide, especially a bromide or iodide. They are formed from the reaction of alkyl and aryl halides with magnesium in the presence of dry ether.

grinding — The mechanical pulverization of solid waste.

grit — Heavy material present in wastewater, including sand, coffee grounds, gravel, sand, etc.

gross alpha-particle activity — Total radioactivity due to alpha-particle emission as inferred from measurements on a dry sample.

gross beta-particle activity — Total radioactivity due to beta-particle emission as inferred from measurements on a dry sample.

gross heating value — Total heat obtained from the complete combustion of fuel or waste at 60°F when combustion starts and the combustion products, which are then cooled back down to 60°F before the quantity of heat released is measured.

gross national product — The sum total of all goods and services produced in a national economy.

gross primary production — The total amount of photosynthesis accomplished in a given period of time.

ground — An electrical line with the same electrical potential as the surrounding earth; used to protect people and equipment from shocks and malfunctions.

ground-based inversion — Temperature inversion at the surface of the ground that forms during the night when radiated losses occur.

ground clutter — A pattern of radar echoes from fixed ground targets such as buildings and hills near the radar; it may hide or confuse radar echoes near a radar antenna.

ground cover — Grasses and low-growing plants that keep soil from being blown or washed away.

ground-fault circuit interrupter — A safety device that monitors the difference between current flowing through the hot and neutral wires of a receptacle; a current imbalance of >5 milliamps results in cutoff in less than a second.

ground itch — A skin rash caused by bacteria introduced by invasive hookworm larvae.

ground-level concentration (GLC) — The concentration of air contaminants at the ground surface.

ground-level ozone — Ozone that occurs near the surface of the Earth and is a serious pollutant.

ground motion — Movement of the surface of the Earth due to earthquakes or explosions; the motion is produced by waves that are generated by a sudden slip on a fault or sudden pressure at the explosive source and travels through the Earth and along its surface.

ground-penetrating radar — A geophysical method that uses high-frequency electromagnetic waves to obtain subsurface information.

ground-source heat pump — Heat pump that uses the Earth or groundwater for heating during the winter and cooling during the summer.

ground track — The inclination of a satellite, together with its orbital altitude and period of its orbit, that creates a track defined by an imaginary line connecting the satellite and the center of the Earth.

ground truth data — Field observations used to check the accuracy of satellite measurements.

grounding — The procedure used to carry an electrical charge to the ground through a conductive path.

groundwater — Water that adheres to the surface of soil particles in thin films and prevents the soil from ever becoming totally dry; the water is held in deposits of gravel, sand, or porous rock below the surface of the Earth in the saturation zone.

Groundwater and Drinking Water — An important program of the U.S. EPA that develops standards for the quality of drinking water supply systems, regulates underground injection of waste and protection of groundwater wellhead areas, and provides information on public water supply systems.

groundwater contamination — Pollution of springs and wells at their source underground, potentially by landfills, agriculture, industry, and urbanization.

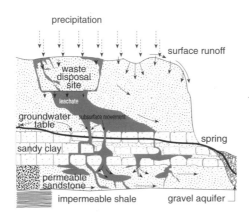

Groundwater Contamination

groundwater discharge — Groundwater near coastal waters contaminated by landfill leachate, well injection of hazardous wastes, septic tanks, and other sources.

Groundwater Disinfection Rule — A 1996 Amendment of the Safe Drinking Water Act requiring the U.S. EPA to promulgate the National Primary Drinking Water Regulations, which require disinfection of all public water systems, including surface water and groundwater.

groundwater hydrology — The branch of hydrology that deals with groundwater, its occurrence and movements, its replenishment, and the properties of rocks that control groundwater movement and storage.

groundwater infiltration — Water that enters treatment plants or pipes from natural springs, aquifers, or runoff from land.

groundwater mining — Removal of groundwater from an aquifer in excess of the rate of natural or artificial recharge.

groundwater recharge — Flow to groundwater storage from precipitation, infiltration from streams, and other sources of water. Inflow to a groundwater

reservoir; the process by which water enters the soil and eventually reaches the saturated zone.

groundwater reservoir — An aquifer or aquifer system in which groundwater is stored.

groundwater runoff — That part of the groundwater that is discharged into a stream channel as a spring or seepage water.

groundwater storage — The storage of water in groundwater reservoirs.

groundwater supply — Water beneath the ground tapped by wells and springs.

groundwater table — The upper boundary of groundwater where water pressure is equal to the atmospheric pressure.

group — See radical.

group A beta-hemolytic streptococcal skin disease — A bacterial skin infection affecting mainly meat packers.

grout — A sealing mixture of cement and water to which sand, sawdust, and other fillers may be added; neat cement, inert natural materials, concrete, heavy drilling mud, heavy bentonite, water slurry impervious to and capable of preventing movement of water.

grouting — The filling of cracks and crevices with a cement mixture.

growth — (*biology*) An irreversible increase in size or volume of a cell, tissue, organ, or organism.

growth factor — An organic compound usually required in trace amounts necessary for growth; an essential cell component or precursor of such components that cannot be synthesized by the organism itself.

growth rate — The rate at which growth occurs.

growth regulator — A chemical used to increase, decrease, or change the normal growth of a plant or animal; does not include fertilizers and other nutrients.

growth yield coefficient (soils) — The quantity of biomass carbon formed per unit of substrate carbon consumed.

grubs — The larvae of certain beetles, wasps, bees, and ants.

GSD — See genetically significant dose.

GT — Acronym for gas turbine.

GTR — Acronym for government transportation request.

GTS — Acronym for Global Telecommunications System.

guaranty — A formal and signed agreement between buyers and sellers in which the latter verifies that the items for sale are not in violation of the Food, Drug, and Cosmetic Act when shipped.

GUI — (*computer science*) See graphical user interface.

guideline — A document suggesting values for maximum concentrations or emissions of a substance for the protection of the environment or human health.

Guillain–Barré Syndrome — An autoimmune reaction of the body that affects the peripheral nerves and causes weakness, paralysis, and occasionally death.

gully erosion — The removal of layers of soil creating channels or ravines too large to be removed by normal tillage operations.

GUP — See general-use pesticide.

gustation — The act of tasting or the sense of taste.

gut — The intestine.

gutter — A trough under an eave to carry off rain water.

GVP — Acronym for gasoline vapor pressure.

GVW — Acronym for gross vehicle weight.

GW — Acronym for groundworking level sampling or groundwater; see gigawatt.

GWDR — Acronym for Groundwater Disinfection Rule.

GWh — See gigawatt-hour.

GWM — Acronym for groundwater monitoring.

GWP — Acronym for global warming potential.

GWPS — Acronym for Groundwater Protection Standard or Groundwater Protection Strategy.

Gy — See gray.

gynecology — Study of female reproductive organs.

gypsum (CaSO$_4$·2H$_2$O) — A monoclinic mineral form of hydrated calcium sulfate; used in the building industry and in the manufacture of cement, rubber, paper, and plaster of Paris.

Gypsum

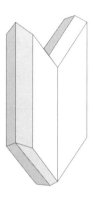

G

H

h — See Planck's constant; SI symbol for the prefix *hecto-*, reflecting a factor of 10^2.

H — See hydrogen, magnetic field intensity, or magnetic field strength.

^{2}H — Scientific notation for deuterium

^{3}H — Scientific notation for tritium

HA — Acronym for health advisory.

HAAP — Acronym for high-altitude aerial photograph.

habitat — The sum of environmental conditions in a specific place occupied by an organism, population, or community.

HACCP — See Hazard Analysis/Critical Control Point Inspection.

HACCP plan — A written document that is based on the principles of HACCP and which describes the procedures to be followed to ensure the control of a specific process or procedure.

HAD — Acronym for health assessment document.

Hadley cells — Circulation patterns of atmospheric convection currents as they sink and rise in several intermediate bands.

Haemophilus — A genus of Gram-negative pathogenic bacteria frequently found in the respiratory tract of humans and other animals.

Haemophilus influenzae — A small, Gram-negative, nonmobile parasitic bacteria that occurs either encapsulated or nonencapsulated and is found in healthy, normal people but can lead to severe disease in debilitated older people.

hafnium (Hf) and compounds — A highly lustrous, ductile, grayish metal; MW: 178.5, BP: 8316°F, Sol: insoluble, sp. gr. 13.31. It is used in the manufacture and fabrication of special alloys; as control rods in water-cooled nuclear reactors; in gas-filled tubes and incandescent lamps as a scavenger for oxygen and nitrogen; in the manufacture of photographic flashbulbs and light-bulb filaments; in electronic equipment as cathodes and capacitors. It is hazardous to the eyes, skin, and mucous membrane and is toxic through inhalation, ingestion, and contact. Symptoms of exposure in animals include liver damage, and eye, skin, and mucous membrane irritation. OSHA exposure limit (TWA): 0.5 mg/m^3 [air].

hail — Precipitation in the form of frozen raindrops produced high in convective clouds, usually in the latitudes between 30° and 60°.

hair — A filament of keratin consisting of a root and shaft formed in a specialized follicle in the epidermis.

hair analysis — Chemical analysis of a hair sample to find possible evidence of exposure to a toxic substance; molecules of lead compounds and other chemicals are absorbed and stored in hair shafts.

hair catcher — A piece of equipment used for straining out lint and other materials before water from a swimming pool precedes to the motor in the filter operation.

half-and-half — A product consisting of a mixture of milk and cream containing not less than 11.5% milk-fat.

half-life — The time required for one half of the atoms of a given amount of radioactive material to undergo radioactive decay.

half-value layer (HVL) — The thickness of absorbing material required to reduce the intensity of a beam of radiation to one half its original intensity.

halfspace — (*geology*) A mathematical model used to approximate the Earth when performing some calculations in seismology.

hallucination — Perception of objects with no basis in reality; usually arises from a disorder of the nervous system or in response to drugs.

hallucinogen — A psychoactive drug capable of altering moods and perceptions of time and space.

halocarbon — (*chemistry*) Any of a group of compounds of carbon and one or more halogens; used as refrigerants, propellant gases, and so on; may be anticipated to contribute to reductions in the concentration of ozone in the stratosphere.

halogen — Any member of the chlorine family containing the elements fluorine, chlorine, bromine, iodine, and astatine.

halogen lamp — A high-pressure incandescent lamp containing halogen gases such as iodine or bromine that allows the filaments to be operated at higher temperatures and higher efficiencies.

halogenated chlorofluorocarbons — A family of refrigerants that, when released into the atmosphere, are destructive to the ozone layer of the Earth.

halogenated hydrocarbon — A chemical substance containing carbon plus one or more of the halogens chlorine, fluorine, bromine, or iodine.

Halon — A compound formed when a halogen, such as fluorine or bromine, attaches to a carbon atom.

Halon-1211 — See bromochlorodifluoromethane.

halophile — An organism that requires a high salt concentration for growth and maintenance.

halothane ($C_2HBrClF_3$) — A colorless volatile liquid with a sweet odor; MW: 197.39, BP: 50.2°C, Sol: soluble in organic solvents, Fl.P: a noncombustible liquid, density: 1.871 at 20°C. It is used as a clinical anesthetic. It is hazardous to the respiratory system and digestive system and, in high concentrations, may cause central nervous system depression, cardiovascular system and liver damage, nausea, vomiting, increased body temperature, excitability, arrhythmia, vasodilation, and severe irritation of the eyes if they come into contact with the liquid. ACGIH exposure limit (TLV-TWA): 50 ppm.

$$F-\underset{\underset{F}{|}}{\overset{\overset{F}{|}}{C}}-\underset{\underset{Cl}{|}}{\overset{\overset{Br}{|}}{C}}-H$$

Halothane

HALSS — Acronym for high-altitude lidar sensing station.

halzoun — A disease resulting from blockage of the nasopharynx by a parasite.

hammermill — A device used for reducing the size of bulk material by means of hammers, which are usually placed on a rotating axle inside a steel cylinder.

handicapped — A person with acquired or congenital mental or physical defects that interfere with normal body functions and participation in activities of life.

hand protection — Specific types of gloves or other hand protectors required to prevent harmful exposure to hazardous materials.

hanger — A support for a pipe.

hantavirus — Viruses of the family Bunyaviridae which are enveloped by lipids and have a negative-stranded RNA genome composed of three unique segments; susceptible to most disinfectants, including dilute hypochlorite solutions, phenolics, detergents, 70% alcohol, and other general household disinfectants. The survival time of the virus in the environment in liquids, aerosols, or the dried state is not known. It is spread by deer mice in the United States and is the cause of hantavirus pulmonary syndrome.

hantavirus infection — See hantavirus pulmonary syndrome.

hantavirus pulmonary syndrome — (*disease*) An acute, viral, flu-like disease with high fever, muscle aches, cough, and headaches; respiratory problems worsen rapidly as lungs fill with fluid, and death may occur because of respiratory failure (over 60% of victims die). Incubation time is 1 to 2 weeks but may range from a few days to 6 weeks. It is caused by hantaviruses and is found worldwide. The reservoir of infection in the United States is the deer mouse, while worldwide it may be in other field rodents. It is transmitted through the inhalation of aerosols from infective saliva or excreta or bites of rodents. The possibility of this being a foodborne infection has not been evaluated. No evidence of person-to-person transmission has been found. It is controlled through elimination of rodents, especially deer mice, in structures where people live or work; by cleaning all surfaces with good household disinfectants; and by removal of dead rodents and contaminated materials. Workers must wear rubber or plastic gloves, and serious infestations require special clothing and breathing apparatus.

HAP — See hazardous air pollution.

HAPEMS — Acronym for Hazardous Air Pollutant Enforcement Management System.

haploid — Having half the number of chromosomes found in the somatic cells of an organism.

HAPPS — Acronym for Hazardous Air Pollutant Prioritization System.

hapten — A molecule of small molecular weight that is immunogenic only when attached to carrier molecules, usually proteins.

haptics — The science of the study of touch.

hard coal — See anthracite.

hard copy — (*computer science*) A paper printout of information shown on a visual display unit.

hard disk — A magnetic disk made of rigid material for high-capacity, random-access storage.

hard swell — (*food*) A tin can bulging so tightly at both ends that no indentation can be made with thumb pressure.

hard water — Water containing dissolved minerals such as calcium, iron, and magnesium that do not lather with soap easily and which form insoluble deposits in boilers; also known as water hardness.

hardness — A measure of the concentration of the multivalent cations in the water equivalent to the approximate amount of calcium and magnesium present.

H

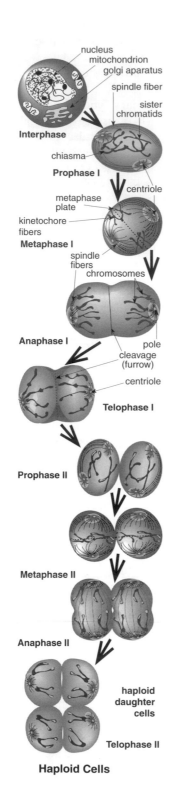

Haploid Cells

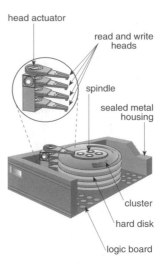

Hard Disk

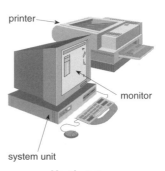

Hardware

hardware — The physical components of a computer.

harmonic — Describes a series of sounds, each with a frequency that is an integral multiple of a fundamental frequency.

harmonic tremor — (*geology*) Continuous-rhythmic earthquakes that can be detected by seismic graphs and that often precede or accompany volcanic eruptions.

harvest — To pick or gather a crop; to gather shellfish.

HATREMS — Acronym for Hazardous and Trace Emissions System.

haustorium — A food-absorbing outgrowth of a plant organ such as a hypha or stem.

HAV — Acronym for hepatitis A.

Haverhill fever — (*disease*) See streptobacillary fever.

hazard — Any biological, chemical, or physical risk to a worker, consumer, or other individual or environment; the probability that a concentration of a substance will cause a significant physiological response.

Hazard Analysis/Critical Control Point Inspection (HACCP) — An in-depth inspection process developed by the Food and Drug Administration

hardpan — A hardened soil layer in the lower A or in the B horizon caused by cementation of soil particles with organic matter or with materials such as silica sesquioxides or calcium carbonate; found in arid and semi-arid regions of the southwestern United States. Also known as caliche or kanker.

used to resolve or prevent disease outbreaks in restaurants and other food establishments by identifying the foods at greatest risk and the critical control points; the approach to use includes: identify the likely health hazards in a given product; identify the critical points in the processing where the hazards may occur; establish control measures to prevent the hazards from occurring; monitor the process to make sure the control measures are working; establish corrective actions if monitoring shows a problem; establish a detailed recordkeeping document monitoring the corrective actions taken; and verify that the system is working

Hazard and Operability Study (HAZOP) — A systematic technique for identifying hazards or operable problems throughout an entire facility.

hazard assessment — An evaluation of the effects of a stressor or the determination of a margin of safety for an organism by comparing the concentration that causes toxic effects with an estimate of exposure to the organism.

hazard classification — The rating for a system based on the potential consequences of failure.

Hazard Communication Standard — An OSHA regulation that requires chemical manufacturers, suppliers, and importers to assess the hazards of the chemicals that they make, supply, or import and to inform employees, customers, and workers of these hazards through MSDS sheets.

hazard criteria — Characteristics of waste that make it hazardous (e.g., ignitable, toxic, corrosive, reactive).

hazard evaluation — A component of risk evaluation that involves gathering and evaluating data on the types of health injuries or diseases that may be produced by a chemical and on the conditions of exposure under which such health effects are produced.

hazard identification — Gathering data on a specific substance, including information about the link between that substance and adverse health effects.

hazard index — The sum of more than one hazard quotient for multiple contaminants or multiple exposure pathways, or both.

hazard information transmission (HIT) — A digital transmission of CHEMTREC emergency data.

hazard minimalization — A reduction in the amount of waste generated by a product or process.

hazard probability — The likelihood that an accident will occur based on an assessment of the location, exposure frequency and duration, and the affected population.

hazard quotient — The ratio of a single contaminant exposure level over a specified time period to a reference dose for that contaminant derived from a similar time period.

Hazard Ranking System — A scoring system used to evaluate potential relative risk to public health and the environment from releases or threatened releases of hazardous substances with a site score of 0 to 100 based on actual or potential releases of the hazardous substances through air, surface water, or groundwater; a score of 28.5 places the site on the National Priorities List.

hazard ratio — Comparison of an animal's daily dietary intake of a pesticide to its LD_{50} value, where a ratio of >1.0 indicates that the animal is likely to consume a dose that would kill 50% of the animals of the same species.

hazard severity — An assessment of the worst potential consequences of a hazard (including degree of injury, occupational illness, health-related performance degradation, or bodily system damage) prior to implementing recommendations to eliminate or minimize the hazard.

hazard warning — Words, pictures, symbols, or a combination thereof presented on a label or other appropriate medium providing information regarding the presence of various hazardous materials.

hazardous air pollutant — An air pollutant to which no ambient air quality standard is applicable and which causes or contributes to air pollution, resulting in an increase in mortality or an increase in serious irreversible or incapacitating reversible illnesses.

hazardous air pollutants engineering program — A program designed to assess various industrial and combustion sources of hazardous air pollutants to determine the magnitude of emissions and the capability of technologies to reduce or eliminate them.

hazardous air pollution — See airborne toxics.

Hazardous and Solid Waste Amendments (HSWA) — A 1984 law banning liquid waste from land disposal; prohibiting certain land disposal practices; establishing minimum technology requirements for landfills, surface impoundments, and incinerators; expanding requirements for groundwater monitoring and cleanup; requiring double-liners in landfills; and expediting permits for new treatment technologies.

hazardous chemical — Any chemical that is explosive, caustic, flammable, poisonous, corrosive, reactive, or radioactive and which requires special care when handling because its presence or use is a physical or health hazard.

hazardous material — Any substance or compound capable of producing adverse health and/or safety effects.

Hazardous Material Information System (HMIS) — A file maintained by the Department of Defense containing Material Safety Data Sheets and trans-

H

H

Hazard Warning

portation data for products purchased by the Department of Defense and the General Services Administration.

hazardous materials manager — An environmental health practitioner responsible for the control and management of materials that are potentially hazardous, from the point of extraction of raw products to their elemental destruction, transformation into nonhazardous materials, or disposal in controlled facilities.

hazardous materials response team — An organized group of employees designated by an employer to perform work to handle and control actual or potential leaks or spills of hazardous substances that could require possible close approach to the substance for the purpose of control or stabilization of the situation.

hazardous substance — A material that poses a threat to human health and/or the environment when discharged into the environment in any quantity and from repeated or prolonged exposure; it may be toxic, corrosive, ignitable, explosive, or chemically reactive.

hazardous waste (HW) — The legal designation for discarded solid or liquid waste containing substances known to be mutagenic, synergetic, or teratogenic to humans or animals and considered to be ignitable, corrosive, reactive, or toxic as defined by Title 40 of the Code of Federal Regulations; excludes domestic, agricultural, and mining wastes.

hazardous waste discharge — Accidental or intentional spilling, leaking, pumping, pouring, emitting, emptying, or dumping of hazardous waste into any land or water.

hazardous waste indicators — Indicators that may signal the presence of hazardous waste including stressed vegetation, lack of vegetation, dead or sick domestic stock, wildlife, birds, fish, unusual coloration of land surfaces, and acrid or other chemical odors.

hazardous waste landfill — An excavated or engineered site where hazardous waste is deposited and covered.

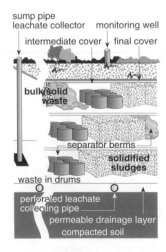

Hazardous Waste Landfill

hazardous waste management (HWM) facility — All contiguous land and structures or other appurtenances and improvements on the land used for treating, storing, or disposing of hazardous waste.

hazardous waste minimization — Reduction of the amount of toxicity or waste produced by a facility through source reduction and environmentally sound recycling.

hazardous waste sites — Any areas containing spills, dumping sites, abandoned or inactive waste disposal sites, or any other areas suspected of containing hazardous materials.

hazards analysis — Procedures used to identify potential sources of release of hazardous materials from fixed facilities or transportation accidents; determines the vulnerability of a geographical area to a release of hazardous materials; compares hazards to determine which presents greater or lesser risks to a community.

hazards identification — Information about which facilities have extremely hazardous substances and what they are, the amount being stored and method of storage, and whether or not they are used at high temperatures.

HazDat — Acronym for Hazardous Substance Release/Health Effects Database.

haze — A phenomenon that results in reduced visibility due to the scattering of light caused by aerosols, many of which are air pollutants and are produced by people.

HAZMAT — Acronym for hazardous materials.

HAZMAT team — See hazardous materials response team.

HAZOP — See Hazard and Operability Study.

hazy — See haze.

Hb — Acronym for hemoglobin.

HB — Acronym for hepatitis B.

HBFC — See hydrobromofluorocarbon.

HBIG — See hepatitis B immune globulin.

HBr — Scientific notation for hydrogen bromide.

HBV — Acronym for hepatitis B virus.

HC — Acronym for hazardous constituents or hydrocarbons; see hazardous chemical.

HCB — See hexachlorobenzene.

HCCPD — See hexachlorocyclopentadiene.

HCFC — Acronym for hydrochlorofluorocarbon; halogenated chlorofluorocarbons.

HCH — See hexachlorocyclohexane.

HCl — See hydrogen chloride.

HCN — See hydrogen cyanide.

HCP — Acronym for hypothermal coal process or hearing conservation program.

HCS — See Hazard Communication Standard.

HCV — Acronym for hepatitis C virus.

HDCV — Acronym for human diploid cell rabies vaccine.

HDD — Acronym for heavy-duty diesel.

HDDT — Acronym for heavy-duty diesel truck.

HDDV — Acronym for heavy-duty diesel vehicle.

HDE — Acronym for heavy-duty engine.

HDL — Acronym for high-density lipoprotein.

HDPE — See high-density polyethylene.

HDT — Acronym for highest dose tested (in a study).

HDTV — Acronym for high-definition television.

HDV — Acronym for hepatitis D virus.

He — See helium.

HE — Acronym for highly erodible.

head — (*physics*) A basic measurement of pressure or resistance in a hydraulic system; equivalent to the height of a column of water capable of exerting a given pressure.

head injury — A traumatic damage to the head resulting from a blunt or penetrating trauma of the skull where blood vessels, nerves, and meninges may be torn.

head loss — The energy loss from a flowing liquid due to friction, transitions, bends, etc.

head pressure — The pressure of the height of a fluid $(P) = yrg$, where y is the the height of the fluid column, r is the fluid density, and g is the the acceleration of gravity.

headache — A pain in the head due to any of a variety of causes.

header — (*computer science*) A prefixed section of a data file that contains metadata.

headspace — Vapor mixture trapped above a solid or liquid in a sealed vessel.

headwater — The source and upper part of a stream.

headworks — The facilities where wastewater enters a wastewater treatment plant consisting of bar screens, comminutors, a wet well, and pumps.

HEAL — Acronym for human exposure assessment location.

health — The avoidance of disease and injury and promotion of normalcy through efficient use of the environment, a properly functioning society, and an inner sense of well-being.

health advisory level — A nonregulatory, health-based reference level of chemical traces (in ppm) in drinking water having no adverse health risks when ingested over 1 day, 10 days, long-term, or lifetime exposure periods; a wide margin of safety exists.

health and safety guides (HSG) — basic information for employers and employees for a safe and healthful work environment.

health assessment — An evaluation of available data on existing or potential risks to human health posed by a Superfund site; ATSDR is required to perform such an assessment on every site on the National Priorities List.

health-based standard — An amount of air pollution that is scientifically determined not to produce such human health effects as asthma, emphysema, and cancer.

H

health consultation — A response to a specific question or request for information pertaining to a hazardous substance or facility, often containing a time-critical element that requires a rapid response.

health economics — Study of the supply and demand of healthcare resources and the effect of health services on a group of people.

health education — A program of activities to promote health and provide information and training about health hazards that will result in the reduction of exposure, illness, and disease and the promotion of wellness.

health hazard — A chemical for which significant evidence exists that acute or chronic health effects may occur in exposed individuals based on at least one study conducted in accordance with established scientific principles; the health hazards may be chemicals that are carcinogens, toxic or highly toxic agents, reproductive toxins, irritants, corrosives, sensitizers, hepatotoxins, nephrotoxins, neurotoxins, agents that act on the hematopoietic system, and agents that damage the lungs, skin, eyes, and mucous membranes of systems of the body.

health hazard assessment — Application of biomedical knowledge and principles to document and quantitatively determine the health hazards of a system.

Health Hazard Assessment Report — Formal documentation for a given system that includes an assessment of health hazards, recommendations for preventive or control actions, and recommendation for training requirements.

health history — The information obtained from a patient and other sources concerning the individual's physical status and psychological, social, and emotional concerns.

health investigation — An investigation of a defined population using epidemiological methods to determine exposures or possible public health impacts by defining health problems requiring further investigation for epidemiological studies, environmental monitoring and/or sampling, and surveillance.

health maintenance — A program or procedure used to prevent illness and to maintain maximum function in order to promote good health.

health maintenance organization (HMO) — A type of group healthcare practice that provides basic and supplemental health maintenance and treatment for a prepaid period that is set without regard to the amount or kind of services received.

health outcomes study — An investigation of exposed persons designed to assist in identifying exposure or effects on public health and defining health problems that require further inquiry by means of a health surveillance or epidemiological study.

health physics — Study of the effects of ionizing radiation on the body and the methods used for protection.

health policy — A statement of the goals for health care and a plan for achieving them.

health professional — A person who has completed a specialized course of study in a specific health field and has either been registered or licensed by a governmental agency or a professional organization.

health professional education — Any activity or activities directed toward public health professionals and the local medical community to improve the knowledge, skill, and behavior of health professionals concerning medical surveillance, screening, and methods of diagnosing, treating, and preventing injury or disease related to exposure to hazardous substances.

health risk — Anything related to a disease that may be associated with higher-than-average morbidity or mortality in the population.

health-risk appraisal — Gathering, analyzing, and comparing various characteristics related to individuals' health within a standard age group to determine the likelihood of disease occurring within that specific group of individuals.

health-risk assessment — (*air pollution*) A document that identifies the risks and number of occurences of possible adverse health effects that may result from exposure to emissions of toxic air contaminants; it cannot predict specific health effects but only describes the increased possibility.

health screening — A program used to evaluate the health status and potential for disease of an individual by use of physical examination, tests, and individual and family history.

health service area — A geographic area designated under the National Health Planning and Resources Development Act of 1974 based on such factors as geography, political boundaries, population, and health resources for effective planning and development of health services.

health standards — Published documents specifying conditions of acceptable risk for individual health hazards including medical exposure limits, health conservation criteria, and material design standards.

health statistics review — An evaluation of information and relevant health outcome data from local, state, or national databases and from private healthcare providers and organizations for an involved population; the data may include reports of injury, disease, or death and come from morbidity and mortality data, tumor and disease registries, birth statistics, and surveillance data.

health supervision — Monitoring the status of a patient's health including appropriate counseling and use of health education techniques.

health surveillance — Periodic medical screening of a defined population for a specific disease or for biological markers of disease for which the population is or is thought to be at significantly increased risk; also called medical monitoring.

health systems agency — Nonprofit private organizations, public regional planning bodies, or local governmental agencies that provide networks of health planning and resources development services in several different areas, as established by the National Health Planning and Resources Development Act of 1974.

healthcare facility — A facility or location where medical, dental, surgical, or nursing attention or treatment is provided to humans or animals.

healthcare industry — The totality of preventive, remedial, and therapeutic services provided by hospitals and other healthcare institutions, nurses, doctors, dentists, physical and occupational therapists, governmental agencies, voluntary agencies, pharmaceutical and medical equipment manufacturers, health insurance companies, and pharmacies.

healthcare institution — Any hospital, convalescent home, or other facility that provides health care, medical treatment, room, board, or other services for the ill or physically or mentally disabled.

healthcare provider — An individual who provides health services to people.

healthcare system — The totality of agencies, facilities, and providers of health care in a specific geographic area.

hearing — The general perceptual behavior and specific responses made in relation to sound stimuli; (*law*) the opportunity for a party to present views in response to a charge.

hearing aid — An electronic device that increases the level of sound for people with impaired hearing.

hearing level — The amount (in decibels) by which the threshold of audibility for an ear differs from a standard audiometric threshold.

hearing loss — Partial or complete loss of the sense of hearing. Conductive hearing loss is associated with impaired transmission of sound waves through the external ear canal to the bones of the middle ear; sensorineural hearing loss is associated with some pathological change in structures within the inner ear or in the acoustic nerve; central hearing loss occurs when a pathological condition is present above the junction of the acoustic nerve and the brain stem.

heart — A muscular, cone-shaped organ that pumps blood throughout the body by coordinated nerve impulses and muscular contractions.

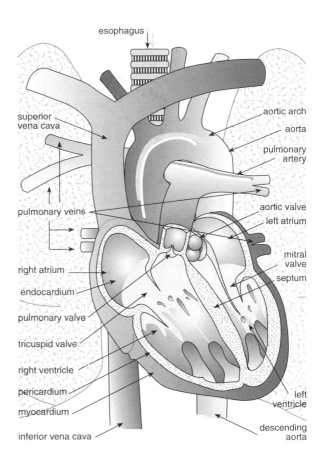

Heart

heart block — Interference with the normal conduction of electrical impulses that control the activity of heart muscle.

heart disease — A disease of the heart having a definite pathological process with a characteristic set of signs and symptoms; it may affect the heart itself or any of its parts and may be of unknown etiology with an unknown prognosis.

heart disease risk factors — The hereditary, lifestyle, and environmental factors that increase the potential for disease and injury.

heart failure — A condition in which the heart cannot pump enough blood to meet the needs of the other body tissues.

heart rate — The number of times the heart beats in one minute; pulse.

heartburn — The burning sensation in the esophagus just below the sternum caused by reflux of the gastric contents.

hearth — The bottom of a furnace where waste materials or fuels are exposed to the flame.

heat — A form of energy that raises the temperature of a body or substance. It is the kinetic energy of the molecules related to the capacity of the substance to do work. Heat will flow from a hot area to a cold area and can be created by mechanical, chemical, electric, or nuclear energy. Heat causes expansion in gases, liquids, and solids.

heat balance — An accounting of the distribution of the heat input and output of an incinerator, usually on an hourly basis.

heat capacity — The quantity of heat in calories needed to raise the temperature of 1 gram of a substance 1°C; also known as thermal capacity.

heat content — The sum total of latent and sensible heat present in a gas, liquid, or solid minus the heat present at an arbitrary set of conditions chosen as the base or zero point.

heat cramps — A condition marked by sudden development of cramps in skeletal muscles resulting from prolonged work in high temperatures, accompanied by profuse perspiration and loss of salt from the body.

heat drying — (*sludge*) The application of heat to evaporate sufficient moisture to render sludge dry to the touch and relatively freeflowing.

heat exchanger — Any device that transfers heat from one fluid to another without allowing them to mix.

heat exhaustion — A condition marked by weakness, nausea, dizziness, and profuse sweating, usually precipitated by physical exertion in a hot environment.

heat flux — The amount of heat passing through any surface per unit time.

heat gain — The heat in an area due to direct radiation, appliances, fireplaces, humans, or animals.

heat island — A reservoir for storing heat through absorption of sunrays during the day and the release of energy at night; typically occurs in cities.

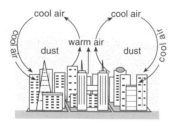

Heat Island

heat island effect — The phenomenon of air circulation found in urban areas with tall buildings; heat and pollution create a haze dome that prevents warm air from rising and being cooled at a normal rate.

heat loss — (*housing*) A decrease in the amount of heat contained in a space resulting from heat flow through walls, windows, roofs, and other building envelope components.

heat of combustion — (*chemistry*) The quantity of heat released per unit mass or unit volume of 1 mole

of a substance when the substance is completely burned at constant pressure.

heat of formation — (*chemistry*) The quantity of heat released when 1 mole of a compound is formed from its elements in the phases in which they exist at ordinary temperatures.

heat of fusion — The heat absorbed and used in the melting process of one unit mass of a solid to a liquid at constant temperature and pressure.

heat of neutralization — The amount of heat evolved when 1 gram equivalent of an acid is neutralized by 1 gram equivalent of a base; for strong acids and bases in dilute solution, the only reaction is the production of water.

heat of reaction — (*chemistry*) The quantity of heat released by a chemical reaction.

heat of vaporization — The heat required to change a liquid into a gas (e.g., water at 100°C does not change into steam until 540 calories of heat have been absorbed by each gram of the boiling water).

heat pump — A mechanical device that transfers heat from the heat source to a heat sink, causing the source to cool and the sink to become warmer; during the summer the heat pump reduces indoor temperatures by transferring heat to the ground, and during the winter the heat pump extracts heat from the ground and transfers it inside the structure.

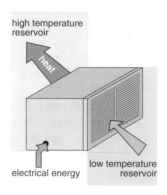

Heat Pump

heat rash — An inflammatory skin condition that occurs in hot, humid environments where sweat is not easily removed from the surface of the skin by evaporation and the skin stays moist; may cause a tingling or prickling sensation. Also known as prickly heat or miliaria.

heat release rate — The amount of heat liberated per second during complete combustion divided by its volume.

heat sensor — A device that opens and closes a switch in response to changes in the temperature.

heat sink — A body that is capable of accepting and storing heat and may be used as a heat source.

heat strain — The natural, physiological reaction of a body to the application of heat stress.

heat stress — Relative amount of thermal stress from the environment.

heat stress index (HSI) — The relation of the amount of evaporation or perspiration required for particular job conditions as related to the maximum evaporative capacity of an average person.

heat survey — A study conducted to determine the measurement of ambient temperatures, air motion, humidity, and radiant heat sources.

heat transfer — The movement of heat from an area of higher temperature to one of lower temperature until an equilibrium is reached; the heat is transferred by conduction, convection, or radiation.

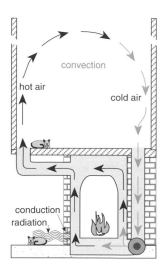

Heat Transfer

heat treatment — (*sewage*) The pressure cooking of sludges in such a manner that little sludge oxidation occurs; any of several processes of metal modification, such as annealing.

heat wave — A period of at least one day of abnormally hot weather.

heating — The transfer of energy to a space or to the air by the existence of a temperature gradient between the source and the space or air.

heating degree days — The number of degrees per day the daily average temperature is below 65°F.

heating value — The amount of heat generated by the complete combustion of a specific amount of fuel.

heatsink — (*physics*) Any substance, body, or region that absorbs or dissipates heat.

heatstroke — A condition marked by the cessation of sweating, extremely high body temperature, and collapse resulting from prolonged exposure to high temperature.

heavy-duty cleaners — Liquid products formulated to emulsify and soften wax and synthetic floor finishes for easy removal; also used to remove heavy soil from surfaces.

heavy hydrogen — Hydrogen isotope having an atomic weight greater than one; also known as deuterium; tritium.

heavy metal — A metallic element such as mercury, chromium, cadmium, arsenic, and lead that has a high molecular weight; can damage living organisms at low concentrations and tends to accumulate in the food chain.

heavy metal poisoning — A poisoning caused by the ingestion, inhalation, or absorption of various toxic heavy metals, including antimony, arsenic, cadmium, lead, and mercury.

heavy water — Water containing significantly more than the natural proportion (one in 6500) of heavy hydrogen (deuterium) atoms to ordinary hydrogen atoms; used as a moderator in a nuclear reactor; also known as deuterium oxide.

hectare — A square kilometer, which is equal to 0.3861 of a square mile.

hecto- (h) — Prefix representing 100 in the metric system.

hectogram — A unit of metric measure equal to 100 grams.

HEEP — Acronym for Health and Environmental Effects Profile.

Heimlich maneuver — An emergency procedure used to dislodge an object obstructing the trachea in order to prevent asphyxiation, accomplished by grasping the choking person from behind and placing the fist, thumb side in, just below the victims sternum. The fist is covered by the other hand in a firm and abrupt manner to force the obstruction up the trachea.

helical — In the shape of a spiral.

helium — A colorless, odorless, rare or inert gas with an atomic number of 2 and an atomic weight of 4; used to lift airships and balloons as well as in respiratory therapy, testing, and the prevention of nitrogen narcosis and decompression sickness in hyperbaric environments.

helium therapy — The use of helium–oxygen gas mixtures to treat patients with airway obstruction; helium has a low density and can move around an object or obstruction more easily.

helix — A coiled spiral-like formation characteristic of many organic molecules such as DNA.

helmet — A device that protects the eyes, face, neck, and other parts of the head.

Helmholtz coils — A pair of flat, circular coils with an equal number of turns and equal diameters arranged with a common axis and connected in series to have a common current. The purpose of the arrangement is to obtain a magnetic field that is more nearly uniform than that of a simple coil.

H

helminth — Any parasitic worm; a nematode or trematode.

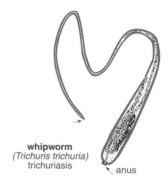

whipworm
(Trichuris trichuria)
trichuriasis

anus

roundworm
(Strongyloides stercoralis)
strongyloidiasis

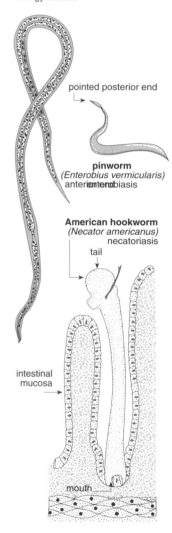

pointed posterior end

pinworm
(Enterobius vermicularis)
anterior end enterobiasis

American hookworm
(Necator americanus)
necatoriasis

tail

intestinal
mucosa

mouth

Some pathogenic **helminths:
nematodes** (not to scale)

helminthemesis — The vomiting of intestinal worms.
helminthiasis — A parasitic infestation of the body by helminths that may be cutaneous, visceral, or intestinal.

helminthic — Relating to worms.

helper/suppressor T cells — White blood cells that are part of the immune system.

HEM — Acronym for human exposure modeling.

hemagglutination — The coagulation of erythrocytes.

hemagglutinin (H) protein — One of two main proteins found on the surface of a virus that are necessary for the virus to attach to the host cell.

hemagglutinin — A type of antibody that promotes clumping of red blood cells.

hemapoiesis — The formation of blood cells.

hematemesis — The vomiting of bright red blood, indicating upper gastrointestinal bleeding.

hematinuria — A dark-colored urine resulting from the presence of hematin or hemoglobin.

hematocyte — A blood cell, especially a red blood cell.

hematogenic shock — A condition of shock caused by the loss of blood or plasma.

hematogenous — Originating in the blood or disseminated by circulation through the bloodstream.

hematologist — A doctor who specializes in treating diseases of the blood.

hematology — The study of anatomy, physiology, pathology, symptomatology, and therapeutics related to the blood and blood-forming tissues.

hematoma — A localized blood clot under the surface of the skin.

hematopoietic system — The blood-forming mechanism of the human body; also known as the reticuloendothelial system.

hematopoietic toxin — Any chemical, including carbon monoxide and cyanide, that decreases hemoglobin function and deprives the body tissues of oxygen, causing symptoms of cyanosis and loss of consciousness.

hematuria — Blood in the urine.

heme ($C_{34}H_{32}O_4N_4Fe$) — The nonprotein, insoluble, iron proroporphyrin constituent of hemoglobin.

hemimetabolism — See incomplete metamorphosis.

hemisphere — One half of the Earth, usually thought of as a division of the globe into two equal parts: north and south or east and west.

hemocyte — A blood corpuscle or formed element of the blood.

hemodialysis — A procedure in which body wastes are removed from the blood by shunting the patient's blood through a machine for diffusion and ultrafiltration and then returning it to the patient's circulatory system; used for treating renal failure and various toxic conditions.

hemoglobin — The iron-containing, oxygen-carrying protein compound in red blood cells that gives them their color.

hemoglobinuria — Bloody urine.

hemolysin — A substance that liberates hemoglobin from erythrocytes by interrupting their structural integrity.

hemolysis — Rupture of erythrocytes with a release of hemoglobin into the plasma.

hemolytic — Pertaining to, characterized by, or producing hemolysis.

hemolytic staphylococci — Strings of cocci that cause the red blood cells to leave the hemoglobin.

hemolytic uremic syndrome — A disease characterized by kidney failure and neurological failure, especially occurring in children under 5 years of age and in the immunocompromised elderly.

hemopoietic — Relating to the process of formation and development of various types of blood cells.

hemoptysis — Bleeding from the lungs, larynx, trachea, or bronchi.

hemorrhage — The escape of blood from the vascular system.

hemorrhagic colitis — An acute disease caused by *Escherichia coli* 015:H7 (enterohemorrhagic strain); symptoms generally include severe abdominal cramping, watery diarrhea that may become grossly bloody, occasional vomiting, and possible low-grade fever. Incubation time is 2 to 9 days. The disease has been found in the Pacific northwest, Canada, and Michigan. The reservoir of infection is cattle, and it is transmitted by raw or improperly cooked ground beef or raw milk. It is not communicable from person to person; general susceptibility, but the very young may develop renal failure and the very old may develop fever and neurologic symptoms that can lead to death. It is controlled by proper refrigeration and cooking of ground beef and using only pasteurized milk. Also known as enterohemorrhagic colitis.

hemorrhagic fever — A group of viral aerosol infections with symptoms of fever, chills, headache, and respiratory or gastrointestinal symptoms, followed by hemorrhaging, kidney failure, hypotension, and possibly death.

hemorrhagic gastritis — A form of acute gastritis usually caused by a toxic agent such as alcohol, aspirin, bacterial toxins, or other chemicals with symptoms of nausea, vomiting, and epigastric distress.

hemorrhagic plague — A form of severe plague with bleeding under the skin.

hemorrhagic shock — A physical collapse and prostration caused by the sudden and rapid loss of significant amounts of blood during severe traumatic injuries.

hemostasis — Stopping of bleeding by mechanical or chemical means or by the coagulation process of the body.

hemothorax — An accumulation of blood and fluid in the pleural cavity, usually as the result of trauma.

hemotoxin — A poison that destroys red blood cells and breaks down the walls of small blood vessels.

hemozoin — The insoluble digestion product of malaria parasites produced from hemoglobin.

Henry's law — At a constant temperature, the solubility of a gas dissolved in a liquid at equilibrium is directly proportional to the partial pressure of the gas.

HEPA — See high-efficiency particulate air filter.

hepatic — Pertaining to the liver.

hepatitis — Inflammation of the liver, commonly of viral origin.

hepatitis A — A form of infectious viral hepatitis caused by the hepatitis A virus with slow onset of signs and symptoms; it is spread by direct contact with food or water contaminated by feces.

hepatitis B — A form of viral hepatitis caused by the hepatitis B virus and transmitted by contaminated serum in blood transfusions, by sexual contact with an infected person, or by use of contaminated needles and instruments, resulting in prolonged illness, destruction of liver cells, cirrhosis, or death.

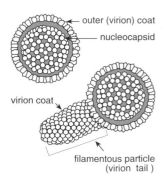

outer (virion) coat
nucleocapsid
virion coat
filamentous particle
(virion tail)

Hepatitis B Virus

hepatitis B immunoglobulin (HBIG) — A preparation that provides some temporary protection following exposure to the hepatitis B virus if given within 7 days after exposure.

hepatitis C — A type of hepatitis usually transmitted by blood transfusion or percutaneous inoculation when intravenous drug users share needles; it progresses to chronic hepatitis in 50% of the cases.

hepatitis D — A form of hepatitis that occurs only in people infected with hepatitis B and usually develops into a chronic state; it is transmitted sexually and through needle sharing.

hepatitis E — A self-limiting type of hepatitis that may occur after a natural disaster because of the consumption of fecal-contaminated water or food.

hepatolenticular degeneration — See Wilson's disease.

hepatoma — A malignant tumor occurring in the liver.

hepatomegaly — Enlargement of the liver.

hepatosis — Any functional disorder of the liver.

hepatotoxin — An agent capable of damaging the liver and causing symptoms such as jaundice and liver enlargement; chemicals include carbon tetrachloride and nitrosamines.

heptachlor ($C_{10}H_5Cl_7$) — White to light-tan crystals with a camphor-like odor; MW: 373.4, BP: 293°F (decomposes), Sol: insoluble, sp. gr. 1.66. It is used in the formulation of pesticides; as an insecticide in seed treatment. In animals, it is hazardous to the liver and central nervous system and is toxic through inhalation, absorption, ingestion, and contact. Symptoms of exposure in animals include tremors, convulsions, liver damage; carcinogenic. OSHA exposure limit (TWA): 0.5 mg/m^3 [skin].

n-heptane ($CH_3(CH_2)_5CH_3$) — A colorless liquid with a gasoline-like odor; MW: 100.2, BP: 209°F, Sol: 0.005% at 60°F, Fl.P: 25°F, sp. gr. 0.68. It is used as a carrier and penetrant solvent for adhesives, in azeotrophic distillations, and in rubber tire manufacture; as a solvent in rubber cements; as an ink solvent in gravure printing; as a solvent in the polymer industry; as a swelling and blowing agent for plastic foams; as a reference fuel for testing gasoline engine knock and for pollution and combustion studies; as a diluent solvent for lacquers during preparation and application; in organic chemical synthesis; as a laboratory solvent for scientific testing. It is hazardous to the skin, respiratory system, and peripheral nervous system and is toxic through inhalation, ingestion, and contact. Symptoms of exposure include lightheadedness, giddiness, stupor, lack of appetite, nausea, dermatitis, chemical pneumonia, and unconsciousness. OSHA exposure limit (TWA): 400 ppm [air] or 1600 mg/m^3.

2-heptanone — See methyl (n-amyl) ketone.

3-heptanone — See ethyl butyl ketone.

herbaceous — Something which is green and leaf-like.

herbicide — A pesticide chemical used to prevent or destroy the growth of plants.

herbicide poisoning — A poisoning caused by the ingestion, inhalation, or absorption of a chemical used as a weed killer or defoliant; symptoms include skin irritation, hypotension, liver and kidney damage, coma, or convulsions.

herd immunity — The resistance of a group to invasion and spread of an infectious agent based on the immunity of a high proportion of individual members of the group.

hereditary — Describes an inborn or inherited characteristic, condition, or disease transmitted from a parent to an offspring.

heredity — The sum of transmission of traits from parent to offspring.

hermaphroditic — An organism that has both male and female reproductive organs.

hermetically sealed container — An airtight container designed and intended to be secure against the entry of microorganisms and to maintain the commercial sterility of its contents after processing.

hernia — An abnormal protrusion of tissue through the wall that contains it.

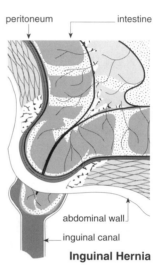

peritoneum   intestine

abdominal wall

inguinal canal

Inguinal Hernia

herniated disk — A rupture of the fibrocartilage surrounding an intervertebral disk that releases a pulpy elastic substance that cushions the surrounding vertebra.

HERS — Acronym for Hyperion Energy Recovery System.

hertz (Hz) — The unit for frequency equal to 1 cycle per second.

heterocyclic — A cyclic or ring-like molecular structure in which one or more of the atoms in the ring are elements other than carbon.

heterogenous — Having dissimilar parts or elements.

heterogonic life cycle — A life cycle involving the alternation of parasitic and free-living generations.

heterotroph — An organism that feeds upon organic compounds produced by other organisms in order to synthesize its own food; see also consumer.

heterotrophic nitrification — Biochemical oxidation of ammonium to nitrite and nitrate by a heterotrophic microorganism.

heterotrophic plate count — The number of colonies of heterotrophic bacteria grown on selected solid media at a given temperature and incubation period, usually expressed in number of bacteria per milliliter of sample.

heterotrophic reducers — Bacteria or fungi that return the complex organic compounds to their original abiotic state and release the remaining chemical energy.

heuristic — Relating to a teaching method in which the student is encouraged to do independent research and study.

HEV — See hybrid electric vehicle.

hexachlorobenzene (HCB; C_6Cl_6) — White or colorless, needle-like crystals that occur as a byproduct during the manufacture of chlorine-containing compounds; MW: 284.79, BP: 322°C (sublimes), Sol: 0.006 mg/l, Fl.P: 242°C. It is not currently in commercial use in the United States. Prior to 1985, it was used as a pesticide and in the production of synthetic rubber and of pyrotechnic and ordinance materials for the military. It is hazardous to the liver, digestive system, central nervous system, and reproductive system and is toxic through ingestion. Symptoms of exposure include skin lesions; weakness; convulsions; osteoporosis, especially of the bones of the hands; and muscle damage. OSHA does not regulate exposure limits. World Health Organization drinking water guidelines: 0.01 μg/l.

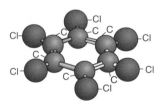

Hexachlorobenzene

hexachlorobutadiene (C_4Cl_6) — A colorless liquid with a turpentine-like odor; MW: 260.76, BP: 215°C, Sol: 2–2.55 mg/l at 20°C in water, density: 1.55 g/cm³ at 20°C. It is used as a chemical intermediate in the manufacture of rubber compounds; as a solvent, a fluid for gyroscopes, a heat-transfer liquid, hydraulic fluid; as a chemical intermediate in the production of chlorofluorocarbons and lubricants. It is hazardous to the respiratory system, kidneys, and liver in animals; may cause death and is toxic through inhalation, ingestion, and contact. Symptoms of exposure in animals include nasal irritation, labored and slow breathing, damaged cortical

and proximal renal tubules. OSHA exposure limit (TWA): 0.002 ppm [air].

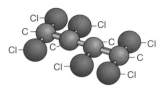

Hexachlorobutadiene

hexachlorocyclohexane (HCH; $C_6H_6Cl_6$) — A white, solid, synthetic chemical existing in eight different isomer forms (γ-HCH is called lindane); MW: 290.83, BP: 60°C at 0.36 mmHg to 323.4°C at 760 mmHg depending on the isomer, Sol: 5 to 17 ppm, density: 1.89 to 1.891 at 19°C. It is used as an insecticide, therapeutic scabicide, pediculicide, and ectoparasiticide for humans and animals. It is hazardous to the respiratory system, neurological system, cardiovascular system, gastrointestinal system, and musculoskeletal system and is toxic through inhalation, ingestion, and contact. Symptoms of exposure include mucous membrane irritation of the nose and throat, electrocardiogram abnormalities, hypochromic anemia, generalized urticaria, paresthesia of the face and extremities, and headache, with occasional deaths. Formerly known as benzene hexachloride (BHC). OSHA exposure limit (TWA): 0.5 mg/m³ [air].

hexachlorocyclopentadiene (C_5Cl_6) — A yellow to yellow-green, dense, oily liquid with a pungent odor: MW: 272.8, BP: 239°C, Sol (water): 0.0002% (reacts), Fl.P: NA (poisonous gas produced in a fire), sp. gr. 1.71. It is used as a chemical intermediate for many insecticides, polyester resins, dyes, pharmaceuticals, and flame retardants. It is hazardous to the respiratory system, digestive system, skin, and eyes. Acute exposure may cause severe irritation of the skin, eyes, and respiratory tract. It also causes sneezing; watery eyes; headache; burning of the eyes and throat; damage to the liver, kidneys, and nervous system; pulmonary edema; and death. Chronic exposure may damage the kidneys, the liver, nervous system, and lungs. OSHA exposure limit (8-hour TWA): 0.1 ppm.

hexachloroethane (Cl_3CCCl_3) — Colorless crystals with a camphor-like odor; MW: 236.7, BP: sublimes, Sol (72°F): 0.005%, sp. gr. 2.09. It is used in the manufacture of pyrotechnics; in lubricants in foundries for mold treatment; in preparation of nitrocellulose esters and camphor substitutes; as a fumigant, insecticide, and fungicide; in the manufacture of synthetic rubber; as a chemical additive; as an additive to fire extinguishers. It is hazardous to the eyes and is toxic through inhalation, absorption, ingestion, and contact. Symptoms of exposure include eye irritation; carcinogenic. OSHA exposure limit (TWA): 1 ppm [skin] or 10 mg/m^3.

Hexachloroethane

H

hexachloronaphthalene (C$_{10}$H$_2$Cl$_6$) — A white to light-yellow solid with an aromatic odor; MW: 334.9, BP: 650 to 730°F, Sol: insoluble, sp. gr. 1.78. It is used in the manufacture of electric equipment as insulating material; as an inert component of resins or polymers for coating and impregnating textiles, wood, and paper for its flameproofing, waterproofing, fungicidal, and insecticidal properties; as an additive to special lubricants and cutting oils; in polymer manufacture as fillers. It is hazardous to the liver and skin and is toxic through inhalation, absorption, ingestion, and contact. Symptoms of exposure include acne-forming dermatitis, nausea, confusion, jaundice, coma. OSHA exposure limit (TWA): 0.2 mg/m^3 [skin].

hexamethylene diisocyanate ((CH$_2$)$_6$(NCO)$_2$) — A colorless liquid; BP: 213°C, Sol: soluble in most organic solvents and decomposes with water, Fl.P: 140°C (open cup), may react violently to strong oxidizers, acids, and bases. It is used in the production of polyurethane foam. It is hazardous to the respiratory tract, skin, and eyes. It causes wheezing, dyspnea, sweating, coughing, breathing difficulty, and insomnia. ACGIH exposure limit (TLV-TWA): 0.005 ppm.

***n*-hexane (C$_6$H$_{14}$)** — A colorless liquid with a faint odor; MW: 86.20; BP: 69°C; Sol: miscible with organic solvents; Fl.P: 30.5°C (closed cup), a highly flammable liquid; density: 0.66 g/nl at 20°C. It is used as an extractant of agricultural products and animal fat; in petroleum ether and gasoline; as a rubber solvent; in the manufacture of polyolefins and certain elastomers as a catalyst carrier; as a solvent

in adhesives; in the pharmaceutical industry as a reaction medium, immiscible solvent, and extraction ergot; in compounding printing inks, lacquers, or stains; as a laboratory reagent and general solvent; in the manufacture of low-temperature thermometers. It is hazardous to the respiratory tract. It causes acute toxicity leading to convulsions and death in animals. In humans, it may cause hallucinations, distorted vision, headaches, dizziness, nausea, and irritation of the eyes and throat. It has reproductive effects in rats and mice. ACGIH exposure limit (TLV-TWA): 50 ppm [air] or 180 mg/m^3.

$$CH_3-CH_2-CH_2-CH_2-CH_2-CH_3$$

***n*-Hexane**

2-hexanone (CH$_3$CO(CH$_2$)$_3$CH$_3$) — A colorless liquid with an acetone-like odor; MW: 100.2, BP: 262°F, Sol: 2%, Fl.P: 77°F, sp. gr. 0.81. It is used as a commercial solvent for nitrocellulose, natural and synthetic resins, oils, waxes, vinyl polymers, and cellulose acetates; as a solvent in the manufacture of varnish removers, vinyl lacquers, and nitrate wood lacquers; as an extractive solvent for paraffin wax; in the separation and purification of certain metals. It is hazardous to the central nervous system, skin, and respiratory system and is toxic through inhalation, absorption, ingestion, and contact. Symptoms of exposure include peripheral neuropathy, weakness, paresthesia, dermatitis, headache, drowsiness, and irritation of the eyes and nose. OSHA exposure limit (TWA): 5 ppm [air] or 20 mg/m^3.

hexavalent — An atom able to bond with six other atoms.

hexone (CH$_3$COCH$_2$CH(CH$_3$)$_2$) — A colorless liquid with a pleasant odor; MW: 100.2, BP: 242°F, Sol: 2%, Fl.P: 64°F, sp. gr. 0.80. It is used as a separating agent for certain inorganic salts; in the extraction and the manufacture of antibiotics and purification of petroleum products; in the manufacture of drycleaning preparations, germicides, fungicides, and electroplating solutions; in blending raw materials for molded plastics; in cleaning and maintaining ketone processing equipment; in the application and drying of lacquers, varnishes, epoxy, acrylic, vinyl, or other cellulose- or resin-based coatings, finishes, and adhesives. It is hazardous to the respiratory system, eyes, skin, and central

$$O=C=N-CH_2-CH_2-CH_2-CH_2-CH_2-CH_2-N=C=O$$

Hexamethylene diisocyanate

nervous system and is toxic through inhalation, ingestion, and contact. Symptoms of exposure include headache, narcosis, coma, dermatitis, and irritation of the eyes and mucous membrane. OSHA exposure limit (TWA): 50 ppm [air] or 205 mg/m³.

sec-**hexylacetate** **($C_8H_{16}O_2$)** — A colorless liquid with a mild, pleasant, fruity odor; MW: 144.2, BP: 236°F, Sol: 0.08%, Fl.P: 113°F, sp. gr. 0.86. It is used in the application of nitrocellulose lacquers and other lacquers; as an inhibitor in handling diacyl peroxide solutions. It is hazardous to the eyes and central nervous system and is toxic through inhalation, ingestion, and contact. Symptoms of exposure include headache; in animals, narcosis and irritation of the eyes, nose, and throat. OSHA exposure limit (TWA): 50 ppm [air] or 300 mg/m³.

HFC — See hydrofluorocarbon.

HFE — Acronym for unit factors engineering.

HFID — Acronym for heated flame ionization detector.

HFO — Acronym for high-frequency oscillation.

HFV — See high-frequency ventilation.

Hg — See mercury.

HGF — Acronym for human growth factor.

HHAR — See Health Hazard Assessment Report.

HHE — Acronym for human health and the environment.

HHV — Acronym for higher heating value.

HI — Acronym for hazard index.

hidosis — The production of sweat and secretions.

high — (*weather*) An area of high-pressure areas that may develop any place where air cools, compresses, and sinks; wind movement that follows a normal flow from high pressure to low pressure.

high altitude — Any elevation over 4000 feet.

high clouds — Types of clouds composed almost entirely of tiny ice crystals, with bases averaging 2000 feet above the surface of the Earth.

high-density polyethylene (HDPE) — A material used to make plastic bottles that produces toxic fumes when burned.

high-efficiency particulate air (HEPA) filter — Designation for a type of filter that is 99.97% efficient in removing particles 0.3 μm in diameter from a body of air.

high-end exposure — A reasonable estimate of an individual's risk to an exposure that is greater than the 90th percentile.

high-flow pump — A pump used for sampling particulates, gases, and vapors where the flow rate is 2 liters per minute of air.

high-frequency loss — A hearing deficit starting at 2000 hertz or greater.

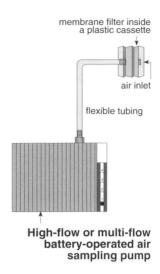

High-flow or multi-flow battery-operated air sampling pump

high-frequency sound waves — A frequency of 3 to 30 megahertz that produces high-pitched or shrill tones.

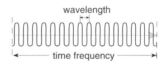

High-Frequency Sound Waves

high-frequency ventilation — A technique of providing ventilation support to patients at a breathing rate of 60 breaths per minute or more with small tidal volume.

high heat value — The BTUs liberated when a pound of solid waste is burned completely and the products of combustion are cooled to the initial temperature of the solid waste.

high-intensity discharge — A generic term for mercury vapor, metal halide, and high-pressure sodium lamps and fixtures.

high level — Having a high level of radioactivity.

high-level radioactive waste — Aqueous waste from the operation of a nuclear reactor.

high-level waste repository — A place where intensely radioactive wastes can be buried deep in the ground and remain unexposed to groundwater and earthquakes for tens of thousands of years.

high-line jumpers — Pipes or hoses connected to fire hydrants and laid on top of the ground to provide emergency water service for an isolated portion of the distribution system.

high-performance liquid chromatography (HPLC) — A technique used to determine the amount of a compound in various media such as the air, water, or blood; samples of the desired media are dissolved in a solution and the components are separated by injecting the solution through a mobile-phase solution and then a stationary-phase solution to separate out different components of the material, which can then be analyzed separately as they pass through the detector.

high-pressure cell — See air mass.

high-quality energy — The intense, concentrated, and high-temperature energy that is considered high quality because of its usefulness in carrying out work.

high-resolution infrared radiation sounder (HRIRS) — A satellite instrument used to determine vertical temperature profiles.

high-resolution — (*computer science*) Describing the quality and accuracy of detail presented by a graphics system or a computer printout.

high-risk community — A population area located near numerous sites or facilities with potential or actual environmental health hazards that may result in high levels of exposure to contaminants or pollutants.

high-risk, high-population areas — Heavily populated areas of the United States particularly susceptible to high-intensity earthquakes for which federal emergency response may be necessary in the event of an earthquake, including Honolulu, HI; San Diego, Los Angeles, and San Francisco, CA; Puget Sound, WA; Anchorage, AK; Salt Lake City, UT; the seven-state area of the central United States that includes Missouri, Kentucky, Tennessee, Mississippi, Arkansas, Indiana, and Illinois; Charleston, SC; Boston, MA; New York; Puerto Rico; and the Virgin Islands.

high-temperature, short-time pasteurization (HTST) — Pasteurization of milk that occurs at 161°F for 15 seconds.

high tide mark — The highest level reached by a rising tide.

high-to-low extrapolation — (*animals*) A mathematical manipulation of data that allows for chemicals to be expressed in terms of dose and cancer risk.

high-volume sampler (Hi-Vol) — A device operating somewhat like a vacuum cleaner and used in the measurement and analysis of suspended particulate pollution; it is operated at an average sampling rate of 72,000 ft³ per 24-hour period.

high-velocity air filter (HVAF) — An air pollution control filtration device for removal of sticky, oily, or liquid aerosol particulate matter from exhaust gas streams.

higher plants — Plant forms traditionally recognized as plants in contrast to simpler plant forms such as fungi and algae.

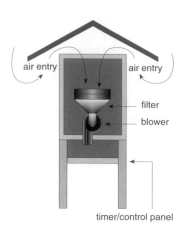

High-Volume (Hi-Vol) Sampler

highly alkaline detergent — A heavy-duty cleaning material used to remove wax and baked-on grease in ranges, ovens, dish washing machines, and power sprays.

highly susceptible population — A group of persons who are more likely than other populations to experience disease because their immune systems have been compromised, such as with the elderly, very young, or those with chronic diseases.

highly toxic — Relating to a chemical with a median lethal dose (LD_{50}) or concentration (LC_{50}) that is fatal to laboratory animals when administered by physical contact, ingestion, or inhalation; relating to a chemical with a median lethal dose (LD_{50}) of 50 mg or less per kilogram of body weight when administered orally to albino rats weighing between 200 and 300 g each; a chemical with a median lethal dose (LD_{50}) of 200 mg or less per kilogram of body weight when administered by continuous contact for 24 hours, or less if death occurs within 24 hours, applied to the bare skin of albino rabbits weighing between 200 and 300 grams each; a chemical that has a median lethal concentration (LC_{50}) in air of 200 ppm by volume or less of gas or vapor, or 2 mg per liter or less of mist, fume, or dust when administered by continuous inhalation for 1 hour, or less if death occurs within 1 hour, to albino rats weighing between 200 and 300 grams each.

highly volatile — A liquid that quickly forms a gas or vapor at room temperature.

Hill reaction — The splitting of a molecule of water during the light reactions of photosynthesis; the photolysis of water.

Hill–Burton Act — Legislation and programs operated under the legislation for federal support to construct and modernize hospitals and other health facilities beginning with public law 79-725, the Hospital Surveying Construction Act of 1946.

HIMSS — Acronym for high-resolution microwave spectrometer sounder.

HIRIS — Acronym for high-resolution imaging spectrometer sounder.

HIRS — See high-resolution infrared radiation sounder.

histamine (C_5HgN_3) — A powerful dilator of the capillaries; also thought to be responsible for many human allergies.

histocyte — A macrophage of connective tissue that is involved in the immune system.

histiotypic growth — The uncontrolled proliferation of cells occuring in bacterial cultures and molds.

histogenesis — The formation and development of tissue.

histogram — (*statistics*) A diagram showing the number of samples and their distribution by category.

histology — Study of the structure of cells and tissues as related to their function.

histopathology — A branch of pathology dealing with tissue changes associated with disease.

histoplasmosis — (*disease*) A systemic infectious mycosis of varying severity with the primary lesion usually in the human lungs; symptoms range from none to chronic pulmonary symptoms. Incubation time is 5 to 18 days, commonly 10 days. It is caused by *Histoplasma capsulatum*, a fungus that grows as mold in soil or as a yeast in animal and human hosts, and is found in many parts of the world including the Americas. The reservoirs of infection include soil around old chicken houses, in caves, around starling roosts, soils with high organic content, and decaying trees. It is transmitted through inhalation and is not communicable from person to person; general susceptibility. It is controlled by disinfection of sputum and articles soiled with sputum and by pharmaceutical treatment.

historical geology — Study of the geologic history of the Earth and its inhabitants.

histotoxin — A substance that is poisonous to the body tissues and usually introduced externally.

histozoic — Dwelling within the tissues of a host.

HIT — See hazard information transmission.

hit space theory — See target theory.

HIV — See human immunodeficiency virus.

hives — A skin condition with elevations and itching resulting from an allergic reaction; also known as urticaria.

Hi-Vol — See high-volume sampler.

HLRW — See high-level radioactive waste.

HM — See hazardous materials.

HMEP — Acronym for hazardous materials emergency preparedness.

HMMP — Acronym for hazardous materials management plan.

HMO — See health maintenance organization.

HMS — Acronym for highway mobile source.

HMTA — Acronym for Hazardous Materials Transportation Act.

HMW — Acronym for high molecular weight.

HNO_3 — See nitric acid.

HNS — Acronym for hazardous and noxious substances.

H_2O — Scientific notation for water.

H_2O_2 — Scientific notation for hydrogen peroxide.

HOC — See halogenated organic carbons; acronym for hydrophobic organic compounds.

HOCl — See hypochlorous acid.

Hodgkin's disease — A lymphatic system cancer that is painless but has progressive enlargement of the lymph nodes, spleen, and general lymphoid tissues.

holding pond — A structure built to contain large volumes of liquid waste to ensure that it meets environmental requirements prior to release.

holding tank — A watertight receptacle designed and constructed to receive and retain sewage for ultimate disposal at another site.

holding time — (*milk pasteurization*) Flow time of the fastest particle of milk at or above 161°F through the holder section, which is a portion of the system outside the influence of the heating medium that slopes continuously upward in the downstream direction and is located upstream from the flow-diversion valve.

holistic medicine — Treatment of the whole entity: the human body, spirit, and mind.

hologram — A realistic, three-dimensional image produced by a laser and photographic plates.

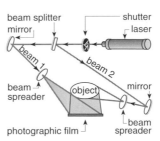

Laser System
in the making of a hologram

Holter monitor — A piece of equipment used to make a 24-hour record of electrocardiographic recordings on a portable tape recorder while the person wearing it conducts normal daily activities.

homeostasis — A steady state that an organism maintains by self-regulating adjustments to its internal environment.

Homestead Act — Legislation passed in 1862 allowing any citizen or applicant for citizenship over 21 years old and head of a family to acquire 160 acres of public land by living on it and cultivating it for 5 years.

homing — The ability of an animal to return to its home site after being displaced.

homogeneous — Pertaining to a substance having uniform composition and structure.

homogenized milk — Milk that has been treated in such a manner as to ensure breakup of the fat globules to such an extent that, after 48 hours of quiescent storage, no visible cream separation occurs in the milk, and the fat percentage of the top 100 ml of milk in a quart bottle does not differ by more than 10% from the fat percentage of the remaining milk.

homogenizing — The process of making a substance more uniform.

homoiothermal — (*animals*) Referring to an organism that maintains a relatively constant internal temperature regardless of the environmental temperature; also known as warm-blooded.

hood — The point of entry into a local exhaust system.

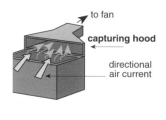

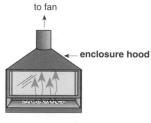

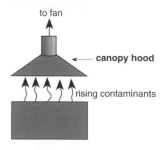

Hoods

hood capture efficiency — The ratio of the emissions captured by a hood and directed into a control or disposal device, expressed as a percent of all emissions.

hood entry loss — The pressure loss from turbulence and friction as air enters a ventilation system.

hood slot — A hood consisting of a narrow slot leading into a plenum chamber under suction to distribute air velocity along the length of the slot.

hood static pressure — The suction or static pressure in a duct near a hood which represents the suction that is available to draw air into the hood.

Hooke's law — The stress within an elastic solid is proportional to the strain responsible for it.

hookworm disease — (*disease*) A common chronic parasitic infection with various symptoms including anemia. Incubation time is a few weeks to many months. It is caused by *Necator americanus*, *Ancylostoma duodenale*, and *A. ceylanicum* and is found in tropical and subtropical climates. The reservoirs of infection include people for *N. americanus* and *A. duodenale* and cats and dogs for *A. ceylanicum*. It is transmitted by eggs in feces that are deposited on the ground and later hatch. Larvae develop and become infective in 7 to 10 days; they penetrate the skin, usually at the foot, and pass through the lymphatics and bloodstream to the lungs where they enter the alveoli. They then migrate up the trachea to the pharynx, are swallowed, and reach the small intestine, where they attach to the intestinal wall. The disease is not transmitted from person to person, but infected people can contaminate soil for several years; general susceptibility. It is controlled through public education, wearing shoes, and proper disposal of sewage. Also known as ancylostomiasis.

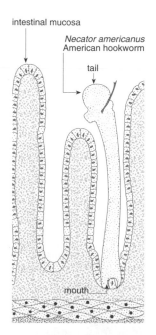

Hookworm Disease

hopper — A storage bin or funnel that is loaded from the top and discharges through a door or chute in the bottom.

horizons — Soil zones from the surface downward that reveal visible horizontal layers of soil.

hormone — A chemical secretion from one of the ductless glands of the body that acts as a catalyst in regulating specific activities of certain cells or organs.

horsepower (hp) — A unit of power equal to 745.7 watts.

hose bib — A location in a water line where a hose is connected.

hospice — A facility or program designed to provide a caring environment for supplying the physical and emotional needs of the terminally ill.

hospital — A healthcare facility where the sick or injured are given medical care.

hospital acquired infection — See nosocomial infection.

host — (*parasitic relationship*) An organism from which a parasite derives its food supply and shelter.

host defense mechanisms — Protective systems of the body including physical barriers and immune responses used against organisms causing infections.

host-mediated assay — An assay evaluating the genotoxicity of a substance to microbial cells introduced by intravenous injection into a host animal.

host specificity — The degree to which a parasite is able to mature in more than one host species.

hot — A relatively high temperature; highly radioactive.

hot spot — See toxic hot spot.

hot tub — A hydrotherapy spa that can be constructed of wood, with separately formed sides and bottoms held together by pressure from the surrounding hoops, veins, or rods.

hot water — Defined as water at 180°F at a pressure of 16 to 25 lb/in² for 30 seconds that serves as an effective physical sanitizing agent for nonselective sanitization of food-contact surfaces where spores may survive.

hot work — Any work involving burning, welding, riveting, or similar fire-producing operations, as well as work that produces a source of ignition such as drilling, abrasive blasting, and space heating.

house sewer — A pipeline connecting a house and drain and septic tank or municipal sewer line.

household — One or more individuals living together in a single dwelling unit and sharing common living, sleeping, cooking, and eating facilities.

household hazardous waste — The hazardous products used and disposed of by residential persons in contrast to industrial sources; may include paints, stains, varnishes, solvents, pesticides, and other materials or products containing volatile chemicals that can catch fire, react, or explode or are corrosive or toxic.

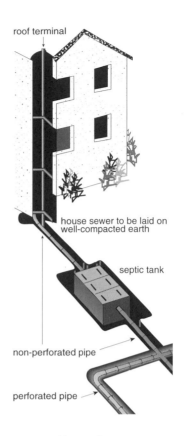

House Sewer

housing maintenance specialist — An environmental health practitioner involved in the regulation of housing, including zoning and occupancy approvals, elimination of nuisances, and regular inspection to ensure that minimum standards of health and safety are met.

hp — See horsepower.

HP — Acronym for health physicist or high pressure.

HPC — See heterotrophic plate count.

HPCC — Acronym for high-performance computing and communications.

HPLC — Acronym for high-performance liquid chromatography.

HPMS — Acronym for Highway Performance Monitoring System.

HPS — See hantavirus pulmonary syndrome.

HPV — Acronym for high production volume or high-priority violator.

hr — notation for hour or hours.

HR — See heart rate; acronym for human resources.

HRA — See health-risk assessment.

HRDI — Acronym for high-resolution Doppler imager.

HRIRS — Acronym for high-resolution infrared radiation sounder.

HROI — Acronym for high-resolution optical instrument.

HRPT — Acronym high-resolution picture transmission.

HRS — See Hazard Ranking System.

HRT — Acronym for hours of retention time.

HRUP — Acronym for high-risk urban problem.

H₂S — Scientific notation for hydrogen sulfide.

HSA — See health systems agency.

HSDB — Acronym for hazardous substance database.

HSG — See health and safety guides.

HSI — See heat stress index.

HSL — Acronym for hazardous substance list.

H₂SO₄ — Scientific notation for sulfuric acid.

HSWA — See Hazardous and Solid Waste Amendments.

HT — Acronym for hypothermally tested.

HTA — Acronym for high threat area.

HTGR — Acronym for high-temperature, gas-cooled reactor.

HTH — Acronym for high-test hypochlorite.

HTML — See Hypertext Markup Language.

HTP — Acronym for high temperature and pressure.

HTST — See high-temperature, short-time pasteurization.

HTTP — See Hypertext Transfer Protocol.

Hubbard tank — A tank containing warm water in which patients perform a series of exercises.

human ecology — Study of the interrelationship between individuals and the environment and among individuals within the environment.

human environment — The natural and physical environment and its relationship to people including all combinations of physical, biological, cultural, social, and economic factors in any given area.

human equivalent dose — A dose that produces an effect in humans equal to that produced by a dose in animals.

human exposure evaluation — A component of risk assessment that involves describing the nature and size of the population exposed to a substance and the magnitude and duration of their exposure. The evaluation can consist of past exposures, current exposures, or anticipated exposures.

human health risk — The likelihood that a given exposure or series of exposures may have damaged or will damage the health of individuals.

human immunodeficiency virus (HIV) — A retrovirus designated as either human T-lymphotropic virus type III (HTLV-III) or lymphadenopathy-associated virus (LAV), both of which compromise the body's ability to deal with various infections and rare forms of cancer by damaging the autoimmune system. It may lie dormant for several years; causes acquired immunodeficiency syndrome (AIDS).

human threadworm — See pinworm.

human waste — Normal excretory waste of the human body.

humecant — A solute that binds free water in a food, reducing the amount of water available to microorganisms.

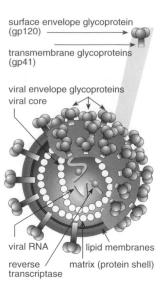

surface envelope glycoprotein (gp120)

transmembrane glycoproteins (gp41)

viral envelope glycoproteins
viral core

viral RNA

lipid membranes

reverse transcriptase

matrix (protein shell)

HIV virus

humidification — The process of transferring a mass of water into the air.

humidifier — A machine that puts moisture into the air.

humidistat — A device used to control systems for maintaining a relative humidity at some set point.

humidity — Atmospheric water vapor content.

humoral immunity — Production of antibodies in response to a foreign antigen.

humpback — See kyphosis.

humus — The normally dark-colored amorphous colloid matter in soil formed by decomposition of plant and animal remains.

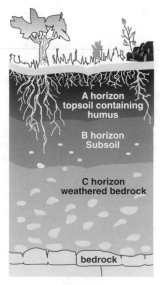

A horizon topsoil containing humus

B horizon Subsoil

C horizon weathered bedrock

bedrock

Humus
in A Horizon

hunchback — See kyphosis.

hurricane — A severe tropical storm with winds of 74 mph or greater that occurs in the western Atlantic and usually is accompanied by rain, thunder, and lightning; it sometimes moves into temperate climates.

HUS — See hemolytic uremic syndrome.

HVAC — Acronym for heating, ventilation, and air conditioning.

HVAF — See high-velocity air filter.

HVIO — Acronym for high-volume industrial organics.

HVL — See half-value layer.

HW — See hazardous waste.

HWHF — Acronym for hazardous waste handling facility.

HWIR — Acronym for Hazardous Waste Identification Rule.

HWLT — Acronym for hazardous waste land treatment.

HWM — Acronym for hazardous waste management.

HWM facility — See hazardous waste management facility.

hybrid electric vehicle — An electric motor vehicle that may operate using electric and gasoline-powered motors.

hydatid — A cyst or cyst-like structure usually filled with a fluid.

hydatid cyst — A cyst in the liver that contains larvae of the tapeworm *Echinococcus granulosus*; the eggs are carried from the intestinal tract to the liver by portal circulation.

hydatid disease — See echinococcosis.

hydrated lime — Limestone that has been burned and treated with water under controlled conditions until the calcium oxide portion has been converted to calcium hydroxide; quicklime combined with water.

hydrates — Compounds containing water, usually salts containing water of crystallization.

hydration — Combining a substance with water.

hydrauger — A horizontal drain installed to stabilize a slope.

hydraulic — The process of water in motion or powered by water.

hydraulic conductivity — (*soils*) The ability of the soil to transmit water in liquid form through pores.

hydraulic efficiency — The efficiency of a pump or turbine to impart energy to or extract energy from water.

hydraulic fill — Fill material transported and deposited using water.

hydraulic fracturing — The physical process that creates fractures in soils to enhance fluid or vapor flow in the subsurface; used especially in petroleum drilling.

hydraulic grade line — The level water would rise to in a small vertical tube connected to a pipe.

hydraulic gradient — The slope of a water surface in an open channel, of the groundwater table, or of water in pipes under pressure; the slope of the water table as measured by the difference in elevation between two points on the slope of the water table and the distance of flow between them.

hydraulic loading — The flow rate (in gal/min/ft²) of liquid applied to a granular bed.

hydraulics — The mechanical properties of water in motion and the application of these properties in engineering.

hydrazine (N₂H₄) — A colorless liquid with an ammonia-like odor; a solid below 36°F; MW: 32.1, BP: 236°F, Sol: miscible, Fl.P: 99°F, sp. gr. 1.01. It is used in the synthesis and handling of high-energy fuels, agricultural chemicals, pharmaceuticals, chemicals for plastics and rubber manufacturing, textile agents and dye intermediates, and photographic chemicals; as an anticorrosion agent. It is hazardous to the skin, eyes, respiratory system, and central nervous system and is toxic through inhalation, absorption, ingestion, and contact. Symptoms of exposure include temporary blindness, dizziness, nausea, dermatitis, skin and eye burns, and irritation of the eyes, nose, and throat; carcinogenic; in animals, bronchitis, pulmonary edema, liver and kidney damage, and convulsions. OSHA exposure limit (TWA): 0.1 ppm [skin] or 0.1 mg/m³.

hydrazine sulfate (H₆N₂O₄S) — An odorless, white or colorless crystalline material; MW: 130.12, BP: decomposes, Sol: 3.415 g/100 ml water at 25°C, Fl.P: NA (poisonous gases produced in a fire include nitrogen oxides and sulfur oxides), sp. gr. 1.378. It is used in refining rare metals, as an antioxidant when soldering flux in light metals, as a reducing agent in the analysis of minerals and slags, in tests for blood, and as a fungicide. It is hazardous to the respiratory system and to the skin. Acute exposure may cause irritation of the eyes, nose, and throat, and it can affect the brain and nervous system to produce symptoms of dizziness, lightheadedness, trembling, and seizures. Chronic exposure may interfere with the ability of the blood to carry oxygen and cause shortness of breath and fatigue. It may also damage the liver and kidneys. It is a suspected carcinogen. OSHA exposure limit (8-hour TWA): NA.

hydrazoic acid (HN₃) — A colorless mobile liquid with a strong, pungent odor; MW: 43.04, BP: 37°C, Sol: soluble in water and organic solvents, Fl.P: a dangerous explosive compound that is unstable and sensitive to heat and shock, density: NA. It is used in making heavy metal azides for detonators. It is hazardous to the respiratory tract and eyes. It is an acute toxin and produces severe irritation, bronchitis, headache, dizziness, weakness, decreased blood pressure, collapse, convulsions, and death. ACGIH exposure limit: 0.1 ppm vapor.

$$H - \ddot{\overset{..}{N}} - \overset{+}{N} \equiv N\mathbf{:}$$

Hydrazoic acid

hydric — Characterized by, or thriving in, an abundance of moisture.

hydric soil — A soil that in an undrained condition is saturated, flooded, or ponded long enough during the growing season to develop anaerobic conditions to support the growth or regeneration of hydrophytic vegetation.

hydriodic acid (HI) — A colorless liquid rapidly becoming yellow or brown when exposed to light and air; MW: 127.91, BP: liquefies at –35°C, Sol: extremely soluble in water and soluble in many organic solvents, Fl.P: a noncombustible gas, density: NA. Used in the manufacture of iodides, as a reducing agent, and in disinfectants and pharmaceuticals. It is hazardous to the skin, eyes, and mucous membranes. It causes severe burns and irritation. ACGIH exposure limit: NA.

hydrobromic acid (HBr) — A colorless and corrosive liquid or gas; MW: 80.92, BP: 126°C, gas liquefies at –66.5°C, Sol: forms an aqueous solution at 25°C, Fl.P: noncombustible gas, density: NA. It is used in the manufacture of bromide, as an alkylation catalyst, and in organic synthesis. It is hazardous to the eyes, nose, respiratory tract, and skin. It causes strong irritation of all mucous membranes. ACGIH exposure limit (TLV-TWA): 3 ppm.

hydrocarbon — One of a very large group of compounds containing hydrogen and carbon atoms that may be saturated or unsaturated. Saturated hydrocarbons have no double or triple bonds. Unsaturated hydrocarbons contain double and/or triple bonds. Aromatic hydrocarbons contain benzene or benzene-like rings. Cyclic hydrocarbons are saturated or unsaturated but contain a ring of carbon atoms.

hydrocarbon radical (R) — (*chemistry*) A group of carbon and hydrogen atoms with one or more free bonds.

hydrochinone — See hydroquinone.

hydrochloric acid — See hydrogen chloride.

hydrochlorination — The application of hypochlorite compounds to water for the purpose of disinfection.

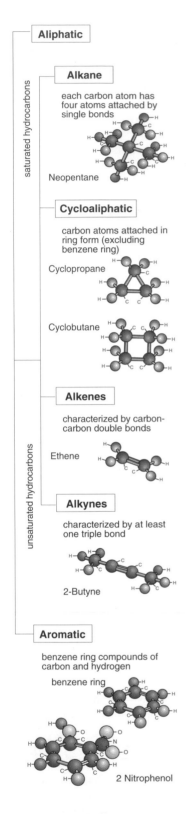

Hydrocarbons

hydrocooling — Spraying refrigerated water at 1°C on products in order to lower the temperature of the food.

hydrocracking — A high-pressure version of catalytic cracking in the presence of hydrogen.

hydrodynamics — The study of fluid motion and fluid boundary interaction.

hydroelectric plant — An electric power plant using flowing water as its primary force.

hydroelectric power — The electrical energy produced by flowing water.

hydrofluoric acid (HF) — A colorless gas or a fuming liquid; MW: 20.01; BP: 19.54°C; Sol: soluble in cold water, highly soluble in alcohol, and slightly soluble in ether; Fl.P: noncombustible as an aqueous solution; density: 0.991 at 19.54°C. Used as a fluorinating agent, as a catalyst, in uranium refining and etching glass, and for pickling stainless steel. It is hazardous to all body tissues and causes extremely corrosive reactions and severe burns; may cause death. ACGIH exposure limit (TWA): 3 ppm.

hydrogen — A gaseous, univalent element with atomic number of 1 and atomic weight of 1.008. It is the simplest and the lightest of the elements and is normally a colorless, odorless, highly inflammable diatomic gas.

hydrogen bond — A relatively weak chemical bond resulting from unequal charge distribution within molecules, in which a hydrogen atom covalently bonded to another atom is attracted to the electronegative portion of another molecule.

hydrogen bromide (HBr) — A colorless gas with a sharp, irritating odor that is often used in an aqueous solution; MW: 80.9, BP: –88°F, Sol: 49%. It is used in the manufacture of organic bromides for use in photography, pharmaceuticals, industrial drying, textile finishing, engraving and lithography, chemical synthesis, and fire retardants; in the manufacture of brominated fluorocarbons for fire extinguishing, refrigeration, and aerosols; in organic synthesis as intermediates for barbiturate manufacture; in the manufacture of synthetic hormones; as a catalyst for alkylations, controlled oxidations, isomerizations, and polymerizations; as a reagent in analytical chemistry; in the etching of germanium crystals, silicon discs, and metal alloys; as a solvent for ore minerals. It is hazardous to the respiratory system, eyes, and skin and is toxic through inhalation, ingestion, and contact. Symptoms of exposure include skin and eye burns and irritation of the eyes, nose, and throat. OSHA exposure limit (TWA): ceiling 3 ppm [air] or 10 mg/m³.

hydrogen chloride (HCl) — A colorless to slightly yellow gas with a pungent, irritating odor that is often used in an aqueous solution; MW: 36.5, BP: –121°F, Sol: 67% at 86°F. It is used during pickling of metals, including stainless steel, iron, and nickel; as a catalyst or chlorinating agent in chemical synthesis; in food processing and manufacture, including sugarcane refining, glucose, and corn sugar and in brewing operations; in industrial chemical cleaning operations; in the production of plastics and resins; in rubber manufacture, including synthesis of chloroprene; as a deliner of hides in leather manufacture; in activation of petroleum wells; in waste treatment operations for neutralization of alkaline wastestreams; in the production of chlorine; in swimming pools. It is hazardous to the respiratory system, skin, and eyes and is toxic through inhalation, ingestion, and contact. Symptoms of exposure include cough, burning throat, choking, burning eyes and skin, dermatitis, and inflammation of the nose, throat, and larynx; in animals, laryngeal spasms and pulmonary edema. OSHA exposure limit (TWA): ceiling 5 ppm [air] or 7 mg/m³. Also known as hydrochloric acid.

hydrogen cyanate (CHNO) — A colorless liquid or gas with an acrid smell; MW: 43.03; BP: 23.5°C; Sol: soluble in water, alcohol, ether, and benzene; Fl.P: NA (the liquid can explode when heated rapidly); density: 1.140 at 20°C. It is used in the preparation of cyanates. It is hazardous to the eyes, skin, respiratory tract, and other mucous membranes and causes severe irritation and injury. ACGIH exposure limit (TLV-TWA): NA.

$$N \equiv C - OH$$

Hydrogen cyanate

hydrogen cyanide (HCN) — A colorless to pale blue liquid or gas (above 78°F) with a bitter, almond-like odor; MW: 27.0, BP: 78°F, Sol: miscible, Fl.P: 0°F, sp. gr. 0.69. It is used in fumigation of structures and agricultural crops; in the production of intermediates in the synthesis of acrylic plastics, nylon 66, chelating agents, dyes, pharmaceuticals, and specialty chemicals. It is hazardous to the central nervous system, cardiovascular system, liver, and kidneys and is toxic through inhalation, absorption, ingestion, and contact. Symptoms of exposure include asphyxia and death at high levels, weakness, headache, confusion, nausea, vomiting, increased rate and depth of respiration, or slow respiration and gasping. OSHA exposure limit (TWA): 4.7 ppm [skin] or 5 mg/m^3.

hydrogen fluoride (HF) — A colorless gas or fuming liquid (below 67°F) with a strong, irritating odor; MW: 20.0, BP: 286°F, Sol: miscible, sp. gr. : 1.00 at 67°. It is shipped in cylinders and is used in the manufacture of chlorofluorohydrocarbons for applications such as refrigeration fluids, aerosol propellants, specialty solvents, high-performance plastics, and foaming agents; in cleaning sandstone and marble; as a pickling agent for stainless steel and other metals; as a cleaner in meat packing; in anhydrous acid in the manufacture of aluminum fluoride and synthetic cryolite; as a catalyst in alkylation of petroleum fractions to produce high-octane fuels; in separation of uranium isotopes; in the manufacture of pharmaceuticals and special dyes; in aqueous acid in etching, frosting, and polishing glassware and ceramics; in the manufacture of insecticides, laundry sours, and stain removers. It is hazardous to the eyes, skin, and respiratory system and is toxic through inhalation, absorption, ingestion, and contact. Symptoms of exposure include pulmonary edema, skin and eye burns, nasal congestion, bronchitis, and irritation of the eyes, nose, and throat. OSHA exposure limit (TWA): 3 ppm [air] or 1.4 mg/m^3.

hydrogen ion — The cation of an acid consisting of a hydrogen atom for which the electron has been transferred to the anion of the acid.

hydrogen ion balance — See acid–base balance.

hydrogen peroxide (H$_2$O$_2$) — A colorless liquid with a slightly sharp odor; the pure compound is a crystalline solid below 12°F and it is often used in aqueous solution; MW: 34.0; BP: 286°F; Sol: miscible; sp. gr. 1.39. It is used in the manufacture of propellants for military and space programs; as a component of explosives; in chemical synthesis as an oxidant in organic and inorganic synthesis; as a polymerization promoter; as a bleaching agent for oils, waxes, fats, and discolored concentrated acids. It is hazardous to the eyes, skin, and respiratory system and is toxic through inhalation, ingestion, and contact. Symptoms of exposure include corneal ulcers, erythema, vesicles on the skin, bleached hair, and irritation of the eyes, nose, and throat. OSHA exposure limit (TWA): 1 ppm [air] or 1.4 mg/m^3.

hydrogen potential — The number of charged hydrogen atoms or protons present; the acidity of the soil; same as pH.

hydrogen selenide (H$_2$Se) — A colorless gas with an odor resembling decayed horseradish; MW: 81.0, BP: –42°F, Sol: 0.9% at 73°F. It is used in the preparation of semiconductor materials; in chemical synthesis for metal selenides and organosele-

nium, lasers, and emulsions. It is hazardous to the respiratory system and eyes and is toxic through inhalation and contact. Symptoms of exposure include nausea, vomiting, diarrhea, metallic taste, garlic breath, dizziness, lassitude, fatigue, and irritation of the eyes, nose, and throat. OSHA exposure limit (TWA): 0.5 ppm [air] or 0.2 mg/m^3.

hydrogen sulfide (H$_2$S) — A colorless gas with a strong odor of rotten eggs; MW: 34.1, BP: –77°F, Sol: 0.4%. Sense of smell becomes rapidly fatigued and cannot be relied on to warn of its continuous presence. It is shipped as a liquefied compressed gas and is used in underground mining operations; in refining of high-sulfur petroleum; in tanneries, glue factories, fat-rendering plants, and fertilizer plants; in the manufacture of viscose rayon; in the production of sulfur dyes, carbon disulfide, sulfur, oleum, and thioprene; in the vulcanization of rubber; in the manufacture of coke from coal with high gypsum content. It is hazardous to the respiratory system and eyes and is toxic through inhalation, ingestion, and contact. Symptoms of exposure include apnea, coma, convulsions, eye irritation, conjunctivitis, pain, lacrimation, photophobia, corneal vesiculation, dizziness, headache, fatigue, irritability, insomnia, gastrointestinal distress, and respiratory system irritation. OSHA exposure limit (TWA): 10 ppm [air] or 14 mg/m^3.

hydrogenation — The process by which hydrogen is made to combine with another substance, usually organic, in the presence of a catalyst, such as nickel.

hydrogeochemistry — The chemistry of groundwater and surface water.

hydrogeologic conditions — Conditions stemming from the interaction of groundwater and the surrounding soil and rock.

hydrogeologic cycle — The natural process of recycling water from the atmosphere to the Earth and back to the atmosphere.

hydrogeologist — A person who studies and works with groundwater.

hydrogeology — Study of the geology of groundwater, with particular emphasis on the chemistry and movement of water.

hydrograph — A chart that measures the amount of water flowing past a point as a function of time.

hydrography — Study of the physical aspects of all water on the surface of the Earth; (*geographical information system*) a database representing the location of water bodies and flow lines.

hydrojet — A device that blends air and water to create a high-velocity turbulent stream of air-enriched water.

hydrokinetic — Of or relating to the motion of fluids or the forces that produce or affect such motions.

hydrologic cycle — Continuous process of water evaporating from the sea, precipitating over land, and eventually returning to the sea; also known as water cycle.

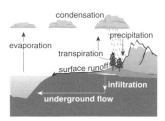

Hydrologic Cycle

hydrology — Science dealing with the properties, distribution, and flow of water on or in the Earth.

hydrolysis — A chemical reaction in which complex molecules are broken down into simpler molecules by combining them with water (e.g., fats into fatty acids and glycerol); (*carbohydrates*) the action of water in the presence of a catalyst upon a carbohydrate to form simpler carbohydrates; (*salts*) a reaction involving the splitting of water into its ions by the formation of a weak acid, base, or both.

hydrolyze — To undergo hydrolysis.

hydrolyzed — A compound that has undergone a chemical reaction with water to form a new chemical compound.

Hydrometeorological Report — A series of reports published by the National Weather Service that discuss issues related mainly to developing estimates of probable maximum precipitation to be used in determination of probable maximum floods for the design of water-control structures.

hydrometer — A direct-reading instrument for indicating the density, specific gravity, or some similar characteristic of liquids.

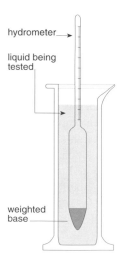

Hydrometer

hydronic — A ventilation system using heated or cooled water pumped through a building.

hydrophilic — Characterized by a tendency to hydrate easily and become quite stable.

hydrophobic — Characterized by an inability to combine with or dissolve in water.

hydrophytic — Characterized by growth in or in close proximity to water.

hydropneumatic — A small water system in which a water pump is automatically controlled by the air pressure in a compressed-air tank.

hydroponics — A commercial technique for growing plants without the use of soil, using nutrient solutions instead.

hydropower — The electric power produced by falling water.

hydroquinone ($C_6H_4(OH)_2$) — Light tan, light gray, or colorless crystals; MW: 110.1, BP: 545°F, Sol: 7%, Fl.P: 329°F (molten), sp. gr. 1.33. It is used in the preparation and use of photographic developers; in the dyeing and fur processing industries; as an antioxidant; as a chemical stabilizer; in the synthesis of hydroquinone ethers and vitamin E. It is hazardous to the eyes, respiratory system, skin, and central nervous system and is toxic through inhalation, ingestion, and contact. Symptoms of exposure include conjunctivitis, keratitis, central nervous system excitement, colored urine, nausea, dizziness, suffocation, rapid breathing, muscle twitches, delirium, collapse, and irritation of the eyes. OSHA exposure limit (TWA): 2 mg/m³ [air]. Also known as hydrochinone.

hydroscopic nuclei — A piece of dust or other particle around which water condenses in the atmosphere.

hydrosol — A dispersion of solid particles in a liquid stream.

hydrosphere — The aqueous or water envelope of the Earth that includes all bodies of water (oceans, lakes, streams, underground waters, and vapor in the atmosphere).

hydrostatic — Of or relating to fluids at rest or the pressures they exert or transmit.

hydrostatic pressure — The pressure exerted by a body of water at rest at a specific elevation; the pressure exerted by a fluid (gas, or liquid), defined as force per unit area. The hydrostatic pressure of 1 atmosphere is 14.7 lb/in.²

hydrotherapy — The use of water in the treatment of various physical disorders.

hydrothermal — The water or steam trapped in fractured or porous rocks beneath the surface of the Earth.

hydrothermal chemistry — The study of aqueous chemistry under high-temperature conditions, such as the chemical composition of water that has been altered by deep, hot rocks; the water tends to come to the surface at fault lines in boiling springs and geysers.

4-hydroxy-4-methyl-2-pentanone — A colorless liquid with a faint pleasant odor; MW: 116.16; BP: 164°C; Sol: soluble in water, alcohol, and ether; Fl.P: 58°C (closed cup), a combustible liquid; density: 0.9387 at 20°C. It is used as a solvent for nitrocellulose, resins, fats, oils, and waxes and is also used in hydraulic fluids and antifreeze solutions. It is hazardous to the eyes, nose, throat, and skin. It is an irritant and a strong narcotic. ACGIH exposure limit (TLV-TWA): 50 ppm.

4-Hydroxy-4-methyl-2-pentanone

hydroxyl — A chemical prefix indicating the presence of an –OH radical in a chemical compound.

hyetometry — The measurement of precipitation; the study of the origin, structure, and other features of all forms of condensation, sublimation, and precipitation.

hygiene — The principles and science of the conservation of health and prevention of disease.

hygrograph — An instrument for recording relative humidity.

hygrometer — An instrument to measure humidity in the atmosphere.

hygrometry — The science of moisture measurements of evaporation, evapotranspiration, condensation, precipitation, and the water vapor content of the atmosphere.

hygroscopic — Characterized by the capacity to take up and readily retain moisture.

hygroscopic particulates — Particles that collect water from the atmosphere and form a mist that may limit visibility.

hygroscopic water — Water adsorbed by dry soil from an atmosphere of high relative humidity.

hymenolepiasis — (*disease*) An intestinal infection with dwarf tapeworms; ranges from asymptomatic to causing enteritis with or without diarrhea, abdominal pain, loss of weight, and weakness. Incubation time is approximately 2 weeks. It is caused by *Hymenolepis nana*, which is found more commonly in warm than cold areas and as the common tapeworm in southeastern United States. Reservoirs of infection include infected people and possibly mice. It is transmitted by the eggs of *H. nana* passing in feces and contaminating food or water

that is then ingested; it is communicable as long as eggs are being passed in the feces, which may be for several years; general susceptibility. It is controlled by good personal hygiene and proper protection of food and water from contamination with human and rodent feces.

hypalon — A widely used synthetic rubber providing exceptional weather, ozone, and sunlight resistance.

hyperbaric chamber — An airtight chamber containing an oxygen atmosphere under high pressure; used for treatment of certain infections, tumors, and cardiovascular diseases.

hyperbaric conditions — Pressure conditions in excess of surface pressure.

hyperbarism — A disorder resulting from exposure to increased ambient pressure.

hypercapnia — Greater than the normal amount of carbon dioxide in the blood.

hypercarbia — See hypercapnia.

hyperemia — An increase of blood in part of the body caused by increased blood flow, an inflammatory response, or obstruction to the outflow of blood from an area.

hyperendemic — Describes a disease or infection that has high, usually seasonal, transmission in a certain geographic area.

hyperflexia — The forcible overextension of a limb.

hyperglycemia — A greater than normal amount of glucose in the blood.

hyperinfection — A condition in *Strongyloides* infections in which filariform juveniles repenetrate the mucosa of the small intestine and proceed with migration.

hyperirritability — An excessive excitability or sensitivity or an exaggerated response to a stimulus.

hyperkeratosis — Hypertrophy of the horny layer of the skin; hypertrophy of the cornea.

hypermenorrhea — See menorrhagia.

hypermetamorphosis — A type of metamorphic development in which different larval instars have markedly dissimilar body forms.

hyperosmotic — Describes a solution for which the osmotic pressure is greater than that of another solution to which it is compared because it contains a greater concentration of dissolved particles and gains water through a selectively permeable membrane from a solution containing fewer particles.

hyperoxemia — Increased oxygen content of the blood.

hyperoxia — An abnormally high oxygen tension in the blood.

hyperparasitism — The parasitism of a parasite by another parasite.

hyperpigmentation — A darkening of the skin that may be a postinflammatory response to chemical photosensitizers such as tar, pitch, and drugs; physical agents such as ultraviolet light, thermal radiation, and ionizing radiation; trauma; and chemicals such as arsenic and certain aromatic hydrocarbons.

hyperplasia — Increase in volume of a tissue or organ caused by the growth of new cells.

hypersensitivity — A state of altered reactivity in which the body reacts with an exaggerated immune response to a foreign agent such as a chemical or antigen; responsible for allergic reactions.

hypersensitivity diseases — Diseases characterized by allergic responses to animal antigens; those most clearly associated with indoor air quality are asthma, rhinitis, and hypersensitivity pneumonitis.

hypersensitivity pneumonitis — An inflammatory form of pneumonia that results from an immunological reaction in a hypersensitive person to a variety of inhaled organic dusts often containing fungal spores, with symptoms of asthma, fever, chills, malaise, and muscle aches developing 4 to 6 hours after exposure.

hypersensitivity reaction — An excessive response of the immune system to a sensitizing agent which may be due to the host response, amount of allergen, kind of allergen, route of entry into the body, timing of the exposure, and site of the allergen–immune reaction.

hyperspectral imagery — (*geographical information system*) A means of classifying types of material on the surface of the Earth used in agriculture and forestry management, mineral exploration, environmental monitoring, and national security activities.

hypertension — Abnormally elevated blood pressure; in adults, it is generally regarded to be a systolic pressure of 140 mmHg and diastolic pressure of 90 mmHg; a diagnosis of hypertension should be based on several readings, as a single reading can be influenced by emotional state or physical activity.

Hypertext Markup Language (HTML) — Standard computer programming language for creating documents on the World Wide Web.

Hypertext Transfer Protocol (HTTP) — Protocol used to provide hypertext links between pages; the standard way of transferring HTML documents between Web servers and browsers; standard computer programming language that computers linked to the World Wide Web use to communicate with each other.

hypertext — Any text in a file containing words, phrases, or graphics that, when clicked, cause another document to be retrieved and displayed; most often such text appears blue and is underlined on the Web pages.

hypertonic — A solution having a greater number of solute particles than another solution to which it is compared.

hypertrophy — Nontumorous increase in the size of an organ as a result of enlargement of constituent cells without increase in the numbers.

hyperventilation — An increase of air in the lungs above the normal amount; leads to fainting, impaired consciousness, tightness of the chest, and a sensation of smothering and apprehension.

hypha — Thread-like structure of the mycelium in a fungus.

hypocenter — The point within the Earth where an earthquake rupture starts; commonly called the focus.

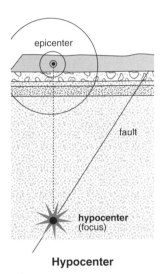

Hypocenter

hypochlorination — Application of hypochlorite compounds to water for the purpose of disinfection.

hypochlorinators — The chlorine pumps, chemical feed pumps, or devices used to dispense chlorine solutions made from hypochlorites such as sodium hypochlorite (bleach) or calcium hydrochloride into the water being treated.

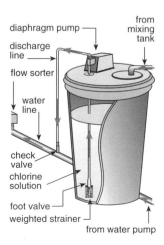

Hypochlorinator

hypochlorite — Any organic compound containing a metal and the OCl⁻ radical.

hypochlorite poisoning — Toxic effects of ingestion or skin contact with household or commercial bleaches or other similar chlorinated products with symptoms of pain and inflammation of the mouth, digestive tract, vomiting, and difficulty breathing.

hypochlorites — A general formula of $M(OCl)^x$, where M is a metal ion having valency x. They are salts of hypochlorous acid (HOCl) which undergo violent reactions with many compounds to cause fires or explosion. They are used as bleaching powders, fungicides, deodorants, disinfectants, and oxidizing agents and are irritants to the skin. Through ingestion, they may cause gastric perforation and corrosion of the mucous membranes.

hypochlorous acid (HOCl) — An unstable acid with excellent bactericidal and algicidal properties.

hypoesthia — Decreased sense of touch.

hypofunction — Diminished or inadequate level of activity of an organ or the body.

hypoglycemia — Low levels of sugar in the blood.

hypolimnion — Region of a body of water that extends from below the thermocline to the bottom of the lake; it is removed from much of the surface influence and is characterized by a uniform temperature that is generally cooler than that of other strata.

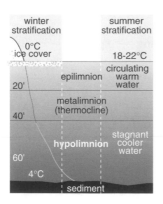

Hypolimnion

hypophysis — See pituitary gland.

hypopigmentation — Loss of pigment in the skin caused by a postinflammatory response due to exposure to thermal radiation, ultraviolet radiation, or chemical burns.

hypostenic — Lack of body tone.

hypotension — Abnormally low blood pressure, generally regarded at levels below 100 systolic and 40 diastolic.

hypothalamus — The area of the brain that controls body temperature, hunger, and thirst.

hypothermia — Subnormal temperature of the body.

hypothermia therapy — Reduction of a person's body temperature to counteract high prolonged fever caused by an infection or neurological disease; also used as an adjunct to anesthesia in heart or brain surgery.

hypothesis — A statement or proposition that can be tested by experiment.

hypothyroidism — A deficiency of thyroid gland activity with underproduction of thyroxin or the condition resulting from it; may be caused by goitrogens.

hypotonic — A solution having a lesser number of solute particles than another solution to which it is compared.

hypovolemia — An abnormally low circulating blood volume.

hypovolemic shock — Physical collapse and prostration caused by massive blood loss, circulatory dysfunction, and inadequate tissue perfusion that occurs when a person loses about one fifth total blood volume; symptoms include low blood pressure, feeble pulse, clammy skin, tachycardia, and rapid breathing.

hypoxemia — Deficient oxygenation of the blood.

hypoxia — Deficiency of oxygen reaching the tissues of the body.

hypoxic — A condition of low oxygen concentration below that considered aerobic.

hypsography — (*geographical information system*) Lines connecting points of equal elevation on the surface of the Earth.

hypsometric diagram — A graph that shows the relative amounts of various elevations of the surface of the Earth with regard to sea level.

hypsometry — (*geography*) Measurement of elevations of surfaces of the Earth with respect to sea level.

hysteresis — A measure of differences in indicated steady-state values for identical conditions when approached from higher and lower states.

hysteria — A behavior showing overwhelming or unimaginable fear or emotional stress.

hysteric paralysis — A loss of movement or muscle weakness that is due to hysteria rather then an identifiable defect or illness.

Hz — See hertz.

I

I — See current.

IA — Acronym for interagency agreement.

I/A — Acronym for innovative/alternative.

IAG — Acronym for interagency agreement.

IAQ — See indoor air quality.

IATDB — Acronym for Interim Air Toxics Database.

IBAD — Acronym for Interim Biological Agent Detector.

IBP — Acronym for initial boiling point.

IC — Acronym for internal combustion or ion chromatography.

ICAD — Acronym for Individual Chemical Agent Detector.

ICB — See immobilized cell bioreactor.

ICD — See International Statistical Classification of Diseases, Injuries, and Causes of Death.

ICDA — Acronym for International Classification of Diseases Adapted for Use in the United States.

ice — A solid form of water.

ICE — Acronym for Industrial Combustion Emissions Model; see internal combustion engine.

Ice Age — A period of time when significant amounts of ice formed on the continents inducing a fall in sea level.

ice fog — An atmospheric suspension of highly reflective ice crystals which affect visibility.

ichthyology — The scientific study of fish.

icon — (*computer science*) A picture used on a computer screen to indicate a purpose or function.

ICP — Acronym for intracranial pressure.

ICR — See information collection request.

ICRE — Acronym for ignitability, corrosivity, reactivity, extraction.

ICSC — Acronym for International Chemical Safety Cards.

ICt_{50} — The inhalation dose of a chemical agent, vapor or aerosol, that produces a given, defined level of incapacitation in 50% of the exposed subjects.

icterus — Yellowing of skin and other organs because of bile pigments in the blood; also called jaundice.

ICU — See intensive care unit.

ID — Notation for inside diameter; acronym for incapacitated dose or infectious diseases.

ideal gas — A gas that strictly obeys Boyle's law ($PV = RT$) at all temperatures and pressures; also known as a perfect gas.

ideal gas law — $PV = nRT$, where the product of pressure (P) and volume (V) is equal to the product of the number of moles of gas (n), temperature (T), and a gas constant (R).

identification code — See EPA I.D. number.

idiopathic — Without a known cause; a morbid state arising without a known cause.

idiopathic disease — A disease that occurs with no apparent or known cause, although it may have a pattern of signs and symptoms that are known.

idiosoma — The posterior of the body of a mite or tick containing the legs and most internal organs.

idiosyncratic reaction — A genetically determined abnormal reaction to a chemical.

idiosyncratic response — An unusual physiologic or metabolic reaction to a chemical that is consistently reproducible; the response observed is qualitatively similar to that observed in all individuals but may take the form of extreme sensitivity to low doses or extreme insensitivity to high doses of the chemical.

IDLH — See immediately dangerous to life or health.

IEUBK — See Integrated Exposure Uptake Biokinetic Model for Lead in Children.

IFB — Acronym for invitation for bid.

IFCAM — Acronym for Industrial Fuel Choice Analysis Model.

IFPP — Acronym for industrial fugitive process particulate.

IGCC — Acronym for integrated gasification combined cycle.

IgE — See immunoglobulin E.

igneous rocks — Rocks formed by the solidification of magma (molten silicate liquid); classified as plutonic, hypabyssal, and volcanic depending on the depth at which they were formed.

Igneous Rock

ignitable — Characterized by the capacity to be set on fire.

ignitable wastes — Liquids with a flash point below 60°C or solids capable of causing fire under standard temperature and pressure.

ignition temperature — The lowest temperature of a fuel at which combustion becomes self-sustaining.

IGR — See insect growth regulator.

IH — Abbreviation for inhalation.

Ikwa fever — See trench fever.

ileum — Third and longest region of the small intestine in mammals; it is the site of digestion and absorption.

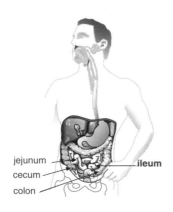

Ileum

illness — A condition marked by a pronounced deviation from the normal health state.

illness prevention — The art devoted to recognition, evaluation, and control of health hazards.

illuminance — The density of light (lumens/area) incident on a surface; measured in footcandles or lux.

illumination — Lighting up a part of the body or an object for the purpose of examining it.

IM — Abbreviation for intramuscular.

I/M — Acronym for inspection/maintenance.

imager — A satellite instrument that measures and maps the Earth and its atmosphere; the data are converted by computer into pictures.

imaging — Procedures that produce pictures of areas inside the body.

imago — The adult and sexually mature insect.

imbibition — The process of physically taking up water.

Imhoff cone — A clear, cone-shaped container marked with graduations and used to measure the volume of settleable solids in a specific volume (1 liter) of water.

Imhoff tank — (*sewage*) A two-story tank used to settle solids from sewage; the liquid flows to the upper story, settling occurs, and the solids fall through a slot into the sludge compartment.

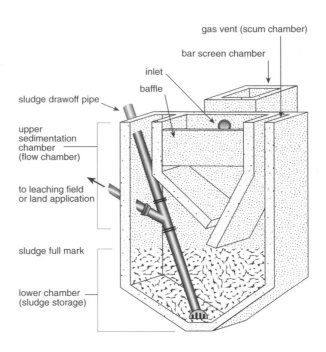

Imhoff Tank

imidazole (C₃H₄N₂) — A colorless solid; MW: 68.09; BP: 257° C; Sol: soluble in water, alcohol, and ether; Fl.P: a noncombustible solid; density: NA. It is used in organic synthesis and as an anti-irradiation agent. In animals, it causes muscle contractions and convulsions. ACGIH exposure limit: NA.

Imidazole

IML — See interface management language.

immediate hypersensitivity — A biological manifestation of an antigen–antibody reaction in which the maximal response is reached in a few minutes or hours. Intravenous injection may produce anaphylactic shock and death.

immediate toxicity — The immediate effects that occur or develop rapidly after a single administration of a substance.

immediately dangerous to life or health (IDLH) — The maximum environmental concentration of a contaminant from which one could escape within 30 minutes without developing escape-impairing symptoms or irreversible health effects.

imminent health hazard — A significant threat or danger to health that is considered to exist when sufficient evidence shows that a product, practice, circumstance, or event creates a situation that requires immediate correction or cessation of an operation to prevent potential injuries and/or severe injuries.

immiscible — Referring to liquids that are not soluble in each other or do not mix.

immobilized cell bioreactor (ICB) — An aerobic, fixed-film bioreactor system designed to remove organic contaminants, including nitrogen-containing compounds and chlorinated solvents, from processed wastewater, contaminated groundwater, and other aqueous systems.

immune — Resistant to disease.

immune body — See antibody.

immune cross reaction — Binding of an antibody or cell receptor site with an antigen other than the one that would provide an exact fit; the antigen is not the same one that stimulated the production of that antibody.

immune gamma-globulin — Passive immunizing agents obtained from pooled human plasma.

immune response — A function of the body that produces antibodies to destroy invading antigens and cancer cells.

immune status — The state of a body's immune system in regard to heredity, age, diet, and physical and mental health.

immune system — A system in the body that helps resist disease-causing microorganisms.

immunity — The ability of an organism to resist disease or toxins by natural or artificial means.

immunization — The process of rendering a subject immune; also known as inoculation or vaccination.

immunoassay — Any of several methods for quantitative determination of the presence of chemical substances such as hormones, drugs, and certain proteins that utilize the highly specific binding between an antigen and an antibody.

immunocompromised — Individuals with a weakened immune system, which makes them susceptible to additional infections.

immunodeficiency — A deficiency of immune response or a disorder characterized by deficient immune response.

immunodeficiency disease — A health condition caused by a defect in the immune system or by chemicals or radiation that results in a susceptibility to infections and chronic diseases.

immunofluorescence — A technique used to visualize and identify specific antibodies and any attached antigens by joining the antibodies to a fluorescent dye.

immunogenic — Any substance that is antigenic and stimulates production of an antibody or cell-mediated immunity.

immunoglobulin — Any of the group of plasma proteins produced by plasma cells that participate in the immune response by combining with the antigen that stimulated its production.

immunoglobulin E — A specific class of anybodies involved in allergic reactions; people who suffer from allergies have elevated levels.

immunohistochemistry — A type of assay in which specific antigens are made visible by the use of fluorescent dye or enzyme markers.

immunologic toxicity — The occurrence of adverse effects on the immune system that may result from exposure to environmental agents such as chemicals.

immunology — Science that deals with the phenomena and causes of immunity and immune responses.

immunopotency — The ability of an antigen to elicit an immune response.

immunosuppression — Suppression of an immune response by the use of drugs or radiation.

immunotherapy — A treatment that attempts to use the body's defenses to control cancer.

immunotoxin — A plant or animal toxin that is attached to an antibody and is used to destroy a specific target cell.

IMPACT — Acronym for Integrated Model of Plumes and Atmosphere in Complex Terrain.

impact injury — Deformation of human body tissues beyond their failure limits which results in damage of anatomic structures or alteration in normal function.

impact mill — A machine that grinds material by throwing it against heavy metal projections rigidly attached to a rapidly rotating shaft.

impact tube — See pitot tube.

impaction — The forcible contact of particles of matter on a surface.

impedance (Z+) — Opposition to the flow of alternating electrical current (measured in ohms); the rate at which a substance absorbs and transmits sound.

impeller — The rotating veins of a centripetal pump, turbine, blower, fan, or mixing apparatus.

imperfect mixing — A combination of two or more substances in which parts of one are unevenly distributed among the parts of another.

impermeable — See impervious.

impermeable layer — A layer of solid material, such as rock or clay, that does not allow water to pass through it.

impervious — Not permitting water or other fluid to pass through; also known as impermeable.

impervious material — A relatively waterproof soil such as clay.

impetigo — A highly contagious streptococcal, staphylococcal, or combined infection of the skin with redness or inflammation of the skin or mucous membranes.

impinge — To collide or strike.

impingement — Removal of liquid droplets from a flowing gas or vapor stream by causing it to collide with a baffle plate at high velocity so that the droplets fall away from the stream; also known as liquid knockout.

impinger — A device used to sample dust in the air; it draws in air and directs it through a jet and onto a wetted glass plate for counting.

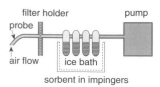

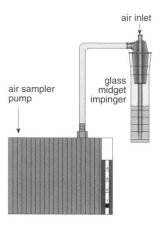

Impinger

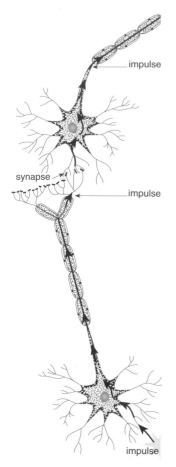

Impulse

implosion — A bursting inward.

impoundment — Confinement of a body of water by a dam, dike, floodgate, or other barrier.

IMPROVE — Acronym for Interagency Monitoring of Protected Visual Environment.

impulse — A type of message triggered by a stimulus, or a response carried by nerve cells through the nervous system.

impulse noise — A noise of short duration, usually less than 1 second, and of high intensity with an abrupt onset and rapid decay.

impurity — A chemical substance that is unintentionally and undesirably present in a pure substance.

IMS — See ion mobility spectrometry.

in. — See inch.

In — See indium

in lieu of — Instead of; in place of.

in rem — (*law*) Against the defendant.

in-service education — A training program that is provided by an agency or institution to its employees to improve their skills and competence in a specific area.

in situ — In the natural or normal place.

in situ **air sampler** — A sensor that makes constant contact with an air sample directly, with or without the aid of an air mover.

in situ **vitrification (ISV)** — A technique using an electric current to melt soil or sludge at extremely high temperatures (1600 to 2000°C) to destroy organic pollutants by pyrolysis.

in vitro — Pertaining to a biological reaction taking place outside the host organism in an artificial environment.

in vivo — Pertaining to a biological reaction taking place within a living organism.

in vivo **studies** — Studies of chemical effects conducted on intact living organisms.

inactivation — A reaction between two chemicals to produce a less toxic product.

inanimate — Lacking signs of life; not alive.

inborn — Acquired or occurring during intrauterine life.

incandescent — Pertaining to an object that has been heated to the point where it becomes hot enough to radiate visible light.

incandescent bulb — The most common and least energy-efficient lamp in which electricity running through

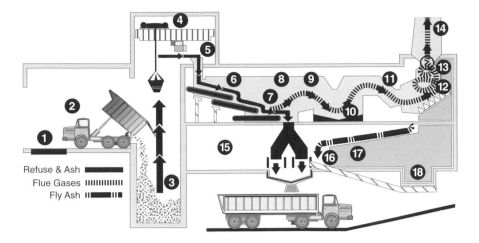

1. Scales
2. Tipping Floor
3. Storage Bin (Pit)
4. Bridge Crane
5. Charging Hopper
6. Drying Grates
7. Burning Grates
8. Primary Combustion Chamber
9. Secondary Combustion Chamber
10. Spray Chamber
11. Breeching
12. Cyclone Dust Collector
13. Induced Draft Fan
14. Stack
15. Garage - Storage
16. Ash Conveyors
17. Forced Draft Fan
18. Fly Ash Settling Chamber

Basic Incinerator Design

a tungsten filament causes it to glow and produce a soft, warm light; much of the energy is lost as heat.

incapacitate — To take away the ability to perform normal activities or tasks.

incapacitated dose — The amount and concentration of chemical that causes an individual to be unable to perform normal activities or tasks.

incapacitating agent — A chemical that produces a temporary, disabling condition that lasts for hours or days after exposure has ceased.

incendiary device — Any mechanical, electrical, or chemical device used intentionally to initiate combustion or start a fire.

incendiary incident — An event in which an incendiary device is used as a terrorist weapon.

inch (in.) — A unit of length equal to 1/12 foot or 2.54 centimeters.

inchoate water right — An unperfected water right.

incidence — (*epidemiology*) The number of cases of disease, infection, or some other event having an onset during a prescribed period of time in relation to the unit of population in which they occur.

incidence rate — The rate at which new cases of a disease occur in a population at risk during a specified period of time; most useful in determining factors associated with the etiology of disease and in evaluating programs of prevention. Also known as incidence ratio.

incidence ratio — See incidence rate.

incident radiation — The quantity of radiant energy striking a surface per unit time and unit area.

incident report — The recording of any accident or deviation from policies or orders.

incidental parasite — An accidental parasite.

incineration — Controlled combustion of solid, liquid, or gaseous combustible wastes, which are ignited and burned to form carbon dioxide, water vapor, and other gaseous products; the solid residues contain little or no combustible material.

incinerator — An engineered piece of equipment used to burn waste substances; all of the factors of combustion, including temperature, retention time, turbulence, and combustion air, can be controlled.

incinerator residue — All solid material remaining after an incineration process is completed

incipient — Describes a symptom or disease that is becoming apparent.

incision — A cut that is surgically produced using a sharp instrument to create an opening into an organ or space in the body.

incisive — Cutting into.

inclusion — A structure within another one.

incoherent waves — Waves for which the crests and troughs are not synchronized.

coherent (in phase) waves

incoherent (out of phase) waves

Incoherent Waves

incompatible — Materials that can cause dangerous reactions when they come into direct contact with one another.

incompatible waste — Waste unsuitable for mixing with another waste or material because it may react to form a hazard.

incomplete metamorphosis — Incomplete form of insect metamorphosis in which the pupal stage is lacking (as in roaches); also known as hemimetabolism.

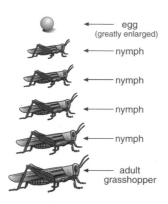

egg (greatly enlarged)

nymph

nymph

nymph

nymph

adult grasshopper

Incomplete Metamorphosis

incompetence — Lack of ability to carry out a task.

incontinence — Inability to control bowel and bladder functions.

incremental reactivity — Formation of additional ozone in the atmosphere due to the incremental addition of a certain amount of a volatile organic compound.

incubation — Provision and maintenance of ideal conditions for growth and development, as in the incubation and growth of bacteria.

incubation period — The period of time after infection before the appearance or development of disease symptoms.

incubator — An apparatus for maintaining optimal conditions of temperature and humidity for growth and development of organisms.

incus — The middle of a chain of three bones of the mammalian middle ear.

indeno(1,2,3-*c*, *d*)pyrene ($C_{22}H_{12}$) — An aromatic hydrocarbon that has been shown to cause lung cancer in animals; MW: 276.34.

Indeno(1,2,3-*c*,*d*)pyrene

indentation — A notch, pit, or depression in the surface of an object.

independent data points — Data points collected for use in the statistical test of a hypothesis independent of every other data point.

independent variables — Variables that define the treatment conditions in an experiment and are manipulated to test their effects.

index case — (*epidemiology*) The first case of a disease.

index of refraction — The refracted or bending ability of light as it passes through various substances.

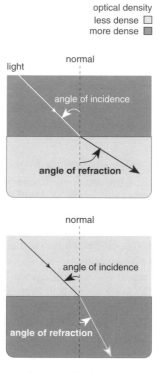

optical density
less dense ☐
more dense ▉

normal

light

angle of incidence

angle of refraction

normal

angle of incidence

angle of refraction

Index of Refraction

Indian Meal Moth (*Plodia interpunctella*) — A moth with a wing expansion of 1/2 to 3/4 inch; the basal one third of each forewing is dull white to cream, the other two thirds, brown to copper. It is a stored food product insect.

Indian tick fever — See Marseille fever.

indicator — (*biology*) A biological entity, process, or community that has characteristics indicating the presence of specific environmental conditions; (*chemistry*) a substance that shows a visible change, usually color, at a desired point in a chemical reaction; a device that indicates the result of a measurement, such as a thermometer.

indicator compounds — Chemical compounds, such as carbon dioxide, whose presence at certain concentrations may be used to estimate air quality conditions in buildings.

indicator instrument — A device that indicates the results of a measurement such as a pressure gauge.

indicator organisms — Microorganisms, such as coliforms, for which their presence is indicative of pollution or more harmful microorganisms.

indicator paper test — A rapid test of limited accuracy in which test papers (usually impregnated with starch iodide) are immersed and the developed color is compared with a standard.

indicator tests — Tests for a specific organism or group of organisms that indicate the presence of contamination or pathogenic bacteria.

indicator tube — A simple and useful field test for detecting a gas or vapor with limited accuracy by utilizing instantaneous samples of air.

indictment — A formal statement of charges against an individual presented by a grand jury, usually utilized in felony cases; a U.S. Attorney or State's Attorney (with witnesses as deemed necessary) appears before a grand jury and attempts to convince this body that an offense has been committed. If the grand jury concurs, it approves, signs, and returns a true bill of indictment to the court, and the matter is then entered on the court docket.

indigenous — Living or occurring naturally in a specific area or environment; native to a particular region.

indigenous species — A species that originally inhabited a particular geographic area.

indigestible — Describes a food substance that cannot be broken down by the digestive system into nutrients that can be used.

indirect association — A valid but noncausal association between a factor and a disease due to some common underlying condition.

indirect discharge — The introduction of pollutants from a non-domestic source into a publicly owned waste treatment system.

indirect source — (*air pollution*) Any facility, building, structure, or installation, or combination of, that generates or attracts mobile source activity that results in emissions of any pollutant subject to a state air quality standard; indirect sources include employment sites, shopping centers, sports facilities, housing developments, airports, commercial and industrial parks, parking lots, and garages.

indirect waste pipe — A waste pipe that does not connect directly with a drainage system but discharges into the drainage system through an air break or air gap into a trap, fixture, receptacle, or interceptor.

indium — A silvery metallic element with some nonmetallic properties that is used in electronic semiconductors; atomic number: 49, atomic weight: 114.82.

individual risk — The probability that an individual person will experience an adverse effect.

individual sewage system — A single system of piping, tanks, or other facilities collecting and disposing of sewage in whole or part into the soil.

individual sewage treatment system — A system or part thereof serving a dwelling, other establishment, or group thereof which utilizes subsurface soil treatment and disposal.

individual susceptibility — Substantial variability in the manner in which individuals are affected by the same exposure to a toxic agent.

indoor air — Air that is breathed inside a habitable structure and often highly polluted with chemicals from solvents, smoke, paints, furniture glues, carpet padding, other synthetic chemicals, and mold and dander because of a lack of exchange with fresh oxygen from outside.

indoor air pollutants — Pollutants within a structure that may contribute to discomfort or disease; these pollutants include radon; environmental tobacco smoke; biological contaminants; contaminants from stoves, heaters, fireplaces, and chimneys; household products; pesticides; formaldehyde; asbestos; lead; and solvents.

indoor air pollution — Pollution occurring indoors from any source, indoors or outdoors.

indoor air pollution strategies — Pollution elimination strategies include source control, removal of the material before it enters the structure, elimination of the individual sources of pollution, or ventilation improvement, including providing a greater flow of air and the use of air cleaners, either tabletop models or whole-house systems.

indoor air quality (IAQ) — A constantly changing interaction of complex factors in the indoor environment including odors; biological, chemical, or physical contaminants; design and/or operation and maintenance of the heating, ventilation, and air-conditioning system; building occupants; and potential for healthful living conditions or disease and discomfort.

indoor air research — The development and testing of monitoring devices and the design and implementation of field studies to identify and quantitate indoor sources of pollutants.

indoor climate — Day-to-day values of physical variables in the building including temperature, humidity, air movement, air quality, concentration of activities and people, which affect the health and/or comfort of the occupants.

induce — To cause or stimulate the beginning of an activity.

induced currents — Currents that flow in an object as a result of exposure to an electric or magnetic field.

induced draft — Negative pressure created by the action of a fan, blower, or ejector located between an incinerator and a stack or at a point where air or gases leave a unit.

induced draft fan — A fan that exhausts hot gases from heat-absorbing equipment, dust collectors, or scrubbers.

induced fever — Deliberate elevation of body temperature to kill heat-sensitive pathogens.

induced radioactivity — Radioactivity produced in a substance after bombardment with neutrons or other particles; also known as artificial radioactivity.

induced vomiting — Vomiting produced by the use of Ipecac syrup, soapy water, or insertion of a finger or blunt instrument into the throat when medically necessary to dispose of ingested noncaustic poisons.

induction — The process of stimulating and determining morphogenic differentiation in a developing embryo through the action of chemical substances transmitted from one to the other embryonic parts.

induction period — The time interval during which a sufficient cause is complete to indicate disease.

induction phase — The time during which a normal cell is transformed into a cancer cell.

indurated — Having become firm or hard.

induration — The process of hardening.

industrial chemical — A chemical developed or manufactured for use in industrial operations or research by industry, government, or academia and which may be hazardous for humans.

industrial gases — Gases produced by industry and classified as industrial, organic solvent vapors, upper respiratory irritants, pulmonary irritants, chemical asphyxiants, and simple asphyxiants. They enter the body through the respiratory system, digestive system, and skin; the gases are primarily inhaled, dissolved in the alveoli, and diffused into the blood. They may form toxic substances that cause acute poisoning, rapid injury, possible death, or chronic poisoning. Gases enter the digestive tract when absorbed in food and saliva; some gases can enter the unbroken skin or may cause dermatitis if they do not penetrate the skin.

industrial health — Health concerns associated with the workplace, such as exposure to chemical agents, physical agents, biological agents, and ergonomics.

industrial hygiene — The science and art of recognition, evaluation, and control of factors or stresses that may lead to disease and injury in the workplace.

industrial hygiene survey — A systematic analysis of a workplace to detect and evaluate health hazards and to determine methods for their control.

industrial hygienist — An environmental health professional who identifies and solves potentially hazardous problems by measuring and assessing harmful chemical, physical, and biological agents in the work environment.

industrial incinerator — An incinerator designed to burn a particular industrial waste.

industrial pollution prevention — A combination of industrial source reduction and toxic chemical use substitution.

industrial property — A parcel of property that is developed and used either in part or in whole for manufacturing purposes.

industrial psychology — The use of psychological principles and techniques to solve problems of business and industry.

industrial site selection process — An environmental study conducted to determine the location of emissions, quantity and quality of pollutants, control of pollutants, impact on the environment, and compliance with regulations prior to purchase of the property.

industrial sludge — Semiliquid residue or slurry remaining from treatment of industrial water and wastewater.

industrial solid waste — Solid waste that results from industrial processes in manufacturing.

industrial source reduction — The practices used to reduce the amount of hazardous substance, pollutant, or contaminant entering any wastestream or released into the environment; the reduction of the threat to public health and the environment associated with releases from industry. The reduction is accomplished through equipment or technology modifications, substitution of raw materials, and improvements in housekeeping, maintenance, training and/or inventory control.

industrial waste — Any liquid, gaseous, or solid waste substance or combination thereof resulting from any process of industry, manufacturing, trade, or business or from the development, processing, or recovery of any natural resources.

industrial waste dam — A specially constructed barrier used to create storage for the disposal of waste products from an industrial process.

industrial water — Water used for industrial purposes in such industries as steel, chemical, paper, and petroleum refining. A large amount of the total water usage is for industry.

industry concentration — An average concentration calculated from results for industry-sourced pollution over a 1-year period.

inedible — Adulterated or not intended for use as human food.

inert — Lacking activity, reactivity, or effect.

inert gas — A rare gas that does not react with other substances under ordinary conditions; in group O of the periodic table of elements; also known as noble gas.

inert ingredient — Inactive ingredient used as a filler or binder; (*pesticides*) any solvent, carrier, dispersant, or surfacant not active against target pests.

inertia — The property of matter that resists a change in motion.

inertial collector — A dry collector mechanism in which a rapidly flowing gas is forced to change direction on contact with an obstacle, causing the particles in the gas to continue in the initial direction. An example is a cyclone collector, where the gas stream moves rapidly into a cylindrical chamber through a tangential inlet duct (set at an angle to the chamber wall) at the top of the cylinder. The gas stream whirls downward with increasing rapidity toward a cone-shaped base; the centripetal force throws the entrained particles out of the spinning gas stream onto the wall of the chamber, where they fall into a collecting hopper.

inertial confinement — A nuclear fusion process in which a small pellet of nuclear fuel is bombarded with extremely high-intensity laser light.

inertial impaction — Deposition of large aerosol particles on the walls of a conduit due to inertia.

inertial separator — Air pollution control equipment that uses the principle of inertia to remove particulate matter from an air or gas stream.

inerting — Displacement of the atmosphere by a nonreactive gas, such as nitrogen, to such an extent that the resulting atmosphere is noncombustible.

in extremis — At the point of death.

in-process control technology — The conservation of chemicals and water throughout production operations to reduce the amount of wastewater discharged.

infant — A liveborne child from the moment of birth through completion of the first year of life.

infant botulism — (*disease*) A distinct clinical form of botulism found only in infants under 1 year of age. Symptoms of exposure start with constipation followed by lethargy, listlessness, poor feeding, sagging or prolapse of an organ or part thereof, difficulty in swallowing, weakness, and sometimes respiratory insufficiency and respiratory arrest. It may result in death. It is caused by colonization of the intestine by bacteria and the production of toxin in the body. Incubation time not known. It is caused by *Clostridiun botulinum* types A, B, or F and is found worldwide. The reservoir of infection is spores in the soil; it is transmitted by ingestion of the spores on food or dust and is not communicable from person to person. Infants 2 weeks to 9 months of age are susceptible; it is controlled by proper handling of soiled diapers, feces, and soiled articles.

infant death — Any death at any time from birth up to but not including 1 year of age.

infant mortality rate — The number of infant deaths per 1000 liveborne infants for a given period of time.

infantile gastroenteritis — See rotavirus.

infarct — The death of a portion of tissue because of a lack of blood supply resulting from obstruction of circulation in the area; also known as infarction.

infarction — See infarct.

infect — To transmit a pathogen that may cause an infectious disease in another person.

infection — The entry and development or multiplication of an infectious agent in the body of a living organism; pathogenic condition resulting from invasion of a pathogen; colonization occurring when an infectious agent has established itself on the host and propagates at a rate sufficient to maintain its numbers with or without disease manifestation.

infection control — Techniques used by a hospital or other health facility to minimize the risk of nosocomial or other infections spreading to patients, staff members, or the community.

infection control committee — A group of healthcare professionals who investigate infection problems and plan and supervise infection control activities.

infection control nurse — A registered nurse who is responsible for surveillance and infection prevention and control activities.

infectious — Being capable of causing an infection.

infectious agent — Any bacterium, parasite, or virus capable of invading body tissues, multiplying, and causing disease.

infectious disease — Any disease that can be transmitted from one person to another or from an animal to a person by direct or indirect contact.

infectious dose — The number of microbial cells required to initiate an infection.

infectious hepatitis — (*disease*) See viral hepatitis A.

infectious isolation — Confining a patient to a specially prepared, isolated room or area and applying particular techniques to prevent the spread of that individual's disease to others.

infectious waste — Equipment, instruments, utensils, and other objects, as well as humans and animal specimens, capable of transmitting infectious material from areas or rooms where persons or animals with suspected or diagnosed communicable disease have been housed or treated; they are not intended for reuse and are designated for disposal. Also includes laboratory wastes that may harbor or transmit infectious material.

infectivity — Ability of a pathogen to spread from one host to another.

inferior — (*anatomy*) A point on the body below a point of reference.

infest — To live on the skin or invade the internal organs of a host.

infestation — The state or condition of having insects or other pests in a place where they can cause annoyance, damage, disease, illness, injury, or death.

infiltrate — A substance that passes through a filter.

infiltration — (*water*) Water entering a sewer system and service connections from the ground through such means as, but not limited to, defective pipes, pipe joints, connections, or manhole walls; (*hydrology*) the movement of water into soil.

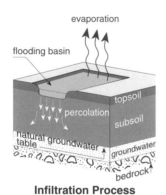

Infiltration Process

infiltration air — Air that leaks into the chambers or ducts of an incinerator.

infiltration gallery — A subsurface groundwater collection system, located close to streams or ponds, typically shallow in depth and constructed with open-jointed or perforated pipes that discharge collected water into a watertight chamber to be pumped into treatment facilities and into the distribution system; a horizontal conduit for intercepting and collecting groundwater by gravity flow.

infiltration rate — The quantity of water usually measured in inches that will enter a particular type of soil per unit time, usually 1 hour.

infirmity — An abnormal, disabling condition of the mind or body.

inflammation — A localized protective response caused by injury or destruction of tissues that serves to destroy or dilute the injurious agent and protect the injured tissue.

inflammatory response — A complex series of interactions between fragments of damaged cells, surrounding tissues, circulating blood cells, and specific antibodies; typical of infections.

inflow — Water discharged into a sewer system, including service connections from such sources as roof leaders, cellars, yard and area drains, foundation drains, pooling water discharges, drains from springs and swampy areas, manhole covers, cross connections from storm sewers and combined sew-

ers, catch basins, storm sewers, surface runoff, street wash waters, or drainage.

influent — The water, wastewater, or other liquid flowing into a reservoir, basin, treatment process, or treatment plant.

influenza — (*disease*) An acute viral disease of the respiratory tract characterized by fever, chills, headache, myalgia, prostration, mild sore throat, and cough. Incubation time is 24 to 72 hours. It is caused by three types of influenza viruses: A, B, and C; type A is associated with widespread epidemics and pandemics; B, with localized epidemics; and type C, with sporadic cases. It is found in all areas. Reservoirs of infection include humans and possibly animals. It is transmitted by direct contact through droplet infection and by airborne spread. It is communicable for 3 days; general susceptibility. It is controlled by active immunization, pharmaceutical treatment, and good personal hygiene.

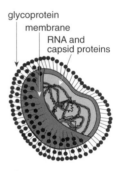

Influenza Virus

information — (*law*) A formal statement of charges against an individual in a misdemeanor; the government may proceed by giving the U.S. Attorney or State's Attorney the information, which is then filed with a clerk of the court.

information collection request — A description of information to be gathered in connection with rules, proposed rules, surveys, and guidance documents that contain information-gathering requirements; it is submitted by the U.S. EPA to the Office of Management and Budget for approval.

Information Collection Rule — A rule promulgated by the U.S. EPA to support future regulation of microbial contaminants, disinfectants, and disinfectant byproducts, intended to provide information about chemical byproducts that form when disinfectants are used for microbial control and react with chemicals already present in the source water, producing disinfection byproducts.

information letter — (*law*) A letter to responsible officials that affirms warning of the existence of a violative practice.

information superhighway — See communications superhighway.

information system — All the means and mechanisms for data receipt, processing, storage, retrieval, and analysis.

informed consent — Consent by an individual to participate in a human subject study after being informed of the nature and implications of the study and its objectives.

infrared — The long rays of radiation beyond the red end of the visible spectrum.

infrared absorption spectrophotometry — Molecular absorption spectrometric technique involving the dispersion of a polychromatic infrared beam of light with a suitable prism or by diffraction grading.

infrared gas analyzer — An instrument used to detect trace gas concentrations by determining the transmission of infrared radiation at a specific absorption frequency through a fixed path length.

infrared light — Light with lower frequencies than red light (less than about 4.3×10^{14} sec^{-1}).

infrared radiation (IR) — Electromagnetic radiation below the visible light spectrum with wavelengths longer than those of red light but shorter than radio waves (0.75 or 0.8 µm to 1000 µm).

infrared radiometer — A sensor that measures the intensity of infrared radiation (720 to 1500 nm) within a specific field of vision.

infrared thermal destruction — A mobile thermal processing system that uses electrically powered silicon-carbide rods to heat organic waste to combustion temperatures.

infrared thermography — The measurement of temperature through detection of infrared radiation emitted from heated tissue.

infusion — Slow administration of drugs into a vein or artery over a period of time.

ingestion — The act or process of taking food and other substances into the body by mouth; a route of entry into the body along with food or water or through inhalation where the materials are eventually swallowed; the materials may cross through the gastrointestinal tract and into the blood.

ingot — A block of iron, steel, or other metal cast in a mold for ease in handling before processing.

ingredient statement — The part of a label on a pesticide container that gives the name and amount by weight of each pesticide chemical plus the amount by weight of the inactive material in the container.

ingredients — (*pesticides*) The active pesticide chemical plus other inactive chemicals in a pesticide formulation or mixture that are listed on a container or package.

inhabit — To occupy as a place of settled residence.

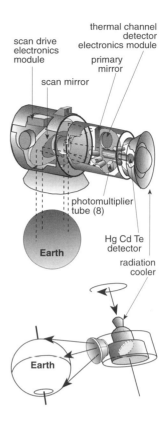

**Visible Infrared Spin Scan
Radiometer (VISSR)**

inhalable particles — All dust capable of entering the human respiratory tract.

inhalant — A medication or volatile chemical introduced into the body by breathing in the chemical.

inhalation — The act or instance of breathing in a substance in the form of a gas, vapor, fume, mist, or dust; a route of entry into the body of microorganisms, chemicals, or physical agents during breathing. Microorganisms can cause infections or chronic disease; chemicals can be absorbed into the lungs, rapidly pass into the bloodstream, and ultimately reach the vital organs; and physical agents can cause mechanical blocking or irritation of the airway.

inhalation LC$_{50}$ — Concentration of a substance (expressed as milligrams per liter or parts per million parts of air) that is lethal to 50% of the test population under specified test conditions.

inhalation reference concentration for noncarcinogens — An estimate of the daily inhalation exposure that is possible without an appreciable risk of adverse non-cancer effects during a lifetime, as set by the U.S. EPA.

inhalation therapy — The therapeutic use of gases or aerosols that are inhaled into the respiratory system.

inhalation valve — A device that allows respirable air to enter the facepiece and prevents exhaled air from leaving the facepiece through the intake opening.

inhalator — A device providing a mixture of oxygen and carbon dioxide for breathing that is used especially in conjunction with artificial respiration.

inhaler — A device containing a medication that is drawn into air passages by inhalation.

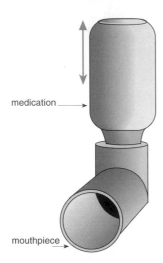

medication

mouthpiece

Pump Style Inhaler

inherent — Belonging by nature or habit.

inHg — Notation for inches of mercury.

inhibitor — Any substance that interferes with a chemical reaction, growth, or other biological activity; also used to prevent or retard rust or corrosion.

inH$_2$O — Notation for inches of water.

initial responders — See first responders.

initial review — (*environmental impact statement*) A preliminary public discussion of the information to be developed, alternatives to be considered, and issues to be discussed.

initiation — (*cancer*) Transformation of a benign or harmless cell to one with the potential for malignant growth.

initiator — A chemical or substance that can cause the initial step in a chain reaction or in the process of carcinogenesis.

injection — The forcing of a liquid into the skin, vessels, muscles, subcutaneous tissue, or any cavity of the body; the forcing of a chemical or other substance into a plant, animal, soil material, or enclosure.

injection well — A well into which fluids are being injected to drive remaining oil into the vicinity of production wells.

injunction — An order issued by the court requiring a defendant to do or refrain from doing a specified act.

injurious — Harmful.

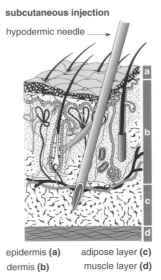

subcutaneous injection

hypodermic needle

epidermis (a)   adipose layer (c)
dermis (b)   muscle layer (d)

Injection

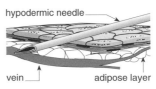

hypodermic needle

vein   adipose layer

intravenous injection

injury — A stress upon an organism that disrupts the structure or function and results in a pathological process.

injury controller — (*nonworkplace*) An environmental health practitioner concerned with the identification and elimination of potentially hazardous construction or conditions and the education of the public in order to prevent unintentional, nonoccupational, and non-motor-vehicle accidents or injuries.

injury prevention — Recognition, evaluation, and control of safety hazards.

inline filtration — The addition of chemical coagulants directly to a filter inlet pipe, where it is mixed with the flowing water; used as a pretreatment method in pressure filter installations.

innage — Space occupied in a product container.

innate — Natural and essential.

inner ear — A cavity in the ear of a vertebrate filled with a liquid and containing many end fibers of the auditory nerve. As sound waves spread through the liquid, they vibrate these nerve endings, and the auditory nerve conveys the impulses to the auditory center of the brain, which produces the sensation of sound.

inner liner — A continuous layer of material placed inside a tank or container to protect the construction materials from the contained waste reagents used to treat the waste.

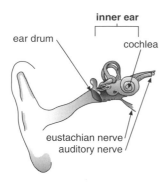

Inner Ear

innervation — The distribution or supply of nerve fibers or nerve impulses to a part of the body.

innocent — Not malignant.

innocuous — Harmless.

innovative control technology — Any system of pollution control technology that has not been adequately demonstrated in practice but would have a substantial likelihood of achieving better results in reducing pollutants than standard control technology.

Innovative Treatment Remediation Demonstration — A program funded by the Department of Energy, Office of Environmental Management, which helps accelerate the adoption and implementation of new and innovative remediation technologies.

inoculate — To inject, introduce, or place fungi, bacteria, nematodes, viruses, or chemicals into a plant or animal to produce a mild form of disease and render the individual immune; see also immunization.

inoculum — Bacteria placed in compost to initiate a biological action.

inorganic — Of mineral origin.

inorganic acid — Compounds of hydrogen and one or more other elements that break down in water and other solvents to produce hydrogen ions; examples include chromic, hydrochloric, nitric, and sulfuric acids. In high concentrations they will destroy body tissue and cause chemical burns. They are especially hazardous to the upper respiratory tract, mucous membrane, and teeth.

inorganic chemistry — The study of the properties and reactions of all chemical elements and compounds other than hydrocarbons.

inorganic compound (IOC) — Compounds that do not contain carbon as the principal element (except carbon dioxide, carbonates, cyanides, and cyanates), as distinguished from organic compounds; organic compounds may be limited to those that contain hydrocarbon groups, often along with oxygen and other elements.

inorganic dust — Dry, finely powdered particles and inorganic substances, especially dust, which can cause abnormal conditions of the lungs if inhaled.

inorganic gases — Oxides of nitrogen, such as nitrogen dioxide, nitric oxide, and nitrous oxide; oxides of sulfur, such as sulfur dioxide and sulfur trioxide; oxides of carbon, such as carbon monoxide and carbon dioxide; and other inorganics, such as hydrogen sulfide, ammonia, and chlorine, which are constituents of nonliving things and are not characteristically biological.

inorganic soil material — Inorganic components consisting of rock fragments and minerals ranging in size from stones and gravel to microscopic pieces of clay.

inorganic solids — (*sewage*) Inert materials not subject to decomposition, including sand, gravel, silt, and mineral salts.

inorganic waste — Materials such as sand, salt, iron, calcium and other mineral materials that are only slightly affected by the action of organisms.

input — The process of entering data.

input bias current — Current that flows into the inputs of a non-ideal amplifier due to leakage current, gate current, or transistor bias current.

input horsepower — The total power used in operating a pump and motor.

input/output — (*computer science*) A data acquisition system monitoring signals through its inputs and sending control signals through its outputs.

insect — An arthropod with three pairs of legs and three body regions (head, thorax, and abdomen) and often, when adult, wings.

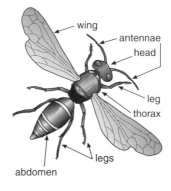

Insect
(Potter Wasp)

insect bite — The bite of an arthropod such as a louse, flea, mite, tick, or arachnid, which may inject a venom that produces poisoning or a severe local reaction or saliva that may contain viruses, bacteria, or substances that produce irritation.

insect growth regulator (IGR) — A chemical substance, such as methoprene; a true analog of a mosquito's own juvenile hormone which causes the death of a mosquito by maintaining high levels of the hormone during the latter instar stages, thus prohibiting the development of physical features necessary for adult emergence from the pupal stage.

insecticide — A pesticide chemical that prevents damage from or destroys insects.

insenescence — The process of aging.

insidious — A development, such as a chronic disease, that is gradual, subtle, or imperceptible.

insight — Ability to comprehend the true nature of a situation.

insipid — Dull, tasteless, lifeless.

insol — See insoluble.

insolation — The magnitude of solar energy that is incident on a particular building element.

insoluble (insol) — Incapable of being dissolved in another material.

insomnia — Inability to sleep.

inspection — (*insurance*) Investigation of certain risks that may be made by independent inspection firms or by an insurance company before issuance or during the term of a policy; an evaluation of a problem or area by visual means.

inspection criteria — The maximum and minimum acceptable values associated with a particular sampling plan.

inspirable particulate mass threshold limit value — Those materials that are hazardous when deposited anywhere in the respiratory tract.

inspiration — Drawing air into the lungs to exchange oxygen for carbon dioxide.

inspirometer — An instrument used to measure the volume, force, and frequency of a person's breathing.

instantaneous field of view — A ground or target area viewed by a sensor in a given period of time.

instar — A stage in the life of an arthropod, such as a mosquito, between two successive molts.

institutional and research laboratory waste — A combination of garbage, refuse, disposables, and dressings that are considered to be contaminated.

institutional environmental health manager — An environmental health professional responsible for the control of biological, physical, and chemical factors that affect the institutional setting.

institutional solid waste — Solid waste originating from educational, healthcare, and research facilities.

institutionalized populations — People who are in schools, hospitals, nursing homes, prisons, and other facilities and who require special care and consideration during emergencies or accidents and in regard to potential infections.

instrument — A device for making observations and measurements of a quantity under investigation.

instrument data processing circuit — A data processing module that transforms the data from the instru-

ment output format into a designed data structure for a computer.

instrument detection limit — The lowest amount of a substance that can be detected by an instrument without correction for the effects of the sample matrix, handling, and preparation.

instrument inlet — The opening through which a sample enters an analyzer, excluding all external sample lines, probes, and manifolds.

instruments system — A grouping of distinct sensors or instruments using the same reporting mechanism.

insulating brick — A firebrick with low thermal conductivity and a bulk density of less than 70 pounds per cubic foot; suitable for insulating industrial furnaces.

insufficiency — Inability to carry out a necessary function adequately.

insulator — A material through which electric charges, heat, or sound are not readily conducted.

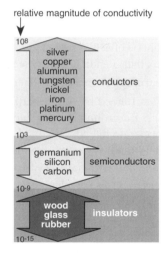

Insulator

insulin — A hormone secretion of the islet of Langerhans in the pancreas that regulates the oxidation of sugar in the tissues and controls its concentration in the blood.

INT — Acronym for intermittent.

intake — The amount of material inhaled, ingested, or absorbed dermally during a specified period of time.

intake area — See recharge zone.

integer — A number without a decimal component.

integrated exposure assessment — The cumulative summation over time of the magnitude of exposure to a toxic chemical in all media.

Integrated Exposure Uptake Biokinetic Model for Lead in Children — A model used to attempt to predict blood-lead concentrations in children exposed to lead in their environment.

integrated performance view — (*indoor air quality*) A study conducted by environmental engineers to analyze building performance from a variety of sources during peak demands, including the pattern of energy use, the levels and kinds of pollutants, and other aspects of air quality.

integrated pest management (IPM) — A systematic, comprehensive approach to pest control that uses the insect's or rodent's own biology and behavior to determine the least toxic control methods at the lowest cost; it includes monitoring, biological control, chemical control, removal of harborage, removal of food, and insect- and rodent-proofing.

integrated sample — The sum of a series of small samples or a continuous flow of samples collected over a defined time period so as to create a large, average sample.

integrated sampling device — An air sampling device that allows estimation of air quality components over a period of time; the laboratory then analyzes the sampler's medium.

Integrated Services Digital Network — A broadband system that allows connection to the Internet at high speeds by standard phone lines.

integrated system — (*computer science*) A group of interconnecting units that form a functioning computer system.

integrated waste management — A variety of practices used to handle municipal solid waste, including source reduction, recycling, incineration, and landfilling.

integrating dose meter — (*radiotherapy*) An ionization chamber designed to be placed on a patient's skin to determine the total radiation administered during an exposure.

integrating nephelometer — An instrument for remote sensing of the vertical and horizontal profile of particle concentration in air in terms of the light-scattering coefficient.

integration — Computing an area within mathematically defined limits; arranging components in a system so they function together in a logical and efficient way.

integrator — A device or meter that continuously measures and calculates total flow in gallons, millions of cubic feet, or some other unit of volume measurement.

integrity — (*global positioning satellite*) The ability of a system to provide timely warnings to users when the system should not be used for navigation as a result of errors or failures in the system.

integument — A covering or skin.

intelligent transportation system — (*geographical information system*) The application of information technologies, especially GIS, to improve the efficiency and safety of the transportation network.

intensity — Amount of energy per unit time passing through a unit area perpendicular to the line of propagation at the point in question; (*geology*) a number describing the severity of an earthquake

in terms of its effects on the surface of the Earth and on humans and their structures.

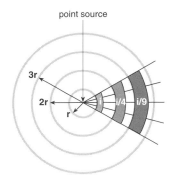

Sound Intensity

Intensity (Mercalli Scale)	Relative Effect(s)	Ergs
I	instrument records only	$<2 \times 10^{10}$
II	barely felt suspended objects swing	$4-9000 \times 10^{10}$
III	felt by some vibration like heavy vehicle	$1-7 \times 10^{15}$
IV–V	felt by most dishes/ windows rattle– may break	$1-30 \times 10^{16}$
VI	felt by all chimneys topple; furniture moves	$1-200 \times 10^{18}$
VII–IX	buildings substantially damaged	$4-230 \times 10^{20}$
X–XI	large landslides; fissures in ground	$4-50 \times 10^{22}$
XII	ground waves; complete devastation	$>1 \times 10^{25}$

Earthquake Intensity

intensive care — A comprehensive, continuous, detailed healthcare program provided for life-threatening conditions by highly experienced and trained individuals.

intensive care unit — A hospital program in which patients requiring close monitoring and specialized care by highly skilled specially trained healthcare workers are housed in a unit as long as necessary.

intentional additives — Substances that are purposely added in the manufacture of food or pharmaceutical products to improve or maintain flavor, color, texture, or consistency or to enhance or conserve nutritional value.

interactive — (*computer science*) A system that allows the operator to receive information from the computer and initiate or modify a program.

interactive graphics system — (*computer science*) A computer system used for the preparation of graphics consisting of a central computer and workstation allowing input from all areas.

intercellular — Between or among cells.

interceptor — A device designed and installed so as to separate and retain hazardous or undesirable material from normal waste while permitting normal sewage or liquid waste to discharge into the drainage system by gravity.

interceptor sewers — A means of control of the flow of sewage to a treatment plant; in the event of a storm, they allow some of the sewage to flow directly into a receiving stream, which protects the treatment plant from being overloaded in case of a sudden surge of water.

intercerebral — Between the cerebral hemispheres.

intercurrent disease — A disease that may alter the course of another disease.

interface — A common boundary layer between two different substances such as water and a solid, between two fluids such as water and a gas, or between a liquid and another liquid.

interface design — (*ergonomics*) The exchange of information between person, machine, and environment to provide necessary data for efficient operation and a safe work place.

interface management language — (*computer science*) A programming language used to communicate with measurement instruments.

interfacial tension — The strength of a film separating two immiscibile fluids, such as oil and water, measured in dynes per centimeter or millidynes per centimeter.

interference — Any undesired energy that tends to disrupt the reception of desired signals; an effect resulting when two series of waves merge into each other.

interferent — A chemical or physical phenomenon that can interfere or disrupt a reaction or process.

interferon — A natural glycoprotein released by cells invaded by viruses.

interflow — Lateral movement of water in the upper layer of soil.

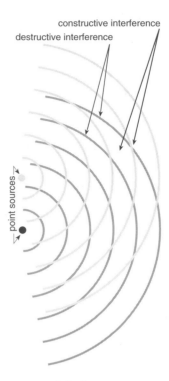

Interference

interleukin — One of several proteins important for lymphocyte proliferation.

interlock — An electrical switch, usually magnetically operated, that interrupts all local power to a panel or device when the door is opened or the circuit is exposed during service.

intermedia transport — Movement from one environmental medium to another.

intermediate — An organic chemical formed during one step of a series of chemical reactions; a precursor to a desired product.

intermediate-care facility — A healthcare facility that provides a level of medically related services to people in an institutional setting that is less than that provided in a hospital or skilled nursing facility.

intermediate exposure — Occupational exposure to a chemical for a duration of 15 to 364 days.

intermediate host — An animal in which a larva spends part of its life cycle.

intermittent — Characterized by alternating periods of activity and inactivity.

intermittent filter — A natural or artificial bed of sand or other fine-grain material; wastewaters are applied intermittently to the surface of the bed in flooding doses to provide filtration and maintain an aerobic condition.

intermittent heating — A means of conserving energy by cutting off the heat supply in any type of building during periods of nonoccupancy.

intermittent positive-pressure breathing (IPPB) — A form of respiratory therapy using a ventilator for the treatment of patients with inadequate pulmonary activity. The treatment involves application of pressure during only the inspiratory phase, forcing adequate oxygen delivery and the removal of carbon dioxide from the lungs.

intermittent sand filter — Specially prepared beds of sand to which effluents from primary treatment, trickling filters, or secondary settling tanks are applied intermittently; the filter bed is a true filter for the solids present, although oxidation may occur at or near the surface of the sand.

intermittent stream — A stream that flows during parts of the year.

internal bleeding — Hemorrhage from an internal organ or tissue.

internal combustion engine — An engine in which fuel burns inside the cylinders rather than in a separate furnace.

internal conversion — One of the possible mechanisms of decay from a metastable state (isomeric transition) in which the transition energy is transferred to an orbital electron, causing its ejection from the atom.

internal dose — The amount of a substance penetrating across absorption barriers by physical or biological means.

internal drainage — The movement of water down through soil to porous aquifers and to surface outlets at lower elevations.

internal ear — Complex inner structure of the ear that contains receptors for hearing and balance.

internal erosion — Formation of voids within soil or soft rock caused by the mechanical or chemical removal of material by seepage.

internal friction — Friction within a fluid such as water due to cohesive forces.

internal housing environment — Heat, light, ventilation, electrical facilities, plumbing, structural soundness, and physical structure of dwellings as they affect people.

internal injury — Any wound or damage to the viscera.

internal medicine — A branch of medicine concerned with the study of the physiology and pathology of internal organs and with medical diagnosis and treatment of disorders of these organs.

internal pressure — Pressure inside a building envelope or space usually expressed with respect to outside or atmospheric pressure.

internal radiation — Exposure to ionizing radiation from radioisotopes deposited inside body tissues through inhalation, ingestion, absorption, or injection.

internal respiration — The exchange of gas between the blood and tissues of an organism.

internal soil drainage — The quality of a soil that permits the downward flow of excess water through the soil.

internal temperature — The temperature within the interior of a food product.

International Classification of Disease Adapted for Use in the United States — A classification system adapted by the U.S. Public Health Service from a master system developed by the World Health Organization for categorizing and indexing hospital records.

International Classification of Diseases — An official list of categories of physical and mental diseases issued by the World Health Organization used primarily for statistical purposes in the classification of morbidity and mortality data.

International Nursing Index — An international index of over 260 nursing journals and selected topics from 2700 allied health and biomedical journals from 1967 to the present.

International Statistical Classification of Diseases, Injuries, and Causes of Death (ICD) — An internationally recognized system for coding problems of health that is revised every 10 years by the World Health Organization.

International System of Units (SI units) — An international system of physical units recommended by the General Conference on Weights and Measures (1960). The fundamental quantities are length, time, mass, electric current, temperature, luminous intensity, and amount of substance; the corresponding units are the meter, second, kilogram, ampere, kelvin, candela, and mole. Also known as *Système International d'Unités*.

International Unit — A unit of measure in the International System of Units.

Internet — An international network of networks; an electronic communications vehicle allowing computers to talk to other computers to send and receive mail, search databases, transfer files, etc.

Internet server — A computer that stores data that can be accessed through the Internet.

Internet service provider — A company that sells access to the Internet.

Internet site — A computer connected to the Internet and containing information that can be accessed using an Internet navigation tool.

interneuron — A neuron located between a sensory neuron and a motor neuron and functioning as an integrating center.

internship — A period of intense, highly supervised apprenticeship for a specified period of time before the person begins professional practice (e.g., a healthcare worker).

interpersonal — Referring to interactions between people.

interphase — The phase of the cell cycle between cell divisions during which much of the synthesis of cellular constituents occurs.

interpolate — To estimate the value of an attribute at an unsampled point from measurements made at surrounding sites.

interpolation — The process of estimating values for missing points.

interrogatories — (*law*) Written questions propounded by one party and served to an individual who must present written answers under oath.

intersection — Nonparallel touching or crossing of fibers with the projection having an aspect ratio of 5:1 or greater.

interspike interval (ISI) — The time between two successive repetitions of a motor unit action potential.

interstate (IS) — Between or among states.

interstate carrier — Any vehicle or transport that conveys passengers or freight in interstate commerce.

interstate carrier water supply — A source of water for drinking and sanitary use on planes, buses, trains, and ships operating in multiple states.

interstice — A gap or break in something generally continuous; space in a rock or soil; see soil pore.

interstitial — Pertaining to or situated in the space between two things.

interstitial fluid — An extracellular fluid that fills the spaces between cells of the body and provides most of the liquid environment of the body.

interstitial monitoring — Continuous surveillance of the space between the walls of an underground storage tank.

interstitial water — Moisture that collects in voids between soil particles; it is relatively mobile, dissolves solids, and moves through the soil.

intertidal — The area that is alternately exposed to the air and to the sea between low and high tides.

intertrigo — An erythemous skin eruption occurring on surfaces of the skin such as in creases of the neck, folds of the groin, and armpits.

interval — Period of time separating two events or the distance between two objects.

intervention — A process or procedure used to prevent harm from occurring to a person or to improve the mental, emotional, or physical functions of an individual.

interview — A communication with a person to gain data.

interzonal airflow — The process of air exchange between internal zones of a building.

intestinal flora — Bacteria normally found in the digestive tract.

intestinal fluke — An internal parasite that enters the body through the mouth as an encysted larva in aquatic vegetation or freshwater fish.

intestinal tract — The segments of small and large intestines between the pyloric valve and rectum.

intestine — Portion of the alimentary canal extending beyond the stomach; it is the site of the final digestion of food matter from the stomach, the absorption of soluble food matter and water, and the production of feces.

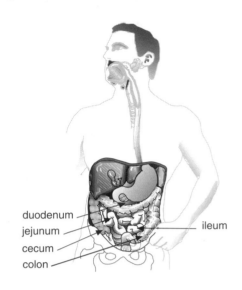

Intestine

intima — The innermost structure.

intolerance — Inability to absorb or metabolize a nutrient or medication.

intoxicant — A chemical that can cause intoxication or poisoning.

intoxication — The state of being poisoned; the state produced by overindulgence in alcohol.

intracellular — Within the cell.

intracelluar fluid — A fluid within the cell membranes throughout the body containing the dissolved solutes required for electrolytic balance and healthy metabolism.

intracerebral — Within the tissues of the cerebrum.

intracranial — Within the cranium.

intracranial pressure — A pressure that occurs within the cranium.

intractable — A symptom or disease that is unrelieved by therapeutic measures.

intractable pain — A chronic and persistent pain that is unrelieved by ordinary medical or surgical techniques.

intracutaneous — Within the substance of a skin.

intradermal — Within the tissue of the skin.

intradermal injection — Injection of a hypodermic needle within the tissue of the skin.

intramedia transport — Movement within an environmental medium.

intraocular — Within the eyeball.

intraperitoneal — Within the abdominal cavity.

intrastate — Within a state.

intrauterine — Within the uterus.

intravenous — Within a vein.

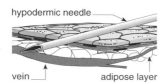

Intravenous Injection

intravenous drug — Drug injected directly into a vein.

intravenous line — A catheter inserted into a vein through which fluids, antibiotics, and other medication can be given directly.

intruding noise level — The total sound level (in decibels) created, caused, maintained, or originating from an alleged offensive source at a specified location while the alleged offensive source is in operation.

intubate — To insert a tube into an organ or body part.

intubation — The insertion of a breathing tube through the mouth or nose or into the trachea for delivery of anesthetic gases and oxygen.

inundate — To cover with impounded waters or flood waters.

invasive — The tendency to spread, especially the tendency to invade healthy tissues.

invasive procedure — A diagnostic or therapeutic technique requiring entry into a body cavity or interruption of normal body functions.

inventory — A specific list of problems or equipment-related problems determined by inspections, field visits, or use of specialized forms prepared by personnel at the source of the potential pollution.

inverse-square law — The strength of a physical quantity is proportional to the reciprocal of the square of the distance from the source of that property; applies to sound, gravity, heat, light, and radio waves.

inversion — The phenomenon of a layer of cool air trapped below a layer of warmer air so that the bottom layer cannot rise as usual; an increase in the temperature of an area with height, as opposed to the usual decrease of temperature with height. Also known as temperature inversion.

invert — The lowest point of the channel inside a pipe, conduit, or canal.

invert emulsion — An emulsion having water suspended as small droplets in oil.

invertebrate — An animal without a backbone or internal skeleton.

inverter — A piece of equipment that converts direct current to alternating current.

invoice — Paperwork that shows the vendor's intent to sell an article but does not prove actual movement.

involuntary muscle — Any muscle that cannot be controlled at will; also known as smooth muscle.

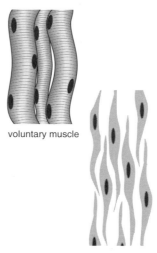

voluntary muscle

Involuntary Muscle

I/O — See input/output.

IOC — See inorganic compound.

iodates — Compounds with the general formula M(IO$_3$)$_x$, where M is a metal ion with valency x or an ammonium ion. They are strong oxidizers that ignite violently with organic or other oxidizable substances when subjected to heat, percussion, or shock. These are salts of iodic acid (HIO$_3$). Potassium iodate and sodium iodate are used as oxidizing agents in volumetric analysis, as dough conditioners, and as antiseptics. They produce mild skin irritations and may cause nausea, vomiting, and diarrhea.

iodide — A chemical compound containing iodine; when potassium or sodium iodide is used with a super oxidizing agent, such as chlorine, iodine will be released in pool water.

iodine (I$_2$) — A violet solid with a sharp, characteristic odor; MW: 253.8, BP: 365°F, Sol: 0.01%, sp. gr. 4.93. It is used in the synthesis of chemical intermediates, pharmaceuticals, photographic chemicals, antiseptics, disinfectants, detergent sanitizers, and organic and inorganic compounds; as a reagent in analytical chemistry; as a catalyst in organic synthesis; in the synthesis of food dyes, food additives, and coloring agents; in the production of intermediates in purification of minerals; as a catalyst in the modification of selenium during the manufacture of photoelectric cells and rectifiers; during the manufacture of specialty lubricants; and in the production of stereospecific polymers. It is hazardous to the respiratory system, eyes, skin, central nervous system, and cardiovascular system and is toxic through inhalation, ingestion, and contact. Symptoms of exposure include lacrimation, headache, tight chest, skin burns, rash, cutaneous hypersensitivity, and irritation of the eyes and nose. OSHA exposure limit (TWA): ceiling 0.1 ppm [air] or 1 mg/m^3.

iodine heptafluoride (IF$_7$) — A colorless gas; MW: 259.91; BP: NA; Sol: soluble in water and absorbed by caustic soda solution; Fl.P: a nonflammable gas that reacts violently with ammonium halides and with organic matter; density: NA. It is used as a fluorinating agent. It is hazardous to the skin and mucous membranes and is highly irritating. ACGIH exposure limit: NA

iodine monobromide (IBr) — A brownish-black crystalline solid; MW: 206.84; BP: 116°C (decomposes); Sol: soluble in water, alcohol, ether, and carbon disulfide; Fl.P: a noncombustible solid that explodes when mixed with potassium and ignites when mixed with aluminum; density: NA. It is hazardous to the skin, eyes, and mucous membranes and is extremely irritating and corrosive. ACGIH exposure limit: NA.

iodine monochloride (ICl) — A black crystalline solid or brown liquid; MW: 162.35; BP: 97°C in the beta form; Sol: soluble in water, alcohol, ether, and carbon disulfide; Fl.P: a noncombustible solid or liquid that explodes in contact with potassium; density: 3.10 at 29°C. It is used to estimate the iodine values of fats and oils. It is hazardous to the skin, eyes, and mucous membranes. It is highly corrosive and causes severe burns. ACGIH exposure limit: NA.

iodine pentafluoride (IF$_5$) — A colorless, viscus liquid; MW: 221.90; BP: 1.5°C; Sol: miscible with water but reacts violently; Fl.P: a noncombustible liquid but highly reactive with water, potassium, molten sodium, arsenic, bismuth, tungsten, silicon, sulfur, or phosphorus; density: 3.190 at 25°C. It is used as a mild fluorinating agent. It is hazardous to the skin, eyes, and mucous membranes and causes severe burns. ACGIH exposure limit (TLV-TWA): 2.5 mg/m^3.

iodine trichloride (ICl$_3$) — A yellow crystalline solid or powder with an irritating pungent odor; MW: 233.39; BP: NA; Sol: soluble in water, alcohol, and ether; Fl.P: noncombustible solid that reacts violently with potassium, sodium, aluminum, phosphorus, or organic matter; density: NA. It is used as a chlorinating and oxidizing agent. It is hazardous to the skin, eyes, and mucous mem-

branes and causes severe irritation and burns. ACGIH exposure limit: NA.

iodized salt — Table salt to which sodium iodide has been added as a preventive measure to protect against goiter.

idoform — A topical antiinfective substance used as an antiseptic.

iodophore — A soluble complex of iodine and a nonionic surface-active agent that releases iodine gradually and creates a bacteriocidal action in cold or hot water.

ion — An atomic particle, atom, or chemical radical bearing an electrical charge that is either negative or positive, depending on whether it has gained or lost one or more electrons; its migration affects the transport of electricity through an electrolyte or to a certain extent through a gas.

ion exchange — A process by which certain ions of given charge may be absorbed from solution and replaced in the solution by other ions from the absorbent with a similar charge; a chemical process involving the reversible interchange of ions between a liquid and a solid but no radical change in structure of the solid.

ion exchange chromatography — Separation and analysis of different substances according to their affinities for chemically stable but reactive synthetic compounds of mostly polystyrene and cellulose.

ion exchange resin — A special type of molecule, such as zeolite, or a synthetic resin containing active groups that give the resin the property of combining with or exchanging ions between the resin and the solution, as in the process of softening hard water.

ion exchange treatment — A water-softening method that reduces some organics and radium by adding calcium oxide or calcium hydroxide to increase the pH to a level where the metals will precipitate out.

ion mobility spectrometry (IMS) — A technique used to detect and characterize organic vapors in air, involving the ionization of molecules and their subsequent temporal drift through an electric field.

ion pair — Two particles of opposite charge, usually referring to the electron and positive atomic or molecular residue left after the interaction of ionizing radiation with the orbital electrons of atoms.

ionic bond — A type of chemical force holding atoms together resulting from differences in electrical charges; also known as electrovalent bond.

ionic concentration — The concentration of an ion in solution (usually expressed in moles per liter).

ionic dissociation — The ions in ionic compounds in an aqueous solution freed from their mutual attractions and distributed uniformly throughout the solvent.

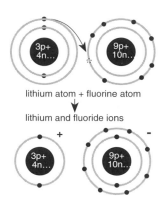

Ionic Bond

lithium atom + fluorine atom

lithium and fluoride ions

ionization — The process by which a neutral atom or molecule acquires a positive or negative charge.

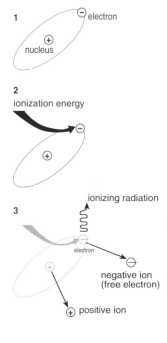

Ionization

ionization chamber — An instrument designed to measure the quantity of ionizing radiation in terms of the charge of electricity associated with ions produced within a defined volume.

ionization constant — At ionic equilibrium, the product of the molar concentration of the ions divided by the molar concentration of the non-ionized molecules.

ionization density — The number of ion pairs per unit volume.

ionization path — The trail of ion pairs produced by ionizing radiation and their passage through matter.

ionize — To change into ions.

ionizing energy — The average energy lost by ionizing radiation producing an ion parent in a gas.

ionizing radiation — High-energy electromagnetic or particulate radiation released by nuclear decay and capable of producing ions directly or indirectly in its passage through matter.

ionosphere — Layer of the atmosphere that reflects radio waves and which is located directly above the stratosphere at 40 or 50 miles; it extends indefinitely.

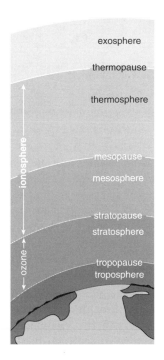

Ionosphere

Iowa vane — A flow deflector that will divert water to attract fish.

IP — Acronym for inhalable particle.

Ipecac — An emetic or expectorant made from the dried rhizome and root of several tropical South American shrubs.

IPM — Acronym for inhalable particulate matter; see integrated pest management.

IPM-TLV — See inspirable particulate mass threshold limit value.

IPPB — See intermittent positive-pressure breathing.

ipso facto — By the act itself.

IPV — See integrated performance view.

IR — See infrared radiation or incremental reactivity.

IRA — Acronym for interim response action.

IRGA — Acronym for infrared gas analyzer.

iridium — A silvery-bluish metallic element used as a radioactive isotope in interstitial brachytherapy; atomic number: 77, atomic weight: 192.2.

iris — Muscular, colored portion of the eye behind the cornea and surrounding the pupil that regulates the amount of light reaching the retina in vertebrates and some cephalopod molluscs.

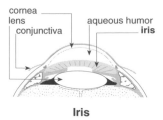

Iris

IRM — Acronym for intermediate remedial measure.

iron bacteria — Those bacteria able to obtain their energy for living by oxidizing iron compounds.

iron oxide dust and fume (Fe_2O_3) — A reddish-brown solid; MW: 159.7, BP: not known, Sol: insoluble, sp. gr. 5.24. Exposure to fume may occur during the arc-welding of iron. It is liberated during the production of steel ingots; processing of iron ore and pig iron; heating and pouring of molten metal in foundry operations; hot rolling of sheet and strip steel; forging of metal items containing iron and steel; pressing of metal items; grinding and polishing of glass, precious metals, stones, and gemstones. It is hazardous to the respiratory system and is toxic through inhalation. Symptoms of exposure include benign pneumoconiosis with x-ray shadows indistinguishable from fibrotic pneumoconiosis. OSHA exposure limit (TWA): 10 mg/m^3 [air].

iron pentacarbonyl ($Fe(CO)_5$) — A colorless oily liquid that turns yellow; MW: 195.90; BP: 103°C; Sol: slightly soluble in alcohol and readily dissolves in most organic solvents; Fl.P: –15°C (closed cup), a highly flammable liquid that ignites in air; density: 1.50 at 20°C. It is used as a catalyst in organic synthesis and as an antiknock agent in motor fuels. It is hazardous to the respiratory tract. It may cause headache, dizziness, somnolence, fever, coughing, and cyanosis and may affect the lungs, liver, and kidneys. ACGIH exposure limit (TLV-TWA): 0.1 ppm.

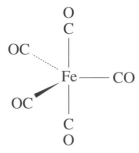

Iron pentacarbonyl

iron sulfide pyrite (FeS) — An ore commonly found in rocks; also known as fool's gold.

Iron Sulfide Pyrite

IRPTC — Acronym for International Register of Potentially Toxic Chemicals.

irradiance — Power per unit area of optical radiation; radiant flux per unit area of a surface.

irradiated food — Food exposed to wavelengths shorter than those of visible light, primarily gamma rays, to kill insects, bacteria, and mold to permit storage without refrigeration and lengthen shelf life.

irradiation — Exposure to any form of radiation.

irreducible — Unable to be returned to the normal position or condition.

irreversible — Permanent or incurable.

irreversible effect — An effect characterized by the inability of the body to partially or fully repair injury caused by a toxic agent.

irreversible toxicity — Toxic effects that cannot be repaired.

irrigation — The controlled application of water for agricultural purposes by transporting it through canals, pipes, and sprinklers to supply water requirements not satisfied by rainfall.

irritability — The ability to detect, interpret, and respond to a stimulus in the environment; a condition of abnormal excitability to slight stimuli.

irritant — A chemical that is not corrosive but which causes a reversible inflammatory effect on living tissue by chemical action at the site of contact; a liquid or solid substance which, upon contact with fire or when exposed to air, gives off dangerous or intensely irritating fumes.

irritation — An aggravation of a tissue at the point of contact with a material.

IRS — Acronym for International Referral Systems.

IRTM — Acronym for infrared thermal mapper.

IS — Acronym for interim status; see interstate.

ISC — Acronym for industrial source complex.

ishemia — A local and temporary deficiency of blood as a result of obstruction of the blood supply; a decreased supply of oxygenated blood to a body organ or part.

ISCST — Acronym for Industrial Source Complex Short-Term Model.

ISDN — See Integrated Services Digital Network.

ISE — Acronym for ion-specific electrode.

ISI — Acronym for International Statistical Institute; see interspike interval.

islets of Langerhans — Cells in the pancreas that secrete insulin.

ISMAP — Acronym for Indirect Source Model for Air Pollution.

isoamyl acetate (CH₃COOCH₂CH₂CH(CH₃)₂) — A colorless liquid with a banana-like odor; MW: 130.2, BP: 288°F, Sol: 0.3%, Fl.P: 77°F, sp. gr. 0.87. It is used in treating natural leathers by tanning; as an extractant in purification of pharmaceuticals including penicillin; in the manufacture of artificial leather; in drycleaning preparations; during the manufacture of artificial silk, rayon, and pearls. It is liberated during application of varnishes and nitrocellulose lacquers as protective and finish coatings for wood, paper, metal, leather, and other surfaces; during application of cellulosic adhesives in shoe manufacturing, book binding, packaging, leather and paper processing; during the manufacture of shoe and furniture polishes; during fermentation of whiskey grains; during the manufacture of cellulosic photographic film; during the preparation of perfumes, foods, and other materials for use as a flavoring and odorant; during the manufacture of bath sponges; during the cleaning and maintenance of acetate-processing equipment. It is hazardous to the eyes, skin, and respiratory system and is toxic through inhalation, ingestion, and contact. Symptoms of exposure include narcosis, dermatitis, and irritation of the eyes, nose, and throat. OSHA exposure limit (TWA): 100 ppm [air] or 525 mg/m³.

isoamyl alcohol (primary: (CH₃)₂CHCH₂CH₂OH; secondary: ((CH₃)₂CHCH(OH)CH₃) — Colorless liquids with a disagreeable odor; MW: 130.2; BP: 270 and 234°F, respectively; Sol: 2% and not known, respectively, at 57°; Fl.P (open cup): 109 and 95°F, respectively; sp. gr. 0.81 and 0.82,

respectively. They are used as a vehicle, latent, or diluent solvent during the application of paints, lacquers, varnishes, thinners, and paint removers; in the synthesis of drugs and medicinals; as a solvent for alkaloids; as an extractant of antibiotics in the pharmaceutical industry; in the manufacture of lacquers, paints, varnishes, thinners, and paint removers; as a chemical intermediate in organic synthesis of photographic chemicals and esters; in the manufacture of printing inks; as a solvent for resins, gums, waxes, oils, perfumes, explosives, and shoe cement; in analytical determination of fat in milk and artificial rubber; in the manufacture of antifoaming agents; in the mining industry as a frothing agent for flotation of nonferrous ores; as a vehicle solvent for celloidin solutions in microscopy. They are hazardous to the eyes, skin, and respiratory system and are toxic through inhalation, ingestion, and contact. Symptoms of exposure include narcosis, headache, dizziness, dyspnea, nausea, vomiting, diarrhea, skin cracking, and irritation of the eyes, nose, and throat. OSHA exposure limit (TWA): 100 ppm [air] or 360 mg/m³.

isobar — (*physiology*) One of two or more chemical species with the same atomic weight but different atomic numbers; (*meteorology*) a line drawn on a map or chart connecting locations of identical atmospheric pressure.

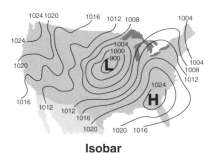

Isobar

isobutane (C₄H₁₀) — A colorless gas; MW: 58.14, BP: NA, Sol: soluble in water and organic solvents, Fl.P: a flammable gas, density: 2.04. It is used as a fuel gas, in organic synthesis, and in liquefied petroleum gas. It is hazardous to the respiratory tract and may cause asphyxia, narcosis, and drowsiness. ACGIH exposure limit: NA.

Isobutane

isobutyl acetate (CH₃COOCH₂CH(CH₃)₂) — A colorless liquid with a fruity, floral odor; MW: 116.2,

BP: 243°F, Sol: 0.6% at 77°F, Fl.P: 64°F, sp. gr. 0.87. It is used in the manufacture of some perfumes, cosmetics, and flavoring agents; in shoe manufacturing, book binding, packaging, leather processing, photographic film manufacturing, and paper processing. It is liberated during the application of varnishes and nitrocellulose lacquers as protective coatings for wood, plastic, metal, leather, and other surfaces; during oven baking of phenolic and epoxy coatings; during the cleaning and maintenance of acetate-processing equipment. It is hazardous to the skin, eyes, and respiratory system and is toxic through inhalation, ingestion, and contact. Symptoms of exposure include headache, drowsiness, anesthesia, and irritation of the eyes, skin, and upper respiratory system. OSHA exposure limit (TWA): 150 ppm [air] or 700 mg/m³.

Isobutyl acetate

isobutyl acetate ($C_6H_{12}O_2$) — A colorless liquid with a fruity smell; MW: 116.18, BP: 118°C, Sol: slightly soluble in water and miscible with most organic solvents, Fl.P: 18°C (closed cup), density: 0.871 at 20°C. It is used as a solvent and as a flavoring agent. It is toxic to the respiratory tract; causes symptoms of headaches, drowsiness, and irritation of the upper respiratory tract; and acts as an anesthesia. ACGIH exposure limit (TLV-TWA): 150 ppm.

isobutyl alcohol (($CH_3)_2CHCH_2OH$) — A colorless, oily liquid with a sweet, musty odor; MW: 74.1, BP: 227°F, Sol: 10%, Fl.P: 82°F, sp. gr. 0.80. It is used in the manufacture of industrial cleaners, nitrocellulose lacquers, paint removers, lubricating oil and hydraulic fluids, amino resins, and plasticizers; during the distillation of whiskey. It is hazardous to the eyes, skin, and respiratory system and is toxic through inhalation, ingestion, and contact. Symptoms of exposure include headache, drowsiness, skin irritation and cracking, and irritation of the eyes and throat. OSHA exposure limit (TWA): 50 ppm [air] or 150 mg/m³.

isobutyraldehyde (C_3H_7CHO) — A colorless liquid with a pungent smell; MW: 72.1; BP: 64.5°C; Sol: soluble in water, ether, acetone, and chloroform; Fl.P: −18°C (closed cup); density: 0.794. It is used in the synthesis of cellulose esters, resins, and plasticizers. It is hazardous to the skin and eyes

and causes severe skin irritation and moderate eye irritation. ACGIH exposure limit: NA

Isobutyraldehyde

isodrin ($C_{12}H_8Cl_6$) — An odorless, crystalline solid; MW: 364.90, BP: decomposes, Sol: NA, Fl.P: NA (a flammable material that may be ignited by heat, sparks, or flames and produces poisonous chlorine gas in a fire), sp. gr. NA. It is used as a chlorinated hydrocarbon insecticide. It is hazardous to the respiratory system, digestive system, skin, and eyes. Acute exposure causes central nervous system, depression, or excitation with tremors and seizures. It also causes stomach pain, nausea, vomiting, diarrhea, respiratory depression, pneumonitis, acute renal failure, weakness of the extremities, confusion, dizziness, amnesia, and death. Chronic exposure may damage the liver and kidneys and cause systemic toxicity, anorexia, nausea, fatigue, memory deficits, personality changes, and loss of coordination. OSHA exposure limit (8-hour TWA): NA.

isoenzyme — See isozyme.

isokinetic sampling — Any technique for collecting airborne particulate matter in which the linear velocity of the gas entering the sampling nozzle is equal to that of the undisturbed gas stream at the sample point.

isolan ($C_{10}H_{17}N_3O_2$) — A colorless liquid; MW: 211.27, BP: 118°C, Sol: miscible in water, sp. gr. 1.07. It is a cholinesterase inhibitor that is used as a pesticide. It is extremely toxic and exhibits acute, delayed, and chronic effects. It may cause trembling, sweating, slurring of speech, nausea, vomiting, loss of bladder control, convulsions, coma, and death.

Isolan

isolate — To separate a pure chemical substance from a contaminating material.

isolation — The separation of infected individuals from those uninfected for the period of communicability of a particular disease; a technique used in environmental control where potential pollutants are removed from substantial population centers.

isolation rate — The rate at which an organism is identified in a culture.

isolation room — A separate room containing the sterilizer and the sterilizer loading area as it relates to the use of ethylene oxide; a room in which an infected patient is treated and housed.

isomer — One of several nuclides having the same number of neutrons and protons but capable of existing for a certain period of time in different states with different energies and physical and chemical properties.

isomeric transition — The process by which a nuclide decays to an isomeric nuclide of lower quantum energy.

isomeric — Two chemical compounds having the same proportion of elements and molecular weight but differing in chemical and physical properties because of the arrangement of atoms in the respective molecules.

isometric — Maintaining or pertaining to the same length.

isometric work — A state of muscular contraction without movement.

isomotic — See isotonic.

isooctane (C_8H_{18}) — A colorless mobile liquid with an odor of gasoline; MW: 114.26; BP: 99.3°C; Sol: mixes readily with ether, chloroform, benzene, toluene, carbon disulfide, and oils; Fl.P: –12.2°C (closed cup), a highly flammable liquid; density: 0.692 at 20°C. It is used as the standard for determining the octane levels of fuels and as a solvent in chemical analysis. It is hazardous to the respiratory tract and may produce irritation. ACGIH exposure limit: NA

$$CH_3-\overset{\overset{\displaystyle CH_3}{|}}{\underset{\underset{\displaystyle CH_3}{|}}{C}}-CH_2-\overset{\overset{\displaystyle CH_3}{|}}{CH}-CH_3$$

Isooctane

isophorone ($C_9H_{14}O$) — A colorless to white liquid with a peppermint-like odor; MW: 138.2, BP: 419°F, Sol: 1%, Fl.P: 184°F, sp. gr. 0.92. It is used as a solvent in the manufacture of vinyl resins, nitrocellulose, fats, chlorinated rubber, herbicides, coatings, roll-coating finishes, paints, adhesives, and inks; in organic synthesis; in the manufacture of lubricating oil additives, fungicides, and tetramethylquanidine. It is hazardous to the respiratory system and skin and is toxic through inhalation, ingestion, and contact. Symptoms of exposure include narcosis, dermatitis, and irritation of the

eyes, nose, and throat. OSHA exposure limit (TWA): 4 ppm [air] or 23 mg/m³.

isophorone diisocyanate ($C_{10}H_{18}(NCO)_2$) — A colorless to slightly yellow liquid; MW: 222.3, BP: 158°C, Sol: soluble in most organic solvents and decomposes in water and alcohol, Fl.P: NA (reacts with compounds containing active hydrogen atoms), density: 1.062 at 20°C. It is used in the production of high-quality coatings, polyurethane paints, and varnishes. It is hazardous to the respiratory tract. It may cause bronchitis, asthma, tightness of the chest, and dyspnea. It is also an irritant to the skin and eyes. ACGIH exposure limit (TLV-TWA): 0.005 ppm.

$$\begin{array}{c} H_3C \\ H_3C \end{array}\!\!\!\diagdown\!\!\!\bigcirc\!\!\!\diagup\!\!\! N=C=O$$
$$H_3C \quad CH_2-N=C=O$$

Isophorone diisocyanate

isopleth — The lines on a graph connecting points of constant value.

isopropyl acetate ($CH_3COOCH(CH_3)_2$) — A colorless liquid with a fruity odor; MW: 102.2, BP: 194°F, Sol: 3%, Fl.P: 36°F, sp. gr. 0.87. It is used as a vehicle solvent during spray applications of cellulose nitrate and ethyl cellulose lacquers; as a solvent in synthetic resins in the plastics industry; in the manufacture of printing and lithography inks, perfumes, and flavoring agents; as a general solvent for waxes, gums, and oils; in organic synthesis. It is hazardous to the eyes, skin, and respiratory system and is toxic through inhalation, ingestion, and contact. Symptoms of exposure include narcosis, dermatitis, and irritation of the eyes, nose, and skin. OSHA exposure limit (TWA): 250 ppm [air] or 950 mg/m³.

isopropyl alcohol ($(CH_3)_2CHOH$) — A colorless liquid with the odor of rubbing alcohol; MW: 60.1, BP: 181°F, Sol: miscible, Fl.P: 53°F, sp. gr. 0.79. It is used as a solvent in spray and heat applications of surface coatings, including stain, varnish, nitrocellulose lacquers, and quick-drying inks and paints; in the manufacture of acetone; as a solvent in the manufacture of surface coatings and thinners; in organic synthesis for isopropyl derivatives; in the manufacture of cosmetics, including liniments, skin lotions, permanent-wave lotions, and hair-coloring rinses; as a disinfectant and sanitizer; in cleaning and degreasing operations; in the extraction and purification of alkaloids, proteins, chlorophyll, vitamins, kelp, pectin, resins, waxes, and gums; in the manufacture of rubber products and adhesives; as an additive in antistalling gasoline, lubricants, denatured ethyl alcohol, hydraulic brake fluids, and rocket fuel. It is hazardous to the eyes, skin, and respiratory system and is toxic through inhalation, ingestion, and contact. Symptoms of exposure include drowsiness, dizziness, headache, dry and cracking skin, and mild irritation of the eyes, nose, and throat. OSHA exposure limit (TWA): 400 ppm [air] or 980 mg/m³.

isopropyl ether ((CH₃)₂CHOCH(CH₃)₂) — A colorless liquid with a sharp, sweet, ether-like odor; MW: 102.2, BP: 154°F, Sol: 0.2%, Fl.P: 92°F, sp. gr. 0.73. It is used as a solvent in extraction processes, rubber adhesives, lacquers, resins, oils, cellulose, pharmaceutical manufacture, smokeless gunpowder, and textile spot cleaning; in organic synthesis as an alkylation agent; as an emulsion breaker in the petroleum industry; as a blending agent for gasoline. It is hazardous to the respiratory system and skin and is toxic through inhalation, ingestion, and contact. Symptoms of exposure include respiratory discomfort, dermatitis, and irritation of the eyes and nose; in animals, drowsiness, dizziness, unconsciousness, and narcosis. OSHA exposure limit (TWA): 500 ppm [air] or 2,100 mg/m³.

isopropyl glycidyl ether (C₆H₁₂O₂) — A colorless liquid; MW: 116.2, BP: 279°F, Sol: 19%, Fl.P: 92°F, sp. gr. 0.92. It is used as a reactive diluent for epoxy resins; as a chemical intermediate for the synthesis of esters and ethers; as a stabilizing agent for organic chemicals. It is hazardous to the eyes, skin, and respiratory system and is toxic through inhalation, ingestion, and contact. Symptoms of exposure include skin sensitization and irritation of the eyes, skin, and upper respiratory system. OSHA exposure limit (TWA): 50 ppm [air] or 240 mg/m³.

isopropylamine-2 — A colorless, volatile, liquid with an ammonia odor; MW: 59.13; BP: 33 to 34°C; Sol: soluble in water, alcohol, and ether; Fl.P: –37°C (closed cup); density: 0.694. It is used as a dehairing agent and in the preparation of many organics. It is hazardous to the eyes, skin, and respiratory system. It can cause irritation to the nose and throat and, with prolonged exposure, can cause pulmonary edema, dermatitis, and skin burns. ACGIH exposure limit (TLV-TWA): 5 ppm

$$CH_3 \diagdown$$
$$CH-NH_2$$
$$CH_3 \diagup$$

Isopropylamine

isopynic — A line on a chart that connects all points of equal or constant density.

isoquinoline (C₉H₇N) — A colorless liquid or solid with a pungent odor; MW: 129.17, BP: 242°C, Sol: soluble in organic solvents, Fl.P: a noncombustible liquid or solid, density: NA. It is used in the manufacture of dyes, pharmaceuticals, and insecticides; as an antimalarial agent. It causes irritation of the skin and eyes of animals. ACGIH exposure limit: NA.

Isoquinoline

isoseismal line — A line on a map joining points of equal earthquake intensity.

isotherm — A line on a map or chart joining locations recording identical temperature.

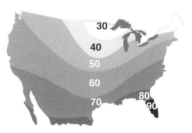

Isotherm

isothermal — Having the same temperature.

isothermal process — Any process that occurs at constant temperature.

isothermal surface — A surface with an even temperature at all its points.

isotone — One of two or more atoms having the same number of neutrons in their nuclei.

isotonic — A solution having the same concentration of solute as another solution, thereby exerting the same amount of osmotic pressure as the original solution.

isotope — One of two or more forms of the same element differing in atomic weight but having the same atomic number and the same chemical properties.

isotopic tracer — The isotope or non-natural mixture of isotopes of an element that may be incorporated into a sample, making possible the observation of the course of that element, alone or in combination, through a chemical, biological, or physical process; also known as label.

isotropic mass — A mass having the same property or properties in all directions.

isozyme — A different chemical form of the same enzyme; thought to be important in adaptation to environmental extremes.

ISP — See Internet service provider.

ISS — Acronym for interim status standard.

IST — Acronym for Instrument Support Terminal.

ISV — See *in situ* vitrification.

IT — See immunotoxin.

itch — A tingling, annoying sensation on an area of the skin.

itch mite — A tiny, eight-legged insect with piercing and sucking mouth parts.

ITRD — See Innovative Treatment Remediation Demonstration.

IU — Acronym for industrial user; see International Unit.

IUP — Acronym for Intended Use Plan.

IUR — Acronym for Inventory Update Rule.

IV — See intravenous line.

IWC — Acronym for in-stream waste concentration or inches of water column.

IWI — Acronym for Index of Watershed Indicator.

IWIC — Acronym for International Waste Identification Code.

IWRP — Acronym for Industry Waste Reduction Plan.

IWS — Acronym for ionizing wet scrubber.

Ixodes — A genus of parasitic hard-shelled ticks associated with the transmission of a variety of arbovirus infections.

I

J

J — See joule.

J curve — A growth curve that depicts exponential growth.

Jackson turbidity unit (JTU) — A measure of the turbidity of water proportional to the parts per million of silica, where 100 ppm of silica equals 21.5 JTU; it has been replaced by the formazin turbidity unit.

jail fever — See epidemic typhus.

Japanese encephalitis — A severe epidemic infection of brain tissue found in East Africa and the South Pacific with symptoms of shaking, chills, paralysis, and weight loss caused by a group of B arboviruses transmitted by mosquitoes; a variety of neurological and psychiatric problems may follow the infection, and mortality rate may be as high as 33%.

Japanese flood fever — See scrub typhus.

jar test — A laboratory procedure that simulates the coagulation/flocculation units of a water treatment plant with differing chemical doses, rapid mix, slow mix, and settling time to estimate a minimum or ideal coagulant dose required to achieve certain water quality goals. Samples of water to be treated are commonly placed in six jars, various amounts of chemicals are added to each jar, and settling of the solids is observed. The dose of chemicals that provides satisfactory settling removal of turbidity and/or color is the amount used to treat the water being taken into the plant at that particular time.

Jason-1 — An oceanography mission to monitor global ocean circulation, discover the tie between the oceans and atmosphere, improve global climate predictions, measure sea level rise, and monitor El Niño conditions.

jaundice — A condition in which the skin and the whites of the eyes become yellow and the urine darkens; occurs when the liver is not working properly due to blockage of the bile duct.

Java — A computer programming language that enables Web pages to use animation and other sophisticated techniques.

jejunum — A section of the small intestine lying between the duodenum and the ileum; its primary function is the absorption of digested material.

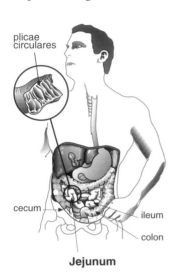

Jejunum

jet — A nozzle of a specific size that limits the flow of a liquid.

jet humidifier — A humidifier that increases the surface area for exposure of water to gas by breaking the water into small aerosol droplets.

jet stream — A swift, high-altitude wind or current of air at the edges of the circumpolar whirls flowing toward the east.

Jet Stream

jetteau — Jet of water.

jetting — A method of compacting soil using a hose or other device with a high-velocity stream of water.

jetting pump — A water pump that develops a very high discharge pressure.

jetty — A structure, such as a pier, extending into a sea, lake, or river.

JIC — See Joint Information Center.

jigging — A process for separating solid materials of different densities by using the periodic pulsation of a liquid, usually water, through a bed of the mixed material to float the lighter solids.

JIISE — See Joint Interagency Intelligence Support Element.

jogging — Frequent starting and stopping of an electric motor.

join — (*geographic information system*) To connect two or more separately digitized maps.

joint — (*anatomy*) The contact point at which two separate bones are joined by ligaments; (*geology*) a fracture or crack in which the rock on either side of the fracture does not exhibit evidence of relative movement.

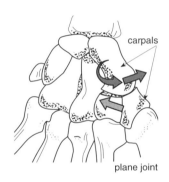
carpals

plane joint

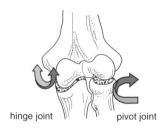
hinge joint   pivot joint

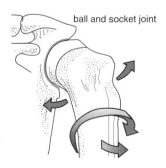
ball and socket joint

Joints

joule (J) — The standard international unit of work and energy equal to the work done when the point of application of a force of 1 newton moves a distance of 1 meter; 1 calorie equals 4.184 joules; the MKS unit of work; a unit of energy equivalent to 10^7 ergs. Also known as newton-meter of energy.

JTU — Acronym for Jackson turbidity unit.

judgment (law) — An award, penalty, or sentence given by a court.

judicial district — Physical perimeters set by Congress or a state legislature designating counties under the jurisdiction of a specific court.

judicial review — A constitutional doctrine that gives a court system the power to annul legislative or executive acts which the judge declares to be unconstitutional.

junction — An interface or meeting place for tissues or structures.

Junin fever — See Argentine hemorrhagic fever.

Junin virus — A South American arenavirus that is rodent-borne and causes Argentine hemorrhagic fever.

junk — Unprocessed materials suitable for reuse or recycling.

jurat — Part of an affidavit or other document where an officer certifies witnessing the sworn testimony.

jurisdiction — The boundary of authorization for a state, county, or city agency.

jurisprudence — The science and philosophy of law.

juvenile hormone — A hormone produced by insects for the maintenance of larval or nymphal characteristics during development.

juxtaposition — The placement of objects side-by-side or end-to-end.

J

K

k — SI symbol for the prefix *kilo-*, representing a factor of 10^3 power; see kilo-.

K — See dissociation constant or ionization constant.

***k*-electron capture** — Electron capture from a *k* shell by the nucleus of the atom; see also electron capture.

K-factor — The thermal conductivity of a material, expressed as BTUs per square foot per hour in degrees Fahrenheit.

K-selected — Describes populations of organisms maintained near the carrying capacity of the environment.

K-strategy — An ecological strategy where organisms depend on physiological adaptations to environmental resources.

Kalman filter — (*global positioning satellite*) A numerical method used to track a time-varying signal in the presence of noise.

kanker — See hardpan.

karst — A geologic formation of irregular limestone deposits; a topography formed over limestone, dolomite, or gypsum and characterized by sinkholes, caves, underground streams, and drainage.

karyotype — Character of a cell as determined by the nature of all its chromosomes.

kb — See kilobase.

KBE — Acronym for keyboard entry.

kbs — Acronym for kilobits per second.

kcal — See kilocalorie.

K_d — See adsorption ratio.

KE — Acronym for kinetic energy.

Kelvin temperature scale — Absolute temperature scale; a scale for measuring temperature that sets 0 degrees at the point at which molecular motion stops and where there is no heat.

kepone ($C_{10}Cl_{10}O$) — A crystalline solid; MW: 490.68; BP: 170°C; Sol: dissolves in acetone, alcohol, and acetic acid; density: NA. It is used as a pesticide. It is highly toxic to the digestive tract, skin, and respiratory tract and may cause tremors, ataxia, hyperactivity, or muscle spasms; it is highly injurious to the liver, kidneys, and central nervous system. It is a teratogenic substance and may possibly be carcinogenic to humans. NIOSH ceiling limit: 0.001 mg/m³/15 minutes.

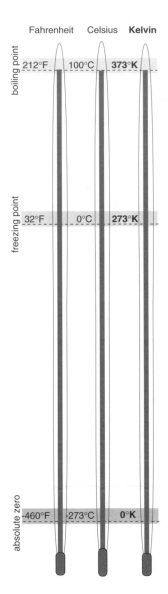

Kelvin Temperature Scale
compared to C and F scales
(not proportionate)

keratin — The horny substance present in the epidermis, hair, and all horny tissues, such as feathers and nails.

keratitis — An inflammation of the cornea of the eye.

kerogen — Hydrocarbon liquid found in sedimentary rock; see also oil shale.

Kepone

kerosene — A flammable hydrocarbon oil usually obtained by distillation of petroleum and used as a fuel and a solvent.

ketene (CH_2CO) — A colorless gas with a penetrating odor; MW: 42.0, BP: –69°F, Sol: decomposes. It is used as an intermediate in the production of acetic anhydride, cellulose, and vinyl acetate resins and plastics; in acrylic resins, dyes, pigments, and pharmaceuticals; in industrial organic synthesis; in laboratory operations as an acetylating agent. It is hazardous to the eyes, skin, and respiratory system and is toxic through inhalation and contact. Symptoms of exposure include pulmonary edema and irritation of the eyes, nose, lungs, and throat. OSHA exposure limit (TWA): 0.5 ppm [air] or 0.9 mg/m^3.

ketone — One of a class of compounds containing a carbonyl group (–CO–) in the molecule attached to two hydrocarbon radicals. Ketones are important intermediates in the synthesis of organic compounds; the simplest ketone is acetone.

keV — See kiloelectronvolt.

Kew Gardens spotted fever — See rickettsialpox.

keyboard — A device for typing alphanumeric characters into the computer.

keyword — (*computer science*) A word or words that can be searched for in documents or menus; words entered into a Web search engine to find particular Websites or special information.

kg — See kilogram.

kidneys — In mammals, a pair of glandular organs that filter waste liquid and secrete urine.

kilkenny coal — See anthracite.

killed vaccine — A vaccine prepared from dead microorganisms used for immunization.

kiln — A furnace in which the heating operations stop just before complete fusion; used for drying, burning, or firing materials.

kilo- (k) — A prefix meaning 1000 in the metric system (e.g., kilogram, which is 1000 grams).

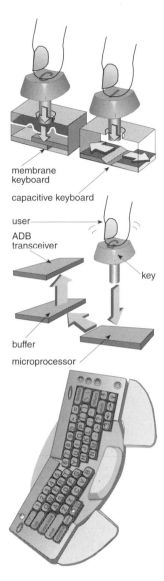

membrane keyboard

capacitive keyboard

user

ADB transceiver

key

buffer

microprocessor

Keyboard
(Showing an
Ergonomic Design)

kilobase (kb) — A unit used when designating the length of a nucleic acid sequence; 1 kilobase equals a sequence of 1000 purine bases.

kilocalorie (kcal) — A unit of heat energy equal to 1000 calories; also known as kilogram calorie.

kilocurie — 1000 curies.

kiloelectronvolt (keV) — 1000 electron volts.

kilogram (kg) — 1000 grams, equal to 2.2046 pounds.

kilogram calorie — See kilocalorie.

kilojoule — 1000 joules.

kilometer (km) — 1000 meters, equal to 0.6214 miles.

kilopascal (kPa) — 1000 pascals; see also pascal.

kilovolt — 1000 volts.

K

kilovolt meter (kV/m) — 1000 volts per meter.

kilowatt (kW) — 1000 watts.

kilowatt-hour (kWh) — A unit of electrical energy equivalent to using 1000 watts of power for 1 hour.

kinematics — A branch of dynamics that deals with aspects of motion apart from consideration of mass and force.

kinesiology — Study of the principles, mechanics, and anatomy of human movement.

kinesis — Movements of an organism in random directions in response to stimulus.

kinetic — Relating to the motion of material bodies and associated forces and energy.

kinetic energy — The energy possessed by a mass because of its motion; equal to one half the mass of the body times the square of its speed ($1/2mv^2$).

kinetic rate coefficient — A number that describes the rate at which a water constituent such as biochemical oxygen demand or dissolved oxygen rises or falls or at which an air pollutant reacts.

kinetics — A branch of science that deals with the effects of forces upon the motions of material bodies or with changes in a physical or chemical system.

kinetochore — See centromere.

kitchenware — All multiuse utensils other than tableware.

kjeldahl method — An analytical method for the determination of nitrogen content, particularly in organic materials.

Klebsiella — A genus of diplococcal bacteria in which small, plump rods with rounded ends are capable of causing infections.

Klebsiella pneumonia — A species of bacteria found in soil, water, cereal grains, and the intestinal tract of humans and other animals associated with pathological conditions including pneumonia.

km — See kilometer.

knapsack sprayer — A sprayer that can be strapped to a person's back and is used to apply liquid pesticide chemicals; the attached hose has a nozzle at the tip that can be aimed at the spots to be treated.

knee — A joint complex that connects the thigh with the lower leg.

knot — (*wind*) A speed unit of 1 nautical mile per hour, equal to approximately 1.15 statute miles per hour.

known and reasonable quality — (*computer science*) The quality of data necessary to meet the needs of the user as judged by their utility.

K_{oc} — See adsorption coefficient.

Koch's bacillus — The *Mycobacterium tuberculosis* organism.

Koch's law — To determine the cause of a given disease by an organism, the following conditions must be met: (1) a microorganism must be present in every case of the disease, (2) it must be isolated and cultivated in pure culture, (3) inoculation of the culture must produce the disease in susceptible animals, and (4) it must be observed in and recovered from the experimentally diseased animals. Also known as Koch's postulates or the law of specificity of bacteria.

Koch's postulates — See Koch's law.

KOH — The scientific notation for potassium hydroxide.

konimeter — An instrument utilized to determine the amount of dust particles in a sample of air.

K_{ow} — See octanol–water partition coefficient.

kPa — See kilopascal.

kraft processing — A method of producing wood pulp by digesting wood chips in an alkaline liquor consisting chiefly of caustic soda together with sodium sulfate.

Krebs cycle — A sequence of enzymatic reactions involving the metabolism of carbon chains of sugars, fatty acids, and amino acids to yield carbon dioxide, water, and high-energy phosphate bonds. The cycle provides a major source of adenosine triphosphate energy and also produces intermediate molecules that serve as starting points for a number of vital metabolic pathways including amino acid synthesis.

Kuru disease — A slow-virus disease spread through cannibalism and ritualistic butchering; causes degenerative changes including gait disturbance, incoordination, and difficulty swallowing; rarely seen today.

kV — See kilovolt.

kV/m — See kilovolt meter.

kW — See kilowatt.

kwashiorkor — A form of malnutrition caused by a diet high in carbohydrates and extremely low in proteins.

kWh — See kilowatt-hour.

Kyasanur forest disease — An arbovirus infection transmitted by the bite of ticks whose hosts are shrews and other forest animals in western tropical India; symptoms include fever, headache, muscle ache, cough, abdominal pain, eye pain, and photophobia.

kyphosis — An abnormal curvature of the spine of the upper back in the anteroposterior plane; also known as humpback or hunchback.

L

l — See liter.

lab — Acronym for laboratory.

label — Notice attached to a container bearing information concerning its contents, proper use, manufacturer, and any cautions or hazards of use; (*nuclear physics*) see isotopic tracer.

labeled compound — A compound consisting in part of labeled molecules, including radionuclides.

labile — Unstable and likely to change under certain influences.

laboratory — Any place organized and operated for the performance of any chemical, microbiological, biochemical, hematological, microscopical, immunological, parasitological, or other tests, examinations, or evaluations; facilities that work with pathogens or animals or that use various biotechnologies that generate infectious waste.

laboratory-acquired infection — Any infection resulting from exposure to biohazardous material in a laboratory environment.

laboratory diagnosis — A diagnosis determined by study of various specimens.

laboratory error — An error made by personnel in carrying out a test or in interpretation, reporting, or recording of results.

laboratory test — A procedure used to detect, identify, or quantify one or more significant substances, evaluate organ functions, or determine the nature of a condition or disease.

laboratory waste — Discarded materials generated by research and analytical activities.

laboratory water — Purified water used in the laboratory as a basis for making up solutions or making dilutions; water devoid of substances.

labradorite — A triclinic feldspar that is an essential constituent of basalts and gabbros.

lacerum — Torn.

lacquer — A colloidal dispersion or solution of nitrocellulose or similar film-forming compounds used as a glossy, protective, and decorative coating for surfaces.

lacrimation — Secretion and discharge of tears.

lacrosse encephalitis — An acute inflammatory disease of short duration involving parts of the brain, spinal cord, and meninges. Symptoms range from

Label
examples of warning placards
required on trucks

being mild to high fevers, meningeal signs, disorientation, coma, tremors, seizures, and possibly death. Incubation time is 5 to 15 days. It is caused by a bunyavirus and is found in the United States and Canada. The reservoir of infection is *Aedes* eggs; it is transmitted by the bite of a mosquito but not from person to person. It is usually a disease of children but can affect others. It is controlled by destroying larvae, eggs, and adult mosquitos and using mechanical means of mosquito control.

lactase — A digestive enzyme of the intestinal fluid that changes lactose to glucose.

lactating animal — Any animal that is producing milk.

lactation — The secretion of milk by the mammary glands.

lactic acid ($C_3H_6O_3$) — A hygroscopic acid produced by anaerobic metabolism of glucose.

Lactobacillus — A nonpathogenic, Gram-positive, rod-shaped bacteria that produces lactic acid from carbohydrates.

Lactobacillus acidophilus — A bacterium found in milk and dairy products and in the feces of adults and bottlefed babies.

lactose — A milk sugar composed of glucose and galactose.

lacustrine — Referring to lake or river habitats.

LADD — Acronym for lifetime daily average dose; lowest acceptable daily dose.

lading — See bill of lading.

LAER — Acronym for lowest achievable emission rate.

lag — A delay in the effect of a changed condition at one point in a system or some other condition to which it is related.

lag phase — The part of the growth curve of bacteria in which, during a given time span, bacteria do not multiply.

lag time — Time between the occurrence of a primary event and occurrence of the symptoms of the event; the initial period in the life of bacteria when cells are adjusting to a new environment before beginning the growth phase.

lagoon — A shallow pond generally near but separated from a larger body of water; often created by humans using rigid specifications in which sunlight, algae, and oxygen interact to restore water to a reasonable state of purity.

lahar — A rock-laden flood, made up of 40% or more by weight volcanic rock, which flows like wet concrete at a speed of 40 mph.

lake — An inland body of water, usually freshwater, formed by glaciers, river drainage, springs, etc.

lamella (*pl.* **lamellae**) — A thin-layer, plate-like arrangement or membrane.

laminar flow — A flow in which no turbulence is present; diffusion in such a flow occurs only by molecular processes; (*air*) the introduction of large volumes of clean air through a very large diffuser or perforated panel that reduces the velocity of the incoming air, thereby preventing agitation and reintroduction of settled contaminants.

laminate — A sheet of material made of many different layers.

LAMP — Acronym for Lake Acidification Mitigation Project.

LAN — See Local Area Network.

land breeze — A nocturnal coastal breeze blowing from land to the sea in the evening when the water may be warmer than the land, causing pressure differences.

land cover — Characteristics of a land surface as determined by its spectral signature.

land farming — The biological degradation involved in the incorporation of waste into soil and utilizing healthy soil microorganisms to metabolize the waste components.

Land Information Systems — Specialized geographical information systems usually found among municipal agencies and used for legal, administrative, and economic spatial analysis.

land planner — An environmental health professional who reviews land-use proposals to assess environmental impacts, evaluate health risks, recommend actions that will protect the public from exposure to disease and other health and safety hazards, and will protect the environment from degradation.

land use — The ultimate use permitted for currently contaminated lands, waters, and structures of each Department of Energy installation.

landfill — A disposal facility or part of a facility where waste is buried in layers below ground, compacted, and covered.

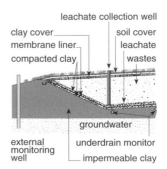

Landfill Cell

landfill blade — A U-blade with an extension on top that increases the volume of solid waste that can be

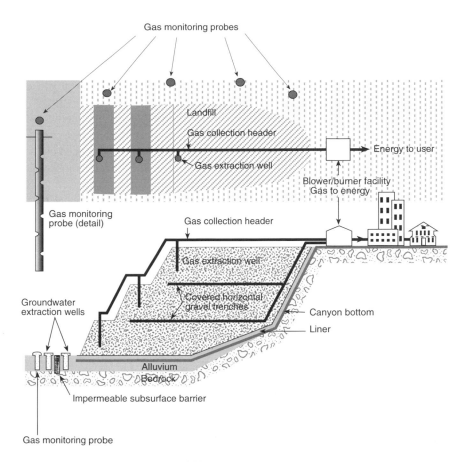

Gas monitoring probes

Landfill

Gas collection header

Gas extraction well

Energy to user

Blower/burner facility
Gas to energy

Gas monitoring
probe (detail)

Gas collection header

Gas extraction well

Covered horizontal
gravel trenches

Canyon bottom

Liner

Groundwater
extraction wells

Alluvium
Bedrock

Impermeable subsurface barrier

Gas monitoring probe

Landfill Gas Collection

pushed and spread and protects the operator from any debris thrown out of the solid waste.

landfill gas — A gas produced in sanitary landfills during anaerobic digestion of the organic contents; contains 40 to 60% methane.

landslide — Mass movement of earth or rock along a definite plane under the influence of gravity.

LANDSTAT — Generic name for a series of Earth resource scanning satellites launched by the United States beginning in 1972; the Land Remote-Sensing Satellite designed to gather data on the resources of Earth in a regular and systematic manner to develop a land-use inventory for geological/and mineralogical exploration, crop and forest assessment, and cartography.

Langelier's index — A mathematically derived factor obtained from the values of calcium hardness, total alkalinity, and pH at a given temperature, where 0 indicates perfect water balance; an index reflecting the equilibrium pH of water with respect to calcium and alkalinity that is used when stabilizing water to control corrosion and scale deposition.

Lantz process — A destructive distillation technique in which the combustible components of solid waste are converted into combustible gases, charcoal, and a variety of distillates.

laptop — A portable computer.

Larder Beetle (*Dermestes lardarius*) — A stored food product insect about 1/3 inch long, dark brown, with a wide yellow band across the front part of the wing cover.

large dyne — See newton.

large intestine — The part of the digestive system between the ileocecal valve of the small intestine and the anus that removes salt and water from undigested food and releases feces through the anus.

large quantity generator — A person or facility generating more than 2200 pounds of hazardous waste per month and subject to all RCRA requirements. These generators produce about 90% of the nation's hazardous waste.

large water system — A water system that serves more than 50,000 people.

larva — Immature worm-like stage of an insect that undergoes complete metamorphosis before assuming characteristic features of the parent.

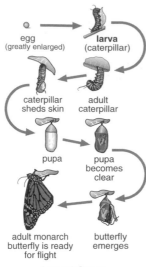

egg
(greatly enlarged)

larva
(caterpillar)

caterpillar
sheds skin

adult
caterpillar

pupa

pupa
becomes
clear

adult monarch
butterfly is ready
for flight

butterfly
emerges

Larval Stage
of *Danaus plexippus*
(Monarch butterfly)

larvacide — A pesticide chemical that will prevent damage from or destroy larvae.

laryngitis — An inflammation of the mucous membrane of the larynx characterized by dryness and soreness of the throat, hoarseness, cough, and dysphagia.

larynx — The anterior portion of the trachea that is made of cartilages and related structures and which contains the vocal chords in amphibians, reptiles, and mammals.

LAS — See linear alkylate sulfonate.

laser — Acronym for light amplification by stimulated emission of radiation; a device that utilizes the natural oscillations of atoms or molecules between energy levels for generating coherent electromagnetic radiation in the ultraviolet visible or infrared regions of the spectrum. Lasers produce monochromatic coherent radiation that travels in a plane wave front with very little spreading. It is amplified, meaning a chain reaction of energy is produced when the highly excitable photon travels between polished or mirror-like ends of the laser. An enormous amount of energy is released simultaneously. Lasers are used in industry, medicine, and defense.

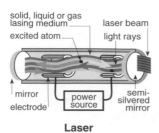

solid, liquid or gas
lasing medium

excited atom

laser beam

light rays

mirror
electrode

power
source

semi-
silvered
mirror

Laser

laser air pollution probe — A device used for measuring the concentration and location of a particulate pollution by use of a laser.

laser beam welding — A process that produces coalescence of materials using heat obtained in the application of a concentrated coherent light beam impinging on the members to be joined.

laser disc — A plastic-coated disc that stores computer data as tiny pits etched on the surface; a laser beam scans the pits and translates the data into computer language.

laser–Doppler velocimeter (LDV) — A mechanism used to measure velocity and concentration levels of pollution by means of a carbon dioxide (CO_2) laser beam.

laser-induced fluorescence — A method for measuring the relative amount of soil and/or groundwater with an *in situ* sensor.

laser light region — A portion of the electromagnetic spectrum that includes ultraviolet, visible, and infrared light.

laser plotter — A plotter for mapping in which the information is written onto a light-sensitive material using a laser.

laser printer — An extremely fast printer that uses a laser to form areas of static electric charge which attract metallic powder to paper.

Laser Printer

laser ranging — The use of lasers to measure distances.

laser system — A device of electrical, mechanical, and optical components including a laser.

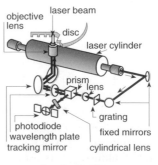

objective
lens

laser beam

disc

laser cylinder

prism

lens

photodiode
wavelength plate
tracking mirror

grating

fixed mirrors

cylindrical lens

Laser System
in a compact disc player

Lasioderma serricorne — See Cigarette Beetle.

Lassa fever — A highly contagious disease caused by a virulent arenavirus. Symptoms include fever, pharyngitis, pleural effusion, edema, renal problems, mental disorientation, confusion, and death from cardiac failure.

lassitude — Weariness; exhaustion.

late effects — Effects that appear 60 days or more following an acute exposure to radiation.

latency — Time from the first exposure to a chemical, microbiological, or physical agent until the appearance of a toxic effect or sickness.

latency period — The presymptomatic interval between initiation of a disease and its becoming clinically apparent.

latent — Referring to something that is dormant.

latent energy — The heat released when water vapor changes back to liquid water.

latent heat — The quantity of heat absorbed or given off per unit weight of material during a change of state, such as ice to water or water to steam.

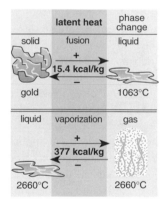

Latent Heat

latent heat of evaporation — The amount of heat energy that is needed to evaporate a substance, to change it from a liquid to a gas. For water, it is 540 calories per gram.

latent heat of fusion — The heat that must be extracted from a liquid to freeze it to a solid at the same temperature. For water, it is 80 calories per gram.

latent heat of melting — The amount of heat energy needed to melt a substance, changing it from a solid to a liquid.

latent heat of vaporization — The quantity of energy required to evaporate a unit mass of liquid at constant temperature and pressure.

latent heat transfer — The heat added or removed during a change of state of a substance from a solid to a liquid to a gas or vice versa, with the temperature remaining constant.

latent malaria — A continuing infection without clinical symptoms, due to the parasite and the person's immune system establishing an equilibrium.

latent period — The time between exposure and the first manifestation of disease or injury.

lateral sewer — A sewer that runs under city streets and into which sewers from homes or businesses empty.

laterite — A group of hard, red soils from tropical areas that show intense weathering and leaching of bases and silica, leaving aluminum hydroxides and iron oxides.

latex — A water emulsion of natural or synthetic rubbers or resins; used in emulsion and paints.

latitude — The angular distance north or south of the equator of a position on the surface of the Earth; measured in degrees.

latrine — A receptacle in the earth used as a toilet.

lattice — A crystalline-like structure caused by the precise orientation of molecules in a solid or liquid.

laughing gas — See nitrous oxide.

laundering weir — A sedimentation basin overflow weir having a plate with V-notches along the top to assure a uniform flow rate and avoid short-circuiting.

launders — Sedimentation basin and filter discharge channels consisting of overflow weir plates in sedimentation basins and conveying troughs.

lava — Fluid, molten igneous rock that flows at the surface of the Earth.

law — A rule of civil conduct prescribed by the supreme power in a state stating what is correct and forbidding what is wrong.

law of conservation of energy — See conservation of energy.

law of conservation of matter — See conservation of matter.

law of definite composition — See law of definite proportion.

law of definite proportion — Every compound always contains the same proportions by weight of the elements that compose it; also known as the law of definite composition.

law of electrostatic attraction — See Coulomb's law.

law of limiting factors — The biological law that a minimum quantity of essentials, such as nutrients, light, heat, moisture, and space, must be available within the ecosystem for survival of the organisms. The ecosystem can be affected by pesticides or other environmental changes.

law of partial pressures — See Dalton's law.

law of specificity of bacteria — See Koch's law.

layer — A spatial dataset containing a common feature type; also referred to as coverages or themes.

lb — See pound.

L-band — The group of radio frequencies extending from 390 to 1550 MHz, with global positioning satellite carrier frequencies at 1227.6 and 1575.42 MHz.

lbd — Acronym for lower back disorder.

lb$_f$ — Acronym for pound-force.

lb$_f$/ft^2 — Acronym for pound-force per square foot.

lb-ft — See pound-foot.

lbp — Acronym for low back pain.

LC — See lethal concentration or liquid chromatography.

LC$_{50}$ — See lethal concentration 50.

LCA — Acronym for lifecycle assessment.

LCCA — See Lead Contamination Control Act.

LCD — See liquid crystal display; acronym for local climatological data.

LCL — Acronym for lower control limit; see low lethal concentration.

LC$_{lo}$ — See lethal concentration lo.

LCM — Acronym for lifecycle management.

LCRS — Acronym for Leachate Collection and Removal System.

LD — Acronym for land disposal; see lethal dose.

LD$_{50}$ — See lethal dose 50.

LDCRS — Acronym for leachate detection, collection, and removal system.

LDIP — Acronym for Laboratory Data Integrity Program.

LDL — See low lethal dose; see lower detection limit.

LD$_{lo}$ — See lethal dose lo.

LDR — Acronym for land disposal restriction.

LDS — Acronym for leak detection system.

LDU — Acronym for land disposal unit.

LDV — See laser–Doppler velocimeter.

Le Chatelier's principle — When an external force is applied to an equilibrium system, the system adjusts to minimize the effect of the force.

leach — Removal of components from the soil by the action of water trickling through it.

leachate — Liquid that has percolated through solid waste or other medium and has extracted, dissolved, or suspended materials from the solid waste; a solution formed by leaching.

leachfield — A subsurface structure built to distribute liquids across a suitable area for disposal.

leaching — The loss of soluble soil minerals as the result of movement of groundwater; the operation by

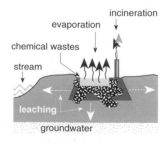

Leaching

which color is partially or wholly removed form a colored material; the dissolving by a liquid solvent of soluble material from its mixture with an insoluble solid; movement of a chemical downward through the soil.

leaching field — A lined or partially lined underground pit or trenches into which sewage is discharged and then seeps into the surrounding soil.

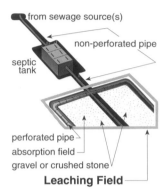

Leaching Field

lead (Pb) — A naturally occurring, bluish-gray metal found in small amounts in the crust of the Earth and having no characteristic taste or smell; MW: 207.20, BP: 1,740°C, Sol: insoluble, Fl.P: not known, density: 11.34 g/cm^3 at 20°C. It is used in the production of batteries, ammunition, sheet lead, solder, and pipes. The tetraethyl lead and tetramethyl lead forms were used as gasoline additives. They are used in paint, ceramic products, roofing, caulking, electrical applications, tubes, or containers. Lead exposure may be due to air, water, food, or soil. Lead in the air is primarily due to lead-based fuels and the combustion of solid waste, coal, oils, and emissions from alkyl lead manufacturers, windblown dust from volcanoes, the burning of lead-painted surfaces, and cigarette smoke. Lead in drinking water comes from leaching from lead pipes, connectors, and solder in both the distribution system and household plumbing. In surface water, lead comes from lead-containing dust that has settled from the atmosphere; from wastewater from industries; from handling lead; from steel, iron, and lead producers; and from urban runoff. Lead in food comes from dust in the atmosphere that settles on crops during fruit processing and from lead-soldered food cans. Lead in the soil comes from dust from the atmosphere, solid waste disposal, paint chips, deterioration of lead-based paints, and fertilizers that become part of sewage sludge. It is hazardous to the digestive system, reproductive system, blood, central nervous system, kidneys, and the mental and physical development of children. Mortality occurs at a higher rate in individuals in the working environment in battery plants and lead production plants, as well as in the ambient environment, especially in children. Symptoms of exposure include cerebrovascular disease, hypertension, nephritis, nephrosis, cancer of the digestive tract, colic, hepatic and renal effects, dullness, irritability, poor attention span, headaches, muscular tremors, memory loss, hallucinations, paralysis, coma, and death; in animals, it causes decreased leukocyte counts, circulating antibodies and antibody-forming cells. It is a suspected causative agent in sudden infant death syndrome and is carcinogenic. Respiratory effects appear to be rare but may occur; colic is a consistent, early gastrointestinal symptom of lead poisoning in occupationally exposed individuals at acutely high levels of lead, especially during the removal of lead-based paint; colic symptoms include abdominal pain, constipation, cramps, nausea, vomiting, anorexia, weight loss, and decreased appetite. Colic is also a symptom of lead poisoning in children. Lead has a profound affect on hemabiosynthesis and inhibits the activity of certain enzymes involved in hemabiosynthesis; children experience hepatic effects with a decreased activity of hepatic mix-function oxygenases. Renal effects of early or acute lead-induced nephropathy include nuclear inclusion bodies, mitochondrial changes and cytomegaly of the proximal tubular epithelial cells, and disfunction of the proximal tubules manifested as aminoaciduria glucosuria phosphaturia with hypophosphatemia. Thyroid function is adversely affected; lead interferes with the conversion of vitamin D to its hormonal form, 1,25-dihydroxyvitamin D. An association exists between blood lead levels and growth in children, and it may have an affect on the cellular component of the immune system. Neurological effects are related to brain function, dullness, irritability, poor attention span, headaches, muscular tremors, loss of memory, and hallucinations; these may worsen to delirium, convulsion, paralysis, coma, and death. Behavioral changes occur in children with a drop of 5 IQ points; developmental effects include reduced birth rate and gestational age and neurobehavioral deficits or delays. Adverse reproductive effects include miscarriages and stillbirths. Because of the indirect effects of lead on maternal nutrition or the hormonal status of the pregnant woman, it has been suggested that lead may have an effect on chromosomes. Immunological effects in animals include a decrease in leukocyte counts, circulating antibodies, and antibody-forming cells. OSHA exposure limits (TWA): 50 μm/m^3 [air].

lead acetate (Pb(C$_2$H$_3$O$_2$)$_2$·3H$_2$O) — An odorless, white, brown, or gray solid; MW: 325.28, BP: decomposes, Sol: 44.3 g/100 ml water at 20°C, Fl.P: NA, sp. gr. 3.25 at 20°C. It is used in a variety of commercial and industrial processes such as production of lead-lined containers for corrosive gases and liquids; as alloys in metallurgy, ceramics, electronic devices, and plastics. It is hazardous to the respiratory system and digestive system. Acute exposure may cause seizures. Chronic exposure may cause tiredness, stomach problems, constipation, headaches, depression, aching and weakness in the arms and legs, and eventually death. OSHA exposure limit (8-hour TWA): 0.05 mg/m^3.

lead additive — Any substance containing lead or lead compounds.

lead agency — The federal department or agency assigned lead responsibility under U.S. law to manage and coordinate the federal response in a specific area.

lead apron — A protective device made of lead and rubber worn by a patient, technician, or radiologist during exposure to x-rays or other diagnostic or therapeutic radiation

lead arsenate (PbHAsO$_4$) — An odorless, heavy white powder; MW: 347.12, BP: decomposes, Sol: insoluble in water, Fl.P: nonflammable, sp. gr. 5.79 at 20°C. It is used in paint pigments, alloy metallurgy, ceramics, electronic devices, plastics, and insecticides. It is hazardous to the respiratory system and digestive system. Acute exposure may cause poor appetite, nausea, vomiting, and muscle cramps. Chronic exposure may cause nerve damage, numbness, weakness, high blood pressure, brain damage, and kidney damage. It is a known carcinogen in humans. OSHA exposure limit (8-hour TWA): 0.05 mg/m^3.

lead-based paint — Any paint with a lead content exceeding 0.06% by weight of the total nonvolatile content of the paint or the weight of the dry paint film; any applied paint film that contains 0.7 mg/cm^2 or more of lead as measured by an *in situ* analyzer device; paint that contains more lead, as determined by chemical analysis, than the appropriate authority considers to be safe.

lead chromate (PbCrO$_4$) — An odorless, yellow, crystalline substance or powder; MW: 323.22, BP: decomposes, Sol: insoluble, Fl.P: nonflammable (poisonous gas produced in a fire), sp. gr. 6.12 at 12°C. It is used as a pigment, in ceramics, and in plastics. It is hazardous to the respiratory system, digestive system, and skin. Acute exposure may cause poor appetite, colic, upset stomach, headache, irritability, muscle or joint pains, and weakness. Chronic exposure may cause permanent

L

damage, memory loss, irritability, anemia, and fatigue. It is a known carcinogen. OSHA exposure limit (8-hour TWA): 0.05 mg/m³.

Lead Contamination Control Act (LCCA) — Legislation passed in 1988 requiring identification of water coolers that are not lead-free, repair or removal of water coolers with lead-lined tanks, identification and resolution of lead problems in schools' drinking water, and banning the manufacture and sale of water coolers that are not lead-free.

lead encephalopathy — A disease of the brain structure and function caused by lead poisoning; symptoms include delirium, convulsions, mania, cortical blindness, and coma.

lead federal agency — The agency designated by the President of the United States to lead and coordinate the overall federal response depending on the type of emergency situation

lead fluoborate (PbB_2F_8) — An odorless, colorless liquid; MW; 380.81, BP: NA, Sol: insoluble in water, Fl.P: nonflammable (produces poisonous gases in a fire), sp. gr. 1.75 at 20°C. It is hazardous to the respiratory system and digestive system. Acute exposure may cause tiredness, mood changes, headaches, stomach problems, sleeping problems, and serious permanent kidney or brain damage. Chronic exposure may cause tiredness, sleeping problems, stomach problems, constipation, headaches, moodiness, weakness in arms and legs, trouble concentrating, and low blood counts. It is considered to be a suspected teratogen. OSHA exposure limit (8-hour TWA): 0.05 mg/m³.

Lead Hazard Control — A major program of the Department of Housing and Urban Development which advises all agencies and health authorities, as well as the housing industry, on lead poisoning prevention.

lead marcasite — See sphalerite.

lead peroxide candle — A method for determining the amount of sulfur dioxide in the atmosphere when exposing lead oxide paste to air in which gaseous sulfur dioxide is present; sulfation occurs at a rate proportional to the amount of sulfur dioxide present.

lead poisoning — A form of poisoning caused by the presence of lead or lead salts in the body with symptoms of weight loss, anemia, stomach cramps, mental depression, irritation, and convulsions affecting the brain, nervous system, blood, and digestive system.

lead service line — A service line made of lead that connects a water main to a building inlet and any lead pigtail, gooseneck, or other fitting that is connected to the lead line.

lead sulfate ($PbSO_4$) — An odorless, white, crystalline powder; MW: 303.25, BP: NA, Sol: soluble in cold water, Fl.P: nonflammable, sp. gr. 6.2 at 20°C. It is used in storage batteries and paint pigments. It is hazardous to the respiratory system, digestive system, and eyes. Acute exposure causes burning of the eyes, poor appetite, upset stomach, colic, headaches, irritability, aching joints and muscles, constipation, disturbed sleep, reduced memory, and brain damage. Chronic exposure causes tiredness, disturbed sleep, stomach problems, constipation, headaches, moodiness, muscle weakness, fatigue, anemia, and permanent kidney damage. It is a known teratogen and is corrosive. OSHA exposure limit (8-hour TWA): 0.05 mg/m³.

lead sulfide (PbS) — An odorless, silvery, metallic, crystalline material or black powder; MW: 239.26, BP: 1281°C, Sol: insoluble in water, Fl.P: nonflammable (produces poisonous gases in a fire), sp. gr. 7.5 at 20°C. It is used in ceramics, infrared radiation detectors, and semiconductors. It is hazardous to the respiratory system and digestive system. Acute exposure may cause poor appetite, colic, upset stomach, headaches, irritability, muscle or joint pain, weakness, and permanent kidney damage. Chronic exposure may cause poor appetite, upset stomach, colic, headaches, irritability, aching joints and muscles, constipation, disturbed sleep, reduced memory, kidney damage, and brain damage. It is a known teratogen. OSHA exposure limit (8-hour TWA): 0.05 mg/m³.

lead(II)azide (PbN_6) — A colorless needle-shaped material or white powder; BP: 350°C (explodes), Sol: slightly soluble in water at 18°C and soluble in acidic acid, Fl.P: detonation temperature 350°C (a primary explosive), density: about 4.0 g/cm³. It is used as a primary explosive in detonators and fuses to initiate the booster or bursting charge. Its aqueous solution is toxic. ACGIH exposure limit: NA.

leader — A conduit used to carry rainwater.

leakage — The undesired entrance or escape of fluid.

leaking underground storage tank (LUST) — Older tanks placed underground as a fire prevention measure that leak into the ground and potentially into the groundwater supply; leaking tanks used to store gasoline, other oil products, or hazardous substances.

leak-protector valve — A valve equipped with a leak-diverting device that will prevent leakage of the fluid past the valve in any closed position.

leavening agent — A substance that provides volume and desired texture to baked goods.

lectins — Any of a variety of plants containing substances that agglutinate red blood cells.

legal — Any actions or conditions that are permitted or authorized by law.

legal status — Type of business enterprise, as allowed by law: corporation, partnership, cooperative, or sole ownership.

legend — (*mapping*) The part of a map explaining the meaning of the symbols used to code the geographical information.

legionellosis — An acute bacterial disease with two distinct manifestations: Legionnaire's disease and Pontiac fever. Symptoms include malaise, anorexia, myalgia, headache, cough, and high temperature. Incubation time for Legionnaire's disease is 2 to 10 days, usually 5 to 6 days. Incubation time for Pontiac fever is 5 to 66 hours, usually 24 to 48 hours. Both are caused by *Legionella pneumophila*, a poor-staining, Gram-negative, rod-shaped bacteria that is difficult to grow outside of the body. It has been found in a variety of places for at least the last 45 years. Reservoirs of infection include possibly water and soil; transmission is possibly by air. It is not communicable from person to person; general susceptibility, but rarely found under age 20. It is controlled by use of antibiotics, appropriate disinfection of cooling tower waters, and proper treatment of water supplies.

Legionnaire's disease — See legionellosis.

legislation — The exercise of power and function of making laws.

Leishmania — A genus of protozoan parasites transmitted to humans by any of several species of sandflies.

leishmaniasis — A disease caused by a protozoan transmitted through the bite of a female sandfly; may be either cutaneous or visceral.

LEL — Acronym for lowest effect level; see lower explosive limit.

L-electron capture — See electron capture.

lens — A transparent disc that directs light rays to the retina of the eye of many invertebrates.

lentic environment — Standing water and its various intergrades such as lakes, ponds, and swamps.

LEPC — See Local Emergency Planning Committee.

Leptophos (C$_{13}$H$_{10}$BrCl$_2$O$_2$PS) — An odorless, white, crystalline solid or tan waxy solid; MW: 412.07, BP: decomposes, Fl.P: NA, sp. gr. 1.53. It is an organophosphate chemical that is used as an insecticide outside of the United States. It is hazardous to the respiratory system, digestive system, skin, and eyes. Acute exposure may cause central nervous system depression, agitation, confusion, delirium, coma, seizures, bradycardia, broncospasm, vomiting, diarrhea, and death. Chronic exposure may cause damage to the nerves, poor coordination in the arms and legs, depression, anxiety, and irritability and may affect the eyes and vision. OSHA exposure limit (8-hour TWA): NA.

Leptospira — A genus of the family Treponemataceae, order Spirochaetales, having tightly coiled spiral microorganisms with hooked ends that live in the urine of infected animals, especially rodents, and cause disease with symptoms of hepatitis, jaundice, skin hemorrhage, fever, renal failure, and muscular pain.

leptospirosis — Zoonotic disease with the ability to assume different forms, including fever, headache, chills, severe malaise, vomiting, sometimes jaundice, renal insufficiency, and hemorrhage. Incubation time is 10 days, with a range of 4 to 19 days. It is caused by Leptospira, especially in the United States, and Icterohaemorrhagie, which is found worldwide. Reservoirs of infection include farm and pet animals as well as rats; it is transmitted through contact with the skin, especially if abraded, or contact with mucous membranes in water. It is rarely transmitted from person to person; may be communicable for 1 to 11 months; general susceptibility. It is controlled by protective clothing, boots for workers, and proper rodent control.

LERC — Acronym for Local Emergency Response Committee.

lesion — Any pathological or traumatic discontinuity of tissue or loss of function due to injury or disease.

Lesser Grain Borer (*Rhyzopertha dominica*) — A brown or black, slender, cylindrical beetle with numerous coarse elevations on the pronatum, about 1/8 inch long and commonly found in the Gulf states, but may occur elsewhere in the country; a stored food product insect.

LET — See linear energy transfer.

lethal — Capable of causing death.

lethal concentration (LC) — A calculated concentration of a chemical in air to which exposure for a specific length of time is expected to cause a death rate of 100% in members of the population affected.

lethal concentration 50 (LC$_{50}$) — A calculated concentration of a chemical in air to which exposure for a specific length of time is expected to cause death in 50% of a defined experimental animal population.

lethal concentration lo (LC$_{lo}$) — The lowest concentration of a chemical in air that has been reported to cause death in humans or animals.

lethal dose (LD) — The quantity of a substance being tested that will kill 100% of the defined experimental population.

lethal dose 50 (LD$_{50}$) — The dose of a chemical that has been calculated to cause death in 50% of a defined animal laboratory population.

lethal dose lo (LD$_{lo}$) — The lowest dose of a chemical introduced by a route other than inhalation that is expected to cause death in humans or animals.

L

lethal time 50 (LT$_{50}$) — A calculated period of time within which a specific concentration of a chemical is expected to cause death in 50% of a defined experimental animal population.

lethargy — A state of deep drowsiness, usually of mental origin but also caused by illness or fatigue.

leukemia — A progressive, malignant disease of the blood-forming organs; a distorted proliferation and development of leukocytes and their precursors in the blood and bone marrow.

leukocyte — A colorless cell with a nucleus found in blood and lymph; also known as white blood cell.

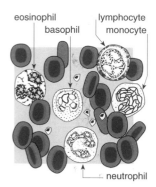

eosinophil   lymphocyte
basophil   monocyte

neutrophil

Leukocytes

leukocytosis — A transient increase in the number of white blood cells in the blood.

leukopenia — Reduction of the number of leukocytes in the blood to 5000 or less.

leukorrhea — A white, pus-like discharge resulting from infection.

LEV — Acronym for low emissions vehicle; see local exhaust ventilation.

levee — A natural or man-made earthen barrier along the edge of a stream, lake, or river.

level controls — A float device or pressure switch that senses changes in a measured variable and opens or closes a switch in response to that change. An example of this would be a floating ball connected mechanically to a switch or valve to stop water flow into a toilet when the tank is full.

level gauging system and alarm — A control that indicates the level of liquid inside of a tank and is installed on tanks to prevent overfilling and spilling of liquid and damage to the tanks.

level of concern — The concentration in air of an extremely hazardous substance above which anyone exposed to it for short periods may experience serious and immediate health effects.

levels of consciousness — The various stages of the response of the mind to stimuli, varying from unconsciousness to total alertness.

lever — A bar used for prying or dislodging something.

LF — Acronym for low frequency.

LFA — See lead federal agency.

LFEM — Acronym for low-frequency electromagnetic field.

LFG — See landfill gas.

LFL — See lower flammable limit.

LFM — See linear feet per minute.

LFT — Acronym for liver function test.

LHA — Acronym for lifetime health advisory.

LI — See Langelier's index.

liaison — An agency official who facilitates interagency communications and coordination with another agency.

libel of information — See complaint for forfeiture.

lice — The plural of louse.

license — A legal document granting the bearer the right to carry forth a given business dependent upon adherence to the rules, regulations, and applicable laws.

licensure — The process by which a state government agency grants permission to a person meeting predetermined qualifications to engage in a given occupation.

LIDAR — Acronym for light detection and ranging; a ruby laser radar used to measure atmospheric backscattering and/or attenuation as a function of the slant range, primary and secondary haze layers, smoke-plume capacity, and particulate clouds; derived from laser infrared radar.

LIDAR ground sensor — A mechanism used to determine the opacity of gases coming from smoke stacks; accurate within 3% opacity for opacities of less than 50%.

LIDAR radar backscatter — A device that measures cross-sections of the plume for particulate emissions by using a laser beam through the plume and comparing the relative backscattering intensity of the beam in front and behind the plume; used in stack transmissometer measurements.

life — The quality that distinguishes a vital and functional being from a dead body; a principal or force that is considered to underlie the distinctive quality of animate beings.

life expectancy — The average length of time an individual may live based on an analysis of actual time until the death of a large group of similar people.

life extension — The process of extending lifespan by the use of preventive medicine and established diagnostic or therapeutic techniques.

life science — Branch of science that deals with living organisms and life processes.

life support — The use of therapeutic techniques or equipment to keep someone alive.

life cycle — The stages of an organism's existence.

lifetime average daily dose — A figure for estimating excess lifetime cancer risk.

L

lifetime exposure — The total amount of exposure to a substance that a human would receive in a lifetime, usually assumed to be 70 years.

lift — (*sanitary landfill*) A compact layer of solid waste and top layer of cover material.

ligament — A flexible, tough strand of connective tissue that holds bones together at a joint.

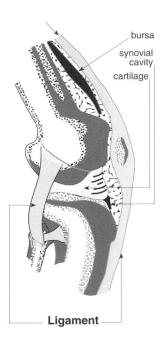

Ligament

light — A form of radiant energy coming from atoms when they are violently disturbed by heat or electricity; electromagnetic waves that travel freely through space in a vacuum at a speed of 299,792 km/sec, where the oscillations are at right angles to the direction in which the light travels; an electromagnetic radiation in a wavelength range including infrared, visible, ultraviolet, and x-ray.

light amplification by stimulated emission of radiation — See laser.

light cream — Coffee cream or cream that contains less than 30% milkfat.

light-emitting diode — A long-lasting illumination technology used for exit signs, requiring very little power.

light intensity — The degree of brightness, depending on the number of photons striking a given area at a point in time.

light meter — A small, portable photometer with light-sensitive, barrier-layer cells that is used to measure illumination directly; usually calibrated in foot candles.

light non-aqueous-phase liquid — A non-aqueous-phase liquid with a specific gravity less than 1.0, causing it to float on top of the water table; petroleum

hydrocarbon fuels and lubricating oils are in this category.

light quantum — See photon.

light therapy — A technique using exposure of the body to electromagnetic waves to treat depression, especially during winter months.

light trap — A device used for collecting or destroying insects that consists of a bright light in association with a trapping or killing medium.

light year — A measure of distance, not time, defined as the distance light travels in one year; roughly equal to 5.88 trillion miles and used for linear measurement of interstellar distances.

lightning — The large spark produced by attraction of unlike electrical charges within a thundercloud or between a thundercloud and the Earth.

Lightning

Lightning Imaging Sensor — A small, highly sophisticated instrument that detects and locates lightning over the tropical regions of the globe and will lead to better weather forecasting.

lignin — An amorphous polymeric substance that together with cellulose forms the woody cell walls of plants.

lignite — A brownish-black solid fuel in the second stage in the development of coal. It has a little over half the heating value of bituminous or anthracite coal.

lignocellulose — Any of several closely related substances constituting the essential part of woody plant cell walls and consisting of cellulose and lignin.

LIMB — Acronym for limestone-injection multistage burner.

limbic system — A system of functionally related neural structures in the brain that are concerned with emotion and motivation.

lime — See calcium oxide.

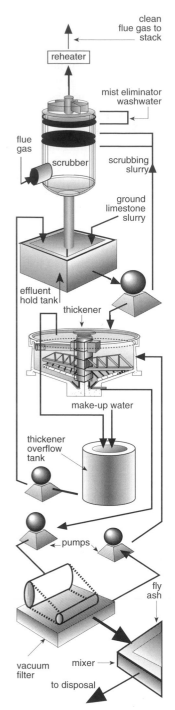

Lime Scrubbing System

lime scrubbing process — A process used on flue gases for control of sulfur dioxide; dried limestone is injected into the firebox, where it becomes reactive lime; this combines with the gas stream and passes into a wet scrubber, where the lime reacts with the water and together with the fly ash forms a slurry. The slurry and the sulfur dioxide react to form sulfite and sulfate salts, which are then disposed of.

lime treatment — (*sewage*) A means of controlling odor and reducing the number of pathogens through the use of lime without significantly reducing sludge solids.

limestone — A sedimentary rock composed mainly of calcite and concentrated shale, coral, algae, and other debris.

limestone scrubbing — The use of a limestone and water solution to remove gaseous stackpipe sulfur before it reaches the atmosphere.

limit of detection — The minimum concentration of a substance that can be detected and identified quantitatively or qualitatively 99% of the time.

limit of quantization (EPA) — The lowest concentration of a chemical that can be accurately and reproducibly quantized.

limiting factor — A condition whose absence or excessive presence exerts some restraining influence upon a population through incompatibility with a species' requirements or tolerance.

limnetic zone — The open-water region of a lake supporting plankton and fish as the principle inhabitants.

limnology — The study of biological, chemical, geographical, and physical features of freshwaters, especially lakes and ponds; used in understanding aquatic ecology.

limonite — A group of hydrated ferric oxide minerals ($Fe_2O_3H_2O$) that commonly occur in many types of rocks; they are generally rusty or blackish with a dull, earthy luster and a yellow-brown streak.

lindane ($C_6H_6CL_6$) — A white to yellow, crystalline powder with a slight musty odor that is the pure gamma-isomer of benzene hexachloride; MW: 290.8, BP: 614°F, Sol: 0.001%, sp. gr. 1.85. It is used in the formulation of insecticides, scabicides, pediculicides, and vermifuges. It is hazardous to the eyes, central nervous system, blood, liver, kidneys, and skin and is toxic through inhalation, absorption, ingestion, and contact. Symptoms of exposure include headache, nausea, clonic convulsions, respiratory difficulty, cyanosis, aplastic anemia, skin irritation, muscle spasms, and irritation of the eyes, nose, and throat; in animals, it causes liver and kidney damage. OSHA exposure limit (TWA): 0.5 mg/m³ [skin].

line — (*mapping*) One of the basic geographical elements defined by at least two pairs of *x,y* coordinates.

linear — Relating to a straight line.

linear accelerator — A device for accelerating charged particles in a straight line through a vacuum tube or series of tubes by means of alternating negative and positive impulses from electric fields; a machine with a rating of 1,000,000 electron volts (1 MeV) or greater which accelerates particles,

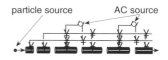

Linear Accelerator

electrons, or protons with high velocities along a straight line.

linear alkylate sulfonate (LAS) — A rapidly biodegradable variety of alkylbenzene sulfonate.

linear energy transfer (LET) — The average amount of energy transferred locally to the medium per unit of particle track length.

linear envelope detection — An electrical circuit assembly designed to detect the shape of the envelope of an amplitude modulated carrier or signal.

linear feet per minute — A unit of measurement for air velocity.

linear flow velocity — The velocity of a particle carried in a moving stream; usually measured in centimeters per second.

linear hypothesis — The assumption that a dose–effect curve in the high-dose and high-dose-rate ranges may be extrapolated through the low-dose and low-dose-rate ranges to zero, implying that theoretically any amount of radiation will cause some damage.

linearity — A determination of how closely an instrument measures actual values of a variable through its effective stage; a measure used to determine the accuracy of an instrument.

liner — Material used on the inside of a furnace wall to ensure the chamber is impervious to escaping gases; (*swimming pools*) a membrane that acts as a container for the water; (*hazardous waste*) a continuous layer of natural or man-made materials beneath or on the sides of a surface impoundment, landfill, or landfill cell that restricts the downward or lateral movement of hazardous waste or leachate.

lingual — Pertaining to the tongue.

lining — Material used on the inside of a furnace wall which is usually a high-grade refractory tile or brick or plastic refractory material; a material used to protect inner surfaces.

linuron ($C_9H_{10}Cl_2N_2O_2$) — A crystalline solid; MW: 249.11, BP: NA, Sol: moderately soluble in acetone

Linuron

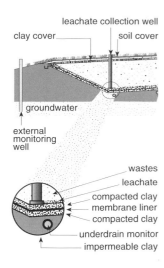

Landfill Liner

and alcohol, density: NA. It is used as an herbicide. It has low to moderate toxicity through ingestion but may be lethal through exposure to its vapors after 4 hours in experimental animals.

lipase — A digestive enzyme of the pancreatic fluid that changes fats to glycerol and fatty acids.

lipid — One of a class of nonpolar organic compounds that is insoluble in water but soluble in certain solvents, including alcohol and ether; includes fatty acids, greasy, oily, and waxy compounds.

lipid solubility — The maximum concentration of a chemical that will dissolve in fatty substances and may be taken up selectively by living tissues.

lipophilic — Having an affinity for fats or other lipids.

liposome — An artificial lipoid particle used to deliver antiparasitic drugs directly to macrophages, which eat the particles.

liquefaction — A change in phase from solid into a liquid form.

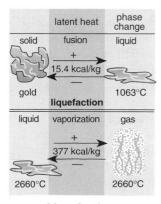

Liquefaction

liquid — The condensed phase of matter between solid and gas in which molecules are very close together but still have enough energy to move over one another, allowing the substance to flow; it conforms to the shape of the confining vessel.

liquid air — Air cooled and compressed until it liquefies; used as a refrigerant.

liquid all-purpose cleaners — Anionic and nonionic synthetic detergents that are low sudsing and will not damage or dull synthetic or wax finishes on floors.

liquid and solid biological treatment (LST) — A process that can be used to remediate soils and sludges contaminated with biodegradable organics.

liquid capacity — (*sewage*) The internal volume of a tank below the invert of the outlet line.

liquid carriage — (*air pollution*) An air control system in which dust particles are forced to separate from a gas stream by striking a liquid surface and are then entrained and removed from the area.

liquid crystal display — A type of digital display made of a material that changes reflectance or transmittance when an electrical field is applied to it.

liquid injection incinerator — A system that uses high pressure to prepare liquid wastes for incineration by breaking them up into tiny droplets to allow easier combustion.

liquid knockout — See impingement.

liquid–liquid extractions — A chemical technique in which two mutually immersible liquids redistribute the desired solute on the basis of differences in solubility.

liquid-media sampler — A device using an absorbing liquid to capture gaseous substances from air for sampling and analysis.

liquid metal fast breeder — A nuclear power plant that converts uranium-238 to plutonium-239, creating more nuclear fuel than it consumes. Because of the extreme heat and density of its core, the breeder uses liquid sodium as its coolant.

liquid oxygen — An intensely cold, transparent liquid formed by putting oxygen under very great pressure and cooling it to –182.97°C.

liquid scintillation counter — An analytical instrument used to quantify tritium, carbon-14, and other beta-emitting radionuclides.

liquified natural gas — A natural gas liquefied either by refrigeration or by pressure.

liquified petroleum gas (LPG) ($C_3H_8/C_3H_6/C_4H_{10}/C_4H_8$) — A mixture of butane, propane, and other light hydrocarbons derived from refining crude oil which can be cooled or subjected to pressure to facilitate storage and transportation. A colorless, noncorrosive, odorless gas when pure (a foul-smelling odorant is usually added) that is shipped as a liquefied compressed gas; MW: 42 to 58, BP: less than –44°F, Sol: not known. It is used in the chemical industry as feedstock for catalytic cracking in the manufacture of petrochemicals and polymers; as an internal combustion engine fuel; in the manufacture of synthetic rubber; as a fuel for industrial space and wall heating, air conditioning, and water heating; as a fuel for drying agricultural products; as a low-sulfur fuel for cutting, soldering, and brazing; as an aerosol propellant for shaving lather, window cleaners, starch sprays, room deodorants, insecticides, and pharmaceuticals; as an agricultural tool in nonselective cultivation and weed control. It is hazardous to the respiratory system and central nervous system and is toxic through inhalation and contact. Symptoms of exposure include lightheadedness and drowsiness. OSHA exposure limit (TWA): 1000 ppm [air] or 1800 mg/m³.

liquor — Any fluid or liquid.

LIS — See Land Information Systems or Lightning Imaging Sensor.

Listeria — A genus of Gram-negative bacteria (Corynebacterium family) found chiefly in the feces of lower animals; produces upper respiratory disease, septicemia, and encephalitic disease in people.

Listeria monocytogenes — A gram-positive motile bacillus that causes listeriosis.

listeriosis — A bacterial disease usually found in the very young or old, during pregnancy, or among people with immunosuppressed diseases. Symptoms include septicemia, fever, meningoencephalitis, intense headaches, nausea, vomiting, delirium, coma, shock, endocarditis, lesions in the liver and other organs, and infections of the fetus. Incubation time is not known, probably a few days to 3 weeks; the fetus is usually infected within several days of the maternal disease. It is caused by the bacterium *Listeria monocytogenes*, types Ia, Ib, IVa, and IVb. It is an unusual infection, sporadic and occurring in all seasons. It is found in institutions and elsewhere. Reservoirs of infection include infected domestic and wild mammals, fowl, and people; it is transmitted from mother to fetus by passage through the infected birth canal and by direct contact from lesions on hands and arms, infectious material, or contaminated soil. It is associated with ingestion of contaminated vegetables and dairy products; also through person-to-person transmission; communicable for mothers of infected newborn infants for 7 to 10 days in vaginal discharges or urine; other person-to-person communicability not known. Fetuses and newborn infants are highly susceptible, children and young adults generally are resistant; older people are more susceptible. It is controlled by avoidance of contact with infected materials or aborted animal fetuses and by pharmaceutical treatment.

liter (l) — The metric unit of volume equal to 1.057 quarts, 2.2046 pounds of pure water at 4°C, 0.0353 cubic feet, or 0.2642 gallons.

liters per minute — A measurement of air flow (in liters) per minute past a structure or into a piece of equipment.

lithium (Li) — A silvery-white metal that turns grayish on exposure to moist air; atomic weight: 6.941, BP: 1340°C, Sol: dissolves in liquid ammonia to form a blue solution, Fl.P: finely divided metal particles ignite spontaneously in air, density: NA. It is used in making alloys and in vacuum tubes. It is hazardous to the respiratory tract, digestive tract, skin, and eyes. It reacts with moisture to produce a corrosive hydroxide that causes severe irritation and burns. It may also cause kidney damage. ACGIH exposure limit: NA.

lithium aluminum hydride (LiAlH$_4$) — A white crystalline solid; MW: 37.96; BP: NA; Sol: soluble in ether, reacts vigorously with water; Fl.P: a flammable solid that ignites in moist or heated air; density: 0.917 g/cm^3. It is used as a powerful reducing agent in organic synthesis. It is hazardous to the skin and respiratory tract; it is highly moisture sensitive and causes extreme burns. ACGIH exposure limit (TLV-TWA): 2 mg Al/m^3.

lithium hydride (LiH) — An odorless, off-white to gray translucent, crystalline mass or white powder; MW: 7.95, BP: decomposes, Sol: reacts, Fl.P: not known, sp. gr. 0.78. It is used in the manufacture of buoyancy devices, fuel cells, and portable field generators; in the manufacture of reducing agents and propellants; in powder metallurgy; as a shielding material for thermal neutrons in the nuclear industry; as a desiccant; as a condensing agent in organic synthesis; in a condensation polymerization; in the manufacture of electronic tubes; in ceramics. It is hazardous to the respiratory system, skin, and eyes and is toxic through inhalation, ingestion, and contact. Symptoms of exposure include nausea, muscle twitches, mental confusion, blurred vision, and eye and skin burns; if ingested, it burns the mouth and esophagus. OSHA exposure limit (TWA): 0.025 mg/m^3 [air].

lithology — The description of rocks on the basis of their physical and chemical characteristics as determined by the eye or a low-power microscope.

lithosphere — The outer part of solid earth composed of rock, essentially like that exposed at the surface; considered to be 60 miles thick.

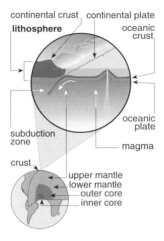

Lithosphere
in a whole earth context

litigation — The act or process of carrying on a lawsuit.

litmus — A water-soluble organic substance obtained from the lichen plant and used as an indicator to determine if a solution is acidic (turns red) or is basic (turns blue).

litmus paper — An absorbent paper containing a blue dye (litmus) that is used to determine the level of pH.

litter — Unwanted, discarded material that is solid waste existing outside of a collection system.

littoral zone — The shallow, shoreward, or coastal region of a body of water having light penetration to the bottom, frequently occupied by rooted plants; the zone between the high- and low-water marks.

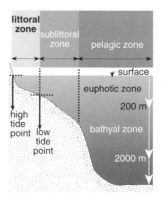

Littoral Zone

liver — The large, dark-red gland located in the upper right portion of the abdomen in the human body that is associated with several vital activities including digestion, sugar metabolism, and removal of toxins.

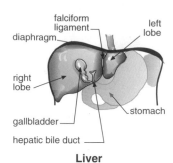

Liver

liver fluke — A parasitic trematode that may infest the liver; usually acquired by eating freshwater fish containing the encysted larvae.

LLD — See low-lethal dose.

LLRW — Acronym for low-level radioactive waste.

LLW — Acronym for low-level waste.

lm — See lumen.

LMFBR — Acronym for liquid metal fast breeder reactor.

LNAPL — Acronym for light non-aqueous-phase liquid.

LNG — See liquified natural gas.

load — The electrical demand of a process expressed in power, current, or resistance.

loading — The quantity of a substance entering the environment (soil, water, or air).

loading rate — The allowable rate of application of septic tank effluent to the soil, expressed in gallons per day per square foot.

LOAEL — See lowest observed adverse effect level.

loam — A soil having an even mixture of sand, silt, clay, and humus.

lobar pneumonia — A severe infection of one or more of the five major lobes of the lungs. Symptoms include fever, chills, cough, rapid shallow breathing, cyanosis, nausea, vomiting, and pleurisy.

LOC — See level of concern

local air exchange index — An index that characterizes conditions at a particular point within a room and may be largely due to the position of the measurement point.

Local Area Network — A private transmission that interconnects computers within a building or among buildings for the purpose of sharing voice, data, facsimile, and/or video.

local emergency planning committee — An organization of state and local officials; police, fire, civil defense, public health professionals; environmental, hospital, and transportation officials; representatives of facilities or community groups, and the media whose primary responsibility is the development of an emergency response plan for a local area; established under the Superfund Amendments and Reauthorization Act of 1988, Title III.

local exhaust ventilation (LEV) — Mechanical ventilation applied at or close to the source of an emission for the purpose of drawing clean, uncontaminated air past the worker, capturing the emission, collecting it in an exhaust hood, and removing it from the building.

local government — A county, city, village, town, district, or political subdivision of any state or Indian tribe.

local infection — An infection involving bacteria that invade the body at a specific point, remain there, and multiply until destroyed.

local mean age of air — The average time it takes for air to travel from an inlet to any point in a room or enclosure.

local runoff — Water running off a local area, such as rainfall draining into a nearby river.

localized — Limited to a given area or part.

location — A coordinate on a globe expressed in longitude and latitude.

lockjaw — See tetanus.

locomotion — The ability to move from one place to another.

locular pneumonia — See bronchopneumonia.

locus — The position a gene occupies on a chromosome.

locus of infection — A place on or in the body where an infection originates.

LOD — Acronym for limit of detection.

loess — A silty, wind-deposited material consisting almost entirely of silt but which may include small amounts of very fine sand, clay, or fossils.

log — A chronological record of a series of events by time, day, value, and physical reference.

log on — To sign on to a computer system.

logarithm — The power to which a base (usually 10) must be raised to produce a given number; for example, base 10 raised by a logarithm of 3 equals 1000.

logbook — A dated, written record of performed operation and maintenance items or observations pertinent to a structure.

logging capacity — The number of data elements that can be stored on a log.

logistic growth — A growth rate regulated by internal and external factors that establishes an equilibrium with environmental resources; the species may grow exponentially when resources are unlimited but slowly as resources are used up.

logistic regression — A statistical method for calculating odds ratios for individual risk factors where a variety of risk factors may be contributing to the occurrence of disease.

long-path monitoring system — A field-deployable, long-path, Fourier-transformed infrared spectrometer that measures the absorption caused by infrared-active molecules.

long-term animal bioassays — Laboratory studies in which animals are exposed to a suspected hazard. The animals are examined for the presence of tumors and other signs of disease throughout the study. At the end of the study, the surviving animals are sacrificed and their tissues are examined.

long-term exposure — The prolonged condition of being exposed to infectious, chemical, or physical agents that may have an adverse effect on the body.

long ton — Measurement equal to 2240 pounds.

longevity — The length or duration of life.

longitude — Location of a position on the surface of the Earth, east or west, measured in degrees by the angle contained between the meridian of a particular place and some prime meridian, such as Greenwich, England.

longitudinal — A measurement in the direction of the long axis of an object.

longitudinal study — See prospective study.

longitudinal wave — A wave in which the particles of a medium move back and forth in the same direction as the wave itself moves.

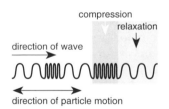

Longitudinal Wave

longwave radiation — The radiation emitted in spectral wavelengths of >4 μm, corresponding to the radiation emitted from the Earth and atmosphere.

loop — See antinode.

LOPE — See low-density polyethylene.

LOQ — Acronym for limit of quantification.

losing stream — A stream or portion of the stream that discharges water into groundwater.

lot — The product produced during a period of time indicated by a specific code; a government lot or sub-division lot representing the boundary of a legally conveyable unit of land identified on a record or document.

lotic environment — The presence of running waters, such as streams or rivers.

lotion — A liquid or cream preparation applied externally to protect the skin or treat a dermatological condition.

loudness — The physiological perception of sound intensity; loudness is dependent on frequency.

louse — The common name for a grayish, wingless, and dorsoventrally flattened insect varying in length from 1/6 to 1/16 inch; a human parasite.

Crab Louse
Phthirus pubis

Head Louse
Pediculus humanus capitis

Body Louse
Pediculus humanus

**The three basic kinds
of human lice**

louse bite — A minute puncture wound produced by a louse that may transmit typhus fever, trench fever, or relapsing fever with secondary infections coming from scratching the affected areas.

louse-borne typhus — See epidemic typhus.

low — (*weather*) An area of low pressure; also known as a depression.

low-acid foods — Any food, other than alcoholic beverages, with a finished equilibrium pH value greater than 4.6 and a water activity greater than 0.85;

includes any normally low-acid fruits, vegetables, or vegetable products in which, for the purpose of thermal processing, the pH value is reduced by acidification.

low altitude — Any elevation less than 4000 feet above sea level.

low back pain — A local or referred pain at the base of the spine that is caused by a sprain, strain, osteoarthritis, neoplasm, or a prolapsed intervertebral disk.

low clouds — Types of clouds whose bases range in height from near the surface of the Earth to 6500 feet; principle clouds in this group include stratocumulus, stratus, and nimbostratus.

low-density material — A material having a low weight per unit volume either as it occurs in its natural state or after being compacted.

low-density polyethylene — A plastic material used for both rigid containers and plastic film applications.

low-emissivity windows — Windows that lose less energy due to inhibition of the transmission of radiant heat while still allowing sufficient light to pass through; also called low-E windows.

low-flow pump — A pump used for gas and vapor sampling where the flow rate of air is 200 ml/min.

low-frequency sound waves — Frequency of sound ranging from 30 to 300 kHz.

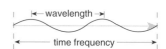

Low-Frequency Sound Waves

low-grade infection — A subacute or chronic infection with a mild fever.

low-head hydropower — A small-scale hydrotechnology that can extract energy from small headwater dams, causing much less ecological damage to the area.

low heat value — The high heat value minus the latent heat of vaporization of water as formed by burning the hydrogen in the fuel; also known as net heating value.

low-LET — See low linear energy transfer.

low linear energy transfer (low-LET) — Radiation characteristic of electrons, x-rays, and gamma-rays.

low lethal concentration (LCL) — The lowest concentration of a gas or vapor capable of killing a specified species.

low lethal dose (LDL) — The lowest administered dose of a material capable of killing a specified species.

low-level nuclear waste — Radioactive wastes coming largely from national defense facilities (e.g., contaminated lab coats, gloves, and laboratory equipment); waste generated by hospitals, research laboratories, and certain industries that are much less hazardous than those coming from nuclear reactors.

low permeability layers — The soil, sediment, or other geological materials that inhibit water movement.

low volatile — A liquid or solid that does not evaporate quickly at normal temperatures.

lower atmosphere — The part of the atmosphere in which most weather occurs; from the ground to approximately 60 km into the air.

lower detection limit — The smallest signal above background noise an instrument can reliably detect.

lower explosive limit (LEL) — The lowest concentration of a vapor or gas at ordinary ambient temperatures expressed in percent of the gas or vapor that will produce a flash of fire when an ignition source is present.

lower flammable limit (LFL) — See lower explosive limit.

lower respiratory tract — A part of the respiratory tract that includes the left and right bronchi and alveoli where the exchange of oxygen and carbon dioxide occurs during the respiratory cycle.

lowest acceptable daily dose — The largest quantity of a chemical that will not cause a toxic effect, as determined by animal studies.

lowest observed adverse effect level (LOAEL) — The lowest dose producing an observable adverse effect.

LOX — See liquid oxygen.

LPG — See liquefied petroleum gas.

LPM — See liters per minute.

LQER — Acronym for lesser quantity emission rate.

LQG — Acronym for large quantity generator.

LRTAP — Acronym for long-range transboundary air pollution.

LSE — Acronym for levels of significant exposure.

LST — See liquid and solid biological treatment.

LT$_{50}$ — See lethal time 50.

LTRA — Acronym for long-term remedial action.

lubricant — A liquid, ointment, or other substance used to reduce friction or make a surface slippery.

LUIS — Acronym for Label Use Information System.

LULUs — Locally unwanted land used for toxic waste dumps, incinerators, smelters, and other sources of environmental, economic, or social degradation.

lumbar puncture — Insertion of a needle into the lower part of the spinal column to collect cerebrospinal fluid for diagnosis.

lumen (lm) — (*anatomy*) A canal or cavity in a tubular organ; (*optics*) the unit of measurement for the rate at which light energy is radiated from a source equal to the amount of light striking 1 square foot of a surface 1 foot away from the light source (1 candle power Å 12.5 lumens).

luminescence — The low-temperature emission of light by a chemical or physiological process.

luminous — Referring to the emission of a steady, suffused light.

luminous flux — The rate or flow of light measured in lumens.

lumpectomy — The surgical removal of a malignant lump in the breast with only a small amount of normal breast tissue around it being taken.

lung — Either of a pair of the respiratory organs for breathing air in vertebrates.

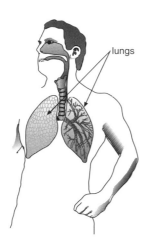

Lungs

lung cancer — A pulmonary malignancy caused by cigarette smoking and a variety of inhaled chemicals, with symptoms of persistent cough, dyspnea, blood-streaked sputum, chest pains, bronchitis, or pneumonia.

lung fluke — A parasitic flatworm, *Paragonimus westermani*, found in Africa, the Orient, and Latin America but rarely in North America. It is transmitted by crabs and crayfish. Symptoms include peribronchiolar distress and blood-streaked sputum.

lung toxins — Chemicals that irritate or damage the pulmonary tissue with symptoms of cough, tightness in the chest, and shortness of breath.

LUST — See leaking underground storage tank.

luster — The manner in which light reflects from the surface of a mineral, describing its quality and intensity.

lux — The SI unit of illumination where 1 lumen is uniformly distributed over an area of 1 square meter.

lye — A common name for sodium hydroxide or potassium hydroxide; used as a strong alkaline solution in industry.

lye poisoning — The toxic effects of ingesting caustic soda or sodium hydroxide resulting in usually irreversible chemical damage to the mouth and throat.

Lyme disease — A tickborne spirochetal zoonotic disease with a distinctive skin lesion, systemic symptoms, polyarthritis, neurological, and cardiac involvement. Incubation time is 3 to 32 days; it is caused by the spirochet *Borrelia burgdorferi* and is found in the United States, along the Atlantic and Pacific coasts and in the south. The reservoir of infection is Ixodid ticks. It is not communicable; general susceptibility. It is controlled by tick repellent, proper clothing, and avoiding areas containing ticks.

lymph — The clear, liquid part of blood that enters the tissue spaces and lymph vessels of the lymphatic system.

lymph node — Glandular centers in which lymphatics converge.

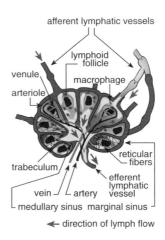

Lymph Node Structure

lymphatic cancer — Cancer of the lymphatic system.

lymphatic system — Tissues and organs that produce, store, and carry white blood cells and fight infection and disease; the system includes the bone marrow, spleen, thymus, lymph nodes, and a network of thin tubes that carry lymph and white blood cells.

lymphocyte — A type of white blood cell that is produced by stem cells in the bone marrow and is a component of the immune system.

lymphocytes — The white blood cells that fight infection and disease.

lymphoid — Resembling lymph tissue.

lymphokine — A molecule secreted by an activated or stimulated lymphocyte that causes physiological changes in certain other cells.

lymphoma — A cancer that arises in cells of the lymphatic system.

lyse — To cause or produce disintegration of a compound, substance, or cell.

lysimeter — A device that measures the quantity or rate of water movement or leaching loss through or from a block of soil usually undisturbed and *in situ* or is used to collect such percolated water for quality analysis.

lysing — A disintegration or breakdown of cells that releases organic matter.

lysis — Destruction or decomposition especially by enzymatic digestion.

lysosome — The digestive system of a cell.

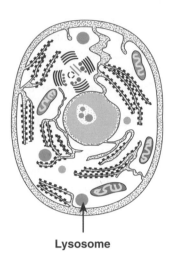

Lysosome

lysozyme — An enzyme that dissolves the cell walls of many bacteria; present in many body secretions such as tears.

L

M

μ/l — See microgram per liter.

μCi — See microcurie.

μg — See microgram.

μg/m³ — See microgram per cubic meter.

m — See meter, milli-, or micro-.

m³ — See cubic meter.

M — SI symbol for the prefix *mega-*, repesenting a factor of 10^6 power; see mega-.

M&E — Acronym for monitoring and enforcement.

M&S — Acronym for modeling and simulation.

M&S R&D — Acronym for modeling and simulation research and development.

mA — See milliampere.

MA — Acronym for mental age.

MAC — See maximum allowable concentration; acronym for mobile air conditioner.

maceration — Softening of a solid by soaking.

machine location — Room, enclosure, space, or area where vending machines are installed and operated.

macroclimate — The climate of a relatively large area.

macroelements — Essential elements for plant growth that are required in relatively large amounts, including carbon, hydrogen, oxygen, phosphorus, potassium, sodium, sulfur, calcium, iron, and magnesium.

macromolecule — A very large molecule; used in reference to carbohydrates, lipids, proteins, and nucleic acids.

macronutrient — A chemical element necessary in large amounts, usually greater than 1 ppm, for the growth and development of plants.

macroparasite — A large parasite that does not multiply in the host; examples are cestodes, trematodes, and most nematodes.

macrophage — Any of the large mononuclear, high phagocytic cells derived by monocytes that occur in the walls of blood vessels and in loose connective tissue.

macrophyte — (*botany*) The larger aquatic plants.

macropores — Secondary soil features such as root holes or desiccation cracks that can create significant conduits for movement of dissolved or vapor-phase contaminants.

macroscopic — Visible without the aid of a microscope.

MACT — See best available control technology or maximum available control technology.

Macrophage

psuedopodia / bacteria

macula — A discolored spot on the skin not elevated above the surface.

mad cow disease — See bovine spongiform encephalopathy.

mad hatter's disease — See mercury poisoning.

made land — Areas in which the natural soil profile has been worked or disturbed to such an extent that the soil cannot be classified in accordance with the standards established for soil classification by the National Cooperative Soil Survey.

MAF — See million acre-feet.

maggot — The larval stage of a fly.

magistrate — A public civil officer having the power to issue a warrant for arrest and charge an individual with a public offense.

magma — Molten silicate source material from which igneous rocks are formed.

Magma

magnesium (Mg) — A silver-white, light, malleable, ductile metallic element that is found abundantly in nature in a combined state and is used in metallurgical and chemical processes such as photography, signaling, and pyrotechnics and produces an intense white light when it burns; atomic number: 12, atomic weight: 24.32.

magnesium hardness — A measure of the magnesium salts dissolved in water; not a factor in water balance.

magnesium oxide fume (MgO) — A finely divided white particulate dispersed in the air; MW: 40.3, BP: 6512°F, Sol (86°F): 0.009%, sp. gr. 3.58. Exposure may occur when magnesium is burned, thermally cut, or welded. It is liberated from the fabrication of alloys for aircraft, ships, automobiles, boats, tools, machinery, and military equipment; from casting of metal and alloys. It is hazardous to the respiratory system and eyes and is toxic through inhalation and contact. Symptoms of exposure include metal fume fever, cough, chest pain, flu-like fever, and irritation of the eyes and nose. OSHA exposure limit (TWA): 10 mg/m³ [air].

magnetic field (*B*) — The space around a magnet or electric current in which a magnetic force is felt.

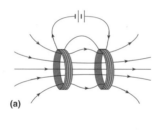

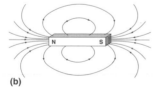

Magnetic Field
(a) Helmholtz coils
(b) bar magnet

magnetic field intensity — See magnetic field strength.

magnetic field strength (*H*) — Vector quantity given by the ratio B/μ, where B is the magnetic flux density and μ is the permeability of the medium; the magnitude of the ratio indicates the strength of the magnetic field at a point in the direction of the line of force at that point. Also known as magnetic field intensity.

magnetic flux (*F*) — The product of a particular area under consideration and the component, normal to the area, of the average magnetic flux density over it.

magnetic resonance — The response of electrons, atoms, molecules, or nuclei to various discrete radiation frequencies as a result of space quantization in a magnetic field.

magnetic resonance imaging (MRI) — A diagnostic imaging technique that employs magnetic and radio-frequency fields to produce cross-sectional images of body tissues and to monitor body chemistry noninvasively; formerly known as nuclear magnetic resonance (NMR).

magnetic separator — Any device that removes ferrous metals by means of magnets.

magnetometer — An instrument to measure the local intensity of a magnetic field, such as that of the Earth.

magnetosphere — Region surrounding a celestial body where a magnetic field controls the motion of charged particles.

magnification — Apparent enlargement of an object by an optical instrument.

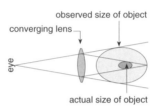

Magnification

magnitude — (*geology*) A number that characterizes the relative size of an earthquake based on measurement of the maximum motion recorded by a seismograph and expressed as ordinary numbers based on a logarithmic scale.

Magnox reactor — A thermal reactor named after the magnesium alloy in which the uranium metal fuel is contained; the moderator is graphite and the coolant is CO_2 gas.

mainframe — (*computer science*) A large computer supporting many users.

maintainability — A measure of the speed with which loss of performance is detected, the problem located, repair carried out and completed, and a check made that the equipment is functioning normally again.

maintenance — The combination of technical actions and corresponding administrative actions undertaken to keep a piece of equipment in a state adequate to fulfill its required function.

maintenance dose — The amount of drug to be administered to maintain a desired concentration in the tissues.

MAIS — Acronym for Major Automated Information Systems.

major disaster — A natural catastrophe (hurricane, tornado, storm, flood, high water, wind-driven water, tidal wave, tsunami, earthquake, volcanic eruption, landslide, mudslide, snowstorm, drought) or,

Magnitude (Richter Scale)	Relative Effect(s)	Ergs
0–1.9	instrument records only	$<2 \times 10^{10}$
2–2.9	barely felt suspended objects swing	$4–9000 \times 10^{10}$
3–3.9	felt by some vibration like heavy vehicle	$1–7 \times 10^{15}$
4–4.9	felt by most dishes/ windows rattle– may break	$1–30 \times 10^{16}$
5–5.9	felt by all chimneys topple; furniture moves	$1–200 \times 10^{18}$
6–6.9	buildings substantially damaged	$4–230 \times 10^{20}$
7–7.9	large landslides; fissures in ground	$4–50 \times 10^{22}$
8–8.6	ground waves; complete devastation	$>1 \times 10^{25}$

Earthquake Magnitude

regardless of cause, any fire, flood, or explosion in any part of the United States that the President determines has caused damage of sufficient severity and magnitude to warrant major assistance under federal law to supplement the efforts and available resources of states, local governments, and disaster relief organizations in alleviating damage, loss, hardship, or suffering of people.

makeup air — Clean, temperature-adjusted, outdoor air supplied to a workplace to replace air removed by exhaust ventilation.

makeup tank — Balance tank; (*swimming pools*) a tank that provides a continuous flow of makeup water for a pool and may be used as a receiving point for overflow or as a receiving point for chemicals in the event they have to be added on an emergency basis.

makeup water — Fresh water used to fill or refill a pool; water added to boiler cooling towers, water tanks, or other systems to maintain the volume of water required.

malabsorption — Poor absorption of nutrients from the gastrointestinal tract.

malady — A disease or illness.

malaise — A vague feeling of discomfort, ill-being, and poor health.

malaria — Four different human diseases: *Plasmodium falciparum* malaria, the most serious form, causes fever, chills, sweats, and headaches that lead to jaundice, coagulation defects, shock, renal and liver failure, and coma; fatality is 10% among children and untreated adults, and incubation time is 12 days. *Plasmodium vivax*, *P. malariae*, and *P. ovale* malaria are usually not life threatening except in the very young and very old. Symptoms include malaise, followed by a shaking chill and rapidly rising temperature, headache, nausea, and profuse sweating. The cycle of chills, fever, and sweating is repeated daily, every other day, or every third day, with relapses occurring irregularly for 2 to 5 years; *P. malariae* relapses may persist for 50 years. Incubation time is 14 days for *Plasmodium vivax*, *P. malariae*, and *P. ovale*. Although not typically found in temperate climates today, malaria is still found in tropical and subtropical areas. The reservoir of infection is people, and it is transmitted by the bite of an infective female anopheline mosquito. The mosquito ingests blood containing the sexual stages of the parasite, the gametocytes. The male and female gametes unite in the mosquito's stomach and enter the stomach wall to form a cyst in which thousands of sporozoites develop in a period of about 8 to 35 days, depending on the parasite and temperature. The sporozoites migrate to various organs of the infected mosquito and some reach the salivary glands, mature, and are infective when injected into humans. In humans, the sporozoites enter hepatocytes and develop into exo-erythrocytic schizonts; the hepatocytes rupture and asexual parasites (tissue merozoites) appear in the bloodstream. They invade the erythrocytes, grow, and multiply. Some develop into asexual forms to mature blood schizonts, which rupture and liberate merozoites and invade other erythrocytes. The clinical symptoms occur with the rupture of the erythrocytic schizonts. The disease is communicable for as long as infective gametocytes are in the blood of patients, which can be 3 years for *P. malariae*, 1 to 2 years for *P. vivax*, and 1 year for *P. falciparum*. It may also be spread by transfusions; general susceptibility. It is controlled by removal of mosquito breeding sites, extermination of larvae and adults, use of residual insecticide, mechanical screening of all sleeping areas, use of insect repellents, and avoidance of blood donors who have a history of malaria.

malarial parasite — One of four species of Plasmodium that may be injected into the human bloodstream by an anopheles mosquito which initiates the cycle of malaria.

M

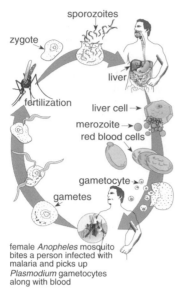

sporozoites

zygote

liver

fertilization

liver cell →

merozoite →

red blood cells

gametocyte →

gametes

female *Anopheles* mosquito
bites a person infected with
malaria and picks up
Plasmodium gametocytes
along with blood

Malaria Cycle

malathion (C$_{10}$H$_{19}$O$_6$PS$_2$) — A deep brown to yellow liquid with a garlic-like odor; a solid below 37°F; MW: 330.4, BP: 140°F (decomposes), Sol: 0.02%, Fl.P (open cup): >325°F, sp. gr. 1.21. It is used in the formulation of pesticide products and as an insecticide for treatment of grain, nut, fruit, and fiber crops. It is hazardous to the respiratory system, liver, blood cholinesterase, the central nervous system, cardiovascular system, and gastrointestinal tract and is toxic through inhalation, absorption, ingestion, and contact. Symptoms of exposure include miosis, aching eyes, blurred vision, lacrimation, salivation, anorexia, nausea, vomiting, abdominal cramps, diarrhea, giddiness, confusion, ataxia, rhinorrhea, headache, tight chest, wheezing, laryngeal spasms, and irritation of the eyes and skin. OSHA exposure limit (TWA): 10 mg/m^3 [skin].

Malathion

maleic anhydride (C$_4$H$_2$O$_3$ — Colorless needles, white lumps, or pellets with an irritating, choking odor; MW: 98.1, BP: 396°F, Sol: reacts, Fl.P: 218°F, sp. gr. 1.48. It is used in the manufacture of polyester resins for automobile bodies, structural building panels, molded boats, and chemical storage; during the manufacture of fumaric acid and polyester; in the manufacture of alkyd resins used as enamels, interior flat finishes, automotive finishes, printing inks, reac-

tive plasticizers, and marine paints and varnishes; in the manufacture of detergents and lubricating additives as dispersants and wetting agents; in the manufacture of drying oils, oil resins used as general coatings and industrial dip coatings; in the manufacture of terpene resins used as shellac substitutes; in aniline inks; as protective coatings on paper; in metal foil, cellulose films, and natural synthetic fibers; in the manufacture of chlorendic anhydride for fire-retardant polyester resins; in the manufacture of extreme pressure lubricants; in organic synthesis; in production of chemical intermediates. It is hazardous to the eyes, skin, and respiratory system and is toxic through inhalation, ingestion, and contact. Symptoms of exposure include conjunctivitis, photophobia, double vision, bronchial asthma, dermatitis, and nasal and upper respiratory system irritation. OSHA exposure limit (TWA): 0.25 ppm [air] or 1 mg/m^3.

malfeasance — Inappropriate action or official misconduct.

malformation — Permanent structural change that may adversely affect survival, development, or function.

malfunction — Any unanticipated or unavoidable failure of pollution-control equipment or processes.

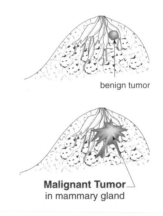

benign tumor

Malignant Tumor
in mammary gland

malignant — An abnormal growth that endangers the life or health of an individual and is resistant to treatment; having the potential to be transmitted from an original site to one or more sites elsewhere in the body if not checked by treatment.

malignant tumor — A cancerous tumor.

malinger — To pretend or exaggerate illness to avoid work.

malleable — Capable of being hammered or pounded into thin sheets without structural failure.

malnutrition — Poor nourishment resulting in improper diet or from some defect in metabolism that prevents the body from using its food properly.

malononitrile (CH$_2$(CN)$_2$) — A colorless solid; MW: 66.07; BP: 218°C; Sol: soluble in water, alcohol, ether, acetone, and benzine; Fl.P: >94°C; density: 1.191 at 20°C. It is used in organic synthesis. It is hazardous to the respiratory tract, digestive tract, and skin and is a highly toxic compound that may cause death. NIOSH exposure limit (TLV-TWA): 3 ppm.

$$N \equiv C - CH_2 - C \equiv N$$

Malononitrile

malpractice — Failure of a professional person to render proper services due to reprehensible ignorance, negligence, or criminal intent.

mammary gland — A highly modified sebaceous gland found in female mammals that secretes milk.

managed care — A healthcare system with control over primary healthcare services in a medical care practice designed to reduce costs.

mandate — A command, order, or direction, written or oral, that the court is authorized to give and a person is bound to obey; this is done upon the decision of an appeal or writ of error.

mandible — The strong, cutting mouthparts of arthropods used for crushing foods; the lower jaw.

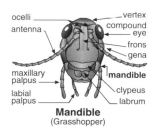

Mandible
(Grasshopper)

manganese (Mn) — A lustrous, brittle, silvery metal; MW: 54.9, BP: 3564°F, Sol: insoluble, Fl.P: not known, sp. gr. 7.20. It is liberated during welding operations; during the casting of molten ferromanganese; during the mixing and pressing of dry battery depolarization; during dumping, weighing, and mixing operations in ceramics and glass manufacture (pigmentation and coloration); during the manufacture of manganese soap and wood preservatives; in the manufacture of safety matches, signal flares, and fireworks. It is hazardous to the respiratory system, central nervous system, blood, and kidneys and is toxic through inhalation and ingestion. Symptoms of exposure include Parkinson's disease, asthenia, insomnia, mental confusion, metal fume fever, dry throat, cough, tight chest, dyspnea, râles, flu-like fever, low-back pain, vomiting, malaise, and fatigue. OSHA exposure limit (TWA): ceiling 5 mg/m³ [air].

manifest — An itemized listing of the cargo of a vessel with other particulars for the facility of customs officers.

manifest system — Tracking hazardous waste from "cradle to grave" that includes any accompanying documents.

manifold — A pipefitting with numerous branches that convey fluids between a large pipe and several smaller pipes or permit the choice of diverting flow from one of several sources or to one of several discharge points.

Manning's roughness coefficient — (*hydraulics*) A coefficient used to describe the relative roughness of natural and artificial channels; used in hydraulic computations.

manometer — A device for measuring pressure using common mercury or another liquid to compare the difference in height of two liquid columns; an instrument used for measuring pressure; U-tube-shaped device partially filled with a liquid (usually water, mercury, or a light oil) constructed so that the amount of displacement of the liquid indicates the pressure being exerted on the instrument.

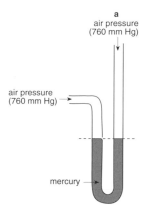

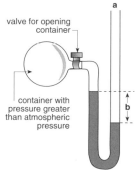

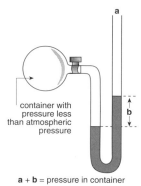

Manometer

mantle — (*geology*) A layer of rock deep within the Earth that separates the Earth's crust from the its metallic outer core.

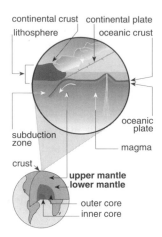

Mantle (geology)

manual immersion cleaners — A blend of anionic and nonionic organic liquids or powders containing special additives for corrosion inhibition and skin protection.

manual materials handling — (*ergonomics*) Any handling task involving the human body as the power source, including lifting, lowering, pushing, pulling, carrying, and holding.

manufacturer's formulation — A list of substances or component parts described by the producer of a coating, pesticide, or other product containing chemicals or other substances.

manure — The excrement of animals found in stables and barnyards; may also contain some spilled feed or bedding.

map — See cartography.

map gallery — A collection of maps sometimes grouped by a common theme.

mapping unit — A soil or combination of soils delineated on a map or showing the taxonomic unit or units included; depicts soil types, phases, associations, or complexes.

MAPS — Acronym for Map Analysis Package System.

Marburg–Ebola virus disease — A devastating febrile disease with symptoms of severe gastrointestinal bleeding, easily transmitted to hospital personnel by improper handling of contaminated needles or the lesions of patients. An epidemic occurred in Marburg, Germany, in 1967, apparently from imported African green monkeys. Explosive epidemics occur periodically in the Ebola River District of Zaire and Sudan with a mortality rate as high as 85%.

margin of exposure — Ratio of the no-observed adverse effect level (NOAEL) to the estimated exposure dose.

margin of safety (MOS) — A factor estimated by dividing the experimental no-observed adverse effect level (NOAEL) by the estimated daily human dose; the difference between an allowable level for a given pollutant and the criteria level at which adverse effects have been noted, assuming the allowable level is lower.

mariculture — Cultivation of fish and shellfish in estuarine and coastal areas.

marine — Referring to life in saltwater.

Marine Protection and Sanctuaries Act — A 1972 law, amended through 1988, stating that unregulated dumping of material into ocean waters endangers human health, welfare, amenities, the marine environment, ecological systems, and economic potential. The policy of the United States is to regulate dumping of all types of materials into ocean waters and prevent or strictly limit dumping of any material that would adversely affect health, welfare, and the ecosystem. The law also regulates transportation of material and dumping at sea.

marine sanitation device — Any equipment or process installed on board a vessel to receive, retain, treat, or discharge sewage.

Mariotte's law — See Boyle's law.

marl — An earthy, unconsolidated deposit formed at freshwater lakes and consisting chiefly of calcium carbonate mixed with clay or other impurities in varying proportions; used as fertilizer.

marrow — Soft tissue in the central cavity of large bones.

Marseille fever — An endemic disease found around the Mediterranean, in Africa, in the Crimea, and in India. It is caused by *Rickettsia conorii* transmitted by the brown dog tick. Symptoms include chills, fever, an ulcer covered with a black crust, and a rash.

marsh — Wet, soft, low-lying land that provides a habitat for many plants and animals and is covered at least part of the time by estuarine or coastal waters.

marsh gas — See methane.

MART — See mean active repair time.

MASER (microwave amplification by stimulated emission of radiation) — An instrument used to amplify radio microwaves.

mask — A protective device usually fitting over the face or mouth and nose to filter poisonous chemicals in gas or vapor form.

masking — Blocking out of a sight, sound, or smell with another.

masking agent — A substance used to cover up or disguise an unpleasant odor.

mass — The definite quantity of matter a substance possesses depending on the gravitational force acting on it; an assigned unit of weight.

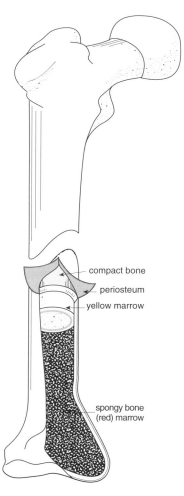

Marrow

mass care center — Facility for providing emergency lodging and care for people temporarily homeless because of an emergency or disaster.

mass flow rate — Volumetric flow rate times density.

mass number — The sum of the numbers of protons and neutrons in the nucleus of an atom; the integer nearest to the atomic mass. Also known as nucleon number.

mass spectrograph (MS) — A device used to determine the relative atomic masses or weights of electrically charged particles by separating them into distinct streams by means of magnetic deflection.

mass spectrometer — A device similar to the mass spectrograph designed in such a way that the beam constituents of a given mass-to-charge ratio are focused on an electrode and detected or measured electrically.

mass storage — A disk that can store large amounts of data easily accessible to the central processing unit of a computer.

massive soil — A soil structure in which the material clings together in large uniform masses, usually found in substratum of the C horizon.

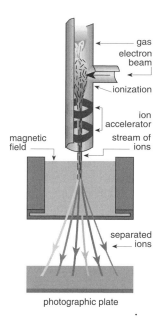

Mass Spectrograph

Massive Soil

MAST pants — See military antishock trousers.

mastectomy — Surgical removal of as much of the breast tissue of one or both breasts as needed to ensure that all cancer cells are gone; usually performed because of a malignant tumor.

mastication — Chewing, tearing, or grinding food with the teeth while mixing it with saliva.

mastitis — (*disease*) Inflammation of the breast occurring in a variety of forms and in varying degrees of severity.

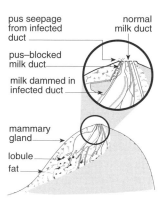

Mastitis

M

MATC — Acronym for maximum acceptable toxic concentration.

material fact — A fact that establishes or refutes an element essential to a charge or defense.

material handling — The movement, storage, control, and protection of materials and products throughout the process of their manufacturer, distribution, consumption, and disposal.

Material Safety Data Sheet (MSDS) — A form used by industry and government to meet specifications of the Occupational Safety and Health Administration Hazard Communication Standard. It contains information regarding product manufacture, identity of hazardous ingredients, physical/chemical characteristics, fire and explosion hazard data, reactivity data, precautions for safe handling and use, and control measures.

maternal death — The death of a woman while pregnant or within 42 days of termination of the pregnancy irrespective of the duration and site of the pregnancy.

maternal death rate — The number of direct and indirect maternal deaths per 100,000 live births for a specified time period.

maternal transmission — Transmission of disease from mother to offspring.

mathematical model — Model used during risk assessment to perform extrapolations.

maticulism — A combination of catabolism and anabolism; the chemical change in living cells by which energy is provided for bioprocesses and activities and new material is assimilated.

matrix — (*EPA definition*) Fiber or fibers with one end free and the other end imbedded in or hidden by a particulate; the exposed fiber must meet the fiber definition; (*NIOSH definition*) one or more fibers attached to or imbedded in a nonasbestos particle.

matte — A mixture of sulfides formed in smelting sulphide ores.

matter — Anything that occupies space and has mass; occurs in three states: solid, liquid, or gas. Solid matter has both a definite size and shape. Liquid matter has a definite volume that takes the shape of its container. Gas has neither definite shape nor definite volume.

maturation — The process or condition of attaining complete development.

maxilla — Mouth part of an arthropod; the upper jaw.

maximally exposed individual — An individual whose location and habits tend to maximize his or her radiation dose, resulting in a dose higher than that received by other individuals in the general population.

maximum acceptable toxic concentration — The range or geometric mean between the no-observable

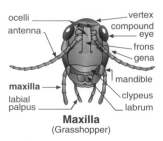

Maxilla
(Grasshopper)

adverse effect level and the lowest observable adverse effects level for any given ecological effects test.

maximum allowable concentration (MAC) — The maximum exposure to a physical or chemical agent allowed in an 8-hour workday to prevent disease or injury.

maximum available control technology — The emission standard for sources of air pollution requiring maximum reduction of hazardous emissions, taking cost and feasibility into account.

maximum contaminant level — The maximum permissible level of a contaminant in water that is delivered to the free-flowing outlet of the consumer; unenforceable concentration of a drinking water contaminant that is set at a level at which no known or anticipated adverse effects on human health occur and which allows an adequate safety margin under the Safe Drinking Water Act.

maximum contaminant level goals — Public drinking water standards that serve as unenforceable goals for selected contaminants contained in drinking water that pose no health risk to people over a lifetime of exposure.

maximum credible earthquake — The largest hypothetical earthquake that may be reasonably expected to occur along a given fault or other seismic source produced under the current tectonic setting.

maximum dosage — The largest amount of a chemical that is safe to use without resultant excess residues or damage.

maximum evaporative capacity — The amount of sweat from a person that can be accepted into the surrounding air.

maximum exposure range — An estimate of exposure or dose level received by an individual in a defined population greater than the 98th percentile dose for all individuals in that population but less than the exposure level received by the person at the highest exposure level.

maximum incremental reactivity — A measure of the increase in ozone formation per unit weight of a hydrocarbon when added to the atmosphere.

maximum operating temperature — The maximum temperature at which an instrument or sensor can be safely used.

maximum permissible concentration (MPC) — The recommended maximum average concentration of radionuclides or chemicals to which a worker can be exposed 8 hours a day, 5 days a week, 50 weeks per year.

maximum permissible dose (MPD) — The maximum dose of radiation that may be received by persons working with ionized radiation that produces no detectable damage over a normal life span.

maximum permissible dose equivalent — The greatest dose equivalent that a person or specified part thereof shall be allowed to receive in a given period of time.

maximum permissible exposure (MPE) — The maximum quantity of a chemical a person can be exposed to without causing short-term or long-term disease or injury.

maximum permissible intake — A limit established by the U.S. EPA (usually expressed as milligrams per day for a 60-kg person) used to determine the level of pesticide residues permitted on crops for human consumption.

maximum permissible level (MPL) — The tolerable dose rate of nuclear radiation to which humans can be exposed.

maximum permissible limit (MPL) — See maximum permissible level.

maximum power rating — The maximum power at which a piece of equipment can safely be used.

maximum tolerated dose — The maximum dose that an animal species can tolerate for a major portion of its lifetime without significant impairment or toxic effect other than carcinogenicity.

maximum total trihalomethane potential — The maximum concentration of total trihalomethanes produced in a given water containing a disinfectant residual after 7 days and 25°C or above.

mbar — See millibar.

MBAS — See methylene blue active substance.

Mbbl — An acronym for 1000 barrels.

MBDRF — Acronym for Medical Biological Defense Research Program.

mbps — See megabit.

MCAT — Acronym for Medical Chemical Advisory Team.

MCBAT — Acronym for National Medical Chem-Bio Advisory Team.

MCBC — Acronym for Management of Chemical and Biological Casualties course.

MCBW — Acronym for mass-casualty biological weapon.

MCE — Acronym for maximum credible event; see maximum credible earthquake.

Mcf — Acronym for 1000 cubic feet.

Mcfe — Acronym for 1000 cubic feet of natural gas equivalent.

MCi — See megacurie.

mCi — See millicurie.

MCL — Acronym for maximum contaminant level.

MCLGs — See maximum contaminant level goals.

MCPA ($C_9H_9Cl_2O_3$) — A crystalline solid; MW: 200.52, BP: NA, Sol: dissolves in most organic solvents. It is used as an herbicide. It is moderately toxic to the digestive tract, respiratory tract, and skin and may cause irritation.

MCRT — See mean cell retention time.

MCS — See multiple chemical sensitivity.

MDL — Acronym for methane detection limit; see minimum detection limit.

MDR — See minimum daily requirement.

MDS — Acronym for Modular Decontamination System.

mean — (*statistics*) See arithmetic mean.

mean active repair time — The average time an item may be expected to be out of service for maintenance and repair when the required tools/parts are available.

mean cell retention time — An expression of the average time in days that a microorganism will spend in an activated sludge process.

mean number — (*statistics*) The average number.

mean radiant temperature (MRT) — The temperature of a uniform black enclosure in which a solid body or person would exchange the same amount of radiant heat as in the existing nonuniform environment.

mean sea level — The elevation of the ocean halfway between high and low tide.

mean temperature — The average of the maximum and minimum temperatures of a process.

mean time between failures — The total number of hours equipment is in use divided by the number of failures.

mean time to failure — The average time an item may be expected to function before failure.

mean time to repair — The average time an item may be expected to be out of service for maintenance and repair.

means of egress — An exit or way and method of passage to free and safe ground.

MEAs — Acronym for multilateral environmental agreements.

measure of exposure — A measurable characteristic of a stressor used to help quantify the exposure of an ecological entity or individual organism.

measured variable — A characteristic or component part that is sensed and quantified by a primary element or sensor.

measurement — The numerical representation, direct or derived, of a physical property.

measurement error — (*statistics*) A statistical error made in a study because of poor selection of populations to be studied or the reliability and validity of the study instrument; basic sources of error are random and systematic.

Measurements of Pollution in the Troposphere (MOPITT) — A scanning radiometer employing gas correlation spectroscopy to measure upwelling and reflected infra red gradients in three absorption bands of carbon monoxide and methane. It flies aboard the NASA Earth Observing System and is designed to enhance knowledge of the lower atmosphere and to observe how the atmosphere interacts with the land and ocean biospheres.

meat — The part of the muscle of any cattle, sheep, swine, or goats that is skeletal or which is found in the tongue, diaphragm, heart, or esophagus with or without the accompanying and overlying fat and the portions of bone, skin, sinew, nerve, and blood vessels that normally accompany the muscle tissue not separated in the process of dressing.

meat and poultry irradiation — The use of x-rays, electronic beams, or gamma rays to damage or destroy living organisms that may be present in food products.

meat byproduct — Any part of cattle, sheep, swine, or goats capable of use as human food, other than meat.

MEC — Acronym for Model Energy Code.

mechanical aeration — The use of mechanical energy to inject air into water, thus causing the waste stream to absorb oxygen from the air.

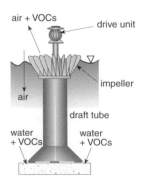

Mechanical Surface Aerator

mechanical composting — A method of composting in which the compost is continuously, mechanically mixed and aerated.

mechanical joint — A flexible device that joins pipes or fittings together by the use of lugs and bolts.

mechanical separation — A mechanical means of separating waste into its various components.

mechanical turbulence — The random irregularities of fluid motion in air caused by buildings or other nonthermal processes.

mechanical ventilation — The air movement caused by a fan or other air moving device.

mechanical weathering — The disintegration of a rock by physical forces, such as freezing and thawing of water in the rock crevices, or the disruption of the rock by plant roots or burrowing animals reducing the rock to small fragments without changing its mineral composition.

mechanism — A process by which something is done or comes into being; a machine or machine-like system.

MED — Acronym for minimal effective dose.

MED$_{50}$ — The dose of a chemical that is minimally effective for mild cognitive impairment in 50% of the exposed population.

MedEvac — Acronym for medical evacuation.

media — A specific environment such as air, water, or soil; soil, water, air, animals, or any other parts of the environment that can contain contaminants.

media filter — A filter containing sand, compost, sand peat, perlite, and zeolite designed to filter particulates, oil, bacteria, or dissolved metals out of storm water runoff as it passes through the filter.

medial tolerance limit (TL$_{50}$) — The concentration of a test material in a suitable diluent (e.g., water) at which 50% of the test animals are able to survive for a specified period of exposure.

median — (*statistics*) The middle value in a population distribution, above and below which lie an equal number of individual values.

median lethal concentration (MLC) — The concentration of a test material that causes the death of 50% of a population within a given period.

median lethal dose (MLD) — The dose of radiation or other material ingested or injected that causes the death of a population of test organisms within a given period. Also known as LD$_{50}$.

median nerve — The nerve that travels through the carpal tunnel of the wrist and services the thumb and first three fingers of the hand; compression of this nerve is called carpal tunnel syndrome.

mediation — The act or process of intervening between conflicting parties to promote reconciliation, settlement, or compromise.

Medicaid — A federally funded, state-operated program of medical assistance to people with low incomes.

medical benefits — Benefits that are usually provided without dollar or time limits in the case of injury on the job or with specific limitations for the normal maintenance of health.

medical care — Provision by a physician of services related to health maintenance, disease prevention, and treatment of illness or injury.

medical center — A healthcare facility.

medical contaminant standards — A level of contaminants that cannot be exceeded.

medical diagnosis — The determination of the cause of a person's illness by using physical examinations, interviews, laboratory tests, review of the person's medical records, and the knowledge of the cause of observed signs and symptoms.

medical engineering — The biomedical engineering and technological concepts used to develop equipment and instruments to improve health care.

medical jurisprudence — The interaction of medicine with criminal and civil law.

medical record — Part of a person's health record that is written by a physician concerning various illnesses or injuries, inoculations, allergies, treatments, prognosis; used by physicians when diagnosing a condition.

medical technologist — A person who under the direction of a physician or medical scientist does specialized chemical, microscopic, and bacteriological tests on blood, tissue, and fluids.

medical waste — Any solid waste that is generated in diagnosis, treatment, research, or production or testing of biologicals in regard to the immunization of human beings or animals.

Medical Waste Tracking Act (MWTA) — An act of Congress amending the Resource Conservation and Recovery Act by adding subtitle J, defining medical waste, and requiring the U.S. EPA to establish a 2-year demonstration program on medical waste identification and disposal in several states.

Medicare — A nationwide health insurance program supported by payroll taxes for people ages 65 and over or for persons under 65 but who have been eligible for Social Security disability payments for over 2 years.

medication — A drug or other substance used as medicine.

medication error — Incorrect administration of a dose of a medication, such as wrong amount, wrong dose, wrong time, outdated medication, overlooked drug allergies or sensitivities, and adverse effects from drug combinations.

medium — The environmental vehicle by which a pollutant is carried to the receptor.

medium texture — (*soils*) Texture exhibited by very fine sandy loams, loams, silt loams, and silts.

medium-size water system — A water system that serves between 3300 and 50,000 people.

MEFR — Acronym for maximal expiratory flow rate.

MEG — A procedure used for checking the insulation resistance on motors, feeders, buss bar systems, grounds, and branch circuit wiring.

mega- (M) — A prefix representing one million.

megabit (Mb) — One million bits; used to measure the transfer rate per second (Mbps) of data.

megacurie (MCi) — A unit of radiation equal to 1 million curies.

megaelectronvolt (MeV) — A unit of energy for expressing the kinetic energy of subatomic particles equal to 1 million electron volts.

megavolt (MV) — A unit of electromotive force equal to 1 million volts.

megavolt per meter (MV/m) — One million volts per meter.

megawatt (MW) — One million watts, or 1000 kilowatts.

megawatt day (MWd) — The amount of energy obtained from 1 megawatt power in one day.

megawatt-hour (MWh) — One million watt-hours of electrical energy.

megohm (MΩ) — One million ohms.

MEI — See maximally exposed individual.

meiosis — The type of cell division in which the chromosomes are reduced to the haploid number during oogenesis and spermatogenesis, producing four reproductive cells.

MEIR — Acronym for medical effects of ionizing radiation.

melanin — A skin pigment that gives the skin its color.

melanoma — A malignant tumor arising from the melanocytic system of the skin, in the eye, or rarely in the mucous membranes of the genitalia, anus, oral cavity, or other sites.

melting — The change of phase of a substance from solid to liquid; see also fusion.

melting point (MP) — The definite temperature at which a solid changes to a liquid.

membrane — A thin layer of tissue that covers a surface, lines a cavity, divides a space, or connects adjacent structures.

membrane filter — A filter medium made from various polymeric materials, such as cellulose, polyethylene, or tetrapolyethylene, and used in collecting microscopic and submicroscopic particles.

membrane microfiltration system — A technique designed to remove solid particles from liquid waste to form filter cakes, typically ranging from 40 to 60% solids.

membrane potential — The difference in electrical polarization or charge between two sides of a membrane or a cell wall.

membranous — Resembling or consisting of a membrane.

M

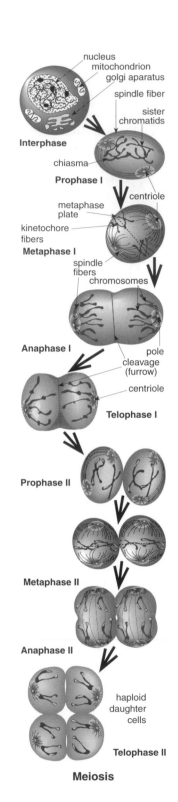

Meiosis

M

memory — (*computer science*) The amount of information a computer is capable of storing.

Ménière's disease — A disorder of the labyrinth of the inner ear with symptoms of tinnitus, heightened sensitivity to loud sounds, progressive loss of hear-

ing, headache, and vertigo. It usually develops after a blow to the head or infection of the middle ear.

meninges — The three membranes covering the brain and spinal cord of vertebrates.

meningitis — Inflammation of the meninges; the term does not refer to a specific disease entity but rather to the pathological condition of inflammation of the tissues of the meninges.

meningococcus — A bacterium, *Neisseria meningitidis*, which is a nonmotile, Gram-negative diplococcus found in the nasopharynx of asymptomatic carriers that may cause septicemia or epidemic cerebrospinal meningitis, usually occurring in crowded conditions.

meniscus — The curved surface of a liquid in a cylinder which is caused by the attraction between a liquid and the walls of a cylinder.

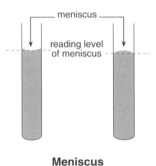

Meniscus

menopause — The period of natural cessation of menstruation occurring usually between the ages of 45 and 50 but varying according to disease or other conditions.

menorrhagia — Excessive menstruation; also known as hypermenorrhea.

menstruation — Periodic discharge of blood from the vagina from a nonpregnant uterus in women from puberty to menopause.

menu — (*computer science*) A set of options that can be used in a computer program.

MEP — Acronym for maximum extent practicable.

meq — See milliequivalent.

meq/l — Scientific notation for milliequivalent per liter.

mercaptoacetic acid ($C_2H_4O_2S$) — A colorless liquid with a disagreeable odor; MW: 92.12, BP: 108°C, Sol: miscible in water and most organic solvents, Fl.P: a noncombustible liquid, density: 1.325 at 20°C. It is used as a reagent for metals analysis and in the manufacture of pharmaceuticals and permanent wave solutions. It is highly toxic and a severe irritant. It can cause inflammation of the cornea and opacity, burns, and necrosis of the skin. ACGIH exposure limit (TLV-TWA): 1 ppm.

$$HS-CH_2-\overset{\overset{\displaystyle O}{\|}}{C}-OH$$

Mercaptoacetic acid

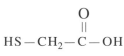

mercurial barometer — A glass tube closed at the top and filled with mercury; the mercury column is supported by the pressure of the air. Units are measured in inches, centimeters, or millibars.

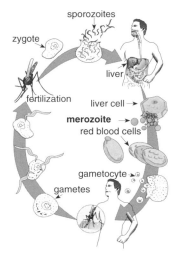

Merozoite

mercuric cyanide (Hg(CN)$_2$) — An odorless, white or colorless crystalline solid; MW: 252.65, BP: NA, Sol: 9.3 g/ml water at 14°C, Fl.P: nonflammable (poisonous gases including cyanide gas and oxides of nitrogen are produced in a fire), sp. gr. 3.996 at 20°C. It is hazardous to the respiratory system, skin, and eyes. Acute exposure may cause eye irritation, skin irritation, irritability, memory loss, personality changes, and brain damage. Chronic exposure may cause skin allergies and kidney damage. OSHA exposure limit (8-hour TWA): 1 mg/m^3 as mercury, 5 mg/m^3 as cyanide.

mercury (Hg) — A silver-white, heavy, odorless liquid; MW: 200.6, BP: 674°F, Sol: insoluble, sp. gr. 13.6. It is used as a liquid cathode in electrolytic production of chlorine and caustic soda from brine; during the manufacture of industrial and medical apparatus; during the manufacture of inorganic and organic compounds for use as pesticides, antiseptics, germicides, and skin preparations; in the preparation of amalgams for use in tooth restorations, chemical processings, and molding operations; during the manufacture of mildew-proof paints and antifouling agents; in the manufacture of batteries, lamps, power tubes, tungsten–molybdenum wire and rods; in the manufacture of inorganic salts for use as catalysts in the production of urethanes, vinyl chloride monomers, anthraquinone derivatives, and other chemicals; as a chemical intermediate and in the manufacture of felt; as a flotation agent in the manufacture of bowling balls; as a laboratory reagent or as a working fluid in instruments; as a conductor during construction of military and nuclear power systems and in air-rectifiers; in the manufacture of explosives; in the preparation of amalgams for use in artificial jewelry; in the manufacture of compounds for pulp and paper industry as controls for biological growths; during the mining and subsequent refining of ore-containing cinnabar. It is hazardous to the skin, respiratory system, central nervous system, kidneys, and eyes and is toxic through inhalation, absorption, and contact. Symptoms of exposure include cough, chest pain, dyspnea, bronchial pneumonitis, tremors, insomnia, irritability, indecision, headache, fatigue, weakness, stomatitis, salivation, gastrointestinal distress, anorexia, weight loss, proteinuria, and irritation of the eyes and skin. OSHA exposure limit (TWA): 0.05 mg/m^3 [skin]. Also known as quicksilver.

mercury–argon calibration lamp — A lamp that produces radiation centered at 253.7 nm, which is then diffused from a diffuser plate. Radiation measurements are made at multiple wavelengths, and possible shifts are noted.

mercury fulminate (Hg(CNO)$_2$) — A white cubic crystal; MW: 284.63, BP: explodes on heating, Sol: slightly soluble in water, Fl.P: powerful explosive that is highly sensitive to impact and heat, density: 3.6 g/cm^3. It is used as a primary explosive to initiate boosters. It is a highly toxic compound. ACGIH exposure limit: NA.

meridian — A great circle of the Earth passing through the poles and any given point on the surface of the Earth.

merozoite — An organism produced from a segmentation of a schizont during the asexual reproductive phase of the lifecycle of a sporozoan.

mesenchyma — A meshwork of embryonic connective tissues in the mesoderm from which the connective tissues of the body and the blood and lymphatic vessels are formed.

mesentery — A double layer of peritoneum connecting the intestines to the posterior abdominal wall.

mesh — One of the openings or spaces in a screen or woven fabric; the number of openings per lineal inch in wire screen.

MESH — An acronym for Medical Subject Headings.

M

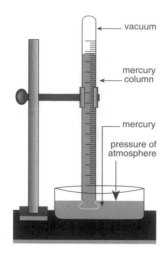

Mercury Barometer

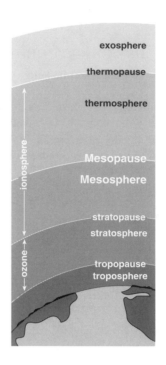

mesityl oxide ((CH₃)₂CCHCOCH₃) — An oily, color-less to light-yellow liquid with a peppermint- or honey-like odor; MW: 98.2, BP: 266°F, Sol: 3%, Fl.P: 87°F, sp. gr. 0.86 at 59°F. It is used as a solvent; as a paint and varnish remover and carbu-retor cleaner; in the preparation of roll-coating inks; during flotation processes for selective bene-faction of ores. It is hazardous to the eyes, skin, respiratory system, and central nervous system and is toxic through inhalation, ingestion, and contact. Symptoms of exposure include narcosis, coma, and irritation of the eyes, skin, and mucous mem-brane; in animals, the central nervous system is affected. OSHA exposure limit (TWA): 15 ppm [air] or 60 mg/m³.

mesocosm — A field-scale model used to understand the interactive relationship of the microbial communi-ties and their roles within their ecosystems.

mesoderm — The middle of the three primary germ lay-ers of an embryo that is the source of bone, muscle, connective tissue, and dermis.

meson — Any unstable, elementary nuclear particle having a mass between that of an electron and a proton.

mesopause — The upper boundary of the mesosphere where the temperature of the atmosphere reaches its lowest point.

mesophiles — A microorganism having an optimal growth temperature between 20 and 45°C.

mesosphere — The region of atmosphere above the stratosphere and below the thermosphere (35 to 60 miles from the Earth) in which temperature decreases with altitude.

mesothelioma — A diffuse, rare form of cancer spread-ing over the surface of the lung and abdominal organs.

mesothorax — Middle portion of the thorax of an insect bearing the second pair of legs and usually a pair of wings.

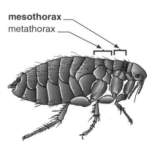

Mesothorax of Flea

mesotrophic — Having a nutrient load resulting in mod-erate productivity.

messenger RNA — The ribonucleic acid produced by deoxyribonucleic acid in the nucleus; functions as a template for protein building in the cytoplasm of a cell.

MET — Acronym for metabolic equivalent.

metabolic balance — An equilibrium between the intake of nutrients and their eventual loss through absorp-tion or excretion.

metabolic disorder — A pathophysiological dysfunction resulting in a loss of metabolic control of homeo-stasis in the body.

metabolic energy — Energy obtained from adenosine triphosphate produced during metabolism.

metabolic enzymes — Protein-based substances that promote change in bodily cells.

metabolic equivalent — The amount of oxygen required while sitting very quietly at rest, which is approximately 3.5 ml of oxygen per kilogram of body weight.

metabolic reaction — The chemical changes in living cells by which energy is provided for vital processes.

metabolism — The sum of all the physical and chemical processes in the buildup and destruction of protoplasm; the transformation during which energy is made available for use by the organism.

metabolite — Any product of metabolism, especially a transformed chemical.

metacycle — Stage in the life cycle of a parasite during which it is infective to its definitive host.

metacyst — The cystic stage of a parasite that is infective to a host.

metadata — Information about data.

metal — A substance with the usual characteristic properties of luster, high strength, and good conductivity of heat and electricity and whose oxide combines with water to form a base. The bonding is provided by a large group of mobile electrons.

metal fume fever — An acute occupational condition caused by a brief high exposure to the freshly generated fumes and particles of metals; also known as fume fever.

metalimnion — The middle layer of a thermally stratified lake or reservoir where there is a rapid decrease in temperature with depth; thermocline.

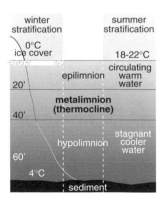

Metalimnion

metallurgy — The theoretical and applied sciences of the nature and uses of metals.

metamorphic rock — Rock that has been changed from its original form (igneous or sedimentary) by great heat, pressure, or chemically active fluids; marble and slate are examples.

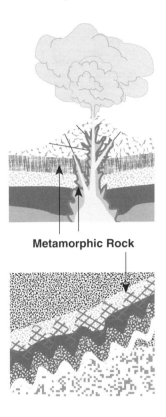

Metamorphic Rock

metamorphism — (*geology*) The process of change that rocks undergo when exposure to increasing temperatures and pressures causes their mineral components to become unstable.

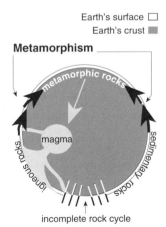

metamorphosis — A marked rapid change in structure of an animal during its growth, as it changes from larva to pupa and pupa to adult in the life cycle of many invertebrates and amphibians.

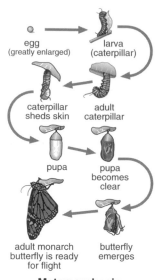

Metamorphosis
of *Danaus plexippus*
(Monarch butterfly)

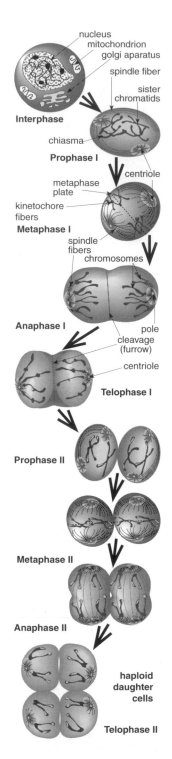

Metaphase

metaphase — The stage of mitosis or meiosis during which the chromosomes are aligned on the equatorial plate.

metastasis — Transfer of disease from one organ or part of the body to another not directly connected with it; may be due to either the transfer of pathogenic microorganisms or the transfer of cells, such as in malignant tumors.

metathorax — The posterior portion of the thorax of an insect bearing the third pair of legs and usually a pair of wings.

metazoan — A group of animals for which the body is composed of cells differentiated into tissues and organs and having a digestive tract.

Meteor-3 — The third in a series of weather satellites launched by the former Soviet Union and carrying a payload, including a Total Ozone Mapping Spectrometer.

meteoric water — Groundwater that originates as precipitation.

meteorology — The science of weather-related phenomena.

meter (m) — The international standard unit of length equal to 39.37 inches, 3.2808 feet, or 1.0936 yards.

meter–kilogram–second system (MKS system) — A metric system of units devised in 1901 that, with some modifications, forms the basis of the International System of Units.

meter–kilogram–second–ampere system — A system of electrical and mechanical units in which length (m), mass (kg), time (sec), and electrical current (A) are the fundamental quantities.

metering pump — (*milk pasteurization*) A pump that is sealed at its maximum speed and wired so it oper-

ates only if the flow diversion valve is fully operable; used to move milk under pressure to the holding tube and the remainder of the high-temperature, short-time pasteurization process.

methacrylic acid (C₄H₆O₂) — A colorless liquid with an acrid odor; MW: 86.10, BP: 163°C, Sol: soluble

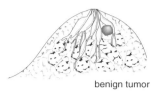

benign tumor

malignant tumor

Metastasis
from mammary gland

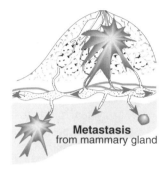

mesothorax
metathorax

Metathorax of Flea

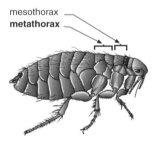

in water and most organic solvents, Fl.P: 76°C (open cup), density: 1.015 at 20°C. Used in the manufacture of methacrylate resins and plastics. It is hazardous to the eyes and skin. It may cause blindness and produce burns of the eyes and skin. ACGIH exposure limit (TLV-TWA): 20 ppm.

$$CH_2 = \overset{\overset{\textstyle CH_3}{|}}{C} - \overset{\overset{\textstyle O}{\|}}{C} - OH$$

Methacrylic Acid
(2–methylacrylic acid)

methane (CH₄) — A colorless, odorless, simple, saturated hydrocarbon that comes from the decomposition of organic material and is the chief component of natural gas; also known as marsh gas.

methane flux — The continuous flow of methane from one place in the atmosphere to another.

methane-producing bacteria — See methanogen.

methanogen — A bacterium that anaerobically oxidizes hydrogen to methane and water using carbon dioxide as the electronic acceptor; found in anaerobic muds, ponds, and sewage sludge.

methanogenesis — The biological production of methane.

methanoic acid — See formic acid.

methanol — See methyl alcohol.

methanotroph — An aerobic bacteria that can use methane as a sole source of carbon.

methanotrophic bioreactor system — An above-ground remedial technology for water contaminated with halogenated hydrocarbons, trichloroethylene, and related compounds.

methemoglobin — A compound formed from hemoglobin by oxidation of the iron atom from the ferrous to ferric state; also known as ferrihemoglobin.

methemoglobinemia — Methemoglobin in the blood usually due to the toxic action of drugs or other agents or to the hemolytic process; especially a problem in children up to age 5. Nitrates are converted in the body to nitrites by bacterial action, and the nitrites combine with hemoglobin, thereby causing the hemoglobin to lose its ability to carry oxygen.

methicillin — An antibiotic that is a variation of penicillin.

methicillin-resistant *Staphylococcus aureus* (MRSA) — A Gram-positive bacterium growing in clusters not affected by the antibiotic methicillin and many other antibiotics.

methiocarb (C₁₁H₁₅NO₂S) — A crystalline or powder solid; MW: 225.33; BP: NA; Sol: dissolves in methylene chloride, chloroform, and toluene; density: NA. It is used as a pesticide. It is hazardous to the digestive tract, skin, and respiratory tract and may cause acute, delayed, or chronic effects including salivation, bradycardia, blurred vision, labored breathing, headache, muscle twitching, tremor, nausea, abdominal pain, diarrhea, convulsions, coma, and death.

Methiocarb

method detection limit (EPA) — The limit at which an agent can be detected, taking into account the sample matrix, reagents used, and preparation steps applied to the sample in the analytical methods.

methomyl (C$_5$H$_{10}$N$_2$O$_2$S) — A white crystalline solid with a slight sulfur odor; MW: 162.23, BP: NA, Sol: dissolves in most organic solvents and has a moderate solubility in water, density: 1.295 at 25°C. It is used as a carbamate insecticide and is available as a wettable powder or liquid. It is hazardous to the respiratory system, digestive system, skin, and eyes. Acute exposure causes central nervous system depression, stupor, coma, seizures, tachycardia, hypertension, urinary incontinence, diarrhea, gastrointestinal cramping, blurred vision, nausea, vomiting, irritation of the eyes and skin, respiratory failure, and death. No chronic exposure health effects are known. OSHA exposure limit (8-hour TWA): 2.5 mg/m^3.

$$CH_3-S=N-O-\overset{\overset{\displaystyle O}{\|}}{C}-NH-CH_3$$

with S—CH$_3$ on the CH$_3$—S group.

Methomyl

methoprene (C$_{19}$H$_{34}$O$_3$) — (*insecticides*) An amber liquid that is an insect growth regulator and true analog of the mosquito's own juvenile hormone; MW: 310, Sol: 1 to 39 ppm in water, density: 8.56 lb/gal. It is used as a larvacide for mosquitos. No toxic effect is noted on non-target organisms. It is a biodegradable compound of very low persistence.

methoxychlor (Cl$_3$CCH(C$_6$H$_4$OCH$_3$)$_2$) — Colorless to yellow crystals with a slight, fruity odor; MW: 345.7, BP: decomposes, Sol: insoluble, Fl.P: not known, sp. gr. 1.41 at 77°F. It is used as an agricultural insecticide and is an analog of DDT. It is not known which organs are harmed by methoxychlor in humans, but it is toxic through inhalation and ingestion in animals. No known symptoms of exposure in humans; symptoms of exposure in animals include fasciculation, trembling, convulsions, kidney, and liver damage; carcinogenic. OSHA exposure limit (TWA): 10 mg/m^3 [air].

Methoxychlor

methyl acetate (CH$_3$COOCH$_3$) — A colorless liquid with a fragrant, fruity odor; MW: 74.1, BP: 135°F, Sol: 30%, Fl.P: 14°F, sp. gr. 0.93. It is used as a solvent for nitrocellulose and cellulose acetate; as a chemical intermediate; in the preparation of artificial leather; during the formulation of lacquers, paints, perfumes, and vinyl resin coatings. It is hazardous to the respiratory system, skin, and eyes and is toxic through inhalation, ingestion, and contact. Symptoms of exposure include headache, drowsiness, optic atrophy, and irritation of the nose and throat. OSHA exposure limit (TWA): 200 ppm [air] or 610 mg/m^3.

methyl acetylene (CH$_3$CCH) — A colorless gas that has a sweet odor and is shipped as a liquefied, compressed gas; MW: 40.1, BP: –10°F, Sol: insoluble. It is used during the synthesis of pharmaceuticals and aromatics and is liberated during high-temperature gas welding operations. It is hazardous to the central nervous system and is toxic through inhalation. Symptoms of exposure include hyperexcitability, tremors, anesthesia, and respiratory system irritation. OSHA exposure limit (TWA): 1000 ppm [air] or 1650 mg/m^3.

methyl acetylene–propadiene mixture (CH$_3$CCH/ CH$_2$CCH$_2$) — A colorless gas that has a strong, characteristic, foul odor and is shipped as a liquefied gas; MW: 40.1, BP: –36 to –4°F, Sol: insoluble. It is liberated from flame hardening, metalizing, brazing, welding, and cutting. It is hazardous to the central nervous system, skin, and eyes and is toxic through inhalation and contact. Symptoms of exposure include frostbite, disorientation, and excitement. OSHA exposure limit (TWA): 1000 ppm [air] or 1800 mg/m^3.

methyl (*n*-amyl) ketone (CH$_3$CO(CH$_2$)$_4$CH$_3$) — A colorless to white liquid with a banana-like, fruity odor; MW: 114.2, BP: 305°F, Sol: 0.4%, Fl.P: 102°F, sp. gr. 0.81. It is used in the preparation of synthetic resins, especially for metal coating, and as a solvent for rubber and nitrocellulose. It is hazardous to the eyes, skin, respiratory system, central nervous system, and peripheral nervous system and is toxic through inhalation, ingestion, and contact. Symptoms of exposure include headache, narcosis, coma, dermatitis, and irritation of the eyes and mucous membrane. OSHA exposure limit (TWA): 100 ppm [air] or 465 mg/m^3. Also known as 2-heptanone.

methyl acrylate (CH$_2$:CHCOOCH$_3$) — A colorless liquid with an acrid odor; MW: 86.1, BP: 176°F, Sol: 6%, Fl.P: 25°F, sp. gr. 0.96. It is used in the preparation of thermoplastic coatings; in the manufacture of acrylic fibers; in the synthesis of higher acrylates; during preparation of adhesives and sealants based on methyl acrylate; during polymerization in aqueous emulsions; during preparation of amphoteric surfactants for use in hair shampoos. It is hazardous to the respiratory system, eyes, and skin and is toxic through inhalation, absorption, and contact. Symptoms of exposure include irritation of the eyes, skin, and upper respiratory system. OSHA exposure limit (TWA): 10 ppm [skin] or 35 mg/m^3.

methyl alcohol (CH$_3$OH) — A colorless liquid with a characteristic pungent odor; MW: 32.1, BP: 147°F, Sol: miscible, Fl.P: 52°F, sp. gr. 0.79. It is used as a solvent for rotogravure inks, aniline dyes, and duplicator fluids; in the plastics industry to produce plasticizers, softening agents, and acrylic resins; as a solvent in the rubber industry. It is liberated during the application of surface coatings. It is hazardous to the eyes, skin, central nervous system, and gastrointestinal tract and is toxic through inhalation, absorption, ingestion, and contact. Symptoms of exposure include headache, drowsiness, lightheadedness, nausea, vomiting, visual disturbances, blindness, eye irritation. Also known as methanol; wood alcohol. OSHA exposure limit (TWA): 200 ppm [skin] or 260 mg/m^3.

methyl bromide (CH$_3$Br) — A colorless gas that has a chloroform-like odor at high concentrations, is a liquid below 38°F, and is shipped as a liquefied compressed gas; MW: 95.0, BP: 38°F, Sol: 2%, Fl.P: not known, sp. gr. 1.73. It is used as a space and soil fumigant in agriculture and industry; in food sterilization for pest control in fruits, vegetables, dairy products, nuts, and grains; in organic synthesis as a methylating agent; for preparation of quaternary ammonium compounds and organotin derivatives; as a selective solvent in aniline dyes; in laboratory procedures. It is hazardous to the central nervous system, skin, eyes, and respiratory system and is toxic through inhalation, absorption, ingestion, and contact. Symptoms of exposure include headache, visual disturbances, vertigo, nausea, vomiting, malaise, hand tremors, convulsions, dyspnea, vesiculation, irritation of the skin and eyes; carcinogenic. OSHA exposure limit (TWA): 5 ppm [skin] or 20 mg/m^3.

methyl Cellosolve® (CH$_3$OCH$_2$CH$_2$OH) — A colorless liquid with a mild, ether-like odor; MW: 76.1, BP: 256°F, Sol: miscible, Fl.P: 102°F, sp. gr. 0.96. It is used in printing on plastic materials, rotogravure printing, and cellulose acetate pigment printing on textiles; in the manufacture of surface coatings and dyeing agents for stains, lacquers, dyes, and inks; as an anti-icing agent in military jet fuel; as an antistall additive in gasoline. It is liberated during the sealing of moisture-proof cellophane wrappers and packaging. It is hazardous to the central nervous system, blood, skin, eyes, and kidneys and is toxic through inhalation, absorption, ingestion, and contact. Symptoms of exposure include headache, drowsiness, weakness, ataxia, tremors, somnolence, anemic pallor, and eye irritation; carcinogenic. OSHA exposure limit (TWA): 25 ppm [skin] or 80 mg/m^3.

methyl Cellosolve® acetate (CH$_3$COOCH$_2$CH$_2$OCH$_3$) — A colorless liquid with a mild, ether-like odor; MW: 118.1, BP: 293°F, Sol: miscible, Fl.P: 120°F, sp. gr. 1.01. It is used as a solvent during the manufacture and spray or heat applications of surface coatings and adhesives; in the manufacture of photographic film; during drycleaning operations. It is hazardous to the kidneys, brain, central nervous system, and peripheral nervous system and is toxic through inhalation, absorption, ingestion, and contact. Symptoms of exposure include kidney and brain damage and eye irritation; in animals, narcosis, and irritation of the eyes, nose, and throat. OSHA exposure limit (TWA): 25 ppm [skin] or 120 mg/m^3.

methyl chloride (CH$_3$Cl) — A colorless gas that is shipped as a liquefied compressed gas and has a faint, sweet odor that is not noticeable at dangerous concentrations; MW: 50.5, BP: –12°F, Sol: 0.5%. It is used in the manufacture of silicone resins and tetramethyl lead; as a methylating and chlorinating agent; as a dewaxing agent in petroleum refining; as a catalyst solvent in the production of butyl rubber; in the synthesis of a variety of other compounds; as an extractant for greases, oils, and resins; in the manufacture of pesticides, pharmaceuticals, and perfumes; as a propellant in aerosols; as a refrigerant. It is hazardous to the central nervous system, liver, kidneys, and skin and is toxic through inhalation and contact. Symptoms of exposure include dizziness, nausea, vomiting, visual disturbances, staggering, slurred speech, convulsions, coma, liver and kidney damage, and frostbite; carcinogenic. OSHA exposure limit (TWA): 50 ppm [air] or 105 mg/m^3. Also known as chloromethane.

methyl chloroform (CH$_3$CCl$_3$) — A colorless liquid with a mild, chloroform-like odor; MW: 133.4, BP: 165°F, Sol: 0.4%, Fl.P: none, sp. gr. 1.34. It is used as a solvent in the cold cleaning of metals and plastics; in vapor degreasing; in ultrasonic

M

cleaning; in the dyeing and cleaning of fabrics and yarns; in organic synthesis; in polymer manufacture; as a primary and carrier solvent in spot cleaners, adhesives, shoe polishes, stain repellents, hair sprays, Mace™, insecticides, resins, inks, lubricants, protective coatings, asphalt extraction, and wastewater treatment; in aerosol manufacture; as a coolant; during printed circuit boards production; in liquid Drano® production; in photographic film processing. It is hazardous to the skin, central nervous system, cardiovascular system, and eyes and is toxic through inhalation, ingestion, and contact. Symptoms of exposure include headache, lassitude, central nervous system depression, poor equilibrium, dermatitis, cardiac arrhythmias, eye irritation. Also known as trichloroethane. OSHA exposure limit (TWA): 350 ppm [air] or 1900 mg/m³.

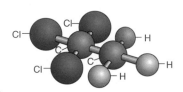

Methyl Chloroform
(1,1,1–Trichloroethane)

3-methyl-4-chlorophenol (C_7H_7ClO) — An odorless, white or slightly pink crystal; MW: 142.58, BP: 235°C, Sol: 0.385 g/100 g water at 20°C, Fl.P: NA (container may explode violently in heat or fire; also volatile with steam), sp. gr. NA. It is used as a preservative for gums, glues, paints, inks, cosmetics, textiles, leather goods; as a topical antiseptic in veterinary medicine. It is hazardous to the respiratory system, digestive system, eyes, and skin. Acute exposure may cause a burning feeling; irritation of the skin, nose, throat, and lungs; shortness of breath; nausea; vomiting; diarrhea; arrhythmias; pulmonary edema; seizures; coma; and liver and renal damage. Chronic exposure may cause damage to the kidneys, liver, brain, pancreas, lungs, and heart muscle; possible human carcinogen. OSHA exposure limit (8-hour TWA): NA.

methyl ethyl ketone (C_4H_8O) — A colorless liquid with an acetone-like odor; MW: 72.11; BP: 79.6°C; Sol: soluble in water, alcohol, ether, acetone, and benzene; Fl.P: –6°C (closed cup), a flammable liquid that can travel to an ignition source and flash back; density: 0.8054 at 20°C. It is used in the manufacture of smokeless powder and colorless synthetic resins; as a solvent and surface coating. It is hazardous to the digestive tract, respiratory tract, eyes, and nose. It causes irritation, headaches, a narcotic

effect, dizziness, and vomiting. ACGIH exposure limit (TLV-TWA): 200 ppm.

$$CH_3-C-CH_2-CH_3$$
$$\underset{O}{\overset{\|}{}}$$

Methyl ethyl ketone

methyl ethyl ketone peroxide ($C_6H_{16}O_4$, monomeric acrylic; $C_8H_{18}O_6$, polymeric acrylic) — A crystalline solid; MW: a mixture of monomeric ($C_8H_{16}O_4$) and polymeric ($C_8H_{18}O_6$) acrylic, BP: NA, Sol: soluble in most organic solvents, Fl.P: an extremely hazardous substance, density: NA. It is used to initiate the polymerization of ethylene and styrene.

methyl formate ($HCOOCH_3$) — A colorless liquid with a pleasant odor; a gas above 89°F; MW: 60.1, BP: 89°F, Sol: 30%, Fl.P: –2°F, sp. gr. 0.98. It is used as an insecticide and larvicide for fumigating dried fruits, nuts, tobacco, cereals, and infected clothing; as a general industrial solvent for greases, fatty acids, cellulose acetate, collodion, and celluloid; in organic synthesis. It is hazardous to the eyes, respiratory system, and central nervous system and is toxic through inhalation, absorption, ingestion, and contact. Symptoms of exposure include chest depression, dyspnea, visual disturbances, central nervous system depression, and irritation of the eyes and nose; in animals, pulmonary edema. OSHA exposure limit (TWA): 100 ppm [air] or 250 mg/m³.

5-methyl-3-heptanone ($CH_3CH_2CO(CH_2)_4CH_3$) — A colorless liquid with a pungent odor; MW: 128.2, BP: 315°F, Sol: insoluble, Fl.P: 138°F, sp. gr. 0.82. It is used in the preparation of perfumes; as a specializing solvent in cleaning operations. It is hazardous to the eyes, skin, respiratory system, and central nervous system and is toxic through inhalation, ingestion, and contact. Symptoms of exposure include headache, narcosis, coma, dermatitis, and irritation of the eyes and mucous membrane. OSHA exposure limit (TWA): 25 ppm [air] or 130 mg/m³.

methyl hydrazine (CH_3NHNH_2) — A fuming, colorless liquid with an ammonia-like odor; MW: 46.1, BP: 190°F, Sol: miscible, Fl.P: 17°F, sp. gr. 0.87 at 77°F. It is used in the preparation of liquid rocket propellants; as a chemical intermediate for the synthesis of pesticides; in the polymerization technology. It is hazardous to the central nervous system, respiratory system, liver, blood, eyes, and cardiovascular system and is toxic through inhalation, absorption, ingestion, and contact. Symptoms of exposure include vomiting, diarrhea, tremors, ataxia, anoxia, cyanosis, convulsions, and irritation of the eyes and respiratory system; carcinogenic. OSHA exposure limit (TWA): ceiling 0.2 ppm [skin] or 0.35 mg/m³.

methyl iodide (CH₃I) — A colorless liquid that has a pungent, ether-like odor and turns yellow, red, or brown on exposure to light and moisture; MW: 141.9, BP: 109°F, Sol: 1%, sp. gr. 2.28. It is used as a methylating agent in organic synthesis; as a laboratory reagent; as an insecticidal fumigant; in analytical chemistry. It is hazardous to the central nervous system, skin, and eyes and is toxic through inhalation, absorption, ingestion, and contact. Symptoms of exposure include nausea, vomiting, vertigo, ataxia, slurred speech, drowsiness, dermatitis, and eye irritation; carcinogenic. OSHA exposure limit (TWA): 2 ppm [skin] or 10 mg/m³.

methyl isoamyl ketone (C₇H₁₄O) — A colorless liquid with a sweet odor; MW: 114.21; BP: 144°C; Sol: soluble in alcohol, ether, acetone, and benzene; Fl.P: 36°C (closed cup), a flammable liquid; density: 0.888 at 20°C. It is used as a solvent for polymers, cellulose, esters, and acrylics. It is hazardous to the respiratory tract and eyes and can cause a narcotic effect and irritation. ACGIH exposure limit (TLV-TWA): 50 ppm.

$$CH_3-\underset{\underset{O}{\overset{\|}{}}}{C}-CH_2-CH_2-CH\Big\backslash{\overset{CH_3}{\underset{CH_3}{}}}$$

Methyl isoamyl ketone

methyl isobutyl carbinol ((CH₃)₂CHCH₂CH(OH) CH₃) — A colorless liquid with a mild odor; MW: 102.2, BP: 271°F, Sol: 2%, Fl.P: 106°F, sp. gr. 0.81. It is used as an extractant in dewaxing of mineral oils; in froth flotation of various minerals; as an extractant in the manufacture of antibiotics; in the synthesis of surfactants; in the preparation of lubricating oil additives. It is hazardous to the eyes and skin and is toxic through inhalation, absorption, ingestion, and contact. Symptoms of exposure include headache, drowsiness, dermatitis, and irritation of the eyes. OSHA exposure limit (TWA): 25 ppm [skin] or 100 mg/m³.

methyl isobutyl ketone (CH₃COCH₂CH(CH₃)₂) — A colorless liquid with a pleasant odor; MW:100.16, BP: 116.8°C, Sol: 1.91 g/100 g water at 20°C, Fl.P: 24°C, sp. gr. 0.798. It is used as a solvent for coating systems; in rare-metal extraction; as a solvent and denaturant for other miscellaneous appli-

cations. It is hazardous to the respiratory system, skin, and eyes. Acute exposure may irritate the eyes, nose, throat, skin and cause loss of appetite, nausea, vomiting, diarrhea, dizziness, lightheadedness, and unconsciousness. Chronic exposure may cause skin irritation, and damage to the liver and kidneys. OSHA exposure limit (8-hour TWA): 205 mg/m³.

methyl isocyanate (CH₃NCO) — A colorless liquid with a sharp, pungent odor; MW: 57.1, BP: 139°F, Sol: 10% at 59°F, Fl.P: 19°F, sp. gr. 0.96. It is used as a cross-linking agent; as an additive in polymer technology; in organic synthesis. It is hazardous to the respiratory system, eyes, and skin and is toxic through inhalation, absorption, ingestion, and contact. Symptoms of exposure include cough, secretions, chest pain, dyspnea, asthma, eye and skin injury, and irritation of the eyes, nose, and throat; in animals, pulmonary edema. OSHA exposure limit (TWA): 0.02 ppm [skin] or 0.05 mg/m³.

methyl mercaptan (CH₃SH) — A colorless gas that has a disagreeable odor like garlic or rotten cabbage, is a liquid below 43°F, and is shipped as a liquefied compressed gas; MW: 48.1, BP: 43°F, Sol: 2%, Fl.P (open cup): 0°F, sp. gr. 0.90. It is used as a catalyst and activator; in wood processing; in the synthesis of chemical intermediates for the manufacture of resins, plastics, insecticides, and pressure-sensitive and oil-resistant adhesives; as an odorant and warning agent in natural gas; in jet fuels. It is hazardous to the respiratory system and central nervous system and is toxic through inhalation and contact. Symptoms of exposure include narcosis, cyanosis, convulsions, and pulmonary irritation. OSHA exposure limit (TWA): 0.5 ppm [air] or 1 mg/m³.

methyl methacrylate (CH₂C(CH₃)COOCH₃) — A colorless liquid with an acrid, fruity odor; MW: 100.1, BP: 330°F, Sol: slight, Fl.P (open cup): 50°F, sp. gr. 0.94. It is used during the casting of acrylic sheets; during polymerization to produce molding and extruding powders; during the manufacture of unsaturated polyester resins; during production of emulsion polymers for use in adhesives, sealants, fabrics, sizes, leather finishes, paper coatings, and polishes. It is hazardous to the eyes, upper respiratory system, and skin and is toxic through inhalation, ingestion, and contact. Symptoms of exposure include dermatitis and irritation of the eyes, nose, and throat. OSHA exposure limit (TWA): 100 ppm [air] or 410 mg/m³.

methyl orange alkalinity — A measure of the total alkalinity in a water sample in which the color of methyl orange reflects the change in level.

M

methyl parathion (C$_8$H$_{10}$NO$_5$PS) — White solid crystals or a brownish liquid with the odor of rotten eggs; MW: 263.23, BP: not known (decomposes above ambient temperature), decomposes violently at 120°C, Sol: 50 ppm at 25°C, Fl.P: not known, density: 1.358 at 20°C. It is used as a broad-spectrum, nonsystemic contact and stomach insecticide with some fumigant action and is used to control insects on a wide variety of crops. It is hazardous to the lungs, pulmonary system, liver, kidneys, and heart and is toxic through inhalation and contact. Symptoms of exposure include pulmonary edema, bronchoconstriction, and hypersecretion of bronchial glands, heart, liver, kidneys, blood vessel lesions, cholinesterase, depression, lethargy, miosis, depressed plasma, and chromosomal abberations. OSHA exposure limit (TWA): 0.2mg/m^3 [air].

methyl propyl ketone (C$_5$H$_{10}$O) — A colorless liquid with a characteristic pungent odor; MW: 86.13; BP: 102.2°C; Sol: soluble in alcohol and ether but slightly soluble in water; Fl.P: 7°C (closed cup), a flammable liquid; density: 0.809 at 20°C. It is used as a solvent in organic synthesis and as a flavoring agent. It is hazardous to the eyes and respiratory tract and can cause narcosis and irritation. ACGIH exposure limit (TLV-TWA): 200 ppm.

$$CH_3 - \underset{\underset{O}{\|}}{C} - CH_2 - CH_2 - CH_3$$

Methyl propyl ketone

methyl salicylate (C$_8$H$_8$O$_3$) — A colorless to yellowish oily liquid; MW: 152.16, BP: 220 to 224°C, Sol: slightly soluble in water and soluble in organic solvents, Fl.P: noncombustible liquid, density: 1.184 at 25°C. It is used in perfume and as a flavoring agent. It is a highly toxic compound and hazardous to the digestive tract and respiratory tract. It causes symptoms of nausea, vomiting, gastritis, diarrhea, respiratory stimulation, pulmonary edema, convulsions, and coma. Ingestion of 15 to 25 ml may be fatal to humans.

Methyl salicylate

α-methyl styrene (C$_6$H$_5$C(CH$_3$)CH$_2$) — A colorless liquid with a characteristic odor; MW: 118.2, BP: 330°F, Sol: insoluble, Fl.P: 129°F, sp. gr. 0.91. It is used in the manufacture of specialized α-methyl styrene polyester and alkyd surface-coating resins; in the manufacture of certain plasticizers in varnishes, adhesives, and plastics; in the manufacture of carbolic acid with cumene peroxidation process. It is hazardous to the eyes, respiratory system, and skin and is toxic through inhalation, ingestion, and contact. Symptoms of exposure include drowsiness, dermatitis, irritation of the eyes, nose, and throat. OSHA exposure limit (TWA): 50 ppm [air] or 240 mg/m^3.

methylacrylonitrile (C$_3$H$_5$CN) — A colorless liquid; MW: 67.10; BP: 90°C; Sol: soluble in common organic solvents; Fl.P: 13°C (open cup), the vapor forms an explosive mixture with air; density: 0.8001 at 20°C. It is used in the preparation of acids, amines, amides, and esters. It is hazardous to the respiratory tract, digestive tract, and skin. Exposure to high concentrations can result in asphyxia and death. ACGIH exposure limit (TLV-TWA): 1 ppm [skin].

$$H_2C \equiv C - C \equiv N$$
$$|$$
$$CH_3$$

Methylacrylonitrile

methylal (CH$_3$OCH$_2$OCH$_3$) — A colorless liquid with a chloroform-like odor; MW: 32.1, BP: 111°F, Sol: 33%, Fl.P: −26°F (open cup), sp. gr. 0.86. It is used as a solvent for adhesives, resins, gums, waxes, and protective coatings; as a solvent for the extraction of alkaloids, barbiturates, organic acids, and hydroxyacids; in the manufacture of artificial resins; as a gasoline and diesel fuel additive; as a special fuel for rocket and jet engines; in the manufacture of perfume; as a methylating agent or chemical intermediate. It is hazardous to the skin, respiratory system, and central nervous system and is toxic through inhalation, ingestion, and contact. Symptoms of exposure include anesthesia, irritation of the skin, and mild irritation of the eyes and upper respiratory system. OSHA exposure limit (TWA): 1000 ppm [air] or 3100 mg/m^3.

methylamine (CH$_3$NH$_2$) — A colorless gas that has a fish- or ammonia-like odor, is a liquid below 21°F, and is shipped as liquefied compressed gas; MW: 31.1, BP: 21°F, Sol: soluble, sp. gr. 0.70. It is used as a chemical intermediate in the production of insecticides, herbicides, and fungicides; in the production of surfactants; in the production of rocket fuels and explosives; in the production of pharmaceuticals and photographic chemicals; as an intermediate for dyes, textiles, dye assists, rubber, and anticorrosive chemicals; as a polymerization inhibitor of hydrocarbons; in paint removers; to prevent webbing in natural and synthetic latex. It is hazardous to the respiratory system, skin, and eyes and is toxic through inhalation, absorption, ingestion, and contact. Symptoms of exposure include cough, dermatitis, conjunctivitis, skin and mucous membrane burns, and irritation of the eyes and respiratory system. OSHA exposure limit (TWA): 10 ppm [air] or 12 mg/m³.

methylation — A chemical process for the combination of a methyl radical with an organic compound.

methylbutyl ketone ($C_6H_{12}O$) — A colorless liquid with a pungent odor; MW: 100; BP: 127°C; Sol: soluble in alcohol, ether, and benzene; Fl.P: 25°C (closed cup), a flammable liquid that may travel some distance from an ignition source and flash back; density: 0.821 at 20°C. It is used as a solvent for nitrocellulose, resins, lacquers, oils, fats, and waxes. It is hazardous to the respiratory tract and skin. It causes peripheral nervous system and neuropathic diseases, muscle weakness, difficulty in grasping heavy objects, and severe irritation of the eyes and nose. ACGIH exposure limit: NA.

$$CH_3-C-CH_2-CH_2-CH_2-CH_3$$
$$O$$
Methyl butyl ketone

methylcyclohexane ($CH_3C_6H_{11}$) — A colorless liquid with a faint, benzene-like odor; MW: 98.2, BP: 214°F, Sol: insoluble, Fl.P: 25°F, sp. gr. 0.77. It is used as a diluent solvent for cellulose during spray application of lacquers; during the manufacture of rotogravure inks; as a solvent for oils, fats, waxes, and rubber; in industrial organic synthesis in the hydro-reforming process; as a degreasing agent. It is hazardous to the respiratory system and skin and is toxic through inhalation, ingestion, and contact. Symptoms of exposure include lightheadedness, drowsiness, and irritation of the skin, nose, and throat. OSHA exposure limit (TWA): 400 ppm [air] or 1600 mg/m³.

methylcyclohexanol ($CH_3C_6H_{10}OH$) — A straw-colored liquid with a weak odor like coconut oil; MW: 114.2, BP: 311 to 356°F, Sol: 4%, Fl.P: 154°F, sp. gr. 0.92. It is used as a solvent in lacquers, oils,

gums, waxes, and resins; as an auxiliary solvent for drycleaning; in soap manufacture; in the textile industry as a blending agent in textile soaps; as a degreaser; as an antioxidant in lubricants. It is hazardous to the respiratory system, skin, and eyes in humans; in animals, it is hazardous to the central nervous system, liver, and kidneys. It is toxic through inhalation, absorption, ingestion, and contact. Symptoms of exposure include headache and irritation of the eyes and upper respiratory system; in animals, narcosis and liver and kidney damage. OSHA exposure limit (TWA): 50 ppm [air] or 235 mg/m³.

o-methylcyclohexanone ($CH_3C_6H_9O$) — A colorless liquid with a weak peppermint-like odor; MW: 112.2, BP: 325°F, Sol: insoluble, Fl.P: 118°F, sp. gr. 0.93. It is used as a solvent in the plastics industry; in the manufacture of lacquers and varnishes; as a co-solvent with cyclohexanone; in the leather industry; as a rust remover. In animals, it is hazardous to the respiratory system, liver, kidneys, and skin and is toxic through inhalation, absorption, ingestion, and contact. Symptoms of exposure in animals include narcosis, dermatitis, and irritation of the eyes and mucous membrane. OSHA exposure limit (TWA): 50 ppm [skin] or 230 mg/m³.

methylene-bis-(4-cyclohexylisocyanate) ($(C_6H_{10})_2CH_2$ (NCO)$_2$) — A colorless liquid; MW: 262.39, BP: NA, Sol: soluble in most organic solvents and decomposes in water and ethanol, Fl.P: noncombustible, 100°C, density: 1.07. It is used to produce urethane foam with color stability. It is hazardous to the respiratory tract and skin and may cause irritation of the respiratory tract, tremors, convulsions, congestion of lungs, and edema in rats. ACGIH exposure limit (TLV-TWA): 0.0535 mg/m³.

N=C=O

CH$_2$

N=C=O

Methylene bis(4-cyclohexylisocyanate)

methylene bisphenyl isocyanate (CH$_2$(C$_6$H$_4$NCO)$_2$) —
White to light-yellow, odorless flakes; a liquid above 99°F; MW: 250.3, BP: 342°F, Sol: 0.2%, Fl.P: 396°F (open cup), sp. gr. (122°F): 1.19. It is liberated during the manufacture of lacquer; the production of component chemicals for foam systems; the casting of high-density polyurethane elastomers; the application of lacquer and sealant finishes; flame lamination of fabrics. It is hazardous to the respiratory system and eyes and is toxic through inhalation, ingestion, and contact. Symptoms of exposure include cough, pulmonary secretions, chest pain, dyspnea, asthma, and irritation of the eyes, nose, and throat. OSHA exposure limit (TWA): ceiling 0.2 mg/m^3 [air].

methylene blue (C$_{16}$H$_{18}$N$_3$SCl·3H$_2$O) — A dark-green crystalline compound used as a dye, as a stain in bacteriology, in medicine as an antidote for cyanide poisoning, and as a reagent in oxidation-reduction processes.

methylene blue active substance — Substance used in surfactants or detergents.

methylene blue reduction test — A test designed to give a rough estimate of the bacterial load in food; in milk, decolorization in 1/2 hour indicates a high bacterial load, whereas decolorization in 8 hours indicates a low bacterial load.

methylene chloride (CH$_2$Cl$_2$) — A colorless liquid that has a chloroform-like odor and is a gas above 104°F; MW: 84.9, BP: 104°F, Sol: 2%, Fl.P: not known. It is used as a solvent in paint and varnish removers; in the manufacture of aerosols; as an extraction solvent for foods and furniture processing; as a cooling solvent in the manufacture of cellulose acetate; in organic synthesis; in plastics processing; as a solvent in vapor degreasing of thermal switches and thermometers; as a secondary refrigerant in air conditioning and scientific testing; as an extraction solvent for edible fats, cocoa, butter, beer flavoring, decaffeinated coffee, oleoresin manufacture, oils, waxes, perfumes, flavorings, and drugs; as a solvent for paints and lacquers, varnishes, enamels, adhesives, rubber cements; as a carrier for pharmaceutical tablet coatings; in the dyeing of synthetic fibers. It is hazardous to the skin, cardiovascular system, central nervous system, and eyes and is toxic through inhalation, ingestion, and contact. Symptoms of exposure include fatigue, weakness, sleepiness, lightheadedness, tingling and numb limbs, nausea, and eye and skin irritation; carcinogenic. OSHA exposure limit (TWA): 500 ppm [air].

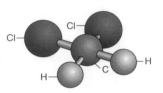

Methylene Chloride

2-methylnaphthalene (C$_{11}$H$_{10}$) — A crystalline solid; MW: 142.2, BP 241°C, Sol: NA, Fl. P: NA, sp. gr. 1.0058. It is used in the manufacture of alkyd and polyester resins, dyes, pigments, pharmaceuticals, insecticides, and as a dye carrier. It is hazardous to the respiratory system, digestive system, skin, and eyes. Acute exposure may irritate the eyes, nose, and throat and cause headaches, restlessness, lethargy, nausea, vomiting, anorexia, fever, liver and kidney damage, renal failure, tachycardia, seizures, coma, shock, and death. Chronic exposure may damage vision and cause headaches, fatigue, nausea, and damage to the developing fetus. OSHA exposure limit (8-hour TWA): NA.

4-methylphenol — See *p*-cresol.

metolcarb (C$_9$H$_{11}$NO$_2$) — A colorless crystalline solid with no odor; MW: 165.21, BP: NA, Sol: soluble in water and organic solvents, density: NA. It is used as a pesticide. It is hazardous to the skin, digestive tract, and respiratory tract and may cause acute, delayed, and chronic effects, including blurred vision, headache, breathing difficulty, loss of reflexes, weakness, sweating, nausea, vomiting, diarrhea, convulsions, coma, and death.

$$\text{NH}-\overset{\overset{\displaystyle O}{\|}}{\text{C}}-\text{OCH}_3$$

Metolcarb

CH$_3$

metribuzin (C$_8$H$_{14}$N$_4$OS) — A white crystalline solid; MW: 214.32, BP: NA, Sol: soluble in alcohol, density: NA. It is used as a herbicide. It has low to moderate acute toxicity that causes depression of the central nervous system. ACGIH exposure limit (TLV-TWA): 5 mg/m^3.

Metribuzin

metric system — A system of scientific units based on the decimal system; its units of length, time, and mass are the meter, second, and kilogram, respectively.

metric ton — A unit of weight equal to 1000 kg or 2204.6 pounds; international measurement for the quantity of greenhouse gas emissions equivalent to 2205 pounds.

metrics — A method of measuring quality or performance.

Metropolitan Statistical Area (MSA) — Except in New England, as prescribed by federal rules and regulations, a county or group of contiguous counties that contains at least one city with a population of 50,000 or twin cities with a population of at least 50,000; they are essentially metropolitan in character and are socially and economically integrated with the central city.

MeV — See megaelectronvolt.

mexacarbate ($C_{12}H_{18}N_2O_2$) — An odorless, white crystalline solid; MW: 222.29, BP: decomposes, Sol: soluble in organic solvents, density: NA, Fl.P: NA (poisonous gases produced in a fire include nitrogen oxides), sp. gr. NA. It is a highly toxic carbamate pesticide no longer used in the United States. It is used as a pesticide and is hazardous to the skin, gastrointestinal tract, and respiratory tract. It is extremely toxic, and acute exposure may cause central nervous system depression, stupor, coma, seizures, hypertension, tachycardia, cardiorespiratory depression, urinary incontinence, diarrhea, gastrointestinal cramping, nausea, vomiting, irritation of the eyes and skin, pulmonary edema, and death. No chronic effects are known. OSHA exposure limit (8-hour TWA): NA.

$$(CH_3)_2N - \text{benzene ring with } CH_3 \text{ top and } CH_3 \text{ bottom} - NH - \overset{\overset{O}{\|}}{C} - O - CH_3$$

Mexacarbate

mg — See milligram.

mg/l — See milligrams per liter; a unit of the concentration of a substance in water or wastewater.

mg/m³ — See milligrams per cubic meter.

mgd — Notation for million gallons per day.

MHAPPS — Acronym for Modified Hazardous Air Pollutant Prioritization System.

mho — A unit of electrical conductance equal to the conductance of a body through which 1 ampere of current flows when the potential difference is 1 volt; it is reciprocal to the ohm; see siemen.

MHz — A unit of frequency equal to 1 million cycles per second.

miasmatist — A person who has made a special study of infectious particles or microorganisms floating in the air.

mica ($K_2Al_4(Al_2Si_6O_{20}) (OH)_4$) — Colorless, odorless flakes or sheets of hydrous silicates containing less than 1% quartz; MW: 797, BP: not known, Sol: insoluble in water, sp. gr. 2.6 to 3.2. It is used in the manufacture of asphalt shingles and roll roofing; in the manufacture of paint, wall paper, and bituminized cardboard; in the manufacture of molded rubber products, plastics, special greases; in the fabrication of windows and diaphragms; in the manufacture of electrical insulation for low thermal conductivity and high dielectrical strength. It is hazardous to the lungs and is toxic through inhalation. Symptoms of exposure include pneumoconiosis, cough, dyspnea, weakness, and weight loss. OSHA exposure limit (TWA): 3 mg/m³ [resp].

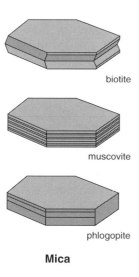

biotite

muscovite

phlogopite

Mica

micelle — An aggregation or cluster of molecules, ions, or minute submicroscopic particles.

micro- (μ) — A prefix meaning one millionth or 10^{-6}.

microaerophile — An organism that requires a low concentration of oxygen for growth.

microaerophilic — An environment that is low in oxygen but is not anaerobic.

microaerophillic environment — An environment with reduced oxygen concentrations often below 5% and with carbon dioxide levels that may approach 10%.

microaerotonometer — An instrument used for measuring the volume of gases in blood or other fluids.

microamp — One millionth of an ampere.

microanalysis — An analysis performed on a relatively small sample and requiring the use of specialized instruments.

microbar — A unit of pressure used in acoustics equal to 1 dyne/cm²; also known as barye.

microbe — A microorganism, especially a bacterium of a pathogenic nature.

microbial biomass — The total mass of microorganisms alive in a given mass of soil.

microbial digestion — The breakdown and use of a substance by microorganisms.

microbial growth — Amplification or multiplication of microorganisms such as bacteria, algae, diatoms, plankton, viruses, and fungi.

microbial insecticide — Bacteria or other tiny plants or animals used to prevent damage by insects or to destroy insects.

microbicide — Any drug, chemical, or other agent that kills microorganisms.

microbiological hazard — The presence of or introduction of pathogenic organisms into food that causes outbreaks of disease.

microbiology — The study of microorganisms, including algae, bacteria, fungi, viruses, and protozoa.

microclimate — Local climatic patterns caused by geographical, biological, and artificial factors.

microcomputer — A very small computer that uses a microprocessor to handle information.

microcosm — A community or other unit that is representative of a larger group; a laboratory-scale model used to understand the interactive relationship of the microbial communities and their roles within their ecosystems.

microcurie (μCi) — A unit of measurement equal to 1 millionth of a curie.

microcytic — Smaller than normal cells.

microelectronics — A branch of electronics concerned with the design, production, and application of electronic components, circuits, and devices of extremely small dimensions.

microenvironment — A well-defined place, such as a home, office, or car, in which a chemical or biological agent is present in a uniform manner.

microfauna — Protozoa, nematodes, and arthropods generally less than 200 μm long.

microfiche — A sheet of microfilm containing rows of microimages of pages of printed matter.

microfilaria — Partially developed juveniles born alive by filarial worms.

microfilm — A strip of 16-mm or 35-mm film that contains photographic reproductions of pages of books, documents, and records, greatly reduced in size.

microflora — Bacteria, fungi, algae, and viruses.

microgram (μg) — A unit of mass equal to 1 millionth of a gram.

microgram per cubic meter (μg/m³) — A measurement of weight per unit volume equal to 1 part per billion; micrograms of particulates per cubic meter of air sampled.

microgram per liter (μ/l) — A unit of weight per volume equal to 1 part per billion.

microhm — A unit of electrical resistance equal to 1 millionth of an ohm.

microhydrogenerator — A small power generator that can be used in low-level rivers to provide economical power for 4 to 6 homes.

microliter — A unit of liquid volume equal to one millionth of a liter.

micrometer — See micron.

micromicro- — See pico.

micromillimeter — A nanometer.

micron — A metric unit of measurement equal to 1/1000 of a milliliter, 10^{-6} meters, and approximately 1/25,000 of an inch; now replaced by the micrometer (μm).

micronutrient — A chemical element necessary in only small amounts for growth and development; also known as trace element.

microorganism — An organism, including bacteria, protozoans, viruses, rickettsiae, and some algae and fungi, that can be observed only with the aid of a microscope.

microphage — A neutrophile that ingests small things, such as bacteria.

microphone — An instrument that is actuated by electricity and transmits sound.

microphyte — (*botany*) A microscopic plant or plant-like organism, especially a fungus.

microplankton — Plankton with a size range of 20 to 200 μm.

micropores — The spaces between soil particles that hold water by means of capillary forces.

microscope — An instrument with different lenses used for looking at very small objects.

microscopic — A substance that is only visible when magnified and illuminated by a microscope.

microsite — A small volume of soil where biological or chemical processes differ from those of the soil at large.

microtubule — Abundant organella in eukaryotic cell cytoplasm consisting of aggregates of globular proteins arranged in strands of protofilaments.

microvolt — One millionth of a volt.

microwave — A high-frequency electromagnetic wave usually 1 mm to 10 cm (sometimes 30 cm) in wavelength; causes structural or functional changes to the eyes, testicles, bone marrow, and cardiovascular and central nervous systems due to the heating caused by the microwave absorption.

microwave amplification by stimulated emission of radiation — See MASER.

microwave radiometer — A sensor that measures the intensity of microwave radiation (0.3 to 30 cm), within a specific field of view.

microwave thermography — Measurement of temperature through the detection of microwave radiation emitted from heated tissue.

midbody — The middle of the body or mid-region of the trunk.

middle clouds — Types of clouds, such as stratus or cumulus, with bases averaging 10,000 feet above the Earth.

middle ear — A small, membrane-lined cavity that is separated from the outer ear by the eardrum and transmits sound waves from the eardrum to the partition between the middle and inner ears through a chain of tiny bones.

middle infrared — Electromagnetic radiation between the near-infrared and thermal infrared (about 2 to 5 μm).

midline — An imaginary line that divides the body into right and left halves.

midpoint — See median.

Mie scattering — Atmospheric scattering caused by large particles such as dust, pollen, smoke, and water droplets; more prevalent in the lower atmosphere, from 0 to 5 km.

migrant — A seasonal farm worker who performs agricultural labor and must travel from job to job.

migration — Moving from place to place.

mil — A unit of length equal to 0.001 inch.

mil gal — Notation for a unit of measurement equal to 1 million gallons.

mild — Low intensity.

mildly alkaline detergent — A general-purpose cleaning material used for manual cleaning of floors, walls, ceilings, equipment, and utensils.

mile — A unit of measurement equal to 5280 feet or 1.6093 kilometers.

miliaria — See heat rash.

milieu — The surrounding environment.

military antishock trousers — A specially designed pair of pants used to produce pressure on the lower part of the body to prevent the pooling of blood in the legs and abdomen.

milk — Fluid secretion of the mammary glands that is practically free of cholestrum, containing a mini-

mum of 8.25% nonfat milk solids and 3.25% milk fat.

milk and food specialist — An environmental health practitioner concerned with the protection of the milk and food supply against contamination from biological, chemical, and physical hazards from source to consumer.

milk products — Products made from milk, including cream, sour cream, half-and-half, whipped cream, concentrated milk, flavored milk, and cottage cheese.

milkfat — The fat of milk; also called butterfat.

milking — A technique used to express the contents of a duct or tube.

mill — A unit of monetary cost equal to 1/1000 of a U.S. dollar that is used by jurisdictions to determine the level of taxes; also used by utilities for billing purposes.

milli- (m) — A prefix meaning 1/1000 or 10^{-3}.

milliamperage — (*radiography*) The x-ray-tube current during an exposure measured in milliamperes.

milliampere (mA) — A unit of current equal to 1/1000 of an ampere.

millibar (mbar) — A unit of pressure used in meteorology and international weather observation equal to 1000 dynes per square centimeter.

millicoulomb — A unit of electric charge that is 1/1000 of a coulomb.

millicurie (mCi) — A unit of radiation equal to 1/1000 of a curie.

millidyne — A unit of force equal to 1/1000 of a dyne.

milliequivalent (meq) — One thousandth of a chemical equivalent.

milligram (mg) — A unit of mass equal to 1/1000 of a gram.

milligrams per cubic meter (mg/m³) — A unit of measurement of air concentrations of dusts, gases, mists, and fumes.

milliliter (ml) — A unit of volume equal to 1/1000 of a liter.

millimeter (mm) — A unit of length equal to 1/1000 of a meter.

millimeter of mercury (mmHg) — The unit of pressure equal to the pressure exerted by a column of liquid mercury 1 millimeter high at a standard pressure.

millimicron (mμ) — A unit of length equal to 1/1000 of a micron, or one billionth of a meter.

millimole — A unit of metric measurement of mass equal to 1/1000 of a mole.

million acre-feet — The volume of water that would cover 1 million acres to a depth of one foot.

million gallons per day — A measure of water flow; a rate of flow of water equal to 133,680.56 cubic feet per day, 1.5472 cubic feet per second, or 3.0689 acre-feet per day.

millipore filter — Trademark for a device used to filter solutions as they are administered.

millirad (mrad) — A unit of absorbed ionizing radiation equal to 1/1000 of a rad.

millirem (mrem) — A unit of radiation equal to 1/1000 of a rem.

milliroentgen (mR) — A unit of radioactive dose of electromagnetic radiation equal to 1/1000 of a roentgen.

millisecond (ms) — A unit of time equal to 1/1000 of a second.

millivolt (mV) — A unit of electromotive force equal to 1/1000 of a volt.

milliwatt (mW) — A unit of power equal to 1/1000 of a watt.

min — See minute.

Minamata disease — (*disease*) A severe neurological disorder due to alkyl mercury poisoning that leads to permanent neurologic and mental disabilities or death; prevalent among those who eat seafood contaminated with mercury.

mine drainage — Any drainage or water pumped or siphoned from an active mining area or a post-mining area.

mineral — A naturally occurring substance that has a characteristic chemical composition and, in general, a crystalline structure; identified by the properties of its crystal system, hardness, relative density, luster, color, cleavage, and fracture.

mineral soil — A soil consisting predominantly of or having its properties determined by mineral matter.

mineral spirits — Flammable petroleum distillates that boil at temperatures lower than kerosene and are used as solvents and thinners, especially in paints and varnishes.

mineralization — The conversion of an element from an organic form to an inorganic state as a result of microbial decomposition.

minimal dose — The smallest dose of a drug or other agent needed to produce a desired effect.

minimization — The measures or techniques that reduce the amount of waste generated during industrial production processes.

minimum daily requirement — The daily human requirements of nutrients necessary for good health and to prevent a deficiency disease.

minimum detection limit — The lowest level to which an analytical parameter can be measured with certainty by the analytical laboratory performing the measurement.

minimum lethal dose — The smallest dose of a drug compared to body weight that will kill an experimental animal.

minimum risk level — An estimate of daily human exposure to a chemical that is likely to be without appreciable risk of deleterious or noncancerous effects over a specified duration of exposure.

mining — The removal of soil or rock because of its chemical composition.

mining wastes — Residues that result from the extraction of raw materials from the Earth.

mining water use — The water used during quarrying rocks and extracting minerals from the land.

minute (min) — A unit of time equal to 1/60 of an hour.

miosis — Excessive contraction of the pupil of the eye.

MIR — See maximum incremental reactivity.

Mirex ($C_{10}Cl_{12}$) — An odorless, snow-white crystal; MW: NA, BP: decomposes, Sol: insoluble in water, Fl.P: supports combustion, sp. gr. NA. It is used as an insecticide and as a fire retardant for plastics, rubber, paint, paper, and electrical goods. It is hazardous to the respiratory system, digestive system, skin, and eyes. Acute exposure may cause cough, dyspnea, chest discomfort, pulmonary edema, necrotizing bronchitis, and degenerative changes in the central nervous system, heart, liver, kidneys, and adrenal glands. It is highly irritating to the skin and eyes. Chronic exposure may cause degenerative changes in the central nervous system, heart, liver, kidneys, adrenal glands and may also cause dermatitis. OSHA exposure limit (8-hour TWA): NA.

mirror image (chemistry) — Chemical molecules with the same composition but asymmetrical arrangement of the atoms.

misbranded food — A food for which the labeling is false or misleading.

miscible — The capability of liquids to be soluble in each other.

miscibile liquids — Two or more liquids that can be mixed and will remain mixed under normal conditions.

misdemeanor — A crime punishable by a year or less in prison.

misfeasance — Doing wrongfully and injuriously an act that might otherwise be done in a lawful manner.

MISR — See Multiangle Imaging Spectro-Radiometer.

mission — A stated self-imposed duty.

mist — Liquid particles (up to 100 μm in diameter) suspended in the air and caused by condensation of gases in the liquid state due to splashing, foaming, or atomizing. It may cause irritation to the nose and throat, ulceration of the nasal passages, damage to the lungs, and damage to other internal tissues, depending on the chemical present, and may also cause dermatitis.

mite — Any of numerous small to very minute arachnids that often infest animals, plants, or stored food.

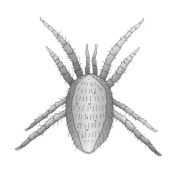

Mite
Order Acarina

miticide — A pesticide chemical or agent that prevents damage from or destroys mites and ticks; see also acaricide.

mitigation — Alleviation, abatement, or diminution of a penalty or punishment imposed by law.

mitochondria — Minute, rod-shaped bodies in the cytoplasm known to be centers of cellular respiration.

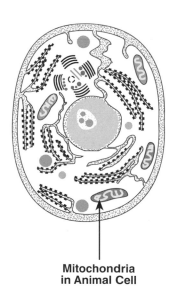

**Mitochondria
in Animal Cell**

mitosis — Cell division from which two identical cells result, each containing the same number and kind of chromosomes as the parent cell.

mix — (*food*) The unfrozen combination of all ingredients of a frozen dessert (e.g., ice cream) with or without fruit, fruit juices, candy, nut meats, flavor, or harmless color.

mixed culture — A laboratory culture that contains two or more different types of organisms.

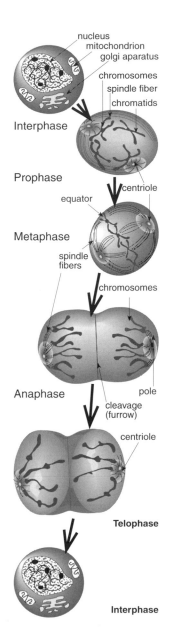

Mitosis

mixed infection — An infection by several different types of microorganisms; see polymicrobic infections.

mixed liquor — The combination of raw influent and returned activated sludge.

mixed-liquor suspended solids (MLSS) — (*wastewater*) Suspended solids in a mixture of activated sludge and organic matter undergoing activated sludge treatment in the aeration tank.

mixed-liquor volatile suspended solids (MLVSS) — Organic solids that can evaporate at relatively low temperatures from the mixed liquor of an aeration tank and are used as a measure or indication of microorganisms present.

mixed waste — Wastes that contains both a hazardous waste component and a radioactive component

mixing chamber — A chamber usually placed between the primary and secondary combustion chambers of an incinerator; the products of combustion are thoroughly mixed there by the turbulence created by an increased velocity of gases or turns in the direction of the gas flow.

mixing depth — The atmospheric layer near the surface of a region in which warmer air rises, mixes with colder air, and creates turbulence that disperses pollutants.

mixing ratio — The relative number of molecules of a specific type in a given volume of air.

mixotroph — An organism able to assimilate organic compounds as carbon sources while using inorganic compounds as electron donors.

mixture — An association of substances not chemically combined and able to be separated by mechanical means.

MKS system — A system of measurement based on the meter, kilogram, and second.

MKS units — See meter–kilogram–second system.

ml — See milliliter.

ML — See mixed liquor.

MLC — See median lethal concentration.

MLD — See median lethal dose; acronym for minimum lethal dose.

MLSS — See mixed-liquor suspended solids.

MLVSS — See mixed-liquor volatile suspended solids.

mm — See millimeter.

MMBbl — Abbreviation for 1 million barrels.

MMBtu — Abbreviation for 1 million Btu.

MMcf — Abbreviation for 1 million cubic feet.

MMcfe — Abbreviation for one million cubic feet of natural gas equivalent.

MMH — See manual materials handling.

mmHg — See millimeter of mercury.

MMTCDE — Acronym for million metric tons of carbon dioxide equivalents.

MMTCE — Acronym for million metric tons of carbon equivalents.

mobile environmental monitor — A field transportable analytical instrument designed to identify and measure organic pollutants in various environmental media by use of a spectrometer.

mobile food unit — A vehicle-mounted food establishment that is readily movable.

mobile home — A factory-assembled structure equipped with the necessary service connections and that is readily movable as a unit; designed to be used as a dwelling without a permanent foundation.

mobile incinerator system — A hazardous waste incinerator that can be transported from one site to another.

mobile noise source — Any noise source other than a fixed noise source.

mobile source — A nonstationary source of air pollution found in cars, trucks, motorcycles, buses, airplanes, and locomotives.

mock lead — See sphalerite.

mock ore — See sphalerite.

model — A mathematical or physical system used to simulate real events and processes.

modeling — The use of mathematical equations to simulate and predict real events and processes.

modem — (*computer science*) A device for the conversion of digital and analog signals to allow data transmission over telephone lines.

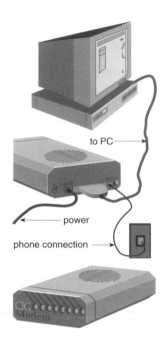

to PC

power

phone connection

Modem

moderate limitations — (*soils*) Limitations that usually can be overcome or modified with direct planning and careful design.

moderately well drained — (*soils*) Soil in which the water is removed somewhat slowly so that the profile is wet for a small but significant part of the time.

moderator — A material used to slow neutrons in a reactor.

modular — A piece of hardware or software that is constructed with standardized units or dimensions for rapid change.

modular incinerator — A small-scale waste combustion unit prefabricated at a manufacturing facility and transported to a waste site.

modulate — (*physics*) To vary the amplitude, frequency, or phase of electromagnetic waves, especially carrier waves.

modus operandi — Manner of working.

MOE — Acronym for margin of exposure.

Mohs scale — A ten-point scale of mineral hardness.

moist soil — Loose, very friable, friable, firm, very firm, or extremely firm soil.

moisture content — (*solid waste*) Measured by the weight loss (expressed in percent) in a sample of solid waste that is dried to a constant weight and a temperature of 100 to 105°C.

moisture penetration — The depth to which irrigation water or rain penetrates soil before the rate of downward movement becomes negligible.

mol — See mole.

mol. wt. — See molecular weight.

molar solution — A solution consisting of 1 gram molecular weight of a compound dissolved in enough water to make 1 liter of solution; of or pertaining to the gram molecular weight of a substance.

molarity — See molar.

mold — A fungus of the order Mucorale; a growth of fungi.

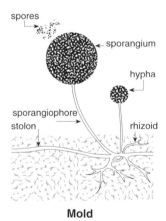

Mold

mole (mol) — A unit of amount of substance, molecule, etc.; it replaces the term "gram formula."

molecular absorption spectrophotometry — A technique using absorption of radiation by molecular species.

molecular diffusion — The process whereby molecules of various gases tend to intermingle and eventually become uniformly dispersed.

molecular epidemiology — Measurement of the interaction of a toxic chemical or its derivative with a tissue constituent or a tissue alteration resulting from exposure to a chemical; provides an indirect measure of individual exposure.

molecular fluorescence spectrophotometry — A technique used for special measurements of fluorescent radiation from luminescent compounds upon excitation by incident radiation.

molecular weight (MW) — The sum of the atomic weight of all the atoms in a molecule of a substance; also known as relative molecular mass.

molecule — The smallest particle of a substance that can exist in a free state and retain all its chemical properties.

molluscicide — A pesticide chemical or agent that prevents damage by or kills slugs or snails.

mollusk — Any of a large phylum of the invertebrates Mollusca, such as snails or clams, with soft unsegmented bodies usually enclosed in a calciphoric shell. Mollusks are indicators of environmental contamination and environmental quality because they live in the mud or sandbottoms of aquatic ecosystems, feed by taking water through their bodies, and therefore reflect the chemical or microbiological conditions of the water.

molten salt reactor — A thermal treatment unit that rapidly heats waste in a heat-conducting fluid bath of carbonate salt.

molting — Shedding an exoskeleton and secreting a new exoskeleton during periods of growth.

molybdenum and insoluble molybdenum (Mo) — Appearance, odor, and properties vary depending upon the specific soluble compound or insoluble metal: dark gray or black powder with a metallic luster; MW: 95.9, BP: 8717°F, Sol: insoluble, sp. gr. 10.28. They are used as lubricants in greases, oil dispersions, resin-bonded films, and dry powders; as catalysts in petroleum refining and chemical processing; in pigment mixture for labeling glass bottles; for printing inks, paints, plastics; as reagents in analytical chemistry laboratories; as catalysts in organic synthesis in medicinals; in decorative and protective coatings for metal; as feed additives; as a chemical intermediate and plating agent for mirrors; in the manufacture of cutting tools. Soluble molybdenum is hazardous to the respiratory system; in animals, it is hazardous to kidneys and blood. There is no known harm to humans from metal Mo but they are toxic by inhalation, ingestion, and contact. Symptoms of exposure in animals include anorexia; incoordination; dyspnea; anemia; irritation of the eyes, nose, and throat; diarrhea; weight loss; listlessness; and damage to the liver and kidneys. OSHA exposure limit

M

(TWA): 5 mg/m³ (molybdenum) and 10 mg/m³ (insoluble molybdenum).

molybdenum hexacarbonyl (Mo(CO)₆) — A white crystalline solid; MW: 264.00, BP: NA, Sol: soluble in most organic solvents, Fl.P: reacts strongly with strong oxidizers and produces a violent reaction, density: 1.96 g/cm³. It is used as a catalyst in many organic, synthetic reactions. It is hazardous to the respiratory tract, digestive tract, and skin. It is highly toxic and may cause headaches, dizziness, nausea, vomiting, fever, severe irritation, and death. ACGIH exposure limit: NA.

Molybdenum hexacarbonyl

moniliasis — See candidiasis.

monition — A summons to appear and answer issues listed on a complaint.

monitor — To observe, collect, check, and measure data to determine the capability of a given condition with predetermined parameters; a planned sequence of observations or measurements of critical limits designed to produce an accurate record and intended to ensure that the critical limit maintains product safety; a means of measuring concentrations and behavior of substances in environmental media or in human or other biological tissues.

monitor stations — (*global positioning satellite*) Worldwide groups of stations used in the GPS control segment to track satellite clock and orbital parameters; collected data are linked to a master control station, where corrections are calculated and from which correction date is uploaded to the satellites as needed.

monitored retrievable storage — The holding of wastes in underground mines or secure surface facilities such as dry casks, where they can be watched and repackaged, if necessary.

monitoring well — A well that collects groundwater for evaluation of water quality, evaluation of groundwater flow and elevation, effectiveness of treatment systems, and effectiveness of administrative or engineering controls.

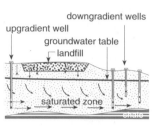

Monitoring Wells

monochromatic light — Light consisting of only one color.

monochromator — A spectrometer that operates within a narrow range of the electromagnetic spectrum.

monoclonal — A group of identical cells or organisms derived from a single cell.

monoclonal antibodies — Clones of a molecule that are produced in quantity for medical or research purposes; the molecules of living organisms that selectively find and attach to other molecules to which their structure conforms exactly.

monocyte — A mononuclear, phagocytic leukocyte, between 13 and 25 μm in diameter, having an ovoid- or kidney-shaped nucleus and azurophilic cytoplasmic granules.

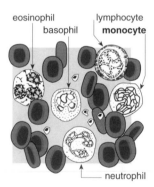

Monocyte

monocytosis — An excess of monocytes in the blood.

monoethanolamine (C₂H₄(OH)NH₂) — A colorless, viscous liquid with a mild ammonia odor: MW: 61.10; BP: 171°C; Sol: miscible with water, alcohol, and acetone; Fl.P: 85°C (closed cup); density: 1.018 at 20°C. It is used as a dispersing agent for agricultural chemicals, in the synthesis of surface active agents, and in emulsifiers and polishes. It is hazardous to the eyes and the skin and may have

$$HO - CH_2 - CH_2 - NH_2$$

Monoethanolamine

reproductive toxicity in animals. ACGIH exposure limit (TLV-TWA): 3 ppm.

monomer — A compound of relatively low molecular weight that, under certain conditions and either alone or with another monomer, forms various types and lengths of molecular chains called polymers or copolymers of high molecular weight. An example of this would be styrene, which readily becomes the polymer polystyrene when polymerized.

monomethyl aniline ($C_6H_5NHCH_3$) — A yellow to light-brown liquid with a weak, ammonia-like odor; MW: 107.2, BP: 384°F, Sol: insoluble, Fl.P: 175°F, sp. gr. 0.99. It is used in organic synthesis and in dye intermediates. It is hazardous to the respiratory system, liver, kidneys, and blood and is toxic through inhalation, absorption, ingestion, and contact. Symptoms of exposure include weakness, dizziness, headache, dyspnea, cyanosis, methemoglobinemia, pulmonary edema, and liver and kidney damage. OSHA exposure limit (TWA): 0.5 ppm [skin] or 2 mg/m³.

monophasic — Having one phase, stage, or aspect.

monosaccharide — Simple sugars, such as $C_6H_{12}O_6$, which cannot be hydrolyzed to simpler carbohydrates.

monotone — See monotonic.

monotonic — (*mathematics*) Either never increasing or never decreasing as a function; also known as monotone.

monoxenous — Living within a single host during a parasite's life cycle.

monthly catalog of U.S. government publications — A series of catalogs of all United States government publications, available from 1895 to the present.

montmorillonite — An aluminosilicate clay mineral with an expanding structure; that is, it has two silicon tetrahedral layers enclosing an aluminum octahedral layer. Considerable expansion may be caused by water moving between silica layers of contiguous units.

Montreal Protocol — A treaty signed in 1987 that governs stratospheric ozone protection and research, production, and use of ozone-depleting substances, including the end of the production of CFCs and other chemicals affecting the ozone layer.

monuron ($C_9H_{11}ClN_2O$) — A crystalline solid with a faint odor; MW: 198.67, BP: NA, Sol: moderately soluble in alcohol and acetone, density: NA. It is used as an herbicide. It shows low toxicity although it is a carcinogen in experimental animals.

moor — See bog.

MOPITT — See Measurements of Pollution in the Troposphere.

Monuron

morbidity — Illness, injury, or disability in a defined population.

morbidity rate — The number of cases of a specific disease occurring in a single year for a specified population unit, such as cases per 10,000.

morbidity statistics — A set of statistics concerned with a disease rate of a specific population or geographic region.

mordant — A chemical that fixes a dye in or on a substance by combining with the dye to form an insoluble compound.

moribund — In the act of dying.

Morgan — A unit of measure used in mapping the relative distances of genes on a chromosome.

morpholine (C_4H_8ONH) — A colorless liquid with a weak, ammonia- or fish-like odor; a solid below 23°F; MW: 87.1, BP: 264°F, Sol: miscible, Fl.P: 98°F (open cup), sp. gr. 1.00. It is used in the manufacture of chemicals for rubber accelerators, catalysts, plasticizers, curing agents, stabilizers of halogenated butyl rubber, and emulsifying agents; as a corrosion inhibitor; in the manufacture of optical brightening agents in bleaches and detergents; in the compounding of waxes and polishes; as a chemical intermediate for the textile industry; in the pharmaceutical industry; as solvents for dyes, waxes, resins, and casein. It is hazardous to the respiratory system, eyes, and skin and is toxic through inhalation, absorption, ingestion, and contact. Symptoms of exposure include visual disturbances, cough, liver and kidney damage, and irritation of the nose, eyes, skin, and respiratory system. OSHA exposure limit (TWA): 20 ppm [skin] or 70 mg/m³.

Morpholine

morphology — Branch of biological science that deals with the study of the structure and form of living organisms.

mortality — Death.

mortality rate — Ratio of the total number of deaths to the total population of a given community at a given time, usually expressed as deaths per 1000, 10,000, or 100,000; also known as death rate.

mortar — A hard mass made from calcium hydroxide, sand, and water that is used in masonry.

M

MOS — See margin of safety; (*physics*) abbreviation for metal oxide semiconductor.

mosquito — Small, long-legged, two-winged insect belonging to the order Diptera and family Culicidae; differs from other flies by having an elongated proboscis and scales on the wing veins and the wing margins.

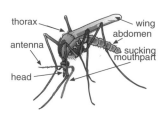

Mosquito

most probable number — A statistical term inferring a particular microbial contamination level on the basis of a sample tested.

motile — Capable of spontaneous movement.

motion — (*law*) An application to a court, either written or oral, for rule or order; (*mechanics*) a continuous change of position of a body.

motor nerve — A nerve containing only motor fibers.

motor neuron — A neuron that carries impulses from the brain or spinal cord to a muscle or gland to initiate a physiological response.

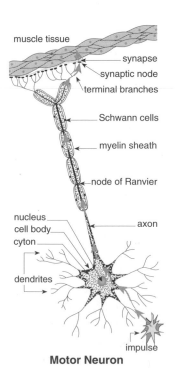

Motor Neuron

motor unit — An anterior horn cell, its axon, and all of the muscle fibers innervated by the axon.

Motor Vehicle Air Pollution Control Act — A law passed in 1965 recognizing feasibility of setting automobile emission standards, establishing national standards for automobile emissions, and applying California state emission standards for hydrocarbons and carbon monoxide nationally.

mottled — Spotty or uneven in appearance; (*soils*) a zone of chemical oxidation and reduction activity appearing as patches of red, brown, orange, or dull gray.

MOU — Acronym for memorandum of understanding.

mouse — A small rodent with a pointed snout, rather small ears, elongated body, and slender tail; (*computer science*) a hand-steered device for controlling the position of a cursor on a video display screen that allows relaying computer commands by pointing and clicking.

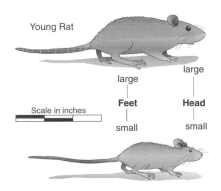

House Mouse *Mus musculus*

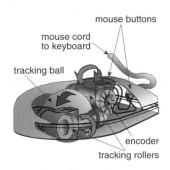

Mouse

mousse — A thick, foamy oil and water mixture formed when petroleum products are mixed with water by the action of waves and wind.

movable grate — A grate with moving parts designed to feed solid fuel or solid waste to a furnace.

MP — Acronym for Montreal Protocol; see melting point.

MPC — See maximum permissible concentration.

MPCA — Acronym for microbial pest control agent.

MPD — See maximum permissible dose.

MPE — See maximum permissible exposure.

MPI — Acronym for maximum permitted intake.

MPL — See maximum permissible level.

MPN — Acronym for maximum possible number.

mppcf — Acronym for million particles per cubic foot.

MPWC — Acronym for multiprocess wet cleaning.

mR — See milliroentgen.

MR — See magnetic resonance; acronym for methane recovery.

mrad — See millirad.

mrem — See millirem.

MRF — Acronym for materials recovery facility.

MRI — See magnetic resonance imaging.

MRID — Acronym for master record identification number.

MRIR — Acronym for medium-resolution infrared radiometer.

MRL — Acronym for maximum residue limit (pesticide tolerance).

MRSA — See methicillin resistant *Staphylococcus aureus*.

MRT — See mean radiant temperature.

MS — See mass spectrograph.

MS4 — See municipal separate storm sewer system.

MSA — See Metropolitan Statistical Area.

MSD — See musculoskeletal disorder.

MSD hazard — The presence of risk factors for a job that occur at a magnitude, duration, or frequency that is reasonably likely to cause musculoskeletal disorders that can result in work restrictions or medical treatment

MSD symptoms — The physical indications that an employee may be developing a musculoskeletal injury, including symptoms of pain, numbness, tingling, burning, cramping, and stiffness.

MSDS — See Material Safety Data Sheet.

msec — See millisecond.

MSGP — See multisector general permit.

MSL — See mean sea level.

MSS — See municipal sewage sludge; multispectral scanner system.

MSTCDE — Acronym for million short tons of carbon dioxide equivalents.

MSW — See municipal solid waste.

MSWI — See municipal solid waste incinerator.

Mt — Abbreviation for millions of tons.

MTBF — See mean time between failures.

MTD — Acronym for maximum tolerated dose or million tons per day.

MTF — Acronym for medical treatment facility.

MTP — Notation for maximum total trihalomethane potential.

MTTF — See mean time to failure.

MTTP — See maximum total trihalomethane potential.

MTTR — See mean time to repair.

muck soil — Earth made from decaying plant materials.

mucociliary clearance — Removal of materials from the respiratory tract via ciliary action.

mucosa — The mucous membrane.

mucous — Of or pertaining to the cells of the mucous membrane; secreting mucous.

mucous membrane — The thin layer of protective mucous-secreting tissue that lines the digestive tract and other organs that have contact with air.

mucous — The fluid of the mucous membranes composed of secretion of the glands, various salts, desquamated cells, and leukocytes.

mud — Any soil containing enough water to make it soft.

mulch — A mixture of wood chips, dry leaves, straw, hay, or other material placed on the soil around plants to hold moisture in the ground, keep weeds from growing, soak up rain, or maintain soil temperature.

Multiangle Imaging Spectro-Radiometer — An instrument that is part of the NASA Earth Observing System that monitors the monthly, seasonal, and long-term trends in the amount and type of atmospheric aerosol particles caused by natural sources and by human activities.

multiclone — See multicyclone.

Multiclone

multicyclone — A combination of cyclone collectors; also known as multiple cyclone or multiclone.

multimedia exposure — The exposure to a toxic substance from multiple pathways including air, water, soil, food, and breast milk.

multiphasic screening — The simultaneous use of multiple tests to detect severe pathological conditions.

multiple chamber incinerator — An incinerator consisting of two or more chambers arranged as in-line or retort types, interconnected by gas passage ports or ducts.

multiple chemical sensitivity — A condition of the modern age in which a person is considered to be sensitive to a number of everyday household, industrial, or agricultural chemicals at very low concentrations; a diagnostic label for people who suffer multisystem illnesses as a result of contact with, or proximity to, a variety of airborne agents and other substances.

multiple dwelling — Any dwelling containing more than two dwelling units.

multiple source — A source of pollution similar in character and widespread throughout the community, such as an automobile, home heating, or incinerators found in apartment houses.

multiple use — Refers to land with more than one purpose.

multiplecyclone — See multicyclone.

multiple-family dwelling — Any shelter containing two or more dwellings, units, or both.

multisector general permit — An NPDES permit that regulates stormwater discharges from 11 categories of industrial activities.

multispectral remote sensing — The science of detection and analysis of an object at a distance.

multispectral scanner system (MSS) — A device found in airplanes or satellites for recording radiation in several wavebands at the same time.

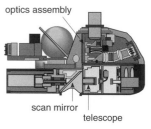

optics assembly

scan mirror
telescope

Multispectral Scanner

multistage pump — A pump that has more than one impeller.

multivalent — The capacity of an atom to combine with three or more univalent atoms.

municipal collection — Collection of solid waste by public employees and equipment under the supervision and direction of a municipal department or office.

municipal discharge — The release of effluent into a drainage basin from wastewater treatment plants that receive wastewater from households, commercial establishments, and industries.

municipal separate storm sewer system — A publicly owned system of conveyances that discharge into the waters of the United States and is designed or used for collecting or conveying stormwater and is not a combined sewer or part of a publicly owned treatment works.

municipal sewage sludge (MSS) — Sludge produced during the operation of a municipal or public sewage plant operation.

municipal solid waste (MSW) — The normal, residential, and commercial solid waste generated within a community and usually collected at the structures.

municipal solid waste incinerator (MSWI) — A privately or publicly owned incinerator primarily designed and used to burn residential and commercial solid waste.

municipal waste — The combination of street litter, discarded parts of automobiles, power plant and incinerator ashes and residue, sludge from sewage, dead animals, and abandoned cars and trucks generated by a community.

municipal waste treatment plant (MWTP) — A municipal facility that treats sewage and releases the effluent into a body of water.

municipal water system — A water system that has at least five service connections or which regularly serves 25 individuals for 60 days; also called a public water system.

muriatic acid — An acid used to reduce pH, alkalinity, stain, and scale.

murine model — A model of disease using mice as the animals being infected.

murine typhus fever — (*disease*) A mild form of typhus, producing symptoms of headache, fever, and some pain. Incubation time is 1 to 2 weeks, usually 12 days. It is caused by *Rickettsia typhi* and is found worldwide. The reservoir of infection is rats. It is transmitted by fleas, usually *Xenopsylla cheopis*, whose feces contaminate the bite site of the person; the fleas remain contaminated for life. The disease is not transmitted from person to person; general susceptibility. It is controlled by using pesticides where rats travel from area to area and avoiding exposure to fleas that might have been on rats.

Musca domestica — The house fly; a fly about 1/4 inch long, gray or buff in color; the male is a little

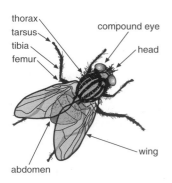

thorax
tarsus
tibia
femur
compound eye
head
wing
abdomen

House Fly
(Musca domestica)

smaller than the female. The body and legs are covered with hairs that readily hold microorganisms; the life cycle includes egg, larva, pupa, and adult.

muscarinism — Poisoning due to ingestion of poisonous varieties of mushrooms.

muscle — Type of tissue composed of fibers that are able to contract, causing or allowing movement of the parts and organs of the body.

muscle fiber — The contractile unit of a muscle.

muscle poison — A poison that causes a tingling of the neck and face, colic, vomiting, diarrhea, elevation of blood pressure, and death; includes barium salts, benzene, digitalis, and potassium salts.

muscle relaxant — Chemical agent that reduces the contractility of muscle fibers.

muscle spindle — Specialized proprioceptive sensory organ composed of a bundle of fine striated muscle fibers innervated by gamma nerve fibers.

muscle sprain — A torn muscle fiber, typically microscopic.

muscle tissue — Type of tissue that allows movement, including skeletal, smooth, and cardiac, which may be found in bundles of long cells.

musculoskeletal disorder — A disorder of the muscles, nerves, tendons, ligaments, joints, cartilage, blood vessels, or spinal discs resulting in muscle strains and tears, ligament sprains, joint and tendon inflammation, pinched nerves, and spinal disk degeneration.

musculoskeletal system — A system composed of or pertaining to the muscles and the skeleton.

MUSE — Acronym for monitor of ultraviolet solar energy.

mushroom — Common name for a group of fungi (Basidiomycetes) that produce an aboveground reproductive structure.

mushroom poisoning — Toxic condition caused by the ingestion of certain mushrooms, especially *Amanita muscaria*; it produces an intoxication within a few minutes to 2 hours. Symptoms include lacrimation, salivation, sweating, vomiting, difficulty in breathing, abdominal cramps, diarrhea, and, potentially, convulsions, coma, and circulatory failure. Intoxication by *A. phalloides* and *A. verna* causes similar symptoms, as well as liver damage, renal failure, and death in 30 to 50% of the cases. *Psilocybe mexicana* produces hallucinatory effects.

mussel poisoning — Poisoning due to the ingestion of mussels that have consumed contaminated plankton.

mustard gas ($C_4H_8Cl_2S$) — A general term referring to several chemicals, most commonly sulfur mustard. A colorless, odorless, oily liquid that becomes brown with a slight garlic scent when mixed with other chemicals. MW: 159.08; BP: 215 to 217°C; Sol: soluble in common organic solvents, fats, and lipids; Fl.P: 221°F; density: 1.274 at 20°C. It is used as a vesicant chemical warfare agent. It is hazardous to the respiratory system, skin, and possibly the gastrointestinal system and is toxic through inhalation and contact. Symptoms of exposure include painful coughing, vomiting, burning eyes, shock, chest tightness, sneezing, rhinorrhea, and sore throat. Delayed symptoms include chronic bronchitis, recurrent pneumonia, skin burns, nausea, anoxia, abdominal pain, diarrhea, headache, and lassitude. It is potentially carcinogenic. Department of Defense exposure limit (TWA): 3 mg/m^3 [air] or 0.1 mg/m^3 for the general population. Also known as *bis*-(2-chloroethyl) sulfide.

$$Cl-CH_2-CH_2-S-CH_2-CH_2-Cl$$

Mustard gas

MUTA — Acronym for mutagenicty.

mutagen — A chemical or physical agent that induces a permanent change in the genetic material.

mutagenesis — Induction of changes in genetic material that are transmitted during cell division.

mutagent — An agent, such as a chemical or radiation, that damages or alters genetic material in cells.

mutant — An organism that contains a trait that is not inherited but is a result of a change in its genes.

mutation — Permanent change in the genetic makeup of an organism that results in a new characteristic which may be passed on to offspring.

mutualism — Two species that live together in a relationship in which both benefit.

mV — See millivolt.

MV — See megavolt.

MV/m — See megavolt per meter.

MVV — Acronym for maximal voluntary ventilation.

mW — See milliwatt.

MW — See molecular weight; megawatt.

MWC — Acronym for machine wet cleaning.

MWd — See megawatt-day.

MWh — See megawatt- hour.

MWL — Acronym for municipal waste leachate.

MWTA — See Medical Waste Tracking Act.

MWTP — See municipal waste treatment plant.

MX gene — A human gene that helps the body resist viral infections.

myalgia — A diffuse muscle pain usually accompanied by malaise.

mycelium — The mass of interwoven filamentous hyphae that forms the vegetative portion or thallus

M

of a fungus and is often submerged in the tissues of a host.

Mycobacterium — A genus of Gram-positive, aerobic, acid-fast bacteria occurring as slightly curved or straight rods and containing many species, including the highly pathogenic organisms that cause tuberculosis and leprosy.

mycologist — A person who studies fungi.

mycology — Branch of biology dealing with fungi.

mycosis — Any infection or disease caused by fungi.

mycotoxin — A toxic substance produced by a fungus, especially a mold.

mycovirus — A virus that infects fungi.

mydriasis — Excessive dilation of the pupil of the eye.

myelin — A fatty material forming the medullary sheath of nerve fibers throughout the body.

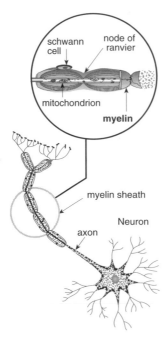

Myelin Sheath

myocardial — Pertaining to the muscular tissue of the heart.

myocardial infarction — Local arrest or sudden insufficiency of arterial or venous blood supply of an area of the heart muscle.

myocardium — A thick, contractile, middle layer of uniquely constructed and arranged muscle cells that forms the bulk of the heart wall.

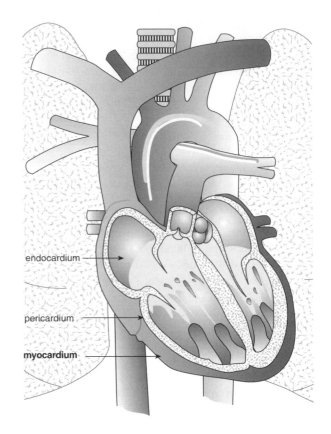

Myocardium

myoclonic — Shock-like, as with contractions of a muscle or a group of muscles.

myoglobin — The oxygen-transporting pigment of muscle tissue.

N

n — See nano- or flow exponent; (*hydraulics*) see Manning's roughness coefficient.

N — Notation for number of neutrons; see Newton or nitrogen.

N_2O — See nitrous oxide.

NAA — Acronym for nonattainment area.

NAAEC — See North American Agreement on Environmental Cooperation.

NAAQS — See National Ambient Air Quality Standards.

NAAS — Acronym for nerve agent antidote system.

NAD — Acronym for no appreciable disease; see nicotinamide adenine dinucleotide.

nadir — The point on Earth directly beneath a satellite.

NADP — Acronym for National Atmospheric Deposition Process Network.

NAERG — See *North America Emergency Response Guidebook.*

NAFTA — Acronym for North American Free Trade Agreement.

NAIC Code — See North American Industrial Classification Codes.

NAMS — Acronym for National Air Monitoring Station.

NANB — Acronym for non-A, non-B hepatitis.

nano- (n) — A prefix in the metric system representing 10^{-9}.

nanocurie (nCi) — A unit of radiation equal to one billionth (10^{-9}) of a curie.

nanogram (ng) — A unit of weight equal to one billionth (10^{-9}) of a gram.

nanograms per cubic meter — A unit of weight equal to one billionth (10^{-9}) of a gram of a substance in a cubic meter of air, soil, or water.

nanometer — A unit of length equal to one billionth (10^{-9}) of a meter; also known as nanon.

nanon — See nanometer.

nanopore — A soil pore having dimensions measured in nanometers.

nanosecond — A unit of time equal to one billionth (10^{-9}) of a second.

nanotesla — A unit of magnetic flux density equal to 1×10^{-9} weber per square centimeter.

NAO — Acronym for North Atlantic Oscillation.

napalm — An aluminum soap made from coprecipitating aluminum hydroxide naphehenic acid and palmitic acid or jelling gasoline using hydroxyaluminum-*bis*-(2-ethylhexanoate). It is used in chemical warfare as a fire bomb or flame land mine. It is also used with incendiary devices to burn targets that are difficult to ignite for a short duration. It will stick to the target when it burns.

NAPAP — Acronym for National Acid Precipitation Assessment Program.

naphtha (coal tar) — A reddish-brown, mobile liquid with an aromatic odor; MW: 110 (approx.), BP: 320 to 428°F, Sol: insoluble, Fl.P: 100 to 109°F, sp. gr. 0.89 to 0.97. It is used in the preparation of coal-tar paints; in the preparation of coumarone and indene; as a solvent in the rubber industry; in the manufacture of waterproof cloth, shoe adhesives, and rubber tires; as a solvent, diluent, or thinner in the paint, varnish, and lacquer industries; in the formulation of nitrocellulose and ethylcellulose; as a solvent for polymerized styrol, phenolic varnishes, urea, resins, melamine, and other synthetic resins; as a solvent for DDT. It is hazardous to the respiratory system, eyes, and skin and is toxic through inhalation, ingestion, and contact. Symptoms of exposure include lightheadedness, drowsiness, dermatitis, and irritation of the eyes, nose, and skin. OSHA exposure limit (TWA): 100 ppm [air] or 400 mg/m³.

naphthalene ($C_{10}H_8$) — A colorless to brown solid with an odor of mothballs; shipped as a molten solid; MW: 128.2, BP: 424°F, Sol: 0.003%, Fl.P: 174°F, sp. gr. 1.15. It is used in the formulation of insecticides and moth repellent as flakes, powder, balls, or cakes; as a fumigant; in the manufacture of chemical intermediates for production of pharmaceuticals, resins, dyes, plasticizers, solvents, coatings, insecticides, pigments, rubber chemicals, tanning agents, surfactants, waxes, cable coatings, textile spinning lubricants, rodenticides, and storage batteries. It is hazardous to the eyes, blood, liver, kidneys, skin, central nervous system, and red blood cells and is toxic through inhalation, absorption, ingestion, and contact. Symptoms of exposure include confusion, headache, excitement, malaise, nausea, vomiting, abdominal pain, bladder irritation, profuse sweating, jaundice, hemoglobinuria, renal shutdown, and dermatitis. OSHA exposure limit (TWA): 10 ppm [air] or 50 mg/m³. Also known as tar camphor.

Naphthalene

2-naphthol (C$_{10}$H$_8$O) — A crystalline solid that darkens on exposure to light and has a slight phenolic odor; MW: 144.18, BP: 285°C, Sol: soluble in organic solvents, Fl.P: a noncombustible solid, density: NA. It is used in the manufacture of dyes, perfumes, medicinal organics; in the production of antioxidants for synthetic rubber. It is hazardous to the skin and digestive tract and respiratory tracts. It may cause nausea, vomiting, diarrhea, abdominal pain, convulsions, respiratory failure, and death.

2-Naphthol

α-naphthyl thiourea (ANTU; C$_{10}$H$_7$NHC(NH$_2$)S) — A white crystalline or gray odorless powder; MW: 202.3, BP: decomposes, Sol: 0.06%, Fl.P: N/A. It is used as a rodenticide. It is hazardous to the respiratory system and is toxic through inhalation and ingestion. Symptoms of exposure after ingestion of large doses include vomiting, dyspnea, cyanosis, coarse pulmonary râles, and mild liver damage. OSHA exposure limit (TWA): 0.3 mg/m³ [air].

1-naphthylamine (C$_{10}$H$_7$NH$_2$) — A colorless crystal with a disagreeable odor that turns red with exposure to air and light; MW: 143.20, BP: 301°C, Sol: dissolves readily in alcohol and ether, Fl.P: a noncombustible solid. It is used in the manufacture of dyes. It is hazardous to the digestive system and skin and causes acute hemorrhagic cystitis, dyspnea,

1-Naphthylamine

ataxia, dysuria, hematuria; it is a cancer-causing substance. ACGIH exposure limit: NA.

2-naphthylamine (C$_{10}$H$_7$NH$_2$) — A white to reddish crystalline solid; MW: 143.20; BP: 396°C; Sol: soluble in hot water, alcohol, and ether; Fl.P: noncombustible solid. It used in the manufacture of dyes and rubber. It is hazardous to the digestive tract, skin, and respiratory tract. Symptoms include tumors in the kidney, bladder, liver, lungs, skin, and blood tissues. It produces severe acute toxic effects resulting in respiratory distress and hypoxia. It is carcinogenic. ACGIH exposure limit: NA

2-Naphthylamine

α-naphthylamine (C$_{10}$H$_7$NH$_2$) — A colorless crystal with an ammonia-like odor; darkens in air to a reddish-purple color; MW: 143.2, BP: 583°F, Sol: 0.002%, Fl.P: 315°F, sp. gr. (77°F): 1.12. It is used in the manufacture of dyes, herbicides, and rubber antioxidants. It is hazardous to the bladder and skin and is toxic through inhalation, absorption, ingestion, and contact. Symptoms of exposure include dermatitis, hemorrhagic cystitis, dyspnea, ataxia, methemoglobinemia, and dysuria; carcinogenic. OSHA exposure limit (TWA): regulated use.

NAPL — Acronym for non-aqueous-phase liquid.

NAPRI — Acronym for North American Pollutant Release Inventory.

NAPS — Acronym for National Air Pollution Surveillance.

narcosis — A state of stupor, unconsciousness, or arrested activity produced by the influence of narcotics or other chemicals.

narcotic — A chemical agent that diminishes awareness of sensory impulses and in large doses can cause coma.

narel — Acronym for National Air and Radiation Environmental Laboratory.

nares — The anterior and posterior openings in the nose that allow the passage of air from the nose to the pharynx and the lungs during respiration.

NARSTO — Acronym for North American Research Strategy for Tropospheric Ozone.

nasal — The nose and nasal cavity.

nasal cannula — A rubber or vinyl tube that extends around a user's face; it has curved prongs that fit into the nostrils for delivery of oxygen at low flow rates.

nascent — Pertaining to something that is just forming.

nasopharynx — The uppermost of the three regions of the throat situated behind the nose extending from the posterior nares to the level of the soft palate.

natal rate — The ratio of total number of births to the total population of a given community at a given time; also known as birth rate.

natality — Birth rate.

National Ambient Air Quality Standards (NAAQS) — Standards established by the federal government for regulated pollutants in ambient air.

Pollutant	Averaging Period	Primary Standard	Secondary Standard
O$_3$	1 hr	0.125 ppm	0.125 ppm
	8 hr	0.085 ppm	0.085 ppm
CO	1 hr	35 ppm	
	8 hr	9 ppm	
SO$_2$	3 hr		0.5 ppm
	24 hr	0.145 ppm	
	annual	0.03 ppm	
NO$_3$	annual	0.053 ppm	0.053 ppm
Respirable Particulate Matter PM$_{10}$	24 hr	150 µg/m^3	150 µg/m^3
	annual	50 µg/m^3	50 µg/m^3
Respirable Particulate Matter PM$_{2.5}$	24 hr	65 µg/m^3	65 µg/m^3
	annual	15.1 µg/m^3	15.1 µg/m^3
Lead	quarter	1.55 µg/m^3	1.55 µg/m^3

National Ambient Air Quality Standards (NAAQS)

National Containment Occurrence Database — A database required by the Safe Drinking Water Act and which provides a repository and access to data on the occurrence of contaminants in drinking water that would support decision making for future drinking water contaminant regulations.

National Cooperative Soil Survey — A system for classifying soil material based on textural classes and the relative proportions of sand, silt, and clay in the soil fraction.

National Defense Authorization Act — A federal law enacted in 1994, and amended in 1995, requiring the Secretary of Energy to prepare the Baseline Report.

National Drinking Water Contaminant Occurrence Database — A database of the U.S. EPA developed to satisfy the statutory requirements set by Congress and the 1996 Safe Drinking Water Act amendments to help in making decisions related to identifying contaminants for regulation and subsequent regulation development.

National Emission Standards for Hazardous Air Pollutants (NESHAPS) — Standards set by the Environmental Protection Agency in compliance with the Clean Air Act for pollutants that could contribute to an increase in mortality or serious illness.

National Environmental Data Index — An index developed by the NOAA which provides access to environmental and descriptive data to facilitate their use and promotes protection of human health and safety and avoidance of disease.

National Environmental Policy Act (NEPA) — An act of Congress passed in 1969 and setting forth the responsibilities of federal agencies declaring a national policy that will encourage productive and enjoyable harmony between people and the environment; to promote efforts that will prevent or eliminate damage to the environment and biosphere; to stimulate the health and welfare of people; to enrich the understanding of ecological systems and natural resources important to the nation; and to establish the Council on Environmental Quality.

National Fire Codes — Sixteen volumes of codes, standards, recommended practices, and manuals developed by the National Fire Protection Association technical committees.

National Geologic Mapping Act — A law mandating that the U.S. Geological Service develop a National Geologic Map Database to be used as a national archives containing geologic maps and related databases.

National Health Planning and Resources Development Act of 1974 — A federal law that establishes a nationwide network of health system agencies for the coordination and direction of national health policy through state and regional regulatory agencies.

National Municipal Plan — A policy created in 1984 by the U.S. EPA and states to bring all publicly owned treatment works into compliance with Clean Water Act requirements.

National Occupational Exposure Survey (NOES) — A survey conducted in a sample of nearly 5000 establishments from 1981 to 1983 to compile data on the types of potential exposure agents found at the workplace and the kinds of safety and health programs implemented at the plant level.

National Ocean Pollution Planning Act — A law passed in 1978 and updated through 1988; states that the activities of people in marine environment can have profound short- and long-term impact on the environment and affect ocean and coastal resources; calls for a federal plan for ocean pollu-

N

tion research, monitoring of materials that have been dumped, fate of material, and effects of pollutants on marine environment.

National Oil and Hazardous Substances Contingency Plan — A federal regulation that guides the determination of sites to be corrected under the Superfund program and the program to prevent or control spills into surface waters or elsewhere.

National Pollutant Inventory — An Australian Internet database of pollutant emissions designed to provide the community, industry, and government with information on the types and amounts of certain substances being emitted to the environment.

National Pollutant Release Inventory — A source of information on environmental releases of toxic chemicals and waste management of those chemicals.

National Pollution Discharge Elimination System (NPDES) — A national program for issuing, modifying, revoking, reissuing, terminating, monitoring, and enforcing permits and imposing and enforcing pretreatment requirements under the Clean Water Act by the U.S. EPA; establishes effluent limitation discharge standards for industrial categories and sewage treatment plants based on the best available technology economically feasible.

National Primary Drinking Water Regulations (NPDWR) — Drinking water regulations establishing maximum levels of chemical, physical, and biological agents permitted in drinking water.

National Priorities List (NPL) — The U.S. EPA's official list of hazardous waste sites that need to be corrected under Superfund.

National Response Team (NRT) — A group consisting of and representative of 14 government agencies prepared for a response in the event of an emergency.

National Strike Force — A highly trained group of individuals directed by the U.S. Coast Guard to back up federal on-scene coordinators in the event of a major oil spill.

National Warning System — A dedicated, commercially leased nationwide telephone warning system operated on a 24-hour basis and designed and maintained to warn the public of a nuclear attack or a natural or person-made disaster.

National Water Data Exchange Program — A program that used to provide water data; now replaced by the USGS Water Resources Web site and the U.S. EPA STORET.

National Water Quality Assessment Program — A program designed to describe current water quality conditions for a large part of U.S. freshwater streams, rivers, and aquifers.

natural convection — Movement of heat through a fluid in a body that occurs when warm, less dense fluid rises and cold, dense fluid sinks because of the influence of gravity.

natural draft — The gas flow created by the height of a stack or chimney and the difference in temperature between flue gases and the atmosphere.

natural gas — A combustible gas composed largely of methane and other hydrocarbons with variable amounts of nitrogen and noncombustible gases; obtained from natural earth fissures or wells; used as a fuel in the manufacture of carbon black and in chemical synthesis.

natural gas liquids — Liquid hydrocarbons that have been extracted from natural gas, such as at ethane, butane, and propane.

natural graphite — A crystalline form of carbon used to make lead pencils; also used in electrodes and electrical equipment. High concentrations may cause pneumoconiosis and anthracosilicosis. ACGIH exposure limit (TLV-TWA): 2.5 mg/m^3.

natural history of disease — The history of a disease from when it is acquired, manifests, and terminates; includes information on susceptibility, presymptomatic stage, clinical disease, and stage of disability.

natural immunity — An immunity present in an individual at birth and not artificially acquired.

natural killer cells — Effector cells that have the capacity for spontaneous cytotoxicity toward various target cells.

natural radioactivity — Radioactivity exhibited by more than 50 naturally occurring radionuclides.

natural resource damages — The injury to, destruction of, or loss of natural resources.

natural selection — Survival of the fittest, by which organisms that successfully adapt to their environments survive and those that do not disappear; first propounded by Charles Darwin.

natural soil drainage — The condition of soil in which the removal of water occurs by natural means.

natural sources — (*air pollution*) Natural emission sources, including biological and geological, wildfires, and windblown dust.

natural ventilation — Air movement caused by wind or temperature differences.

nausea — An unpleasant sensation usually felt when nerve endings in the stomach and other parts of the body are irritated; accompanied by a tendency to vomit.

nautical mile — A unit of distance equal to 1852 meters, which is very close to the mean value of 1 minute of latitude, which varies from approximately 1843 m at the equator to 1861.6 m at the pole.

NAWAS — See National Warning System.

NAWDEX — See National Water Data Exchange.

NCA — See Noise Control Act.

NCE — Acronym for net carbon emission.

NCP — Acronym for National Contingency Plan.

NCPDI — Acronym for National Coastal Pollutant Discharge Inventory.

NCWS — See Non-Community Water System.

NDIR — Notation for nondispersed infrared analyzer.

NDVI — Acronym for normalized difference vegetation index.

nebulize — To disperse a liquid in a fine spray.

nebulizer — An atomizer device that sprays liquid medication in aerosol form into the air a patient breathes.

neburon (C_{12}H_{16}Cl_2N_2O) — A crystalline solid; MW: 275.20, BP: NA, Sol: slightly soluble in hydrocarbon solvents, density: NA. It is used as an herbicide and is moderately toxic by the intravenous route.

Neburon

NEC — Acronym for National Electrical Codes.

Necator — A genus of nematode that is an intestinal parasite causing hookworm disease.

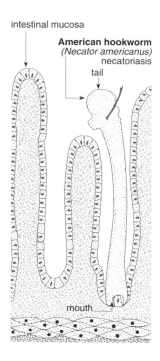

Necator

necatoriasis — See hookworm disease.

Necrobia rufipes — See Red-Legged Ham Beetle.

necrogenic — Capable of causing death of cells or tissues.

necrology — Study of the causes of death, including the preparation and interpretation of mortality statistics.

necrolysis — Exfoliation of dead tissue.

necropsy — See autopsy.

necrosis — The morphological changes indicative of cell death as a result of injury, disease, or other pathogenic state.

necrotic — The death of tissue caused by disease or injury.

necrotizing entertitis — An acute inflammation of the small and large intestine by the bacterium *Clostridium perfringens*; symptoms include severe abdominal pain, bloody diarrhea, and vomiting.

necrotrophic — A nutritional mechanism by which an organism produces a group of hydrolytic enzymes to kill and break down host cells and then absorb nutritional compounds from the dead organic material.

NEDS — Acronym for National Emission Data System.

needle biopsy — The removal of a piece of living tissue by means of inserting a hollow needle through the skin or organ for microscopic examination.

needle valve — (*plumbing*) Any of a family of valves regulating flow through the use of a needle moving into and out of an orifice.

needs assessment — Identification of differences between current or projected future levels of concern and the desired levels.

negative — An indication that a substance or reaction is not present.

negative electrode — A cathode, or the negative pole of an electric current, battery, or dry cell.

negative feedback — An interaction that reduces or dampens the response of a system in which it is incorporated.

negative pressure — A condition that exists when less air is supplied to a space than is exhausted from the space, so the air pressure within the space is less than that in the surrounding area; a vacuum.

negative temperature coefficient — A decrease in resistance with an increase in temperature.

negligence — The act of not taking prudent care.

neighborhood — An area comprising all the public facilities and conditions required by the average family for their comfort and existence.

Neisseria — A genus of Gram-negative aerobic or facultatively anaerobic cocci which are part of the normal flora of the oropharynx, nasopharynx, and genitourinary tract.

NEL — The no-effect level in high-to-low-dose extrapolation models.

nemafos (C_{18}H_{13}N_2O_3PS) — An amber liquid; MW: 248.26, BP: 80°C, Sol: readily mixes with organic

N

solvents, density: NA. It is a cholinesterase inhibitor and is used as a pesticide. It is extremely toxic and hazardous to the skin, digestive tract, and respiratory tract. It exhibits acute, delayed, and chronic effects including blurred vision, headaches, weakness, mental confusion, vomiting, diarrhea, stomach pain, convulsions, coma, respiratory paralysis, and death.

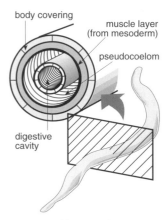

Nemafos

nematocide — A pesticide chemical or agent that prevents injury from or destroys nematodes.

Nematoda — A class of unsegmented round worms with elongated bodies that are pointed at both ends; includes hookworms, trichina, and *Ascaris*, many of which are parasites.

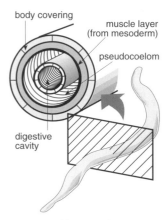

Nematoda

neonatal — Pertaining to the period within 28 days after birth.

neonatal death — The death of a liveborn infant under 28 days of age.

neoplasia — A condition characterized by the presence of new growths or tumors.

neoplasm — Any new and abnormal growth, specifically one in which cell multiplication is uncontrolled and progressive; a benign or malignant tumor.

neoprene — Synthetic rubber closely paralleling natural rubber in flexibility and strength; used in paints, putties, adhesives, the soles of shoes, and tank linings.

neper (Np) — Division of a logarithmic scale.

nephelometer — (*optics*) An instrument for measuring the concentration of suspended matter in a liquid dispersion by measuring the amount of light scattered by the dispersion.

nephelometric — A means of measuring turbidity in a sample by using an instrument called a nephelometer; light passes through a sample and the deflected light, usually at a 90° angle, is then measured.

nephelometric turbidity unit (NTU) — Unit of measurement for water clarity.

nephometry — Measurement of clouds.

nephritis — Inflammation of the kidney.

nephrologist — Licensed physician who specializes in the treatment of kidney disease.

nephron — Structural and functional unit of the kidney; each nephron is capable of forming urine by itself.

nephrosis — Any kidney disease, especially disease marked by purely degenerative lesions of the renal tubules.

nephrotoxin — A substance that causes injury to the kidneys, including halogenated hydrocarbons and uranium.

neritic — Of or pertaining to the part of the ocean that is associated with the continental shelf, from the seacoast to approximately 660 feet out.

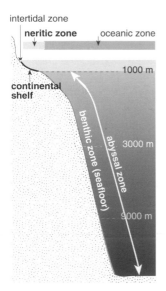

Neritic Zone

nerve — A bundle of nerve fibers bound together by connective tissue leading to or from the central nervous system.

nerve agent — The organic esters of phosphoric acid used as chemical warfare agents because of their extreme toxicity, including Tabun-GA, Sarin-GB, and Soman-GB, -GF, and -VX, which are powerful inhibitors of the enzyme acetylcholinesterase and allow an excess accumulation of acetylcholine, thus presenting severe danger to individuals. The nerve agents are readily absorbed by inhalation and/or through intact skin.

nerve cell — A cell specialized to originate or transmit nerve impulses.

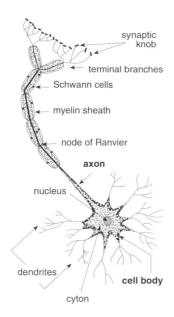

Nerve Cell Body

nerve cell body — The largest part of a neuron that typically contains the nucleus.

nerve compression — Harmful pressure on one or more nerve trunks resulting in nerve damage and muscle weakness or atrophy.

nerve cord — A long, compact bundle of nerve cells that is part of the central nervous system.

nerve excitability — The readiness of a nerve to respond to a stimulus.

nerve fiber — A slender process of a neuron, usually the axon, which may be myelinated or unmyelinated.

nervous system — The entire network of nerve cells that enables an organism to be aware of its environment and react to it; in vertebrates, the system consists of the ganglia, cranial and peripheral nerves, spinal cord, brainstem, and brain.

NES — Acronym for National Eutrophication Survey.

NESHAPS — See National Emission Standards for Hazardous Air Pollutants.

nested design — (*epidemiology*) The combination of two study designs of a problem; used to gather more data at less cost and more quickly.

net area — The portion of a site that can actually be built upon.

net energy yield — Total useful energy produced during the lifetime of an entire energy system minus the energy used, lost, or wasted in making the useful energy available.

net heating value — Gross heating value minus the latent heat of vaporization of the water formed by

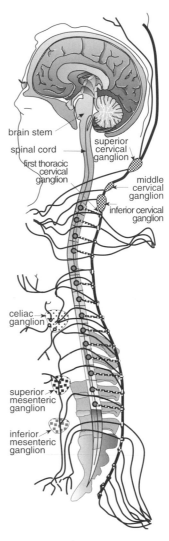

Nervous System

the combustion of the hydrogen in the fuel; also known as low heat value.

net primary production — The net amount of carbon dioxide taken in by vegetation in a particular area as determined by subtracting the carbon dioxide produced by plants (photorespiration) from the carbon dioxide used by plants (photosynthesis).

netting — A concept in which all emission sources in the same area, owned or controlled by a single company, are treated as one large source, allowing flexibility in controlling individual sources to meet a single emission standard.

network — Two or more interconnected computer systems for the implementation of specific functions.

network performance — (*computer science*) The ability of a network to transmit data.

neural — Relating to nerve cells and their processes.

neural loss — Hearing loss.

N

neuralgia — A pain in a nerve or along the course of one or more nerves.

neurasthenia — An abnormal condition with symptoms of nervous exhaustion and vague fatigue that often follows depression.

neuraxon — See axon.

neurite — See axon.

neuritis — Noninflammatory lesions of the peripheral nervous system.

neuroanatomy — The anatomy of the brain, spinal cord, and peripheral nervous system.

neurogenic — The formation of nervous tissue.

neurohormone — A chemical transmitter produced by nervous tissue using the bloodstream or other body fluids for distribution to its target site.

neurologic assessment — Evaluation of a patient's neurologic status and symptoms.

neurological examination — Complete study of the nervous system including mental status, function of each of the cranial nerves, sensory and neuromuscular function, reflexes, and other cerebral functions.

neurologist — A physician who specializes in dealing with the nervous system and its disorders.

neurology — Branch of medical science dealing with the nervous system.

neuromotor — Nerve impulses transmitted to muscles; refers to nerves and muscles.

neuron — The nerve cell and its branches along which nerve impulses travel.

neuropathy — An inflammation or degeneration of the peripheral nerves.

neurosecretions — The hormone-like chemical substances that are produced by specialized neurons and affect various physiological processes in the body.

neurosis — An emotional disorder that can interfere with normal health and daily living; anxiety is the chief symptom.

neurotoxicity — The occurrence of adverse effects on a nervous system following exposure to a chemical.

neurotoxin — A poisonous chemical that produces primary toxic effects on the nervous system, with symptoms of narcosis, behavioral changes, and a decrease in motor functions; chemicals include mercury and carbon disulfide.

neurotransmitter — A chemical substance secreted by the terminal end of an axon that stimulates a muscle fiber contraction or impulse in another neuron.

neurotrauma — Mechanical injury to a nerve.

neutral — Neither basic nor acidic; referring to the absence of a net electrical charge.

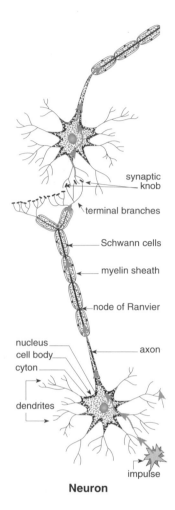

Neuron

neutral posture — Body position that minimizes stresses on the body, typically near the mid-range of the range of motion of any joint.

neutralization — (*chemistry*) Chemical reaction between an acid and a base.

neutralize — (*chemistry*) To make a solution neutral by adding a base to an acidic solution or an acid to a basic solution.

neutrino — A neutral particle having zero rest mass, originally postulated to account for the continuous distribution of energy among particles in the beta-decay process.

neutron — An electrically neutral subatomic particle with approximately the same mass as a proton and

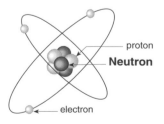

Neutron

that is within the nucleus of an atom; has an approximate mass of greater than 1.

neutron probe — An instrument used to estimate soil moisture, where the rate of attenuation in pulsed neutron emissions is related to soil water content.

neutrophil — A polymorphonuclear granular leukocyte that stains easily with neutral dyes; see polymorphonuclear leukocytes.

New Source Performance Standards — Uniform, national U.S. EPA air emissions standards that limit the amount of pollution allowed from new sources or from modified existing sources.

newborn — A human offspring from the time of birth through the 28th day of life.

newton (N) — An SI unit of force equal to the force required for 1 kilogram to accelerate 1 meter per second squared; formerly known as large dyne.

newton-meter of energy — See joule.

Newton's law of universal gravitation — The force of attraction between two small bodies is proportional to the product of their masses and inversely proportional to the square of the distance between their centers.

Newton's laws of motion — (1) Every body continues in a state of uniform motion in a straight line unless acted upon by some external force; (2) the time rate of change of momentum is proportional to the impressed force; (3) for every force (action) there is an equal and opposite force (reaction).

NEXUS — See numerical examination of urban smog.

NFRAP — Acronym for no further remedial action planned.

NFSAM — Acronym for *National Foods Security Act Manual*.

ng — See nanogram.

niche — See ecological niche.

nickel–cadmium cell — A rechargeable dry cell containing nickel oxide, cadmium, and an electrolyte of potassium hydroxide.

nickel carbonyl (Ni(CO)$_4$) — A colorless to yellow liquid with a musty odor; a gas above 110°F; MW: 170.7, BP: 110°F, Sol: 0.05%, Fl.P: 4°F, sp. gr. 1.32 at 63°F. It is used in the Mond process for nickel refining; in plating operations on foundry patterns, steel, and electronics manufacture; as a reagent in the synthesis of acrylic esters; as a catalyst or reagent in organic synthesis; in petroleum and petrochemical processing. It is hazardous to the nasal cavities, lungs, and skin and is toxic through inhalation, ingestion, and contact. Symptoms of exposure include sensitization dermatitis, allergic asthma, and pneumonitis; carcinogenic.

OSHA exposure limit (TWA): 0.001 ppm [air] or 0.007 mg/m^3.

nickel cyanide (Ni(CN)$_2$) — A yellow-brown powder when dry, greenish when wet with a weak almond odor; MW: 110.75, BP: decomposes, Sol: insoluble in water, Fl.P: nonflammable (poisonous gases, including cyanide, are produced in a fire), sp. gr. 2.41 at 25°C. It is used in metallurgy and electroplating. It is hazardous to the respiratory system, skin, and eyes. Acute exposure may cause eye irritation and a pneumonia-like illness with cough and shortness of breath. Chronic exposure may damage the heart, liver, kidneys, and lungs. It also causes burning, itching, redness, and a rash on the skin.

nickel dermatitis — An allergic contact dermatitis caused by exposure to the metal nickel.

nickel metal (Ni) and other compounds — A lustrous, silvery metal; MW: 58.7, BP: 5139°F, Sol: insoluble, sp. gr. 8.90. It is used in the manufacture of more than 3000 alloys; in electronic tube parts, coins, heavy machinery, tools, instrument parts, magnets, food and chemical processing equipment, flatware, jet engines, automotive parts, zippers, surgical and dental instruments, and cooking utensils; in chemical synthesis; in the textile industry. It is hazardous to the lungs, paranasal sinus, and central nervous system and is toxic through inhalation, ingestion, and contact. Symptoms of exposure include headache, vertigo, nausea, vomiting, epigastric pain, substernal pain, cough, hyperpnea, cyanosis, weakness, leukocytosis, pneumonitis, delirium, and convulsions; carcinogenic. OSHA exposure limit (TWA): 0.1 mg/m^3 for soluble compounds, 1 mg/m^3 for metal and insoluble compounds.

nickel metal hydride battery — An environmentally safe, high-performance battery considered to be the best available electric vehicle technology.

nickel tetracarbonyl (Ni(CO)$_4$) — A colorless, volatile liquid with a characteristic sooty odor; MW: 170.75; BP: 43°C; Sol: soluble in most organic solvents; Fl.P: below −18°C (closed cup), highly flammable and forms explosive mixtures with air; density: 1.318 at 17°C. It is used in the manufacture of nickel powder and nickel-coated metals. It is an extremely toxic substance and hazardous to the respiratory tract, digestive tract, eyes, and skin. It causes dizziness, giddiness, headache, weakness, increased body temperature, irritation, rapid breathing, congestion of the lungs, tightness in the

N

chest, convulsions, hemorrhage, and death. ACGIH exposure limit (TLV-TWA): 0.05 ppm.

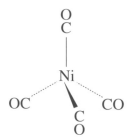

Nickel tetracarbonyl

nicotinamide adenine dinucleotide — A coenzyme that accesses an electron acceptor particularly important in respiration.

nicotine ($C_5H_4NC_4H_7NCH_3$) — A pale-yellow to dark-brown liquid with a fish-like odor when warm; MW: 162.2, BP: 482°F (decomposes), Sol: miscible, Fl.P: 203°F, sp. gr. 1.01. It is used as a pesticide and fumigant on vegetable crops, fruit, grasses and turf, and greenhouse plants. It is hazardous to the central nervous system, cardiovascular system, lungs, and gastrointestinal tract and is toxic through inhalation, absorption, ingestion, and contact. Symptoms of exposure include nausea, salivation, abdominal pain, vomiting, diarrhea, headache, dizziness, hearing and vision disturbance, confusion, weakness, incoordination, paroxysmal atrial fibrillation, convulsions, and dyspnea. OSHA exposure limit (TWA): 0.5 mg/m³ [skin].

NICT — Acronym for National Incident Coordination Team.

NIDCD — Acronym for National Institute on Deafness and Other Communications Disorders.

nidus — Specific locality of a given disease resulting in a unique combination of ecological factors that favor the maintenance and transmission of the disease organism.

nimbostratus — True rain clouds, darker than ordinary stratus and with a base; height from near the Earth to 6500 feet up; diffuse from continuous streaks of rain often extending to the ground.

nimbus — A family of clouds usually producing precipitation; the term is usually used in combination, such as in cumulonimbus.

NIMBUS — A series of research-oriented meteorological satellites used to develop sensor technology for determining vertical atmospheric temperature and moisture profile; uses microwave spectrometer and

electrically scanning microwave radiometer for vertical temperature profiles.

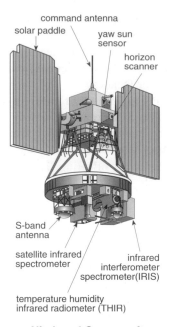

Nimbus 4 Spacecraft

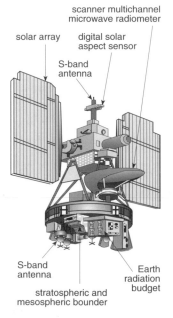

Nimbus 7 Satellite System

NIMBUS I — A satellite that photographs nighttime cloud cover from space.

NIOSH 7400 Method — A standardized method of air sample collection, preparation, and analysis by phase-contrast microscopy to quantify airborne fiber concentrations.

NIOSH equations — National Institute of Occupational Safety and Health guidelines consisting of a series of mathematical equations developed based on historical injury data and related job data to calculate recommended maximum safe weights of lift.

NIPDWR — Acronym for National Interim Primary Drinking Water Regulations.

nipple — (*plumbing*) A short piece of pipe with male threads on each end.

Nissl bodies — Large, granular bodies that stain with basic dyes, forming the substance of the reticulum of the cytoplasm of a nerve cell.

nit — The egg of a parasitic insect.

nitrasamines — See nitrosoamines.

nitrate — A salt or ester of nitric acid.

nitrate respiration — The use of nitrate as a terminal electron acceptor for anaerobic respiration.

nitrate-forming bacteria — Bacteria that converts nitrites into compounds that can be used by green plants to build proteins.

nitrates — General formula of $M(NO_3)_x$; includes potassium nitrate (KNO_3) and sodium nitrate ($NaNO_3$). Explosivity differs significantly among nitrates. They are oxidizing substances that decompose by heat. They are used in fireworks, matches, gunpowder, and freezing mixtures and for pickling meats. Ingestion of these compounds can cause severe gastroenteritis, and sodium nitrate may be lethal to humans.

nitric acid (HNO₃) — A colorless, yellow, or red fuming liquid with an acrid, suffocating odor; MW: 63.0, BP: 181°F, Sol: miscible, sp. gr. 1.50 at 77°F. It is used in metallurgy as a pickling agent; in metal refining, ore recovery, metal etching, and photoengraving; in acidulation of phosphate rock and in the manufacture of nitrogen solutions for use in the fertilizer industry; in the wood pulping industry; during organic synthesis in the manufacture of fertilizers, explosives, herbicides, antibiotics, meat-curing, pickling, ceramics, and pharmaceuticals; during organic synthesis in the manufacture of nitrating and oxidizing agents, nylons, foams, lubricants, insecticides, dyes, explosives, photographic films, lacquers, and celluloids. It is hazardous to the eyes, respiratory system, skin, and teeth and is toxic through inhalation, ingestion, and contact. Symptoms of exposure include delayed pulmonary edema, pneumonitis, bronchitis, dental erosion, and irritation of the eyes, mucous membrane, and skin. OSHA exposure limit (TWA): 2 ppm [air] or 5 mg/m³.

nitric oxide (NO) — A colorless gas; MW: 30.0, BP: –241°F, Sol: 5%. It is used as an intermediate in the synthesis of nitric acid, acrylonitrile, hydroxylamine, nitrosyl halides, nitrosyl hydrogen sulfate, and nitrogen dioxide; as a decomposition agent in certain gaseous products; as an additive to rocket propellants; in the bleaching of rayon. It is hazardous to the respiratory system and is toxic through inhalation and contact. Symptoms of exposure include drowsiness, unconsciousness, and irritation of the eyes, nose, and throat. OSHA exposure limit (TWA): 25 ppm [air] or 30 mg/m³.

nitrification — The biochemical oxidation of ammonium to nitrates by certain bacteria.

nitrifier — A nitrifying bacteria.

nitrifying bacteria — Gram-negative bacteria commonly found in the soil and obtaining energy through the process of nitrification.

nitrilotriacetate — The salt of nitrilotriacetic acid, which has the ability of complex metal ions and is proposed as a builder for detergents.

nitrilotriacetic acid (NTA; N(CH₂COOH)₃) — A white powder; MP: 240°C. It is used as a chelating agent in the laboratory, is hazardous to a developing fetus, and is toxic through ingestion. Also known as TGA.

nitrite — Any salt or ester of nitrous acid.

nitrite-forming bacteria — Bacteria that combine ammonia with oxygen to form nitrites.

nitrites — A general formula of $M(NO_2)_x$; nitrites are oxidizing agents as well as reducing substances and react violently.

m-nitroaniline (C₆H₆N₂O₂) — Yellow, needle-like crystals with a burning sweet odor; MW: 138.12, BP: 581 to 584.6°F (decomposes), Sol: 90 mg/100 ml, Fl.P: no data, sp. gr. 1.43. It is used as a dye-stuff

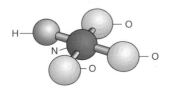

Nitric Acid

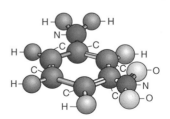

m-Nitroaniline

intermediate; other information is not available. OSHA exposure limit (TLV-TWA): not established.

p-nitroaniline (NO₂C₆H₄NH₂) — A bright-yellow, crystalline powder with a slight ammonia-like odor; MW: 138.1, BP: 630°F, Sol: 0.08%, Fl.P: 390°F, sp. gr. 1.42. It is used as a gasoline-gum inhibitor; as an intermediate in the pharmaceutical industry; in the production of intermediates for dyestuff manufacture; during antioxidation and antiozonation operations in rubber manufacture. It is hazardous to the blood, heart, lungs, and liver and is toxic through inhalation, absorption, ingestion, and contact. Symptoms of exposure include cyanosis, ataxia, tachycardia, tachypnea, dyspnea, irritability, vomiting, diarrhea, convulsions, respiratory arrest, anemia, and methemoglobinemia. OSHA exposure limit (TWA): 3 mg/m³ [skin].

nitrobenzene (C₆H₅NO₂) — A yellow, oily liquid with a pungent odor like paste shoe polish; a solid below 42°F; MW: 123.1, BP: 210.8°C, Sol: 0.2%, Fl.P: 190°F, sp. gr. 1.20. It is used in the production of chemical intermediates; in solvent refining of lubricating oils; in the production of intermediates in the synthesis of rubber chemicals, photographic chemicals, explosives, liquid propellants, and pharmaceuticals; as a solvent in specialized surface coatings; as a solvent in organic synthesis; as perfume in the manufacture of toilet and household soaps; in the synthesis of insecticides and germicides. It is hazardous to the blood, liver, kidneys, cardiovascular system, and skin and is toxic through inhalation, absorption, ingestion, and contact. Symptoms of exposure include anoxia, dermatitis, anemia, and eye irritation; in animals, liver and kidney damage. OSHA exposure limit (TWA): 1 ppm [skin] or 5 mg/m³.

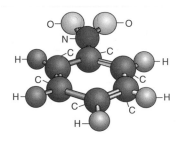

Nitrobenzene

4-nitrobiphenyl (C₆H₅C₆H₄NO₂) — A white to yellow, needle-like crystalline solid with a sweetish odor; MW: 199.2, BP: 644°F, Sol: insoluble, Fl.P: 290°F. It is used in the production of rocket fuel; as an industrial solvent; as an oxidant; as an additive in lubricants. It is hazardous to the bladder and blood and is toxic through inhalation, absorption, inges-

tion, and contact. Symptoms of exposure include headache, lethargy, dizziness, dyspnea, ataxia, weakness, methemoglobinemia, urinary burning, and acute hemorrhagic cystitis; carcinogenic. OSHA exposure limit (TWA): regulated use.

p-nitrochlorobenzene (ClC₆H₄NO₂) — Yellow crystalline solid with a sweet odor; MW: 157.6, BP: 468°F, Sol: insoluble, Fl.P: 261°F, sp. gr. 1.52. It is used in the production of chemical intermediates for the manufacture of pesticides, fungicides, and preservatives; in the manufacture of pharmaceuticals, rubber chemicals, antioxidants, gasoline gum inhibitors, corrosion inhibitors, and photographic chemicals. It is hazardous to the blood, liver, kidneys, and cardiovascular system and is toxic through inhalation, absorption, ingestion, and contact. Symptoms of exposure include anoxia, unpleasant taste, mild anemia; carcinogenic; in animals, hemoglobinuria. OSHA exposure limit (TWA): 1 mg/m³ [skin].

nitroethane (CH₃CH₂NO₂) — A colorless, oily liquid with a mild, fruity odor; MW: 75.1, BP: 237°F, Sol: 5%, Fl.P: 82°F, sp. gr. 1.05. It is used as a solvent in coatings and adhesives on cellulose esters and synthetic resins; as an intermediate in the synthesis of organic dyes, insecticides, pesticides, nitroplasticizers, pharmaceuticals, and other organic chemicals; as a propellant; as a reaction-media fluid; as an extraction solvent in petroleum fractionation; as a recrystallization solvent. It is hazardous to the skin and is toxic through inhalation, ingestion, and contact. Symptoms of exposure include dermatitis; in animals, lacrimation, dyspnea, pulmonary râles, edema, narcosis, and liver and kidney injury. OSHA exposure limit (TWA): 100 ppm [air] or 310 mg/m³.

nitrogen — A gaseous, nonmetallic element; atomic number: 7, atomic weight: 14.008. It constitutes approximately 78% of the atmosphere and is a component of all proteins and a major complement of most organic substances. It is found in mineral compounds and is the 17th most abundant element in the crust of the Earth.

nitrogen balance — The difference between nitrogen intake and nitrogen loss in the body or in soil.

nitrogen base — The basic component of nucleic acids composed of a nitrogen-containing molecule of purine or pyrimidine.

nitrogen cycle — The natural sequence through which nitrogen circulates in the biosphere. Nitrogen compounds in the soil are taken in and stored by plants; when the plants are eaten by animals, these compounds are broken down into a variety of new compounds, and a portion is passed on as waste. Along with nitrogen in the atmosphere, the nitrogen passed into the soil or by the decay of plants or animals is then converted by nitrogen-fixing bacteria into nitrogen compounds used by plants.

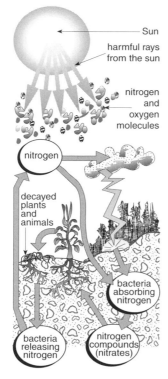

Nitrogen Cycle

nitrogen dioxide (NO₂) — A yellowish-brown liquid or reddish-brown gas (above 70°F) with a pungent acrid odor; in solid form (below 15°F) it is found structurally as N_2O_4; MW: 46.0, BP: 70°F, Sol: reacts, sp. gr. 1.44 (a liquid at 68°F). It is used during metal surface treatment with nitric acid; in the production of intermediates for the manufacture of sulfuric acid, nitric acid, and fertilizers; in chemical synthesis; in the synthesis of dyes; in the manufacture of nitrocellulose paints and lacquers; in the production and handling of rocket propellants. It is hazardous to the respiratory system and cardiovascular system and is toxic through inhalation, absorption, ingestion, and contact. Symptoms of exposure include cough, mucoid frothy sputum, dyspnea, chest pain, pulmonary edema, cyanosis, tachypnea, tachycardia, and eye irritation. Nitrogen dioxide pollution reacts with water vapor in the air, affecting visibility and contributing to acid rain. OSHA exposure limit (TWA): short-term, 1 ppm [air] or 1.8 mg/m³.

nitrogen fixation — The utilization of atmospheric nitrogen in synthesizing amino acids by some bacteria and blue-green algae.

nitrogen-fixing bacteria — Bacteria that convert nitrogen from the atmosphere or soil into ammonia that can then be converted to plant nutrients by nitrite- and nitrate-forming bacteria.

nitrogen oxides (NOₓ) — Gases formed from nitrogen and oxygen during combustion in engines or in industries during conditions of high temperature and pressure.

nitrogen trifluoride (NF₃) — A colorless gas with a moldy odor that is shipped as a liquefied compressed gas; MW: 71.0, BP: –200°F, Sol: insoluble. It is used in the manufacture of flash cubes; in welding, brazing, and cutting of metals using a hydrogen–nitrogen trifluoride torch; in research as an oxidizer for rocket and other high-energy fuels. In animals, it is hazardous to the blood and is toxic through inhalation. Symptoms of exposure in animals include methemoglobin, anoxia, cyanosis, weakness, dizziness, headache, and injury of the liver and kidneys. OSHA exposure limit (TWA): 10 ppm [air] or 29 mg/m³.

nitrogenase — An enzyme of various nitrogen-fixing bacteria that catalyzes the reduction of molecular nitrogen to ammonia.

nitrogenous — Chemical compounds, usually organic, containing nitrogen in combined forms.

nitrogenous waste — Animal or plant residue that contains large amounts of nitrogen.

nitroglycerine (CH₂NO₃CHNO₃CH₂NO₃) — A colorless to pale-yellow, viscous liquid or solid (below 56°F); MW: 227.1, BP: begins to decompose at 122 to 140°F, Sol: 0.1%, Fl.P: explodes, sp. gr. 1.60. It is used during formulation and filling operations in the manufacture of industrial explosives and propellants; in the preparation of dosage forms, including tablets and solutions in the manufacture of pharmaceuticals. It is hazardous to the cardiovascular system, blood, and skin and is toxic through inhalation, absorption, ingestion, and contact. Symptoms of exposure include throbbing headache, dizziness, nausea, vomiting, abdominal pain, hypotension, flush, palpitations, methemoglobin, delirium, depressed central nervous system, angina, and skin irritation. OSHA exposure limit (TWA): short-term, 0.1 mg/m³ [air].

nitromethane (CH₃NO₂) — A colorless, oily liquid with a disagreeable odor; MW: 61.0, BP: 214°F, Sol: 10%, Fl.P: 95°F, sp. gr. 1.14. It is used as a solvent in coatings and adhesives on cellulose esters and synthetic resins; as a propellant or fuel additive; as an intermediate in the synthesis of organic dyes, textiles, surfactants, insecticides, pharmaceuticals, and explosives; as a reaction-media fluid; as a recrystallization solvent; as a stabilizer for halogenated alkanes, aerosol formulations, and paste

N

formulations for inks. It is hazardous to the skin and is toxic through inhalation, ingestion, and contact. Symptoms of exposure include dermatitis. OSHA exposure limit (TWA): 100 ppm [air] or 250 mg/m³.

2-nitrophenol (C₆H₅NO₃) — A yellow, crystalline solid with an aromatic odor; MW: 139.11, BP: 216°C, Sol: 1.08 g/100 g water at 20°C, Fl.P: 101.7°C, sp. gr. 1.495. It is used to make dyestuffs, pesticides, rubber chemicals, lumber preservatives, photographic chemicals, and fungicide agents. It is hazardous to the respiratory tract, digestive tract, skin, and eyes. Acute exposure may cause headaches, blue skin, changes in visual acuity, decreased visual fields, weakness, lethergy, shortness of breath, nausea, vomiting, confusion, drowsiness, irritation to the skin and eyes, respiratory depression, seizures, and death. Chronic exposure may cause central nervous system problems, headaches, dizziness, vertigo, weaknesses, decreased visual acuity, and fatigue. It may pose significant risk to fetuses. OSHA exposure limit (8-hour TWA): NA.

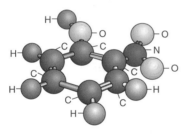

2 Nitrophenol

4-nitrophenol (C₆H₅NO₃ — An odorless, colorless, slightly yellow to brown crystalline solid or flake; MW: 139.11, BP: decomposes, Sol: 1.6 g/100 g water at 25°C, Fl.P: NA (poisonous gases produced in a fire include nitrogen oxides), sp. gr. 1.270. It is used as a fungicide for leather products or as a chemical intermediate in the synthesis of methyl and ethyl parathion. It is hazardous to the respiratory system, digestive system, skin, and eyes. Acute exposure causes headaches, blue skin, changes in visual acuity, weakness, lethergy, shortness of breath, nausea, vomiting, confusion, drowsiness, burning of the skin and eyes, shock, rapid heart beat, respiratory depression, seizures, and death. Chronic exposure causes central nervous system effects, headaches, dizziness, vertigo, weakness, fatigue, and failure of visual acuity. It may also be a potential cause of cancer and reproductive damage in humans and may be a signifi-

cant risk to the fetus. OSHA exposure limit (8-hour TWA): NA.

1-nitropropane (CH₃CH₂CH₂NO₂) — A colorless liquid with a somewhat disagreeable odor; MW: 89.1, BP: 269°F, Sol: 1%, Fl.P: 96°F, sp. gr. 1.00. It is used as a thinner and solvent for cellulose compounds, lacquers, and dopes; in vinyl resins for industrial coatings and printing inks; in synthetic finish removers; as an extraction solvent for purification, separation, recrystallization, and recovery of natural and synthetic resins, tars, coating materials, fats, and oils; as a reaction medium in polymer technology; as a catalyst, initiator, and solvent; in organic chemical synthesis for preparation of amines, nitrated alcohols, acids, and chloronitroparaffins; in the manufacture of explosives. It is hazardous to the eyes and central nervous system and is toxic through inhalation, ingestion, and contact. Symptoms of exposure include headache, nausea, vomiting, diarrhea, and eye irritation. OSHA exposure limit (TWA): 25 ppm [air] or 90 mg/m³.

2-nitropropane (CH₃CH(NO₂)CH₃) — A colorless liquid with a pleasant, fruity odor; MW: 89.1, BP: 249°F, Sol: 2%, Fl.P: 75°F, sp. gr. 0.99. It is used in the manufacture of explosives; as a thinner and solvent; in organic chemical synthesis; as a propellant in rocket motors. It is hazardous to the respiratory system and central nervous system and is toxic through inhalation, ingestion, and contact. Symptoms of exposure include headache, anorexia, nausea, vomiting, diarrhea, and irritation of the respiratory system; carcinogenic. OSHA exposure limit (TWA): 10 ppm [air] or 35 mg/m³.

nitrosation — The reaction between nitrous acid and secondary or tertiary amines.

nitrosoamines — Derivatives of amines containing *N*-nitroso groups. They are formed by the reaction of an amine with a nitrite and are also produced by the action of nitrate-reducing bacteria. They are found in trace quantities in tobacco smoke, processed food, meat products, and salted fish. They are used as gasoline and lubricant additives, antioxidents, stabilizers, and softeners for copolymers. They are noncombustible liquids or solids at ambient temperature. Chronic exposure can cause jaundice and liver damage. Methyl and ethyl nitrosoamines cause nausea, vomiting, ulceration, and an increase in body temperature. They are potent animal and human carcinogens and may cause cancers in the liver, kidneys, lungs, bladder, and pancreas.

nitrotoluene (NO₂C₆H₄CH₃) — Yellow liquids (*ortho-*, *meta-*isomers) or crystalline solid (*para-*isomers) with a weak, aromatic odor; MW: 137.1; BP: 432, 450, 460°F, respectively; Sol: 0.07, 0.05, 0.04%, respectively; Fl.P: 223, 232, 223°F, respectively; sp. gr. 1.16, 1.16, 1.12, respectively. It is used for the production of dyes; as intermediates for vulcanization accelerators, gasoline inhibitors, and flotation agents; as analytical reagents; in the production of explosives; in organic synthesis. It is hazardous to the blood, central nervous system, cardiovascular system, skin, and gastrointestinal tract and is toxic through inhalation, absorption, ingestion, and contact. Symptoms of exposure include anoxia, cyanosis, headache, weakness, dizziness, ataxia, dyspnea, tachycardia, nausea, and vomiting. OSHA exposure limit (TWA): 2 ppm [skin] or 11 mg/m^3.

nitrous — Pertaining to or containing nitrogen in its lowest valence.

nitrous oxide (N$_2$O) — A colorless gas with a sweet odor and taste that is used as an anesthetic; also known as laughing gas.

NK — Natural killer cells.

NMHC — Acronym for non-methane hydrocarbon.

NMOC — Acronym for non-methane organic component.

NMR — Acronym for nuclear magnetic resonance; now known as magnetic resonance imaging.

NMVOC — Acronym for non-methane volatile organic chemical.

N-**nitrosodimethylamine ((CH$_3$)$_2$N$_2$O)** — A yellow, oily liquid with a faint, characteristic odor; MW: 74.1, BP: 306°F, Sol: soluble, Fl.P: not known, sp. gr. 1.00. It is used in the production of rocket fuel; as a solvent in the fibers and plasticizers industries; as an oxidant; as an additive in lubricants. It is hazardous to the liver, kidneys, and lungs and is toxic through inhalation, absorption, ingestion, and contact. Symptoms of exposure include nausea, vomiting, diarrhea, abdominal cramps, headache, fever, enlarged liver, jaundice, reduced function of the liver, kidneys, and lungs; carcinogenic. OSHA exposure limit (TWA): regulated use.

NO — See nitric oxide.

no code — An order written in the patient record and signed by a senior or attending physician instructing the staff not to attempt to resuscitate a special patient in the event of cardiac or respiratory failure.

No Further Remedial Action Plan Database — A U.S. EPA database containing information on sites that have been removed from the inventory of Superfund sites and now put into the archives.

NOAEL — Acronym for no observed adverse effect level.

noble gas — See inert gas.

noble metal — A chemically inactive metal, such as gold, which does not corrode easily.

NOC — See not otherwise classified.

Nocardia — A genus of Gram-positive aerobic bacteria in which some species are pathogenic.

nocardiosis — An infection with *Nocardia asteroides*, an aerobic, Gram-positive species of actinomycetes with symptoms of pneumonia, cavitation, and chronic abscesses in the brain and subcutaneous tissue.

nociceptor — A sensory receptor responding to potentially harmful stimuli, producing a sensation of pain.

nocturnal — Referring to activity during the night.

node — A small mass of tissue in the form of a swelling knot or protuberance, either normal or pathological.

nodule — A small mass of rounded or irregular shaped cells or tissues; a small node.

NOEL — Acronym for no observed effects level.

NOIC — Acronym for Notice of Intent to Cancel.

NOIS — Acronym for Notice of Intent to Suspend.

noise — Unwanted sounds.

Noise Control Act — A law passed in 1972 and amended in 1978 stating that inadequately controlled noise presents a growing danger to the health and welfare of the population of the United States; that major sources of noise include transportation vehicles, equipment, machinery, appliances, and other products used in commerce; that the primary responsibility for noise belongs to state and local governments; and that calls for federal action to deal with major noise sources in commerce.

noise control specialist — An environmental health and safety practitioner who prevents loss of hearing, minimizes the psychological effects of noise, and reduces the nuisance factors associated with noise.

noise exposure — A combination of amount of noise, kind of noise, and duration of noise.

noise-induced hearing loss — The slow, progressive inner ear loss that results from exposure to continuous noise over a period of time as contrasted to loss from a trauma or injury.

noise level — Physical quantity of unwanted sound, usually expressed in decibels.

noise pollution — A level of sound in the environment that is either uncomfortable or leads to disease or injury.

nolle prosequi — Not willing to prosecute, either as the sum of accounts or sum of defendants, or all.

nolo contendere — Literally, "I will not contest it" — an implied confession of guilt in a criminal case having the same legal effect as a guilty plea.

nomenclature — A classified system of technical terms.

nominal size — The approximate dimensions of standard materials.

nonaggressive sampling — An environmental sampling method performed in a quiescent atmosphere.

N

nonattainment area — A geographic area identified by the U.S. EPA as not meeting the standards for a given pollutant.

nonauditory effects of noise — The stress, fatigue, health, work efficiency, and performance effects due to continuous loud noise.

nonbiodegradable — A chemical that cannot be broken down by bacteria or other organisms; considered to be persistent.

noncommunicable disease — A disease for which the causative agents cannot be passed or carried from one person to another directly or indirectly.

noncommunity water system — A public water system that serves less than 15 connections or 25 individuals on a daily basis.

noncompliance — Failure or refusal to conform to or follow rules, regulations, or the advice or wishes of another person.

nonconductor — A substance that does not readily transmit electricity, light, or heat.

non-consumptive use — Use of water in such a way that it does not reduce the supply (e.g., swimming).

nonconventional pollutant — Any pollutant that is not listed by rules or laws or which is poorly understood by the scientific community.

noncritical soils — Those undisturbed soil materials that can support a conventional private sewage disposal system where at least the lower portion of the soil absorption part of the system can be installed in original, uncompacted, or undisturbed soil.

nondirectional — See omnidirectional.

nonelectrolyte — A compound that, when dissolved in water, does not separate into charged particles and is incapable of conducting an electrical current.

nonexpanding clay — Clay particles that do not change appreciably in volume when dried by weather.

nonfat — Skim milk that contains not more than 0.1% milkfat; also known as fat free.

nonfeasance — Failure to carry out a task that should be done as dictated by law.

nonferrous metal — Any metal that contains no iron or iron alloys.

nonflammable — Incapable of being easily ignited or burning rapidly.

non-food-contact surfaces — Exposed surfaces other than food-contact surfaces.

nonfouling — A property of cooling water that allows it to flow over steam condenser surfaces without accumulation of impediments.

nonfriable — A material that cannot be crushed, pulverized, or reduced to powder by normal hand pressure when dry.

non-intact skin — Skin that is chapped, abraded, weeping, or that has rashes or eruptions.

noninvasive — A diagnostic or therapeutic technique that does not invade the body.

non-ionic — Not dependent on a surface-active anion for effect.

non-ionic detergent — A synthetic detergent that produces electrically neutral colloidal particles in solution.

non-ionizing radiation — Electromagnetic radiation, including ultraviolet, infrared, laser, microwave, and radiowave, that does not produce ionization.

nonpersistent — Lasting only for a few weeks or less.

non-point source — Man-made or induced alteration of the chemical, physical, biological, or radiological integrity of water originating from any source other than a point source. Major non-point sources of pollution include excess farm and lawn nutrients, uncontrolled stormwater runoff, forestry operations, animal waste, and pollutants from the atmosphere.

non-point-source water pollution — Polluted water typically coming from rural areas; it enters a receiving body from many small, scattered sources.

nonpolar — Any molecule or liquid that has a reasonable degree of electrical symmetry such that there is low or no separation of charge, such as benzene and carbon tetrachloride.

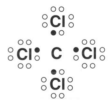

● hydrogen electron
○ oxygen electron

Nonpolar Molecule

nonporous — A substance that does not allow water to pass through.

nonpotable water — Any water not defined as potable.

nonreactive liquid — A simple solvent that dissolves vapor or releases it with no chemical reaction involved, such as water or heavy carbon oil.

nonregenerative absorbent — An absorbent that is irreversibly converted and must be discharged.

nonselective herbicide — A pesticide chemical or agent that will prevent growth or kill most of the plants it contacts.

nonsystemic — A chemical not capable of affecting the entire system.

nontoxic — A substance that is not poisonous.

nonventilated air sampler — An instrument kept constantly in direct contact with an air sample by means of natural ventilation rather than by use of an air mover.

nonvolatile — A pesticide chemical that does not evaporate by turning into a gas or vapor.

norepinephrine ($C_8H_{11}O_3N$) — A hormone that acts as a neurotransmitter of most sympathetic postganglionic neurons and also of certain tracts in the central nervous system; produces vasoconstriction, an increase in heart rate, and elevation of blood pressure.

norm — A numerical or statistical measure of average observed performance.

NORM — Acronym for naturally occurring radioactive material.

normal — (*math*) Perpendicular to; a mean or average.

normal saline solution — A 0.9% in weight per volume of sterile solution of sodium chloride and water that is isotonic with blood and injectable by intravenous methods.

normal solution — A solution that contains 1 gram equivalent of the active reagent in grams in 1 liter of solution.

normal temperature and pressure — See standard temperature and pressure.

normalized difference vegetation index — A model for converting satellite-based measurements into surface vegetation types using a complex ratio of reflectance in the red and near-infrared portions of the spectrum.

North American Emergency Response Guidebook — A guidebook developed by Transport Canada, U.S. Department Of Transportation, Secretariat of Communications and Transportation of Mexico, for use by firefighters, police, and other service personnel, who are first to arrive at the scene of the transportation incident, where dangerous goods is being transported, and where the first responders are involved in protecting the injured individuals, bystanders, and themselves from the chemicals.

North American Industrial Classification Codes — A system of numerical codes that categorizes industrial facilities by the type of activity in which they are engaged.

Norwalk-type disease — Diarrhea caused by Norwalk virus, a 27-nanometer particle that is a parvovirus-like agent. It is a limited, mild disease with symptoms of nausea, vomiting, abdominal pain, malaise, and low-grade fever. The occurrence of the disease is worldwide, with people being the only reservoir of infection; general susceptibility. The mode of transmission is probably the oral–fecal route through food and water. The incubation period is 24 to 48 hours, with communicability occurring during the acute state of disease and shortly thereafter. Also known as epidemic viral gastroenteritis.

nosocomial — Originating in a hospital; describes a disease or infection caused or aggravated by hospitalization.

nosocomial infection — An infection acquired at least 72 hours after hospitalization.

not otherwise classified (NOC) — Risks that do not have a classification definition that precisely describes their operation.

notice — A legal warning from an administrative agency advising an individual or individuals concerning a violation of rules and regulations.

NOVI — Acronym for notice of violation.

NO_x — Scientific notation for nitrogen oxide.

noxious — Something that is harmful, injurious, or detrimental to health.

noy — A unit of perceived noise level.

Np — See neper.

NPDWR — See National Primary Drinking Water Regulations.

NPHAP — Acronym for National Pesticide Hazard Assessment Program.

NPI — See National Pollutant Inventory.

NPL — See National Priorities List.

NPP — See net primary production.

NPRI Chemicals — A list of substances included in the National Pollutant Release Inventory reasonably anticipated to cause acute or chronic adverse human health effects or adverse environmental effects.

NRD — Acronym for natural resource damage.

NRe — See Reynolds number.

NRT — See National Response Team.

NSDWR — Acronym for National Secondary Drinking Water Regulations.

NSEC — Acronym for National System for Emergency Coordination.

NSEP — Acronym for National System for Emergency Preparedness.

NSPS — Acronym for New Source Performance Standards.

NSR — Acronym for New Source Review.

NSWS — Acronym for National Surface Water Survey.

NTA — See nitrilotriacetate.

NTNCWS — Acronym for nontransient, noncommunity water system.

NTU — See nephelometric turbidity unit.

nuclear battery — A device in which the energy emitted by the decay of a radioisotope is converted first to heat and then directly to electricity.

nuclear bombardment — The rapid movement of atomic particles toward nuclei to split the atom or to form a new element.

N

nuclear breeder reactor — A highly efficient nuclear reactor design that uses weapons-grade plutonium-239 as the fuel and is estimated to be 60 times more efficient than uranium-235 reactors.

nuclear change — The process by which an element is converted into another element through a change in the number of protons in its atoms, such as uranium-238, which is radioactive and decays through a series of nuclear reactions to form lead; atomic weight: 206; a stable, nonradioactive isotope.

nuclear disintegration — A spontaneous nuclear transformation of radioactivity characterized by the emission of energy and/or mass in the nucleus.

Nuclear Emergency Search Team — A group of experts assisted by radiation detection systems and associated personnel assigned responsibility to provide technical assistance to law enforcement agencies in nuclear threat emergencies for the search and identification of any ionizing radiation-producing materials that may have been lost or stolen or may be associated with bomb threats or radiation dispersal threats.

nuclear energy — Energy released by spontaneous or artificially produced fission, fusion, or disintegration of the nuclei of atoms; also known as atomic energy.

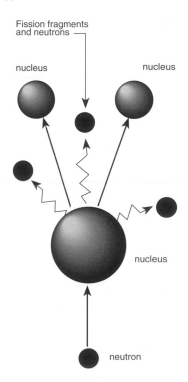

Nuclear Fission

nuclear fission — A nuclear transformation in which a nucleus is divided into at least two nuclei with a release of energy, such as when uranium-235 is bombarded with neutrons and split into barium-144 and krypton-90.

nuclear fuel — A substance that is consumed in a reactor during nuclear fission or fusion.

nuclear fusion — A nuclear transformation in which two smaller atomic nuclei fuse into one larger nucleus and release energy.

nuclear magnetic resonance — See magnetic resonance.

nuclear medicine — The use of radionucleides for diagnosing and treating disease in patients.

nuclear power plant — An energy plant that utilizes uranium-235 in fuel rods to produce heat, which is converted into energy for populations; also known as atomic power.

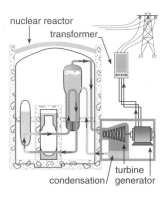

Nuclear Power Plant

nuclear reaction — Any reaction involving a change in nuclear structure, such as fission, fusion, neutron capture, or radioactive decay.

nuclear reactor — A device in which a controlled chain reaction of fissionable material can be produced and sustained.

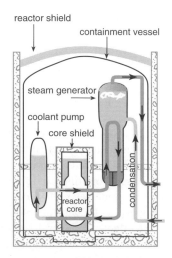

Nuclear Reactor

nuclear transformation — The process by which a nuclide is transformed into a different nuclide by absorbing or emitting a particle; also known as transmutation.

nucleic acid — The large organic molecule made of nucleotides that functions in the transmission of hereditary traits in protein synthesis and in the control of cellular activities.

nucleolus — A rounded refractile body in the nucleus of most cells that is the site of synthesis of ribosomal RNA; it grows larger during periods of synthesis and smaller during quiescent periods.

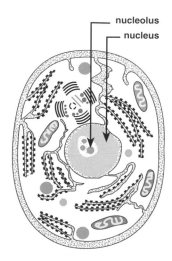

Nucleus & Nucleolus

nucleon — The common name for a proton or neutron, the main constituent particles of the nucleus.

nucleon number — See mass number.

nucleonics — The science that deals with the constituents and changes in the atomic nucleus.

nucleus — (*biology*) The central controlling body within a living cell, usually a spherical unit enclosed in a membrane containing genetic codes for maintaining life systems of the organism and issuing commands for growth and reproduction; (*chemistry*) center of an atom about which electrons rotate.

nuclide — A species of atom characterized by the constituents of its nucleus or by its atomic number, mass number, and atomic mass.

nuisance — An action or situation that is annoying, unpleasant, or potentially hazardous, such as the presence of rodents or sewage on top of the ground.

nuisance contaminant — (*water*) A constituent in water that is not normally harmful to health but may cause offensive taste, odor, color, corrosion, foaming, or staining.

nuisance dust — Dust that has little or no adverse effect on the lungs or health of individuals.

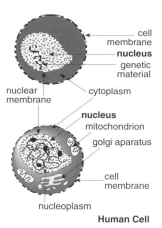

Human Cell

Nucleus

nuisance law — Common law stating that the use of private property is unrestricted only as long as it does not injure another person or his or her property; an injury or potential injury would constitute a nuisance.

null hypothesis — (*statistics*) A hypothesis for a research study involving two or more groups that assumes that no difference between the groups will be demonstrated; if a difference is shown, the null hypothesis is rejected, and the results are said to be statistically significant.

numbness — A partial or total lack of sensation in a part of the body caused by an interruption of the transmission of impulses from the sensory nerve fibers.

numerical examination of urban smog (NEXUS) — Mathematical model of air pollution that uses a computer for simulation of transport and diffusion of pollutants in the atmosphere.

nunc pro tunc — Literally, "now for then" — a phrase applied to acts allowed to be done after the time that they should have been done with a retroactive effect.

N-unit — A measure of radiation due to fast neutrons; a unit of index of refraction.

nurse — A person educated and licensed in the practice of nursing and concerned with the diagnosis and treatment of actual or potential health problems in people.

nursery diarrhea — A potentially life-threatening diarrhea of newborns that is found in hospital nurseries; it is caused by *Escherichia coli*, *Salmonella*, ecoviruses, or adenoviruses, with infection occurring at the time of birth from the mother's stools or later by organisms spread by hands of hospital personnel. It results in severe dehydration and electrolyte imbalance.

nutrient — An organic and inorganic chemical necessary for the growth or reproduction of an organism.

N

nutrient pollution — Contamination of water resources by excessive amounts of nutrients in surface waters which produce excess algae.

nutrition — The science of food; the nutrients and other substances therein; their actions, interactions, and balance in relation to health and disease; and the processes by which an organism ingests, digests, absorbs, transports, utilizes, and excretes food substances.

NWQSS — Acronym for National Water Quality Surveillance System.

nymph — The immature larval stage of various insects, such as grasshoppers and roaches.

nystagmus — An involuntary movement of the eyeball; occurs among workers who subject their eyes to

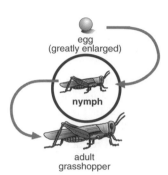

Grasshopper Nymph

abnormal and unaccustomed movements for extended time periods.

N

O

O — See oxygen.

O₂ — Scientific notation for oxygen.

O_2 — Scientific notation for oxygen.

O₃ — See ozone.

O&M — Acronym for operations and maintenance.

O&MP — See operation and maintenance program.

obesity — An abnormal increase in the proportion of fat cells mainly in the viscera and subcutaneous tissues of the body.

obfuscation — A confused, clouded, or obscure fact or situation.

objective — A quantified statement of a desired future state or condition to be achieved within a specified time period.

objective sign — A clinical observation that can be seen, heard, or otherwise measured.

oblate spheroid — A flattened sphere in which the diameter of its equatorial circle exceeds the length of the axis of revolution.

obligate aerobe — An organism that cannot live without atmospheric oxygen.

obligate anaerobe — An organism that cannot grow in the presence of atmospheric oxygen.

obligate symbiont — An organism that is physiologically dependent on establishing a symbiotic relationship with another.

oblique — Having a slanted direction or position.

oblique sensing — Sensing in which the optical axis of the sensor is not perpendicular to the terrain.

obliquity — The angle between the plane of the Earth's orbit and the plane of the Earth's equator; the tilt of the Earth.

obliteration — Removal or loss of function of a part of the body by surgery, disease, injury, or degeneration.

observation — The process of recognizing and recording information.

observation well — A hole used to observe the groundwater surface at atmospheric pressure within soil or rock.

observational study — Research in which the investigator can only observe the occurrence of a disease (or other characteristics) in persons already categorized on the basis of some experience or exposure, the degree to which is not under the investigator's control.

observations — Objective measuring or recording of results when testing a theory by watching and evaluating.

obstruction — Something that blocks or clogs a passageway.

oc — See open cup.

occlude — To close, shut, or stop up a passage.

occluded front — A composite of two fronts formed as a cold front that overtakes a warm front.

occlusion — See occluded front.

occupancy factor — The amount of people in an area adjacent to a source of radiation; used to determine the shielding required in the walls.

occupancy — The time during which people are in a building, expressed in hours per day.

occupant — Any individual over 1 year of age living, sleeping, cooking, eating in, or having possession of a dwelling, dwelling unit, or rooming unit.

occupational accident — An accidental injury to a person that occurs in the workplace.

occupational asthma — An abnormal condition of the respiratory system resulting from exposure in the workplace to allergenic or other irritating substances.

occupational disability — A condition in which a worker is unable to perform necessary functions because of an occupational disease or accident.

occupational disease — A disease that is due to short-term or long-term exposure to specific substances or continuous or repetitive physical acts.

occupational health — The ability of a person to function at an optimum level of productivity in a work situation while enjoying good health and longevity.

occupational health history — Description of a person's health history at work. The history includes previous jobs; duration of the jobs; exposure to specific environmental problems of a microbiological, physical, chemical, and psychological nature resulting diseases and injuries; and current status of the person's health related to previous exposures.

occupational health nurse — A registered nurse who provides acute and rehabilitative care and participates in preventive and health maintenance programs for workers.

occupational health physician — A physician who conducts medical practices and diagnosis and treatment activities, directs medical departments in the workplace, and collaborates with other occupational safety and health personnel to identify, investigate, and prevent the occurrence and recurrence of occupational safety and health problems.

occupational injury — An injury occurring in the course of employment and caused by inherent or related factors arising from the operation of materials of that occupation.

occupational lung disease — A lung disease caused by the inhalation of dust, fumes, gases, or vapors in the occupational environment.

occupational medicine — A field of preventive medicine related to the problems and practices used in protecting health and preventing disease or injury of workers.

Occupational Safety and Health Act — A 1970 law establishing the Occupational Safety and Health Administration (OSHA) and the National Institute for Occupational Safety and Health (NIOSH); the Congressional purpose of the Act was to ensure as far as possible that every working person in the nation has safe and healthful working conditions, thereby preserving our human resources.

occupational safety specialist — An environmental health practitioner who plans, develops, implements, and administers occupational health and injury control programs in industry to identify and correct hazardous conditions that may cause disease and injury to employees, damage to equipment and facilities, or loss of materials. Also known as safety management specialist or environmental safety specialist.

occupational skin diseases or disorders — A series of diseases or disorders including contact dermatitis, eczema, or rash caused by primary irritants, poisonous plants, oil acne, chrome ulcers, chemical burns, or inflammations.

occupational stress — A neurosis associated with a job or work; symptoms include extreme tension and anxiety as well as headaches or cramps.

occupational therapist — An allied health professional who is concerned with the evaluation, diagnosis, and/or treatment of people whose ability to cope with activities of daily living is impaired by physical injury, illness, emotional disorder, congenital or developmental disability, or aging. The occupational therapist must be licensed, registered, certified, or otherwise regulated by law.

occurrence — An accident; includes continuous or repeated exposure to conditions that result in bodily injury or property damage that the owner or operator neither expected nor intended.

OCD — Acronym for offshore and coastal dispersion.

OCE — Acronym for Ocean Color Experiment.

ocean — Any portion of the high sea beyond the contiguous zone and occupying depressions in the surface of the Earth; banded by continents and imaginary lines.

ocean color — The ability of phytoplankton to appear as different colors in certain bands of the electromagnetic spectrum because of their chlorophyll concentrations.

ocean disposal — The deposition of waste into an ocean or estuary.

ocean thermal electric conversion — The energy derived from temperature differentials between warm, surface waters and cold, deep waters of the ocean; the differential can be used to drive turbines attached to electric generators.

oceanic — Pertaining to marine waters seaward of the continental shelf margin.

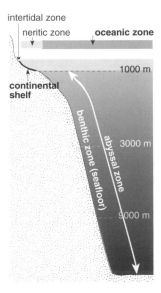

Oceanic Zone

Oceanic Remote Sensing Assembly — A sensor package for a polar platform.

ochronosis — A blue or brownish-blue discoloration of body tissues caused by deposits of alkapton bodies as a result of a metabolic disorder.

OCI — Acronym for Ocean Color Imager.

OCM — Acronym for Ocean Color Monitor.

OCS — Acronym for Ocean Color Scanner.

octachloronaphthalene ($C_{10}Cl_8$) — A waxy, pale-yellow solid with an aromatic odor; MW: 403.7, BP: 824°F, Sol: insoluble, sp. gr. 2.00. It is used as an insulating material; as an inert compound of resins or polymers for coatings or impregnating textiles, wood, and paper to impart flame resistance, waterproofness, and fungicidal and insecticidal properties; as an additive for cutting oil in various operations preformed on metal; as an additive to special lubricants. It is hazardous to the skin and liver and is toxic through inhalation, absorption, ingestion, and contact. Symptoms of exposure include acneforming dermatitis, liver damage, and jaundice. OSHA exposure limit (TWA): 0.1 mg/m^3 [skin].

octane (CH₃(CH₂)₆CH₃) — A colorless liquid with a gasoline-like odor; MW: 114.2, BP: 258°F, Sol: insoluble, Fl.P: 56°F, sp. gr. 0.70. It is used in the preparation of gasoline and rocket fuels; as an industrial solvent; as a lacquer diluent; in phosphate manufacture; in the preparation of liquid soaps and detergents; in printing ink manufacture; as an additive and solvent in polymer manufacture; in the manufacture of benzene, toluene, and xylene aromatics; as a blowing agent for foam rubber used in rocket propellants. It is hazardous to the skin, eyes, and respiratory system and is toxic through inhalation, ingestion, and contact. Symptoms of exposure include drowsiness, dermatitis, chemical pneumonia, and irritation of the eyes and nose. OSHA exposure limit (TWA): 300 ppm [air] or 1800 mg/m^3.

octane number — A conventional rating for gasoline based upon its behavior under standard conditions in a standard internal combustion engine as compared with isooctane, whose octane number is 100.

octanol–water partition coefficient (K$_{ow}$) — The equilibrium ratio of the concentration of a chemical in octanol and water in dilute solution.

octave — The interval between any two frequencies having the ratio 2:1.

octave band — A band of frequency for which the lower frequency is related to the upper frequency by the ratio of 2:1.

octave band analyzer — An instrument used to determine where noise energy lies in the frequency spectrum.

OCTS — Acronym for Ocean Color and Temperature Scanner.

ocular — Pertaining to the eye.

ocular route — A route of entry into the body through the eye.

OD — Notation for outside diameter; see optical density.

odds ratio — Ratio of the probability of a specific event occurring relative to the probability of the event not occurring.

odor — The property of a substance that affects the sense of smell.

odor threshold — The lowest concentration of an airborne odor that a human can detect.

odorant — Material added to odorless gases to impart an odor for safety purposes.

odoriferous — A substance that produces a strong or offensive smell.

Sound level meter and
octave band analyzer

odorous — A substance that has an odor, smell, or fragrance.

ODP — Acronym for ozone-depletion potential.

ODS — Acronym for ozone-depleting substances.

ODTS — See organic dust toxic syndrome.

OF — Acronym for optional form.

offal — Intestines and discarded parts, including manure, remaining after animals have been slaughtered.

off-line — (*computer science*) A situation where information is not available on a computer because part of the operating system is not connected or operating.

offsets — A concept whereby emissions from proposed new or modified stationary sources are balanced by reductions from existing sources to stabilize total emissions.

off-site transfers — Chemicals in waste that are moved off the grounds of a facility, including transfers of waste sent to other facilities or other locations, such as hazardous waste treatment facilities, municipal sewage treatment, or landfills.

ohm (Ω) — An MKS and SI unit of electrical resistance equal to the resistance through which a current of 1 ampere will flow when there is a potential difference of 1 volt across it.

ohmeter — An instrument used to measure electrical resistance.

Ohm's law — Electric current flowing through an electric circuit is equal to the voltage divided by the resistance.

OHMTADS — Acronym for Oil and Hazardous Material Technical Assistance Data System.

OHZORA — Name of the Japanese scientific satellite launched in 1984.

OII — Acronym for optical imaging instrument.

oil — A greasy, liquid substance not miscible in water.

oil mist — A colorless, oily, liquid aerosol dispersed in air that has an odor like burned lubricating oil; MW: varies, BP: 680°F, Sol: insoluble, Fl.P: 275 to 500°F (open cup), sp. gr. 0.90. It is used as a coolant or quenching agent; in steel rolling operations; in printing pressroom operations; as an insecticide. It is hazardous to the respiratory system and skin and is toxic through inhalation. No reported symptoms of exposure. OSHA exposure limit (TWA): 5 mg/m^3 [air].

Oil Pollution Act — A 1924 law preventing oily discharges into coastal waters.

oil retention boom — A floating baffle used to contain and prevent the spread of floating oil on a water surface.

oil shale — A generic name for the sedimentary rock that contains substantial organic materials known as kerogen. The kerogen content of oil shale may vary between 5 and 80 gallons of oil per equivalent ton. Extraction of oil from the oil shale occurs when the shale is heated to 350 to 550°C under an inert atmosphere. The products include oil vapor, hydrocarbon gases, and a carbonaceous residue. This retorting shale process produces 1 ton of spent shale per barrel of oil. The volume of the shale increases by more than 50% during the crushing process, and a considerable amount of alkaline minerals contained in the oil shale may leach into the groundwater supply. A large amount of water is needed for controlling dust and reducing the alkalinity of the spent shale.

oil slick — A smooth area on the surface of water caused by the presence of oil.

oil solution — A pesticide chemical dissolved in oil.

oil spill — An accidental discharge of oil into bodies of water that may be controlled by chemical dispersion, combustion, mechanical containment, or absorption.

Oil Spill Contingency Fund — A revolving fund administered by the U.S. Coast Guard for spill control when the federal government has taken over containment and cleanup operations.

olefiant gas — See ethylene.

olefins (CnH$_{2n}$) — A class of unsaturated hydrocarbons characterized by relatively great chemical activity; butene, ethylene, and propylene are examples. See also alkene.

oleum — An oily, corrosive liquid consisting of a solution of sulfur trioxide and anhydrous sulfuric acid that turns to fume in moist air; also known as sulfuric acid.

olf — A unit used to quantify odorous pollution sources.

olfactory — Pertaining to the sense of smell.

olfactory fatigue — A condition in which a person's nose, after exposure to certain odors, can no longer detect the odors.

oligotrophic lake — A deep, clear lake with low nutrient supply containing little organic matter and a high dissolved-oxygen level.

oliguria — Diminished urine secretion in relation to fluid intake.

OLR — See outgoing longwave radiation.

OLS — Acronym for optical line scanner.

OLTS — Acronym for online tracking system.

ombudsman — A person who investigates and mediates problems and complaints between individuals and institutions.

omission — The intentional or unintentional neglect of duty.

omnidirectional — Relating to signals sent or received in all directions; also known as nondirectional.

onboard processing — Data processing that takes place on board a sensor platform.

onchocerciasis — A form of filariasis common in Central and South America and Africa; symptoms include subcutaneous nodules, pruritic rash, and eye lesions. It is transmitted by the bite of a blackfly that deposits *Onchocerca volvulus* microfilariae under the skin.

onchosphere — The larva of the tapeworm contained within the external embryonic envelope and having six hooks and cilia.

oncogen — Any substance capable of inducing tumors; a carcinogen.

oncogene — A gene found in the chromosomes of tumor cells for which activation is associated with the initial and continuing conversion of normal cells into cancer cells.

oncogenic — Causing tumor formation.

oncogenic virus — One of over 100 viruses able to cause the development of malignant neoplastic disease.

oncology — The study of the causes, development, and treatment of neoplasms.

oncotic — Pertaining to or resulting from a tumor.

oncovirus — A member of a family of viruses associated with leukemia and sarcoma in animals and possibly humans.

one-hundred-year flood — A flood level with a 1% chance of being equaled or exceeded in any given year.

one part per billion — A unit of measure equal to 1 μg per kg.

one part per trillion — A unit of measure equal to 1 ng per kg.

online — Providing access to information or equipment that is part of an operating computer system.

on-site disposal — Any method or process to eliminate or reduce the volume or weight of solid waste on the property of a generator.

on-site incinerator — An incinerator that burns solid waste on property utilized by a generating person or persons.

on-site sewage — Gray- and blackwater containing solids coming from a structure that receives primary settlement in a septic tank and whose effluent is disposed on-site in the ground.

ontogenetic niche — The multitude of inherited ecological and social traits that are passed on from generation to generation and play integral roles along with genetic inheritance to influence the development of behavioral traits.

onus — Some task, duty, or responsibility that involves or necessitates considerable effort or that is a burden.

oocyst — The encysted or encapsulated zygote of some sporozoa.

oogenesis — The process by which an egg cell forms from an oocyte.

ookinete — Fertilized form of the malarial parasite in a mosquito's body formed by fertilization by a microgamete and which develops into an oocyst.

OP — An acronym for an organophosphate pesticide.

OPA — See Oil Pollution Act.

opacity — The degree of obstruction of light.

opacity reading — The apparent obfuscation of an observer's vision that equals the apparent obfuscation of smoke of a given rating on the Ringelmann chart.

opaque — Unable to transmit light; totally absorbent of rays of a specified wavelength and not transparent to the human eye.

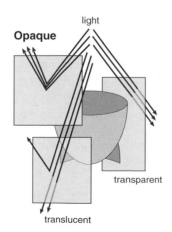

Opaque

light

transparent

translucent

open burning — Uncontrolled burning of waste in an open area, such as a dump.

open cup (oc) — A method for determining the flashpoint of a chemical.

open dump — Any facility or site where solid waste is disposed of that is not a sanitary landfill.

open hearth furnace — A long, wide, shallow furnace used to produce steel from cast or pig iron.

open-loop system — A system of heat exchange where well water at 55°F is used as the condenser; the water is pumped into a house and through a refrigeration process where the heat from the water can be removed and used to heat the property; when cooling the property, the hot air can be passed across the 55°F water to raise the temperature of the water and to decrease the temperature of the air within the property.

open-pit incinerator — A burning device that has an open top and a system of closely spaced nozzles that places a stream of high-velocity air over the burning zone.

open wound — A wound that disrupts the integrity of the skin.

operable — Able to use surgical intervention.

operable unit — The division of a contaminated site into separate areas based on the complexity of the problems associated with it.

operating environment — An environment in which a sensor operates.

operating flow rate — The flow rate at which a facility actually operates as opposed to the design flow rate.

operating humidity range — Range of ambient relative humidity over which a machine or instrument will meet all performance specifications.

operating room (OR) — A specially equipped room in which an operation on a patient is conducted in a sterile manner.

operating temperature range — Range of ambient temperatures over which a machine or instrument will meet all performance specifications.

operation and maintenance program (O&MP) — A program of training, work practices, and periodic surveillance to maintain friable-asbestos-containing building material in good condition, to clean up asbestos fibers previously released, and to prevent further release by minimizing and controlling friable-asbestos-containing building material disturbance or damage.

operator — An environmental health or safety practitioner qualified by education, training, and practice to manage plant and system processes necessary for environmental health or safety control and prevention activities; certification is usually required.

ophthalmic — Pertaining to the eye.

O

ophthalmoscope — An instrument used for examining the interior of the eye through the pupil.

lens selection disk
aperture selection disk
light
18

Ophthalmoscope

opisthotonos — A form of spasm of the back muscles in which the head and heels are bent backward and the body bowed forward.

opportunistic infection — An infection that is usually warded off by a healthy immune system, but which can affect an individual whose immune system has been suppressed.

opportunistic organisms — Usually saprophytic or commensal organisms but which can also be pathogenic under conditions of impaired host resistance; such organisms include systemic fungi and *Pneumocystis carinii* and are important agents of nosocomial infections.

OPS — Acronym for optical scanner.

opsonin — An antibody in blood serum that renders bacteria and other cells susceptible to phagocytosis.

optic nerve — Sensory nerve leading from the retina of the eye to the optic lobe of the brain.

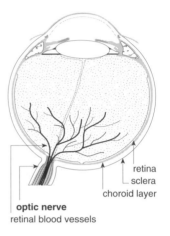

retina
sclera
choroid layer
optic nerve
retinal blood vessels

Optic Nerve

optical density (OD) — A logarithmic expression of the degree of opacity of a translucent medium.

optical depth — See optical thickness.

optical microscopy — A basic system of magnification using light rays reflected from an object which bend as they pass through one or more lenses; used to determine particle size and to identify solid pollutants.

optical pyrometer — A temperature-measuring instrument that matches the intensity of radiation at a single wavelength from a tungsten filament with the intensity of the radiation at the same wavelength emitted by a heat source.

optical radiation — Electromagnetic radiation comprised of ultraviolet, visible, and infrared radiations.

optical spectrum — A portion of the electromagnetic spectrum, from 0.30 to 15 μm, that can be reflected and refracted with mirrors and lenses.

optical system — A collection of mirrors, lenses, prisms, and other devices that reflect, refract, disperse, absorb, polarize, or otherwise act on light.

optical thickness — The mass of an absorbing or emitting material lying in a vertical column of unit cross-sectional area and extending between two specified levels; used in calculating the transfer of radiant energy.

Optical Transient Detector — The world's first space-based sensor capable of detecting and locating lightning events in the daytime as well as during the nighttime with high detection efficiency; built and designed by NASA for the purpose of detecting the full spectrum of lightning flashes, including cloud-to-ground, cloud-to-cloud, and within-cloud lightning events.

optics — The study of phenomena associated with the generation, transmission, and detection of radiation in the spectral range extending from the longwave edge of the x-ray region to the shortwave edge of the radio region.

optoacoustic device — A device making use of the interaction of sound waves and light waves for signal processing.

OR — See operating room.

oral — Pertaining to the mouth, or taken by mouth.

oral cancer — A malignant neoplasm on the lip or in the mouth with predisposing factors of alcoholism, heavy use of tobacco, poor oral hygiene, poorly fitting dentures, syphilis, overexposure to sun and wind, and pipe smoking.

oral route — A route of entry into the body by ingestion.

oral toxicity — The ability of a chemical to sicken or kill an organism if eaten or swallowed; usually measured as LD_{50}.

orb — Something spherical or globe like.

orbit — The act of a spacecraft circling the Earth; the type of orbit describes the path a satellite takes as it circles the Earth. A path one body takes as it revolves around another one.

orbit crossover — The points where the orbital ground tracks of a satellite intersect.

orbivirus — See reovirus.

orchitis — Inflammation of the testes.

order — (*law*) A formal written statement from a court requiring action or stating a ruling.

order of magnitude — A range of magnitudes of a quantity extending from some value of the quantity to some small multiple of the quantity (usually 10).

ordinance — A statute enacted by the legislative body of a local, generally municipal, government.

ordinary summer conditions — A temperature of 10°F below the highest recorded temperature in a locality during the most recent 10-year period.

ordinary winter conditions — A temperature 15°F above the lowest recorded temperature in a locality during the most recent 10-year period.

ore — A naturally occurring mineral substance from which economically valuable elements may be obtained.

organ — An organized collection of tissues that have a specific and recognized function.

organic — Referring, relating, or deriving from living organisms; (*chemistry*) of or relating to any

covalently bonded compound containing carbon atoms.

organic acid — A chemical compound containing one or more carboxyl group (–COOH); examples include formic, acidic, propionic, chloracetic, and trifluoracetic acids. It causes severe damage to skin and mucosal surfaces and eye injury and may cause damage to the enzyme systems of the body.

organic chemistry — Branch of chemistry concerned with the composition, properties, and reactions of chemical compounds containing carbon.

organic compound — Compounds containing carbon that refer, relate, or derive from living organisms; can also be produced by chemical synthesis.

organic disease — A disease associated with a detectable or an observable change in one or more body organs.

organic dust — The dried particles of plants, animals, fungi, or bacteria that are fine enough to become suspended in air.

organic dust toxic syndrome — A febrile illness occurring after heavy organic dust exposure.

organic gas — A gas containing carbon and usually hydrogen and possibly other elements bound to the carbon; examples include hydrocarbons such as methane, butane, octane, benzene, acetylene, ethylene, and butadiene; aldehydes and ketones such as formaldehyde and acetone; other organics such as chlorinated hydrocarbons, benzopyrene, and alcohols.

organic material in soils — The organic matter in soils consisting of decaying tissue, residues, decaying matter, and organic byproducts created by microorganisms that break down plant and animal tissue.

organic matter — The carbonaceous waste contained in plant or animal matter and originating from domestic or industrial sources.

organic molecule — A molecule that contains one or more carbon atoms.

organic nitrogen — Nitrogen combined in organic molecules, such as proteins and amino acids.

organic peroxide — An organic compound that contains the bivalent –O–O structure and may be considered a structural derivative of hydrogen peroxide, as one or both of the hydrogen atoms has been replaced by an organic radical.

organic pesticide — Any chemical containing carbon, which alone, in chemical combination, or in formulation with one or more substances is an "economic poison" as defined by the Federal Insecticide, Fungicide, and Rodenticide Act.

organic pump — Uptake of large quantities of water by plant and tree roots and removal to the atmosphere; helps reduce the flow of water in a given area.

organic soil — A soil that contains a high percentage of organic matter throughout the solum.

organic solid — (*sewage*) Solid matter of animal or vegetable origin including waste products, dead animal matter, plant tissues, organisms, and synthetic organic compounds.

organic solvent — A liquid organic material, including diluents and thinners, used to dissolve solids, reduce viscosity, or clean substances.

organic wastes — Includes saliva, urine, perspiration, and suntan oils introduced into pools by swimmers.

organism — An animal, plant, fungus, or microorganism able to carry out all life functions.

Organization of Economic Cooperation and Development — A Paris-based intergovernmental organization with 29 member countries; it develops common solutions to various social problems, including issues of toxic chemical management.

organogenesis — Secretion of tissues in various organs during embryonic development.

organoleptic — Pertaining to or perceived by a sensory organ.

organophosphate — One of a group of pesticide chemicals containing phosphorous; it interferes with the normal working of the nervous systems of animals and humans.

orientation — (*remote sensing*) Determination of the position of an image or photograph with respect to the altitude of the sensor.

orifice — An opening through which a fluid can pass; (*pesticides*) an opening or hole in a nozzle through which liquid material is forced out and broken up into a spray.

orifice meter — A flowmeter used to measure upstream and downstream differential pressures for a restriction within a pipe or duct.

original container — The package (e.g., can, bag, or bottle) in which a company sells a chemical.

O-ring — A rubber seal used around the stems of some valves to prevent water from leaking past.

ORM — Acronym for other regulated material.

orogeny — The period of mountain building.

orographic precipitation — A precipitation caused by topographic barriers, such as mountains that force moisture-laden air to rise and cool.

ORP — Acronym for oxidation–reduction potential; see redox.

ORS — Acronym for optical remote sensing.

ORSA — See Oceanic Remote Sensing Assembly.

Orsat analyzer — An apparatus used to analyze flue gases volumetrically by measuring the amount of carbon dioxide, oxygen, and carbon monoxide present.

orthoimage — An image derived from a conventional perspective image by simple or differential recti-

fication so that image displacements caused by a sensor tilt and the relief of the terrain are removed.

orthopedics — A branch of medicine concerned with the prevention and correction of problems related to the skeleton, muscles, joints, and other related tissues.

orthophoto — (*remote sensing*) A photograph derived from a conventional-perspective photograph by simple or differential rectification so that image displacements caused by camera tilt and the relief of the terrain are removed.

orthophotography — (*geographical information system*) Digital imagery in which distortion from the camera angle and topography have been removed, equalizing the distances represented on the image; (*remote sensing*) the process of making an orthophoto.

orthotolidine — An organic test agent that turns yellowish-green in the presence of chlorine, bromine, or iodine.

orthotolidine colorimetric comparison — A test in which a colorless solution of orthotolidine is added to a chlorine solution, producing an orange-brown to pale-yellow color representative of the concentration of chlorine; the color is then compared to a standardized color chart.

Oryzaephilus surinamensis — See Saw-Toothed Grain Beetle.

OSC — See federal on-scene commander.

oscillate — To swing backward and forward like a pendulum; to move or travel periodically back and forth between two points.

oscillation — Variation with time of the magnitude of a quantity with respect to a designated reference; the magnitude is alternately greater and smaller than the reference.

oscillator — A radiant tube or circuit that produces steady alternating current, usually at high frequency.

oscilloscope — A type of cathode-ray tube in which sound waves are changed to electrical impulses and are displayed on a screen; used to make visible

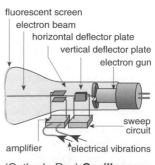

(Cathode Ray) **Oscilloscope**

the pattern of electrical vibrations, sound waves. Also known as cathode ray oscilloscope.

osmium tetroxide (OsO$_4$) — A colorless, crystalline solid or pale-yellow mass with an unpleasant, acrid, chlorine-like odor; a liquid above 105°F; MW: 254.2, BP: 266°F, Sol: 6% at 77°F, sp. gr. 5.10. It is used as an intermediate for separation of metals and ores of osmium from platinum; as a catalyst in organic synthesis; as a staining and fixing agent in pathological and histological analysis; as a laboratory reagent. It is hazardous to the eyes, respiratory system, and skin and is toxic through inhalation, ingestion, and contact. Symptoms of exposure include lacrimation, visual disturbances, conjunctivitis, headache, cough, dyspnea, dermatitis. OSHA exposure limit (TWA): 0.0002 ppm [air] or 0.002 mg/m^3.

osmol — See osmole

osmole — The molecular weight of the solute, in grams, divided by the number of ions or particles into which it disassociates in solution.

osmology — Science of the sense of smell and the production and composition of odors.

osmoregulator — A body mechanism concerned with the maintenance of constant osmotic pressure relationships; regulates internal osmotic concentration of body fluids.

osmosis — The diffusion of a solvent through a semipermeable membrane from a region of greater concentration to a region of lesser concentration.

osmotic potential — See osmotic pressure.

osmotic pressure — In hypoosmotic conditions, it is the internal fluid pressure that develops from the osmotic inflow of water.

OSPS — See Outreach and Special Projects Staff.

ossification — Formation of bone.

ostemia — An abnormal congestion of blood in the bone.

osteogenesis — Formation or development of bones.

osteoma — A tumor of bone tissue.

osteoporosis — An abnormal loss of bone substance.

OT — See occupational therapist.

OTC — Notation for over-the-counter drug.

OTD — See Optical Transient Detector.

OTH radar — Acronym for over the horizon radar.

other waste — Substances, not including sewage or industrial waste, that may cause or tend to cause pollution such as garbage, refuse, wood, residues, sand, lime, cinders, ashes, offal, oil, tar, dye stuffs, acids, and chemicals.

otitis — An inflammation or infection of any part of the ear.

otitis media — Inflammation of the middle ear.

otocariasis — An infestation of the external ear canal by ticks or mites.

otology — Diagnosis and treatment of diseases and disorders of the ear.

otosclerosis — Formation of spongy bone just inside the inner ear, causing progressive deafness.

otoscope — An instrument for inspecting or auscultating the ear and rendering the tympanic membrane visible.

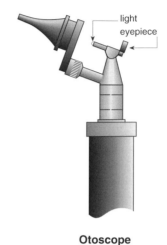

light
eyepiece

Otoscope

OU — See operable unit.

ounce (oz.) — A unit of mass in avoirdupois measure equal to 1/16 of a pound or approximately 28.3495 grams.

outbreak data — Data on foodborne disease outbreaks, where an outbreak is defined as an incident in which two or more persons experience a similar illness after ingestion of a common food and epidemiological analysis implicates the food as being the source of the illness; the two exceptions to this are botulism and chemical poisoning, in which one case constitutes an outbreak.

outcome — A final consequence, result, or effect.

outdoor air — Air taken from external surroundings and not previously circulated through a system.

outfall — The point at which an effluent is discharged into receiving waters.

outgoing longwave radiation — Thermal infrared radiation leaving the Earth's atmosphere and moving into space.

outlet — An opening through which water can be freely discharged from a reservoir to the river for a particular purpose.

outlet capacity — The amount of water that can be safely released through an outlet system.

outwash — A deposit of sand and gravel formed by streams of meltwater flowing from a glacier.

OVA — Acronym for organic vapor analyzer.

O

ovary — The female sex gland in which ova are formed in vertebrates.

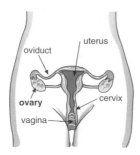

Ovary

over-the-horizon radar — Long-range radar in which the transmitted and reflected beans are bounced off the ionosphere layers to achieve ranges far beyond the line of sight.

overburden — The overlying layer of sediment that must be removed to reach a mineral or coal deposit; (*mineral engineering*) to charge in a furnace too much ore and flux in proportion to the amount of fuel.

overdraft — Water from a groundwater basin or aquifer pumped in excess of the supply flowing into the basin.

overexposure — Exposure beyond the specified limits.

overfire air — Air introduced above or beyond a fuel bed by a natural, induced, or forced draft; also known as secondary combustion air.

overfire air fan — A fan used to provide air above a fuel bed.

overflow system — (*swimming pools*) A series of devices or channels used to remove overflow water from a pool, including surface skimmers and surface-water collection systems of various design and manufacture.

overhead — The general costs of operating an entity which are distributed to all the revenue-producing operations but which are not directly attributable to a single activity.

overlay — (*mapping*) The process of stacking digital representations of various spatial data on top of one another so each position in the area covered can be analyzed in terms of these data.

overload — A burden greater than the capacity of a system designed to move or process it.

overpressure — An induced pressure above ambient atmospheric pressure or other given reference pressure.

overturn — The almost spontaneous mixing of all layers of water in a reservoir or lake when the water temperature becomes similar from top to bottom; the sinking of surface water that has become more dense than the water below.

ovicide — A pesticide chemical or agent used to destroy insect, mite, or nematode eggs.

oviparous — The production of eggs that develop and hatch outside the maternal body.

ovoid — Shaped like an egg.

ovulation — Discharge of an ovum from an ovary.

ovum — An egg; a female gamete.

OX — See photochemical oxidant.

oxalic acid (HOOCCOOH•2H$_2$O) — A colorless, odorless powder or granular solid (the anhydrous form is an odorless, white solid); MW: 126.1, BP: sublimes, Sol: 14%, Fl.P: not known, sp. gr. 1.65. It is used in metal cleaning and polishing operations to remove carbonaceous deposits from steel; in textile cleaning and bleaching operations to remove rust, ink, and stains; in bleaching of cotton and other fabrics; as a sour in treatment of woolen and other pieces of goods; in bleaching and tanning of hides during leather manufacture; in the manufacture of household and industrial cleaning and bleaching agents; in metallurgical processing of tin; in the synthesis of other chemicals for the manufacture of dyes and blue-print photography; in extraction and purification operations; in lithography and photoengraving operations; as an analytical reagent; in the manufacture of inks and lacquers; in the synthesis of pharmaceuticals. It is hazardous to the respiratory system, skin, kidneys, and eyes and is toxic through inhalation, absorption, ingestion, and contact. Symptoms of exposure include eye burns, local pain, cyanosis, shock, collapse, convulsions, and irritation of the eyes, mucous membrane, and skin. OSHA exposure limit (TWA): 1 mg/m^3 [air].

$$HO - C - C - OH$$

Oxalic Acid

oxamyl (C$_7$H$_{13}$N$_3$O$_3$S) — A white, crystalline solid with a slight sulfur odor; MW: 219.25, BP: 100°C, Sol: 28 g/100 g water at 25°C, Fl.P: NA (poisonous gases produced in a fire), sp. gr. NA. It is a potent carbamate chemical used as an insecticide, nematicide, and acaricide. It is hazardous to the respiratory system, digestive system, skin, and eyes. Acute exposure may cause central nervous system depression, stupor, coma, seizures, hypertension, tachycardia, cardiorespiratory depression, incontinence, diarrhea, blurred vision, sweating, nausea, vomiting, irritation to the eyes and the skin, pulmonary edema, and death. No chronic health effects are known. OSHA exposure limit (8-hour TWA): NA.

oxidant — See oxidizing agent.

O

oxidase — An enzyme that induces biological oxidation by activating the oxygen in molecules containing the element, such as hydrogen peroxide.

oxidation — A chemical union of oxygen with any substance; a chemical reaction in which an element loses electrons and thus increases in oxidation number.

oxidation number — The charge that an ion has (or which an atom appears to have) when its electrons are counted.

oxidation pond — A shallow lake or other body of water in which wastes are consumed by bacteria; also known as a sewage lagoon.

oxidation potential — A measure in volts of the tendency of an element to oxidize or lose electrons; used to predict how the element will react with another substance.

oxidation–reduction — A chemical reaction in which one or more electrons are transferred from one atom or molecule to another.

oxidation–reduction potential — The electromotive force resulting from oxidation–reduction, usually measured relative to a standard hydrogen electrode.

oxidation state — The number of electrons lost or gained by an element or atom in a compound; expressed as a positive or negative number indicating the ionic charge and the number of electrons lost or gained is equal to the valence.

oxidative phosphorylation — The transfer of phosphate ion onto an acceptor molecule, usually adenosine diphosphate, to establish a more energy-rich molecule driven by the coupling of hydrogen transport down an electron transport chain.

oxide — A compound of two elements, one of which is oxygen.

oxidize — To combine with oxygen.

oxidizer — A chemical, other than a blasting agent or explosive, that initiates or supports combustion in another material, causing fire either by itself or through the release of oxygen or other gases.

oxidizing agent — A substance that gives up its oxygen readily, removes hydrogen from a compound, or attracts electrons from elements; also known as an oxidant.

oximeter — A noninvasive device used for measuring continuously the estimated degree of oxygen saturation of the circulating blood and the heart rate.

oxirane — See ethylene oxide (ETO).

oxyacetylene welding — An oxyfuel gas welding process that produces coalescence of metals by heating them with a gas flame obtained from the combustion of acetylene with oxygen; also known as acetylene welding.

oxyfuel gas welding — A process that produces coalescence by heating materials with an oxyfuel gas flame, with or without the application of pressure and with or without the use of filler metal.

oxygen — A tasteless, odorless, colorless gas essential for human respiration; atomic number: 8, atomic weight: 15.9994.

oxygen cycle — Circulation and reutilization of oxygen in the biosphere.

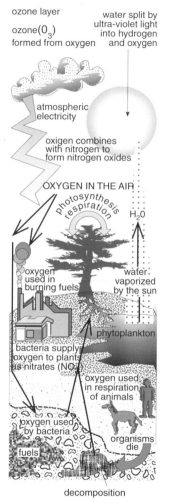

Oxygen Cycle

oxygen debt — Cumulative deficiency of oxygen that develops in the body during periods of intense activity and must be made up when the body activity returns to a normal level.

oxygen deficient — Pertaining to an atmosphere with a partial pressure of oxygen (PO_2) less than 132 mmHg; normal air at sea level contains approximately 21% oxygen at a PO_2 of 160 mmHg.

oxygen demand — The molecular oxygen needed for biological and chemical processes in water.

oxygen-demanding waste — Organic water pollutants that are usually degraded by bacteria if sufficient dissolved oxygen is present.

oxygen depletion — Reduction of the dissolved-oxygen level in a water body.

oxygen difluoride (OF$_2$) — A colorless gas with a peculiar, foul odor; MW: 54.0, BP: –230°F, Sol: 0.02%. It is used as an oxidizer in rocket propellants. It is hazardous to the lungs and eyes and is toxic through inhalation and contact. Symptoms of exposure include intractable headache, respiratory system irritation, pulmonary edema, and eye and skin burns. OSHA exposure limit (TWA): ceiling 0.05 ppm [air] or 0.1 mg/m^3.

oxygen-enriched atmosphere — Any oxygen concentration greater than 25%.

oxygen sag — A declining oxygen concentration downstream from a pollution source having a high biological oxygen demand.

oxygen therapy — The administration for medical reasons of any gas that contains more than 21% oxygen.

oxygen tolerance — An increased capacity to withstand the toxic effects of hyperoxia because the organism has adapted.

oxygen toxicity — The damage or injury caused by inhaling too much oxygen; occurs where the oxygen concentration or oxygen pressure is too high, causing potential seizures.

oxygen transport — The process of oxygen being absorbed in the lungs by hemoglobin in circulating deoxygenated red cells and carried to the peripheral tissues.

oxygen trim — An energy management function in which an attempt is made to reduce heat losses from the exhaust gases of a boiler by adjusting the air/fuel ratio.

oxygen uptake — The amount of oxygen an organism takes from the environment.

oxygenase — An enzyme that catalyzes a reaction in which one or both atoms of molecular oxygen are incorporated into a molecule or substrate; the first step in degradation of straight-chain and aromatic hydrocarbons.

oxygenated fuel — A type of gasoline that has been blended with alcohol or ethers that contain oxygen in order to reduce carbon monoxide and other emissions.

oxygenated hydrocarbons — Compounds that contain carbon, hydrogen, and oxygen (e.g., alcohols).

oxygenates — See oxygenated hydrocarbons.

oxygenation — The process whereby the blood is supplied with oxygen from the lungs; the process of treating, infusing, or combining with oxygen.

oxyhemoglobin (HHbO$_2$) — Oxygen in blood combined with hemoglobin in red blood cells.

oz. — See ounce.

ozonation — The application of ozone to water for disinfection or for taste and odor control.

ozone (O$_3$) — A colorless to blue gas with a very pungent odor; MW: 48.0, BP: –169°F, Sol: insoluble. It is liberated during welding operations; during the oxidizing process of fine organic chemicals; during operations involving high-intensity ultraviolet light; operations involving high-voltage electrical equipment; bleaching operations; food preserving operations for mold and bacteria control. It is hazardous to the eyes and respiratory system and is toxic through inhalation. Symptoms of exposure include pulmonary edema, chronic respiratory disease, irritation of the eyes and mucous membrane, reduced resistance to infection, and possibly premature aging of the lung tissue. OSHA exposure limit (TWA): ceiling 0.1 ppm [air] or 0.2 mg/m^3.

ozone depletion — Destruction of the stratospheric ozone layer, which shields the Earth from ultraviolet

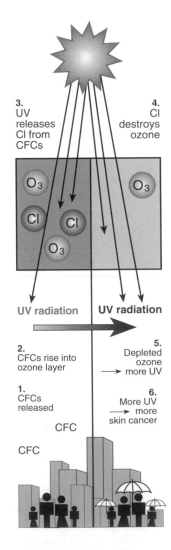

Ozone Depletion Process

radiation harmful to life; caused by the breakdown of certain chlorine and/or bromine-containing compounds that catalytically destroy ozone molecules.

ozone-depletion potential — A measure of the capacity of a particular chemical to destroy ozone compared to a standard (chlorofluorocarbon-11, which has an ozone depletion potential 1.0).

ozone hole — A thinning break in the stratospheric ozone layer where the depletion exceeds 50%. A large area of intense stratospheric ozone depletion over the Antarctic continent occurs particularly between late August and early October each year and generally ends in mid-November. This severe ozone thinning has increased since the late 1970s and early 1980s; it is a result of chemical mechanisms initiated by anthropogenic chlorofluorocarbons.

ozone layer — The protective area in the atmosphere, about 15 miles above the ground, that absorbs some of the ultraviolet rays of the Sun, thus reducing the amount of potentially harmful radiation that reaches the surface of the Earth; ozonosphere.

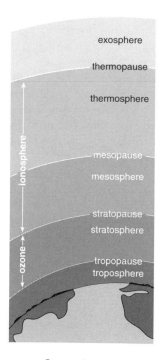

Ozone Layer

ozone mini-hole — A rapid, transient, polar-ozone depletion usually within a 50-km^2 area. It is caused by weather patterns in the upper troposphere that produce an ozone decrease with no chemical depletion of the ozone; however, the cold stratospheric temperatures associated with the weather systems can cause clouds to form, which can lead to conversion of chlorine compounds from inactive to reactive forms, resulting in longer-term problems of ozone depletion.

ozone precursors — Chemicals such as non-methane hydrocarbons and oxides of nitrogen occurring either naturally or as a result of human activities contributing to the formation of ozone, a major component of smog.

ozone shield — The layer of ozone that is 20 to 40 miles above the surface of the Earth and protects the Earth from excessive ultraviolet radiation.

ozone sickness — An abnormal condition caused by the inhalation of ozone at altitudes over 40,000 feet; symptoms include headaches, chest pains, itchy eyes, and sleepiness.

ozone-measuring satellite instruments — Instruments that can measure ozone by looking at the amount of ultraviolet absorption reflected from the surface of the Earth and clouds.

ozonosphere — The region in the upper atmosphere where most atmospheric ozone is concentrated, from about 8 to 30 miles above the Earth, with maximum ozone occurring about 12 miles above the Earth; also known as ozone layer.

P

p — See pico.

P — SI symbol for the prefix *peta-*, representing a factor of 10^{15}; scientific notation for phosphorus.

***p* value** — See probability value.

P2 — Acronym for pollution prevention.

pA — Scientific notation for picoampere.

Pa — See pascal.

PA — Notation for permanent abeyance.

PA/SS — See preliminary assessment and site selection.

pacemaker — An electrical apparatus used for maintaining a normal heart rhythm of myocardial contraction by electrically stimulating the heart muscle.

pacemaker of the heart — See sinoatrial node.

package — (*computer science*) A set of computer programs that can be used for various operations.

package treatment plant — A small wastewater treatment facility serving from a few homes to large institutions.

packaged foods — Candy bars, for example, and other foods stored in specific containers.

packed bed scrubber — An air pollution control device in which emissions pass through alkaline water to neutralize hydrogen chloride gas.

packed tower — A control device in which a gaseous pollutant is removed from a rising air stream by absorption in a liquid flowing downward through a closely packed, chemically inert substance.

PADI — See Provisional Acceptable Daily Intake.

PAH — See polycyclic aromatic hydrocarbon; polynuclear aromatic hydrocarbon.

PAI — Acronym for pure active ingredient.

pain — A feeling of distress, suffering, or agony caused by stimulation of specialized nerve endings.

pain intervention — Techniques used to relieve the pain from disease or trauma.

pain mechanism — A network of nerves that transmit unpleasant sensations to the body because of physical disease and trauma involving tissue damage.

pain receptor — A modified nerve ending that, when stimulated, helps produce the sensation of pain.

pain threshold — The point at which a stimulus of pressure or temperature activates pain receptors and produces a sensation of pain.

pair production — An absorption process for x- and gamma-radiation in which the incident photon is annihilated in the vicinity of the nucleus of the absorbing atom with subsequent production of an electron and positron pair.

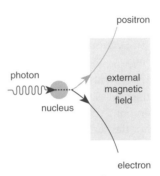

Pair Production

pallid — Lacking color.

pallor — Paleness of the skin.

palpitation — An abnormally rapid heartbeat of which a person is acutely aware; usually over 120 beats per minute.

PAMS — Acronym for Photochemical Assessment Monitoring Stations.

PAN — See peroxyacylnitrates.

panchromatic — Imagery taken of all wavelengths within the visible spectrum, although not uniformly.

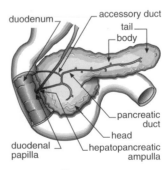

Pancreas

pancreas — A digestive and endocrine gland located between the stomach and intestine in most vertebrates.

pandemic — A disease widespread throughout a geographic area.

panic — An intense, sudden, and overwhelming fear or anxiety that produces immediate physiological and psychological effects.

paper — A thin, felted matrix of interlacing cellulose fibers.

paperboard — A paper product that is heavier, more rigid, and thicker than paper.

paperstock — Paper waste that is recovered and reused.

papilloma — A small growth or tumor of the skin or mucous membrane.

papovavirus — A group of relatively small ether-resistant deoxyribonucleic acid viruses, including papilloma and polyoma viruses, many of which are oncogenic or potentially oncogenic.

PAPR — Acronym for powered air-purifying respirator.

papule — A small, circumscribed, solid, elevated lesion of the skin.

PAR — Acronym for population at risk.

parabolic mirror — A piece of equipment with a large, shiny, curved surface that focuses solar radiation on a specific point.

paraffin series — Any of the methane series of hydrocarbons.

parallel circuit — An arrangement of circuit elements such that each has the same voltage.

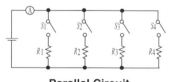

Parallel Circuit
S=Switch   A=Ammeter
R_X =Resistor

paralysis — Loss or impairment of motor or sensory function, in part due to lesions of neural or muscular mechanisms.

paralytic shellfish poisoning — A toxic, neurologic condition that results from eating clams, oysters, or mussels that have ingested the poisonous protozoa called red tide; symptoms, within a few minutes, include nausea, light-headedness, vomiting, and tingling and numbness around the mouth, followed by paralysis of the extremities and possibly respiratory paralysis.

paramedic — A health professional who has completed a course in advance life support care and has been certified by the Emergency Medical Services Commission.

parameter — The property of a system that can be measured numerically; (*statistics*) an arbitrary constant, such as a population mean or standard deviation.

paraplasma — An abnormal growth or malformation.

paraquat ($C_{12}H_{14}Cl_2N_2$) — A yellow solid with a faint ammonia-like odor; may also be found commercially as a methyl sulfate salt; MW: 257.2, BP: 347 to 356°F (decomposes), Sol: miscible, sp. gr. 1.24. It is used as an agricultural herbicide, desiccant, and defoliation agent; to control aquatic weeds. It is hazardous to the eyes, respiratory system, heart, liver, gastrointestinal tract, and kidneys and is toxic through inhalation, absorption, ingestion, and contact. Symptoms of exposure include dermatitis; fingernail damage; damage to the heart, liver, and kidneys; acute pulmonary inflammation; irritation of the gastrointestinal tract; and irritation of the eyes, nose, and epistaxis. OSHA exposure limit (TWA): 0.1 mg/m³ [resp. and skin].

parasite — An organism living in or on another organism from whose body the parasite takes nutrients.

parasitic hemoptysis — Spitting out bright red blood because of a parasitic infection.

parasitemia — Parasites in the blood.

parasympathetic nervous system — A division of the autonomic nervous system.

paratenic host — A host in which a parasite survives without undergoing further development; also known as a transport host.

parathion ($C_{10}H_{14}NO_5PS$) — A pale-yellow to dark-brown liquid with a garlic-like odor (solid below 43°F); MW: 291.27, BP: 375°C, Sol: soluble in most organic solvents, density: NA. It is used as a pesticide on agricultural crops, vegetables, and ornamentals. It is extremely toxic through the respiratory tract and digestive tract. It is an acetylcholinesterase inhibitor and may cause nausea, vomiting, diarrhea, constriction of the pupils, miosis, rhinorrhea, bronchoconstriction, headache, tight chest, wheezing, laryngeal spasm, salivation, cyanosis, anorexia, nausea, vomiting, abdominal cramps, diarrhea, sweat, muscle fasciculation, weakness, paralysis, giddiness, confusion, ataxia, convulsions, coma, low blood pressure, cardiac irregularities, irritation of the eyes and skin, convulsions, coma, respira-

P

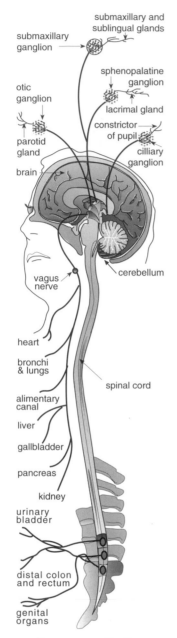

**Parasympathetic
Nervous System**

tory failure, and death. OSHA exposure limit (TWA): 0.1 mg/m³ [skin].

$$C_2H_5O \diagdown \underset{\underset{C_2H_5O \diagup}{\overset{\overset{S}{\|}}{P}}}{} - O - \bigcirc - NO_2$$

Parathion

parathyroid gland — One of the small glands located within a lobe of the thyroid gland.

paratyphoid fever — (*disease*) A bacterial enteric infection in humans with abrupt onset of fever, malaise,

headache, enlargement of the spleen, diarrhea, and involvement of lymphoid tissues. Incubation time is 1 to 3 weeks for enteric fever, 1 to 10 days for gastroenteritis. It is caused by *Salmonella paratyphi* A, B, and C and is found in the United States and Canada. The reservoir of infection is people. It is transmitted by direct or indirect contact with feces or carrier and is spread through food, milk, milk products, shellfish, flies, water, and contaminated individuals' hands. It is communicable as long as the infectious agent is in the excreta, which may range from weeks to months or permanently; general susceptibility. It is controlled by treatment of the disease, good health habits, elimination of flies and feces, and protection of food and water from the microorganisms.

parent — A radionuclide that, upon disintegration, yields a specified nuclide (the daughter), either directly or as a later member of a radioactive series; also known as precursor.

parent material — (*soils*) The undecomposed mineral particles and unweathered rock fragments beneath the subsoil.

parenteral route — A route of entry into the body through injection.

paresis — A slight paralysis.

paresthesia — A feeling of numbness burning, prickling, or tingling sensation of the skin.

paries — The wall of an organ or cavity.

Parshall flume — A calibrated device for measuring the flow of liquid in an open conduit; consists of a contracting length, a throat, and an expanding length. At the throat, the flow passes a sill opening at critical depths, and the upper and lower heads are each measured at a definite distance from the sill. The lower head need not be measured unless the sill is submerged more than 67%.

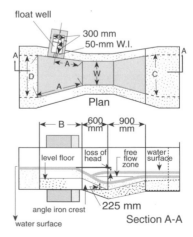

Parshall Flume

partial pressure — The pressure exerted by a gas in a mixture of gases or a liquid; the pressure is directly related to the concentration of that gas to the total pressure of the mixture.

partial pressure carbon dioxide — The pressure exerted by carbon dioxide in any mixture of gases or dissolved in a liquid.

partial pressure nitrogen — The pressure exerted by nitrogen in any mixture of gases or dissolved in a liquid.

partial pressure of arterial oxygen — The part of total blood gas pressure exerted by oxygen gas.

partial pressure of carbon dioxide — The part of total blood gas pressure exerted by carbon dioxide.

partial pressure oxygen — The pressure exerted by oxygen in any mixture of gases or dissolved in a liquid.

particle — A small, discrete mass of solid or liquid matter.

particle accelerator — (*nuclear physics*) A device for speeding up charged subatomic particles to high enough energies to smash the nuclei of the target atoms; used to produce radioisotopes.

particle conditioning — Increasing the effective size of small particles so they can be collected more efficiently by a variety of mechanisms.

particle count — The results of a microscopic examination of treated water with a special instrument that classifies suspended particles by number and size.

particle size — The diameter in millimeters of suspended sediment or bed materials. Clay equals 0.00024 to 0.004 mm; silt, 0.004 to 0.062 mm; sand, 0.062 to 2.0 mm; and gravel, 2.0 to 64.0 mm.

particle-size distribution — Amounts of the various soil types separated in a soil sample; usually expressed as weight percentage.

particleboard — Board product composed of distinct particles of wood or other lignocellulosic materials not reduced to fibers but bonded together with an organic or inorganic resin binder and then cured.

particulate — Unburned carbon usually present when coal or residual oil is used as fuel; appears in the form of black smoke.

particulate loading — Mass of particulates per unit volume of air or water.

particulate matter (PM) — A suspension of fine solid or liquid particles in air (e.g., dust, fog, fume, mist, smoke, spray); also known as aerosol.

particulate tracer — Solid particles and aerosol or bubbles used as tracers for measuring the rate of air movement; the particles have a diameter between 2 and 3 μm and are detected by using a fluorescent light scattering detector, a photomultiplier detector, or a phosphorescence detector.

partition coefficient — A measure of the sorption phenomenon when a pesticide is divided between the soil and water phase; also known as the adsorption partition coefficient.

parts per billion (ppb) — The number of parts of a chemical or substance found in 1 billion parts of a particular gas, liquid, or solid.

parts per million (ppm) — The number of parts of a chemical or substance found in 1 million parts of a particular gas, liquid, or solid.

parts per thousand — The number of parts of a chemical or substance found in 1000 parts of a particular gas, liquid, or solid.

parts per trillion — The number of parts of a chemical or substance found in 1 trillion parts of a particular gas, liquid, or solid.

parvovirus — (*disease*) A virus belonging to the family Parvoviridae and containing linear, single-stranded DNA surrounded by a protein capsid about 22 nm in diameter; may cause viral gastroenteritis in all age groups with symptoms of diarrhea, nausea, vomiting, malaise, abdominal pain, and fever. The mode of transmission is the oral–fecal route, from person to person, by ingestion of contaminated food (especially shellfish), and by drinking contaminated water. The incubation period is 10 to 70 hours. The disease has been reported in England, Australia, and the United States. The reservoir of infection is people; general susceptibility. Also known as acute nonbacterial infectious gastroenteritis or viral gastroenteritis.

pascal (Pa) — A unit of pressure equal to the force of 1 newton per square meter.

Pascal's principle (physics) — A confined liquid transmits pressure applied to it from an external source equally in all directions.

passage — Inoculation of an infection into an animal and then harvesting the infective agent from the same animal.

passive adsorption — A process by which a gas or vapor is condensed out of the air and held on the surface of a piece of solid material by natural forces only.

passive carrier — A healthy person who carries the organisms of an infectious disease although exhibiting no symptoms of the disease.

passive heat absorption — The use of natural materials or absorptive structures without moving parts to gather and hold heat; the simplest and oldest use of solar energy.

passive immunity — Immunity acquired by the injection or transfer of antibodies in another organism.

passive motion — Involuntary motion caused by an external stress.

passive sampling — A method of sampling tracer gas in a building by the process of passive adsorption.

passive smoking — Involuntary second-hand inhalation of tobacco smoke by nonsmokers.

P

Pasteurella — A genus of Gram-negative, nonmotile, facultatively anaerobic, ovoid to rod-shaped bacteria sometimes causing infection in humans.

pasteurization — The process of heating milk or other liquids for a specified time to destroy microorganisms that would cause spoilage and/or disease; the process of heating every particle of milk or milk products to at least 145°F for 30 minutes, or 161°F for 15 seconds and holding it at such temperature continuously.

patella — Knee bone.

patent — The stage in an infection at which infectious agents produce evidence of their presence, such as eggs or cysts.

pathogen — An organism that produces disease in a host and may alter the response of the host to other diseases' processes.

pathogenesis — The cellular events and reactions and other pathologic mechanisms occurring in development of a disease.

pathogenic — Disease causing.

pathological — Abnormal or diseased.

pathological diagnosis — Examination of the substance and function of tissues of the body especially when abnormal changes have occurred.

pathological histology — The study of the effects of disease on the structure, composition, and function of tissues.

pathologist — A licensed physician whose practice of medicine is limited to or specialized in the study of the origin, nature, and course of disease.

pathology — Study of the essential nature, cause, and effects of diseases and their abnormalities.

pathway — A history of the flow of a pollutant from source to receptor, including qualitative descriptions of emission-type transport medium and exposure route.

patient — An individual receiving diagnosis and/or treatment for disease or injury.

payload — Instruments that are carried on a spacecraft.

Pb — See lead.

PBB — See polybrominated biphenyl.

PbS — See galena.

PBZ — See personal breathing zone.

PC — Acronym for percutaneous.

PCA — Acronym for principal component analysis.

PCB — See polychlorinated biphenyl.

PCDD — Acronym for polychlorinated dibenzo-*p*-dioxin; see 2,3,7,8-tetrachlorodibenzo-*p*-dioxin.

PCDF — Acronym for polychlorinated dibenzofuran.

PCE — Acronym for perchloroethylene; notation for pentachloroethane.

pCi — See picocurie.

PCM — See phase-contrast microscope.

PCO — Acronym for pest control operator.

PCO$_2$ — See partial pressure carbon dioxide.

PCP — See pentachlorophenol.

pct — See percent.

PCV valve — A valve that regulates the flow rate of unburned crankcase vapors back to the combustion chamber in order to control their release from automobiles and becoming pollutants.

PD — See potential difference.

PDA — Acronym for portable decontamination apparatus.

PDP–CVS — Acronym for positive displacement pump–constant volume sampler.

PDS — Acronym for personal decontamination station.

PE — Acronym for Professional Engineer.

pea gravel — A uniformly graded gravel with a particle size of approximately 3/16 inch.

Pea Weevil (*Vruchus pisorum*) — Similar to the Bean Weevil but 1/5 inches long, brownish flecked with white, and with black to gray patches or scales; a stored food product insect.

peak — The amount of chemical in a body at its highest level.

peak exposure standard — Maximum concentration of a particular airborne substance determined over the shortest analytically practicable period of time, not exceeding 15 minutes.

peak flow — The maximum instantaneous discharge of a stream or river at a given location; (*wastewater plant*) the highest flow expected to be encountered under conditions of high rainfall and prolonged periods of wet weather.

peak levels — The levels of airborne pollutants much higher than average or occurring over a short period of time in response to sudden releases or due to a longer term buildup over several days.

peak optical response — The wavelength of maximum sensitivity of an instrument.

peat — Partially decomposed dark-brown or black plant material found in marshes and other wet places.

ped — (*soils*) A natural soil aggregate.

pediculosis — (*disease*) An infestation of the head and hairy parts of a body with lice, causing severe itching; incubation time is 15 to 17 days. It is caused by *Pediculus capitis*, the head louse; *P. humanus*, the body louse; and *Phthirus pubis*, the crab louse. It is found worldwide, with head lice being especially common among children in schools and institutions. The reservoir of infection is infested people; it is transmitted by direct contact or contact with clothing, combs, or towels. It is communicable as long as lice or eggs remain alive on the person or clothing; general susceptibility. It is controlled by avoidance of physical contact with infested people, their clothing, and bedding; good health habits; inspection of children for head lice; and appropriate treatment.

pedology — (*soils*) The scientific studies of soils, also known as soil science; (*medicine*) study of the physiological and psychological aspects of childhood.

pedon — (*soils*) The smallest volume in a soil body that displays normal variations in properties of a soil profile.

pedosphere — (*soils*) The soil layers of the crust of the Earth.

peer review — The evaluation by practitioners or other professionals of the effectiveness and efficiency of services ordered, services performed, or professional studies completed and prepared for publication.

peer-reviewed literature — A referenced published document that is reviewed by a minimum of two technical reviewers located outside of the author's organization.

PEIS — See Programmatic Environmental Impact Statement.

PEL — See permissible exposure limit.

pel — See pixel.

pelagic — Of, relating to, living in, or occurring in the open sea.

pellagra — A dietary deficiency disease due to a lack of niacin; characterized by skin lesions, inflammation of the soft tissues of the mouth, diarrhea, and central nervous system disorders.

pen plotter — (*mapping*) A device for drawing maps and figures using a computer-guided pen.

pendente lite (**pending litigation**) — (*law*) Litigation that continues (or is pending).

penetration — Action of a liquid entering into porous materials from cracks and pinholes; (*protective clothing*) flow of a chemical through closures, porous materials, seams, pinholes, or other imperfections in a protective clothing material on a nonmolecular level; (*virus*) stage of viral replication at which the virus genome enters the cell.

penicillin — Collective name for an antibiotic produced by the mold *Penicillium* which is toxic to a number of bacteria, both pathogenic and nonpathogenic.

pentaborane (B$_5$H$_9$) — A colorless liquid with a pungent odor like sour milk; MW: 63.1, BP: 140°F, Sol: reacts, Fl.P: 86°F, sp. gr. 0.62. It is used in chemical research as jet and rocket fuels, catalysts, corrosion-inhibitor/fluxing agents, and oxygen scavengers; as a chemical intermediate. It is hazardous to the central nervous system, eyes, and skin and is toxic through inhalation, absorption, ingestion, and contact. Symptoms of exposure include dizziness; headache; drowsiness; lightheadedness; incoordination; tremors; spasms of the face, neck, and limbs; convulsions; and irritation of the eyes and skin. OSHA exposure limit (TWA): 0.005 ppm [air] or 0.01 mg/m^3.

pentachloronaphthalene (C$_{10}$H$_3$Cl$_5$) — A pale-yellow or white solid or powder with an aromatic odor; MW: 300.4, BP: 636°F, Sol: insoluble, Fl.P: 180°F, sp. gr. 1.73. It is used as an insulating material; as an inert compound of resins or polymers for coating or impregnating textiles, wood, and paper to impart flame resistance, waterproofness, and insecticidal properties; as an additive for cutting oil in various operations performed on metals; as an additive to special lubricants. It is hazardous to the skin, liver, and central nervous system and is toxic through inhalation, absorption, ingestion, and contact. Symptoms of exposure include headache, fatigue, vertigo, anorexia, pruritus, acne-form skin eruptions, jaundice, and liver necrosis. OSHA exposure limit (TWA): 0.5 mg/m^3 [skin].

pentachlorophenol (PCP; C$_6$Cl$_5$OH) — A colorless to white, crystalline solid with a benzene-like odor; MW: 266.4, BP: 588°F, Sol: 0.001%, sp. gr. 1.98. It is used as a preservative for wood, starch, paint, adhesives, leather, latex, and oils; in slime-algae control; as a pesticide, herbicide, and snail control agent. It is hazardous to the cardiovascular system, respiratory system, eyes, liver, kidneys, skin, and central nervous system and is toxic through inhalation, absorption, ingestion, and contact. Symptoms of exposure include sneezing, cough, weakness, anorexia, weight loss, sweating, headache, dizziness, nausea, vomiting, dyspnea, chest pain, high fever, dermatitis, and irritation of the eyes, nose, and throat. OSHA exposure limit (TWA): 0.5 mg/m^3 [skin].

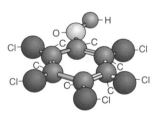

Pentachlorophenol

***n*-pentane (C$_5$H$_{12}$)** — A colorless, volatile liquid with a gasoline-like odor (gas above 97°F); MW: 72.17; BP: 36.1°C; Sol: miscible with organic solvents; Fl.P: –49.5°C (closed cup), a highly flammable liquid; density: 0.6264 at 20°C. It is used as a gasoline additive; in the synthesis of amyl chlorides for intermediates; in the synthesis of polychlorocylopentanes as intermediates; as a solvent; as a component of lighter fluids and blow-torch fuel; as a heat-exchange medium in the manufacture of artificial ice; as a blowing agent for plastics. It is hazardous to the skin, respiratory system, and eyes and is toxic through inhalation, ingestion, and contact.

P

Symptoms of exposure include drowsiness, dermatitis, chemical pneumonia, and irritation of the eyes and nose. It may cause narcosis. ACGIH exposure limit (TLV-TWA): 600 ppm [air] or 1800 mg/m³.

$$CH_3 - CH_2 - CH_2 - CH_2 - CH_3$$
n-Pentane

2-pentanone (CH₃COCH₂CH₂CH₃) — A colorless to water-white liquid with a characteristic acetone-like odor; MW: 86.1, BP: 215°F, Sol: 6%, Fl.P: 45°F, sp. gr. 0.81. It is used in the manufacture of pharmaceuticals and some flavorings; as an extractant in dewaxing petroleum products. It is hazardous to the respiratory system, eyes, skin, and central nervous system and is toxic through inhalation, ingestion, and contact. Symptoms of exposure include headache, dermatitis, narcosis, coma, and irritation of the eyes and mucous membrane. OSHA exposure limit (TWA): 200 ppm [air] or 700 mg/m³.

2–Pentanone

pentavalent — A chemical radical or element that has a valence of 5.

pepsin — A digestive enzyme that changes proteins to peptones and proteoses in mammals, birds, reptiles, and fish.

peptide — A chain of less than 20 amino acids.

peptization — Breaking down materials into particles of colloidal size by chemical action.

peptone — A stage in protein digestion prior to the formation of amino acids.

***per capita* use** — The average amount of a substance used per person during a standard time period, usually a day.

percent (pct or %) — One part per 100 parts.

percent by weight — The amount of a substance in a mixture based on its weight compared to the total weight of the mixture.

percent saturation — The amount of a substance that is dissolved in a solution compared to the amount that could be dissolved in it.

percent solution — Relationship of a solute to a solvent expressed in terms of weight of solute per weight of solution.

percent volatile — A measurement of the percentage of a liquid or solid by volume that will evaporate at an ambient temperature of 70°F.

percentile — The 100th part of a statistical distribution; one of the values of a variable that divides the distribution of the variable into 100 groups having equal frequencies.

perception — The awareness of the effects of the stimuli.

perched water table — The water table of a discontinuous saturated zone in a soil.

perchlorates — M salts of perchloric acid (HClO₄); an example would be potassium perchlorate (KClO₄). These compounds are used in explosives, pyrotechniques, photography; as desiccants; in analytical chemistry. They are low to moderately soluble in water and decompose at around 400°C. They are hazardous to the skin and eyes and cause irritation of the respiratory tract. They are powerful oxidizing substances that explode when mixed with combustible, organic, or other easily oxidizable compounds and are exposed to heat or friction.

perchloroethylene — See tetrachloroethylene.

perchloromethyl mercaptan (Cl₃CSCl) — A pale-yellow, oily liquid with an unbearable acrid odor; MW: 185.9, BP: 297°F (decomposes), Sol: insoluble, sp. gr. 1.69. It is used in chemical synthesis for agricultural and dye chemicals; as an organic intermediate. It is hazardous to the eyes, respiratory system, liver, kidneys, and skin and is toxic through inhalation, absorption, ingestion, and contact. Symptoms of exposure include lacrimation, eye inflammation, cough, dyspnea, deep breathing pain, coarse râles, vomiting, pallor, tachycardia, acidosis, anuria, and irritation of the nose and throat. OSHA exposure limit (TWA): 0.1 ppm [air] or 0.8 mg/m³.

perchloryl fluoride (ClFO₃) — A colorless gas with a characteristic, sweet odor that is shipped as a liquefied compressed gas; MW: 102.5, BP: –52°F, Sol: 0.06%. It is used as a liquid oxidizer in rocket fuels; in chemical synthesis; in flame photometry for analysis of different elements. It is hazardous to the respiratory system, skin, and blood and is toxic through inhalation and contact. Symptoms of exposure include respiratory irritation and skin burns; in animals, methemoglobin, anoxia, cyanosis, weakness, dizziness, headache, pulmonary edema, and pneumonitis. OSHA exposure limit (TWA): 3 ppm [air] or 14 mg/m³.

percolation — The flow or trickling of a liquid downward toward contact with a filtering medium; (*soil*) the flow of water down through the soil.

percutaneous absorption — The absorption of a substance into the skin.

percutaneous exposure — Absorption of a contaminant through unbroken skin.

percutaneously — Pertaining to entry through the skin.

perennial stream — A stream that flows throughout the year.

perfect gas — See ideal gas.

P

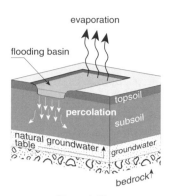

Percolation

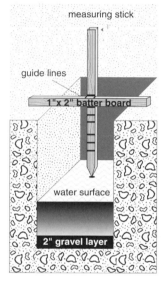

Percolation test methods

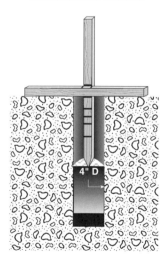

perfect mixing — See uniform mixing.

perfluorocarbons — A group of anthropogenic chemicals composed of carbon and fluorine used as alternatives to ozone-depleting substances but which are powerful greenhouse gases.

perforate — To puncture or pierce.

perforated pipe — A pipe designed to discharge water or sewage through small, multiple, closely spaced orifices for liquid distribution purposes.

performance-based methods system — A system of performance criteria, such as precision, sensitivity, specificity, detection limit, and any other appropriate measurement methods, that demonstrate the ability to meet established performance criteria and comply with specified data-quality needs.

perfusion — The process of flooding fluid through the artery to saturate the surrounding tissue; the passage of blood through the lungs.

pericardial cavity — The area of the body in which the heart lies.

pericardial fluid — The clear fluid contained in the pericardial cavity.

pericarditis — An inflammation of the pericardium associated with trauma, malignant disease, infection, uremia, myocardial infarction, or idiopathic causes.

pericardium — Membranous sack surrounding the heart.

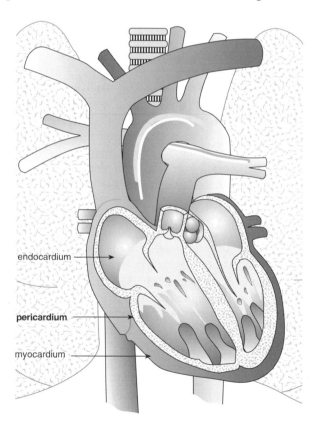

Pericardium

perigee — The point nearest a planet or a satellite reached by an object orbiting it.

perimeter — The circumference, outer edge, or periphery of an object; the boundary of a closed plane figure.

perinatal — Pertaining to an event happening just before, during, or immediately after birth.

perinatal mortality — The statistical rate of fetal and infant deaths, including stillbirths, from 28 weeks of gestation to the end of a neonatal period of 4 weeks after birth.

period — Time for one complete cycle, vibration, or oscillation; time required for a single wave to pass a given point; the time required for a satellite to make one complete orbit; punctuation mark at the end of a sentence.

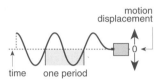

Period

period decay — The tendency of a satellite to lose orbital velocity because of atmospheric drag and gravitational forces.

periodates — Metal salts of periodic acid, $M(IO_4)_x$, where M is a metal ion having valency x. The compounds are highly irritating to the skin, eyes, and mucous membranes. They explode when heated and ignite when mixed with combustible materials, although they are noncombustible substances.

periodic table — Systematic classification of the elements according to atomic numbers or atomic weights and by physical and chemical properties.

periodicity — Events or episodes that tend to repeat at predictable intervals.

periosteum — A fibrous vascular membrane covering the bones.

periostitis — An inflammation of the periosteum caused by disease or injury.

peripheral nervous system (PNS) — Nerve cells lying outside the skull and the vertebral column, including the somatic nervous system and the autonomic nervous system.

periphery — Parts or areas near or outside a perimeter or boundary.

peritoneal cavity — The potential space between the parietal and the visceral layers of the peritoneum.

periphyton — Microscopic underwater plants and animals that are firmly attached to solid surfaces such as rocks, logs, and pilings.

perishable food — Any food of such type or in such condition as it may spoil unless hermetically sealed in containers processed by heat to prevent spoilage, properly packed or dehydrated, or dried to so low a moisture content as to preclude the development of microorganisms.

peristalsis — A series of circular contractions by which the digestive tract propels its contents.

peristaltic — Of or pertaining to peristalsis.

peritoneal fluid — Fluid contained in the membrane lining of the abdominal cavity.

peritoneum — Membrane lining the abdominal cavity.

peritonitis — Inflammation of the peritoneum.

perlite — A volcanic glass that is expanded by heat and forms a lightweight aggregate.

permafrost — Permanently frozen soil in the polar regions.

permameter — A device used for containing a soil sample and subjecting it to fluid flow to measure permeability or hydraulic conductivity.

permanent injunction — A decree that, when signed by the court, perpetually restrains the defendant from engaging in specified violative practices and remains in force until final termination.

permanent parasite — A parasite that lives its entire adult life within or on a host.

permanent partial disability — As defined by workers' compensation insurance, a condition that actually or presumptively results in partial loss of earning power.

permanent total disability — As defined by workers' compensation insurance, a condition that actually or presumptively is considered to be the equivalent of complete and permanent loss of earning power.

permanganates — General formula of $M(MnO_4)_x$, where M can be ammonium or a metal ion with valency x; the salts of permanganic acid ($HMnO_4$). Permanganates are powerful oxidizers and may cause explosions with concentrated sulfuric acid. They are used for bleaching resins, waxes, oils, silk, cotton, and other fibers; in purifying water; as analytical reagents. Potassium permanganate is a moderately toxic compound causing nausea, vomiting, and dyspnea and may be fatal if ingested.

permeability — The ease with which liquids move through a soil; the ability of substances to pass through certain membranes; the capacity of a porous medium to conduct or transmit fluids.

permeable — Describes a substance or soil that has pores or openings that permit liquids or gases to pass through.

permeation — The process by which a chemical moves through protective clothing on a molecular level that involves sorption of a chemical into the contacted material, diffusion of the chemical molecules within the material, and desorption from the opposite surface of the material; (*chemistry*) the movement of atoms, molecules, or ions into or through a porous or permeable substance.

permethrin — (*insecticides*) A synthetic pyrethroid used for control of adult mosquitos in residential and recreational areas.

permissible dose — The amount of radiation that may be safely received by an individual within a specified period; formerly known as tolerance level.

permissible exposure limit (PEL) — An Occupational Safety and Health Administration standard designating the maximum occupational exposure permitted as an 8-hour, time-weighted average (TWA).

permit — A means of registering a source of air, water, or land pollution.

permittivity ($\hat{I}$) — The ability of a dielectric to store electrical energy.

pernicious — Something that is potentially injurious, destructive, or fatal unless treated.

peroxide — A compound in which oxygen is visualized as joined to an oxygen.

peroxides — For example, sodium peroxide (Na_2O_2). Peroxides are used for bleaching fibers, waxes, and woods; for dying and printing textiles; for purifying air. They are powerful oxidizing substances and are corrosive to the skin. They may react violently when exposed to air or water

peroxyacetic acid ($C_2H_4O_3$) — A colorless liquid with an acrid odor; MW:76.05; BP: 105°C; Sol: soluble in water, alcohol, ether, and sulfuric acid; Fl.P: 40.5°C (open cup), density: 1.150 at 20°C. It is used as an epoxidizing agent, for bleaching, as a germicide, and in the synthesis of pharmaceuticals. It is hazardous to the skin, eyes, respiratory tract, and digestive tract. It causes severe irritation and severe acid burns. It causes skin tumors in mice at the site of the contamination. ACGIH exposure limit: NA.

$$H_3 - \overset{\overset{\textstyle O}{\|}}{C} - O - O - H$$

Peroxyacetic Acid

peroxyacylnitrates (PAN) — Any of a group of organic compounds that can come from several sources and are formed by the chemical reactions of other pollutants; these reactions proceed more rapidly in intense sunlight. They are extremely reactive oxidizing agents that cause eye irritation and respiratory distress.

peroxybenzoic acid ($C_7H_6O_3$) — A volatile solid with a pungent odor that sublimes; MW: 138.12, BP: NA, Sol: slightly soluble in water and mixes readily with most organic solvents, Fl.P: NA (violent

decomposition may occur at high temperatures). Used in organic analysis to measure the degree of unsaturation; in the production of epoxides. It is mostly nontoxic in humans but can cause skin tumors in mice with prolonged contact. ACGIH exposure limit: NA.

$$\overset{\overset{\textstyle O}{\|}}{\underset{\textstyle}{C}} - O - O - H$$

Peroxybenzoic Acid

peroxyformic acid (CH_2O_3) — A colorless acid that is unstable in the concentrated form; MW: 62.03; BP: NA; Sol: soluble in water, alcohol, ether, and chloroform; Fl.P: the solutions are shock and heat sensitive, highly reactive, and decompose violently when exposed to heat; density: NA. It is used as an epoxidizing agent. It is hazardous to the skin and eyes and causes skin and eye irritation. ACGIH exposure limit: NA.

$$H - \overset{\overset{\textstyle O}{\|}}{C} - O - O - H$$

Peroxyformic Acid

persist — To stay or to remain.

persistent — Describes a pesticide that remains in the soil or on the crops after it has been applied and is degraded only slowly by the environment.

persistent agent — A chemical agent that does not hydrolyze or volatilize readily, such as VX and HD.

persistent infection — An infection in which ongoing virus replication occurs, but the virus adjusts its replication and pathogenicity to avoid killing the host. This differs from chronic infection in that the virus may continue to be present and to replicate in the host for its entire lifetime instead of eventually being cleared out by the host.

persistent organic pollutants — The chemicals, chiefly compounds of carbon, that persist in the environment, accumulate through the food chain, and pose a risk of causing adverse effects to human health and the environment.

persistent pesticides — Pesticides that do not break down chemically or that break down very slowly and remain in the environment for a long period of time.

P

person-rem — (*radiation*) Product of the average individual dose in a population and the number of individuals in the population.

person-year — The activity or exposure of an individual during the course of an entire year; used in epidemiological study design.

personal air samples — Air samples taken with a pump that is directly attached to the worker; the collecting filter and cassette are placed in the worker's breathing zone.

personal breathing zone (PBZ) — The region of air inhaled by a worker in the normal execution of duties; samples are collected by a device secured to the worker's lapel.

personal measurement — A measurement collected from an individual's immediate environment.

personal noise dosimeter — The portion of a personal noise dosimeter set that must actually be worn by the worker, including the microphone.

personal protection equipment (PPE) — Devices worn to protect against environmental hazards.

personnel monitoring — The monitoring of radioactive contamination based on measurements from any part of an individual, including breath, excretions, or clothing; the monitoring of hazardous chemicals in the breathing zone of a person.

perspiration — The excretion of moisture through the pores of the skin; also known as sweat.

persulfates — Examples include sodium persulfate ($Na_2S_2O_7$) and potassium persulfate ($K_2S_2O_7$), which are strong oxidizers that decompose violently upon heating. They are used for bleaching fabrics, in photography and as reagents in chemical analysis. They are strong irritants to the skin and mucous membranes.

PERT — See program evaluation and review technique.

Perthane ($C_{16}H_2OCl_2$) — An odorless, cream- to tan-colored semisolid in the technical product state; MW: 307.26, BP: decomposes at 52°C, Sol: insoluble in water, Fl.P: NA (flammable material that may be ignited by heat, sparks, or flames and may travel to a source of ignition and flash back), sp. gr. NA. It is used as a chlorinated hydrocarbon insecticide but is no longer permitted in the United States. It is hazardous to the respiratory system, digestive system, skin, and eyes. Acute exposure may cause dizziness, abdominal pain, headache, nausea, vomiting, diarrhea, mental confusion, a sense of apprehension, weakness, loss of muscle control, tremors, severe seizures, and death. Chronic exposure may cause depressed sperm counts, anorexia, nausea, fatigue, headaches, memory deficits, personality changes, disorientation, loss of coordination, dry eyes, seizures, damage to the liver and kidneys, and damage to the developing fetus. OSHA exposure limit (8-hour TWA): NA.

respirator

helmet

goggle

Ear Plugs
(3–flange design)

Personal Protective Devices

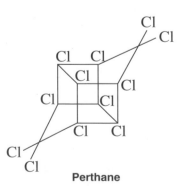

Perthane

perturbation — A disturbance of motion, course, arrangement, or state of equilibrium.

perturbed electric field — Modification of an electric field by the presence of a conducting object.

pervious — See permeable.

pest — An unwanted plant or animal.

pest control operator — A person or company that applies pesticides as a business, usually for households.

pest resurgence — A rebound of pest populations due to acquired resistance to chemicals and nonspecific destruction of natural predators and competitors by broad-scale pesticides.

pesticide — Any agent that destroys pests.

pesticide formulation — The substance or mixture of substances comprised of all active and inert ingredients (if any) of a pesticide product.

pesticide poisoning — A toxic condition caused by the ingestion or inhalation of a substance used to kill insects or rodents.

pesticide rain — Long-range transport of pesticides by air currents and deposition through precipitation in sites far from its origin; similar to acid rain.

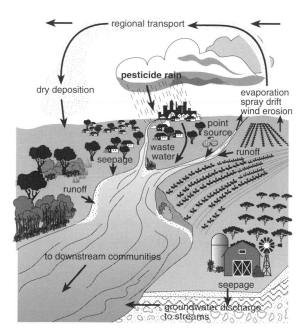

Pesticide Rain

Pesticide Regulation Notice — A formal notice to pesticide registrants about important changes in regulatory policy, procedures, and regulations.

pesticide solid waste — Residue resulting from the manufacturing, handling, or use of chemical pesticides.

pesticide tolerance — The amount of a pesticide chemical that can safely remain on or in food for human consumption or for consumption by livestock.

pesticide treadmill — A need for constantly increasing doses or new pesticides to prevent pest resurgence.

pet — A domesticated dog, cat, fish, bird, or snake, for example.

PET — See positron emission topography.

PET scan — See positron emission tomography scan.

petcock — A small drain valve.

Petri dish — A flat-bottom, transparent dish with cover used when culturing microorganisms and tissues.

petrochemicals — Chemicals isolated or derived from petroleum or natural gas.

petroleum — A complex, naturally occurring green to black mixture of gaseous, liquid, and solid hydrocarbons; commercial petroleum products are obtained from a crude petroleum distillation cracking chemical treatment.

petroleum derivatives — Chemicals formed when gasoline breaks down while in contact with groundwater.

petroleum distillate (naphtha) — A colorless liquid with a gasoline- or kerosene-like odor and which is a mixture of paraffins that may contain a small amount of aromatic hydrocarbons; MW: 99 (approx.), BP: 86 to 460°F, Sol: insoluble, Fl.P: –40 to –86°F, sp. gr. 0.63 to 0.66. It is used as a solvent in the rubber industry; during the manufacture of waterproof cloth, shoe adhesives, and rubber tires; as extractants; in the preparation of paint, varnish, and lacquer as solvents, diluents, and thinners; as solvents in pesticides; for dry-cleaning operations. It is hazardous to the skin, eyes, respiratory system, and central nervous system and is toxic through inhalation, ingestion, and contact. Symptoms of exposure include dizziness, drowsiness, headache, nausea, dry and cracked skin, and irritation of the eyes, nose, and throat. OSHA exposure limit (TWA): 400 ppm [air] or 1600 mg/m^3.

PF — See protection factor.

PFBC — Acronym for pressurized fluidized bed combustor.

PFC — Acronym for perfluorated carbon.

PFCRA — Acronym for Program Fraud Civil Remedies Act.

PFCs — See perfluorocarbons.

PFE — Acronym for pneumatic fracturing extraction.

PFT — See pulmonary function test.

pg — See picogram.

pH — Scientific notation for potential hydrogen.

pH adjustment — A means of maintaining the optimum pH through the use of a chemical additive.

pH scale — Scale that measures acidity and alkalinity, ranging from 0 for most acidic to 14 for most basic; pH stands for potential hydrogen and refers to the concentration of hydrogen ions.

pH value — Measure of the acidity or alkalinity of a solution on a scale from 0 to 14, with 7 being neutral, <7 being acid, and >7 being alkaline.

P

PHA — Acronym for Preliminary Hazard Analysis.

phage typing — Identification of bacteria by testing vulnerability to bacterial viruses.

phagocyte — A cell that engulfs foreign material and consumes debris and foreign bodies.

phagocytic cell — A cell that engulfs bacteria and digests them with enzymes, including lysozymes.

phagocytize — To engulf and destroy bacteria or other foreign materials.

phagocytosis — The engulfing of microorganisms or other cells and foreign particles by phagocytes.

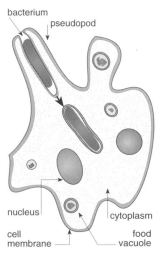

Phagocytosis

phagotrophic — Describes a form of feeding where animals, such as protozoans, engulf particulate nutrients, including bacterial cells or debris.

phalanges — Bones of the fingers and toes in most vertebrates.

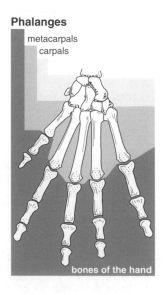

Phalanges

pharmaceuticals — Drugs and related chemicals produced by companies for medicinal purposes.

pharmacodynamics — Study of how a drug acts on a living organism.

pharmacogenetics — Study of the effects of genetic factors on the response of an individual or group to certain drugs.

pharmacokinetics — Quantitative analysis of interactions between an organism and a drug; the exposure to a drug or chemical that can lead to a variety of chemical and biological reactions in the body, including absorption into the body, metabolism into other substances, distribution to other organs and tissues, or removal from the body, as well as the various biochemical pathways in the body that activate, metabolize, detoxify, transport, and excrete chemicals and the relationship between the administered doses and the effective doses.

pharmacological agent — An oral, parenteral, or topical substance used to relieve symptoms and treat or control disease.

pharmacology — A science dealing with drugs, their sources, appearance, chemistry, actions, uses, and potential reactions.

pharmacopeia — A compendium of descriptions, strengths, standards or purity, and dosage levels for drugs.

pharynx — The muscular throat cavity extending up over the soft palate and to the nasal cavity in vertebrates.

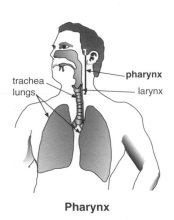

Pharynx

phase — A measure of the progression of a cyclic wave or motion form in time or space from a chosen instant or position.

phase contrast microscope (PCM) — A magnifying instrument that uses the differences in phases of light passing through or reflected by the object being examined.

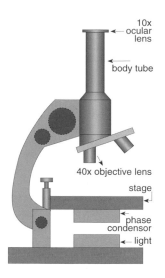

Phase Contrast Microscope

PHC — Acronym for principal hazardous constituent.

phenanthrene ($C_{14}H_{10}$) — A monoclinic plate crystalized from alcohol; MW: 178.24; BP:340°C; Sol: soluble in alcohol and dissolves readily in benzene, toluene, and chloroform; density, NA. A polycyclic aromatic hydrocarbon formed during incomplete combustion of coal, oil, gas, garbage, or other organic substances and found especially at hazardous waste sites; see polycyclic aromatic hydrocarbons. May cause tumors in the skin of animals.

Phenanthrene

phenol (C_6H_5OH) — A colorless to light-pink, crystalline solid with a sweet, acrid odor that liquefies by mixing with about 8% water; MW: 94.1, BP: 359°F, Sol: 9% at 77°F, Fl.P: 175°F, sp. gr. 1.06. It is used in plywood manufacture; in the manufacture of molded articles; in industrial coatings; in the synthesis of thermosetting phenolic resins, epoxy, polycarbonate, phenoxy, and polysulfone; in the synthesis of aprolactam; in the synthesis of agricultural chemicals and intermediates; in the synthesis of pharmaceuticals; in the synthesis of stabilizers and preservatives for dyes, perfumes, and fungicides; in the synthesis of intermediates in polyester production; in the synthesis of surface-active agents and detergent intermediates; in the

synthesis of explosives; in the manufacture of disinfectant agents and products; in the synthesis of synthetic cresols and xyenols. It is hazardous to the liver, kidneys, and skin and is toxic through inhalation, absorption, ingestion, and contact. Symptoms of exposure include anorexia, weight loss, weakness, muscle aches, pain, dark urine, cyanosis, liver and kidney damage, skin burns, dermatitis, ochronosis, tremors, convulsions, twitches, and irritation of the eyes, nose, and throat. OSHA exposure limit (TWA): 5 ppm [skin] or 19 mg/m³.

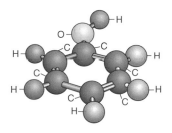

Phenol

phenol-red — An organic dye used in swimming pool testing; it is yellow at a pH of 6.8 and turns progressively deeper red as the pH increases to 8.4.

phenolic resins — A class of resins produced as the condensation product of phenol, formaldehyde, or other aldehydes.

phenolphthalein (($C_6H_4OH)_2COC_6H_4CO$) — Pale-yellow crystals used as an organic indicator in volumetric analysis; colorless in acid solutions below pH 8 and red in the presence of ⁻OH ions above pH 9.6.

phenotype — The visible properties of an organism that are produced by the interaction of the genotype and the environment.

phenyl ether (vapor) ($C_6H_5OC_6H_5$) — A colorless, crystalline, solid or liquid (above 82°F) with a geranium-like odor; MW: 170.2, BP: 498°F, Sol: insoluble, Fl.P: 234°F, sp. gr. 1.08. It is used as a high-temperature, heat-transfer medium; as an intermediate in organic synthesis; as an odorant in the manufacture of soaps and perfumes; as a flavoring agent in the manufacture of food products. It is hazardous to the eyes, skin, and respiratory system and is toxic through inhalation, ingestion, and contact. Symptoms of exposure include nausea, irritation of the eyes, nose, and skin. OSHA exposure limit (TWA): 1 ppm [air] or 7 mg/m³.

phenyl ether–biphenyl mixture (vapor) ($C_6H_5OC_6H_5$/ $C_6H_5C_6H_5$) — A colorless to straw-colored liquid or solid (below 54°) with a disagreeable, aromatic odor. A mixture typically contains 75% phenyl

P

ether and 25% biphenyl; MW: 166 (approx.), BP: 495°F, Sol: insoluble, Fl.P: 255°F (open cup), sp. gr. 1.06 at 77°F. It is used in the chemical, petroleum, and nuclear industries as a high-transfer medium; in heating and cooling; as a dye carrier for printing and dyeing polyester textiles. It is hazardous to the eyes, skin, and respiratory system and is toxic through inhalation, ingestion, and contact. Symptoms of exposure include nausea and irritation of the eyes, nose, and skin. OSHA exposure limit (TWA): 1 ppm [air] or 7 mg/m³.

phenyl glycidyl ether (C₉H₁₀O₂) — A colorless liquid; a solid below 38°F; MW: 150.1, BP: 473°F, Sol: 0.2%, Fl.P: 248°F, sp. gr. 1.11. It is used as a coupling agent, catalyst, or reactive diluent in curing rubber, thermostable epoxy resins, tire cord, and electric insulating material; as a stabilizer of halogenated compounds; in the treatment of fabric; as a copolymer in the production of epoxy polymers; as a chemical intermediate. It is hazardous to the skin, eyes, and central nervous system and is toxic through inhalation, ingestion, and contact. Symptoms of exposure include narcosis and irritation of the skin, eyes, and upper respiratory system; carcinogenic. OSHA exposure limit (TWA): 1 ppm [air] or 6 mg/m³.

m-phenylenediamine (C₆H₈N₂) — A white crystalline solid that turns red when exposed to air; MW: 108.16; BP: 284 to 287°C; Sol: soluble in water, alcohol, acetone, and chloroform. It is used in the manufacture of a variety of dyes, including hair dye, and is a rubber curing agent. It is hazardous to the respiratory tract and digestive tract. It creates symptoms of tremor, excitement, convulsions, cyanosis, and death. ACGIH exposure limit (TLV-TWA): 0.1 mg/m³.

NH₂

m-Phenylenediamine

o-phenylenediamine (C₆H₈N₂) — A brownish-yellow crystal; MW: 108.16, BP: NA, Sol: slightly soluble

NH₂—NH₂

o-Phenylenediamine

in water and dissolves in organic solvents, Fl.P: noncombustible solid but may react violently to strong oxidizers. It is used as an intermediate in the manufacture of dyes, pigments, and fungicides. It is hazardous to the respiratory tract and digestive tract and causes symptoms of excitement, tremors, convulsions, and respiratory depression. It is a suspected human carcinogen. ACGIH exposure limit(TLV-TWA): 0.1 mg/m³.

p-phenylenediamine (C₆H₄(NH₂)₂) — A white to slightly red crystalline solid; MW: 108.2, BP: 513°F, Sol: 5%, Fl.P: 312°F. It is used in dye and dyestuff intermediates for fur, hair, leather, cotton, and synthetics; in photographic application as a fine-grain developing agent, for preparation of color developers, and for photochemical measurements; in polymer technology; as a catalyst in the preparation of epoxy resins, synthetic fibers, heat-resistant polymers, and coatings for leather, paper, and textiles; in rubber technology for natural and synthetic rubber for vulcanization; in organic synthesis; as an analytical reagent; as a special solvent. It is hazardous to the respiratory system and skin and is toxic through inhalation, absorption, ingestion, and contact. Symptoms of exposure include sensitization dermatitis, bronchial asthma, and irritation of the pharynx and larynx. OSHA exposure limit (TWA): 0.1 mg/m³ [skin].

p-Phenylenediamine

phenylhydrazine (C₆H₅NHNH₂) — A colorless to pale-yellow liquid or solid (below 67°F) with a faint, aromatic odor; MW: 108.1, BP: 470°F (decomposes), Sol: slight, Fl.P: 190°F, sp. gr. 1.10. It is used in the manufacture of pharmaceuticals, photographic developers, and polymethene dyes; as an analytical reagent; as an intermediate in organic synthesis; as a reducing agent. It is hazardous to the blood, respiratory system, liver, kidneys, and skin and is toxic through inhalation, absorption, ingestion, and contact. Symptoms of exposure include skin sensitization, hemolytic anemia, dyspnea, cyanosis, jaundice, kidney damage, and vascular thrombosis; carcinogenic. OSHA exposure limit (TWA): 5 ppm [skin] or 20 mg/m³.

phenotype — Complete observable characteristics of an organism as determined by the interaction of genetic traits and environmental factors.

pheromone — A chemical substance produced by a body and acting as a stimulus to individuals of the same species to elicit a behavioral response.

Philadelphia rod — A leveling rod in which the hundredths of feet or eighths of inches are marked by alternate bars of color the width of the measurement.

PHL — Acronym for Preliminary Hazards List.

phlebotomus fever — An acute mild infection caused by one of five distinct arboviruses transmitted to people by the bite of an infected sandfly with symptoms of rapidly developing fever, headache, eye pain, conjunctivitis, myalgia, and an occasional rash.

phlegm — Thick mucous secreted by tissues lining the airways of the lungs.

PHN — Acronym for Center for Population, Health, and Nutrition.

phobia — An anxiety disorder with an obsessive, irrational, and intense fear of specific things; symptoms include faintness, fatigue, palpitations, perspiration, nausea, tremor, and panic.

phon — Unit of loudness level.

phonic — Pertaining to voice, sound, or speech.

phosdrin ($C_7H_{13}O_4PO_2$) — A pale-yellow to orange liquid with a weak odor; MW: 224.2, BP: decomposes, Sol: miscible, Fl.P: 347°F (open cup), sp. gr. 1.25. It is used as an insecticide on agricultural crops, fruits, and vegetables and in sewage treatment plants. It is hazardous to the respiratory system, central nervous system, cardiovascular system, skin, and blood cholinesterase and is toxic through inhalation, absorption, ingestion, and contact. Symptoms of exposure include miosis, rhinorrhea, headache, tight chest, wheezing, laryngeal spasms, salivation, cyanosis, anorexia, nausea, vomiting, abdominal cramps, diarrhea, paralysis, ataxia, convulsions, low blood pressure, cardiac irregularities, and irritation of the skin and eyes. OSHA exposure limit (TWA): 0.01 ppm [skin] or 0.1 mg/m³.

phosgene ($COCl_2$) — A colorless gas with a suffocating odor like musty hay; a fuming liquid below 47°F; shipped as a liquefied compressed gas; MW: 98.9, BP: 47°F, Sol: slight, sp. gr. 1.43. It is used in sandbleaching operations during glass manufacture; in the synthesis of isocyanates for the manufacture of polyurethane plastics. It is liberated during fire extinguishing with agents containing chlorinated hydrocarbons. It is hazardous to the respiratory system, skin, and eyes and is toxic through inhalation, absorption, ingestion, and contact. Symptoms of exposure include eye irritation, dry and burning throat, vomiting, cough, foamy sputum, dyspnea, chest pain, cyanosis, and skin burns. OSHA exposure limit (TWA): 0.1 ppm [air] or 0.4 mg/m³.

phosphatase — Any of a group of enzymes capable of catalyzing the hydrolysis of esterified phosphoric acid with liberation of inorganic phosphate; found in practically all tissues, body fluids, and cells, including erythrocytes and leukocytes.

phosphatase test — A test used to determine the efficiency of pasteurization and quality of milk.

phosphate (PO_4^{-3}) — A generic term for any compound containing a phosphate group.

phosphine (PH_3) — A colorless gas with a fish- or garlic-like odor (pure compound is odorless) that is shipped as a liquefied compressed gas; MW: 34.0, BP: –126°F, Sol: slight. It is used as a doping agent in the manufacture of solid-state components for electronic circuits and in the manufacture of lasers; in preparations for textile treatments; in organic intermediates; as a fumigant for stored grain; in the manufacture of safety matches and pyrotechnics. It is liberated during the use of acetylene. It is hazardous to the respiratory system and is toxic through inhalation. Symptoms of exposure include nausea, vomiting, abdominal pain, diarrhea, thirst, chest pressure, dyspnea, muscle pain, chills, stupor, or syncope. OSHA exposure limit (TWA): 0.3 ppm [air] or 0.4 mg/m³.

phosphobacterium — A bacterium that is especially good at making soluble the inorganic phosphate in soil.

phosphorescence — The capacity of a substance to glow for more than 0.1 nanoseconds after removal of a light source.

phosphoric acid (H_3PO_4) — A thick, colorless, odorless, crystalline solid often used in an aqueous solution; MW: 98.0, BP: 415°F, Sol: miscible, sp. gr. (77°F): 1.87. It is used in the manufacture of aluminum products; in the synthesis of intermediates for the manufacture of soil fertilizers; in the manufacture of poultry and livestock feed; during synthesis of detergent and soap builders and water-treatment chemicals; as an acidulant and flavor agent in the manufacture of carbonated beverages and jellies and preserves; in the manufacture of cleaning preparations and disinfectants; as a bonding agent in the manufacture of refractory bricks; during lithography and photoengraving operations; as a catalyst in the synthesis of other chemicals; in the synthesis of textile and leather processing chemicals, clays, ceramics, cements, and clay-thinning agents; in the synthesis of pharmaceuticals; as a laboratory reagent; during the manufacture of opal glass; in the manufacture of dental cements and dentrifice adhesives; in the manufacture of electric lights. It is hazardous to the respiratory system, skin, and eyes and is toxic through inhalation, ingestion, and contact. Symptoms of exposure include dermatitis;

irritation of the upper respiratory tract, eyes, and skin; and burns of the skin and eyes. OSHA exposure limit (TWA): 1 mg/m^3 [air].

phosphorus, yellow (P$_4$) — A white to yellow, soft, waxy allotrope with acrid fumes in air that is usually shipped or stored in water; MW: 124.0, BP: 536°F, Sol: 0.0003%, Fl.P: not known, sp. gr. 1.82. It is used in the synthesis of high-purity phosphoric acid salts for use as fertilizers, water treatment chemicals, food products, beverages, and dentrifices; in the synthesis of inorganic phosphorous compounds for use as pesticides, flame retardants, and gasoline and lube oil additives; in the synthesis of inorganic and organic compounds; in the manufacture of phosphorus alloys; in the manufacture of explosives, munitions, and pyrotechnics for military use; as an ingredient of rat poison and roach powders. It is hazardous to the respiratory system, liver, kidneys, jaw, teeth, blood, eyes, and skin and is toxic through inhalation, ingestion, and contact. Symptoms of exposure include abdominal pain, nausea, jaundice, anemia, cachexia, dental pain, excess salivation, jaw pain, swelling, skin and eye burns, and irritation of the eyes and respiratory tract. OSHA exposure limit (TWA): 0.1 mg/m^3 [air].

phosphorus cycle — Circulation of phosphorus in the biosphere; phosphorus is found in sediment and bones and is taken up by plants, is consumed by animals, is used to produce and fortify teeth and bones, and is returned to the soil upon the animals' deaths.

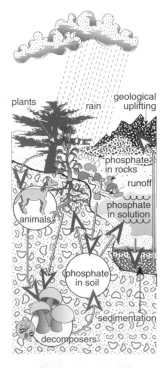

Phosphorus Cycle

phosphorus oxychloride (POCl$_3$) — A colorless liquid with a pungent odor; MW: 153.32; BP: 106°C; Sol: mixes with water and alcohol; Fl.P: a nonflammable solid that reacts with water, alcohols, and amines, resulting in a violent heat release and splattering; density: 1.645. It is used to produce hydraulic fluids, plasticizers, and fire-retarding agents. It is hazardous to the respiratory tract, producing acute and chronic toxicity. It may cause headaches, dizziness, weakness, nausea, vomiting, coughing, chest pains, bronchitis, and pulmonary edema. ACGIH exposure limit (TLV-TWA): 0.1 ppm.

$$O \equiv P\!\!\begin{array}{l} \diagup Cl \\ -Cl \\ \diagdown Cl \end{array}$$

Phosphorus oxychloride

phosphorus pentachloride (PCl$_5$) — A white to pale-yellow, crystalline solid with a pungent, unpleasant odor; MW: 208.3, BP: sublimes, Sol: reacts. It is used as a chlorinating agent in organic synthesis or chemical reactant; as a catalyst in organic synthesis in the production of polyethylene from ethylene. It is hazardous to the respiratory system, skin, and eyes and is toxic through inhalation, ingestion, and contact. Symptoms of exposure include bronchitis, dermatitis, and irritation of the eyes and respiratory system. OSHA exposure limit (TWA): 1 mg/m^3 [air].

phosphorus pentafluoride (PF$_5$) — A colorless gas that fumes strongly in air; MW: 125.97, BP: NA, Sol: reacts with water and is soluble in carbon disulfide and carbon tetrachloride, Fl.P: a nonflammable gas that is decomposed by water, density: 5.8 g/l. It is used as a catalyst in polymerization reactions. It is hazardous to the skin, eyes, and respiratory tract and is highly irritating; may cause pulmonary edema and lung injury. ACGIH exposure limit (TLV-TWA): 2.5 mg/m^3.

phosphorus pentasulfide (P$_2$S$_5$) — A light-yellow to greenish-yellow solid with the odor of rotten eggs; MW: 222.24, BP: 513°C, Sol: soluble in carbon disulfide and aqueous solutions of caustic alkalies and decomposes violently in water, Fl.P: a flammable solid that burns slowly, density: 2.09. It is used in the manufacture of lubricant additives, pesticides, safety matches, and flotation agents. It is hazardous to the skin, eyes, and respiratory tract and causes severe irritation and intoxication when it produces hydrogen sulfide in the presence of moisture. ACGIH exposure limit (TLV-TWA): air 1 mg/m^3.

P

phosphorus pentoxide (P₂O₅) — A white deliquescent crystal; MW: 141.94, BP: NA, Sol: readily absorbs moisture and mixes with water to form phosphoric acid, Fl.P: a nonflammable compound that does not support combustion but reacts violently with water and alcohol, density: 2.3. It is a highly corrosive substance that is an irritant to the eyes, skin, and mucous membranes and is hazardous to the skin and respiratory tract. It may cause chronic pulmonary edema, injury to lungs, and hemorrhage in experimental animals.

phosphorus sesquisulfide (P₄S₃) — A yellowish-green crystal; MW: 220.06; BP: 407 to 408°C; Sol: soluble in carbon disulfide, benzene, and toluene and reacts with water; Fl.P: a highly flammable solid that can ignite by friction; density: 2.03 at 20°C. It is used in making safety matches. It is hazardous to the skin, causing a mild irritation.

phosphorus trichloride (PCl₃) — A colorless to yellow, fuming liquid with an odor like hydrochloric acid; MW: 137.4, BP: 169°F, Sol: reacts, sp. gr. 1.58. It is used during the synthesis of plasticizers and intermediates; in the production of intermediates and during chemical synthesis of dyes, pharmaceuticals, other chlorinating agents, and other organic chemicals; in the treatment of polypropylene before drying; in the manufacture of knitted fabrics. It is hazardous to the respiratory system, skin, and eyes and is toxic through inhalation, ingestion, and contact. Symptoms of exposure include pulmonary edema, eye and nose burns, and irritation of the eyes, nose, and throat. OSHA exposure limit (TWA): 0.2 ppm [air] or 1.5 mg/m³.

phosphorylation — The addition of a phosphate group to a molecule.

photoallergic — Delayed hypersensitivity reaction after being exposed to light.

photoallergic contact dermatitis — Oozing of fluid and other materials from cells and tissues due to an inflammation or injury occurring 24 to 48 hours after exposure to light of a previously sensitized person.

photoautotroph — An organism that synthesizes organic matter using the energy of light.

photobiologic effect — A photosensitive effect associated with specific chlorinated hydrocarbons, particularly the chlorobenzols, diphenyls, and triphenyls.

photochemical — Influenced or initiated by light.

photochemical oxidant (Oₓ) — Any of the products of secondary atmospheric reactions that occur during smog.

photochemical process — Chemical changes caused by the radiant energy of the sun acting upon various polluting substances to produce photochemical smog.

photochemical reaction — A chemical reaction between different air pollutants involving absorption or emission of radiation, particularly ultraviolet light.

photochemical smog — A mixture of secondary air pollutants including ozone, organic nitrates, and others produced by photochemical reactions from primary pollutants, such as nitrogen oxides and hydrocarbons.

photochemistry — Study of the impact of light on certain chemical molecules.

photoconductive cell — A photoelectric cell in which the electrical resistance varies inversely with the intensity of light that strikes its active material.

photocurrent — An electric current induced by radiant energy.

photodegradable — A chemically degradable action due to exposure to light.

photodisintegration — A nuclear reaction involving the interaction of a gamma ray with the nucleus of an atom.

photodissociation — See photolysis.

photoelectric effect — The ejection of electrons in a substance by incident electromagnetic radiation.

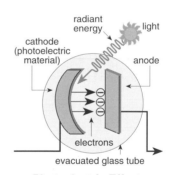

Photoelectric Effect
in a Photocell

photoelectron — An electron that is discharged when light strikes a metal surface.

photogrammetry — The science of making reliable measurements by the use of photographs, especially aerial photographs, as in surveying.

photographic dosimeter — A photographic film or badge used in personal radiation monitoring to measure the cumulative dosage of radiation to which an individual is exposed.

photoheterotroph — An organism able to use light as a source of energy and organic materials as carbon sources.

photokinetic — Related to any movement stimulated by light rays.

photolysis — Degradation of a chemical caused by exposure to light or ultraviolet radiation.

P

photolysis — Chemical decomposition by the actions of radiant energy.

photolyze — To cause to undergo photolysis.

photometer — An instrument for measuring luminous intensity, luminous flux, illumination, or brightness; a direct-reading physical instrument using the relative radiant power of a beam of radiant energy measured as it passes through a gas-air mixture or a suspension of solid or liquid particles.

photometry — The branch of physics dealing with measurements of the intensity of light, light distribution, elimination, and luminous flux.

photon — An invisible unit of electromagnetic radiation having zero rest mass, yet traveling at the speed of light; the energy (E) of each photon in radiation of frequency (v) is given by the formula $E = hv$, where h is Planck's constant. Also known as quantum.

photooxidation — A change in the structure of a molecule due to exposure to light; sometimes called bleaching.

photophobia — Abnormal sensitivity to and discomfort from light.

photophosphorylation — The synthesis of high-energy phosphate bonds as adenosine triphosphate using light energy.

photoreaction — A chemical reaction that is stimulated by light.

photoreceptor — A sensor sensitive to light.

photorespiration — Production of carbon dioxide when plants are exposed to light and take up oxygen from the air; oxidation involving production of carbon dioxide during photosynthesis.

photosensitivity — The capacity of an organ, organism, or certain chemicals in plants to be stimulated into activity by light or to react to light.

photosensitize — To make sensitive to the influence of radiant energy, especially light.

photosensitizer — A substance that increases the response of a material to electromagnetic irradiation.

photosynthesis — The biochemical process by which green plants and some bacteria in the presence of chlorophyll use light energy to produce chemical bonds, resulting in the consumption of carbon dioxide and water and the production of oxygen and simple sugars; symbolized by the equation $CO_2 + 2H_2O \rightarrow (CH_2O) + H_2O + O_2$. Virtually all atmospheric oxygen originates from oxygen released during photosynthesis.

photosynthetically active radiation — Electromagnetic radiation in the part of the spectrum used by plants for photosynthesis.

phototaxis — Movement toward light.

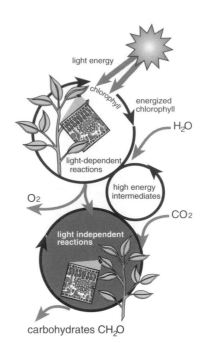

Photosynthesis

phototoxicity — Toxicity resulting from sequential exposure to a photosensitizing agent and sunlight.

phototrophs — Microorganisms that are capable of using light energy for metabolism.

photovoltaic cell — An energy-conservation device that captures solar energy and converts it directly into electrical energy. Also known as photovoltaic solar cell.

photovoltaic solar cell — See photovoltaic cell.

PHSA — Acronym for Public Health Services Act.

phthalic anhydride ($C_6H_4(CO)_2O$) — A white solid with a characteristic, acrid odor; MW: 148.1, BP: 563°F, Sol: 0.6%, Fl.P: 305°F, sp. gr. 1.53. It is used in the manufacture of plasticizers; in the manufacture of unsaturated polyester resins for use in structural building parts, swimming pools, automotive parts, and luggage; in the synthesis of dyes; in the manufacture of chemicals and chemical intermediates; in the manufacture of pharmaceuticals and pharmaceutical intermediates; in the manufacture of metallic and acid salts; in the manufacture of epoxy resins; in the manufacture of fire retardants. It is hazardous to the respiratory system, skin, eyes, liver, and kidneys and is toxic through inhalation, ingestion, and contact. Symptoms of exposure include conjunctivitis, nasal ulcer bleeding, upper respiratory irritation, bronchitis, bronchial asthma, and dermatitis. OSHA exposure limit (TWA): 1 ppm [air] or 6 mg/m³.

Phthirus — The genus of bloodsucking lice that includes the pubic louse or crab louse.

PHWC — See project hazardous waste coordinator.

phycologist — A specialist in the study of algae.

phylum — A major classification of the plant and animal kingdoms that has one or more classes; the second-highest taxonomic classification for the kingdom Animalia between kingdom level and class level.

physiatrist — A physician specializing in physical medicine and rehabilitation and who has been certified by the American Board of Physical Medicine and Rehabilitation after completing specialized training and other requirements.

physical agents — Energy-related agents such as heat, cold, vibration, noise, and electromagnetic radiation of all types and their associated fields.

physical allergy — An allergic reaction to cold, heat, light, or trauma.

physical change — A change that does not produce a new substance.

physical chemistry — A natural science dealing with the relationship between the chemical and physical properties of matter.

physical climate system — The system of processes that regulates climate, including atmospheric and ocean circulation, evaporation, and precipitation.

physical fitness — The ability to carry out daily assignments without undue fatigue and with enough energy reserve to handle emergencies and/or enjoy leisure time.

physical geology — Overall study of the Earth, including its composition and the physical changes occurring in it.

physical hazard — (*hazardous chemicals*) A chemical classified or described as a combustible liquid, as a compressed gas, as an explosive, as being flammable, as an organic peroxide, as an oxidizer, as being pyrophoric, as being unstably reactive, or as being water reactive; the particles or fragments of items not supposed to be in foods.

physical medicine — The use of physical therapy techniques to help individuals return to a useful life after disease or injury.

physical therapist — A person who is licensed to assist in the examination, testing, and treatment of physically disabled or handicapped people through the use of special exercise, application of heat or cold, use of sonar waves, and other techniques.

physical therapy — The treatment of individuals with massage, manipulation, therapeutic exercises, cold, heat, hydrotherapy, electrical stimulation, and light to assist in rehabilitation and to restore normal functions after illness or injury.

physics — Study of the laws and phenomena of nature, especially of forces and general properties of matter and energy.

physiochemical — A branch of science dealing with the chemical aspects of physiological and biological systems.

physiological age — Age of a body as determined by its stage of development or deterioration.

physiological chemistry — See biochemistry.

physiological ecology — Study of biophysical, biochemical, and physiological processes used by living organisms to deal with factors of the physical environment and which are used during ecological interactions with other organisms.

physiological half-life — The rate at which pollutants are eliminated from the body.

physiological limits — The tolerance range of a species for temperature-induced rates of metabolic activity.

physiology — The basic, biomedical science dealing with vital functions of living organisms such as nutrition, respiration, reproduction, and excretion.

physiotherapy — See physical therapy.

phytoaccumulation — See phytoextraction.

phytodegradation — Process by which plants are able to break down organic pollutants through their metabolic processes.

phytoextraction — The use of plants to extract contaminants (such as metals) from the environment.

phytogenesis — The origin an evolution of plant organisms.

phytomining — The use of plans to extract inorganic substances of economic value such as precious metals.

phytoplankton — Microscopic, free-floating, autotrophic organisms that function as producers in aquatic ecosystems.

phytoremediation — The use of plants to remediate contaminated soil or groundwater.

phytostabilization — The use of soil amendments and plants to reduce bioavailability and off-site migration of contaminants.

phytotoxic — Of or pertaining to a substance or state that is injurious to a plant.

phytovolatilization — The use of plants to volatilize contaminants such as solvents from soil or water.

PI — Acronym for preliminary injunction, public information, or point of intersection.

PIC — Acronym for product of incomplete combustion.

pica — The consumption of nonfood items; an abnormal appetite as seen in some children and pregnant women.

pickling — (*air pollution/industry*) Using various industrial baths for cleaning or processing, commonly concentrated with sulfuric acid; also known as acid pickling.

pico (p) — A prefix meaning 10^{-12}; also known as micro-micro-.

P

picocurie (pCi) — A measurement equal to 1 millionth of a microcurie.

picocurie per liter (pCi/l) — A measurement equal to 2.2 radioactive disintegrations per minute in 1 liter of air.

picogram (pg) — A measurement equal to 10^{-12} grams.

picograms per cubic meter — A unit of weight equal to 1000 billionths (10^{-12}) of a gram of a substance in a cubic meter of air, soil, or water.

picomole (pmol) — A unit of quantity equal to 10^{-12} mole.

picornavirus — An extremely small, ether-resistant RNA virus.

picosecond — A unit of measurement equal to 1/1,000,000,000,000 of a second.

picric acid ($C_6H_2OH(NO_2)_3$) — A yellow, odorless solid; usually used as an aqueous solution; MW: 229.1, BP: explodes above 572°F, Sol: 1%, Fl.P: 302°F, sp. gr. 1.76. It is used in munitions and explosives manufacture; in the synthesis of dye and dye intermediates in the textile industry; in the manufacture of medicinals; in the manufacture of pyrotechnics and compounds for pyrotechnics as color intensifiers; as oxidizers in matches; as chemical reagents; in the manufacture of colored glass; in the manufacture of electric batteries. It is hazardous to the kidneys, liver, blood, skin, and eyes and is toxic through inhalation, absorption, and contact. Symptoms of exposure include sensitization dermatitis, eye irritation, yellow-stained hair and skin, weakness, myalgia, anuria, polyuria, bitter taste, gastrointestinal disturbances, hepatitis, hematuria, albuminuria, and nephritis. OSHA exposure limit (TWA): 0.1 mg/m³ [skin].

PICs — See products of incomplete combustion.

PID — Acronym for photoionization detector.

piezochemistry — A branch of chemistry concerned with reactions that occur under pressure.

piezoelectric — The property of certain crystals having the ability to produce a voltage when subjected to a mechanical stress, such as bending, or to produce a mechanical force when voltage is applied.

piezoelectric accelerometer — A transducer that produces an electrical charge in direct proportion to the acceleration of vibration.

piezoelectricity — The electricity of certain asymmetric crystalline materials that relates the electric field to the mechanical strain.

piezometer — An instrument used to measure the pressure or compressibility of a fluid; an instrument for measuring the change of pressure of a material subjected to hydrostatic pressure.

piezometric — Referring to a piezometer.

piezoresistance — A resistance that changes with stress.

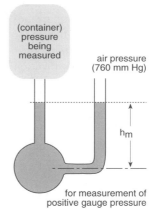

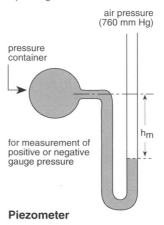

Piezometer

pig — A lead container used to ship radioactive material.

pigment — A chemical that has color because it reflects light of only certain wavelengths; any coloring matter in plant or animal cells.

pigmentosa — A rare inherited skin disease making the skin very sensitive to ultraviolet light.

PIGS — Acronym for Pesticides in Groundwater Strategy.

pigtail — (*plumbing*) See gooseneck.

pile — (*nuclear physics*) A nuclear reactor; (*solid waste*) any non-containerized accumulation of solid, non-flowing hazardous waste that is used for treatment or storage.

pili — Thread-like structures present on some bacteria that are shorter than flagella and are used to adhere bacteria to one another during mating and to adhere to animal cells.

PILOT — A computer interpreter language similar to BASIC and used in computer-assisted instruction.

pilot health study — Any investigation of exposed individuals, using epidemiological methods, that would assist in determining exposures or possible public health impacts by defining health problems requiring further investigation through epidemiological studies, environmental monitoring or sampling, surveillance, or registries.

pilot plant — A small-scale industrial process unit operated to test the application of a chemical or other manufacturing process under conditions that will yield information useful in design and operation of full-scale manufacturing equipment and/or processes.

pilot program — A program that is initiated on a limited basis for the purpose of facilitating a future full-scale program.

pilot test — Evaluation of a cleanup technology under actual site conditions to identify potential problems prior to full-scale implementation.

pindone ($C_9H_5O_2C(O)C(CH_3)_3$) — A bright-yellow powder with almost no odor; MW: 230.3, BP: decomposes, Sol (77°F): 0.002%, Fl.P: not known, sp. gr. 1.06. It is used in pesticide formulations; as an intermediate in pharmaceutical synthesis. It is hazardous to blood prothrombin and is toxic through inhalation. Symptoms of exposure include epistaxis, excess bleeding from minor cuts and bruises, smoky urine, black tarry stools, and pain of the abdomen and back. OSHA exposure limit (TWA): 0.1 mg/m³ [air].

pineal gland — An endocrine gland located near the midline of the brain that produces melatonin, a hormone involved in biological rhythms, especially in annual cycles.

pinkeye — A bacterial conjunctivitis sometimes transmitted by flies.

pint (pt) — A unit of volume equivalent to 2 cups or 1/8 gallon.

pinworm — The nematode that causes enterobiasis; also known as human threadworm or seatworm.

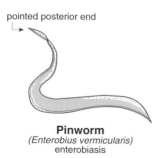

pointed posterior end

Pinworm
(Enterobius vermicularis)
enterobiasis

pions — A family of particles created in nuclear reactions that are unstable but can survive long enough to be formed into beams and used in certain types of medical therapy, such as the treatment of brain tumors.

pipe — A circular conduit constructed of any one of a number of materials that conveys water or other fluids by gravity or under pressure.

pipe gauge — A number that defines the thickness of the sheet used to make steel pipe; the larger the number, the thinner the pipe wall.

pipe lagging — Asbestos-containing material used to insulate pipes carrying heated or refrigerated liquids or vapors.

piperazine ($C_4H_{10}N$) — A crystalline solid with a salty taste; MW: 86.16, BP: 146°C, Sol: soluble in water and alcohol, Fl.P: 81°C (open cup), density: NA. It is used as an intermediate in the manufacture of dyes, pharmaceuticals, polymers, surfactants, and rubber accelerators. It is hazardous to the digestive tract, respiratory tract, skin, and eyes and causes nausea, vomiting, excitement, changes in the motor activity, muscle contraction, and severe irritation. ACGIH exposure limit: NA.

Piperazine

pipette — A calibrated, transparent, open-ended tube of glass or plastic used for measuring or transferring small quantities of a liquid.

PIR — See proportional incidence ratio.

pirimcarb ($C_{11}H_{18}N_4O_2$) — A colorless, crystalline solid; MW: 238.33, BP: NA, Sol: soluble in water and polar organic solvents, density: NA. It is used as a pesticide. It is a highly toxic substance and is hazardous to the skin and gastrointestinal and respiratory tracts. It may cause a slow heartbeat, blurred vision, headache, tremor, convulsions, vomiting, nausea, abdominal pain, diarrhea, and death.

Pirimicarb

P

piscicide — A pesticide specifically designed to kill unwanted fish.

pit — Any mine, quarry, or excavation area worked by the open-cut method to obtain material of value.

PIT — Acronym for Permit Improvement Team.

pitch — A characteristic of auditory sensation in which noises are ordered on a scale that extends from low to high, depending on the frequency of the sound stimulus, the sound pressure, and wave form of the stimulant.

pitchblend — The main component of high-grade African or domestic uranium ore and also containing other oxides and sulfides including radium, thorium, and lead; uranium oxide.

pitless installation device — An assembly of parts that permits water to pass through the casing or extension thereof, provides access to the well and to the parts of the water system within the well, and provides for the transportation of the water and the protection of the water therein from surface or near-surface contaminants.

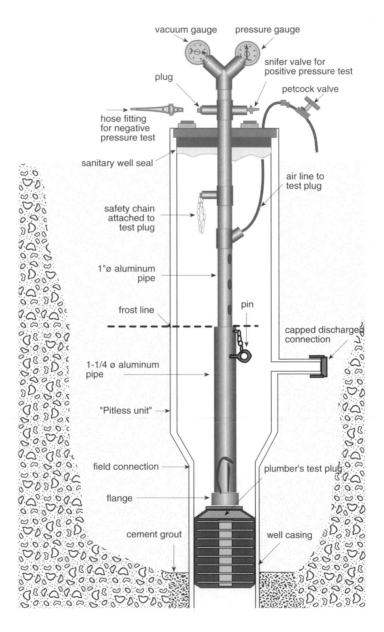

Pitless Installation Device

pitot tube — An instrument for measuring the total static and dynamic pressure of a fluid stream that consists of a small-bore tube for which one end is connected to a manometer and the other end is open and pointing upstream; also known as impact tube.

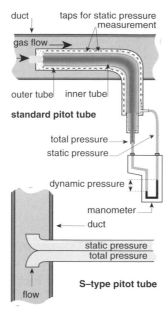

Pitot Tube

pituitary gland — An endocrine gland in the head of higher vertebrates that helps coordinate the work of other endocrine glands; also known as hypophysis.

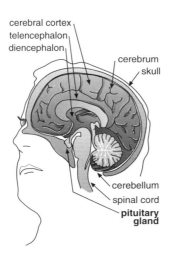

Pituitary Gland

pival ($C_{14}H_{14}O_3$) — A yellow powder with a mild, moldy, acrid odor suggesting marigolds. It is used as an anticoagulant. It causes internal bleeding in rats or mice and therefore prevents development of avoidance of the chemical. Also known as tertiary butyl valone.

pixel (pel) — (*mapping*) A contraction for "picture element"; the smallest unit of information in a grid, cell, map, or scanner image.

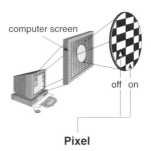

Pixel

PL — Acronym for public law.

placard — A legal notice posted in a public place.

placebo — A drug or treatment that has no known pharmacological effect on disease but works because the patient believes in its efficacy.

placenta — A large, thin membrane in the uterus of most mammals that transports substances between the mother and developing fetus by means of the umbilical cord.

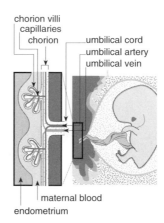

Placenta

plague — An infectious bacterial disease of rodents and humans caused by *Yersinia pestis* or *Pasterella pestis*. It is transmitted to humans by the bite of an infected flea (*Xenopsylla cheopis*) or by inhalation. Symptoms of exposure include inflammation and tenderness of the lymph nodes, fever, septicemia, and pneumonia. Incubation time is 2 to 6 days for plague, and 1 to 6 days for plague pneumonia. It is found in various parts of the world, including among the wild rodent population of the western United States. The reservoir of infection is wild rodents. It is not usually transmitted from person to person, but pneumonic plague is highly communicable; general susceptibility. It is controlled by checking the rodent population for fleas, rat control on ships and docks, and immunization where necessary. Also known as black death or bubonic plague.

Xenopsylla cheopis
(Rat Flea)
a vector for bubonic plague

plaintiff — The party initiating a suit or action in a civil case; opposite of the defendant.

plan — A detailed formulation of a program of action; a design of desired future states.

plan of action — A written document that consolidates all of the operational actions to be taken by various personnel in order to stabilize an incident.

planaria — Non-parasitic flatworm.

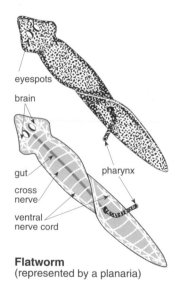

eyespots
brain
gut
pharynx
cross nerve
ventral nerve cord

Flatworm
(represented by a planaria)

Planck's constant (*h*) — A universal constant having the value 6.626076×10^{-34} J·s; see also Planck's law.

Planck's law — The energy of electromagnetic radiation is confined to small indivisible packets or photons, each of which has an energy of hf, where h is the Planck constant and f is the frequency of radiation; this law forms the basis of quantum theory.

plane — A flat surface determined by three points in space.

planimeter — A mechanical or electronic device that calculates the area of a map feature.

plankton — Microscopic plants and animals that drift in water, usually because of currents and tides.

plant (physical) — A station where mechanical, chemical, and/or nuclear energy is converted into electrical energy.

plant classification — A system of identification of plants and their natural relationships based on embryology, structure, or physiological chemistry. The system of classification, in descending order, is kingdom, division, class, order, genus, and species.

plant regulator — Any substance or mixture of substances that, through physiological action, accelerates or retards the rate of growth or rate of maturation or otherwise alters the behavior of plants; does not include plant nutrients, trace elements, nutritional chemicals, plant inoculants, and soil amendments.

plant toxin — Any poisonous substance derived from a plant.

plaque — A flat, often raised patch on the skin or any other organ of the body.

plasma — The fluid portion of the lymph or blood containing approximately 92% water, 7% proteins, and less than 1% inorganic salts, organic substances other than proteins, dissolved gases, hormones, antibodies, and enzymes; a state of matter that exists in space in which atoms are positively charged and share space with free negatively charged electrons able to conduct electricity and interact strongly with electric and magnetic fields.

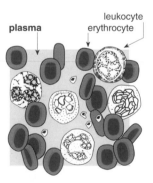

leukocyte
erythrocyte
plasma

Plasma

plasma arc cutting — A process that severs metal by melting a localized area with a constricted arc and removes the molten material with a high-velocity jet of high-temperature ionized gas issuing from the orifice.

plasma arc reactor — An incinerator that operates at extremely high temperatures and is used for treating highly toxic wastes that do not burn easily.

plasma arc welding — A process that produces coalescence of metals by heating them with a constricted arc between an electrode and the work piece or between an electrode and the constricting nozzle.

plasma state — A phase of matter consisting of nuclei and free electrons in contrast to liquid, solid, and gas states.

plasmid — A circular piece of DNA that exists apart from the chromosome and replicates independently of it; often used in genetic engineering to carry desired genes into organisms. Bacterial plasmids carry information that allows the bacteria to be resistant to antibodies.

Plasmodium — A genus of protozoa, several species of which cause malaria; transmitted to humans by the bite of an infected *Anopheles* mosquito.

plastic — A generic name for any one of a large and varied group of materials that consists of or contains as an essential ingredient an organic substance of large molecular weight and which, while solid in the finished state or at some stage in its manufacture, has been or can be cast, molded, etc. into various shapes by application of heat and/or pressure. Each plastic has its own individual physical, chemical, and electrical properties. Adaptability, uniformity of composition, lightness, and good electrical properties give plastic substances wide application, although relatively low resistance to heat, strain, and weather is a limiting factor in their use.

plastic limit — (*soils*) A soil moisture content at the point of transition between being plastic and semisolid.

plastic soil — (*soils*) A soil capable of being molded or deformed continuously and permanently by relatively moderate pressure into various shapes.

plasticity — The property of a soil or rock that allows it to be deformed beyond the point of recovery without cracking or appreciable volume change.

plasticity index — The difference, expressed numerically, between the liquid limit and the plastic limit.

plasticizers — High-boiling, organic, liquid chemicals used in modifying plastics, synthetic rubber, and other materials to give them special properties such as elongation, flexibility, and toughness.

plate — (*geology*) Any of the huge moveable segments into which the crust of the Earth is divided and which float on or travel on the mantle.

plate count — A bacteriological technique that involves taking a sample, diluting it, transferring it to nutrient agar in a Petri dish, incubating it, and then counting the colonies.

plate tectonics — A theory supported by considerable evidence that considers the crust and upper mantle of the Earth to be composed of several large, thin, relatively rigid plates that move relative to one another; when a slip or fault occurs it results in an earthquake. Also, collective geologic processes that move the crustal plates of the Earth and cause continental drifting and seafloor spreading.

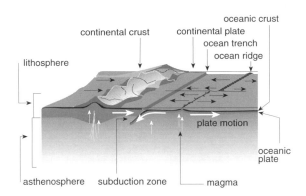

Plate Tectonics

plate tower — A vertical shell in which are mounted a large number of equally spaced, circular, perforated sieve plates; the gases or vapors bubble upward through the liquid seal above each plate.

plateau — (*dose–response assessment*) A point in the dose–response curve at which higher levels of the drug will not produce any greater response.

platelet — An irregularly shaped disk that is found in blood, contains granules in the central part and clear protoplasm peripherally, and has no definitive nucleus; it is important in clotting. Also known as thrombocyte.

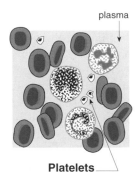

Platelets

platform — A satellite that can carry instruments.

platinum (Pt) (soluble salts) — Appearance, odor, and properties vary depending upon the specific soluble salt. They are used as catalysts in the production of high-octane gasoline, nitric and sulfuric acids, vinyl esters, petrochemicals, and pharmaceuticals; as a photographic paper sensitizer. They are hazardous to the respiratory system, skin, and eyes and are toxic through inhalation, ingestion, and contact. Symptoms of exposure include cough, dyspnea, wheezing, cyanosis, skin sensitization,

P

lymphocytosis, and irritation of the eyes and nose. OSHA exposure limit (TWA): 0.002 mg/m³ [air].

platy soil — (*soils*) A soil structure in which the aggregates are flat or plate like with horizontal dimensions greater than vertical. The plates overlap and cause slow permeability; they are usually found in the subsurface or lower A horizon of timber and claypan soil.

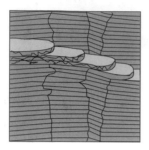

Platy Soil

pleading — (*law*) Process performed by parties to a suit or action, each presenting written statements of their contention and each serving to narrow the field of controversy until there revolves a single point affirmed on one side and denied on the other (called the "issue") upon which the parties go to trial.

plenum — A receiving enclosure for gases in which the static pressure at all points is relatively uniform.

Plesiomonas shigelloides — (*disease*) Gastroenteritis caused by *Plesiomonas shigelloides*, a Gram-negative, rod-shaped bacterium that has been isolated from freshwater fish, shellfish, and many animals; it is usually a mild, self-limiting disease with fever, chills, abdominal pain, nausea, diarrhea, or vomiting. In severe cases, diarrhea may be greenish yellow, foamy, and blood tinged. Although this condition cannot yet be considered a definite cause of human disease, the organism has been isolated in the stools of patients with diarrhea but also is sometimes isolated from healthy individuals. The disease occurs primarily in tropical or subtropical areas, with rare infections reported in the United States or Europe. The mode of transmission is probably from contaminated water (drinking and recreational) or water used to rinse foods that are uncooked or unheated. The mode of transmission is probably through the oral–fecal route. The incubation period is 20 to 24 hours. The reservoir of infection is not clear but is possibly people and/or animals; general susceptibility among children under 15 and people who are immunocompromised.

pleura — The serous membrane covering the lungs and lining the walls of the thoracic cavity.

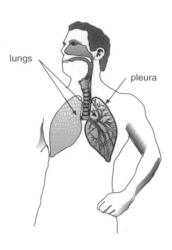

Pleura

pleural fluid — Fluid contained in the membrane covering the lung and lining the chest cavity.

pleurisy — Inflammation of the pleura caused by infection, injury, or tumor; a condition where the visceral and parietal pleura lose their lubricating properties, resulting in friction and causing pain and irritation.

plexus — A network, especially of nerves or blood vessels.

PLIRRA — Acronym for Pollution in Liability Insurance and Risk Retention Act.

PLM — Acronym for polarized light microscopy.

Plodia interpunctella — See Indian Meal Moth.

plug — Cement, grout, or other material used to seal a hole drilled for a water well.

plumb — The state of being exactly perpendicular.

plumb line — A line of cord that has a weight at one end; usually used to determine vertical alignment of distance.

plumbago — See graphite.

plumbing — Procedures, materials, and fixtures used in installing, maintaining, and altering all pipes, fixtures, appliances, and appurtenances that connect with sanitary or storm drainage, a venting system, and a public or individual water supply system.

plume — The column of smoke emitted into the atmosphere through a smoke stack or chimney.

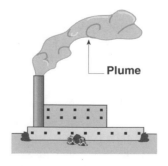

Plume

plutonium (Pu) — Found in five different states: plutonium dioxide (PuO_2), plutonium nitride (PuN), plutonium hexafluoride (PuF_6), plutonium oxalate ($Pu(C_2O_4)_2 \cdot 6H_2O$), and plutonium tetrafluoride (PuF_4); also found as a variety of isotopes (plutonium-236 to -243). Plutonium is a silvery-white radioactive metal that exists as a solid under normal conditions. It is produced when uranium absorbs an atomic particle; small amounts occur naturally, although larger amounts have been produced in nuclear reactors; MW: 242.00, BP: 3232°C, Sol: not known, Fl.P: not known, density: 19.84 at 20°C. Plutonium dioxide is yellowish green; MW: 274.00, BP: not known, Sol: not known, Fl.P: not known, density: 11.46 at 20°C. Plutonium nitride is black in color; MW: 256.01, BP: not known, Sol: soluble in water at 20°C and hydrolyzes in cold water, Fl.P: not known, density: 14.25 at 20°C. Plutonium hexafluoride is reddish brown; MW: 355.99, BP: 62.3°C, Sol: soluble in water at 20°C and decomposes in cold water, Fl.P: not known, density: no data. Plutonium oxalate is a yellowish-green solid; MW: 526.13, BP: not known, Sol: insoluble in water at 20°C, Fl.P: not known, density: not known. Plutonium tetrafluoride is pale brown; MW: 317.99, BP: not known, Sol: insoluble in water at 20°C, Fl.P: not known, density: 7.0 at 20°C. It is used primarily in the form of plutonium-239 as an ingredient in nuclear weapons; plutonium-238 is a heat source in thermoelectric power devices, such as in satellites and for powering artificial hearts. Adverse health effects and death may occur in occupational settings; in treated animals such as rats, mice, hamsters, dogs, and baboons from a single acute inhalation. Mortality in people is uncertain as people working with it are extremely healthy and therefore are not a typical population. Radiation pneumonitis with alveolar edema, fibrosis pulmonary hyperplasia, and metaplasia have been observed in dogs, mice, rats, hamsters, and baboons; biological effects occur in the hematopoietic system of animals; increases in liver enzymes occur in dogs. No conclusive information is available regarding genotoxic effects, although chromosomal aberrations have been observed in rhesus monkeys and Chinese hamsters. A causal link between plutonium exposure and cancer has not been demonstrated in workers. In areas surrounding production plants, a somewhat higher incidence than normal for all cancers is observed. Radiation-induced cancer occurs in experimentally exposed dogs in the lungs; also, osteosarcomas are found. Gastrointestinal effects are observed in neonatal rats, and liver tumors have been observed in dogs. OSHA exposure limits do not exist; Nuclear Regulator Commission limits in air for accumulative annual dose for the general population from nuclear power plant operations: 0.5 rem/yr.

PLUVUE — Acronym for Plume Visibility Model.

ply — One of several layers of fabric, wood, or other strength-contributing material.

plywood — A building material consisting of two or more thin sheets or wood bonded together.

PM — See preventive maintenance; acronym for particulate matter or program manager.

PM_{10} — Acronym for particulate matter less than 10 μm in diameter; indicates particles under 10 μm in diameter that have the greatest adverse effect on human health.

$PM_{2.5}$ — Acronym for particulate matter smaller than 2.5 μm in diameter.

PMAS — Acronym for Photochemical Assessment Monitoring Station.

PMI — Acronym for Pesticide Monitoring Inventory.

PMN — Acronym for Premanufacturer Notification.

pmol — See picomole.

PMR — Acronym for pollutant mass rate or proportionate mortality ratio.

PMRS — Acronym for Performance Management and Recognition System.

PMS — Acronym for Program Management System.

PN_2 — See partial pressure nitrogen.

PNA — See polynuclear aromatic hydrocarbons.

pneumatic — Powered or inflated by compressed air.

pneumococcal meningitis — Meningitis caused by a pneumococcal infection.

pneumococcal — Referring to a bacteria of the genus *Pneumococcus*.

pneumoconiosis — A pneumonia-like disease characterized by the development of fibrotic tissue formation in the alveoli of the lungs caused by dust inhalation; also known as black lung disease or dust diseases of the lungs.

pneumocystosis — An infection with the parasite *Pneumocystis carinii* that is found among people who are debilitated, immunosuppressed, have AIDS, or infants; symptoms include fever, cough, tachypnea, and frequently cyanosis. An almost 100% mortality occurs if untreated.

pneumonia — Inflammation of the lung with solidification of a fluid with a high content of protein and cellular debris that has escaped from blood vessels and has been deposited in tissues or on tissue surfaces (called exudate).

pneumonic plague — A highly virulent and rapidly fatal form of plague characterized by bronchopneumonia; see also plague.

pneumonitis — Inflammation of lung tissue.

P

pneumothorax — A collection of air or gas in the pleural space that causes the lung to collapse; symptoms include sudden, sharp chest pains followed by difficult and rapid breathing, cessation of normal chest movements on the affected side, tachycardia, a weak pulse, hypotension, diaphoresis, an elevated temperature, pallor, dizziness, and anxiety.

PNS — See peripheral nervous system.

PO — Acronym for project officer.

PO/GO — Acronym for privately owned/government-operated.

PO$_2$ — See partial pressure oxygen.

POC — Acronym for point of compliance.

pocket dosimeter — A direct-reading portable dosimeter shaped like a pen with a pocket clip used to measure x- and gamma-radiation. It consists of a quartz fiber, a scale, a lens to observe the movement of the fiber across the scale, and an ionization chamber. The fiber is charged electrostatically until it reaches zero on the scale; as the fiber is exposed to radiation, some of the air atoms in the chamber become ionized, which allows the static electricity charge to leak from the quartz fiber in direct relationship to the amount of radiation present. The fiber then moves to a new position on the scale, indicating the amount of radiation exposure.

podiatrist — A health professional who diagnoses and treats diseases and other disorders of the feet.

podzol — (*soils*) Any of a group of zonal soils that develop in a moist climate under coniferous or mixed forests and have an organic mat and a thin organic-mineral layer.

POE — Acronym for point of exposure or point-of-entry treatment.

POES — See Polar-Orbiting Operational Environmental Satellite.

POHC — Acronym for principle organic hazardous constituent.

POI — Acronym for point of interception.

point mutations — A change in the structure of a gene usually arising from the addition, deletion, or substitution of one or more nitrogenous bases.

point-of-contact measurement — A measurement of the contact of a chemical and person over time while the exposure is taking place.

point-of-entry treatment — The treatment of all water entering a facility regardless of its intended use; anion exchange is used to remove nitrates.

point of supply — Location where water is obtained from a specific source.

point of use — Location where water is actually used in a process or incorporated into a product.

point source — A stationary source that emits a given pollutant.

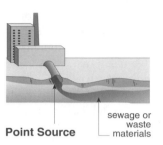

Point Source

sewage or waste materials

point-source water pollution — The water pollutants from urban areas or from industries and entering a receiving body from a single pipe.

poison — Any highly toxic chemical or agent that through ingestion, injection, absorption, or inhalation can cause injury, harm, or destruction to organs, tissue, or life.

poison control center — One of a large network of specialty centers that provides information concerning all aspects of poisoning.

poisoning — Morbid condition produced by a poison that may be swallowed, inhaled, injected, or absorbed through the skin.

polar — A molecule that has a strong overall charge or polarity.

● hydrogen electron
○ oxygen electron

H :O:

H

Polar Molecule

polar ice cap — Portions of the globe closest to the poles, which are permanently covered with ice.

Polar Operational Environmental Satellite — A satellite system of the National Oceanic and Atmospheric Administration that provides daily global coverage by making nearly 14.1 polar orbits daily. It is able to collect global data on a variety of land, ocean, and atmospheric applications by monitoring weather, climate research, global sea surface temperature, atmospheric temperature and humidity, volcanic eruptions, forest fires, global vegetation, and search and rescue activities.

Polar-Orbiting Operational Environmental Satellite — A third-generation, polar-orbiting environmental spacecraft operated by the National Oceanic and Atmospheric Administration and included in the current series of TIROS-N satellites.

polar solvents — Solvents, such as alcohols and ketones, that contain oxygen.

polar stratospheric clouds — High, thin clouds composed of nitric acid and water that form in the coldest regions of the stratosphere when temperatures drop below –80°C. The ice crystal surfaces within these clouds are efficient in converting inert chlorine into reactive chlorine compounds.

polarity — The quality of having two oppositely charged poles, one positive and one negative.

polarization — (*electromagnetic waves*) Direction of the electric field vector.

Polarization

polarization microscope — A microscope that uses polarized light for special diagnostic purposes, such as examining crystals of chemicals found in people.

polarized light — Light that vibrates in only one plane.

polarized light microscope — A light microscope equipped with two crossed polarizing filters, where minerals between the filters can be identified from their refractive index, color, size, shape, and various optical properties in different directions relative to the axis of the fiber.

polarography — A method of quantitative or qualitative analysis based on current–voltage curves obtained during electrolysis of a solution with a steadily increasing electromotive force.

policy — A statement of the principles that guide and govern the activities, procedures, and operations of a program.

pollen — The male microspores of seed plants produced by the anther of a flower or by the male cone of a conifer.

pollutant — A polluting agent or medium; a contaminant.

pollutant migration — The movement of indoor air pollutants throughout a building between rooms (zones).

pollutant source — Any object, usually within a building, that produces a substance that will contaminate the internal environment; includes human bioeffluents or pollutants from carpets or furniture.

Pollutant Standards Index — A numerical index previously used for reporting the severity of pollutant air levels to the general public based on the levels of five pollutants (ozone, PM_{10}, carbon monoxide, sulfur dioxide, and nitrogen dioxide) combined into a single index based on the 1-hour ozone standard. The levels range from 0 for good air quality to 500 for hazardous air quality.

pollution — The presence of a foreign substance — organic, inorganic, radiological, or biological — that tends to degrade the quality of the environment so as to create a health hazard.

pollution audit — A thorough assessment of a company's product and processes that may be contributing to environmental degradation.

pollution fee — A charge for the amount of waste or pollution produced.

pollution prevention — Elimination or minimization of hazardous materials through source reduction using techniques such as substitution and process control; identification of areas, processes, and activities that create excessive waste products or pollutants in order to reduce or prevent them by altering or eliminating a process.

Pollution Prevention Assessment Team — A group of individuals selected on the basis of their expertise and knowledge of process operations to conduct waste-reduction assessments.

Pollution Prevention Coordinator — A person selected to facilitate a pollution prevention program by resolving conflicts.

pollution tax — See pollution fee.

polybrominated biphenyl (PBB) — A chemical substance, the composition of which, without regard to impurities, consists of brominated biphenyl molecules having the molecular formula $C_{12}H_xBr_y$.

polychlorinated biphenyls (PCBs) — A class of 209 colorless, synthetic liquid compounds with the trade name Aroclor®. They are used as an insulating fluid in electrical equipment; as a plasticizer; in surface coatings, inks, adhesives, pesticide extenders, and carbonless duplicating paper. They are hazardous to the upper respiratory tract, the digestive system, liver, blood, eyes, and skin and are toxic through inhalation, ingestion, and skin absorption. Symptoms of exposure include eye and upper respiratory tract irritation; cough; tightness of the chest; chest pain; loss of appetite; anorexia; weight loss; nausea; vomiting; abdominal pain after eating; epigastric distress and pain; liver enlargement; increased cholesterol; chloracne and skin rashes; increased urinary excretion; headache, dizziness, depression, fatigue, and nervousness; possibly carcinogenic. OSHA exposure limit (TWA): chlorodiphenyl (42% chlorine), 1 mg/m³ [air]; (54% chlorine), 0.5 mg/m³.

polychromatic light — Light composed of several colors.

polycyclic aromatic hydrocarbons (PAHs) — A class of 15 colorless, white, or pale-yellow-green solids, including acenaphthene, acenaphthylene, anthracene, benz(*a*)anthracene, benzo(*a*)pyrene,

P

benzo(*b*)fluoranthene, benzo(*g,h,i*)perylene, benzo(*k*)fluoranthene, chrysene, didenzo(*a,h*)anthracene, fluorine, indeno(1,2,3-*c,d*)pyrene, phenanthrene, and pyrene; formed during the incomplete combustion of coal, oil, gas, garbage, or other organic substances and found especially at hazardous waste sites. Only anthracene is used for more than research: as an intermediate in dye production; in smoke screens; in scintillation canter crystals; for inorganic semiconductor research; as an insecticide and fungicide; in the manufacture of plastics. Exposure to emissions containing PAHs may increase mortality due to lung cancers. Animal research indicates exposure may cause respiratory tract tumors induced in the nasal cavity, pharynx, larynx, and trachea of hamsters; decreased survival time due to benzo(*a*)pyrene has been observed in mice. Preneoplastic hepatocytes in animals are correlated with cancer promotion. Also observed are renal effects in rats and developmental effects in mice, as well as skin disorders such as regressive verruca, local villus eruptions, and skin lesions. OSHA exposure limits (PEL): 0.2 mg/m^3 [air]. Also known as polynuclear aromatic hydrocarbon.

polyelectrolyte — A natural or synthetic chemical used to speed the removal of solids from sewage; it causes solids to flocculate or clump together more rapidly than chemicals such as alum or lime.

polyethylene — A thermoplastic material with good flexibility, tensile strength, and resistance to solvents.

polyethylene sheeting structure — A microscopic bundle, cluster, fiber, or matrix of polyethylene that may contain asbestos.

polygon — An area bounded by a closed loop of three or more straight sides.

polymer — Material of high molecular weight formed by the joining together of many simple molecules.

polymerization — A chemical reaction in which two or more similar molecules combine to form larger, very stable compounds with the same chemical proportions but greater molecular weight; used extensively in the production of plastics.

polymicrobic infections — An infection involving more than one type of pathogen.

polymorphic — The ability to assume several different forms.

polymorphonuclear leukocytes — The white blood cells that respond quickly, phagocytose, and destroy foreign antigens.

polynominal — (*mathematics*) An expression having a finite number of terms of the form $a + bx + cx^2$; a mathematical expression of one or more algebraic terms, each of which consists of a constant multiplied by one or more variables raised to a positive integral power.

polynuclear aromatic hydrocarbon (PAH) — See polycyclic aromatic hydrocarbon.

polypeptide — A compound formed by the peptide bonds of two or more amino acids.

polypropylene ((C$_3$H$_6$)$_x$) — A thermoplastic material that is a polymer of propylene; it is prone to ozone and ultraviolet attack but is tolerant to chemicals and extremes of temperatures and has good tensile strength and low permeability to water and solvents.

polysaccharide — A group of large, complex carbohydrates, such as starch and cellulose, with the general formula (C$_6$H$_{10}$O5)$_n$.

polystyrene resins — Synthetic resins formed by polymerization of styrene.

polythemia — A condition of excess numbers of red corpuscles in the blood.

polyvinyl chloride (PVC; H$_2$CCHCl$_x$) — A white thermoplastic substance with good mechanical and electrical properties and a high resistance to chemicals; the inhalation of vinyl chloride may be a health hazard. It is produced by the polymerization of vinyl chloride in the presence of initiators such as benzoyl peroxide.

POM — Acronym for polycyclic organic matter.

pond — A body of water usually smaller than a lake and larger than a pool, either naturally or artificially confined.

pondcrete — The process of mixing materials from ponds with concrete to mobilize waste constituents in the material.

ponding — (*sewage*) A condition where the spaces between a stone become clogged with solids and air cannot pass through the filter due to excessive growths on the filter, plugging up the filter with primary tank solids, or the gradual breaking down and breaking up of filter stones.

pons — A portion of the brain stem above the medulla oblongata and below the midbrain.

Pontiac fever — See Legionnaires' disease.

poor soil suitability — A condition in which one or more of the soil properties are somewhat unfavorable for cover material.

poorly drained soil — Soil in which the water is removed so slowly that the soil remains wet for a large part of the time.

POP — Acronym for persistent organic pollutant.

population — Number of humans or other living creatures in a designated area.

population at risk — A group of people who share a common characteristic that causes each member to be vulnerable to a particular event.

population attributal risk — (*epidemiology*) The attributal risk multiplied by the prevalence of exposure to a factor in a total population; the number of new cases of the diseases in a total population that can be attributed to the factor.

population crash — A sudden population decline caused by waste accumulation or resource depletion.

population density — The number per unit area of individuals in any given species at any given time.

population equivalent — An average waste loading equivalent to that produced by one person (equal to 100 gallons per day in volume) or any lesser amount containing a biochemical oxygen demand of 0.17 pounds per day.

population explosion — The growth of a population at exponential rates to a size that exceeds environmental carrying capacity.

population risk — (*epidemiology*) Excess number of cases of disease in an exposed population above the expected number of cases.

pore — A small to minute opening in a rock or soil.

pore spaces — The open areas or spaces in soil, sediment, and rocks, that are filled by air or water.

pork tapeworm disease — (*disease*) See taeniasis.

porosity — Ratio of the volume in any porous material that is not filled with solid matter to the total volume occupied.

porphyria — A genetic, hereditary disorder characterized by a disturbance in porphyrin metabolism with a resultant increase in the formation and excretion of porphyrins or their precursors; characterized by photosensitivity and porphyrinuria.

porphyrin — Any of a group of iron- or magnesium-free cyclic tetrapyrrole derivatives, occurring universally in protoplasm and forming the basis of respiratory red pigments of animals and plants; in combination with iron, it forms hemes.

Port and Tanker Safety Act — A 1978 law, updated through 1986, in which navigation, vessel safety, and protection of marine environment are matters of major national importance; the law states that the handling of dangerous articles and substances immediately adjacent to navigable waters will be subjected to appropriate care and advanced planning for critical areas.

port of entry effect — A local effect produced in the tissue or organ of first contact between a toxicant and the biological systems.

portability — (*computer science*) The property of computer software that allows its use in a variety of compatible systems.

portal — Place of entrance.

portal circulation — The pathway of blood flow from the gastrointestinal tract and spleen to the liver through the portal vein and its tributaries.

Portland cement — A gray, odorless powder containing less than 1% crystalline silica and composed of dicalcium silicate, tricalcium silicate, and small amounts of alumina, iron oxide, and tricalcium aluminate; Sol: insoluble in water. It is used in the manufacture of mortar for building blocks, bricks, stone, and precast items; as a moisture sealant for exterior concrete blocks; in concrete for highway paving and domestic and commercial building construction. It is a nuisance particulate that is hazardous to the respiratory system, eyes, and skin and is toxic through inhalation, ingestion, and contact. Symptoms of exposure include eye and nose irritation, cough, expectoration, exertional dyspnea, wheezing, chronic bronchitis, and dermatitis. ACGIH exposure limit (TLV-TWA): 10 mg/m^3 [resp].

positional error — (*geographical information system*) The difference between the true location and map location of a point.

positive-displacement pump — A pump that forces or displaces a liquid under pressure through a pumping mechanism.

positive pressure — A condition when more air is supplied to a space than is exhausted, so the air pressure within that space is greater than that in surrounding areas.

positive-pressure fabric filter — A fabric filter with the fans on the upstream side of the filter bags.

positive temperature coefficient — An increase in resistance due to an increase in temperature.

positron — A positively charged particle of electricity with about the same weight but opposite charge as the electron.

positron emission topography (PET) — A computerized radiographic technique using radioactive substances to examine the metabolic activity of various body structures; used to determine the presence of disease.

possible carcinogen — A compound that has shown evidence of being a carcinogen in animals but for which no human data are available.

posterior — Referring to the tail or back end of an organism.

postictal — After a convulsion.

postmortem — See autopsy.

postnatal — Occurring after birth.

postpartum — The period following childbirth during which generative organs are returning to their normal, nonpregnant state.

P

post-processor — (*computer science*) A computer program that is used to convert the results of another operation into a standard format ready for further analysis.

posttraumatic — Any emotional, mental, or physiological consequence that follows a major illness or injury.

postviral fatigue syndrome — A condition of chronic muscle fatigue, even with rest, after a virus infection.

potable water — Fresh water of a quality for drinking, culinary, and domestic purposes not less than that prescribed in international standards for drinking water, especially as concerns bacteriological requirements and chemical and physical requirements.

potassium (K) — A silvery-white metal that loses its luster when exposed to air; atomic weight: 39.098, BP: 765.5°C, Sol: soluble in liquid ammonia and aniline, Fl.P: a very reactive metal that ignites in air or oxygen at room temperature and has a violent reaction when mixed with water, density: NA. It is used in the manufacture of many reactive potassium salts, in organic synthesis, and as a heat exchange fluid when alloyed with sodium. It is hazardous to the skin, eyes, and respiratory tract. It reacts with moisture to cause severe burns. ACGIH exposure limit: NA.

potassium alum (KAl(SO$_4$)$_2$) — White, odorless crystals used as a flocculate in sand filters; in medicines and baking powder; in dyeing, papermaking, and tanning. Also known as alum, aluminum potassium sulfate, and potassium aluminum sulfate.

potassium aluminum sulfate — See potassium alum.

potassium cyanide (KCN) — A white, crystalline solid; MW: 65.12, BP: NA, Sol: soluble in water and hydroxylamine, Fl.P: a noncombustible solid, density: 1.553 at 20°C. It is used for electrolytic refining of platinum and for the separation of gold, silver, and copper from platinum. It is a dangerous poison that is hazardous to the skin, eyes, digestive tract, and respiratory tract. It causes nausea, vomiting, headache, confusion, muscle weakness, collapse, and cessation of breathing. ACGIH exposure limit (TLV-TWA): 5 mg/m^3 [skin].

potassium methoxide (CH$_3$KO) — A yellowish-white, free-flowing powder; MW: 70.14, BP: NA, Sol: soluble in alcohol, Fl.P: a flammable solid igniting in moist air, density: 1.0 g/ml. It is used as a catalyst and an intermediate in organic synthesis. ACGIH exposure limit: NA.

potassium-*tert*-butoxide ((CH$_3$)$_3$COK) — A white, crystalline powder; MW: 112.2, BP: NA, Sol: soluble in tetrahydrofuran and *tert*-butanol, Fl.P: a flammable solid that ignites on heating, density: 0.50 g/ml. It is used as a catalyst and in organic synthesis. ACGIH exposure limit: NA.

$$K - O - C(C_4H_9)_3$$

Potassium-*tert*-butoxide

potency — The amount of a chemical that will produce an effect.

potent — Powerful or strong.

potential — An expression of the energy involved in transferring a unit of electric charge.

potential difference (PD) — A measure of force (expressed in volts) produced between charged objects that move free electrons; also known as voltage or electromotive force.

potential dose — The amount of a chemical contained in material ingested, inhaled, or absorbed by the skin.

potential energy — The energy due to the position of a body or to the configuration of its particles; also known as stored energy.

potential evapotranspiration — The rate at which water, if available, would be removed from soil and plant surfaces.

potential ionization — The ionization necessary to separate one electron from an atom, resulting in the formation of an ion pair.

potential occupational carcinogen — Any substance, combination, or mixture of substances that causes an increased incidence of benign and/or malignant neoplasm or a substantial decrease in the latency period between exposure and onset of neoplasms in human or in one or more experimental species in a million; results from any oral, respiratory, dermal, or other exposure that causes induction of tumors at a site other than the site of administration.

potential spill — An accident or other circumstance that threatens to result in the discharge of oil or other hazardous substance.

potentially exposed — The condition where valid information, usually analytical environmental data, indicates the presence of contaminants of public health concern in one or more environmental media in contact with people and an identified route of exposure.

potentially hazardous food — Any natural or synthetic perishable food or ingredient, such as milk, eggs, meat, poultry, or shellfish, in a form capable of supporting the rapid and progressive growth of infectious or toxigenic microorganisms or the slower growth of *Clostridium botulinum*.

potentiation — The process of a nontoxic or relatively nontoxic substance increasing the effect of another toxic substance.

potentiometer — An instrument used to measure voltage.

potentiometry — A technique of analysis of a gas that causes a change in the hydrogen-ion concentration (pH) when the gas reacts with a reagent, which is sensed by a galvanic cell; the change in pH is a measure of the concentration of the gas.

POTW — Acronym for publicly owned treatment works (municipal treatment plant).

POU — See point of use treatment.

poultry — Domesticated birds, such as chickens, turkeys, ducks, geese, and pigeons, that are raised for food consumption.

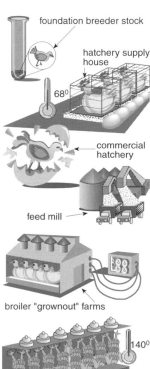

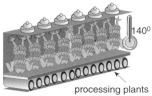

Poultry Factory

poultry processing plant — Any food establishment or portion thereof in which poultry are killed, dressed, processed, stored for sale, or offered for live sale.

pound (lb) — A unit of mass equal to 0.4536 kilograms or 16 ounces.

pound-foot (lb-ft) — A unit of energy that represents the force required to move a substance weighing 1 pound 1 foot; see also foot-pound of torque.

pour plate — A method for performing a plate count of microorganisms using a known amount of a serial dilution placed in a sterile Petri dish to which a melted agar medium is added; the two are mixed together by gently swirling them. After growth appears, the number of colony forming units is counted.

pour point — The ability of crude oil to flow at low temperatures.

power — The rate at which work is done. Electrical energy generated, transferred, or used, usually expressed in kilowatts.

power density — The intensity of electromagnetic radiation per unit area, expressed in watts per centimeter squared.

PP — Acronym for program planning.

ppb — See parts per billion.

ppbv — Acronym for parts per billion by volume.

PPE — Acronym for personal protection equipment.

PPFD — Acronym for photosynthetic photon flux density.

ppm — A unit of measure equal to 1 mg/kg or 1 μg/g; see also parts per million.

PPMAP — Acronym for Power Planning Modeling Application Procedure.

PPO — See Preferred Provider Organization.

PPSP — Acronym for Powerplant Siting Program.

ppt — See parts per trillion.

ppth — See parts per thousand.

pptv — Acronym for parts per trillion by volume.

PQUA — Acronym for preliminary quantitative usage analysis.

PR — Acronym for Pesticide Regulation Notice or preliminary review.

PRA — Acronym for Paperwork Reduction Act or planned regulatory action.

PRATS — Acronym for Pesticides Regulatory Action Tracking System.

preassessment — See preliminary assessment.

precancerous — Pertaining to a pathologic process that tends to become cancer.

preceptorship — The position of teacher or instructor.

prechlorination — The addition of chlorine at the beginning of a treatment prior to the other treatment processes. It is used for disinfection; control of taste, odors, and aquatic growths; and promotion of coagulation and settling.

precious metal — Any relatively scarce, valuable metal, such as gold, silver, or platinum group metals, and the principal alloys of those metals.

P

precipitate — An insoluble compound formed in the chemical reaction between two or more substances in solution; to form a precipitate.

precipitation — The physical settling of particles; any form of water particles falling from the atmosphere to the ground.

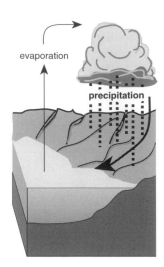

Precipitation

precipitator — A device using mechanical, electrical, or chemical means to collect particulates.

precipitin — An antibody that specifically aggregates the macromolecular antigen to give a visible precipitate.

precision — Degree of accuracy of agreement of repeated measurements of the same property; (*mathematics*) number of significant digits to the right of the decimal point; (*statistics*) degree of variation about the mean.

precision of measurement — (*statistics*) See random error.

preclinical — Early stage of a disease when a specific diagnosis cannot be made because signs and symptoms have not yet fully developed.

precoat — A layer of diatomaceous earth deposited on a filter at the start of a filter run.

precoat feeder — A chemical feeder designed to inject diatomaceous earth into a filter in sufficient quantity to coat the filter septa at the start of a filter run.

precordia — The region over the heart and lower thorax.

precursor — A pollutant that takes part in a chemical reaction resulting in the formation of one or more new pollutants; see also parent.

predator — An insect or other animal that attacks, feeds on, or destroys other insects or animals.

predictive control — A control system that attempts to predict the effects of a disturbance on future output.

predisaster response — The deployment of resources necessary for immediate response and initial recovery operations based on the potential or known threat of a natural disaster.

predisposing cause — A condition such as heredity, lifestyle, or environment that enhances the potential for disease.

preexisting condition — Any injury, disease, or disability that predisposes an individual to limited health in the future.

Preferred Provider Organization — A group of physicians, hospitals, and pharmacists who discount their healthcare services to subscribers.

preformed water — Water that is contained in food.

preliminary assessment — A facility survey performed early in the development of a pollution prevention program for the purpose of determining which areas present opportunities for pollution prevention.

preliminary assessment and site selection — A technique used by the U.S. EPA to evaluate the potential for release of hazardous substances from a site; information collected is used to calculate a score on the Hazardous Ranking System; a score of 28.50 or greater puts the site on the National Priorities List.

preliminary injunction — (*law*) The intermediate phase of an injunction process, which is initiated by the filing of a motion of preliminary injunction. If granted and signed by the court, the defendant is preliminarily restrained from committing future violations of the act. It is not a final process and is an unadjudicated civil case as far as the court is concerned. See also temporary restraining order.

preliminary treatment — The removal of metals, rocks, rags, sand, and other materials that may cause a problem in the operation of a treatment plant by the use of racks, bar screens, comminutors, and grit removal systems.

premature — Not fully formed or developed.

premature death — A death that occurs before its statistical expectation in an average population; usually attributable to a specific cause.

premature ventricular contractions — An irregular cardiac rhythm with ectopic beats followed by full compensatory pauses; caused by irritable focus within the ventricle commonly associated with myocardial infarction and resulting in increased ventricular irritability.

premise — A proposition that is laid down as the basis for an argument.

premises — A lot, parcel, or plot of land either occupied or unoccupied by any dwelling or non-dwelling structure, including any such buildings, accessory structures, or other structures thereon.

P

premium — An amount of money paid to an insurance company in return for insurance protection.

prenatal — Preceding birth.

presbycusis — A condition of hearing loss due primarily to aging.

prescreen — To evaluate a person or group to determine who is at greatest risk for a potential injury or disease.

prescribed burning — The planned application of fire to vegetation to achieve certain specific objectives.

prescription (R$_x$) — A written directive for the compounding, dispensing, or administration of drugs or other services to a particular patient.

presentence investigation — An investigation of parties convicted in criminal cases prior to imposition of a sentence; it is conducted by a probation officer.

preservative — Any substance added to a product to destroy or inhibit multiplication of microorganisms in a foodstuff.

pressed wood product — Materials used in building and furniture construction that are made from wood veneers, particles, or fibers bonded together with an adhesive under heat and pressure; may lead to indoor air quality problems.

pressure — Force per unit area.

pressure bandage — A bandage applied to stop bleeding or prevent edema.

pressure coefficient — A dimensionless coefficient relating the velocity pressure on the outer surface of a building to the velocity pressure derived from the main wind velocity at a reference point.

pressure difference — The difference in pressure of the volume of air enclosed by the house envelope and the air surrounding the envelope.

pressure differential — Difference in pressure between two points in a hydraulic system.

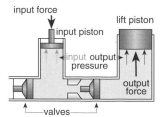

Pressure Differential
in a hydraulic lift

pressure drop — Difference in static pressure measured at two locations in a ventilation system due to friction or turbulence.

pressure head — The height of a column of fluid required to produce a given pressure at its base.

pressure loss — Energy lost from a pipe or duct through friction or turbulence.

pressure point — A place over an artery where a pulse may be felt; pressure can be applied here to stop the flow of blood from a wound distal to that point.

pressure potential — The water force created by real pressure against a membrane.

pressure sewer — A small-diameter pipeline shallowly buried and following the profile of the ground.

pressure sore — See decubitus ulcer.

pressurization — (*indoor air quality*) A method of testing air leakage of a building or components by installing a fan in the building envelope through a door or window and creating a static pressure excess inside the building.

pressurized water reactor — A thermal reactor using water as both a moderator and coolant and utilizing enriched uranium oxide fuel.

presymptomatic disease — An early phase of the disease process when physiological changes have begun but signs or symptoms are still not showing.

pretibial disease — An acute infection caused by *Leptospira autumnalis*; symptoms include headache, chills, fever, enlarged spleen, myalgia, low white blood cell count, and a rash on the anterior surface of the legs.

pretreatment — Reduction of the amount of pollutants, elimination of pollutants, or alteration of the nature of pollutant properties in wastewater to a less harmful state prior to treatment facilities.

pretreatment facility — An industrial wastewater treatment plant consisting of one or more treatment systems used to remove sufficient pollutants from the wastewater to allow the industry to comply with effluent limits established by U.S. EPA, local, or state regulations.

pretrial conference — A conference of parties in a suit or case called by the court to consider various actions in order to limit the various issues for trial; discussion between an agency attorney and witnesses before a trial.

prevalence — Total number of cases of a disease existing in a population at a certain time in a designated area.

prevalence rate — The number of cases of a disease that exists in a population at some point in time; expressed per unit of population.

prevalence study — An epidemiological study that examines the relationship between diseases and exposures as they exist in a defined population at a particular point in time.

prevention — A means of control where proper planning, maintenance of equipment, location of sources of

P

pollutants, and an interruption in the spread of the pollutant occurs; a means of avoidance of disease.

preventive maintenance (PM) — A management-initiated process of inspection and corrective action undertaken prior to any actual failure of the facility assets, including the physical structure and related equipment.

preventive medicine — A branch of medicine that is concerned with the prevention of disease and development and use of techniques to improve the ability of the individual and the community to resist disease, promote health, and prolong life.

PRF laser — See pulsed-recurrence frequency laser.

PRI — Acronym for periodic reinvestigation.

prickly heat — See heat rash.

prima facie — A fact presumed to be true, unless it is proved by evidence to the contrary.

primacy — The state of being first in importance.

primary air pollutants — Pollutants emitted to the atmosphere as the result of some heating or industrial process.

primary burner — A burner that dries out and ignites material in the primary combustion chamber.

primary care physician — The initial physician who examines a patient, determines necessary treatment, and recommends medical or surgical specialists when needed.

primary care — The initial evaluation and treatment of a person's illness.

primary combustion air — Air admitted to a combustion system at the point where the fuel is first oxidized.

primary combustion chamber — The chamber of an incinerator where waste is ignited and burned.

primary drinking water standard — A regulation that applies to public systems and specifies contaminants that may have adverse effects on human health.

primary health care — The level of routine outpatient care provided by a supervised primary healthcare physician.

primary ionization — (*collision theory*) Ionization produced by the primary particle.

primary irritant — A chemical that causes dermatitis by direct action on normal skin at the site of contact if permitted to act in sufficient intensity or quantity for a sufficient length of time.

primary irritation dermatitis — An inflammation and irritation of the skin due to brief contact with a concentrated chemical such as an acid, alkali, or irritant gas; may also be caused by extended exposure to a dilute chemical, friction, cold, or heat.

primary metals — Ferrous or nonferrous metals produced by the smelting of ore in industry; the smelting process may cause metallic oxide pollution as well as the emission of carbon monoxide, smoke, dust, sulfur dioxide, lead, and other contaminants.

primary particles — Particles that are directly emitted from combustion and fugitive dust sources.

primary pollutant — A pollutant directly emitted from a polluting source.

primary sedimentation — The first major process in wastewater treatment plants in which solids settle by gravity.

primary septicemia — See *Vibrio vulnificus*.

primary sewage treatment — A system in which preliminary treatment consisting of the use of screens and grit chambers, comminuting devices, preaeration tanks, sedimentation tanks, chemical treatment, and discharge of the effluent to the receiving stream may occur; the effluent may go on to secondary treatment, and the sludge is removed to a sludge digester.

primary sludge — The portion of raw wastewater solids contained in effluent that is directly captured and removed in the primary sedimentation process.

primary standard — A substance with a known property that can be defined, calculated, or measured and is readily reproducible; see health-based standard.

primary treatment — The first stage in wastewater treatment where substantially all floating or setting solids are removed by flotation and/or sedimentation.

primary voltaic cell — A pair of plates in an electrolyte from which electricity is derived by chemical action; also known as voltaic cell.

prime — The action of filling a pump casing with water to remove the air.

principle component analysis — (*geographic information system*) A technique of analyzing multivariate data and expressing their variation in small numbers of principle components or linear combinations of the original data.

printout — A printed copy of information from a computer.

prion — A protein particle that lacks nucleic acid and is thought to be the cause of various infectious diseases of the nervous system.

prion protein — A normally occurring protein found on the surface of particular cell types; accumulations of these proteins are found in a diseased brain in Creutzfeldt–Jacob disease.

prior appropriation doctrine — A system for allocating water to private individuals, used in most western states and originating with early settlers and miners who developed the land.

priority — A rational assignment of a preferential rating to something meriting attention among competing alternatives.

prism — A 3-, 4-, 6-, 8-, or 12-sided piece of glass that separates light into its colors and is used in a number of optical instruments.

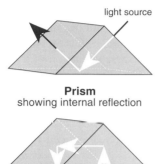

Prism
showing internal reflection

prismatic soil — A soil structure that is without rounded caps and is prism like; the vertical axis is greater than the horizontal, and it is usually found in subsoil or the B horizon.

Prismatic Soil

privacy — The existence of conditions that permit an individual or individuals to be without interruption or interference from the sight or sound of unwanted individuals.

private sector — A division of people as related to business to produce products or services using private capital and the profit motive.

private water system — Any water system intended for the provision of water for human consumption, has fewer than 15 service connections, and does not readily serve an average of at least 25 people per day at least 60 days out of the year.

privy — Any sanitary, waterless device for the collection and storage of human excreta; includes chemical commodes or other portable receptacles.

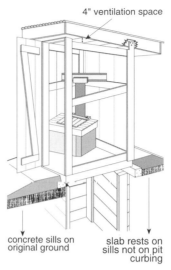

4" ventilation space

concrete sills on original ground   slab rests on sills not on pit curbing

Sanitary pit privy

PRN — Acronym for Pesticide Registration Notice.

prn — Acronym for *pro re nata*, Latin for "as needed".

pro confesso — (*law*) As if conceded as confessed.

pro forma — As a matter of form.

probability — (*statistics*) The relative frequency with which a specific event occurs; expressed as the number of times the event occurs divided by the total number of trials.

probability value — The probability that an index of effect is as extreme or more extreme than that observed, even if no effect exists (i.e., the null hypothesis is false).

probable carcinogen — A compound that has shown some evidence of carcinogenicity in humans or, lacking adequate data in humans, sufficient evidence exists of carcinogenicity in animals.

probe — A tube used for sampling or measuring pressure or temperature at a distance from the actual collection or measuring apparatus; (*soils*) a tube used for taking a soil sample.

proboscis — A tubular mouth part in certain insects and mammals.

procarcinogen — A chemical substance that becomes carcinogenic only after it is altered by metabolic processes.

procaryotic organism — A microorganism that does not have an organized nucleus surrounded by a nuclear membrane.

procedure — A sequence of actions that collectively produce a desired outcome.

P

procedure in admiralty — Seizure actions taken pursuant to the Food, Drug, and Cosmetic Act, which states that the government may initiate confiscation of articles it deems to be contraband without first having to show proof of the allegation.

proceeding — (*law*) Any rulemaking, adjudication, or licensing conducted by a governmental agency under applicable statutes or under regulations used to implement them.

process — A particular method of doing something that generally involves a number of steps or operations.

process emission — Particulate matter that is collected by a capture system.

process study — An organized systematic investigation of a particular process designed to identify all of the variables involved and to establish the relationship among them.

process variable — A physical or chemical quantity that is usually measured and controlled in the operation of a water treatment plant or industrial plant.

process verification — Verification of the quantity and quality of pollutants contained in discharges by analyzing the raw materials, water usage, water treatment processes, production rate, and other facts relative to the process, as outlined in a U.S. EPA permit application.

process wastewater — Water that has been used for cooling purposes and may have become contaminated through contact with raw materials or final products.

processed meat — Meat plus additives, spices, and other ingredients that receives special treatment, such as curing, smoking, or canning.

processed water — Water that comes in contact with an end-product or with materials incorporated in an end-product.

processing — (*solid waste*) Any method, system, or other treatment designed to change the physical form or chemical content of solid waste.

processor — Any person who processes a chemical substance or mixture.

procyazine ($C_{10}H_{13}ClN_6$) — A crystalline solid; MW: 252.74, BP: NA, Sol: dissolves in organic solvents,

Procyazine

density: NA. It is used as an herbicide. It has moderate oral and skin toxicity. It has been found to be teratogenic in experimental animals.

prodromal stage — A premonitory symptom; a symptom indicating the onset of a disease; (*radiation*) the initial period for acute radiation syndrome.

producer — (*food chain*) An autotrophic member of the food chain able to make its food from inorganic substances and upon which the other plants and animals in the chain depend.

product of combustion — Gases, vapors, and solids that result from the combustion of a fuel.

productivity — The efficiency with which a person performs a specific function or the output of a worker under specific environments and conditions.

productivity loss — The monetary value of output that would have been produced in the absence of an illness, disability, injury, morbidity, or premature mortality.

products of incomplete combustion — Organic compounds formed by combustion and usually generated in small, sometimes toxic amounts; heat-altered versions of the original material fed into an incinerator, such as charcoal from burning wood.

profile — A sectional view showing grades and distances usually taken along a center line.

profundal — Referring to the deepest part of the ocean or lake where light does not penetrate.

profundal zone — The deepwater region of a lake that is not penetrated by sunlight.

profuse sweat — Excessive perspiration.

progeny — Decay products produced after radioactive decay; descendants.

proglottids — Segment of a tapeworm containing both male and female reproductive organs.

prognosis — The probable outcome or course of a disease; the chance of recovery.

program — An organized activity with a definable purpose; (*computer science*) a precise, sequential set of instructions directing a computer to perform a task.

program evaluation and review technique (PERT) — Objective analysis using an evaluation of project study and periodic reviews to enhance the effectiveness of management, supervision, and control.

Programmatic Environmental Impact Statement — An environmental impact statement that addresses a proposal to implement a specific policy, to adopt a plan for a group of related actions, or to implement a specific statutory program.

project — A task, that upon completion, will require concerted efforts and will yield data, material, and products.

project hazardous waste coordinator — An individual designated to coordinate records and maintain proper management of hazardous wastes on each project.

projection (*GIS*) — (*geographical information system*) A mathematical calculation transforming the three-dimensional surface of the Earth into a two-dimensional plane.

prokaryote — A cellular organism that does not have a distinct nucleus.

prokaryotic — An adjective for prokaryote.

prolapse — Downward displacement of a part or organ.

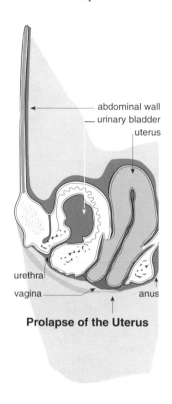

abdominal wall
urinary bladder
uterus

urethra
vagina   anus

Prolapse of the Uterus

prolate — Elongated or cigar shaped.

proliferation — Reproduction or multiplication of cells.

prolific — Highly productive.

prolonged exposure — The continuous exposure of workers to a chemical agent or chemical hazard over a long period of time; may lead to adverse health effects.

promoter — A chemical that, when administered after an initiator has been given, promotes the change of an initiated cell into a cancerous cell.

promotion — (*cancer*) An event that causes a cell to grow rapidly and to acquire the capacity to metastasize.

promulgate — To publish or make known officially.

propagation of flame — The spread of flame through the entire volume of a flammable vapor–air mixture from a single source of ignition.

propagation — The process of increasing numbers.

propane (CH₃CH₂CH₃) — A colorless, odorless gas (a foul-smelling odorant is often added when used for fuel purposes) that is shipped as a liquefied compressed gas; MW: 44.1, BP: –44°F, Sol: 0.01%. It is used as a basic material in chemical synthesis for oxidation, alkylation, nitration, and chlorination; as a solvent and extractant in deasphalting and degreasing of crude oils; as feedstock in the cracking process for the production of ethylene and propylene and motor gasoline; as a refrigerant; in petroleum refining and gas processing operations; in low-temperature crystallizers; for helium recovery from natural gas; as a fuel in welding and cutting operations; in the desalination of water. It is hazardous to the central nervous system and is toxic through inhalation and contact. Symptoms of exposure include dizziness, disorientation, excitation, frostbite. OSHA exposure limit (TWA): 1000 ppm [air] or 1800 mg/m³.

propanol (C₃H₇OH) — A colorless liquid with a smell of alcohol; MW: 60.09; BP: 97.4°C; Sol: soluble in water, alcohol, acetone, and ether; Fl.P: 29°C (open cup); density: 0.8035. It is used as a solvent for waxes, resins, and vegetable oils. It is hazardous to the skin, eyes, gastrointestinal tract, and respiratory system. It causes headaches, drowsiness, abdominal cramps, gastrointestinal pain, ataxia, nausea, diarrhea, eye irritation, dermatitis, and at a high concentration a narcotic effect. ACGIH exposure limit (TLV-TWA): 200 ppm.

2-propanol (C₃H₇OH) — A colorless liquid with an alcohol-like odor and a bitter taste; MW: 6.09; BP: 82.3°C; Sol: soluble in water, ether, acetone, and chloroform; Fl.P: 17.2°C (Tag open cup); density: 0.7849 at 20°C. It is used as an industrial solvent for paints, polishes, and insecticides; as an antiseptic. It is hazardous to the eyes, skin, and respiratory system. It produces mild irritation in the eyes and nose and through ingestion can cause drowsiness, dizziness, nausea, coma, and death. ACGIH exposure limit (TLV-TWA): 400 ppm.

P

$$CH_3$$
$$\diagdown$$
$$CH-OH$$
$$\diagup$$
$$CH_3$$

2-Propanol

propellant — A chemical or other agent used in aerosol bombs to force or push a pesticide chemical out through the nozzle.

properly connected — The condition of a connection or installation being in accordance with all applicable codes and ordinances.

properties — The characteristics by which a substance may be identified, physically or chemically.

property — A characteristic by which a substance may be identified; physical properties include state of matter, color, odor, and density, and chemical properties include behavior in reaction with other materials; (*law*) a parcel of land for which legal title has been recorded.

property right — (*law*) A legal right in or against specific property.

property tax — A tax levied on real or personal property.

property transfer assessment — A determination of those areas of potential legal and economic liability that a buyer may assume by purchasing a business or industry that has actual or potential environmental problems.

propham (C₁₀H₁₃NO₂) — A crystalline solid; MW: 179.21, BP: NA, Sol: soluble in water and readily dissolves in most organic solvents, density: 1.09 at 20°C. It is used as an herbicide. It is hazardous to the digestive tract and skin and is a cholinesterase inhibitor that may be lethal when taken in large amounts.

$$\text{C}_6\text{H}_5-\text{NH}-\overset{\overset{\displaystyle O}{\|}}{\text{C}}-\text{O}-\text{CH(CH}_3)_2$$

Propham

prophase — The first stage of nuclear division by mitosis or meiosis.

prophylactic — An agent that is used in the prevention of the spread of disease.

prophylaxis — Any substance or steps taken to prevent an event.

proportional counter — A gas-filled radiation-detection tube in which the pulse produced is proportional to the number of ions formed in the gas by the primary ionizing particle.

proportional incidence ratio (PIR) — A measure of risk that compares the observed numbers of events with the expected number of events.

proportional mortality ratio — The ratio of deaths for a specific disease among all deaths in an exposed group to the ratio of deaths for that disease among all deaths in an unexposed group.

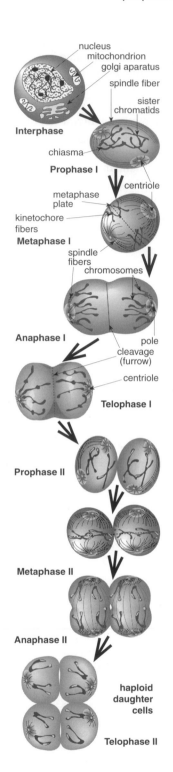

Prophase

proportional sample — See composite sample.

propoxur (C₁₁H₁₅NO₃) — A white crystalline solid or powder; MW: 209.27, BP: NA, Sol: soluble in organic solvents, density: 1.12 at 20°C. It is used as a pesticide. It is hazardous to the digestive tract and respiratory tract and causes slow heart rate, blurred vision, lack of coordination, nausea, weakness, diarrhea, abdominal pain, and death. It is a teratogenic substance. ACGIH exposure limit (TLV-TWA): 0.5 mg/m^3.

Propoxur

proprietary — Of or pertaining to possessing, owning, or holding exclusive rights to something.

proprioceptive — The sensations of body movements and awareness of posture that help the body to orient itself in space without having visual clues.

propulsion — The process of moving forward by means of a force.

_n_-propyl acetate (CH₃COOCH₂CH₂CH₃) — A colorless liquid with a mild, fruity odor; MW: 102.2, BP: 215°F, Sol: 2%, Fl.P: 55°F, sp. gr. 0.84. It is used in the manufacture of lacquers and adhesives; as a solvent for rubber; in the preparation of flavoring agents and perfumes. It is hazardous to the respiratory system, eyes, skin, and central nervous system and is toxic through inhalation, ingestion, and contact. Symptoms of exposure include narcosis, dermatitis, and irritation of the eyes, nose, and throat. OSHA exposure limit (TWA): 200 ppm [air] or 840 mg/m^3.

_n_-propyl alcohol (CH₃CH₂CH₂OH) — A colorless liquid with a mild, alcohol-like odor; MW: 60.1, BP: 207°F, Sol: miscible, Fl.P: 72°F, sp. gr. 0.81. It is used during printing on plastic film and sheeting with polyamide-based inks; during textile leather processing; in the manufacture of surface coatings as a solvent; in the manufacture of cleaning preparations and polishing agents; in the extraction of vegetable oils, castor oil, and pharmaceuticals; during cellulose processing; in the manufacture of brake fluids. It is hazardous to the skin, eyes, respiratory system, and gastrointestinal tract and is toxic through inhalation, absorption, ingestion, and contact. Symptoms of exposure include dry and cracking skin, drowsiness, headache, ataxia, gastrointestinal pain, abdominal cramps, nausea,

vomiting, diarrhea, and mild irritation of the eyes, nose, and throat. OSHA exposure limit (TWA): 200 ppm [air] or 500 mg/m^3.

_n_-propyl nitrate (CH₃CH₂CH₂NO₃) — A colorless to straw-colored liquid with an ether-like odor; MW: 105.1, BP: 231°F, Sol: insoluble, Fl.P: 68°F, sp. gr. 1.07. It is used in liquid rocket propellants. No hazards to humans are known; in animals, it is toxic by inhalation, ingestion, and contact. Symptoms of exposure in animals include methemoglobin, anoxia, cyanosis, dyspnea, weakness, dizziness, headache, and irritation of the eyes and skin. OSHA exposure limit (TWA): 25 ppm [air] or 105 mg/m^3.

_n_-propylamine (C₃H₇NH₂) — A colorless liquid with a strong ammonia odor; MW:59.13; BP: 48°C; Sol: soluble in water, alcohol, and ether; Fl.P. –37°C (closed cup), a highly flammable liquid; density: 0.719 at 20°C. It is used as an intermediate in many organic reactions. It is hazardous to the respiratory tract, digestive tract, skin, and eyes. It causes burns, and possible skin sensitization, severe irritation to the eyes, and labored breathing. ACGIH exposure limit(TLV-TWA): 10 ppm

$$CH_3 - CH_2 - CH_2 - NH_2$$

_n_-Propylamine

propylene (C₃H₆) — A colorless gas; MW: 42.09, BP: NA, Sol: soluble in acetone and benzene, Fl.P: a flammable gas that can travel a considerable distance to a source of ignition and flash back, density: 1.46. It is used to produce polypropylene and in the manufacture of acetone and propylene oxide. It is an asphyxiate. Exposure to high concentrations can cause narcosis and unconsciousness. ACGIH exposure limit: NA.

$$CH_3 - CH = CH$$

Propylene

propylene dichloride (CH₃CHClCH₂Cl) — A colorless liquid with a chloroform-like odor; MW: 113.0, BP: 206°F, Sol: 0.3%, Fl.P: 60°F, sp. gr. 1.16. It is used as a soil fumigant for protection of fruit and nut crops, field crops, beets, and tobacco against nematodes; in cleaning, degreasing, and spot removal operations, including paint and varnish removal; during rubber compounding and vulcanizing operations; during extraction processing of fats, oils, lactic acid, and petroleum waxes; in

P

the manufacture of tetrachloroethylene and propylene oxide; as an additive and lead scavenger in antiknock fluids. It is hazardous to the skin, eyes, respiratory system, liver, and kidneys and is toxic through inhalation, ingestion, and contact. Symptoms of exposure include drowsiness, eye irritation, lightheadedness, skin irritation; carcinogenic; in animals: liver and kidney disease. OSHA exposure limit (TWA): 75 ppm [air] or 350 mg/m^3.

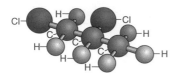

**Propylene Dichloride
(1,2–Dichloropropane)**

propylene glycol monomethyl ether (C₄H₁₀O₂) — A colorless liquid; MW: 90.14, BP: 120°C, Sol: soluble in water and organic solvents, Fl.P: 36°C (closed cup), a flammable liquid, density: 0.931. It is used as a solvent for cellulose, acrylics, dyes, and inks. It is hazardous to the respiratory tract and digestive tract. It causes nausea, vomiting, and a general anesthetic effect. ACGIH exposure limit (TLV-TWA): 100 ppm.

$$\overset{\displaystyle OH}{\underset{\displaystyle |}{CH_3-O-CH_2-CH-CH_3}}$$

**Propylene glycol
monomethyl ether**

propylene imine (C₃H₇N) — A colorless, oily liquid with an ammonia-like odor; MW: 57.1, BP: 152°F, Sol: miscible, Fl.P: 25°F, sp. gr. 0.80. It is used as a polymer modifier; in the manufacture of polypropylene imines and polymers for coating materials, adhesives, chelating agents, emulsifying agents, and fireproofing agents in the textile, rubber, paint, ink, and paper industries; as a chemical intermediate in the pharmaceutical and chemical industries. It is hazardous to the eyes and skin and is toxic through inhalation, absorption, ingestion, and contact. Symptoms of exposure include eye and skin burns; carcinogenic. OSHA exposure limit (TWA): 2 ppm [skin] or 5 mg/m^3.

propylene oxide (C₃H₆O) — A colorless liquid (gas above 94°F) with a benzene-like odor; MW: 58.1, BP: 94°F, Sol: 41%, Fl.P: –35°F, sp. gr. 0.83. It is used in the production of polyurethane foam; in

the manufacture of propylene glycol for use as a solvent, emulsifier, and mold inhibitor; in the manufacture of dipropylene glycol for use as a solvent; in ink formulations; as a chemical intermediate in the production of lubricants, surfactants, and miscellaneous chemicals for the pharmaceutical, petroleum, textile, rubber, and soap industries; as a low-boiling solvent for cellulose derivatives, hydrocarbons, commercial gums, and various resins; as a fumigant, herbicide, germicide, and insecticide; as a stabilizer of vinyl resin lacquers and discoloration preventer of hydrocarbons; as a food preservative; during hydropropylating of wheat flour. It is hazardous to the eyes, skin, and respiratory system and is toxic through inhalation, ingestion, and contact. Symptoms of exposure include irritation, blistering, and burning of the skin and irritation of the eyes, upper respiratory system, and lungs; carcinogenic. OSHA exposure limit (TWA): 20 ppm [air] or 50 mg/m^3.

propylene-β-chlorohydrin (C₃H₇OCl) — A colorless liquid with a pleasant smell; MW: 94.54; BP: 133 to 134°C; Sol: soluble in water, alcohol, and other organic solvents; Fl.P: 44°C (close cup); density: 1.103 at 20°C. It is used in the production of propylene oxide. It is hazardous to the skin, eyes, digestive tract, and respiratory tract. It targets the central nervous system, gastrointestinal system, liver, and kidneys. ACGIH exposure limit (TLV-TWA): 10 ppm [air].

$$\underset{\displaystyle \underset{\textstyle Cl}{|}}{H_3C-CH-CH_2-OH}$$

Propylene β-Chlorohydrin

2-propyn-1-ol (C₃H₃OH) — A colorless and moderately volatile liquid with a mild odor; MW: 56.06; BP: NA; Sol: soluble in water, alcohol, ether, chloroform, and acetone; Fl.P: 33°C (open cup), density: 0.948. It is used in metal plating and pickling and as a corrosion inhibitor of steel. It is moderately toxic and causes depression of the central nervous system and irritation of the eyes and skin. ACGIH exposure limit (TLV-TWA): NA (however, 1 ppm should be appropriate).

$$HC \equiv C-CH_2-OH$$

2-Propyn-1-ol

P

prosecution — The party initiating action in a criminal case; the institution of a criminal proceeding against an individual.

prospective study — (*epidemiology*) A study of susceptible exposed and unexposed individuals; also known as longitudinal study.

prosthesis — An artificial replacement for a missing body part.

prostration — A complete physical or mental breakdown.

protease — An enzyme that digests proteins.

protection factor (PF) — The ratio of the ambient, airborne concentration of a contaminant to the concentration inside the facepiece of respiratory protective equipment.

protective barrier — A barrier of radiation-absorbing material such as lead, concrete, plaster, and plastic used to reduce or eliminate radiation exposure.

protective clothing — Clothing that protects a person against injury, exposure, or death while handling toxic chemicals.

protective equipment — Equipment that protects a person against injury, exposure, or death while handling toxic chemicals.

protein — Any of a class of complex nitrogen-containing compounds made by living organisms from amino acids; may consist of one or more polypeptide chains.

proteinaceous infectious agent — See prion.

proteinuria — An excess of serum proteins in the urine.

proteolysis — Hydrolysis of proteins or peptides with the formation of simpler and soluble products.

proteolytic enzyme — An enzyme that helps break down protein; either rennet or casease.

proteose — A secondary protein derivative formed by hydrolytic cleavage of the protein molecule.

Proteus — A genus of Gram-negative, facultatively anaerobic, motile, rod-shaped bacteria found in fecal material, especially in patients treated with oral antibiotics.

Proteus mirabilis — A species of anaerobic, motile, rod-shaped bacteria found in decomposing meat, abscesses, and fecal material; a leading cause of urinary tract infections.

Proteus morgani — A species of bacteria associated with infectious diarrhea in infants.

Proteus vulgaris — A species of bacteria found in the feces, water, and soil; a frequent cause of urinary tract infections.

protist — An unicellular, colonial, or multicellular organism (i.e., protozoa and most algae).

protocol — The plan and procedures to be followed in conducting a test, performing a service, or setting up a program.

protofilament — The filament extending along the axoneme of a cilium and flagellum.

proton — A positively charged elementary particle that together with the neutron forms atomic nuclei.

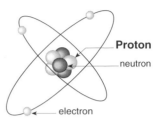

protoplasm — An organized colloidal complex system of substances that constitutes the living material of a cell.

protoplasmic poisons — Poisons that cause damage or death to cells, resulting in an inflammatory change altering cell membranes and inhibiting enzymes; they include ammonia, formaldehyde, dimethyl sulfate, arsenic, and cyanide.

protoplast — A membrane-bound cell from which the outer wall has been partially or completely removed.

protozoan — Single-celled microorganisms belonging to the phylum protozoa.

protozoologist — A person who studies protozoa.

protraction dose — A measure of administering radiation by delivering it continuously over a relatively long period at a low dose rate.

proven reserves — The estimated quantities of oil and gas that geological and engineering data demonstrate with reasonable certainty will be recoverable in future years from known reservoirs under existing economic and operating conditions.

provirus — A form of a virus that is integrated into the genetic material of a host cell; by replicating with it, it can be transmitted from one cell generation to the next.

provisional acceptable daily intake — The maximum dose of a substance that is anticipated to be without health risk to humans when taken daily over the course of a lifetime.

proximal — Situated toward or near the point of attachment.

proximate — The nearest place to a point of origin or attachment.

proximate analysis — Analysis of a solid fuel to determine on a percentage basis how much moisture, volatile matter, fixed carbon, and ash the sample contains; this also helps establish the heat value of the fuel.

P

proximate cause — Factors that underlie the functioning of a biological system at a particular place and time, including those responsible for metabolic, physiological, and behavioral functions at the molecular, cellular, organ, and population levels.

PrP — See prion protein.

PRP — Acronym for potentially responsible party.

pruigo — A group of chronic inflammatory conditions of the skin causing severe itching.

pruritus — An unpleasant itching sensation due to irritation of sensory nerve endings.

PRZM — Acronym for Pesticide Root Zone Model.

PS — Acronym for point source.

PS equation — A means of predicting the air velocity that will be chosen by a person exposed to a certain air temperature when the person has control of the air velocity source.

PSAM — Acronym for point-source ambient monitoring.

PSC — Acronym for program site coordinator.

PSD — Acronym for prevention of significant deterioration.

PSES — Acronym for pretreatment standards for existing sources.

Pseudomonas aeruginosa — A species of Gram-negative, non-sporing, motile bacilli isolated from wounds, burns, and urinary tract infections that may cause a variety of nosocomial infections and purulent meningitis.

pseudomyiasis — The presence within a host of a fly that is not normally parasitic.

pseudopodium — A temporary protrusion or retractile process of the protoplasm of a cell that serves as a locomotor or food-gathering function.

psi — Acronym for pounds per square inch.

PSI — Acronym for pollution standards index.

psia — Acronym for pounds per square inch absolute, measured in a vacuum.

psid — Acronym for the pounds per square inch differential between two points.

psig — Acronym for the pressure per square inch gauge; see gauge pressure.

psis — Acronym for pounds per square inch standard, using standard atmosphere.

psittacosis — (*disease*) An acute, generalized chlamydial disease of humans and birds; human symptoms include fever, headache, myalgia, chills, and upper or lower respiratory tract disease. Incubation time is 4 to 15 days (usually 10 days). It is caused by *Chlamydia psittaci* and is found worldwide, especially around various types of birds. Reservoirs of infection include parakeets, parrots, pigeons, turkeys, ducks, and other birds. It is transmitted by inhaling the chlamydia from dry droppings and secretions of infected birds in an enclosed space. Communicability may last for weeks or months from diseased or healthy birds. It is rarely transmitted from person to person; general susceptibility. It is controlled by regulating the importing, raising, and selling of birds of the parrot family, surveillance of pet shops and bird areas, and educating the public of the potential for the disease.

PSM — Acronym for point-source monitoring.

psoriasis — A usually chronic, recurrent skin disease of unknown origin marked by discrete, bright-red macules, papules, or patches covered with lamellated silvery scales.

PSS-TLVs — Acronym for particulate size-selected threshold limit values.

PSU — Acronym for primary sampling unit.

psychobiology — Study of the biochemical foundations of thought, mood, emotion, and behavior.

psychogenic deafness — Loss of hearing originating in or produced by the mental reaction of an individual to the physical or social environment.

psychogenic illness — A syndrome defined as a group of symptoms developing in an individual or a group of individuals in the same indoor environment who are under some type of physical or emotional stress.

psychoneuroses — A mental disorder of psychogenic origin but presenting the symptoms of a functional nervous disease.

psychosis — A fundamental mental derangement characterized by a defect or loss of contact with reality.

psychosocial factors — Psychological, organizational, and personal stressors that could produce symptoms similar to poor indoor air quality, such as shortness of breath.

psychrometer — A hygrometer consisting of two similar thermometers; the bulb of one is kept wet so that the cooling that results from evaporation makes it register a lower temperature than the dry one, with the difference between the readings constituting a measure of the dryness of the atmosphere. It is used to measure relative humidity.

psychrometric chart — A graphical representation of the thermodynamic properties of moist air.

psychrophile — An organism that thrives in temperatures of 0 to 5°C and that causes spoilage in refrigerated foods; found on uncultivated soil in lakes and streams, on meats, and in ice creams.

psychrophilic — Of or pertaining to an organism that thrives at relatively low temperatures, usually at or below 15°C.

pt — See pint.

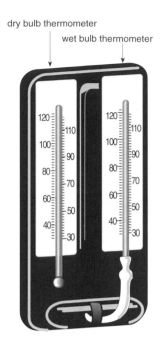

dry bulb thermometer

wet bulb thermometer

Psychrometer

PTB — Acronym for persistent, toxic, and bioaccumulative substance or potential to emit.

PTFE — Acronym for polytetrafluoroethylene (Teflon®).

p-**toluidine ($C_7H_7NH_2$)** — A white, crystalline solid; MW: 107.17, BP: 200°C, Sol: dissolves readily in organic solvents and dilute acids, Fl.P: 86.6°C (closed cup). It is used as an intermediate in the manufacture of various dyes. It is hazardous to the skin and is a suspected human carcinogen. A 1-hour exposure may result in severe poisoning. ACGIH exposure limit (TLV-TWA): 2 ppm [skin].

ptosis — Prolapse of an organ or part.

public health — A part of the health field that deals with the physical and mental health of the community in areas of the environment, nursing, dentistry, education, medicine, disease control, statistics, preventive medicine, and maternal and child health care.

public health assessment — Evaluation of data and information on the release of hazardous substances into the environment in order to assess any current or future impact on public health, develop health advisories or other recommendations, and identify studies or actions needed to evaluate and mitigate or prevent human health effects; also, the document resulting from the evaluation.

public health hazard — A site that poses a public health hazard as a result of long-term exposures to hazardous substances.

public health law — A body of statutes, regulations, and precedence that protects and promotes individual and community health; founded on the preamble of the Constitution and Section 8, Article 1, of the Constitution for promotion of the general welfare.

public health nursing — A field of nursing that is concerned with the health care of individuals in the community.

public health statement — The first chapter of an ATSDR toxicological profile intended to be a health effects summary written in lay language for a target audience, especially people living in the vicinity of a hazardous waste site or chemical release.

public hearing — A public session in which witnesses are heard and testimony is taken.

public involvement — The process of obtaining citizen input into each stage of development of planning documents.

public market — A space, stall, or enclosure designated for selling farm products.

public sewage disposal — A community sewage disposal system in which sewage from assorted homes, businesses, and possibly industries is collected and transported through a piping system to a sewage treatment plant, where it may be treated in a conventional way through settling and biological oxidation or through advanced wastewater treatment methods.

public water supply — Water distributed from a public water system.

public water system — A system for the provision of piped water for public human consumption that has at least 15 service connections and serves at least 25 people.

public water use — Water supplied from a public-water source and used for firefighting, street washing, and municipal parks and swimming pools.

puck — A hand-held device for entering data from a digitizer.

pulldown menu — A list of options that opens up when a menu is selected at the top of a window.

pulmonary — Pertaining to the lung.

pulmonary circuit — A system of blood vessels from the right ventricle of the heart to the lungs; it transports deoxygenated blood to the lungs and returns oxygenated blood from the lungs to the left atrium of the heart.

pulmonary congestion — An excessive accumulation of fluid in the lungs associated with inflammation or congestive heart failure.

pulmonary edema — The presence of fluid in the lung tissue.

P

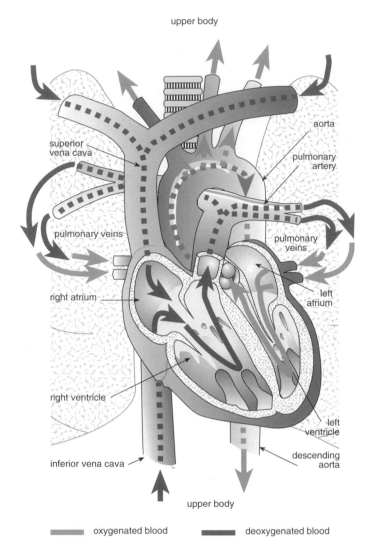

Pulmonary Circuit

pulmonary emphysema — A pathological accumulation of air in tissues or organs in which the terminal bronchials become plugged with mucous and the aveoli lose their elasticity, making breathing difficult.

pulmonary function — Performance of the respiratory system in supplying oxygen to and removing carbon dioxide from the body by way of circulating blood and moving air in and out of the alveoli.

pulmonary function test — A procedure for determining the capacity of the lungs to efficiently exchange oxygen and carbon dioxide.

pulmonary function tests — A set of tests to evaluate the mechanical properties of the lung by studying lung volumes and capacities.

pulmonary ventilation — Inhalation and exhalation of air through the lungs.

pulp — Fiber material produced by chemical or mechanical means from such raw materials as virgin wood, secondary fibers, and rags that are used in the manufacture of paper and paperboard.

pulse — A regular throbbing in the arteries caused by the contractions of the heart.

pulse rate — The number of beats per minute measured at an artery.

pulsed laser — A class of laser where emission occurs in one or more flashes of short duration.

pulsed recurrence frequency laser (PRF laser) — A laser with properties similar to a continuous wave laser.

pulse–purge — The part of the sterilizer cycle after evacuation–exhaust that consists of repeated cycles of operation of the vacuum pump followed by vacuum relief.

pulsus alternans — Latin for "alternating pulse"; regular alternation of weak and strong beats without changes in cycle length.

pulverization — Crushing or grinding material into small pieces; also known as comminution.

pumice — A natural silicate of volcanic ash or lava used as an abrasive.

pump — A device that moves, compresses, or alters the pressure of a fluid such as water or air; a piece of equipment used to move fluids or gases by suction or by positive pressure.

pump and treat — A conventional system of cleaning the polluted aquifers by pumping out water, stripping the pollutant using non-biological methods (such as air stripping or chemical treatment), and returning the water to the aquifer.

pump curve — A graph of performance characteristics of a given pump under varying power flow resistance factors.

pump strainer — A device containing a removable strainer basket designed to protect a pump from debris in the water flow when installed in the pump action line.

pumping — The mechanical transfer of fluids.

pumping station — A machine installed on sewers to pull sewage uphill.

pumping water level — The water level in a well when the pump is operating and water is being removed.

pumplift — The vertical distance that a pump will raise water.

puncture — The act of piercing or penetrating with a pointed object or instrument.

puncture wound — A traumatic injury caused by the penetration of the skin by a narrow, sharp object.

pupa — The resting stage of an insect that has completed metamorphosis; it follows the larva stage and precedes the adult stage.

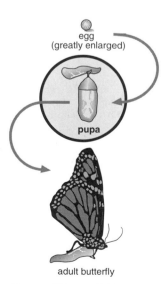

egg
(greatly enlarged)

pupa

adult butterfly

Pupa of Monarch Butterfly

pupil — The opening in front of the vertebrate eyeball, the size of which is controlled by the iris.

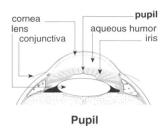

cornea
lens
conjunctiva

pupil
aqueous humor
iris

Pupil

pure culture — A population of microorganisms composed of a single strain.

pure tone — Sound wave in which the instantaneous sound pressure is a simple sinusoidal function of time; a sound sensation that is characterized or signaled by a single pitch.

purging — The method by which gases, vapors, or other airborne impurities are displaced from a confined space.

purine — The parent substance of adenine, chignon, and other naturally occurring purine "bases" not known to exist as such in the body.

purpura — Extensive hemorrhage into the skin, joints, or mucous membrane.

purulent — Consisting of or containing pus.

purulent diarrhea — A diarrhea in which the stools contain pus.

purulent inflammation — An inflammation that contains pus.

pus — A collection of dead bacteria or white blood corpuscles at the site of an infection.

pustule — An elevation on the skin containing pus.

putrefaction — Decomposition of organic matter by microorganisms producing disagreeable odors.

putrescible — Of or pertaining to organic matter capable of being decomposed by microorganisms.

putrid — Something that is decomposing.

PV — Acronym for project verification.

PVC — See polyvinyl chloride.

PWSS — Acronym for public water supply system.

pycnocline — A layer of water that exhibits rapid change in density with depth.

pyogenic — Pus-producing.

pyogenic infection — An infection that results in pus production.

pyogenic microorganism — A microorganism that produces pus.

pyorrhea — A discharge of pus.

pyranometer — An instrument that measures total global radiation.

pyrazole (C₃H₄N₂) — A crystalline solid with a bitter taste; MW: 68.01, BP: 187°C, Sol: soluble in water

P

and organic solvents, Fl.P: a noncombustible solid, density: NA. It is used in organic synthesis and as a chelating agent. It is hazardous to the digestive tract and respiratory tract. In experimental animals, it causes ataxia, muscle weakness, respiratory depression, and reproductive toxicity. ACGIH exposure limit: NA.

Pyrazole

pyrene — A polycyclic aromatic hydrocarbon formed during incomplete combustion of coal, oil, gas, garbage, or other organic substances and found especially at hazardous waste sites; see polycyclic aromatic hydrocarbons.

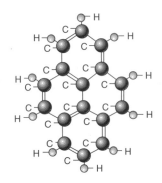

Pyrene

pyrethrin — One of two oily liquid esters ($C_{21}H_{28}O_3$ and $C_{22}H_{28}O_5$) that have insecticidal properties and that occur naturally in old-world chrysanthemums; an excellent knock-down agent.

pyrethrum ($C_{20}H_{28}O_3$/$C_{21}H_{28}O_3$/$C_{22}H_{28}O_5$) — A brown, viscous oil or solid; MW: 316 to 372, BP: not known, Sol: insoluble, Fl.P: 180 to 190°F. It is used as an insecticide on pre- and post-harvest agricultural crops, cattle, and poultry; on food cartons; and in confined areas. It is hazardous to the respiratory system, skin, and central nervous system and is toxic through inhalation, ingestion, and contact. Symptoms of exposure include erythema, dermatitis, papules, pruritus, rhinorrhea, sneezing, and asthma. OSHA exposure limit (TWA): 5 mg/m³ [air].

pyrexia — See fever.

pyrgeometer — An instrument that measures radiation from the surface of the Earth into space.

pyridine (C_5H_5N) — A colorless to yellow liquid with a nauseating, fish-like odor; MW: 79.1, BP: 240°F, Sol: miscible, Fl.P: 68°F, sp. gr. 0.98. It is used in the manufacture of pharmaceuticals; as a solvent in the manufacture of polycarbonate resins used in hand tools, small appliances, camera parts, safety helmets, and electrical connectors; as a starting material used in the manufacture of chemical intermediates and products; in the manufacture of rubber accelerators, epoxy resins, and pharmaceuticals; as a solvent reaction medium or catalyst in paint manufacture, drug manufacture, and rubber manufacture; as a reagent in chemical analysis; in textile treatment as a waterproofing agent; as a denaturant for ethyl alcohol; as a coupling assistant in azo dye manufacture; in purification of mercury fulminate in explosives manufacture; during thermal decomposition of flexible polyurethane foams. It is hazardous to the central nervous system, liver, kidneys, skin, and gastrointestinal tract and is toxic through inhalation, absorption, ingestion, and contact. Symptoms of exposure include headache, nervousness, dizziness, insomnia, nausea, anorexia, frequent urination, eye irritation, dermatitis, and liver and kidney damage. OSHA exposure limit (TWA): 5 ppm [air] or 15 mg/m³.

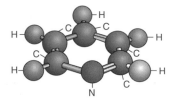

Pyridine

pyrimidine ($C_4H_2N_2$) — An organic base that is a constituent of deoxyribonucleic acid and ribonucleic acid.

pyrite (FeS_2) — A mineral compound of sulfur and a metal, usually iron, that is often found in coal; also known as fool's gold.

pyrocatechol ($C_6H_6O_2$) — A colorless crystalline solid that turns brown on exposure to air or light when in an aqueous solution; MW: 110.12, BP: 245.5°C, Sol: soluble in water and most organic solvents, Fl.P: a noncombustible solid, density: NA. It is used in photography, in dying fur, and as a topical antiseptic. It is hazardous to the skin and respiratory tract and may cause weakness, muscular pain, dark urine, tremors, dyspnea, convulsions, and death. ACGIH exposure limit (TLV-TWA): 5 ppm.

pyrogallol ($C_6H_6O_3$) — A white crystalline solid that turns gray on exposure to air and light; MW:

P

126.12; BP: 309°C; Sol: soluble in water, alcohol, and ether; Fl.P: a noncombustible solid; density: NA. It is used in the manufacture of various dyes; in the dyeing of furs, hair, and feathers; in engraving; as a developer in photography. It is hazardous to the skin and digestive tract. It may cause nausea, vomiting, gastritis, kidney and liver damage, convulsions, congestion of the lungs, and death.

Pyrogallol

pyrogen — (*bacteria*) A fever-producing substance of bacterial origin (believed to be complex polysaccharides) attached to another radical containing nitrogen or phosphorous.

pyrogenic — A substance that causes a rise in body temperature.

pyroligneous acid — A combination of acetic acid, methyl alcohol, and acetone that comes from the destructive distillation of wood.

pyrolysis — An endothermic reaction causing a destructive distillation of solid material in the presence of heat and the absence of air or oxygen.

pyrometer — An instrument used for measuring or recording temperatures.

pyrophoric — Pertaining to a chemical that ignites spontaneously in air at temperatures of 13°F or below.

pyroxene — (*mineralogy*) Any of a group of silicate materials, usually calcium, magnesium, and iron silicate, often found in igneous rocks.

pyrrole (C_4H_5N) — A colorless liquid with a chloroform-like odor; MW: 67.10; BP: 130°C; Sol: slightly soluble in water and miscible in organic solvents; Fl.P: 39°C (closed cup), the vapor forms explosive mixtures with air; density: 0.969 at 20°C. It is used in organic synthesis. It is hazardous to the respiratory tract, digestive tract, and skin and causes irritation. Very few toxicity data are available. ACGIH exposure limit: NA.

Pyrrole

pyrrolidine (C_4H_9N) — A colorless liquid with an ammonia-like odor; MW: 71.14; BP: 89°C; Sol: soluble in water and organic solvents; Fl.P: 3°C (closed cup), the vapor forms explosive mixtures with air; density: 0.892 at 20°C. It is used infrequently but does occur in tobacco and carrot leaves. It is hazardous to the eyes, respiratory tract, and skin and is irritating and corrosive. ACGIH exposure limit: NA.

Pyrrolidine

P

pyuria — The presence of white blood cells in the urine.

Q

Q fever — (*disease*) An acute, febrile, rickettsial disease with sudden chills, headache, weakness, malaise, and severe sweats. Pneumonitis may occur along with chest pains, acute pericarditis, acute hepatitis, and generalized infections. Incubation time is 2 to 3 weeks. It is caused by *Coxiella burnetii* (*Rickettsia burneti*) and is found in all parts of the world. Reservoirs of infection include cattle, sheep, goats, ticks, and some wild animals; it is transmitted by airborne dissemination of rickettsiae in dust. Transmission from person to person is rare; general susceptibility. It is controlled by understanding how the disease occurs, pasteurization of milk, and immunization.

q1* — Upper-bound estimate of the low-dose slope of the dose–response curve as determined by the multistage procedure; can be used to calculate an estimate of carcinogenic potency, the incremental excess cancer risk per unit of exposure (usually μg/l for water, mg/kg/day for food, and μg/m^3 for air).

Q10 — A relative increase in reaction rate with temperature; expressed as the increase over a 10°C interval.

QA — Acronym for quality assurance.

QAC — Acronym for Quality Assurance Coordinator.

QAMIS — Acronym for Quality Assurance Management and Information System.

QAO — Acronym for Quality Assurance Officer.

QAP — Acronym for Quality Assurance Program.

QAPP — Acronym for Quality Assurance Program or Project Plan.

QA/QC — Acronym for quality assistance/quality control.

QAT — Acronym for Quality Action Team.

QBtu — Acronym for quadrillion British thermal units.

QC — See quality control.

QCA — Acronym for Quiet Communities Act.

QCI — Acronym for Quality Control Index.

QF — See quality factor.

QL — Acronym for quantification limit.

QNCR — Acronym for Quarterly Noncompliance Report.

Q-switched laser — A pulsed laser capable of extremely high peak power for a very short duration.

qt — See quart.

QUA — Acronym for qualitative use assessment.

quad — A unit of heat energy equal to one quadrillion or a million billion British thermal units.

quadrant — (*mathematics*) A quarter of a circle; an arc of 90°.

quadrat — A sampling area, most commonly 1 square meter, used for analyzing vegetation.

quadratic polynominal — (*mathematics*) One in which the highest degree of terms is 2.

quadrillion — Equal to 10^{15} (or 1,000,000,000,000,000).

quagmire — See bog.

qualified — A professional or organization that meets the standards of performance of professional competence as recognized by the appropriate agency or organization of peers.

qualitative — Describing size, magnitude, or degree.

qualitative analysis — A chemical analysis designed to identify the components of a substance or mixture.

qualitative test — A test that determines the presence or absence of a substance.

quality — The degree to which techniques contribute to the improvement or maintenance of a product, service, or situation.

quality assessment — The process of evaluating and measuring the effectiveness of an improvement or maintenance of a product, service, or situation.

quality assurance (QA) — A system of actions designated to promote confidence that a product or service will satisfy needs.

quality control (QC) — A group of activities designed to ensure adequate quality, especially in manufactured products; consists of random sampling and inspection techniques.

quality factor (QF) — The linear-energy-transfer-dependent factor by which doses of absorbed radiation are multiplied to obtain, for radiation protection purposes, a quantity that expresses the effectiveness of the absorbed dose on a common scale for all ionizing radiation.

quality measurement experiment — A derived product that compares measurements to assess quality or otherwise analyze data.

quality of life — A subjective sense of well-being.

quantitative — Refers to a measurement of quantity or amount.

quantitative analysis — The determination of relative amounts of significant components present in a substance or mixture of substances.

quantum — The smallest indivisible quantity of radiant energy; a photon.

quantum mechanics — Branch of physics dealing with matter and electromagnetic radiation and the behavior of particles whose specific properties are given by quantum numbers.

quantum number — A number used to describe a specific property of a subatomic particle in mathematical terms.

quantum theory — A theory based on the concept of the subdivision of radiant energy into finite quanta and applied to numerous processes involving transference or transformation of energy at an atomic or molecular scale,

quarantine — The isolation of people with a communicable disease or those exposed to one during the contagious period to prevent the spread of the illness.

quark — A subatomic particle found in the nucleus of the atom.

quarry — A rock pit.

quarry sanitary landfill method — A variation of the area landfill method in which the waste is spread and compacted in a depression and cover material is brought in from elsewhere.

quarry water — The moisture content of freshly quarried stone.

quart (qt) — A unit of liquid volume equal to 32 fluid ounces or 2 pints.

quartan malaria — A form of malaria with fever reoccurring every 72 hours; caused by *Plasmodium malariae.*

quartz (SiO$_2$) — A colorless, transparent, very hard mineral composed of silica that is found in many different types of rocks, such as sandstone and granite; the most abundant and widespread mineral.

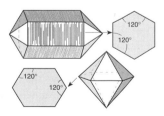

Quartz

quash — To vacate, annul, or make void.

quaternary ammonium compounds — Any of numerous strong bases and their salts derived from ammonium by replacement of the hydrogen atoms with organic radicals; important as surface-active agents and disinfectants.

quench tank — A water-filled tank used to cool incinerator residue or hot materials from industrial processes.

quench trough — A water-filled trough into which burning residue drops from an incinerator furnace.

quenching — Sudden cooling.

quicklime — A substance that is mostly calcium oxide or calcium oxide in natural association with a smaller amount of magnesium oxide; capable of combining with water to form hydrated lime.

quicksilver — See mercury.

quicksilver water — A solution of mercury nitrate used in gilding.

quickwater — The part of a stream that has a strong current.

QuikSCAT — A NASA satellite that provides climatologists, meteorologists, and oceanographers with daily detailed snapshots of the winds swirling above the oceans of the Earth using a state-of-the-art radar instrument, the scatterometer.

quinoline (C$_9$H$_7$N) — A colorless liquid with a characteristic odor; MW: 129.17, BP: 237.5°C, Sol: soluble in water and miscible with organic solvents, Fl.P: a noncombustible liquid, density: NA. It is used in the manufacture of dyes, as an antimalarial agent, and as a solvent for resins and terpenes. It exhibits carcinogenicity in rats and mice (liver cancer). ACGIH exposure limit: NA

Quinoline

quinone (C$_6$H$_4$O$_2$) — A pale-yellow solid with an acrid, chlorine-like odor; MW: 108.1, BP: sublimes, Sol: slight, Fl.P: 100 to 200°F, sp. gr. 1.32. It is used as an oxidizing agent in the synthesis of organic chemicals and intermediates; as a photographic developer (hyperquinone), toner, and intensifier; in the production of insecticides and fungicides; in the production of cortisone and barbiturates; in the polymer and resin industries as an inhibitor and retarder–antioxidant, curing agent, and catalyst; as a leather tanning agent; in the manufacture of quinhydrone electrodes for use in pH determinations. It is hazardous to the eyes and skin and is toxic through inhalation, ingestion, and contact. Symptoms of exposure include conjunctivitis, keratitis, and eye and skin irritation. OSHA exposure limit (TWA): 0.4 mg/m³ [air].

quotidian malaria — A form of malaria with fevers reoccurring every 24 hours; found in cases of overlapping infections.

Q-value — The energy liberated or absorbed in a nuclear reaction.

Q

R

R — Abbreviation or symbol for hydrocarbon radical, radius, Rankine, resistance, roentgen, or Rhydberg constant.

R horizon — Bedrock, such as limestone, sandstone, or shale.

R&D — See research and development.

R/O — Acronym for rule out or reverse osmosis.

R/O unit — A reverse osmosis unit used for water purification.

RA — Acronym for remedial action; see risk assessment.

RAATS — Acronym for RCRA Administrative Action Tracking System.

rabbit fever — (*disease*) See tularemia.

rabies — (*disease*) Viral encephalomyelitis with headache, fever, malaise, sense of apprehension, paralysis, muscle spasm, delirium, and convulsions after the bite of an infected animal. Incubation time is 2 to 8 weeks or as short as 10 days. It is caused by the rabies virus, which is a rhabdovirus, and is found worldwide. Reservoirs of infection include wild and domestic dogs, foxes, coyotes, wolves, cats, raccoons, and other biting mammals; it is also found in bats. It is transmitted in the saliva of a rabid animal and is communicable in dogs and cats for 3 to 5 days before the onset of clinical signs and during the course of the disease; general susceptibility. It is controlled by vaccinations of all pets for rabies, 10-day detention for observation of dogs and cats who have bitten people, immediate evaluation of the heads of animals suspected to be rabid for virus isolation in the brain, and immediate immunization of an individual bitten by a rabid animal.

rack — A set of evenly spaced, parallel metal bars or rods located in an influent channel to remove rags, rocks, cans, and other inert materials from wastewater.

RACM — Acronym for reasonably available control measure.

RACT — See reasonable available control technology.

rad — See radiation absorbed dose.

radar — Acronym for radio detection and ranging; a system for locating natural and artificial objects, navigating, measuring distance or altitude, homing, or bombing by means of radio signals.

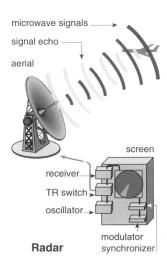

Radar

radar interferometry — The study of interference patterns caused by radar signals which allows scientists to generate three-dimensional images of the surface of the Earth.

radial — A spoke or ray; lines converging at a single center.

radian — An arc of a circle equal in length to the radius.

radiant — (*optics*) The point or object from which light proceeds; (*geometry*) a straight line proceeding from a given point or fixed pole; (*astronomy*) the point in space from which a shower of meteors seems to proceed.

radiant barrier — A reflective material used to block radiant heat especially in the roof of a building.

radiant energy — See radiation.

radiant heat transfer — Energy transfer occurring when a large difference exists between the temperatures of two surfaces exposed to each other but not touching.

radiant temperature — The temperature of the body resulting from absorbed radiant energy.

radiation — The emission and propagation of energy in space or through a material medium in the form of waves; the emission and propagation of the electromagnetic waves of sound and elastic waves;

the transfer of heat or light by waves of energy; a stream of alpha- or beta-particles from a radioactive source or neutrons from a nuclear reactor; usually refers to electromagnetic radiation but can also be applied to sound waves and emitted particles. Also known as radiant energy.

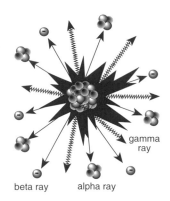

Radiation

radiation absorbed dose (rad) — A measure of the energy imparted to matter by ionizing radiation per unit mass of irradiated material; equal to 100 ERGs per gram; equal to 1 gray (Gy).

radiation budget — A measure of all the inputs and outputs of radiative energy relative to a system such as the Earth.

radiation burn — A burn resulting from exposure to radiant energy in the form of sunlight, x-rays, nuclear emissions, or explosion.

radiation dermatitis — An acute or chronic inflammation of the skin due to exposure to ionizing radiation.

radiation detector — An instrument used for converting radiant energy into a form that will register on a monitor.

radiation equilibrium — A condition occurring in a radiation field when the energy of the radiation entering a volume equals the energy of the radiation leaving that volume.

radiation exposure — A determination of the ionization produced in air by x-rays or gamma rays.

radiation gas flow counter — A counter in which an appropriate atmosphere is maintained in the counter tube by allowing a suitable gas to flow slowly through the sensitive volume.

radiation illness — An illness caused by the harmful effect of radiation and which predisposes people to the development of symptoms of general malaise, nausea, and vomiting.

radiation protection — The use of equipment, distance, and barriers to reduce the risk of exposure to ionizing radiation by individuals.

radiation pyrometer — A device that determines temperature by measuring the intensity of radiation at all wavelengths emitted by material having a high temperature.

radiation sensitivity — The response of tissue to ionizing radiation

radiation sickness — The self-limited syndrome of radiation characterized by nausea, vomiting, diarrhea, and psychic depression following exposure to appreciable doses of ionizing radiation, particularly to the abdominal region; its mechanism is unknown, and no satisfactory treatment is known; it usually appears a few hours after irradiation and may subside within a day but may be more intense and, depending on exposure, cause death.

radiation source — An apparatus or material emitting or capable of emitting ionizing radiation.

radiation standards — Regulations that govern exposure to permissible concentrations of radioactive materials as well as their transportation.

radiation therapy — The use of high-energy radiation from x-rays, neutrons, and other sources to kill cancer cells and shrink tumors.

radiative cooling — Cooling of the surface of the Earth and adjacent air that occurs when infrared heat energy radiates from the surface of the Earth upward through the atmosphere into space, causing a cooling of the temperature of both the surface and the lowest part of the atmosphere.

radiative forcing — A change in the balance between incoming solar radiation and outgoing infrared radiation caused by greenhouse gases; it increases the fraction of infrared radiation reradiating toward the surface, thus creating a warming influence.

radical — A group of atoms of different elements acting as a single unit in a chemical reaction. They are normally incapable of existing separately. A radical may be negatively charged, positively charged, or without a charge. Also known as group.

radio detection and ranging — See radar.

radio spectrum — The complete range of frequencies or wavelengths of electromagnetic waves especially those used in radio and television.

radio wave — An electrical impulse sent through the atmosphere at a radiofrequency.

radioactive — Referring to an unstable isotope that decays spontaneously and releases subatomic particles or energy.

radioactive carbon dating — See carbon dating.

R

radioactive contamination — Deposition of radioactive material in any place where it is not desired, particularly where its presence may be harmful.

radioactive decay — A change in the nuclei of radioactive isotopes so they spontaneously emit high-energy electromagnetic radiation and/or subatomic particles while gradually changing into different isotopes or elements.

radioactive element — An atom subjected to spontaneous degeneration of its nucleus accompanied by the emission of alpha-particles, beta-particles, or gamma rays.

radioactive equilibrium — In a radioactive series, the state that prevails when the rates of decay of two or more successive members of a series remain constant.

radioactive half-life — The time required for a radioactive substance to lose 50% of its activity by decay.

radioactive iodine — A radioactive isotope of iodine used as a tracer in biology and medicine.

radioactive isotope — An unstable form of an element that spontaneously emits either high-energy electromagnetic radiation or subatomic particles, or both. The decay rate describes the time necessary for a radioactive isotope to change to a stable element.

radioactive series — A succession of nuclides, each of which transforms by radioactive disintegration into the next until a stable nuclide results.

radioactive tracer — A substance to which a radioactive tag has been attached allowing it to be followed through a physiological system using radiation detectors.

radioactive waste — Waste generated by the emission of particulate or electromagnetic radiation resulting from the decay of the nuclei of unstable elements.

radioactivity — The quality of emitting or the emission of particulate or electromagnetic radiation as a consequence of the decay of the nuclei of unstable elements.

radiobiology — The study of the principles, mechanisms, and effect of radiation on living organisms.

radiocarbon dating — See carbon dating.

radiochemical — Any compound or mixture containing a sufficient portion of radioactive elements to be detected by a Geiger counter.

radiochemistry — The phase of chemistry concerned with the properties and behavior of radioactive materials.

radiodiagnosis — A means of diagnosis that uses x-ray examination.

radioecology — Study of the effects of radiation on plants and animals in natural communities.

radiofrequency (rf) — Electromagnetic radiations produced by very rapid reverses of current in a conductor; suitable for radio transmission above approximately 10^4 Hz and below 3×10^{12} Hz.

radiofrequency interference — Electromagnetic waves between the frequencies of 10 kHz and 300 Ghz that can affect susceptible systems by conduction through sensor or power input lines and by radiation in space.

radiography — The making of film records of internal structures of the body by exposure of film specially sensitized to x-rays or gamma rays.

radioimmunoassay — A radiobiological technique used to determine the concentration of an antigen, antibody, or other protein in the serum.

radioisotope — A radioactive atomic species of an element with the same atomic number and usually identical chemical properties where the substances are both unstable and decaying; used as biological tracers, in industrial control operations, and in the diagnosis and treatment of disease.

radiological health specialist — An environmental health practitioner responsible for protection of workers and the general public from ionizing and nonionizing radiation hazards.

radiological monitoring — Periodic or continuous determination of the amount of ionizing radiation or radioactive contamination present in an occupied region as a safety measure for health protection.

radiological survey — An evaluation of the radiation hazards in an area that may occur as a result of production, use, or existence of radioactive materials or other sources of radiation under certain conditions.

radiology — A branch of medicine concerned with the use of radiant energy in the diagnosis and treatment of disease.

radiometer — An instrument for measuring the intensity of radiant energy by twisting on its own axis of suspended waves that are blackened on one side and exposed to a source of radiant energy; an instrument for measuring electromagnetic or acoustical energy.

radionuclide — A radioactive nuclide that has the capability of spontaneously emitting radiation; an isotope of an element that is unstable and exhibits radioactive decay; a cancer-causing, hazardous air pollutant that may be naturally occurring or produced by accelerators, reactors, or fission byproducts.

radiopaque — Being opaque to various forms of radiation.

radiophotoluminescence — Luminescence exhibited by certain materials after exposure to ionizing radiation.

radiopoison — A radioactive poison, such as strontium-90.

radiosensitive — Relative susceptibility of cells, tissues, organs, organisms, or any living substance to the injurious action of radiation.

radiosonde — An instrument carried on a balloon that is used to measure and transmit temperature, pressure, and humidity at various heights in the atmosphere.

radiotherapy — Treatment of disease with the application of relatively high roentgen doses.

radiotoxicity — The potential of an isotope to cause damage to living tissue by absorption of energy from the disintegration of radioactive material introduced into a body.

radium (Ra) — A naturally occurring, silvery-white, radioactive metal formed by the decay of uranium and isotopic forms, the most abundant being radium-226. It is used in medicine, in industrial radiography, and as a source of neutrons and radon. Previously used in luminescent paints for watches and clock faces, instrument panels in airplanes, and military instruments and compasses. It is hazardous to the blood, bones, liver, eyes, hematopoietic tissues, and immunological system and is toxic through inhalation, absorption, ingestion, and contact. Symptoms of exposure include anemia, necrosis of the jaw, osteogenic sarcoma, bone marrow failure, chronic liver disease, cataracts, breast and liver cancer, and dental disease. OSHA does not regulate radium. U.S. EPA limit for radium-228 and radium-226: $5(2 \times 10^{-1})$ picocuries per liter of water.

radius (R) — Any line segment going straight from the center to the outside of a circle or sphere.

radius of influence — Radial distance from the center of a well bore to the point where there is no lowering of the water table or to the edge of the cone of depression; the radial distance from an extraction well that has adequate airflow for effective removal of contaminants when a vacuum is applied to the extraction well.

radius of vulnerability zone — The maximum distance from a point of release of a hazardous substance in which the airborne concentration could reach a level of concern under specified weather conditions.

RADM — Acronym for Regional Acid Deposition Model.

radon (Rn) — A naturally occurring, colorless, odorless, tasteless, radioactive gas formed by the decay of radium and which may be found in rocks, soils, and uranium ore. Radon produces a series of daughter products through rapid radioactive decay which are characterized by relatively short half-lives and are easily adsorbed on the dust particles present in the air. The particles in the air containing the adsorbed gases are deposited on the surface of the respiratory tract, emitting alpha-radiation doses to the epithelium and diffusing from the lungs into the blood. It was used from 1930 to 1950 in the medical treatment of malignancies, obliterative arthritis, and atherosclerosis of the lower extremities. It is hazardous to the respiratory system, kidneys, and blood and is toxic through inhalation. Symptoms of exposure include increased mortality due to cancer and nonneoplastic diseases, silicosis, chronic obstructive pulmonary disease, fibrosis, emphysema, chronic nephritis and renal sclerosis, hematological effects, lung cancer, and genotoxic effects with chromosomal aberrations. Radon is released from soils and can enter buildings through cracks in foundations and other openings. U.S. EPA's annual atmospheric release rate for environmental and indoor air from residual radioactive material from inactive uranium processing sites: 20 pCi/m^2/sec [air].

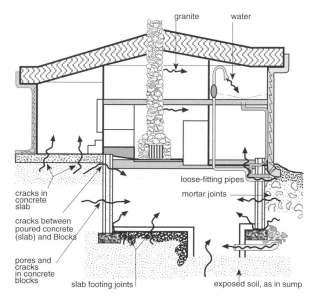

granite   water

cracks in concrete slab

cracks between poured concrete (slab) and Blocks

pores and cracks in concrete blocks

slab footing joints

loose-fitting pipes

mortar joints

exposed soil, as in sump

Radon access routes into house

radon daughters — The short-lived radioactive decay products of radon that decay into longer-lived isotopes that can attach themselves to airborne dust or other particles and when inhaled cause damage to the lining of the lung.

radon decay products — The immediate products of the radon decay chain, including Po-218, Pb-214, Bi-214, and Po-214, with an average combined half-life of about 30 minutes.

Radon Gas and Indoor Air Quality Research Act of 1986 — A federal law enacted because of the potential health effects of high levels of radon gas in structures in certain areas of the country and because existing federal radon and indoor air pollution research programs were fragmented, under-funded, and did not provide adequate information concerning exposure to radon and indoor air pollutants.

radon progeny — See radon daughters.

radura symbol — A circular symbol that must appear on all irradiated food unless the food is used as an ingredient in a processed food or served in a restaurant.

Radura symbol

ragweed — Any of a variety of weeds that produce highly allergenic pollen and potential allergic reactions.

RAI — See radioactive iodine.

rain — Water falling in drops condensed in vapor in the atmosphere.

rain forest — A forest with high humidity, constant temperature, and a large rainfall of approximately 150 inches per year; an evergreen woodland of the tropics distinguished by a continuous leaf canopy and an average rainfall of about 100 inches per year, thus playing an important role in the global environment.

rain gauge — An instrument used for recording and measuring time, distribution, and the amount of rainfall.

râle — A sound or noise in the lungs accompanying normal sounds of respiration.

RAM — See random access memory.

ramp sanitary landfill method — Variation of area landfill methods in which a cover material is obtained by excavation in front of the working face.

random access memory (RAM) — Readable and writeable memory that acts as a storage area while the computer is on and is erased each time the computer is turned off; used to store data and help execute programs when the computer is used.

random allocation — Experimental procedure in which allocation of subjects to treatment groups is controlled by a table of random numbers; an essential part of valid clinical trials.

random error — (*statistics*) The measure of random variations around a specific value, such as taking 10 samples of a gaseous air pollutant in a confined area so the samples range around a specific value; an error that can only be predicted on a statistical basis.

random incident field — A sound field in which the angle of arrival of sound at a given point in space is random in time.

random motion — The molecular movement that is always present but nondirectional.

random noise — Oscillation whose instantaneous magnitude cannot be specified for a given point in time.

random sampling — (*air*) A process in which grab or integrated samples are collected at random intervals; (*statistics*) a sample drawn from a group in such a way that every item in the group has an equal chance of being chosen.

range of motion — Difference between the maximum extension and maximum flexion of a joint in degrees of a circle.

range — (*statistics*) The interval between the lowest and the highest values in a series of data; the area between two limits in which a quantity or value is measured.

Rankine (R) — See Rankine temperature scale.

Rankine cycle — An ideal steam-engine cycle because it is theoretically reversible; also known as a steam cycle.

Rankine temperature scale (R) — An absolute temperature scale on which the unit of measurement equals a Fahrenheit degree; 0°R is the temperature at which molecular energy is a minimum and it corresponds to a temperature of –459.67°F; the freezing point of water is 491.69° and the boiling point is 671.69°.

RAP — Acronym for Radon Action Program or Response Action Plan.

rapid sand filter — A closed, pressurized tank containing filter media made up of filter sand and free of carbonates or other foreign material; used to filter swimming pool water and/or drinking water.

RAPS — Acronym for Regional Air Pollution Study.

RAS — See return-activated sludge; acronym for routine analytical service.

rash — A general term for any abnormal reddish coloring or blotching on some part of the skin.

rasper — A grinding machine in the form of a large vertical drum containing heavy, hinged arms that rotate horizontally over a rasp-and-sieve floor.

raster — A regular grid of cells covering an area.

raster database — A database containing all mapped, spatial information in the form of regular grid cells.

raster display — A device for displaying information in pixels on a visual display unit.

raster map — A map encoded in the form of a regular array of cells.

raster-to-vector — The process of converting an image made up of cells into one described by lines and polygons.

RASVSS — Acronym for return-activated sludge volatile suspended solids.

rat — A rodent considerably larger in size and structural details than a mouse.

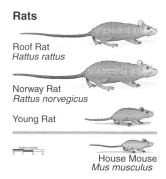

Rats

Roof Rat
Rattus rattus

Norway Rat
Rattus norvegicus

Young Rat

Scale in inches
0   1   2

House Mouse
Mus musculus

RAT — Acronym for relative accuracy test.

rat-bite fever — One of two diseases, streptobacillus fever and spirillary fever, transmitted by the bite of a rat; see streptobacillary fever and spirillary fever.

rat typhus — See murine typhus.

rate — (*pesticides*) The actual amount of a pesticide chemical put in or on a plant, animal, or part of a building per unit of time; the cost of a unit of insurance through a specified period of time; the measure of probability of an event in a population.

rate coding — The process of controlling muscular force by regulating the firing rate of motor neurons.

rate of flow — The quantity of water flowing past a given point in a unit of time; usually measured in gallons per minute.

rate of flow indicator — A flow meter that measures differentials across a calibrated orifice and indicates the rate of flow at that point, usually in gallons per minute.

rate of refrigeration — A measure of temperature control influenced by the heat-transfer properties of the food, volume of food to be refrigerated, kind of containers used, heat conductivity of the containers, agitation of the food, and temperature difference between the food and the refrigerator unit.

rated capacity — The number of tons of solid waste that can be processed at an incinerator in a 24-hour period.

rating curve — A drawn curve showing the relationship between gauge height and discharge of a stream at a given station.

ratio — Relationship in quantity, amount, or size between two or more things.

rationale — A discussion of a concern, some issues involved, and some methods for dealing with the concern.

raw cotten fiber — Untreated cotton fibers before any processing; can cause byssinosis, bronchitis, and chest tightness. ACGIH exposure limit (TLV-TWA): 0.2 mg/m^3 for fibers <15 μm in length.

raw sewage — Untreated wastewater and its contents.

raw water — Water in its natural state prior to any treatment and usually first entering the water treatment plant.

ray — A line of light that appears to radiate from a bright object.

Raynaud's disease — A primary or idiopathic vasospastic disorder characterized by bilateral and symmetrical pallor and cyanosis of the fingers with or without local gangrene; caused by exposure to cold or emotional upset.

RB — Acronym for request for bid.

RBC — Red blood cell; See red blood count, erythrocyte, or rotating biological contactor.

RBE — See relative biologic effectiveness.

RCDO — Acronym for regional case development officer.

RCO — Acronym for regional compliance officer.

RCRA — See Resource Conservation and Recovery Act.

RD — Acronym for registered dietitian.

RD&D — Acronym for research, development, and demonstration.

RD/RA — Acronym for remedial design/remedial action.

RDF — See refuse-derived fuel.

rDNA — Acronym for recombinant DNA.

RDS — Acronym for respiratory distress syndrome.

RDT — Acronym for remote data transmission.

RDU — Acronym for regional decision unit.

RDV — Acronym for reference dose value.

RE — Acronym for reasonable effort or reportable event.

reaction — (*chemistry*) A change involving the rearrangement of the atoms and molecules of one or more substances that often have different properties; chemical transformation or change.

reaction time — The time between application of a stimulus and initiation of a response.

R

reactivation — The treatment of activated carbon to remove adsorbed organic material and restore its adsorption capabilities; also known as regeneration.

reactive bond — (*chemistry*) Readily given to or involving a reaction; (*physiology*) readily responding to a stimulus.

reactive liquid — Any liquid that removes a gaseous pollutant from a chemical reaction, thus transforming the pollutant to a less offensive form.

reactive organic gas — A photochemically reactive chemical gas composed of non-methane hydrocarbons that may contribute to the formation of smog.

reactive wastes — Wastes that are normally unstable, react violently with air or water, or form potentially explosive mixtures with water, including those that emit toxic fumes when mixed with water and materials capable of detonation.

reactivity — (*chemical*) The susceptibility of a substance to undergo a chemical reaction or change that may result in explosion or burning and release of corrosive or toxic emissions.

reactor — A nuclear reactor.

reactor vessel — A piece of equipment in which nuclear fission may be sustained and controlled in a self-supporting nuclear reaction; it houses the core, which is made up of fuel rods, control rods, and instruments.

read — (*computer science*) The retrieval or transfer of data from a storage location to a disk.

read-only memory (ROM) — Part of the memory of a computer where information is permanently stored; users have access to it only for purposes of reading the contents.

ready-to-eat food — Food that is in edible form, including potentially hazardous food that has been cooked; raw, washed, cut fruits and vegetables; whole fruits and vegetables; and other foods that are presented to customers without further washing, cooking, or additional preparation required.

reaeration — The introduction of air into the lower layers of a body of water, thus causing air bubbles to rise through the water, which replenishes dissolved oxygen; also causes the lower waters to rise to the surface, where they take on oxygen from the atmosphere.

reagent — Any substance used in a chemical reaction to produce, measure, examine, or detect another substance.

real conductivity — A property of material which when multiplied by the electric field gives the free current density.

real number — A number that has both an integer and a decimal component.

real time — Tasks or functions executed so rapidly that the user gets an impression of continuous visual feedback.

REAP — Acronym for Regional Enforcement Activities Plan.

reasonable available control technology (RACT) — The level of air pollution emissions control required to be imposed on all existing sources in nonattainment areas, pursuant to the Clean Air Act.

rebar — A reinforcing steel bar.

rebuttal — An answer or response to a statement; refuting, opposing, or contradicting what has been stated.

recalcitrant — A resistance to biodegradation.

recall — Removal of goods from the market by the producer through voluntary action in lieu of seizure.

recarbonation — The diffusion of carbon dioxide gas through water to restore the carbon dioxide removed by adding lime to water in water softeners.

receiver — The person to whom a message is sent.

receiving waters — Rivers, lakes, oceans, or other watercourses that receive treated or untreated wastewaters.

receptor — A special molecule that recognizes and binds to a foreign chemical and is involved in the initial steps in a toxic response; a sense organ.

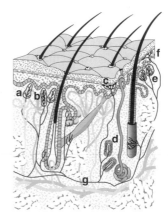

(a) heat (Ruffini's end organ)
(b) touch, pressure
　　(Meissner's corpuscle)
(c) touch, pressure (Merkel's discs)
(d) deep pressure
　　(Pacinian corpuscle)
(e) cold (end bulb of Krause)
(f) pain (naked nerve endings)
(g) cutaneous nerve

Receptors

receptor antagonism — Process that occurs when a chemical blocks the effect of a second chemical.

recharge — To add water to the zone of saturation, as in recharging of an aquifer.

R

recharge area — The area where water predominantly flows downward through the unsaturated zone to become groundwater.

recharge zone — The area through which water is added to an aquifer; also known as an intake area.

recidivism — A tendency to relapse into a previous condition or mode of behavior.

reciprocating internal combustion engine — An engine in which air and fuel are introduced into cylinders, compressed by pistons, and ignited by a spark plug or by compression, resulting in combustion in the cylinders, which pushes the pistons sequentially and transfers energy to the crankcase, causing it to rotate.

recirculated air — Extracted air that is resupplied to a space, normally blended with outside air and reconditioned; used for ventilation, heating, humidification, or dehumidification.

recirculating system — The entire system of pipes, pumps, and filters that allows water to be taken from a pool, to be filter treated, and to be returned to the pool.

RECLAIM — Acronym for Regional Clean Air Initiatives Marker.

reclaimed wastewater — Treated wastewater that can be used for beneficial purposes such as irrigating certain plants.

reclaimed water system — A system of pipelines, pumps, and storage basins for storage and distribution of reclaimed wastewater.

reclamation — Restoration to a better or more useful state, such as land reclamation by sanitary land filling or obtaining useful materials from solid waste.

recombinant bacteria — Microorganisms for which the genetic makeup has been altered by deliberate introduction of new genetic elements.

recombinant DNA — Deoxyribonucleic acid that is artificially introduced into a cell, which alters the genotype and phenotype of the cell and is replicated along with the natural DNA.

recommended action — A statement of activity that, if implemented, is expected to contribute to the achievement of an objective.

recommended exposure limit (REL) — The National Institute of Occupational Safety and Health recommendation for maximum occupational exposure to a hazardous chemical.

recondition — (*law*) The process of reworking a lot of goods under seizure pursuant to a consent decree of condemnation and the attempt to bring the goods into compliance with the act.

reconstitute — To recombine dehydrated food products with water or other liquids.

reconstituted milk — A product resulting from the recombining of milk constituents with water and which complies with the standards for milkfat and solids-not-fat of milk.

record — (*computer science*) A set of related contiguous data in a computer file.

recorder controller — Equipment that records milk pasteurization temperatures and time and automatically controls the position of the flow diversion valve.

recoverable resource — The capacity and likelihood of a valuable substance, such as metal, being recovered from solid waste for a commercial or industrial use.

recovery plan — A plan developed by each state with the assistance of the federal agencies to restore affected areas in the event of an emergency or disaster.

recovery rate — (*solid waste*) The percentage of usable recycled materials removed from the total amount of municipal solid waste generated in a specific area or by a specific business.

recovery — Activities traditionally associated with providing federal supplemental disaster recovery assistance under a presidential major disaster declaration; types of emergency actions used to protect the public and to promote the resumption of normal activities.

recreational benefit — The value of recreational activity to a person, usually measured in dollars above the cost of participating in the recreational activity.

recreational vehicle — A vehicle that may or may not be motor propelled and is used primarily for recreational purposes; includes motor homes, travel trailers, vans, etc.

recreational vehicle park — Any tract of land used for parking five or more self-contained recreational vehicles, including any roadway, building, structure, or enclosure used or intended for use as part of the park facilities and any tract of land subdivided or leased for individual lots.

recruitment — Successive activation of motor units with increasing strength of voluntary muscle contractions.

rectification — The noun for rectify (to correct).

rectifier — An electrical device that converts alternating current to direct current.

rectify — To correct.

recuperation — The process of regaining a former state of health.

recurrence interval — The average amount of time between events of a given magnitude.

recurrence — The reappearance of a sign or symptom of the disease.

R

recycle — To separate waste material from the waste stream and process it so that it can be used again.

recycled water — Water that is used more than one time before being returned to the natural hydrological system.

RED — Acronym for reregistration eligibility decision document.

red blood cell — See erythrocyte.

red blood count (RBC) — The number of erythrocytes in a cubic millimeter of blood.

red bug — See chigger.

red corpuscles — Cells in blood containing hemoglobin.

Red Flower Beetle (*Tribolium castaneum*) — An elongate, reddish-brown flying beetle about 1/7 inch long that has a distinct joint between the thorax and abdomen, with the last three segments being abruptly enlarged; a stored food product insect.

Red-Legged Ham Beetle (*Necrobia rufipes*) — A shiny blue to green beetle, 1/7 to 1/4 inch long, with reddish legs; a stored food product insect especially problematic in the mid-Atlantic states.

red mite — See chigger.

red tide — A proliferation of reddish ocean plankton that may kill large numbers of fish; a natural phenomenon that may be stimulated by the addition of nutrients. Also known as red water.

red water — See red tide.

REDA — Acronym for Recycling Economic Development Advocate.

redevelopment — (*housing*) The elimination of underutilized areas and development of land for other purposes, such as businesses, industries, highways, or new housing.

redox — (*chemistry*) The processes of reduction (a gain of electrons in a chemical reaction) and oxidation (a loss of electrons in a chemical reaction).

reduced hemoglobin — Hemoglobin that has given up its oxygen.

reduced oxygen packaging — (*food*) The reduction of the amount of oxygen in a package by mechanical evacuation, by displacement with another gas or gases, or by controlling the oxygen level to keep it under 21%.

reducing agent — A substance that loses its valence electrons to another element and is readily oxidized.

reduction — A chemical reaction in which an element gains electrons and thereby decreases in valence; the addition of hydrogen or removal of oxygen in a substance; (*geology*) the extraction of a metal from its ore.

reduction control system — An emission control system that reduces emissions from sulfur-recovering plants by converting these emissions to hydrogen sulfide.

reduction potential — The inherent tendency of a compound to act as an electron donor or electron acceptor; measured in millivolts.

reductive sealing method — A method of determining the leakage of specific building components by pressurizing the building and recording the leakage changes as the components are successively sealed.

REE — Acronym for rare earth element.

reentrainment — The process of exhausting air from a building and immediately bringing it back into the system through air intakes and other openings in the building envelope so contaminants are reintroduced into the building air.

reentry interval — (*pesticides*) The period of time immediately following the application of a pesticide to an area during which unprotected workers should not enter the area.

refereed journal — A professional or literary journal for which experts in the field review articles and papers submitted for publication and then select those that are pertinent to the topic and scientifically accurate.

reference concentration — An estimate of the daily inhalation dose expressed in terms of an ambient concentration that can be taken over a lifetime without appreciable risk.

reference dose (RfD) — An estimate of the daily exposure of the human population to a potential hazard that is likely to be without risk of deleterious effects during a lifetime.

reference exposure concentration — An estimate of the daily exposure to the human population, including sensitive subgroups, that is likely to be without appreciable risk of health effects during a lifetime of exposure.

reference exposure level — The concentration at or below which no adverse health effects are anticipated for a specified exposure period.

reference method — A method of sampling and analyzing ambient air for an air pollutant that is specified in the Code of Federal Regulations.

referred pain — A pain felt at a site different from where the injury or disease has occurred.

refinery — A plant at which gasoline or other petroleum-derived products are produced.

refinery gas — Any form or mixture of gases gathered from equipment in a petroleum refinery.

ReFIT — Acronym for reinvention for innovative technologies.

reflectance — See reflectivity.

reflection — The return of a wave or ray after striking and bouncing off of a surface.

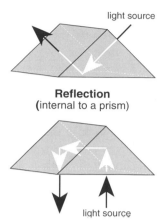

Reflection
(internal to a prism)

reflective spectrum — Portion of the optical spectrum from approximately 0.38 to 15 μm that defines the direct solar radiation used in remote sensing.

reflectivity — The ratio of intensity of the total radiation reflected from a surface to the total radiation incident on the surface.

reflector — A piece of equipment used to reflect and focus solar radiation.

reforming — The refinery process for improving gasoline quality by changing chemical characteristics rather than breaking up molecules, as in cracking.

reformulated gasoline — A fuel with a different composition from conventional gasoline (lower aromatics content) that reduces air pollutants.

refraction — The bending of light waves or rays as they pass at an angle from one material to another.

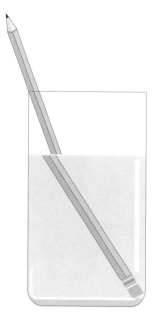

Refraction

refractive index — Ratio of the velocity of light rays in a vacuum to that of a specified medium; concentration and measurement of the bending of light rays by a refractometer.

refractometer — An instrument that measures the refraction of light rays.

refractory — A substance with a high melting point and good physical resistance to the effects of high temperature; resisting ordinary treatment.

refractory erosion — The wearing away of refractory surfaces by the washing action of moving liquids, such as molten slags, metals, or the action of moving gases.

refrigerant — A volatile substance used as a working fluid in a cooling system.

refrigeration — The reduction of temperature of substances to a level below the ambient temperature.

refuse — All putrescible and nonputrescible solid wastes, except sewage and body waste; includes garbage, rubbish, ashes, dead animals, discarded appliances and vehicles, trash, junk, and similar materials.

refuse burner — A device for either central or on-site volume reduction of solid waste by burning.

refuse chute — A pipe, duct, or trough through which solid waste is conveyed pneumatically or by gravity to a central storage area.

refuse container — A watertight container that is constructed of metal or other durable material, impervious to rodents, and capable of being serviced without creating unsanitary conditions; all openings to the container, including covers and doors, must be tight fitting.

refuse-derived fuel (RDF) — Energy produced by burning the remainder of processed solid wastes to remove metal, glass, and other unburned materials.

regeneration — See reactivation.

regenerative absorbent — A liquid that releases a gaseous pollutant due to the application of heat or steam.

regenerative process — The replacement of damaged cells by new cells.

regenerator — A piece of equipment used in milk processing in which heat exchange occurs while the pasteurized milk is always under positive pressure on one side of the plate, and the unpasteurized milk is always under negative pressure on the other side of the plate.

regime — A strictly regulated therapeutic program.

regional haze — The haze produced by a multitude of sources and activities that emit fine particles and their precursors across a broad geographic area; national regulations require states to develop plans to reduce the regional haze that impairs visibility in national parks and wilderness areas.

R

regional response center (RRC) — The physical location of a regional response team.

regional response team (RRT) — A group of official federal, regional, and state representatives who review the plans of the local emergency planning committee and offer necessary advice; also responds to large emergencies.

registered environmental health specialist (REHS) — A person who meets the educational and experience criteria and passes appropriate examinations to be qualified to do the work of an environmental health specialist.

registered pesticide — A pesticide chemical accepted by the U.S. Department of Agriculture for use as stated on the label of the container.

registered trademark — A registered name, symbol, or mark used to identify particular products and distinguish them from others.

registrant — (*pesticides*) Any manufacturer or formulator who obtains registration for a pesticide-active ingredient or product.

registration — The process by which a qualified individual is listed on an official roster maintained by a governmental or nongovernmental agency or board; registration may also require minimum practice standards.

registration standard — A document produced by the U.S. EPA detailing the various conditions that chemical registrants must meet in order to register pesticide products containing active ingredients.

registry — A system for collecting and maintaining in a structured record information on specific persons from a defined population.

regression — A return to previous conditions, signs, or symptoms.

regulated medical waste — Any solid waste generated in the diagnosis, treatment, or immunization of human beings or animals or produced during research or testing of biologicals, including cultures and stocks of infectious agents, human blood and blood products, human pathological body waste from surgery and autopsy, contaminated animal carcasses from medical research, waste from patients with commutable diseases, used sharp implements such as needles and scalpels, and certain unused sharps.

regulation — The legal mechanism for ensuring that the directives of a statute, law, or act are to be carried out; the process of holding constant a property or characteristic.

regulatory letter — A letter to the management of a firm warning of violation of a specific law or regulation and the statement that, unless immediate remedy is initiated, regulatory action will be taken by the governmental agency.

rehabilitation — (*housing*) Elimination of those properties that cannot be brought up to a standard and rebuilding of properties that can be brought up to standard, a less expensive alternative to redevelopment; (*health care*) the process by which physical sensory and mental functional capacities are restored or developed after damage.

reheat — Use of a heat exchanger or reintroduction of some flue gas downstream from a gas scrubber to raise the temperature of the gas to prevent condensation of water vapor in the stack.

REHS — See registered environmental health specialist.

rehydration — Restoring the normal water balance in a person by giving fluids orally or intravenously.

REI — Acronym for restricted entry interval.

Reid vapor pressure (RVP) — The absolute vapor pressure of volatile crude oil and volatile nonviscous petroleum liquids, except liquified petroleum gases.

reinfestation — The return of insects or other pests after extermination or migration.

REL — See recommended exposure limit or reference exposure level.

relabeling — The application of different labels to a product.

relapse — The return of signs and symptoms of a disease after improvement.

relapsing fever — (*disease*) A systemic spirochetal disease in which periods of fever, lasting 2 to 9 days, alternate with no fever for 2 to 4 days; relapses vary from 1 to 10 times or more. The louse-borne disease averages 13 to 16 days, and the tick-borne disease lasts longer. The incubation period 5 to 15 days, usually 8 days. The louse-born disease is caused by *Borrelia recurrentis*, a spirochete; the tick-borne disease is caused by various strains of the spirochete. It is found in the epidemic state when spread by lice and the endemic state when spread by ticks. The tick-borne disease may be found in the United States. The reservoir of infection for the louse-borne disease is people. Reservoirs of

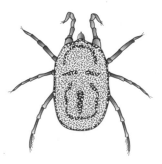

**Relapsing Fever Tick
(*Ornithodorus*)**

infection for the tick-borne disease include rodents and ticks in transovarian transmission; it is not transmitted from person to person. The epidemic relapsing fever occurs when an infective louse is crushed, thus allowing the organisms to enter the wound; communicability for the louse is 4 to 5 days after ingestion of blood from an infected person, and remains so for the life of an infected tick; general susceptibility. It is controlled through the control of lice and ticks.

relational database — A method of structuring data in the form of sets of records so relations between different entities and attributes can be used for data access and transformation.

relative atomic mass — See atomic weight.

relative biologic effectiveness (RBE) — A term used in radiobiology that refers to the relative effectiveness of a given kind of ionizing radiation and producing a particular biological response, as compared with 250-keV x-rays or gamma rays; for any living organism or part of the organism, the RBE is the ratio of the absorbed dose of a reference radiation that produces a specified biological effect to the absorbed dose of the radiation of the substance under evaluation that produces the same biological effect.

relative density — See specific gravity.

relative humidity — A measure of the water vapor content of the atmosphere calculated as the ratio of the vapor pressure in the atmosphere to the saturation vapor pressure at the existing temperature.

relative molecular mass — See molecular weight.

relative risk — A measure of the association between exposure to a factor and occurrence of a disease expressed as a ratio of the incidence rate in exposed persons to the incidence rate in unexposed persons.

relative ventilation efficiency — A quantity describing how the ventilation ability of a system varies between different parts of a room.

release — A controlled or uncontrolled escape of chemical agents into the environment by spilling, leaking, pumping, pouring, emitting, emptying, discharging, injecting, escaping, leaching, dumping, or disposing of a pollutant into the environment; excludes release into the workplace, emissions from engine exhaust, and release from a nuclear incident.

release mechanism — The means by which a hazardous material enters a transporting medium.

release source — Any hazardous material or substance contaminated with hazardous material that can release the contaminant in any medium.

reliability — A measure of the consistency of a method in producing the same results with each trial.

relief valve — A device that vents a sterilizer in the event that pressure exceeds the designed pressure of the chamber.

relocation — Temporary or permanent removal of a population or community in response to an emergency or disaster; to vacate a contaminated area to avoid chronic exposure.

rem — A unit of absorbed energy that takes into account the varying relative biological effectiveness of different types of radiation; a unit of dose equivalent that is equal to the absorbed dose in rads multiplied by the quality factor, distribution factor, and any other necessary modifying factors. See also sievert.

REM/FIT — Acronym for remedial/field investigation team.

remedial actions — Actions taken to restore a contaminated site to its pre-contaminated condition.

remedial investigation — The Comprehensive Environmental Response, Compensation, and Liability Act process of gathering the data necessary to determine the nature and extent of contamination at a site, establishing criteria for cleanup, identifying preliminary alternatives for remedial action, and supporting the technical and cost analyses of the alternatives.

remediation — (*law*) The act or process of recovering a right or preventing or readdressing a wrong.

remediation level — The concentration of a contaminant and applicable controls that are used to protect human health and the environment.

remission — Disappearance of the signs and symptoms of a disease.

remote optical sensing of emissions (ROSE) — A mobile spectrometer system using long-path, high-resolution infrared wavelengths to analyze gaseous air pollutants.

remote sensing — Quantitative or qualitative determination of air pollutants or of meteorological parameters by means of instruments not in physical contact with the sample being examined.

remote sensing air sampler — A sampler used to measure pollution concentration by means of a sensor that detects the absorbed or scattered electrooptical beam as affected by pollutants at a distance.

remotely sensed data — Data collected from a distance.

removal actions — Immediate short-term response actions for cleanup and removal of hazardous materials, assessment of the release, and actions to protect the public, such as temporary relocation.

REMS — Acronym for RCRA Enforcement Management System.

renal — Referring to the kidney.

renal anuria — A cessation of urine production due to disease or injury.

R

rendering — The process of recovering fatty substances from animal parts by heat treatment, extraction, and distillation.

renewable energy — Energy produced from materials that can be regenerated or are virtually inexhaustible, including biomass, solor radiation, wind, water, or heat from the interior of the Earth.

reovirus — A group of animal viruses spread by mosquitos and ticks and causing fevers; also known as orbivirus.

REP — Acronym for Reasonable Efforts Program.

rep — See roentgen equivalent physical.

repair — Specific and limited intervention for corrective maintenance after a failure.

repellent — A pesticide chemical or agent that drives or keeps insects or other pests away from a plant, animal, product, or treated area.

replication — The act or process of duplicating or reproducing.

report of disease — An official report, usually by a doctor, of the occurrence of a communicable or other disease of humans or animals to departments of health, agriculture, or both as locally required; reports include those diseases requiring epidemiological investigation or initiation of special control measures.

reportable quantity (RQ) — Quantity of a hazardous substance that is considered reportable under the Comprehensive Environmental Response, Compensation, and Liability Act.

repository — Any system licensed by the Nuclear Regulatory Commission and is intended to be used for or may be used for the permanent deep geologic disposal of high-level radioactive waste and spent nuclear fuel.

representative sample — A large sample of a universe or whole which can be expected to exhibit the average properties of that universe or whole.

repression — The process by which the synthesis of an enzyme is inhibited by the presence of an external substance.

reprocessing — The action of changing the condition of a scrap material, such as turning scrap aluminum into chairs and other products.

reproductive effect — Toxic effect of a substance that is evident in the second or third generation of the organism.

reproductive effects — Changes that may occur during the reproductive process, including mutagenesis, teratogenesis, diminished fertility, death, growth retardation, functional disorders, and prematurity or death of offspring.

reproductive toxicant — An agent that affects postpubertal reproductive or sexual function.

reproductive toxicity — The occurrence of adverse effects on the reproductive system that may result from exposure to a chemical.

reproductive toxin — A substance that affects the reproductive system and may impair fertility or cause chromosomal damage to a fetus.

repulsion — A force that separates two bodies or things.

request for proposal — A notice issued by a funding agency for proposals to conduct specific research or carry out specific projects.

reregistration — (*pesticides*) The reevaluation and relicensing of existing pesticidal active ingredients originally registered prior to current scientific and regulatory standards.

res — Latin for "things"; includes real and personal property.

res judicata — Matters that have been decided in court and are not to be decided again. In food, for example, a product may be seized for misbranding and adjudicated by order of the court; if the same article is again found to be misbranded, a rejudgment can be obtained without a lawsuit under this rule.

resazurin test — A test in which resazurin dye is added to fresh milk. Reduction of the dye is indicated by a gradual change through various shades of purple and mauve to fading pink, and the longer the time required to reduce the dye, the better the quality of the milk; similar to the methylene blue test.

rescission — Enacted legislation canceling budget authority previously provided by the legislative authority.

research — A studious inquiry or examination of a specific subject using investigative and experimentation techniques to reaffirm, discover, and interpret facts; revise theories or develop new theories based on the new facts; or find practical applications for the information.

research and development (R&D) — Studious inquiry, examination, investigation, and experimentation aimed at the discovery of new information, products, and their practical implementation.

research instrument — A testing device such as a questionnaire, evaluation form, interview technique, or a set of guidelines for observations of specific types of information, used to gather data and determine the level of accuracy by means of statistical tools.

research measurement — An evaluation of the quantity or incidence of a given variable obtained through the use of research tools.

reservoir — Any human being, animal, arthropod, plant, or inanimate matter in which an infectious agent usually lives, also known as reservoir of infection; a body of water impounded in a dam in which it can be stored.

R

reservoir of infection — See reservoir.

resident bacteria — Bacteria living in a specific area of the body.

residential burner — A device used to burn the solid waste generated in an individual dwelling.

residential care facility — A facility that provides custodial care for people unable to live independently because of physical, mental, or emotional problems.

residential drain — Horizontal piping in a house drainage system that receives the discharge from soil waste and drainage pipes inside the walls of the house and conveys it to the residential sewer.

residential sewage disposal system — All equipment and devices necessary for proper conduction, collection, storage, treatment, and on-site disposal of sewage from a one- or two-family dwelling.

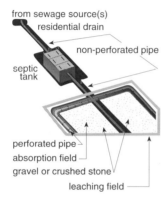

from sewage source(s)
residential drain
non-perforated pipe
septic tank
perforated pipe
absorption field
gravel or crushed stone
leaching field

**Residential Sewage
Disposal System**

residential sewer — Horizontal piping beginning 2 feet outside of a house that carries discharges from the residential drain to the sanitary sewerage system or a residential sewage disposal system.

residential solid waste — All solid waste that normally originates in a residential environment.

residential waste — A combination of garbage, refuse, ashes, and bulky materials coming from residences.

residual — The portion of a substance that remains after the bulk of the material has been removed.

residual chlorine — Chlorine remaining in water, sewage, or industrial waste at the end of a specified contact period and which will react chemically and biologically with materials present.

residual contamination standards — The amount and concentration of contaminants in soil, water, and other media that will remain following environmental management activities.

residual hazard — The hazards that remain after corrective actions are taken.

residual oil — The heavy oil left after refining processes have separated out most of the lower molecular weight compounds in crude oil.

residual pesticide — The amount of a pesticide chemical still effective for more than a few hours after application.

residual risk — The quantity of health risk remaining after application of emission controls.

residual volume (RV) — The amount of air remaining in the lungs after maximum expiratory effort.

residual waste — The solid, liquid, or sludge waste from human activities in urban, agricultural, mining, and industrial environments remaining after collection and necessary treatment.

residue — Material that remains after gases, liquids, or solids have been removed; (*pesticides*) the amount of a pesticide chemical that remains after it has been applied to a plant, animal, product, or area.

residuum — Residue from the weathering of sedimentary rock.

resilience — The ability of any system to resist or recover from any stress or hardship.

resin — A solid or semisolid, amorphous, organic compound or mixture of such compounds that is soluble only in organic solvents, has no definite melting point, and has no tendency to crystallize. It may be of vegetable, animal, or synthetic origin and has distinctive physical and chemical properties; it may be molded, cast, or extruded and used in adhesives or protective coatings.

resistance — (*medicine*) The sum total of body mechanisms placing barriers to the progress or invasion of pathogenic organisms; (R) (*electricity*) the ratio of voltage across an object to the current flowing through it; the ability of all conductors of electricity to resist the flow of current, turning some of it into heat. The smaller the cross-section of a conductor, the greater the resistance; the hotter the cross-section of the conductor, the greater the resistance.

resistance value — A metric measurement of the ability of a material to resist heat transfer; the R-value.

resistance welding — A process that produces coalescence of metals by the application of pressure and with the heat obtained from resistance of the work to electric current by circuit that includes the work.

resistant — Able to survive in conditions that are harmful to other organisms through existing immunity or built-in immunity.

resistant species or strain — A type of plant, animal, or microorganism that has developed a tolerance for a pesticide chemical.

resolution — The smallest spacing between two display elements.

R

resonance — Oscillation of molecules between two or more structures, each possessing identical atoms but different arrangements of electrons; a marked increase occurring in the amplitude of oscillation of a system when a system is subjected to an oscillating force whose frequency is the same as or very close to the natural frequency of the system.

resorb — To absorb again.

resorbent — A substance that is used to absorb blood or other substances.

resorcinol ($C_6H_6O_2$) — A white crystalline solid turning pink when exposed to air or light and having a sweet taste; MW: 110.12; BP: 281°C; Sol: soluble in water, alcohol, and ether; Fl.P: a noncombustible solid, the vapors of which can form explosive mixtures with air at high temperatures; density: NA. It is used in the manufacture of resin adhesives, dyes, drugs, explosives, and cosmetics. It is hazardous to the skin, eyes, digestive tract, and respiratory tract. It may cause cyanosis, convulsions, irritation to the mucous membranes, and strong irritation to the skin and eyes. ACGIH exposure limit (TLV-TWA): 10 ppm.

Resorcinol

resource conservation — Reduction of the amounts of solid waste generated by reduction of overall resource consumption and utilization of recovered resources.

Resource Conservation and Recovery Act (RCRA) — A federal act passed in 1976 and updated through 1988 authorizing the U.S. EPA to identify hazardous wastes and regulate their generation, transportation, treatment, storage, and disposal.

resource recovery — The recovery of materials or energy from waste; the extraction of useful materials or energy from solid waste.

resp — See respiratory route.

respirable — Of a size small enough to be inhaled deeply into the lungs.

respirable dust monitor — A device using beta-attenuation by respirable dust where the aerosols are drawn through a cyclone to remove non-respirable dust and are impacted on a surface positioned between a beta source and a counter; attenuation of the beta-radiation is directly related to the amount of collector particulates.

respirable particle — A particle of a size smaller than 10.0 μm and most likely to be deposited in the pulmonary portion of the respiratory tract.

respirable particulate mass threshold limit values (RPM-TLVs) — Materials that are hazardous when deposited in the gas-exchange region.

respiration — The exchange of oxygen and carbon dioxide between the atmosphere and the body cells, including inspiration and expiration, diffusion of oxygen from the pulmonary alveoli to the blood and of carbon dioxide from the blood to the alveoli, and the transport of oxygen to and carbon dioxide from the cells; breathe.

respiration rate — The weight of oxygen utilized by the total weight of mixed liquor suspended solids in a given time.

respirator — A face mask used to filter out poisonous gases and dust particles from the air so that a person can breathe and work safely.

respiratory depressant — A substance that diminishes normal breathing functions.

respiratory depression — A slowing of respiration below 12 inspirations per minute; failing to provide full ventilation and perfusion of the lungs.

respiratory fibrotic agents — A special class of irritants, including chemical reagents and particulate materials, that damage the lungs and cause scar tissue formation, which lowers respiratory capacity.

respiratory insufficiency — A failure of the respiratory system to maintain adequate ventilation and perfusion of the lungs.

respiratory protection — Any of a group of devices that protect the respiratory system from exposure to airborne contaminants; usually a mask with a fitting to cover the nose and mouth.

respiratory quotient — The total exchange of oxygen for carbon dioxide in the human body expressed as a ratio of the volume of carbon dioxide produced to the volume of oxygen consumed over a specific period of time at specific conditions.

respiratory route (resp) — A route of entry into the body through the breathing process.

respiratory system — A system consisting of the nose, mouth, nasal passages, nasal pharynx, pharynx, larynx, trachea, bronchi, bronchioles, and muscles of respiration involved with the intake, exchange, and exhalation of oxygen in vertebrates.

respiratory toxicity — Ability of a chemical to sicken or kill an animal or human by entering the lungs or other breathing organs as dust particles, gases, fine droplets, smoke, or vapors.

respiratory toxin — A chemical, such as silica or asbestos, that irritates or damages the pulmonary tissue; symptoms include coughing, tightness in the chest, and shortness of breath.

respiratory tract infection — Any infection of the upper or lower respiratory tract.

respiratory tract — The organs and structures that carry out pulmonary ventilation of the body and the exchange of oxygen and carbon dioxide between

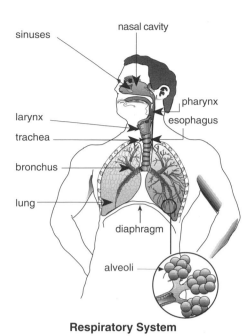

Respiratory System

the surrounding air and blood circulating through the lungs.

respirometer — An instrument used to analyze the quality of a person's respirations.

respondent — (*law*) One who answers a notice of a hearing.

response — Efforts to minimize the risks created in an emergency through protection of people, the environment, and property in an effort to return to normal pre-emergency conditions.

response indicator — An environmental indicator measured to provide evidence of the biological condition of a resource at the organism, population, community, or ecosystem level of organization.

response phase — The activities or immediate actions taken in response to an actual or potential chemical agent release, including elimination of the source of the release and taking life-saving measures for affected personnel, safety measures for potentially affected personnel, and security measures to preclude the exposure of additional personnel.

response time — (*computer science*) The time that elapses between sending a command to a computer and receiving the results at the work station.

resting spore — A cell in a period of reorganization and temporary dormancy; often initiated by adverse or changing environmental conditions.

restoration — Measures taken to return a site to precontamination conditions; (*chemical release*) the removal and decontamination of all chemical warfare agents, removal of any rubble, and emergency repair of structures and facilities.

restricted area — See restricted zone.

restricted use pesticide — A pesticide that is classified for restricted use under the provisions of the Federal Insecticide, Fungicide, and Rodenticide Act, such as fluoroacetamide (1081).

restricted zone — An area with controlled access and from which the population has been evacuated or relocated.

resuscitation — Restoring or sustaining the vital functions of a person in respiratory or cardiac failure using techniques of artificial respiration and cardiac massage.

resuscitator — A piece of equipment used to pump air into the lungs.

retail food processing establishment — Any food establishment or portion thereof undertaking any of the activities of such establishment for sale or delivery of retail, including grocery stores, meat or poultry markets, fish markets, fruit or vegetable markets, bakeries, confectionery stores, restaurants, delicatessens, and soft-drink stands.

retail food store — Any establishment or section of an establishment where food and food products are offered to the consumer and intended for off-premises consumption.

retardation — The retention of a contaminant in the subsurface resulting from adsorptive processes or solubility differences.

retch — Making an effort to vomit.

retention — Abnormal retaining of a fluid or secretion in a body cavity.

reticuloendothelial system — See hematopoietic system.

reticulum — A small network, especially a protoplasmic network, in cells.

retina — The inner, photoreceptive layer of the vertebrate eye formed from the expanded end of the optic nerve.

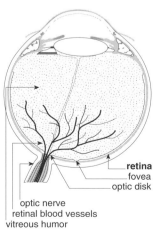

Retina

R

retirement center — A facility with organized programs to provide social services and activities for senior citizens.

retort — A vessel within which substances are subjected to distillation or undergo decomposition.

retort-type incinerator — A multiple-chamber incinerator in which the gases travel from the end of the ignition chamber and pass through the mixing and combustion chamber.

retract — To draw back.

retrofit — A modification to equipment or a facility added to existing equipment .

retrospective study — An epidemiological study that collects information about past events that may be related to the current distribution of the disease.

retrovirus — An RNA virus containing reverse transcriptase in the viron, where during replication the viral DNA becomes integrated into the DNA of the host cell, as occurs in the HIV virus, the causative agent of AIDS.

return — The reoccupation of areas cleared for unrestricted residence or use by previously evacuated populations.

return activated sludge — A settled activated sludge that is collected in secondary clarifiers and returned to an aeration basin to be mixed with incoming raw settled wastewater.

return date — (*law*) The expiration date of the interval between seizure by a U.S. Marshall and when a claimant may appear before the court. If no one appears as a claimant before expiration of the return date and proceedings, the U.S. Attorney will secure a default decree of condemnation, forfeiting and disposing of the article seized.

return of service — (*law*) Proof of service to the court by a person serving the process.

reuse — The reintroduction of a waste material or product into the economic stream without any chemical or physical change.

revenue bond — A bond issued by a public agency authorized to build, acquire, or improve a property and payable out of revenue from that property.

reverberation — Continuation of a sound in an enclosed area resulting from multiple reflections after the sound source has ended.

reverberation time — The time that would be required for the mean-square sound pressure level originally in a steady state to fall 60 decibels after the source is stopped.

reversal — (*law*) The act of reversing a court order or verdict; an appellate court may reverse a decision rendered by a district court.

reverse isolation — An isolation procedure used to protect a vulnerable patient from microorganisms carried by staff, patients, visitors, equipment, materials, and the surrounding environment.

reverse osmosis — A membrane process in which water is forced to flow from a solution of high salt concentration to one of lower concentration.

reverse transcriptase — An enzyme that is present in the viron of retroviruses occurring in leukoviruses and RNA tumor viruses.

reversible effect — An effect that is not permanent.

reversible reaction — Any reaction that reaches an equilibrium or which can be made to proceed in either direction.

reversible toxicity — Toxic effects that can be repaired, usually by the ability of a specific tissue to regenerate or correct itself after chemical exposure.

revocation of probation — Return to the original penalty imposed on a person convicted of a criminal charge because the person violated the law during the period of probation.

revolution — (*geology*) Motion of the Earth around the Sun in a 600-million-mile orbit; the motion of a body around a closed orbit.

revolutions per minute (rpm) — The number of revolutions of an object per minute of time.

Reye's syndrome — An often fatal encephalopathy of childhood with symptoms of fever, vomiting, fatty infiltration of the liver, and swelling of the kidneys and brain.

Reynolds number (NRe) — A dimensionless parameter computed by dividing the product of pipe diameter, average velocity, and fluid density by the fluid viscosity; the ratio of inertial and viscous forces in a fluid defined by the formula $Re = rVD/\mu$, where r is the density of the fluid, V is the velocity, D is the inside diameter of the pipe, and μ is the viscosity.

rf — See radiofrequency.

RF — Acronym for response factor.

RF energy — Notation for radiofrequency energy.

RFA — Acronym for Regulatory Flexibility Act.

RFB — Acronym for request for bid.

RfC — Acronym for reference concentration; see reference exposure concentration.

RfD — See reference dose.

RfDs — Acronym for subchronic reference dose.

RFI — acronym for remedial field investigation; see radiofrequency interference.

RFP — See request for proposal.

RH — Acronym for relative humidity.

rheostat — A resistor for regulating a current by means of varible resistance.

rheumatoid arthritis — A chronic systemic disease with inflammatory changes occurring throughout the body's connective tissues.

normal configuration of hand

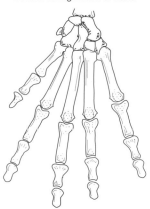

Rheumatoid Arthritis

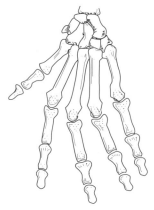

some results of deformity and disintegration of joints and bones

rhinitis — An inflammation of the mucous membrane of the nose.

rhinorrhea — The free discharge of a thick nasal mucous.

rhinovirus — A small RNA virus, one of about 100 serologically distinct types, that causes about 40% of accute respiratory illnesses; symptoms include a dry scratchy throat, nasal congestion, malaise, and headache.

rhizobacteria — Bacteria that aggressively colonize roots.

rhizofiltration — The uptake of contaminants by the roots of plants immersed in water.

rhizoplane — The surface of plant roots.

rhizosphere — Soil in the area surrounding a plant that is influenced by the plant root.

Rhodesian trypanosomiasis — An acute form of African trypanosomiasis caused by the parasite *Trypanosoma brucei rhodesiense*, which causes a disease that may progress rapidly to encephalitis, coma, and then death in only a few weeks.

rhodium (Rh) metal fume and insoluble compounds — A white, hard, ductile, malleable metal with a bluish-gray luster; MW: 102.9, BP: 6741°F, Sol: insoluble, sp. gr. 12.41. It is used in the manufacture of platinum alloys for use in thermocouples, windings for resistance furnaces, laboratory crucibles, and catalysts for chemical processing; in the preparation of dental castings; in applying thin reflective coatings by sublimation to scientific instruments, radio and radar equipment, searchlight reflectors, cinema projectors, and headlight reflectors. No hazards to humans are known; toxic by inhalation. No known symptoms of exposure. OSHA exposure limit (TWA): 0.1 mg/m^3 [air].

rhodium (Rh) soluble compounds — Appearance and odor vary depending upon the specific soluble compound; properties vary depending upon the specific soluble compound. They are used during refining and extraction of metal from platinum; in electroplating baths for finishing scientific instruments, camera fittings, radio equipment, and jewelry. They are hazardous to the eyes and are toxic through inhalation, ingestion, and contact. Symptoms of exposure in animals include mild eye irritation and central nervous system damage. OSHA exposure limit (TWA): 0.001 mg/m^3 [air].

RHRS — Acronym for Revised Hazard Ranking System.

RHS — Acronym for Rural Housing Service.

Rhydberg constant (R) — (*physics*) An atomic constant that appears in the formulas of wave numbers in all atomic spectra.

rhythm — A regular reoccurring quantitative change in a variable biological process.

Rhyzopertha dominica — See Lesser Grain Borer.

RI — Acronym for remedial investigation.

RI/FS — Acronym for remedial information/feasibility study.

RIA — Acronym for regulatory impact analysis or regulatory impact assessment; see radioimmunoassay.

ribonucleic acid (RNA) — A nucleic acid used for transportation and translation of genetic code found on DNA molecules.

ribose ($C_5H_{10}O_5$) — A five-carbon sugar present in ribonucleic acid.

R

ribosomes — Tiny, dense granules attached to the endoplasmic reticulum and lying between its folds; contain RNA and protein-synthesizing enzymes.

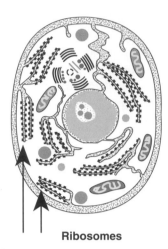

Ribosomes

Rice Weevil (*Sitophilus oryza*) — A small reddish-brown to brown snout beetle, 1/8 to 1/6 inch in length, with small brown pits on the thorax and two reddish or yellowish spots on each wing cover; a stored food product insect.

Richter scale — A logarithmic scale for expressing the magnitude of a seismic disturbance in terms of the energy dissipated; 4.5 on the scale is an earthquake causing slight damage, and 8.5 on the scale is a severe earthquake with considerable damage.

ricin — A white, toxic powder derived from castor beans.

Rickettsia — A group of organisms intermediate in size between viruses and bacteria and which cause disease.

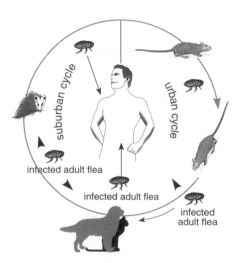

Urban and suburban life cycles
of ***Rickettsia*** and mammalian hosts

rickettsial pox — (*disease*) A rickettsial infection with a skin lesion occurring at the site of a bite of a mite followed by fever and skin rash; caused by *Rickettsia akari*.

Rift Valley fever — An arbovirus infection of Egypt and East Africa that is spread by mosquitoes or by handling infected sheep and/or cattle; symptoms include abrupt fever, chills, headache, and generalized aching followed by epigastric pain, anorexia, loss of taste, and photophobia.

right-of-way — A land area providing legal right of passage.

rim — An outer edge, which may be curved or circular.

RIM — Acronym for regulatory interpretation memorandum.

RIN — Acronym for regulatory identifier number.

ring of fire — The zone surrounding the Pacific Ocean where about 90% of the world's earthquakes occur.

ringbuoy — A ring-shaped floating buoy capable of supporting a drowning person.

Ringelmann chart — A series of charts numbered from 0 to 5 that simulate various smoke densities ranging from white to black; a measure of 1 is equal to 20% black, and a measure of 5 is equal to 100% black.

Ringelmann number — A rating from 0 to 5 on the Ringelmann chart.

RIP — Acronym for RCRA Implementation Plan.

riparian rights — Rights of a land owner to water on or bordering his or her property, including the right to prevent upstream water from being diverted or misused.

risk — A measure of the probability of an adverse or untoward outcome and the severity of the result and harm to the health of individuals in a defined population associated with a specific type of situation; (*emergencies*) the probability that damage to life, property, and/or the environment will occur if a hazard manifests itself.

risk assessment — The evaluation of short- and long-term risks by hazard identification, dose–response assessment, exposure assessment, and risk characterization; the process of integrating available information in the inherent toxicity of the chemical with information on how much of the chemical may come in contact with the individual.

risk assessment policy — The guidelines for value judgment and policy choices that may need to be applied to specific decision points in the risk assessment process and which are used to protect its scientific integrity.

risk characterization — Exposure assessment and the dose–response assessment combined to estimate some measure of the risk of toxicity and any uncertainties in the assessment of risk; a summary of the strengths and weaknesses of each component of the assessment is presented along with major assumptions, scientific judgment, and, to the extent possible, estimates of the uncertainties.

risk communication — The exchange of information about health or environmental risks among risk assessors, managers, news media, interest groups, and the general public.

risk estimate — A description of the probability that organisms exposed to a specific dose of a chemical will develop an adverse response.

risk factor — Characteristics, such as race, sex, age, weight, or variables, such as smoking or occupational exposure level, associated with increased probability of a toxic effect.

risk factor for carcinogens — The extra risk of getting cancer due to exposure to a particular level of a carcinogenic substance as determined by the U.S. EPA or the World Health Organization.

risk management — The development of regulatory options and evaluation of public health, economic, social, and political consequences of the regulatory options leading to agency decisions and actions.

risk perception — The magnitude of risk as perceived by an individual or population.

risk premium — The increased wages necessary to attract workers to riskier jobs.

risk profile — A description of a safety problem and its context to identify elements of a hazard or risk relevant to various risk management decisions, including prioritizing, setting policy, determining safety standards, and management choices.

risk–benefit analysis — An evaluation of the amount of potential problems created by a process or program compared to the promotion of well-being when a specific action is instituted in a community.

RISKIND — A computer code developed by the Department of Energy for analyzing radiological consequences and health risks to individuals and the collective population from exposures associated with the transportation of spent nuclear fuel.

river — A natural stream of water of large volume.

river basin — Land area drained by a river and its tributaries.

Rivers and Harbors Act — A law passed in 1899 prohibiting the illegal discharge or deposit of refuse or sewage into navigable U.S. waters to keep them free for boat traffic; industrial polluters can be subject to prosecution under this act.

RLL — Acronym for rapid and large leakage.

RMCL — Acronym for recommended maximum contaminant level.

RMDHS — Acronym for Regional Model Data Handling System.

RMIS — Acronym for Resources Management Information System.

RMP — Acronym for risk management program.

rms — See root mean square.

rms value — See root mean square value.

rmse/RMSE — Acronym for root mean square error.

RMSF — See Rocky Mountain spotted fever.

RNA — See ribonucleic acid.

RO — Symbol representing a hydrocarbon radical generated from an organic emission, such as formaldehyde or an aromatic derivative.

roach — A broad, flattened, dark- or light-brown or black insect; usually active only at night to seek food; also called a cockroach.

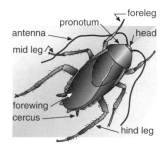

Roach

Robert T. Stafford Disaster Relief and Emergency Assistance Act, Public Law 93-2 88 — A law authorizing the federal government to respond to disasters and emergencies in order to help state and local governments save lives and to protect public health, safety, and property.

ROC — Acronym for record of communication.

rock — The hard, firm, and stable parts of the crust of the Earth.

rock cycle — The process whereby rocks are broken down by chemical and physical forces; moved by wind, water, and gravity; reformed into rock; and then crushed, folded, melted, and recrystallized into new forms.

rock fever — See brucellosis.

rocket — A piece of equipment, consisting essentially of a combustion chamber and an exhaust nozzle; it carries liquid or solid propellant, which provides the fuel and oxygen needed for combustion. It is used to propel a vehicle, missile, or bomb.

rocketsounde — An instrument carried aboard a rocket and used for measurement and transmission of upper air meterological data in the lower 250,000 feet of the atmosphere.

R

Rocky Mountain spotted fever — (*disease*) A rickettsial disease with sudden onset of moderate to high fever lasting for 2 to 3 weeks; symptoms include malaise, deep muscle pain, severe headache, chills, and rash. Incubation time is 3 to 14 days. It is caused by *Rickettsia rickettsii* and is found throughout the United States. Reservoirs of infection include nature and transmission of the organism by transovarial and trans-stadial passage. *Rickettsia* can be transmitted to dogs, rodents, and other animals by the bite of an infected tick; the tick remains infected for life. The disease is not directly transmitted from person to person; general susceptibility. It is controlled by avoidance of tick-infested areas.

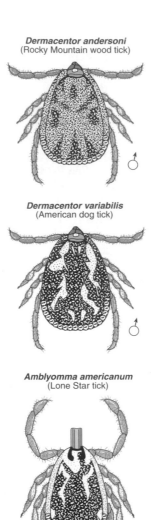

Dermacentor andersoni
(Rocky Mountain wood tick)

Dermacentor variabilis
(American dog tick)

Amblyomma americanum
(Lone Star tick)

Rocky Mountain Spotted Fever
(most common vectors)

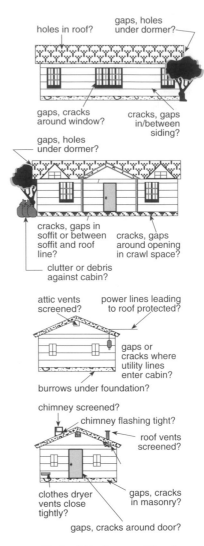

Rodent Harborage Check

rodent — A member of the order Rodentia; a relatively small gnawing mammal such as a mouse or rat.

rodent harborage — Any location where rodents can live, nest, or seek shelter.

rodenticide — A pesticide chemical or agent that prevents damage by or destroys rodents; a class of pesticides that kills rodents, especially rats and mice.

rodenticide poisoning — A toxic condition caused by the ingestion of a substance used for control of rodents.

rodentproof — A type of construction that prevents the ingress or egress of rodents from a given space or building or their access to food, water, or harborage.

RODS — Acronym for Records of Decision System.

roentgen (R) — The amount of ionization made by a beam of gamma- or x-radiation in air that produces 2.082×10^9 ion pairs per cm^3 of standard air; 1 roentgen of exposure will usually produce about 1 rad of absorbed energy in soft tissue. See also exposure unit.

roentgen equivalent physical (rep) — A measurement of ionizing radiation absorbed in human tissue.

roentgen ray — See x-ray.

ROG — Acronym for reactive organic cases.

ROM — Acronym for regional oxidant model; see range of motion or read-only memory.

ronnel ($(CH_3O)_2P(S)OC_6H_2Cl_3$) — A white to light-tan crystalline solid; a liquid above 95°F; MW: 321.6, BP: decomposes, Sol (77°F): 0.004%, sp. gr. 1.49 at 77°F. It is used as an insecticide and pest control agent for agricultural and livestock operations. It is hazardous to the skin, liver, kidneys, and blood plasma and is toxic through inhalation, ingestion, and contact. Symptoms of exposure in animals include cholinesterase inhibition, eye irritation, and liver and kidney damage. OSHA exposure limit (TWA): 10 mg/m³ [air].

R

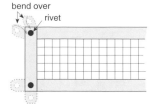

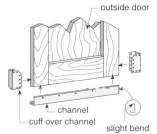

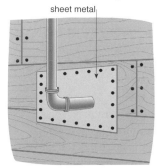

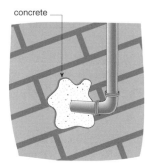

Rodent-proofing Devices

rooming house — Any dwelling or part thereof in which a specified portion may be used separately as a single habitable rooming unit.

rooming unit — Any room or group of rooms forming a single habitable unit used or intended to be used for living and sleeping, but not for dining and/or cooking.

room temperature — Air temperature measured in a specific part of a room.

root mean square — The square root of the sum of the squares of a set of quantities divided by the total number of quantities.

root mean square error — A determination made by calculating the deviations of points from their true position, summing up the measurements, and then taking the square root of the sum.

root mean square of uncertainty — The square root of the sum of the squares of the uncertainties.

root mean square value (rms value) — The square root of the arithmetic mean of the squares of a set of related values.

ROSA — Acronym for Regional Ozone Study Area.

ROSE — See remote optical sensing of emissions.

rosin — Any of a large class of synthetic products having some of the physical properties of natural resins but differing chemically and used primarily in plastics; any of various products made from a natural resin or a natural polymer.

rotameter — A flow-measuring device that operates at a constant pressure and consists of a weight or float in a tapered tube; as the flow rate increases, the float moves to a region of larger area and seeks a new position of equilibrium in the tapered tube related to flow rate.

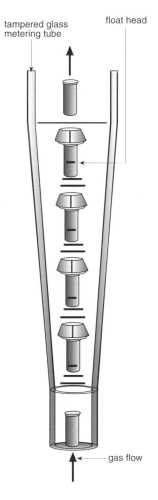

Rotameter

rotary kiln incinerator — An incinerator with a rotating combustion chamber that keeps waste moving, allowing it to vaporize for easier burning.

rotary kiln stoker — A cylindrical, slightly inclined kiln that rotates, causing the solid waste to move in a slow cascading motion forward.

rotary lime kiln — A kiln with an inclined, rotating drum used to produce a lime product from limestone by calcination.

rotary screen — An inclined, meshed cylinder that rotates on its axis and screens material placed at its upper end.

rotating biological contactor — An attached culture wastewater treatment system.

rotation — (*geology*) A complete turn of the Earth on its own axis once every 24 hours (approx.) at a speed of about 1000 miles per hour at the equator; motion about a fixed point.

rotavirus — (*disease*) A virus belonging to the family Reoviridae and containing spherical RNA 65 to 75 nm in diameter; it has a wheel-like appearance. Symptoms include gastroenteritis with vomiting followed by diarrhea, malaise, fever, abdominal pain, and dehydration. Death may occur from dehydration or aspiration of vomitus. The mode of transmission is person to person or possibly through contaminated food. Identical or similar viruses are found in various animals. The disease has been reported in the United States and is the most common cause of diarrhea in infants and young children. The reservoir of infection is humans and possibly animals; general susceptibility. Also known as infantile gastroenteritis.

rotenone ($C_{23}H_{22}O_6$) — A colorless to red, odorless, crystalline solid; MW: 394.4, BP: decomposes, Sol: insoluble, Fl.P: not known, sp. gr. 1.27. It is used as a pesticide, ascaricide, and insecticide on agricultural crops and livestock. It is hazardous to the central nervous system, eyes, and respiratory system and is toxic through inhalation, ingestion, and contact. Symptoms of exposure include eye irritation, numbness of the mucous membrane, nausea, vomiting, abdominal pain, muscle tremors, incoordination, clonic convulsions, stupor, pulmonary irritation, and skin irritation. OSHA exposure limit (TWA): 5 mg/m^3 [air].

rough fish — Species not prized for the purpose of fishing sports or for human consumption; more tolerant of changing environmental conditions than game fish.

roughing filter — A filter used to remove large particulate matter from a wastewater stream prior to treatment by means of ion exchange, adsorption, or other treatment techniques.

round off — To shorten a number to a specific decimal place; also known as truncating.

roundworm — A phylum of round, unsegmented nematodes, including trichina and hookworms.

Roundworm
(*Ascaris lumbricoides*)

route of entry — See route of exposure.

route of exposure — The avenue by which a microorganism or chemical comes in contact with an organism, such as by inhalation, ingestion, dermal contact, or injection; also known as route of entry.

RP — Acronym for respirable particle.

rpm — See revolutions per minute.

RPM-TLVs — See respirable particulate mass threshold limit values.

RQ — See reportable quantity or respiratory quotient.

RR — See respiration rate.

RRC — See regional response center.

Rrfd — Acronym for risk reference dose.

RRT — See regional response team.

RS — Acronym for registration standard.

RSD — Acronym for relative standard deviation; acronym for risk-specific dose.

RSE — Acronym for removal site evaluation.

r-strategy — An ecological strategy where organisms rely on high reproductive rates for continued survival within the community.

RT — See reverse transcriptase.

RTM — Acronym for Regional Transport Model.

RTP — Acronym for Research Triangle Park.

rub — To move one surface over another, causing friction.

rubbish — A general term for solid waste, excluding food waste and ashes taken from residences, commercial establishments, and institutions; nonputrescible solid waste, excluding ashes consisting of either combustible waste including paper, cardboard, plastic containers, vehicle tires, yard clippings, wood and similar materials or noncombustible waste including tin cans, glass crockery, and similar materials.

rubble — Broken pieces of masonry and concrete.

RUD — Acronym for reflectance unit density.

run — The net period during which an emission sample is collected.

runoff — The portion of precipitation or irrigation water that drains from an area as surface flow.

RUP — Acronym for restricted use pesticide.

rupture — A tear or break in the continuity of an organ or body tissue.

rural area — A predominantly agricultural, prairie, forest, range, or undeveloped land where the population is small.

rust — A type of corrosion in which iron is converted to hydrated Fe_2O_3 by the combined action of atmospheric oxygen and water.

RV — See residual volume.

R-value — A unit of thermal resistance use for comparing the insulation values of different materials; a measure of the effectiveness of insulation in stopping heat flow.

RVP — See Reid vapor pressure.

RW — Acronym for radiological/nuclear warfare.

RWC — Acronym for residential wood combustion.

R$_x$ — See prescription.

R

S

s — See second.

S — See siemen.

S&A — Acronym for sampling and analysis; see surveillance and analysis.

S&M — Acronym for surveillance and maintenance.

S&T — Acronym for science and technology.

SA — Acronum for specific activity or *Staphylococcus aureus*.

SAAQS — See Standard Ambient Air Quality Standards.

SAC — Acronym for suspended and canceled pesticides.

saccharomycosis — An infection with yeast fungi.

sacrificial anode — An easily corroded material deliberately installed in a pipe or intake to be corroded while the rest of the water supply facility remains relatively corrosion-free.

SADT — See structure analysis and design technique.

SAE viscosity number — A system established by the Society of Automotive Engineers for classifying crankcase oils, automotive transmission, and differential lubricants according to their viscosities.

SAED — Acronym for selected area electron defraction.

safe — A condition of being reasonably free from danger and hazards that may cause unintentional injury or disease.

Safe Drinking Water Act (SDWA) — A law passed in 1974 and updated through 1988 to ensure a safe drinking water supply by regulating water provided by public water systems and hazardous substances, including carcinogens, in drinking water. It directs the U.S. EPA to prescribe national drinking water standards, allows states to enforce requirements for water purification, establishes a system for emergency allocation of water purification, and provides protection for underground sources of drinking water.

safe temperature — (*food*) A temperature of 41°F or below and 140°F or above as applied to potentially hazardous foods.

safe water — Water that is free of harmful bacteria, toxic materials, or chemicals even if it has taste, odor, color, or other mineral problems.

safe yield — The annual quantity of water that can be taken from a source of supply, over a period of years, without depleting this source beyond its ability to be replenished naturally, in the wet years.

safety — The probability that harm will not occur under specified conditions.

safety assessment — The process of evaluating the level of safety of a chemical in the environment based upon its toxicity and current levels of human exposure.

safety assessment report — A formal summary of safety data collected during the design and development of a system, including risk assessment and recommendations to correct hazardous situations to reduce potential problems to an acceptable level.

safety belt — A device usually worn around the waist or as a harness that secures a person in a vehicle or a worker to a structure.

safety cabinet — Enclosure used for processing of biological materials.

safety can — An approved, closed container of not more than 5 gallons and having a flash-arresting screen, spring-closing lid, and spot cover; designed to safely relieve internal pressure when exposed to fire.

safety director — A specially trained and educated professional whose activities are related to safety functions, such as fire prevention, environmental safety, and disaster planning and coordination.

safety factor — A factor that is used to provide a margin of error when extrapolating from animal experimentation to estimate human risk.

safety glass — A transparent material that is prepared by laminating a sheet of transparent plastic between sheets of glass so that it resists shattering; especially used for car windows.

safety guard — An enclosure designed to restrain the pieces of a grinding wheel and furnish all possible protection in the event that the wheel is broken in operation.

safety management specialist — See occupational safety specialist.

safety objectives — Criteria for comparing and judging measures for adequacy.

safety shoe — A steel-toed shoe or boot that is waterproof and impervious to chemicals.

safety solvent — A solvent that has relatively low toxicity and low flammability, such as inhibited 1,1,1-trichloroethane.

SAGE — See Stratospheric Aerosol and Gas Experiment.

sagittal — The longitudinal median plane of a body or any plane parallel to it.

saline — The condition of containing dissolved or soluble salts.

saline water — Water that contains significant amounts of dissolved solids and has high salinity (10,000 to 35,000 ppm), moderate salinity (3000 to 10,000 ppm), and slight salinity (1000 to 3000 ppm).

salinity — The degree of dissolved salts in water measured by weight in parts per thousand.

salinization — The accumulation of salt in soil.

saliva — A fluid secreted into the mouth by the salivary glands to lubricate the passage of food and sometimes to carry out part of its digestion.

salivary gland — An exocrine gland in the mouth that secretes saliva, which keeps the mouth moist and aids in digestion.

Salmonella — A complex genus of Gram-negative, facultatively aerobic, usually motile, rod-shaped, pathogenic bacteria for humans and animals.

Salmonella enteritidis — A species of salmonella causing foodborne disease and gastroenteritis in humans.

salmonellosis — (*disease*) A bacterial disease with acute enterocolitis, sudden onset of headache, abdominal pain, diarrhea, nausea and sometimes vomiting, severe dehydration, and fever in infants; septicemia may occur. Incubation time is 6 to 72 hours, usually 12 to 36 hours. It is caused by varying serotypes of *Salmonella* that are pathogenic to animals and people. It is found worldwide in small or large outbreaks and is related to institutions, food, or water. Reservoirs of infection include domestic and wild animals, poultry, pigs, cattle, rodents, turtles, chickens, dogs, and cats, as well as people who may be carriers. It is transmitted by ingestion of the microorganisms in food contaminated by feces of infected animals or people; also may be transmitted through water that is contaminated. It is communicable throughout the time of the infection, which can be several days to several weeks, and the carrier state may last for more than a year; general susceptibility. It is controlled by proper cooking of foods, avoidance of raw foods, prevention of contamination of the foods by sick food handlers, and good personal hygiene.

salt — A usually crystalline compound containing positive ions from a base and negative ions from an acid in which the hydrogen of the acid has been replaced by metal or other positive ions; the minerals that water picks up as it passes through the air, over and under the ground, and from household and industrial use.

salt depletion — The loss of salt from the body through excessive elimination of body fluids by perspiration, diarrhea, vomiting, or urination.

salt effect — The effect on the activity coefficient due to salts in the solution.

saltcake — A cake of dry crystals of nuclear waste found in high-level waste tanks.

saltmarsh — A maritime community of salt-tolerant plants growing on intertidal mud and brackish conditions in sheltered estuaries and bays where excess sodium chloride is the primary environmental feature.

salvage — The recovery and utilization of waste materials for reuse or refabrication.

saltwater intrusion — The invasion or displacement of fresh water or groundwater by saltwater because of its greater density.

SAM — Acronym for stratospheric aerosol measurement.

sample — A representative portion or specimen of an entity presented for inspection or analysis.

sample error — A random variation reflecting the inherent variability within a population being counted.

sample quantization limit — A limit of analysis of small quantities accounting for sample characteristics, sample preparation, and analytic adjustments.

sampler — A device used with or without flow measurement to obtain an adequate portion of water, air, or waste for analytical purposes.

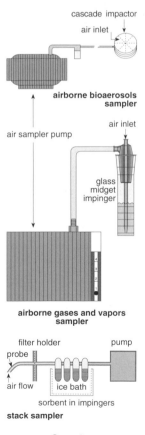

Sampler

sampling frequency — Interval between the collection of successive samples.

sand — Small rock or mineral fragments ranging from 0.05 to 2.0 mm in diameter and distinguishable by the naked eye.

sand filter — A water filter using sand or sand and gravel as a filter medium; a system that removes some suspended solids from sewage.

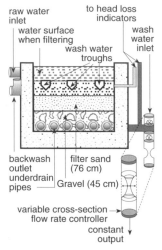

Sand Filter

sandstone — A sedimentary rock consisting of sand-size particles, usually a quartz.

sandstone aquifers — Geological formations found in many areas along with shale formations that supply a free-flowing spring and well water; shale serves as an aquiclude, in effect forming a conduit through which the water flows.

sandy loam — A soil that consists largely of sand but has sufficient quantities of silt and clay to impart stability.

SANE — Acronym for sulfur and nitrogen emissions.

sanitarian — See environmental health practitioner.

sanitary landfill — A site where solid waste is disposed of using sanitary landfilling techniques.

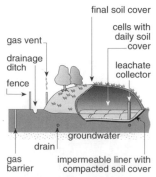

Sanitary Landfill

sanitary landfill liner — An impermeable barrier that is manufactured, constructed, or existing in a natural condition and is utilized to collect leachate.

sanitary landfilling — An engineered method of disposing of solid waste on land in a manner that protects the environment by spreading the waste in thin layers, compacting it to the smallest compactible volume, and covering it with soil by the end of each working day.

sanitary sewer — A sewer that carries sewage and to which storm, surface, and groundwaters are not intentionally admitted.

sanitary sewerage system — A sewer or a system of sewers that conveys sewage away from its origin to a wastewater treatment facility owned and operated by an incorporated city or town, conservation district, regional sewer district, or private utility.

sanitary survey — An on-site review of facilities, equipment, operation, and maintenance of water sources and sewage system problems.

sanitation — Control of physical factors in the human environment that can harm development, health, or survival.

sanitization — The act of reducing microbial organisms on cleaned food-contact surfaces to a safe level.

sanitize — To reduce the number of pathogenic microorganisms by physical and/or chemical means.

sanitizer — Any product used for the effective bactericidal treatment of clean surfaces of equipment and utensils by a process that provides sufficient accumulated heat or concentration of chemicals for a proper amount of time to reduce the bacterial count, including pathogens, to an acceptable level by public health standards.

SANSS — Acronym for Structure and Nomenclature Search System.

saponification — The hydrolysis of esters into acids and alcohols by the action of alkalis or acids, by boiling with water, or by the action of superheated steam.

saponin — Any of numerous plants producing toxins that hemolyze red blood cells.

saprophyte — An organism living on dead organic matter.

SAR — See safety assessment report, structure–activity relationship, supplied air respirator, specific energy absorption rate, sodium adsorption ratio, or synthetic aperture radar.

SARA — See Superfund Amendments and Reauthorization Act.

SARA Title III — Superfund and Amendments and Reauthorization Act of 1986; also known as the Emergency Planning and Community Right-To-Know Act of 1986.

sarcoma — A malignant tumor of mesenchymal derivation.

Sarcoptes scabiei — The itch mite that causes scabies.

sarin (C₄H₁₀FPO₂) — A colorless liquid; MW: 140.09, BP: 147°C, Sol: miscible with organic solvents and water, density: 1.10 at 20°C. It is used as a chemical warfare agent and is one of the most toxic compounds synthesized. It is hazardous to the skin, eyes, digestive tract, and respiratory tract. The poisoning effects come essentially from deactivation of cholinesterase. At an extremely low dose, symptoms of nausea, vomiting, muscle weakness, bronchiolar constriction, asthma, and death occur. ACGIH exposure limit: NA.

$$(CH_3)_2CH-O \underset{H_3C}{\overset{O}{\underset{|}{\overset{\|}{P}}}} -F$$

Sarin

SAS — Acronym for special analytical service or Statistical Analysis System.

SASS — Acronym for Source Assessment Sampling System.

SAT — Acronym for surface air temperature.

satellite image — A digitally recorded image of the surface of the Earth from a satellite in space; the images captured from measurements made in the visible and infrared spectrums create colors not necessarily true to real life.

satellite vehicle — A small collection vehicle that transfers its load into a larger vehicle operating in conjunction with it.

satellite zenith angle — The angle between the position of the satellite and the zenith, which is the point directly over the observed target.

saturated air — Air that contains as much moisture as it is able to hold under existing conditions.

saturated organic compound — An organic compound that cannot take additional hydrogen atoms into its structure; compounds that are less reactive than unsaturated organic compounds.

saturated solution — A solution containing the maximum equilibrium amount of solute at a given temperature; the dissolved substance is in equilibrium with the undissolved substance.

saturated vapor pressure — See vapor pressure.

saturated zone — See zone of saturation.

saturation index — See Langelier's index.

saturation point — The maximum concentration of water vapor the air can hold at a given temperature.

sauna bath — A bath in which hot water is used to induce sweating.

sausage — A highly seasoned mixture of ground meat enclosed in a tube casing of animal intestine.

savanna — An Earth biome characterized by an extensive cover of grasses with scattered trees and associated with climates having seasonal precipitation accompanied with a seasonal drought.

Saw-Toothed Grain Beetle (*Oryzaephilus surinamensis*) — A small, active, brown beetle, 1/10 inch long, with a flattened body and six saw-toothed projections on each side of the thorax; a major stored food product insect found throughout the world.

SBC — Acronym for single-breath canister.

SBS — See sick building syndrome.

SBUV — See solar backscatter ultraviolet radiometer.

scabies — (*disease*) An infectious disease of the skin caused by a mite that penetrates the skin and lays eggs. Incubation time is 2 to 6 weeks before onset of intense itching; if previously infested, 1 to 4 days after re-exposure. It is caused by the mite *Sarcoptes scabiei*. It is widespread, especially in areas with poor hygiene, during war, and during economic disturbances. The reservoir of infection is people; it is transmitted by direct skin-to-skin contact, clothes, or infected individuals and is communicable until the mites and eggs are destroyed, usually 1 to 2 weeks. Some individuals are resistant. It is controlled by isolating and treating infested individuals, their underclothing, clothing, and bed clothing.

scaffold — Any temporary elevated platform and structure used for supporting workers, materials, or both.

scald — A severe burn caused by exposure of skin to a hot liquid.

scale — Relationship between the size of an object on a map and its real size.

scan — To take pictures of organs in the body to be used in diagnosis, treatment, and monitoring.

scanner — A device for automatically converting images into digital form; (*geographic information systems*) used to convert images from maps, photographs, or views of the real world into digital form.

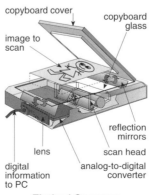

Flatbed Scanner

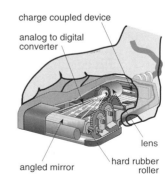

Hand-held Scanner

scanning — A method of studying an area, organ, or system of the body by use of specialized equipment that displays an image of the area.

scanning electron microscope — A microscope in which an electron beam, instead of light, forms a three-dimensional image for viewing, allowing much greater magnification and resolution.

scanning radiometer — An imaging system consisting of lenses, moving mirrors, and solid-state image sensors; used to obtain observations of the Earth and its atmosphere.

SCANSCAT — Acronym for scanning scatterometer.

SCAP — Acronym for Superfund Consolidated Accomplishments Plan.

scattered radiation — Radiation that, during its passage through a substance, has experienced a deviation in direction and may also have been modified by a decrease in energy.

scattering — The process by which electromagnetic radiation interacts with and is redirected by molecules of the atmosphere, ocean, or land surface.

scatterometer — A high-frequency radar instrument that transmits pulses of energy toward the ocean and measures the backscatter from the ocean surface.

scavenger — (*solid waste*) A person who participates in the uncontrolled removal of materials at any point in the solid waste stream; (*chemistry*) a substance added to a system to remove impurities.

scavenger cell — A phagocytic cell that removes tissue debris and some pathogens.

SCBA — See self-contained breathing apparatus.

Schick test — A skin test to determine immunity to diphtheria.

Schistosoma — A genus of blood flukes that may cause urinary, gastrointestinal, or liver disease in humans and requires freshwater snails as intermediate hosts.

schistosome — Any of a genus of elongated trematode worms having separate sexes and parasitizing the blood vessels of birds and mammals; also known as blood fluke.

schistosomiasis — (*disease*) A trematode infection of humans during which parasitic adult worms inhabit mesenteric or vesical veins of the host, resulting in symptoms of diarrhea, abdominal pain, urinary problems, liver fibrosis, hypertension, bacterial infection, and possibly bladder cancer. Incubation time is 2 to 6 weeks after exposure. It is caused by three major species of blood flukes: *Schistosoma mansoni*, *S. haematobium*, and *S. japonicum*. It is is found in Africa, South America, the Caribbean, the Middle East, and Asia. The reservoir of infection for *S. haematobium* and *S. mansoni* is humans. The reservoirs of infection for *S. japonicum* are humans, dogs, cats, pigs, cat-

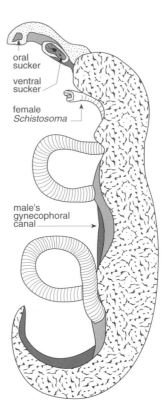

Schistosoma mansoni

tle, and horses. It is transmitted through water when free-swimming larval forms develop in freshwater snails and then leave the snails to penetrate the skin of mammals. It is not communicable from person to person, but an infected individual can excrete the organisms for many years; general susceptibility. It is controlled by the proper disposal of excreta, by destroying the snails, and by

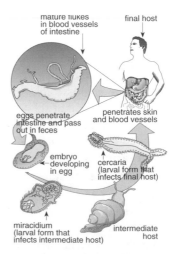

Schistosoma mansoni
**schistosomiasis infection
cycle**

S

wearing proper clothing to prevent the organism from penetrating the skin.

science — A system of knowledge covering general truths or the operation of general laws obtained through testing and observation.

scientific method — An orderly method used in scientific research generally consisting of identifying the problem, gathering data, formulating hypotheses, performing experiments, interpreting results, and reaching a conclusion.

scientific name — The accepted name used throughout the world by scientists for each animal and plant; it is made up of a genus and species name.

scientific notation — A short form of mathematical notation used by scientists in which a number is expressed as a decimal number between 1 and 10 multiplied by a power of 10; also known as standard notation.

scientific rationale — The reasons supported by scientific evidence for why a particular action has been chosen.

scintillation cocktail — An organic chemical solution that produces light when bombarded with radiation; a major component of institutional low-level waste.

scintillation counter — A counter in which light flashes produced on a scintillator by ionizing radiation are converted into electrical pulses by a photomultiplier tube.

scintillation detector — A detector operating on the principle of energy being transferred from radiation to a substance which in turn produces visible or near-visible light that may be picked up on a photosensitive vacuum tube and developed into an electrical pulse.

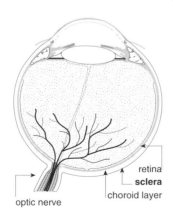

Sclera

sclera — The hard, white, outer coating of the eye.

sclerosis — Hardening of tissue.

scolex — The head-like segment or organ of an adult tapeworm that has hooks, grooves, or suckers and that allows the tapeworm to attach itself to the wall of the intestine.

Scombroid poisoning — (*disease*) Poisoning caused by eating scombroid fish or fish of the family Mahimahi; poisoning is due to a histamine-like substance produced by several species of *Proteus* bacteria or other bacteria.

scooter — A three-wheeled satellite vehicle equipped with a flat bed, dump box, or packer body used to gather solid waste.

scotomo — An area of lost or depressed vision within the visual field surrounded by an area of less depressed or normal vision.

scour — The erosion in a streambed, particularly if caused or increased by channel changes.

SCR — Acronym for selective catalytic reduction.

scrap — Any discarded or rejected material or parts of material that result from manufacturing or fabricating operations and may be recycled.

SCRC — Acronym for Superfund Community Relations Coordinator.

screen — (*solid waste*) A device for separating material according to size by passing it through variously sized mesh and retaining sorted materials on the surfaces.

screening — Checking for a disease when no symptoms are apparent.

screening level — Information about the toxicity of a chemical or its exposure potential, derived from readily available information, that does not require extensive analysis to support preliminary findings.

screening risk assessment — A risk assessment performed using available data to identify toxic chemical releases that have a higher probability of posing health risks.

scroll — (*computer science*) To move all or part of the screen material up or down, left or right.

scrub typhus — (*disease*) A rickettsial disease causing a skin ulcer from an infected mite. Symptoms include fever, headache, profuse sweating, red eruptions on the body, cough, and pneumonitis. Incubation time is 10 to 12 days, varying from 6 to 21 days. It is caused by *Rickettsia tsutsugamushi* and is found in Asia and Japan. The reservoir of infection is the infected larval stages of mites transmitted by the infected larval mite. It is not communicable from person to person; general susceptibility. It is controlled by preventing contact with infected mites and not using contaminated bedding and clothing. Also known as Tsutsugamushi disease.

scrubber — A device using a liquid spray to remove aerosols and gaseous pollutants from an airstream by making the particles heavier than the surround-

S

ing gas, aiding in separation of the pollutants; also known as a wet collector.

Type "D" scrubber

scrubbing — The washing of any impurities from a process gas stream.

S-curve — A curve that depicts logistic growth.

SCW — Acronym for supercritical water oxidation.

SD — See standard deviation.

SDC — Acronym for sample data collection.

SDK — Acronym for skin decontamination kit.

SDR — Acronym for standard dimension ratio.

SDTS — See Special Data Transfers Standard.

SDWA — See Safe Drinking Water Act.

SE — See seasonal efficiency.

SEA — Acronym for state enforcement agreement.

sea level — The datum against which land elevations and sea depth are measured, with mean sea level being the average of high and low tide.

sea surface temperature — The temperature of the layer of seawater (approximately 0.5 meters deep) nearest the atmosphere.

Sea-Viewing Wide Field-of-View Sensor — An ocean color sensor used to study ocean productivity and interactions between the ocean ecosystems and the atmosphere.

seafood — Marine and freshwater animal food products.

seal — Impermeable material, such as cement grout and bentonite, placed in the space between the bore-

hole wall and the casing of a water well to prevent the downward movement of surface water.

sealed — The condition of being free of cracks or other openings that permit the passage or entry of unwanted elements.

sealing — The complete filling of a well (with grout) in order to protect the aquifer from contamination.

SEAM — Acronym for surface, environmental, and mining.

search engine — A giant database on the Internet that stores data on Web sites.

seasonal — Fluctuations in environmental factors such as temperature over an annual cycle.

seasonal efficiency — A measure of the percentage of heat from the combustion of gas and from associated electrical equipment that is transferred to the space being heated during a year under specified conditions.

seasonal energy efficiency ratio — The total cooling output of a central air-conditioning unit (expressed in Btu) during its normal usage period for cooling divided by the total electrical energy input (expressed in Wh) during the same period as determined by using specified federal test procedures.

seatworm — See pinworm.

SeaWiFS — See Sea-Viewing Wide Field-of-View Sensor.

sebaceous — Oily or greasy; relating to a hair follicle that produces sebum.

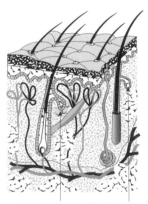

Sebaceous Oil Gland

sebum — Oily secretion of the sebaceous glands of the skin that contains keratin, fat, and cellular debris; when mixed with sweat, it forms a moist, oily, acidic film that is mildly antibacterial and antifungal and also protects the skin from drying out.

sec — See second.

sec-ft — See second-foot.

secator — A separating device that throws mixed material onto a rotating shaft where heavy, resilient materials are propelled to one side and light, elastic materials are propelled to the other side.

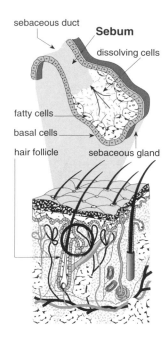

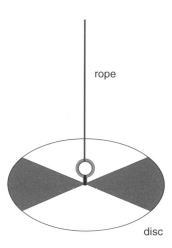

Secchi disc

Secchi depth — A measure of water clarity.

second (s, sec) — A 60th of a minute of time; the 60th part of a minute of angular measure.

second-foot — A shortened term for cubic foot per second.

second-hand smoke — Tobacco smoke inhaled by non-smokers.

second law of thermodynamics — Heat will flow of its own accord from a hot body to a cooler body; all physical processes, natural and technological, proceed in such a way that the availability of the energy involved decreases.

secondary air pollutant — Any pollutant produced by the reaction of two or more primary pollutants in the presence of catalysts and sunlight.

secondary attack rate — The number of cases of a disease developing in a stated period of time among members of a closed group at risk.

secondary burner — A burner installed in a secondary combustion chamber of an incinerator to maintain a minimum temperature and to complete the combustion of incompletely burned gases; also known as an afterburner.

secondary combustion air — See overfire air.

secondary combustion chamber — The chamber of an incinerator where combustible solids, vapors, and gases from the primary chamber are burned and where fly ash settles.

secondary health care — An intermediate level of health care given in a hospital which features specialized equipment and laboratory facilities for diagnosis and treatment.

secondary immune response — The immune response that follows a second exposure to a specific antigen.

secondary infection — Infection acquired by person-to-person transfer from a primary case or from subsequent secondary cases.

secondary material — Material utilized in place of a primary or raw material in manufacturing.

secondary maximum contaminant level (SMCL) — Maximum permissible level of contaminant allowed in a free-flowing public water supply where the water can affect public welfare.

secondary particles — Particles that are formed in the atmosphere and are products of the chemical reactions between gases such as nitrates, sulfur oxides, ammonia, and organic products.

secondary pollutant — A pollutant formed in the atmosphere by chemical changes taking place between primary pollutants and sometimes other substances found in the air, such as acid rain.

secondary radiation — Radiation that results from an absorption of other radiation in matter; may be either electromagnetic or particulate.

secondary sewage treatment — A system that uses primary sewage treatment followed by some type of biological oxidation, such as a trickling filter, activated sludge, or stabilization pond, to reduce the solids present and biochemical oxygen demand of the effluent before it enters a receiving stream.

secondary standard — A unit having a property that is calibrated against a primary standard to a known accuracy.

secondary treatment — The second step in most waste treatment systems in which bacteria consume the organic parts of the waste, which is accomplished by bringing the sewage and bacteria together in trickling filters or in the activated sludge process.

section — An area equal to 640 acres or 1 square mile.

secure landfill — A solid waste disposal site lined and capped with an impermeable barrier to prevent leakage or leaching; also includes drain tiles, sampling wells, and vent systems that provide monitoring and pollution control.

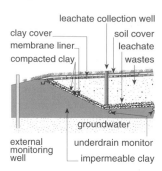

Secure Landfilll

sedative — A treatment used to calm, moderate, or tranquilize a person who is nervous or excited.

sediment — The fine particles of soil produced by weathering and which become suspended in water, air, or ice and finally settle.

sediment concentration — The quantity of sediment relative to the quantity of transporting fluid.

sediment discharge — The rate at which sediment passes a stream cross-section in a given period of time (expressed in millions of tons per day).

sediment load — The mass of sediment passing through a stream cross section in a specified period of time (expressed in millions of tons).

sedimentary rock — A rock formed as the result of the weathering of preexisting rocks, erosion, and deposition.

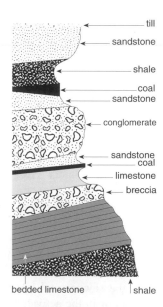

Sedimentary Rock

sedimentation — The deposition of solids by gravity during wastewater treatment.

sedimentation tanks — Tanks in which sewage flows slowly and solids settle to the bottom or float on the top as scum; the scum is skimmed off the top and solids on the bottom are pumped to digestors, with the effluent moving on for further treatment.

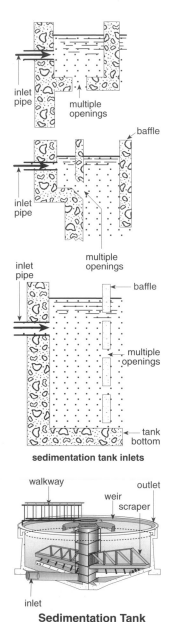

sedimentation tank inlets

Sedimentation Tank
(Circular)

SEDS — Acronym for State Energy Data System.

seeding — Inoculation with cultured organisms of a biological oxidation unit to help minimize the time required to build a biological sludge.

seep — A spot where groundwater oozes slowly to the surface, usually forming a pool.

seepage — The slow movement of water or gas through soil or other medium without forming definite channels.

SEER — See seasonal energy efficiency ratio; acronym for Surveillance, Epidemiology, and End Results.

segregation — See recondition; separation.

SEIA — Acronym for socioeconomic impact analysis.

Seiche wave — A wave generated by either a landslide into a reservoir or a sudden displacement or deformation of a fault line in a reservoir floor during a major earthquake.

seismic — Movement in the crust of the Earth caused by natural relief of rock stresses.

Seismic Wave

seismic intensity — The subjective measurement of the degree of shaking at a specified place by an experienced observer using a descriptive scale.

seizure — Confiscation by the U.S. Marshall of goods through a court order pending determination of the case.

seizures — Sudden involuntary movements of the muscles; also called convulsions.

selective absorption — The trapping of radiant energy from the sun by the carbon dioxide layer and the conversion of that energy into heat.

selectively permeable membrane — A membrane that is permeable to small molecules, usually water,

oxygen, and carbon dioxide, but not permeable to large molecules or ions.

selenium compounds (Se) — An amorphous or crystalline red to gray element that occurs as an impurity in most sulfide ores; MW: 79.0, BP: 1265°F, Sol: insoluble, Fl.P: not known, sp. gr. 4.28. It is used in the glassware industry for decolorization of fiberglass, scientific glassware, vehicular tail lights, traffic and other signal lenses, and infrared equipment; in the manufacture of electrical components; in the manufacture of photography and photocopy devices; in the manufacture of dyes, pigments, and colored glazes; in the manufacture of lubricating oils; in the rubber industry; in the manufacture of pharmaceuticals, fungicides, and dermatitis control; as a catalyst for hardening fats; in the manufacture of insecticides, parasiticides, bactericides, and herbicides; in the manufacture of flame-proofing agents for textiles and electric cables; in the manufacture of delayed-action blasting caps; as solvents in paint and varnish removers; in the refining of copper, silver, gold, or nickel ores; in miscellaneous operations in the manufacture of insect repellents, activators, hardeners, special ceramic materials, plasticizers, and mercury vapor detectors; for preparation of feed additives for poultry and swine. It is hazardous to the upper respiratory system, eyes, skin, liver, kidneys, and blood and is toxic through inhalation, absorption, ingestion, and contact. Symptoms of exposure include eye, nose, and throat irritation; visual disturbance; headache; chills; fever; dyspnea; bronchitis; metallic taste; garlic breath; gastrointestinal disturbances; dermatitis; and eye and skin burns. In animals, symptoms include anemia and liver and kidney damage. OSHA exposure limit (TWA): 0.2 mg/m³ [air].

selenium hexafluoride (SeF$_6$) — A colorless gas; MW: 193.0, BP: –30°F, Sol: insoluble. It is used as a gaseous electric insulator. It is toxic through inhalation and contact. Symptoms of exposure in animals include pulmonary irritation and edema. OSHA exposure limit (TWA): 0.05 ppm [air] or 0.4 mg/m³.

self-calibrating — Referring to an instrument or computer that is able to recalibrate automatically.

self-contained breathing apparatus (SCBA) — A portable respiratory protection device carried by the wearer that consists of a supply of air, oxygen, or oxygen-generating material and includes a mask and hood; a device that provides complete respiratory protection against toxic gases and oxygen deficiency.

self-contained recreational vehicle — Any recreational vehicle that can operate without connections to

S

sewer and water; the plumbing fixtures or appliances are connected to sewage holding tanks located within the vehicle.

self-insure — To assume liability for workers' compensation and avoid administrative costs associated with insurance policies.

self-purification — A process in which the stream proceeds by physical, chemical, and biological means to dispose of materials that may include suspended solids or other organic material.

self-supplied water — Water withdrawn from a surface or groundwater source by a user rather than being obtained from a public water supply.

SEM — Acronym for scanning electron microscopy, standard error of the means, or space environment monitor; see scanning electron microscope.

semiconductor — A material, such as silicon, that has an electrical resistance that is somewhere between that of conductors and insulators.

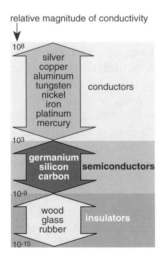

Semiconductor

semipermeable membrane — A membrane through which solvent molecules diffuse easily but through which dissolved substances diffuse slightly or not at all.

semivolatile organic compound — Organic compound that volatilizes slowly at a standard temperature of 20°C and 1 atmospheric pressure.

senescence — The aging process, usually characterized by the loss of some functional capacity.

senility — The general state of reduced mental and physical vigor associated with aging.

sensation — An impression produced by impulses conveyed by an afferent nerve to the sensorium.

sensation of sound — Sensation created when sound waves pass through the outer ear, which is shaped to collect the waves and channel them down the ear canal to the ear drum.

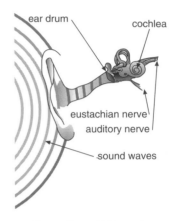

Sensation Of Sound

sense — A sensation, feeling, or mechanism of perception.

sense organ — A body organ sensitive to a particular kind of sensitivity such as vision, hearing, smell, taste, and touch.

sensible — Capable of being perceived by the sense organs.

sensible heat — The heat generated or absorbed by a substance during a change in temperature that is not accompanied by a change of state.

sensible heat transfer — Heat absorbed or evolved by a substance during a change of temperature that is not accompanied by a change of state.

sensing element — A device or component that measures the value of a variable.

sensitive ingredient — Any ingredient historically associated with a known microbiological hazard.

sensitivity — The ratio of the number of persons testing positive for a disease divided by the total number of persons tested; the minimum amount of contamination that can be repeatedly detected by an instrument.

sensitization — The process of making an individual sensitive to the action of a chemical.

sensitization dermatitis — An allergic reaction of the skin to certain chemicals after repeated exposure with production of antibodies that produce skin reactions in sensitized people.

sensitizer — A chemical that causes a substantial proportion of exposed people or animals to develop an allergic reaction in normal tissue after repeated exposure.

sensor — A device that measures a physical quantity or the change in a physical quantity, such as temperature, pressure, flow rate, pH, or liquid level; see also detector.

sensorium — The part of the cerebral cortex that receives and coordinates all the impulses sent to individual nerve centers.

sensory — Relating to the conveying of impulses from the sense organs, such as those of sight, hearing, smell, taste, touch, and pressure, to the nerve centers.

sensory deficit — A defect in the function of one or more of the senses.

sensory deprivation — An involuntary loss of awareness caused by the removal of external sensory stimuli that results in psychological disorders such as panic, mental confusion, depression, and hallucinations.

sensory nerve — A nerve composed only of the fibers of sensory neurons.

sensory neurons — Neurons that carry impulses from a receptor to the central nervous system.

sensory overload — A situation in which the central nervous system receives more stimuli than it can process effectively from sound, visual activities, or other environmental sources.

sensory pathway — The route followed by a sensory nerve impulse via an impulse from an end organ to a reflex center in the brain or spinal cord.

sentence — The formal judgment pronounced by the court or judge upon a defendant after conviction in a criminal prosecution.

SEP — Acronym for supplementary environmental project.

separation — The segregation of solid waste into designated categories as general as paper, metals, and glass or as specific as colors of glass. Also known as segregation.

sepsis — The bacteriological process of decay.

septage — The liquid content, including sludge and scum, from a septic tank periodically pumped out and transported to another site for disposal.

septic fever — A rise in body temperature occurring along with an infection by a pathogenic microorganism or in response to a toxin.

septic filter field use rating — Determination of the limitations of a soil to absorb water.

septic infarct — An infected portion of dead tissue.

septic shock — A form of shock accompanying septicemia following severe infection that is caused by decreased vascular resistance and results in a drastic fall of blood pressure due to the production of endotoxins or exotoxins, with symptoms of fever, tachycardia, increased respirations, and confusion or coma.

septic system — An individual sewer system employing a septic tank and soil treatment seepage trenches that are partially or totally in original soil material; also known as conventional septic system.

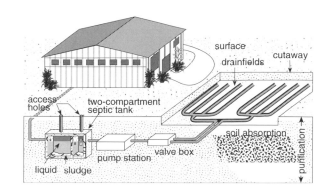

Large-capacity septic system

septic tank — A watertight underground receptacle that receives sewage and is designed and constructed to provide for sludge storage.

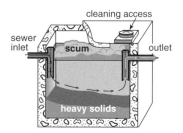

Septic Tank

septicemia — A disease condition due to the presence of pathogenic bacteria and their toxins in the blood; also known as blood poisoning.

septicemic plague — A rapidly fatal form of bubonic plague in which septicemia with meningitis occurs before the lymph nodes have an opportunity to become enlarged and inflamed.

septicity — The condition in which organic matter decomposes to form poor-smelling products associated with the absence of free oxygen.

S

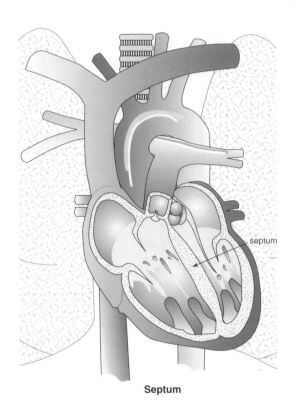

Septum

septum — A membrane between bodily spaces or masses of soft tissue.

sequela — A harmful consequence persisting after recovery from a disease.

sequential sampling — A process in which individual grab or integrated samples are collected, one after another, at regular, predetermined intervals, accumulating results until a determination is made.

sequestering agent — A substance that removes a metal ion from a solution system by forming a complex ion that does not have the chemical reactions of the ion that is removed.

sequestrants — Chelates used to deactivate undesirable properties of metal ions without the necessity for removing these ions from solutions.

sequestration — The process of surrounding or tying up metal atoms in large, ring-shaped molecules.

series — A chain of objects or events arranged in a predictable manner.

series circuit — An arrangement of circuit elements such that a single path is formed for the current.

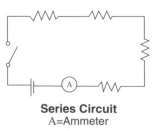

Series Circuit
A=Ammeter

serious injury — Any injury requiring an emergency service response or medical treatment as determined by emergency medical response personnel and/or resulting in medical attention at a hospital emergency room or admittance to a hospital.

serious level — The level of single pollutant combinations likely to lead to chronic disease or significant alteration of physiological function.

seroconversion — The process by which a person previously known to be antibody negative converts to testing positive for human immunodeficiency virus antibodies.

serologic test — Any of a group of tests performed on blood.

serological — Pertaining to serums and their study.

serology — A branch of laboratory science that studies blood serum for evidence of infection by evaluating antigen–antibody reactions.

seropositive — A condition in which antibodies to a disease-causing agent are found in the blood.

serotype — A group of related microorganisms distinguished by the composition of their antigens.

serous membrane — A thin sheet of tissue that lines a closed cavity of the body.

Serratia — A genus of motile, Gram-negative bacilli capable of causing infections in humans, including bacteremia, pneumonia, and urinary tract infections; frequently acquired in hospitals.

serum — A substance, usually an extract of blood, containing antibodies taken from an animal inoculated with bacteria or their toxins that is used to immunize people or animals; the clear liquid of the blood.

serum albumin — The principal blood protein necessary for absorption.

serum globulin — A globular blood protein containing antibodies.

serum hepatitis — (*disease*) See viral hepatitis B.

server — (*computer science*) A computer or software package that provides a specific service to client software running on other computers.

service — (*law*) Delivery of a legal document by an authorized person to a person who is thereby officially notified of some action or proceeding in which he or she is concerned and is thereby advised or warned of some action that he or she is commanded to take or forebear.

service connection — (*water*) The point at which a water system enters any structure, building, or dwelling.

SES — Acronym for secondary emissions standard.

sesquicarbonate — A salt with a composition that is between that of a carbonate and a bicarbonate.

sesquioxide — A compound of oxygen and a metal element in the proportion of three atoms of oxygen to two of the other; it is a nearly obsolete term.

SET — See standard effect of temperature.

seta — The erect, aerial part of the spore-producing structure of masses or liver warts; a bristle-like structure in annelid worms.

setpoint — The value on a controller scale at which the controller is set.

SETS — Acronym for Site Enforcement Tracking System.

settable solids — The matter in wastewater that will not stay in suspension during a preselected settling period but either settles to the bottom or floats to the top.

settlement — Sinking of the surface of a sanitary landfill or other engineered structure because of decomposition, consolidation, drainage, or underground failure.

settling chamber — A dry collector air pollution control device consisting of an enclosed compartment in which the velocity of the carrier gas is reduced sufficiently to allow the particles to settle by gravity.

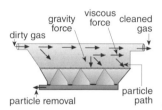

Settling Chamber

settling tank — A container that gravimetrically separates oil, grease, and dirt from petroleum solvent; a holding area for wastewater where heavier particles sink to the bottom and are siphoned off.

settling velocity — Velocity at which a given dust will fall out of a dust-laden gas under the influence of gravity only.

seven-day average — The arithmetic mean of pollutant parameter values for samples collected over seven consecutive days.

severe effects — The effects of nerve agents, including systemic effects such as vomiting, involuntary urination, and/or defecation, tremors, collapse, or convulsions.

severe limitations — Limitations that are difficult and usually costly to overcome or modify for subsurface seepage fields.

severe thunderstorm — A thunderstorm that produces tornadoes, hail that is 0.75 inches or more in diameter, or winds of 58 mph or more.

severity — The degree to which an effect changes and impairs the functional capacity of an organ system.

Sevin® — See carbaryl.

sewage — The water-carried human or animal wastes from residences, buildings, industrial establishments, or other places together with such groundwater infiltration and surface water as may be

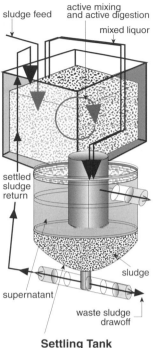

aerobic digester

Settling Tank

present. When fresh, it is gray in color with a musty odor; when old, it is black in color with a foul odor.

sewage disposal system — See sewage treatment system.

sewage lagoon — See oxidation pond.

sewage sludge — A semiliquid substance consisting of settled sewage solids combined with varying amounts of water and dissolved materials; it may be raw or fresh, digested, elutriated, dewatered, or dry.

sewage treatment residue — Coarse screening grit or sludge resulting from wastewater treatment units.

sewage treatment system — Any sewage handling or treatment facility receiving domestic sewage and having a ground surface discharge, or any sewage handling or treatment facility receiving domestic sewage and having no ground surface discharge. Also known as sewage disposal system.

sewage treatment works — Municipal or domestic waste treatment facilities of any type that are publicly owned or regulated.

sewer — An underground pipe or conduit used for carrying sewage.

sewer tile — Glazed waterproof clay pipe with bell joints.

sewerage — The entire system of sewage collection, treatment, and disposal.

sewerage system — Any community or individual system publicly or privately owned for the collection and disposal of sewage or industrial waste of a

S

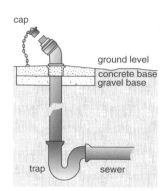

Sewer Connection

liquid nature, including treatment facilities for sewage or industrial waste.

SF — Acronym for standard form; see Superfund Act.

SF$_6$ — Acronym for sulfur hexafluoride, which is a physiologically inert gas used as a tracer in building investigations.

SFA — Acronym for spectral flame analyzer.

SFDS — Acronym for Sanitary Facility Data System.

shale — A soft, sedimentary rock formed by consolidated clay or silt.

shale oil — Any oil derived by retorting crushed oil-bearing rock and characterized by a large proportion of unsaturated hydrocarbons, alkenes, and dialkenes.

sharps — Hypodermic needles, syringes, pipettes, scalpel blades, blood vials, needles, culture dishes, broken or unbroken glassware contaminated with infectious agents, suture needles, slides, and coverslips used in animal or human patient care or treatment or in medical, research, or industrial laboratories.

shear — A strain resulting in a change in shape; variation in wind speed and/or direction over a short distance.

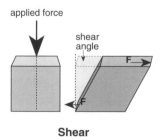

Shear

shear shredder — A size reduction machine that cuts material between two large blades or between a blade and a stationary ledge; used in solid waste disposal.

shear wave — See S-wave.

sheen — An iridescence on the surface of water caused by the addition of a dye that is used as a tracking agent to determine if the water is being contaminated by a particular source.

shell — A pattern of electrons surrounding a nucleus.

shellfish — A group of mollusks, including oysters, mussels, and clams, usually enclosed in a self-secreted shell or exoskeleton.

shellfish poisoning — A toxic neurological condition resulting from eating clams, oysters, or mussels that have ingested poisonous protozoa. Symptoms include nausea, lightheadedness, vomiting, and tingling or numbness around the mouth followed by paralysis of the extremities and possibly respiratory paralysis, occurring within a few minutes after consuming the causative agent, saxitoxin, which is not destroyed by cooking.

shellstock — Raw, in-shell molluscan shellfish.

shield volcano — See basaltic domes.

shielded metal arc welding — A process that produces coalescence of metals by heating them with an arc between a covered metal electrode and the work.

shielding — A control technique in which equipment or substances are physically kept away from people by barriers.

Shigella — A genus of Gram-negative pathogenic bacteria that causes gastroenteritis and bacterial dysentery.

shigellosis — (*disease*) An acute bacterial disease involving the large and small intestine with diarrhea, fever, nausea, sometimes toxemia, vomiting, and cramps; may cause convulsions in young children. Also, blood, mucous, and pus are present in the feces due to an enterotoxin produced by the bacteria. Incubation time is 1 to 7 days, usually 1 to 3 days. It is caused by four species of *Shigella*: *S. dysenteriae*, *S. flexneri*, *S. boydii*, and *S. sonnei*. It is found worldwide, with most cases occurring in individuals under 10 years of age, usually in overcrowded or poor environmental conditions, institutions for children, daycare centers, jails, mental hospitals, and ships. The reservoir of infection is people. It is transmitted directly or indirectly through the oral–fecal route from person to person; the carrier state may exist. It is also transmitted through water, food, flies, or a person's hands. It is communicable for at least 4 weeks and in the carrier state for many months; general susceptibility. It is controlled through proper supervision of institutions (especially children's centers) and food preparation, by using clean water and good personal hygiene practices, by control of feces and contaminated articles, and by proper sewage disposal.

shipper — The party whose name and address appear on a bill of lading or freight bill as the one who introduced an article into interstate commerce.

shock — A condition of acute peripheral circulatory failure due to derangement of circulatory control or loss of circulating fluid.

shock chlorination — The removal of organic waste through oxidation by adding significant quantities of a sanitizer; also called superchlorination or breakpoint coronation. Chlorine is added to disinfect swimming pools or water supply systems, including the wells and all distribution pipelines, when coliform bacteria are detected or after system repairs. The concentration of chlorine in the treated water should be at least 200 ppm for at least 24 hours.

shock load — The raw water arriving at a water treatment plant containing unusual amounts of algae, colloidal matter, color, suspended solids, turbidity, or other pollutants that will affect the normal treatment process.

shock trousers — A pair of pneumatic trousers designed to counteract hypotension associated with internal or external bleeding and hypovolemia.

Shore Protection Act — A 1988 law stating that a vessel may not transport municipal or commercial waste in coastal waters without a permit from the Secretary of the Department of Transportation and without displaying a number or marking on the vessel as prescribed by the secretary; the secretary has the right to enforce regulations concerning loading, securing, offloading, and clean-up.

short circuit — Reduction to zero resistance or impedance across a voltage source, usually resulting in damage if the circuit is not opened elsewhere; the condition occurring when the supply air flows to exhaust registers before entering the breathing zone.

short-term concentration (STC) — Concentration measurements derived from a combination of site monitoring and modeling information averaged over a relatively short period of time, usually 10 to 90 days; the information is used to evaluate potential subchronic effects.

short-term exposure — Multiple or continuous exposures occurring over approximately 1 week.

short-term exposure limit (STEL) — The maximum concentration to which workers can be exposed continually for up to 15 minutes; no more than 4 periods are allowed per day, and at least 60 minutes must expire between exposure periods.

short-term memory — Memory of recent events.

short-term test — Any test that can be completed in a short time that is used to examine genetic changes in laboratory cultures of cells in humans, animals, or lower organisms.

short ton — A measure of weight of a substance equal to 2000 pounds.

shortcuts — (*computer science*) Keystrokes that duplicate the commands available in pull-down menus of a program.

shredder — A machine that reduces discarded automobiles and other low-grade sheet and coated metal to small pieces in a continuous operation.

shrink–swell potential — The amount of shrinkage or swelling that will occur in a soil when wet or dry.

shucked shellfish — Shellfish that have been removed from their shells.

shute-fed incinerator — An incinerator that is charged through a shute that extends two or more floors above it; also spelled chute-fed incinerator.

SHWL — Acronym for seasonal high water level.

SI — Acronym for site inspection; see surface impoundment.

SI units — Système International d'Unités; see International System of Units.

SICEA — Acronym for Steel Industry Compliance Extension Act.

sick building syndrome (SBS) — A generalized term used to describe situations in which building occupants experience acute health and/or comfort effects that appear to be linked to time spent in a particular building, but where no specific illness or cause can be identified. Symptoms include acute discomfort; headaches; eye, nose, or throat irritation; dry cough; dry or itchy skin; dizziness; nausea; difficulty in concentrating; fatigue; and sensitivity to odors. Typically, the cause of the symptoms are not known, but the complainants report relief of the symptoms when they leave the building.

side effects — Symptoms that occur when treatment affects healthy cells.

SIDS — Acronym for screening information dataset; see sudden infant death syndrome.

siemen (S) — In the MKSA system of measurement, it is a unit of conductance; formerly known as mho (W).

sievert (sv) — A unit of radiation dose equal to approximately 8.38 roentgens; 1 sv = 1 joule/kg = 100 rem; formerly known as rem.

siftings — The fine materials that fall from a fuel bed through its grate openings during incineration.

sight — The process, power, or function of seeing shape, size, position, and color of objects.

sign — An objective finding as determined by a competent observer.

signal — An electrical transmission that carries information.

signal power — signal voltage divided by the source impedance.

signal processor — A processor located within the electronic system that consists of multiple voltage-to-

S

frequency converters that are responsible for converting an incoming signal from optical to digital.

signal word caution — A slightly toxic or relatively nontoxic pesticide.

signal word danger — A highly toxic pesticide.

signal word warning — A moderately toxic pesticide.

significance — The probability that a given finding is very unlikely to have occurred by chance alone.

significant deterioration — Describes the generation of pollution from a new source in a previously unpolluted area.

significant discharge — Any point-source discharge for which timely management action must be taken to meet water quality objectives within the period of the operative water quality management plan.

significant hazard to public health — Any level of contaminant that causes or may cause potable water to exceed the maximum contaminant level set forth in any promulgated National Primary Drinking Water Standard or which may otherwise adversely affect health or may require installation of additional public water system treatments to prevent such adverse effects.

significant health risk — Circumstances in which people are being or could be exposed to hazardous substances at levels that pose an urgent public health hazard; public health advisories are generally issued when urgent public health hazards have been identified.

significant impairment — Impairment of visibility that interferes with the management, protection, preservation, or enjoyment of a visual experience by visitors at a national park or forest.

significant risk — Describes a situation posing a moderate likelihood of causing an unacceptable health risk.

SIL — See speech interference level.

silent disease — See subclinical.

silane (SiH_4) — A colorless gas with a disagreeable odor; MW: 32.13, BP: NA, Sol: decomposes slowly in water, Fl.P: a flammable gas that ignites spontaneously in air and chlorine, density: 1.44 g/l at 20°C. It is used for preparing semiconducting silicone for the electronics industry. It is hazardous to the respiratory tract and can cause irritation. ACGIH exposure limit (TLV-TWA): 5 ppm.

silica (SiO_2) (amorphous) — A transparent to gray, odorless powder; MW: 60.1, BP: 4046°F, Sol: insoluble, sp. gr. 2.20. It is used in the manufacture of insulating materials; as a filter medium in food and beverage manufacture; in the manufacture of construction bonding materials; in the manufacture of abrasive cleaning and polishing agents; in the manufacture of surface coatings; as a filler agent in paints, lacquers, and varnishes; in the manufac-

ture of pharmaceuticals and as a constituent of pill masses, dentrifices, and salves; in pottery manufacture; in water treatment; as a carrier for nickel catalysts in the petroleum and petrochemical industries. It is hazardous to the respiratory system and is toxic through inhalation. Symptoms of exposure include pneumoconiosis. OSHA exposure limit (TWA): 6 mg/m³ [air].

silica (SiO_2) (crystalline, as respirable dust) — A colorless, odorless solid that is a component of many mineral dusts; MW: 60.1, BP: 4046°F, Sol: insoluble, sp. gr. 2.66. It is used in the metallurgy industry for foundry molds, iron and steel casting, flux in smelting basic ores; in the manufacture of fiberglass; in the ceramics industry; as an abrasive for scouring and polishing soaps and powders, flint sandpaper, metal polishes, and sandblast work; in the processing of synthetic quartz; in the manufacture of refractories and building products; in grading and classification of electronic and optical grade quartz; in the manufacture of optical equipment in prisms, wedges, and lenses; in a variety of processes, such as in dental composition, in rocket engines and spacecraft, as a paint extender, and in graining lithographic plates. It is hazardous to the respiratory system and is toxic through inhalation. Symptoms of exposure include cough, dyspnea, wheezing, impaired pulmonary function, and various progressive symptoms; carcinogenic. OSHA exposure limit (TWA): 0.1 to 0.05 mg/m³ [air].

silica gel — A regenerative absorbent consisting of the amorphous silica manufactured by the action of hydrochloric acid on sodium silicate; used as a dehumidifying and dehydrating agent.

silicate — The largest group of minerals for which the basic structure is a product of the joining of silicon, oxygen, and one or more metals; may contain hydrogen. The dust causes nonspecific dust reaction.

silicon (Si) — A nonmetallic element that is one of the primary constituents of the crust of the Earth.

silicon carbide (SiC) — A bluish-black refractory material that is very dense, resists abrasion, and has a high melting point.

silicon diode — A device made of a silicon compound in which current flows when exposed to ionizing radiation; the current is converted to electrical pulses and counted.

silicone — A group of compounds (SiR_2O) made by molecular combination of the element silicon or certain of its compounds with organic chemicals. It is produced in a variety of forms, including silicone fluids, resins, and rubber, and has special properties, such as water repellency, wide temperature resistance, and durability.

silicosis — A pneumoconiosis caused by the prolonged inhalation of silica dust, usually for 10 years or more. Symptoms include progressive shortness of breath and steady, dry, unproductive coughing in the early stages, later followed by mucous tinged with blood, loss of appetite, chest pain, and general weakness. The silica produces a nodular fibrotic reaction that scars the lungs and makes them receptive to the further complications of bronchitis, emphysema, and increased susceptibility to tuberculosis.

sill — (*hydraulics*) The top level of a weir or the lowest level of a notch; (*construction*) a beam or threshold.

silt — (*soils*) Soil particles in an intermediate size range between clay particles and sand grains, ranging from 0.002 to 0.05 mm in diameter; sediment carried or deposited by water.

silt loam — (*soils*) Soil having a moderate amount of fine grades of sand (0 to 50%), a small amount of clay (0 to 27%), and a large amount of silt particles (50 to 88%).

siltation — The deposition of finely divided soil and rock particles on the bottom of a stream or riverbed.

silver (Ag) (metal dust and soluble compounds) — A white, lustrous metal; MW: 107.9, BP: 3632°F, Sol: insoluble, sp. gr. 10.49. It is used in the manufacture of silver nitrate for use in photography, mirrors, plating, inks, dyes, and porcelain; as a germicide, antiseptic, caustic, and analytical reagent; in the manufacture of silver salts as catalysts; in chemical synthesis; in the manufacture of glass; in silver plating; in photography; in medicine. It is hazardous to the nasal septum, skin, and eyes and is toxic through inhalation, ingestion, and contact. Symptoms of exposure include skin irritation, ocular burns, ulceration, and gastrointestinal distress. OSHA exposure limit (TWA): 0.01 mg/m^3 [air].

silver acetylide (Ag$_2$C$_2$) — A white, precipitated solid; MW: 239.76, BP: explodes on heating, Sol: slightly soluble in alcohol and soluble in acid, Fl.P: it is extremely sensitive to shock and can explode at a slight touch when dry, density: NA. It is used only as a detonator. ACGIH exposure limit: NA.

$$Ag - C \equiv C - Ag$$

Silver acetylide

silver azide (AgN$_3$) — A white rhombic prism; MW: 149.80, BP: NA, Sol: insoluble in water, Fl.P: it is a primary explosive that explodes violently with thermal or mechanic shock, density: 5.1 g/cm^3. It is used as a primary explosive. It is a highly toxic substance. ACGIH exposure limit: NA.

silver fulminate (Ag$_2$C$_2$N$_2$O$_2$) — A white crystal with a needle shape; MW: 299.77, BP: explodes on heating, Sol: slightly soluble in water, Fl.P: it is extremely sensitive to heat and detonates, density: NA. It is used as a detonator. ACGIH exposure limit: NA.

$$Ag(C \equiv N \longrightarrow O)_2$$

Silver fulminate

silver nitrate — A topical antiinfective agent.

silvex (C$_9$H$_7$Cl$_3$O$_3$) — A crystalline odorless solid; MW: 269.51; BP: NA; Sol: soluble in acetone, methanol, and ether; density: NA. It was formerly used as an herbicide but has been prohibited by the EPA. It is moderately toxic to test animals and may cause embryo toxicity and fetal death.

$$
\begin{array}{c}
CH_3 \\
| \\
O - CH - COOH \\
\end{array}
$$

Silvex

silvicide — A pesticide chemical used to destroy woody shrubs and trees.

silviculture — Management of forest land for timber; may sometimes contribute to water pollution.

simple asphyxiant — A physiologically inert gas that dilutes or displaces atmospheric oxygen below amounts needed to maintain blood levels adequate for normal tissue respiration. Examples include carbon dioxide, ethane, helium, hydrogen, methane, and nitrogen.

simple harmonic motion — Periodic motion from a fixed point in which a particle goes equal distances and opposite directions in the equilibrium position. The particle completes oscillatory motion about a point in the same amount of time. The force exerted on the particle and the resulting acceleration are

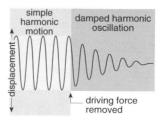

Simple Harmonic Motion

directly proportional to the displacement from the equilibrium position.

SIMS — Acronym for secondary ion-mass spectrometry.

simulated workplace protection factor (SWPF) — A surrogate measure of workplace protection provided by a respirator.

simulation — A mock accident or release designed to test emergency response methods or for use as a training tool; the determination of possible outcomes of situations expressed in the form of mathematical models.

single sources — Industrial and/or municipal sources that release contaminants into the air.

single tracer gas technique — A method for determining the air change rate within a room or zone using one tracer gas.

single-blind design — An experiment in which participants are unaware of their allocation to treatments.

single-family dwelling — An unattached dwelling unit inhabited by an adult person plus one or more related persons.

single-grain soil — A soil in which structure is lacking, and individual soil particles exist separately and do not form aggregates; usually found in substratum or the C horizon.

Single Grain Soil

single-service article — Tableware, flatware, hollowware, carryout utensils, and other items such as bags, containers, stirrers, straws, toothpicks, and wrappers that are designed, fabricated, and intended by the manufacturer for one-time use.

sinkhole — A closed, funnel-shaped cavity formed in limestone regions by the removal of rock through the action of water.

sinks — Any natural or artificial means of absorbing or removing a substance or a form of energy from a system; a place in the environment where a compound or material collects.

sinoatrial node — A collection of atypical muscle fibers in the wall of the right atrium where the rhythm of cardiac contraction is usually established; also known as pacemaker of the heart.

sintering — A heat treatment that causes adjacent particles of a material to cohere below a temperature that would cause them to melt; used in powder metallurgy.

sinus — An air cavity in the bones of the skull; a cavity, recess, or depression in part of an animal's body.

sinus bradycardia — A cardiac rhythm with a heart rate of <60 beats per minute that may be due to a normal response to sleep, being a well-conditioned athlete, diminished blood flow to the SA node, vagal stimulation, or pharmacological agents. It has no clinical significance unless associated with signs of impaired cardiac output and symptoms of dizziness, syncope, and chest pain.

sinus dysrhythmia — A cardiac irregular rhythm with a heart rate of 60 to 100 beats per minute, slowing during inspiration and increasing with expiration. It commonly occurs in children, young adults, and older adults but disappears as heart rate increases. It has no clinical significance unless heart rate decreases and symptoms of dizziness occur with the decreasing heart rate.

sinus tachycardia — A cardiac rhythm with a heart rate between 100 and 180 beats per minute that may be a normal response to exercise, emotion, or stressors such as pain, fever, pump failure, and certain drugs, resulting in a potentially damaged heart.

sinusitis — Inflammation of a sinus.

SIP — See state implementation plans.

siphon — A closed conduit, a portion of which lies above the hydraulic grade line, resulting in a pressure less than atmospheric and acquiring a vacuum within the conduit to start a flow. A siphon utilizes atmospheric pressure to affect or increase the flow of water through the conduit.

siphonage — A partial vacuum created by the flow of liquids in pipes.

SIR — See standardized incidence ratio.

site — A land or water area before activity is physically located or conducted, including adjacent land use in connection with the facility of activity; any location where acutely toxic chemicals are manufactured, processed, stored, handled, used, or disposed.

site assessment program — A means of evaluating hazardous waste sites through preliminary assessments and site inspections to develop a Hazard Ranking System score.

site characterization — An onsite investigation at a known or suspected contaminated waste or release site to determine the extent and types of contamination.

site inspection — The collection of information from a Superfund site to determine the extent and severity of hazards and to determine if it presents an immediate threat that requires prompt removal of the substances.

site-specific surveillance — An epidemiological surveillance activity designed to assess the specific occurrence of one or more defined health conditions

S

among a specific population potentially exposed to hazardous substances in the environment.

siting — The process of choosing a location for a facility.

Sitophilus granarius — See Granary Weevil.

Sitophilus oryza — See Rice Weevil.

Sitotroga cerealella — See Angoumois Grain Moth.

SIVE — See steam injection and vacuum extraction.

skeletal fluorosis — Fluoride deposition in the bones.

skeletal muscle — Striated, voluntary muscle attached to a bone.

skeleton — The supporting framework for the body. The human body is made up of 206 bones that provide attachment for muscles, allow body movement, serve as major reservoirs of blood, and produce red blood cells.

skew — Deviation from a line or symmetric pattern, such as data in a research study that do not follow the expected statistical curve of distribution because of the introduction of another variable.

skim milk — Milk from which a sufficient portion of milkfat has been removed to reduce its milkfat content to less than 3.25%; milk that contains 1% buttcrfat.

skimmer — A device other than an overflow trough for continuous removal of surface water and floating debris from the pool.

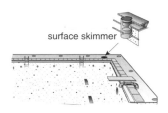

surface skimmer

Skimmer

skimmer weir — Part of a skimmer that adjusts automatically to small changes of water level to ensure a continuous flow of water to the skimmer.

skimming — The use of a machine to remove oil or scum from the surface of water

skin — The outer integument or covering of an animal body consisting of the dermis and epidermis and resting on the subcutaneous tissues.

skin absorption — The process by which some hazardous chemicals pass directly through the skin and enter the bloodstream.

skin cancer — A cutaneous neoplasm caused by ionizing radiation, genetic defects, chemical carcinogens, or overexposure to the sun or other sources of ultraviolet light.

skin prep — A procedure of skin cleansing used to remove pathological organisms to reduce the risk of infection prior to surgery or vena puncture.

slag — The top-layer, nonmetallic waste product formed when flux reacts with the impurities of an ore in a metallurgical process, such as smelting and refining; also known as dross.

slake — The mixing of a chemical with water in a true chemical combination such as hydrolysis.

slaking — Adding water to quick lime (CaO) to produce hydrated lime (CaOH)$_2$.

SLAMS — Acronym for state or local air monitoring station.

slaughter — To kill an animal for food.

slaughterhouse — Any food establishment or portion thereof in which cattle, sheep, swine, goats, or horses are slaughtered for transportation, sale, or processing as food for human consumption.

sleeping sickness — See African trypanosomiasis.

sleet — Transparent or translucent beads of ice occurring when rain dropping from upper warm air falls through a layer of freezing air; the rain drops first become freezing rain and then turn into sleet or ice pellets.

slight limitations — In soils, limitations relating to water permeability that are easy to overcome.

slime layer — A diffuse layer of polysaccharide exterior to the cell wall in some bacteria.

SLN — Acronym for special local need.

slope — (*dose–response*) Linear portion of a curve that defines the potency of an agent; (*mathematics*) deviation of a surface from the horizontal expressed as a percentage by a ratio or in degrees.

slope factor — An upper bound (approximately 95% confidence limit) of the increased cancer risk from a lifetime exposure to an agent.

slope stability — The ability of the slope of soil or rock materials to resist moving downhill.

slot velocity — The linear flow rate of air through a slot.

slough — A wet or marshy area; a place of deep mud or mire.

sloughings — Trickling-filter slimes that have been washed off of filter media; they are usually high in BOD and lower effluent quality unless removed.

slow sand filtration — The passage of raw water through a bed of sand at low velocity resulting in substantial removal of chemical and biological contaminants.

slow virus — A virus that remains dormant for many years in the body after an initial infection and may eventually cause degenerative diseases.

sludge — A semisolid mixture of organic and inorganic materials that settle out of wastewater at a sewage treatment plant; the digested or partially digested solid material accumulated in a septic tank.

sludge age — A measure of the length of time a particle of suspended solids has been retained in the activated sludge process; see cell residence time.

sludge bank — The accumulation of solid, sewage, or industrial waste deposits on the bed of a waterway.

sludge conditioning — Pretreatment of a sludge to facilitate removal of water in a thickening or dewatering process by means of chemical, inorganic or organic, elutriation, or heat treatment.

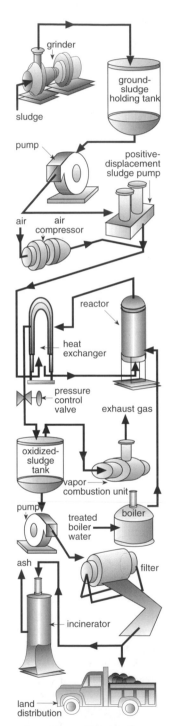

**Thermal Sludge
Conditioning Process
Schematic**

sludge digester — A tank in which complex organic substances such as sewage sludge are biologically degraded, resulting in the release of energy and production of methane, carbon dioxide, water, and stable matter.

sludge reduction — Any of the processes including incineration, wet air oxidation, and pyrolysis that primarily yield a major reduction in volatile sludge solids.

sludge stabilization — A technique for converting raw, untreated sludge into a less offensive form with regard to odor, putrescibility, weight, and pathogenic organism content through anaerobic digestion, aerobic digestion, lime treatment, chlorine oxidation, heat treatment, and composting.

sludge thickening — A technique for increasing solid concentration by gravity, flotation, or centrifugation.

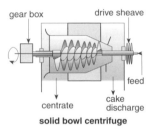

solid bowl centrifuge

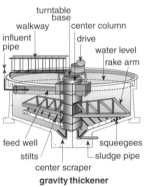

gravity thickener

Sludge Thickening

sludge volume — A settling test using a 2-liter settleometer to measure sludge quality; it is expressed in percent and is related to time.

sludge volume index — A settling test used to measure sludge quality.

slugs — Intermittent releases or discharges of industrial wastes.

sluice — An opening for releasing water from below the static head elevation.

sluicing — The use of low-pressure, high showers of water to mobilize waste.

slurry — A mixture of liquid and insoluble material, such as Portland cement.

slurry feed — A feed of diatomaceous earth introduced as a liquid slurry onto a piece of equipment to aid in filtration.

s/m² — Notation for structures per square millimeter.

small intestine — The part of the digestive system consisting of the duodenum, jejunum, and ileum.

small-quantity generator (SQG) — A generator that produces less than 1000 kg of hazardous waste per month, accumulates less than 1000 kg at any one time, produces less than 1 kg of acutely hazardous waste per month, or accumulates less than 1 kg of acutely hazardous waste at one time.

SMCL — See secondary maximum contaminant level.

SMCRA — Acronym for Surface Mining Control and Reclamation Act.

smear — A laboratory specimen prepared for microscopic examination; a thin film of tissue is spread on a glass slide and a dye, stain, reagent, diluting agent, or lysing agent is used to aid in differentiation.

smell — The special sense concerned with the perception of odor.

smelter — A facility that melts or uses ore, often with an accompanying chemical change to separate its metal content.

smelting — The treatment of an ore by heat to separate out the desired metal.

smog — A hazy mixture of smoke and fog resulting from actions of the Sun on certain pollutants in the air, especially those from automobile exhaust and factories; may also be any air pollution problem that reduces visibility.

smoke — Solid or liquid particles under 1 μm in diameter and dispersed in a gaseous medium; (*incineration*) an aerosol consisting of all the dispersible particulates produced by incomplete combustion of carbonaceous materials entrained in flue gas.

smoke alarm — An instrument that continuously measures and records the density of smoke by determining how much light is obscured when a beam shines through the smoke.

smoke density — The amount of solid matter contained in smoke, often measured by a system of grayness of the smoke compared to an established standard.

smoke detector — A device that can detect the presence of products of combustion in the air even in minute concentrations.

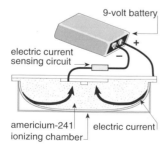

Smoke Detector

smoke inhalation — The inhalation of toxic fumes and/or irritating particulate matter that may cause severe pulmonary damage, chemical pneumonitis, and asphyxiation.

smoke number (SN) — A dimensionless term quantifying smoke emissions.

smokestack — A vertical pipe or flue designed to exhaust gases and any particulate matter suspended in the gases.

smooth muscle — The involuntary muscle found lining the walls of the intestines, stomach, and arteries.

SMP — Acronym for State Management Plan.

SMR — See standardized mortality ratio.

SMS — Acronym for synchronous meteorological satellite.

SMSA — See standard metropolitan statistical area.

SN — See smoke number.

SNAAQS — Acronym for Secondary National Ambient Air Quality Standards.

SNARL — Acronym for suggested no adverse response level.

SNC — Acronym for significant noncompliers.

Snellen chart — A vision-measuring chart consisting of block letters in diminishing sizes; the most common industrial test for vision acuity.

snow — Precipitation in the form of small, white ice crystals formed directly from the water vapor of air at a temperature less than 32°F.

SNUR — Acronym for significant new use rule.

SO₂ — See sulfur dioxide.

SO₃ — See sulfur trioxide.

SOA — Acronym for state of the art.

soap — A type of detergent that can be cast into bars; ordinarily, it is a metal salt of a fatty acid, usually sodium stearate, sodium oleate, or sodium palmitate or some combination of these materials.

soap-bubble meter — A device used to calibrate high-flow and low-flow air-sampling pumps.

soapstone (containing less than 1% quartz) — An odorless, white-gray powder; MW: 379.3, BP: not known, Sol: insoluble, sp. gr. 2.7 to 2.8. It is used in the manufacture of acid-proof coverings in floors, tables, sinks; in switchboard panels for high-electrical resistance; in the Kraft process of pulp manufacture; in fume cupboards and fireless cookers; in crayons for marking cloth, metal, and glass. It is hazardous to the lungs and cardiovascular system and is toxic through inhalation and contact. Symptoms of exposure include cough, dyspnea, digital clubbing, cyanosis, and basal crackles. OSHA exposure limit (TWA): 6 mg/m³ [resp.].

SOB — Acronym for short of breath.

SOC — Acronym for synthetic organic chemical or soil organic carbon.

S

socioeconomic status — A person's position in society based on education, income, type of occupation, place of residence, and, in some areas, heritage and religion.

soda ash (Na$_2$CO$_3$) — Sodium carbonate powder used in glass manufacturing and petroleum refining and for soaps and detergents.

sodium (Na) — A light-silvery-white metal when freshly cut but becoming gray and tarnished when exposed to air; BP: 881°C, Sol: dissolves in liquid ammonia, Fl.P: reacts violently with water and may ignite or explode hydrogen that is released from the water, density: NA. It is used in the manufacture of highly reactive sodium compounds, as a reducing agent in organic synthesis, and as a catalyst of synthetic rubber. It is highly corrosive and can cause severe burns to the skin, eyes, and mucous membranes. ACGIH exposure limit: NA.

sodium adsorption ratio — The relative activity of sodium ions in exchange reactions with the soil.

sodium azide (NaN$_3$) — A colorless crystalline solid; MW: 65.02, BP: NA, Sol: soluble in alcohol and soluble in water, Fl.P: it is inert to shock but has a violent decomposition when heated to 275°C, density: 1.846. It is used in making other metal azides, in chemicals to control blood pressure, as a preservative for laboratory agents, and in organic synthesis. It is a highly toxic compound that is hazardous to the digestive and respiratory tracts. It causes hypotension, headache, tachypnea, hypothermia, convulsions, and diarrhea and may be fatal because of its effect on the central nervous system and brain. It is strongly mutagenic. ACGIH exposure limit: 0.3 mg/m^3 in air.

$$\overset{+}{Na} : \overset{-}{\underset{\cdot\cdot}{N}} \equiv \overset{+}{N} \equiv \overset{-}{\underset{\cdot\cdot}{N}} :$$

Sodium azide

sodium bicarbonate (NaHCO$_3$) — A white crystalline chemical used to raise the total alkalinity of a pool with no change in pH; also known as baking soda.

sodium bisulfate (NaHSO) — A dry, white powder that produces an acid solution when dissolved in water; used as a laundry-rinse neutralizer, preservative, and antiseptic; in glass etching and tin plating.

sodium borohydride (NaBH$_4$) — A white solid; MW: 37.83, BP: NA, Sol: NA, Fl.P: burns in air and when mixed rapidly with acids can cause a dangerous explosion, density: 1.074. It is used as a reducing agent in many organic synthetic reactions. It is mildly corrosive to the skin. In animals,

oral intake produces high levels of toxicity that may be lethal. ACGIH exposure limit: NA.

sodium chloride (NaCl) — Table salt.

sodium cyanide (NaCN) — A white, granular or crystalline solid or powder with a faint almond-like odor; MW: 49.02, BP: 1496°C, Sol: 82 g/100 ml water at 35°C, Fl.P: nonflammable (poisonous gases produced in fires), sp. gr. 1.6 at 25°C. It is used as a solid or solution to extract metal ores; in electroplating or metal cleaning baths, metal hardening, and insecticides. It is hazardous to the respiratory system, digestive system, and skin. Acute exposure may cause irritation of the skin, nose, and throat; headache; dizziness; pounding heartbeat; and sudden death. Chronic exposure may cause irritation to the nose, nosebleeds, skin rash, damage to the thyroid gland, and damage to the nervous system. It is a possible teratogen. OSHA exposure limit (8-hour TWA): 5mg/m^3 as cyanide.

sodium fluoroacetate (FCH$_2$C(O)ONa) — A fluffy, colorless to white (sometimes dyed black), odorless powder; a liquid above 95°F; MW: 100.0, BP: 332°F, Sol: miscible. It is used in the formulation of pesticides. It is hazardous to the cardiovascular system, lungs, kidneys, and central nervous system and is toxic through inhalation, absorption, ingestion, and contact. Symptoms of exposure include vomiting, apprehension, auditory hallucinations, facial paresthesia, twitching of the facial muscles, pulsus altenans, ectopic heartbeat, tachycardia, ventricle fibrillation, pulmonary edema, nystagmus, and convulsions. OSHA exposure limit (TWA): 0.05 mg/m^3 [skin]. Also known as 1080.

sodium hydride (NaH) — A white crystalline powder or silver needle-like structure; MW: 24.00, BP: NA, Sol: reacts violently with water and is soluble in fused salt mixtures, Fl.P: a flammable solid that ignites spontaneously on contact with moist air and violently with water, density: 1.36 g/cm^3. It is used as a reducing agent in organic synthesis. It is hazardous to the respiratory tract and skin. It is highly corrosive and may cause severe irritation and burns. ACGIH exposure limit: NA.

sodium hydroxide (NaOH) — A colorless to white, odorless solid (flakes, beads, granular form); MW: 40.0, BP: 2534°F, Sol: 11%, sp. gr. 2.13. It is used in chemical manufacture; in the explosives industry; in boiler water and as a laboratory reagent; in pH control in the textile, paper, and chemical industries; in the manufacture of pulp and paper; in the pulping kraft process; in the manufacture of insulating board; in metal processing and refining; in petroleum refining; as a flotation reagent; in the manufacture of soaps and detergents; in food processing; in glass manufacture. It is hazardous to

S

the eyes, skin, and respiratory system and is toxic through inhalation, ingestion, and contact. Symptoms of exposure include nose irritation, pneumonitis, eye and skin burns, temporary loss of hair. OSHA exposure limit (TWA): ceiling 2 mg/m^3 [air].

sodium hypochlorite (NaOCL) — Pale-green crystals with a sweet aroma and containing 12 to 15% available chlorine; used as a bleaching agent for paper pulp and textiles, as a chemical intermediate, and in disinfection.

sodium methoxide (CH$_3$ONa) — A white, free-flowing powder; MW: 54.03, BP: NA, Sol: soluble in methanol, Fl.P: a flammable solid that ignites spontaneously in air when heated to 70 to 80°C, density: 0.45 g/ml. It is used as a catalyst for treatment for edible fats and oils and as a reagent in chemical analysis. ACGIH exposure limit: NA.

sodium paratoluene sulfonchloramide — See chloramine T.

sodium thiosulfate (Na$_2$S$_2$O$_3$) — A chemical solution used to remove all chlorine from a test sample to avoid false pH test readings; in photography, as it dissolves silver halides; in tanning, dyeing, and manufacture of chemicals.

SOFC — Acronym for solid oxide fuel cell.

soft detergent — A cleaning agent that breaks down in nature.

soft swell — Describes a can bulged at both ends but not so tightly that the ends cannot be pushed in somewhat with thumb pressure.

software — Programs that control a computer and allow it to perform specific functions.

softwater — Water with very few minerals or other dissolved chemicals.

soil — Unconsolidated particles from weathered rock, water, air, and humus over bedrock from which plants obtain essential materials.

soil absorption — A process that utilizes the soil to treat and dispose of effluent from a septic tank.

soil absorption system — Pipes laid in a system of trenches or elevated beds into which the effluent from the septic tank is discharged for soil absorption.

soil aggregate — A group of soil particles cohering so as to behave mechanically as a unit.

soil air — Air and other gases found in the voids between soil particles; similar to atmospheric air but not mobile.

soil application — The process of applying a pesticide chemical on the soil.

soil atmosphere — Gases occupying the pore space in soil, usually with a greater percentage of carbon dioxide and a lesser percentage of oxygen than the overlying air.

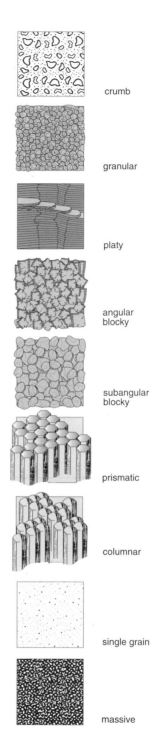

crumb

granular

platy

angular blocky

subangular blocky

prismatic

columnar

single grain

massive

Soil Structures

soil biochemistry — Branch of soil science concerned with enzymes and reactions, activities, and products of soil microorganisms.

soil boring — A soil core taken intact and undisturbed by a probe.

soil bulk density — The mass of dry soil per unit of bulk volume determined before drying to constant weight at 105°C.

soil characteristic limitation — Those limits that preclude installation of an on-site sewage system because of certain soil characteristics, such as a high water table, high clay, or high silt levels.

soil classification — A systematic arrangement of soils into classes of one or more categories; levels of classification for a specific objective.

soil cohesion — The mutual attraction exerted on soil particles by molecular forces and moisture films.

soil conditioner — A synthetic chemical or natural material added in small quantities to improve the structure of soil.

soil conservation — The protection of soil against physical loss by erosion and chemical deterioration by the application of management and land-use methods that safeguard the soil against natural and human-induced factors.

soil consistence — The combined properties of soil material that cause the aggregates to hold together or fall apart, resisting deformation.

soil deposition — The movement of material removed from soil and laid down as sediments.

soil description — A listing of soil properties, both site and profile, specific to a geographical location.

soil drainage — The ability of the soil pores to avoid saturation and therefore drain freely.

soil erosion — The wearing away or removal of soil by the action of water or wind.

soil fumigant — A pesticide chemical used to kill pests in the soil; evaporates quickly.

soil gas — Any gas such as radon, volatile organics, and pesticides entering a building from the surrounding ground.

soil health — See soil quality.

soil horizon — A layer of soil or soil material in a soil profile differing from adjacent related layers in physical, chemical, and biological properties or characteristics such as color, structure, texture, consistency, kind, and number of organisms present and degree of acidity or alkalinity.

soil injection — Placing a pesticide chemical below the soil surface with little or no soil mixing.

soil interpretation — Description of the soil horizon organized and presented to provide an understanding for soil use and management.

soil limitations — See soil characteristic limitation.

soil map — A map showing the distribution of soil, soil types, or other soil mapping units in relation to the prominent physical and cultural features of the surface of the Earth.

soil map unit — An area of the landscape shown on a soil map that consists of one or more soils.

soil microbiology — Branch of soil science concerned with soil-inhabiting microorganisms and their functions and activities.

soil mineralogy — Study of the kinds and proportions of minerals present in soil.

soil moisture — The water contained in the pore space of the unsaturated zone of the soil.

soil morphology — The physical constitution, particularly the structural properties or soil profile, as exhibited by the kinds, thickness, and arrangement of horizons in a profile and by the texture, structure, consistency, and porosity of each horizon.

soil organic matter — Plant and animal residue that decomposes and becomes a part of the soil.

soil permeability — The quality of a soil that enables water and air to move through it; a measure of this quality is the rate at which soil will transmit water under saturated conditions.

soil pipe — A pipe that conveys sewage containing fecal matter to the building drain or building sewer.

soil plasticity — The property of a soil that allows it to be deformed or molded in a moist condition without cracking or falling apart.

soil population — All of the organisms living in the soil, including plants and animals.

soil pore — The part of the bulk volume of the soil not occupied by soil particles.

soil profile — A vertical cross-section of soil horizons to a depth of 2 to 5 feet.

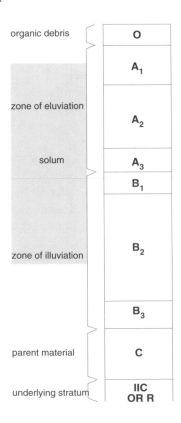

**Soil Profile
showing all the soil horizons**

soil profile analysis — Observation and evaluation of the physical characteristics of a soil horizon or layers to a depth of at least 5 feet or to a shallower layer that cannot be readily penetrated.

soil quality — The continued ability of soil to function as a vital living system to sustain biological productivity, maintain the quality of the environment, and promote plant, animal, and human health.

soil salinity — The amount of soluble salts in a soil as measured by the electrical conductivity of a saturation extract.

soil saturation — The state when all the pores in a soil are filled with water.

soil science — The study of the principles of soil classification and mapping; see also pedology.

soil sealant — A chemical or physical agent used to plug porous soils to prevent leaching or percolation.

soil separate — Any of a group of mineral particles separated on the basis of arrangement and sizes; the principle separates are sand, silt, and clay.

soil series — A basic unit of soil classification consisting of soils that are essentially alike in all major profile characteristics.

soil slope — The incline of the surface of soil area.

soil solution — The aqueous liquid phase of a soil and its solutes consisting of ions associated from the surfaces of the soil particles and other soluble materials.

soil stack — The vertical main of a system of soil, waste, or event piping.

soil sterilant — A pesticide chemical or agent that destroys all plants and animals in the treated soil for extended periods of time.

soil structure — The combination or arrangement of individual soil particles into definable aggregates that are characterized and classified on the basis of size, shape, and degree of distinctiveness.

soil subsidence — Lowering of the normal level of the ground, commonly due to overpumping of water from wells.

soil suction — A measure of the force of water retention in unsaturated soil; equal to a force per unit area that must be exceeded by an externally applied suction to initiate water flow from the soil.

soil suitability — The determination of soil properties to assess their various uses.

soil survey — The systematic examination, description, classification, and mapping of soils in an area.

soil textural class — Soil grouped on the basis of 12 specified ranges and textural classes.

soil textural classification — Soil particles, sizes, or textures that correspond to soil classifications in the Department of Agriculture's *Soil Survey Manual.*

soil texture — The relative proportions of size groups in sand, silt, and clay of individual soil particles.

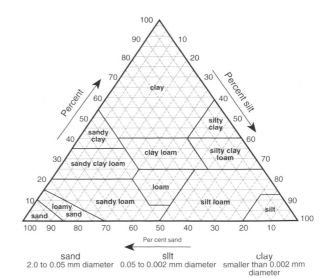

Soil Textural Class

soil type — Subdivision of a soil series based on differences in texture of the A horizon.

soil vapor extraction — An in-place method of extracting volatile organic chemicals from soil by applying a vacuum to the soil, collecting the air, and treating the air to remove chemicals before discharging it to the atmosphere.

soil washing — A technique used to remove contaminants concentrated in a fine-size fraction of soil including silt, clay, and soil organic matter.

soil water — The moisture trapped between and around soil particles that contains nutrients available to plants and may contain pesticides, fertilizers, and other contaminants.

soil water tension — The expression in positive terms of the negative hydraulic pressure of soil water.

Sol — See solubility.

solar backscatter ultraviolet radiometer — An instrument used to measure the vertical distribution and total ozone in the atmosphere of the Earth and to continuously monitor ozone distribution to estimate long-term trends.

solar battery — A battery that is charged through photovoltaic cells.

solar cell — A photovoltaic cell that can convert light directly into electricity typically using semiconductors made from silicon,

solar collector — A piece of equipment that gathers and accumulates solar radiation to produce heat. The solar collectors have a layer of glazing on top to trap that heat, which passes into a collector and an absorber plate, which transfers the heat to a heat transfer medium, such as air, water, antifreeze, or some other substance.

S

solar collector — A surface or device that absorbs solar heat and transfers it into a fluid; a component of an active or passive solar system that absorbs solar radiation to heat a transfer medium which in turn supplies heat energy to a space or water heating system

solar collector efficiency — The total solar radiation that is incident on a collector during a specific time period.

solar concentrator — The part of a solar collector that focuses sunlight onto an absorber surface.

solar constant — A constant expressing the amount of solar radiation reaching the Earth from the Sun at approximately 1370 W/m^2; not truly constant as variations are detectable.

solar cooling — A solar system used to lower the temperature in a room or piece of equipment.

solar cycle — An 11-year cycle of sun spots and solar flares that affects the magnetic field, temperature, and ozone levels of the Earth.

solar energy — Energy produced by the Sun and absorbed by the Earth or trapped in various mechanical pieces of equipment to produce thermal, electrical, mechanical, or chemical energy, which is eventually reradiated into space as heat.

solar heat gain — The heat added to a space due to transmitted and absorbed solar energy.

solar heat gain factor — An estimate used when calculating cooling loads of heat gain due to transmitted and absorbed solar energy through 1/8-inch thick, clear glass at a specific latitude, time, and orientation.

solar heating and hot water systems — The collection and delivery subsystems that capture the radiant energy of the sun, convert it into heat energy, store it in insulated storage tanks, and deliver the stored energy as needed to domestic hot water and/or heating systems.

solar power — Electricity generated from solar radiation.

solar radiation — The increased flow of heat through windows, walls, and roofs by their absorption of radiant heat; radiation received from the sun and emitted in spectral wavelengths of less than 4 μm.

solar satellite power — A proposed process of using satellites in geosynchronous orbit above the Earth to capture solar energy with photovoltaic cells, convert it to microwave energy, and beam the microwaves to Earth, where they would be received by large antennas; the microwave energy would then be changed into usable electricity.

solar thermal — The process of concentrating sunlight on a relatively small area to create the high temperatures needed to vaporized water or other fluids to drive a turbine for generation of electric power.

solar thermal power plant — A thermal power plant in which 75% or more of the total energy output is from solar energy, and backup fuels such as oil, natural gas, and coal do not exceed 25% of the total energy input of the facility during any calendar year.

solar water heater — A water heater that uses radiant energy from the Sun as its source of power.

solarization — A method of controlling pathogens and weeds; moistened soil in hot climates is covered with transparent polyethylene plastic sheets that trap incoming radiation.

solder — A general term for a low-melting metal or alloy used to join to adjacent surfaces of less fusible metals or alloys; the principle types are soft solder and brazing solder.

sole-source aquifer — An aquifer that supplies 50% or more of the drinking water of an area.

solenoid — A coil of wire, commonly in the form of a long cylinder; when carrying a current, it resembles a bar magnet so a movable core is drawn into the coil when a current flows.

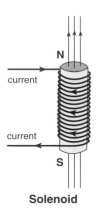

current

current

Solenoid

solid — An object without an internal cavity.

solid flammable — Any solid, other than a blasting agent or explosive, that is liable to cause fire, friction, absorption of moisture, or spontaneous chemical change or to retain heat from manufacturing or processing, or any solid that can be readily ignited and burns so vigorously and persistently as to create a serious hazard.

solid waste — Any garbage, refuse, or sludge from a waste treatment plant, water supply treatment plant, or air pollution control facility; discarded material including solid, liquid, semisolid, or contained gaseous material resulting from industrial, commercial, mining, and agriculture operations and from domestic activities.

solid waste desorption process — A technique using superheated steam (up to 900°F) as a continuous conveying and stripping gas in a pneumatic system to treat contaminated solids.

solid waste disposal — The final disposal of refuse from which it cannot be salvaged or recycled.

solid waste management — The purposeful, systematic control, generation, storage, collection, transport, separation, processing, recycling, recovery, and disposal of solid waste.

solid waste manager — An environmental health practitioner responsible for the planning, administration, and regulation of storage, transportation, and disposal of solid waste.

solid waste storage container — A receptacle used for the temporary storage of solid waste before collection.

solid–gas extraction — A chemical technique used to remove a desired gas from a mixture of gases.

solid–liquid extraction — A chemical technique used to remove impurities from a mixture that involves dissolving the substances from the desired precipitate and the reaction of a solvent with a component to make it either soluble or insoluble.

solid-sorbent tube — A device used to capture insoluble or nonreactive gas and vapor contaminants from air through tubes filled with granular sorbents, such as activated charcoal.

solids-not-fat — The solids in milk such as protein, lactose, and minerals that are not fat.

solids retention time (SRT) — The average residence time of suspended solids in a biological waste treatment system equal to the total weight of suspended solids in the system divided by the total weight of suspended solids leaving the system per unit time.

solubility (Sol) — The extent to which one substance will dissolve in another; usually expressed as the weight in grams of a substance required to saturate 100 grams of a solvent at a given temperature.

soluble powder — A dust that will dissolve in water or other liquid.

solum — (*soils*) The upper part of a soil profile comprising the A and B horizons.

solute — The substance dissolved in a solvent to produce a solution.

solution — A uniform mixture of a solute in a solvent; a preparation made by dissolving a solid, liquid, or gaseous substance into another substance without a chemical change taking place.

solvent — A substance, usually a liquid, that can dissolve other substances.

solvent solubility — Refers to the concentration of a material that dissolves in a given solvent.

SOM — Acronym for sensitivity of method or soil organic matter.

soman ($C_7H_{16}FO_2P$) — A colorless liquid; MW: 182.19, BP: NA, Sol: miscible with organic solvents and hydrolyzes in water, density: 1.15. It is used as a chemical warfare agent. It is an extremely toxic organophosphorus compound that is a highly potent cholinesterase inhibitor. It may cause convulsions, cardiovascular toxicity, difficulty in breathing, bronchial constriction, tremor, and death. ACGIH exposure limit: NA.

$$(CH_3)_3C-CH-O \qquad P-F$$

Soman

somatic cell — Any nonreproductive body cell that has two sets of chromosomes and is not a germ cell.

somewhat poorly drained — A classification describing soil in which the water is removed slowly enough to keep it wet for significant but not constant periods.

sonar — Acronym for sound navigation and ranging; an electronic method of ranging using ultrasonic waves for underwater surveying, detection, navigation, and communication. A narrow pulsed beam of ultrasonic waves is sent through water and reflected back to the receiver; the length of time it takes for the reflected waves to return is a measure of the distance of the reflecting surface.

sonic boom — The thunderous noise made when shock waves reach the ground from a jet airplane or other craft exceeding the speed of sonic velocity.

soot — Agglomeration of very finely divided, tar-impregnated carbon particles that form from the incomplete combustion of carbonaceous material; see also carbon particulate matter.

soot-blowing — Periodic cleaning of heat-transfer equipment required because, during combustion, particulate matter sticks to the surface of the equipment and reduces its heat-exchanging ability; it is done by means of a jet of air or steam while the equipment is in use.

SOP — See standard operating procedure.

sorbent — An inert and insoluble material used to remove oil and hazardous substances from water.

sorbent barrier — A partition or an obstruction that prevents the attraction and retaining of substances by absorption or adsorption.

sorbitol ($C_6H_8(OH)_6$) — A crystalline, higher alcohol produced by the reduction of sugar with hydrogen. It is used in the manufacture of paper, tobacco, textiles, adhesives, pharmaceuticals, and cosmetics and in food.

sorption — A general term including absorption, adsorption, chemisorption, and persorption.

S

sorption coefficient — The rate at which chemicals adhere to solids.

SOTDAT — Acronym for source test data.

sound — Auditory sensation produced by the oscillations, stress, pressure, particle displacement, and particle velocity in a medium with internal forces; pressure variation that the human ear can detect.

sound absorption — The process of removing sound energy by use of special materials.

sound absorption coefficient — The ratio of the sound absorbed by a surface exposed to a specific sound field to the sound energy on the surface.

sound analyzer — A device for measuring the band, pressure level, or pressure-spectrum level of a sound as a function of frequency.

sound energy — Energy added to the medium in which sound travels; consists of potential energy in the form of deviations from static pressure and kinetic energy in the form of particle velocity.

sound energy density — Sound energy per unit volume, usually measured in ergs per cubic centimeter.

sound exposure level — The level of sound accumulated over a given time or event; the time integrated mean square A-weighted sound for a stated time interval or event, with a reference time of 1 second.

sound intensity — The rate at which sound energy flows through a unit area.

sound level — The weighted sound pressure level (measured in decibels) using a metering characteristic; the weighting is in the use of scale A, B, or C.

sound level meter — A sensitive electronic voltmeter that measures the electric signal from a microphone, which is ordinarily supplied with and attached to the instrument; the alternating-current electric signal from the microphone is amplified sufficiently so that after conversion to direct current by means of a rectifier, the signal can deflect a needle on an indicating meter. An attenuator controls the overall amplification of the instrument. The instrument contains A-, B-, and C-scale networks; the high-frequency noise passed by the A-weighting network correlates well with annoyance effects and hearing damage effects on people; most of the studies that are done are read on the A-scale and the meter reading is known as the dBA value.

sound navigation and ranging — See sonar.

sound power — The rate at which acoustic energy is radiated.

sound pressure — The difference between the actual pressure at any point in the field of a sound wave at any instant and the average pressure at that point.

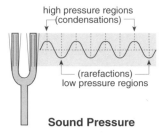

Sound Pressure

sound pressure level (SPL) — The value in decibels equal to 20 times the logarithm to the base 10 of the ratio of the pressure of the sound to a reference pressure, which is explicitly stated.

sound transmission class (STC) — Single-figure rating system designed to yield a preliminary estimate of the sound insulation properties of a partition or preliminary rank ordering of a series of partitions.

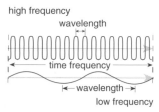

Sound Waves

sound waves — Longitudinal waves causing particles of a gas, liquid, or solid elastic medium to vibrate along the direction of motion.

sounder — A special kind of radiometer that measures changes in atmospheric temperature with height, as well as the content of various chemical species in the atmosphere at various levels.

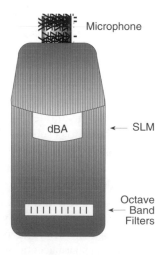

Sound level meter and octave band analyzer

sour — Describes a gasoline or refined oil that contains hydrogen sulfide or mercaptans and has an offensive odor.

sour cream — A cream with an acidity of more than 0.20%, expressed as lactic acid.

source — (*wells*) The site from which water is extracted for the purpose of supplying water to a private water system.

source energy — All of the energy used in delivering energy to a site, including power generation, transmission, and distribution losses, to perform a specific function, such as space conditioning, lighting, or water heating; it takes approximately 3 watts of energy to deliver 1 watt of usable electricity.

source of emissions — An activity or process that can lead to the release of a pollutant.

source reduction — Any practice that reduces the amount of any hazardous substance, pollutant, or contaminant, including fugitive emissions, entering any waste stream or otherwise released into the environment prior to recycling, treatment, and disposal and which reduces the hazards to public health and the environment associated with the release of these substances.

source separation — The separation of waste materials at the source by the public, business, or industry.

source specific — A source of pollution that can be identified in a community.

source term — The release rate of hazardous agents from a facility or activity.

South American trypanosomiasis — See Chagas disease.

SOW — Acronym for scope of work.

SO$_x$ — See sulfur oxides.

sp. gr. — See specific gravity.

spa — A hydrotherapy unit of irregular or geometric shell design.

space bomb — A container that has a pesticide chemical plus a chemical aerosol under pressure which forces the pesticide out as a spray or a mist into a room or building.

space heater — A self-contained heating device of either the convection or radiant type that is intended primarily to heat only one room, two adjoining rooms, or some other limited space or area.

space spray — A pesticide chemical forced out of an aerosol container or sprayer as tiny droplets that form as a mist.

space telescope — An instrument launched into space from a spacecraft; the high-resolution and faint object spectographics provide a wide range of spectral resolutions, while the photometer measures the brightness of stars and the galactic background.

spallation — The process by which a high-energy particle strikes a nucleus, causing fragments to be

Space Telescope

ejected from the nucleus and resulting in a radioactive atom.

SPAR — Acronym for Status of Permit Application Report.

sparge — Injection of air below the water table to strip dissolved volatile organic compounds and/or oxygenate groundwater to facilitate aerobic biodegradation of organic compounds.

spasm — Tightening or contraction of any set of muscles.

spatial — A characteristic that refers to a location relative to an arbitrary point or a specific location on the surface of the Earth.

spatial relationships — The relative locations of various personnel and equipment in an occupational setting.

spatial resolution — The ability to resolve the spatial distribution of air pollution concentrations in air by modeling or monitoring; affected by distance to the source, number of sources, emission height, and emission strength.

SPCC — Acronym for spill prevention, containment, and countermeasure.

SPCS — See state plane coordinate system.

SPDES — See State Pollution Discharge Elimination System.

SPE — Acronym for secondary particulate emissions; see Society of Plastics Engineers.

Special Data Transfers Standard — A comprehensive transfer standard for Earth-referenced data that may be used to transfer all types of spatial data between dissimilar computer systems.

special operation — The use of an area or building, such as a school, hospital, nursing home, or shopping center, for a special purpose or during certain circumstances.

special review — (*EPA*) An intensive analysis of all the data on a chemical including its risks and benefits.

special waste — Waste that requires extra treatment before entering into the normal plumbing system.

specialist — A person who through education, training, and experience is recognized by his peers as one who specializes in a particular occupation, practice, or branch of learning.

speciation — The subdivision of classes of chemicals into groups with closely related properties.

species — One of the smaller taxonomic subgroupings in a classification system of plants or animals consisting of interbreeding organisms of a single kind showing continuous morphological variations within the group but distinct from other such groups.

species barrier — The naturally occurring barrier between different species of animals that makes transfer of a disease from one to the other difficult.

species diversity — A number that relates the density of organisms of each type present in a habitat.

specific activity (SA) — The activity or decay rate of a radioisotope per unit mass of a sample.

specific capacity — The productivity of a well obtained by dividing the well discharge rate by the well drawdown while pumping.

specific chemical identity — The chemical name, the Chemical Abstract Service registry number, or any precise chemical designation of a substance.

specific conductance — A measure of the ability of water to conduct an electrical current as measured using a 1-cm cell expressed in units of electrical conductance (siemens per centimeter at 25°C); a rapid method of estimating the dissolved solid content of a water supply by testing its capacity to carry an electrical current.

specific energy — The actual energy per unit mass deposited per unit volume in a given event.

specific energy absorption rate — The rate at which energy is absorbed by unit mass of tissue in an electromagnetic field.

specific flow — The total volumetric supply-air flow rate per unit volume of a room measured by the equation $a = Q/V$, where Q is the ventilation-air flow rate (m^3/sec), V is the total volume of the room (m^3) and a is the specific flow; often called the air-exchange rate.

specific gravity (sp. gr.) — (*weight*) A number expressing the ratio between the weight of certain volumes of a substance and the weight of an equal volume of some standard material, such as water or air, at standard temperature and pressure; also known as relative density.

specific heat — Ratio between the amount of heat required to raise a given weight of a substance

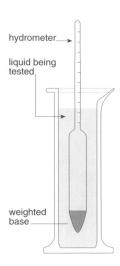

hydrometer

liquid being tested

weighted base

Instrument for finding
specific gravity of a liquid

1 degree in temperature and the amount of heat required to raise the same amount of a standard material, such as water, 1 degree in temperature.

specific humidity — The weight of water vapor per unit weight of dry air.

specific ionization — The number of ion pairs per unit length of the path of ionizing radiation in a medium per centimeter of air or per micrometer of tissue.

specific rates — A ratio that refers to some particulate subgroup of events counted in the numerator and/or to some particular subgroup of a population counted in the denominator.

specific response — Response of an air pollution analyzer to a single pollutant species without interference from other pollutants.

specific volume — The volume occupied by 1 pound of a substance under any specified condition of temperature and pressure.

specific weight — The weight per unit volume of a substance; see density.

specific yield — Ratio of the volume of water that will drain from a unit volume of an aquifer by gravity flow.

specificity — The ability of a screening test to identify persons who do not have a disease; it is measured by the ratio of persons that test negative to the total number of persons testing negative and testing positive.

specimen — An individual, item, or part considered typical of a group, class, or whole.

spectral band — A finite segment of wavelengths in the electromagnetic spectrum; shortwave radiation. See solar radiation.

spectral filter — A filter that allows only a specific bandwidth of the electromagnetic spectrum to pass.

spectral signature — The unique way in which a given type of land cover reflects and absorbs light.

spectrograph — A type of spectroscope that makes a photographic record of the spectrum of an object viewed through a telescope.

spectrometer — An instrument for measuring wavelengths of rays of the spectrum, the deviation of refracted rays, and the angles between faces of a prism; used to determine the distribution of energy within a spectrum of wavelengths.

spectrophotometry — A direct-reading procedure used to measure the wavelength range of energy absorbed by a sample under analysis; similar to photometry, except that a prism made of glass is used in the visible range, quartz is used in the ultraviolet range, and sodium chloride or potassium chloride is used in the infrared range.

spectroscope — An optical instrument that separates a beam of light into its various component colors using a prism or grating and a telescope, camera, or counter for observing the dispersed radiation.

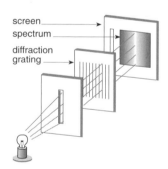

Grating Spectroscope

spectrum — (*physics*) The band of colors formed when a beam of white light breaks up by passing through a prism or by other means; description of a function of time of its resolution into components, each having a different frequency, amplitude, and phase.

spectrum analysis — The use of the frequency components of a vibration signal to determine the source and cause of vibration.

speech — The use of particular vocal sounds that form words to give expression to a person's thoughts or ideas.

speech interference level (SIL) — Average of the sound-pressure levels of a sound expressed in deci-

bels in three octave bands that have center frequencies of 500, 1000, and 2000 Hz.

speech perception test — A test designed to measure hearing acuity by administering a carefully controlled list of words.

speed — The rate of change of position with time.

speed of response — A measurement of how quickly a system responds to a change in the input.

SPEGL — Acronym for short-term public emergency guidance level.

spend fuel — Fuel rods that no longer have enough fissionable uranium in them to be efficiently used to produce power.

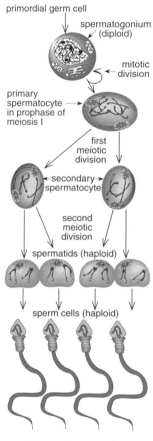

Spermatogenesis

spermatogenesis — The process of male gamete formation, including formation of a primary spermatocyte from a spermatogonium, meiotic division of the spermatocyte, and transformation of the four resulting spermatids into spermatozoa.

SPF — Acronym for structured programming facility; see sun protection factor.

sphalerite ((Zn,Fe)S) — An ore of zinc sulfide, a brownish, yellowish, or black mineral; also known as blende, false galena, lead marcasite, mock lead, or mock ore.

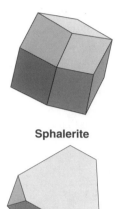

Sphalerite

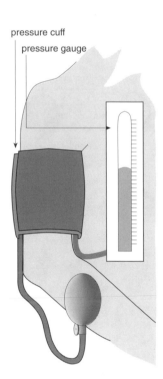

Sphygmomanometer

sphygmomanometer — An apparatus for measuring blood pressure.

SPI — Acronym for strategic planning initiative.

spill event — A discharge of oil into or upon the navigable waters of the United States or adjoining shorelines in harmful quantities, causing a film or sheen upon the water and violating applicable water quality standards.

spillway — The channel or passageway around or over a dam through which excess water is diverted.

spinal cord — The main dorsal nerve of the central nervous system in vertebrates extending down the back from the medulla.

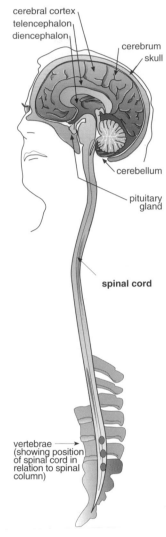

Spinal Cord

spinal cord injury — A dramatic disruption of the spinal cord often associated with extensive musculoskeletal involvement and resulting in a variety of paralyses and loss of bodily functions.

spinal tap — The test in which a fluid sample is removed from the spinal column with a thin needle.

spindle — (*genetics*) A microtubular apparatus that appears in many eukaryotic cells at the beginning of nuclear division and is responsible for the ordered separation of the chromosomes.

spirillary fever — (*disease*) A sporadic rat-bite fever caused by *Spirillum minor*, *S. minus*, or *Spirochaeta morsus muris*; a common form of rat-bite fever in Japan and Asia with fever rash of red or purple plaques. Incubation time is 1 to 3 weeks, with healed wounds reactivating when symptoms appear. The reservoir of infection is rats. It is transmitted by rat bite and is not communicable from person to person. It is controlled by destroying rats and by treatment of the patient.

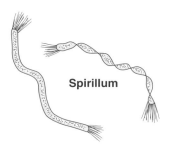

Spirillum

spirit — Any volatile liquid that has been distilled.

spirochete — Any of the slender, spiraling, undulating bacteria of the order Spirochaetales.

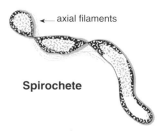

Spirochete

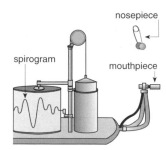

Spirometer

spirometer — An instrument that measures the flow of air in and out of lungs during respiration.

SPL — See sound pressure level.

splash-proof goggles — Eye protection made of a noncorrosive material that fits snugly against the face and has indirect ventilation ports.

spleen — A highly vascular ductless organ located in the left abdominal region near the stomach or intestine that is involved in the final destruction of red blood cells, filtration and storage of blood, and production of lymphocytes.

SPLMD — Acronym for soil-pore liquid monitoring device.

SPOC — Acronym for single point of contact.

spoil — Soil or rock that has been removed from its original location, such as in mining.

spontaneous combustion — The process in which slow oxidation produces enough heat to raise the temperature of a substance to its kindling temperature without direct application of flame, as with the autoignition of rags soaked in flammable liquids.

spontaneously combustible — Pertaining to a material that ignites as a result of retained heat from processing, will oxidize to generate heat and ignite, or will absorb moisture to generate heat and ignite.

sporadic — A number of events occurring at scattered, intermittent, and apparently random intervals.

spore — A tiny single- or multicelled asexual reproductive structure produced by fungi or bacteria; it is often able to survive drought, cold, and other unfavorable environmental conditions.

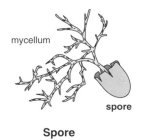

Spore

spore formation — Refers to the production of unicellular, often environmentally resistant, dormant, or reproductive bodies by some microbes.

sporocyst — A structure containing spores or reproductive cells.

sporophyte — The stage that produces spores in an organism having alterations of generations.

Sporozoea — A class of parasitic protozoa with both sexual and asexual phases.

sporozoite — A cell resulting from the sexual union of spores during the life cycle of a sporozoan.

SPOT — An Earth resource satellite with high-resolution sensors launched by France in 1986.

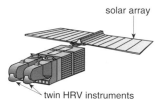

SPOT (French Satellite)

spot check — Supplemental test performed on a random basis.

sprain — A traumatic injury to the tendons, muscles, or ligaments around a joint with symptoms of pain, swelling, and discoloration of the skin over the joint.

sprawl — The unplanned development of open land.

spray — A mixture of water or other liquid plus a pesticide chemical applied in small droplets.

spray drift — The movement of airborne spray particles away from the intended application area.

spray tower — A vertical column for spraying liquid through gases in a variety of chambers of various designs; each liquid droplet contacts a volume of gas equal to the droplet cross-sectional area so as to create maximum contact between the gas and liquid.

spread plate — A method for performing a plate count of microorganisms where a known amount of a serial dilution is spread over the surface of an agar plate and after growth the number of colony-forming units is counted.

spring — A source of running water produced by the intersection of the water table with the surface of the ground.

spring melt — The process of warming temperatures melting snow and ice; may raise the amount of acidity entering streams and rivers in areas where acid deposition has been stored in the frozen water.

spring thaw — See spring melt.

spring turnover — A phenomenon in lakes occurring during the spring when surface ice melts and the surface water temperature warms to its greatest density and then the water sinks, causing it to literally turn over.

springer — A can with one end permanently bulged.

SPS — Acronym for State Permit System.

spurious — A falsified or erroneously attributed origin.

sq. ft. — See square foot.

sq. in. — See square inch.

SQBE — Acronym for small-quantity burner exemption.

SQG — See small-quantity generator.

squall line — An unbroken line (typically 12 to 30 miles) of black, ominous clouds towering 40,000 feet or more into the sky, causing thunderstorms of great severity and occasionally tornadoes.

squamous — An epithelial tissue that has scales.

Squamous Epithelial Tissue

square foot (sq. ft.) — An area that is 144 square inches.

square inch (sq. in.) — An area that is 1 inch wide by 1 inch long.

square mile — An area that is 1 mile wide by 1 mile long and is equal to 640 acres.

square wave — A time-dependent function consisting of a series of discontinuous changes.

square yard — An area that is 1 yard wide by 1 yard long and is equal to 9 square feet.

SR — Acronym for special review.

SRC — Acronym for solvent-refined coal.

SRF — Acronym for State Revolving Fund.

SRM — Acronym for Standard Reference Method.

SRP — Acronym for Special Review Procedure.

SRRP — Acronym for Source Reduction Review Project.

SRT — Acronym for solids retention time.

SRTS — Acronym for Special Requests Tracking System.

SS — Acronym for settleable solids, suspended solids, or stainless steel.

SSA — Acronym for sole-source aquifer.

SSAC — Acronym for soil site assimilated capacity.

SSC — Acronym for State Superfund Contracts.

SSD — Acronym for Standards Support Document.

SSE — See System Safety Engineering.

SSEIS — Acronym for Standard Support and Environmental Impact Statement or Stationary Source Emissions and Inventory System.

SSG — Acronym for special study group.

SSI — Acronym for size-selective inlet.

SSMS — Acronym for sparked source mass spectrometry.

SSO — Acronym for sanitary sewer overflow.

SSSD — Acronym for state soil survey database.

SST — Acronym for sea-surface temperature.

SSTS — Acronym for Section Seven Tracking System.

SSURO — Acronym for stop sale, use, and removal order.

St — See stoke.

St. Louis encephalitis — (*disease*) An acute inflammatory disease of short duration involving parts of the brain, spinal cord, and meninges; symptoms range from mild cases to high fever, disorientation, coma, convulsions, and possible death. Incubation time is usually 5 to 15 days. It is caused by a flavivirus and is found in the United States, Canada, and parts of South America. Possible reservoirs of infection include birds, rodents, bats, reptiles, amphibians, and mosquito adults or eggs. It is transmitted by the bite of a mosquito but not from person to person; it usually affects young children but can affect others and be more severe. It is controlled by destroying mosquito larvae, eggs, breeding places, and adult mosquitos and by using mechanical control techniques.

stability — (*meteorology*) The atmospheric condition existing when the temperature in the air rises rather than falls with altitude, allowing for little or no vertical air movement; (*matter*) the ability of a material to remain unchanged.

stabilization — The treatment of active organic matter in sludge to produce inert, harmless material.

stabilization lagoon — See stabilization pond.

stabilization pond — A shallow oxidation pond with a continuous current used for oxidation of organic material present in the sewage; also known as stabilization lagoon.

stabilizer–conditioner — A chemical compound, normally cyanuric acid, used to fortify and prolong the chlorine residual in pool water by reducing chlorine consumption lost by evaporation, dissipation, and oxidation.

stable — Refers to a substance that does not change without the application of an external force.

stable air — An air mass that remains stationary rather than moving in its normal horizontal and vertical directions, not allowing for the dispersion of pollutants; also known as stagnation.

stable condition — A state of health in which the prognosis indicates little if any immediate change.

stable element — A nonradioactive element not subject to spontaneous nuclear degeneration.

stable isotope — A nonradioactive isotope of an element.

stack — (*construction*) A vertical passage rising above a roof through which products of combustion are conducted into the atmosphere; any chimney, flue, vent, roof monitor, conduit, or duct arranged to vent emissions to the ambient air; (*plumbing*) a general term for any vertical line of soil, waste, vent, or inside conductor piping.

stack effect — Pressure-driven air flow produced by convection as heated air rises, creating a positive pressure area at the top of the building and a negative pressure area at the bottom; the stack effect can overpower the mechanical system and disrupt ventilation and circulation in the building.

stack emissions — Particulate matter captured and released to the atmosphere through a stack, chimney, or flue.

stack gas — Air emitted to the atmosphere through a chimney after a production process or combustion takes place; also called flue gas.

stack pressure — See stack effect.

stack sampling — Collecting representative samples of gases and particulate matter flowing through a duct or stack.

stack testing — A monitoring system that determines emission characteristics and quantities at one particular point in time.

stack transmissometer — A monitoring device used to determine the opacity of gases leaving a smoke stack.

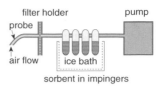

Stack Sampler

Stafford Act — A law passed by Congress to coordinate overall consequence management response and recovery activities related to a disaster or emergency, as directed by the Federal Emergency Management Agency.

stage — The elevation or depth of water.

stagnant water — Polluted water that is motionless or unable to drain off into the streams or into the ground.

stagnation — See stable air.

stagnation pressure — Sum of the static and dynamic pressure.

stain — A dye used to color microorganisms.

stainless steel (SS) — Corrosion-resisting steel containing 15% or more chromium; used for cutlery, furnace parts, chemical plant equipment, valves, stills, turbine blades, and ball bearings.

stakeholder — Any organization, governmental entity, or individual that has a stake in or may be impacted by a given approach to environmental regulation, pollution prevention, energy conservation, etc.

standard — A desired or prescribed level of achievement or reference with respect to a criterion representing acceptable performance; a set value or reference point from which measurements or calibrations are made.

standard air — Dry air at 29.92 inHg absolute pressure and 70°F and weighing 0.075 lb/ft³.

Standard Ambient Air Quality Standards (SAAQS) — Standards established by the U.S. EPA that describe the concentration limits that a given pollutant may reach annually and on a 24-hour basis.

standard candle — A standard measure of light brightness; 1 standard candle radiates about 12.5 lumens. The term has been replaced by the candela.

standard conditions — See standard temperature and pressure.

standard deviation (SD) — A measure of a dispersion or variability in a frequency distribution equal to the square root of the arithmetic mean of the squares of the deviations from the arithmetic mean of the distribution.

standard effective temperature — A temperature index that accounts for radiative and latent heat transfers.

S

standard error — An estimate of the standard deviation of the means of many samples; calculated as the standard deviations divided by the square root of the number of individuals in a sample.

standard gravity — A gravitational force that will produce an acceleration equal to 32.17 ft/sec.

Standard Industrial Classification Code (SIC) — A classification system used to identify places of employment by major types of activity.

standard metropolitan statistical area (SMSA) — An urbanized region with at least 100,000 inhabitants and strong economic and social ties to a central city of at least 50,000 people.

standard notation — See scientific notation.

standard operating procedure (SOP) — A document that describes in detail an operation, analysis, or action commonly accepted as the preferred method for performing certain routine or repetitive tasks; a routine procedure that has been designated by a manager or physician to be used when a specific type of problem occurs.

standard pressure — Pressure equivalent to 760 mmHg at sea level at a temperature of 0°C.

standard rate — A rate calculated to permit a more accurate comparison of the state of health of two or more populations that differ substantially in some important respects, such as distribution by age or sex; also called adjusted rate.

standard sample — The aliquot of finished drinking water that is examined for the presence of coliform bacteria.

standard solution — A solution in which the concentration is known.

standard temperature — The measurement arbitrarily set at 0°C at a pressure of 760 mmHg.

standard temperature and pressure (STP) — A temperature of 0°C at one atmosphere of pressure; also known as standard conditions or normal temperature and pressure.

standardized incidence ratio (SIR) — A ratio of the crude rate of an exposed population to a weighted average of the category-specific rate in the unexposed population, weighted by a distribution of the exposed population.

standardized mortality ratio (SMR) — The ratio of the number of deaths in a specific population to the number of deaths that would have been expected when the population is compared with the same mortality experienced in a standard population.

standby loss — A measure of the losses from a water heater tank; the ratio of heat loss per hour to the heat content of the stored water above room temperature; heat loss per hour per square foot of the surface area of a tank (expressed in watts).

standing orders — Written documents that contain rules, policies, procedures, regulations, and orders for the conduct of patient care in various situations.

standing wave — A periodic wave that has a fixed distribution in space; also known as a stationary wave.

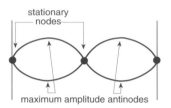

Standing Waves

staphyloccal food poisoning — (*disease*) An intoxication of abrupt and sometimes violent onset with severe nausea, cramps, vomiting, and prostration; it is frequently accompanied by diarrhea for 1 to 2 days. Incubation time is usually 2 to 4 hours but may vary up to 7 hours. It is caused by *Staphylococcus aureus* enterotoxins and is found worldwide. The reservoir of infection is people. It is transmitted by ingestion of food containing the enterotoxin and is not communicable from person to person; general susceptibility. It is controlled by proper heating and cooling of food, educating food handlers, proper hygiene, and exclusion from food handling of people with boils, abscesses, runny noses, or other drainage conditions.

staphylococcal infection disease — (*disease*) Any of a variety of diseases caused by a bacterium; clinical symptoms range from a small boil to septicemia and death. Skin lesions such as impetigo, boils, carbuncles, and abscesses may occur; cellulitis, pneumonia, osteomyelitis, and endocarditis may also occur. Incubation time varies but is commonly 4 to 10 days. It is caused by coagulase-positive strains of *Staphylococcus aureus*, which may have different phage types, antibiotic resistance, or serological agglutination; it is found worldwide. The reservoir of infection is people. It is transmitted from the anterior nares, skin, and sites of infection in people and is communicable as long as the lesions on the skin are present, when there is drainage from the nose, or while the carrier state persists. Susceptibility is highest among newborns and chronically ill, elderly, or debilitated people. It is controlled by good personal hygiene (especially handwashing), disinfection, proper placement and removal of all dressings and discharges in the hospital, treatment of the patient, and isolation in the hospital when it is a hospital-borne infection.

staphylococcal pneumonia — A pneumonia caused by a staphylococcus infection.

Staphylococcus — A genus of Gram-positive bacteria made up of spherical microorganisms tending to occur in grape-like clusters; they are constantly present on the skin and in the upper respiratory tract and are the most common cause of localized suppurating infections.

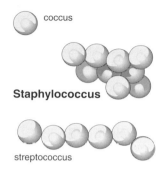

coccus

Staphylococcus

streptococcus

Staphylococcus aureus — A species of *Staphylococcus* commonly found on the skin, in the nose and throat, and in the environment; some serotypes are highly virulent, causing pyogenic infections and especially nosocomial infections.

starch ($(C_6H_{10}O_5)_x$) — Any of a mixture of polymers made up of the sugar glucose; found in all assimilating (green) plants.

starch iodide method (iodometric) — A titration test in which chlorine displaces iodine from potassium iodide in an acid solution and forms a blue color with starch; it is decolorized by the addition of standard sodium thiosulfate; the test is generally used to measure high residuals of chlorine.

STARS — Acronym for Strategic Targeted Activities for Results System.

stasis — The stagnation or inactivity of the life processes within organisms.

statcoulomb — See electrostatic unit of charge.

state implementation plans — U.S. EPA-approved state plans for the establishment, regulation, and enforcement of air pollution standards

state plane coordinate system — A system for specifying positions of geodetic stations using plane rectangular coordinates.

State Pollution Discharge Elimination System — A permit issued by the state regulating the discharge of wastewater and specifying the maximum discharge limits for the parameters present in the particular discharge.

static electric charge — A positive or negative stationary electric charge resulting from rubbing unlike bodies together.

static electricity — An accumulation of positive or negative charges on an object.

static head — The difference in elevation between the pumping source and point of delivery.

static load — The stresses on a body that increase as a function of the body parts remaining immobile for extended periods.

static pressure — (*acoustics*) Pressure of the medium on which the alternating sound pressure is superimposed; (*fluid mechanics*) the pressure exerted by a fluid in a direction normal to the direction of flow; for a fluid at rest it is exerted uniformly in all directions; (*ventilation*) usually the difference between the absolute pressure in an exhaust system and atmospheric pressure such that static pressure less than atmospheric pressure is termed negative static pressure.

static water level — The elevation or level of the water table in a well if the pump is not operating; the level or elevation to which water would rise in a tube connected to an artesian aquifer or basin in a conduit under pressure.

stationary front — A front that moves little or not at all, bringing mild conditions with some rain.

stationary grate — See fixed grate.

stationary phase — A period during the growth cycle of a population in which growth rate equals the death rate.

stationary source — Any building, structure, facility, or installation that emits or may emit any air pollutant subject to regulation under the Clean Air Act.

stationary wave — See standing wave.

statistic — A single term that describes a property of a set of data.

statistical power — The probability that one can detect an effect if one really exists.

statistical significance — Refers to when the probability of observing a certain result or class of results by chance alone is less than some predetermined value.

statistical technique — The use of information gained from an appropriate sample of a larger group to make inferences concerning the group in a specific matter.

statistics — A branch of mathematics dealing with the collection, analysis, interpretation, and presentation of large quantities of numerical data.

statute of limitation — The period during which any criminal prosecution contemplated must be brought.

statutory law — A rule passed by a state or national legislature.

STC — See short-term concentration or sound transmission class.

STD — Acronym for sexually transmitted disease.

steady flow — A flow rate of a substance that does not vary significantly with time.

steady-state vibration — A vibration brought on by an unchanging, continuing, periodic force.

steam — Hot water vapor often used as a physical sanitizing agent for treating food equipment.

steam cycle — See Rankine cycle.

S

steam injection and vacuum extraction (SIVE) — A process used to remove volatile organic compounds and semivolatile organic compounds from contaminated soils.

steam-injection flares — Flares that mix steam with exhaust gases at the top of the stack, separating the hydrocarbon molecules and reducing unwanted polymerization.

steam pasteurization — A technique for killing bacteria on the surface of cattle and hog carcasses by using steam to briefly raise the carcass surface temperature, which kills pathogens occurring after the evisceration stage or before final cooling.

steam sterilization — A process conducted in an enclosed vessel at 121 to 123°C for 30 minutes to kill all bacteria and spores.

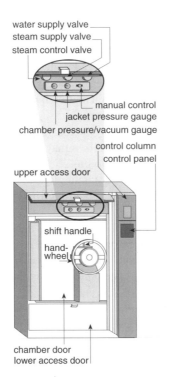

water supply valve
steam supply valve
steam control valve

manual control
jacket pressure gauge
chamber pressure/vacuum gauge

control column
control panel

upper access door

shift handle
hand-wheel

chamber door
lower access door

Steam Sterilizer

steel — An iron-based alloy containing about 2% carbon; it is malleable under certain conditions.

STEL — See short-term exposure limit.

STEM — Acronym for scanning transmission electron microscope.

stenosis — Narrowing of a body passage or an opening such as a blood vessel.

stereoisomer — One of two or more chemical compounds containing the same atoms linked in the same way but organized differently in space.

stereoscopic microscope — A microscope that produces three-dimensional images through the use of double eyepieces in double objectives.

sterile — Free of living microorganisms and spores.

sterile field — An area immediately around a patient that has been prepared for a surgical procedure.

sterility — The inability to conceive or reproduce the species.

sterilization — A process that kills all living organisms, including spores, on and in an object.

sterilize — The destruction of all living organisms, including spores in a material.

sterilizer — A piece of equipment used to destroy all forms of bacteria, viruses, fungi, and spores.

Sterling engine — An external combustion engine that converts heat into usable mechanical energy via the heating or expansion and cooling or contraction of a captive gas such as helium or hydrogen, thereby driving a piston up and down to utilize the energy.

stibine (SbH_3) — A colorless gas with a disagreeable odor similar to hydrogen sulfide; MW: 124.8, BP: –1°F, Sol: slight. It is used as a chemical intermediate and in chemical synthesis; it is liberated from purification of antimony. It is hazardous to the blood, liver, kidneys, and lungs and is toxic through inhalation. Symptoms of exposure include headache, weakness, nausea, abdominal pain, lumbar pain, hemoglobinuria, hemolytic anemia, jaundice, and lung irritation. OSHA exposure limit (TWA): 0.1 ppm [air] or 0.5 mg/m³.

still — A closed chamber in which heat is applied to vaporize a liquid, and the vapor is then condensed.

still air — Air in a building with a velocity of 25 feet per minute or less.

STO — Acronym for science technology objectives.

stochastic — Pertaining to the assumption that the actions of a chemical substance result from probabilistic events.

Stockholm Convention on Persistent Organic Pollutants — A global treaty to protect human health and the environment from persistent organic pollutants, which are chemicals remaining intact in the environment for long periods, becoming widely distributed geographically and accumulating in the fatty tissue of living organisms. They are toxic to humans and wildlife.

Stoddard solvent — A colorless liquid with a kerosene-like odor; MW: varies, BP: 428 to 572°F, Sol: insoluble, Fl.P: 110°F, sp. gr. 0.78. It is used as a solvent in the drycleaning industry; in the paint and varnish industries; as a solvent for printing inks; in textile printing; in the manufacture of aerosol sprays; in the manufacture of sprays for pesticides, herbicides, household cleaners, and silicone compounds; as a solvent and thinner in protective coating materials; in metal cleaning and degreasing; in leather degreasing; as a general solvent in fabric waterproofing; in processing synthetic yarns; in rubber cements. It is hazardous to the skin, eyes, and respiratory and central nervous systems and is toxic through inhalation, ingestion, and contact. Symptoms of exposure include dizziness, dermatitis, and irritation of the eyes, nose, and throat. OSHA exposure limit (TWA): 100 ppm [air] or 525 mg/m³.

stoichiometric — Refers to the set of chemical and physical principles applied to determine the relationships between reactants and products in a chemical process, where the mass relationships existing between the chemical reactants and products are of primary interest; the quantities of reactants and products as determined by the reaction equation for a specific chemical reaction which produces no byproducts.

stoichiometric air — See theoretical air.

stoichiometry — The mass relationship of elements and compounds in a chemical reaction.

stoke (St) — A unit of kinematic viscosity.

stoker — A mechanical device used to feed solid fuel or solid waste to a furnace.

stoma — A tiny pore-like opening on the underside of a leaf through which a plant exchanges gases.

stomach — Expansion of the alimentary canal between the esophagus and the duodenum.

stomach poison — A pesticide that must be eaten in order to kill or poison an organism.

stomatitis — An inflammatory condition of the mouth resulting from an infection by bacteria, viruses, or fungi; exposure to chemicals or drugs; a vitamin deficiency; or a systemic inflammatory disease.

stone coal — See anthracite.

stool — Fecal matter.

stopping power — The average rate of energy loss of a charged particle per unit thickness of a material or per unit mass of material transversed; a measure of the effect of a substance upon the kinetic energy of a charged particle passing through it.

storage — (*hazardous materials*) Any method of keeping raw materials, finished goods, or products while awaiting use, shipment, or consumption; (*computer science*) parts of a computer system used for storing data and programs.

storage pit — A hole below ground in which solid waste is held before processing.

storage tube — A visual display device used for displaying maps and graphic information that continuously retains the image.

stored energy — See potential energy.

storm drain system — A system of inlets and pipes that removes surface rain water or melted snow to a creek, pond, lake, river, or the ocean.

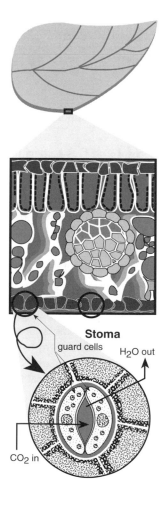

Stoma

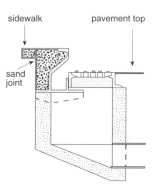

Storm drain system

S

storm sewer — A sewer used for conveying rainwater, surface water, condensate, cooling water, or liquid waste off of a surface to a storm drain system.

storm surge — An increase in water level above the normal water level on the open coast due to the action of wind stress and atmospheric pressure on the surface of the water.

storm track — The path followed by a center of low atmospheric pressure. In many cases, multiple storms follow the same storm track.

storm water — Any water resulting from precipitation mixed with the accumulation of dirt, soil, and other precipitation that has fallen or which flows.

stormwater pollution prevention plan — A plan used to describe a process that a facility uses to evaluate potential pollutant sources and implements appropriate measures to prevent or control the discharge of pollutants into stormwater runoff.

STP — Acronym for sewage treatment plant; see sewage treatment system; see standard temperature and pressure.

STP flow rate — The rate of flow of fluid by volume corrected to standard temperature and pressure.

STP volume — The volume that a quantity of gas or air would occupy at standard temperature and pressure.

straight-chain hydrocarbon — A hydrocarbon in which the carbon atoms are linked together in a long, straight chain.

strain — Muscular damage due to excessive physical effort; a population of cells, all descended from a single pure isolate.

S-trap — Configuration of a sink drain line in which piping beyond the trap runs vertically instead of horizontally; this can cause the water in the trap to be siphoned out, allowing sewer gas to enter.

strata — The distinct layers of stratified, or sedimentary, rock.

strategic petroleum reserve — A reserve consisting of government-owned and -controlled crude oil stockpiles stored at various locations in the Gulf Coast region of the United States.

strategy — The science and art of employing political and psychological pressure by organizations to obtain maximum support for adoption of specific policies.

stratification — The separation of an object into layers; (*hydrology*) a phenomenon occurring when a body of water becomes divided into distinguishable layers; (*geology*) the deposition of sedimentary materials in layers; (*meteorology*) a condition occurring when the air at different levels in the atmosphere is at a different temperature and little mixing occurs between the layers.

stratocumulus — Irregular mass of clouds from near the surface of the Earth to 6500 feet up. It spreads out in a rolling or puffy layer and is gray with darker shadings; it does not produce rain but the clouds sometimes change into rain clouds.

stratosphere — Region extending from the tropopause to a height of approximately 33 miles from the Earth, where the temperature reaches a maximum and where clouds of water never form.

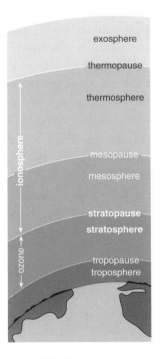

Stratosphere

Stratospheric Aerosol and Gas Experiment — A NASA experiment used to determine the vertical distribution of stratospheric aerosols, ozone, nitrogen oxide, and water vapor on a global scale and to develop a viable, satellite-based, remote sensing technique to measure these gases.

stratum — A single bed or layer, regardless of thickness, that consists generally of the same kind of rock or other material.

stratus — Low clouds that are quite uniform, like fog, with an obvious base above the ground; when they are dull-gray, the sky looks heavy and ready to rain. A fine drizzle can fall from these clouds. They are formed when a layer of air is cooled below the saturation point without vertical movement and are sheet like or layered.

stream — A natural watercourse containing water at least part of the year.

street refuse — Material picked up when streets and sidewalks are swept manually and mechanically.

S

streptobacillary fever — (*disease*) A bacterial disease caused by the bite of a rat with an onset of symptoms within 10 days after the bite has healed; it is characterized by chills, fever, headache, muscle pain, and rash. Joints become swollen, red, and painful. Incubation time is 3 to 10 days. It is caused by *Streptobacillus moniliformis* and is found worldwide but not commonly in North and South America. The reservoir of infection is the infected rat; transmission occurs through secretion of the mouth, nose, or conjunctival sac of an infected animal, usually introduced by biting. It is not communicable from person to person; susceptibility is not known. It is controlled through specific pharmaceutical treatment. Also known as Haverhill fever.

Streptobacillus moniliforms — A species of necklace-shaped bacteria that can cause rat bite fever in humans.

streptococcal infection — (*disease*) Disease or infection caused by *Streptococcus pyogenes* group A streptococci of approximately 75 serological types. It produces strep throat, scarlet fever, impetigo, erysipelas, puerperal fever, tonsillitis, and pharyngitis. The incubation period is 1 to 3 days. It is common in temperate climates and is seen in semitropical climates but less commonly than in tropical climates. The reservoir of infection is people. It is transmitted by direct or close contact with a patient or carrier's nose, skin, hands, clothing, dust, or contaminated food. It is communicable for 10 to 21 days if untreated and 24 to 48 hours if treated with penicillin; general susceptibility. It is controlled by proper health habits that incorporate an understanding of its transmission, by pasteurizing milk, and by protecting and refrigerating foods.

Streptococcus — A genus of Gram-positive facultatively aerobic cocci occurring in pairs or chains.

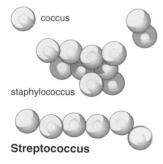

Streptococcus

Streptococcus pneumoniae — One of 70 antigenic types of pneumococci that cause pneumonia and other diseases in humans.

Streptococcus pyogenes — A species of *Streptococcus* with many strains that are pathogenic to humans

and cause such diseases as scarlet fever and strep throat.

stress — The nonspecific response of the body to any demand made upon it. It may result in a fight-or-flight response to potentially dangerous situations. The body reacts through elevated heart rate and blood pressure and a redistribution of blood flow to the brain and major muscle groups and away from the distal body parts.

stress-related disease — See stress shock.

stress shock — A loose set of physical, psychological, and/or behavioral changes thought to result from the stress of excess competition and other members of the same species being in extremely close proximity.

stress test — A test that measures the functions of a system of the body subjected to carefully controlled amounts of physiological stress.

stressor — Any agent causing a condition of stress.

stressor indicator — A characteristic measured to quantify a natural process, an environmental hazard, or a management action that results in changes in exposure and habitat.

striate — Refers to something that is striped, is marked by parallel lines, or has structural lines.

strike — The deposition of fly eggs or larvae on a living host.

strip — To remove friable asbestos materials from any part of a facility.

strip mining — A process that uses machines to scrape soil or rock away from mineral deposits just under the surface of the Earth; surface mining.

stripping — The process of removing one substance from another by heating it until it is loosened and then subjecting it to a gas stream that carries it away.

stroke — See cerebrovascular accident.

stroke counter — A flowrate meter that measures air flow by counting the number of strokes of the piston and equating it to air flow.

strong acid — An acid capable of a high degree of ionization in water solution.

strong base — A base capable of a high degree of ionization in water solution.

strongyloidiasis — (*disease*) A helminthic infection of the duodenum and upper jejunum that is asymptomatic in most cases. Dermatitis may occur when the larvae penetrate the skin; also, cough, râles, and pneumonitis may occur as the larvae pass through the lungs; abdominal symptoms may occur when the adult female is in the mucosa of the intestine; the chronic infection may be either mild or severe, with symptoms of diarrhea, weight loss, vomiting, weakness, or constipation. Incubation

S

time is approximately 2 weeks from penetration of the skin until the larvae are found in the feces; the symptoms vary. It is caused by *Strongyloides stercoralis*, a nematode found in tropical and temperate areas. Reservoirs of infection include people and possibly dogs. It is transmitted from feces or moist soil contaminated with feces; when the larvae enter a penetration in the skin, they are carried to the lungs and then to the stomach. It is communicable for up to 35 years; general susceptibility. It is controlled by proper disposal of human feces, good health habits, and public education.

Strongyloides stercoralis
(parasitic roundworm causing
strongyloidiasis)

strontium (Sr) — A metallic element with an atomic number of 38 and atomic weight of 87.62.

structural chemistry — Science dealing with a molecular structure of chemical substances.

structural formula — A schematic representation of the arrangement of molecules of organic compounds.

structural geology — Study of the structures of the Earth and their relationship to the forces that produce these structures.

structural isomers — Molecules that contain the same number and types of atoms but have a different arrangement of bonds.

structure analysis and design technique (SADT) — A tool for graphic representation of a method of analysis for the simplification of problem solving of complex systems.

structure–activity relationship (SAR) — The association between a chemical structure and carcinogenicity.

strychnine ($C_{21}H_{22}N_2O_2$) — A colorless to white, odorless, crystalline solid; MW: 334.4, BP: decomposes, Sol: 0.02%, Fl.P: not known, sp. gr. 1.36. It is used in the formulation of medicinals and pesticides. It is hazardous to the central nervous system and is toxic through inhalation, ingestion, and contact. Symptoms of exposure include stiff neck and facial muscles, restlessness, apprehension, increased acuity of perception, increased reflex excitability, cyanosis, and tetanic convulsions with opisthotonos. OSHA exposure limit (TWA): 0.15 mg/m^3 [air].

stupor — Partial unconsciousness or nearly complete unconsciousness.

styrene ($C_6H_5CHCH_2$) — A colorless to yellow, oily liquid with a sweet, floral odor; MW: 104.2, BP: 293°F, Sol: slight, Fl.P: 88°F, sp. gr. 0.91. It is used in the manufacture of tires and other rubber goods; in the manufacture of glass fiber; in the manufacture of surface coatings; as an intermediate material. It is hazardous to the central nervous system, respiratory system, eyes, and skin and is toxic through inhalation, ingestion, and contact. Symptoms of exposure include drowsiness, weakness, unsteady gait, narcosis, defatting dermatitis, and irritation of the eyes and nose. OSHA exposure limit (TWA): 50 ppm [air] or 215 mg/m^3.

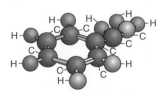

Styrene

styrene oxide (C_8H_8O) — A colorless liquid; MW: 120.15; BP: 194.1°C; Sol: soluble in alcohol, ether, and benzene; Fl.P: 79°C; density: 1.0523 at 16°C. It is used in organic synthesis. It is hazardous to the skin and is lethal to rats at high levels through inhalation. It is a mutigen, teratogen, and carcinogen. ACGIH exposure limit: (TLV-TWA): NA.

$$H_2C\!\!-\!\!CH_2\!\!-\!\!\bigcirc$$

Styrene oxide

subacute — Referring to a stimulus not intense enough to bring about rapid response.

subacute infection — An infection that is less severe or of shorter duration than an acute infection.

subacute toxicity — The property of a substance or mixture of substances to cause adverse effects in an organism upon repeated or continuous exposure within less than one half of the lifetime of that organism.

subangular blocky — A soil structure that looks like a pile of different nuts, where the sides of the aggregates form obtuse angles and the corners are rounded. It is usually more permeable than blocky soil. It is nearly block like, with six or more sides and all three dimensions about the same. It is usually found in subsoil or B horizon.

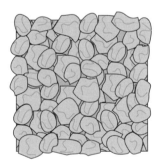

Subangular Blocky Soil

subchronic — Of intermediate duration; usually used to describe studies on levels of exposure between 5 and 90 days.

subchronic exposure — Multiple or continuous exposures occurring usually over a period of 2 weeks to several years.

subclinical — Without clinical manifestations; the early stages or a very mild form of a disease.

subcutaneous — Beneath the layers of the skin.

subdivision ordinance — An ordinance adopted by a local entity for the appropriate establishment of rules relating to tracts of land.

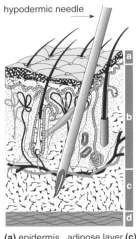

(a) epidermis   adipose layer (c)
(b) dermis   muscle layer (d)

Subcutaneous Injection

subduction — The action or process of one crustal plate descending below the edge of another.

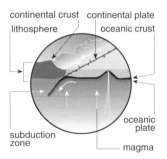

Subduction Zone

subjective data — Data related to a person that were collected from that individual.

sublethal — Referring to a stimulus below the level that causes death.

sublethal dose — A dose of a potentially lethal substance that is not large enough to cause death.

sublimation — The process by which matter passes from a solid to a gas without passing through a liquid phase.

sublime — To vaporize directly from the solid to the gaseous state.

sublittoral — The region in a lake between the deepest-growing rooted vegetation and the part of the lake below the thermocline.

submerged arc welding — A process that produces coalescence of metals by heating them with an arc between a bare metal electrode and the work.

submersible pump — A centrifugal pump driven by a close-coupled electrical motor constructed for submerged operation as a single unit.

Submersible Wastewater Pump Association — An organization of manufacturers of submersible wastewater pumps and systems for municipal and industrial applications and manufacturers of component parts and accessory items for those pumps and systems.

subpoena — A court order requiring a witness to appear and give testimony.

subpoena duces tecum — A court order requiring a witness to attend a proceeding and bring certain documents.

subsidence — A settling or sinking of the surface of the Earth caused by natural means, such as chemical weathering in limestone areas, or by artificial means, such as mining.

subsoil — A layer of soil beneath the topsoil with a lower organic content and higher concentration of fine mineral particles.

substance — Any drug, chemical, or biological entity.

substance abuse — Overindulgence in and dependence on a stimulant, depressant, or other chemical substance leading to effects that are detrimental to the individual's physical or mental health or the welfare of others.

substandard — A deviation from the standard or norm.

substation — A facility that steps up or steps down the voltage in utility power lines.

substitution — A technique used to reduce pollutants by changing or improving the process or altering the fuels used in the process.

substrate — A layer of material on which an organism can grow and multiply; any surface on which a layer of a different material can be deposited.

subsurface soil injection — The injection of hazardous waste under the surface of the ground and away from water supplies.

suction — An aspiration of a gas or fluid by reducing air pressure.

suction filter — See vacuum filter.

sudden death — A death that occurs unexpectedly within 1 hour after the onset of symptoms.

sudden infant death syndrome (SIDS) — The sudden and unexpected death of an apparently healthy infant that typically occurs between the ages of 3 weeks and 5 months and is not explained by postmortem studies.

sugar — The common term for a water-soluble, crystalline mono- or oligosaccharide; examples are sucrose ($C_{12}H_{22}O_{11}$) or cane sugar. See also glucose.

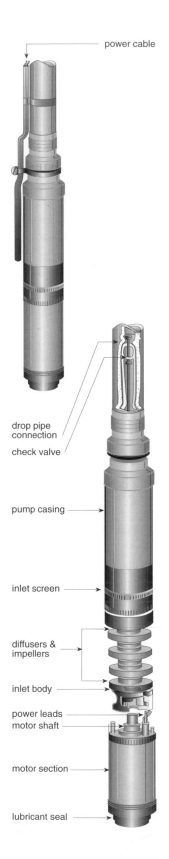

power cable

drop pipe connection

check valve

pump casing

inlet screen

diffusers & impellers

inlet body

power leads

motor shaft

motor section

lubricant seal

Submersible Pump

sulfate — A compound containing the –SO₄ group in which the hydrogen of sulfuric acid is replaced by either a metal or organic radical to become a sulfate salt or ester.

sulfate attack — Damage to concrete caused by the effects of chemical reactions between sulfates in soils or groundwater and hydrated lime and hydrated calcium aluminate in cement paste.

sulfhemoglobin — The reaction product of oxyhemoglobin and hydrogen sulfide.

sulfiting agent — A food preservative composed of potassium or sodium bisulfite or potassium metabisulfite used in processing beer, wine, baked goods, soup mixes, and some imported seafoods and also to give a fresh appearance to salads and vegetables. Symptoms include flushing, faintness, hives, headaches, gastrointestinal distress, breathing difficulties, and, in extreme cases, loss of consciousness and death.

sulfur — A nonmetallic, multivalent, tasteless, odorless chemical element that occurs in a yellow crystalline form, especially in volcanic areas; atomic number: 16, atomic weight: 32.06.

sulfur dioxide (SO_2) — A colorless gas (liquid below 14°F) with a characteristic, irritating, pungent odor that is shipped as a liquefied compressed gas; MW: 64.1, BP: 14°F, Sol: 10%. It is used as a bleaching agent and preservative agent in the wood and pulp, sugar, and food preparation industries; as a solvent in oil refineries; during synthesis in the chemical, petroleum, textiles, pharmaceutical, tanning, photography, metal, and rubber industries; during extraction, enrichment, and recovery processes in mining and metallurgy; in miscellaneous processes during leather tanning, special glass manufacture, water treatment, textile processing, and chrome waste treatment. It is hazardous to the respiratory system, skin, and eyes and is toxic through inhalation and contact. Symptoms of exposure include rhinorrhea, choking, cough, reflex bronchoconstriction, eye and skin burns, and irritation of the eyes, nose, and throat. OSHA exposure limit (TWA): 2 ppm [air] or 215 mg/m³.

sulfur hexafluoride (SF_6) — A very powerful greenhouse gas used primarily in electrical transmission and distribution systems.

sulfur monochloride (S_2Cl_2) — A light-amber to yellow-red, oily liquid with a pungent, nauseating, irritating odor; MW: 135.0, BP: 280°F, Sol: decomposes, Fl.P: 245°F, sp. gr. 1.68. It is used for the production of white vulcanized oils; as a natural or synthetic rubber extender; as a cross-linking catalyst; in chemical synthesis for the pro-duction of intermediates for dyes, pharmaceuticals, insecticides, and war gases; for treatment of drying oils; for cold vulcanizing; as a solvent for sulfur and sulfur compounds. It is hazardous to the respiratory system, skin, and eyes and is toxic through inhalation, ingestion, and contact. Symptoms of exposure include lacrimation, cough, pulmonary edema, and eye and skin burns. OSHA exposure limit (TWA): ceiling 1 ppm [air] or 6 mg/m³.

sulfur oxides (SO_x) — Oxides of sulfur; any of several pungent, colorless gases formed primarily from the combustion of fossil fuels; they are major air pollutants and cause damage to the respiratory tract as well as vegetation.

sulfur pentafluoride (S_2F_{10}) — A colorless liquid or gas (above 84°F) with an odor like sulfur dioxide; MW: 254.1, BP: 84°F, Sol: insoluble, sp. gr. 2.08 at 32°F. It is used during synthesis of sulfur hexafluoride and is not produced commercially. It is hazardous to the respiratory system and central nervous system and is toxic through inhalation and contact. Symptoms of exposure in animals include pulmonary edema and hemorrhage. OSHA exposure limit (TWA): ceiling 0.01 ppm [air] or 0.1 mg/m³.

sulfur recovery plant — Any plant that recovers elemental sulfur from a gas stream.

sulfur trioxide (SO_3) — An oxide of sulfur produced by the action of oxygen on sulfur dioxide in the presence of a catalyst such as iron oxide.

sulfuric acid (H_2SO_4) — A colorless to dark-brown, oily, odorless liquid (pure compound is a solid below 51°F) that is often used in an aqueous solution; MW: 98.1, BP: 554°F, Sol: miscible, sp. gr. 1.84. It is used in the manufacture of phosphoric acid and fertilizers; in petroleum refining; during the manufacture of pigments and dyes, and dyestuff intermediates; in the manufacture of industrial and military explosives; in the production of alcohols, phenols, and inorganic sulfates; in ore leaching and processing; in metal cleaning and plating; in the manufacture of detergents; in coke-oven gas refining; in the plastics industry for the manufacture of rayon, cellophane, cellulose, acetate, and others; in food processing; for the preparation of insecticides; in the manufacture of natural and synthetic rubber; in the treatment of industrial water for pH control; in the manufacture of textiles and leather; as a laboratory reagent (solvent) for chemical analysis; in chemical synthesis. It is hazardous to the respiratory system, eyes, skin, and teeth and is toxic through inhalation,

S

ingestion, and contact. Symptoms of exposure include pulmonary edema, bronchitis, emphysema, conjunctivitis, stomatitis, dental erosion, tracheobronchitis, skin and eye burns, dermatitis, and irritation of the eyes, nose, and throat. See also oleum. OSHA exposure limit (TWA): 1 mg/m³ [air].

sulfuryl fluoride (SO₂F₂) — A colorless, odorless gas that is shipped as a liquefied compressed gas; MW: 102.1, BP: –68°F, Sol: 0.2% at 32°F. It is used as an insecticidal fumigant for control of dry wood termites and other structural pests. It is hazardous to the respiratory system and central nervous system and is toxic through inhalation and contact. Symptoms of exposure include conjunctivitis, rhinorrhea, pharyngeal irritation, and paresthesia; in animals, narcosis, tremor, convulsions, pulmonary edema, and kidney injury. OSHA exposure limit (TWA): 5 ppm [air] or 20 mg/m³.

summary judgment — A judgment by the court made without a trial and based on written interrogatories, depositions, and sworn affidavits; the motion for summary judgment is filed when prosecution can substantiate a lack of genuine issues in the dispute.

summons — A court order requiring the person named in an action to appear before the court and answer the complaint.

sump — A tank or pit that receives sewage liquid or liquid waste located below the normal grade of the gravity system that must be emptied by mechanical means.

sump pump — A pump used for removing collected water from a sump.

sun — A star that is the source of almost all energy on Earth; the energy is transmitted as waves, some of which are visible and others that are invisible. It is a globe of glowing gas, 8.7×10^5 miles in diameter and held together by its own gravity. It is located 93 million miles away from the Earth.

sun protection factor — A scale for rating sunscreens; a rating of 15 or higher provides the best protection from harmful rays of the sun.

sunphotometer — An instrument that measures the properties of light emanating from the sun.

sunspot — A relatively dark, sharply defined region on the sun associated with an intense magnetic field.

SUP — Acronym for standard unit of processing.

superchlorination — Chlorination where the doses are large enough to complete all chlorination reactions and to produce a free chlorine residual, as in a swimming pool.

superconductivity — The pairing of electrons in certain materials when cooled below a critical temperature, causing the material to lose all resistance to

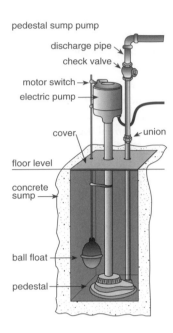

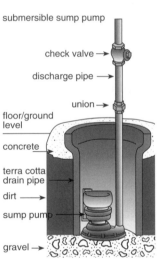

Sump Pump

electricity flow and therefore eliminating any energy losses.

superconductor — A conductor that has no resistance to electrical current; examples include lead, tin, and mercury.

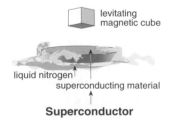

Superconductor

supercooling — The cooling of a liquid below its freezing temperature without forming a solid.

supercritical water — A type of thermal treatment using moderate temperatures and high pressures to enhance the ability of water to break down large, organic molecules into smaller, less toxic ones.

superficial — Refers to something that is not grave or dangerous.

Superfund Act — See Comprehensive Environmental Response Compensation and Liability Act.

Superfund Amendments and Reauthorization Act (SARA) — A 1986 law amending and reauthorizing the Comprehensive Environmental Response Compensation, and Liability Act of 1980; it provided $8.5 billion for the cleanup of Superfund sites and $0.5 billion dollars for the cleanup of underground tank leaks and establishment of alternative technologies for measuring, monitoring, and treatment of hazardous waste.

Superfund Chemical Data Matrix — A source of factor values and benchmark values applied when evaluating National Priorities List sites using the Hazard Ranking System, where factor values are part of the mathematical equation for determining the relative threat posed by a hazardous waste site and reflect hazardous substance characteristics such as toxicity and persistence in the environment, substance mobility, and potential for bioaccumulation.

superheating — The heating of a liquid above its boiling temperature without forming a gas.

superior — Above a point of reference.

supernatant liquid — (*sewage*) Liquid remaining above a layer of settled solids after the solids have collected at the bottom of a storage tank or container.

supersaturated solution — A solution that contains a greater quantity of solute than is normally possible at that temperature.

supersaturation — A condition in which air reaches more than 100% relative humidity.

supersonic — See ultrasonic.

supertanker — A very large ship designed to transport more than 500,000 dead weight tonnage of oil.

superventricular tachycardia — A cardiac rhythm featuring a sudden rapid onset of tachycardia with the stimulus originating above the AV node and a heart rate of 150 to 250 beats per minute that may begin or end spontaneously or be precipitated by excitement, fatigue, caffeine, smoking, or alcohol; usually no significant impairment occurs unless the individual complains of palpitations and shortness of breath.

supplied air respirator — An air-supplied respirator or self-contained breathing apparatus.

supply — (*economics*) The number of qualified personnel available to practice a given occupation, including those employed (or self-employed) and those seeking employment in the occupation.

supply air — Air delivered to a conditioned space and used for the purpose of ventilation, heating, cooling, humidification, or dehumidification.

suppuration — Formation or discharge of pus.

SUR — Acronym for Safe Use Rule.

SURE — Acronym for Sulfate Regional Experiment Program.

surface-active agent — Any of a group of compounds added to a liquid to modify surface or interface tension; used in synthetic detergents to reduce the interfacial tension providing cleansing action. Also known as surfactant.

surface air temperature — The temperature of the air near the surface of the Earth; usually determined by a thermometer in an instrument shelter about 2 meters above the ground.

surface area — The area of the solid particles in a given quantity of material or porous medium.

surface coating — Coating used to cover paint, lacquer, varnish, and other chemical compositions used to protect and/or decorate surfaces.

surface condenser — A heat-transfer device used to condense vapor by contact with a cooling surface behind which the cooling liquid is located.

surface impoundment (SI) — A facility or part of a facility that is a natural topographic depression, an excavation, or dike area formed primarily of earthen materials, which is designed to hold an accumulation of liquid waste or waste containing free liquids.

surface mining — The practice of removing coal by excavation of the surface without an underground mine; also known as strip mining.

surface moisture — Water that is not chemically bound to a metallic mineral or metallic mineral concentrate.

surface pressure — The pressure at an observation point on the surface of the Earth.

surface pump — A mechanism for removing water or wastewater from a sump or wet well.

surface runoff — Any material that is dissolved or suspended in free liquids, usually water, which can leach or flow across the surface of the land from the point of application to the point of discharge.

surface skimmer — A box in the side of a pool that contains a floating weir and a basket through which overflow water from the pool passes in order to clean the surface of the water.

surface soil — The top 10 to 20 cm of soil.

surface tension — A property possessed by liquids in which the water surface in contact with the air acts

S

like an elastic skin; the phenomenon is due to unbalanced molecular cohesive forces near the surface.

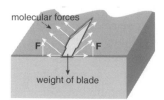

Surface Tension

surface water — Any water on the surface of the Earth.

surface weather maps — Maps used to plot wind direction, speed, pressure, temperature, dew point, visibility, current weather conditions, amount and types of clouds and their heights, pressure changes in the past 3 hours, weather in the past 6 hours, and the amount and kind of precipitation at a given weather station.

surfactant — See surface-active agent.

surgical asepsis — Destruction of organisms before they enter the body, such as in the care of open wounds and in surgical procedures.

surgical waste — The waste generated by surgical techniques in the treatment of disease, injury, or deformity.

surveillance activities — Activities that evaluate exposure or trends in adverse health effects at a rate of a specified period of time; systematic collection, analysis, and interpretation of health data are utilized.

surveillance and analysis (S&A) — Determination of facts related to environmental situations and the analysis of the data to determine a means of prevention and control.

surveillance data — Data on individual cases of foodborne illness cultured in a laboratory and reported to the CDC surveillance system.

surveillance system — A series of monitoring devices designed to determine environmental quality.

survey — A rapid means of gathering information concerning potential problems relating to the environment or of accurately determining position, contour, etc. of an area.

susceptance — The ratio of current-to-voltage in a nondissipative circuit; that part of the admittance that is related to the storage of electric energy.

susceptibility — The degree to which an organism is affected by a chemical or microorganism at a particular level of exposure.

suspended sediment concentration — The ratio of the mass of a dry sediment in a water sediment mixture to the mass of the water sediment mixture; expressed in milligrams of dry sediment per liter of water sediment mixture.

suspended sediment discharge — The quantity of suspended sediment passing a point in a stream over a specified period of time; expressed in tons per day.

suspended sediment — The very fine soil particles that remain in suspension in water for a considerable period of time without contact with the bottom caused by the upward components of turbulence and currents.

suspended solids (SS) — Solids physically suspended in water, sewage, or other liquids and which resist separation from the water by conventional means.

suspension — A mixture of finely divided solid material in a liquid from which the solid settles out on standing; the action of holding insoluble particles in a solution; (*pesticides*) the U.S. EPA's prohibition of the use of a pesticide to prevent the potential hazards of continued use of the pesticide.

sv — See sievert.

SV — Acronym for sampling visit or significant violator; see sludge volume.

SVI — See sludge volume index.

SVOC — Acronym for semivolatile organic compounds.

SVR — The volume of a sludge blanket divided by the daily volume of sludge pumped from the thickener.

SW — Acronym for slow wave.

swath — The amount of ground covered lengthwise in the passing of the satellite.

S wave — A seismic body wave that shakes the ground back and forth perpendicular to the direction in which the wave is moving.

SWDA — See Solid Waste Disposal Act.

sweat — See perspiration.

sweet crude oil — Crude oil with a low sulfur content.

sweetening — The process of improving the odor and color of petroleum products by the oxidation or removal of unsaturated, sulfur-containing compounds.

swill — Semiliquid waste material consisting of food scraps and free liquids; usually fed to hogs.

swimmer load — The number of persons in a pool area at any given moment or during any related period of time.

swimmer's itch — See cutaneous larva migrans.

swimming pool — A body of water of such size in relation to the bathing load that the quality and quantity of the water confined must be mechanically controlled for the purpose of purification and contained in an impervious structure.

SWMU — Acronym for solid waste management unit.

SWPA — Acronym for source water protection area; see Submersible Wastewater Pump Association.

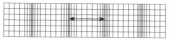

body waves

P waves (primary or compressional wave)

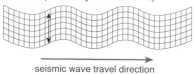

S waves (secondary or shear wave)

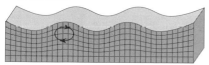

seismic wave travel direction

◄——► = particle motion

surface waves

Rayleigh waves

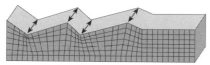

Love waves

S Wave

SWPF — See simulated workplace protection factor.

SWPPP — See stormwater pollution prevention plan.

SWQPPP — Acronym for Source Water Quality Protection Partnership Petitions.

SWTR — Acronym for Surface Water Treatment Rule.

sylvatic plague — An endemic disease of wild rodents caused by *Yersinia pestis* and transmitted to humans by the bite of an infected flea.

symbiosis — A relationship in which two organisms live together for the mutual advantage of each organism; also known as consortism.

symbol — An image, object, action, or notation that represents something else.

symclosene ($C_3Cl_3N_3O_3$) — A white crystalline solid with an odor of chlorine; MW: 232.41; BP: NA; Sol: dissolves in chloroform, alcohol, and acetone; density: NA. It is used as an herbicide. It is hazardous to the digestive tract and skin and may cause

Symclosene

lethargy, weakness, bleeding from the stomach, injury to the liver and kidneys, and death.

symmetrical — Having a definite shape.

sympathetic nervous system — A division of the autonomic nervous system controlling the smooth muscles and glands of the body; it accelerates heart rate, constricts blood vessels, and raises blood pressure.

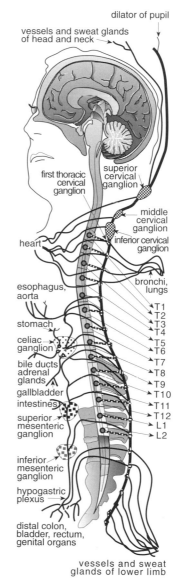

Sympathetic Nervous System

symptom — A warning that something is abnormal; an indication of a disease, infection, or poisoning.

synapse — The space between nerve endings where impulses can pass from one to another.

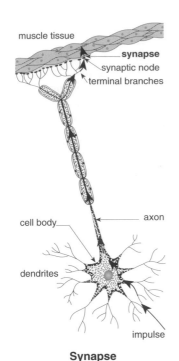

Synapse

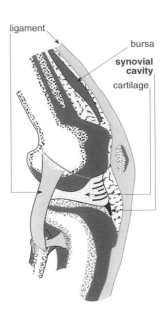

Synovial Cavity

syncope — A temporary suspension of consciousness due to cerebral anemia; also known as faint.

syncrude — A synthetic crude oil made from coal or from oil shale.

syndrome — A group of symptoms that together are characteristic of a specific condition or disease.

synecology — The branch of ecology dealing with communities rather than individual species.

synergetic — See synergism.

synergism — The cooperative action of separate substances so that the total effect is greater than the sum of the effects of the substances acting independently.

synergist — A chemical that produces greater effects when mixed with another chemical than when the added effects of each are used alone.

synergistic effect — A combined effect of two or more substances or organisms that is greater than the sum of the individual effect of each.

synfuels — Synthetic petroleum products produced from coal or natural gas.

syngas — A synthetic gas made from coal.

synopsis — A brief review, condensation, summary, or abridgement.

synoptic chart — A chart showing meteorological conditions over a region in any given time; weather mapping.

synoptic view — The ability to see large areas at the same time.

synovial fluid — A secretion of cartilage that lubricates a joint.

syntax — (*computer science*) A set of rules governing the way statements can be used in a computer language.

synthesis — The reaction or series of reactions by which a complex compound is obtained from simpler compounds or elements.

synthesize — To form a material by combining parts of elements.

synthetic — Referring to chemicals or other substances made artificially.

synthetic aperture radar — A high-resolution ground-mapping technique that effectively combines diverse information into a coherent whole through the use of a large receiving antenna and by processing the phase of the reflected radar return.

synthetic detergent — A liquid or solid chemical that is used as a cleaning agent, is soluble in water, and does not form insoluble precipitates.

synthetic fiber — Fiber derived from petrochemicals or coal chemicals, such as polyester or nylon.

synthetic graphite — A crystalline form used as an adsorbent and in water purification; also called activated carbon. Inhalation causes cough, dyspnea, black sputum, and fibrosis. ACGIH exposure limit (TLV-TWA): 2.5 mg/m^3 respirable dust.

synthetic pyrethrum — See allethrin.

SYSOP — Acronym for systems operator.

system — (*anatomy*) A group of separate organs performing a similar function (e.g., the digestive system).

System Safety Engineering — The application of system safety management and engineering principles throughout the life cycle of a system.

system sensitivity — A measure of the lowest concentration of a chemical that can be detected in a testing procedure.

systematic acidosis — A condition of raised acidity in the blood or body tissues.

systematic error — (*statistics*) See bias.

systematic maintenance — Preventative maintenance carried out on a schedule established according to a time period or number of units in use.

Système International d'Unités — See International System of Units.

systemic — Refers to a condition or process affecting the whole body, such as when the effects of a chemical agent, radiological agent, or microorganism spread throughout the body.

systemic pesticide — A pesticide chemical that is carried to other parts of a plant or animal after it is injected or taken up from the soil or body surface.

systemic poison — A poison that spreads throughout the body and affects all body systems and organs.

systemic toxicity — Adverse effects caused by a substance which affect the body in a general rather than local manner.

systems design — (*ergonomics*) Design based upon the principle of an individual being an integral part of a total person–machine system that allocates the functions carried out by machines and people.

S

T

1080 — See sodium fluoroacetate.

1081 — See fluoroacetamide.

T — The SI symbol for the prefix *tera-*, representing a factor of 10^{12}; see tesla or tritium.

2,4,5-T — See 2,4,5-trichlorophenoxyacetic acid.

T cell — A type of lymphocyte derived from bone marrow stem cells that matures into an immunologically competent cell under the influence of the thymus.

TA — See temporary abeyance.

tablespoon (Tbsp) — A measurement of volume equal to 3 teaspoons.

tableware — All multiuse eating and drinking utensils.

TAC — See toxic air contaminant.

tachycardia — A cardiac rhythm in which the myocardium contracts at a rate >100 beats per minute.

tachypnea — Very rapid, abnormal respirations.

tactile receptor — Sensory receptor in the skin that detects a light pressure.

TAD — Acronym for technical assistance document.

Taenia saginata — A species of tapeworm that is found in the tissues of cattle during its larval stage and which infests the intestines of humans in its adult form.

Taenia solium — A species of tapeworm that is found in the tissues of pigs during its larval stage and infects the intestine of humans in its adult form.

Taenia spiralis — A large tapeworm species found in the striated muscle of various animals; a common cause of infection in people as a result of ingestion of poorly cooked pork.

taeniasis — A human intestinal infection characterized by nervousness, insomnia, weight loss, abdominal pain, and digestive disturbances. Incubation time is 8 to 14 weeks. It is caused by *Taenia saginata* (beef tapeworm) or *Taenia solium* (pork tapeworm). It is found worldwide, and the reservoir of infection is humans, with the immediate hosts being cattle or pigs. It is spread by the consumption of raw or inadequately cooked beef or pork or the consumption of water containing the eggs of adult worms; *T. saginata* is not transmitted from person to person, while *T. solium* may be found in the feces for 30 to 40 years; general susceptibility. It is controlled by the proper cooking of beef and pork, proper sewage disposal, inspection of livestock, and proper water treatment. Also known as beef or pork tapeworm disease.

TAG — Acronym for technical assistance grant.

tail water — The runoff of irrigation water from the lower end of an irrigated field.

tailings — The remaining mining waste materials after a substance, such as uranium or iron, has been mechanically or chemically extracted from ore.

tailpipe standards — Emissions limitations applicable to mobile source engine exhausts.

talc ($Mg_3H_2(SiO_3)_4$) (non-asbestos form with less than 1% quartz) — An odorless white powder; MW: varies, BP: not known, Sol: insoluble, sp. gr. 2.70 to 2.80. It is used in the cosmetics industry as face, body, or dusting powder and in soap and toilet preparations; in the pharmaceutical industry in tablets, pills, salves, and lotions; as a filler and selective absorbent in the paper industry; in the ceramics industry; to clean up oil spills. It is hazardous to the lungs and cardiovascular system and is toxic through inhalation and contact. Symptoms of exposure include fibrotic pneumoconiosis. OSHA exposure limit (TWA): 2 mg/m^3 [resp.].

tamp — To firmly compact earth during backfilling.

tamper — A tool for compacting soil in spots not accessible to rollers.

tamping — A method of compacting soil using the impact of a power or hand tamper on the surface of the soil.

TAMS — Acronym for Toxic Air Monitoring System.

tandem — A pair in which one part accompanies the other part.

tangent — A line that touches a circle and is perpendicular to its radius at the point of contact.

tank — A stationary device designed to contain an accumulation of hazardous waste or other chemicals or fuels; it is constructed primarily of nonearthen materials, such as wood, concrete, steel, or plastic, and provides structural support for the chemicals.

tantalum (Ta) metal and oxide dust — A steel-blue to gray metal or a black powder; MW: 180.9, BP: 9797°F, Sol: insoluble, sp. gr. 16.65 (metal) and 14.40 (powder). It is used in the manufacture of electronic equipment; in the fabrication of metals;

in the fabrication of refractory nonferrous alloys for nuclear and aerospace applications; in the manufacture of surgical metals, mesh, and clips; in the manufacture of special optical glass. It is toxic through inhalation. Symptoms of exposure in animals include pulmonary irritation. OSHA exposure limit (TWA): 5 mg/m^3 [air].

TAPDS — Acronym for Toxic Air Pollutant Data System.

tape sampler — A device used in the measurement of fine particulates allowing air sampling to be done automatically at predetermined times by drawing air across the filter strip and determining what is on the strip.

tapeworm — A parasitic cestode worm of the class of Cestoidea (phylum Platyhelminthes) having a flattened band-like form that lodges in the intestines of animals and humans; they are transmitted to people in larval form imbedded in cysts of improperly cooked meat or fish; in humans they develop to maturity and attach themselves to the wall of the intestine, where they grow and release eggs.

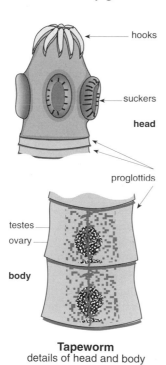

Tapeworm
details of head and body

tapeworm infection — An intestinal infection by one of several different species of parasitic worms caused by eating raw or undercooked meat containing tapeworms or larva.

taproot — A large root that grows downward from the base of a tree.

tar — A viscous, complex mixture of chemicals of top fractional distillation systems.

tar balls — The dense, black, sticky spheres of hydrocarbons formed from weathered oil.

tar camphor — See naphthalene.

tar sand — A sand coated with bitumen.

tar sand deposit — A geological formation that can be used to convert bitumen into crude oil.

tare — A deduction from gross weight made to allow for the weight of an empty container.

target hazard quotient — A value that is combined with exposure and toxicity information to calculate a risk-based concentration for noncarcinogenic contaminants.

target organ — An organ such as the liver or kidney that is specifically affected by a toxic chemical.

target organ effect — The health effect on a specific organ, such as the liver, caused by a particular chemical or substance.

target organ toxicity — A broad range of adverse effects on target organs or physiological systems coming from a single exposure or limited lifetime exposure to a chemical.

target organ toxin — A toxic substance that attacks a specific organ of the body.

target species — An organism that a pesticide is intended to control.

target theory — A theory explaining some biological effects of radiation on the basis that ionization occurring in a discrete (target) volume within a cell causes a lesion, which subsequently results in a physiological response to the damage at that location; also known as hit space theory.

TAS — Acronym for Tolerance Assessment System.

taste — One of the special senses that allows the individual to distinguish among sweet, sour, bitter, or salty qualities of a dissolved substance, as determined by the taste buds on the tongue.

taxonomy — The science of classification.

TB — Acronym for technical bulletin.

T-BACT — See toxic best available control technology.

TBM — Acronym for transportation of biomedical materials.

TBW — Acronym for total body water.

Tbsp — See tablespoon.

TC — Acronym for target concentration, toxicity characteristics, or toxic concentration.

TCA — Acronym for trichloroacetic acid.

TCBD — Notation for tretrachlorodibenzo-*t*-dioxin.

TCCL — Acronym for Toxic Chemicals Control Law.

TCDD — Notation for tretrachlorodibenzo-*p*-dioxin.

TCDF — Acronym for tetrachlorodibenzofuran.

T

TCE — See trichloroethylene.

TCF — Acronym for total chlorine free.

TCL — See toxic concentration low.

TCLP — Acronym for total concentrate leachate procedure or toxicity characteristic leachate procedure.

TCM — Acronym for transportation control measure.

TCP — Acronym for transportation control plan; see trichloropropane.

T cytotoxic cells — A subset of T lymphocytes able to directly kill foreign cells, especially virally infected host cells.

TD — notation for toxic dose.

TD$_{50}$ — The dose that produces 50% of the toxic effect of a chemical.

TDH — See total dynamic head.

TDI — Notation for toluene diisocyanate.

TDL — See toxic dose low.

TDRSS — See Tracking and Data Relay Satellite System.

TDS — See total dissolved solids.

TEAM — Acronym for Total Exposure Assessment Model.

tear — To rip, rend, or pull apart by force.

teaspoon (tsp) — Measurement equal to 50 drops.

Technical Assistance Grant — Money provided for activities that help a community participate in the decision-making process of site-specific cleanup strategies at eligible Superfund sites.

technical grade — Refers to a high level of concentration and purity for a chemical.

technical material — A pesticide chemical as it is manufactured by the chemical company.

technician — An environmental health or safety practitioner qualified by education, training, and practice to operate sampling, monitoring, and other data-gathering equipment in the conduct of environmental health or safety control and prevention activities. The minimum education level is specialized training.

technological hazard — A range of hazards coming from the manufacture, transportation, and use of such substances as radioactive materials, chemicals, explosives, flammables, agricultural pesticides, herbicides, and disease agents; oil spills on land, in coastal waters, or in inland water systems; and debris from space.

technology assessment — Systematic analysis of the anticipated impact of a particular technology in regard to its safety and efficacy as well as its social, political, economic, and ethical consequences.

technology-based limitations — Industry-specific effluent limitations based on best available preventive technology and applied to a discharge when it will not cause a violation of water quality standards at low stream flows; usually applies to discharges into large rivers.

technology-based standards — Industry-specific limitations applicable to direct and indirect sources and which are decided on a category-by-category basis using statutory factors, not including water-quality effects.

tectonics — The science that deals with the structure of the crust of the Earth.

tee — A T-shaped pipe fitting that has a threaded opening at right angles to the other two threaded openings.

teepee burner — See conical burner.

tegument — The covering of a living body or some part of an organ.

TEL — See tetraethyllead.

teleconnection — A strong statistical relationship between weather in different parts of the globe; particularly true of the tropics and North America during El Niño.

telemetry — A space-to-ground data system of measured values that does not include command, tracking, computer memory transfer, audio, or video signals.

television and infrared observation satellite (TIROS) — A satellite sent into orbit in 1960 to determine weather patterns.

telluric — A natural electric current flowing near the surface of the Earth.

tellurium (Te) and compounds — An odorless, dark-gray to brown, amorphous powder or grayish-white, brittle metal; MW: 127.6, BP: 1814°F, Sol: insoluble, Fl.P: not known, sp. gr. 6.24. It is used as an insecticide, germicide, fungicide, and photographic print toner. It is hazardous to the skin and central nervous system and is toxic through inhalation, absorption, ingestion, and contact. Symptoms of exposure include garlic odor on the breath, sweating, dry mouth, metal taste, somnolence, anorexia, dermatitis. OSHA exposure limit (TWA): 0.1 mg/m^3 [air].

tellurium hexafluoride (TeF$_6$) — A colorless gas with a repulsive odor; MW: 241.6, BP: sublimes, Sol: decomposes. It is used in scientific studies on physical and chemical properties. It is hazardous to the respiratory system and is toxic through inhalation. Symptoms of exposure include headache, dyspnea, and garlic odor on the breath; in animals, pulmonary edema. OSHA exposure limit (TWA): 0.02 ppm [air] or 0.2 mg/m^3.

Telnet — A system that allows users to access computers and their data at thousands of places around the world, usually in libraries, universities, and government agencies.

telomer — A reduced polymer formed by the reaction between a substance capable of being polymerized and an agent that arrests the growth of the chain of atoms.

telophase — The phase of nuclear division by mitosis or meiosis that begins as chromosomes reach the centrioles.

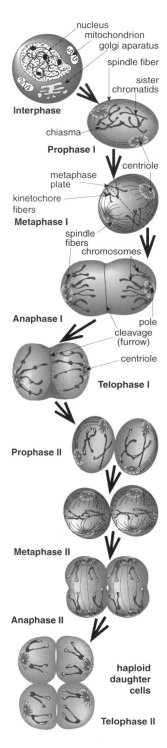

Telophase

TEM — See transmission electron microscope.

temperate — A mild or moderate temperature.

temperature — A measure of the tendency of objects to transfer heat to or absorb heat from other objects; the unit of measurement is the degree, and it can be read directly on a thermometer.

temperature danger zone — The temperatures between 41 and 140°F (5 and 60°C) at which bacteria grow best.

temperature extremes — Environmental conditions that will cause adverse effects on individuals due to excessive heat or cold.

temperature gradient — The change of temperature along a given line in space.

temperature inversion — See inversion.

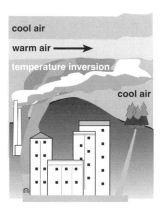

Temperature Inversion

temperature measuring device — A thermometer, thermocouple, thermistor, or other device for measuring the temperature of food, air, or water.

tempered air — The supply of air that has been heated sufficiently to prevent cold drafts; also known as conditioned air.

tempering — The preheating and rapid cooling of a metal to increase its hardness.

temporal resolution — Resolution of the temporal distribution of air pollution concentrations in a study area by modeling or monitoring. It is affected by sources of air pollution that seldom emit material into the air at a constant and continuous rate; changing wind direction, which has a profound effect on the distribution of air pollutants; changing wind speed, which has a major effect on the distribution of pollutants; changing atmospheric stability, which can produce fluctuations in the ambient air pollution levels; and topography, which can affect the distribution of air pollutants by affecting air movements that might not otherwise occur.

T

temporary abeyance (TA) — The holding of sample or samples in abeyance for possible future regulatory action.

temporary food establishment — A food establishment that operates in one location for not more than 14 consecutive days in conjunction with a single event.

temporary housing — Any tent, trailer, mobile home, or any other structure used for human shelter that is designed to be transportable and not attached to the ground, another structure, or any utility system on the same premises for more than 30 consecutive days.

temporary partial disability — A condition under workers' compensation insurance that results in partial loss of earning power but from which recovery can be expected.

temporary restraining order (TRO) — The first step of most injunctions in which the government or plaintiff requests the imposition of immediate temporary restraint of the defendant lasting for 10 days; see also preliminary injunction.

temporary total disability — A condition under workers' compensation insurance that disables the employee from working but from which a complete or partial recovery can be expected.

tendency — A local rate of change of the vector or scalar quantity with time at a given point in space.

tendon — A fibrous mass connecting muscle to bone.

tendonitis — Inflammation of a tendon, which connects muscle to bone, that usually results from a strain.

tenosynovitis — Swelling and inflammation of the sheath that produces a lubricating fluid and surrounds the tendons; results in a decreased capacity to produce this fluid.

tensile strength — The capacity of a metal to withstand a stretching force.

tensiometer — A device for measuring a negative hydraulic pressure or tension of water in soil involving a porous, permeable ceramic cup connected through a tube to a manometer or vacuum gauge.

TEOR — See thermally enhanced oil recovery.

TEPP (ethyl pyrophosphate; $(C_2H_5)_4P_2O_7$) — A colorless to amber liquid with a faint, fruity odor; a solid below 32°F; MW: 290.2, BP: decomposes, Sol: miscible, sp. gr. 1.19. It is used as a contact insecticide on agricultural and ornamental crops. It is hazardous to the central nervous system, respiratory system, cardiovascular system, and gastrointestinal tract and is toxic through inhalation, absorption, ingestion, and contact. Symptoms of exposure include eye pain, blurred vision, lacri-mation, rhinorrhea, headache, tight chest, cyanosis, anorexia, nausea, vomiting, diarrhea, weakness, twitch, paralysis, Cheyne–Stokes respiration, convulsions, low blood pressure, cardiac irregularities, and local sweating. OSHA exposure limit (TWA): 0.05 mg/m³ [skin].

teratogen — A chemical that causes structural defects that affect the development of a fetus.

teratogenesis — The study of malformation induced during development of an animal or human from the time of conception.

teratogenic agent — See teratogen.

teratology — Study of the causes and effects of congenital malformations and developmental abnormalities.

terminal — (*computer science*) A device that usually includes a cathode-ray tube and a keyboard and is used for communicating with a computer.

terminal illness — An advanced stage of a disease leading to death.

terminated — Ended or adjudicated.

terpenes — A class of volatile organic compounds.

terphenyls ($C_6H_5C_6H_4C_6H_5$) — A colorless or light-yellow solid that has three different isomers; MW: 230.3, BP: 630 to 761°F, Sol: insoluble, Fl.P: 325 to 405°F (open cup), sp. gr. 1.10 to 1.23. It is used in the formulation of waxes, polishes, and resin body paints; in the manufacture of plastic scintillators. It is hazardous to the skin, respiratory system, and eyes and is toxic through inhalation, ingestion, and contact. Symptoms of exposure include thermal skin burns, headache, sore throat, and irritation of eyes and skin. OSHA exposure limit (TWA): ceiling 0.5 ppm [air] or 5 mg/m³.

terracing — Dikes built along the contour of agricultural land to hold runoff and sediment and reduce water pollution.

terrorism — The unlawful use of force against persons or property to intimidate or coerce governments or the civilian population in the furtherance of political or social objectives.

tertatogenic — The ability to produce birth defects.

tertian malaria — A form of malaria in which fevers reoccur every 48 hours. It is caused by *Plasmodium vivax, P. ovale,* and *P. falciparum.*

tertiary butyl valone — See pival.

tertiary treatment — See advanced wastewater treatment.

tertiary wastewater treatment — See advanced wastewater treatment.

TES — Acronym for technical enforcement support; see Tropospheric Emission Spectrometer.

tesla (T) — The SI unit for magnetic field; a unit of magnetic flux density is equal to 1 weber per square centimeter.

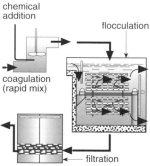

complete treatment

direct filtration

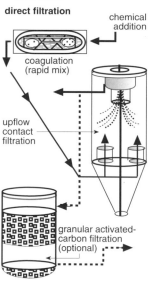

**contact filtration
with optional granular
activated carbon adsorption**

Tertiary Treatment

test — A critical examination, observation, or evaluation.

test hole — Any excavation, regardless of design or method of construction, done for the purpose of determining the most suitable site for further excavation — for example, determining where to obtain groundwater from an aquifer for use in a private water system.

tetanic — Pertaining to tetanus.

tetanus — An acute disease caused by the bacterium *Clostridium tetani*; the spore of the organism is introduced into the body in an anaerobic state and then produces the exotoxin, which causes painful muscle contraction of the neck and trunk. Abdominal rigidity occurs, as well as generalized spasms. Incubation time is 3 to 21 days, and it is found worldwide. The reservoirs of infection include intestines of animals, including people, or soil contaminated with animal feces. It is transmitted through a puncture wound and is not transmitted from person to person; general susceptibility. It is controlled through active immunization with tetanus toxoid at least once every 10 years. Also known as lockjaw.

tetanus antitoxin — An equine antitoxin used against the toxins of *Clostridium tetani*.

2,3,7,8-tetrachlorodibenzo-*p*-dioxin (PCDD; C$_{12}$H$_4$Cl$_4$O$_2$) — A colorless solid with no distinguishable odor; MW 321.97, BP 412.2°C, Sol: 7.91 mg/l at 20 to 22°C in water, Fl.P: not known, density: 1.827 g/ml (estimated). It is used in Germany for flameproofing polyesters; as a control against insects and wood-destroying fungi. It is hazardous to the skin, liver, digestive system, immune system, and fetus and is toxic through inhalation, ingestion, and contact. Symptoms of exposure include chloracne, hepatotoxicity and immunotoxicity, hyperpigmentation, hyperkeratosis, hirsutism of the skin, hypertriglyceride hypercholesterolemia, insomnia, aching muscles, weight loss, loss of appetite, digestive disorders, headaches, neuropathy, sensory changes, loss of libido, developmental disorders; probable carcinogen. U.S. EPA drinking water advisory: 3.5×10^{-8} milligrams per liter.

1,1,1,2-tetrachloro-2,2-difluoroethane (CCl$_3$CClF$_2$) — A colorless solid with a slight, ether-like odor; a liquid above 105°F; MW: 203.8, BP: 197°F, Sol: 0.01%. It is used for drycleaning in combination with other solvents; in the polymer and plastics industries; as a dye solvent; as a solvent extractant for the separation and purification of biological material; as a corrosion inhibitor in brake fluid and surface coatings. It is hazardous to the respiratory system and skin and is toxic through inhalation, ingestion, and contact. Symptoms of exposure include central nervous system depression, pulmonary edema, drowsiness, dyspnea, and irritation of the skin and eyes. OSHA exposure limit (TWA): 500 ppm [air] or 4170 mg/m^3.

T

1,1,2,2-tetrachloroethane (CHCl₂CHCl₂) — A colorless to pale yellow liquid with a pungent, chloroform-like odor; MW: 167.9, BP: 296°F, Sol: 0.3%, sp. gr. (77°F): 1.59. It is used as a chemical intermediate; in cleaning and extraction processes; as a fumigant in greenhouses; in the manufacture of lacquers, varnishes, and paint and varnish removers; in cleaning and degreasing metals; as a solvent in the preparation of adhesives; in refining waxes and resins. It is hazardous to the liver, kidneys, and central nervous system and is toxic through inhalation, absorption, ingestion, and contact. Symptoms of exposure include nausea, vomiting, abdominal pain, tremor of the fingers, jaundice, tenderness and enlargement of the liver, dermatitis, monocytosis, and kidney damage. OSHA exposure limit (TWA): 1 ppm [skin] or 7 mg/m³.

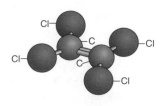

1,1,2,2–Tetrachloroethane

tetrachloroethylene (Cl₂CCCl₂) — A colorless liquid with a mild, chloroform-like odor; MW: 165.8, BP: 250°F, Sol: 0.02% at 77°F, sp. gr. 1.62. It is used as a drycleaning solvent; as a degreasing and metal-cleaning agent; as a chemical intermediate; as a scouring, sizing, desizing, solvent, and grease remover in the processing and finishing of textiles; as a general industrial solvent; as an extraction agent; as a drying medium in the metal and wood industries. It is hazardous to the liver, kidneys, eyes, upper respiratory system, and central nervous system and is toxic through inhalation, ingestion, and contact. Symptoms of exposure include nausea, flushed face and neck, vertigo, dizziness, incoordination, headache, somnolence, skin erythema, liver damage, and irritation of the eyes, nose, and throat. OSHA exposure limit (TWA): 25 ppm [air] or 170 mg/m³.

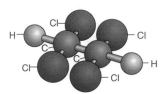

Tetrachloroethylene
(Tetrachlorethane)

tetrachloronaphthalene (C₁₀H₄Cl₄) — A colorless to pale yellow solid with an aromatic odor; MW: 265.9, BP: 593 to 680°F, Sol: insoluble, Fl.P: 410°F (open cup), sp. gr. 1.59 to 1.65. It is used during the manufacture of electric equipment; as an inert compound of resins or polymers for coating or impregnating textiles, wood, and paper; as an additive for cutting oil; as an additive to special lubricants. It is hazardous to the liver and skin and is toxic through inhalation, absorption, ingestion, and contact. Symptoms of exposure include acne-form dermatitis, headache, fatigue, anorexia, vertigo, jaundice, and liver injury. OSHA exposure limit (TWA): 2 mg/m³ [skin].

tetraethyllead (TEL; (C₂H₅)₄Pb) — A colorless liquid; MW: 323.45; BP: 200°C; Sol: miscible with benzene, toluene, petroleum ether, and gasoline; Fl.P: 163°F; density: 1.65 g/ml at 20°C. It is used as an additive to gasoline to prevent knocking. It is a highly toxic substance that is hazardous to the digestive tract, skin, and respiratory tract. It is toxic to the central nervous system and causes insomnia, hypotension, tremor, hypothermia, pallor, weight loss, hallucinations, nausea, convulsions, and coma. ACGIH exposure limit (TLV-TWA): 0.1 mg Pb/m³.

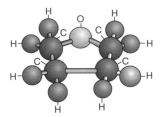

Tetrahydrofuran

tetrahydrofuran (C₄H₈O) — A colorless liquid with an ether-like odor; MW: 72.1, BP: 151°F, Sol: miscible, Fl.P: 6°F, sp. gr. 0.89. It is used as a solvent in the preparation of printing inks, adhesives, lacquers, and other coatings; as a Grignard reagent in the synthesis of motor fuels, vitamins, hormones, pharmaceuticals, synthetic perfumes, organometallic compounds, and insecticides; as an intermediate in the preparation of various chemicals. It is hazardous to the eyes, skin, respiratory system, and central nervous system and is toxic through inhalation, ingestion, and contact. Symptoms of exposure include nausea, dizziness, headache, and irritation of the eyes and upper respiratory system. OSHA exposure limit (TWA): 200 ppm [air] or 590 mg/m³.

tetramethyl succinonitrile ((CH₃)₂C(CN)C(CN) (CH₃)₂) — A colorless, odorless solid; MW: 136.2, BP: sublimes, Sol: insoluble, Fl.P: not known, sp. gr. 1.07. It is used in the manufacture of polymers by and in the processing of products expanded with azo-*bis*-isobutyronitrile. It is hazardous to the central nervous system and is toxic through inhalation, absorption, ingestion, and contact. Symptoms of exposure include headache, nausea, convulsions, and coma. OSHA exposure limit (TWA): 0.5 ppm [skin] or 3 mg/m^3.

tetramethylene chlorohydrin — A colorless, oily liquid; MW: 108.58, BP: 84.5°C, Sol: soluble in water and in organic solvents, Fl.P: 36°C (vapor forms an explosive mixture with air), density: 1.0083 at 20°C. It is used in organic synthesis. It is hazardous to the digestive tract and causes symptoms of muscle contraction, gastrointestinal pain, and ulceration. ACGIH exposure limit: NA.

$$Cl-CH_2-CH_2-CH_2-CH_2-OH$$

Tetramethylene chlorohydrin

tetramethyllead (Pb(CH$_3$)$_4$) — A colorless liquid (unless dyed red, orange, or blue) with a fruity odor; MW: 267.3, BP: 212°F (decomposes), Sol: insoluble, Fl.P: 100°F, sp. gr. 2.00. It is used as an antiknock additive for gasoline. It is hazardous to the central nervous system, cardiovascular system, and kidneys and is toxic through inhalation, absorption, ingestion, and contact. Symptoms of exposure include insomnia, bad dreams, restlessness, anxiousness, hypotension, nausea, anorexia, delirium, mania, convulsions, and coma. OSHA exposure limit (TWA): 0.075 mg/m^3 [skin].

tetranitromethane (C(NO$_2$)$_4$) — A colorless to pale-yellow liquid or solid (below 57°F) with a pungent odor; MW: 196.0, BP: 259°F, Sol: insoluble, Fl.P: not known, sp. gr. 1.62. It is used as an ingredient in the manufacture of liquid explosives; as an octane number improver in diesel fuels; as a laboratory analytical reagent; in research into rocket propellants. It is hazardous to the respiratory system, eyes, skin, blood, and central nervous system and is toxic through inhalation, ingestion, and contact. Symptoms of exposure include dizziness, headache, chest pain, dyspnea, methemoglobinuria, cyanosis, skin burns, and irritation of the eyes, nose, and throat. OSHA exposure limit (TWA): 1 ppm [air] or 8 mg/m^3.

1,1,2,2-tetrochloro-1,2-difluoroethane (CCl$_2$FCCl$_2$F) — A colorless solid or liquid (above 77°F) with a slight, ether-like odor; MW: 203.8, BP: 199°F, Sol: 0.01% at 77°F, sp. gr. 1.65. It is used in the dry-cleaning industry in combination with other sol-

vents; in the polymer and plastics industries; as a dye solvent; as a solvent extractant for the separation and purification of biological material; as a corrosion inhibitor in brake fluid and surface coatings. It is hazardous to the lungs and skin and is toxic through inhalation, ingestion, and contact. Symptoms of exposure include conjunctivitis, pulmonary edema, and skin irritation. OSHA exposure limit (TWA): 500 ppm [air] or 4170 mg/m^3.

tetryl ((NO$_2$)$_3$C$_6$H$_2$N(NO$_2$)CH$_3$) — A colorless to yellow, odorless, crystalline solid; MW: 287.2, BP: 356 to 374°F (explodes), Sol: insoluble, Fl.P: explodes, sp. gr. 1.57. It is used in the manufacture of explosives; as a pH indicator. It is hazardous to the respiratory system, eyes, central nervous system, and skin; in animals, it damages the liver and kidneys; it is toxic to humans and animals through inhalation, absorption, ingestion, and contact. Symptoms of exposure include sensitization dermatitis, itching, erythema, keratitis, sneezing, anemia, fatigue, cough, coryza, irritability, malaise, headache, lassitude, insomnia, nausea, vomiting, and edema on nasal folds, cheeks, and neck. OSHA exposure limit (TWA): 1.5 mg/m^3 [skin].

texture — See soil texture.

Tfx — See toxic effects.

TGA — See nitrilotriacetic acid.

TG — Acronym for trace gas.

TGO — Acronym for total gross output.

Th — See thorium.

thalamus — One of a pair of large oval organs forming most of the lateral walls of the third ventricle of the brain and part of the diencephalon; functions in the integration of sensory information.

thalidomide (C$_{13}$H$_{10}$N$_2$O$_4$) — A chemical substance used extensively as a sedative in sleeping pills in Europe and the United States in the early 1960s; its use was discontinued when it was determined that the drug caused phocomelia, or failure of limbs to develop in fetuses exposed to the drug at a certain time during pregnancy.

thallium (Tl) soluble compounds — Appearance and odor vary depending upon the specific soluble solution; properties vary depending upon the specific soluble solution. It is used in the manufacture of special lenses, plates and prisms, medicinals, pyrotechnic products, and fuel additives; as a rodenticidal agent. It is hazardous to the eyes, central nervous system, lungs, liver, kidneys, gastrointestinal tract, and body hair and is toxic through inhalation, absorption, ingestion, and contact. Symptoms of exposure include nausea, diarrhea, abdominal pain, vomiting, ptosis, strabismus, peripheral neuritis, tremors, chest pain, pulmonary

T

edema, seizures, chorea, psychosis, liver and kidney damage, alopecia, and paresthesia of the legs. OSHA exposure limit (TWA): 0.1 mg/m³ [skin].

thallium sulfate (Tl₂SO₄) — Toxic, tasteless, and odorless crystal used as rat poison; acts similarly to arsenic.

thallus — A plant body that lacks differentiation into distinct members, such as stem, leaves, and roots, and does not grow from an apical point.

THC — Acronym for total hydrocarbons.

T helper cell — A type of T lymphocyte that normally orchestrates the immune response by signaling other cells in the immune system to perform their special functions.

thematic map — A map showing special information such as soil, land use, population, density, hazardous waste sites, and sewerage systems.

thematic mapper — A Landsat multispectral scanner designed to acquire data to categorize the surface of the Earth, with special emphasis on agricultural applications and identification of land use.

theme — (*geographical information system*) A spatial dataset containing a common feature type.

theoretical combustion air — See theoretical air.

theoretical maximum residue contribution — The amount of a pesticide present in the 1.5-kg average daily diet of an average person; expressed in milligrams of pesticide per kilogram of body weight per day.

theory — A plausible or scientifically acceptable general principle or body of principles used to explain certain phenomena.

therapeutic dose — A dose of a drug that may be required to produce a desired effect.

therapeutic equivalent — A drug that has the same effect in the treatment of a disease or condition as another drug.

therapeutic exercise — An exercise program used to create specific physical benefits.

therapeutic index — The ratio of the dose required to produce toxic or lethal effects to the dose required to produce nonadverse or therapeutic response.

therapeutic recreation — The use of recreational programs supervised by specialists to improve upon physical and mental health.

therapeutics — A healthcare field concerned with the treatment of disease to relieve symptoms or cure the individual.

therapist — A specially trained individual working in one or more areas of health care.

therapy — The treatment of disease.

therm — A quantity of heat energy equivalent to 100,000 British thermal units.

thermal — The production, application, or maintenance of heat.

thermal barrier — See thermal break.

thermal break — An element of low heat conductivity placed in such a way as to reduce or prevent the flow of heat.

thermal burn — A tissue injury usually of the skin caused by exposure to extreme heat.

thermal capacity — See heat capacity.

thermal comfort — The condition of being satisfied with the thermal environment.

thermal conductivity — A specific rate of heat flow per unit time through refractories or other substances, expressed in British thermal units per square foot of area for a temperature difference of 1°F and for a thickness of one inch.

thermal efficiency — The ratio of actual heat used to total heat generated.

thermal energy storage — A technology that lowers the amount of electricity needed for comfort conditioning during utility peak load periods.

thermal expansion — An increase in size due to an increase in temperature expressed in units of 11; increase in length or increase in size per degree.

thermal gradient — The distribution of a differential temperature through a body or across a surface.

thermal infrared — Electromagnetic radiation with wavelengths between about 3 and 25 μm.

thermal mass — A material such as concrete, brick, masonry, tile and mortar, water, and rock that is used to store heat, thereby slowing the temperature variation within a space.

thermal neutrons — Neutrons that have been slowed to the degree that they have the same average thermal energy as the atoms or molecules through which they are passing.

thermal pollution — A discharge of heat into bodies of water to the point that increased warmth activates organic material, depleting necessary oxygen and eventually destroying some fish and other aquatic organisms.

thermal power plant — A stationary or floating electrical generating facility using any source of thermal energy with a capacity of 50 MW or more.

thermal processing — See thermal treatment.

thermal radiation — The transmission of energy by means of electromagnetic waves longer than visible light.

thermal regenerative system — An incineration system that relies on the heat radiation capability of an inert ceramic.

thermal sand — Sand used to dissipate heat away from buried electrical cables.

thermal shock resistance — The ability of a material to withstand sudden heating or cooling, or both, without cracking.

thermal staple — A material that is resistant to change induced by heat.

thermal stratification — Layers of different temperatures of water or air caused by different densities, with the less dense floating on the more dense layers.

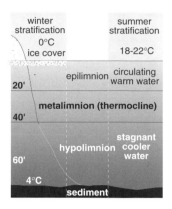

**Thermal stratification
in a reservoir**

thermal system insulation — Asbestos-containing material applied to pipes, fittings, boilers, tanks, ducts, or other interior structural components to prevent heat loss or gain or water condensation.

thermal transmittance — The heat flow transmitted through a unit area of a given structure divided by the difference between the effective ambient temperature on either side of the structure under steady-state conditions.

thermal treatment — The treatment of a hazardous waste in a device that uses elevated temperatures as the primary means to change the chemical, physical, or biological character or composition of the hazardous waste; also known as thermal processing.

thermal turbulence — Turbulence caused by heated air rising from the surface of the Earth.

thermal zone — The region in a building where the temperature is assumed to be more or less the same.

thermalgesia — Pain caused by exposure to high temperatures.

thermally enhanced oil recovery — The injection of steam to increase the amount of oil that can be recovered from a well.

thermistor — An electrical resistor making use of a semiconductor for which the resistance varies sharply in a specific manner along with a change in temperature.

thermochemistry — A branch of chemistry concerned with the heat changes involved in chemical reactions.

thermocline — A layer in a thermally stratified body of water in which the temperature changes rapidly relative to other layers of the water.

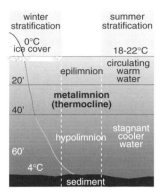

Thermocline

thermoconductivity — The specific heat of conduction of a gas or vapor in a carrier gas, such as air, is a measure of its concentration; two unlike materials, one with extra electrons and one with missing electrons are placed between a temperature gradient (a difference in termperaure). The electrons will move from one material to the other generating a small electric current.

thermocouple — A device composed of dissimilar metals joined to form a circuit in which current flows when heat is applied at a junction of the metals; an electromotive force develops when one junction is at a different temperature than the other.

thermoduric — Bacteria that can survive the pasteurization process but do not reproduce or grow at high temperatures.

thermodynamics — The relationship between heat and other properties such as temperature, pressure, density, etc.; usually referring to the distribution of temperature and moisture as related to atmospheric instability.

thermoelectric power water use — Water used in the process of generation of thermoelectric power, primarily in plants that burn coal and oil.

thermogenesis — The production of heat especially by the cells of the body.

thermograph — A self-recording thermometer.

thermography — The process of converting the heat emitted from an object into visible pictures used to indicate the temperature distribution of part of the building envelope and for locating infiltration flow paths.

thermohaline — The combined effects of temperature and salinity that contribute to density variations in the oceans.

thermoluminescence detector (TLD) — A personal dosimeter made of small chips of lithium fluoride; the absorbed ionizing radiation energy displaces electrons from their balanced state. The electrons are then trapped in a metastable state but can be returned to the balanced state by heating. When this occurs, light is emitted relative to the absorbed radiation dose.

thermoluminescent material — An irradiated material that releases light in proportion to the ionizing radiation adsorbed when it is heated.

thermometer — An instrument used for measuring temperature.

thermonuclear — The transformation in the nucleus of atoms of isotopes of hydrogen fused at extremely high temperatures into heavier nuclei of helium atoms.

thermonuclear reaction — A fusion reaction in which two light nuclei combine to form a heavier atom, releasing a large amount of energy.

thermophile — An organism that thrives at temperatures exceeding 55°C; causes spoilage of pasteurized milk.

thermopile — A piece of equipment that combines a number of thermocouples to multiply the effect; used for generating electric currents or for determining the intensities of radiation.

thermoplastic — A plastic that can be softened by heat and hardened when cooled and remolded; examples include vinyls, acrylics, and polyethylenes.

thermoregulation — The control of heat production and heat loss in the body through physiological mechanisms activated by the hypothalamus.

thermosetting material — A plastic that is heat-set in its final processing to a permanently hard state and cannot be remolded; examples are phenolics, ureas, and melamines.

thermosphere — The outermost layer of the atmosphere of the Earth extending from about 60 miles to several hundred miles, with the temperature varying from many hundreds to thousands of degrees Celsius.

thermostable — Resistant to changes by heat.

thermostat — A device used to measure and control the temperature within a given space; a device that controls a heating or cooling system by responding to changes in temperature.

thermostatic radiator valve — A control valve that combines sensor, controller, and actuator functions.

thermotaxis — Normal adjustment and regulation of body temperature.

Thévenin equivalent circuit — Representation of a circuit as a single voltage source and series resistance.

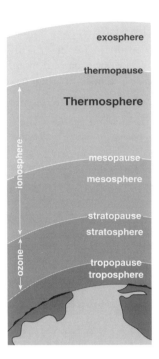

thinking — The cognitive process of forming mental images or concepts.

thinner — A liquid used to increase the fluidity of paints, varnishes, shellacs, and other materials.

thiophenol (C_6H_6S) — A colorless liquid with a penetrating garlic odor; MW: 110.18; BP: 168°C; Sol: soluble in alcohol, ether, benzene, and toluene; Fl.P: 56°C (closed cup), a combustible liquid; density: 1.0728 at 25°C. It is used as a mosquito larvicide and as an intermediate in organic synthesis. It is hazardous to the digestive tract, skin, and respiratory tract. It may cause restlessness, incoordination, muscle weakness, headache, dizziness, cyanosis, lethergy, respiratory depression, coma, and death. ACGIH exposure limit (TLV-TWA): 0.5 ppm.

thiourea (CH_4N_2S) — A crystalline solid; MW: 76.13, BP: NA, Sol: soluble in water and alcohol, Fl.P: a noncombustible solid, density: 1.405. It is used in the manufacture of resins and as a vulcanization accelerator and is a photographic fixing agent. It is carcinogenic to animals.

thiram ($(CH_3)_2NC(S)SSC–(S)N(CH_3)_2$) — A colorless to yellow, crystalline solid (commercial pesticide products may be dyed blue); MW: 240.4, BP: decomposes, Sol: insoluble, Fl.P: not known, sp. gr. 1.29. It is used in the rubber industry as an accelerator, peptizing agent, and vulcanizing agent; as a bacteriostat in commercial and surgical soap, antiseptics, sunburn oils, and fats; as an animal repellent on plants and trees. It is hazardous to the respiratory system and skin and is toxic through inhalation, ingestion, and contact. Symptoms of exposure include dermatitis and irritation of the mucous membrane; for ethanol consumption, symptoms include flush, erythema, pruritus, urticaria, headache, nausea, vomiting, diarrhea, weakness, dizziness, and dyspnea. OSHA exposure limit (TWA): 5 mg/m³ [air].

thirty-day average — The arithmetic mean of pollutant parameter values of samples collected over a period of 30 consecutive days.

thixotropy — A property of materials that appear and act as solids when undisturbed but will change to a semiliquid when agitated.

THM — See trihalomethane.

thoracic — Pertaining to the chest cavity.

thoracic cavity — The space that lies above the diaphragm and below the neck and is enclosed within the walls of the thorax; it contains the heart and lungs.

thoracic particulate mass threshold limit values (TPM-TLVs) — Materials that are hazardous when deposited anywhere within the lung airways and the gas-exchange region.

thoracodynia — Chest pain.

thorax — The middle region of the body of some insects between the head and the abdomen.

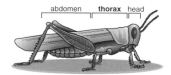

Grasshopper Thorax

thorium (Th) — A naturally occurring, radioactive metal; MW: 232.04, BP: 4500°C, Sol: insoluble, density: 9.7 to 11.7 g/cm³. It is used in the generation of nuclear energy; in refractory applications, lamp mantles, aerospace alloys, welding electrodes, nuclear weapon production; in ceramics; in sun lamps. It is hazardous to the respiratory system, blood, pancreas, and bone marrow and is toxic through inhalation. Symptoms of exposure in animals include cirrhosis of the lungs, hematological effects, hepatic effects, and pancreatic cancer. Nuclear Regulatory Commission maximum permissible concentrations in air per 40-hour week: 9×10^{-12} μCi/ml.

threadworm — Any of a group of nematode worms.

threshold — The lowest dose of a chemical at which a specific measurable effect is observed; the point at which a stimulus is of sufficient intensity to begin to produce an effect.

threshold dose — The minimum absorbed dose that will produce a detectable degree of any given effect.

threshold level — Time-weighted average pollutant concentration values beyond which an exposure is likely to cause adverse human health effects.

threshold limit value (TLV) — The average airborne concentration of a toxic substance to which nearly all workers may be repeatedly exposed 8 hours a day without adverse effects; also can be stated as any amount of a toxic substance regardless of how it makes its way into a living organism.

threshold limit value ceiling (TLV-C) — The average concentration of a toxic substance that should not be exceeded during any working exposure.

threshold limit value short-term exposure limit (TLV-STEL) — A concentration to which workers can be exposed continuously for a short period of time without suffering from irritation, chronic or irreversible tissue damage, narcosis, or accidental injury and will not impair self-rescue or materially reduce work efficiency, provided that the daily threshold limit value time-weighted average is not exceeded.

threshold limit value time-waited average (TLV-TWA) — A concentration based on a time-weighted average for a normal 8-hour work day and 40-hour work week to which nearly all workers may be repeatedly exposed without adverse effects.

threshold of audibility — The minimum effective sound pressure level of a signal capable of generating an auditory sensation in a specified condition; it is assigned a value of 0 decibels (1000 hertz).

threshold of consciousness — The lowest limit of perception of a stimulus.

threshold of pain — The upper intensity level for audible sounds that cause intense discomfort in the average human listener, usually between 130 and 140 decibels.

threshold planning quantity — The amount of an extremely hazardous substance present at a facility above which the facilitiy owner/operator must give emergency planning notification to the Local Emergency Planning Committee.

threshold stimulus — A stimulus that is just adequate to produce a response.

throb — A deep, pulsating kind of discomfort or pain.

T

thrombin — A substance formed in blood clotting as a result of the reaction of prothrombin, thromboplastin, and calcium.

thrombocyte — See platelet.

thromboplastin — A substance essential in blood clotting and formed by the disintegration of blood platelets.

thrombus — A plug or clot in a blood vessel remaining at the point of its formation and causing damage to the area.

thrush — (*disease*) See candidiasis.

thumb separation — A technique that employs the tendency of certain surface active solutes, such as alkyl benzene sulfonate, to collect at a gas–liquid interface.

thunder — Sudden, tremendous heat from lightning that causes the audible compression of shock waves in the air.

thunderheads — See cumulonimbus clouds.

thunderstorm — A violent vertical movement of air with clouds sometimes reaching in excess of 75,000 feet accompanied by periodic lightning, thunder, rain, and sometimes hail.

thymus — A glandular structure of largely lymphoid tissue that functions in development of the body's immune system.

thymus gland — A ductless mass of flattened lymphoid tissue situated behind the top of the sternum that forms antibodies in the newborn and is involved in the development of the immune system.

thyroid carcinoma — A malignant tumor of the thyroid gland, an endrocrine gland that lies in front of the trachea.

thyroid gland — An endocrine gland that produces, secretes, and stores thyroxin; its basic function is to control the rate of metabolism in the body. Insufficient secretions cause sluggish, apathetic, and emotionally depressed behavior; excessive secretions cause a thyroid crisis where all of the bodily processes are accelerated to dangerous levels which may cause death. See also thyrotoxicosis.

thyrotoxicosis — A condition resulting from excessive quantities of thyroid hormones that may cause an increase of the pulse to 200 beats per minute, a sharp rise in respiration, and a sharp rise in body temperature because of loss of control by the temperature control center of the brain.

TI — Acronym for thermal imagery.

TIA — Acronym for transient ischemic attack.

TIC — Acronym for total inorganic carbon.

tick — A bloodsucking arachnid parasite of either the hard or soft type; important vectors of significant arthropod-borne disease.

Dermacentor andersoni
(Rocky Mountain wood tick)

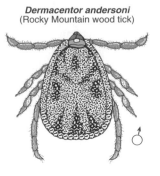

Amblyomma americanum
(Lone Star tick)

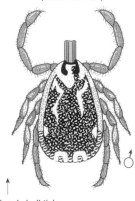

hard shell tick

soft shell tick

Ornithodoros sp.

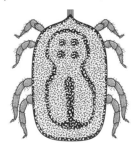

Ticks

tick fever — See relapsing fever.

tick paralysis — A rare, progressive, reversible disorder caused by the bite of a tick, which injects a neurotoxin found in its saliva into the body, causing weakness, incoordination, and paralysis.

tidal marsh — Low, flat, marshland traversed by interlaced channels and tidal sloughs and subject to tidal inundation.

tidal volume — The amount of air inhaled and exhaled during normal ventilation.

tide — The periodic rhythmic rise and fall of the sea surface.

Tier I screening — An evaluation of contaminant concentrations from a site, for all media of concern, compared to those from background samples collected from nearby areas that have not been affected by the substances of concern.

Tier II screening — An evaluation of contaminant concentrations from a site, for all media of concern, compared to medium-specific values obtained from published sources, such as the U.S. EPA National Primary Drinking Water Regulations.

Tier III screening — A site-specific analysis weighing current and potential exposure scenarios for the populations of concern and characteristics of the affected media.

tight soil — A compact, relatively impervious and tenacious soil or subsoil that may or may not be plastic.

tile — A pipe made of baked clay; (*geographic information system*) a part of the database in a geographic information system representing a discrete part of the surface of the Earth.

TIM — Acronym for toxic industrial material.

timbre — Characteristic of hearing in which the listener is capable of distinguishing between two sounds, even though the two sounds are of equal pressure and pitch.

time-weighted average — The average value of a parameter; concentration of a chemical in air that varies over time.

time-weighted average concentration — A concentration of contaminant that has been weighted for the time duration of the sample.

time-weighted average exposure (TWA) — The average exposure over a given working period used to determine the contaminants within a given sample during that period.

tin (Sn) organic compounds — Appearance and odor vary depending upon the specific organic compound; properties vary depending upon the specific organic compound. It is used as a stabilizing agent; as a catalyst in organic and inorganic synthesis; in the manufacture of insecticides, bactericides, fungicides, and molluscides; on agricultural crops; in the manufacture of rodent repellent for wire cable; as a fuel additive and lubricant; in veterinary medicine as poultry anthelmintic. It is hazardous to the central nervous system, eyes, liver, urinary tract, skin, and blood and is toxic through inhalation, absorption, ingestion, and contact. Symptoms of exposure include headache, vertigo, eye irritation, psycho-neurologic disturbances, sore throat, cough, abdominal pain, vomiting, urine retention, paresis, focal anesthesia, skin burns, and pruritus; in animals, symptoms include hemolysis and hepatic necrosis. OSHA exposure limit (TWA): 0.1 mg/m³ [skin].

tin (Sn) inorganic compounds (except oxides) — A gray to almost silver-white, ductile, malleable, lustrous metal; MW: 118.7, BP: 4545°F, Sol: insoluble, sp. gr. 7.28. It is used in the manufacture of blueprint paper; as a perfume stabilizer in soap; in the preparation of lubricating oil additives; in treating glass surfaces; as a catalyst in the manufacture of pharmaceuticals; as a bleaching agent for sugar; as a food additive; in dyeing silk; in metallurgy. It is hazardous to the eyes, skin, and respiratory system and is toxic through inhalation and contact. Symptoms of exposure include skin and eye irritation. OSHA exposure limit (TWA): 2 mg/m³ [air].

TIN — See triangulated irregular network.

tine test — A tuberculin test in which a small, disposable disk with multiple tines carrying tuberculin antigen is used to puncture the skin to determine if there had been a previous exposure to tuberculosis or the actual active disease.

tingling — A prickly sensation in the skin or a part of the body along with diminished sensitivity to stimulation of the sensory nerves.

tinnitus — A ringing sound in one or both ears.

tipping — Unloading refuse from a collection truck.

tipping fee — A fee charged at a disposal or recycling facility for accepting wastes.

TIROS — See Television and Infrared Observation Satellite.

Tiros

TIROS-N — A group of TIROS satellites designed to carry high-resolution radiometers to map clouds, sea-surface temperatures, and the surface of the Earth by day and night; these satellites contain sensors to obtain vertical temperature soundings, moisture, and carbon dioxide measurements.

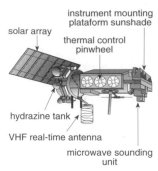

TIROS-N Spacecraft

TIROS-N/NOAA satellites — NOAA satellites that continuously orbit the Earth from the North to the South Pole at an altitude of approximately 870.44 km to collect visible and infrared imagery and provide atmospheric-sounding data and meteorological data relay and collection.

TIS — Acronym for tolerance index system.

tissue — A group of cells having the same or related functions and structure.

tissue dose — The absorbed dose received by tissue in the region of interest; expressed in rads.

tissue fluid — Fluid that bathes the cells of the body and is called lymph when contained in vessels.

tissue response — A reaction or change in a tissue exposed to a disease, toxin, or other stimulus.

titanium (Ti) — A dark-gray metal; atomic weight: 47.88, BP: 3277°C, Sol: NA, Fl.P: flammable in a very finely divided state and will explode when combined with liquid oxygen, density: NA. It is added to steel and aluminum to enhance their tensile strength and acid resistance and is also alloyed with copper and iron in titanium bronze. It is hazardous to the respiratory tract and may cause coughing, irritation, and dyspnea. ACGIH exposure limit: NA.

titanium dioxide (TiO₂) — A white, odorless powder; MW: 79.9, BP: 4532 to 5432°F, Sol: insoluble, sp. gr. 4.26. It is used as a pigment in the manufacture of paints, varnishes, and lacquers to impart whiteness, opacity, and brightness; in the manufacture of paper, photographic paper, and cellophane coatings; in the manufacture of elastomers; in the manufacture of floor coverings; in the manufacture of ceramics and glass for capacitors; in the manufacture of printing inks; in the manufacture of coated fabrics and textiles; in the manufacture of building materials in roofing granules, ceiling tiles, cement-curing aids, and titanium carbide cutting tools; in

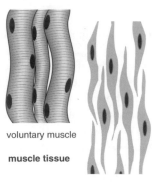

voluntary muscle

muscle tissue

involuntary muscle

Tissue

simple squamous

epithelial tissue

simple columnar

the manufacture of cosmetics, food color additives, and synthetic diamonds. It is hazardous to the lungs and is toxic through inhalation. Symptoms of exposure include slight lung fibrosis; carcinogenic. OSHA exposure limit (TWA): 10 mg/m³ [air].

titanium tetrachloride (TiCl₄) — A colorless or light-yellow liquid with a fuming, penetrating odor; MW: 189.71, BP: 136.4°C, Sol: soluble in cold water, Fl.P: nonflammable (poisonous gases produced in a fire include titanium oxides and hydrochloric acid), sp. gr. 1.726. It is used along with potassium bitartrate as a mordant in the textile industry, in dying leather, as a smoke-producing screen, and in the manufacture of iridescent glass and artificial pearls. It is hazardous to the respiratory system, skin, and eyes. Acute exposure may cause burning and scarring of the skin and eyes, irritation of the throat and air passages, cough, and pulmonary edema. Chronic exposure may cause emphysema, cough, and shortness of breath. OSHA exposure limit (8-hour TWA): NA.

titer — The concentration of a substance in a solution as determined by titration.

titrant — A standard solution of known concentration and composition used during titration.

titration — The process of adding one reactant to another a drop at a time until the two are present in exactly equivalent amounts or a given reaction occurs.

titrimetric procedure — Procedure in which a sample is reacted with a titrant solution containing a known concentration of a reactive chemical.

TL$_{50}$ — See medial tolerance limit.

TLC — See total lung capacity.

TLD — See thermoluminescence detector.

TLM$_{96}$ — The concentration of a substance that would kill 50% of exposed aquatic organisms within a 96-hour period.

TLV — See threshold limit value.

TLV-C — See threshold limit value ceiling.

TLV-STEL — See threshold limit value short-term exposure limit.

TLV-TWA — See threshold limit value time-waited average.

TM — See thematic mapper.

TMDL — Acronym for total maximum daily limit; total maximum daily load.

TMRC — See theoretical maximum residue contribution.

TNCWS — Acronym for transient noncommunity water system.

TNT — See trinitrotoluene.

tntc — Acronym for too numerous to count.

TOA — Acronym for trace organic analysis.

TOC — See total organic carbon.

TOD — See total oxygen demand.

toe — The bottom of the working face of a sanitary landfill where deposited solid waste is in contact with virgin ground.

TOG — See total organic gases.

TOGA — Acronym for tropical ocean–global atmosphere.

togavirus — An arbovirus spread by mosquitos; may cause fever, encephalitis, and rashes.

tolerance — The allowable concentration of a chemical on or in any food, plant, or animal meant for human or animal consumption considered to be safe by the Food and Drug Administration; (*physiology*) the ability of an organism to withstand environmental stress; (*geographical information system*) a numerical value representing the acceptable error range for a feature with respect to its actual point found on Earth.

tolerance level — See permissible dose.

toluene (C$_6$H$_5$CH$_3$) — A colorless liquid with a sweet, pungent, benzene-like odor; MW: 92.1, BP: 484°F, Sol: 0.05% at 61°F, Fl.P: 40°F, sp. gr. 0.87. It is used as a solvent in the pharmaceutical, chemical, rubber, and plastics industries; as a thinner for paints, lacquers, coatings, and dyes; as a paint remover; in insecticides; as a starting material and intermediate in organic chemical and chemical synthesis industries; in the manufacture of artificial leather; in photogravure ink production; as a constituent in automotive and aviation fuels. It is hazardous to the central nervous system, liver, kidneys, and skin and is toxic through inhalation, absorption, ingestion, and contact. Symptoms of exposure include fatigue, weakness, confusion, euphoria, dizziness, headache, dilated pupils, lacrimation, nervousness, muscle fatigue, insomnia, paresthesia, and dermatitis. OSHA exposure limit (TWA): 100 ppm [air] or 375 mg/m^3.

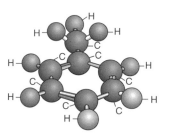

Toluene

2,4-toluenediamine (C$_7$H$_{10}$O$_2$) — A crystalline solid; MW: 122.19, BP: 284°C, Sol: slightly soluble in water but dissolves in organic solvents, Fl.P: a noncombustible solid that can react violently to strong oxidizers. It is used as an intermediate in the manufacture of dyes. It is a carcinogen and is hazardous to the digestive tract and skin. ACGIH exposure limit: NA.

2,4-Toluenediamine

toluene-2,4-diisocyanate (CH$_3$C$_6$H$_3$(NCO)$_2$) — A colorless to pale-yellow solid (liquid above 71°F) with a sharp, pungent odor; MW: 174.2, BP: 484°F, Sol: insoluble, Fl.P: 260°F, sp. gr. 1.22. It is used in the manufacture of surface coatings and finishes, polyurethane paints, and electrical and thermal insulation; in the manufacture of flexible polyurethane foams and elastoplastics, adhesives, and sealants. It is hazardous to the respiratory system and skin and is toxic through inhalation, ingestion,

Toluene Diisocyanate

and contact. Symptoms of exposure include nose and throat irritation, choking, paroxysmal cough, chest pain, nausea, vomiting, abdominal pain, bronchial spasms, pulmonary edema, dyspnea, asthma, conjunctivitis, lacrimation, dermatitis, and skin sensitization; carcinogenic. OSHA exposure limit (TWA): 0.005 ppm [air] or 0.04 mg/m³.

***m*-toluidine (C₇H₇NH₂)** — A light yellow liquid; MW: 107.17, BP: 203°C, Sol: dissolves easily in organic solvents and dilute acids, Fl.P: 85.5°C (closed cup), density: 0.990 at 25°C. It is used as an intermediate in the manufacture of dyes and other chemicals. It is hazardous to the digestive tract, respiratory tract, and skin. It causes anemia and hematuria and exposure to it has the same symptoms as *o*-toluidine. It has not caused any cancer in tests subjects. ACGIH exposure limit (TLV-TWA: 2 ppm [skin].

NH₂

CH₃

***m*-Toluidine**

***o*-toluidine (C₇H₇NH₂)** — A colorless to light-yellow liquid becoming reddish brown on exposure to air or light with a weak aromatic odor; MW: 107.17; BP: 200°C; Sol: soluble in water and very soluble in organic solvents and dilute acids; Fl.P: 85°C (closed cup), combustible liquid; density: 1.008 at 20°C. It is used in the manufacture of dyes; as an intermediate in rubber chemicals, pesticides, and pharmaceuticals. It is a cancer-causing compound. It is hazardous to the respiratory system, digestive system, and skin. It causes severe effects to the kidneys, liver, blood, cardiovascular system, skin, and eyes. It produces anoxia, cyanosis, headache,

weakness, drowsiness, and dizziness. ACGIH exposure limit (TLV-TWA): 2 ppm.

***o*-Toluidine**

tomography — An x-ray technique that produces a film representing a detailed cross-section of tissue at a predetermined depth.

TOMS — Acronym for the Total Ozone Mapping Spectrometer, which was launched on the NASA NIM-BUS-7 satellite and used to monitor the Earth between October 31, 1978, and May 6, 1993.

ton — A unit or weight measurement equal to 2000 pounds; also known as short ton.

ton of refrigeration — A measure of the rate of refrigeration equal to 12,000 Btu per hour.

tone — Sound sensation having pitch and capable of causing an auditory sensation.

tonic spasm — A slight, continuous contraction of a muscle.

tonnage — The amount of waste that a landfill accepts; expressed in tons per month.

tonne — One metric ton; equivalent to 1000 kilograms.

tonometer — An instrument used to measure pressure in the eyeball; used in glaucoma testing.

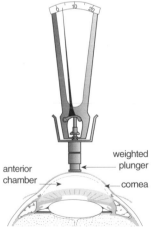

(Schiötz) Tonometer
(not proportionate to eye)

topical — Meant for local application on a body surface.

topographic map — A map showing topography in great accuracy and detail relative to the scale of the map.

topography — The detailed mapping or charting of the surface features of a relatively small area, district, or locality to show as much accurate detail as possible.

topology — The way in which geographical elements are linked together.

topsoil — The uppermost few inches of soil consisting of organic matter and plant nutrients; usually the A horizon.

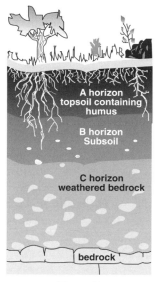

Topsoil
in the A horizon

torch brazing — A brazing process in which the heat required is furnished by a fuel gas flame.

tornado — A violent storm characterized by strong swirling winds and updrafts formed when a strong cold front pushes under a warm, moist air mass over the land.

torpor — A time of decreased metabolism and lowered body temperature that occurs in daily activity cycles.

torque — A force that produces or tends to produce rotation or torsion.

torr — A unit of pressure equal to 1/760 atmosphere.

torsion — The process of twisting in a positive or negative direction.

tort — A civil wrong, other than a breach of contract, where the court is asked to provide a remedy in the form of damages for actions taken against the individual bringing the tort.

TOSC — See Technical Outreach for Communities.

total alkalinity — A measure of the sum total of all alkaline materials dissolved in water.

total body burden — Concentration of pollutants remaining in the body from previous exposures.

total body radiation — An exposure of the entire body, where all the cells receive the same amount of radiation.

total column ozone — The amount of ozone measured from the surface of the Earth to the top of the atmosphere over a given surface area.

total dissolved phosphorus — The total phosphorus content of all material, as orthophosphate, that will pass through a filter without prior digestion or hydrolysis; also called soluable P or ortho P.

total dissolved solids — A water quality parameter defining the concentration of dissolved organic and inorganic chemicals in water; the total dissolved (filterable) solids in wastewater.

total dynamic head — Vertical distance from the elevation of an energy grade line on the suction side of a pump to the elevation of the energy grade line on the discharge side of the pump.

total exchange — See cation-exchange capacity.

total hazard value — A quantitative value representing the total hazard of a chemical substance derived from integrating the human health effects, ecological effects, and exposure potential of the chemical.

total ionization — The total positive or negative charge on ions produced by radiation in the process of losing its kinetic energy.

total lung capacity (TLC) — The volume of air left in the lungs after maximum voluntary inspiration.

total maximum daily load — The total allowable pollutant load to a receiving water within a designated period that will not produce a violation of water-quality standards.

total maximum daily loads — Estimates of the amount of specific pollutants that a body of water can safely take without threatening beneficial uses.

total organic carbon (TOC) — A measure of the kind, type, and amount of organic pollution occurring in a body of water.

total organic gases — Gaseous organic compounds, including reactive organic gases and relatively unreactive organic gases such as methane.

total oxygen demand (TOD) — A combination of the biochemical oxygen demand and chemical oxygen demand of a body of water or a quantity of sewage or organic waste.

Total Ozone Mapping Spectrometer — An instrument flown on NASA's NIMBUS-7 satellite to continue the high-resolution global mapping of total ozone on a daily basis.

total pressure — The algebraic sum of static pressure and velocity pressure.

T

total recovered petroleum hydrocarbon — A method for measuring petroleum hydrocarbons in samples of soil or water.

total solar irradiance — See solar constant.

total solids — All solid constituents of sewage; the total of organic and inorganic solids suspended and dissolved.

total suspended nonfilterable solids (TSS) — Total suspended solids as measured by a standard technique using glass fiber disks.

total suspended particulates (TSP) — All suspended particles of varying sizes.

totalizer — See integrator.

TOVS — Acronym for TIROS Operational Vertical Sounder.

TOX — Acronym for tetradichloroxylene.

toxaphene ($C_{10}H_{10}Cl_8$) — An amber, waxy solid having an odor similar to chlorine or camphor; MW: a complex mixture of a large number of chloroderivatives of camphene; BP: dechlorinates at 155°C; Sol: 0.0003 g/100 cc water, soluble in benzene, toluene, and other aromatic solvents; density: 1.65 gm/cm³. It was one of the most widely used insecticides in the Unites States until it was banned by the U.S. EPA in 1982. It is hazardous to the respiratory system, liver, adrenal gland, and central nervous system and is toxic through inhalation and ingestion. Symptoms of exposure include acute pulmonary insufficiency with extensive miliary shadows, high serum globulin, liver damage, damage to the adrenal gland, salivation, restlessness, hyperexcitability, muscle tremors, convulsions, and loss of consciousness. It may be fatal to human adults and causes liver cancer in animals. OSHA exposure limit (PEL-TWA): 0.5 mg/m³ [air].

toxemia — A condition of poisoning by toxins or bacterial products in the blood.

toxic — Relating to poisonous or deadly effects on the body caused by inhalation, ingestion, absorption, or contact with a toxin.

toxic air contaminant — An air pollutant that may cause or contribute to an increase in deaths or serious illness or which may pose a present or potential hazard to human health.

toxic best available control technology — The most effective emission limitation or control technique that has been achieved in practice for a particular class of pollutants using new processes and/or equipment changes.

toxic chemical — A substance that can cause severe illness, poisoning, birth defects, disease, or death when ingested, inhaled, or absorbed by living organisms; a chemical listed in U.S. EPA rules as "Toxic Chemicals Subject to Section 313 of the Emergency Planning and Community Right-to-Know Act of 1986."

toxic chemical use substitution — The replacing of toxic chemicals with less-harmful chemicals in industrial processes.

toxic cloud — An airborne plume of gases, vapors, fumes, or aerosols containing toxic materials.

toxic concentration low (TCL) — The lowest concentration of a gas or vapor capable of producing a defined toxic effect in a specified test species over a specified time.

toxic concentration — The concentration at which a substance produces a toxic effect.

toxic dose (TD50) — The calculated dose of a chemical introduced by a route other than inhalation that is expected to cause a specific toxic effect in 50% of a defined experimental animal population.

toxic dose low (TDL) — The lowest administered dose of a material capable of producing a defined toxic effect in a specified test species.

toxic effects (Tfx) — The health effects that occur due to exposure to a toxic substance; any poisonous effect on the body, either reversible or irreversible, resulting in an undesirable disturbance of physiological function.

toxic equivalency potential — Adverse human health effects of a chemical compared to those of other chemicals after being converted into a common unit of comparison.

toxic hot spot — A location where emissions from specific sources may expose individuals and population groups to elevated risk of adverse health effects including but not limited to cancer and may contribute to the cumulative health risks of emissions from other sources in the area.

toxic pollutants — Materials contaminating the environment that cause death, disease, or birth defects in organisms that ingest or absorb them.

toxic shock syndrome (TSS) — A severe illness caused by *Staphylococcus aureus* and characterized by a sudden onset of high fever, vomiting, profuse watery diarrhea, and hypotension. Several body organs are typically involved; a rash appears during the acute phase and then the scaly skin on the palms and soles of the feet peels off. It can be prevented by not using tampons during the menstrual cycle, or the risk can be reduced by using tampons for short periods of time only.

toxic substance — Any substance that can cause acute or chronic injury to the human body or which is suspected of being able to cause disease or injury under some conditions; examples include convulsant poisons, central nervous system depressants, peripherally acting nerve poisons, muscle poisons, protoplasmic poisons, poisons of the blood, and blood forming poisons.

Toxic Substances Control Act (TSCA) — A 1976 law amended through 1986 that allows for the regulation of chemicals in commerce as well as before they enter commerce. It establishes a policy in which chemical manufacturers and processors are responsible for developing data about the health and environmental effect of their chemicals. It allows for government regulation of chemical substances that pose an unreasonable risk of injury to health or the environment.

toxic symptom — Any feeling or sign indicating the presence of a poison in the system.

toxic waste — A waste that can produce injury if inhaled, swallowed, or absorbed through the skin.

toxicant — Any harmful substance or agent that can injure or kill an exposed organism: humans, animals, or plants; a poison.

toxicity — The degree to which a substance is injurious or poisonous.

toxicity assessment — The process of characterizing the inherent toxicity of a chemical.

Toxicity Characteristic Leachate Procedure — A method for determining the toxicity of a contaminant; used to measure the impact of a contaminant on groundwater.

toxicity reduction evaluation — A study conducted to determine the sources of toxicity in a discharge effluent so that a discharger can comply with permit limits.

toxicity test — A way of determining the toxicity of a chemical or an effluent on living organisms.

toxicity value — The numerical expression of the dose–response relationship of a substance to be used in risk assessments.

toxicokinetics — Study of the action of toxins in the body, including the routes and mechanisms of absorption and excretion, the rate at which the toxic action begins, the biotransformation of the substance in the body, the metabolites produced, and the duration of the effect.

toxicological endpoint — The biological effect judged to be the most meaningful and specific in relation to the toxic chemical being studied.

toxicological profile — An examination, summary, and interpretation of a hazardous substance to determine levels of exposure and associated health effects.

toxicologist — A specialist in toxicology.

toxicology — The study of how natural or synthetic poisons cause undesirable effects in living organisms.

toxicosis — A disease condition caused by the absorption of metabolic or bacterial poisons.

toxics use reduction — A source reduction program used to reduce, avoid, or eliminate the use of toxics in processes or products to reduce overall risks to the health of workers, consumers, and the environment.

toxin — Any poisonous substance of microbial, vegetable, animal, or synthetic chemical origin that reacts with specific cellular components to kill cells, alter growth or development, or kill the organism.

toxoid — A modified, microbial, vegetable, or animal toxin used for immunization.

Toxoplasma — A genus of sporozoan parasites in mammals and some birds.

Toxoplasma gondii — A protozoan parasite that causes toxoplasmosis with symptoms of mild flu-like symptoms.

toxoplasmosis — A systemic protozoal disease that may be asymptomatic or produce fever with an increase in the number of lymphocytes in the blood. Incubation time may vary from 5 to 23 days. It is caused by *Toxoplasma gondii* and is found worldwide in mammals and birds. Reservoirs of infection include rodents, pigs, cattle, sheep, goats, and birds. It is transmitted by eating raw or undercooked infected meat containing the cysts or ingestion of the oocysts in food, water, or dust that is contaminated with cat feces. It is not directly transmitted from person to person; general susceptibility. It is controlled through proper cooking of meats, proper feeding of cats and disposal of their feces and litter, and thorough hand washing.

TP — Acronym for total particulates or technical product.

2,4,5-TP ($C_9H_7Cl_3O_3$) — A white powder with a faint odor; MW: 269.51, BP: NA, Sol: 0.02 g/100 g water at 25°C, Fl.P: NA (some formulations contain petroleum solvents that can ignite and containers may explode violently in the heat of a fire), sp. gr. 1.2085. It has been used as a plant hormone and an herbicide but has been limited because it is teratogenic and carcinogenic in test animals. It is hazardous to the respiratory tract, digestive tract, skin, and eyes. Acute exposure may cause skin, nose, throat, and bronchial tube irritation; cough; fatigue; weakness; poor appetite; seizures; and kidney or liver damage. Chronic exposure may cause liver and kidney damage and a possible reproductive risk for cancer. OSHA exposure limit (8-hour TWA): NA.

TPH — Acronym for total petroleum hydrocarbons.

TPI — Acronym for technical proposal instructions.

TPM-TLVs — See thoracic particulate mass threshold limit values.

TPQ — See threshold planning quantity.

TPR — Acronym for temperature, pulse, respiration.

TPY — Acronym for tons per year.

TQM — Acronym for Total Quality Management.

trace analysis — Determination of the presence of a substance in a sample at a relatively low concentration.

trace element — See micronutrient.

trace gas — A small quantity of gas leaking into the ambient environment that may cause adverse health effects on personnel in the vicinity of the equipment; any one of the less common gases found in the atmosphere of the Earth that account for about 1% of the atmosphere, including carbon dioxide, water vapor, methane, oxides of nitrogen, ozone, and ammonia.

tracer — A minute quantity of radioactive isotope used in medicine or biology to study the chemical changes within living tissues.

tracer gas analyzer — An instrument used to evaluate the concentration of tracer gas in a sample of air over time.

tracer gas technique — A method using tracer gases to evaluate air infiltration and ventilation rates.

tracer gases — Compounds, such as sulfur hexafluoride, that are used to identify suspected pollutant pathways and to quantify ventilation rates.

trachea — In vertebrates, the windpipe, which is used for conducting air; in insects and spiders, an air tube.

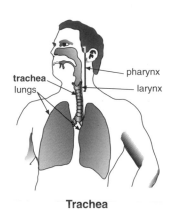

Trachea

tracheitis — An inflammation of the air passage (the trachea) between the larynx and lungs.

tracheotomy — An incision made into the trachea through the neck below the larynx that is used to gain access to the airway below a blockage caused by a foreign body, tumor, or edema of the glottis.

Tracking and Data Relay Satellite System — A NASA-developed orbiting communications satellite used to relay data from satellite sensors to ground stations and to track the satellites in orbit.

trade name — A trademarked name or commercial name for a material or product.

transboundary pollutant — An air pollutant that travels from one jurisdiction to another, often crossing state or international boundaries.

transcutaneous — Performed through the skin.

transducer — Any system or device that is actuated by waves from one or more transmission systems or which converts an input signal into an output signal of a different form.

transect — To sever or cut across.

transfer — The loading and unloading of chemicals between transport vehicles and storage vessels and the sending of chemicals by way of pipes between storage vessels and process reactors.

transfer air — The movement of indoor air from one space to another.

transfer DNA (tDNA) — DNA transferred from its original source and present in transform cells.

transfer efficiency — A measure of the percent of the total amount of coating used that is transferred to a unit surface by spray gun or other device.

transfer facility — Any transportation-related facility, including loading docks, parking areas, storage areas, and other similar areas, in which shipments of hazardous waste or solid waste are held during the normal course of transportation.

transfer RNA (tRNA) — A form of RNA that transfers the genetic code from messenger RNA for the production of a specific amino acid.

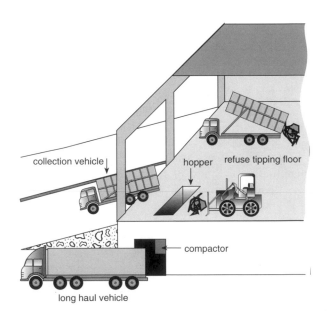

Solid Waste Transfer Station

transfer station — A site in which solid waste is concentrated and then taken to a processing facility or a sanitary landfill.

transferring — Moving a substance to another location.

transform — (*mapping*) Changing the scale, projection, or orientation of a mapped image.

transformation — The chemical alteration of a compound by a process such as biodegradation, hydrolysis, photodegradation, or reaction with other compounds.

transformer — A piece of equipment that converts the low-voltage electricity of a generator to higher voltage levels for transmission to the load center.

transfusion — Adding whole blood or blood components to the bloodstream.

transgenic plant — A genetically engineered plant.

transient — Referring to a condition that is temporary.

transient equilibrium — The state reached if the half-life of the parent is short enough so the quantity present decreases appreciably during the period under consideration but is still longer than that of successive members of the series; a stage of equilibrium will be reached after the activity of all members of the series decreases exponentially with the period of the parent.

transient heat fatigue — A temporary condition of discomfort and mental or psychological strain resulting from prolonged heat exposure.

transient water system — A non-community water system that does not serve 25 of the same non-resident persons per day for more than 6 months per year.

transillumination — Passage of light through a solid or liquid substance.

transistor — An electronic device used in place of a vacuum tube to control electric current.

transit — A surveying instrument that can measure vertical and horizontal angles.

transition — A change in the cross-sectional shape or area of a duct or hood.

translocation — The conduction of soluble material from one part of a plant to another; (*toxicology*) the movement of a toxicant from a given part of an animal or plant to a different part of the organism.

translucent — The ability to transmit light so that objects beyond are entirely visible.

transmissibility — (*groundwater*) The capacity of a rock to transmit water under pressure.

transmission — (*radiation*) The process by which radiation passes through a body or medium.

transmission electron microscope (TEM) — An electron microscope that routinely can resolve fibers 0.0025 μm in diameter; it is 100 times more sensitive than phase contrast microscopy.

transmission loss — The ratio (expressed in decibels) of the sound energy on a structure to the sound energy transmitted.

transmissivity — The capacity of an aquifer to transmit water, which is dependent on the water-transmitting characteristics of the saturated formation (hydraulic conductivity) and the saturated thickness.

transmissometer — An instrument used to measure visibility that consists of a fixed light projector and a receiver. The light projected at a fixed intensity either reaches the receiver as projected or decreases in intensity in proportion to the visibility-restricting particles present in the air between the projector and the receiver; the receiver translates the light it receives in visibility readings expressed in hundreds of feet.

transmittance — The fraction of incident light that is transmitted through an optical medium.

transmutation — A process in which the nuclei of a radioactive element decays at a fixed rate by emitting nuclear particles; the element is changed to another element, such as when radioactive radium becomes radon. See also nuclear transformation.

transovarial — Pertaining to passing an infection from one generation to another through the egg.

transparent — Refers to the ability of a substance to transmit light so objects can be seen clearly through it.

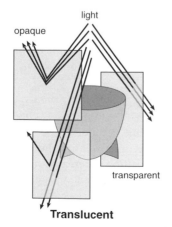

Translucent

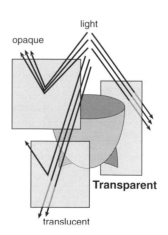

Transparent

transpiration — The process in which green plants lose water by evaporation through their stomata; the passage of a gas through a membrane.

Transpiration

transplacental — Pertaining to an agent that causes physical defects in a developing embryo.

transponder — A unit that receives pulses from a radar set or interrogator and automatically responds to the received pulse by transmitting a pulse or sequence of pulses that can be recognized by the interrogating station.

transport — The hydrologic, atmospheric, or other physical processes that convey the mass of a pollutant through and across media from source to receptor; (*engineering*) to carry or convey goods from one place to another using ships, trucks, trains, pipelines, or planes; (*solid waste*) the movement of solid waste subsequent to collection.

transport mode — Any method of transportation, such as highway, rail, water, pipelines, or air.

transport velocity — The velocity required to prevent the settling of a contaminant from an air stream, usually related to the flow of air in a duct.

transportation control measure — Any technique used to reduce vehicle trips, vehicle use, vehicle miles traveled, vehicle idling, or traffic congestion that results in a smaller amount of motor vehicle emissions.

transstadial — Pertaining to the transmission of organisms from infected adult ticks through the eggs to the larval, nymphal, and adult stages.

transverse — At right angles to the long axis of the body.

transverse wave — A wave in which the particles of a medium meet at right angles to the direction of the motion of the wave.

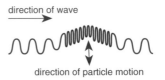

Transverse Wave

trap — (*plumbing*) A fitting that provides a liquid seal to prevent the back passage of gases without materially affecting the flow of sewage or water.

trapezoidal weir — See Cipolletti weir.

trauma — An injury caused by accident or violence.

traumatic meningitis — A form of meningitis that develops as the result of an injury to the skull or spinal column.

traumatic myelitis — A spinal cord inflammation coming from an injury.

traumatic myositis — Inflammation of the muscles resulting from a wound or other trauma.

traumatic neuritis — Inflammation of a nerve caused by an injury.

traumatic psychosis — A major mental disorder caused by an injury to the head.

traumatic shock — An emotional or psychological state induced by an injury that may produce abnormal behavior.

traumatic thrombosis — An intravascular coagulation of a vein or other blood vessel after injury or irritation.

traumatology — The study of wounds and injuries.

traumatopathy — A pathological condition resulting from a wound or injury.

traveler's diarrhea — A diarrheal disorder usually seen in people visiting different parts of the world other than their own; caused by some strains of *Escherichia coli* that produce a powerful exotoxin; also caused by *Giardia lamblia* and species of *Salmonella* and *Shigella*. Symptoms include abdominal cramps, nausea, vomiting, some fever, and watery stools.

TRC — Acronym for technical review committee.

TRD — Acronym for technical review document.

TRE — See toxicity reduction evaluation.

treated wastewater — Wastewater that has been subjected to one or more physical, chemical, and biological processes to reduce the pollutants that cause health hazards.

treatment — Any method, technique, or process designed to change the physical, chemical, or biological character or composition of any hazardous waste so as to neutralize or render it nonhazardous or less hazardous; to recover it; to make it safer to transport, store, and dispose of; to make it amenable for recovery, storage, or volume reduction. Methods include neutralization, precipitation, evaporation, incineration, filtration, and ion exchange. Also, a technique used to improve a medical problem.

Trechoria — Genus of spiders whose bite is toxic and irritating to humans.

trematodes — See flukes.

tremie — A device used to place concrete or grout underwater.

tremor — Involuntary shaking, trembling, or quivering.

trench fever — A nonfatal, febrile, louse-borne bacterial disease characterized by headache, malaise, pain, and tenderness ranging in severity. Incubation time is 7 to 30 days. It is caused by *Rickettsia quintana* and was found during World War I and World War II among troops and prisoners of war and in other occasional conditions of crowded living. The reservoir of infection is people, with the body louse *Pediculus humanus* being the intermediate host. It is not directly transmitted from person to person but through the feces of the louse via a break in the skin. It is communicable as long as the organisms circulate in the blood, which may be weeks to years; general susceptibility. It is controlled through delousing procedures and by dusting clothing and bodies.

trench foot — A painful foot disorder resembling frostbite and resulting from exposure to cold and wet.

trench sanitary landfill method — A landfill method in which the waste is spread and compacted in a trench; the excavated material is spread and compacted over the waste to form the basic cell structure.

triage — The sorting out or screening of patients seeking care to determine which service is initially required and with what priority.

trial burn — An incinerator test in which emissions are monitored for the presence of specific organic compounds, such as trichloroethylene, particulates, and hydrogen chloride.

triangulated irregular network — (*geographical information system*) A surface representation derived from irregularly spaced points and break-line features with each sample point having an x,y coordinate and a z-value or surface value.

Tribolium castaneum — See Red Flower Beetle.

Tribolium confusum — See Confused Flower Beetle.

tributary — A small river or stream that flows into a larger river or stream with a number of these units merging to form a river.

tributyl phosphate ($(C_4H_9)PO_4$) — A colorless, odorless liquid; MW: 266.32, BP: 289°C, Sol: soluble in most organic solvents, Fl.P: a noncombustible solid, density: 0.976 at 25°C. It is used as a plasticizer for cellulose esters, vinyl resins, and lacquers and in making fire retardants, biocides, defoamers, and catylists. It is a neurotoxic compound and cholinesterase inhibitor. It is hazardous to the skin, eyes, and respiratory tract. It may cause paralysis, depression of the central nervous system, and irritation. ACGIH exposure limit (TLV-TWA): 0.2 ppm.

$$C_4H_9O - \overset{\overset{\displaystyle O}{\displaystyle \|}}{\underset{\underset{\displaystyle OC_4H_9}{\displaystyle |}}{P}} - OC_4H_9$$

Tributyl phosphate

Trichinella — A genus of nematode parasites.

trichinosis — An infection caused by *Trichinella spiralis*; an intestinal nematode whose larvae migrate and become encapsulated in muscles. It may be asymptomatic or have symptoms of thirst, profuse sweating, chills, weakness, and prostration. It may become a fulminating fatal disease. Incubation time is usually 10 to 14 days with a range of 5 to 45 days. It is found worldwide in pork or wild meat. Reservoirs of infection include pigs, dogs, cats, rats, and wild animals. It is transmitted by eating raw or improperly cooked meat containing larvae. It is not transmitted from person to person; general susceptibility. It is controlled by cooking meat thoroughly.

T

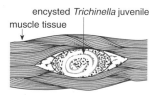

encysted *Trichinella* juvenile
muscle tissue

Trichinosis

1,2,4-trichlorobenzene ($C_6H_3Cl_3$) — A colorless liquid; MW: 181.46, BP: 415.4°F, Sol: not available, Fl.P: 222°F, sp. gr. 1.463. Other information not available. ACGIH ceiling exposure limit (TLV-C): 5 ppm or 40 mg/m³.

1,2,4 Trichlorobenzene

1,1,2-trichloroethane ($CHCl_2CH_2Cl$) — A colorless liquid with a sweet, chloroform-like odor; MW: 133.4, BP: 237°F, Sol: 0.4%, sp. gr. 1.44. It is used in organic synthesis for the production of vinylidene chloride. It is hazardous to the central nervous system, eyes, nose, liver, and kidneys and is toxic through inhalation, absorption, ingestion, and contact. Symptoms of exposure include central nervous system depression, liver and kidney damage, and irritation of the nose and eyes; carcinogenic. OSHA exposure limit (TWA): 10 ppm [skin] or 45 mg/m³. See also methyl chloroform.

1,1,2–Trichloroethane

trichloroethylene ($ClCHCCl_2$) — A colorless liquid (unless dyed blue) with a chloroform-like odor; MW: 131.4, BP: 189°F, Sol: 0.1% at 77°F, Fl.P: 90°F, sp. gr. 1.46. It is used as a cleaning solvent in cold cleaning and vapor degreasing operations; as a scouring and cleaning agent in textile processing; in the extraction and purification of animal and vegetable oils in the food and pharmaceutical industries; in the manufacture of adhesives, anesthetics, and analgesics; as a fumigant and disinfectant for seeds and grains. It is hazardous to the respiratory system, heart, liver, kidneys, central nervous system, and skin and is toxic through inhalation, ingestion, and contact. Symptoms of expo-

sure include headache, vertigo, visual disturbances, tremors, somnolence, nausea, vomiting, eye irritation, dermatitis, cardiac arrhythmia, and paresthesia; carcinogenic. OSHA exposure limit (TWA): 50 ppm [air] or 270 mg/m³.

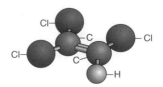

Trichloroethylene
(Trichloroethene)

trichlorofluorocarbon — A halocarbon compound used in aerosols that photolyzes at high altitudes to release chlorine atoms, resulting in ozone depletion.

trichlorofluoromethane (CCl_3F) — A colorless liquid with an ether-like odor; MW: 137.38, BP: 23.7°C, Sol: 0.1 g/100 g water at 20°C, Fl.P: NA (poisonous gases are produced when heated to decomposition), sp. gr. 1.494. It is used in refrigeration and air conditioning; as a foam-blowing agent in fire extinguishers; as a polymer precursor, a solvent, and aerosol propellant; to sterilize medical equipment. It is hazardous to the respiratory system, digestive system, skin, and eyes. Acute exposure can cause irregular heartbeat, irritation of the skin and eyes, dizziness, headache, disorientation, cerebral edema, pulmonary edema, and death. Chronic exposure can cause permanent neurological damage, impaired memory and learning, and emotional instability and may damage lung function. OSHA exposure limit (8-hour TWA): 5600 mg/m³.

trichloromethane — See chloroform.

trichloronaphthalene ($C_{10}H_5Cl_3$) — A colorless to pale-yellow solid with an aromatic odor; MW: 231.5, BP: 579 to 669°F, Sol: insoluble, Fl.P: 392°F (open cup), sp. gr. 1.58. It is used as an insulating material; as an inert component for resins or polymers; as a flame-resistant, waterproofing, and insecticidal/fungicidal agent in coatings for wood, textiles, and paper; in catalytic chlorination of naphthalene; in the manufacture of special lubricants; as an additive for cutting oil. It is hazardous to the skin and liver and is toxic through inhalation, absorption, ingestion, and contact. Symptoms of exposure include acneform dermatitis, anorexia, nausea, vertigo, jaundice, and liver injury. OSHA exposure limit (TWA): 5 mg/m³ [skin].

2,4,5-trichlorophenol ($C_6H_3Cl_3O$) — A gray flake or needle-shaped solid with a strong phenolic odor; MW: 197.46, BP: 253°C, Sol: 0.119 g/100 g water at 25°C, Fl.P: NA (when heated to decomposition it can produce toxic fumes of chloride or explode), sp. gr. 1.678. It is used as a preservative for adhesives, synthetic textiles, rubber, wood, and paints and in paper manufacturing. It is hazardous to the respiratory system, digestive system, skin, and eyes. Acute exposure may cause inflammation and swelling, coughing, wheezing, headaches, nausea, vomiting, diarrhea, and pulmonary edema. Chronic exposure may affect the nervous system, causing headaches, dizziness, nausea, vomiting, weakness, and coma. OSHA exposure limit (8-hour TWA): NA.

2,4,6-trichlorophenol (C$_6$H$_3$Cl$_3$O) — A synthetic yellow solid with a strong, sweet smell; MW: 197.46, BP: 246°C, Sol: 800 mg/l in water at 25°C, density: 1.4901 g/cm^3. It is used in the preservation of wood and glue; as a disinfectant, sanitizer, bactericide, germicide; in pesticides; as an anti-mildew treatment for textiles; as a fungicide, herbicide, and defoliant. It is hazardous to the respiratory system and is toxic through inhalation. Symptoms of exposure include cough, chronic bronchitis, chest wheezing, and reduced expiratory flow; may be carcinogenic for soft tissues. U.S. EPA guidelines: 2×10^{-2} mg/kg/day [air].

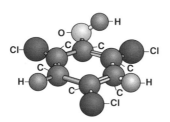

2,4,6–Trichlorophenol

2,4,5-trichlorophenoxyacetic acid (2,4,5-T; C$_6$H$_2$Cl$_3$OCH$_2$COOH) — A colorless to tan, crystalline solid; MW: 255.5, BP: decomposes, Sol: 0.03%, Fl.P: not known, sp. gr. 1.80. It is used in the formulation of herbicides and plant hormones. It is hazardous to the skin, liver, and gastrointestinal tract and is toxic through inhalation, ingestion, and contact. Symptoms of exposure in animals include ataxia, skin irritation, and acne-like rash. OSHA exposure limit (TWA): 10 mg/m^3 [air].

1,2,3-trichloropropane (C$_3$H$_5$Cl$_3$) — A colorless liquid with a chloroform-like odor; MW: 147.43; BP: 156°C; Sol: miscible with organic solvents; Fl.P: 73°C (closed cup), forms explosive mixtures with air; density: 1.389 at 20°C. It is used as a solvent and extractant for resins, oils, fats, waxes, and chlorinated rubber; as a commercial solvent for degreasing metal parts; in organic synthesis for the manufacture of other chemicals; as a copolymer, telomere, or cross-linking agent for sealing compounds. It is hazardous to the eyes, skin, respiratory system, central nervous system, and liver and is toxic through inhalation, absorption, ingestion, and contact. Symptoms of exposure include central nervous system depression, liver injury, skin irritation, irritation of the eyes and throat, narcosis, convulsions, and, at high enough levels, death; carcinogenic. ACGIH exposure limit (TLV-TWA): 10 ppm. [air] or 60 mg/m^3.

1,2,3 Trichloropropane

1,1,2-trichloro-1,2,2-trifluoroethane (C$_2$Cl$_3$F$_3$) — A colorless to water-white liquid (gas above 118°F) with an odor like carbon tetrachloride at high concentrations; MW: 187.37, BP: 47.6°C, Sol: miscible with organic solvents, Fl.P: a noncombustible liquid, density: 1.563 at 25°C. It is used as a selective solvent in degreasing electrical equipment, photographic films, magnetic tapes, precision instruments, plastics, glass, elastomers, or metal components; as a drycleaning solvent; as a refrigerant in commercial/industrial air conditioning and industrial process cooling; as a chemical intermediate for dechloronization of chemicals in the manufacture of polymers and copolymers; as a solvent in the textile industry. It is hazardous to the skin and heart and is toxic through inhalation, ingestion, and contact. It may cause a weak narcotic effect, cardiac sensitization, irritation, nausea, lethargy, and nervousness. ACGIH exposure limit (TLV-TWA): 1000 ppm. [air] or 7,600 mg/m^3.

$$Cl - \overset{\displaystyle \overset{Cl}{|}}{C} - \overset{\displaystyle \overset{Cl}{|}}{\underset{\displaystyle \underset{F}{|}}{C}} - F$$

**1,1,2-Trichloro-
1,2,2-trifluoroethane**

Trichomonas — A genus of parasitic protozoan occurring in the digestive and reproductive system of humans and other animals.

trichuriasis — A nematode infection of the large intestine, usually without symptoms; can have bloody, mucous stools and diarrhea. Incubation time is indefinite. It is caused by *Trichuris trichiuria* and is found worldwide. The reservoir of infection is people. It is transmitted through the eggs of the whipworm found in feces and hatches into larvae in contaminated soil. It is communicable for several years; general susceptibility. It is controlled by proper disposal of feces and good health habits, especially handwashing. Also known as whipworm.

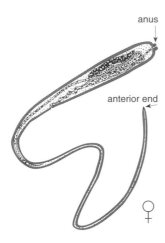

Trichuriasis
(*Trichuris trichuria*–whipworm
which causes trichuriasis)

trickling filter — A treatment system in which wastewater is trickled over stones or other material covered with bacteria that break down the organic waste and produce cleaner water.

trickling filter system — A system of secondary sewage treatment that is similar to the self-purification action of streams; it is more accurately a biological oxidizing bed. The effluent is placed on stones in the bed and microorganisms present consume the solids as a food supply.

triethylaluminum (Al(C$_2$H$_5$)$_3$) — A colorless liquid; MW: 114.17; BP: 194°C; Sol: soluble in ether, hexane, and isooctane but reacts violently with water, alcohol, chloroform, carbon tetrachloride, and many other solvents; Fl.P: ignites spontaneously in air and violently with water, alcohol, halogenated hydrocarbons, and oxidizing substances; density: 0.83 g/ml at 25°C. It is used as an intermediate in organic synthesis. It is hazardous to the skin, eyes, and respiratory tract. It causes violent reactions with moisture on the skin and dangerous burns to the eyes and respiratory tract. ACGIH exposure limit: NA.

triethylamine (C$_2$H$_5$)$_3$N) — A colorless liquid with a strong, ammonia-like odor; MW: 101.2, BP: 193°F, Sol: 2%, Fl.P (open cup): 16°F, sp. gr. 0.73. It is used in polymer technology as a catalyst; as a stabilizing agent and chain-transfer agent; in the preparation of film-forming resins; as a solvent pH stabilizer and to improve water solubility in polymers; in extraction and purification of antibiotics; in herbicides and pesticides; in organic synthesis; as a corrosion inhibitor; as a dye assist; as an ingredient of photographic development; in carpet cleaners. It is hazardous to the respiratory system, eyes, and skin and is toxic through inhalation, absorption, ingestion, and contact. Symptoms of exposure include irritation of the respiratory system, skin, and eyes. OSHA exposure limit (TWA): 10 ppm [air] or 40 mg/m^3.

triethylsilane ((C$_2$H$_5$)$_3$SiH) — A colorless liquid; BP: 109°C, MW: 116.28, Sol: insoluble in water, Fl.P: –4°C (can explode on heating), density: 0.73 g/ml. It is used as a reducing agent in many organic synthetic reactions. ACGIH exposure limit: NA.

trifluoromonobromomethane (CBrF$_3$) — A colorless, odorless gas that is shipped as a liquefied compressed gas; MW: 148.9, BP: –72°F, Sol: 0.03%. It is used as a commercial and military fire extinguishant; as a refrigerant for food processing and storage; in organic synthesis; during the manufacture of hydraulic fluids; for special purposes in bubble chambers for ionization studies, in quark detection, and in radiation counters. It is hazardous to the heart and central nervous system and is toxic through inhalation and contact. Symptoms of exposure include lightheadedness and cardiac arrhythmias. OSHA exposure limit (TWA): 1000 ppm [air] or 6100 mg/m^3.

trigger — A substance, object, or agent that initiates or stimulates an action.

trigger finger — Swelling of the tendons in finger joints due to overuse, locking the finger into a fixed position.

trihalomethane (THM) — Any member of the family of organic compounds that are derivatives of methane, wherein three of the four hydrogen atoms are substituted by halogen atoms in the molecular structure.

triisobutylaluminum ([(CH$_3$)$_2$CHCH$_2$]$_3$Al) — A colorless liquid; MW: 198.33; BP: NA; Sol: soluble in ether, hexane, and toluene but reacts explosively with water; Fl.P: –18°C (ignites spontaneously in air). It is used as a reducing agent. It is hazardous to the skin, eyes, and respiratory tract and causes severe burns because of its reaction with the moisture on or in the body. ACGIH exposure limit: NA.

trimester — A period of time divided into three sections.

trimethacarb (C$_{11}$H$_{15}$NO$_2$) — A crystalline solid; MW: 193.27, BP: NA, Sol: soluble in most organic solvents, density: NA. It is used as a pesticide. It is moderately toxic to the digestive tract and respiratory tract. It is a cholinesterase inhibitor that may cause headaches, weakness, blurred vision, nausea, vomiting, and convulsions.

$$CH_3 \quad O-C-NH-CH_3$$
$$CH_3 \quad \| \quad O$$
$$CH_3$$

Trimethacarb

trimethyl phosphate ((CH$_3$)$_3$PO$_4$) — A colorless liquid; MW: 140.08, BP: 197°C, Sol: soluble in most organic solvents, Fl.P: a noncombustible liquid, density: 1.197 at 20°C. It is a mutagen, teratogen, and cancer-causing compound.

$$H_3C-O-P-O-CH_3$$
$$\underset{O-CH_3}{\overset{O}{\|}}$$

Trimethyl phosphate

trimethyl phosphite ((CH$_3$)$_3$PO$_3$) — A colorless liquid with a characteristic pungent odor; MW: 124.09; BP: 112°C; Sol: mixes with most organic solvents and decomposed by water; Fl.P: 37.8°C (closed cup), a flammable liquid; density: 1.052 at 20°C. It is hazardous to the skin, eyes, and respiratory tract. Can produce irritation and lung inflammation in experimental animals. It shows a teratogenic effect in newborn rats. ACGIH exposure limit (TLV-TWA): 2 ppm.

$$CH_3O-P-OCH_3$$
$$OCH_3$$

Trimethyl phosphite

trimethylaluminum (Al(CH$_3$)$_3$) — A colorless liquid with a corrosive odor; MW: 72.08; BP: 125°C; Sol: soluble in ether, toluene, and hexane and reacts explosively with water; Fl.P: extremely reactive and volatile and ignites spontaneously in air, density: 0.75 g/ml. It is used as a highly reactive reducing and alkylating agent. It is hazardous to the skin, eyes, and respiratory tract. It reacts

explosively with moisture on the skin, can cause blindness, and severe damage to the respiratory tract. ACGIH (TLV-TWA): 2 mg (Al)/m^3.

trimethylene chlorohydrin (ClCH$_2$CH$_2$CH$_2$OH) — A colorless, oily liquid with an agreeable odor; MW: 94.55; BP: 165°C; Sol: soluble in water, alcohol, and other organic solvents; Fl.P: 73°C (vapor may form an explosive mixture with air), density: 1.1318 at 20°C. It is used in organic synthesis to produce cyclopropane and trimethylene oxide. It is hazardous to the digestive tract and skin. It causes central nervous system depression, muscle contraction, gastrointestinal pain, stomach bleeding, and liver injury. ACGIH exposure limit (TLV-TWA): 100 ppm [air].

$$Cl-CH_2-CH_2-CH_2-OH$$

Trimethylene chlorohydrin

trinitrobenzene (C$_6$H$_3$N$_3$O) — A pale-yellow crystalline solid, MW: 213.12, BP: NA, Sol: soluble in carbon disulfide and petroleum ether, Fl.P: a highly explosive substance extremely sensitive to shock and heat, density: NA. It is used as an explosive. It is hazardous to the digestive tract and respiratory tract and may cause irritation, headache, dyspnea, and cyanosis. ACGIH exposure limit: NA.

$$\underset{O_2N \quad\quad NO_2}{NO_2}$$

Trinitrobenzene

2,4,6-trinitrotoluene ((C$_6$H$_2$(CH$_3$)(NO$_2$)$_3$)) — A colorless to pale-yellow, odorless solid or crushed flakes; MW: 227.1, BP: 464°F (explodes), Sol (77°F): 0.01%, Fl.P: not known (explodes), sp. gr. 1.65. It is used in the manufacture of shells, bombs, grenades, and mines; in commercial explosives; in propellant compositions; in the production of intermediates for synthesis of dyestuffs and photographic chemicals. It is hazardous to the blood, liver, kidneys, eyes, cardiovascular system, central nervous system, and skin and is toxic through inhalation, absorption, ingestion, and contact. Symptoms of exposure include liver damage, jaundice, cyanosis, sneezing, cough, sore throat, peripheral neuropathy, muscular pain, kidney damage, cataracts, sensitization dermatitis, leukocytosis, anemia,

T

and cardiac irregularities. OSHA exposure limit (TWA): 0.5 mg/m³ [skin].

2,4,6 Trinitrotoluene

triorthocresyl phosphate (CH₃C₆H₄₀)₃PO) — A colorless to pale-yellow, odorless liquid or solid (below 52°F); MW: 368.4, BP: 770°F (decomposes), Sol: slight, Fl.P: 437°F, sp. gr. 1.20. It is used as a flame retarder and plasticizer; in hot extrusion molding or bulk forming of plasticized polyvinylchloride; in coatings and adhesives; as a gasoline additive to control preignition; as a hydraulic fuel and a heat exchange medium; as a synthetic lubricant; as a waterproofing agent; as a solvent mixture for nitrocellulose and other natural resins; as an extraction solvent in recovery of phenol from gas-plant effluents and coke-oven wastewaters; in grinding media for pigments; as an intermediate in the synthesis of pharmaceuticals. It is hazardous to the central and peripheral nervous systems and is toxic through inhalation, absorption, ingestion, and contact. Symptoms of exposure include gastrointestinal distress, peripheral neuropathy, cramps in calves, paresthesia in feet or hands, weak feet, wrist drop, and paralysis. OSHA exposure limit (TWA): 0.1 mg/m³ [skin].

triphasic — Something that has three phases.

$$
\begin{array}{c}
O \\
\parallel \\
\bigcirc -O-P-O-\bigcirc \\
\mid \\
O \\
\mid \\
\bigcirc
\end{array}
$$

Triphenyl phosphate

triphenyl phosphate ((C₆H₅)₃PO₄) — A colorless solid with a faint aromatic odor; MW: 326.30, BP: 370°C, Sol: soluble in most organic solvents, Fl.P: a noncombustible solid, density: NA. It is used in fireproofing, in impregnating roofing paper, as a plasticizer, and in lacquers and varnishes. It is a neurotoxic agent and a cholinesterase inhibitor. It causes tremor, diarrhea, muscle weakness, and paralysis in experimental animals. ACGIH exposure limit (TLV-TWA): 3 mg/m³ [air].

tripod — A three-legged support for a surveying instrument.

tritium (T; ³H) — A heavy hydrogen isotope with one proton and two neutrons in the nucleus.

trivalent — An atom able to bond with up to three other atoms.

TRMM — See Tropical Rainfall Measuring Mission.

Tropical Rainfall Measuring Mission — A joint NASA/NASDA mission launched in 1997 to obtain a minimum of 3 years of climatologically significant observations of rainfall in the tropics and to help provide accurate estimates of vertical distributions of heat in the atmosphere for use with cloud models.

TRO — Acronym for temporary restraining order.

trophic action — The stimulation of cell production and enlargement by nurturing and causing growth.

trophic accumulation — The passing of a substance through a food chain such that each organism retains all or a portion of its food and eventually acquires a higher concentration in its flesh than that which was present in the original food.

trophic level — A system of categorizing organisms by the way they obtain food; examples are producers or organic detritus.

trophozoite — The active feeding stage of a protozoan in contrast to a cyst; also called the vegetative stage.

tropical rain forest — An ecosystem characterized by very high rainfall and temperatures between 20 and 25°C, with very diverse life forms; broadleaf, nondeciduous trees; and highly stratified forest.

tropical storm — A tropical cyclone with strong winds of less than hurricane intensity.

tropism — The ability of a virus to infect specific cells or tissue types.

tropopause — The boundary between the troposphere and stratosphere, usually characterized by an abrupt change of lapse rate.

troposphere — The layer of the atmosphere extending from the surface of the Earth outward about 5 miles at the poles and 10 miles at the equator; most weather activities occur in this layer.

Tropospheric Emission Spectrometer — A high-resolution infrared spectrometer for monitoring minor components of the lower atmosphere.

T

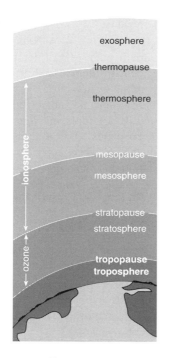

Tropopause
Troposphere

tropospheric ozone — The ozone that is located in the troposphere and plays a significant role in the greenhouse gas effect and urban smog.

trough — An elongated area of low atmospheric pressure either at the surface or in the upper atmosphere.

true color — The color of water resulting from substances that are totally in solution.

true negative — A negative test result for an individual who does not have that particular disease.

true positive — A positive test result for an individual who does have that particular disease.

true respiration — The use of oxygen by the cells with the release of carbon dioxide.

truncate — See round off.

TRV — See Thermostatic Radiator Valve.

TSA — Acronym for technical systems audit.

TSCA — See Toxic Substances Control Act.

TSCATS — Acronym for TSCA Test Submissions Database.

TSD — Acronym for technical support document; notation for treatment storage for disposal facility.

TSDF — Acronym for treatment, storage, and disposal facility.

TSE — Acronym for transmissibile spongiform encephalopathy.

tsetse fly — A bloodsucking fly of the genus *Glossina*.

TSI — Acronym for thermal system insulation or total solar irradiance.

TSM — Acronym for transportation system management.

tsp — See teaspoon.

TSP — See total suspended particulates.

TSS — See total suspended nonfilterable solids or toxic shock syndrome.

tsunami — An unusually large sea wave produced by a sea quake or undersea volcanic eruption; not of tidal origin.

tsutsugamushi disease — See scrub typhus.

TTHM — Acronym for total trihalomethane.

TTO — Acronym for total toxic organics.

tuberculosis — A mycobacterial disease with lesions and involvement of lymph nodes, pleura, pericardium, kidneys, bones, joints, and larynx. Incubation time is 4 to 12 weeks before a skin test may be positive; disease symptoms may appear within 1 to 2 years after infection. It is caused by *Mycobacterium tuberculosis* and is found worldwide. Reservoirs of infection include people and diseased cattle. It is transmitted by airborne droplet nuclei from sputum of people with infectious tuberculosis or from raw, unpasteurized milk or dairy products from contaminated cattle. It is communicable as long as the tubercle bacilli are being discharged in the sputum; general susceptibility. It is controlled by limiting overcrowding, the use of diagnostic medical tests and x-rays, and proper control of cattle.

tularemia — An infectious zoonotic disease with a variety of symptoms, including swelling of the regional lymph nodes, intestinal pain, diarrhea, and vomiting. Incubation time is 2 to 10 days, usually 3 days. It is caused by *Francisella tularensis* and is found throughout North America, Europe, and parts of Asia. Reservoirs of infection include wild animals (including rabbits and hard ticks). It is transmitted when blood or tissue that has been infected enters a break in the skin or by the bite of infected ticks or deer flies. It is not communicable from person to person; general susceptibility. It is controlled through proper use of gloves when skinning or handling animals, especially rabbits, and by avoiding the bites of deer flies, mosquitos, and ticks. Also known as rabbit fever.

tumescence — A state of swelling or edema.

tumor — An abnormal swelling or growth; see also neoplasm.

tumor registry — A set of data concerning the incidence of cancers, personal history, treatment, and treatment results of people diagnosed with cancer.

tumorigenic — Able to produce a tumor.

tundra — An area, including the Arctic and Antarctic zones, that can support vegetation such as lichens, grasses, and dwarf woody plants; characterized by long, severe winters and short, mild summers.

tuple — See database.

TUR — Acronym for Toxics Use Reduction.

turbid — Having a cloudy or muddy appearance.

turbidimeter — A device that measures the amount of suspended solids in a liquid.

turbidity — An expression of the optical property of a sample that causes light to be scattered and absorbed rather than transmitted in straight lines through the sample.

turbine — An engine that forces a stream of liquid through jets at high pressure against the curved blades of a wheel, forcing the blades to turn.

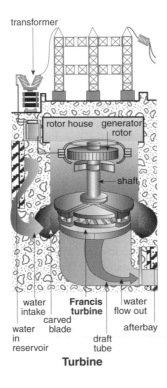

transformer

rotor house   generator rotor

shaft

water intake
Francis turbine
water flow out

water in reservoir
carved blade
draft tube
afterbay

Turbine

turbine generator — A piece of equipment that uses steam, heated gases, water flow, or wind to cause a spinning motion that activates electromagnetic forces and generates electricity.

turbulent flow — A state of fluid flow in which the instantaneous velocities exhibit irregular and apparently random fluctuations so, in practice, only statistical properties can be recognized and subjected to analysis.

turbulent loss — The pressure or energy lost from a ventilation system through air turbulence.

turgid — Swollen, hard, and congested, usually as a result of the accumulation of fluid.

turnover rate — A period of time, usually in hours, required to circulate a volume of water equal to the pool capacity.

turpentine ($C_{10}H_{16}$ approx.) — A colorless liquid with a characteristic odor; MW: 136 (approx.), BP: 309 to 338°F, Sol: insoluble, Fl.P: 95°F, sp. gr. 0.86. It is used in the manufacture of synthetic pine oil; in insecticides, flavors, and perfumes; in the preparation of polishes; in the manufacture of synthetic camphor; in paints. It is hazardous to the skin, eyes, kidneys, and respiratory system and is toxic through inhalation, absorption, ingestion, and contact. Symptoms of exposure include headache, vertigo, albuminuria, skin irritation and sensitization, and irritation of the eyes, nose, and throat. OSHA exposure limit (TWA): 100 ppm [air] or 560 mg/m³.

tutorial — A set of materials in which instruction is given to the student to help the individual discover the correct answers to problems.

tuyere — An opening or port in a grate through which air can be directed to improve combustion.

TV — See tidal volume.

TVOC — Notation for total volatile organic compound.

TWA — See time-weighted average exposure.

twitch — A contraction of a small muscle.

TWS — Acronym for transient water system.

tympanic membrane — A thin, semitransparent membrane in the middle ear that transmits sound vibrations to the internal ear by means of small auditory bones.

typical — A representative sample.

typhoid carrier — A person who lacks signs or symptoms of typhoid fever but who has the organisms and sheds them.

typhoid fever — A highly infectious, systemic, bacterial disease with sustained fever, headache, malaise, anorexia, bradycardia, nonproductive cough, constipation, occasionally diarrhea, and involvement of the lymphoid tissues. Incubation time is 1 to 3 weeks. It is caused by 106 types of *Salmonella typhi* and is found worldwide. The reservoir of infection is people. It is transmitted by food or water contaminated by feces or urine from an infected person or carrier, shellfish, sewage-contaminated waters, raw fruits, vegetables, and milk products. It is spread by flies and is communicable for 3 months or more; general susceptibility. It is controlled through the proper disposal of human feces, fly control, protection of private and public water supplies, appropriate food handling, and removal of any infected persons or carriers from food-handling situations.

typhoon — A hurricane in the western Pacific Ocean.

typhus fever — Any of three epidemic louse-borne human diseases transmitted by lice to people, with symptoms of headaches, chills, prostration, fever, and general pains. Incubation time is 1 to 2 weeks, usually 12 days. It is caused by *Rickettsia prowazekii* and is found in the United States, Mexico, and Central and South America. The reservoir of infection is people. It is transmitted by the body louse *Pediculus humanus*; general susceptibility. It is controlled by proper use of insecticides on clothes and people to kill lice and by immunization of special groups.

TZ — Acronym for treatment zone.

T

U

U value — See thermal transmittance.

ULV — See ultra-low volume.

UA — A measure of the amount of heat that would be transferred through a given surface or enclosure with a 1° temperature difference between the two sides.

UAM — Acronym for urban airshed model.

UAO — See upper airway obstruction.

UAQI — Acronym for uniform air quality index.

UARS — See Upper Atmosphere Research Satellite.

U-blade — A bulldozer blade with an extension on each side to increase moving capacity.

UC — Notation for uniformity coefficient.

UCC — Acronym for ultra-clean coal.

UCL — Acronym for upper control limit.

UEL — See upper explosive limit.

UF — See uncertainty factor.

UFFI — Acronym for urea-formaldehyde foam insulation.

UFL — See upper flammable limit.

UHF — See ultra-high-frequency.

UHV — See ultra-high-voltage.

UIC — Acronym for underground injection control.

ulcer — Destruction of skin or mucous membrane with or without infection or pain.

ulceration — Formation of an open sore or ulcer.

ULEV — Acronym for ultra-low-emission vehicle.

ultimate analysis — Chemical analysis of a solid, liquid, or gaseous fuel to determine the amounts of carbon, hydrogen, sulfur, nitrogen, oxygen, and ash it contains.

ultimate strain — Strain at the point of failure.

ultimate stress — The highest load that can be sustained by a material at the point of failure.

ultracentrifuge — A high-speed centrifuge able to produce sedimentation of viruses; used in many kinds of biochemical analyses.

ultra-clean coal — Coal that is washed, ground into fine particles, and chemically treated to remove sulfur, ash, silicone, and other substances; usually in briquette form and coated with a sealant made from coal.

ultrafiltrate — Low-molecular-weight solutes that have passed through a semipermeable membrane with very small pores.

ultra-high-frequency (UHF) — The band of frequencies between 300 and 3000 megahertz in the radio spectrum.

ultra-high-temperature pasteurization — Pasteurization accomplished by steam injection or by use of vacuum to heat milk to any of the following combinations of temperature and time: 191°F for 1 second, 194°F for 0.5 second, 201°F for 0.1 second, 204°F for 0.05 second, or 212°F for 0.01 second.

ultra-high-voltage transmission — Transportation of electricity over a bulk-power line at voltages greater than 800 kilovolts.

ultra-low volume (ULV) — (*pesticides*) A technique used in mosquito larvaciding where adulticiding equipment is set at a low level, typically 2 pints per acre, and the product is allowed to drift down into otherwise inaccessible areas such as wetlands; appears to be a successful technique for control of the Asian tiger mosquito.

ultrasonic — Sounds for which the frequencies are above the normal range of human hearing, about 20,000 hertz; also known as supersonic.

ultrasonic nebulizer — A humidifier in which high-frequency vibrations break a fluid into aerosol particles.

ultrasonics — The science of sound having frequencies above the audible range.

ultrasonography — A test in which sound waves are bounced off tissues and the echoes are converted into a picture (sonogram).

ultrasound — Mechanical radiant energy of a frequency greater than 20,000 cycles per second; the upper limit of human hearing.

ultraviolet — Radiations above the visible light spectrum with wavelengths shorter than those of violet light.

ultraviolet absorption spectrophotometry — Study of the spectra produced by absorption of ultraviolet radiant energy during the transformation of an electron.

ultraviolet light — See ultraviolet radiation.

ultraviolet (UV) radiation — The invisible part of the light spectrum where the rays have wavelengths shorter than those of the bowed end of the visible spectrum and longer than those of x-rays; the ultraviolet spectrum is usually considered to extend from about 50 to 380 nanometers; the maximum reddening of the skin occurs at 260 nm; bacteria and viruses may be killed at ultraviolet rays shorter than 320 nm. Also known as ultraviolet light.

UMTRCA — Acronym for Uranium Mill Tailing Radiation Control Act of 1978.

uncertainty factor (UF) — A factor used when deriving a reference dose from experimental data; intended to account for variation in sensitivity among members of the human population, uncertainty in extrapolating animal data to humans, uncertainty in extrapolating from data obtained in a study that is of less than lifetime exposure, and uncertainty in using lowest observed adverse effect level data rather than no observed adverse effect level.

unconfined aquifer — A geological formation that can be readily contaminated because no protective stratum or aquitard is above it.

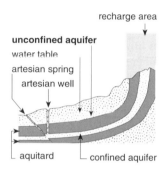

Unconfined Aquifer

unconscious — Being unaware of the surrounding environment.

uncoupling — The process of prohibiting a phosphate from joining adenosine diphosphate (ADP) to form adenosine triphosphate (ATP).

underdrain — A distribution system at the bottom of a sand filter to collect filtered water during a filter run and to distribute the backwash water.

underfire air — Any forced or induced air under control as the quantity and direction is supplied beneath a grate and passes through a fuel bed.

underground injection wells — Shafts that are encased in steel and concrete and into which hazardous waste is deposited by force and under pressure.

underground storage tank (UST) — A tank placed underground as a fire-preventive measure; used to store gasoline, other oil products, or hazardous substances.

underwater light — A lighting fixture designed to illuminate a pool from beneath the water surface; may be "wet niched" (located in the pool water) or

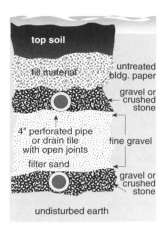

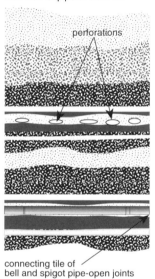

side view of pipe or drain tile

Underdrained sand-filter trench

"dry niched" (located in the pool side wall beyond or behind a waterproof window and serviced from outside the pool).

Underwater Light

underwrite — To insure; to scrutinize risk and decide if something can be insured.

undulant — A wave-like action.

unicellular — A life-form in which all functions are carried out within the confines of a single plasma membrane.

unified classification — A system of classification of soils according to particle size, distribution, plasticity, liquid limit, and organic matter.

uniform corrosion — Deterioration over the entire metal surface at the same rate.

Uniform Resource Locator (URL) — The address of a Web site.

unipolar — An analog signal range that is always positive through zero.

unit — A single item.

unit risk — The excess lifetime risk due to continuous constant lifetime exposure to one unit of carcinogen concentration.

United Soil Classification System — A method of grouping and describing soils according to their engineering properties.

univalent — A chemical valency of one.

universal antidote — A mixture of 50% activated charcoal, 25% magnesium oxide, and 25% tannic acid, previously thought to be useful as an antidote for numerous poisons but now considered to be no more effective than activated charcoal taken with water.

Universal Transfer Mercator — An international plane (rectangular) coordinate system developed by the U.S. Army; it divides the world into 60 zones of 6° longitude, with each zone extending 3° east and west from its central meridian and numbered consecutively west to east from the 180° meridian.

Unix — A computer operating system.

unleaded gasoline — Gasoline containing not more than 0.05 grams of lead per gallon and not more than 0.005 grams of phosphorus per gallon.

unrestricted use — The designation of acceptable future use for a site at which the remediation levels based on either background or standard residential exposure factors have been attained throughout the site in all media.

unsaturated — A solution that is capable of dissolving more of the solute.

unsaturated compound — A chemical compound that contains double or triple bonds.

unsaturated flow — The movement of water in a soil that is not filled to capacity with water.

unsaturated organic compound — An organic compound having some of the carbon atoms in its molecule linked by two or three covalent bonds.

unsaturated zone — See zone of aeration.

unstable — Tending toward decomposition or other unwanted chemical change during normal handling or storage.

unstable reactive — A chemical that in the pure state or as produced or when transported vigorously will polymerize, decompose, condense, or become self-reactive under conditions of shock, pressure, or temperature.

updraft — A relatively small-scale current of air with marked upward vertical motion.

upload — To copy or send data or documents from one computer to another.

upper airway obstruction — An abnormal condition of the mouth, nose, or larynx that interferes with breathing when the rest of the respiratory system is functioning properly.

upper atmosphere — A general term applied to the atmosphere 60 kilometers and higher above the surface of the Earth.

Upper Atmosphere Research Satellite — A NASA satellite launched in September 1991 to serve as a platform for instruments to measure temperature, wind, and composition of the upper atmosphere, including the stratosphere.

upper detection limit — The largest concentration that an instrument can reliably detect.

upper explosive limit (UEL) — The highest concentration of a vapor or gas in air that will produce a flash of fire when an ignition source, heat arc, or flame is present; also known as upper flammable limit.

upper flammable limit (UFL) — See upper explosive limit.

upper respiratory infection — See respiratory tract infection.

uptake — Drawing up a substance.

upwelling — The process of rising or appearing to rise to the surface and flowing outward; the upward motion of sub-surface water toward the surface of the ocean.

uracil ($C_4H_4N_2O_2$) — A pyrimidine base; one of four bases coding genetic information in the polynucleotide chain of ribonucleic acid.

uranium (U) (insoluble compounds) — A silver-white, malleable, ductile, lustrous metal that is weakly radioactive; MW: 238.0, BP: 6895°F, Sol: insoluble, Fl.P: not known, sp. gr. 19.05 at 77°F, atomic number: 92. It is used as a chemical intermediate in the preparation of uranium compounds; for nuclear technology; in nuclear reactors as fuel to pack nuclear fuel rods; in the ceramics industry for pigments, coloring porcelain, painting on porcelain, and enameling; as a catalyst for many reactions; in the production of fluorescent glass. It is hazardous to the skin, bone marrow, lymphatics and is toxic through inhalation, ingestion, and contact. Symptoms of exposure include dermatitis; it is carcinogenic. In animals, dermatitis and lung and lymph node damage occur. OSHA exposure limit (TWA): 0.2 mg/m³ [air].

U

Uranium Mill Tailings Radiation Control Act of 1978 — An act that directs the Department of Energy to provide for stabilization and control of uranium mill tailings from inactive sites in a safe and environmentally sound manner to minimize radiation health hazards to the public.

uranium mill tailings — Sand-like materials left over from the separation of uranium from its ore, with more than 99% of the ore becoming tailings.

urban renewal — Programs used to revitalize old and blighted sections of inner cities.

urban runoff — Stormwater from city streets, usually carrying litter and organic wastes.

urbanization — An increasing concentration of population in cities and transformation of land use to an urban pattern of organization.

urea ($CO(HN_2)_2$) — A nitrogenous waste substance found chiefly in the urine of mammals; it is formed in the liver from metabolized proteins and is synthesized as white crystals or powder.

urea–formaldehyde foam insulation — A material that has been used in the past to conserve energy by sealing crawl spaces, attics, etc.; it can cause a specific health hazard.

uremia — Accumulation in the blood of excessive amounts of waste products normally eliminated in the urine; a condition resulting from kidney failure.

urethra — The passageway through which urine leaves the bladder in most mammals.

urgent public health hazard — A site that poses a serious risk to public health as the result of short-term exposure to hazardous substances at the site.

uricosuria — Excretion of uric acid in the urine.

urinalysis — Physical, microscopic, or chemical examination of urine.

urinary tract — Organs and ducts involved in the secretion and elimination of urine from the body.

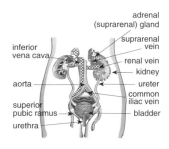

Urinary system

urinary tract infection — An infection of one or more structures in the urinary tract, most frequently caused by Gram-negative bacteria such as *Escherichia coli* and species of *Klebsiella*, *Proteus*, *Pseudomonas*, or *Enterobacter*.

urine — Liquid waste filtered from blood in the kidneys and excreted by the bladder.

urinogenital — Pertaining to the urinary tract.

URL — See Uniform Resource Locator.

urticaria — Elevated itchy patches of skin; also known as hives.

U.S. Code — Codification of all laws of the United States.

USC — See U.S. Code; acronym for United Soil Classification.

USDW — Acronym for underground source of drinking water.

use-dilution — Dilution, as specified on labeling, that produces the concentration of the pesticide for a particular effect.

user friendly — Refers to computer hardware or software programs that are designed to help users by presenting information or instructions that are easy to understand..

UST — See underground storage tank.

utensil — Any food-contact implement used in the storage, preparation, transportation, or dispensing of food.

uterus — A muscular organ that holds and nourishes a growing fetus during gestation in mammals.

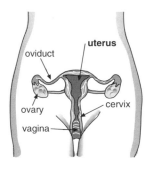

Uterus

UTI — See urinary tract infection,

UTM — See Universal Transverse Mercator.

UTP — Acronym for urban transportation planning.

UV — See ultraviolet radiation.

UVA — Acronym for an ultraviolet radiation band.

U value — The number of British thermal units that flow through 1 square foot of a material in 1 hour when there is a 1° difference in temperature between the inside and outside air under steady-state conditions; the number of watts that flow through 1 square meter of a material in 1 hour when there is a 1° difference in temperature between the inside and outside air under steady-state conditions; the reciprocal of the resistance, or R, value.

UVB — Acronym for an ultraviolet radiation band.

UZM — Acronym for unsaturated zone monitoring.

µg/m³ — Notation for micrograms per cubic meter.

V

V — See volt.

V/m — See voltmeter.

vaccination — See immunization.

vaccine — A preparation made from a killed or attenuated pathogen designed to protect individuals from the disease through their production of active immunity.

vaccinia — A contagious disease of cattle believed to be a form of small pox and which is transmissible to humans; also known as cow pox.

vacuole — A clear or fluid-filled space or cavity within a cell.

vacuum — A space that contains no gas, liquid, or solid; devoid of matter; any pressure less than that exerted by the atmosphere.

vacuum cleaning — (*swimming pools*) A system of lines leading from the main drain and skimmers throttled or shut off until the vacuum connection is opened, thus permitting the pump to pull dirt from the pool through the lines.

vacuum filter — A filter that operates under a vacuum from the suction of a pump; also known as a suction filter.

vadose — The unsaturated soil zone.

vadose zone — The unsaturated soil zone between the water table and the surface of the ground; also known as the zone of aeration.

vagabond's disease — Darkened, thickened skin caused by years of infestation with body lice.

valence — The chemical-combining power of an atom; indicates the number of electrons that can be lost, gained, or shared by an atom in a compound. Also known as bond or chemical bond.

valence electron — An outermost orbiting electron of an atom that bonds the atom into crystals, molecules, and compounds.

***n*-valeraldehyde (C_4H_9CHO)** — A colorless liquid; MW: 86.1, BP: 103°C, Sol: slightly soluble in water and readily soluble in alcohol and ether, Fl.P: 12°C (open cup), density: 0.8095 at 20°C. It

$$H_3C-CH_2-CH_2-\overset{\overset{\textstyle H}{\textstyle |}}{C}=O$$

***n*-Valeraldehyde**

is used for food flavoring and in resin and rubber products. It is a moderate skin and eye irritant. ACGIH exposure limit (TLV-TWA): 50 ppm.

validate — To test the accuracy of mathematical models used to predict pollution concentrations by comparing predictive concentrations to measured concentrations under equivalent conditions.

validity — A measure of the extent to which an observed situation reflects the actual situation.

value — A personal belief about the worth of an idea or behavior.

valve — A device that controls the direction of fluid flow, rate, and pressure.

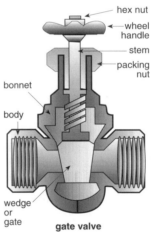

gate valve

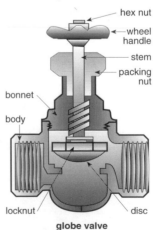

globe valve

Valves

valve authority — The rate of pressure across a fully open valve to the pressure drop across the remainder of the circuit.

valve regulation — The ratio of maximum controlled flow to minimum controlled flow.

valve stop — A guide that permits turning the valve plug to but not beyond the fully closed position.

Van Allen belt — A zone of intense, ionizing radiation circling the Earth and containing charged particles trapped in the magnetic field of the Earth.

Van de Graaff accelerator — An electrostatic generator.

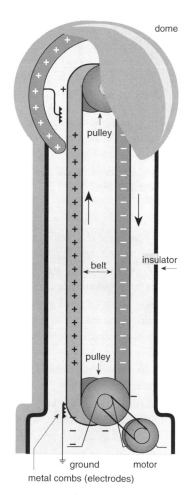

Van de Graaff Accelerator

van der Waals force — Intermolecular and interatomic forces that are electrostatic in origin; also known as dispersion force.

vanadium hexacarbonyl (V(CO)₆) — A bluish-green crystalline solid; MW: 291.01, BP: NA, Sol: soluble in most organic solvents with the solution turning yellow, Fl.P: may explode on heating, density: NA. It is used as a catalyst in isomerization and hydrogenation reactions. It should be treated as a potential health hazard as it can produce highly toxic compounds such as carbon monoxide and vanadium oxides. It causes irritation to the skin. ACGIH exposure limit: NA.

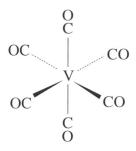

Vanadium hexacarbonyl

vanadium pentoxide respirable dust (V₂O₅) — A yellow-orange powder or dark gray, odorless flakes dispersed in air; MW: 181.9, BP: 3182°F (decomposes), Sol: 0.8%, sp. gr. 3.36. It is used as a catalyst in the preparation of organic and inorganic compounds; in chemical synthesis; in the manufacture of ultraviolet filter glass to prevent radiation injury and fading of fabrics; in the manufacture of afterburners for automobiles; in the textile industry as a catalyst to yield intensive black dyes; in the printing industry; in the manufacture of ceramic pigments; as a component of special steels in electric furnace steels, welding rods, and permanent magnets. It is hazardous to the respiratory system, skin, and eyes and is toxic through inhalation, ingestion, and contact. Symptoms of exposure include eye irritation, green tongue, metallic taste, eczema, cough, fine râles, wheezing, bronchitis, dyspnea, and throat irritation. OSHA exposure limit (TWA): 0.05 mg/m³ [air].

vanadium pentoxide respirable fume (V₂O₅) — A finely divided particulate dispersed in air; MW: 181.9, BP: 3182°F (decomposes), Sol: 0.1%, sp. gr. 3.36. It is liberated from the production of pellets from electric furnaces and from the manufacture of semiconductors fused with sodium oxide. It is hazardous to the respiratory system, skin, and eyes and is toxic through inhalation and contact. Symptoms of exposure include eye irritation, green tongue, metallic taste, throat irritation, cough, fine râles, wheezing, bronchitis, and eczema. OSHA exposure limit (TWA): 0.05 mg/m³ [air].

vapor — The gaseous form of a substance that is a liquid at ordinary temperatures and pressure.

vapor barrier — A moisture-impervious layer applied to surfaces enclosing a space or to the surface of thermal insulation to limit moisture migration through the surface; (*housing*) a material such as plastic or paint applied to a wall, floor, or ceiling to prevent the passage of moisture.

vapor bath — The exposure of the body to vapor.

vapor capture system — A combination of hoods and ventilation systems that captures or contains organic vapors for abatement or recovery.

vapor density — The weight of a vapor or gas compared to the weight of any equal volume of air.

vapor diffusion — The movement of water vapor between two areas caused by a difference in vapor pressure independent of air movement; the rate of diffusion is determined by the difference in vapor pressure, the distance the vapor must travel, and the permeability of the material to water vapor.

vapor dispersion — The movement of vapor clouds in air due to wind, gravity, spreading, and mixing.

vapor/hazard ratio — The relationship between the toxicity of a substance in its vapor phase and in its vapor pressure.

vapor plumes — Flue gases that are visible because they contain water droplets.

vapor pressure — The pressure at any given temperature of a vapor in equilibrium with its liquid or solid form; (*soils*) the pressure exerted by a vapor in a confined space as a function of the temperature; (*meteorology*) the partial pressure of water vapor in the atmosphere; (*matter*) the partial pressure of any liquid, also known as saturated vapor pressure.

vapor recovery system — A closed system of pipes, valves, and compressors in which vapors that might otherwise escape into the atmosphere are compressed into liquid form and returned to their source.

vaporization — The change of a substance from a liquid to a gaseous state.

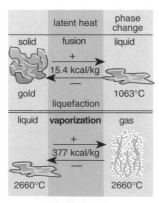

Vaporization

variables — (*statistics*) Any of a set of physical properties for which the values determine the characteristics or behavior of a population.

variable air volume (VAV) system — An air handling system that conditions the air to a constant temperature and varies the outside air flow to ensure thermal comfort.

variability — The degree of divergence of an object from a standard.

variance — Difference between what is required or specified and what is permitted.

variant — The differences between individuals of the species.

varicose — Pertaining to a blood vessel that is unnaturally swollen.

variegated — Having characteristics that vary, especially as to color.

vas — A duct or vessel.

vascular — Relating to a channel for the conveyance of a body fluid, such as blood, or to a system of such channels.

vasoconstriction — A decrease in the cross-sectional area of blood vessels.

vasodilator — An agent that causes dilation of blood vessels.

vasogenic shock — A shock resulting from peripheral vascular dilation caused by toxins that directly affect the blood vessels.

vasopressin — See antidiuretic hormone.

vasospastic — A spasm in a blood vessel.

vat — Large container or tank.

vat pasteurization — Pasteurization of milk at 145°F for 30 minutes.

VAV — See variable air volume system.

VC — See vital capacity.

VCM — Acronym for vinyl chloride monomer.

VCO$_2$ — Symbol for carbon dioxide output per unit time.

VCP — Acronym for voluntary cleanup program.

VDT — See visual display terminal.

VDU — See visual display unit.

VE — Acronym for visual emission.

vector — (*mathematics*) A quantity that has both direction and magnitude; (*medicine*) an agent, such as an insect, capable of transmitting an infection or disease from one organism to another.

vector control specialist — An environmental health practitioner responsible for surveillance and control of rodents, insects, and other vectors of public health and economic significance.

vector graphics structure — (*geographic information system*) A means of coding line and area information in the form of units of data expressing magnitude, direction, and connectivity.

vegetative cell — A form of bacteria in which an organism is able to grow and divide continuously.

vegetative controls — Non-point-source pollution control practices that involve vegetative cover to reduce erosion and minimize the loss of chemicals.

vehicle — Water, food, milk, or any other substance or particle serving as an intermediate means by which a pathogenic agent is transported from a reservoir and introduced into a susceptible host through ingestion, inhalation, inoculation, skin contact, or mucous membrane.

vein — A vessel that conveys blood to the heart in vertebrates.

Thermal (hot wire) Velometer

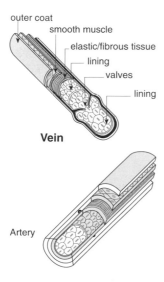

Vein

Artery

velocity — A vector specifying the time/rate-of-change of displacement with respect to some reference plane; see also potential difference.

velocity of ultrasound — The speed of ultrasound energy (measured in meters per second) in a specific medium.

velocity pressure — The algebraic difference between total pressure and static pressure; the kinetic pressure equal to that required to bring a fluid at rest to flow at a given velocity; the velocity pressure is always positive and in the direction of the flow.

velometer — A device for measuring air velocity.

vena cava — The major vein returning blood to the heart of vertebrates.

vending machine — Any self-service device, which, upon insertion of a coin, paper currency, token, card, or key, dispenses unit servings, either in bulk

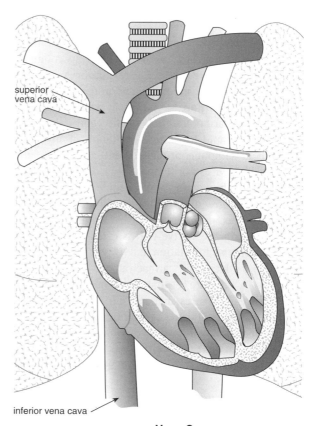

Vena Cava

or in packages, without the necessity of replenishing the device between each vending operation.

venom — A toxic or poisonous agent of animal origin.

ventilate — To change the quality of air by using a fan or opening windows or vents.

ventilated air sampler — An instrument constantly in contact with the air sample.

ventilated cell composting — A composting method in which compost is mixed and aerated by being dropped through a vertical series of ventilated cells.

ventilation — Circulation of air in a room; (*medical*) the movement of gas in and out of the lungs to facilitate blood oxygenation and carbon dioxide removal.

ventilation air — The total air or combination of air brought into a system from the outdoors and the air that is recirculated within the building.

ventilation effectiveness — The ability of a mechanical ventilation system to remove pollution originating in a space.

ventilation efficiency — A measure of how quickly a contaminant is removed from a space.

ventilation rate — The rate at which outside air enters and leaves a building, expressed as the number of changes of outdoor air per hour or the number of cubic feet per minute of air entering the building.

ventilator — A machine used in operating rooms and intensive care units for respiratory support of patients who cannot breathe on their own.

ventricular fibrillation — A cardiac arrhythmia marked by fibrillar contractions of the ventricular muscle due to rapid, repetitive excitation of myocardial fibers without coordinated ventricular contraction; also known as fibrillation. See also atrial fibrillation.

ventricular tachycardia — A cardiac rhythm that is slightly irregular, with a heart rate of 100 to 200 beats per minute; caused by irritable ventricular foci firing repetitively. It is often a forerunner of ventricular fibrillation and, if persistent and rapid, causes decreased cardiac output.

venturi — A pressure jet that draws in and mixes air.

venturi flares — Flares that mix air with a gas stream before it is ignited; usually used at ground level.

venturi meter — A fluid-flow measuring device consisting of a short tube with a tapering constriction in the middle that causes an increase in the velocity of the fluid flow and a corresponding decrease in fluid pressure; also known as venturi tube.

venturi scrubber — A scrubber system in which dust-laden gas passes through a duct incorporating a venturi section at the throat, which causes an increase in the velocity of the gas flow and a decrease in gas pressure, resulting in removal of dust from the gas.

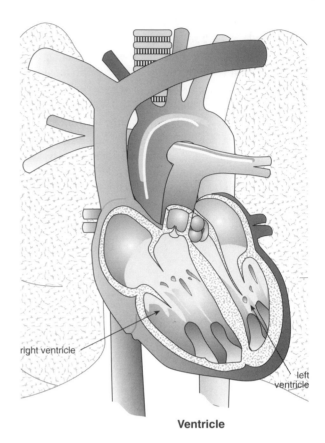

Ventricle

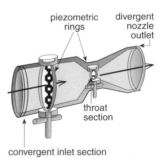

Venturi Meter

venturi tube — See venturi meter.

venturi wet scrubber — A collection device where an aqueous stream or slurry is used to remove particulate matter and/or gaseous pollutants.

venue — Place of trial.

venule — A small blood vessel that collects blood from a capillary bed and joins a vein.

VEO — Acronym for visible emission observation.

verdict — The ruling of the judge or jury in a criminal case.

verification — Methods, procedures, and tests used to determine that a proposed system is in compliance with specific rules and regulations.

vermicide — An agent that kills worms, especially those in the intestine.

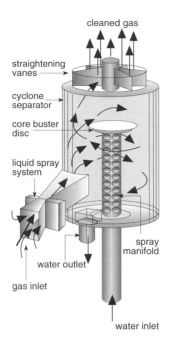

Venturi Scrubber

vermicomposting — A type of indoor composting that uses red wigglers to break down organic waste, producing a rich, sweet-smelling compost.

vermiculite ((Mg,Fe,Al)$_3$(Al,Sl)$_4$O$_{10}$(OH)$_2$·4H$_2$O) — An expanded mica used as a sorbent for spill control and cleanup.

vermiform — Resembling a worm.

vernier — A device that permits finer measurement or control than standard markings or adjustments.

vernon globe — A thin-walled, flattened, black globe made up of a copper sphere, 15 cm in diameter, with a temperature sensing device at its center; used to measure radiant energy transferred between surrounding surfaces and the convective heat exchange of the surrounding air. See also globe thermometer.

verruca — A wart-like elevation on the endocardium of a plant or animal.

version — (*computer science*) An integer value assigned to an application or file name; the larger the integer, the more recent the application.

vertebra — Each of the bones making up the spinal column.

vertebral disc — The discs that separate the bones that make up the spinal column; composed of fiberous structures filled with a pulpy, gelatinous matter. They function as shock absorbers for the spine and can be damaged readily, resulting in bulging and rupturing.

vertical — Perpendicular or at a right angle to the plane of the horizon.

vertical and horizontal spread (VHS) — The transport of landfill wastes to nearby receptors.

vertical transmission — The transmission of illness from parents to offspring.

vertigo — A sensation of dizziness.

very high frequency (VHF) — The band of frequencies from 30 to 300 megahertz in the radio spectrum.

very poorly drained soil — Soil in which the water is removed so slowly that the water table often remains at or on the surface.

very severe limitations — (*soils*) Limitations that preclude development of a tract of land without the use of sanitary sewer systems.

veryllium compounds (VE) — A hard, grayish element that does not occur naturally but as a chemical component of certain soils, volcanic dust, and rocks, such as bertrandite and beryl; MW: 9.012, BP: 2970°C, Sol: insoluble, density 1.8469/cm^3. It is used in aircraft disc brakes, x-ray transmission, windows, and space vehicles and instruments; in satellite structures, missile parts, nuclear reactors, and weapons; in fuel containers, precision instruments, rocket propellants, navigational systems, heat shields, and mirrors. Veryllium oxide is used in high-technology ceramics, electronic heat sinks, electrical insulators, microwave oven components, gyroscopes, military vehicle armor, rocket nozzles, crucibles, thermocouple tubing, and laser structural components. Veryllium alloys are used as electrical connectors and relays, springs, precision instruments, aircraft engine parts, nonsparking tools, submarine cable housings and pivots, wheels and pinons, automotive electronics and molds for injection-molded plastics, and automotive, industrial, and consumer applications. Veryllium is hazardous to the respiratory system and the endocrine system and is toxic by inhalation. Symptoms of exposure include non-neoplastic respiratory disease, acute pneumonitis, weight loss, and effects on the endocrine system. Veryllium workers have a death rate higher than the national average for cardiovascular and pulmonary diseases and increased incidence of lung cancer. OSHA exposure limits (TWA): 0.002 mg/m^3 [air].

vesical — Pertaining to either the bladder or gallbladder.

vesicant — Any substance or material that produces blisters on the skin.

vesicants — Chemical agents, also called blister agents, that cause severe burns to eyes, skin, and tissues of the respiratory tract.

vesicating agent — An agent that acts on the eyes and lungs and blisters the skin.

vesication — The process of blistering.

vesicle — A small blister on the skin; a bladder-like cavity.

vesicular disease — Any disease of the urinary bladder.

vesiculation — Formation of vesicles.

vesiculitis — Inflammation of a vesicle.

VHAP — See volatile hazardous air pollutant.

VHF — See very high frequency.

VHS — See vertical and horizontal spread; refers to the transport of landfill wastes to nearby receptors.

viability — Refers to the period of time an organism remains alive.

viable — Living; capable of growing or evolving.

vial — A glass container with a metal-enclosed rubber seal.

vibrating screen — An inclined sizing screen that is vibrated mechanically.

vibration — Periodic motion of the particles of an elastic body or medium in alternately opposite directions from the position of equilibrium when that equilibrium has been disturbed; any movement of the earth, ground, other surface, or equipment by temporal and special oscillation of displacement, velocity, or acceleration in any mechanical device or equipment located upon, attached to, affixed to, or in conjunction with that surface.

vibration white finger — A condition where the blood vessels in the hand constrict due to long-term use of vibrating tools, resulting in decreased blood flow; also called Raynaud's phenomenon.

Vibrio — A genus of short, rigid, motile bacteria typically shaped like a comma or an S.

vibrio gastroenteritis — See *Vibrio parahaemolyticus* food poisoning.

***Vibrio parahaemolyticus* food poisoning** — (*disease*) An intestinal disorder with watery diarrhea, abdominal cramps, and sometimes nausea, vomiting, fever, and headache. Incubation time is 12 to 24 hours but can range from 4 to 96 hours. It is caused by *Vibrio parahaemolyticus*, a halophilic vibrio; approximately 72 antigen types have been identified. Outbreaks have occurred in Japan, Southeast Asia, and the United States, primarily in warm months. Reservoirs of infection include marine coastal environments, where the organisms congregate in marine silt in the cold season and are free swimming in coastal waters and in fish and shellfish in the warm season. It is transmitted by ingestion of raw or improperly cooked seafood or through the handling of raw seafood. It is not communicable from person to person; general susceptibility. It is controlled by proper cooking of all seafood and handling of the cooked seafood.

Vibrio vulnificus — (*disease*) A bacterium that causes gastroenteritis, primary septicemia, or wound infections, generally with symptoms of diarrhea and bullous skin lesions. Consumption of the microorganisms in raw seafood by people with chronic diseases (especially liver disease) may result in septic shock, followed rapidly by death. Incubation time is 16 hours for gastroenteritis; the disease is found sporadically in the United States. The reservoir of infection is humans and other primates. It is transmitted through water, sediment, plankton, and consumption of raw shellfish; people with open wounds or lacerations can become contaminated in seawater. It is not communicable from person to person; general susceptibility. It is controlled by avoiding contact of open cuts with seawater, the proper disposal of feces, and cooking all shellfish thoroughly.

villi — Microscopic finger-like projections of the intestinal mucosa.

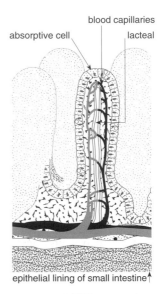

Villi

vinyl — A general term applied to a class of resins such as polyvinyl chloride, acetate, and butyryl.

vinyl acetate (CH₃COOCH:CH₂) — A clear, colorless liquid with a sweet, pleasant, fruity smell; MW: 86.09, BP 72 to 73°C, Sol: 2 g/100 ml water at 20°C, Fl.P: –8°C (closed cup) and –1.1°C (open cup), density: 0.932 at 20°C. It is used in the production of polyvinyl acetate and polyvinyl alcohol; with polyvinyl chloride to produce copolymers for adhesives; in architectural painting, nonwoven textile fibers, and textile sizing and finishing; in adhesives, as a raw material for polyvinyl butyryl; as a barrier in packaging paint and in coating applications; in plastic floor coverings; in phonograph records; in flexible film, sheeting, and molding; in coextrusion compounds. It is hazardous to the respiratory system and is toxic through inhalation. Symptoms of exposure include irritation of the nose, throat, and mucous membrane; respiratory damage in animals causes death due to lung congestion, hemorrhage froth in the trachea, and excess pleural fluid. OSHA exposure limits (PEL-TWA): 10 ppm [air].

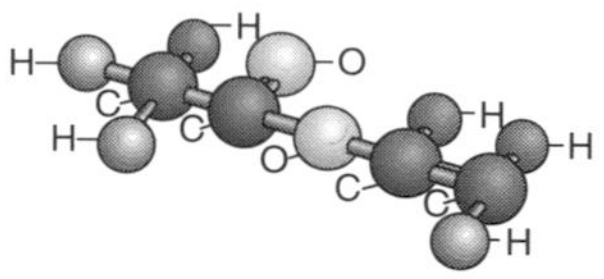

Vinyl Acetate

vinyl bromide (C_2H_3Br) — A colorless gas with a characteristic odor; MW: 106.96, BP: NA, Sol: miscible with organic solvents, Fl.P: the gas can travel a considerable distance to a source of ignition and flash back, density: 1.493. It is used as a fire retardant in plastics. It is hazardous to the respiratory system and can produce an anesthesia effect and kidney damage. It has shown carcinogenic properties in laboratory animals and is suspected to be a human carcinogen. ACGIH exposure limit (TLV-TWA): 5 ppm.

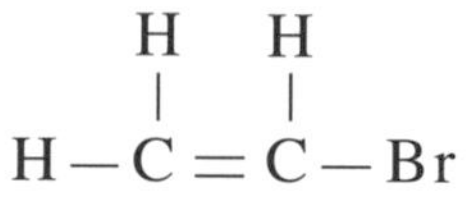

Vinyl bromide

vinyl chloride (CH_2=CHCl) — A colorless gas or vapor with a mild, sweet odor; MW: 62.5, BP: –13.4°C, Sol: 2763 mg/l water at 25°C, density: 0.9106 g/cm^3 at 20°C. It is used in organic synthesis and in adhesives; in automotive parts, furniture, packaging materials, pipes, wall coverings, wire coatings; in copolymer products, such as films and resins. It is hazardous to the pulmonary system, hepatic system, liver, kidneys, reproductive system, gastrointestinal system, central nervous system, and fetal development and is toxic through inhalation. Symptoms of exposure include dizziness, disorientation, a burning sensation in the soles of the feet, headaches, gastritis, ulcers, upper gastrointestinal bleeding, lack of blood clotting, hepatic abnormalities, burns on the conjunctiva and cornea, angiosarcoma of the liver, muscle pain, scleroderm-like skin changes, ataxia, visual and/or hearing disturbance, nausea, developmental toxicity, birth defects, and decreased sexual function. OSHA exposure limits (PEL-TWA): 1 ppm [air].

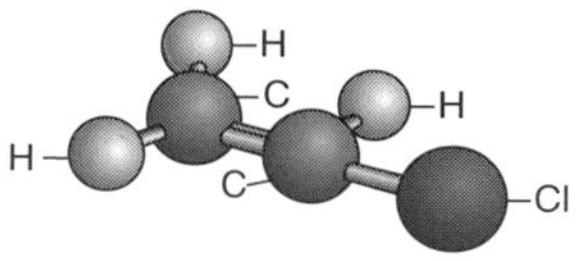

Vinyl Chloride

vinyl ether ($(C_2H_3)_2O$) — A clear, colorless liquid with a characteristic odor; MW: 70.10, BP: 28.5°C, Sol: soluble in most organic solvents, Fl.P: less than –30°C (closed cup), density 0.774 at 20°C. It is used as an anesthetic and in organic synthesis. It is hazardous to the respiratory tract and can cause unconsciousness in humans. ACGIH exposure limit: NA.

$$H_2C=CH-O-CH=CH_2$$

Vinyl ether

vinyl toluene ($CH_2CHC_6H_4CH_3$) — A colorless liquid with a strong, disagreeable odor; MW: 118.2, BP: 339°F, Sol: 0.009%, Fl.P: 120°F, sp. gr. 0.89. It is used in spray applications of vinyl toluene polyester surface coatings; during the preparation of unsaturated polyester resins and alkyd coatings; in the manufacture of thermoplastic moldings via extrusion, injection, or other processes. It is hazardous to the eyes, skin, and respiratory system and is toxic through inhalation, ingestion, and contact. Symptoms of exposure include drowsiness and irritation of the eyes, skin, and upper respiratory system. OSHA exposure limit (TWA): 100 ppm [air] or 480 mg/m^3.

violation — (*pollution*) Any incident of excess emissions, regardless of the circumstances of the occurrence; (*law*) an infringement of rules or laws.

viral disease — See viral infection.

viral dysentery — A form of dysentery caused by a virus with symptoms of acute watery diarrhea.

viral gastroenteritis — A generic term for diseases caused by astrovirus, calicivirus, enteric adenovirus, parovirus, and acute nonbacterial infectious gastroenteritis.

viral hepatitis A — (*disease*) A type of hepatitis with sudden onset; fever, malaise, anorexia, nausea and abdominal discomfort are followed in a few days by jaundice. It varies from being a mild condition of 1 to 2 weeks' duration to being severely disabling and lasting many months. Incubation time is 15 to 50 days, with an average of 28 to 30 days. It is caused by hepatitis A virus, a 27-nm picornavirus with the characteristics of an enterovirus. It is found worldwide, especially in areas with poor environmental control and in institutions, daycare centers,

V

low-cost housing, rural areas, and the military. It is transmitted from person to person by the oral–fecal route and through food and water. It is communicable until a few days after the onset of jaundice; general susceptibility. It is controlled through public health education, good personal hygiene, proper management of daycare centers and other institutions, proper sterilization of syringes and needles, and vaccines. Also known as infectious hepatitis.

viral hepatitis B — (*disease*) A type of hepatitis with a gradual and cumulative effect; anorexia, nausea, and vomiting lead to jaundice. Incubation time is 45 to 180 days, with an average of 60 to 90 days. It is caused by hepatitis B virus, a 42-nm, double-stranded deoxyribonucleic acid virus composed of a 27-nm nucleocapsid core surrounded by an outer lipoprotein coat containing the surface antigen. It is found worldwide, commonly in high-risk groups such as parenteral drug abusers, homosexual men, institutional staff and patients, and hemodialysis individuals. The reservoir of infection is people. It is transmitted by all body secretions and excretions; blood, saliva, semen, and vaginal fluids are the only fluids to have been shown to be infectious. It is communicable for many weeks, and a chronic carrier state may exist for years; general susceptibility. It is controlled by vaccines, good personal hygiene, not sharing reusable needles, and protecting against infected bodily fluids. Also known as serum hepatitis.

viral infection — Approximately 200 viruses are pathogenic to people and may cause mild or extremely dangerous disease. The viruses may be introduced into the body through a break in the skin, transfusion, inhalation, or ingestion. The process of viral infection includes entry into the body; attachment of the virus to a susceptible cell and absorption of the virus; penetration of the viral nucleic acid into the cell; maturation of the virus within the cell, including replication; using the chemical building blocks and energy available in the cell; and movement of the viruses outside of the cell, ready to attack new cells.

viremia — Presence of a virus in the blood.

virgin material — Raw material.

virino — A small substance that may be the infective agent of transmissible spongiform encephalopathy (TSE)-containing protein and nucleic acid.

virions — Structurally mature, extracellular virus particles.

viroid — A small infective segment of nucleic acid.

virologist — A specialist who studies viruses and diseases caused by viruses.

virtual — (*computer science*) Referring to a computer-generated environment.

virucide — Any agent that destroys or inactivates a virus.

virulence — A relative measure of the frequency with which a microorganism will produce disease once it has established itself in the host; infectiousness.

virulent — Capable of overwhelming bodily defense mechanisms.

virus — The smallest organism known, ranging in size from about 0.025 to 0.25 μm, characterized by a lack of independent metabolism, and is a disease-producing organism that requires living cells to grow; (*computer science*) a computer program that is transferred by means of a computer network without the user's knowledge and can infect and damage computer files.

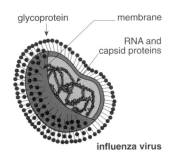

influenza virus

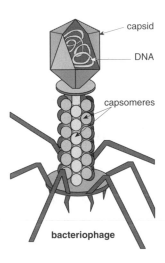

bacteriophage

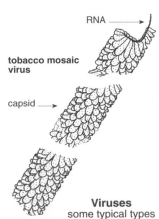

Viruses
some typical types

VIS — Abbreviation for visible.

viscera — Internal organs of the body.

visceral — Of the organs of the thoracic, abdominal, and pelvic cavities.

viscosity — The property of a fluid or gas that offers resistance to flow; the resistance to flow due to the friction between molecules.

viscous — Having a heavy or gluey consistency.

visible — Pertaining to objects that are perceptible to the eye.

visibility — A measurement of the ability to see and identify objects at various distances; for example, visibility reduction in the air is caused by air pollution, such as the presence of sulfur and nitrogen oxides as well as particulate matter.

visibility-reducing particles — Any particles in the atmosphere that obstruct the range of visibility.

Visible Infrared Spin–Scan Radiometer (VISSR) — An instrument used on the synchronous meteorological satellite to provide both day and night images of the Earth and cloud cover.

visible light — Light that can be detected by the human eye, ranging from extreme red (with a frequency of 4.3×10^{14} sec^{-1}), through orange, yellow, green, blue, and indigo to violet (with a frequency of 7.5×10^{14} sec^{-1}); see also spectrum.

visible radiation — Wavelengths of the electromagnetic spectrum between 10^{-4} and 10^{-5} cm.

visible spectrum — The portion of radiation of the sun that can be perceived as light by humans; between approximately 380 and 750 nm in wavelength.

vision — The special sense by which the eye can observe the color, shape, and size of objects.

VISSR — See Visible Infrared Spin–Scan Radiometer.

VISTTA — Acronym for Visibility Impairment from Sulfur Transformation and Transport in the Atmosphere.

visual — Pertaining to the sense of sight.

visual discomfort — A situation in which an impairment of vision is caused by an excessive range of brightness in the visual field, called glare.

visual display terminal (VDT) — See cathode-ray tube.

visual display unit (VDU) — A terminal with a cathode-ray tube.

visual environment — Area having a spatial distribution of illuminance, luminance ratios, color, and glare that might affect the person.

vital — Pertaining to or contributing to life forces.

vital capacity (VC) — The largest volume of air measured on complete expiration after taking the deepest inspiration without a forced or rapid effort.

vital records — Records relating to birth, death, marriage, health, and disease.

vital signs — Pulse rate, respiration rate, and body temperature; blood pressure is typically taken when measuring these vital signs.

vital statistics — Data collected from the registration of vital events such as births, deaths, fetal deaths, marriages, divorces, etc.

vitamin — An organic substance found in natural foodstuffs and essential for normal body activity.

vitamin D milk — Milk with added vitamin D content, which has been increased by the approved method to at least 400 USP units per quart.

vitiated air — Spoiled, impure, or polluted air.

vitiligo — A condition in which destruction of melanocytes occurs in small or large areas, resulting in patches of depigmentation.

vitrification — A process whereby high temperatures create permanent chemical and physical changes in a ceramic body, converting it into glass or a glassy substance.

vitrification technology — A system of converting contaminated soils, sediments, and sludges into oxide glasses, rendering them nontoxic and suitable for landfilling as a nonhazardous material.

viviparous — An animal whose eggs develop inside the female while the embryo derives nutrition from the mother.

V-notch weir — A weir that is V-shaped, with its apex pointed downward; used to accurately measure small rates of flow.

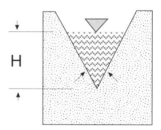

V-notch Weir

VOC — See volatile organic compound.

void — To empty or evacuate.

voids — Spaces in or between particles or fibers of a filtering medium; (*environmental engineering*) spaces found in an existing landfill. See soil pore.

vol — See volume.

volatile — Referring to a substance that evaporates rapidly and explodes readily; the ability to evaporate readily and go to a gas phase from a liquid or solid.

volatile hazardous air pollutant (VHAP) — Volatile organic compounds released into the air, such as benzene, perchlorethylene, and carbon tetrachloride, and are potential health hazards.

volatile liquids — Liquids that easily vaporize or evaporate at room temperature.

volatile matter — The gaseous constituents of solid substances.

volatile oils — Terpenes, such as lemon and peppermint, composed of two to four isoprene units; also known as essential oils.

volatile organic compound (VOC) — Organic chemical that evaporates readily and exists as a gas in the air.

volatile solid — The quantity of a solid in water, sewage, or other liquid lost on ignition of total solids.

volatility — The quality of a solid or liquid to evaporate quickly at ordinary temperatures when exposed to the air.

volatilization — The process of evaporation; to pass into the atmosphere as a gas.

volcanic ash — A fine, white, porous powder similar to diatomite but lighter in weight and having a particle diameter of <4 mm; it is used as a filter medium or filter aid.

volcanic glass — Natural obsidian formed by the cooling of molten lava too rapidly to allow crystalization.

volcanism — Phenomenon and activities associated with the origin of volcanoes and the movement of both lava and other volcanic material.

volt (V) — The MKSA unit of voltage potential equal to the potential difference across a conductor when 1 ampere of current in it dissipates 1 watt of power.

voltage — See potential difference.

voltage gain — Ratio of the output signal level with respect to the input level in decibels.

voltage–voltage difference — The electric potential; the work required to transport a unit of electric charge from one point to another.

voltaic cell — A device that changes chemical energy into electrical energy by the action of two dissimilar metals immersed in an electrolyte; see also primary voltaic cell.

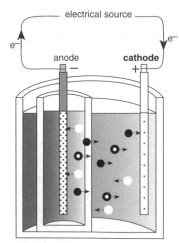

Voltaic Cell showing Cathode

voltammetry — The measurement of electric current as a function of potential.

voltmeter — An instrument used to measure in volts the difference of potential between two points in an electrical circuit.

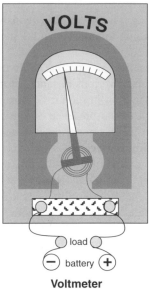

Voltmeter
showing galvanometer

volume flow rate — The quantity of a unit flowing per unit time.

volume reduction program — (*solid waste*) A technique developed by a generator to identify waste minimization activities and goals; includes a series of steps from data gathering and analysis to identification of deficiencies, recommendations for a schedule, and its incorporation into the ongoing corporate management program.

volume — The amount of space occupied by a substance.

volumetric analysis — A technique of quantitative analysis in which the volume of a liquid reagent of known concentration required to react completely with a substance whose concentration is being determined is measured by titration.

volumetric flow rate — The amount of fluid flowing past a designated point; usually measured in liters per second.

volumetric tank test — A test that determines the volume of fluid in a tank measured directly or calculated from product-level changes; a sharp drop in volume indicates a leak.

voluntary agency — A service agency that is legally controlled by a board of volunteers but may hire paid staff to carry out specific functions.

voluntary muscle — Muscle tissues, including all the skeletal muscles, that appear microscopically to consist of striped myofibrils and are composed of bundles of parallel, striated fibers under voluntary control; the heart, which is a striated muscle, is under involuntary control.

Voluntary Muscle

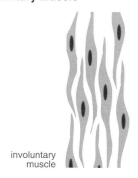

involuntary muscle

voluntary nervous system — The part of the nervous system that relays commands to skeletal muscles.

voluntary residents tracking system — A collection of people who are contacted periodically, for a limited period of time, to disseminate information or coordinate other health-related services.

vomiting — Expelling the contents of the stomach through the esophagus and out of the mouth.

vomitus — Material expelled from the stomach during vomiting.

vortex — A revolving mass of water that forms a whirlpool.

VOST — Acronym for volatile organic sampling train.

VP — See vapor pressure.

vps — Acronym for vibrations per second.

VRP — See visibility-reducing particle.

Vruchus pisorum — See Pea Weevil.

VSD — Acronym for virtually safe dose.

VSI — Acronym for visual slate inspection.

VSS — Acronym for volatile suspended solid.

VT — See tidal volume.

vulcanism — All of the processes associated with the movement of molten rock material.

vulcanization — The process of combining natural, synthetic, or latex rubber with sulfur and various other additives, usually under heat and pressure to eliminate tackiness when warm and brittleness when cool, and to change permanently the material from a thermoplastic to a thermosetting composition; it improves the strength, elasticity, and abrasive resistance of the material.

vulnerable — Being susceptible to infection or injury because of a situation in the environment.

vulnerable zone — An area over which the airborne concentration of a chemical accidentally released could reach a level of concern.

VX ($C_{11}H_{26}NO_2S$) — A colorless and odorless liquid nerve agent, ethyl-*S*-dimethylaminoethyl methylphosphonothiolate; MW: 236.44, BP: NA, Sol: soluble in organic solvents and water, density: NA. It is used as a chemical warfare agent. It is an extremely toxic organophosphorus compound and is a highly potent cholinesterase inhibitor. It may cause convulsions, cardiovascular toxicity, difficulty in breathing, bronchial constriction, tremor, and death. ACGIH exposure limit: NA.

$$H_3C\diagdown \overset{\overset{\textstyle O}{\|}}{P} - S - CH_2 - CH_2 - N \diagup^{CH(CH_3)_2}_{\diagdown CH(CH_3)_2}$$

$$CH_3CH_2O$$

VX

V

W

W — See watt.

wading pool — An impervious structure less than 3 feet deep that contains water and is mechanically controlled for purification purposes; usually designed for children's use or wading.

WAIS — (*computer science*) See Wide Area Information Server.

waiting period — A period that must elapse to show that a person is disabled during which income benefits are not paid; refers only to compensation, not medical or hospital care, which must be provided immediately.

WAP — Acronym for Waste and Analysis Plan.

warewashing — (*food*) Cleaning and sanitizing food-contact surfaces of equipment and utensils; formerly dishwashing.

warfarin ($C_{19}H_{16}O_4$) — A colorless, odorless crystalline powder; MW: 308.3, BP: decomposes, Sol: 0.002%, Fl.P: not known. It is used as a rodenticide. It is hazardous to the blood, cardiovascular system, and is toxic through inhalation, absorption, ingestion, and contact. Symptoms of exposure include back pain, hematoma of the arms and legs, epistaxis, bleeding lips, mucous membrane hemorrhage, abdominal pain, vomiting, fecal blood, petechial rash, and abnormal hematologic indices. OSHA exposure limit (TWA): 0.1 mg/m³ [air].

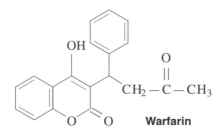

Warfarin

warm air mass — A large mass of stable air that produces steady winds and creates poor surface visibility with stratus clouds and drizzle; it is warmer than the surrounding air.

warm-blooded — See homoiothermal.

warm front — Advancing edge of a warm air mass.

warning — (*meteorology*) A report issued by National Weather Service local offices indicating that a par-

Warm Front

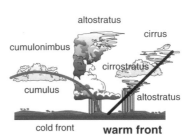

ticular weather hazard is either imminent or has been reported; it indicates the need to take action to protect life and property.

warning device — A sound-emitting device used to alert people to the presence of a dangerous situation.

warrant — A court order directed to a marshal or other proper officer requiring the arrest and appearance before the court or U.S. magistrate of the defendant named in the order.

waste — Unwanted materials left over from a manufacturing process or from places of human or animal habitation.

waste characterization — Identification of chemical and microbiological constituents of a waste material.

waste disposal — The orderly process of discarding useless or unwanted material.

waste exchange — An arrangement in which companies exchange their waste for the benefit of both.

waste feed — The continuous or intermittent flow of wastes into an incinerator.

waste generation — The weight or volume of materials or products that enter the waste stream before recycling, composting, landfilling, or combustion.

waste heat — Heat that remains after useful energy generation or use.

waste heat boiler — A boiler that uses the heat of exhaust or processed gas to generate steam or to heat water.

waste load allocation — The maximum load of pollutants that each discharger of waste is allowed to release into a particular waterway.

waste minimization audit — A survey conducted by in-plant personnel or consultants that identifies and evaluates opportunities to reduce hazardous waste generation.

waste products — The products of metabolic activity after oxygen and nutrients have been supplied to a cell.

waste reduction — A policy mandated by the U.S. Congress in the 1984 Hazardous and Solid Wastes Amendments to the Resource Conservation and Recovery Act (RCRA); a part of waste management that includes source reduction and recycling.

waste stabilization pond — A large, shallow basin, usually 2 to 4 feet deep, where the biological oxygen demand of wastewater is reduced; the organics are converted to inorganics or stabilized because of the metabolic activity of bacteria, algae, and surface aeration.

waste stream — The total amount of solid waste from homes, businesses, institutions, and manufacturing plants that are recycled, burned, or disposed of in landfills.

waste water — Water used to carry waste from homes, businesses, and industries and which contains dissolved or suspended solids; it is not intended for reuse unless it is treated..

wastewater drain — A receptor and appurtenances for the disposal of liquid waste.

wastewater specialist — An environmental health practitioner who controls water pollution from municipal and industrial sources in order to ensure water quality that fosters the health, welfare, and comfort of the community.

wastewater treatment plant (WWTP) — A facility designed and utilized to remove impurities from water through a series of aerobic processes and the production of an effluent released to a stream that will not be polluting in nature.

wastewater treatment return flow — Water returned to the environment by wastewater treatment facilities.

water — A chemical compound made up of one atom of oxygen and two atoms of hydrogen.

water activity (A_w) — A measure of free moisture in a food, determined by dividing the water vapor pressure of the substance by the vapor pressure of pure water at the same temperature.

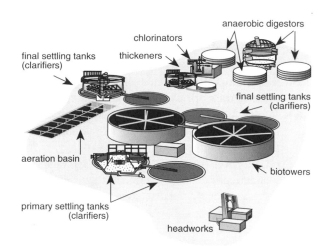

Wastewater Treatment Plant

water base — Water used as a solvent for coatings such as paint.

water column — A unit used in measuring pressure; the vertical column of seawater that extends from the surface to the bottom.

water cycle — (*geology*) The cycle by which water evaporates from oceans, lakes, and other bodies of water; forms clouds; and is returned to those bodies of water through rain and snow, the runoff from rain and snow, or groundwater. See also hydrologic cycle.

water droplet coalescence — A mechanism of condensation that occurs in clouds and causes ice crystal formation.

water hardness — A characteristic of water caused mainly by the salts of calcium and magnesium, such as bicarbonate, carbonate, sulfate, chloride, and nitrate, with excessive amounts being undesirable because of the formation of soap curds; hard water leads to increased use of soap, deposition of scale in boilers, damage of some industrial processes, and sometimes the production of objectionable tastes in drinking water. See hard water.

water ice — A food that is prepared by freezing water while stirring a mix composed of one or more optional fruit ingredients, optional sweetener ingredients, sugar, dextrose, or corn syrup.

water mass — A body of water that can be identified by its temperature and salinity.

water miscible — Referring to the quality of a substance to mix readily with water.

water of hydration — Water present as a structural part of certain crystals.

water pollution — The addition of harmful or objectionable material, causing an alteration of water quality.

Water Pollution Control Act — A 1948 law funding state water pollution control agencies and providing technical assistance to states containing limited provisions for legal actions against polluters.

water potential — The relevant ability of water molecules to do work by interacting with each other.

water quality — The chemical, physical, and biological characteristics of water and its suitability for a particular purpose.

Water Quality Act — A 1970 law updated through 1987 mandating that states establish water quality standards and regulating discharges into river systems.

water quality criteria — Acceptable levels of pollutants in bodies of water that are consistent with uses such as drinking, fishing, industry, and swimming.

water quality standard — The combination of a specific use and the maximum concentration of a pollutant that will protect that use for any given body of water.

water-reactive — A chemical that reacts with water to release a gas that is either flammable or presents a health hazard.

water recreational facility (WRF) — Any artificial basin or other structure containing water used or intended to be used for recreation, bathing, relaxation, or swimming where body contact with the water occurs or is intended to occur; includes auxiliary building and appurtenances.

Water Resources Act — A 1984 law that declares the necessity of the existence of an adequate supply of quality water for production of materials and energy for the nation's needs and efficient use of the nation's energy and water resources that are essential to economic stability, national growth, and citizens' well-being.

Water Resources Development Act — A 1999 law that provides amended authority for the U.S. Army Corps of Engineers to support states', tribes', and local governments' water resources development, management, projection, and restoration.

Water Resources Planning Act — A 1965 law updated through 1988 that declares the policy of Congress to encourage conservation, development, and utilization of water and related land resources of the United States on a comprehensive and coordinated basis by the federal government, states, counties, and private enterprises.

water saturation — The point at which a material will no longer absorb water.

water softening — The removal of minerals, primarily calcium and magnesium, to reduce the amount of soap needed, scale on cooking utensils or laundry basins, and materials deposited on interior pipes; improves the heat-transfer efficiency through walls

of heating elements or exchange units of water tanks.

water solubility — The maximum concentration of a chemical that dissolves in pure water at a specific temperature, pressure, and pH; the ability of a substance to remain dissolved in water and be less likely to volatilize, thereby reducing potential air pollution problems.

water supply engineering — Study of the sources of water supply, including water treatment, storage, transmission, and distribution facilities.

water supply specialist — An environmental health practitioner responsible for the supply of safe and acceptable water to the public to prevent disease.

water table — The upper boundary of groundwater below which all spaces within the rock are completely filled with water; the area between the zone of saturation and the zone of aeration.

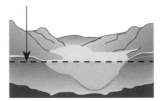

Water Table

water turbine — A piece of equipment that converts the energy of falling water into rotating mechanical energy.

water use — Human interaction with and influence on the hydrological cycle, including elements of water withdrawal from surface and groundwater sources, water delivery to homes and businesses, water usage, water released from wastewater treatment plants, and instream uses such as water for producing hydroelectric power.

water wheel — A wheel of buckets or blades turned by the weight or velocity of falling water or by water moving beneath it.

waterborne — Something carried by waters such as a disease organism.

waterborne-disease outbreak — A significant occurrence of acute illness due to drinking water that is deficient in treatment, as determined by local or state agencies.

waterlogging — Water saturation of soil that fills all airspaces and causes plant roots to die from lack of oxygen.

watershed — A region or area bounded by a divide and draining into a water course or body of water.

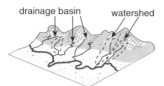

Watershed

watt (W) — An SI unit of power equal to 1 joule per second.

watt/cm² — See watts per centimeter squared.

watt-hour — An electrical energy unit of measure equal to 1 watt of power supplied to or taken from an electrical circuit steadily for 1 hour.

watts per centimeter squared (watt/cm²) — A unit of power density used when measuring the amount of power per absorbing area.

wave — Undulation that forms as a disturbance and moves along the surface of the water.

wave period — The time it takes for two successive crests of a wave to pass a given point.

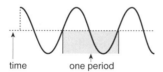

Wave Period

wavelength — The distance between any two corresponding points of two consecutive waves.

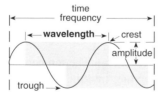

Wavelength

wax — A lipid material that has a high melting point and is relatively impermeable to water.

way bill — A record accompanying a shipment during transport.

WB — Acronym for wet bulb.

WBC — Acronym for white blood cell.

WBGT — See wet-bulb globe temperature index.

WBT — Acronym for wet-bulb thermometer.

WC — Acronym for water column.

weak electrolyte — An electrolyte only slightly ionized in moderately concentrated solutions.

weapon of mass destruction — Any device, material, or substance used in adequate quantities and in such a manner as to cause death or serious injury to people or significant damage to property.

weather — Description of the short-term physical conditions of the atmosphere including the moisture, temperature, pressure, and wind.

weather radar — A radar system designed to detect the physical characteristics of precipitation or clouds, their distribution, and movement.

weathering — A physical disintegration of rocks that are slowly decomposed by exposure to either physical or chemical agents.

Web — See World Wide Web.

web of causation — An interrelationship of multiple events contributing to causation of a disease or injury.

weber — Named after Wilhelm E. Weber, a German physicist; the practical MKS unit of magnetic flux equal to the flux which, when linking a circuit of one turn, produces in it an electromotive force of 1 volt as the flux is reduced to zero at a uniform rate of 1 ampere per second.

Webmaster — The person responsible for managing a Web site.

weed — Any plant growing where it is not wanted or growing to the detriment of some crop; vegetation that has attained a height of 12 inches or more and that constitutes a potential rodent harborage or any other health or safety hazard.

weed eradication — Destruction or removal of all weeds.

weep hole — A drain embedded in a concrete or masonry structure intended to relieve pressure caused by seepage behind the structure.

Weevils — Various beetles, usually having a long snout and whose larvae destroy crops.

weight (wt) — The gravitational force with which a body is attracted toward the Earth.

weight of evidence — The extent to which available information supports the hypothesis that a substance causes an effect in humans. Information used for this determination includes the number of tissue sites affected by the agent, the number of animal species tested, and statistical significance in the occurrence of the adverse effect in treated subjects compared to untreated controls.

weighted moving average — An average value of an attribute computed for a point from the values for surrounding data points, taking into account their distance or importance.

weighting networks — (*sound*) Electrical networks associated with sound level meters.

Weil's disease — (*disease*) See leptospirosis.

weir — A device in a waterway or swimming pool surface skimmer used to control the water level or divert the flow.

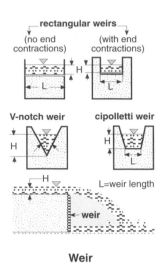

Weir

weld — To build up or fasten together metals by bonding molten metal.

well — Any excavation regardless of design or method of construction done to remove groundwater from an aquifer or to determine the quality, quantity, or level of groundwater.

well-drained soil — (*soils*) Soil in which the water is removed readily but not rapidly.

well field — An area containing one or more wells that produce usable amounts of water or oil.

well injection — Subsurface placement of fluids into a well.

well monitoring — The measurement by on-site instruments or laboratory methods of well water quality.

well plug — A watertight, gastight seal installed in a bore hole to prevent movement of fluids.

well point — A hollow, vertical tube, rod, or pipe terminating in a perforated driving point and covered by a fine-mesh screen; used to remove underground water.

wellhead protection area — A protected surface and subsurface zone surrounding a well or well field and supplying a public water system to keep contaminants from reaching the well water.

WENDB — Acronym for Water Enforcement National Database.

western equine encephalitis — (*disease*) An acute inflammatory encephalitis of short duration involving parts of the brain, spinal cord, and meninges. It is caused by an alpha virus and is found chiefly in western and central United States, Canada, and Argentina. Possible reservoirs of infection include birds, rodents, bats, reptiles, amphibians, and surviving mosquito eggs or adults. It is transmitted

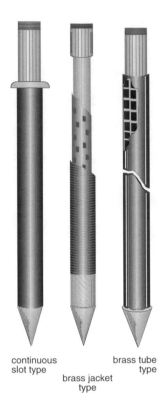

continuous slot type   brass tube type
                  brass jacket type

Drive Well Points

by the bite of infected mosquitos, not from person to person; it is usually a disease of children but may affect others. It is controlled by destroying mosquito larvae, eggs, breeding places, and adult mosquitos and by using mechanical mosquito control techniques.

WET — Acronym for whole effluent toxicity test.

wet-bulb globe temperature index (WBGT) — A heat index recognized by the National Institute of Occupational Safety and Health to establish threshold limit values for heat stress, utilizing wet-bulb temperature, dry-bulb air temperature, and globe temperature.

wet-bulb temperature — The temperature at which liquid evaporating into the air can bring the air into saturation.

wet-bulb thermometer — A thermometer in which the bulb is covered with a water-saturated cloth or other material.

wet chemistry — Chemical analysis of pollutants dissolved in a chemical solution.

wet collector — One of a variety of devices used for cleaning, cooling, and deodorizing gas particulates and vapor emissions to the atmosphere; see also scrubber.

wet digestion — A solid waste stabilization process in that mixed solid organic wastes are placed in an open digestion pond to decompose anaerobically.

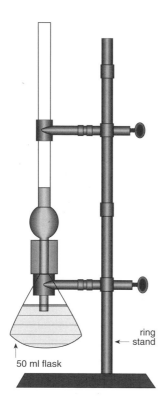

Wet-Bulb Thermometer

wet-gas meter — A meter that measures the total gas volume by entrapping the gas in inverted cups or veins under a liquid; the buoyancy of a gas causes a rotation of the cups or veins that is proportional to the volume indicated on a precalibrated meter.

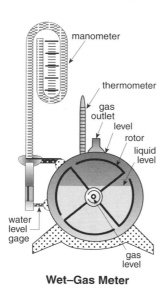

Wet–Gas Meter

wet milling — Mechanical size reduction of solid waste that has been wetted to soften the paper and cardboard constituents.

wet soil — Nonsticky to very sticky; nonplastic to very plastic soil.

wet weather flows — The water entering storm drains during rainstorms and other wet weather events.

wetlands — A piece of land or area, such as a tidal flat or swamp, that contains much soil moisture because of its location at the land/water interface; provides ecological, biological, economic, scenic, and recreational resources; among the most productive ecosystems on the Earth, especially estuaries and mangrove swamps. At least one half of the biological production of the oceans occurs in the coastal wetlands; 60 to 80% of the world's commercially important marine fish either spend time in the estuary or feed upon the nutrients produced there. Coastal wetlands help protect inland areas from erosion, flood, waves, and hurricanes and provide cash crops such as timber, marsh, hay, wild rice, cranberries, blueberries, fish, and shellfish.

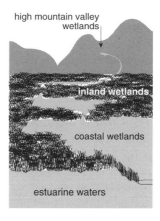

Wetlands

wettable powder — A chemical in dust form that can be mixed with water to form a suspension.

wetting — A means of reducing surface tension of soil and allowing detergent solutions to penetrate and spread out.

wetting agent — A material that decreases the surface tension of a liquid, causing it to spread out easily across a surface.

Wh — See watt-hour.

wheal — A localized area of edema on the body surface; often associated with severe itching.

Wheatstone bridge — (*electricity*) A bridge used for measuring an unknown resistance by comparing it with a known resistance.

whey — A liquid or semiliquid material remaining after the removal of fat and casein from milk or cream in the process of making cheese.

whiplash injury — Injury to the cervical vertebrae or their supporting ligaments and muscles with symptoms of pain and stiffness; usually caused by a sudden acceleration or deceleration.

whipped cream — Cream containing sugar and flavoring to which a harmless gas has also been added to cause whipping of the product.

whipping cream — Cream that contains not less than 30% milkfat.

whipworm — See trichuriasis.

white blood cell — See white corpuscle; leukocyte.

white corpuscle — A colorless cell of the blood; also known as leukocyte or white blood cell.

white noise — A random-like signal with a flat frequency spectrum described as a relative power density in volts squared per hertz, varying directly with bandwidth, having twice as much power in a higher octave than in the current one; the random energy containing all frequency components in equal proportions. The introduction of a white noise audio signal can destroy high-frequency loudspeakers.

wholesale food processing establishment — A food establishment or portion thereof undertaking any of the activities of an establishment for the purpose of sale or delivery to another person for the purpose of resale; includes warehouses, food manufacturing plants, beverage manufacturing plants, bakeries, etc.

wholesome — Food that is nutritious, clean, free from adulteration, and otherwise suitable for human consumption.

WHPA — Acronym for wellhead protection area.

Wide Area Information Server — Servers that allow users to conduct full-text keyword searches in documents, databases, and libraries connected to the Internet.

Wilson's disease — A rare, progressive disease inherited as an autosomal recessive trait. It causes a defect in the metabolism of copper leading to the accumulation of copper in the liver, brain, kidney, cornea, and other tissues. Also known as hepatolenticular degeneration.

wind — The natural horizontal movement of air over the surface of the Earth for which the speed may be given in meters per second, miles per hour, or knots; movement of the atmosphere brought about not only by rotation of the Earth but also by unequal temperatures on the Earth.

wind chill — Loss of heat from the body when it is exposed to wind of a given speed at a given temperature and humidity.

wind chill factor — The amount of chilling of the body, beyond that caused from a cold ambient temperature, because of exposure to cool air currents.

wind chill index — A chart that compares temperatures of the atmosphere with various wind speeds, thus allowing calculation of the wind chill factor.

wind energy — Energy that is captured from the natural movement of the air.

wind farms — A large number of windmills concentrated in a single area to produce energy.

Wind Farm

wind power — Power available from the wind that is used by various types of wind machines; expressed as: $P = E(.5)(DAV3)$, where P is the power (kW), E is the efficiency of the piece of equipment (%), D is the air density (kg/m^3), A is the swept area (m^2), and V is the wind velocity (m/sec).

wind power plant — A set of wind turbines that are interconnected to a common utility system through a system of transformers, distribution lines, and a substation.

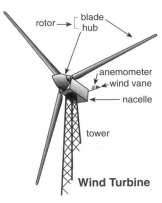

Wind Turbine

propellor model
(horizontal axis–HAWT–wind turbine)

Darrieus model
(vertical axis–VAWT–wind turbine)

wind shear — The change in wind speed with height.

wind vane — An instrument used to measure the direction of the wind.

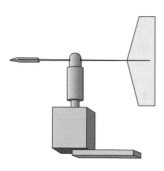

Wind Vane

wind velocity — The speed of air movement in miles per hour or meters per second; the amount of power available is based on velocity.

windbreak — Rows of trees or shrubs planted to block wind flow, reduce soil erosion, and protect sensitive crops from high winds; also used to reduce noise.

window — (*computer science*) An area on a computer screen that is used to view specific data.

windrow — A low-technology operation using long rows of shredded refuse, normally 10 feet wide at the base and 4 to 6 feet high, that allows sufficient moisture and oxygen to support aerobic life.

windrow composting — An open-air composting method where compostable material is placed in windrows, piles, or ventilated bins or pits and is occasionally turned or mixed; the process may be anaerobic or aerobic.

Winterbottom's sign — Swollen lymph nodes at the base of the skull; symptomatic of African sleeping sickness.

witness — The party having personal knowledge of the facts who appears to present that knowledge as testimony under oath in any proceeding in a lawsuit. An expert witness is one who is skilled in some art, science, trade, profession, or other human activity and possesses particular knowledge concerning it.

WL — Acronym for warning letter; (*radon*) acronym for working level.

WLA/TMDL — Acronym for waste load allocation/total maximum daily load.

WLM — See working level month.

WMD — See weapon of mass destruction.

wood — Secondary xylem.

wood alcohol — See methyl alcohol.

wood pulp waste — Wood or paper fiber residue resulting from a manufacturing process.

word — (*computer science*) A set of bits, usually 16 or 32, that occupies a single storage location and is treated by the computer as a unit of information.

work — A force acting against resistance to produce motion in a body.

work area — A service area or areas where major activities are conducted; a room or defined space in a workplace where hazardous chemicals are produced or used while employees are present.

work situation design — (*ergonomics*) The assignment of appropriate work hours, rest periods, training, supervision, and management to provide a good working environment and reduce the potential for disease and injury.

workers' compensation — Coverage required by state law for compensation to workers who have been injured on the job, regardless of the employer's negligence.

Workers' Compensation Law — Rules of conduct or action prescribed and enforced by a controlling authority governing employer and employee relations and handling occupational disabilities.

working face — The portion of a sanitary landfill where waste is discharged by collection trucks and is compacted primary to placement of cover material.

working level (WL) — Any combination of short-lived radon daughters in 1 liter of air that will result in the ultimate emission of 1.3×10^5 MeV of potential alpha energy.

working level month (WLM) — The air concentration in working levels multiplied by exposure duration in multiples of the 170-hour occupational month; one working level is equal to 200 picocuries per liter.

workplace — An establishment at one geographical location that contains one or more work areas.

workplace protection factor (WPF) — A measure of protection provided in the workplace by a properly functioning respirator when correctly worn and used.

worksheet — A document on which an analyst records analytical work and findings.

workspace design — Design of a physical environment to fit the needs of workers to perform their tasks without excessive effort and within a range of healthy postures while standing, sitting, or exerting force.

workstation — (*computer science*) Desk, keyboard, digitizing tablet, and cathode-ray monitor connected together and working as a single unit; (*ergonomics*) the entire area accessed by a worker when performing a specific task or job cycle.

World Geodetic System — A consistent set of parameters describing the size and shape of the Earth; the positions of a network of points with respect to the center of mass of the Earth, transformations in major geodetic datums, and the potential of the Earth in terms of harmonic coefficients.

wound — A physical injury causing a break in the skin, usually due to an accident.

WP — See wettable powder.

WPF — See workplace protection factor.

WQS — Acronym for water quality standard.

WRDA — See Water Resources Development Act.

WRF — See water recreational facility.

writ — A court order commanding the performance of some act by the person to whom it is directed.

writ of mandamus — Literally "we order"; a court order to a public official or board to do their duty by virtue of office or position, such as compelling the issuance of a license or the payment of salaries or expenses improperly withheld.

WS — Acronym for work status.

WSF — Acronym for water-soluble fraction.

WSRF — Acronym for Wild and Scenic Rivers Act.

WSTP — Acronym for wastewater sewage treatment plants; see wastewater treatment plant.

wt — See weight.

WWTP — See wastewater treatment plant.

WWTU — Acronym for Wastewater Treatment Unit.

WWW — Acronym for World Weather Watch; see World Wide Web.

X

x-axis — The horizontal axis of any type of graph.

xenobiote — Any biotum displaced from its normal habitat; a chemical foreign to a biological system.

xenobiotic — Pertaining to a chemical compound, such as a drug, pesticide, or carcinogen, that is formed in a living organism.

xenodiagnosis — The diagnosis of a disease by infecting a test animal.

xenon — A heavy gas used in specialized electric lamps.

xeric — Referring to a very dry environment.

xeriscaping — A method of landscaping that uses plants that are well adapted to a local area and are drought resistant.

xerogram — An x-ray of soft tissue.

xerophile — An organism adapted to grow at a low water potential and in very dry habitats.

x-radiation — See x-ray.

x-ray — A penetrating, electromagnetic radiation for which the wavelength is shorter than that of visible light; usually produced by bombarding a metallic target with fast electrons in a high vacuum. Also known as x-radiation or roentgen ray.

x-ray crystallography — A technique used for studying the molecular atomic structure of a substance.

x-ray diffraction (XRD) — The scattering of x-rays by matter, causing varied intensity due to interference.

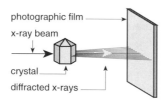

X-Ray Diffraction
in crystallography

x-ray film — A radiograph made by projecting x-rays through organs or structures of the body onto photographic film.

x-ray fluorescence analyzer (XRF) — An instrument containing a small, radioisotopic source which, when placed against a painted surface, stimulates any lead atoms that are present in the paint to emit characteristic x-rays.

x-ray fluoroscopy — Real-time imaging using an x-ray source to project an image through the patient onto a fluorescence screen.

x-ray microscope — A microscope that produces images by using x-rays and records them on film or projects them on screens.

XRD — See x-ray diffraction.

XRF — See x-ray fluorescence analyzer.

xylem — Tissue that conducts water and minerals in vascular plants.

xylene ($C_6H_4(CH_3)_2$) — A colorless liquid with an aromatic odor and which has three isomeres (*ortho-*, *meta-*, *para-*); MW: 106.2; BP: 292, 269, 281°F, respectively; Sol: insoluble; Fl.P: 63, 84, 81°F, respectively; sp. gr. 0.88, 0.86, 0.86, respectively. They are present in coal and wood, tar, naphtha, etc. They are used in the manufacture of dyes and as raw materials for polyester fibers. They are hazardous to the eyes, skin, respiratory system, and digestive system. Symptoms of exposure include dizziness; excitement; incoherence; staggering gait; irritation of the eyes, nose, and/or throat; vomiting; abdominal pain; and nausea. OSHA exposure limit (TWA): 100 ppm [air].

xylidine (($CH_3)_2C_6H_3NH_2$) — A pale-yellow to brown liquid with a weak, aromatic, amine-like odor; MW: 121.2, BP: 415 to 439°F, Sol: slight, Fl.P: 206°F, sp. gr. 0.98. It is used in the manufacture of dyes; in the manufacture of curing agents, antioxidants, and antiozonants; in polymer production; as a gasoline additive; in the manufacture of pharmaceuticals; in organic synthesis for the preparation of wood preservatives; as wetting agents in textiles. It is hazardous to the blood, lungs, liver, kidneys, and cardiovascular system and is toxic through inhalation, absorption, ingestion, and contact. Symptoms of exposure include anoxia, cyanosis, and damage to the lungs, liver, and kidneys. OSHA exposure limit (TWA): 2 ppm [skin] or 10 mg/m³.

xyloid coal — A brown coal or lignite mostly derived from wood.

Y

yard (yd) — A unit of length measurement equal to 3 feet or 0.9144 meters.

yard waste — Plant clippings, prunings, and other discarded materials from yards and gardens.

yaws — A bacterial disease caused by the spirochete *Treponema pertenue* and transmitted by direct contact; it is characterized by chronic ulcerating sores anywhere on the body with eventual tissue and bone destruction. It is often transmitted by flies.

y-axis — The axis perpendicular to and in the horizontal plane through the *x*-axis of any type of graph.

yd — See yard.

yeast — Any unicellular, usually oval, nucleated fungus that reproduces by budding.

yellow fever — (*disease*) An acute, viral, infectious disease of short duration and various severities with jaundice, sudden onset of fever, headache, backache, prostration, nausea, and vomiting. Incubation time is 3 to 6 days; it is caused by the flavivirus of yellow fever and is found in Africa, South America, and Central America (formerly found in the United States). It is transmitted by the bite of the infected *Aedes aegypti* mosquito. Reservoirs of infection include people and *A. aegypti* mosquitos. It is communicable during initial stages of the disease when a mosquito bites an infected person and then bites another person; this is usually about 9 to 12 days. Some immunity exists in infants, and lifelong immunity exists in individuals who have had yellow fever. It is controlled by immunization and/or control of *A. aegypti* mosquitos.

yellowcake — The concentrate of 70 to 90% uranium oxide extracted from crushed ore.

Yersinia pestis — A small, Gram-negative bacillus that causes plague.

yersiniosis — (*disease*) An acute, enteric disease with watery diarrhea, enterocolitis, false appendicitis, fever, headache, pharyngitis, anorexia, and vomiting. Incubation time is 3 to 7 days, usually under 10 days. It is caused by *Yersinia pseudotuberculosis* and *Y. enterocolitica*, with numerous serotypes and biotypes, and is found worldwide, especially in infants and children. It is associated with household pets, especially sick puppies and kittens. Reservoirs of infection include domestic animals, birds, and mammals. It is transmitted by the oral–fecal route from contact with infected persons or animals or by eating or drinking food or water contaminated with feces. It is communicable as long as symptoms last, which may be 2 to 3 months, or it may persist in a chronic-carrier state; general susceptibility. It is controlled by proper disposal of human, cat, and dog feces; protection of water supplies; proper handwashing; proper preparation of food; and pasteurization of milk.

yield — The quantity of water (expressed as a rate of flow or total amount per year) that can be collected for a given use from surface or groundwater sources.

yttrium compounds (Y) — A dark-gray to black metal; MW: 88.9, BP: 5301°F, Sol: not known, sp. gr. 4.47. It is used in the manufacture of red phosphorus, which is used in color picture tubes and fluorescent and mercury vapor lamps; as simulated diamonds; in the manufacture of ceramics for use in high-temperature furnaces and high-intensity incandescent lamps; in the manufacture of refractories for stable, high-temperature, high-strength compositions; in the manufacture of lasers; in the manufacture of electronic components in telephones, radar, and space communications networks. It is hazardous to the eyes and lungs and is toxic through inhalation, ingestion, and contact. Symptoms of exposure include eye irritation; in animals, pulmonary irritation, eye injury, and possible liver damage. OSHA exposure limit (TWA): 1 mg/m^3 [air].

Y

Z

Z — A symbol indicating the number of protons.

Z+ — See impedance.

Z value — The elevation value of a surface at a particular *x,y* location.

z-axis — The vertical axis in any three-dimensional coordinate system.

zeolite — Any one of a group of hydrated aluminum, calcium, or sodium silicates used to soften water; frequently made synthetically.

zero — (*mapping*) The origin of all coordinates defined in an absolute system.

zero air — Atmospheric air purified to contain less than 0.1 ppm total hydrocarbons.

zero emission vehicle — A vehicle that produces no emissions from the on-board source of power; an electric vehicle.

zero energy band — An energy conservation technique that allows temperatures to float between selected settings, thereby preventing the consumption of heating or cooling energy while the temperature is in this range.

zero fluid balance — A state in which the amount of fluid input equals the amount of fluid output.

zero gas — A gas containing less than 1 ppm sulfur dioxide.

zero population growth (ZPG) — The maintenance of a given population at a constant level, without growth.

ZEV — Acronym for zero emissions vehicle.

zinc (Zn) — A bluish-white metal with luster, for which exposure to moist air produces a white carbonate coating; atomic weight: 65.37, BP: 908°C, Sol: reacts with water, Fl.P: zinc dusts form explosive mixtures with air and in the presence of moisture may heat spontaneously and ignite, density: NA. It is used as an ingredient in alloys such as brass, bronze, and German silver, and it is used as a protective coating to prevent corrosion of other metals, for galvanizing sheet iron, in gold extraction, in dry cell batteries, and as a reducing agent in organic synthesis. It is hazardous to the respiratory tract and may cause irritation, coughing, sweating, dyspnea, fever, nausea, and vomiting. ACGIH exposure limit: NA.

zinc chloride fume (ZnCl$_2$) — A white particulate dispersed in air; MW: 136.3, BP: 1350°F, Sol: 432% at 77°F, sp. gr. 2.91 at 77°F. It is liberated from arc welding of galvanized iron and steel pipes; from fluxing iron/steel prior to galvanizing; from vulcanizing and reclaiming processes for rubber; from solutions in glass and metal etching; from the manufacture of dry-cell batteries; from petroleum refining operations. It is hazardous to the respiratory system, skin, and eyes and is toxic through inhalation and contact. Symptoms of exposure include conjunctivitis, nose and throat irritation, cough, copious sputum, dyspnea, chest pain, pulmonary edema, bronchopneumonia, pulmonary fibrosis, fever, cyanosis, tachypnea, burning skin, and eye and skin irritation. OSHA exposure limit (TWA): 1 mg/m^3 [air].

zinc oxide fume (ZnO) — A fine, white, odorless particulate dispersed in air; MW: 81.4, BP: not known, Sol: insoluble, sp. gr. 5.61. It is liberated during brazing, welding, burning, and cutting of zinc and galvanized metals; during use as ceramic flux; from the manufacture of glass, where ZnO is used to increase the brilliance and luster of glass; from use as an intermediate in the manufacture of other zinc compounds; in the manufacture of electronic devices. It is hazardous to the respiratory system and is toxic through inhalation. Symptoms of exposure include sweet metallic taste, dry throat, cough, chills, fever, tight chest, dyspnea, râles, reduced pulmonary function, headache, blurred vision, muscle cramps, low back pain, nausea, vomiting, fatigue, lassitude, and malaise. OSHA exposure limit (TWA): 5 mg/m^3 [air].

zirconium compounds (Zr) — A soft, malleable, ductile metal or a gray to gold, amorphous powder; MW: 91.2, BP: 6471°F, Sol: insoluble, sp. gr. 6.51. It is used in the manufacture of ceramics, glass, and porcelains; in the synthesis of pigments, dyes, and water repellents; in tanning operations; as abrasives and polishing materials; as an igniter in the manufacture of munitions and other items; as detonators, photoflash bulbs, and lighter flints; in the manufacture of skin ointments and antiperspirants; as a gas getter in the manufacture of high vacuum tubes and radio valves; in chemical synthesis. It is hazardous to the respiratory system and skin and is toxic through inhalation and contact. Symptoms of exposure include skin granulomas, with some evidence of retention in lungs and irritation of the skin and mucous membranes in animals. OSHA exposure limit (TWA): 5 mg/m³ [air].

ZOI — Acronym for zone of incorporation.

zone — An area with specific boundaries and characteristics.

zone of aeration — The area above the water table and below the ground surface where the interstices of rocks and soil are filled with water and air; also known as vadose zone or unsaturated zone.

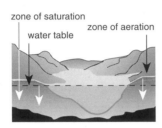

**Zone of Aeration
Zone of Saturation**

zone of capillarity — The area above a water table where some or all of the pores are filled with water that is held there by capillary action.

zone of decomposition — A zone in a stream where the dissolved oxygen is exhausted, allowing for anaerobic decomposition putrefaction.

zone of degradation — Part of the self-purification process in a stream where visible evidence of pollution, including floating solids, pieces of garbage, sticks, papers, and fecal materials, are seen.

zone of leaching — The layer of soil just beneath the top soil where water percolates, removing soluble nutrients that accumulate in the subsoil.

zone of recovery — A zone in a stream where the dissolved oxygen appears to increase and the organic solids decrease, causing the microorganisms, especially anaerobes, to die.

zone of saturation — A region beneath the surface of the soil in which all spaces between rocks and soil are filled with water; also known as saturated zone.

zoning — A means of planning to ensure that community land will be used in the best possible way for the health and general safety of the community.

zoning ordinance — An ordinance adopted by a local entity to prohibit or approve certain uses of land.

zoobiology — Study of the biology of animals.

zoochemistry — The biochemistry of animals.

zoogenous — Something that is acquired from, or originates in, animals.

zoogleal film — A complex population of organisms that form a slime growth on a trickling-filter media and break down the organic matter in wastewater.

zoogleal mass — A jelly-like mass of bacteria found in both the trickling filter and activated sludge processes.

zoom — (*computer science*) The ability to proportionately enlarge or reduce the scale of a figure or map on a visual display terminal or other optical instrument.

zoonoses — A disease adapted to and normally found in animals that can also affect humans.

zooparasite — A parasitic animal organism.

zooplankton — Microscopic aquatic animals that fish passively feed on.

zootoxin — A poisonous substance from an animal.

ZPG — See zero population growth.

ZRL — Acronym for zero risk level.

zygomycosis — Any infection caused by fungi of the class Zygomycetes.

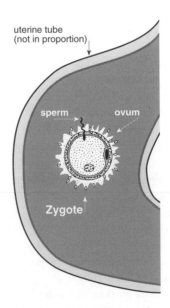

zygote — A cell produced by the joining of two gametes that are either sex or germ cells.

zymosis — See fermentation.

Resource Directory*

3-A Sanitary Standards Committees — A series of committees of the International Association of Food Industry Suppliers, made up of manufacturers and users of dairy and food equipment, state and local environmental health personnel, the U.S. Public Health Service, and the U.S. Department of Agriculture, who provide the dairy and food industries with voluntary sanitary standards, on a national level.

3-A SSC — See 3-A Sanitary Standards Committees.

AAAC — See American Association of Analytical Chemists.

AAAI — See American Academy of Allergy and Immunology.

AAAR — See American Association for Aerosol Research.

AAAS — See American Association for the Advancement of Science.

AACN — See American Association of Colleges of Nursing.

AACR — See American Association for Cancer Research.

AACT — See American Academy of Clinical Toxicology.

AAEE — See American Academy of Environmental Engineers.

AAEM — See American Academy of Environmental Medicine.

AAFP — Acronym for American Academy of Family Practice; see American Academy of Family Physicians.

AAG — See Association of American Geographers.

AAGP — Acronym for American Academy of General Practice; see American Academy of General Physicians.

AAHE — See American Association for Health Education.

AAIH — See American Academy of Industrial Hygiene.

AAIN — See American Association of Industrial Nurses.

AAM — See American Academy of Microbiology.

AAMC — See Association of American Medical Colleges.

AAMI — See Association for the Advancement of Medical Instrumentation.

AAN — See American Academy of Nursing; acronym for American Academy of Neurology.

AAOHN — See American Association of Occupational Health Nurses.

AAOM — See American Academy of Occupational Medicine.

AAP — Acronym for Asbestos Action Program.

AAPB — See American Association of Pathologists and Bacteriologists.

AAPCO — See Association of American Pesticide Control Officials.

AAPHP — See American Association of Public Health Physicians.

AAPMR — See American Academy of Physical Medicine and Rehabilitation.

AAR/BOE — Acronym for Association of American Railroads/Bureau of Explosives.

AARC — Acronym for Alliance for Acid Rain Control; see Arctic and Antarctic Research Center.

AARP — Acronym for American Association of Retired Persons.

AART — Acronym for American Association for Respiratory Therapy.

AAS — See American Academy of Sanitarians.

AASHTO — See American Association of State Highway and Transportation Officials.

ABA — Acronym for American Bar Association, American Bankers Association, or American Bakers Association; see American Bioenergy Association.

ABC — See Association of Boards of Certification.

ABES — Acronym for Alliance for Balanced Environmental Solutions.

ABET — See Accreditation Board for Engineering and Technology.

ABIH — See American Board of Industrial Hygiene.

ABPA — See American Backflow Prevention Association.

ABPM — See American Board of Preventive Medicine.

ABSA — See American Biological Safety Association.

ACA — Acronym for American Conservation Association; see American Camping Association or American Correctional Association.

ACAA — See American Coal Ash Association.

ACCCL — Acronym for American Coke and Coal Chemicals Institute.

Accreditation Board for Engineering and Technology — An organization of professional societies representing over 800,000 engineers that is responsible for the quality control of engineering education

---

* The Resource Directory identifies available environmental, occupational, and engineering resources. To obtain more in-depth information, please consult the Internet.

and accrediting college programs in engineering, engineering technology, and engineering-related areas.

ACDP — See Advisory Committee on Dangerous Pathogens.

ACE — See Agriculture in Concert with the Environment; acronym for Alliance for Clean Energy, Advisory Committee on the Environment, Aerosol Characterization Experiment, or Atmospheric Chemistry Education in Global Change.

ACEEE — See American Council for an Energy Efficient Economy.

ACEO — See Association of Corporate Environmental Officers.

ACFSA — See American Correctional Food Service Association.

ACGIH — See American Conference of Governmental Industrial Hygienists.

ACHE — See American College of Healthcare Executives.

ACHSA — See American Correctional Health Services Association.

ACOEM — See American College of Occupational and Environmental Medicine.

ACP — See American College of Physicians; acronym for Agricultural Control Program, Air Carcinogen Policy, American College of Pathologists, Agricultural Conservation Program, Atmospheric Chemistry Program, or Area Contingency Plan.

ACPA — See American Concrete Pipe Association; acronym for American Crop Protection Association.

ACQUIRE — Acronym for Aquatic Information Retrieval System.

ACR — Acronym for American College of Radiology.

ACS — See American Chemical Society; acronym for American College of Surgeons.

ACSH — See American Council on Science and Health.

ACSM — See American Congress on Surveying and Mapping.

ACT — See American College of Toxicology.

ACWA — Acronym for American Clean Water Association.

ADA — See American Dietetic Association, acronym for American Dairy Association.

ADAMHA — See Alcohol, Drug Abuse, and Mental Health Administration.

ADDNET — Acronym for Acid Deposition Data Network.

Administration on Aging — The principal agency of the Department of Health and Human Services designated to carry out provisions of the Older Americans Act of 1965.

ADSA — See American Dairy Science Association.

Advisory Committee on Dangerous Pathogens — A government committee producing information regarding the handling of dangerous materials.

AEA — Acronym for Atomic Energy Act.

AEC — See Atomic Energy Commission.

AEE — See Alliance for Environmental Education; acronym for Alliance for Environmental Education or Association of Energy Engineers.

AEHS — See Association for the Environmental Health of Soils.

AEPI — See Army Environmental Policy Institute.

AERE — See Association of Environmental and Resource Economists.

AFBF — Acronym for American Farm Bureau Federation (Farm Bureau).

AFCEE — See Air Force Center for Environmental Excellence.

AFDO — See Association of Food and Drug Officials.

AFFI — See American Frozen Food Institute.

AFGL — Acronym for Air Force Geophysics Laboratory.

AFPA — Acronym for American Forest and Paper Association.

AFRRI — See Armed Forces Radiobiology Research Institute.

AFS — Acronym for American Fisheries Society.

AFSCME — See American Federation of State, County, and Municipal Employees.

Agency for Toxic Substances and Disease Registry (ATSDR) — An agency within the Public Health Service whose mission is to carry out health-related responsibilities of the Comprehensive Environmental Response Compensation and Liability Act of 1980 and its amendments by collecting, maintaining, analyzing, and disseminating information relating to serious diseases, mortality, and human exposure to toxic or hazardous substances.

AGI — See American Geological Institute.

AGIS — Acronym for Association for Geoscientists for International Development.

Agricultural Quarantine Inspection — A program administered by the USDA's Animal and Plant Health Inspection Service consisting of the inspection of incoming passengers, luggage, and cargo at U.S. ports of entry to protect U.S. agriculture from foreign animal and plant pests and diseases.

Agricultural Research Service — A USDA agency using federal scientists to conduct basic, applied, and developmental research in livestock, plants, soil, water and air quality, energy, food safety and quality, nutrition, food processing, storage and distribution efficiency, nonfood agricultural products, and international development.

Agricultural Retailers Association — A national trade association, providing a national voice for the retail sector of the agricultural industry and focusing on issues and programs that affect the professionalism of the group.

Agriculture in Concert with the Environment — A U.S. EPA program administered cooperatively with the

USDA's Sustainable Agriculture Research and Education program to fund research projects to reduce the risk of pollution from pesticides and soluble fertilizers.

AGU — See American Geophysical Union.

AGWT — See American Ground Water Trust.

AHA — See American Hospital Association or American Hydrogen Association.

AHERA — See Asbestos Hazard Emergency Response Act.

AHI — Acronym for Animal Health Institute.

AHRI — See American Heritage Rivers Initiative.

AIAA — Acronym for American Institute of Aeronautics and Astronautics.

AIB — See American Institute of Baking.

AIBC — See American Institute of Biomedical Climatology.

AIBS — See American Institute of Biological Sciences.

AIC — See American Institute of Chemists.

AIChE — See American Institute of Chemical Engineers.

AID — Acronym for Agency for International Development.

AIDIS-USA — An organization of sanitary engineers in government, private business, and educational institutions throughout the Western Hemisphere which promotes study and solution of sanitary engineering and environmental problems as a necessary condition for economic and social development in the Americas.

AIHA — See American Industrial Hygiene Association.

AIHC — See American Industrial Health Council.

AIME — See American Institute of Mining, Metallurgical, and Petroleum Engineers.

Air and Waste Management Association (AWMA) — An association of industrialists, researchers, equipment manufacturers, governmental personnel, educators, and meteorologists sponsoring continuing education and a library to seek economical answers to air pollution and waste management problems. Formerly known as Air Pollution Control Association (APCA).

Air Force Center for Environmental Excellence — An operating agency of the Air Force Civil Engineer, which provides leaders with a comprehensive expertise needed to protect, preserve, restore, develop and sustain the nation's environmental and installation resources.

Air Pollution Control Association (APCA) — See Air and Waste Management Association (AWMA).

Air Pollution Training Institute — A U.S. EPA division that provides training on an annual basis to approximately 20,000 air pollution specialists in various parts of the country, consistent with the 1990 Amendments to the Clean Air Act and other pertinent rules and regulations, using appropriate scientific and technological material.

Air Resources Laboratory — A program of the National Oceanic and Atmospheric Administration that conducts research on processes that relate to air quality and climate, concentrating on the transport, dispersion, transformation, and removal of trace gases and aerosols; their climatic and ecological influences; and exchanges between the atmosphere and biological and non-biological surfaces.

AIRMON — Acronym for Atmospheric Integrated Research Monitoring Network.

AIS — Acronym for International Association of the Soap and Detergent Industry.

ALA — See American Lung Association; acronym for American Library Association.

ALAPCO — Acronym for Association of Local Air Pollution Control Officers.

Alcohol, Drug Abuse, and Mental Health Administration — An agency of the U.S. Department of Health and Human Services that conducts and supports research on the biological, psychological, epidemiological, and behavioral aspects of alcoholism, drug abuse, and mental health and illness.

Alliance for Environmental Education — A group of organizations, corporations, and government agencies that promote the science and art of environmental education, formally and informally.

Alliance for Environmental Technology — An organization of U.S. and Canadian chemical manufacturers and forest products companies trying to improve the environmental performance of the pulp and paper industry, especially in wastewater disposal.

Alliance for Responsible Atmospheric Policy — An organization of users and producers of chlorofluorocarbons and their alternatives seeking to ensure scientifically sound, economically and socially effective government policies pertaining to ozone protection.

Alliance of Foam Packaging Recyclers — An organization of companies that recycle foam packaging material and coordinate a national network of collection centers for postconsumer foam packaging products.

Alliance to Save Energy — A coalition of government, business, consumer, and labor leaders concerned with increasing efficiency of energy use; they conduct research, demonstration projects, and public education programs.

AMA — See American Medical Association; acronym for Adhesives Manufacturers Association.

AMC — See Army Material Command.

American Academy of Neurology — A worldwide professional association of medical specialists who are committed to improving the care of patients with neurological diseases. It provides practice guidelines, continuing education, research activities, and professional opportunities to share experiences and technical knowledge.

American Academy of Allergy and Immunology — A national organization of physicians specializing in the diagnosis and treatment of allergies and immune system disorders.

American Academy of Clinical Toxicology — An organization of physicians, veterinarians, pharmacists, nurses, research scientists, and chemists who facilitate the exchange of information on therapeutic methods and techniques and conduct professional training regarding poisons for emergency service personnel.

American Academy of Environmental Engineers — An organization of state-licensed environmental engineers who have passed examinations in environmental engineering specialties, including general environment, air pollution control, solid waste management, hazardous waste management, industrial hygiene, radiation protection, water supply, and waste water.

American Academy of Environmental Medicine — An organization of physicians and other people interested in the clinical aspects of environmental medicine who promote a better understanding of illness due to environmental factors.

American Academy of Family Physicians — A national medical organization representing more than 93,000 family doctors with family practices, whose mission is to preserve and promote the science and art of family medicine and to ensure high-quality, cost-effective health care for patients of all ages. It provides educational opportunities for family practitioners to integrate the biological, clinical, and behavioral sciences into the practice of family medicine.

American Academy of General Physicians — An association whose mission is to further the interests of general medicine and general practitioners and to contribute to the field of medicine in the areas of professional development, biomedical ethics, medical scientific research, medical education, and the betterment of patient care.

American Academy of Industrial Hygiene — A professional society of industrial hygienists.

American Academy of Microbiology — A select group of leaders component honored by the American Society for Microbiology that encourages the exchange of information among members.

American Academy of Nursing — An honorary organization of the American Nurses Association recog-

nizing superior achievement in nursing to further nursing practice and education.

American Academy of Occupational Medicine — A professional association of physicians working in occupational medicine full time to promote maintenance and improvement of the health of industrial workers.

American Academy of Physical Medicine and Rehabilitation — A national association of professional healthcare workers who diagnosis physical problems and develop therapies to improve physical function.

American Academy of Sanitarians — An organization of registered environmental health specialists who possess a minimum of a master's degree in public health, environmental health, or environmental management plus have 10 years of progressive practical experience, including supervision and management, and who have completed successfully examination by a jury of peers to achieve the status of diplomate.

American Association for Aerosol Research — An organization of scientists and engineers who promote and communicate technical advances in aerosol research.

American Association for Cancer Research — An organization of research scientists dedicated to the study of cancer and its etiology, prevention, and treatment throughout the world.

American Association for Clinical Chemistry — An international society of chemists, physicians, and other scientists specializing in clinical chemistry.

American Association for Health Education — An organization of professional health educators who have responsibility for health education in schools, colleges, communities, hospitals, clinics, and industries and who advance health education through program activities, federal legislation, encouragement of relationships between health educators and health service organizations, and use of computer services, as well as conferences.

American Association for the Advancement of Science — The world's largest general scientific society, which seeks to advance science and innovation throughout the world; foster communication among nations, scientists, and the public; enhance international cooperation in science and its applications; and foster education in science and technology for everyone.

American Association of Analytical Chemists — See AOAC International.

American Association of Colleges of Nursing — An organization that represents college nursing programs and trains nurses to promote education, research, data collection, publications, and other programs to establish standards for college nursing programs, to improve health care, and to promote public support for higher education and research.

American Association of Engineering Societies — A federation of engineering societies, whose members work in industry, construction, government, academia, and private practice and who advance the knowledge, understanding, and practice of engineering.

American Association of Industrial Nurses — A national professional association of nurses working in industry and concerned with occupational health.

American Association of Medical Colleges — A national organization of faculty members and deans of medical schools and colleges who deal with issues in medical education.

American Association of Occupational Health Nurses — A professional association of registered nurses that advances the profession of occupational and environmental health by creating the pursuit of excellence in areas of health, safety, productivity, and disability management for worker populations.

American Association of Pathologists and Bacteriologists — A national professional organization of specialists in pathology and bacteriology.

American Association of Poison Control Centers — An association that keeps up-to-date information on the ingredients and potential acute toxicity of substances that may cause accidental poisoning and the proper management of a poisoned person.

American Association of Public Health Physicians — An organization of physicians engaged in public health practices who promote and support leadership in the field of public health, appropriate health legislation at various levels, and a variety of public health efforts.

American Association of State Highway and Transportation Officials — An association of public officials in state agencies responsible for highways and transportation.

American Backflow Prevention Association — An organization whose members have a common interest in protecting drinking water from contamination through cross-connections and who are involved in education and technical assistance in all areas of backflow prevention and water consumption.

American Bioenergy Association — An association whose mission is to represent ethanol fuel producers, biomass power producers, chemical companies, utilities, farmers, equipment manufacturers, environmental groups, and the forest products industry to promote the economic and environmental benefits of biomass utilization.

American Biological Safety Association — An association that promotes biosafety as a scientific discipline and services the growing needs of biosafety professionals and practitioners throughout the world.

American Board of Industrial Hygiene (ABIH) — A specialty board authorized to certify properly qualified industrial hygienists.

American Board of Preventive Medicine (ABPM) — A medical specialty board authorized to certify properly qualified specialists in preventive medicine.

American Boiler Manufacturers Association — An organization of manufacturers of boiler systems, boiler-related products, and fuel-burning systems who are concerned with environmental issues.

American Camping Association — An association of caring professionals, who share their knowledge and experience to ensure the quality of camp programs for children and adults and provide safe camp communities, competent adult role models, service to the community, opportunities for leadership and personal growth, experiential education, and excellent continuous self-improvement.

American Chemical Council — An organization of manufacturers of chemicals who conduct and administer research in areas of pollution prevention and other special problems.

American Chemistry Council — An organization of manufacturers of basic industrial chemicals who provide members with technical research, communications services, legal aid, and counseling in environmental safety and health, transportation, and energy.

American Chemical Society — A scientific, educational, and professional society of chemists and chemical engineers who evaluate college chemistry departments and high-school chemistry curricula and give grants for basic research.

American Coal Ash Association — An organization of electric utility companies; coal and transportation companies; engineering, research, and development organizations; and ash marketing companies who provide technical assistance and information to coal ash producers, transporters, and other groups to stimulate research in the use of ash and its proper means of transport and disposal.

American Coke and Coal Chemicals Institute — An organization of producers of coke, metallurgical coal, and chemicals interested in environmental, safety, and health issues.

American College of Healthcare Executives — An organization of healthcare executives who conduct credentialing and educational programs, research and development programs, and public policy programs and who publish various books, journals, and textbooks.

American College of Occupational and Environmental Medicine — An organization of physicians specializing in occupational and environmental medicine who promote maintenance and improvement of the health of workers, awareness of occupational medicine as a medical specialty, educational programs, computer services, and publications.

American College of Physicians — A national professional organization of physicians.

American College of Toxicology — An organization of individuals interested in toxicology and related disciplines such as analytical chemistry, biochemistry, pathology, teratology, and immunology who disseminate information and provide a forum for discussion of approaches to the resolution of problems in order to advance the science of toxicology and protect society.

American Concrete Pipe Association — A nonprofit organization composed of manufacturers of concrete piping, providers of products and services, and other conveyance product manufacturers located in the United States, Canada, and other foreign countries; it provides research, technical, and marketing support to promote and advance the use of concrete piping for drainage and to control pollution.

American Conference of Governmental Industrial Hygienists — A professional society of government and university employees engaged in full-time programs of industrial hygiene.

American Congress on Surveying and Mapping — A worldwide association of surveyors, cartographers, geodesists, and other spatial data information professionals working in the public and private sectors who establish standards; prepare papers, reports, and other materials; hold conferences; promote education; conduct research; and use information to develop a better understanding of spatial relationships of both actual objects and planned concepts.

American Correctional Association — The oldest international correctional association in the world involving all disciplines and providing professional development and certification to meet standards and accreditation as well as provide research and publications in the correctional field.

American Correctional Food Service Association — An organization of employees of food services of federal, state, county, and city correctional detention institutions who work to professionalize and improve correctional food service through education, information, and networking.

American Correctional Health Services Association — An organization of healthcare providers, individuals, or organizations interested in improving the quality of correctional health services and who promote the provision of health services to incarcerated persons consistent in quality and quantity with acceptable healthcare practices, encourage continuing education, and provide technical professional guidance for correctional healthcare personnel.

American Council for an Energy Efficient Economy — An independent research organization concerned with energy policy, technologies, and conservation.

American Council on Science and Health — An independent association that provides consumer education while promoting scientifically balanced evaluations of food, chemicals, the environment, and their relationship to human health.

American Dairy Science Association — A professional society of dairy educators, scientists, researchers, extension workers, dairy equipment manufacturers and suppliers, commercial dairy plants, breeding associations, farmers, and others who promote the dairy industry by stimulating scientific research, improving educational methods, and encouraging worthy industry cooperative endeavors.

American Dietetic Association — An organization of food and nutrition professionals who promote optimal nutrition and well-being for people by controlling obesity, conducting research in dietary supplements, conducting genetic research, evaluating biotechnology, and improving the area of environmental concerns about the safety of food.

American Federation of State, County, and Municipal Employees — A labor union of federal, state, county, and municipal employees.

American Frozen Food Institute — A national trade association representing all aspects of the frozen food industry supply chain, from manufacturers to distributors, suppliers, and packagers, who are concerned with legislative affairs, product distribution, regulatory compliance, research and technology, and environmental matters.

American Geological Institute — An organization of earth science societies and associations that maintains a computerized database containing worldwide information on geology, engineering, environmental geology, oceanography, and other geological fields.

American Geophysical Union — A scientific society that provides information to the public on what is known about the Earth and other planets, as well as space, and are involved in the atmospheric and ocean sciences, solid earth sciences, hydrologic sciences, and space sciences.

American Ground Water Trust — An organization that disseminates public education information about groundwater to industry, government, educational institutions, and the public.

American Heritage Rivers Initiative — A program that identifies historic American rivers that are in need of restoration and conservation.

American Hospital Association — An association of hospitals, specialists, and individuals dedicated to the promotion of patient care through education, lobbying, publications, and provision of consultative services.

American Hydrogen Association — An organization of individuals interested in renewable natural resources and advocating a transition from fossil and nuclear energy sources to solar or hydrogen technologies in order to help resolve environmental problems such as global warming, acid rain, ozone depletion, and urban air pollution.

American Industrial Health Council — An organization of people who advocate the importance of sound science in regulatory decision making concerning chronic human health hazards, and promote the sound use of scientific principles and procedures in the assessment and regulation of risks of chronic human health effects and ecological health effects.

American Industrial Hygiene Association — An association promoting the study and control of environmental stresses arising in or from the workplace or its products in relation to the health or well-being of workers and the public.

American Institute of Baking — A baking research and educational center that conducts basic and applied research and hands-on training regarding in-plant sanitation and worker safety.

American Institute of Biological Sciences — An organization of professional members and biological associations, laboratories, and museums having an interest in the life sciences to promote the effectiveness of individuals involved in biological research, education, environment, and medicine.

American Institute of Biomedical Climatology — An organization of meteorologists, biologists, epidemiologists, physicians, atmospheric physicists, engineers, architects, physiologists, climatologists, and other professionals interested in investigating the influence of the outdoor and indoor environments on the creation of disease and the health of people.

American Institute of Chemical Engineers (AIChE) — A member of the Accreditation Board for Engineering and Technology, the American National Standards Institute, and related organizations. It produces numerous publications and conducts annual educational meetings.

American Institute of Chemists — An organization of chemists and chemical engineers who promote advancement of chemical professions in the United States and establish and enforce high practice standards for professionals.

American Institute of Inspectors — An organization of certified home inspectors who work to set standards for impartial evaluations of residential properties.

American Institute of Mining, Metallurgical, and Petroleum Engineers — An organization of professional engineers who advance, record, and disseminate significant knowledge of engineering, and the arts and sciences involved in the production and use of minerals, metals, energy sources, and materials for the benefit of people.

American Journal of Nursing — The professional journal of the American Nurses Association.

American Lung Association — The oldest voluntary health organization in the United States, it was founded in 1904 to fight tuberculosis. Today the organization is involved in asthma, tobacco control, and environmental health; formally known as the Tuberculosis Society.

American Meat Institute — An organization founded in 1906 to assist member companies in adjusting to federal inspection. It is currently involved in scientific research, industry education, and a variety of environmental issues in order to promote food safety.

American Meat Institute Research — An organization of meat and poultry packers, processors, sausage manufacturers, meat and poultry suppliers, canners, and others related to the industry, who work in the areas of research, foodservice, improving operating methods and products, worker safety and environment, inspection, and education.

American Medical Association — A professional association of licensed physicians in the United States that includes practitioners in all recognized medical specialties.

American Meteorological Society — An organization that promotes the development and dissemination of information and education on the atmospheric and related oceanic and hydrologic sciences and is composed of professionals, professors, students, and people interested in the weather.

American National Standards Institute — A privately funded, voluntary membership organization that identifies industrial and public needs for National Consensus Standards and coordinates development of such standards.

RESOURCE DIRECTORY

American Nuclear Society — A non-profit international, scientific, and educational organization composed of engineers, scientists, administrators, educators, corporations, educational institutions, and governmental agencies who develop and safely apply nuclear science and technology for the public's benefit and provide an exchange of knowledge and professional development to enhance public understanding of nuclear energy.

American Nurses Association — A national professional association of registered nurses in the United States founded in 1896 to improve the standards of health and the availability of health care in order to improve standards for nursing, promote professional development, and advance the economic and general welfare of nurses.

American Oceans Campaign — An organization committed to the use of scientific information to develop sound public policy on protecting the environment, health, food, and coastal recreation.

American Petroleum Institute — An organization of corporations in petroleum and allied industries, including producers, refiners, and transporters of crude oil, lubricating oil, gasoline, and natural gas, who provide public policy development, advocacy, research, and technical services to enhance the ability of the petroleum industry to fill its mission of meeting the nation's energy needs while enhancing the environmental and health and safety performance of the industry.

American Physical Therapy Association — A professional association whose goal is to foster advancements in physical therapy practice, research, and education.

American Plastics Council — An organization affiliated with the Society of the Plastics Industry that tries to increase plastics recycling, conducts research on disposal of plastic products, and supports alternative waste management technologies.

American Public Health Association — An association representing health professionals in over 40 disciplines in the development of health standards and policies.

American Public Works Association — An international educational and research association of public agencies, private sector companies, and individuals dedicated to providing high-quality public works goods and services, and that provides accreditation services, special task forces, and technical committees relating to all aspects of water and the environment.

American Red Cross (ARC) — A voluntary, humanitarian organization that provides relief to victims of disasters and helps people prevent, prepare for, and respond to emergencies.

American Registry of Radiologic Technologists — A credentialing organization that seeks to ensure high-quality patient care in radiologic technology through testing and certifying technologists and conducting continuing education, annual registration, and special seminars.

American School Food Service Association — An organization of professionals in the school food service area, the purpose of which is to advance the availability, quality, and acceptance of food nutrition programs and to promote food safety. It provides education and training, establishes standards through certification and credentialing, and gathers and transmits regulatory, legislative, industry, and nutritional information to schools with the goal of obtaining the best nutritional status for all children.

American School Health Association — An organization of school physicians, school nurses, dentists, nurses, nutritionists, health educators, dental hygienists, school-based professionals, and public health workers who promote coordinated school health programs that include health education, health services, a healthful school environment, physical education, nutrition services, and psychosocial health services for students, families, and other members of the community.

American Slow Sand Association — An organization of water districts, corporations, public authorities, and political subdivisions with a primary interest in providing and distributing potable water or treatment of wastewater using slow sand filtration systems.

American Society of Agricultural Engineers — A professional and technical organization dedicated to the advancement of engineering applicable to agriculture, food, and biological systems and whose members serve in industry, academia, and public service. The organization is concerned with the efficient and environmentally safe methods of cultivating food and fiber for an increasing population. It has areas of biological engineering, food and process engineering, information and electrical technologies, power and machinery, soil and water, and structures and environment.

American Society for Cell Biology — An organization of scientists with educational or research experience in cell biology or allied field who provide computer services and educational programs.

American Society for Engineering Education — A professional educational society of college and university engineering deans, administrators, and professors; practicing engineers; corporate executives; persons interested in engineering education; engineering colleges; technical colleges and institutes; junior colleges; governmental agencies; and industrial organizations who seek to advance education and research in engineering, science, and related fields.

American Society for Environmental History — An organization of teachers and researchers in environmental studies, primarily in higher education, government, and business, who have a forum for sharing information and ideas.

American Society for Healthcare Engineering of the American Hospital Association — An organization of the American Hospital Association composed of hospital engineers, facility managers, directors of buildings and grounds, assistant administrators, directors of maintenance, directors of construction, and safety officials who promote better patient care by encouraging and assisting members to develop their knowledge in the field of facilities management and safety practices.

American Society for Healthcare Environmental Services of the American Hospital Association — An organization of the American Hospital Association that includes managers and directors of institutional environmental services, laundry departments, and housekeeping departments who provide educational opportunities, career advancement, and forums for discussion among members.

American Society for Healthcare Food Service Administrators — An organization of healthcare food service managers who promote excellence in healthcare food service, educational programs, and research and development programs.

American Society for Quality — An international organization made up of individuals and corporations whose mission is to initiate quality methods used throughout the world, including statistical process control, quality cost measurement control, Total Quality Management, failure analysis, and zero defects. It publishes books and standards in a variety of quality areas and provides in-depth educational programs taught by highly qualified instructors from business, industry, and academia. It awards certifications to professionals through the various programs.

American Society for Testing and Materials — A nonprofit organization that develops standard testing methods by the consensus of volunteers from manufacturers, users, and others.

American Society of Anesthesiologists — An organization of physicians who are involved in education, research, and scientific investigation to raise and maintain the standards of medical practice of anesthesiologists and improve the care of patients.

American Society of Civil Engineers — An organization of professionals and students in civil engineering who develop and produce consensus standards for construction documents and building codes.

American Society of Clinical Pathologists — A professional society of clinical pathologists, clinical scientists, chemists, microbiologists, medical technologists, and medical technicians whose purpose is to promote a wider application of pathology and laboratory medicine to the diagnosis and treatment of disease, to conduct education programs and publish educational materials in the field of clinical and anatomic pathology, and to conduct an examination and certification program of medical laboratory personnel.

American Society of Heating, Refrigerating, and Air Conditioning Engineers — A society of professionals dedicated to the establishment of standards in refrigeration, heating, and air conditioning, and to the promotion of education through research and publications.

American Society of Home Inspectors — An organization of professional home inspectors whose goals are to establish home inspector qualifications, set standards of practice for home inspections, adhere to a code of ethics, and inform members of the most advanced methods and techniques.

American Society of Limnology and Oceanography — An organization of persons involved in scientific study of freshwater and oceans, including their physical, chemical, and biological properties.

American Society of Mechanical Engineers — A technical society with extensive programs in the development of safety codes, equipment standards, and educational guidance for students.

American Society of Plumbing Engineers — An international organization for professionals skilled in the design, specification, and inspection of plumbing systems that is dedicated to the advancement of the science of plumbing engineering, to the professional growth and advancement of its members, and the health, welfare, and safety of the public. It provides technical data and information, sponsors research and education, and provides numerous training sessions.

American Society of Safety Engineers — A society of safety-related professionals dedicated to improving the occupational area through the promotion of standards, education, and professionalism.

American Society of Sanitary Engineering — An organization of sanitary engineers, plumbing officials, building officials, architects, engineers, design engineers, physicians, and others interested in health and the environment who conduct research on plumbing and sanitation and develop performance standards for components of the plumbing system as well as conduct studies of waterborne epidemics.

American Water Resources Association — An organization of professionals, students, academia, corporations, engineers, and librarians whose mission is to advance multidisciplinary water resources management and research, information exchange, professional development, and education about water resources and related issues. Its major efforts are concerned with improving water resources through educational programs, professional meetings, and publications.

American Water Works Association — An organization of water utility managers, superintendents, engineers, chemists, bacteriologists, public health departments, and other individuals interested in public water supply who develop standards; support research programs in waterworks design, construction, operation, and management; and conduct in-service training schools.

American Wind Energy Association — An association of individuals, companies, and other organizations involved with wind energy that provides statistics and information on development of domestic and international markets, wind energy conferences on the latest industry trends, technological developments, and renewable energy policy developments. Its Web site features research and educational materials on wind energy.

American Wood Preservers Institute — A national industry trade association representing the pressure-treated wood industry throughout the United States and working to conserve forest resources, preserve the environment, and extend the life of wood products through the manufacture of pressure-treated wood.

Ames Research Center — A research center located at Moffett Field, CA, that is active in aeronautical research, life sciences, space science, and technology research; it houses the world's largest wind tunnel and the world's most powerful supercomputer system.

AMI — See American Meat Institute.

AMPS — Acronym for Association of Metropolitan Sewer Agencies.

AMS — See American Meteorological Society; acronym for American Mathematical Society.

AMSA — See Association of Metropolitan Sewerage Agencies.

AMWA — See Association of Metropolitan Water Agencies.

ANA — See American Nurses Association.

Animal and Plant Health Inspection Service — A USDA agency established to conduct inspections and regulatory and control programs to protect animal and plant health through inspections, pre-

vention activities, disease control, quarantine, and eradication programs.

ANL — Acronym for Argonne National Laboratory.

ANRHRD — Acronym for Air, Noise, and Radiation Health Research Division/ORD.

ANS — See American Nuclear Society.

ANSI — See American National Standards Institute.

ANSIR — Acronym for Awareness of National Security Issues and Response.

AOA — See Administration on Aging.

AOAC — See Association of Official Analytical Chemists.

AOAC International — An international association of analytical science professionals, companies, government agencies, nongovernmental organizations, and institutions who promote methods of validation and quality measurements in the analytical sciences.

AOEC — See Association of Occupational and Environmental Clinics.

AOTA — Acronym for American Occupational Therapy Association.

APA — Acronym for American Plastics Association.

APCA — See Air Pollution Control Association.

APCO — See Association of Public Safety Communications Officials.

APHA — See American Public Health Association.

APHIS — See Animal and Plant Health Inspection Service.

API — See American Petroleum Institute.

APIC — See Association for Practitioners of Infection Control and Epidemiology.

APICE — See Association for Professionals in Infection Control and Epidemiology.

APMA — Acronym for American Podiatric Medical Society.

Applied Science and Technology Index — A reference that indexes over 300 journals by subject from 1958 to the present.

APTA — See American Physical Therapy Association.

APTI — See Air Pollution Training Institute.

APWA — See American Public Works Association.

AQI — See Agricultural Quarantine Inspection.

ARA — See Agricultural Retailers Association.

ARAC — Acronym for Acid Rain Advisory Committee.

ARB — Acronym for Air Resources Board.

ARC — See American Red Cross or Ames Research Center.

ARCC — Acronym for American Rivers Conservation Council.

Arctic and Antarctic Research Center — A research center that maintains an archive of more than 175,000 satellite passes of the polar regions of the Earth, provides data processing services to a wide variety of polar researchers, and supplies data to the National Snow and Ice Data Center in collaboration with the National Science Foundation's Office of Polar Programs.

ARIP — Acronym for Accidental Release Information Program.

ARL — See Air Resources Laboratory.

ARM — See Atmospheric Radiation Measurements Program.

ARMA — Acronym for the American Roofing Manufacturers Association.

Armed Forces Radiobiology Research Institute — A program of the Department of Defense that serves as the principal ionizing radiation radiobiology research laboratory under the jurisdiction of the Uniformed Services University of the Health Sciences.

Army Corps of Engineers — The branch of the U.S. Army responsible for maintaining and regulating inland waterways.

Army Environmental Policy Institute — A program area within the U.S. Army involved in environmental issues and which provides support on policy matters to the Secretary of the Army. Its mission is to provide recommendations for policy options, to conduct policy analysis and studies on the full range of environmental issues facing the Army, and to develop solutions to emerging and potential environmental problems. It gathers its data through questionnaires, surveys, and review.

ARRA — See Asphalt Recycling and Reclaiming Association.

ARRP — Acronym for Acid Rain Research Program.

ARRT — See American Registry of Radiologic Technologists.

ARS — See Agricultural Research Service.

ASA — See American Society of Anesthesiologists; acronym for American Statistical Association.

ASAE — See American Society of Agricultural Engineers.

ASAHP — See Association of Schools of Allied Health Professionals; acronym for American Society of Allied Health Professionals.

Asbestos Analysts Registry — A registry of individual asbestos analysts who have met asbestos air sample analysis criteria set by the American Industrial Hygiene Association.

Asbestos Information Association/North America — An organization of companies that manufacture, sell, and use products containing asbestos fiber and those that mine, mill, and sell asbestos; it provides information on health and safety in the industry.

ASC — Acronym for the Adhesives and Sealant Council, Inc.

ASCB — See American Society for Cell Biology.

ASCE — See American Society of Civil Engineers.

ASCP — See American Society of Clinical Pathologists.

ASDWA — Acronym for Association of State Drinking Water Administrators.

ASEE — See American Society for Engineering Education.

ASEH — See American Society for Environmental History.

ASFE — An organization of consulting geotechnical and geoenvironmental engineering firms that conducts seminars and peer review programs on quality control policies and procedures in geotechnical engineering; formerly the Association of Soil and Foundation Engineers.

ASFPM — See Association of State Floodplain Managers.

ASFSA — See American School Food Service Association.

ASHE — See American Society for Healthcare Engineering of the American Hospital Association.

ASHES — See American Society for Healthcare Environmental Services of the American Hospital Association.

ASHFSA — See American Society for Healthcare Food Service Administrators.

ASHI — See American Society of Home Inspectors.

ASHRAE — See American Society of Heating, Refrigerating, and Air Conditioning Engineers.

ASIWCPA — See Association of State and Interstate Water Pollution Control Administrators.

ASLO — See American Society of Limnology and Oceanography.

ASMD — See Atmospheric Sciences Modeling Division.

ASME — See American Society of Mechanical Engineers.

ASPE — See American Society of Plumbing Engineers.

Asphalt Recycling and Reclaiming Association — An organization of contractors and engineers engaged in the reworking of asphalt as part of a recycling process.

ASPPT — Acronym for Association for Suppliers of Printing and Publishing Technologies.

ASPRS — Acronym for American Society for Photogrammetry and Remote Sensing.

ASQ — Acronym for American Society for Quality.

ASSE — See American Society of Safety Engineers or American Society of Sanitary Engineering.

Associated Laboratories — An organization of regional water treatment companies that conducts studies on water treatment efficiency and tests new equipment.

Association for Practitioners of Infection Control — A national professional organization of nurses who work in the field of infection control.

Association for Professionals in Infection Control and Epidemiology — A nonprofit international organization of physicians, microbiologists, nurses, epidemiologists, medical technicians, environmental health specialists, and pharmacists who enhance patient care by improving the profession of infection control through the development of educational programs, quality research, standardization of practices and procedures, and implementation of standards.

Association for the Advancement of Medical Instrumentation — An organization of clinical engineers, biomedical equipment specialists, physicians, hospital administrators, consultants, engineers, manufacturers of medical devices, nurse researchers, and others interested in medical instrumentation who work to improve the quality of medical care through the application of medical technology and management of knowledge.

Association for the Care of Children's Health — An international, interdisciplinary organization concerned with the psychosocial needs of children and their families in healthcare settings.

Association for the Environmental Health of Soils — An organization of individuals interested in the contamination, analysis, assessment, remediation, and regulation of soils.

Association of American Geographers — A scientific and educational society of professionals who share interests in the theory, methods, and practice of geography and conduct research, annual conferences, and specialized educational programs.

Association of American Medical Colleges — A non-profit organization that works to reform and upgrade medical education. It represents accredited medical schools, teaching hospitals, academic and scientific societies, medical students and residents, administrative leadership, and professional development groups. Its purpose is to improve the nation's health through the advancement of medical schools and teaching hospitals by means of better research, administration, facilities, and educational programs.

Association of American Pesticide Control Officials — An organization of state agencies controlling the sale, use, and distribution of pesticides that promotes uniform laws, regulations, and policies of enforcement.

Association of Boards of Certification — An international organization representing over 150 boards that certify the operators of waterworks and wastewater facilities.

Association of Corporate Environmental Officers — An organization of corporate environmental officers and managers in business, industry, and government who conduct research and develop solutions to environmental problems facing the business community and an understanding of environmental laws.

Association of Energy Engineers — An organization of engineers, architects, manufacturers, industries, and other professionals with an interest in energy management and cogeneration who promote the advancement of the profession and contribute to the professional development of its members.

Association of Environmental and Resource Economists — An organization of professionals, economists, individuals, universities and governmental services interested in resource and environmental issues, as well as air, water, and land pollution.

Association of Food and Drug Officials — An organization of officials who enforce federal, state, district, county, and municipal laws and regulations relating to food, drugs, cosmetics, consumer product safety, and similar areas who promote uniform laws and administrative procedures concerning the enforcement of food and drug programs.

Association of Local Air Pollution Control Officials — An organization of local representatives of air pollution control programs nationwide that are responsible for implementing the various programs related to the Clean Air Act.

Association of Metropolitan Sewerage Agencies — An association that represents over 300 public agencies and organizations and emphasizes scientifically based, technically sound, and cost-effective laws and regulations, as well as implementation of environmental programs.

Association of Metropolitan Water Agencies — An organization of municipal public water supply agencies representing the members' interests before the U.S. EPA and other federal bodies.

Association of Occupational and Environmental Clinics — A non-profit organization committed to improving the practice of occupational and environmental health, through information sharing and collaborative research; identifying, reporting, and preventing occupational and environmental health hazards and their effects; and providing high-quality clinical services.

Association of Official Analytical Chemists — An independent international association devoted to the development, testing, validation, and publication of methods of analysis for foods, drugs, feeds, fertilizers, pesticides, water, forensic materials, and other substances.

Association of Public Safety Communications Officials — An organization of employees of municipal, county, state, and federal public safety agencies who foster the development of programs on the art of public safety communications and ensure greater cooperation in the work and activities of local agencies, counties, states, and federal agencies.

Association of Schools of Allied Health Professions — An organization of 2- and 4-year colleges, academic health science centers with allied health professional training programs, administrators, educators, practitioners, and professional societies that works with the Department of Health and Human Services to conduct surveys of allied health education programs; it is concerned with the areas of health information and disease prevention, ethics in health care, disabled people, and allied health programs.

Association of Schools of Public Health — An organization of accredited graduate schools of public health that promotes improved education and training of professional public health personnel.

Association of State and Interstate Water Pollution Control Administrators — An organization of administrators of state water pollution agencies and related associations representing their states' concerns on implementation, funding, and reauthorization of the Clean Water Act.

Association of State and Territorial Health Officials — An organization of executive officers of state and territorial health departments serving as a legislative review agency and providing information for members.

Association of State and Territorial Solid Waste Management Officials — An organization of individuals who are state and territorial solid waste management officials and who work with the U.S. EPA to develop policy on solid and hazardous waste.

Association of State Floodplain Managers — An organization of corporations, agencies, and individuals interested in and responsible for floodplain management, including coastal policies, floodplain regulations, mapping and engineering standards, stormwater management, and local environmental protection for floodplain areas.

Association of Teachers of Preventive Medicine — A professional organization of medical educators, practitioners, administrators, students, and health-care agencies who work to advance education in preventive medicine, public health, international health, clinical prevention, and aerospace and occupational medicine.

Association of Technical and Supervisory Professionals — An organization of technical and supervisory personnel who are involved in USDA poultry and meat inspection programs and who monitor legislation affecting the meat processing industry and inform members of pending regulatory agency policy revisions.

Association of Water Technologies — An organization that provides regional water treatment companies with technical education, industry communication, access to information, and management skills.

ASTHO — See Association of State and Territorial Health Officers.

ASTM — See American Society for Testing and Materials.

ASTSWMO — See Association of State and Territorial Solid Waste Management Officials.

ASUPA — Acronym for Aluminum Sulfate Producers Association.

ATDD — See Atmospheric Turbulence and Diffusion Division.

Atmospheric Radiation Measurements Program — A U.S. Department of Energy program for the continual, ground-based measurement of atmospheric and meteorological parameters over approximately a 10-year period.

Atmospheric Sciences Modeling Division — A division of the NOAA's Air Resources Laboratory established to collaborate with the U.S. EPA in developing advanced air quality models that can simulate the transport and fate of pollutants in the atmosphere.

Atmospheric Turbulence and Diffusion Division — A division of the NOAA's Air Resources Laboratory that began as a weather bureau special projects research office but has now expanded to perform air-quality-related research in addition to its original mission. It develops better methods for predicting transport, dispersion, and air-surface exchanges of air pollutants under realistic situations.

Atomic Energy Commission — The agency now known as the Nuclear Regulatory Commission in the Department of Energy.

ATSDR — See Agency for Toxic Substances and Disease Registry.

ATSP — See Association of Technical and Supervisory Professionals.

Audio Visuals On-Line — A database maintained by the National Library of Medicine of educational aids in the health sciences.

AVLINE — See Audio Visuals On-Line.

AWEA — See American Wind Energy Association.

AWMA — See Air and Waste Management Association.

AWPI — See The American Wood Preservers Institute.

AWRA — See American Water Resources Association.

AWWA — Acronym for American Water Works association.

Baking Industry Sanitation Standards Committee — An industry association representing 135 companies that seeks to establish standards of sanitation in bakery food processing equipment, receives advisory assistance from national and international public health and food sanitation groups, and develops and publishes sanitation standards for the baking industry.

Baseline Environmental Management Report — A report mandated by Congress and prepared by the Secretary of Energy to estimate the cost and schedule of cleaning up the nation's nuclear weapons complex.

RESOURCE DIRECTORY

Battery Council International — An organization of manufacturers, suppliers of materials, and national distributors of lead-acid storage batteries who recommend industry standards, compile statistics, and are concerned with environmental and safety issues.

BBSRC — Acronym for Biotechnology and Biological Sciences Research Council.

BCHCM — See Board of Certified Hazard Control Management.

BCI — See Battery Council International.

BCPSM — See Board of Certified Product Safety Management.

BCSP — See Board of Certified Safety Professionals.

BEMS — See Bioelectromagnetics Society.

BERA — See Biomass Energy Research Association.

BFRL — See Building and Fire Research Laboratory.

BHPr — Acronym for Bureau of Health Professions.

BIA — See Bureau of Indian Affairs.

BIDS — Acronym for Biological Integrated Detection System.

BIO — See Biotechnology Industry Organization.

Bioelectromagnetics Society — An independent organization of biophysical scientists, physicians, and engineers interested in the interactions of non-ionizing radiation with human systems; an international resource for excellence in scientific research, knowledge, and understanding of the interactions of electromagnetic fields with biological systems.

Biological Abstracts — A bibliography of available journals of biology from 1926 to the present.

Biological and Agricultural Index — An index of available journal articles of biology and agriculture from 1916 to the present.

Biomass Energy Research Association — An organization of corporations and individuals from universities, public utilities, industries, and research laboratories with a common interest in encouraging biomass and waste-to-energy and fuels research.

Biomedical Engineering Society — An organization of biomedical, chemical, electrical, and mechanical engineers; physicians; managers; university professors representing all fields of biomedical engineering; students; and corporations who encourage the development, dissemination, integration, and utilization of knowledge of biomedical engineering.

Biosis Previews — A database covering 1969 to the present and including biological abstracts, biological abstract reports, reviews, and meetings, as well as coverage of environmental biology.

Biotechnology Industry Organization — An organization of biotechnology companies, academic institutions, and state biotechnology centers in 46 states and 25 nations who are involved in the research and development of healthcare, agricultural, and environmental biotechnology products.

BISSC — See Baking Industry Sanitation Standards Committee.

BLM — See Bureau of Land Management.

BLS — See Bureau of Labor Statistics.

BMES — See Biomedical Engineering Society.

BNL — Acronym for Brookhaven National Laboratory.

Board of Certified Hazard Control Management — An organization of safety managers who establish professional standards, evaluate and certify safety programs, and provide advice to those who wish to acquire skills in administration and technical safety areas.

Board of Certified Product Safety Management — An organization of professional managers and engineers involved in safety and liability who evaluate qualifications of personnel in the administration of product safety management programs.

Board of Certified Safety Professionals — An organization that conducts research in the evaluation of competencies in safety practice areas and grants certification to occupational health and safety technologists, construction health and safety technicians, and safety-trained supervisors in construction.

BOM — Acronym for the Bureau of Mines (U.S. Department of the Interior).

BOP — Acronym for Federal Bureau of Prisons.

BOR — See Bureau of Reclamation.

BPA — See Bonneville Power Administration.

BPHC — See Bureau of Primary Health Care.

BSRN — Acronym for Baseline Surface Radiation Network.

Building and Fire Research Laboratory — A division of the National Institute of Standards and Technology that conducts research on high-performance building materials to provide a means for evaluating and predicting the performance of materials and products; on fire loss reduction with respect to engineering fire safety for people, products, and facilities; on enhancing firefighter effectiveness, with a 50 percent reduction in fatalities; on enhancing building performance, to ensure buildings work better throughout their useful lives; and on advanced construction technology innovations to ensure the safety and performance of construction. It also has available a Fire Research Information Service, Fire Dynamics Simulator, and Virtual Cement and Concrete Testing Laboratory.

Bureau of Indian Affairs — A federal government agency within the U.S. Department of the Interior whose mission is to enhance the quality of life, to promote economic opportunity, and carry out the responsibility of protecting and improving the true assets of American Indians, Indian tribes, and Alaskan natives.

Bureau of Land Management — A major program of the U.S. Department of the Interior that evaluates, classifies, develops, and administers regulations for natural gas, geothermal resources, and all solid energy and mineral resources, including coal and uranium, on federal lands, in addition to overseeing implementation of the Mining Law of 1872 and the Mineral Materials Act of 1955. It has total jurisdiction over about 268 million acres of federally owned lands, with approximately one third of them in Alaska and a majority of the remainder in the western states. Its function is to sustain the health, diversity, and productivity of these lands for the use and enjoyment of current and future generations.

Bureau of Primary Health Care — A major department within the Health Resources and Services Administration that assists states and local communities in providing health care to medically underserved populations through special grants and helps in meeting the health needs of such populations as migrant workers, black lung disease victims, the homeless, and high-risk pregnant women. It also administers the National Health Services Corps, which helps states and communities arrange for physicians and other healthcare professionals to provide care in areas with a shortage of health care.

Bureau of Reclamation — A major division of the U.S. Department of the Interior that constructs dams, power plants, and canals in 17 western states and promotes economic development of the west through water projects. It has constructed more than 600 dams and reservoirs and is today the largest wholesaler of water in the country, as well as the second largest producer of hydroelectric power in the western United States.

Business Periodicals Index — Indices of business literature from 1958 to the present.

CAAHEP — See Commission on Accreditation of Allied Health Education Programs.

CAC — Acronym for Committee on Atmospheric Chemistry.

CACC — See Citizens for Alternatives to Chemical Contamination.

CACFP — Acronym for Child and Adult Care Food Program.

CACGP — Acronym for International Commission on Atmospheric Chemistry and Global Pollution.

CAER — See Community Awareness and Emergency Response Program.

CAG — Acronym for Carcinogen Assessment Group.

CAHEA — Acronym for Committee on Allied Health Education and Accreditation.

Canadian Council for Human Resources in the Environment Industry — A nonprofit corporation, that assists the Canadian environment sectors in implementing good human resource development policies, national occupational standards for skills and training, employment opportunities for the highly skilled workforce, labor market projections, alliances between business and labor, and dialog between industry and the academic community.

Cancer Literature — A database containing more than 475,000 citations with abstracts to worldwide literature on oncology epidemiology, pathology, treatment, and research.

CANCERLIT — See Cancer Literature.

CAR — See Corrective Action Report.

CARAT — Acronym for Committee to Advise on Reassessment and Transition.

CARDS — Acronym for Chemical Agent Remote Detection System.

CARF — Acronym for Commission on Accreditation of Rehabilitation Facilities.

CAS — See Chemical Abstract Services.

CASAC — Acronym for Clean Air Scientific Advisory Committee.

CASTNeT — Acronym for Clean Air Status and Trends Network.

Catastrophic Disaster Response Group — A national level group of representatives from federal departments and agencies who serve as a centralized coordinating committee to support on-scene federal response and recovery efforts in a catastrophic disaster.

CATS — Acronym for Corrective Action Tracking System.

CBDCOM — Acronym for Chemical Biological Defense Command (of the U.S. Army).

CBDP — Acronym for Chemical/Biological Defense Program.

CBE — See Citizens for a Better Environment.

CBEP — See Community Based Environmental Protection.

CBER — Acronym for Center for Biologics Evaluation and Research.

CBIAC — Acronym for Chemical and Biological Defense Information Analysis Center.

CBIRF — Acronym for Chemical Biological Incident Response Force.

CBO — Acronym for Congressional Budget Office.

CB-RRT — Acronym for Chemical Biological Rapid Response Team.

CC/RTS — Acronym for Chemical Collections/Request Tracking System.

CCAE — Acronym for Council of Chemical Association Executives.

RESOURCE DIRECTORY

CCAP — Acronym for Center for Clean Air Policy or Climate Change Action Plan.

CCC — See Chlorine Chemistry Council.

CCHEST — See Council on Certification of Health, Environmental, and Safety Technologists.

CCHREI — See Canadian Council for Human Resources in the Environmental Industry.

CCHW — Acronym for Citizens Clearinghouse for Hazardous Wastes.

CCPS — See Center for Chemical Process Safety.

CCR — See Council for Chemical Research.

CCTP — Acronym for Clean Coal Technology Program.

CDC — See Centers for Disease Control and Prevention.

CDRG — See Catastrophic Disaster Response Group.

CDRH — Acronym for Center for Devices and Radiological Health.

CDRP — Acronym for Carbon Dioxide Research Program.

CDTF — Acronym for Chemical Defense Training Facility (at the U.S. Army Chemical School).

CEC — See Center for Environmental Citizenship or Committee on Energy and Commerce; acronym for Commission for Environmental Cooperation.

CEDR — See Consortium for Environmental Education in Medicine; acronym for Comprehensive Epidemiological Data Resource.

CEE — Acronym for Center for Environmental Education.

CEEC — See Corporate Environmental Enforcement Council.

CEEM — Acronym for Center for Energy and Environmental Management.

CEF — Acronym for Chemical Educational Foundation.

CEI — See Center for Environmental Information.

CEMCS — Acronym for Comprehensive Energy Management and Control System.

Cement Kiln Recycling Coalition — An organization of cement companies involved in the use of hazardous-waste-derived fuel as well as companies involved in the collection, processing, managing, and marketing of such fuel who have developed a mandate that its members must be responsible and conduct their businesses to protect human health and environment.

Census Bureau — A major program of the Department of Commerce that provides information on imports and exports of energy commodities, including coal, oil, and natural gas.

Center for Auto Safety — A public interest organization that conducts research on air pollution caused by automobile emissions and monitors fuel economy regulations.

Center for Chemical Process Safety — An organization of chemical and hydrocarbon manufacturers and engineering firms who study process safety issues in the chemical and hydrocarbon industries related to manufacture, handling, and storage of toxic and reactive materials and the potential release of hazardous materials.

Center for Clean Air Policy — An organization of governors, corporations, environmentalists, and academicians who analyze economic and environmental effects of air pollution and related environmental problems and work with each of the members to resolve the problems.

Center for Devices and Radiological Health — A major program of the Food and Drug Administration that administers national programs to control exposure to radiation, establishes standards for emissions from consumer and medical products, and conducts factory inspections and research, training and educational programs.

Center for Energy Policy and Research — A facility of the New York Institute of Technology designed to disseminate information and conduct research into energy utilization and conservation and to assist public, quasi-public, and private-sector organizations in the practical use of current and future findings in the energy field.

Center for Environmental Citizenship — A national nonpartisan organization that encourages college students to be environmental citizens and is dedicated to educating, training, and organizing a diverse national network of young leaders to protect the environment.

Center for Environmental Information — An organization that disseminates information on environmental issues and conducts annual climate issues conferences.

Center for Environmental Study, The — An organization that seeks to increase public awareness of environmental processes and issues through development and implementation of educational and informational programs and materials with a regional focus while still providing research materials nationwide.

Center for Food Safety and Applied Nutrition — An agency within the Food and Drug Administration responsible for regulating the food processing industry.

Center for Health, Environment, and Justice — An organization that provides citizens' groups, individuals, and municipalities with support and information on solid and hazardous waste.

Center for International Environment — An organization of environmental attorneys who seek to strengthen national, international, and comparative international law and public policy worldwide.

Center for Marine Conservation — An organization that protects the health of oceans and seas by advocating policies that restrict discharge of pollutants harmful to marine ecosystems.

Center for Study of Responsive Law — A nonprofit organization that supports and conducts a wide variety of research and educational projects to encourage political, economic, and social institutions of the United States to be more responsive to the needs of the citizen–consumer.

Center for Tropical and Subtropical Aquaculture — One of five regional aquaculture centers in the United States established by the USDA to integrate individual and institutional expertise and resources in the support of commercial aquaculture. Its function is to support aquaculture research, development, demonstration, and extension education to enhance viable profitable U.S. aquaculture.

Centers for Disease Control and Prevention — A federal health agency and branch of the Department of Health and Human Services located in Atlanta, GA; it provides national health and safety guidelines, statistical data, and specialized training in control of infectious diseases.

CEPP — See Chemical Emergency Preparedness Program.

CEPPO — See the Chemical Emergency Preparedness and Prevention Office.

CEPR — See Center for Energy Policy and Research.

CEPS — Acronym for Center for Earth and Planetary Studies.

CEPW — See Committee on Environment and Public Works.

CEQ — See Council on Environmental Quality.

CERCLIS — See Comprehensive Environmental Response, Compensation, and Liability Act Information System.

CERES — Acronym for Clouds and the Earths Radiant Energy System.

CERF — See Civil Engineering Research Foundation.

CES — See Center for Environmental Study.

CESB — Acronym for Council of Engineering and Scientific Specialty Boards.

CFDC — See Clean Fuels Development Coalition.

CFSAN — See Center for Food Safety and Applied Nutrition.

CGA — Acronym for Compressed Gas Association, Inc., or Community Advisory Group.

CGES — Acronym for the Center for Global Environmental Studies.

CHAMP — Acronym for Community Health Air Monitoring Program.

CHAMPUS — Acronym for Civilian Health and Medical Programs for Uniformed Services.

Chartered Institution of Building Services Engineers — An international body that represents and provides services to the building services profession by producing information, good-practice publications, and guides for operating a wide range of facilities, in addition to providing extensive networking activities for members to share their experiences.

CHC — See Clean Harbors Cooperative.

ChemAlliance — A computer site funded by the U.S. EPA and used as a source of up-to-date information concerning the environmental regulations affecting the chemical industry; it was designed to be a way for industry, regulators, and technical assistance providers to work together to improve both the environment and the profitability of corporations.

Chemical Abstract Services — An organization of the American Chemical Society.

Chemical Abstracts — Abstracts of chemical-related articles available in journals from 1907 to the present.

Chemical Biological Rapid Response Team — A program of the Department of Defense that provides chemical and biological defense support to civil authorities by supplying and establishing an integrated capability to coordinate and synchronize all medical and nonmedical technical assistance; it is the lead federal agency in the event of a major crisis. It is focused on domestic concerns but is responsive worldwide.

Chemical Control — A program of the U.S. EPA that selects and implements control measures for new and existing chemicals that pose a risk to human health and the environment.

Chemical Dictionary On-Line — The National Library of Medicine's on-line, interactive chemical dictionary file created by the Specialized Information Services in conjunction with the Chemical Abstracts Service.

Chemical Emergency Preparedness and Prevention — A major program of the U.S. EPA that develops and administers chemical emergency preparedness and prevention programs, reviews the effectiveness of existing programs, and prepares community right-to-know regulations.

Chemical Emergency Preparedness Program — A program developed by the U.S. EPA to address accidental releases of acutely toxic substances.

Chemical Exposure — A database from 1974 to the present providing comprehensive coverage of chemicals identified in both human tissues and body fluids and in feral and food animals; the service receives information on chemical properties, analytical methods, and toxic effects as well as references to selected journal articles, conferences, and reports.

Chemical Hazards Response Information System/Hazard Assessment Computer System — A program developed by the U.S. Coast Guard to provide chemical-specific data and determine how to handle chemical spills.

Chemical Industry Centers for Health Research —
An internationally known center for biomedical sciences and high-tech engineering located in Research Triangle Park, NC, that is funded by various industry groups as well as the federal government. It conducts state-of-the-art biochemical research studies, provides training programs to advance the quality of science in resolving environmental and public health issues, emphasizes an interdisciplinary approach to excellence and innovation, and helps the public make informed decisions concerning the assessment of risk and benefits of exposure to chemicals.

Chemical Industry Data Exchange — A trade association and standards body that is focused on creating transactional efficiency throughout the global chemical industry supply chain and on improving the speed and cost-effectiveness of electronic business transactions between chemical companies and their trading partners, resulting in improved data accuracy, cost-effectiveness, and safety.

Chemical Information Database — A database available covering journal articles with information on pure chemicals, natural substances, and mixtures that result from industrial processes.

Chemical Management and Resources Association —
An association of professionals within the chemical and allied industries who are involved in marketing, research, planning, product management, sales, consulting, and commercial development and who promote the growth and development of the chemical and allied process industries, provide continuing education, contribute to public information, and promote high standards and knowledge, skills, and ethics.

Chemical Manufacturers' Association — An organization of chemical companies conducting specific research on chemicals germane to industrial activity.

Chemical Producers and Distributors Association —
A voluntary, nonprofit membership organization consisting of about 100 member companies involved in the manufacture, formulation, distribution, and sale of generic products used on food, feed, and fiber crops and in the care and maintenance of lawns, gardens, and turf.

Chemical Specialties Manufacturers Association —
An organization of manufacturers, marketers, formulators, and suppliers of household, industrial, and personal care chemical specialty products, such as pesticides, cleaning products, disinfectants, sanitizers, and polishes, who develop consumer education information.

Chemical Transportation Emergency Center — A center provided by the Chemical Manufacturers'
Association to give information and/or assistance to emergency responders.

CHEMLINE — See Chemical Dictionary On-Line.

CHEMNET — A mutual-aid network of chemical shippers and contractors.

CHEMTREC — See Chemical Transportation Emergency Center.

CHESS — Acronym for Community Health and Environmental Surveillance System.

CHID — Acronym for Combined Health Information Database.

CHIP — See Chemical Hazardous Information Profile.

CHLOREP — See Chlorine Emergency Plan.

Chlorine Chemistry Council — A business council comprised of chlorine and chlorinated product manufacturers who work toward achieving policies that promote the continued responsible use of chlorine-based products and conduct research concerning the potential health and environmental effects of chlorine.

Chlorine Emergency Plan — A 24-hour, mutual-aid program operated by the Chlorine Institute.

Chlorine Institute, Inc. — An organization of industries concerned with the safety, health, and environmental protection in regard to chlor-alkali products.

CHRICS/HACS — See Chemical Hazards Response Information System/Hazard Assessment Computer System.

CHRIS — Acronym for Chemical Hazard Response Information System.

CHy — Acronym for Commission for Hydrology.

CI — Acronym for the Chlorine Institute.

CIAQ — Acronym for Council on Indoor Air Quality.

CIB — See Current Intelligence Bulletins.

CIBSE — Acronym for the Chartered Institution of Building Services Engineers.

CICIS — Acronym for Chemicals in Commerce Information System.

CIDRS — Acronym for Cascade Impactor Data Reduction System.

CIDX — See Chemical Industry Data Exchange.

CIEL — See Center for International Environmental Law.

CIIT — See Chemical Industry Institute of Technology.

CIS — Acronym for Cancer Information Service.

CIT — Acronym for Center for Information Technology.

Citizens for a Better Environment — An organization that works to reduce exposure to toxic substances in air, water, and land by focusing on research, public information, and advocacy, including formal and informal interaction with policymaking bodies on state, regional, and national levels.

Citizens for a Sound Economy — An education and research organization that seeks market-oriented solutions to environmental problems.

Citizens for Alternatives to Chemical Contamination — A grass-roots environmental education and advocacy organization dedicated to the principles of social justice, pollution prevention, and protection of the Great Lakes human and natural ecosystems to increase public awareness of toxic chemical threats to the environment.

Civil Engineering Research Foundation — An organization of civil engineers who facilitate, coordinate, and integrate research results more quickly into practice, revitalize the infrastructure, and enhance the environment.

CKRC — See Cement Kiln Recycling Coalition.

Clean Fuels Development Coalition — An organization that advocates the development of a national energy policy that addresses environmental concerns and provides for increased production and use of nonpetroleum motor fuels.

Clean Harbors Cooperative — An organization of major petroleum and energy companies who make available oil-spill cleanup equipment and trained equipment operators to member companies in the greater New York Harbor area.

Clean Water Action — A citizens' organization interested in clean, safe, and affordable water that works to influence public policy through education, technical assistance, and grass-roots organization in the areas of toxins, pollution, drinking water, water conservation, sewage treatment, pesticides, mass burn incineration, estuary protection, and consumer water issues.

Clean Water Council — An organization of engineers, manufacturers, distributors, and general business associations who promote federal funding for water quality infrastructure and educational programs to raise public awareness of water quality issues.

Clean Water Fund — An organization of scientific and policy experts, writers of environmental legislation, politicians, and grass-roots organizers who conduct research and educational programs to promote public interest and involvement in issues related to water, toxic materials, and natural resources.

CLF — Acronym for Conservation Law Foundation.

Climate Institute — An organization that educates the public and policymakers on climate change and on depletion of the ozone layer.

CLIPS — Acronym for Chemical List Index and Processing System.

CLP — See Contract Laboratory Program.

CMA — See Chemical Manufacturers' Association.

CMCHS — Acronym for Civilian–Military Contingency Hospital System.

CMRA — See Chemical Management Resources Association.

CNPP — Acronym for Center for Nutrition Policy and Promotion.

Coalition for Responsible Waste Incineration — An organization of manufacturing companies, academic institutions, and interested individuals and organizations that promotes responsible incineration of industrial waste as part of an overall waste management strategy.

Coalition to End Childhood Lead Poisoning — An organization that works to prevent lead poisoning in children and promotes a lead-free environment.

COE — See Army Corps of Engineers.

Cold Regions Research and Engineering Laboratory — A department of the U.S. Army Corps of Engineers whose mission is to gain knowledge of cold regions through scientific and engineering research and to put the knowledge to work for the Corps of Engineers, the Army, Department of Defense, and the nation.

Commission on Accreditation of Allied Health Education Programs — An association of approximately 70 member organizations, including the American Medical Association, that accredits educational programs for health professionals in 18 different disciplinary areas, accreditation being a major step in ensuring that the public has a supply of qualified healthcare professionals.

Committee on Energy and Commerce — A committee of the U.S. House of Representatives that oversees commerce, trade, and consumer protection; energy and air quality; environment and hazardous materials; public health, quarantine, hospital construction, mental health, and biomedical programs; health protection; and drugs and drug abuse. It has a subcommittee on oversight and investigation of all agencies, departments, and programs within the jurisdiction of the full committee.

Committee on Environment and Public Works — A major committee of the U.S. Senate whose responsibility is in areas of clean air, wetlands, climate change, transportation, infrastructure, nuclear safety, fisheries, wildlife, water, Superfund, toxics, and risk and waste management.

Community Advisory Group — An organization made up of members of a community related to a Superfund site and serving as a focal point for the exchange of information among local leaders, the U.S. EPA, the state regulatory agency, and other pertinent federal agencies involved in the cleanup of a Superfund site.

Community Awareness and Emergency Response Program — A program designed by the Chemical Manufacturers Association to help plant managers

RESOURCE DIRECTORY

take the initiative to work with local communities to develop integrated, hazardous material response plans.

Community Development Block Grant — A federal program under the Department of Housing and Urban Development that provides grants to carry out a wide range of community development activities directed toward neighborhood revitalization, economic development, and improved community facilities and services.

Community Viability Program — An area of the Department of Housing and Urban Development that develops policies promoting energy efficiency, conservation, and renewable sources of supply in housing and community development programs.

Compressed Gas Association — An organization of companies producing and distributing compressed, liquefied, and cryogenic gases and manufacturers of related equipment who act as consultants and make recommendations to appropriate government agencies to improve safety standards and methods of handling, transporting, and storing gases. It emphasizes the areas of atmospheric gases and equipment, carbon dioxide, environmental controls, hazard communications, hazard materials, hydrogen sulfide, medical gases, technology, safety, health, and environmental pollutants.

Computer-Stored Ambulatory Record System — An on-line interactive computerized information system accessible through a minicomputer and used in the public health field.

Concern — An environmental education organization interested in sustainable communities, global warming, energy, agriculture, pesticides, water resources, and waste reduction.

Conference of Radiation Control Program Directors — An organization of local and state radiological program directors and individuals from related federal protection agencies who serve as a forum for the exchange of experiences, concerns, new developments, and recommendations among radiation control programs and related agency programs.

Conservation and Research Foundation — A group of conservation professionals who seek to encourage biological research, promote conservation of natural resources, increase the understanding of the relationship between humans and the environment, and encourage publication of original materials.

Conservation Fund, The — An organization that creates partnerships among the private sector, nonprofit organizations, and public agencies to promote land and water conservation.

Consortium for Environmental Education in Medicine — An organization that includes the faculty from five Massachusetts and Rhode Island medical schools. Responses to a survey it conducted showed that the average medical student receives only 6 total hours, not credit hours, of training in occupational and environmental medicine, so it plans to implement faculty development, curriculum dissemination, and student outreach programs in the area of occupational and environmental medicine.

Consumer Energy Council of America Research Foundation — An organization of individuals who analyze the economic and social effects of energy policy, develop long-range conservation projects, evaluate air pollution emissions trading, etc.

Consumer Product Safety Commission — An agency created in 1970 to evaluate and regulate chemicals that may be carcinogenic and also to evaluate a number of products that may be hazardous to individuals.

Container Recycling Institute — An organization that studies alternatives for reducing container and packaging waste, including reuse and recycling.

Contract Laboratory Program — A national network of U.S. EPA personnel, commercial laboratories, and support contractors whose fundamental mission is to provide data of known and documented quality.

Co-Op America — An organization that teaches consumers and businesses about social and environmental responsibility and publishes a directory of environmentally responsible businesses.

Cooperative State Research, Education, and Extension Service — An agency of the USDA that administers federal funds approved for agriculture and forest research, extension, and education programs in the food and agricultural science areas.

COP — Acronym for Coastal Ocean Program.

Corporate Environmental Enforcement Council — An organization made up of corporate counsel and environmental professionals from a wide range of industrial sectors that focuses on civil and criminal environmental enforcement issues.

Corps of Engineers — See Army Corps of Engineers.

COSCDA — See Council of State Community Development Agencies.

COSTAR — See Computer-Stored Ambulatory Record System.

Council for Chemical Research — An organization of universities that grant advanced degrees in chemistry or chemical engineering, chemical companies, government laboratories, and independent research laboratories that employ chemists and chemical engineers in research and development; it promotes more effective interactions between university chemistry and chemical engineering departments and the research function of industry and government and supports basic research in chemistry and chemical engineering.

Council for Textile Recycling — An organization of textile and clothing manufacturers; national, state, and local recycling organizations; trade associations; state and local government agencies; and interested individuals who work together to create greater awareness of the benefits of textile recycling.

Council of Engineering and Scientific Specialty Boards — An independent, voluntary membership body that accredits a variety of programs in engineering, science, and related fields.

Council of State Community Development Agencies — A nonprofit organization made up of state agencies that manage state and federal grant and loan programs in housing, community development, and economic development.

Council on Certification of Health, Environmental, and Safety Technologists — Through this council, the American Board of Industrial Hygiene and the Board of Certified Safety Professionals operate certification programs for a variety of technical personnel in the safety and health fields.

Council on Education for Public Health — An organization that accredits schools of public health, graduate programs, community health education, and community health preventive medicine.

Council on Environmental Quality — An office within the White House that coordinates federal environmental efforts and works closely with agencies and other White House offices in the development of environmental policies and initiatives. It reports annually to the President on the state of the environment, oversees federal agency implementation of environmental impact assessment processes, and acts as a referee when agencies disagree on the adequacies of the assessments. Congress passed the national Environmental Policy Act in 1969 and the Environmental Quality Improvement Act of 1970, thereby recognizing that nearly all federal activities affect the environment in some way, and mandated that before federal agencies could make decisions they must consider the effects of their actions on the quality of the human environment.

CPDA — See Chemical Producers and Distributors Association.

CPMA — Acronym for Color Pigments Manufacturers Association, Inc.

CPSC — See Consumer Product Safety Commission.

CRCPD — See Conference of Radiation Control Program Directors.

CRF — See Conservation and Research Foundation.

CRISP — Acronym for Computer Retrieval of Information on Scientific Products.

CRO — Acronym for Contract Research Organization.

CRP — Acronym for Conservation Reserve Program.

CRR — Acronym for Center for Renewable Resources.

CRREL — See Cold Regions Research and Engineering Laboratory.

CRWI — See Coalition for Responsible Waste Incineration.

CSB — Acronym for Chemical Safety and Hazard Investigation Board.

CSCS — See Comprehensive Soil Classification System.

CSFP — Acronym for Commodity Supplemental Food Program.

CSGWPP — Acronym for Comprehensive State Ground Water Protection Program.

CSHIB — Acronym for U.S. Chemical Safety and Hazard Investigation Board.

CSIN — Acronym for Chemical Substances Information Network.

CSMA — Acronym for Chemical Specialties Manufacturers Association.

CSPA — Acronym for Council of State Planning Agencies.

CSPI — Acronym for Center for Science in the Public Interest.

CSR — Acronym for Center for Scientific Review.

CSRL — See Center for the Study of Responsive Law.

CSRS — Acronym for Cooperative State Research Service.

CTARC — Acronym for Chemical Testing and Assessment Research Commission.

CTR — See Council for Textile Recycling.

CTSA — See Center for Tropical and Subtropical Aquaculture.

Cumulative Index to Nursing and Allied Health Literature — A reference index of approximately 300 nursing, allied health, and health-related journals from 1956 to the present.

Current Intelligence Bulletins — Relay important public health information and recommended protective measures to industry, labor, public interest groups, and academia concerning occupational health problems.

CVM — Acronym for Center for Veterinary Medicine.

CWC — See Clean Water Council.

CWF — See Clean Water Fund.

CWS — Acronym for Chemical Warfare Service.

CWTC — Acronym for Chemical Waste Transportation Council.

DAAC — See Distributed Active Archive Center.

Dangerous Goods Advisory Council — An international, nonprofit, educational organization devoted to promoting safety in domestic and international transportation of hazardous materials and dangerous goods.

Data Element Dictionary — A dictionary that enables users to find specific data that are characterized under tables listed as site, action, operable unit, and financial.

Data Web — An electronic presentation of the Bureau of Reclamation's *Project Data Book* containing historical, statistical, and technical information on the projects of the Bureau of Reclamation.

DCAT — Acronym for Drugs, Chemical, and Allied Trades Association, Inc.

DEA — Acronym for Drug Enforcement Agency.

Defense Meteorological Satellite Program — A U.S. Air Force meteorological satellite program. Satellites circle in Sun-synchronous orbit to gather imagery collected in the visible to near-infrared band (0.4 to 1.1 μm) and in the thermal-infrared band (8 to 13 μm) at a resolution of about 3 km.

Defense Nuclear Facilities Safety Board — An independent board created by Congress and appointed by the President to provide external oversight of Department of Energy and defense nuclear facilities and to make recommendations concerning public health and safety.

Defense Nuclear Nonproliferation — A major program of the Department of Energy that provides the intelligence community with technical and operational expertise on foreign nuclear and energy issues.

DENR — Acronym for Department of Environmental and Natural Resources.

Department of Defense — The federal agency responsible for providing the military forces needed to deter war and protect the security of the United States.

Department of Energy — The federal agency responsible for providing the framework for a comprehensive and balanced national energy plan through the coordination and administration of the energy functions of the federal government.

Department of Health And Human Services — A federal department carrying out the functions of health and social programs such as Social Security, human development, family support, healthcare financing, and public health. The Department of Health and Human Services includes the Centers for Disease Control and Prevention, the U.S. Public Health Service, the Agency for Toxic Substances and Disease Registry, the Food and Drug Administration, and the National Institutes of Health.

Department of Housing and Urban Development — The federal agency principally responsible for programs concerned with the nation's housing needs, development and preservation of the nation's communities, and provision of equal housing opportunities for all individuals.

Department of Justice — The federal agency that serves as counsel to citizens of the United States by representing them in enforcing the law in the public interest; it protects citizens against criminals and subversion, ensures healthy competition of business, safeguards the consumer, and works to enforce drug, immigration, civil rights, and naturalization laws.

Department of Labor — The federal agency responsible for fostering, promoting, and developing the welfare of the wage earners of the United States to improve their working conditions and to advance their opportunities for profitable and safe employment.

Department of Transportation — The federal agency responsible for establishing the nation's overall transportation policy. It deals with highway planning, development, and construction; urban mass transit; railroads; and aviation, as well as the safety of waterways, ports, highways, and oil and gas pipelines.

DEST — Acronym for Domestic Emergency Response Team.

DGAC — See Dangerous Goods Advisory Council.

DHHS — See Department of Health and Human Services.

Distributed Active Archive Center — Eight centers located around the United States that process, store, and distribute satellite remote sensing data for NASA and other agencies.

DMSP — See Defense Meteorological Satellite Program.

DOC — See U.S. Department of Commerce.

DOD — See Department of Defense.

DOE — See Department of Energy.

DOE Energy — A database from 1974 to the present covering all aspects of energy and related topics, including the environment and conservation, as well as indexing journal articles, reports, conference papers, books, patents, dissertations, and translations.

DOI — See U.S. Department of the Interior.

DOJ — See Department of Justice.

DOL — See Department of Labor.

DOT — See Department of Transportation.

DSM — Acronym for *Diagnostic and Statistical Manual of Mental Disorders*.

EAA — See Environmental Assessment Association.

EADS — Acronym for Environmental Assessment Data System.

Earth Ecology Foundation — An organization of individuals interested in earth science and human ecology who study the interrelationship between people, technology, and nature to provide modern solutions to ecological problems, to develop management plans relating to climate change and weather disasters, and to promote scientific and humane use of the Earth's ecology through natural and technological means.

Earth Science Enterprise — An international research program designed to understand the environment of the planet as a system by observing, understanding, modeling, assessing, and eventually predicting global change due to the impact of human activity and nature by utilizing ground-based measurements, satellite measurements, and aircraft measurements.

Earth Share — An organization of environmental and conservation groups who work with government and private payroll deduction programs to solicit contributions for environmental research, education, and community programs.

Earthjustice — A nonprofit law firm for the environment representing hundreds of public interest clients, large and small, who work through the courts to safeguard public lands, national forests, parks, and wilderness areas; to reduce air and water pollution; to prevent toxic contamination; to preserve endangered species; and to achieve environmental justice. It was founded in 1971 as the Sierra Club Legal Defense Fund and has played a leading role in shaping the development of environmental law.

EBA — See Environmental Bankers Association.

EBI — See Environmental Business International.

EC — Acronym for Environment Canada or European Community.

ECL — Acronym for Environmental Chemical Laboratory.

ECO — See Environmental Careers Organization.

Ecological and Toxicological Association of Dyes and Organic Pigments Manufacturers — An international organization that represents the interests of the industry in areas relating to health and the environment and bases their decisions on sound science and practical state-of-the-art materials.

Ecological Society of America — An organization of educators, professional ecologists, and scientists interested in the study of plants, animals, and people in relationship to their environment and who are seeking to develop a better understanding of biological processes and their contributions to agriculture, forestry, wildlife, range management, fisheries, industry, conservation, and public health.

ECOS — See Environmental Council of the States.

ECP — Acronym for Emergency Conservation Program.

ECS — See Electrochemical Society.

EDF — See Environmental Defense Fund.

Edison Electric Institute — An association of electric companies formed in 1933 to exchange information on industry developments and to act as an advocate for utilities on subjects of national interest.

EDRS — Acronym for Enforcement Document Retrieval System.

EDS — Acronym for Electronic Data System or Energy Data System.

EDSTAC — Acronym for Endocrine Disruptors Screening and Testing Advisory Committee.

Educational Resources Information Center — A national information system designed to provide ready access to an extensive body of education-related literature, established and supported by the U.S. Department of Education's Office of Educational Research and Improvement and administered by the National Library of Education. It is the largest education database in the world, containing more than 1 million records of journal articles, research reports, curriculum and teaching guides, conference papers, and books.

EEC — Acronym for European Economic Community.

EEF — See Earth Ecology Foundation.

EEGS — See Environmental and Engineering Geophysical Society.

EEI — See Edison Electric Institute.

EEOC — See Equal Employment Opportunity Commission.

EESI — See Environmental and Energy Study Institute.

EESL — Acronym for Environmental, Ecological, and Support Laboratory.

EETFC — Acronym for Environmental Effects, Transport, and Fate Committee.

EFNEP — See Expanded Food and Nutrition Education Program.

EHAC — See Environmental Health Accreditation Council.

EHC — See Environmental Health Center; acronym for Environmental Health Committee.

EHP-online — Acronym for Environmental Health Perspectives-online.

EIA — See Energy Information Administration, Environmental Investigation Agency, or Environmental Industry Association.

EIC — Acronym for Environmental Industry Council.

EII — Acronym for Environmental Industry Interactive.

EIS — Acronym for Epidemic Intelligence Service.

ELAW — See Environmental Law Alliance Worldwide.

ELC — See Environmental Literacy Council.

Electrochemical Society — A technical society of electrochemists, chemists, chemical engineers, electrochemical engineers, metallurgists, metallurgical engineers, physical chemists, physicists, electrical engineers, research engineers, teachers, technical sales representatives, and patent attorneys who seek to advance the science and technology of chemistry, electronics, electrothermics, electrometallurgy, and applied subjects.

Electronic Industries Association — An organization that specializes in the electrical and functional characteristics of interface equipment and sets standards for the industry.

Electronic Radiology Laboratory — One of nine research laboratories within the Division of Radiologic Sciences, Mallinckrodt Institute of Radiology, Washington University School of Medicine. It was established to investigate digital imaging technologies important to the distributed radiology department of the future, and to be responsive for the delivery of clinical image information to the medical decision maker.

ELI — See Environmental Law Institute.

ELR — Acronym for Environmental Law Reporter.

ELS — See Harvard Environmental Law Society.

EM — Acronym for Office of Environmental Management.

EMAP — Acronym for Environmental Mapping and Assessment Program.

EMAS — Acronym for Enforcement Management and Accountability System.

EMB — Acronym for Environmental Management Bureau.

EMBASE — A database from 1974 to present abstracting the world's biomedical literature, including environmental health and pollution control.

Emergency Management Institute — The part of the federal Environmental Management Agency's National Emergency Training Center that conducts resident and nonresident training activities for federal, state, and local government officials; managers of the private sector; and members of professional and volunteer organizations on subjects from civil nuclear preparedness systems to domestic emergencies caused by natural and technological hazards.

Emergency Nurses Association — A professional organization of emergency room nurses who promote continuing education, certification, and the enhancement of emergency skills.

Emergency Response Notification System — A U.S. EPA database used to store information on notification of oil discharges and hazardous substance releases.

Emergency Response Program — A U.S. EPA program providing adequate and timely response measures in communities affected by hazardous substances and oil releases where state and local first-responder capabilities have been exceeded or additional support is needed.

Emergency Watershed Protection — A program of the USDA's Natural Resources Conservation Service that provides technical and financial assistance to protect lives and property threatened by natural disasters, excessive erosion, and flooding.

Emergency Wetlands Reserve Program — A federal law passed in 1993 authorizing emergency supplemental appropriations to respond to widespread floods in the Midwest by providing payments to purchase easements and partial financial assistance to landowners to permanently restore wetlands at sites where the restoration cost exceeds the fair market value of the land.

EMI — See Emergency Management Institute.

EML — Acronym for Environmental Measurements Laboratory.

EMSL — Acronym for Environmental Monitoring Support Systems Laboratory.

ENA — See Emergency Nurses Association.

Energy Bar Association — A group of lawyers interested in all areas of energy law, including production, development, conservation, transmission, and economic regulation.

Energy Efficiency and Renewable Energy Clearinghouse — A project of the U.S. Department of Energy that aids in the transfer of technology to the public by answering inquiries, providing referrals, and disseminating information concerning the use of renewable energy technologies, conservation, and energy efficiency techniques for residential and commercial needs.

Energy Information Administration — A statistical agency of the U.S. Department of Energy that provides policy, in-depth data, forecasts, and analysis to promote sound policymaking, efficient markets, and public understanding regarding energy and its interaction with the economy and the environment.

Energy Research Institute — An organization of individuals and companies interested in alternative energy sources and the development of projects concerning wind generators, methane digesters, solar cells and collectors, steam generators, and hydrogen production with special focus on alcohol production for fuel purposes.

Enforcement and Compliance Assurance — A program of the U.S. EPA that enforces laws that protect public health and the environment from hazardous materials, pesticides, and toxic substances.

ENRIC — Acronym for Environment and Natural Resources Information Center.

Enviroline — A database from 1971 to present that covers all aspects of the environment and provides coverage of more than 5000 international sources.

Environment and Plastics Industry Council — An organization responsible for the use and recovery of plastics resources and that promotes an integrated approach that selectively utilizes source reduction, reuse, recycling, recovery of energy, and retention in landfills of plastics.

Environment, Safety, and Health — A major program of the Department of Energy that ensures that energy programs comply with federal policies and standards designed to protect the environment and government property; evaluates and establishes standards related to radiation, industrial hygiene, occupational medicine; and oversees epidemiological studies.

Environmental Abstract Annual — An index of available environmental articles from 1972 to present; originally known as the *Environmental Index.*

Environmental Action Foundation — An organization that promotes environmental research and education and serves as a resource for concerned citizens and other organizations in the areas of energy policy, toxic substances, and solid waste reduction.

Environmental and Energy Study Institute — An organization that promotes environmentally sustainable societies by working to educate the general public on related issues of groundwater protection, water efficiency, and climate change.

Environmental and Engineering Geophysical Society — An organization of corporations, students, and others interested in geophysics or related sciences seeking to promote the application of geophysical techniques for environmental and engineering use and to promote education and research in these areas.

Environmental Assessment Association — An organization of certified environmental inspectors, certified environmental specialists, certified remediation/testing specialists, and other environmental professionals dedicated to providing members with information and education in the environmental industry concerning environmental inspections, testing and hazardous material removal, and working with governmental and other environmental agencies to conduct educational programs.

Environmental Bankers Association — An organization of banks and financial services organizations interested in environmental risk management and liability issues.

Environmental Bibliography — A database from 1973 to present indexing periodical articles on general human ecology, land and water resources, pollution, health hazards, solid waste management, and other related topics.

Environmental Business International — A strategic business information service that provides newsletters, comprehensive research reports, and consulting services to business, technology, and regulatory corporations faced with environmental problems.

Environmental Careers Organization — An organization that seeks to protect and enhance the environment through development of professionals and promotion of careers by offering paid internships, career development educational programs, and publications. The participants are mostly upper-level undergraduates, graduates, and doctoral students or recent graduates seeking professional experience relevant to careers in the environmental fields of biology, chemistry, community development, hazardous wastes, natural resources, pollution, public health, occupational health, and transportation.

Environmental Council of the States — An organization that works to improve the environment by providing for the exchange of ideas and experiences among states and territories and promotes cooperation and coordination among environmental management professionals.

Environmental Defense Fund (EDF) — A national association of individuals concerned about the environment who lobby, bring about legal action, and educate the public as well as legislators on a variety of environmental issues.

Environmental Health Accreditation Council — An organization of highly skilled and educated environmental health personnel from academia, government, business, and industry who develop criteria for accreditation of undergraduate and graduate environmental health programs, establish a systematic means of accreditation, establish the appropriate curriculum for each of the different types of degree programs, provide professionals for peer evaluation of the academic programs, and finally determine whether the academic programs being evaluated meet the appropriate standards for accreditation.

Environmental Index — See Environmental Abstract Annual.

Environmental Industry Associations — An organization of trade associations from the waste services and environmental technology industries representing the National Solid Waste Management Association and the Waste Equipment Technology Association; an organization that compiles statistics, conducts research, and promotes educational programs in the environmental and safety industries.

Environmental Industry Council — An organization of manufacturers of pollution control equipment and systems involved in pollution control and promoting consistent national environmental policies.

Environmental Information Association — An organization of individuals and corporations concerned with environmental management and control who collect and disseminate information concerning environmental issues and solutions.

Environmental Investigation Agency — An organization involved in international resolution of environmental crimes, illegal trade in wildlife, and distruction of the natural environment.

Environmental Law Alliance Worldwide — A global alliance of public interest attorneys and scientists

RESOURCE DIRECTORY

in 50 countries who defend the environment through the use of law.

Environmental Law Institute — A research and education organization with an interdisciplinary staff of lawyers, economists, scientists, and journalists who publish materials on environmental issues, sponsor education and training courses, issue policy recommendations, and co-sponsor conferences on environmental law.

Environmental Literacy Council — An organization dedicated to helping citizens, especially young people, participate wisely in the area of environmental issues by giving teachers the tools to help students develop environmental literacy, which requires a fundamental understanding of the systems of the natural world and the relationships and interactions between living and non-living environments, as well as the ability to deal sensibly with problems that involve scientific evidence, uncertainty, economic, aesthetic, and ethical considerations.

Environmental Management — A program of the Department of Energy that provides policy guidance for and oversees waste management operations.

Environmental Management Program — An office of the Department of Energy that was created in 1989 to oversee the department's waste management and environmental cleanup efforts; originally called the Office of Environmental Restoration and Waste Management.

Environmental Media Services — An organization that advocates expanded and improved coverage of environmental issues in the nation's media.

Environmental Policy Center — An organization that operates the Global Cities Project, which provides environmental information services to cities and other communities to assist in building a sustainable future and working with public policy issues with respect to local environmental concerns.

Environmental Quality Incentives Program — A federal effort established in the 1996 Farm Bill to provide a voluntary conservation program for farmers and ranchers who face serious threats to soil, water, and related natural resources.

Environmental Radiation Ambient Monitoring System — A national network of monitoring stations operated by the U.S. EPA's National Air and Radiation Environmental Laboratory that regularly samples major pathways through which the U.S. population could be exposed to radiation by air, drinking water, surface water, and pasteurized milk.

Environmental Resource Council — A nonprofit group that works to improve and safeguard the public health, involved in child protection research, anti-tobacco messages and programs, issues related to

waste management, and promotion of general environmental health and safety.

Environmental Safety — An advisory committee of former officials of the U.S. EPA, lawyers, public health officials, and environmental specialists, whose primary concern is to ensure that the U.S. EPA and other federal agencies take active steps toward implementing their environmental protection responsibilities, particularly with respect to the dangers of toxic chemicals in the air, water, food, the workplace, and waste streams.

Environmental Security — A program of the Department of Defense that integrates environmental, safety, and occupational health considerations into U.S. defense and economic policies.

Environmental Technology and Regulatory Affairs — A program of the U.S. Chamber of Commerce that develops policy on all issues affecting the production, use, and conservation of natural resources, including fuel and nonfuel minerals, timber, water, public lands, energy, wet lands, and endangered species.

Environmental Technology Council — An organization of environmental service firms interested in recycling, detoxification, disposal of hazardous and industrial waste, and cleanup of contaminated industrial sites.

Environmental Working Group — A research and advocacy organization that studies and reports on the presence of herbicides and pesticides in food and drinking water.

EOIC — See Ethylene Oxide Industry Council.

EPA — See U.S. Environmental Protection Agency.

EPAA — Acronym for Environmental Programs Assistance Act.

EPC — See Environmental Policy Center.

EPI — Acronym for Environmental Policy Institute.

EPIC — See Environment and Plastics Industry Council.

EPR2 — Acronym for Electronic Product Recovery and Recycling.

EPRI — Acronym for Electric Power Research Institute.

EPSRC — Acronym for Engineering and Physical Sciences Research Council.

EPW — See Committee on Environment and Public Works.

Equal Employment Opportunity Commission — A federal program that focuses on the elimination of illegal discrimination in the workplace and accomplishes these goals by using an integrated approach that strategically links all activities to new and innovative programs such as alternative dispute resolution methods (e.g., mediation).

ERAMS — See Environmental Radiation Ambient Monitoring System.

ERC — See Environmental Resource Council.

EREC — See Energy Efficiency and Renewable Energy Clearinghouse.

EREN — Acronym for Energy Efficiency and Renewable Energy Network.

ERI — See Energy Research Institute.

ERIC — See Educational Resource Center.

EREN — Acronym for Energy Efficiency and Renewable Energy Network.

ERL — See Electronic Radiology Laboratory; acronym for Environmental Research Laboratory.

ERNS — Acronym for Emergency Response Notification System.

ESA — Acronym for Ecological Society of America.

ESE — See Earth Science Enterprise.

ESRC — Acronym for Economic and Social Research Council or Environmental Science Research Center.

ETA — See Electronic Industries Association.

ETAD — See Ecological and Toxicological Association of the Dyes and Pigments Manufacturers Industry.

Ethylene Oxide Industry Council — An organization of producers and users of ethylene oxide used in the manufacture of consumer products such as antifreeze and polyester fibers, and in health care as a sterilizing agent; they work to develop scientific, technological, and safe manufacture, use, and handling of the chemical.

Expanded Food and Nutrition Education Program — A program of the USDA designed to assist limited-resource audiences in acquiring the knowledge, skills, attitudes, and change of behavior necessary for nutritionally sound diets and to contribute to their personal development and the improvement of the total family diet and nutritional well-being.

FACA — Acronym for Federal Advisory Committee Act.

FACOSH — See Federal Advisory Council for Occupational Safety and Health.

FACSM — Acronym for Fellow of the American College of Sports Medicine.

FAO — See Food and Agricultural Organization of the United Nations.

FAOTA — Acronym for Fellow of the American Occupational Therapy Association.

FAPTA — Acronym for Fellow of the American Physical Therapy Association.

FBI — Acronym for Federal Bureau of Investigation.

FCC — See Federal Communications Commission.

FCCC — Acronym for Framework Convention on Climate Change.

FDA — See Food and Drug Administration.

Federal Advisory Council for Occupational Safety and Health — A joint management–labor council that advises the Secretary of Labor on matters relating to the occupational safety and health of federal employees.

Federal Aviation Administration — A division of the U.S. Department of Transportation that inspects and rates civilian aircraft and airmen and enforces the rules of air safety.

Federal Communications Commission — A federal agency that regulates interstate and foreign communications by radio, television, wire, and cable.

Federal Emergency Management Agency — A federal agency serving as a single point of accountability for all federal emergency preparedness mitigation and response activities and chartered to enhance the multiple use of emergency preparedness and response resources at the federal, state, and local levels.

Federal Energy Regulatory Commission — An independent, five-member commission within the Department of Energy that sets rates and charges for the transportation and sale of natural gas, for the transmission and sale of electricity, and for the licensing of hydroelectric power projects.

Federal Facilities Restoration and Reuse Office — A U.S. EPA office facilitating faster, more effective, and less costly cleanup and reuse of federal facilities.

Federal Housing Administration — An agency of the U.S. government whose function has been transferred to the Department of Housing and Urban Development.

Federal Mine Safety and Health Review Commission — An independent agency established by the Federal Mine Safety and Health Act of 1977 that holds hearings and issues orders concerning the Secretary of Labor's enforcement actions regarding mine safety and health.

Federal Motor Vehicle Control Program — The federal actions used to control pollution from motor vehicles by establishing and enforcing tailpipe and evaporative emissions standards for new vehicles, testing methods development, and guidance to states operating inspection and maintenance programs.

Federal Power Commission — The predecessor agency of the Federal Energy Regulatory Commission, first created by an Act of Congress under the Federal Water Power Act of June 10, 1920.

Federal Radiological Monitoring and Assessment Center — A facility established by the Department of Energy, usually at an airport near the scene of a radiological emergency, from which the offsite technical director conducts the Federal Radiological Monitoring and Assessment Plan.

Federal Register — A bulletin published daily by the federal government containing executive branch proposed or final regulations, proclamations, and schedules of hearings before Congress and federal agency committees.

Federal Trade Commission — An agency of the federal government that enforces a variety of federal antitrust and consumer protection laws seeking to ensure that the nation's markets function competitively and are vigorous, efficient, and free of undue restrictions.

FEDMAP — Acronym for Federal Geologic Mapping Project.

FEDRIP — Acronym for Federal Research in Progress Database.

FEDS — Acronym for Federal Energy Data System.

FEMA — See Federal Emergency Management Agency.

FERC — See Federal Energy Regulatory Commission.

FFIS — Acronym for Federal Facilities Information System.

FFRRO — See Federal Facilities Restoration and Reuse Office.

FFSHC — See Field, Federal, Safety, and Health Councils.

FHA — See Federal Housing Administration.

FHCP — Acronym for Federal Hazard Communication Program.

FIELD — Acronym for Foundation for International Environmental Law and Development.

Field, Federal, Safety, and Health Councils — Councils organized throughout the country to improve federal safety and health programs at the field level and within a geographic location.

FIPS — Acronym for Federal Information Procedure System.

Fish and Wildlife Service — The principal federal agency within the U.S. Department of the Interior responsible for conserving, protecting, and enhancing fish, wildlife, and plants and their habitats for the continuing benefit of the American people.

Fish, Wildlife, and Parks — A program of the U.S. Department of the Interior responsible for development, conservation, and use of fish, wildlife, recreational, historical, and national park system resources.

FLC — Acronym for Federal Laboratory Consortium.

FLECT — Acronym for Federal Law Enforcement Training Center.

FMCSA — Acronym for Federal Motor Carrier Safety Administration.

FMSHRC — Acronym for Federal Mine Safety and Health Review Commission.

FMVCP — Acronym for Federal Motor Vehicle Control Program.

FNCS — Acronym for Food, Nutrition, and Consumer Services.

FNII — See Food and Nutrition Index.

FNS — Acronym for Food and Nutrition Service.

FOE — See Friends of the Earth.

Food and Agricultural Organization — The agency of the United Nations that institutes and administers programs, especially in underdeveloped countries, for improving farming methods and increasing food production.

Food and Drug Administration — A federal agency within the Department of Health and Human Services authorized to regulate food additives, contaminants, drugs, cosmetics, and potential carcinogens related to food additives, drugs, and cosmetics.

Food and Nutrition Internet Index — A searchable Web site describing and indexing food and nutrition resources available on the Internet, with its main focus being on food science, food technology, and human nutrition.

Food Safety Consortium — A group consisting of researchers from the University of Arkansas, Iowa State University, and Kansas State University and established by Congress in 1988 as a special Cooperative State Research Service Grant to conduct extensive investigation into all areas of poultry, beef, and pork meat production from the farm to the consumer's table.

Food, Science, and Technology Abstracts — A database from 1969 to the present that provides access to research and new development literature; also covers allied disciplines such as agriculture, chemistry, biochemistry, and physics and related disciplines such as engineering and home economics when relevant to food science. It covers over 1200 journals from over 50 countries, patents from 20 countries, and books in any language.

Foodborne Diseases Active Surveillance Network — The principal foodborne disease component of the CDC's Emergency Infections Program, a collaborative project of 19 states, the U.S. Department of Agriculture, and the U.S. Food and Drug Administration consisting of active surveillance for foodborne diseases and related epidemiological studies to help public health officials better understand the epidemiology of foodborne diseases in the United States.

FoodNet — See Foodborne Diseases Active Surveillance Network.

Foodservice and Packaging Institute — An organization of manufacturers, raw materials suppliers, machinery suppliers, and distributors who strive to expand and enhance acceptance of all foodservice disposables as an integral part of serving food and beverages to the public in a safe and sanitary manner.

Forecast Systems Laboratory — A division of the National Oceanic and Atmospheric Administration Research Laboratories that conducts applied meteorological research and development to improve and create short-term warning and weather fore-cast systems, models, and observing technology.

Forest Service — See USDA Forest Service.

FPC — See Federal Power Commission.

FPI — See Foodservice and Packaging Institute.

FRES — Acronym for Forest Range Environmental Study.

Friends of the Earth — An environmental advocacy group involved in conservation, renewable energy resources, air and water pollution, and groundwater protection and the financial resources related to these areas.

FSAC — Acronym for Food Safety Advisory Committee.

FSL — See Forecast Systems Laboratory.

FTC — See Federal Trade Commission.

FTS — Acronym for Federal Technology Service.

FURICS — Acronym for Federal Underground Injection Control Reporting System.

FVMP — Acronym for Federal Visibility Monitoring Programs.

FWS — See Fish and Wildlife Service.

GAO — See General Accounting Office.

Gas Research Institute — An organization of gas producers who plan and manage research and technology development related to supply of natural gas, efficient gas-fueled appliances and equipment, safety, and environmental impacts.

Gasification Technologies Council — An organization of corporations involved in the gasification of coal, petroleum, coke, and heavy oils and that gathers and disseminates information on gasification methods and technologies.

GAW — Acronym for Global Atmosphere Watch.

GC — See General Council.

GCC — See Global Climate Coalition.

GCF — See Greenhouse Crisis Foundation.

GCRP — See Global Change Research Program.

GEF — See Global Environmental Facility or Global Environmental Forum.

GELPI — See Georgetown Environmental and Policy Institute.

GEMI — Acronym for Global Environmental Management Initiative.

GEN — See Geneva Environment Network.

General Accounting Office — A federal accounting office established by the Budget and Accounting Act of 1921 to independently audit government offices; under the control and direction of the Comptroller General of the United States.

General Council — A group of attorneys employed by the U.S. Department of Health and Human Services and assigned to handle litigation of lawsuits brought by the U.S. Food and Drug Administration to ensure the legality of proposed regulations.

General Science Index — A bibliographical index available for journals or general science papers from 1978 to the present.

Geneva Environment Network — A cooperative partnership between over 40 environment and sustainable development organizations in units based in the International Environment House, including United Nations offices and programs and specialized agencies.

Geological Society of America — A broad, unifying, scientific society of professionals from academia, government, business, and industry whose mission is to advance the geosciences, to enhance the professional growth of its members, and to promote the geosciences in the service of humankind through promoting a human understanding of the Earth, planets, and life and catalyzing new scientific ways of thinking about natural systems.

Geospatial and Statistical Data Center — A data center located at the University of Virginia Library that provides archives, map collections, and a data clearinghouse; retrieves and creates custom datasets and maps; and provides information from the *Mapping Census 2000: The Geography of U.S. Diversity*, a synthesis of the basic patterns and changes in U.S. population distribution in the last decade, including color maps, race, diversity today, age, population density, and demographic shifts.

GEOSTAT — See Geospatial and Statistical Data Center.

GESAMP — See Joint Group of Experts on the Scientific Aspects of Marine Pollution.

GFDL — Acronym for Goddard Fluid Dynamics Laboratory.

GIFAP — Acronym for International Group of National Associations of Agrochemical Manufacturers.

GIPSA — Acronym for Grain Inspection, Packers, and Stockyards Administration.

GISS — Acronym for Goddard Institute for Space Studies.

GLERL — See Great Lakes Environmental Research Laboratory.

GLNPO — Acronym for Great Lakes National Program Office.

Global Change Research Program — A government-wide program with the goal to establish a scientific basis for national and international policymaking relating to natural and human-induced changes in the global Earth system.

Global Climate Coalition — An organization that works to coordinate participation in a scientific and policy debate relating to global climate change issues.

Global Environment Facility — An organization that helps developing countries fund projects and programs that protect the global environment.

Global Environmental Form — An organization that protects and nurtures the environment of the Earth for present and future generations by engaging in scientific and policy research and promoting a wide range of environmental activities at the local and international levels through information dissemination and networking.

Global Environmental Management Initiative — An organization of individuals and environmental companies who work to improve their operations through comprehensive self-assessment tools, planning, and understanding sustainable development, environmental impact, and economic development. They create a variety of strategies, using case studies to explain them, to improve various areas of the environment including water.

Global Investigation of Pollution in the Marine Environment — An international cooperative program of scientific investigations focused on marine contamination and pollution and established in response to recommendations of the United Nations Conference on the Human Environment, held in Stockholm, Sweden, in 1972.

Global Programs Division — A division of the U.S. EPA that implements programs to protect the ozone layer, including requirements under the Montreal Protocol and Title 6 of the Clean Air Act.

Global Reference Network — Program of geochemical data against which existing local and national geochemical data can be compared.

Global Water Partnership — A working partnership of various groups involved in water management, including government agencies, public institutions, private companies, professional organizations, and a variety of other agencies, that identifies critical knowledge needs at global, regional, and national levels; helps design programs for meeting these needs; and serves as an alliance for disseminating information on integrated water resources management.

GLU — See Great Lakes United.

GMCC — Acronym for Global Monitoring for Climatic Change.

GMENAC — Acronym for Graduate Medical Educational National Advisory Committee.

GPO — Acronym for Government Printing Office.

GPO Monthly Catalog — A database published by the Government Printing Office from 1976 to the present that covers all types of federal government publications.

GRCDA — Acronym for Government Refuse Collection and Disposal Association.

Great Lakes Environmental Research Laboratory — An NOAA laboratory that conducts integrated interdisciplinary environmental research and monitoring and provides supportive resource management and environmental services in coastal and estuarine waters, with a special emphasis on the Great Lakes.

Great Lakes United — An international conservation coalition formed by representatives of environmental, sports, union, community, and business groups that promote the conservation and enhancement of the Great Lakes ecosystem and target issues including hazardous and toxic substances, biodiversity, habitat protection, and water pollution.

Greenhouse Crisis Foundation — An organization that attempts to create global awareness of the greenhouse effect and resulting environmental issues through educational programs and litigation.

GRI — See Gas Research Institute.

GRN — See Global Reference Network.

Ground Water Protection Council — An organization of professionals and corporations involved in groundwater protection and underground disposal of drilling products and other hazardous materials who work to protect the groundwater supply of the United States.

GSA — See Geological Society of America; acronym for Gerontologic Society of America or General Services Administration.

GTC — See Gasification Technologies Council.

GTN — Acronym for Global Trend Network.

GWP — Acronym for Global Water Partnership.

GWPC — See Ground Water Protection Council.

HARRPA — Acronym for Hydrocarbon and Rosin Resins Producers Association.

Harvard Environmental Law Society — An organization of Harvard law students united to protect the environment through legal research and educational programs in the areas of nuclear power, toxic wastes, land-use planning, and wilderness preservation.

Hazardous Materials Advisory Council — An organization of shippers, carriers, container manufacturers, emergency response personnel, and cleanup companies that promotes safety in the domestic and international transportation of hazardous materials; see Dangerous Goods Advisory Council.

Hazardous Substance Release/Health Effects Database — A scientific and administrative database that provides access to information on the release of hazardous substances from Superfund sites or from emergency events and on the effects of hazardous substances on the health of human populations; it is produced and maintained by the Agency for Toxic Substances and Disease Registry.

Hazardous Substance Research Centers — Five multi-university research centers overseeing basic and applied research, technology transfer, and training involving problems relating to hazardous substance management.

HazMat Safety — A computer database of the Research and Special Programs Administration Office of Hazardous Material Safety, U.S. Department of Transportation, that presents a variety of information on the transportation of hazardous materials by air, rail, highways, and water.

HDRL — Acronym for High-Dose Reference Laboratory.

Health and Safety Science Abstracts — An abstract service covering health and safety science articles from 1977 to the present.

Health Effects Institute — An independent nonprofit corporation that provides high-quality, impartial, and relevant science on the health effects of pollutants from motor vehicles and from other sources in the environment and is supported by the U.S. EPA and industry.

HECC — Acronym for House Energy and Commerce Committee.

HEI — See Health Effects Institute.

HEW — Acronym for Department of Health, Education, and Welfare.

HFM — See National Society for Healthcare Foodservice Management.

HHS — Acronym for Department of Health and Human Services.

HMAC — Acronym for Hazardous Materials Advisory Council.

HMIS — See Hazardous Material Information System.

HMR — See Hydrometeorological Report.

HMTC — Acronym for Hazardous Materials Technical Center.

HPCRC — Acronym for High-Performance Computing Research Center.

HSRC — See Hazardous Substance Research Centers.

HST — Acronym for Hubble Space Telescope.

HUD — See Department of Housing and Urban Development.

Human Factors Society — A professional society of psychologists, engineers, physiologists, and other related scientists concerned with the use of human factors in the development of systems and devices.

HWDMS — Acronym for Hazardous Waste Data Management System.

HWGTF — Acronym for Hazardous Waste Groundwater Task Force or Hazardous Waste Groundwater Test Facility.

HWRTF — Acronym for Hazardous Waste Restrictions Task Force.

HWTC — Acronym for Hazardous Waste Treatment Council.

Hydromet — A system of data collection platforms that gathers hydrometeorological data and transmits it by satellite to a computer downlink.

IAA — Acronym for International Academy of Astronautics.

IAAC — Acronym for Interagency Assessment Advisory Committee.

IABO — Acronym for International Association of Biological Oceanography.

IACGEC — See Inter-Agency Committee on Global Environmental Change.

IADN — Acronym for Integrated Atmospheric Deposition Network.

IAE — See International Association for Ecology.

IAEA — See International Atomic Energy Agency.

IAEG — See International Association for Engineering Geology.

IAEMS — See International Association of Environmental Mutagen Societies.

IAF — Acronym for International Astronomical Federation.

IAFF — Acronym for International Association of Firefighters.

IAFP — See International Association for Food Protection.

IAGA — See International Association of Geomagnetism and Aeronomy.

IAGC — Acronym for International Association of Geochemistry and Cosmochemistry.

IAH — See International Association of Hydrogeologists.

IAHR — Acronym for International Association for Hydraulic Research.

IAHS — Acronym for International Association of Hydrological Sciences.

IAIGCR — Acronym for Inter-American Institute for Global Change Research.

IAPSO — See International Association for Physical Sciences of the Ocean.

IARC — Acronym for International Agency for Research on Cancer.

IAWPRC — Acronym for International Association on Water Pollution Research and Control.

IBEM — See International Board of Environmental Medicine.

IBSRAM — Acronym for International Board for Soil Research and Management.

IBT — Acronym for Industrial Biotest Laboratory.

IBT Reference Laboratory — A national research and specialty clinical lab that provides a wide range of tests and services in the area of allergies, clinical immunology, molecular biology, latex allergy, and pharmaceutical research, with a staff of doctoral scientists who are involved in a variety of research projects and clinical studies.

IBWA — See International Bottled Water Association.

ICAC — See Institute of Clean Air Companies.

ICAIR — See International Center for Advanced Internet Research.

ICBEN — Acronym for International Commission on the Biological Effects of Noise.

ICC — See Interstate Commerce Commission.

ICCA — Acronym for International Council of Chemical Associations.

ICCP — Acronym for International Climate Change Partnership.

ICCS — Acronym for International Conference on Chemical Safety.

ICGEB — Acronym for International Centre for Genetic Engineering and Biotechnology.

ICMASA — See Intersociety Committee on Methods for Air Sampling and Analysis.

ICME — Acronym for International Council on Metals and the Environment.

ICMM — See International Council on Mining and Metals.

ICOH — See International Commission on Occupational Health.

ICPEMC — See International Commission for Protection against Environmental Mutagens and Carcinogens.

ICRP — See International Commission on Radiation Protection.

ICRU — See International Commission on Radiation Units and Measurements.

ICS — See International Cytokine Society; acronym for Institute for Chemical Studies.

ICWM — Acronym for Institute for Chemical Waste Management.

ICWP — See Interstate Council on Water Policy.

IDAC — Acronym for International Disaster Advisory Committee.

Idaho National Engineering and Environmental Laboratory — A laboratory that is part of the Department of Energy and is involved in complex missions in the delivery of science-based engineering solutions, environmental cleanup programs, Department of Energy environmental management programs, and scientific programs for the northwest region of the United States.

IEA — See International Energy Agency or International Ergonomics Association.

IEB — Acronym for International Environmental Bureau.

IECA — Acronym for International Erosion Control Association.

IEEE — See Institute of Electrical and Electronics Engineers.

IEMP — Acronym for Integrated Environmental Management Project.

IEMS — Acronym for Integrated Emergency Management System.

IES — Acronym for Institute for Environmental Studies.

IFCB — See International Federation of Cell Biology.

IFCE — See International Federation of Consulting Engineers

IFCS — Acronym for International Forum on Chemical Safety.

IFIC — See International Food Information Council.

IFIS — Acronym for Industry File Information System or International Food Information Service.

IGCI — Acronym for Industrial Gas Cleaning Institute.

IGIS — Acronym for Imaging and Geospatial Information Society.

IHS — Acronym for Indian Health Service.

IHSPCB — See International Healthcare Safety Professional Certification Board.

IJC — See International Joint Commission on Great Lakes or *Internet Journal of Chemistry*.

ILO — See International Labor Organization.

ILSI — See International Life Sciences Institute.

ILSI Risk Sciences Institute — An organization established to advance and improve the scientific basis of ecological and human health risk assessment and working toward this goal through an international program of research, working groups, conferences, workshops, publications, seminars, and training programs.

IMA — See International Mycological Association; acronym for Industrial Medical Association.

IME — See Institute of Makers of Explosives.

IMLS — Acronym for Institute of Museum and Library Services.

IMPCA — Acronym for International Methanol Producers and Consumers Association.

INCE — See Institute of Noise Control Engineering.

Independent Petroleum Association of America — An organization of thousands of independent oil and gas producers and service companies across the United States that acts as an advocate for the exploration and production of oil and who presents its views before the U.S. Congress, the administration, and federal agencies.

Index Medicus — A monthly index of the world's leading biomedical literature that is indexed by subject and author and published by the National Library of Medicine.

Index to United States Government Periodicals — An index of government periodicals from 1974 to the present.

Indian Health Service — A bureau within the Department of Health and Human Services providing public health and medical services to Native Americans in the United States.

Industrial Medical Association — A professional organization whose members are concerned with the identification, prevention, diagnosis, and treatment of diseases and injuries associated with technology and industry.

INEEL — See Idaho National Engineering and Environmental Laboratory.

INFOTERRE — See International Environmental Referral and Research Network.

Innovative Treatment Remediation Demonstration — A program funded by the Department of Energy, Office of Environmental Management, that helps accelerate the adoption and implementation of new and innovative remediation technologies.

Institute for Local Self-Reliance — An organization that conducts research and provides technical assistance on environmentally sound economic development for government, small businesses, and community organizations.

Institute for Science and International Security — An organization that analyzes scientific and policy issues that affect national and international security, including the problems of war, spread of nuclear weapons, and environmental, health, and safety hazards of nuclear weapons production.

Institute of Clean Air Companies — An organization of companies that manufacture industrial gas-cleaning, air pollution control, and monitoring equipment to encourage general improvement of engineering and technical standards in the manufacture, installation, operation, and performance of the equipment for stationary sources to reduce environmental pollutants and protect the health and safety of the public.

Institute of Electrical and Electronics Engineers — An American society that establishes international standards in the computing, electronics, and telecommunications fields.

Institute of Makers of Explosives — An organization of manufacturers of commercial explosives and blasting supplies concerned with the safety and protection of their employees, users, the public, and the environment.

Institute of Noise Control Engineering — An organization of professionals in the field of noise control engineering who seek to develop the technology of noise control with emphasis on engineering solutions to environmental noise problems.

Institute of Scrap Recycling Industries — An organization representing processors, brokers, and consumers of scrap paper, glass, plastic, textiles, rubber, and ferrous and nonferrous metals who are interested in environmental control and recycling.

Instructional Resources Information System — An electronic database containing the U.S. EPA's latest descriptive and quantitative regulatory information on chemical constituents.

Integrated Waste Services Association — An organization of companies that design, build, and operate resource recovery facilities and promote integrated solutions to municipal solid waste management issues, including the use of waste-to-energy technology.

Inter-Agency Committee on Global Environmental Change — An organization in the United Kingdom, established to review national and international research activities directed at or related to global environmental research issues and to prepare a national strategy for data management, coordination, and development of appropriate environmental activities.

Intergovernmental Panel on Climate Change — An organization, established in 1988 by the United Nations Environmental Programme and the World Meteorological Organization to assess information in the scientific and technical literature related to all significant components of the issue of climate change.

International Agency for Cancer — An expert international agency of the World Health Organization that publishes evaluations of evidence on the carcinogenicity of a wide range of chemicals.

International Association for Ecology — An association of member libraries and institutions, national and international ecological associations, students, and other individuals, who promote and communicate the science of ecology and the application of ecological principles to global needs by encouraging public awareness of ecological problems, collecting and disseminating information, and acting as a clearinghouse and center for coordination of ecological projects.

International Association for Engineering Geology and the Environment — A worldwide scientific society with individual members and national groups devoted to the investigation, study, and solution of engineering and environmental problems that may arise as a result of the interaction between geology and activities of people as well as to the prediction and the development of techniques for prevention or remediation of geological hazards through research, teaching, and in-service education.

International Association for Food Protection — An organization of food safety professionals who provide educational programs and services to members, function as a clearinghouse on food safety, and serve as a forum for the exchange of information among members.

International Association for the Physical Sciences of the Oceans — An international association that participates in scientific research of the oceans along with other international associations.

International Association of Environmental Mutagen Societies — An organization of chemical, biological, and medical scientists in 50 countries who work in research, teaching, and administration in the fields of environmental mutagens, carcinogens, and their control and who organize training courses and workshops to test methods and interpretation of data; affiliated with the World Health Organization.

International Association of Geomagnetism and Aeronomy — An international scientific association promoting the study of magnetism and space physics.

International Association of Hydrogeologists — A scientific and educational organization whose mission is to promote research and an understanding of the proper management and protection of groundwater for the common good throughout the world.

International Atomic Energy Agency — An independent, intergovernmental science- and technology-based organization that is part of the United Nations and serves as the global focal point for nuclear cooperation in planning for and using nuclear science and technology for various peaceful purposes, including the generation of electricity.

International Board of Environmental Medicine — An accrediting agency for physicians, osteopaths, and persons working in related environmental professions who examine licensed practitioners, facilities, and relevant training programs and offer programs to evaluate qualifications of healthcare professionals, their training programs, and the facilities offering special types of treatment.

International Bottled Water Association — An association of bottled water plants, manufacturers of bottled water supplies, and international bottlers, distributors, and suppliers who conduct technical research and provide seminars.

International Center for Advanced Internet Research — An organization dedicated to accelerating innovation and enhancing global communications through advanced Internet technologies in partnership with the international community and using high-performance communication of digital information, innovative Internet technologies, and high-quality experimentation.

International Commission for Protection Against Environmental Mutagens and Carcinogens — An organization of scientists from academic and industrial institutes working in the areas of genetics, mutagensis, cancer, the epidemiology of cancer, genetic toxicology, and related sciences to identify and promote scientific principles and to determine guidelines and regulations for the purpose of preventing or minimizing the deleterious effects of chemicals on human genetic material.

International Commission on Occupational Health — An international non-governmental society whose goal is to foster scientific progress, knowledge, and development of occupational health and safety programs, in all its aspects.

International Commission on Radiation Protection — A nongovernmental organization founded in 1928 to provide general guidance on the safe use of radiation in medical radiology.

International Commission on Radiation Units and Measurements — An organization of senior advisers, consultants, and representatives of 12 countries who develop internationally acceptable recommendations regarding quantities and units of radiation and radionuclides, procedures suitable for measurement and application of these quantities in clinical settings, data needed for application of the procedures, and techniques to be used for uniformity in reporting.

International Council on Mining and Metals — An international organization that initiates, conducts, promotes, and communicates research and analysis in the world's mining, minerals, and metals industries. It seeks to lead changes within these industries by stimulating discussion and coordinating activities of those involved; develops and communicates a clear position on mining, minerals, metals, and the environment; promotes global best practice performance standards; maintains a high-level dialog with governmental, non-governmental, and community organizations; and works at all levels with academia, organizations, industry, and government to promote sustainable development while protecting the environment.

International Cytokine Society — An organization of persons who have research, clinical, or educational experience in the field of cytokines or an allied discipline and work together to combat a variety of serious diseases through the exchange of scientific information.

International Energy Agency — A major program of the Organization for Economic Cooperation and Development that promotes cooperation in energy research among developed nations, helps developing countries negotiate energy programs, and prepares plans for international emergency energy allocation.

International Environment House — A large structure in Geneva, Switzerland, where a group of United Nations and non-governmental organizations act together in the field of environment and sustainable development to foster synergies and encourage partnership between the different organizations and governmental agencies.

International Environmental Referral and Research Network — An international organization made up of 177 countries coordinated by the United Nations Environment Programme responding to requests from the international community for environmental information through document delivery, database searching, bibliographical products, purchasing information, and referrals to experts.

International Ergonomics Association — A federation of ergonomics and human factors societies around the world whose mission is to elaborate and advance ergonomics science and practical applications to improve the quality of life by expanding the scope of application and contributions of ergonomics to society.

International Federation of Cell Biology — An organization of seven national and 14 regional associations of cell biologists who work together to promote international cooperation among scientists working in cell biology and related fields and contribute to the advancement of cell biology in all of its branches.

International Federation of Consulting Engineers — An association of national member associations of engineering-based consulting companies that arranges seminars and other events; develops publications and international ethical and professional standards; works to strengthen the consulting industry in developing countries; and provides resources for international businesses.

International Food Information Council — An organization that serves as an information and educational resource on nutrition and food safety and provides science-based information to journalists, health professionals, educators, government officials, and other opinion leaders who communicate with the public.

International Healthcare Safety Professional Certification Board — An organization of individuals who work in hospitals or healthcare facilities and are specialists in the handling and control of hazardous materials; biological, chemical, and physical hazards; fire prevention and protection; maintenance and engineering; and personal protective equipment and who are certified to do the work by education and experience.

International Institute for Energy Conservation — An organization that works with developing nations to establish sustainable growth through efficient use of energy while reducing air and water pollution and the threat of global warming.

International Joint Commission — A governmental commission of members from Canada and the United States created to protect and enhance the lakes and river systems along the border of the two countries.

International Labor Organization — A permanent international organization created by the Paris Peace Conference of 1919 to promote the improvement of working and living conditions as an essential contribution to safeguarding peace throughout the world.

International Life Sciences Institute — A nonprofit worldwide foundation that seeks to improve the well-being of the general public by their pursuit of balanced science and to further the understanding of scientific issues related to nutrition, food safety, toxicology, risk assessment, and the environment by bringing together scientists from academia, government, and industry.

International Mycological Association — An international society representing 20,000 mycologists from 80 countries that promotes the study of mycology in all its aspects.

International Occupational Hygiene Association — An organization of occupational hygienists worldwide and their member organizations who promote and develop occupational hygiene standards, the exchange of occupational hygiene information, and the development of occupational hygiene at a professional level, in addition to maintaining and promoting a high standard of ethical practice.

International Organization for Standards — A worldwide federation of national standards bodies from more than 140 countries, whose mission is to promote the development of standards and related activities in the world in an attempt to facilitate the international exchange of goods and services and to develop cooperation in the areas of intellectual, scientific, technological, and economic activity.

International Packaged Ice Association — An organization of manufacturers and distributors of packaged ice, ice-making equipment, and suppliers who are involved in improving technology, sanitation, and food service.

International Paint and Printing Council — An organization that provides a forum for information exchange and cooperation on the major issues and priorities facing paint and printing industries worldwide, acts as a focal point for monitoring and communicating specified international issues, and develops recommendations and analysis of problems.

International Programme on Chemical Safety — A World Health Organization program that is involved in evaluation of risks to human health and the environment, methodologies for risk assessment, prevention and management of toxic exposures and chemical emergency use, exchange of information on chemical safety and communication of related hazards and risks, and capacity building for sound management of chemicals and risk reduction.

RESOURCE DIRECTORY

International Red Cross Society — An international organization based in Geneva, Switzerland, concerned primarily with the humane treatment and welfare of victims of war and disaster and with neutrality of hospitals and medical personnel in times of war.

International Society for Analytical Cytology — An organization of researchers and academics from government and private sectors interested in using high-power technical equipment in the study of the structure, function, multiplication, pathology, and life history of cells.

International Society for Environmental Toxicology and Cancer — An organization of clinicians and researchers working in the fields of environmental toxicology and oncology who promote research and information exchange.

International Society for Soil Mechanics and Geotechnical Engineering — An international association that fosters worldwide technology transfer, contributes to advancing the state of engineering and construction practice, and accelerates the reliable use of innovative ground improvement geosystems for a variety of engineering applications. It emphasizes reinforcement and grouting.

International Society for the Study of Harmful Algae — An international association that promotes and fosters research and training programs on harmful algae and co-sponsors meetings on this topic at the national, regional, and international levels.

International Society of Soil Science — A non-governmental, nonprofit scientific society whose purpose is to promote all branches of soil science and its applications, to promote contacts among scientists and other persons engaged in the study and the application of soil science, to stimulate scientific research, and to further the application of research for the benefit of people.

International Test and Evaluation Association — An organization of engineers, scientists, managers, and other industry, government, and academic professionals interested in testing and evaluating products and complex systems.

International Union of Geological Sciences — A non-govermental, nonpolitical, and nonprofit scientific organization that addresses earth science problems of a broad international scope and encourages the highest levels of international cooperation and participation in making the Earth and human welfare a priority.

International Union of Pure and Applied Chemistry — An international organization of chemists from industry and academia who recognize the need for international standardization in chemistry, nomen-clature of inorganic chemicals, atomic weights, physical constants, names, and symbols.

Internet Journal of Chemistry — A journal dedicated to promoting the use of the Internet and development of network resources to enable chemists to better communicate by providing a mechanism for chemists to publish their research in developing new techniques, new resources, new databases, etc. on the Internet for use by other chemists.

Inter-Organizational Programme for the Sound Management of Chemicals — A division of the United Nations that coordinates and promotes the joint planning of other international organizations in the area of chemistry to ensure the global sound management of hazardous chemicals and to protect human health and the environment from the impact of toxic chemicals.

Intersociety Committee on Methods for Air Sampling and Analysis — A committee made up of members of societies of environmental engineers, chemists, biologists, and physicists to prepare, publish, and apply recommended methods of air sampling.

Interstate Commerce Commission — A federal agency regulating interstate surface transportation, including trains, trucks, buses, water carriers, freight forwarders, transportation brokers, and coal slurry pipelines.

Interstate Council on Water Policy — An organization of state, interstate, and intrastate officials with responsibilities for all water quantity and quality programs and who work with individuals, businesses, universities, and other governmental agencies with primary interest in water.

Interstate Oil and Gas Compact Commission — An organization of states promoting conservation and efficient recovery of domestic oil and natural gas resources while protecting health, safety, and the environment.

Interstate Professional Applicators Association — An organization of companies working in the application of horticultural spraying whose goal is to protect the health and safety of the public and the environment through proper chemical usage.

Interstate Technology Regulatory Council — A state-led coalition working together with industry and stakeholders to achieve regulatory acceptance of environmental technologies, break down barriers against these technologies and reduce compliance costs, deepen technical knowledge, and streamline the regulation of new environmental technologies.

IOC — Acronym for International Oceanographic Commission.

IOCC — See Inter-Organizational Programme for the Sound Management of Chemicals.

IOGCC — See Interstate Oil and Gas Compact Commission.

IOHA — See International Occupational Hygiene Association.

IOMC — Acronym for International Organization Committee on the Sound Management of Chemicals.

IPAA — See Interstate Professional Applicators Association or Independent Petroleum Association of America.

IPCC — See Intergovernmental Panel on Climate Change.

IPCS — See International Programme on Chemical Safety.

IPCSINCHEM — A computer searching program that offers quick electronic access to thousands of searchable full-text documents on chemical risk and the sound management of chemicals. It is produced through the cooperation of the International Programme on Chemical Safety and the Canadian Centre for Occupational Health and Safety.

IPPIC — See International Paint and Printing Ink Council.

IRG — Acronym for Interagency Review Group.

IRIS — See Instructional Resources Information System.

IRLG — Acronym for International Regulatory Liaison Group.

IRMC — Acronym for Inter-Regulatory Risk Management Council.

IRPTC — Acronym for International Register of Potentially Toxic Chemicals.

IRR — Acronym for Institute of Resource Recovery.

ISA — Acronym for Instrument Society of America.

ISAC — See International Society for Analytical Cytology; acronym for Industry Sector Advisory Committee.

ISC — Acronym for Intersociety Committee on Methods (for Air Sampling and Analysis).

ISETC — See International Society for Environmental Toxicology and Cancer.

ISO — See International Organization for Standardization.

ISSA — Acronym for International Sanitary Supply Association, Inc.

ISSHA — Acronym for International Society for Study of Harmful Algae.

ISSMFE — See International Society of Soil Mechanics and Foundation Engineering.

ISSMGE — Acronym for International Society of Soil Mechanics and Geotechnical Engineering.

ISSS — See International Society of Soil Science.

ITC — Acronym for Innovative Technology Council, Interagency Testing Committee, or International Trade Commission.

ITRC — See Interstate Technology Regulatory Coordination.

ITRI — Acronym for Inhalation Toxicology Research Institute.

IUGS — See International Union of Geological Sciences.

IUPAC — See International Union of Pure and Applied Chemistry.

IWSA — See Integrated Waste Services; acronym for International Water Supply Association.

JAMA — Acronym for *Journal of the American Medical Association.*

JAPCA — Acronym for *Journal of Air Pollution Control Association.*

JAPHA — Acronym for *Journal of the American Public Health Association.*

JCAH — See Joint Commission on Accreditation of Hospitals.

JCAHO — See Joint Commission on Accreditation of Healthcare Organizations.

JEC — See Joint Economic Committee.

JECFA — See Joint Expert Committee on Food Additives.

JEI — See Joint Education Initiative.

JEIOG — Acronym for Emissions Inventory Oversight Group.

JIFSAN — See Joint Institute for Food Safety and Applied Nutrition.

JIFSR — See Joint Institute for Food Safety Research.

JMPR — Acronym for Joint Meeting on Pesticide Residues.

Joint Commission on Accreditation of Healthcare Organizations — A private, nongovernmental agency that establishes guidelines for the operation of hospitals and other healthcare facilities, conducts accreditation programs, and surveys and promotes high standards of institutional care.

Joint Commission on Accreditation of Hospitals — A private, nonprofit organization whose purpose is to encourage the attainment of uniformly high standards of institutional medical care; it is comprised of representatives of the American Hospital Association, American Medical Association, American College of Physicians, and American College of Surgeons.

Joint Economic Committee — Created by Congress to be one of only four joint committees made up of members of the U.S. Senate and the House of Representatives whose function is to review economic conditions and to recommend improvements in economic policy.

Joint Education Initiative — A project developed by the USGS, NOAA, NASA, industry, and teachers to enable teachers and students to explore the massive quantities of earth science data published by the U.S. government and to encourage research and analysis in science education.

Joint FAO/WHO Expert Committee on Food Additives — An international expert scientific committee administered jointly by the Food and Agricultural Organization of the United Nations and the World Health Organization that evaluates the safety of food additives, contaminants, naturally occurring toxicants, and residues of veterinary drugs in food.

Joint Group of Experts on the Scientific Aspects of Marine Environmental Protection — A program of the United Nations that utilizes the skills of a variety of United Nations organizations to work on the problems of all scientific aspects of the prevention, reduction, and control of the degradation of the marine environment and sustaining life-support systems, resources, and amenities.

Joint Information Center — The primary field location for the coordination of federal and state media relations.

Joint Institute for Food Safety and Applied Nutrition — An institute established in April 1996 by the U.S. Food and Drug Administration and the University of Maryland to conduct and improve the quantity and quality of multidisciplinary research and educational programs to provide the basis for sound public health policy; promote food safety, human nutrition, animal health; and to integrate academic and regulatory science programs.

Joint Institute for Food Safety Research — An institute of the U.S. Department of Health and Human Services and the U.S. Department of Agriculture created to coordinate planning and priority setting for food safety research for the two departments, other governmental agencies, and the private sector in order to use research results in a more effective manner to improve the practice of food safety from the farm to the table.

Joint Interagency Intelligence Support Element — An interagency intelligence group used to merge intelligence information from the various agencies involved in a response to a threat or incident of a weapon of mass destruction.

JPL — Acronym for Jet Propulsion Laboratory.

JSC — Acronym for Johnson Space Center.

Land and Minerals Management — A major program of the U.S. Department of the Interior that directs and supervises the Bureau of Land Management, the Minerals Management Service, and the Surface Mining and Reclamation Enforcement units.

Landfill Methane and Outreach Program — A program of the U.S. EPA that encourages the use of methane gas from landfill decomposition as an energy source instead of allowing it to go to waste.

LANL — Acronym for Los Alamos National Laboratory.

LDC — See Linguistic Data Consortium; acronym for London Dumping Convention.

LDRTP — Acronym for Land Disposal Restrictions Task Force.

LEAF — See Legal Environmental Assistance Foundation.

Legal Environmental Assistance Foundation — An organization that promotes the protection of the environment and health of the community by enforcing environmental regulations, discouraging harmful toxic and hazardous waste dumping, and encouraging energy efficiency.

Legal Information Institute — A nonprofit activity of the Cornell University Law School whose mission is to carry out applied research on the use of digital information technology in the distribution of legal information, the delivery of legal information, and the practice of law to make the law more accessible to legal professionals, students, teachers, and the general public.

LEP — Acronym for Laboratory Evaluation Program.

Library of Congress — The national library of the United States offering diverse materials for research, including the world's most extensive collection in many areas, such as American history, music, and art.

LII — See Legal Information Institute.

Linguistic Data Consortium — A Consortium of universities, companies, and government research laboratories hosted by the University of Pennsylvania that creates, collects, and distributes speech and text databases, lexicons, and other resources for research and development purposes.

LMOP — See Landfill Methane Outreach Program.

LOC — See Library of Congress.

LOCIS — Acronym for Library of Congress Information System.

Manufacturers of Emission Controls Association — An organization of manufacturers of motor vehicle and stationary-source emission control equipment who provide information on emission technology and industry capabilities.

MARAD — Acronym for Maritime Administration.

Marine Technology Society — An organization of scientists, engineers, technologists, and others interested in marine science and technology.

Materials Technology Institute of the Chemical Process Industries — An organization of chemical processors and their suppliers who conduct studies on the deterioration of materials and equipment used in the industry and develops projects to solve these problems.

MEDEX — An educational program accredited by the American Medical Association for training military personnel with medical experience to become physician's assistants.

Medical Literature Analysis and Retrieval System — A computerized bibliographic system of the National Library of Medicine from which the *Index Medicus* is produced; see also MEDLINE.

Medical Mycological Society of the Americas — An organization of medical professionals interested in fungi and fungal diseases who seek to exchange professional information, promote continuing education, and increase knowledge of diseases due to fungi.

MEDLARS — See Medical Literature Analysis and Retrieval System

MEDLINE — Acronym for MEDLARS on-line, a computerized bibliographical computer system; see also Medical Literature Analysis and Retrieval System.

MESA — See Mining Enforcement and Safety Administration.

Meteorological Service of Canada — A source for meteorological information that monitors water quantities, provides information, and conducts research on climate, atmospheric science, air quality, ice, other environmental issues, and provides expertise in these areas.

MFSA — Acronym for Metal Finishing Suppliers Association.

Mine Safety and Health Administration — A program of the Department of Labor that administers and enforces the health and safety provisions of the Federal Mine Safety and Health Act of 1977 and assists in rescue operations following mining accidents. It regulates coal mines, metal and nonmetal mines, and constant exposure to asbestos and various carcinogens found in mines.

Minerals Management Service — A program of the U.S. Department of the Interior that administers the Outer Continental Shelf Lands Act and evaluates, classifies, and supervises oil, gas, and other mineral reserves and operations on outer Continental shelf lands as well as submerged lands.

Mining Enforcement and Safety Administration — A now defunct administration whose functions have been transferred to the Mine Safety and Health Administration.

MMSA — See Medical Mycological Society of the Americas.

MMWR — See *Morbidity and Mortality Weekly Reports.*

Morbidity and Mortality Weekly Reports — A Centers for Disease Control weekly publication that gives information on current trends in the nation's problems of disease.

MRC — Acronym for Medical Research Council.

MSC — See Meteorological Service of Canada.

MSEL — Acronym for Materials Science and Engineering Library.

MSHA — See Mine Safety and Health Administration.

MTI — See Materials Technology Institute of the Chemical Process Industries.

Municipal Waste Management Association — An organization composed of mayors of cities with populations of 30,000 or more who belong to the U.S. Conference of Mayors and join together with other organizations with a common interest in the management of solid waste and broader environmental issues related to environmental protection in the urban setting.

NABIE — See National Academy of Building Inspection Engineers.

NACA — Acronym for National Agricultural Chemicals Association.

NACCA — Acronym for North American Council of Chemical Associations.

NACCED — See National Association for County Community and Economic Development.

NACCHO — See National Association of County and City Health Officials.

NACD — See National Association of Chemical Distributors.

NACEC — See North American Commission for Environmental Cooperation.

NACEPT — Acronym for National Advisory Council for Environmental Policy and Technology.

NACo — See National Association of Counties.

NAECA — See National Automotive Environmental Compliance Assistance.

NAELS — See National Association of Environmental Law Societies.

NAEP — See National Association of Environmental Professionals.

NAFEC — Acronym for North American Fund for Environmental Cooperation.

NAFEM — See North American Association of Food Equipment Manufacturers.

NAFSMA — See National Association of Flood and Stormwater Management Agencies.

NAHRO — See National Association of Housing and Redevelopment Officials.

NAHIT — See National Affordable Housing Training Institute.

NAL — Acronym for National Agricultural Library.

NALBOH — See National Association of Local Boards of Health.

NALGEP — See National Association of Local Government Environmental Professionals.

NALHFA — See National Association of Local Housing Finance Agencies.

NAM — See National Association of Manufacturers.

NAMH — Acronym for National Association for Mental Health.

NANCO — See National Association of Noise Control Officials.

NAPT — Acronym for National Association of Physical Therapists.

NAPE — See National Association of Physicians for the Environment.

NAPHSIS — See National Association for Public Health Statistics and Information Systems.

NAPIM — See National Association of Printing Ink Manufacturers.

NARA — Acronym for National Agrichemical Retailers Association.

NARUC — See National Association of Regulatory Utility Commissioners.

NAS — See National Academy of Sciences.

NASA — See National Aeronautics and Space Administration.

NASBO — See National Association of State Budget Officers.

NASDA — Acronym for National Association of State Departments of Agriculture.

NASEO — See National Association of State Energy Officials.

NASQUAN — Acronym for National Streamwater Quality Accounting Network.

NASS — Acronym for National Agricultural Statistical Service.

NAST — Acronym for National Academy of Science and Technology.

NASUCA — See National Association of Utility Consumer Advocates.

NATCOL — Acronym for Natural Food Colors Association.

National Academy of Building Inspection Engineers — An organization of state-registered professional engineers who strive to lead and advance the integrity, value, and understanding of the practice of professional engineering as it applies to the inspection, investigation, and evaluation of buildings and homes.

National Academy of Engineering — An independent society whose members are elected in recognition of important contributions to the field of engineering and technology and who share responsibility with the National Academy of Sciences for examining questions of science and technology at the request of the federal government.

National Academy of Sciences — A private, honorary organization of scholars in scientific and engineering research chartered by an act of Congress in 1863 to serve as an advisory agency to the federal government on questions of science and technology.

National Acid Precipitation Assessment Program — An interagency program of the NOAA, EPA, DOE, DOI, USDA, and NASA that conducts research, monitoring, and assessment programs on the causes, effects, and controls used for acid rain.

National Aeronautics and Space Administration — A federal government agency whose function is to conduct research for the solution of problems of flight within and outside of the atmosphere of the Earth and to develop, construct, test, and operate aeronautical and space vehicles.

National Affordable Housing Training Institute — A nonprofit organization composed of eight national public interest groups that provides technical assistance and training to its members in the area of affordable housing and is currently funded through a cooperative agreement with the Department of Housing and Urban Development.

National Agricultural Library — A division of the USDA that is a major international source for agricultural and related information.

National Association for County Community and Economic Development — An affiliate of the National Association of Counties that aids county governments in developing the technical capacities needed in a variety of areas, such as housing, community and economic development, and other programs.

National Association for Public Health Statistics and Information Systems — An organization that advocates, creates, and maintains comprehensive public health information systems that integrate vital records registration, public health statistics, and other health information.

National Association of Chemical Distributors — An organization of member companies who are in full compliance with the responsible distribution process of chemicals in all areas. It serves the public interest by conducting research and educational programs for proper distribution practices and the safe handling, use, storage, transportation, disposal, and recycling of chemical products.

National Association of Chemical Recyclers — An organization of commercial chemical recyclers and others interested in the industry that promotes recovery and reuse of spent solvent as an alternative to waste disposal.

National Association of Church Food Service — An organization of individuals involved in production of food for churches that offers certification to become a certified food-service director.

National Association of Conservation Districts — An organization of conservation districts that works to promote the conservation of land, forest, and other natural resources and is interested in erosion and sediment control, water quality, floodplains, and rural development.

National Association of Counties — The only national organization that represents county governments in the United States and provides services including legislative, research, technical, and public affairs assistance, as well as enterprise services, for its members. It acts as a liaison to other levels of government, works to improve public understanding of counties, serves as a national advocate for counties, and provides the counties with resources to help to find innovative solutions to meet the challenges that occur.

National Association of County and City Health Officials — An organization of city, county, and district health officers who work to develop the technical competence, managerial capacity, and leadership potential of local public health officials.

National Association of Energy Service Companies — An organization of energy service companies, equipment manufacturers, affiliates of utilities, financial institutions, and governmental and other organizations involved in energy conservation and alternate energy projects.

National Association of Environmental Law Societies — An organization of environmental law societies that promotes the presentation of environmental issues among law schools and in the law profession as well as the public at large.

National Association of Environmental Professionals — An organization that promotes ethical practice, technical competency, and professional standards in the environmental field through environmental research, technology, law, and policy study.

National Association of Flood and Stormwater Management Agencies — An organization of county and local governments concerned with the management of water resources and the ability to reduce or eliminate flooding as well as provide stormwater management and conservation of watersheds.

National Association of Housing and Redevelopment Officials — An organization of individuals and public agencies involved in community rebuilding through community development, public housing, large-scale private or cooperative housing rehabilitation, and conservation of existing neighborhoods through housing code enforcement, voluntary citizen action, and government actions.

National Association of Local Boards of Health — An organization representing the interests of local boards of health throughout the United States who relate their concerns to individuals responsible for developing public health policy at the national level.

National Association of Local Government Environmental Professionals — An organization of environmental officials responsible for municipal environmental compliance and promoting effective administrative performance in municipal environmental quality projects.

National Association of Local Housing Finance Agencies — A national association of professionals working together at the local level to finance affordable housing in a broader community development context.

National Association of Manufacturers — The nation's largest industrial trade association of people that make things in America and whose mission is to enhance the competitiveness of manufacturers and improve American living standards by shaping a legislative and regulatory environment conducive to U.S. economic growth; also, to increase understanding among policymakers, the media, and the general public about the importance of manufacturing American goods.

National Association of Noise Control Officials — An organization of state and local officials working to control environmental and industrial noise and interested individuals from government, science, industry, education, and citizens groups to promote laws to control noise, sponsor use of effective noise measurement and analysis devices, and promote a reduction of excessive and unnecessary noise in the community.

National Association of Physicians for the Environment — An organization of physicians, medical and environmental organizations, and corporations who promote improved understanding of the impacts of environmental pollutants on the organs, systems, and disease processes of the human body.

National Association of Printing Ink Manufacturers — A trade association whose purpose is to represent the printing ink industry in the United States and provide direction of managment in the areas of environmental issues, business management, governmental regulations, and regulatory compliance.

National Association of Regulatory Utility Commissioners — An organization of people from the federal, state, municipal, and Canadian regulatory commissions that has jurisdiction over utilities and carriers and is involved with electricity, natural gas, and nuclear power.

National Association of State Budget Officers — A nonprofit, nonpartisan, professional membership research organization of state finance officers, whose mission is to improve the quality and availability of information for state budgeting, provide opportunities to share practices across states, provide training and research publications and educational meetings, and assist the National Governors Association in the development of fiscal policies.

National Association of State Energy Officials — An organization of state energy officials who promote establishment and enforcement of effective state

RESOURCE DIRECTORY

laws and regulations regulating the energy industries.

National Association of Utility Consumer Advocates — An organization of members from 38 states and the District of Columbia who exchange information and take positions on issues affecting utility rates before federal agencies, Congress, and the courts.

National Atmospheric Deposition Process Network — A major monitoring network involving cooperative effort among numerous federal, state, and private research agencies in the area of acid rain.

National Automotive Environmental Compliance Assistance Center — A 24-hour automobile service center available to help persons engaged in automotive services, collision repairs, and other sectors of the automotive industry to better understand their environmental responsibilities and help them achieve compliance with environmental program requirements.

National Bioenergy Industries Association — An organization of private corporations interested in or involved with the improvement and productivity of biomass resources focusing on the commercial, utility, and industrial use of biomass for energy as well as encouraging public and private participation in the development of biomass as a renewable energy source through the use of responsible forest, agricultural, and land management practices.

National Biomedical Research Foundation — A nonprofit institution affiliated with Georgetown University Medical Center, dedicated to carrying out scientific research in advanced fields, including the application of computers and modern technology to medical research.

National Bureau of Standards — An agency in the U.S. Department of Commerce that sets accurate measurement standards for commerce, industry, and science in the United States.

National Cancer Institute — One of the National Institutes of Health whose function is to expand existing scientific knowledge on cancer causes and prevention as well as on the diagnosis, treatment, and rehabilitation of cancer patients.

National Center for Atmospheric Research — An organization whose mission is to plan, organize, and conduct atmospheric and related research programs in collaboration with universities, to provide state-of-art research tools and facilities to the entire atmosphere sciences community, to support and enhance university atmospheric research and education, and to facilitate the transfer of technology to the public and private sectors.

National Center for Environmental Assessment — A program of the U.S. EPA that conducts research to help ensure that efforts to reduce environmental

risks are based on the best available scientific information in order to protect human health and to safeguard the air, water, and land.

National Center for Environmental Health Strategies — An organization of persons with environmental illnesses (including those with chemical sensitivity disorders); medical, legal, and scientific professionals; governmental agencies; environmentalists; and interested others who promote public awareness of health problems caused by chemical and environmental pollutants, with a focus on chemical sensitivity disorders.

National Center for Geographic Information and Analysis — An independent research consortium dedicated to basic research and education in the geographic information science and its related technologies, including geographic information systems; members include the University of California, Santa Barbara, University of Buffalo, and University of Maine.

National Center for Health Statistics — A part of the Centers for Disease Control whose function is to collect, maintain, analyze, and disseminate data on health status and services.

National Center for Toxicological Research — A program within the Department of Health and Human Services whose function is to study the biological effects of potentially toxic chemical substances found in the environment, emphasizing determination of health effects resulting from long-term, low-level exposure to chemical toxicants.

National Centers for Environmental Prediction — An agency of the NOAA's National Weather Service composed of nine centers working together to save lives and property and create economic opportunity; formerly known as the National Meteorological Center.

National Climatic Data Center — A division of the NOAA that works with scientists and researchers worldwide, serves as a national resource for climate information, provides information on the trends and anomalies of weather and climate, and provides a Web site for easy distribution of the material.

National Coal Association — An organization of coal companies representing the coal industry in all matters except labor relations.

National Coal Council — An organization of individuals appointed by the Secretary of Energy and representing coal producers, transporters, manufacturers, and minorities who make recommendations to the Secretary on issues involving coal.

National Committee on Radiation Protection — An advisory group of scientists and professionals that makes recommendations for radiation protection in the United States.

National Conference of Local Environmental Health Administrators — An organization of professional environmental health personnel engaged in or officially concerned with city, county, or district environmental health administration or teaching of environmental health who promote improvement and greater use of science in the practice of environmental health in community life.

National Conference of State Legislatures — A national advisory council that provides services to state legislatures by bringing together information from all the states to determine workable answers to complex policy questions.

National Conference of States on Building Codes and Standards — An organization composed of individuals appointed by governors of the states and territories as well as from organizations concerned with building standards to promote the updating and adoption of model energy conservation codes for new and existing buildings.

National Cooperative Geologic Mapping Program — A program that is part of the 1999 FEDMAP Geologic Mapping project and is involved in mapping areas regions of the United States and providing the information for use by professionals and the public.

National Coordination Office for Computing Information, and Communications — An agency of the federal government that coordinates multi-agency research and development projects that involve computing, information, and communications, including the High-Performance Computing and Communications Program.

National Council for Air and Stream Improvement — A council originally established to assist the pulp and paper industry in addressing wastewater treatment issues and which is now involved in extensive research in all areas of the environment.

National Council for Science and the Environment — An organization of over 500 academic, scientific, environmental, and business organizations seeking to improve the scientific basis for environmental decision making; it uses an integrated, interdisciplinary, research, scientific, and educational approach to deal with environmental issues and is deeply involved in full implementation of the recommendations of the National Science Foundation's report, *Environmental Science and Engineering for the 21st Century: The Role of the National Science Foundation.*

National Council of State Housing Agencies — A national, nonprofit organization created by the nation's state housing finance agencies to assist them in increasing housing opportunities for lower income and underserved people through the financing, development, and preservation of affordable housing.

National Council on Radiation Protection and Measurements — An organization of nationally recognized scientists who share the belief that significant advances in radiation protection and measurement can be achieved through cooperative effort and by conducting research focusing on safe occupational exposure levels as well as disseminating the information chartered by Congress.

National Disaster Medical System (NDMS) — An emergency medical system housed at the U.S. Public Health Service that is based on the concept of the Civilian–Military Contingency Hospital System.

National Drinking Water Advisory Council — An organization that advises the U.S. EPA administrator on activities, functions, and policies related to implementation of the Safe Drinking Water Act.

National Environmental Balancing Bureau — An organization of qualified heating, cooling, and air conditioning contractors specializing in the fields of systems balancing, sound vibration, mechanical heating and cooling systems, and testing of clean rooms; it maintains industry standards and procedures for testing, adjusting, and balancing systems to improve the indoor environment.

National Environmental Development Association — An organization of corporations and individuals who provide information on balancing environmental and economic needs.

National Environmental Education and Training Foundation — An organization dedicated to environmental learning to connect people to the solutions of issues such as health care, educational excellence, consumers' "right-to-know", promotion of individual responsibility, and community participation.

National Environmental Health Association — A professional association of environmental health and protection specialists that provides certification, training, continuing education, consulting, and professional liaisons; all members are in the environmental health field. It also provides databases, mailing lists, educational materials, publications, and credentials; meetings are available to all members and nonmember professionals who strive to improve the environment. Special sections include air, land, water quality, counter-bioterrorism, emerging pathogens, environmental management, food, geographic information systems, hazardous waste/toxic substances, injury prevention and occupational health, international environmental health, noise control, on-site wastewater management, vector control, and zoonotic diseases. Members include professionals in environmental health protection, including registered sanitarians, registered environmental health specialists, registered environmental technicians, certified environmental health technicians, registered hazardous substances professionals, and registered hazardous substances specialists.

National Environmental Performance Partnership Agreements — A system that allows states to assume greater responsibility for environmental programs based on their relative ability to execute them.

National Environmental Publications Information System — A division of the U.S. EPA that provides 9000 full-text documents.

National Environmental Satellite, Data, and Information Service — A program of the NOAA whose mission is to provide and ensure timely access to global environmental data from satellites and other sources to promote, protect, and enhance the nation's economy, security, environment, and quality of life.

National Environmental Training Association — A professional association of environmental trainers in air pollution, noise pollution, hazardous waste, water pollution, and wastewater treatment.

National Environmental Trust — An organization that identifies and publicizes environmental issues related to climate change, endangered species, hazardous chemicals, and weakening of environmental laws at the national and local levels.

National Estuary Program — A program established under the Clean Water Act Amendments of 1987 to develop and implement conservation and management plans for protecting estuaries and restoring and maintaining their chemical, physical, and biological integrity, as well as controlling point and nonpoint pollution sources.

National Exposure Registry — A listing of people exposed to hazardous substances; this database is used to facilitate epidemiological research in determining adverse health effects of persons exposed to low levels of chemicals over a long period of time.

National Eye Institute — A division of the National Institutes of Health established in 1968 to support research on the normal functioning of the human eye, the pathology of visual disorders, and rehabilitation of the visually handicapped.

National Fire Academy — A component of the Federal Environmental Management Agency's National Emergency Training Center. It provides fire protection and control training for fire services and allied services.

National Fire Protection Association — An international organization that promotes improved fire protec-

tion and prevention and establishes safeguards against loss of life and property by fire.

National Foundation for the Chemically Sensitive — An organization of individuals suffering from chemical hypersensitivity, their families and friends, healthcare professionals, and other interested people who promote public awareness concerning chemical hypersensitivity disorders such as multiple chemical sensitivities, environmental illness, food intolerance, total allergy syndrome, and chronic fatigue.

National Geodetic Survey — A part of the NOAA that defines and manages the National Spatial Reference System.

National Geologic Map Database — A database project that serves as a central archive or point-of-contact for users searching for earth science information.

National Geophysical Data Center — A division of the NOAA whose mission is to provide and ensure timely access to global environmental data from satellites and other sources to promote, protect, and enhance the nation's economy, security, environment, and quality of life.

National Ground Water Association — An organization of groundwater drilling contractors; manufacturers and suppliers of drilling equipment; groundwater scientists, such as geologists and engineers; public health officials; and others interested in the process of locating, developing, preserving, and using groundwater supplies that encourages scientific educational research and development of standards.

National Hazardous Materials Information Exchange — An organization providing information on hazardous materials (HazMat) training courses, planning techniques, events and conferences, and emergency response experiences for the handling of hazardous materials.

National Health Services Corps — A unit within the Public Health Service that is responsible for providing physicians for areas that are underserved.

National Health Survey — An ongoing health survey by the National Center for Health Statistics that includes studies to determine the extent of illness and disability in the population of the United States; describes the use of health services by Americans and gathers related information.

National Highway Traffic Safety Administration — An agency of the federal government whose function is to carry out programs related to the safety performance of motor vehicles and related equipment, motor vehicle drivers, and pedestrians and to promote a uniform nationwide speed limit.

National Hydropower Association — An organization of hydrodevelopers, dam site owners, manufacturers, utilities, municipalities, financial institutions, contracting firms, architects, engineering firms, and others actively involved in the promotion and development of hydropower that promotes the development of hydroelectric energy.

National Independent Study Accreditation Council — An organization that has been created to provide an accreditation process and a self-governing standard for public and private schools offering independent study and/or home education.

National Institute for Chemical Studies — An organization that helps communities manage safety, health, and environmental risks assessment in the manufacture, storage, transportation, and disposal of chemicals, by providing research, training, education, and consulting services.

National Institute for Environmental Studies — A major environmental research institute in Japan that allows scientists from the basic sciences, engineering, agriculture, and fisheries sciences, medicine, pharmacology, and economics to work together on problems of industrial pollution, global warming, and resolution of waste and hazardous waste substances disposal.

National Institute for Global Environmental Change — An agency of the Department of Energy's Biological and Environmental Research Program whose purpose is to contribute to the knowledge base of global climate change, the reduction of key scientific uncertainties inherent in projections of future climatic states, and the understanding of perturbations for climate system attributed to human activities.

National Institute for Occupational Safety and Health — A federal agency whose function is to identify substances that pose potential health problems and recommend exposure levels to the Occupational Safety and Health Administration. NIOSH does in-depth research on potential hazards in industry and carries out a variety of field studies.

National Institute of Allergy and Infectious Diseases — A component of the National Institutes of Health that conducts and supports research that strives to understand, treat, and ultimately prevent infectious, immunologic, and allergic diseases that threaten hundreds of millions of people; its mission is to conduct basic research in the fields of immunology, microbiology, and infectious diseases.

National Institute of Child Health and Human Development — A branch of the National Institutes of Health that is responsible for all aspects of the growth, development, and health of children in the United States.

National Institute of Environmental Health Sciences — An institute of the National Institute of Health that conducts and supports fundamental research concerned with defining, measuring, and understanding the effects of chemical, biological, and physical factors in the environment on the health and well-being of people.

National Institute of General Medical Sciences — A part of the National Institutes of Health that is involved in understanding fundamental life processes and answering basic research questions by developing data and techniques on analytical and separation methods, biomedical instrumentation, cell organization, lipid biochemistry, membrane structure and function, molecular biophysics, structural biology, etc.; it supports basic biomedical research that is not targeted to specific diseases or disorders.

National Institutes of Health — An important component of the Department of Health and Human Services composed of several specialized institutes in various critical health areas. Their objective is to improve the general health by conducting and supporting biomedical research into the causes, prevention, and cure of disease; research, training, and the development of research sources; and the provision of biomedical information.

National Institute of Mental Health — A branch of the National Institutes of Health responsible for federal research and educational programs dealing with mental health.

National Institute of Neurological Disorders and Stroke — A part of the National Institutes of Health whose mission is to reduce the burden of neurological disease by conducting, promoting, coordinating, and guiding research in the prevention, diagnosis, and treatment of neurological disorders and stroke, providing grants-in-aid to public and private institutions and individuals, operating a program of contracts for the funding of research, providing individuals institutional fellowships to increase scientific expertise in the neurological fields, conducting a diversified program of collaborative research, and collecting and disseminating research information related to neurological disorders.

National Institute of Science, Law, and Public Policy — An organization that develops public policies on food production, sustainable agriculture, food safety, and nutrition and promotes agricultural techniques involving the proper use of chemical fertilizers or pesticides.

National Institute of Standards and Technology — A major program of the Department of Commerce that provides the focus for research, development, and applications in the fields of electrical, electronic, quantum electric, electromagnetic materials, industrial, and mechanical engineering.

National Institute on Aging — A part of the National Institutes of Health that leads a broad scientific effort to understand the nature of aging and to extend the healthy, active years of life of people through research, training, health information dissemination, and other programs relevant to aging and older people.

National Insulation Association — An organization made up of commercial and industrial insulation companies as well as asbestos abatement industries that monitors legislation and regulations.

National Juice Products Association — An organization of juice processors and representatives of related businesses, such as container and equipment manufacturers, suppliers, and brokers, that promotes research, technology, and communications and monitors Food and Drug Administration regulatory action and pending legislation.

National League for Nursing — An organization of professional nurses concerned with the improvement of nursing education, nursing service, accreditation of nursing programs, testing for nursing students, statistical data on nursing personnel, and trends in healthcare delivery.

National League of Cities — A federation of states and cities who develop and pursue a national municipal policy that can meet the future needs of cities and help cities solve critical problems they have in common, including areas of environmental health and safety.

National Library for the Environment — An agency of the National Council for Science and the Environment that provides an online library, including directories of academic environmental programs, journals, funding sources, meetings, job or news sources, laws and treaties, reports, reference materials, and more.

National Library of Medicine — The nation's chief medical information source, authorized to provide medical library services and on-line bibliographic searching capabilities, such as MEDLINE, TOXLINE, and others, to public and private agencies, organizations, and individuals.

National Lime Association — An organization of manufacturers of lime products, including quick lime and hydrated lime, used in steel manufacture, soil stabilization, building construction, air pollution control, and water treatment; it conducts research and operates educational programs.

National Ocean Industries Association — An organization including manufacturers, producers, suppliers, and support and service companies involved in marine, offshore, and ocean work that is focused on offshore oil and gas supplies, production, deep-sea mining, ocean thermal energy, and new energy sources.

National Ocean Service — A part of the NOAA that helps promote coastal awareness and provide an ecosystem of robust health and usefulness to the public by encouraging appropriate management responses to problems, restoration of areas, and navigation conducted to avoid environmental disasters.

National Oceanic and Atmospheric Administration — A major organization within the Department of Commerce whose objective is to explore, map, and chart the global ocean and its living resources; to manage, use, and conserve these resources; and to describe, monitor, and predict conditions in the atmosphere, ocean, sun, and space environment.

National Oceanographic Data Center — One of the national environmental data centers, operated by the National Oceanic and Atmospheric Administration, whose function is to acquire, process, preserve, and disseminate oceanographic data in an easily accessible manner to the world's science community and to other users.

National Office Paper Recycling Project — A program of the U.S. Conference of Mayors that promotes increased recycling of office paper and use of recycled products.

National Oil Recyclers Association — An organization of recycling companies, plant managers, environmental engineers, laboratory managers, environmental consultants, and other individuals and companies with an interest in the recycling of oil that promotes reductions of emissions and pollution associated with the production and use of oil.

National Park Service — A major program of the U.S. Department of the Interior that administers national parks, monuments, historic sites, and recreation areas and preserves the natural and cultural resources and values of the national park system.

National Pest Control Association — An organization of companies engaged in control of insects, rodents, birds, and other pests in and around structures through the use of proper chemical and non-chemical methods while protecting the health and safety of the public and the environment.

National Petroleum Council — An advisory council to the Secretary of Energy on matters relating to oil and gas.

National Pollutant Release Inventory — A source of information on environmental releases of toxic chemicals and waste management of those chemicals.

National Pollution Prevention Roundtable — An organization of state agencies, academic institutions, private industry, consultants, and trade associations who work to promote pollution prevention at its source.

National Recycling Coalition — An organization of public officials; community recycling groups; local, state, and national agencies; environmentalists; waste haulers; solid waste disposal consultants; and private recycling companies that encourages recycling to reduce waste, preserve resources, and promote economic development.

National Registry of Environmental Professionals — An organization that certifies auditors, property assessors, lending analysts, indoor air quality specialists, hazardous and chemical material managers, program administrators, environmental managers, engineers, technologists, scientists, and technicians and provides lists of qualified environmental professionals to governmental agencies.

National Research Council — A council established by the National Academy of Sciences in 1916 to serve as the operating arm of the Academy and the National Academy of Engineering by providing scientific and technical advice to the government, the public, and the scientific and engineering communities.

National Response Center — A communication center for activities related to response actions; it is located at Coast Guard headquarters.

National Restaurant Association Quality Assurance Study Group — An organization of professional quality assurance and quality control personnel who are environmental health specialists belonging to the National Restaurant Association and work in food service facilities.

National Safety Council — An independent, nonprofit organization with the goal of reducing the number and severity of accidents and industrial illnesses by collecting and distributing information about the causes and means of prevention of accidents and illnesses.

National Safety Council Environmental Health Center — An organization that develops a variety of programs to help resolve environmental problems. These programs include on-line search engines for collecting information on chemical safety and issues for industry, government, media, and the public; risk management programs; climate change updates; electronic product recovery and recycling programs; radon education and testing programs; and lead-based paint removal programs.

National Sanitation Foundation — An organization consisting of representatives of industry and government to determine standards for a variety of equipment used in the environmental health science field; it runs elaborate research and testing laboratories and conducts training and education programs.

National Science Foundation — An independent agency created by the National Science Foundation Act of 1950 as amended that supports and promotes the progress of science and engineering through research and educational programs with major emphasis on high-quality, merit-selected research to improve the understanding of the fundamental laws of nature and to promote increased understanding of science and engineering at all educational levels.

National Shellfish Sanitation Program — A voluntary system by which regulatory authorities for shellfish-harvesting waters and shellfish processing and representatives of the transportation and shellfish industries implement specific controls to ensure that raw and frozen shellfish are safe.

National Snow and Ice Data Center — An information and referral center that supports polar and ice data research; distributes data and maintains information about snow cover, avalanches, glaciers, ice sheets, freshwater ice, sea ice, ground ice, permafrost, atmospheric ice, and ice cores; and publishes reports and quarterly newsletters.

National Society for Healthcare Foodservice Management — An organization dedicated to advocacy, support, and education to improve the healthcare environment.

National Society of Professional Engineers — A professional organization of U.S. licensed professional engineers from all disciplines that sponsors seminars and an information center to provide additional in-service education.

National Solid Waste Management Association — A nonprofit trade association that promotes the management of waste in an environmentally responsible, efficient, profitable, and ethical operating format that benefits the public as well as protecting employees from accidents, injuries, and health problems.

National Spa and Pool Institute — An organization of builders, dealers, designers, service companies, retail stores, engineers, manufacturers, and public officials who are concerned with the health and safety of people using public and residential swimming pools, spas, and hot tubs.

National Space Science Data Center — A program sponsored by the Information Systems Office of NASA's Office of Space Sciences providing on-line and off-line access to a wide variety of astrophysics, space plasma, solar physics, lunar, planetary, and Earth science data from NASA's space flight missions as well as selected other data, models, and software.

National Spatial Reference System — The framework for latitude, longitude, height, scale, gravity, orientation and shorelines throughout the United States, providing the foundation for transportation, communication, and defense systems; boundary and property surveys; land records systems; mapping and charting; and a multitude of scientific and engineering applications.

National Swim and Recreation Association — An organization of operators of swim clubs and recreation areas that provides a forum for the exchange of ideas and information concerning areas of interest, including health and safety and the environment.

National Swimming Pool Foundation — An organization that initiates and supports education and research for development and improvement of design, construction, operation, management, and safety of aquatic facilities.

National Technical Information Service — A federal agency involved in the development and publishing of advanced information, products, and services for the achievement of productivity and innovative goals.

National Telecommunications and Information Administration — An agency of the U.S. Department of Commerce whose principal function is dealing with domestic and international telecommunications and information technology issues.

National Toxicology Program — A program in operation since 1961 that has developed testing techniques for determining the carcinogenicity of chemicals. The program runs long-term animal studies and other tests determining carcinogenic activity.

National Transportation Safety Board — An independent safety board created by Congress, consisting of five members appointed by the President with the advice and consent of the Senate and whose function is to ensure that all types of transportation in the United States are conducted safely. The board is responsible for investigating, determining probable cause, making safety recommendations, and reporting the facts and circumstances of aircraft accidents, railroad accidents, pipeline accidents, highway accidents, major marine casualties and accidents, and hazardous materials accidents.

National Weather Service — A federal agency that provides weather, hydrologic, and climate forecasts and warnings for the United States, its territories, adjacent waters, and ocean areas for the protection of life and property and enhancement of the national economy. It is a major program of the NOAA that issues warnings of hurricanes, severe storms, and floods and provides weather forecasts and services for the general public and aviation and marine interests. See also National Oceanic and Atmospheric Administration.

National Wetlands Coalition — An organization of local governments, port authorities, water agencies, communities, agricultural groups, electrical utilities, oil and gas producers, the mining industry, banks, environmental and engineering consultants, and Native American groups who work together for legislative reform of the regulatory improvements to the federal wetlands program.

National Wildlife Control Operators Association — A nonprofit trade association that assists persons and organizations in providing commercial wildlife damage management and control activities, training, educating, promoting competencies, and service.

National World Network — A group of nongovernmental organizations that are rural stakeholders who can provide Congress, the administration, and the public with the full breadth of issues that confront the new rural reality in America by providing research and information necessary for creating awareness and understanding of the effect of public policies on the viability of rural America.

Natural Environment Research Council — A research council of the United Kingdom that provides independent research and training in the environmental sciences to gather and apply knowledge, improve understanding, and predict the behavior of the natural environment and its resources.

Natural Hazards Research and Applications Information Center — A clearinghouse for research information relating to the economic, social, behavioral, and political aspects of natural disasters and their mitigation; it assists others in preparing workshops, symposia, and meetings dealing with natural hazards issues and is located at the University of Colorado.

Natural Resources — A program of the League of Women Voters Education Fund that conducts a national project concerning community drinking water systems and groundwater.

Natural Resources and Environment — A program of the USDA that formulates and promulgates policy relating to environmental activities and management of natural resources.

Natural Resources Defense Council — An organization of lawyers, scientists, public health specialists, transportation, energy, land use, and economic planners who promote the wise management of natural resources through research, public education and development of public policies in areas of land use, coastal protection, air and water pollution, nuclear safety and energy production, toxic substances, and protection of wilderness and wildlife.

Natural Resources Law Center — A center at the University of Colorado at Boulder that promotes sustainability in the rapidly changing American west by informing and influencing natural resources policies and decisions by means of a comprehensive program of research, education, and advice on policy decisions for western natural resources.

Natural Resources Section — A program of the National Governors Association that develops governors' recommendations on energy and environmental issues and presents these policies to Congress and federal agencies.

Naval Research Laboratory — A branch of the U.S. Navy that conducts a broadly based multidisciplinary program of scientific research and advanced technological development that is directed toward maritime applications of new and improved materials, techniques, equipment, systems, and ocean, atmospheric, and space sciences and related technologies.

NBIA — See National Bioenergy Industries Association.

NBRF — See National Biomedical Research Foundation.

NBS — See National Biological Survey, National Bureau of Standards.

NCA — Acronym for National Coal Association.

NCAB — Acronym for National Cancer Advisory Board.

NCAMP — Acronym for National Coalition against the Misuse of Pesticides.

NCAR — See National Center for Atmospheric Research.

NCASI — See National Council for Air and Stream Improvement.

NCBI — Acronym for National Center for Biotechnology Information.

NCCDPHP — Acronym for National Center for Chronic Disease Prevention and Health Promotion.

NCD — Acronym for National Council on Disability.

NCDC — The National Climatic Data Center (a program of the National Oceanic and Atmospheric Administration).

NCEA — See National Center for Environmental Assessment.

NCEHS — See National Center for Environmental Health Strategies.

NCEP — See National Centers for Environmental Prediction.

NCEPI — Acronym for National Center for Environmental Publications and Information.

NCGIA — See National Center for Geographic Information and Analysis.

NCGMP — See National Cooperative Geologic Mapping Project.

NCHS — See National Center for Health Statistics.

NCI — See National Cancer Institute.

NCLEHA — See National Conference of Local Environmental Health Administrators.

NCLIS — Acronym for National Commission on Libraries and Information Science.

NCRP — See National Committee on Radiation Protection.

NCRPM — See National Council on Radiation Protection and Measurements.

NCS&T — Acronym for National Counsel for Science and Technology.

NCSE — See National Council for Science and the Environment.

NCSHA — See National Council of State Housing Agencies.

NCSL — See National Conference of State Legislatures.

NCTR — See National Center for Toxicological Research.

NCVHS — Acronym for National Committee on Vital and Health Statistics.

NDMS — See National Disaster Medical System.

NEA — See Nuclear Energy Agency.

NEB — Acronym for National Environmental Board.

NEBB — See National Environmental Balancing Bureau.

NEETF — See National Environmental Education and Training Foundation.

NEHA — See National Environmental Health Association.

NEI — See National Eye Institute or Nuclear Energy Institute.

NEIC — Acronym for National and Enforcement Investigation Center.

NEJM — See *New England Journal of Medicine.*

NEPA — See National Environmental Policy Act.

NEPI — Acronym for National Environmental Policy Institute.

NEPIS — Acronym for National Environmental Publications Information System.

NEPMU — Acronym for Navy Environmental and Preventive Medicine Unit.

NEPPS — Acronym for National Environmental Performance Partnerships System.

NERC — Acronym for Natural Environmental Research Council.

NERI — Acronym for National Environmental Research Institute.

NETA — See National Environmental Training Association.

NETL — Acronym for National Energy Technology Laboratory.

New England Journal of Medicine — A weekly professional medical journal that publishes findings of medical research and articles about political and ethical issues in the practice of medicine.

NFA — See National Fire Academy.

NFCH — See National Foundation for the Chemically Sensitive.

NFPA — Acronym for National Fire Protection Agency or National Fire Protection Association.

NGDC — See National Geophysical Data Center.

NGS — See National Geodetic Survey.

NGWA — See National Ground Water Association.

NHA — See National Hydropower Association.

NHI — Acronym for National Highway Institute.

NHIC — Acronym for National Health Information Center.

NHLBI — Acronym for National Heart, Lung, and Blood Institute.

NHMIE — See National Hazardous Materials Information Exchange.

NHRAIC — See Natural Hazards Research and Applications Information Center.

NHSC — See National Health Services Corps.

NHTSA — See National Highway Traffic Safety Administration.

NIA — See National Institute on Aging.

NIAID — See National Institute of Allergy and Infectious Diseases.

NICHHD — See National Institute of Child Health and Human Development.

NICNAS — Acronym for National Industrial Chemicals Notification and Assessment Scheme.

NICS — See National Institute for Chemical Studies.

NIDCD — Acronym for National Institute on Deafness and Other Communications Disorders.

NIDRR — Acronym for National Institute on Disability and Rehabilitation Research.

NIE — Acronym for National Institute for the Environment.

NIEHS — See National Institute of Environmental Health Sciences.

NIES — See National Institute of Environmental Studies.

NIFC — Acronym for National Interagency Fire Center.

NIGEC — See National Institute for Global Environmental Change.

NIGMS — See National Institute of General Medical Science.

NIH — See National Institutes of Health.

NIJ — Acronym for National Institute of Justice.

NIMA — Acronym for National Imagery and Mapping Agency.

NIMH — See National Institute of Mental Health.

NINDS — See National Institute of Neurological Disorders and Stroke.

NINR — Acronym for National Institute of Nursing Research.

NIO — Acronym for National Institute of Oceanography.

NIOSH — See National Institute for Occupational Safety and Health.

NIRP — Acronym for National Institute of Radiation Protection.

NISAC — Acronym for National Industrial Security Advisory Committee or National Independent Study Accreditation Council.

NISLAPP — See National Institute of Science, Law, and Public Policy.

NIST — See National Institute of Standards and Technology.

NJPA — See National Juice Products Association.

NLA — Acronym for National Lime Association.

NLC — See National League of Cities.

NLE — Acronym for National Library for the Environment.

NLM — See National Library of Medicine.

NLN — See National League for Nursing.

NMC — Acronym for that National Meteorological Center.

NMFS — Acronym for National Marine Fisheries Service.

NNSA — Acronym for National Nuclear Security Administration.

NOAA — See National Oceanic and Atmospheric Administration.

NODC — See National Oceanographic Data Center.

NOES — See National Occupational Exposure Survey.

NOHS — Acronym for National Occupational Hazard Survey.

NOHSC — Acronym for The National Occupational Health and Safety Commission.

NORA — See National Oil Recyclers Association.

North American Agreement on Environmental Cooperation — An agreement on the environment that is part of the North American Free Trade Agreement and promotes sustainable development through mutually supportive environmental and economic policies. The U.S. EPA is the lead agency in the United States in this area.

North American Association of Food Equipment Manufacturers — An organization of manufacturers of food equipment who work together to develop and utilize effective standards, protect the environment, and reduce the potential for foodborne disease.

North American Commission for Environmental Cooperation — An international organization created by Canada, Mexico, and the United States, established to address regional environmental concerns and help prevent potential trade and environmental conflicts.

North American Insulation Manufacturers Association — An organization made up of manufacturers of insulation products for use in homes, commercial buildings, and industrial facilities interested in thermal efficiency, sound control, and fire safety.

NOS — See National Ocean Service.

NPC — See National Petroleum Council.

NPCA — Acronym for National Paint and Coatings Association, Inc., or National Pest Control Association.

NPDES — See National Pollution Discharge Elimination System.

NPIRS — Acronym for National Pesticide Information Retrieval System.

NPMA — See National Pest Management Association.

NPPH — Acronym for *National Planning Procedures Handbook.*

NPPR — See National Pollution Prevention Roundtable.

NPRI — See National Pollutant Release Inventory.

NPS — See National Park Service.

NPTN — Acronym for National Pesticide Telecommunications Network.

NPTO — Acronym for National Petroleum Technology Office.

NRAQASG — See National Restaurant Association Quality Assurance Study Group.

NRC — See National Research Council, Nuclear Regulatory Commission, or National Response Center.

NRCS — Acronym for Natural Resources Conservation Service.

NRDC — See Natural Resources Defense Council.

NREP — See National Registry of Environmental Professionals.

NREVSS — Acronym for National Respiratory and Enteric Virus Surveillance System.

NRL — See Naval Research Laboratory.

NRLC — See Natural Resources Law Center.

NRN — See National Rural Network.

NSC — See National Safety Council.

NSF — See National Sanitation Foundation, National Science Foundation, or National Safety Council; acronym for national strike force.

NSFNET — Acronym for National Science Foundation NETwork.

NSIDC — See National Snow and Ice Data Center.

NSPF — See National Swimming Pool Foundation.

NSPI — Acronym for National Swimming Pool Institute or National Spa and Pool Institute.

NSRA — See National Swim and Recreation Association.

NSSDC — See National Space Science Data Center.

NSSP — See National Shellfish Sanitation Program.

NSTL — Acronym for National Space Technology Laboratories.

NSWMA — See National Solid Waste Management Association.

NTC — Acronym for National Training Center.

NTI — Acronym for National Toxics Inventory.

NTIA — See National Telecommunications and Information Administration.

NTIS — See National Technical Information Service.

NTN — Acronym for National Trains Network.

NTOF — Acronym for National Traumatic Occupational Fatality Database.

NTP — See National Toxicology Program.

NTSB — See National Transportation Safety Board.

Nuclear Energy Agency — An international organization whose mission is to assist member countries in maintaining and developing international cooperation on the scientific, technological, and legal basis required for the safe, environmentally friendly, and economic use of nuclear energy for peaceful purposes by sharing information and experience and promoting international cooperation.

Nuclear Energy Institute — An organization of electric utilities, manufacturers, industrial firms, research and service organizations, educational institutions, labor groups, and governmental agencies engaged in development and utilization of nuclear energy, especially nuclear produced electricity, and other energy matters.

Nuclear Energy Research Mission — A program of the Department of Energy that addresses and helps resolve the principal technical and scientific issues affecting the future use of nuclear energy in United States; that helps preserve the nuclear science and engineering infrastructure within universities and industry laboratories; and that advances the state of nuclear energy technology to maintain a competitive position worldwide.

Nuclear Regulatory Commission — A federal agency that regulates the civilian uses of nuclear energy to protect the public health and safety and the environment.

Nuclear Waste Technical Review Board — An independent board of scientists and engineers appointed by the President to review, evaluate, and report on Department of Energy waste disposal systems and repositories for spent fuel and high-level radioactive waste.

NWC — See National Wetlands Coalition.

NWCDA — See National Wildlife Control Operators Association.

NWS — See National Weather Service.

NWTRB — See Nuclear Waste Technical Review Board.

NWWA — Acronym for National Water Well Association.

OAQPS — Acronym for Office of Air Quality Planning Standards.

OASDHI — Acronym for Old Age, Survivors, Disability and Health Insurance Program.

OAT — Acronym for Office for the Advancement of Telehealth.

Occupational Safety and Health — A database from 1972 to present covering virtually all aspects of occupational health and safety including hazardous agents, unsafe workplace, environment, and

RESOURCE DIRECTORY

toxicology; it includes coverage of over 150 journals plus government publications.

Occupational Safety and Health Administration — An organization formed in 1971 as a result of the Occupational Safety and Health Act of 1970; it establishes and enforces regulations and standards to control occupational health and safety hazards, such as exposure to carcinogens and diseases or disorders related to exposure to hazardous chemicals or agents.

Occupational Safety and Health Review Commission — An independent, quasi-judicial agency established by the Occupational Safety and Health Act of 1970 and charged with ruling on cases forwarded to it by the Department of Labor when disagreements arise from the results of safety and health inspections performed by occupational safety and health personnel.

Oceanic Abstracts — A database from 1964 to the present, indexing technical literature published worldwide on marine-related subjects (including marine pollution) from journals, books, technical reports, conference proceedings, and government publications.

OCRS — Acronym for Ontario Center for Remote Sensing.

OEC — Acronym for Office of Environmental Compliance.

OECA — See Office of Enforcement and Compliance Assurance.

OECD — Acronym for Organization for Economic Cooperation and Development.

OEE — See Office of Environment and Energy.

OEG — Acronym for Office of Environmental Guidance.

OEJ — See Office of Environmental Justice.

OER — Acronym for Office of Energy Research.

OERR — See Office of Emergency and Remedial Response.

OERR-GIS — See Office of Emergency and Remedial Response Geographic Information System Workgroup.

OES — Acronym for Bureau of Oceans and International Environmental and Scientific Affairs.

OFAP — See Office of Federal Agency Programs.

Office of Emergency and Remedial Response — A U.S. EPA office responsible for managing the Superfund program.

Office of Emergency and Remedial Response Geographic Information System Workgroup — A U.S. EPA office that coordinates and shares information on GIS projects related to the Superfund and oil programs.

Office of Enforcement and Compliance Assurance — A U.S. EPA office working in partnership with EPA regional offices, state, tribal, and other federal agencies to ensure compliance with the nation's environmental laws.

Office of Environmental Justice — A program of the U.S. EPA that has a broad mandate to serve as a focal point for ensuring that communities comprised predominantly of people of color or low-income populations receive protection under environmental laws.

Office of Federal Agency Programs — The organizational unit of the Occupational Safety and Health Administration that provides federal agencies with guidance to develop and implement occupational safety and health programs for federal employees.

Office of Management and Budget — A federal government agency within the Executive Branch whose function is to evaluate, formulate, and coordinate management procedures and program objectives within and among federal departments and agencies.

Office of Pesticide Programs — A U.S. EPA office regulating the use of all pesticides in the United States and establishing maximum levels for pesticide residues in food to safeguard the nation's food supply.

Office of Pesticides and Toxic Substances (OTS) — A branch of the U.S. EPA whose function is to develop national strategies for control of toxic substances, direct enforcement activities, develop criteria, and assess the impact of existing chemicals and new chemicals.

Office of Pollution Prevention and Toxics — A US EPA office focusing on the promotion of pollution prevention efforts for controlling industrial pollution; providing safer chemicals through a combination of regulatory and voluntary efforts; improving risk reduction to minimize exposure to existing substances such as lead, asbestos, dioxin and polychlorinated biphenyls; and assisting the public to understand the risk involved in the use of chemicals.

Office of Prevention, Pesticides, and Toxics Substances — A U.S. EPA office promoting pollution prevention and the public's right-to-know about chemical risks.

Office of Research and Development — A branch of the U.S. EPA whose function is national research in pursuit of technological controls of all forms of pollution.

Office of Response and Restoration — A major program of the NOAA that responds to spills of oil and other hazardous materials, helps emergency planners prepare for potential accidents, creates remedies for environmental damage, assesses injury to coastal resources, and creates software, databases, and other tools to help people respond to hazardous materials accidents.

Office of Science — A major program of the Department of Energy that advises the Secretary of Energy in the areas of physical science and energy research, the use of multipurpose laboratories, education and training, and other applied research activities and which coordinates energy research, science, and technology programs among producing and consuming nations.

Office of Science and Technology Policy — A major program of the Executive Office of the President that provides the President with policy analyses of scientific and technological issues including energy.

Office of Science Coordination and Policy — A U.S. EPA office providing coordination, leadership, peer review, and synthesis of science policy within the Office of Prevention, Pesticides, and Toxics Substances; it currently is working in such areas as biotechnology and endocrine disruptors and with the Federal Insecticide, Fungicide, and Rodenticide Act Scientific Advisory Panel.

Office of Solid Waste — A branch of the U.S. EPA whose function is to provide policy guidance and direction for the agency's hazardous waste and emergency response programs.

Office of Solid Waste and Emergency Response — A federal government agency within the U.S. EPA whose function is to provide policy, guidance, and direction for the agency's hazardous waste and emergency response program.

Office of Sustainable Ecosystems and Communities — A U.S. EPA office helping implement integrated, geographic approaches to environmental protection with an emphasis on ecological integrity, economic sustainability, and quality of life.

Office of Technology Assessment — An agency of Congress whose function is to provide objective analysis of major public policy issues related to scientific and technological change.

Office of the Environment and Energy — A program of the Federal Aviation Authority that coordinates national aviation policy relating to the environment and energy and provides instructions, guidance, oversight, and technical assistance to the Federal Aviation Authority to comply with applicable environmental, occupational safety and health, and energy statutes and regulations prescribing federal environmental protection, worker protection, and energy conservation policies.

Office of Workers' Compensation Program — A federal program within the Department of Labor whose function is to administer federal laws that compensate workers in a variety of special areas.

OFR — Acronym for Office of the Federal Register.

OGWP — Acronym for Office of Ground Water Protection.

OHASIS — Acronym for Office of Health and Safety Information System.

OIEA — Acronym for Office of Integrated Environmental Analysis.

OMAR — Acronym for Office of Medical Applications of Research.

OMB — See Office of Management and Budget.

ONEB — Acronym for Office of National Environmental Board.

ONR — Acronym for Office of Naval Research.

ONWI — Acronym for Office of Nuclear Waste Isolation.

OPEC — Acronym for Organization of Petroleum Exporting Countries.

OPHS — Acronym for Office of Public Health and Science.

OPP — See Office of Pesticide Programs.

OPPT — See Office of Pollution Prevention and Toxics.

OPPTS — See Office of Prevention, Pesticides, and Toxics Substances.

ORACBA — Acronym for Office of Risk Assessment and Cost–Benefit Analysis.

ORD — See Office of Research and Development.

Organization of Economic Cooperation and Development — A Paris-based intergovernmental organization with 29 member countries that develops common solutions to various social problems, including issues of toxic chemical management.

ORNL — Acronym for Oak Ridge National Laboratory.

ORWH — Acronym for Office of Research on Women's Health.

OSCP — See Office of Science Coordination and Policy.

OSEC — See Office of Sustainable Ecosystems and Communities.

OSHA — See Occupational Safety and Health Administration.

OSHRC — Acronym for Occupational Safety and Health Review Commission.

OSM — Acronym for Office of Surface Mining.

OSO — See Office of Satellite Operations.

OSPS — See Outreach and Special Projects Staff.

OST — Acronym for Office of Science and Technology.

OSTI — Acronym for Office of Scientific and Technical Information.

OSW — See Office of Solid Waste.

OSWER — See Office of Solid Waste and Emergency Response.

OTA — See Office of Technology Assessment.

OTAG — Acronym for Ozone Transport Assessment Group.

OTC — Acronym for Ozone Transport Commission.

OTR — Acronym for Ozone Transport Region.

OTS — See Office of Pesticides and Toxic Substances.

Outreach and Special Projects Staff — A U.S. EPA office coordinating and implementing the Office of Solid Wastes and Emergency Response's new initiatives, such as Brownfields and Environmental Justice.

OWCP — See Office of Workers' Compensation Program.

Pacific Disaster Center — A federal information processing center designed to provide information and support to federal, state, local, and regional emergency managers to support mitigation, preparation, response, and recovery within the Pacific region.

PAG — Acronym for Pesticide Assignment Guidelines.

PAHO — Acronym for Pan-American Health Organization.

PAIS Bulletin — See *Public Affairs Information Services Bulletin*.

PAM — Acronym for *Pesticide Analytical Manual*.

PATS — Acronym for Pesticide Action Tracking System or Pesticides Analytical Transport Solution.

PCA — Acronym for Pulp Chemicals Association, Inc.

PCAST — See President's Council of Advisors on Science and Technology.

PCI — Acronym for Powder Coating Institute.

PCSD — Acronym for President's Council on Sustainable Development.

PDI — See Plumbing and Drainage Institute.

PDR — See *Physicians' Desk Reference*.

PEA — Acronym for Plastics Environmental Council.

PestWorld — An Internet site of the National Pest Management Association that provides to the public a wide variety of information designed to help solve pest problems in homes and businesses.

PHA — Acronym for Public Housing Agency.

PhARMA — Acronym for Pharmaceutical Research and Manufacturers Association of America.

Pharmaceutical Manufacturers Association — A trade association made up of producers of ethical pharmaceuticals and biological products sold under their own labels.

PHPPO — Acronym for Public Health Practice Program Office.

PHPS — Acronym for Public Health Prevention Service.

PHS — Acronym for Public Health Service.

PHTN — Acronym for Publlic Health Training Network.

Physicians' Desk Reference — A text produced annually that contains information primarily about prescription drugs and products used for diagnostic procedures in the United States; the information is supplied by the manufacturers.

PIH — Acronym for Office of Public and Indian Housing.

PIMS — Acronym for Pesticide Incident Monitoring System.

PIN — Acronym for Pesticide Information Network.

PIP — Acronym for Public Involvement Program.

Pipeline and Hazardous Material Safety — A program of the National Transportation Safety Board that investigates accidents involving the transportation of hazardous materials.

PIRG — Acronym for Public Interest Research Group.

PIRT — Acronym for Pretreatment Implementation Review Task Force.

PITRI — Acronym for Petroleum Industry Technology and Research Institute, Inc.

PITS — Acronym for Project Information Tracking System.

Plant Science and Water Conservation Research Laboratory — A research laboratory that is part of the U.S. Department of Agriculture and conducts research on alternatives to chemical pesticides to protect plants while focusing on developing safe and effective hydraulic structures to control surface water runoff.

Plumbing and Drainage Institute — An organization of manufacturers of engineered plumbing and drainage products that distributes publications to architects, engineers, contractors, and other plumbing industry representatives on codes and standards for plumbing drainage products.

PMA — See Pharmaceutical Manufacturers Association.

PMEL — Acronym for Pacific Marina Environmental Laboratory.

Pollution Abstracts — A database covering articles on air, noise, and water pollution; environmental quality; pesticides; and solid waste from 1970 to the present.

Pollution Prevention and Toxics — A program of the U.S. EPA that assesses the health and environmental hazards of existing chemical substances and mixtures; collects information on chemical use, exposure, and effects; reviews new chemicals; and regulates the manufacture, distribution, use, and disposal of harmful chemicals.

Pollution Prevention Program — A program established by the Office of Environmental Management in 1991 to minimize pollution generation and release by implementing cost-effective technologies, practices, and policies in the use of materials, processes, and work methods, including recycling activities to reduce pollutants, contaminants, and hazardous substances released to land, water, and air.

Pollution Prevention Task Force — A group of individuals responsible for instituting a pollution prevention program, for performing a preliminary assessment, and for guiding the program through the developmental stages.

POLREP — Acronym for Pollution Report.

PPDC — Acronym for Pesticide Program Dialog Committee.

PPIC — Acronym for Pesticide Programs Information Center.

PPIS — Acronym for Pesticide Product Information System or Pollution Prevention Incentives for States.

PRC — Acronym for Planning Research Corporation.

President's Council of Advisers on Science and Technology — An advisory council established under the Federal Advisory Committee Act as amended whose function is to advise the President on matters involving science and technology policy.

Prevention, Pesticides, and Toxic Substances — A program of the U.S. EPA that makes studies and recommendations for regulating chemical substances under the Toxic Substances Control Act.

Prior Informed Consent Procedure — A program associated with the Rotterdam Convention that encourages governments to address problems related to toxic pesticides and other hazardous chemicals and makes it mandatory for countries to be given the tools and information they need to identify potential hazards and exclude chemicals they cannot manage safely.

PRM — Acronym for *Prevention Reference Manual*.

Professional Pest Management Alliance — An organization of pest control operators established to increase awareness among consumers of the value of professional pest management services, protect the pest control industry, and increase the size of the consumer market.

PRTPs — Acronym for Pollutant Release and Transfer Registers.

PSRO — Acronym for Professional Standards Review Organization.

PSWCRL — See Plant Science and Water Conservation Research Laboratory.

Public Affairs Information Services Bulletin — A subject index published in quarterly paperbacks and compiled in annual hardcover volumes; it covers areas of economics, political science, public administration, international law, sociology, and demography.

Public Utilities — An agency within the General Services Administration whose mission is to provide leadership within the federal government with regard to developing programs that enable customers to procure utility services at the lowest cost to the taxpayer and the greatest value to the American citizens.

Public Utility Commission — A governmental agency whose mission is to ensure all residential and business customers access to adequate, quick, safe, and reliable utility services at fair prices while facilitating an environment that provides competitive choices.

PUC — See Public Utility Commission.

QCP — Acronym for Quiet Communities Program.

QUIPE — Acronym for *Quarterly Update for Inspector in Pesticide Enforcement*.

RAC — See Risk Assessment Consortium; acronym for Radiation Advisory Committee or Response Action Coordinator.

Radiation and Indoor Air — A major program of the U.S. EPA that establishes standards to regulate the amount of radiation discharged into the environment from uranium mining and milling projects and other activities that result in radioactive emissions and to ensure safe disposal of radioactive waste.

Rainbow Report — A comprehensive document giving the status of all pesticides now or ever registered or having special reviews.

RAMP — Acronym for Rural Abandoned Mine Program.

RAMS — Acronym for Regional Air Monitoring System.

RARG — Acronym for Regulatory Analysis Review Group.

RCC — Acronym for Radiation Coordinating Council.

RCMA — Acronym for Roof Coatings Manufacturers Association.

RCP — Acronym for Research Centers Program.

RCRA/Superfund/EPCRA hotline — A hotline of the U.S. EPA that provides information on regulations and programs and responds to queries about waste minimization programs, source reduction, and hazardous wastes reduction.

RCRIS — Acronym for Resource Conservation and Recovery Information System.

REEP — Acronym for Review of Environmental Effects of Pollutants.

Renewable Fuels Association — An organization of engineering and financial firms, marketers, producers, and state governments working toward developments in biomass fuel technology, mainly alcohol fuels.

Renewable Natural Resources Foundation — A consortium of professional, scientific, and education organizations working to advance scientific and public education in renewable natural resources.

Resources for the Future — A research organization that conducts studies on economic and policy aspects of energy, conservation, and development of natural resources, including effects on the environment; it is also interested in hazardous wastes, the Superfund, and biodiversity.

Responsible Industry for a Sound Environment — An organization of manufacturers, formulators, distributors, and representatives of the specialty pesticides industry that promotes the environmental and health and safety benefits of the proper use of specialty pesticides.

RFA — See Renewable Fuels Association.

RESOURCE DIRECTORY

RESOURCE DIRECTORY

RIC — Acronym for Radon Information Center.

RICHS — Acronym for Rural Information Center Health Service.

RISC — Acronym for Regulatory Information Service Center.

RISE — See Responsible Industry for a Sound Environment.

Risk Assessment Consortium — A group of representatives from all of the government agencies with any food safety responsibilities, including the FDA, USDA, EPA, CDC, NMFS, NIH, and DOD, who work together to enhance communication and coordination among the member agencies to promote the conduct of scientific research that will facilitate risk assessments.

Risk Reduction Engineering Laboratory — A division of the U.S. EPA responsible for planning, implementing, and managing research, development, and demonstration programs to provide an authoritative, defensible engineering basis in support of the policies, programs, and regulations of the U.S. EPA with respect to drinking water, wastewater, pesticides, toxic substances, solid waste, hazardous waste, and Superfund-related activities.

RMA — See Rubber Manufacturers Association; acronym for Risk Management Agency.

RMMC — See Rocky Mountain Mapping Center.

Rocky Mountain Mapping Center — A division of the U.S. Geological Survey that conducts mapping activities in the Western United States and produces digital elevation, graphic maps, and new mapping techniques.

ROMCOE — Acronym for Rocky Mountain Center on Environment.

RREL — See Risk Reduction Engineering Laboratory.

RSCC — Acronym for Regional Sample Control Center.

RTECS — Registry of Toxic Effects of Chemical Substances (produced by NIOSH)

Rubber Manufacturers Association — The primary national trade association for the finished rubber products industry in the United States and whose major function is to provide safe, reliable, and essential rubber products and to be involved in a tire safety education campaign.

SAB — Acronym for Science Advisory Board or Scientific Advisory Board.

SAE — Acronym for Society of Automotive Engineers.

SAEWG — Acronym for Staining Air Emissions Work Group.

Safe Drinking Water Information System — A U.S. EPA-designed database used to help states run their drinking water programs.

SAFed — See Safety Assessment Federation.

Safety Assessment Federation — An independent inspection industry association that plays a key role in maintaining high standards of safety within workplaces in the United Kingdom and presents a focal point for all issues and concerns relating to the safe use and operation of plants, machinery, and equipment.

SAFSR — See Society for the Advancement of Food Service Research.

SAMA — See Scientific Apparatus Makers Association.

SAMSA — See Substance Abuse and Mental Health Services Administration.

SAP — Acronym for Scientific Advisory Panel and Scientific Advisory Board.

SAPL — See Seacoast Anti-Pollution League.

SBCCOM — See U.S. Army Soldier and Biological Chemical Command.

SC — See Sierra Club.

SCC — Acronym for Source Classification Code.

SCD — Acronym for Soil Conservation District.

SCDM — Acronym for Superfund Chemical Data Matrix.

SCHC — Acronym for Society for Chemical Hazard Communication.

SCI — See Society of Chemical Industry.

Science Advisory Board — A group of external scientists who advise the U.S. EPA on science and policy.

Scientific Apparatus Makers Association — An association that issues standards covering platinum, nickel, and copper resistance elements.

SCLDF — Acronym for Sierra Club Legal Defense Fund.

SCS — See Soil Conservation Service; acronym for Supplementary Control Strategy/System.

SCSA — Acronym for Soil Conservation Society of America.

SCSP — Acronym for Storm and Combined Sewer Program.

SDA — See Soap and Detergent Association.

SDWIS — See Safe Drinking Water Information System.

Seacoast Anti-Pollution League — A citizen-based environmental organization concerned with health and safety issues affecting the regional seacoast environment and its population and seeking to prevent ecological, economic, and public health damage from the Seabrook nuclear reactor; it monitors the cleanup efforts at the Portsmouth Naval Shipyard.

SEAS — Acronym for Strategic Environmental Assessment System.

SEC — See Securities and Exchange Commission.

Securities and Exchange Commission — A quasi-judicial commission formed by an act of Congress to provide the fullest possible disclosure to the investment public, to regulate investments, and to protect the public and investors against malpractice in the securities and financial markets.

SEGIP — Acronym for State Environmental Goals and Improvement Project.

SEHSC — See Silicones Environmental Health and Safety Council.

Senate Energy and Natural Resources Committee — A committee of the U.S. Senate that has jurisdiction over mineral conservation and over energy conservation measures such as emergency fuel allocation, gasoline rationing, and coal conversion.

Senate Environment and Public Works Committee — The committee of the U.S. Senate that has jurisdiction over the U.S. EPA research and development programs in air, water, noise, solid wastes, hazardous materials, and toxic substances, as well as jurisdiction over the many environmental laws enforced by the agency.

SEPWC — Acronym for Senate Environmental and Public Works Committee.

SER — See Society for Ecological Restoration.

SERC — See State Emergency Response Commission; acronym for Smithsonian Environmental Research Center.

SERI — See Solar Energy Research Institute.

SETAC — See Society of Environmental Toxicology and Chemistry.

SIAM — See Society for Industrial and Applied Mathematics.

SIC — See Standard Industrial Classification Code.

Sierra Club — A citizens' interest group that promotes protection and responsible use of the Earth's ecosystems and its natural resources and focuses on global warming and the greenhouse effect through energy conservation and efficient use of renewable energy resources.

Silicones Environmental Health and Safety Council — An organization of organosilicones manufacturers that coordinates programs dealing with health, environmental, and safety issues of interest to the industry.

SITE — See Superfund Innovative Technology Evaluation.

SLA — See Special Libraries Association.

Smithsonian Environmental Research Center — A program under the Smithsonian Institution that performs laboratory and field research that measures physical, chemical, and biological interactions to determine the mechanisms of environmental responses to humans' use of the air, land, and water.

SNA — Acronym for System Network Architecture.

SNL — Acronym for Sandia National Laboratories.

Soap and Detergent Association — An organization of manufacturers of cleaning products, their ingredients, and finished packaging that is involved in consumer education and environmental and human safety research.

Social Security Administration — A major division of the Department of Health and Human Services that operates a national program of contributory social insurance to provide funds for retirement, death, and disability benefits and for health benefits under Medicare.

Society for Ecological Restoration — An organization of scientists, researchers, educators, academics, environmental consultants, governmental agencies, and other interested individuals that promotes ecological restoration as a scientific and technical discipline and provides a strategy for environmental conservation and a technique for ecological research that result in a beneficial relationship between humans and nature.

Society for Industrial and Applied Mathematics — An organization of professional mathematicians and engineers that utilizes applied mathematics to solve many real-world problems, from modeling physical, chemical, and biological phenomena to designing engineered parts and structures in systems; to optimize performance; to understand and optimize manufacturing processes; and to promote safety.

Society for Occupational and Environmental Health — An organization of scientists, academicians, and industry and labor representatives that seeks to improve the quality of both working and living places by operating as a neutral forum for conferences involving all aspects of the occupational health environment and focuses public attention on scientific, social, and regulatory problems.

Society for Public Health Education — An organization of researchers and practitioners in health education that is concerned with promotion of personal and community health programs and contributing to the advancement of health through continuing research and the improvement of public health practices.

Society for Risk Analysis — A professional association that brings together individuals from diverse disciplines and from different countries and provides the opportunity to exchange information, ideas, and methodologies for risk analysis and risk problem solving.

Society for the Advancement of Food Service Research — An organization of restaurant operators, institutional food service directors, dietitians, university professors and consultants, environmental health officials, engineering and research personnel from the food and equipment industries, and others interested in studying problems and identifying areas of research in the food service field. It is concerned with the chemistry and other basic properties of food; nutrition; physiological attributes of foodstuffs; problems of handling, storage, and production; design, manufacture, and purchase of food-related equipment; and administration and management functions.

Society for Vector Ecology — An organization of individuals with bachelor's or master's degrees from an accredited college or university plus 2 or more years' experience in vector ecology or related fields of education and research, or 7 years of progressive technical or administrative service in the field; it provides educational meetings and facilitates the exchange of information among members concerning vectors and vector control.

Society of Chemical Industry — An international network of chemistry professionals who interact to share knowledge, research, and practical applications, in the complex and diverse sectors of applied science and this science based industry.

Society of Environmental Toxicology and Chemistry — An organization of professionals in the fields of chemistry, toxicology, biology, ecology, atmospheric science, health science, earth science, and environmental engineering who promote the use of multidisciplinary approaches to examine the impacts of chemicals and technology on the environment.

Society of Plastics Engineers — A professional society that acts as a means of communication between scientists and engineers engaged in the development, conversion, and applications of plastics and promotes scientific and engineering knowledge relating to plastics.

Society of the Plastics Industry — A trade association representing all facets of the manufacturing industry, including processors, machinery and equipment manufacturers, and raw materials suppliers.

SOCMA — See Synthetic Organic Chemical Manufacturers Association.

SOCMI — Acronym for Synthetic Organic Chemicals Manufacturing Industry.

SOD — See Superintendent of Documents.

SOEH — See Society for Occupational and Environmental Health.

Soil Conservation Service — An agency of the USDA whose function is to develop and carry out national soil and water conservation programs in cooperation with landowners, land users, developers, and governmental agencies.

Soil Science Society of America — A worldwide professional association of soil scientists who advance the discipline and practice of soil science by acquiring and disseminating information about soils in relation to crop production, environmental quality, ecosystems sustainability, bioremediation, waste management, and recycling.

Solar Energy Research Institute — An institute established in 1974 and funded by the federal govern-

ment to support the U.S. Department of Energy's solar energy program and promote the widespread use of all aspects of solar technology, including photovoltaics, solar heating and cooling, solar thermal power generation, wind ocean thermal conversion, and biomass conversion.

Solid Waste Association of North America — An organization of government and private industry officials that manages municipal solid waste programs and is interested in waste reduction, collection, recycling, combustion, and disposal.

SOPHE — See Society for Public Health Education.

SPC — See Storm Prediction Center.

SPCP — See Spill Prevention, Containment, and Countermeasures Plan.

Special Libraries Association — An international association representing the interests of thousands of information professionals in over 70 countries; special librarians are information resource experts who collect, analyze, evaluate, package, and disseminate information to facilitate accurate decision making in corporate, academic, and government settings.

SPI — See Society for the Plastics Industry.

Spill Prevention, Control, and Countermeasures — A U.S. EPA oil-spill program to aid in the prevention, assessment, control, and treatment of oil spills.

Spill Prevention, Containment, and Countermeasures Plan — A plan developed by the U.S. EPA to cover the release of hazardous substances as defined in the Clean Water Act.

SPMS — Acronym for Strategic Planning and Management System.

SPSS — See Statistical Package for the Social Sciences.

SPUR — Acronym for Software Package for Unique Reports.

SRA — See Society for Risk Analysis.

SRAP — Acronym for Superfund Remedial Accomplishment Plan.

SRI — See Steel Recycling Institute.

SRI International — A research organization that conducts policy-related energy studies and scientific research, including surveys of energy supply and demand; analysis of fossil fuel, solar, and nuclear energy; environmental effects of advanced energy technology; and energy management.

SRRB — See Surface Radiation Research Branch.

SSA — Acronym for Social Security Administration.

SSPMA — See Sump and Sewage Pump Manufacturers Association.

SSSA — See Soil Science Society of America.

STALAPCO — Acronym for State and Local Air Pollution Control Officials.

Standing Committee on Environmental Law — A committee of the American Bar Association that conducts domestic and international projects in environmental law and policy and coordinates environmental law activities.

STAPPA — See State and Territorial Air Pollution Program Administrators.

STAR — Acronym for State Acid Rain Projects.

State and Territorial Air Pollution Program Administrators — An organization of state, territorial, and local officials responsible for implementing programs established under state and local legislation and the Clean Air Act.

State Drinking Water Information/Federal Version — A U.S. EPA national database storing routine information about the nation's drinking water to monitor approximately 175,000 public water systems.

State Emergency Response Commission — A committee of various state agencies, public and private groups, and associations appointed by the governor to assist in emergency preparedness and response capabilities through better coordination and planning.

Statistical Package for the Social Sciences — A computer program used in research for the analysis of complex data from large samples.

Steel Recycling Institute — An organization that provides information and assistance to communities on methods of collection, preparation, and transportation of steel scrap and which provides ideas and processes for recycling.

STN — Acronym for Scientific and Technical Information Network.

Storage and Retrieval System — A U.S. EPA data bank of a variety of environmental information.

STORET — See Storage and Retrieval System.

Storm Prediction Center — A National Forecast Center in Norman, OK, responsible for providing short-term forecasts for severe and excessive rainfall and severe winter weather over the contiguous United States.

Subcommittee on Energy — A committee of the House Science Committee that has jurisdiction over legislation on research and development of energy sources and over the Department of Energy basic research programs.

Subcommittee on Energy and Air Quality — A committee of the House Committee on Energy and Commerce that has jurisdiction over legislation on proposals to label appliances to indicate energy consumption and on emergency fuel allocation.

Subcommittee on Energy and Water Development — A major subcommittee of the Senate Appropriations Committee that has jurisdiction over legislation to appropriate funds for the Energy Department, the Federal Energy Regulatory Commission, the Nuclear Regulatory Commission, and the Tennessee Valley Authority.

Subcommittee on Interior — A major subcommittee of the House Appropriations Committee that has jurisdiction over legislation to appropriate funds for the U.S. Department of the Interior, strategic petroleum reserves, naval petroleum and oil shale reserves, clean coal technology, fossil energy research and development, energy conservation, and alternate fuels production.

Subcommittee on Water and Power — A major subcommittee of the House Resources Committee that has jurisdiction over water-related programs of the U.S. Geological Survey, saline water research and development, water resources research programs, and matters related to the Water Resources Planning Act.

Submersible Wastewater Pump Association — An organization of manufacturers of submersible wastewater pumps and systems for municipal and industrial applications and manufacturers of component parts and accessory items for those pumps and systems.

Substance Abuse and Mental Health Services Administration — A major division of the U.S. Department of Health and Human Services that is involved in mental health and drug and alcohol abuse and has prepared a number of studies to be used by state and federal agencies, including the *2000 National Household Survey on Drug Abuse* and *Changing the Conversation: A National Plan to Improve Substance Abuse Treatment*.

Substance-Specific Applied Research — A program of research designed to fill data needs, including laboratory and other studies to determine short-term, intermediate, and long-term health effects from human exposure to a given substance; organ-, site-, and system-specific acute and chronic toxicity, as well as metabolites and their effects, are measured.

Sump and Sewage Pump Manufacturers Association — An organization of manufacturers of residential sump pumps and sewage pumps seeking to develop and promulgate quality standards, implement certification and labeling, and improve provisions of building codes.

Superfund Innovative Technology Evaluation — A U.S. EPA program in which research is used to determine state-of-the-art cleanup methods for hazardous waste sites around the nation and techniques for removing hazardous substances from soils.

Superintendent of Documents — An agency of the federal government that disseminates the largest amount of U.S. government publications and information in the world.

Surface Mining Reclamation and Enforcement Section — A program of the U.S. Department of the Interior that administers the Surface Mining Control and Reclamation Act of 1977 by establishing and enforcing national standards for the regulation and reclamation of surface and underground mining.

Surface Radiation Research Branch — A program of the NOAA that is staffed by national and international scientists conducting research into many aspects of solar radiation at the surface of the Earth and the effects of surface ultraviolet radiation on people.

SVE — See Society for Vector Ecology.

SWAP — Acronym for Source Water Assessment Program.

SWCD — Acronym for Soil and Water Conservation District.

SWIE — Acronym for Southern Waste Information Exchange.

Synthetic Organic Chemical Manufacturers Association — An organization of manufacturers of synthetic organic chemicals that are manufactured from coal, natural gas, and crude petroleum and certain natural substances such as vegetable oils, fats, proteins, carbohydrates, rosin, grains, and their derivatives; it is concerned with air pollution, hazardous wastes, occupational safety and health, toxic substances control, and water pollution.

TAB — See Technical Assistance to Brownfields Communities.

TAMTAC — Acronym for Toxic Air Monitoring Technical Advisory Committee.

TAP — Acronym for Technical Assistance Program.

TAPPI — See Technical Association of the Pulp and Paper Industry.

TCRI — Acronym for Toxic Chemical Release Inventory.

TDMA — Acronym for Titanium Dioxide Manufacturers Association.

TEC — Acronym for Technical Evaluation Committee.

Technical Assistance to Brownfields Communities — A U.S. EPA program that helps communities clean and redevelop properties that have been damaged or undervalued by environmental contamination.

Technical Association of the Pulp, Paper, and Converting Industry — An international group of technically experienced people who have a comprehensive collection of reliable technical information and knowledge in the industry and strive to produce the highest quality products and services and to resolve technical problems.

Technical Outreach for Communities — A U.S. EPA program using university, educational, and technical resources to help community groups understand the technical issues of hazardous waste sites in their area.

Technology Innovation Office — A U.S. EPA office acting as an advocate for new technologies and working to increase the application of innovative treatment technologies to contaminated waste sites, soils, and groundwater.

Teratology Society — An organization of individuals from academia, government, private industry, and the professions that stimulates scientific interest and promotes an exchange of ideas and information about problems of abnormal biological cells and malformations at the fundamental or clinical level.

TIO — See Technology Innovation Office.

TIPS — Acronym for Terrorism Information and Prevention System.

TITC — Acronym for Toxic Substance Control Act Interagency Testing Committee.

ToxFAQs — A series of fact sheets about hazardous substances developed by the Division of Technology of the Agency for Toxic Substances and Disease Registry.

Toxic Release Inventory — A database of annual toxic releases from certain manufacturers who must report to the U.S. EPA that states the amounts of approximately 350 toxic chemicals and 22 chemical categories that they directly release to air, water, or land; inject underground; or transfer to off-site facilities.

Toxicology Information On-Line — The National Library of Medicine's collection of computerized references on human and animal toxicologic studies, toxicology, effects of environmental chemicals and pollutants, adverse drug reactions, and analytical methods.

TOXLINE — See Toxicology Information On-Line.

TPSIS — Acronym for Transportation Planning Support Information System.

TRANSPORT — A comprehensive database of international transportation information consisting of databases of a variety of bibliographies produced by the world's leading transportation research organizations, Organization for Economic Cooperation and Development, and the Transportation Research Board of the United States.

Transportation Research Information Service — A database of transportation research information from 1968 on air, highway, rail, transport, mass transit, and other transportation modes; information is also included on environmental and safety concerns.

TRI — See toxic release inventory.

TRIP — Acronym for Toxic Release Inventory Program.

TRIS — See Transportation Research Information Service.

TRLN — Acronym for Triangle Research Library Network.

TS — See Teratology Society.

TSCC — Acronym for Toxic Substances Coordinating Committee.

TSDG — Acronym for Toxic Substances Dialog Group.

TSI — Acronym for Transportation Safety Institute.

TTN — Acronym for Technology Transfer Network.

TVA — Acronym for Tennessee Valley Authority.

UAPSP — Acronym for Utility Acid Precipitation Study Program.

UARG — Acronym for utility air regulatory group.

UL — See Underwriters Laboratories.

UMWA — Acronym for United Mine Workers of America.

UNAMAP — Acronym for Users Network for Applied Modeling of Air Pollution.

UNCEAU — Acronym for United Nations Centre for Urgent Environmental Assistance.

UNCED — Acronym for United Nations Conference on Environment and Development.

Underwriters Laboratories — A product safety certification laboratory that establishes and operates product safety certification programs to ascertain that items produced are safeguarded against reasonable forseeable risk.

UNECE — Acronym for United Nations Economic Commission for Europe.

UNEP — See United Nations Environmental Programme.

UNFCCC — Acronym for United Nations Framework Convention on Climate Change.

Union of Concerned Scientists — An independent group of scientists and others that advocates sustainable international, national, and state energy policies.

United Nations Environment Programme Chemicals — The center for all chemical related activities of the United Nations whose mission is to make the world a safer place from toxic chemicals by helping governments take global action for sound management of chemicals, by promoting the exchange of information on chemicals, and by helping to build the capacities in countries around the world for chemical safety.

U.S. Army Chemical and Biological Defense Command — An agency of the U.S. Army that conducts research and concept exploration, demonstration, and validation, in addition to the manufacturing, development, and internal production of chemical defense systems.

U.S. Army Materiel Command — A section of the U.S. Army that conducts research in a world-class research laboratory, stores and maintains the U.S. chemical weapons stockpile, provides training to help emergency responders in 125 U.S. cities and will respond to chemical or biological weapons. It conducts support programs for executive officers in research and development and assigns worldwide science advisers to on-scene commanders.

U.S. Army Soldier and Biological Chemical Command — An organization of the U.S. Army that provides support in three main areas of defense: research, development, and acquisition; emergency preparedness and response; and safe, secure chemical weapons storage, remediation, and demilitarization.

U.S. Chemical Safety and Hazard Investigation Board — An independent, scientific investigative agency, not a regulatory or enforcement body, created by the Clean Air Act Amendments of 1990 but not funded until 1998; its members are appointed by the President and confirmed by the Senate and are involved in the promotion and prevention of major chemical accidents at fixed facilities.

U.S. Coast Guard — A branch of the Armed Forces of the United States at all times and a service within the Department of Transportation except during wartime or when the President directs it to become part of the Navy. It is now a part of the Department of Homeland Security.

U.S. Conference of Mayors — An organization of mayors of cities with populations over 30,000 who promote improved municipal government by cooperation between the cities and the federal government.

U.S. Department of Agriculture — A department of the U.S. government that works to improve and maintain farm income and to develop and expand markets abroad for agricultural products; helps to curb and cure poverty, hunger, and malnutrition; and works to enhance the environment and to maintain production capacity by helping landowners protect soil, water, forest, and other natural resources.

U.S. Department of Commerce — A department of the U.S. government that promotes job creation, economic growth, sustainable development, and improved living standards for all Americans by working in partnership with businesses, universities, communities, and the public.

U.S. Department of the Interior — The federal agency whose function is to act as the nation's principle conservation agency; it has the responsibility for most of the nationally owned public lands and natural resources.

U.S. Environmental Protection Agency — An agency of the federal government with headquarters, program offices, 10 regional offices, and 17 laboratories spread across the United States that employs a highly educated, technically trained staff, including engineers, scientists, and environmental protection specialists, as well as legal, public affairs, financial, and computer specialists. Its mission is to protect human health and to safeguard the natural environment (air, water, and land).

RESOURCE DIRECTORY

U.S. Geological Survey — A major program of the U.S. Department of the Interior whose function is to classify public lands and examine geological structure, mineral resources, and products of the national domain.

U.S. Global Change Research Program — A comprehensive and multidisciplinary scientific research agenda addressing significant uncertainties concerning the natural and human-induced changes to the environment of the Earth.

U.S. Green Building Council — A coalition of leaders across the building industry working to promote buildings that are environmentally responsible, profitable, and healthy places to live and work; accomplished by industry and government joining together to develop and utilize practices, technologies, policies, and standards to achieve these goals.

U.S. Public Health Service — Originated in 1798 by an Act of Congress authorizing marine hospitals to care for American merchant seamen; reorganized in 1944 and expanded subsequently to support and protect the nation's physical and mental health.

U.S. Public Interest Research Group — An organization that coordinates grass-roots efforts to advance environmental and consumer protection laws, conducts research on environmental issues, and works in areas of toxic and solid wastes, air and water pollution, pesticides, energy, and endangered species.

University of Iowa Injury Prevention Research Center — A university-based medical research center that seeks to reduce and prevent injuries, provide educational research programs, and make available libraries and databases.

UOMA — See Used Oil Management Association.

USACMLS — Acronym for U.S. Army Chemical School.

USANCA — Acronym for U.S. Army Nuclear and Chemical Agency.

USBR — See Bureau of Reclamation.

USCC — Acronym for U.S. Chamber of Commerce.

USCG — See U.S. Coast Guard.

USCGS — Acronym for U.S. Coast and Geodetic Survey.

USCIEP — Acronym for U.S. Council for International Engineering Practice.

USCM — See U.S. Conference of Mayors.

USDA — See U.S. Department of Agriculture.

USDA Forest Service — An agency of the USDA whose mission is to sustain the health, diversity, and productivity of the nation's forests and grasslands to meet the needs of present and future generations.

USDOC — See U.S. Department of Commerce.

USDOI — Acronym for U.S. Department of the Interior.

Used-Oil Management Association — An organization of manufacturers and distributors of appliances used to convert waste oil in a safe manner and to protect the health of the public and the environment.

USFS — Acronym for U.S. Forest Service.

USG — Acronym for U.S. Government.

USGBC — See U.S. Green Building Council.

USGCRP — See U.S. Global Change Research Program.

USGS — See U.S. Geological Survey.

USP — See *U.S. Pharmacopeia*.

USPHS — See U.S. Public Health Service.

USPIRG — See U.S. Public Interest Research Group.

Vessel Sanitation Program (VSP) — A cooperative activity between the cruise ship industry and the U.S. Public Health Service for evaluation of food, water, sewage disposal, and housing and to help prevent disease and injury.

VSP — See Vessel Sanitation Program.

WADEM — See World Association for Disaster and Emergency Medicine.

WADTF — Acronym for Western Atmospheric Deposition Task Force.

Walter Reed Army Institute of Research — The largest, most diverse, and oldest laboratory in the U.S. Army Medical Research and Materials Command that conducts research in a range of militarily relevant issues, including naturally occurring infectious diseases, combat casualties care, operational health hazards, and medical defense against biological and chemical weapons.

Waste Equipment Technology Association — An organization of manufacturers, designers, and distributors of waste collection, treatment, storage equipment, and waste handling consultants who promote effective processing of solid and hazardous wastes and more extensive use of recycling.

Waste Management Information Systems — A system that allows users at multiple military installations worldwide to have 24-hour access to obtain waste profiles, to print waste manifests, for transportation planning, to view regulatory databases, and to archive waste inventory data.

Waste Pollution Abstracts — An index of water pollution articles from 1971 to the present.

WASTEC — See Waste Equipment Technology Association.

Water Environment Federation (WEF) — Formerly known as the Water Pollution Control Federation; a group of autonomous regional associations concerned with the treatment and disposal of domestic and industrial wastewater, groundwater, and toxic and hazardous waste.

Water Science and Technology Board — A board established by the National Research Council to provide a focal point for studies related to water resources;

accomplished under the control of the National Academy of Sciences and the National Academy of Engineering.

Water Systems Council — An organization of manufacturers and installers of pitless adapters who promote sound principles of equipment construction and installation while maintaining appropriate standards.

WCC — Acronym for World Chlorine Council.

WCED — Acronym for World Commission on Environment and Development.

WCRP — See World Climate Research Program.

WDROP — Acronym for Distribution Register of Organic Pollutants in Water.

WEC — See World and Environment Center.

WEF — See Water Environment Federation.

WENDB — Acronym for Water Enforcement National Database.

WERL — Acronym for Water Engineering Research Laboratory.

WEVS — Acronym for Wetland Ecosystem Vulnerability Study.

WHA — Acronym for World Health Assembly.

WHO — See World Health Organization.

WHP — Acronym for Wellhead Protection Program.

WHWT — Acronym for Water and Hazardous Waste Team.

WIC — Acronym for Women, Infants, and Children.

WICE — Acronym for World Industry Council for the Environment.

WICEM — Acronym for World Industry Conference on Environmental Management.

WIN — Acronym for Watershed Information Network.

WMIS — See Waste Management Information System.

WMO — See World Meteorological Organization.

World Association for Disaster and Emergency Medicine — An international humanitarian association dedicated to the improvement of disaster and emergency medicine by improving international collaboration, facilitating academic and research-based education and training, and facilitating data collection.

World Climate Research Program — A program established under the joint sponsorship of the International Council for Science and the World Meteorological Organization whose function is to develop the fundamental scientific understanding of the physical climate system and climate processes necessary to determine to what extent climate can be predicted and the extent to which humans influence climate.

World Environment Center — An organization that works to strengthen industrial and urban environmental health and safety policies by establishing and promoting partnerships among industry, the government, and nongovernmental organizations; encourages corporate environmental leadership and responsibility and provides training and technical cooperation programs utilizing volunteer experts.

World Health Organization — An international organization of countries whose objective is the attainment by all people of the highest possible level of health (physical, mental, and social); chartered by the United Nations.

World Meteorological Organization — Successor to the International Meteorological Committee and International Meteorological Organization; helps explain long-term global acid precipitation trends.

World Resources Institute — An environmental think tank that goes beyond research to find practical ways to protect the Earth and improve the lives of people while working with organizations and individuals worldwide.

World Wildlife Fund — A world federation whose mission is to stop the degradation of the planet's natural environment; to build a future in which humans live in harmony with nature by conserving the world's biological diversity; to ensure that the use of renewable natural resources is sustainable; and to promote the reduction of pollution and wasteful consumption.

Worldwatch Institute — A research organization that focuses on interdisciplinary approaches to solving global environmental problems.

WQA — Acronym for Water Quality Associaation.

WRAIR — See Walter Reed Army Institute of Research.

WRC — Acronym for Water Resources Council.

WRI — See World Resources Institute.

WRRC — Acronym for Water Resources Research Center.

WSTB — See Water Science and Technology Board.

WWEMA — Acronym for Waste and Wastewater Equipment Manufacturers Association.

WWF — See World Wildlife Fund.

ZBA — See Zoning Board of Appeals.

Zoning Board of Appeals (ZBA) — A board established by a local government to hear appeals concerning a change of zoning for a parcel of land or structure to allow usage different than originally approved by law.

RESOURCE DIRECTORY